Name Reactions for Carbocyclic Ring Formations

Wiley Series on Comprehensive Name Reactions

Jie Jack Li, *Series Editor*

Name Reactions in Heterocyclic Chemistry
Edited by Jie Jack Li

Name Reactions of Functional Group Transformations
Edited by Jie Jack Li

Name Reactions for Homologation, Part 1 and Part 2
Edited by Jie Jack Li

Name Reactions for Carbocyclic Ring Formations
Edited by Jie Jack Li

Name Reactions for Carbocyclic Ring Formations

Edited by

Jie Jack Li
Bristol-Myers Squibb Company

Foreword by

E. J. Corey
Harvard University

WILEY

A JOHN WILEY & SONS, INC., PUBLICATION

Published by John Wiley & Sons, Inc., Hoboken, New Jersey.

Published simultaneously in Canada.

Library of Congress Cataloging-in-Publication Data:

Name reactions for carbocyclic ring formations / edited by Jie Jack Li ; foreword by E.J. Corey.
p. cm.
Includes index.
ISBN 978-0-470-08506-6 (cloth)
1. Ring formation (Chemistry) I. Li, Jie Jack.
QD281.R5N36 2010
547'.2—dc22 2010008429

Printed in the United States of America.

10 9 8 7 6 5 4 3 2 1

Dedicated to Professor Keith R. Fagnou

June 27, 1971–November 11, 2009

Contents

Foreword

Part of the charm of synthetic organic chemistry derives from the vastness and multidimensionality of the intellectual landscape. First, there is the almost infinite variety and number of possible target structures that lurk in the darkness waiting to be made. Then there is the vast body of organic reactions that serve to transform one substance into another, now so large in number as to be beyond credibility to a nonchemist. There is the staggering range of reagents, reaction conditions, catalysts, elements, and techniques that must be mobilized in order to tame these reactions for synthetic purposes. Finally, it seems that new information is being added to that landscape at a rate that exceeds the ability of a normal person to keep up with it. In such a challenging setting, any author or group of authors must be regarded as heroic if through their efforts, the task of the synthetic chemist is eased.

This volume on methods for formation of carbon rings brings to the attention of practicing synthetic chemists and students of chemistry a wide array of tools for the formation of such rings by synthesis. Since cyclic structures are among the most useful molecules, it is a valuable addition to the literature that will prove its merit for years to come. The new knowledge that arises with its help will prove to be of great benefit to humankind.

E. J. Corey
February 1, 2010

Preface

This book is the fifth volume of the series *Comprehensive Name Reactions*, an ambitious project conceived by Professor E. J. Corey of Harvard University in the summer of 2002. Volume 1, *Name Reactions in Heterocyclic Chemistry*, was published in 2005. Volume 2, *Name Reactions for Functional Group Transformations* was published in 2007. Volumes 3 and 4 on homologations were both published in 2009. They have been warmly received by the organic chemistry community. After this Volume 5, *Name Reactions on Carbocyclic Ring Formations* is out in 2010, we will roll out the final volume, Volume 6 on *Name Reactions in Heterocyclic Chemistry—Part II*, in 2011. Continuing the traditions of the first four volumes, each name reaction in Volume 5 is reviewed in seven sections:

1. *Description,*
2. *Historical Perspective,*
3. *Mechanism,*
4. *Variations and Improvements,*
5. *Synthetic Utility,*
6. *Experimental, and*
7. *References.*

I also introduced a symbol [R] to highlight review articles, book chapters, and books dedicated to the respective name reactions.

I have incurred many debts of gratitude to Professor E. J. Corey. What he once told me — "The desire to learn is the greatest gift from God" — has been a true inspiration. Furthermore, it has been my great privilege and a pleasure to work with a collection of stellar contributing authors from both academia and industry. Some of them are world-renowned scholars in the field, some of them have worked intimately with the name reactions that they have reviewed, some of them even discovered the name reactions that they authored in this series. As a consequence, this volume truly represents the state-of-the-art for *Name Reactions for Carbocyclic Ring Formations*.

I welcome your critique.

Jack Li
February 1, 2010

Jie Jack Li and E. J. Corey, circa 2002

Contributing Authors:

Dr. Nadia M. Ahmad
Takeda Cambridge
418 Cambridge Science Park
Cambridge
CB4 0PA
United Kingdom

Dr. Jeffrey A. Campbell
Department of Chemistry
Lehigh University
Bethlehem, PA 18015

Dr. Louis S. Chupak
Discovery Chemistry
Bristol-Myers Squibb Company
5 Research Parkway
Wallingford, CT 06492

Dr. Timothy T. Curran
Chemical Development
Vertex Pharmaceuticals
130 Waverly Street
Cambridge, MA 02139

Professor Roman Dembinski
Department of Chemistry
Oakland University
2200 North Squirrel Road
Rochester, MI 48309

Dr. Matthew J. Fuchter
Department of Chemistry
Imperial College London
London SW7 2AZ

Dr. Paul Galatsis
Medicinal Chemistry
Pfizer Global Research & Development
Eastern Point Road
Groton, CT 06340

Professor Brian Goess
Department of Chemistry
Furman University
3300 Poinsett Highway
Greenville, SC 29613

Dr. Martin E. Hayes
Medicinal Chemistry
Abbott Bioresearch Center
381 Plantation Street
Worcester, MA, 01605

Professor Nessan Kerrigan
Department of Chemistry
Oakland University
2200 North Squirrel Road
Rochester, MI

Dr. Ewa Krawczyk
Department of Heteroorganic Chemistry
Centre of Molecular & Macromolecular Studies
Polish Academy of Sciences
Sienkiewicza 112
90-363 Łódź, Poland

Dr. Jie Jack Li
Discovery Chemistry
Bristol-Myers Squibb Company
5 Research Parkway
Wallingford, CT 06492

Noha S. Maklad
Medicinal Chemistry
Pfizer Global Research & Development
Eastern Point Road
Groton, CT 06340

Professor Richard J. Mullins
Department of Chemistry
Xavier University
3800 Victory Parkway
Cincinnati, OH 45207-4221

Dr. Kevin M. Peese
Discovery Chemistry
Bristol-Myers Squibb Company
5 Research Parkway
Wallingford, CT 06492

Dr. Frank Rong
ChemPartner
No. 2 Building, 998 Halei Road
Zhangjiang Hi-Tech Park, Pudong
Shanghai, China 201203

Professor Kevin M. Shea
Department of Chemistry
Clark Science Center
Smith College
Northampton, MA 01063

Professor Nicole L. Snyder
Hamilton College
198 College Hill Road
Clinton, NY 13323

Dr. Stephen W. Wright
Medicinal Chemistry
Pfizer Global Research & Development
Eastern Point Road
Groton, CT 06340

Dr. Yong-Jin Wu
Discovery Chemistry
Bristol-Myers Squibb Company
5 Research Parkway
Wallingford, CT 06492

Chapter 1 Three-Membered Carbocycles 1

1.1 Freund Reaction

Frank Rong

1.1.1 Description

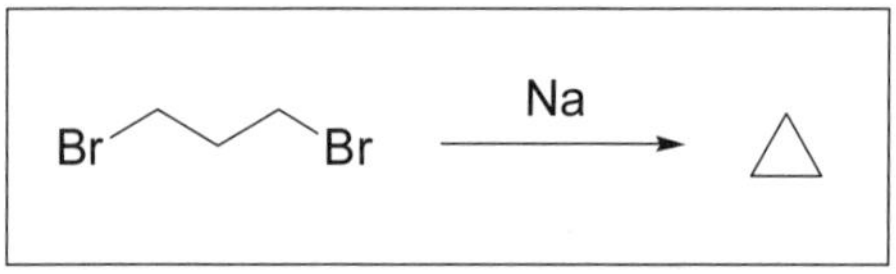

The Freund reaction refers to the formation of alicyclic hydrocarbons by the reaction of sodium on open chain dihalo compounds.[1]

1.1.2 Historical Perspective

In 1882 Freund reported that treating trimethylene glycol with hydrobromic acid gave trimethylene dibromide, which was further treated with sodium in reflux temperature. As a result the sodium dissolved, the sodium bromide was precipited, and a gas from the reaction was collected. What is the gas? By treacting with bromine it went back to trimethylene dibromide. By treacting with hydriodic acid it gave iodopropane. Therefore, the gas was concluded to be cyclopropane for the first time.[1]

This reaction has been called the Freund reaction on occasion. However, reference to the original literature shows that although Freund was the first to make cyclopropane itself, he used an extension of the Wurtz reaction and therefore had no claim to the method of ring closure that employs zinc in the presence of protonic solvent. Gustavson published in 1887 a paper titled "Concerning a New Method of Preparation of Trimethylene."[2] Gustavson and Popper extended this method to the preparation of substituted cyclopropanes; using zinc dust-treated trimethylene dibromide gave cyclopropane.[2,3] In 1936 Hass reported addition of sodium iodide to the zinc dust and 1,3-dichloropropane reaction mixture, both the yield of cyclopropane and the conversion rate were changed singnificantly.[4]

1.1.3 Mechanism

The mechanism of Freund reaction is more likely as same as the Wurtz reaction, a free-radical mechanism. In the presence of iodide ions, the pathways might be a combination of substitution (S_N1 or S_N2) with a free-radical mechanism.[3]

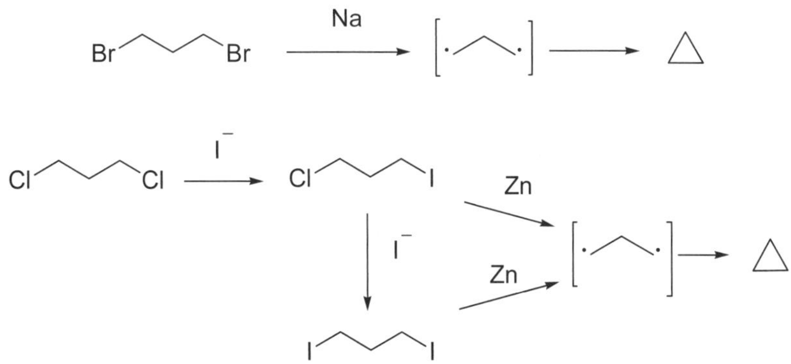

1.1.4 *Variations and Improvements*

Gustavson reported a new method in the preparation of cyclopropane.[2] Treating trimethylene dibromide (10 g) with zinc dust (12 g) suspended in aqueous alcohol at 50–60 °C gave cyclopropane. He tried different ratios of alcohol to water and found without water the reaction was very slow and at least 2% water was needed.

Hass has further modified the Gustavson reaction condition.[4] By using a large excess of zinc dust and by raising the temperature of the reaction mixture with high-boiling solvents, the rate of conversion was increased materially. When sodium iodide was added to a refluxing mixture of zinc dust, ethanol, and 1,3-dichloropropane, a marked acceleration of the reaction rate occurred. For example, using 1 mole of anhydrous sodium carbonate for each mole of 1,3-dichloropropane, a 100% excess of zinc dust, and 1/6 mole of sodium iodide in a solvent consisting of 75% ethanol and 25% water, a 95% yield of crude cyclopropane was obtained in 12 h in the same apparatus as before. A better yield and purer product were obtained if both sodium carbonate and acetamide were employed. The 1,3-dichloropropane was prepared by the chlorination of propane obtained from natural gas. This is called as Hass cyclopropane process.

1.1.5 *Synthetic Utility*

The cyclopropane was an important anesthetic in 1930s. Galasso said: "The safest anesthetic agent—the one which presents all the good qualities and none of the objectional side effects of the agents we have on hand—cyclopane."[5] This drug has been manufactured execusively by the following reaction sequence.

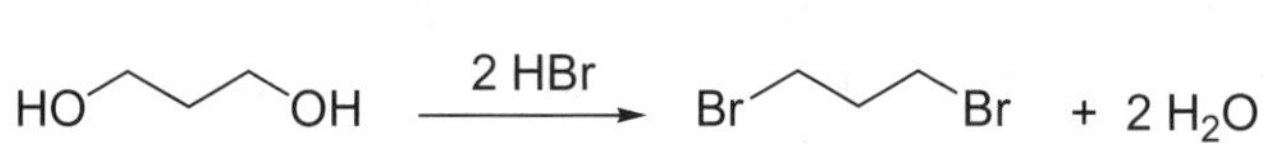
(1)

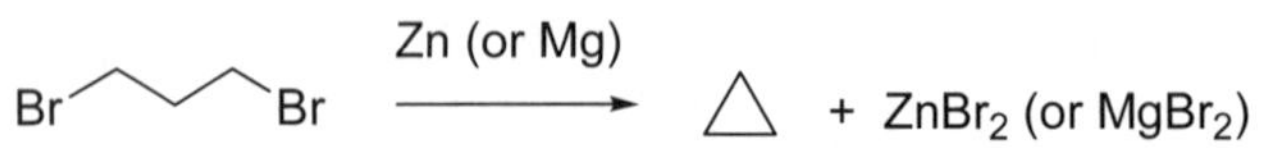

(2)

The 1,3-propanediol (trimethylene glycol) was obtained as a by-product of the soap industry, where it exists as a minor impurity in the glycerol. However, both 1,3-propanediol and hydrobromiic acid are relatively expensive compared to propane and chlorine.[4] Hass process made the production of cyclopropane more cost effective.

Shortridge and co-workers reported an extension of the Gustavson method for the synthesis of cyclopropane and its derivatives.[6] They successfully prepared spiranes containing a cyclopropane ring and provide an easy, straightforward way of producing this type of hydrocarbon in quantity and in a good state of purity. The corresponding dibromide **1** was cyclized by zinc in aqueous ethanol to give spirane **2** in excellent yield.

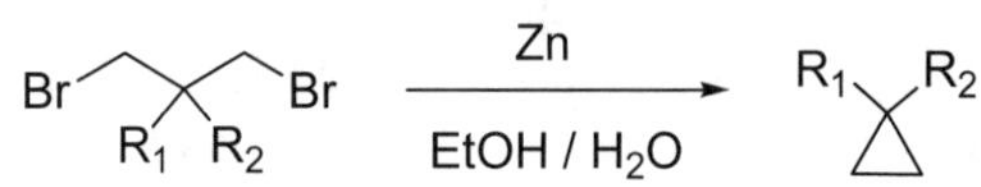

1 **2**

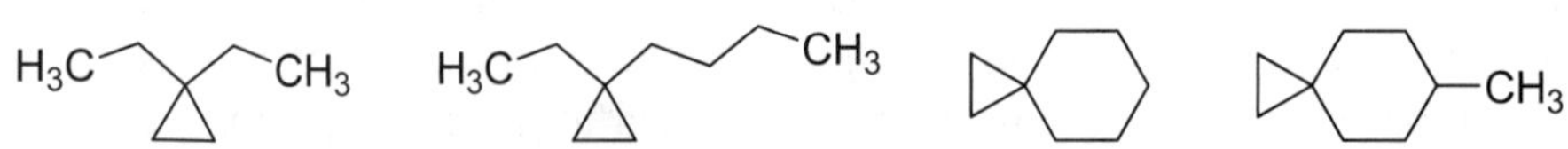

Yield: **3**: 92% **4**: 94% **5**: 91% **6**: 89%

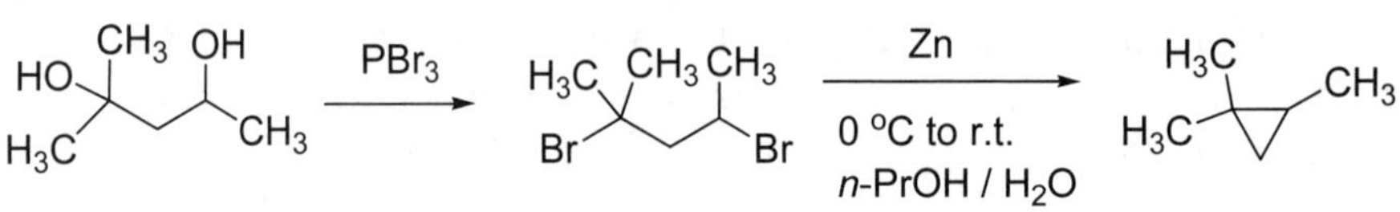

7 **8** **9**

The Freund reaction for the preparation of cyclopropane derivatives has in certain cases been unsatisfied due largely to the formation of olefins as the principal product. In general primary–primary 1,3-dibromides give high yields, primary–secondary dibromides give good yields. Secondary–secondary dibromides give fair yields, and all condensations involving a tertiary bromide give products containing an olefin as the principal or sole product. Bartleson and co-workers found that this problem can be solved at low terperature for the ring closure reaction.[7] The 1,1,2-trimethylcyclopropane **9** was prepared from 2-methyl-2,4-dibromopentane **8** by the Freund reaction at low temperature. The yield of crude product was

86% and the purity can reach 95% by fractional distillation in a high efficiency distilling column.

The 1,2-dimethyl-3-ethylcyclopropane **12** was similarly synthesized from 3-methyl-2,4-dibromohexane **11**.[7] A yield of 90% was obtained in the ring closure step. The secondary–tertiary and secondary–secondary 1,3-dibromides, **8** and **11**, were prepared in 90% yields by the reaction of phosphorus tribromide with the diols, **7** and **10**, at low temperature (–24 °C). The low temperature prevents the loss of hydrogen bromide from the reaction mixture.

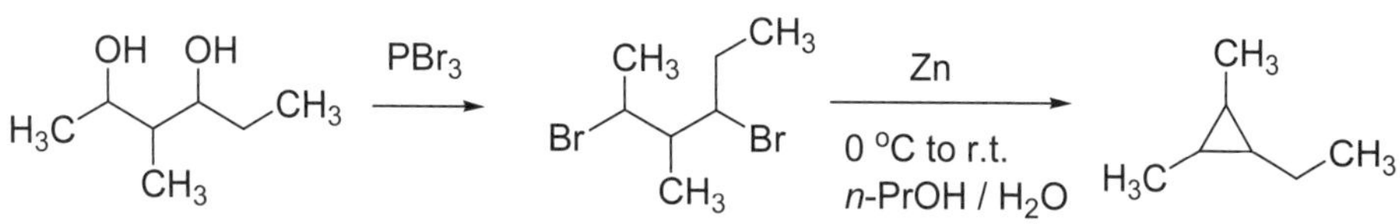

10 **11** **12**

1.1.6 Experimental

Preparation of 1,1-dimethylcyclopropane (14)[6]

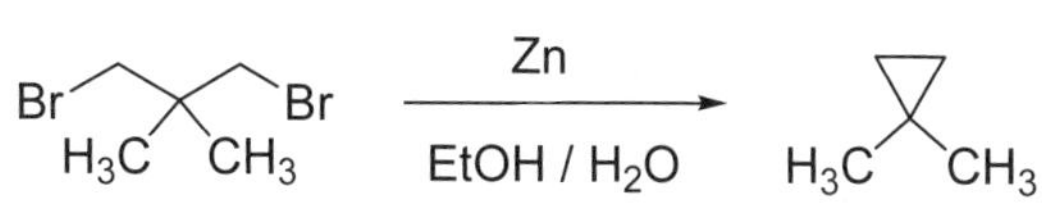

13 **14**

In a 2 L three-necked flask equipped with a dropping funnel, mercury-sealed stirrer and reflux condenser (connected to a trap surrounded by a dry ice–acetone bath) were placed 900 mL 95% ethanol, 90 mL distilled water and 628 g (9.6 mol) zinc dust; it was necessary to maintain vigorous stirring at all times to prevent caking of the zinc. The mixture was brought to gentle reflux, and 562 g (2.4 mol) of 1,3-dibromo-2,2-dimethyl-propane **13** was added dropwise at the temperature. Heating and stirring were continued for 24 h after the last of the dibromide had been added; the bulk of the hydrocarbon collected in the trap during this period. The remaining 1,1-dimethylcyclopropane (along with some alcohol) was then distilled from the reaction flask and was collected in the trap. The crude product **14** (162 g) was washed with ice water and dried. The product **12** was obtained in 96% yield (based on distilled dibromide) with these physical properties: b.p. 20.63 °C (760 mm) and n^{20}_{D} 1.3668.

Preparation of 1,1,2-trimethylcyclopropane (9)[7]

The reaction was carried out in a 1 L, three-necked flask fitted with a reflux condenser, thermometer, dropping funnel and mercury sealed stirrer. To the flask was added 100 mL water, 300 mL *n*-propyl alcohol and 196 g oxygen-free zinc dust prepared from commercial-grade zinc dust. The flask was placed in an ice-bath and 244 g (1 mol) of freshly distilled 2-methyl-2,4-dibromopentane was added dropwise with efficient stirring over a period of about 90 min. The icebath was then removed and the mixture was stirred at room temperature for about 32 h. After about 10 h an immiscible layer of hydrocarbon had formed. At the end of the reaction the hydrocarbon product was separated by distillation. The crude product **9** was collected over a temperature range of 49–51 °C and weighed 78.1 g, a yield of 86%. The refractive index of the crude product was $n^{20}{}_{D}$ 1.3847. The crude product was further purified by fractional distillation in a high-efficiency distilling column to give 95% pure product **9** with these physical properties: b.p. 52.1 °C (736 mm) and $n^{20}{}_{D}$ 1.3850.

1.1.7 References

1. Freund, A. *Monatsh.* **1882**, *3*, 625–635.
2. Gustavson, G. *J. Prakt. Chem. (2)* **1887**, *36*, 300–303.
3. Gustavson, G.; Popper; J. *J. Prakt. Chem. (2)* **1898**, *58*, 458.
4. Hass, H. B.; McBee E. T.; Hinds, G. E.; Gluesenkamp, E. W. *Ind. Eng. Chem.* **1936**, *28*, 1178–1181.
5. Galasso, *Anesth. and Analges.* **1936**, *15*, 32.
6. Shortridge, R. W.; Craig, R. A.; Greenlee, K. W.; Derfer, J. M.; Boord, C. E. *J. Am. Chem. Soc.* **1948**, *70*, 946–949
7. Bartleson, J. D.; Burk, R. E.; Lankelma, H. P. *J. Am. Chem. Soc.* **1946**, *68*, 2513–2518.

1.2 Kishner Cyclopropane Synthesis

Frank Rong

1.2.1 Description

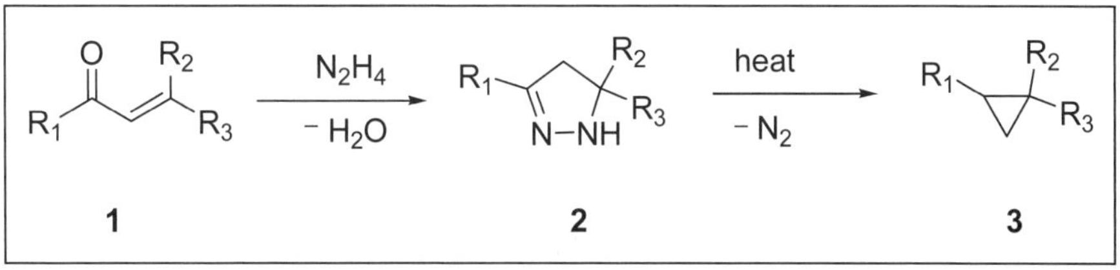

Kishner cyclopropane synthesis refers to the formation of cyclopropane derivatives **3** by decomposition of pyrazolines **2** formed by reacting α,β-unsaturated ketones or aldehydes with hydrazine.[1]

1.2.2 Historical Perspective

In 1912 Kishner and Zavadovskii reported the synthesis of phenylcyclopropane by heating decomposition of 5-phenyl-3-pyrazoline.[1] The Kishner cyclopropane synthesis has become wellknown due to its unique and the smallest cyclic core structure.[2]

1.2.3 Mechanism

It is believable that the pyrazoline, **4** or **5**, undergoes thermolytic decomposition and gives the diradical **6** first. Then, the diradical formed a bond quickly to give the cyclopropane **7**.[2,3] This could be a reversible reaction between the diradical **6** and the cyclized product **7**.

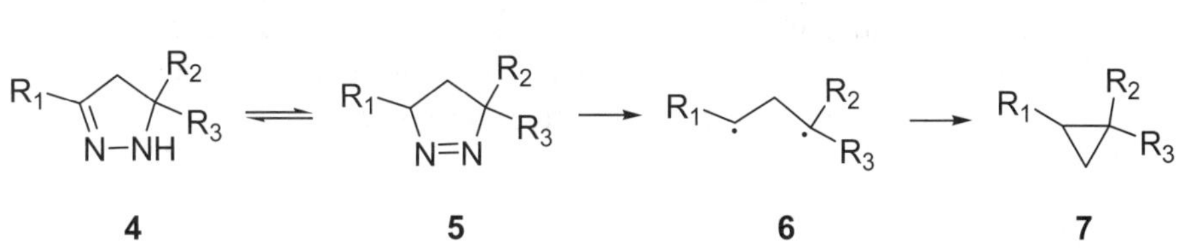

Stereochemical crossover in the pyrolysis of 3,5-disubstituted pyrazolines was proposed.[3,4] The observation of a stereochemical crossover phenomenon stimulated a consideration of the mechanism from a different viewpoint. The loss of molecular nitrogen in the pyrolysis of a *cis*-3,5-disubstituted pyrazoline, *cis*-**8**, might be expected to give a trimethylene intermediate that could cyclized to a *cis*-disubstituted cyclopropane **10**, if internal rotations were slow, or to a mixture of *cis*- and *trans*-cyclopropanes, if internal rotations were fast. In the later case, the *cis*- and *trans*-pyrazoline, **8** and **9,** should give identical mixtures of cyclopropanes **10** and **11**. The experimental facts, however, are inconsistent with either of these models since the stereochemistry of the cyclopropane product in each case is predominantly (3:1) opposite to that of the pyrazoline. Obviously, a stereorandomized trimethylene cannot be the sole intermediate. In fact, the stereochemistry of deazetation of pyrazolines is still not completely understood.

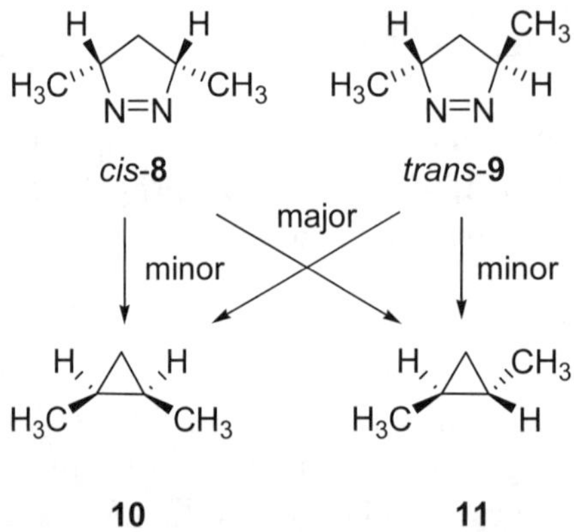

The research groups of McGreer[5,6] and Crawford[7–11] have done comprehensive investigation on the cyclic azo compounds thermal decomposition. Crawford's group investigated the stereochemistry problem in the thermal decomposition of *cis*- and *trans*-3,5-dimethylpyrazolines (**12** and **13**).[7] The major products of these decompositions are the stereoisomeric dimethylcyclopropane, and the major pathway is apparent single inversion of stereochemistry in each case. Crawford and Mishra rationalized these observations by assuming that the pyrazolines decompose in the envelope conformation leading directly to 0,0 intermediates. Predominant conrotatory closure then leads to overall single inversion of stereochemistry.[7]

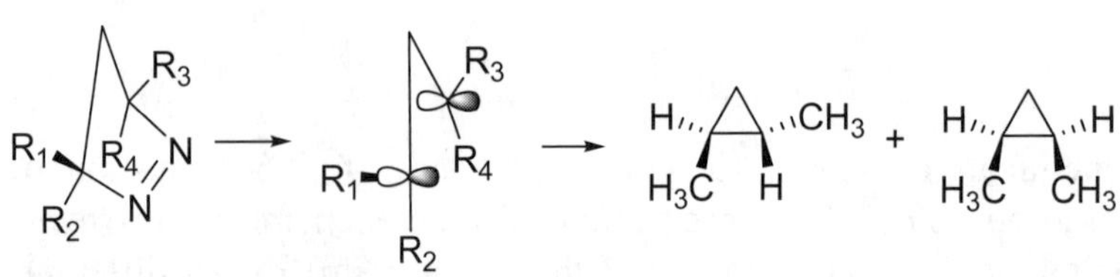

12) $R_1 = R_3 = CH_3$ and $R_2 = R_4 = H$ *trans*- 66.1% *cis*- 33.2%
13) $R_1 = R_4 = CH_3$ and $R_2 = R_3 = H$ *trans*- 25.4% *cis*- 72.6%

One of the most difficult mechanisms to rule out rigorously involves the possibility than only one C–N bond breaks initially, leading (in the case of *trans*-pyrazoline **14** or **15**) to diradical **16**. If the radical center at C-2 is now required to carry out a backside displacement of N_2 at C-4, a product of the correct stereochemistry is produced.[12] However, Crawford and his co-workers have carried out a number of elegant studied that provide support for a mechanism that involves simultaneous cleavage of both C–N bonds.[7–11]

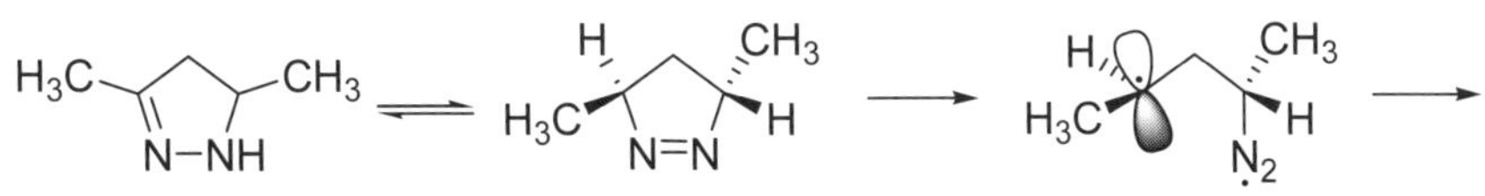

14 **15** **16**

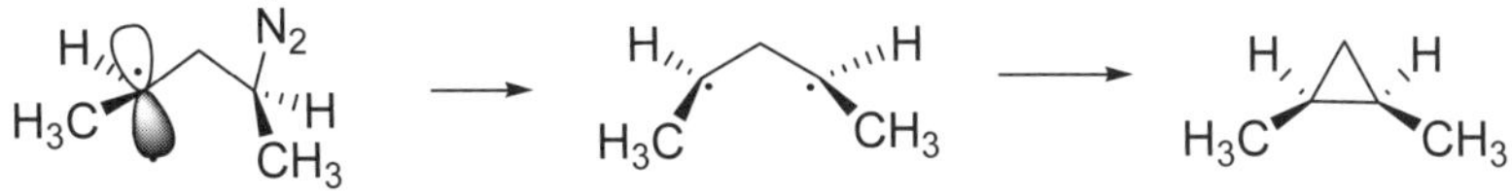

17 **18** **19**

1.2.4 Variations and Improvements

A couple of different approaches in the synthesis of pyrazolines were shown. Crawford and Ohno synthesized a 40:60 mixture of *cis*- and *trans*-3,5-divinyl-1-pyrazoline, **21** and **22,** by adding a concentrated solution of vinyldiazomethane **20** in diethyl ether, purified by distillation, to a large excess of 1,3-butadiene maintained as a liquid in a pressure bottle at low temperature.[8] The intermolecular 1,3-cycloaddition of the diazoalkane proceeded more rapidly than the intramolecular cyclization to pyrazole. The kinetic of the thermolysis of these compounds in diphenyl ether at 35–65 °C producing divinylcyclopropane **23** were studied by measuring the rate of nitrogen evolution. They concluded that both carbon–nitrogen bonds are being broken in the rate determining step.[8]

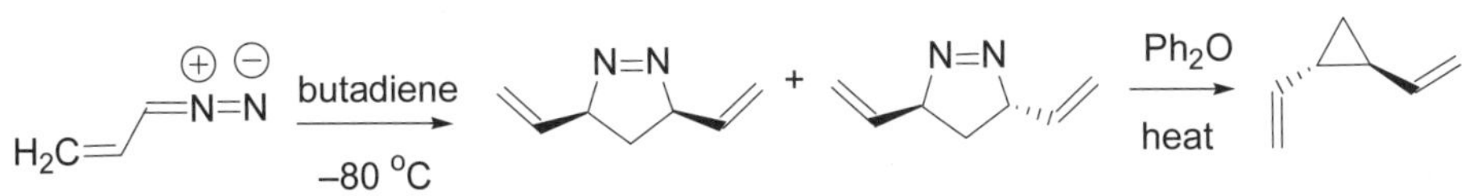

20 **21** **22** **23**

The *cis*- and *trans*-3-ethyl-5-vinyl-1-pyrazoline, **25** and **26,** were similarly prepared by the cycloaddition of 1-diazopropane **24** to 1,3-butadiene.[8] The mixture of *cis*-**25** and *trans*-**26** (18% *cis*, 82% *trans*, by NMR) was heated at 120 °C for 2 h. The product proportions were determined by GC to be 57% *cis*- and 43% *trans*-ethyl-2-vinylcyclopro-pane, **27** and **28**, respectively.

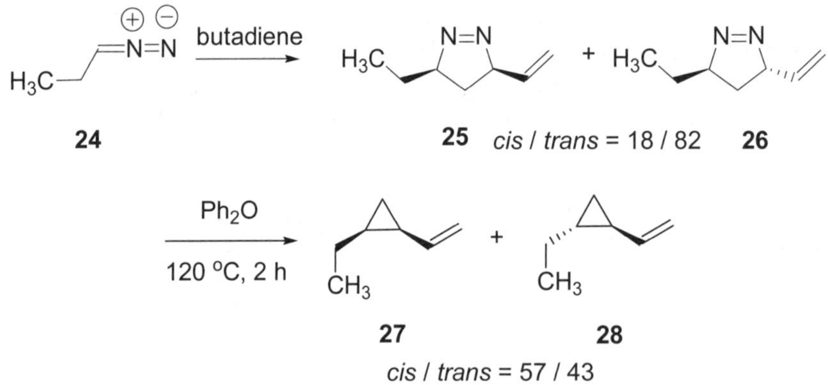

24 **25** **26** **27** **28**

1.2.5 Synthetic Utility

A great attention has been paid on the cyclopropane derivatives after Kishner reported his discovery due to the similarities between olefins and cyclopropane with respect to both their chemical and physical properties.[2,3] Cyclopropane resembles ethylene in some respects, and both systems can enter into conjugation with other unsaturated groups such as a carbonyl group, a phenyl group, or a pyridyl group. Smith and Rogier synthesized the 2-phenylbicyclopropyl by the Kishner method.[13] The styryl cyclopropyl ketone **29** was converted to pyrazoline **30**, which was decomposed at 220 °C to give the 2-phenylbicyclopropyl **31** in 74% yield. The physical and chemical properties of this compound were also studied. The results indicated that this compound does not exhibit any of the conjugative effect shown by phenylcyclopeopane.

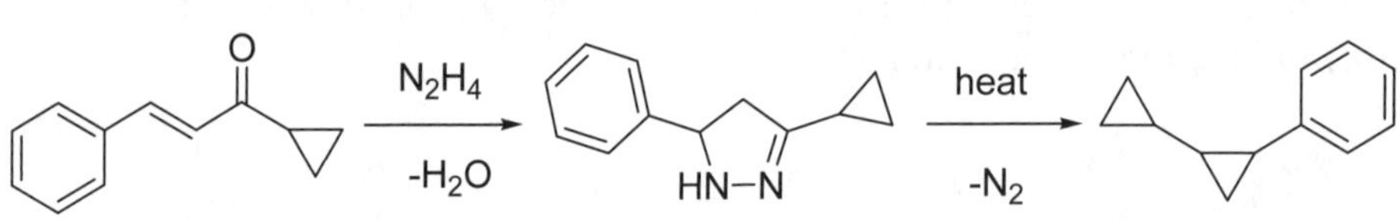

29 **30** **31**

Mishra and Crawford reported the synthesis of (*3R,5R*)-(+)-*trans*-3,5-dimethyl-1-pyrazoline **35** by different approach starting from alcohol **32**.[7] The pyrazoline **35** undergoes thermolysis, producing 25.6% of *trans*-1,2-dimethyl-cyclopropane **36**, processing 23% optical purity, and having the *S:S* configuration.

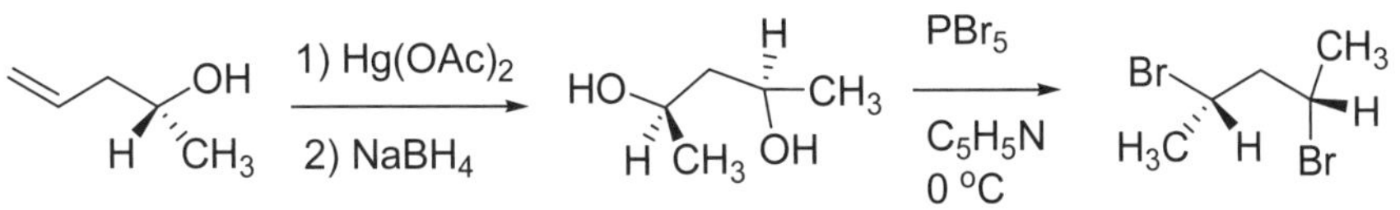

32 **33** (*2R* : *4R*) **34** (*2S* : *4S*)

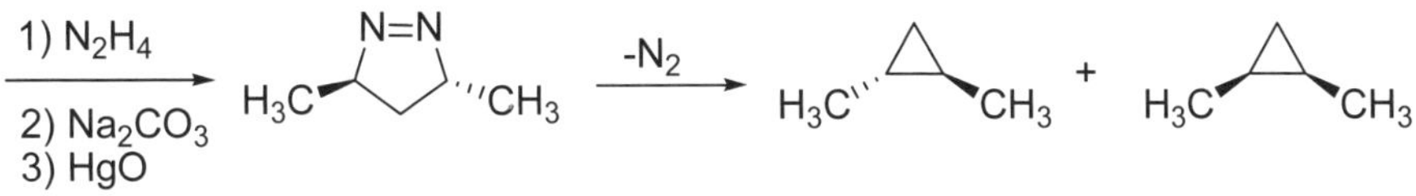

35 (*3R* : *5R*) **36** (*1S* : *2S*) **37** (*1R* : *2S*)

Crawford and Tokunaga synthesized 3,3,5,5-tetramethyl-4-methylene-1-pyrazoline **41** by a different approach.[10] The tetramethyl-4-pyrazolidone **39** was prepared by the reaction of hydrazine with α,α′-dibromodiisopropylketone **38**. Then oxidizing **39** by MnO_2 gives intermediate **40**, which was converted to **41** by following Mock's procedure.[14]

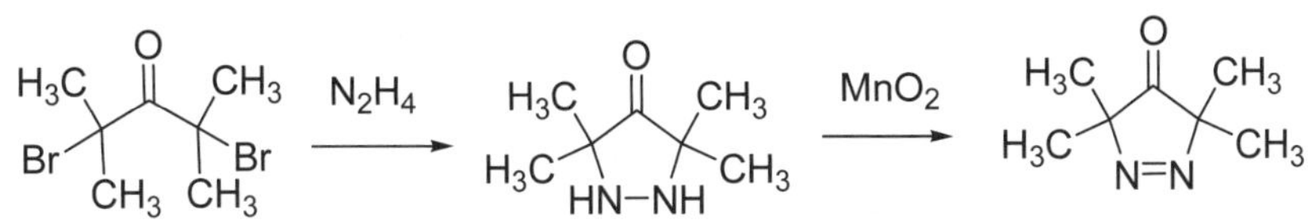

38 **39** **40**

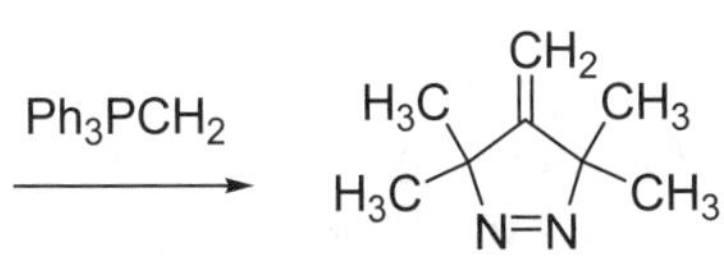

41

They also studied the thermolysis of the compound **41**.[10] They found the thermolysis of **41** proceeds at 1/63 the rate of 4-methylene-1-pyrazoline. The **41** is undergoing thermolysis by a mechanism different from that for 4-methylene-1-pyrazoline. The 2,2,3,3-tetramethylmethylenecyclopropane **42** produced rapidly isomerizes under the reaction conditions to 2,2-dimethylisopropylidenecyclopropane **43**. The four opposed methyl groups of **42** have created sufficient ground state destabilization as to cause its isomerization to be 147 times faster than the conversion of 2,2-dimethylmethylenecyclopropane **44** to isopropylidenecyclopropane **45**.

41

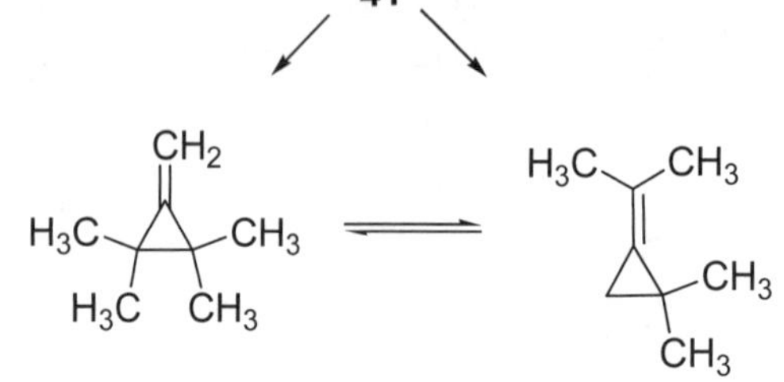

42 **43**

44 **45**

1.2.6 Experimental

Preparation of phenylcyclopropane 47.[15]

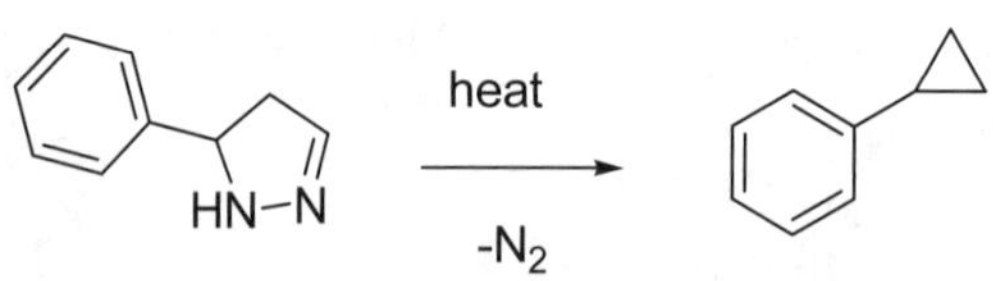

46 **47**

A mixture of 118 g 5-phenyl-3-pyrazoline **46**, prepared by published procedure; 30 g pulverized potassium hydroxide; and 2.5 g platinized asbestos was heated in a 1 L, three-necked flask equipped with a stirrer and a Claisen distillation head. The temperature was raised slowly and the heat was shut off at the first sign of reaction. When the exothermic reaction ceased the temperature was again raised and the product was distilled. Both the distillate and the residue were steam distilled and the steam distillate was taken up in ether and dried first with sodium sulfate and then with sodium and redistilled. The product **47** was collected at 60–63 °C (11 mm Hg) and was finally redistilled giving a colorless oil, 11.5 g (12%), b.p. 173.5 °C (740 mm Hg), and n^{20}_{D} 1.5320.

Preparation of 2-phenylbicyclopropyl (31)[13]

Preparation of 3-cyclopropyl-5-phenyl-2-pyrazoline ***(30)****:* Styryl cyclopropyl ketone **29** (42 g, 0.245 mol) was added to a solution of aqueous hydrazine (25 mL, 0.42 mol) in ethanol (95%, 70 mL); the mixture became warm and acquired a green color. It was allowed to stand for 45 min., then warmed on the steambath for 1 h, after which excess hydrazine and solvent were removed under reduced pressure. Distillation of the residue gave **30** as a light green liquid (37.8 g, 86%), which boiled at 164 °C (1 mm Hg).

Preparation of 2-phenylbicyclopyl (31): Powdered potassium hydroxide (7.2 g) and platinized asbestos (3.2 g) were placed in a 100 mL round-bottomed flask arranged for distillation, and immersed in a metal bath. The bath was heated to 220 °C (thermometer in the bath), and the pyrazoline **30** (38.8 g, 0.2 mol) was slowly added. After the rapid evolution of nitrogen cased, the product **31** was distilled from the reaction mixture; it distilled at 92–96 °C (0.8 mm Hg), and weighed 22.5 g (74%). The products from several runs were combined and fractionated through a column (15 × 1.5 cm) packed with glass helices. A center cut was taken for analysis. This boiled at 57 °C (0.12 mm), and had n^{20}_{D} 1.5352, d^{20}_{4} 0.9587, and molar refraction 51.40.

1.2.7 References

1. Kishner, N. M.; Zavadovskii, A. *J. Russ. Phys. Chem. Soc.* **1912**, *43*, 1132.
2. [R] Berson, J. A. in *Rearrangements in Ground and Excited States* vol. 1, P. de Mayo, Academic Press, New York, 1980, pp 326–329.
3. [R] Bergmann, R. G. in *Free Radicals,* vol. 1, J. Kochi, Wiley, New York, 1973. Pp 191–230.
4. Crawford, R. J.; Mishra, A. *J. Am. Chem. Soc.* **1965**, *87*, 3768.
5. McGreer, D. E.; Chiu, N. W. K.; Vinje, M. G.; Wong, K. C. K. *Can. J. Chem.* **1965**, *43*, 1407.
6. McGreer, D. E.; McDaniel, R. S.; Vinje, M. G. *Can. J. Chem.* **1965**, *43*, 1389–1397.
7. Mishra, A.; Crawford, R. J. *Can. J. Chem.* **1969**, *47*, 1515–1519.
8. Crawford, R. J.; Ohno, M. *Can. J. Chem.* **1974**, *52*, 3134–3139.
9. Crawford, R. J.; Erickson, G. L. *J. Am. Chem. Soc.* **1967**, *89*, 3907.
10. Crawford, R. J.; Tokunaga, H. *Can. J. Chem.* **1974**, *52*, 4033-4039.
11. Crawford, R. J.; Mishra, A. *J. Am. Chem. Soc.* **1966**, *88*, 3963.
12. Roth, W. R.; Martin, M. *Ann.* **1967**, *702*, 1.
13. Smith, L. I.; Rogier, E. R. *J. Am. Chem. Soc.* **1951**, *73*, 3840–3842.
14. Mock, W. L. Ph.D. thesis, Harvard University, Cambridge, MA, 1964.
15. Hammond, G. S.; Todd, R. W. *J. Am. Chem. Soc.* **1954**, *76*, 4081–4083.

1.3 Kulinkovich Cyclopropanol Synthesis

Jie Jack Li

1.3.1 Description

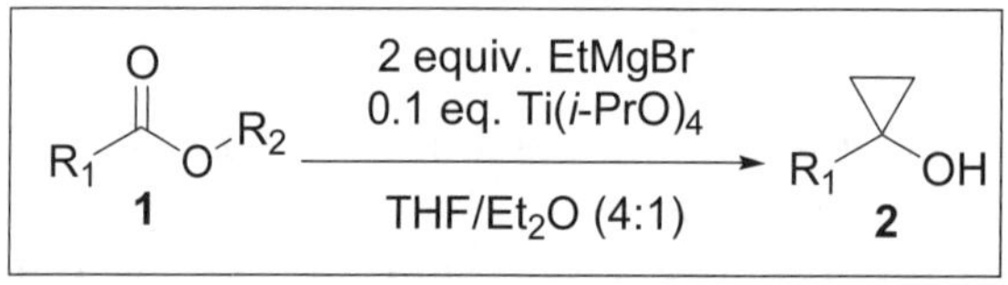

Kulinkovich cyclopropanol synthesis, also known as Kulinkovich reaction or Kulinkovich cyclopropanation, is titanium-catalyzed transformation of esters **1** to its corresponding cyclopropanols **2**.[1–9] The EtMgBr/Ti(*i*-OPr)$_4$ mixture resulting in bis-ethoxytitanacyclopropane is known as the Kulinkovich reagent, which is considered a synthetic equivalent of a two-carbon-1,2-dianion synthon ($^{\ominus}CH_2CH_2^{\ominus}$).

1.3.2 Historical Perspective

Professor Oleg Grigor'evich Kulinkovich,[10] a student of I. G. Tischenko of the Tischenko reaction fame, discovered the titled reaction in 1989 at Belorussian State University.[11–14] The unprecedented transformation has received great attention and utility, as testified by the references cited herein.

1.3.3 Mechanism

Kulinkovich himself proposed that the dialkoxytitanacyclopropanes as the key intermediate in the Kulinkovich cyclopropanation.[12] Extensive theoretical study on mechanism was published in 2001.[15] Eisch also provided detailed exploration of the mechanism for the Kulinkovich reaction in 2003.[16] In 2007, Kulinkovich proposed a modified "ate" complex mechanism for titanium-mediated cyclopropanation of carboxylic esters with Grignard reagents.[17]

Summing up the state-of-the-art understanding, the mechanism may be described as the following: When Ti(O*i*-Pr)$_4$ was mixed with the ethyl Grignard reagent, they react to provide diethyl titanium intermediate **3**, which immediately undergoes β-elimination with formation of the titanacyclopropane **4** and with release of ethane gas. Next, a nucleophilic attack at the ester carbonyl furnishes the titanoxacyclopentane **5**. Rearrangement to the homoenolate with concomitant activation of the carbonyl group allows for an intramolecular attack of the titanium–carbon bond to give the titanium cyclopropane alkoxide **6**. Metal exchange reaction with excess of Grignard reagent liberates the product as the magnesium alkoxide **7** and regenerates the catalytically active species **3**.

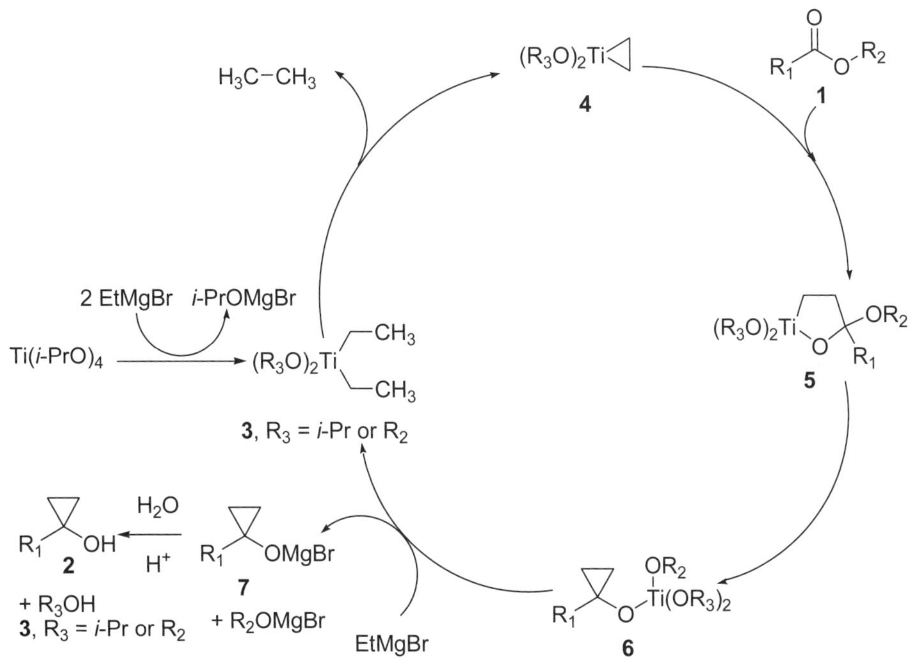

1.3.4 Variations and Improvements

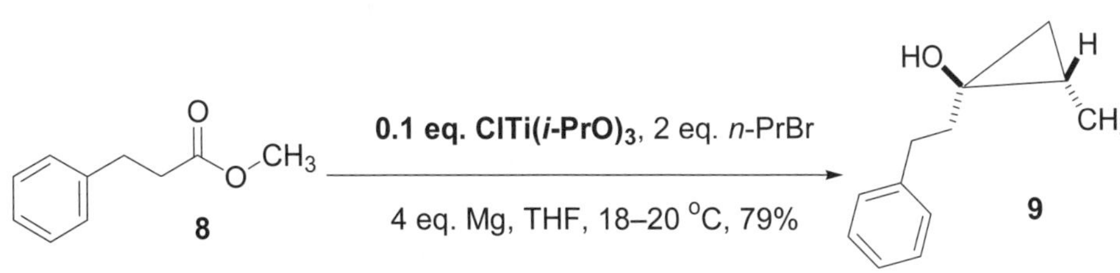

Initially, the ethyl Grignard reagent was successfully employed in the prototypical Kulinkovich reaction. In 1994, Corey demonstrated that when chlorotitanium(IV) triisopropoxide was better suited for higher substituted Grignard reagents, such as *n*-butylmagnesium bromide, can be used.[18] As exemplified by transformation **8** to **9**, the reaction was completely diastereoselective to give the *cis*-1,2-dialkylated cyclopropanol **9**. Furthermore, Corey also carried out the preliminary studies of an enantioselective version of the cyclopropanol synthesis with promising results. Employing a chiral TADDOL-based titanium reagent, 85:12 to 89:11 enantio-selectivity was achieved.

The other significant variation of the prototypical Kulinkovich reaction is the so-called Kulinkovich–de Meijere reaction, where de Meijere extended the substrates from esters to amides.[19,20] Other carboxylic acid derivatives including (cyclic) carbonate, imides, and nitriles also react with the key Kulinkovich intermediate. Szymoniak[21] developed an efficient new synthesis of cyclopropanes via hydrozirconation of allylic ethers (e.g., using $Cp_2Zr(H)Cl$) followed by addition of a Lewis acid (e.g., $BF_3{\bullet}OEt_2$). Casey et al. further investigated the stereochemistry of this interesting cyclopropanation reaction using deuterated allylic ethers.[22]

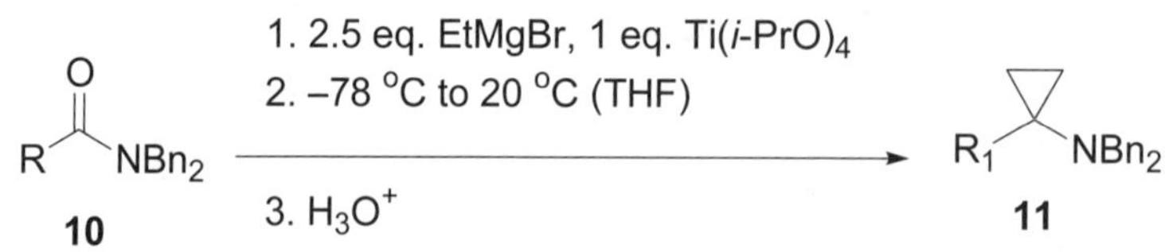

1.3.5 Synthetic Utility

1.3.5.1 General Utility

In general, esters, acid chlorides, and anhydrides are most reactive toward the Kulinkovich reagent. Carbonates and thioesters are of moderate reactivity, whereas carbonamides are least reactive. Case in point was made by chemoselective Kulinkovich reaction of succinic ester–amide **12**.[23] Cha observed that only the ester portion underwent the Kulinkovich reaction to afford cyclopropanol **13**. Szymoniak[24] demonstarted that nitriles are more reactive than ester and amides.

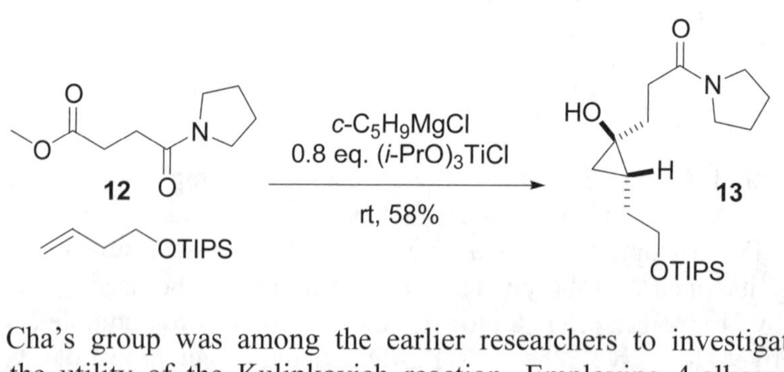

Cha's group was among the earlier researchers to investigate and extend the utility of the Kulinkovich reaction. Employing 4-alkoxybutyyl Grignard reagents such as **15** (several other Grignard reagents such as 2-phenethylmagnesium bromide did not work) and chlorotitanium(IV) triiso-propoxide converted cyclopropanol **16** as a single diastereomer.[25]

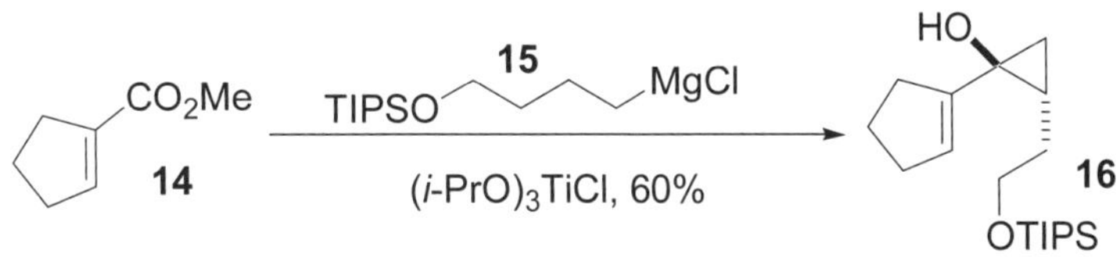

An important contribution from Sato[26] and Cha[27] was their successful extension of the Kulinkovich reaction to the intramolecular version. For instance, reductive coupling of carboxylic ester **17** with a terminal olefin provided bicyclo[3.1.0]hexan-1-ol (**18**) in 71% yield.[27] Cha also extended the low-valent titanium-mediated cyclopropanation to vinylogous esters as substrates. An interesting application is an intramolecular version of the methodology that transformed vinylogous ester **19** with a pendent terminal olefin to tricycle **20**.[28]

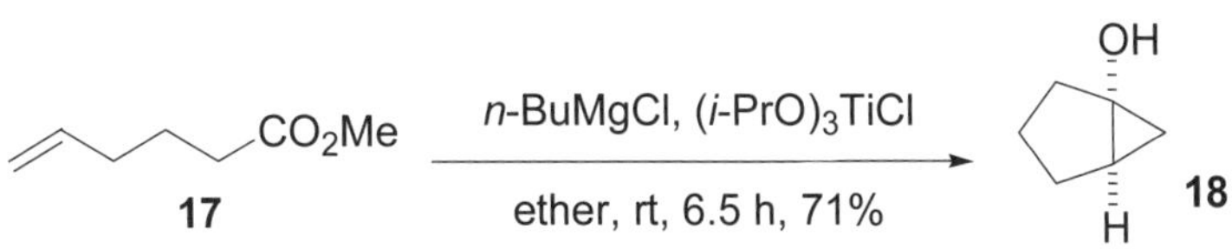

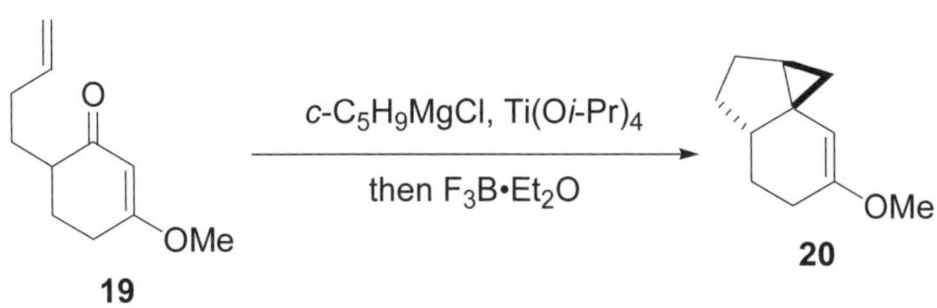

Cha also explored substrate directed asymmetric synthesis using the Kulinkovich reaction. Sequential treatment of homoallylic alcohol **21** with Ti(O*i*-Pr)$_4$ and *c*-C_5H_9MgCl furnished the putative intermediate **22**, which upon exposure to ethyl acetate produced *trans*-1,2dialkylcyclopropanol **23** in 12.2 : 1 *dr*.[29]

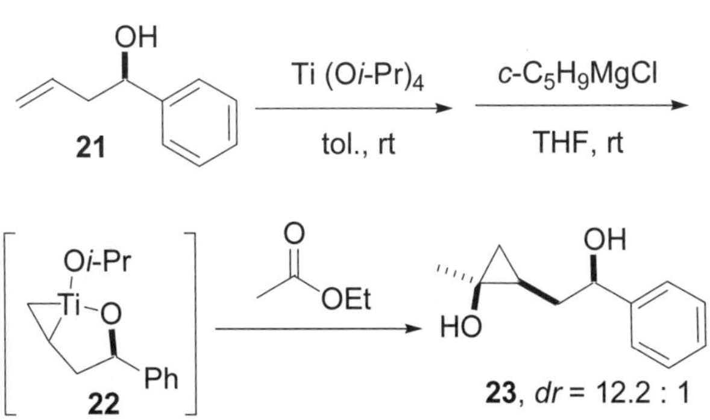

Sato and colleagues discovered a new titanium complex, (η^2-propene)Ti(O*i*-Pr)$_2$ (**24**), generated in situ, by treatment of Ti(O*i*-Pr)$_4$ with 2 equiv of *i*-PrMgX.[3] Compond **24** was proven to a versatile reagent in synthetic reactions involving alkynes. At the onset of the investigation, the Sato group soon discovered that use of *i*-PrMgX as the Grignard reagent was very important for generating a titanium compound that afforded alkyne–titanium complexes by the reaction with alkynes.[30] When silyl alkyne **25** was treated with **24**, the putative complex **26** was formed. Exposure of **26** to cyclopentyl aldehyde was followed by iodine to afford adduct **27** in 71% yield. An intramolecular version of the aforementioned transformation has been developed. Therefore, intramolecular coupling of diyne **28** provided [5,6]-bicyclic cyclopentadienol **29**.[31]

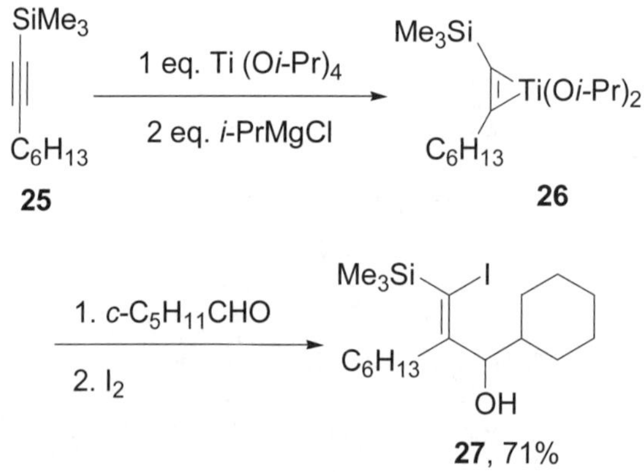

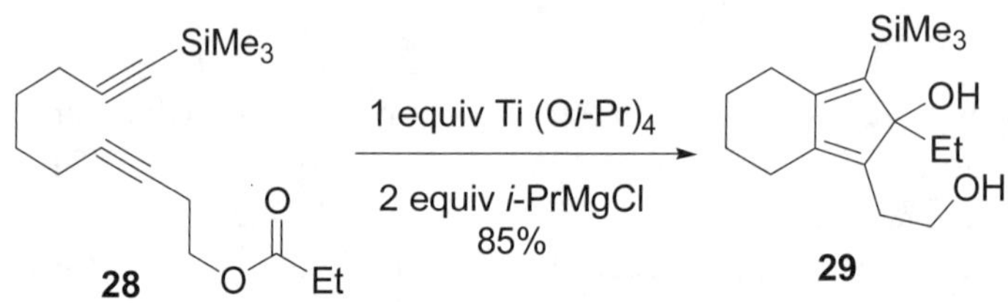

In addition, Sato et al. developed an interesting enantioselective synthesis of bicyclic cyclopropanols **31** from *N*-acylcamphorsultam (the Oppolzer's chiral auxiliary) derivatives **30**.[32]

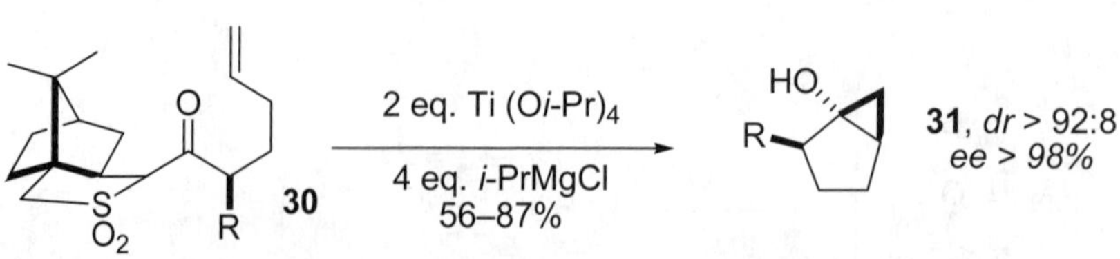

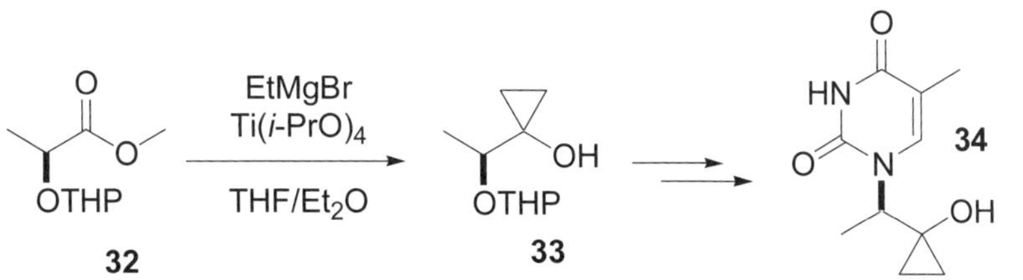

In the realm of medicinal chemistry, cyclopropanol provides a unique rigidity for the side chain. For instance, cyclopropanol **33**, from ester **32**, was incorporated into nucleoside **34**, an analogue of acyclovir.[33] Similarly, ester **35** was converted to cyclopropanol **36**, which was assembled onto guanine **37**, an anti-HBV agent.[34]

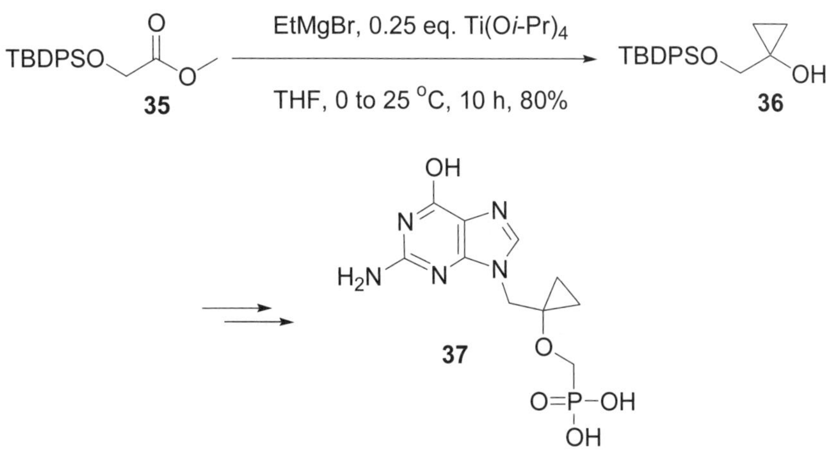

1.3.5.2 Applications in the total synthesis of natural products

Ollivier and co-workers used the Kulinkovich reaction in their total synthesis of heliannuols K and L.[35,36] Thus, conversion of ester **38** with pendent olefin to cyclopropanol **39** was achieved using **bromo**titanium(IV) triisopropoxide and cyclohexyl Grignard reagent. Cyclopropanol **39** then underwent oxidation using $FeCl_3$ and dechlorination to yield benzoxocinone **40**, an intermediate for both heliannuols K (**41**) and L.

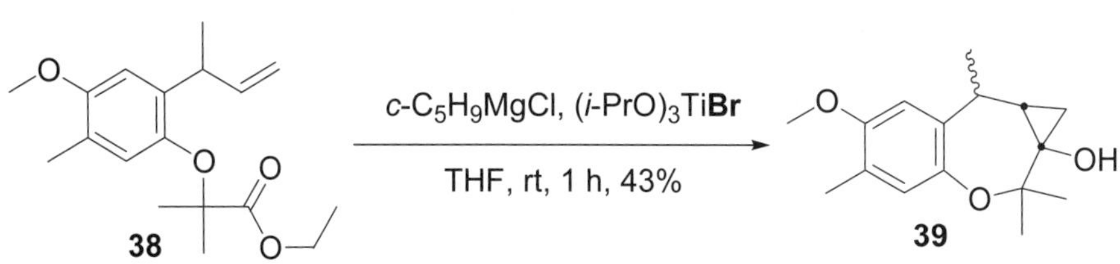

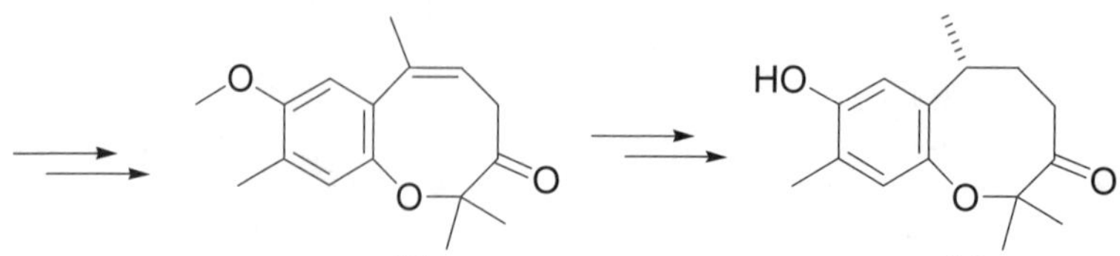

40 **41**

Singh's group converted ester **42** to its corresponding cyclopropanol **43**, which was treated with NBS and then Et_3N to deliver enone **44** via the intermediacy of β-bromoketone.[37] Enone **44** was transformed to tarchonanthuslactone **45**.

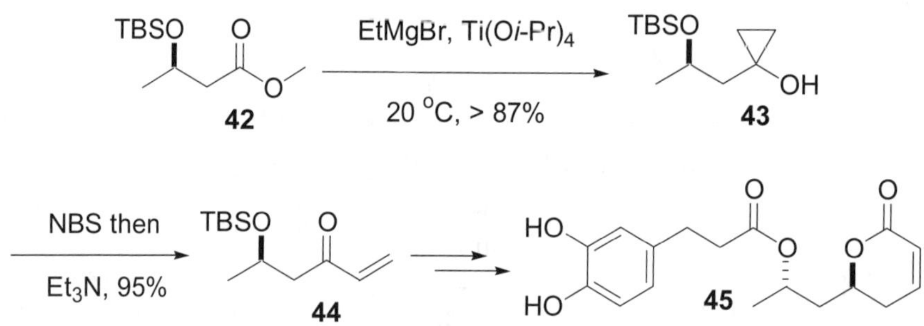

42 **43** **44** **45**

Corey employed the Kulinkovich reaction in the total synthesis of β-arianeosene (**48**).[38] Ester **46** was converted to cyclopropanol **47**, which served as the substrate to make cyclobutanone by treating **47** with $Al(Me)_3$.

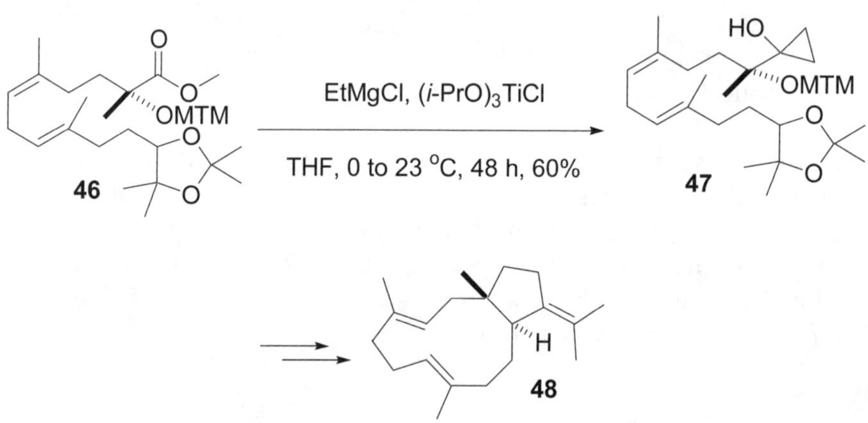

46 **47** **48**

1.3.5.3 Utility of the Kulinkovich–de Meijere reaction

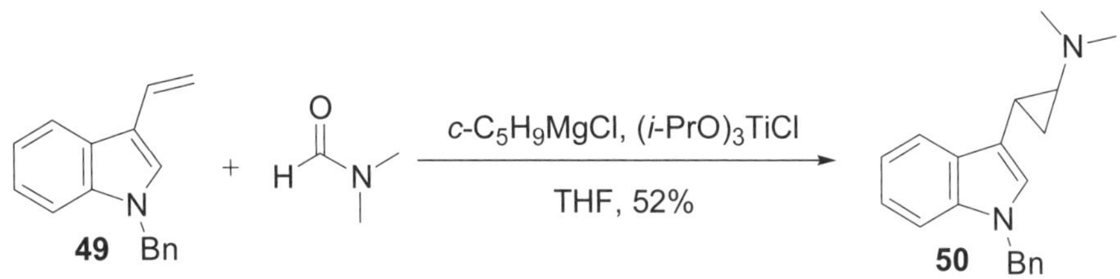

Joullié took advantage of the Kulinkovich–de Meijere reaction and synthesized a series of constrained *N,N*-dialkyl neurotransmitter analogues.[39] For instance, indolylcyclopropylamine **50** was assembled from indole-olefin **49** and DMF in 52% yield.

While intermolecular Kulinkovich–de Meijere reaction assembles cyclopropylamine, many have taken advantage of a pendent olefin at the substrates to synthesize bicycles. Cha converted olefinyl amide **51** to bicyclic amine **52** in excellent yields.[40] Six and co-workers prepared eight bicyclic aminocyclopropanes including **54** (from substrate **53**) with yields ranging from 26 to 87%.[41,42]

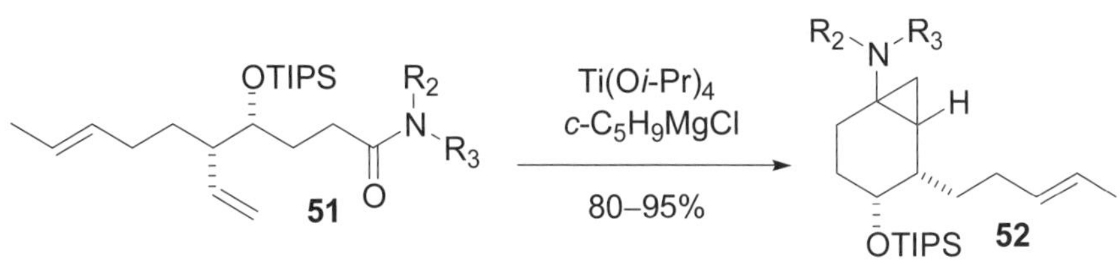

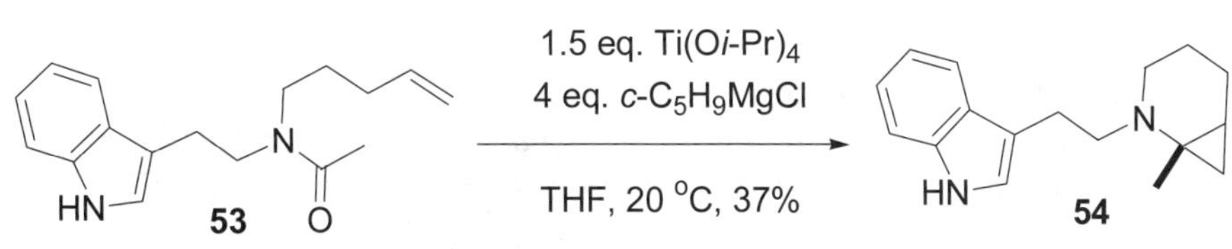

1.3.6 Experimental

1.3.6.1 Intramolecular Kulinkovich cyclopropanation reaction of carboxylic ester with olefin: bicyclo[3.1.0]hexan-1-ol (2)[43]

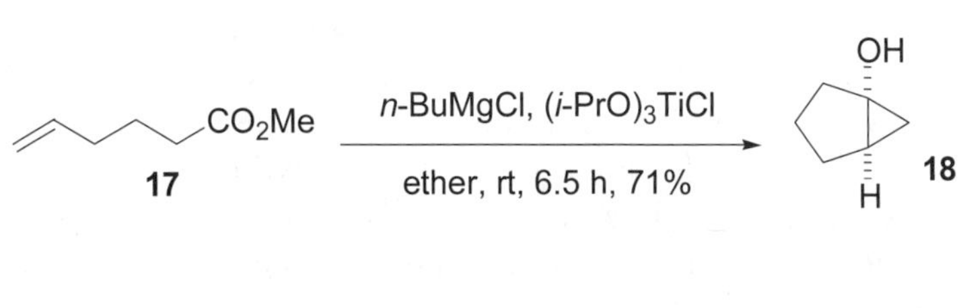

To a 500-mL, round-bottomed flask, equipped with a magnetic stirring bar and rubber septum, is added at room temperature a mixture of 2.0 g (15.6 mmol) of methyl 5-hexenoate (**17**, 11.2 mL, 11.2 mmol) of a 1 M solution of chlorotitanium triisopropoxide in hexane, and 54 mL anhydrous ether under a nitrogen atmosphere. A 1 M solution of *n*-butylmagnesium chloride in ether (52 mL, 52 mmol) is added over a period of 6.5 hr via a syringe pump at room temperature. After the addition is complete, the resulting black reaction mixture is stirred for an additional 20 min. The mixture is cooled to 0 °C with an ice bath, diluted with 50 mL ether and then quenched by slow addition of water (14 mL). The resulting mixture is stirred for an additional 3 h at room temperature. The organic phase is separated and the aqueous phase is extracted with ether (3 × 100 mL). The combined organic extracts are washed with brine (2 × 50 mL), dried over anhydrous magnesium sulfate, filtered, and concentrated under reduced pressure using a rotary evaporator. Purification of the crude product by column chromatography on 40 g silica gel using a gradient of 5% to 10% ether/pentane as the eluent provides 1.09 g (71%) of bicyclo[3.1.0]hexan-1-ol (**18**) as a colorless oil.

1.3.6.2 Intermolecular Kulinkovich cyclopropanation reaction of carboxylic ester with olefin: *trans*-2-benzyl-1-methylcyclopropan-1-ol (56)[43]

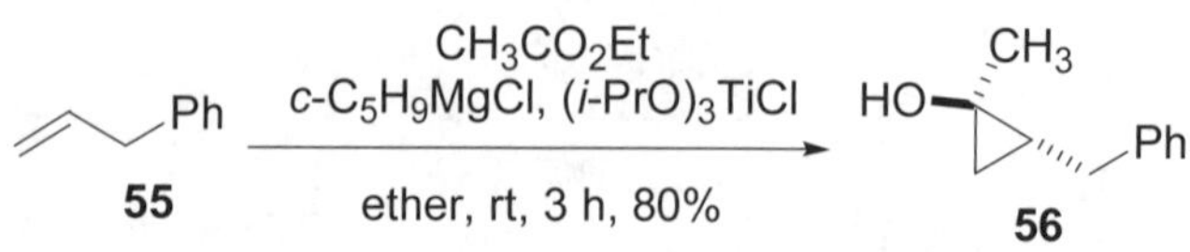

To a 500 mL, round-bottomed flask, equipped with a magnetic stirring bar and rubber septum, is added a mixture of 2.5 g (21 mmol) allylbenzene (**55**, 2 mL, 20 mmol) ethyl acetate, 20 mL of a 1 M solution of chlorotitanium triisopropoxide in hexane, and 160 mL anhydrous tetrahydrofuran (THF). After the mixture has been cooled to 0 °C with an ice bath under a nitrogen atmosphere, a 1 M solution of cyclopentylmagnesium chloride in ether (80 mL, 80 mmol) is added over a period of 2.5 h via a syringe pump. After the addition is complete, the resulting black reaction mixture is stirred for 30 min at 0 °C, then is quenched by the cautious addition of water (15 mL). The resulting mixture is stirred for an additional 1 h at room temperature and filtered through a pad of Celite, which is rinsed thoroughly with ether (4 × 50 mL). The combined filtrate and rinsings are poured into a separatory funnel containing 50 mL water and shaken thoroughly. The organic phase is separated, washed with brine (50 mL), dried over anhydrous magnesium sulfate, filtered, and concentrated under reduced pressure using a rotary evaporator. Purification of the crude product (obtained as a pale yellow oil)

by column chromatography on 80 g silica gel using 1:20 ethyl acetate:hexane as the eluent provides 2.6 g (80%) *trans*-2-benzyl-1-methylcyclopropan-1-ol (**56**) as a colorless oil.

1.3.7 References

1. [R] Sato, F.; Urabe, H.; Okamoto, S. *Pure Appl. Chem.* **1999**, *71*, 1511–1519.
2. [R] (a) Kulinkovich, O. G.; de Meijere, A. *Chem. Rev.* **2000**, *100*, 789–2834. (b) Kulinkovich, O. G. *Pure Appl. Chem.* **2000**, *72*, 1715–1719.
3. [R] Sato, F. Urabe, H.; Okamoto, S. *Synlett* **2000**, 753–775.
4. [R] Sato, F.; Urabe, H.; Okamoto, S. *Chem. Rev.* **2000**, *100*, 2835–2886.
5. [R] Breit, B. *J. Prak.Chem.* **2000**, *342*, 211–214.
6. [R] Eisch, J. J. *J. Organomet. Chem.* **2001**, *617–618*, 148–157.
7. [R] Kulinkovich, O. G. *Chem. Rev.* **2003**, *103*, 2597–2632.
8. [R] Oestreich, M. *Nac. Chem.* **2004**, *52*, 805–808.
9. [R] Tyvorskii, V. I.; Epstein, O. L. *ARKIVOC* **2008**, 1–5.
10. Kulinkovich, O. G.; Sviridov, S. V.; Vasilevski, D. A.; Pritytskaya, T. S. *Zh. Org. Khim.* **1989**, *25*, 2244–2245.
11. Kulinkovich, O. G.; Sviridov, S. V.; Vasilevskii, D. A.; Savchenko, A. I.; Pritytskaya, T. S. *Zh. Org. Khim.* **1991**, *27*, 294–298.
12. Kulinkovich, O. G.; Sviridov, S. V.; Vasilevskii, D. A. *Synthesis* **1991**, 234.
13. Kulinkovich, O. G.; Sorokin, V. L.; Kel'in, A. V. *Zh. Org. Khim.* **1993**, *29*, 66–69.
14. Kulinkovich, O. G.; Savchenko, A. I.; Sviridov, S. V.; Vasilevskii, D. A. *Mendeleev Commun.* **1993**, 230–231.
15. Wu, Y.-D.; Yu, Z.-X. *J. Am. Chem Soc.* **2001**, *123*, 5777–5786.
16. Eisch, J. J.; Adeosun, A. A.; Gitua, J. N. *Eur. J. Org. Chem.* **2003**, 4721–4727.
17. Kulinkovich, O. G.; Kananovich, D. G. *Eur. J. Org. Chem.* **2007**, 2121–2132.
18. Corey, E. J.; Rao, S. A.; Noe, M. C. *J. Am. Chem. Soc.* **1994**, *116*, 9345–9346.
19. Chaplinski, V.; de Meijere, A. *Angew. Chem., Int. Ed. Engl.* **1996**, *35*, 413–414.
20. de Meijere, A.; Williams, C. M.; *Chem. Eur. .* **2002**, *8*, 3789–3801.
21. Gandon, V.; Szymoniak, J. *Chem. Commun.* **2002**, 1308–1309.
22. Casey, C. P.; Strotman, N. A. *J. Am. Chem. Soc.* 126, 1699–1704.
23. Lee, J.; Cha, J. K. *J. Org. Chem.* **1997**, *62*, 1584–1585.
24. Bertus, P.; Menant, C.; Tanguy, C.; Szymoniak, J. *Org. Lett.* **2008**, *10*, 777–780.
25. Lee, J.; Kim, H.; Cha, J. K. *J. Am. Chem. Soc.* **1995**, *117*, 9919–9920.
26. Kasatkin, A.; Sato, F. *Tetrahedron Lett.* **1995**, *36*, 6079–6082.
27. Lee, J.; Kim, H.; Cha, J. K. *J. Am. Chem. Soc.* **1996**, *118*, 4198–4199.
28. Masalov, N.; Feng, W.; Cha, J. K. *Org. Lett.* **2004**, *6*, 2365–2368.
29. Takayanagi, Y.; Yamashita, K.; Yoshida, Y.; Sato, F. *Chem. Commun.* **1996**, 1725–1726.
30. Harada, K.; Urabe, H.; Sato, F. *Tetrahedron Lett.* **1995**, *36*, 3203–3206.
31. Ollero, L.; Mentink, G.; Ruyjes, F. P. J. T.; Speckamp, W. N.; Hiemstra, H. *Org. Lett.* **1999**, *1*, 1331–1334.
32. Mizojiri, R.; Urabe, H.; Sato, F. *Angew. Chem., Int. Ed. Engl.* **1998**, *37*, 2666–2668.
33. Esposito, A.; Taddei, M. *J. Org. Chem.* **2000**, *65*, 9245–9248.
34. Choi, J.-R.; Cho, D.-G.; Roh, K. Y.; Hwang, J.-T.; Ahn, S.; Jang, H. S.; Cho, W.-Y.; Kim, K. W.; Cho, Y.-G.; Kim, J.; Kim, Y.-Z. *J. Med. Chem.* **2004**, *47*, 2864–2869.
35. Lecornué, F.; Ollivier, J. *Org. Biomol. Chem.* **2003**, *1*, 3600–3604.
36. Lecornué, F.; Paugam, R.; Ollivier, J. *Eur. J. Org. Chem.* **2005**, 2589–2598.
37. Baktharaman, B.; Selvakumar, S.; Singh, V. K. *Tetrahedron Lett.* **2005**, *46*, 7527–7529.
38. Kingsbury, J. S.; Corey, E. J. *J. Am. Chem. Soc.* **2005**, *127*, 13813–13815.
39. Faler, C. A.; Joullié, M. M. *Org. Lett.* **2007**, *9*, 1987–1990.
40. Lee, J. C.; Sung, M. J.; Cha, J. K. *J. Am. Chem. Soc.* **2001**, *42*, 11322–11324.
41. Larquetoux, L.; Ouhamou, N.; Chiaroni, A.; Six, Y. *Eur. J. Org. Chem.* **2005**, 4654–4662.
42. Chiaroni, A.; Six, Y. *Org. Biomol. Chem.* **2003**, *1*, 3007–3009.
43. Kim, S.-H.; Sung, M. J.; Jin, K. C. *Org. Synth.* **2003**, *80*, 111–119.

1.4 Simmons–Smith Cyclopropanation

Matthew J. Fuchter

1.4.1 Description

The stereospecific addition of a metal carbenoid (mainly zinc based) to a double bond is known as the Simmons–Smith cyclopropanation.[1–6] It is one of the most powerful methods of converting olefins to cyclopropanes.

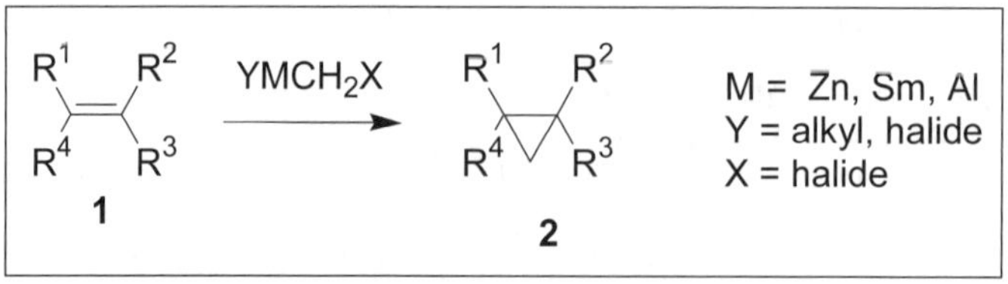

The classical conditions use the zinc-copper couple (Zn–Cu) and diiodomethane to prepare the active carbenoid species, although there are alternative conditions (see 1.4.5.1), the most important being the Furukawa modification, which uses diiodomethane in the presence of diethylzinc. Samarium and aluminium carbenoids are also effective as cyclopropanating reagents (see 1.4.5.2). The reaction can be performed in a range of solvents, however, increased rates are observed in noncoordinating solvents such as dichloromethane. A wide range of alkenes can be used, including simple olefins, α,β-unsaturated systems, and electron-rich alkenes such as enol ethers. In general, the reaction conditions are highly tolerant of most functional groups. The reaction takes place stereospecifically whereby the stereochemistry of the double bond is preserved in the product. If a substituted methylene group is added to the double bond, in the majority of cases a *syn* product is observed. Many polar functional groups (OH, OAc, OMe, NHR) have a directing effect on the cyclopropanation either enabling regioselective reactions for substrates containing multiple double bonds, or stereoselectivity for chiral substrates. Asymmetric Simmons–Smith cyclopropanations are possible by either using stoichiometric chiral auxiliaries, or by the use of chiral catalysts (see 1.4.4.3.3).

1.4.2 Historical Perspective

In 1958, Howard Simmons and Ronald Smith reported a general method for preparing cyclopropane compounds from olefin substrates.[7,8] For example, cyclohexene (**3**) was converted to cyclopropane **4** in moderate yield. The development of this method was built on earlier work by Emschwiller in

1929, who reported the preparation of diiodomethane and its reaction with the zinc-copper couple to form iodomethyl zinc iodide.[9] In fact, even Emschwiller's studies were preceded by extensive work from numerous other chemists, with the reaction of diiodomethane and a variety of metals attracting attention ever since the 1860s. In the field of cyclopropane synthesis, the procedure developed by Simmons and Smith was particularly notable for its broad generality. At the time, the only other viable options were the classical addition of diazo compounds to olefins,[10] or the production and use of dihalocarbenes.[11]

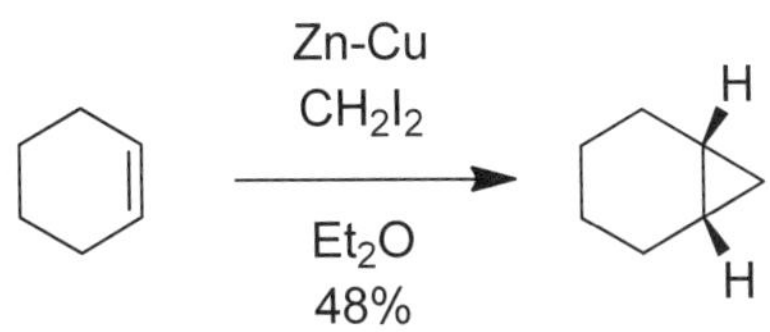

3 4

The only downside of this method was the irreproducibility of generating the active reagent from the zinc-couple and diiodomethane. Furukawa and co-workers provided one solution in 1966 when they showed that a mixture of diethyl zinc and diiodomethane gives very reproducible results in generating the active reagent (see 1.4.5.1 for this and other methods).[12] It has subsequently been shown that other carbenoid species, including samarium[13] and aluminium,[14] are also effective reagents in the cyclopropanation reaction, and in some cases demonstrate interesting chemoselectivity. Nowadays, the Simmons–Smith reaction has developed into one of the most powerful methods of cyclopropane formation available to synthetic chemists. Indeed, in their 2001 review, Charette and Beauchemin documented all the examples of the Simmons–Smith reaction to appear in the literature between 1973 and 1999, and this totaled more than 1500 olefin substrates. Since the 1990s, the major developments in the Simmons–Smith reaction have focused on asymmetric methods to stereoselectively prepare chiral cyclopropanes. One of the most widely used methods in this regard originates from a report by Charette and co-workers in 1994 on the use of chiral dioxaborolane auxiliaries (see 1.4.4.3.2).[15]

1.4.3 Mechanism

Simmons–Smith cyclopropanation proceeds via the addition of a zinc carbenoid (**6**/**8**) to an olefin. There are three classes of reactions, however, that can generate the reactive zinc species, each with it is own mechanistic pathway.[5] The oxidative addition of an activated form of zinc metal into a C–X bond is by far the most widely used method for the formation of **6**

(Pathway A). Indeed, it was this method that Simmons and Smith first used in their seminal study,[7,8] with many methods of zinc activation available (see 1.4.5.1). Furukawa's modification of the reaction uses diethyl zinc, and proceeds via alkyl ligand exchange to give **8** (Pathway B). Finally, Wittig reported the insertion of diazomethane (**7**) into zinc iodide to give **6**,[16] although this method is not widely used.

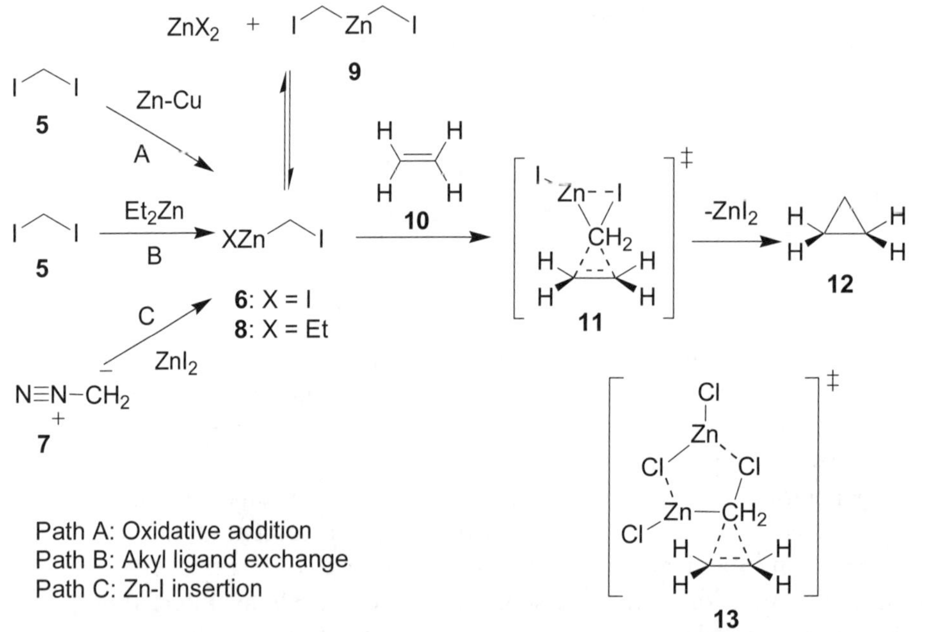

Zinc carbenoids **6** and **8** are in Schlenk equilibria with dialkyl zinc **9** and ZnX_2, although the position of the equilibrium depends on the counter ligand X. Spectroscopic studies have shown that in the case of iodomethyl zinc iodide (**6**), the equilibrium lies strongly in favour of the active species **6**.[17,18] For ethyl-substituted carbenoid **8**, however, Et_2Zn and dialkyl zinc **9** are far more prevalent, and an additional self-destructive pathway is apparent, giving rise to PrZnI.[17] Since this destructive reaction can be competitive with cyclopropanation, in certain cases it may be advantageous to add additional diiodomethane (to convert PrZnI to the active species).

After generation of the active species, the reaction proceeds via concerted addition of the methylene group to the olefin substrate with retention of configuration. This process was postulated to proceed via a three-centre "butterfly-type" transition state **11** on the basis of experimental observations, and numerous theoretical calculations are in agreement with this postulate.[5] However this transition state may not be the favoured reaction pathway under all relevant experimental conditions. For example, theoretical studies on the cyclopropanation of ethylene with chloromethylzinc chloride, in the presence of zinc chloride indicated the five-centered complex **13** to be the kinetically favoured transition state.[19]

1.4.4 Synthetic Utility

1.4.4.1 Substrate Scope

The Simmons–Smith reaction is a broadly applicable procedure and has a wide-ranging functional group tolerance. In general, higher reactivity is observed with electron-rich alkenes, due to their increased nucleophilicity. Enol ethers are just one example of substrates whose cyclopropanation under the reaction conditions is generally facile.[5] Silyl enol ethers are the most widely used in this regard since they are readily available from enolization of the corresponding ketone. For example, substrate **14** was converted to **15** in excellent yield using Furukawa's conditions (see 1.4.5.1).[20]

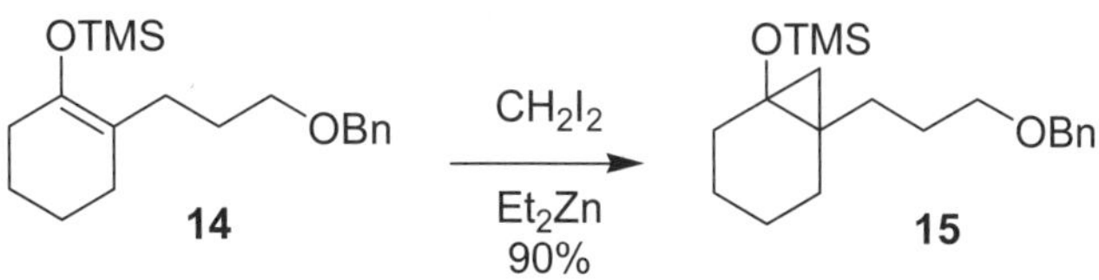

The cyclopropanation of vinyl halides is also effective, with fluoro-, bromo- and iodo-substituted olefins all being suitable substrates. For example, vinyl bromide **16** underwent the reaction in good yield, using Denmark's modification of the reaction (see 1.4.5.1).[21]

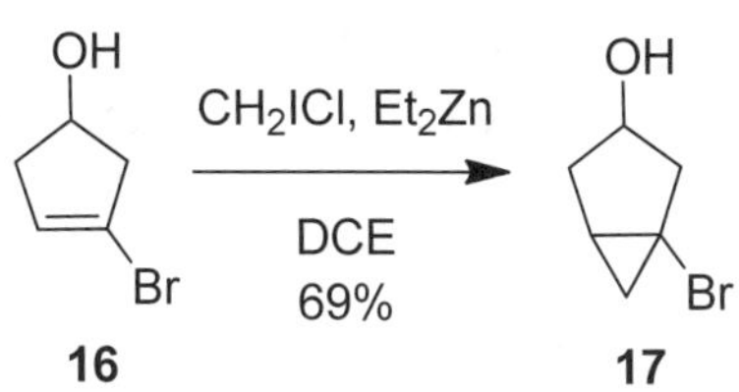

While the classical Simmons–Smith regent (Zn/Cu) is often used, Furukawa's conditions (Et_2Zn/CH_2I_2, see 1.4.5.1) are the preferred choice for less reactive olefin substrates. This is since generation of the active carbenoid under Furukawa's conditions can be performed in noncomplexing solvents, which in turn, leads to a reagent with a higher electrophilicity.[5] Taber and co-worker's synthesis of *trans*-dihydroconfertifolin (**20**) employed such conditions in their endgame strategy, cyclopropanating substrate **18** in excellent yield.[22] The resulting cyclopropane was subjected to catalytic hydrogenolysis to give *trans*-dihydroconfertifoin (**20**).

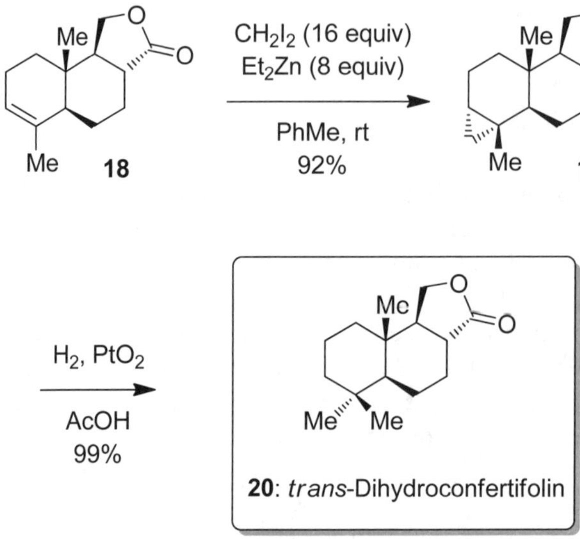

1.4.4.2 Regioselectivity and Stereoselectivity

When more than one double bond is available to react within a substrate, a combination of both steric and electronic effects control the regioselectivity of cyclopropanation.[5] In light of the electrophilic nature of the reagent, highly regioselective Simmons–Smith reactions are observed when one double bond is significantly more nucleophilic than another. For example, the high reactivity of enol ethers can be used to obtain excellent regioselectivity. Winkler et al. capitalized on such high regioselectivity in their conversion of substrate **21**, into cyclopropane **22**, a key derivative in the synthesis of taxol analogs.[23]

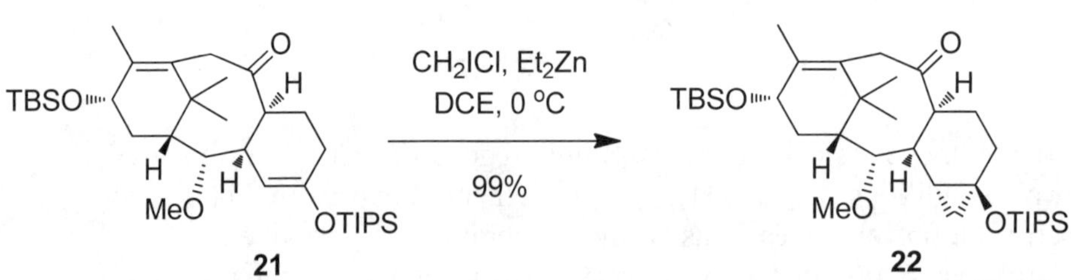

Polar functionality remote from the double bond can also be used to direct both the regioselectivity and stereoselectivity of the Simmons–Smith reaction. Much work has been performed on the directing effect of basic functional groups in the allylic position of cyclic alkenes.[5] Indeed, it was recognized early on that proximal hydroxyl groups can direct the cyclopropanation reaction.[24] In general, cyclopropanation of five-, six-, and seven-membered ring 1-cycloalken-3-ols proceed with high levels of *syn* selectivity.[6] Such diastereoselectivity has been used to good effect, even with more complex substrates, such as the use of derivative **23** by Corey and co-workers.[25] It is interesting that a switch in selectivity is observed with the analogous eight- and nine-membered ring systems. In these cases, high *anti* selectivity is often observed, a fact that can be readily rationalized by the conformation of the ground state molecule.[6]

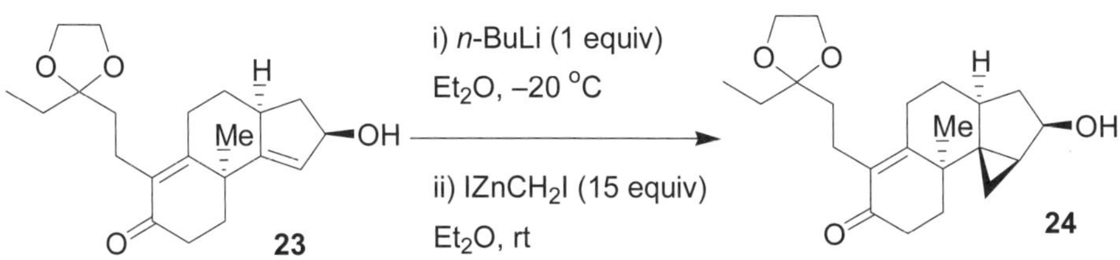

While the hydroxyl group has been most extensively used to direct the Simmons–Smith reaction, other basic groups such as ethers, esters and acetamides are also able to direct the zinc reagent in certain cases.[5] For example, in the case of substrate **25**, the unsubstituted acetamide (R = H) is a better directing group than the hydroxyl group, giving cyclopropane **27** as the sole stereoisomer. Once benzoylated however (**25**, R = Bz), the selectivity is reversed, resulting in the selective formation of compound **26**.[26]

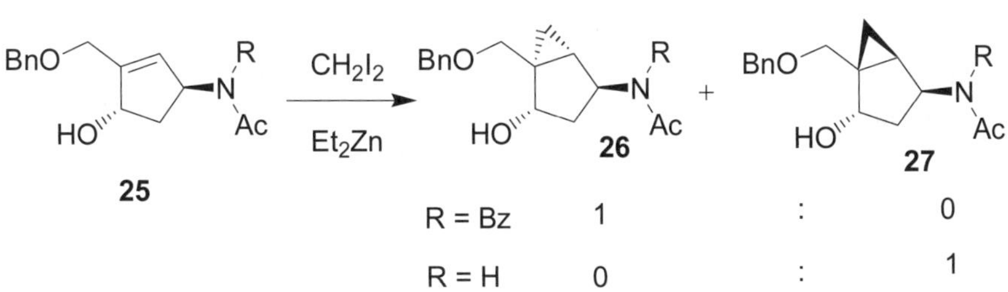

In the absence of a directing group, the cyclopropanation of cyclic olefins is generally under steric control. The stereochemical preference can be predicted from the ground state conformation of the molecule and often high levels of stereocontrol are observed. For example, Paquette and co-workers used a Simmons–Smith reaction in their total synthesis of the secondary marine metabolite (+)-acetoxycrenulide (**30**), whereby high β-face selectivity was observed.[27]

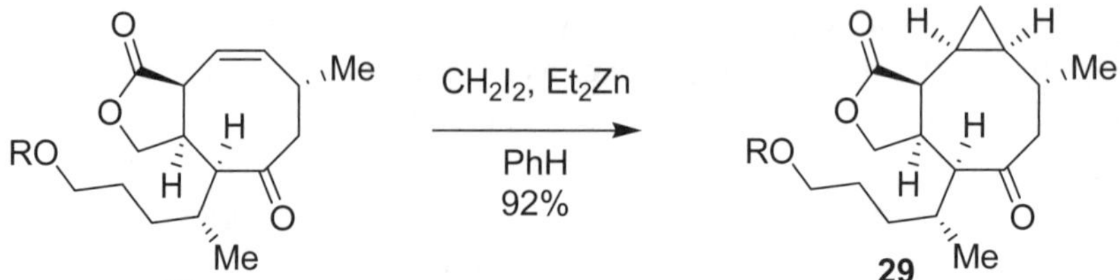

28 **29**

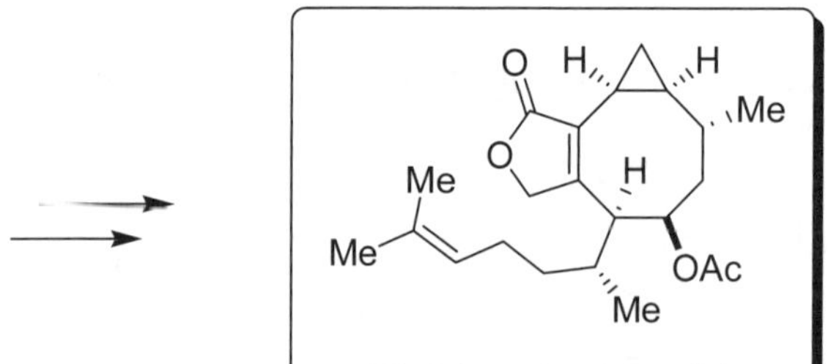

30: (+)-acetoxycrenulide

The stereoselective Simmons–Smith reaction of a chiral acyclic allylic alcohol was first reported by Pereyre and co-workers in 1978.[28] They observed very high *syn* selectivities (> 200:1) when (*Z*)-disubstituted olefins such as **31** were treated under classical Simmons–Smith conditions. The analogous reaction of (*E*)-olefins however was reported to only give modest selectivity (< 2:1). In depth studies by Charette and co-workers however has revealed that the nature zinc carbenoid used in these reactions is key to obtaining high selectivities.[29,6] While in the case of simple (*E*)-disubstituted olefins, the classical Simmons–Smith conditions (Zn–Cu, CH_2I_2, Et_2O) gives only modest selectivity, the use of Furukawa's reagent (Et_2Zn, CH_2I_2) in excess produces a much higher degree of selectivity, especially when dichloromethane is used as the solvent.[29,6] Indeed, these trends are generally maintained for more complex acyclic systems, although it should be stated that stereoelectronic effects also play an important role.[6]

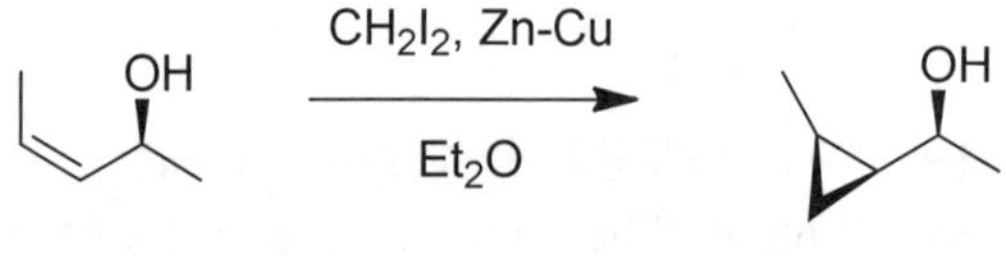

31 **32** *syn*:*anti* >99:1

Impressive regioselectivity and stereoselectivity has been reported for a number of complex acyclic systems, including the synthesis of halicholactone (**35**) by Takemoto and co-workers. In their total synthesis, substrate **33** was converted into the desired product **34** in good yield and total selectivity.[30,31] It is interesting that a hydroxyl group was required to direct the reaction (an acetal being ineffective), and the protecting groups used were crucial to its success.

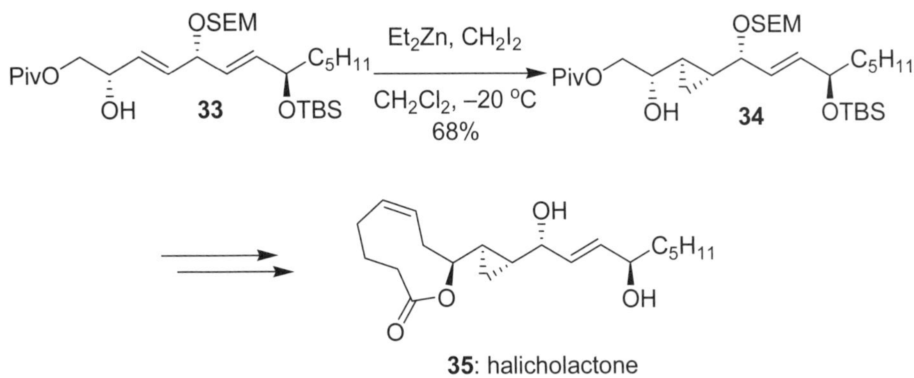

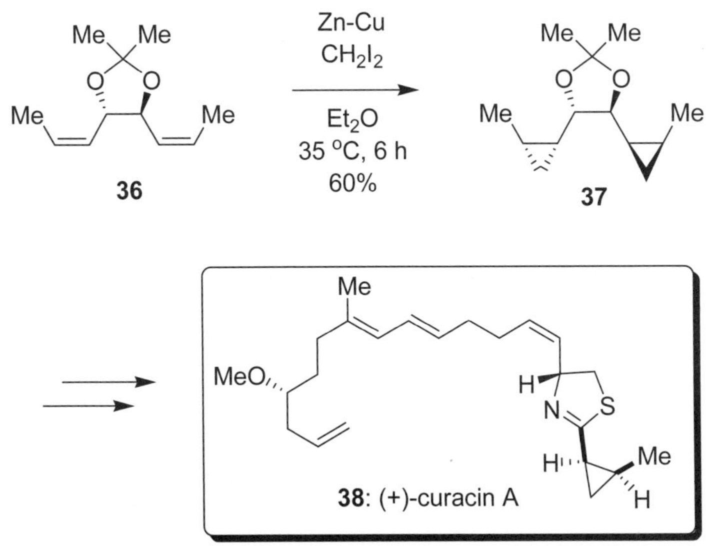

As is the case with cyclic substrates, other basic functional groups are able to effectively direct the reaction in certain cases. An impressive example was reported by Iwasaki and co-workers in their synthesis of the antimitotic agent (+)-curacin A (**38**), whereby an acetal oxygen was used to direct the reaction.[32] In a two-directional approach, substrate **36** was converted into product **37** as a single diastereoisomer in good yield. This compound was subsequently deprotected and subjected to oxidative cleavage to give the desired 2-methylcyclopropane carboxylic acid, which was required to form the thiazoline portion of curacin A. Indeed, such a stereoselective two-directional cyclopropanation of tartrate derived substrates had been previously reported by Barrett and co-workers and used in their total synthesis of FR-900848.[33,34]

The stereocontrol in acyclic chiral olefins in which the basic directing group is not the stereogenic centre is generally quite poor.[6] One exception to this rule however, was reported by Panek et al., who demonstrated that a stereogenic bulky silicon group in the allylic position of acyclic substrates can induce good diastereoselectivities.[35]

In terms of the synthesis of substituted cyclopropane derivatives using either halo- or alkyl-substituted reagents, *endo*:*exo* selectivity of the product is often poor.[5] For example, cyclohexene (**39**) was converted into an approximately 2:1 ratio of **40** and **41** on exposure to bromoform and diethylzinc.[14]

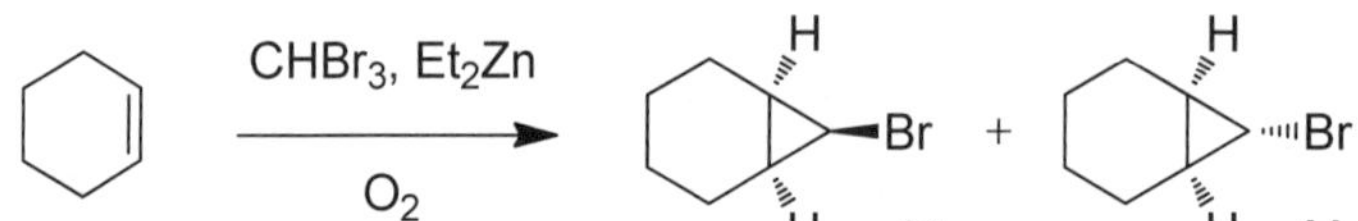

39 **40** **41**

1.9 : 1

The use of directing groups is effective at controlling the relative stereochemistry of the cyclopropane and the directing functionality, however once again, *endo*:*exo* selectivity is usually modest. For example, relatively poor selectivity was observed in the conversion of substrate **42** into derivative **44**, although the samarium-derived reagent (see 1.4.5.2) provided slightly better results.[36]

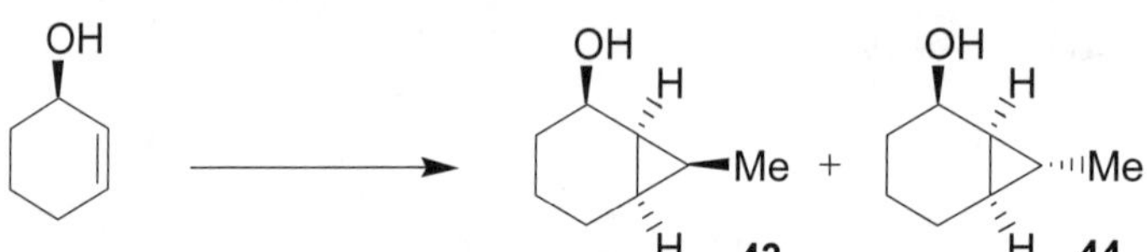

42 **43** **44**

	43		44
Sm, $MeCHI_2$ (> 99%)	17	:	83
Et_2Zn, $MeCHI_2$ (64%)	37	:	63
Zn-Cu, $MeCHI_2$ (84%)	28	:	72

As is usually the case, however, specific results are dependent on the substrate in question, and the selectivity observed a case of both steric and electronic factors. For example, in their enantioselective synthesis of (–)-pinidine, Momose et al. reported a highly diastereoselective reaction employing the reagent derived from 1,1-diiodoethane and diethyl zinc.[37] Substrate **45** was selectively converted to stereoisomer **46** in excellent yield.

It is important to note, however, that the nitrogen-protecting group was crucial to the success of this reaction.

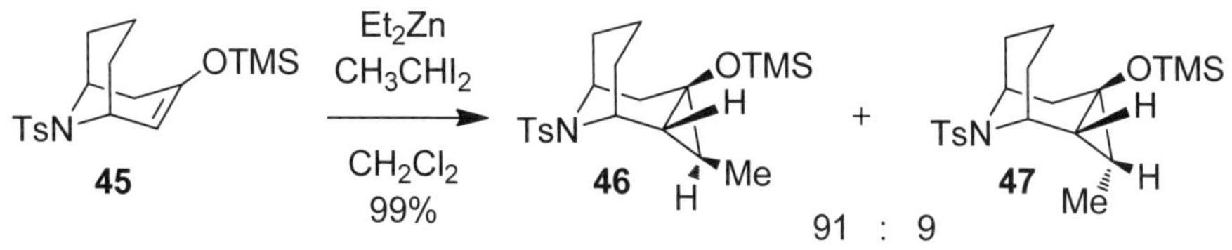

1.4.4.3 Asymmetric Simmons–Smith Reactions

1.4.4.3.1 Chiral Auxiliaries

The use of a chiral auxiliary is one strategy of preparing enantiomerically pure cyclopropyl derivatives after cleavage of the auxiliary. There are several classes of auxiliary that have been used for different substrates, and these can be divided into chiral allylic ethers, acetals, α,β-unsaturated carbonyl derivatives, enamines, and enol ethers.[6] In the case of chiral allylic ethers, carbohydrate derivatives have proved particularly effective, inducing excellent levels of diastereoselectivity in the reaction. For example, substrates of the general class **48**, were cyclopropanated asymmetrically in excellent yield and very high levels of diastereoselectivity.[38] It is believed that the chiral auxiliary acts as a bidentate ligand for the zinc reagent, and indeed, structurally simplified auxiliaries are almost as effective.[39]

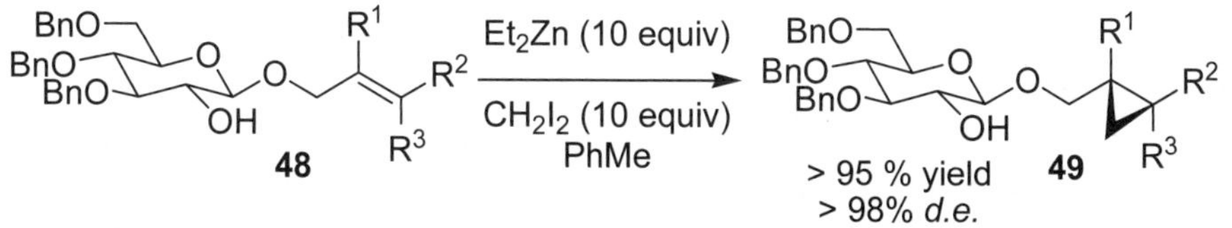

A number of chiral acetal derivatives have also proved effective in asymmetric cyclopropanation reactions, with auxiliaries based on tartaric acid proving to be particularly useful.[6] In the case of cyclic α,β-unsaturated compounds, di-*O*-benylthreitol derivatives (see **51**) undergo efficient and diastereoselective Simmons–Smith reactions to give the cyclopropanated products **53**.[40]

The configuration of the products can be rationalized by model **52**, whereby coordination of the zinc reagent occurs to the least sterically hindered dioxolane oxygen atom proximal to the olefin.

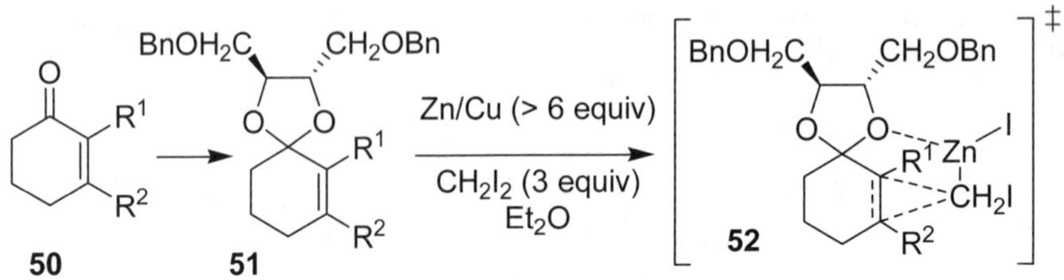

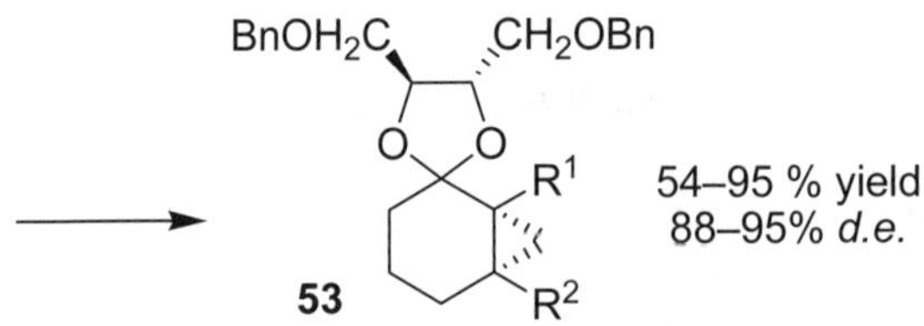

1.4.4.3.2 Stoichiometric Chiral Ligands

As early as 1968, the addition of chiral ligands to the reaction was performed in an attempt to induce asymmetry.[41] Despite several early attempts, however,[6] only modest enantioselectivities were obtained. Fujisawa and co-workers reported the first moderate levels of asymmetric induction by adding stoichiometric amounts of diethyl tartrate to the Furukawa's Simmons–Smith conditions.[42] In 1994, however, a major breakthrough was reported by Charette and co-workers, who demonstrated that bifunctional non-racemic chiral ligands induced good levels of enantioselectivity in the reaction.[15] These ligands contained both acidic and basic sites that allowed simultaneous chelation of the acidic halomethylzinc reagent and the basic zinc alkoxide. In particular, dioxaborolane **55**, prepared from *N,N,N′,N′*-tetramethyltartaric acid diamide and butyl boronic acid, was a particularly useful chiral controller. This stoichiometric ligand shows good substrate scope, and the products (**56**) are usually isolated in good yield and high enantiomeric purities.[6]

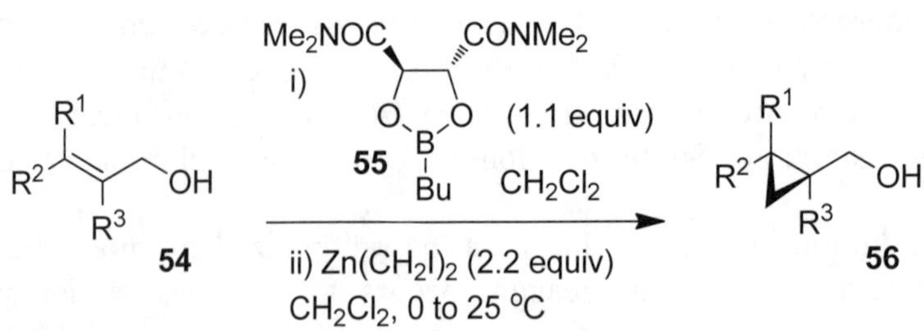

Charette's procedure is so efficient that it has been used in numerous syntheses of complex molecules. For example, en route to (+)-ambruticin,

Jacobsen and co-workers used the Charette ligand to mediate asymmetry in the cyclopropanation of substrate **57**.[43] This reaction is particularly notable since a substituted cyclopropane is installed with high diastereoselectivity.

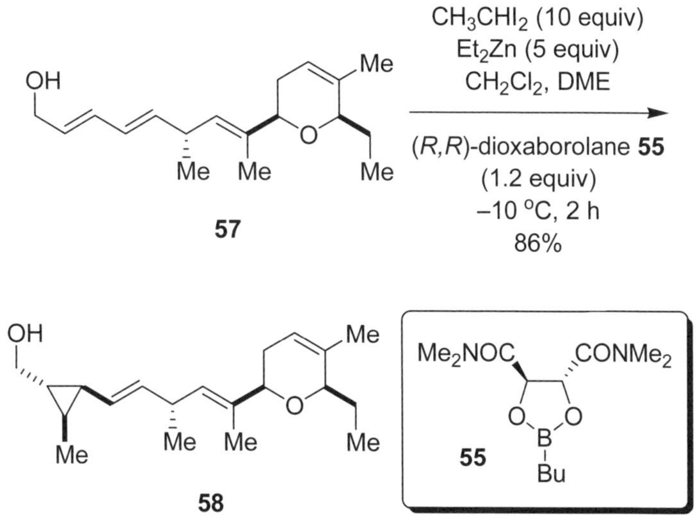

Although other stoichiometric mediators of the Simmons–Smith reaction have been reported, such as biaryl alcohols[44] and dipeptides,[45] none have to date shown such broad applicability as the Charette ligand.

1.4.4.3.3 Sub-Stoichiometric Chiral Ligands

Several examples have been reported of asymmetric Simmons–Smith reactions whereby the chiral, nonracemic ligand is added in sub-stoichiometric quantities. Kobayashi and co-workers were the first to report such a system, and showed that catalytic quantities of disulfonamide ligand **60** could result in isolation of the product (for example **62**) in good enantioselectivity.[46] This method has broad applicability and results in consistently high enantioselectivities for a wide range of substrates.[6] Denmark and co-workers subsequently reported an in-depth study of this reaction and highlighted that the rate and selectivity of the catalytic cyclopropanation greatly depends on the order of addition of the reagents.[18] Preformation of the ethylzinc alkoxide and bis(iodomethyl)zinc was crucial and the reaction was shown to be autocatalytic due to the generation of zinc iodide. These and other observations led to the proposed transition state assembly **61**, in which three zinc atoms are involved in the methylene delivery process.

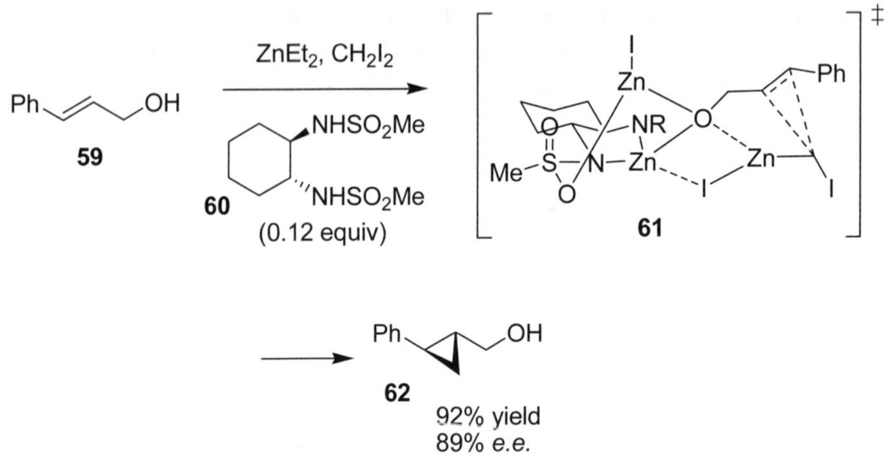

Charette and co-workers reported a chiral Lewis acid-catalysed Simmons–Smith reaction, using a titanium TADDOL complex, although in general this system shows limited substrate scope compared to the Kobayashi system.[47]

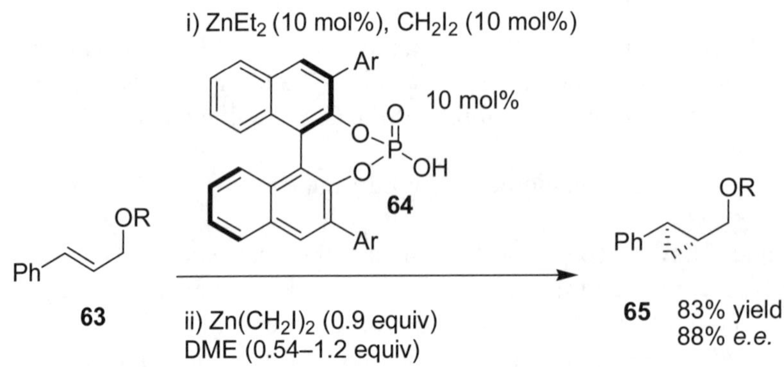

More recently, however, the group have developed chiral zinc phosphate reagents as mediators of the asymmetric Simmons–Smith reaction. A chiral, nonracemic zinc reagent derived from phosphoric acid **64** was shown to be effective for the enantioselective cyclopropanation of substrates **63** when used in stoichiometric quantities. After significant optimization, it was shown that modified conditions could allow the use of just 10 mol % of **64**, resulting in the production of the product **65** with good levels of enantiomeric excess.[48] Charette and co-workers have extended this work, developing alternative ligands such as a TADDOL derived phosphoric acid.[49]

1.4.5 Variations and Improvements

1.4.5.1 Methods of generating active species

There are three classes of reaction that can generate the reactive haloalkylzinc species: (1) Oxidative addition of zinc metal into a carbon-halogen bond, (2) alkyl group exchange between and organozinc reagent and a dihaloalkane, and (3) the insertion of a diazoalkane into a zinc iodide bond.

Class 1, oxidative addition

The oxidative addition of activated zinc metal into a carbon-halogen bond is still one of the most widely used methods for the cyclopropanation of simple olefins. Indeed, it is this method that Simmons and Smith used in their seminal publications, whereby they favoured the use of a zinc-copper couple.[7,8] While their procedure involved heating a mixture of zinc dust and cupric oxide under a hydrogen atmosphere, this has been replaced by more convenient methods, including treatment of zinc powder with a cupric sulphate solution, treatment of zinc dust with a hot solution of cupric acetate in acetic acid, and mixing zinc dust with cuprous chloride under nitrogen.[5] While these procedures have remained the mainstay of Simmons–Smith reactions over the past 25 years, related activation procedures exist, including the use of the zinc-silver couple.[50] Despite the wide use of these procedures, irreproducible results are occasionally observed as a result of inconsistencies in forming the active zinc reagent. The other major disadvantage is that an ethereal solvent must be used for the activation process. Under such conditions the electrophilicity of the active zinc reagent is reduced, thus lowering its reactivity. Also, the majority of stereoselective Simmons–Smith reactions (see 1.4.4.3) require noncomplexing solvents to maximize stereoselectivity and so this method is not applicable.[5]

Class 2, alkyl group exchange

Many of the downsides highlighted above were overcome by Furukawa and co-workers, who showed that a mixture of diethyl zinc and diiodomethane gives very reproducible results in generating the active reagent via alkyl group exchange.[12] This procedure can be performed in non-coordinating solvent and thus is highly useful in stereoselective Simmons–Smith reactions (see 1.4.4.3). Subsequent work by Denmark and co-workers showed that in certain cases (especially for deactivated olefin substrates) it is advantageous to use bis(chloromethyl zinc as the active species, which is prepared from $ZnEt_2$ and CH_2ICl.[51] Another underused method for preparing $IZnCH_2I$ involves the treatment of EtZnI with CH_2I_2.[52] This method is particularly useful on large scale because it avoids the use of pyrophoric Et_2Zn. More recently, several other highly effective reagents have been reported for use in Simmons–Smith reagents, prepared via alkyl group exchange. Iodomethylzinc trifluoroacetate, prepared by mixing tifluoroacetic acid, diethyl zinc and diiodomethane is a very effective cyclopropanating reagent.[53] Likewise, substituted iodomethylzinc aryloxides (for example 2,4,6-$Cl_3C_6H_2OZnCH_2I$) are very useful in the cyclopropanation of unfunctionalized olefins.[54]

Class 3, insertion of a diazoalkane into a zinc iodide bond

Despite the fact that diazoalkane derived reagents were some of the first examined by Wittig and co-workers for the Simmons–Smith reaction,[10] and the huge growth of diazo compound usage in other cyclopropanating methods, this reagent preparation procedure has only appeared sporadically in the literature for Simmons–Smith reactions. A very recent publication by Charette and co-workers may draw more attention to this method however.[55] In an attempt to enantioselectively prepare aryl-substituted cyclopropanes, they showed that exposure of allylic alcohol substrates to a reagent formed from EtZnI, phenyldiazomethane, and their chiral ligand (**55**, see 1.4.4.3.2) resulted in the formation of the product in good yield and excellent diastereoselectivity and enantioselectivity.[55] Of particular note, however, was the fact that consideration of the mechanism led the team to consider the possibility of a Simmons–Smith reaction catalytic in zinc. Indeed, they found that exposure of nonracemic chiral substrate **66** to just 5 mol % of zinc iodide along with stoichiometric NaH and excess phenyldiazomethane, resulted in the formation of product **67** in excellent yield, good diastereoselectivity and excellent enantioselectivity. This is the first example of an asymmetric cyclopropanation catalytic in a zinc salt.

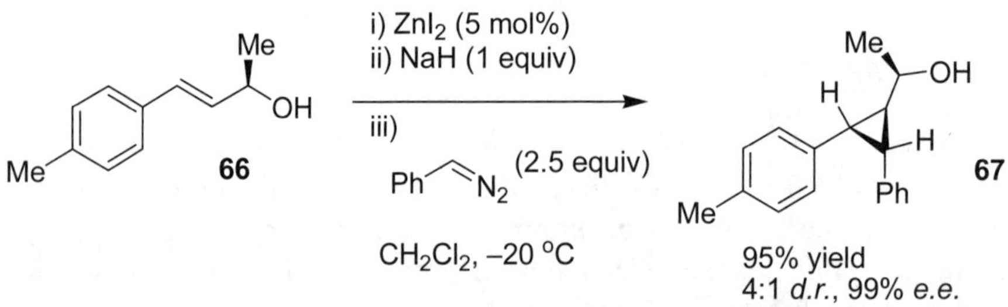

Mechanistically the reaction is hypothesized to proceed via reaction of zinc iodide with phenyldiazomethane to form a zinc carbenoid, which in turn reacts with the sodium alkoxide formed in situ (from the alcohol and

NaH) to produce the cyclopropanated product, regenerating the zinc iodide salt.[55]

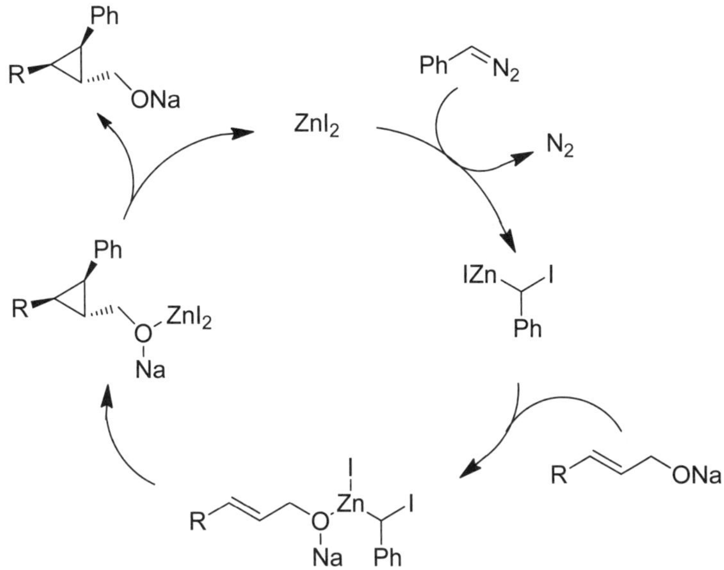

1.4.5.2 Other Metal Carbenoids: Samarium and Aluminium

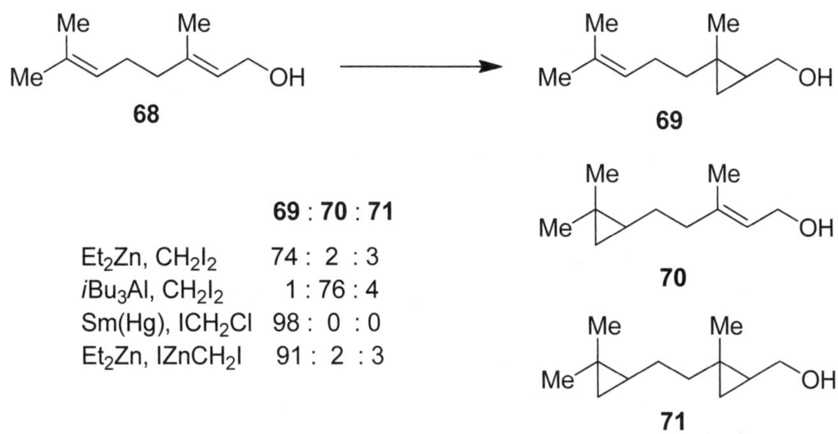

	69 : **70** : **71**
Et_2Zn, CH_2I_2	74 : 2 : 3
iBu_3Al, CH_2I_2	1 : 76 : 4
Sm(Hg), ICH_2Cl	98 : 0 : 0
Et_2Zn, $IZnCH_2I$	91 : 2 : 3

In addition to zinc-based carbenoids, other potential active agents of the general structure "MCH_2X" have been proposed. For example in 1985, Yamamoto and co-workers described the preparation and use of aluminium based carbenoids (R_2AlCH_2I).[14] Subsequently, in 1987 Molander and co-workers reported the use of a samarium/mercury amalgam and CH_2I_2 to generate samarium carbenoids.[13] While these species are less well characterized than their zinc counterparts and their use has not been so widely adopted, they do show some interesting chemoselectivity. This is clearly demonstrated in the cyclopropanation of geraniol (**68**). The allylic alcohol functionality is selectively cyclopropanated (see **69**) in the presence of the isolated olefin for the zinc and samarium derived reagents, whereas it is the terminal double bond that selectively reacts (see **70**) in the presence of the aluminium carbenoid.[6] It is interesting that if the alcohol is protected as a benzyl ether, all three reagents cyclopropanate the allylic position.

Other interesting selectivities are observed in the stereoselective cyclopropanation of acyclic chiral nonracemic allylic alcohols. For example, cyclopropanation of substrate **72** gave the *syn* isomer **73** as the major product in the case of the zinc carbenoid and the *anti* isomer **74** in the case of the samarium reagent.[6]

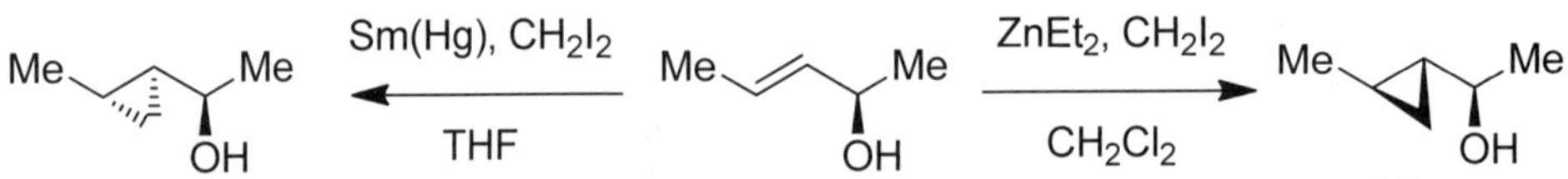

74 **72** **73**

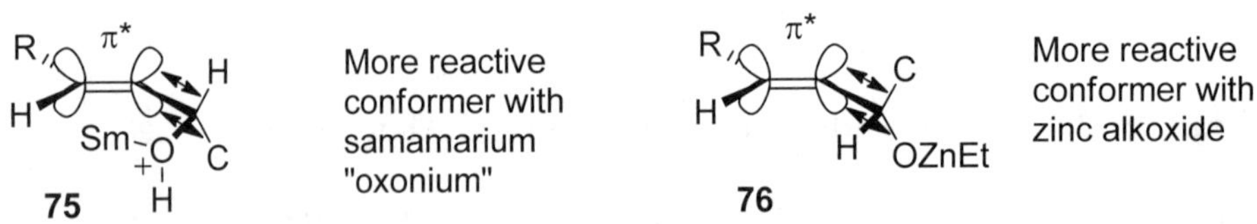

As stated in section 1.4.4.2, a variety of factors, including substrate ratios, solvent, and stereoelectronic effects, play important roles in the selectivity of these reactions, however, in general, the stereochemical outcome can be qualitatively predicted by assuming an oxygen group-assisted delivery of the methylene group from a conformation that minimizes A(1,3) strain.[6] The fact that the samarium reagent gives the *anti* isomer as the major product for substrate **72** suggests a different conformer is involved in the cyclopropanation reaction. One possibility is that deprotonation of the alcohol does not occur with the less basic samarium reagent, and the most reactive conformer is, therefore, the one in which the C–O(H)Sm is orthogonal to the π system (see **75**) to maximize the nucleophilicity of the alkene.[6] Delivery of the methylene group from the face away from the alkyl group would then lead to the *anti* isomer. This is in contrast to the proposed favoured conformer of the zinc alkoxide (see **76**).

1.4.6 Experimental

1.4.6.1 Standard conditions

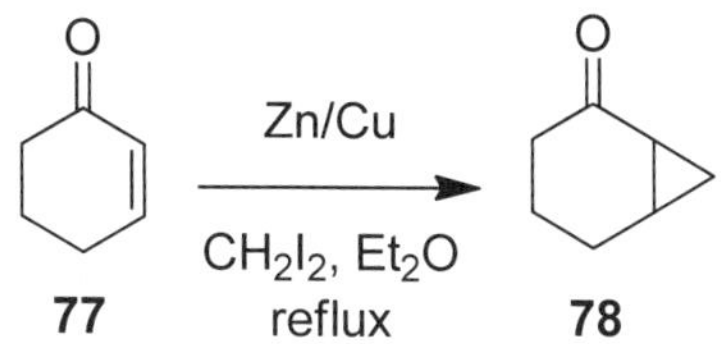

Bicyclo[4.1.0]heptan-2-one (78)[56]

Cupric acetate monohydrate (0.16 g, 0.8 mmol) was dissolved in hot glacial acetic acid (5 mL). Zinc powder (2.8 g, 42.8 mmol) was added to this stirred solution, and after 30–60 s the green colouration disappeared and metallic red copper was deposited on the zinc. The supernatant liquid was decanted and replaced by fresh acetic acid (5 mL). The suspension was stirred, and then the supernatant liquid was once again decanted and replaced by Et_2O (10 mL). The couple was washed in the same fashion with Et_2O (3 × 10 mL). Finally, the couple was covered with Et_2O (20 mL). A few drops of CH_2I_2 was added, and an exothermic reaction occurred. A mixture of cyclohexen-2-one (**77**, 0.96 g, 0.01 mol) and CH_2I_2 (7.5 g, 28 mmol) was then added dropwise, inducing a gentle reflux for 30 min. to 1 h. The mixture was then heated to reflux for 36 hours, during which time a white precipitate appeared. After cooling, H_2O (2 mL) was added dropwise, and the mixture was separated by centrifugation. The ether phase was decanted and washed with 10% aqueous HCl and then three times with H_2O. The solution was dried over Na_2SO_4, filtered, and the solvent removed in vacuo. This gave bicyclo[4.1.0]heptan-2-one (**78**, 1.0 g, 90%) as a colorless liquid.

1.4.6.2 Furukawa modification

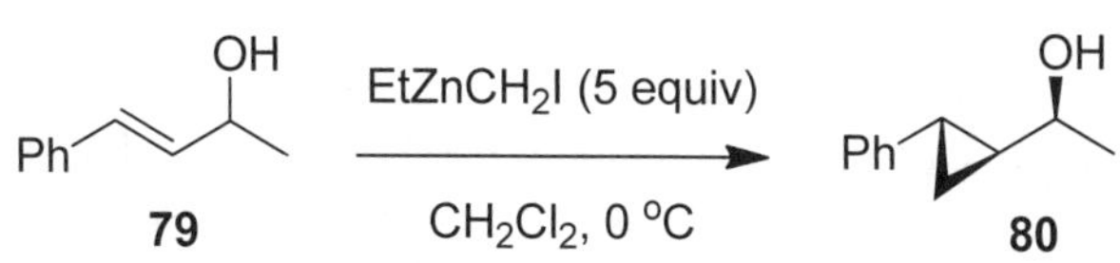

(*αR,1R,2R*)-α-Methyl-2-phenylcyclopropanemethanol (80)[29]

To a solution of alcohol **79** (296 mg, 2.0 mmol) in anhydrous CH_2Cl_2 (20 mL) at –10 °C was added dropwise diethylzinc (1.0 mL, 10 mmol) followed by CH_2I_2 (810 μL, 10 mmol). The mixture was then allowed to warm to room temperature over 3 h. Saturated aqueous NH_4Cl (10 mL) was added, and the mixture diluted with Et_2O (80 mL) and 10% aqueous HCl (10 mL). The organic layer was successively washed with saturated aqueous Na_2SO_3 (20 mL), saturated aqueous $NaHCO_3$ (20 mL), and brine (20 mL). The organic layer was dried over anhydrous $MgSO_4$, and filtered; the solvent was removed in vacuo. Silica gel chromatography (EtOAc:hexane, 15:85) gave the *syn* product **80** (280 mg, 86%) as the major isomer. The less polar anti isomer (40 mg, 12%) was also isolated.

1.4.6.3 Asymmetric Simmons–Smith Using the Charette Auxiliary

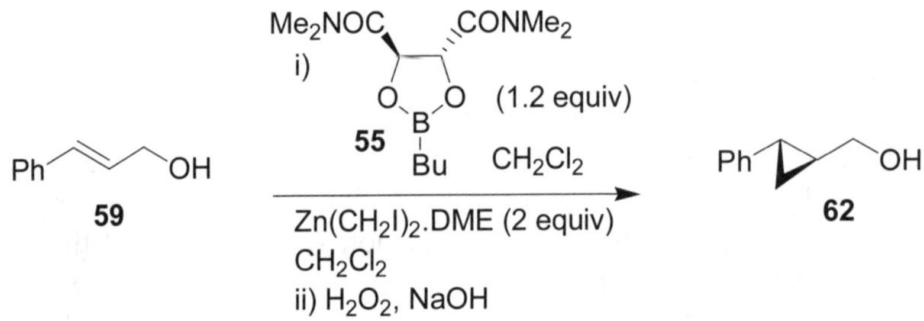

(+)-(*1S,2S*)-2-Phenylcyclopropanemethanol (62)[57]

To a solution of dry DME (1.60 mL, 14.0 mmol) in anhydrous CH_2Cl_2 (45 mL) cooled at –10 °C (internal temperature) was added diethylzinc (1.50 mL, 14.9 mmol). Then to this mixture was added CH_2I_2 (2.40 mL, 29.8 mmol) over 15–20 min while maintaining an internal temperature between –8 and –12 °C. After the addition, the resulting clear solution was stirred for an additional 10 min at –10 °C. A solution of dioxaborolane **55** (2.41 g, 8.94 mmol) in anhydrous CH_2Cl_2 (10 mL) was then added via cannula over 5 min, followed by immediate addition of cinnamyl alcohol (**59**, 1.00 g, 7.45 mmol) in anhydrous CH_2Cl_2 (10 mL) via cannula over a further 5 min, maintaining the internal temperature below –5 °C. The mixture was then allowed to warm to room temperature and stirred for 8 hours at this temperature. The reaction was quenched with saturated aqueous NH_4Cl (10 mL) and 10% aqueous HCl solution (10 mL). The mixture was diluted with Et_2O (60 mL), and the phases were separated. The reaction flask was further washed with Et_2O (15 mL) and 10% aqueous HCl solution, and these washings were combined with the extracts. The aqueous layer was further extracted with Et_2O (20 mL). A solution of 2 N aqueous NaOH (60 mL) and 30% aqueous H_2O_2 (10 mL) was added in one portion to the combined organic extracts. The biphasic mixture was stirred vigorously for 5 min. The two layers were the separated, and the organic layer was washed successively with 10% aqueous HCl solution (50 mL), saturated aqueous Na_2SO_3 (50 mL), saturated aqueous $NaHCO_3$ (50 mL), and brine (50 mL). The organic layer was dried over anhydrous $MgSO_4$, and filtered; the solvent removed in vacuo. Further drying of the product in vacuo was performed overnight to remove residual *n*-butanol from the oxidative workup. The product **62** was purified by Kugelrohr distillation (90 °C, 0.8 mm Hg) to after alcohol **62** (1.05 g, 95%, 94% *ee*. as determined by GC analysis of a chiral nonracemic trifluoroacetate ester derivative) as a colorless oil.

1.4.7 References

1. [R] Simmons, H. E.; Cairns, T. L.; Vladuchick, S. A.; Hoiness, C. M. *Org. React.* **1973**, *20*, 1–131.
2. [R] Helquist, P. M. In *Comprehensive Organic Synthesis*; Vol. 4, Trost, B. M., Fleming, I., Eds.; Pergamon Press: Oxford, 1991; pp 951–998.
3. [R] Charette, A. B. *Organozinc Reagents* **1999**, 263–285.
4. [R] Boche, G.; Lohrenz, J. C. W. *Chem. Rev.* **2001**, *101*, 697–756.
5. [R] Charette, A. B.; Beauchemin, A. *Org. React.* **2001**, *58*, 1–415.
6. [R] Lebel, H.; Marcoux, J.-F.; Molinaro, C.; Charette, A. B. *Chem. Rev.* **2003**, *103*, 977–1050.
7. Simmons, H. E.; Smith, R. D. *J. Am. Chem. Soc.* **1958**, *80*, 5323–5324.
8. Simmons, H. E.; Smith, R. D. *J. Am. Chem. Soc.* **1959**, *81*, 4256–4264.
9. Emschwiller, G. *Compt. rend.* **1929**, *188*, 1555–1557.
10. Huisgen, R. *Angew, Chem.* **1955**, *67*, 439–463.
11. Doering, W. von E.; Hoffmann, A. K. *J. Am. Chem. Soc.* **1954**, *76*, 6162–6165.
12. Furukawa, J.; Kawabata, N.; Nishimura, J. *Tetrahedron Lett.* **1966**, 3353–3354.
13. Molander, G. A.; Etter, J. B. *J. Org. Chem.* **1987**, *52*, 3942–3944.
14. Maruoka, K.; Fukutani, Y.; Yamamoto, H. *J. Org. Chem.* **1985**, *50*, 4412–4414.
15. Charette, A. B.; Juteau, H. *J. Am. Chem. Soc.* **1994**, *116*, 2651–2652.
16. Wittig, G.; Schwarzenbach, K. *Angew. Chem.* **1959**, *71*, 652–652.
17. Charette, A. B.; Marcoux, J.-F. *J. Am. Chem. Soc.* **1996**, *118*, 4539–4549.
18. Denmark, S. E.; O'Connor, S. P. *J. Org. Chem.* **1997**, *62*, 3390–3401 and references therein.
19. Nakamura, E.; Hirai, A.; Nakamura, M. *J. Am. Chem. Soc.* **1998**, *120*, 5844–5845.
20. Morrison, V.; Barnier, J. P.; Blanco, L. *Tetrahedron* **1998**, *54*, 7749–7764.
21. Yong, W.; Vandewalle, M. *Synlett* **1996**, 911–912.
22. Taber, D. F.; Nakajima, K.; Xu, M.; Reingold, A. L. *J. Org. Chem.* **2002**, *67*, 4501–4504.
23. Winkler,J. D.; Bhattacharya, S. K.; Batey, R. A. *Tetrahedron Lett.* **1996**, *37*, 8069–8072.
24. Winstein, S.; Sonnenberg, J.; De Vries, L. *J. Am. Chem. Soc.* **1959**, *81*, 6523–6524.
25. Corey, E. J.; Virgil, S. C. *J. Am. Chem. Soc.* **1990**, *112*, 6429–6431.
26. Russ, P.; Ezzitouni, A.; Marquez, V. E. *Tetrahedron Lett.* **1997**, *38*, 723–726.
27. Paquette, L. A.; Wang, T.-Z.; Pinard, E. *J. Am. Chem. Soc.* **1995**, *117*, 1455–1456.
28. Ratier, M.; Castaing, M.; Godet, J.-Y.; Pereyre, M. *J. Chem. Res. (M)* **1978**, 2309–2318.
29. Charette, A. B.; Lebel, H. *J. Org. Chem.* **1995**, *60*, 2966–2967
30. Takemoto, Y.; Baba, Y.; Saha, G.; Nakao, S.; Iwata, C.; Tanaka, T.; Ibuka, T. *Tetrahedron Lett.* **2000**, *41*, 3653–3656.
31. Baba, Y.; Saha, G.; Nakao, S.; Iwata, C.; Tanaka, T.; Ibuka, T.; Ohishi, H.; Takemoto, Y. *J. Org. Chem.* **2001**, *66*, 81–88.
32. Onoda, T.; Shirai, R.; Koiso, Y.; Iwasaki, S. *Tetrahedron Lett.* **1996**, *37*, 4397–4400.
33. Barrett, A. G. M.; Kasdorf, K.; Williams, D. J. *J. Chem. Soc., Chem. Commun.* **1994**, 1781–1782.
34. Barrett, A. G. M.; Doubleday, W. W.; Kasdorf, K.; Tustin, G. J. *J. Org. Chem.* **1996**, *61*, 3280–3288.
35. Panek, J. S.; Garbaccio, R. M.; Jain, N. F. *Tetrahedron Lett.* **1994**, *35*, 6453–6456.

36. (a) Molander, G. A.; Etter, J. B. *J. Org. Chem.* **1989**, *54*, 3525–3532. (b) Kawabata, N.; Nakagawa, T.; Nakao, T.; Yamashita, S. *J. Org. Chem.* **1977**, *42*, 3031–3035. (c) Friedrich, E. C.; Biresaw, G. *J. Org. Chem.* **1982**, *47*, 2426–2429.
37. Momose, T.; Nishio, T.; Kirihara, M. *Tetrahedron Lett.* **1996**, *37*, 4987–4990.
38. Charette, A. B.; Côté, B.; Marcoux, J. F. *J. Am. Chem. Soc.* **1991**, *113*, 8166–8167.
39. Charette, A. B.; Marcoux, J.-F. *Tetrahedron Lett.* **1993**, *34*, 7157–7160.
40. Mash, E. A.; Nelson, K. A. *Tetrahedron* **1987**, *43*, 679–692.
41. Sawada, S.; Takehana, K.; Inouye, Y. *J. Org. Chem.* **1968**, *33*, 1767–1770.
42. Ukaji, Y.; Nishimura, M.; Fujisawa, T. *Chem. Lett.* **1992**, 61–64.
43. Liu, P.; Jacobsen, E. N. *J. Am. Chem. Soc.* **2001**, *123*, 10772–10773.
44. Kitajima, H.; Aoki, Y.; Ito, K.; Katsuki, T. *Chem. Lett.* **1995**, 1113–1114.
45. Long, J.; Yuan, Y.; Shi, Y. *J. Am. Chem. Soc.* **2003**, *125*, 13632–13633.
46. Takahashi, H.; Yoshioka, M.; Ohno, M.; Kobayashi, S. *Tetrahedron Lett.* **1992**, *33*, 2575–2578.
47. Charette, A. B.; Brochu, C. *J. Am. Chem. Soc.* **1995**, *117*, 11367–11368.
48. Lacasse, M.-C.; Poulard, C.; Charette, A. B. *J. Am. Chem. Soc.* **2005**, *127*, 12440–12441.
49. Voituriez, A.; Charette, A. B.; *Adv. Synth. Catal.* **2006**, *348*, 2363–2370.
50. Denis, J. M.; Girard, C.; Conia, J. M. *Synthesis* **1972**, 549–551.
51. Denmark, S. E.; Edwards, J. P. *J. Org. Chem.* **1991**, *56*, 6974–6981.
52. Sawada, S.; Inouye, Y. *Bull. Chem. Soc. Jpn.* **1969**, *42*, 2669–2672.
53. Yang, Z.; Lorenz, J. C.; Shi, Y. *Tetrahedron Lett.* **1998**, *39*, 8621–8624.
54. Charette, A. B.; Francoeur, S.; Martel, J.; Wilb, N. *Angew. Chem., Int. Ed.* **2000**, *39*, 4539–4542.
55. Goudreau, S.; Charette, A. B. *J. Am. Chem. Soc.* **2009**, *131*, 15633–15635.
56. Limasset, J.-C.; Amice, P.; Conia, J.-M. *Bull. Soc. Chim. Fr.* **1969**, 3981–3990.
57. Charette, A. B.; Juteau, H.; Lebel, H.; Molinaro, C. *J. Am. Chem. Soc.* **1998**, *120*, 11943–11952.

Chapter 2 Four-Membered Carbocycles

2.1 Staudinger Ketene–Imine Cycloaddition

Stephen W. Wright

2.1.1 Description

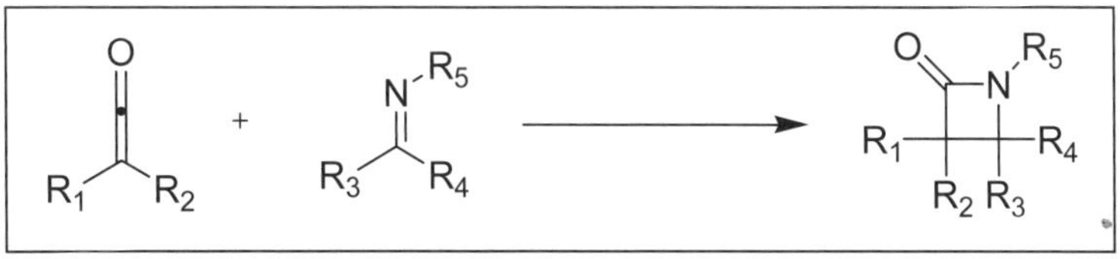

The Staudinger ketene cycloaddition is the nonphotochemical [2 + 2] cycloaddition of a ketene and an imine to form a β-lactam. Related ketene cycloaddition reactions include the cycloaddition of a keten with an olefin to afford a cyclobutanone, with a carbonyl to give a β-lactone, and with carbodiimides to form 4-imino β-lactams.

2.1.2 Historical Perspective

Professor Hermann Staudinger first reported the reaction in 1907, as part of an overall study on the chemistry of ketenes carried out at the University of Strassburg.[1] His discovery remained largely overlooked until the emergence of penicillin and other β-lactams as antibiotics during World War II. Since then, other classes of β-lactam antibiotics have been developed, and the β-lactams are still widely used to treat infection.[2] After over 100 years, this general reaction yielding β-lactams remains one of the key methods for the synthesis of these strained heterocycles,[3] which remain important in both medicinal and synthetic chemistry.[4]

2.1.3 Mechanism

The mechanism of the Staudinger ketene imine cycloaddition reaction has been the subject of much debate and has recently been reviewed.[5] The mechanism has been studied both computationally and experimentally.[6] Experimental evidence gathered on solution phase reactions supports a two-step mechanism, in which addition of the imine nitrogen to the ketene carbonyl group occurs to generate an intermediate zwitterion. Subsequent cyclization of the zwitterion results in formation of the key C3–C4 σ-bond.

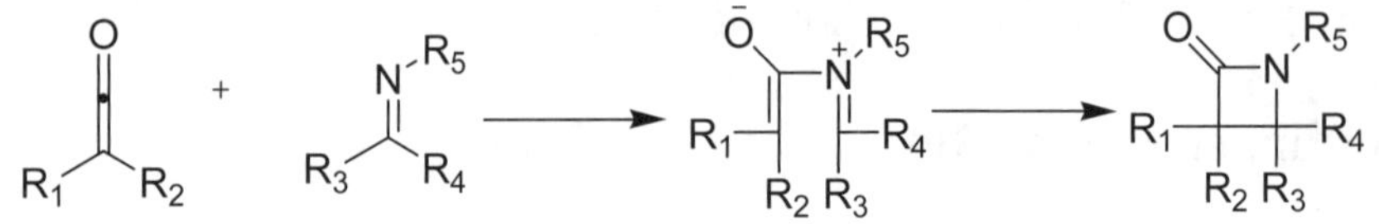

1 2 3 4

The existence of the zwitterionic intermediates has been inferred from the results of trapping experiments and by direct observation of zwitterionic intermediates that are unable to complete the cyclization to a β-lactam. Generation of the zwitterion **7** from ketene **5** and imine **6**, followed by in situ trapping with sulfur dioxide, afforded the thiazolidine-4-one 1,1-dioxide **8**.[7] Similarly, reaction of the formamidate **10** with ketene **9** generated in situ by pyrolysis of the acylal afforded the formamide diethylacetal **12** upon trapping with ethanol, which suggests the intermediacy of the zwitterion **11** in this reaction pathway.[8]

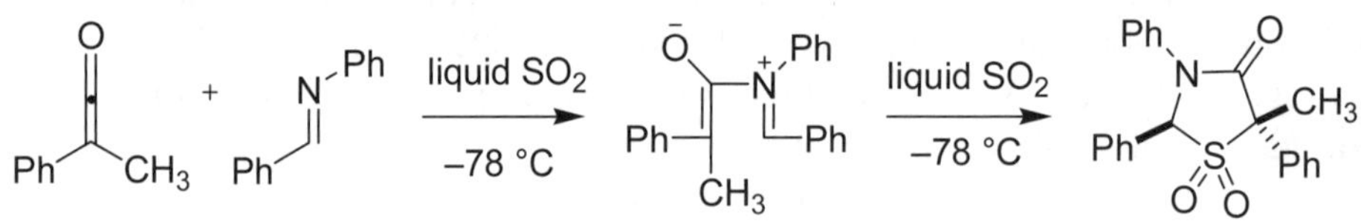

5 6 7 8

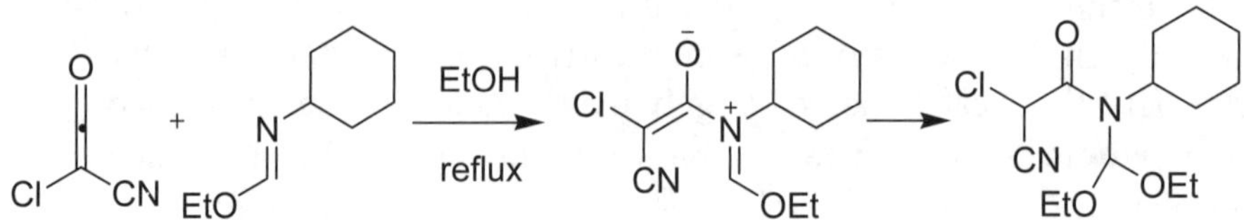

9 10 11 12

Direct observation of zwitterionic intermediates has been made by the addition of azomethine heterocycles, such as pyridine to ketenes.[9] In these cases, the cyclization of the zwitterion to a β-lactam is energetically unfavorable because of the loss of aromaticity of the pyridine ring that would result.

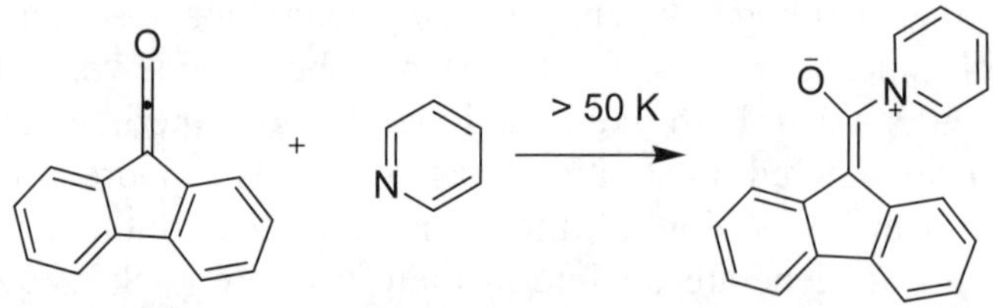

The long-standing observation that the presence of electron-donating groups on the imine facilitate the addition of the imine to the ketene while electron-withdrawing groups retard the reaction is consistent with the nucleophilic attack shown above.[10]

The cyclization step of the Staudinger reaction has been the subject of much study. It is has been generally thought that the ring closure step of the Staudinger reaction is an electrocyclic process, due to its obvious similarity to the well-studied cyclobutene system. The zwitterionic intermediate of the Staudinger reaction is regarded as a 4π-electron system and has been believed to obey the Woodward–Hoffmann rules and undergo a conrotatory ring closure. However, when monosubstituted ketenes are allowed to react with aldimines, the *cis/trans* ratio of the products obtained have been found to vary in rather unpredictable ways. For example, the electron-donating or electron-withdrawing nature of the ketene substituent has significant effects that are difficult to explain if the ring closure is an electrocyclic reaction.[6] In addition, it has been shown that, unlike substituted 1,3-butadienes, the zwitterionic intermediates in photochemical Staudinger reactions do not undergo disrotatory ring closure as expected by application of the Woodward–Hoffmann rules.[11] Recently, evidence has been developed to provide a general explanation for the various stereochemical outcomes of Staudinger reactions in which the ring closure step occurs as an intramolecular nucleophilic addition of the enolate to the iminium ion.[12]

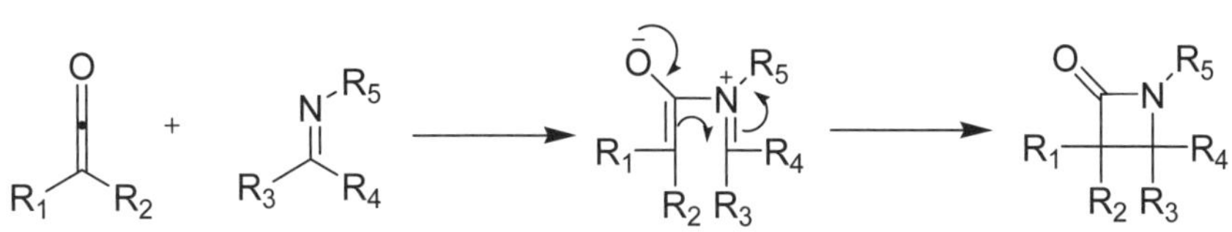

2.1.4 Stereochemical Outcome

The reaction of a monosubstituted ketene with an aldimine produces two new stereocenters, C3 and C4, in the β-lactam ring. The substituents on these two carbon atoms may therefore be *cis* or *trans* to each other, and the reaction of any particular pair of ketene and imine may afford the *cis* β-lactam, the *trans*-β-lactam, or a mixture of the two. The importance of β-lactams as antibacterial agents has resulted in extensive effort to understand the diastereoselectivity of the Staudinger reaction.[13] The two-step nature of the reaction pathway has made the interpretation of experimental results more difficult and increased the number of possible contributing factors.

The earliest hypothesis was the simplest; that the stereochemical outcome of the Staudinger reaction was determined by the *Z* or *E* configuration of the aldimine.[14] *Z*-Aldimines would be expected to lead to *trans*-β-lactam products while *E*-aldimines would lead to *cis*-β-lactams.

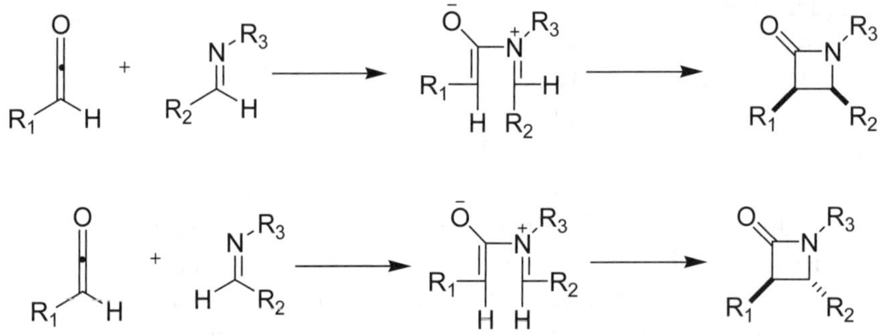

However, very few acyclic aldimines exist in the *Z*-conformation; therefore, the aldimine geometry cannot provide a complete explanation for the stereochemical outcome of the Staudinger reaction.

Another possibility is that the initial addition of the aldimine to the ketene determines the product stereochemistry. When a monosubstituted ketene is treated with an aldimine, the attack of the aldimine may occur from the less-hindered side of the ketene bearing the smaller substituent, most often H (*exo* attack), or from the side of the ketene bearing the larger substituent (*endo* attack). According to this model, *exo* attack would be expected to lead to a *cis*-β-lactam product, whereas *endo* attack would lead to the *trans*-β-lactam.[15]

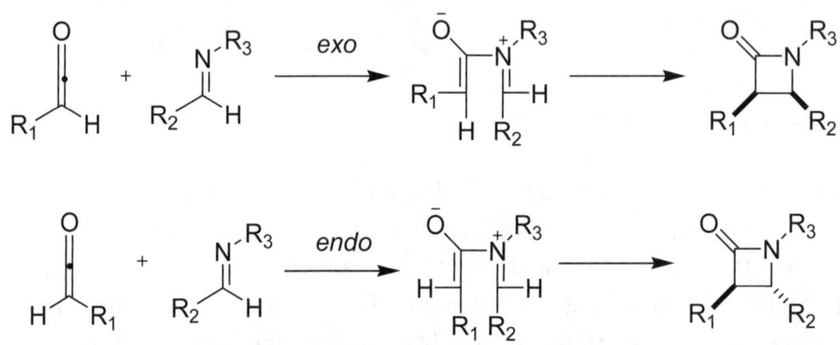

One would expect that the aldimine would add to the less-hindered face of the ketene. Computational studies support this conclusion, showing that *exo* attack leads to lower energy transition states. Further proof that *exo* attack is the exclusive pathway of aldimine addition has been developed using cyclic *Z*-imines substituted with various electron-donating and electron-withdrawing substituents.[6]

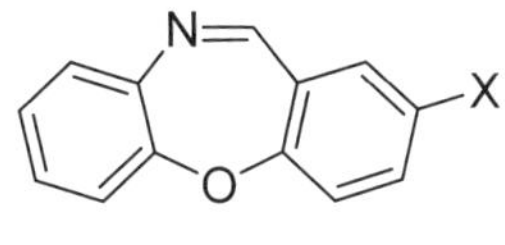

These afforded exclusively the *trans*-β-lactams in practically quantitative yields, indicating that the aldimine approaches exclusively from the less-hindered side of the ketene, and further that the electronic nature of the aldimine does not influence the direction of aldimine attack.

The third possibility is that isomerization of the aldiminium group in the intermediate zwitterion may occur at a rate that is competitive with the direct closure of the β-lactam ring. If direct ring closure occurs at a rate significantly greater than that of aldimine isomerisation in the zwitterion, then the *cis*-β-lactam will be formed. However, if the aldimine isomerizes at a rate greater than that of cyclization, then the *trans*-β-lactam may be formed. Isomerization of the aldimine would be expected to relieve steric crowding in the cyclization transition state; therefore, it might be expected that the zwitterion may favor isomerization to the *Z*-aldimine, in which the aldimine substituent is moved away from the ketene substituent.

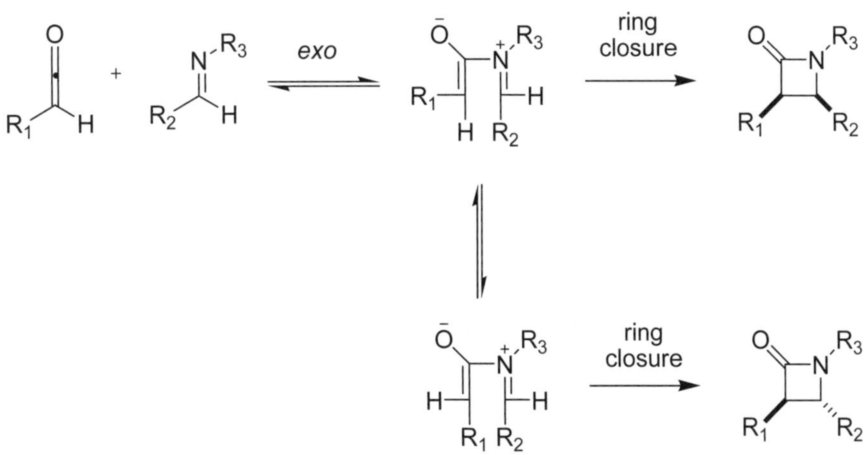

The possibility of aldimine isomerization before cyclization is consistent with experimental observation when the cyclization is viewed as occurring by nucleophilic attack of the ketene enolate on the aldiminium ion in the zwitterionic intermediate. The effects of electron-donating and electron-withdrawing substituents in the ketene and aldimine as well as steric effects may be understood in terms of this model. Thus the stereochemical outcome of the Staudinger reaction is determined by the competition between direct ring closure of the zwitterion and isomerization of the aldimine.

Further studies into the stereochemical outcome of the Staudinger reaction continue to generate new information.[12]

The selection of experimental conditions can influence the stereochemical outcome of the Staudinger reaction as well, further complicating matters. This can be especially important when the ketene is generated in situ, as is often the case. Many different experimental factors, such as reaction temperature, solvent,[16] and the presence of potential auxiliary nucleophiles (particularly the base used to generate a ketene from an acid chloride[17] and the by-product chloride ion[18]) or electrophiles (such as metal ions or excess acid chloride) may affect the ratio of *cis-* to *trans*-β-lactam obtained in any particular experiment. Even the order of addition of reagents has been shown to affect the stereochemical outcome. For example, in the example shown below, the addition of the acid chloride to a mixture of aldimine and triethylamine afforded predominantly the *cis*-β-lactam (*cis/trans* = 3:1). However, the addition of a mixture of the aldimine and triethylamine to the acid chloride gave the reverse stereochemical outcome, favoring the formation of the *trans*-β-lactam (*cis/trans* = 1:3).[19]

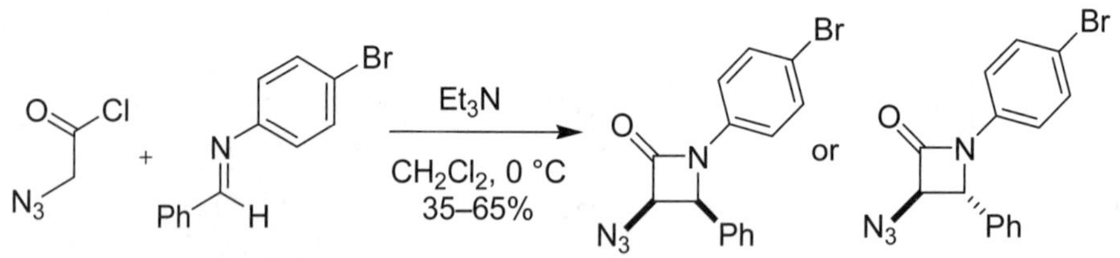

2.1.5 *Periselectivity*

The reaction of a ketene with an α,β-unsaturated imine may be expected to afford either a β-lactam, resulting from [2 + 2] addition, or a δ-lactam, resulting from [4 + 2] addition.

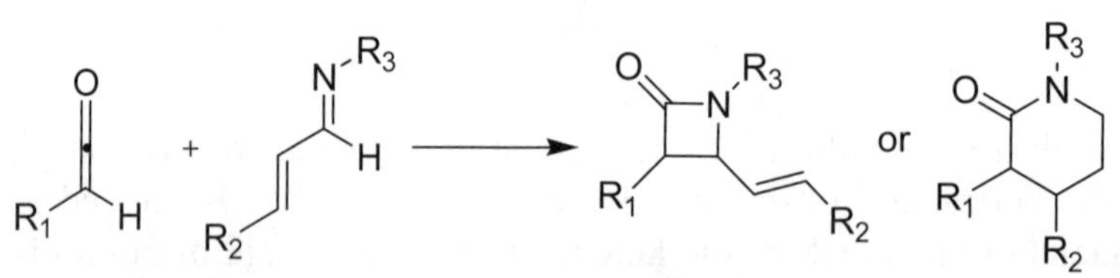

The formation of both types of products has been observed experimentally.[20] While the influence of steric and electronic effects on the outcome of the reaction have been analyzed,[5] the reaction of a ketene with an α,β-unsaturated imine still proceeds by the two-step mechanism and the periselectivity is determined in the second step. Thus it should be recognized that the periselectivity of any particular reaction may be influenced by the selection of reaction conditions and the presence of additional reagents or reaction by-products, as noted previously.

2.1.6 Variations and Improvements

The oldest method for the formation of a ketene, used by Staudinger in his studies on diphenylketene, is the reduction of an α-haloacyl halide with activated zinc.[21] Most often, the ketene components used in the Staudinger reaction are usually produced by either of two ways: the elimination of an acyl chloride (or less frequently another activated carboxyl derivative) in the presence of a base,[22] or the Wolff rearrangement of α-diazocarbonyl compounds.[23] The ketene is usually generated in situ in the presence of the imine; however, if the ketene is stable enough, it may be prepared separately and then introduced into reaction with the imine. Other methods to produce ketenes have been used less often in the Staudinger reaction due to incompatibility with the imine component or β-lactam product or due to the harsh conditions required, such as the high temperatures employed in the pyrolysis of acid anhydrides or ketone acylals.

The ease of preparation of ketenes and their use in the Staudinger reaction depends on their reactivity. Most ketenes dimerize readily, are hydrolyzed easily, and are sensitive to oxidation by oxygen. However, some ketenes, such as trimethylsilylketene and diphenylketene, may be isolated as pure compounds.[24]

The elimination of an acyl chloride by a tertiary amine base has been widely used to generate the ketene component in situ due to its convenience and the ready availability of the starting acyl chlorides. The tertiary amine must be a nucleophilic tertiary amine, and triethylamine is generally used. The tertiary amine forms an intermediate acylammonium salt that undergoes decomposition to the ketene.[25]

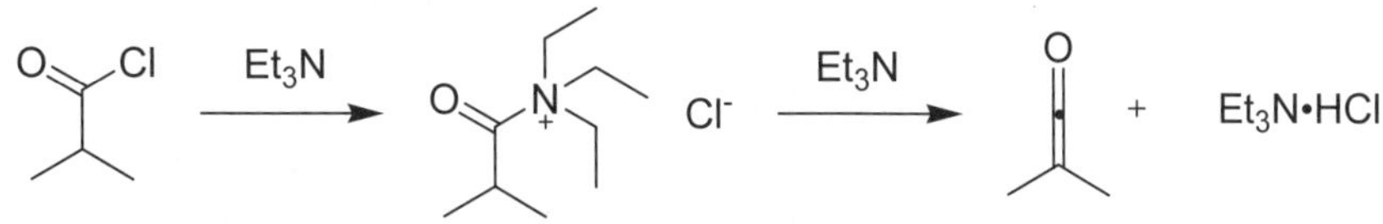

Despite the method's convenience, the presence of the acyl chloride, the tertiary amine, and the tertiary amine hydrochloride may all introduce further complications.[26]

The Wolff rearrangement of α-diazocarbonyl compounds offers perhaps the "cleanest" means of generating a ketene in situ without the presence of additional reactants or by-products. The rearrangement may be induced by photolysis or heat.[27] The ready availability of α-diazocarbonyl compounds by diazo transfer from sulfonyl azide reagents to ketone and ester enolates,[28] and the ease with which the Wolff rearrangement may be induced make this a valuable method for the generation of the ketene component.

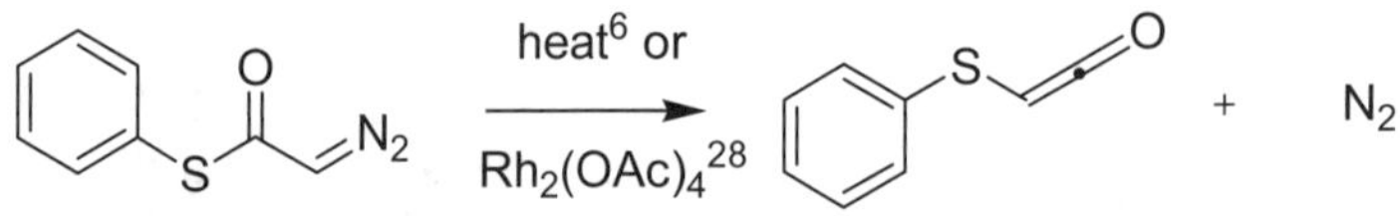

Other methods for ketene generation that are occasionally used are conceptually similar to the elimination of acyl chlorides but use different carboxyl activating groups. Activation of a carboxylic acid by Mukaiyama's reagent, for example, followed by treatment with triethylamine to generate a ketene in situ, has been used on occasion.[29] A very mild method for ketene formation involves treatment of the carboxylic acid with triphenylphosphine and carbon tetrabromide in the presence of the imine.[30] The photolysis of metal–carbene complexes, particularly chromium carbonyl carbenes, has been used but this necessarily involves more effort in the preparation of the necessary ketene precursor.[31]

2.1.7 Enantioselective methods

Until recently, attempts to conduct the Staudinger reaction in such a way as to favor the formation of one *cis* (or much less frequently, one *trans* diastereomer) over the other possible *cis* or *trans* diastereomer have made use of a fixed chiral center in one of the reactants to influence the stereochemical outcome of the reaction. Thus asymmetric induction may be brought about by the presence of one or more chiral centers on the imine, or the ketene, or both. Further, a chiral imine may be derived either from a chiral aldehyde and an achiral amine or from a chiral amine and an achiral aldehyde. The asymmetric synthesis of β-lactams using the Staudinger reaction has been recently reviewed.[5]

The use of a chiral aldehyde to generate a chiral imine is somewhat more difficult than it might first appear. The chiral center should be close to the aldehyde carbonyl, but these are notorious for their facile enolization. Further, it is imperative that the derived imine does not also enolize to form the corresponding enamine. In general, aldehydes substituted with α-nitrogen or α-oxygen substituents have been found suitable, while successful applications of α-alkyl aldehydes have been less common. Aldehydes with chiral α-nitrogen substituents are readily derived from α-amino acids, and *N*-

BOC imines, derived from *N*-BOC α-amino aldehydes or Garner's aldehyde[32] often afford β-lactams with high diastereoselectivity.[33]

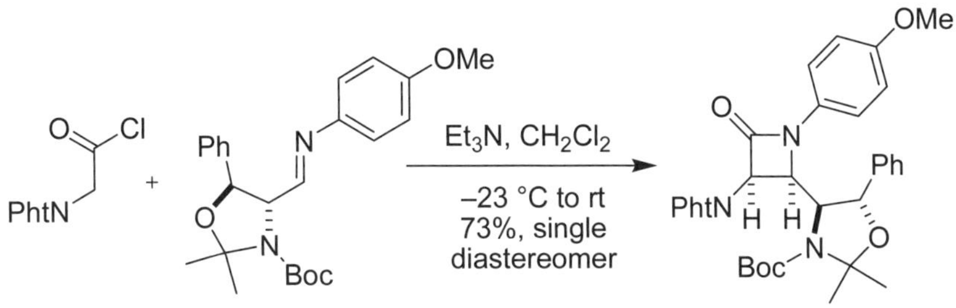

Aldehydes with chiral α-alkoxy substituents may be derived from α-hydroxy acids, sugars,[34] and other chiral pool fragments such as sodium erythorbate.[35] Highly diastereoselective Staudinger reactions have been reported with these imines as well. Glyceraldehyde imines have also been used to provide β-lactam products with high diastereoselectivity, as shown below.[36] Cyclic imines, such as enantiomerically pure dihydropyrazinones[37] and benzoxazepines,[38] have been used in highly enantioselective Staudinger reactions.

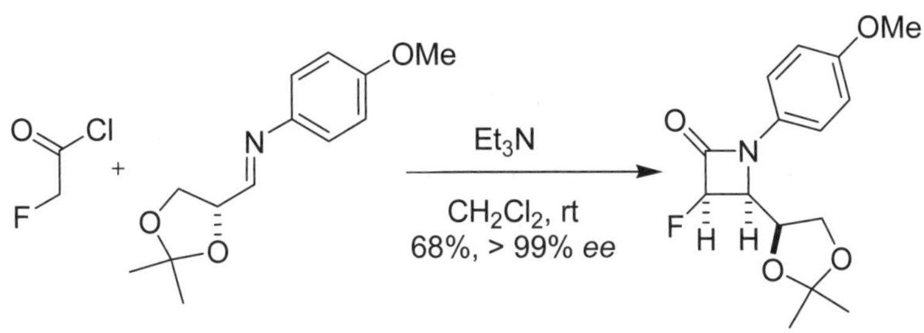

In contrast, imines derived from alkyl aldehydes generally perform poorly in the Staudinger reaction under most circumstances, and examples of their successful use to cause asymmetric induction in the Staudinger reaction are unusual. However, enolization may be suppressed by appropriate design of the aldehyde, such as the example shown in which the formation of an enol or an enamine is disfavored.[39]

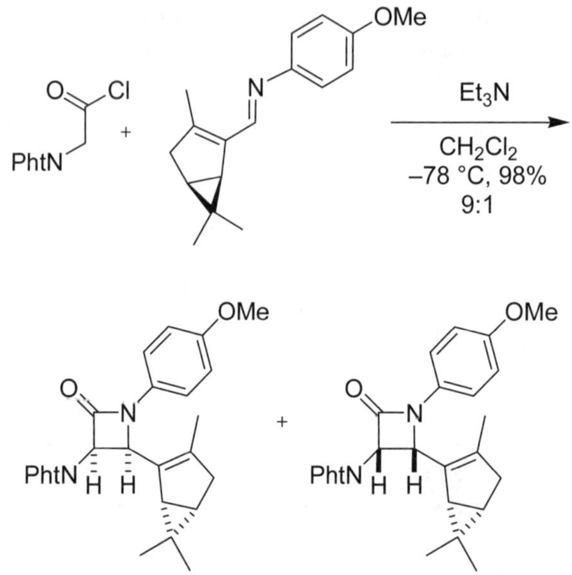

The use of a chiral amine to generate a chiral imine is obviously attractive. Chiral amines (and amino alcohols) are readily available, and they have been used successfully in numerous enantioselective methodologies. However, imines derived from chiral amines and achiral aldehydes have generally afforded disappointing levels of asymmetric induction.[40] This is perhaps to be expected because the imine *N*-substituent is relatively remote from the C3 and C4 substituents in the assembled transition state. For example, modest diastereoselectivity was observed with the use of the imine **13**, which is derived from a commercially available aminodiol.[40g]

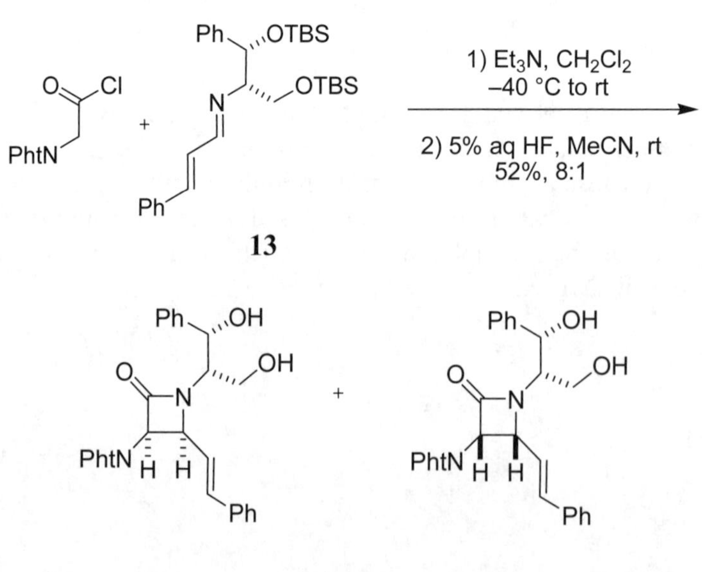

More selective examples have used imines derived from amino acids, including phenylalanine[41] and, more generally useful, *O*-protected threonine esters.[42]

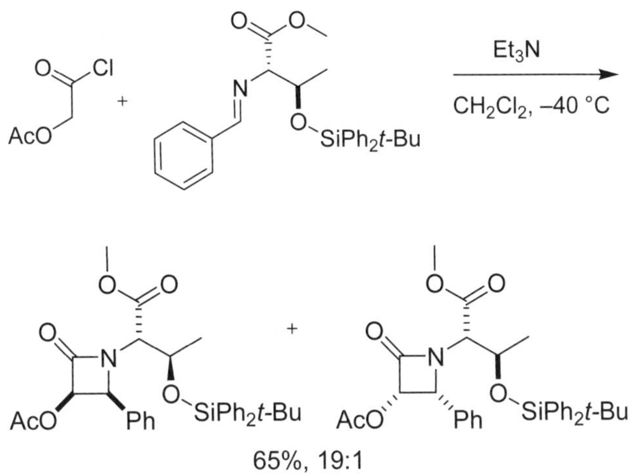

By contrast, very high diastereoselectivities have been observed by the use of chiral hydrazones prepared from the C2 symmetric hydrazine 2,5-dimethyl-pyrrolidin-1-ylamine.[43] The asymmetric induction comes at a price, however: The hydrazine is not commercially available, and it cannot be recovered as such following oxidative cleavage from the β-lactam. More readily available, and therefore more expendable, hydrazines are SAMP analogs derived from L-proline.[44]

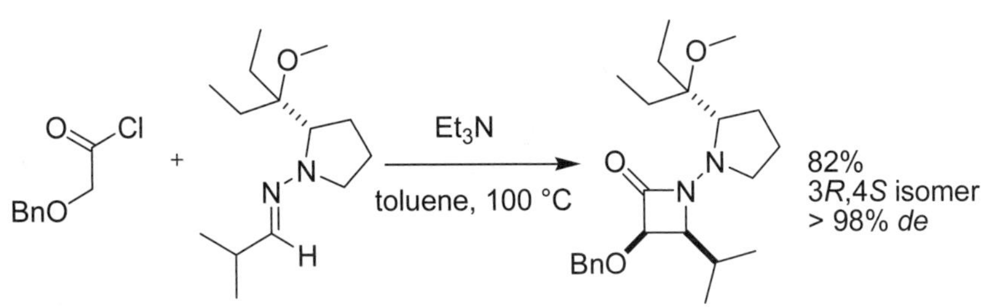

Chiral ketene fragments have proven to offer more general utility in the quest for enantioselective Staudinger reactions. In particular, ketenes containing the Evans oxazolidinone have proven to offer high diastereoselectivities in the synthesis of 3-amino-β-lactams.[45] The ease of preparation of the carboxylic acid precursor, the low cost of the auxiliary, and the facile unmasking of the 3-amino group from the auxiliary all contribute to the utility of this approach. The following is an example in which the β-lactam was obtained as a single enantiomer in good yield.[45f]

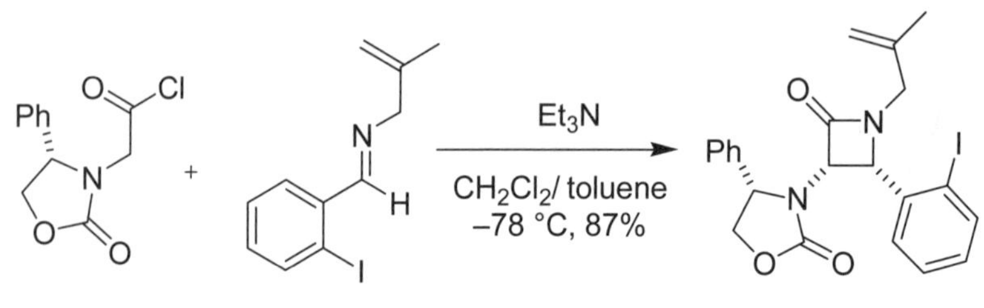

Thiazolidine-substituted ketenes derived from cysteine and other chiral aminothiols have also been used successfully to generate spiro-β-lactams with excellent diastereoselectivity and, with a large enough substituent, high enantioselectivity.[46] Ketenes derived from proline have been used similarly.[47]

Solutions to the synthesis of β-lactams substituted with functionality other than an amino group at C3 are less general. Carbon and oxygen substituents bearing chiral groups have been employed as potential chiral directors at the ketene 2-position, often with disappointing levels of diastereoselectivity.[48] A rigid and more sterically demanding bicyclic ketene has afforded excellent diastereoselectivity (*dr* 7:1 to 50:1).[49] Recently, ethyl L-(+)-tartrate has been used to generate a ketene with a chiral alkoxy substituent.[50] Diastereoselectivity was again very modest; however this provides access to 3-keto-β-lactams, which offer many options for further functionalization.

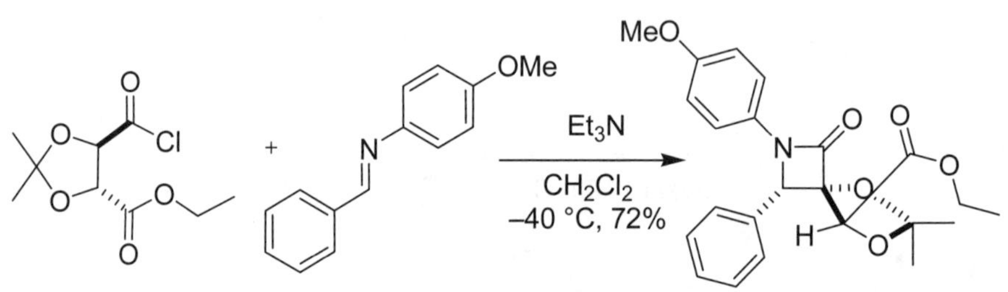

The two-component nature of the Staudinger reaction suggests that double asymmetric induction may afford improved diastereoselectivity. The chiral groups may be paired in any of three ways: (1) a chiral aldehyde with a chiral amine to place two defined stereocenters on the imine, (2) an imine with a chiral *N*-substituent with a chiral ketene substituent, and (3) an imine with a chiral *C*-substituent with a chiral ketene. Some examples of these possibilities have been reported.

The first pairing, that of a chiral aldehyde with a chiral amine, has been used to further develop the threonine imine method. The use of a threonine-derived imine of (*S*)-glyceraldehyde acetonide afforded a single diastereomer in 61% yield.[51]

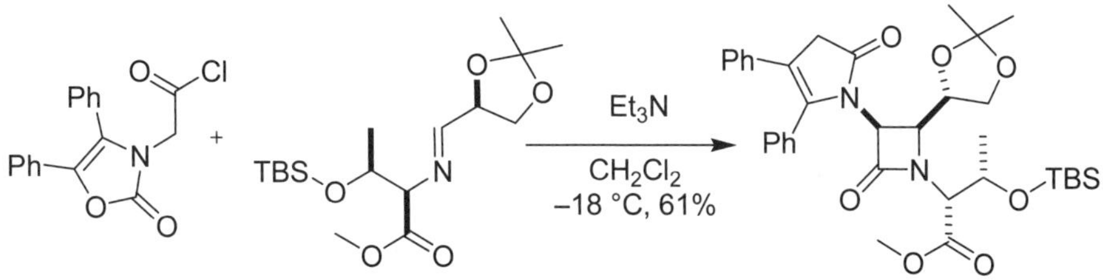

The second pairing, that of an imine with a chiral *N*-substituent with a chiral ketene substituent, presents an attractive synthetic design because both chiral auxiliaries may be removed subsequent to β-lactam formation. The Evans–Sjogren chiral auxiliary on the ketene and imines derived from amino acids are clear choices for such a combination. An example is the combination of the Evans–Sjogren chiral auxiliary with an imine derived from (*S*)-valine, which gave a single diastereomer in excellent yield.[52]

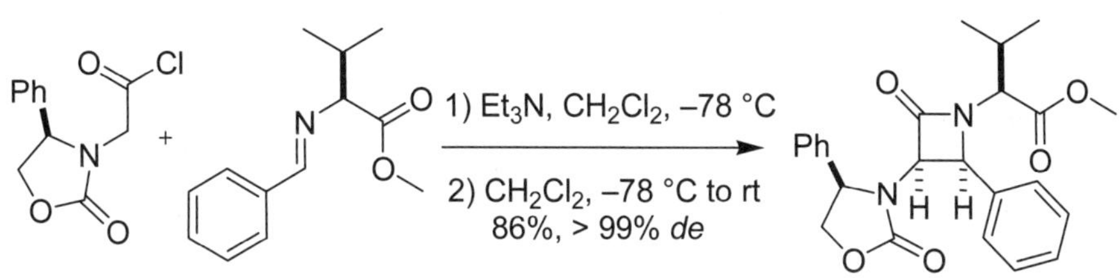

(3*R*, 4*S*)

The third pairing, that of an imine with a chiral *C*-substituent with a chiral ketene, has also been studied. Again, the Evans–Sjogren chiral auxiliary was used as the directing group on the ketene, while imines derived from (*R*)-glyceraldehyde acetonide[53] and lactic acid[54] have been used.

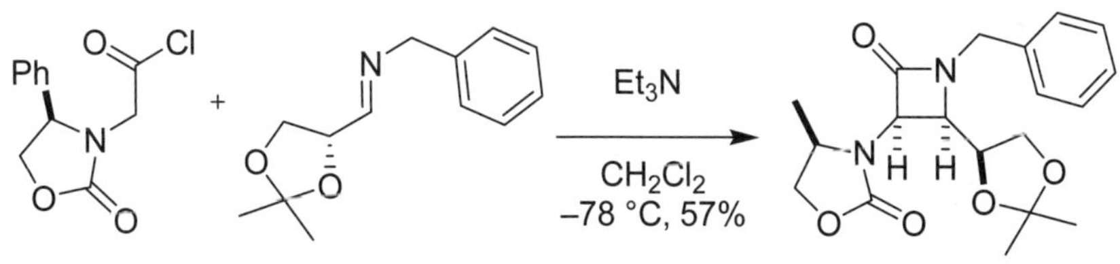

In addition, asymmetric induction from chiral centers on all three possible groups—the imine *C*-substituent, the imine *N*-substituent, and the ketene—has been described.[53]

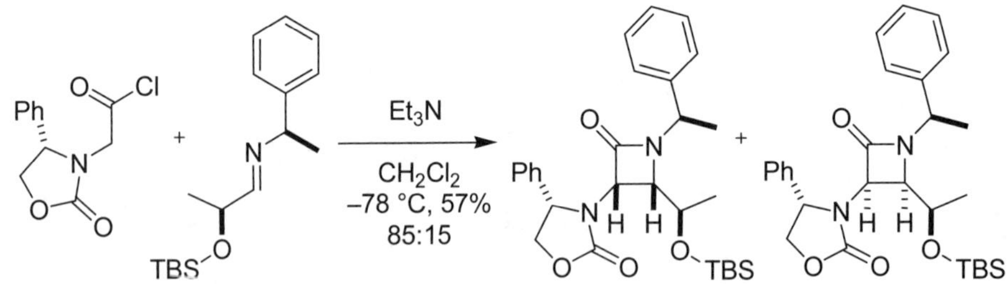

The emergence of catalytic asymmetric methods to effect the Staudinger reaction appears to have largely displaced further efforts to identify new methodology based on the use of chiral auxiliaries. These methods rely on the nucleophilic activation of the ketene to form a zwitterionic enolate, which then undergoes nucleophilic addition to the imine, followed by cyclization. While the assembly of enantiodifferentiated transition states using Lewis acid catalysis has been well developed, the use of Lewis base catalysts to accomplish the same purpose is a relatively recent development and well suited to the Staudinger reaction.

The first catalytic asymmetric Staudinger reaction to be described used chiral tertiary amines **14** and **15** derived from the *Cinchona* alkaloids as the nucleophile to activate the ketene via zwitterion formation.[55] The ketene was conveniently generated in situ from the acid chloride. Because the HCl generated in the elimination would consume the chiral tertiary amine catalyst, a nonnucleophilic strong base (e.g., Proton Sponge) was included to remove the HCl formed. Yields of β-lactams were on the order of 60% in 99% *ee*.

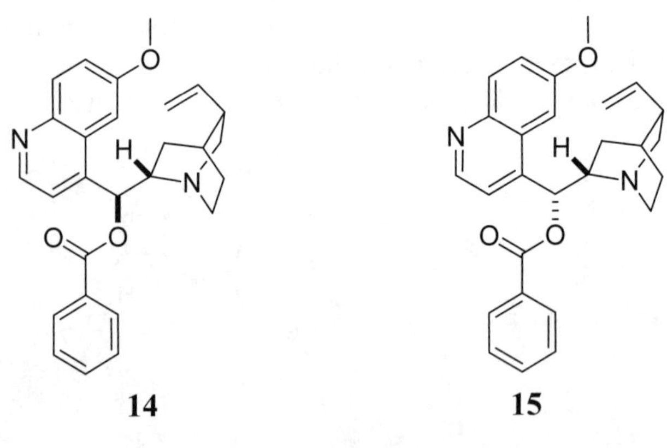

14 **15**

The next catalyst to be described for the catalytic asymmetric Staudinger reaction was the planar chiral dialkylaminopyridine derivative **16**.[56] In this case, the ketene was prepared separately before reaction with the imine, and therefore, an auxiliary base was not necessary.

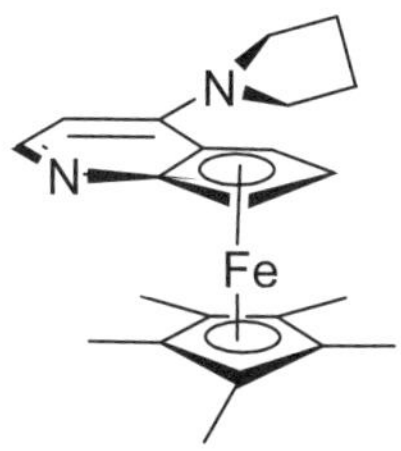

16

In each of these approaches, it is necessary that the imine be a relatively electrophilic *N*-tosyl imine. Further work by Fu showed that *trans*-β-lactams could be formed in good yields and *ee*, if the corresponding triflyl imines were employed.[57]

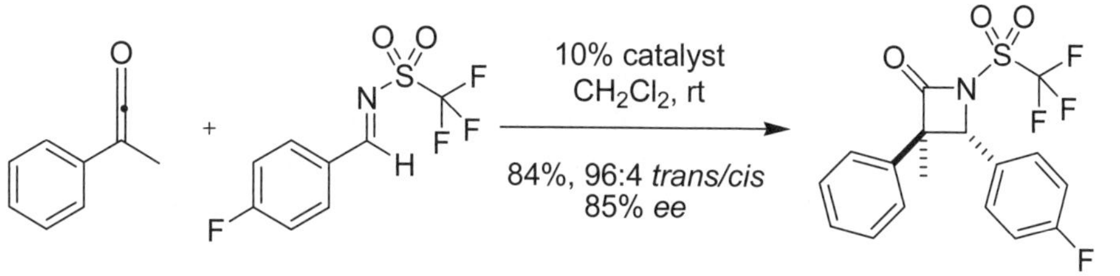

Both of these approaches have been developed and refined further,[58] although no truly general solution has yet been developed. Lectka has described less expensive bases to consume the HCl formed during ketene generation, with sodium hydride/15-crown-5 being the most promising early candidate.[59] It should be noted that diisopropylethylamine (Hunig's base) was specifically described as being unsuitable, due to competitive catalysis of the Staudinger reaction by this amine. Subsequently, conditions were found in which sodium bicarbonate performed well as the terminal base.[60] Further refinements of this method include the use of enantiomerically pure cyclophane[61] and indium[62] cocatalysts. Recently, an anionic co-catalyst has been described that favors the formation of *trans* diastereomers in high *ee*,[63] which have been difficult to access previously. Similarly, sodium bis(trimethylsilyl)amide has been used to catalyze the Staudinger reaction of relatively unreactive ketenes with imines in a nonenantioselective manner.[64]

Chiral *N*-heterocyclic carbenes have recently been described as nucleophilic catalysts for the Staudinger reaction. The carbenes may be conveniently prepared enantiomerically pure from either pyroglutamic acid[65] or *trans*-1,2-diaminocyclohexane.[66] The ketenes were generated separately from the Staudinger reaction mixture, suggesting that further improvements in convenience and reaction scope may be identified.

2.1.8 Synthetic Utility

2.1.8.1 General Utility

In general, ketenes substituted with electronegative atoms such as O, N, F, or Cl, as well as those bearing by SO_2 or aryl groups, perform satisfactorily in the Staudinger reaction. Alkyl ketenes remain problematic. It is not clear whether the difficulties are entirely due to the inherent reactivity alkyl ketenes, particularly monoalkyl ketenes, which dimerize within minutes at room temperature, or whether the difficulties are due at least in part to the conditions generally used to generate the ketenes in situ from the acid chlorides. Some evidence that the latter may be at least partially responsible is the successful use of alkyl ketenes generated by the Wolff rearrangement in the Staudinger reaction.[67]

Acetyl chloride has been used as a ketene precursor to prepare β-lactams unsubstituted at C3.[68] Unsubstituted imine fragments pose special difficulties in the Staudinger reaction. Formaldehyde imines are difficult to prepare and handle, and generally β-lactams unsubstituted at C4 are not readily accessible. By contrast, formaldehyde hydrazones are more stable and can be used in place of the unstable imines in the Staudinger reaction to afford 4-unsubstituted β-lactams.[69]

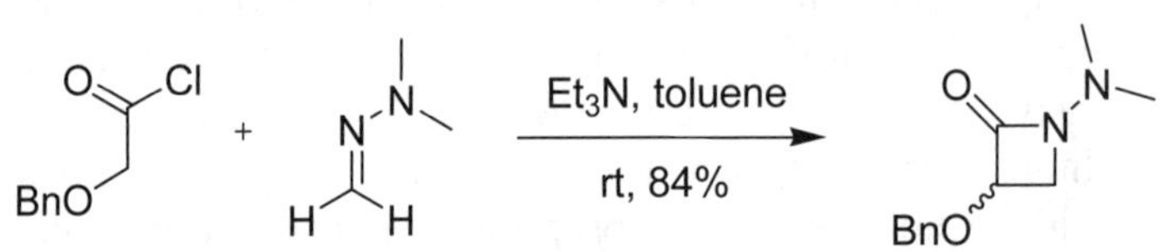

The synthesis of β-lactams unsubstituted on nitrogen also cannot be accomplished directly due to the instability of most imines derived from ammonia. However, imines derived from 4-methoxyaniline and 4-ethoxyaniline readily afford *N*-aryl β-lactams; cleavage of the *N*-aryl bond is accomplished by oxidation with ceric ammonium nitrate.[70] *N*-Trimethylsilyl imines have also been used to provide NH β-lactams.[54]

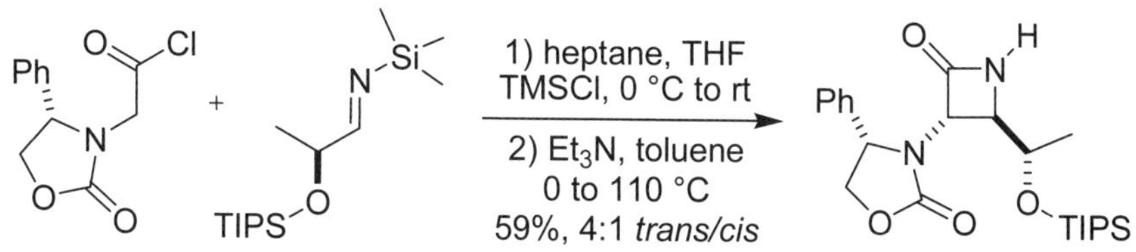

The catalytic enantioselective synthesis of *trans*-β-lactams has also been problematic because the nucleophilic catalysts employed facilitate the cyclization reaction, which leads to *cis*-β-lactam products from *E*-imines. A workaround has been developed by Lectka, in which an anionic nucleophilic catalyst is used.[63]

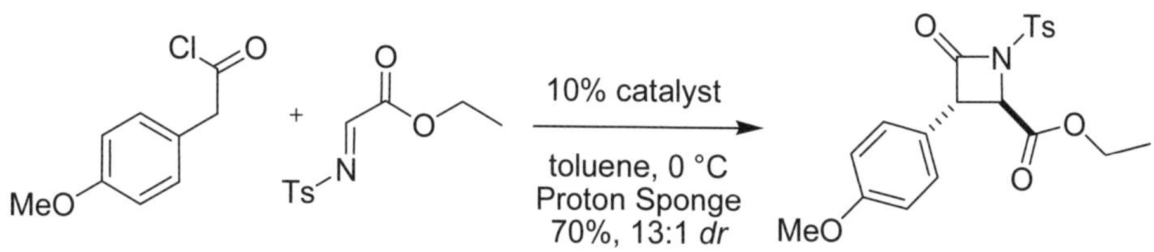

The catalyst is a dihydro-1*H*-imidazol-2-yl-benzenesulfonic acid; which is used as the tetraheptylammonium salt to suppress the formation of ion pairs as much as possible.

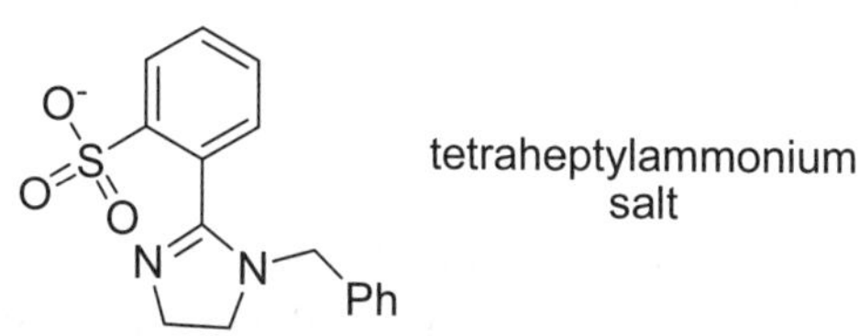

Remarkably, both 3-keto and 3-aza β-lactams are accessible. The 3-keto β-lactams may be prepared ultimately from diethyl tartrate.[50] Conversion to the acetonide, followed by saponification and treatment with oxalyl chloride, gave the acid chloride, which was treated with triethylamine and an imine to give the spiro β-lactam. Removal of the acetonide followed by oxidative cleavage of the diol afforded the 3-keto-β-lactam.

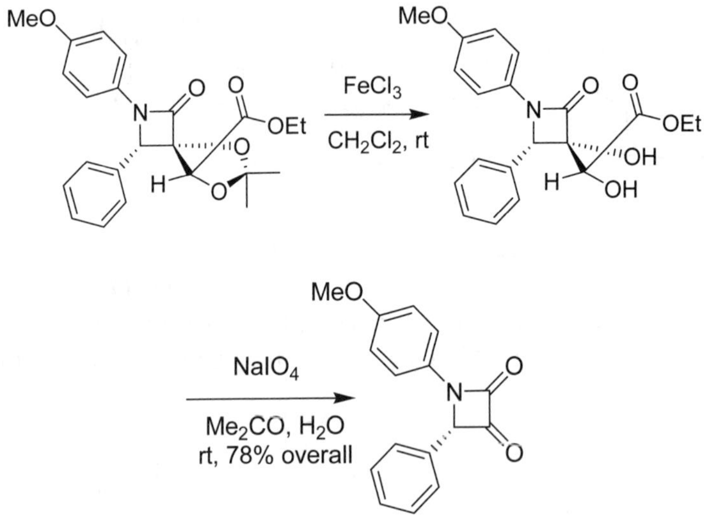

An alternate approach to 3-keto-β-lactams involves formation of the thiazolidine-derived spiro-β-lactam **17**, followed by cleavage of the thiazolidine moiety by oxidation.[71]

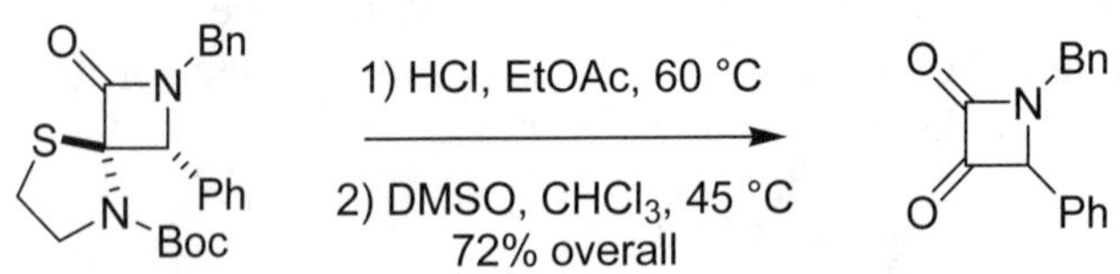

17

3-Aza-β-lactams may be prepared by the Staudinger reaction of a ketene with an azodicarboxylate ester under catalysis by a planar chiral nucleophile.[72] Dimethyl azodicarboxylate and diethyl azodicarboxylate performed well in this reaction. Higher azodicarboxylate esters afforded lower yields and lower enantioselectivities, while an azodicarboxamide failed to undergo reaction.

2.1.8.2 Applications in synthesis

The 2-azetidinone system has emerged as a useful, densely functionalized fragment that may be elaborated in a variety of different ways. Thus β-lactams have emerged as interesting synthetic intermediates, besides their obvious utility as antibacterial agents. In particular, ring opening of β-lactams has been shown to provide stereocontrolled synthesis of both α- and β-amino acids, as well as functional group derivatives of these, such as β-amino amides, alcohols, esters, and ketones. The intense interest generated by the discovery of the antitumor agent paclitaxel spurred interest in the Staudinger reaction to generate a synthetic equivalent to the phenylpropionyl side chain fragment **18**.[73] For example, the β-lactam **19** was appended to the paclitaxel core by acylation of the core 13-position alcohol by the β-lactam carbonyl.[73b]

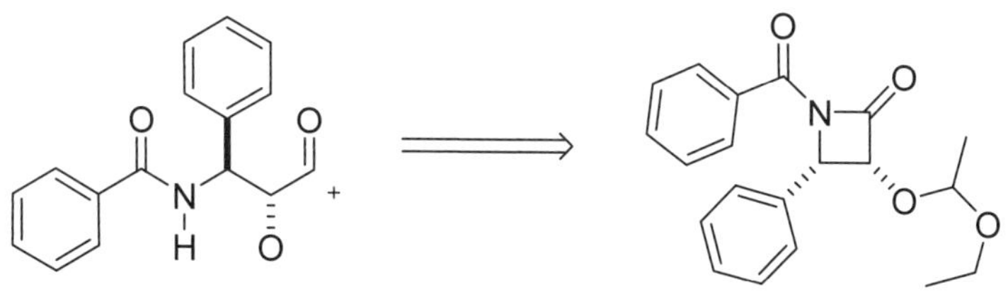

18 **19**

Ring opening of β-lactams by amines affords β-amino amides, which are fragments of interest from both a medicinal chemistry perspective and for the synthesis of natural products. Intramolecular aminolysis of a β-lactam affords a 1,4-diazepin-5-one.[74]

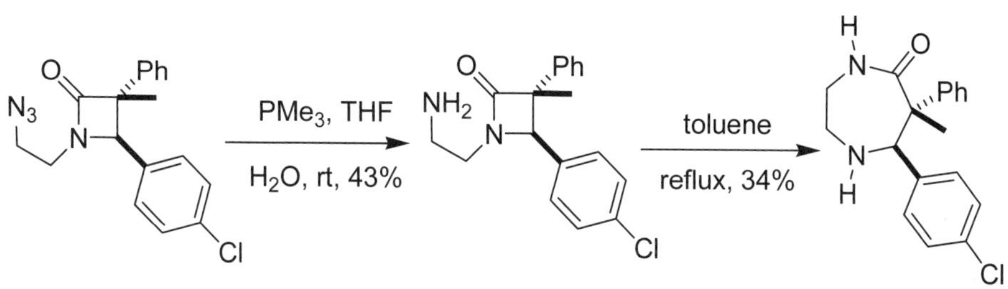

Two recent examples of the use of the Staudinger reaction to construct key intermediates in the total synthesis of natural products are the syntheses of oseltamivir and (–)-cribrostatin.

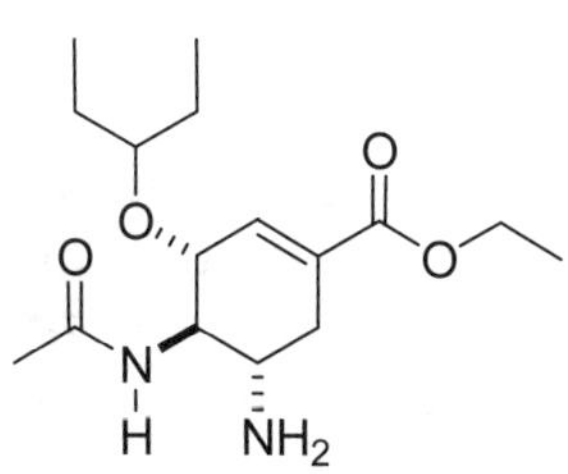

20

The key step in the synthesis of oseltamivir **20** was the use of the Staudinger reaction to set all three contiguous chiral centers using the chiral center of (*S*)-methionine to induce the required asymmetry at C3 and C4 of the β-lactam.[75]

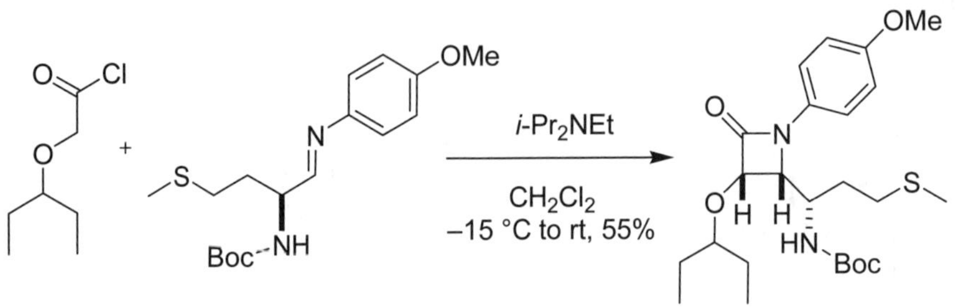

Similarly, a key strategy in an elegant synthesis of (–)-cribostatin was the use of a chiral auxiliary to fix the stereochemistry of C3 and C4 of the β-lactam, which subsequently determined the stereochemical outcome of a Pictet–Spengler reaction on the aryl ether. Subsequently, the β-lactam was used to set the stereochemistry of the pentacyclic framework of (–)-cribrostatin by reduction with lithium borohydride and spontaneous cyclization.[76]

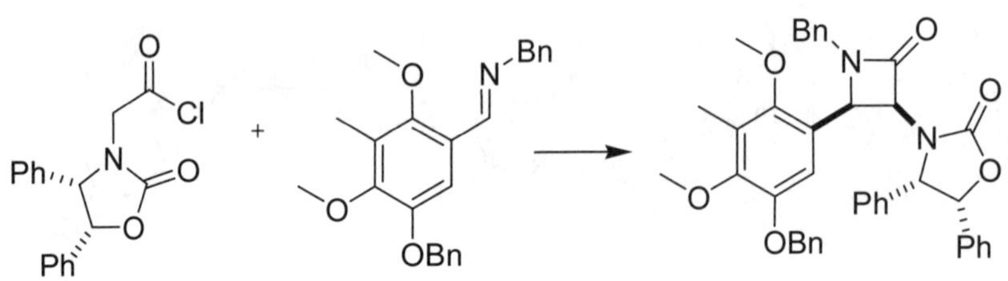

2.1.9 *Experimental*

Staudinger cycloaddition using the Wolff rearrangement to generate a ketene[28]

Caution! Diazo compounds are presumed to be toxic and potentially explosive and, therefore, should be handled with caution in a fume hood. Although in carrying out this reaction numerous times we have never observed an explosion, we recommend that these reactions be conducted behind a safety shield.

N-Benzylidene-*p*-anisidine

A 100 mL, three-necked, round-bottomed flask is equipped with an argon inlet adapter, rubber septum, glass stopper, and a magnetic stirring bar. The flask is charged with 45 mL of CH_2Cl_2 and 3.00 mL (0.030 mol) benzaldehyde, and then is cooled in an ice–water bath while a solution of 3.50 g (0.028 mol) *p*-anisidine in 5 mL CH_2Cl_2 is added dropwise via syringe over 15 min. After 30 min, 7.5 g anhydrous magnesium sulfate is added in one portion. The ice-water bath is removed, and the reaction mixture is stirred at room temperature for 2 h. The resulting mixture is then filtered through a sintered glass funnel with the aid of two 5-mL portions of CH_2Cl_2, and the filtrate is concentrated at reduced pressure by rotary evaporation at room temperature to afford a pale brown powder. This material is dissolved in 150 mL ethanol heated in an 80 °C water bath while 270 mL hot water is added with stirring. The resulting solution is allowed to cool to room temperature and then is cooled in an ice-water bath for 2 h. Filtration provides 5.31 g (88%) of *N*-benzylidene-*p*-anisidine as brown metallic plates.

trans-1-(4-Methoxyphenyl)-4-phenyl-3-(phenylthio)azetidin-2-one

A 500-mL, three-necked, round-bottomed flask is equipped with a magnetic stirring bar, reflux condenser fitted with an argon inlet adapter, glass stopper, and 50-mL pressure-equalizing dropping funnel fitted with a glass stopper. The flask is charged with 3.50 g (0.016 mol) *N*-benzylidene-*p*-anisidine, 200 mL CH_2Cl_2, and 0.045 g (0.10 mmol) rhodium(II) acetate dimer, and the resulting green solution is heated at reflux while a solution of 4.43 g (0.025 mol) *S*-phenyl diazothioacetate in 40 mL of CH_2Cl_2 is added via the dropping funnel over 1 h (the dropping funnel is rinsed with 2 mL CH_2Cl_2). The reaction mixture is further heated at reflux for 5 min, then allowed to cool to room temperature. After transfer to a 500-mL, round-bottomed flask, the mixture is concentrated by rotary evaporation at reduced pressure to provide a brown oil. This material is filtered through a column of 50 g of silica gel (elution with 750 mL CH_2Cl_2) to remove polar impurities and is concentrated under reduced pressure to afford a brown solid, which is washed on a sintered glass funnel with 10 mL ethyl acetate and then 20 mL hexane. The resulting pale yellow powder is dissolved in 40 mL ethyl acetate at 80 °C, and 400 mL hexane (pre-heated in a water bath at 80 °C) is then added in one portion. The resulting solution is allowed to cool to room temperature and then cooled further at −20 °C for 2 h to afford 5.50 g (91%) *trans*-1-(4-methoxyphenyl)-4-phenyl-3-(phenylthio)azetidin-2-one as off-white crystals.

Catalytic asymmetric Staudinger cycloaddition using benzoylquinine as nucleophilic catalyst and sodium bicarbonate as base[60]

To a vigorously stirred solution of $NaHCO_3$ (350 mg, 4.01 mmol), benzoylquinine (6 mg, 0.0129 mmol), and 15-crown-5 (3 mg, 0.0129 mmol) in toluene (6 mL) at –40 °C, phenylacetyl chloride **1** (20 mg, 0.129 mmol) in toluene (1 mL) at –40 °C was added dropwise, followed by α-imino ester **2** (33 mg, 0.129 mmol) in toluene (2 mL). The reaction was allowed to stir for 5 h as it slowly warmed to room temperature. The reaction mixture was washed with 1 M HCl, extracted with CH_2Cl_2 (3 ×). The organics were combined and dried with $MgSO_4$. The solvent was removed under reduced pressure, and the crude mixture was subjected to column chromatography (15% EtOAc–hexanes) on a plug of silica gel to yield **5a** (58% yield, 28 mg).

2.1.10 References

1. (a) Staudinger, H. *Chem. Ber.* **1907**, *40*, 1145–1148, (b) Staudinger, H. *Liebigs Ann. Chem.* **1907**, *356*, 51–123. These experiments were facilitated by the relative stability of diphenylketene (**1**), which was stored and handled as a solution in diethyl ether. The concentration of the ether solution could be accurately determined by titration after hydrolysis of the ketene to diphenylacetic acid with water. Staudinger believed that the structural assignment of his product as a β-lactam was precedented by the work of Meyer (Meyer, H. *Monatshefte fuer Chemie* **1900**, *21*, 965–980), which he cited. However, Meyer's structural assignments of the cantharidine ring and its derivatives were subsequently shown to be erroneous.
2. Kattan, J. N.; Villegas, M. V.; Quinn, J. P. *Clin. Microbiol. Inf.* **2008**, *14*, 1102–1111.
3. (a) Palomo, C.; Aizpurua, J. M.; Ganboa, I.; Oiarbide, M. *Eur. J. Org. Chem.* **1999**, 3223–3235; (b) Palomo, C.; Aizpurua, J. M.; Ganboa, I.; Oiarbide, M. *Curr. Med. Chem.* **2004**, *11*, 1837–1872.
4. [R] *Chemistry and Biology of β-Lactam Antibiotics*; Morin, R. B.; Gorman, M., Eds.; Academic Press: 1982; Vols. 1–3.
5. Cossio, F. P.; Arrieta, A.; Sierra, M. A. *Acc. Chem. Res.* **2008**, *41*, 925–936.
6. Jiao, L.; Liang, Y.; Xu, J. *J. Am. Chem. Soc.* **2006**, *128*, 6060–6069.
7. (a) de Souza Gomes, A.; Joullik, M. M. *J. Heterocycl. Chem.***1969**, *6*, 729–734; (b) Decazes, J. M.; Luche, J. L.; Kagan, H. B.; Parthasarathy, R.; Ohrt, J. T. *Tetrahedron Lett.* 1972, *13*, 3633–3636; (c) Bellus, D. *Helv. Chim. Acta* **1975**, *58*, 2509–2511.
8. (a) Moore, H. W.; Hernandez Jr., L.; Kunert, D. M.; Mercer, F.; Sing, A. *J. Am. Chem. Soc.* **1981**, *103*, 1769–1777; (b) Moore, H. W.; Hernandez Jr., L.; Chambers, R. *J. Am. Chem. Soc.* **1978**, *100*, 2245–2247.
9. [R] Quiao, G. H.; Andraos, J.; Wentrup, C. *J. Am. Chem. Soc.* **1996**, *118*, 5634–5638.
10. [R] Moore, J. A. In *Heterocyclic Compounds with Three and Four Membered Rings, Part 1* Weissberger, A., Ed.; John Wiley & Sons: New York, 1964; pp 929–939.
11. Liang, Y.; Jiao, L.; Zhang, S. W.; Xu, J. X. *J. Org. Chem.* **2005**, *70*, 334–337.
12. Liang, Y.; Jiao, L.; Zhang, S. W.; Yu, Z.-X.; Xu, J. X. *J. Am. Chem. Soc.* **2009**, *131*, 1542–1549.
13. See reference 12 and references cited therein, particularly references 8–10.
14. Hegedus, L. S.; Montgomery, J.; Narukawa, Y.; Snustad, D. C. *J. Am. Chem. Soc.* **1991**, *113*, 5784–5791.
15. (a) Brady, W. T.; Gu, Y. Q. *J. Org. Chem.* **1989**, *54*, 2838–2842; (b) Dumas, S.; Hegedus, L. S. *J. Org. Chem.* **1994**, *59*, 4967–4971.
16. Palomo, C.; Cossio, F. P.; Odriozola, J. M.; Oiarbide, M.; Ontoria, J. M. *Tetrahedron Lett.* **1989**, *30*, 4577–4580.

17. Browne, M.; Burnett, D. A.; Caplen, M. A.; Chen, L. Y.; Clader, J. W.; Domalski, M.; Dugar, S.; Pushpavanam, P.; Sher, R.; Vaccaro, W.; Viziano, M.; Zhao, H. *Tetrahedron Lett.* **1995**, *36*, 2555–2558.

18. (a) Bose, A. K.; Spiegelman, G.; Manhas, M. S. *Tetrahedron Lett.* **1971**, *12*, 31673170; (b) Arrieta, A.; Lecea, B.; Cossio, F. P. *J. Org. Chem.* **1998**, *63*, 5869–5876. See also reference 15b.

19. Bose, A. K.; Anjaneyulu, B.; Bhattacharya, S. K.; Manhas, M. S. *Tetrahedron* **1967**, *23*, 4769–4776.

20. (a) Brady, W. T.; Shieh, C. H. *J. Org. Chem.* **1983**, *48*, 2499–2502; (b) Moore, H. W.; Hughes, G.; Srinivasachar, K.; Fernandez, M.; Nguyen, N. V.; Schoon, D.; Tranne, A. *J. Org. Chem.* **1985**, *50*, 4231–4238; (c) Lecea, B.; Arrastia, I.; Arrieta, A.; Roa, G.; Lopez, X.; Arriortua, M. I.; Ugalde, J. M.; Cossıo, F. P. *J. Org. Chem.* **1996**, *61*, 3070–3079.

21. Staudinger, H. *Chem. Ber.*, **1905**, *38*, 1735–1739.

22. [R] Ward, R. S. In *The Chemistry of Ketenes, Allenes, and Related Compounds, Part 1*; Patai, S. , Ed.; John Wiley and Sons: New York, 1980; pp 231–232.

23. [R] Ward, R. S. In *The Chemistry of Ketenes, Allenes, and Related Compounds, Part 1*; Patai, S. , Ed.; John Wiley and Sons: New York, 1980; pp 227–239.

24. (a) Pelotier, B.; Rajzmann, M.; Pons, J.-M.; Campomanes, P.; López, R.; Sordo, T. L. *Eur. J. Org. Chem.* **2005**, 2599–2606; (b) Darling, S. D. ; Kidwell, R. L. *J. Org. Chem.* **1968**, *33*, 3974–3975.

25. Brady, W. T.; Scherubel, G. A. *J. Am. Chem. Soc.* **1973**, *95*, 7447–7449.

26. (a) Brady, W. T.; Waters, O. H. *J. Org. Chem.* **1967**, *32*, 3703–3705; (b) Brady, W. T.; Smith, L. *J. Org. Chem.* **1971**, *36*, 1637–1640.

27. The Wolff rearrangement may also be brought about by catalysts such as rhodium, copper, or silver salts.

28. Danheiser, R. L.; Okamoto, I.; Lawlor, M. D.; Lee, T. W. *Org. Synth.* Vol. *80*, 160–171.

29. (a) Manhas, M. S.; Chawla, H. P. S.; Amin, S. G.; Bose, Ajay K. *Synthesis* **1977**, 407–409; (b) Amin, S. G.; Glazer, R. D.; Manhas, M. S. *Synthesis*, **1979**, 210–213.

30. Manhas, M. S.; Amin, S. G.; Ram, Bhagat; Bose, Ajay K. *Synthesis* **1976**, 689–690.

31. (a) McGuire, M. A.; Hegedus, L. S. *J. Am. Chem. Soc.* **1982**, *104*, 5538–5540; (b) Hegedus, L. S.; Imwinkelried, R.; Alarid-Sargent, M.; Dvorak, D.; Satoh, Y. *J. Am. Chem. Soc.* **1990**, *112*, 1109–1117; (c) Hegedus, L. S. *Tetrahedron* **1997**, *53*, 4105–4128. See also reference 14.

32. Garner, P.; Park, J. M. *Org. Synth.* **1991**, *70*, 18–28.

33. (a) Jayaraman, M.; Deshmukh, A. R. A. S.; Bhawal, B. M. *Tetrahedron* **1996**, *52*, 8989–9004; (b) Jayaraman, M.; Puranik, V. G.; Bhawal, B. M. *Tetrahedron* **1996**, *52*, 9005–9016.

34. (a) Palomo, C.; Cossio, F. P.; Cuevas, C.; Lecea, B.; Mielgo, A.; Roman, P.; Luque, A.; Martinez-Ripoll, M. *J. Am. Chem. Soc.* **1992**, *114*, 9360–9369; (b) Shaikh, A. L.; Kale, A. S.; Shaikh, M. A.; Puranik, V. G.; Deshmukh, A. R. A. S. *Tetrahedron* **2007**, *63*, 3380–3388; (c) Shin, D. G.; Heo, H. J.; Jun, J.-G. *Synth. Commun.* **2005**, *35*, 845–855.

35. Frazier, J. W.; Staszak, M. A.; Weigel, L. O. *Tetrahedron Lett.* **1992**, *33*, 857–860.

36. (a) Araki, K.; O'Toole, J. C.; Weilch, J. T. *Bioorg. Med. Chem. Lett.*, **1993**, *3*, 2457–2460. (b) Welch, J. T.; Araki, K.; Kawcki, R.; Wichtowski, J. A. *J. Org. Chem.* **1993**, *58*, 2454–2462.

37. Viso, A.; Fernandez de la Pradilla, R.; Flores, R. *Tetrahedron Lett.* **2006**, *47*, 8911–8915.

38. Del Buttero, P.; Molteni, G.; Papagni, A.; Miozzo, L. *Tetrahedron: Asymmetry* **2004**, *15*, 2555–2559.

39. Jayaraman, M.; Deshmukh, A. R. A. S.; Bhawal, B. M. *Tetrahedron* **1996**, *52*, 3741–3756.

40. (a) Van der Steen, F. H.; Van Koten, G. *Tetrahedron* **1991**, *47*, 7503–7524; (b) G. I. Georg, Z. Wu, *Tetrahedron Lett.* **1991**, *35*, 381–384; (c) Bourzat, J. D.; Commerçon, A. *Tetrahedron Lett.*, **1993**, *34*, 6049–6052; (d) Hashimoto, Y.; Kai, A.; Saigo, K. *Tetrahedron Lett.*, **1995**, *36*, 8821–8224; (e) Srirajan, V.; Deshmukh, A. R. A. S.; Puranik, V. G.; Bhawal, B. M. *Tetrahedron: Asymmetry* **1996**, *7*, 2733–2738; (f) Abouabdellah, A.; Begue, J. P.; Bonnet-Delpon, D.; Nga, T. T. T. *J. Org. Chem.* **1997**, *62*, 8826–8833; (g) Gunda, T. E.; Sztaricskai, F. *Tetrahedron* **1997**, *53*, 7985–7998; (h) Brown, S.; Jordan, A. M.; Lawrence, N. J.; Pritchard, R. G.; McGown, A. T. *Tetrahedron Lett.* **1998**, *39*, 3559–3562.

41. Jarrahpour, A. A.; Shekarriz, M.; Taslimi, A. *Molecules* **2004**, *9*, 939–948.

42. (a) Bose, A. K.; Manhas, M. S.; van der Veen, J. M.; Bari, S. S.; Wagle, D. R. *Tetrahedron* **1992**, *48*, 4831–4844; (b) Farina, V.; Hauck, S. I.; Walker, D. G. *Synlett* **1992**, 761–763.
43. Martin-Zamora, E.; Ferrete, A.; Llera, J. M.; Munoz, J. M.; Pappalardo, R. R.; Fernandez, R.; Lassaletta, J. M. *Chem. Eur. J.* **2004**, *10*, 6111–6129.
44. Fernandez, R.; Ferrete, A.; Lassaletta, J. M.; Llera, J. M.; Martin-Zamora, E. *Angew. Chem., Int. Ed. Engl.* **2002**, *41*, 831–833.
45. (a) Evans, D. A.; Sjögren, E. B. *Tetrahedron Lett.* **1985**, *27*, 3783–3786; (b) Boger, D. L.; Myers, J. B. Jr. *J. Org. Chem.* **1991**, *56*, 5385–5390; (c) Muller, M.; Bur, D.; Tschamber, T.; Streith, J. *Helv. Chim. Acta* **1991**, *74*, 767–773; (d) Duczek, W.; Jähnisch, K.; Kunath, A.; Reck, G.; Winter, G.; Schulz, B. *Liebigs Ann. Chem.* **1992**, 781–787; (e) Alcaide, B.; Martin-Cantalejo, Y.; Perez-Castells, J.; Rodriguez-Lopez, J.; Sierra, M. A.; Monge, A.; Perez-Garcia, V. *J. Org. Chem.* **1992**, *57*, 5921–5931; (f) Burwood, M.; Davies, D.; Diaz, I.; Grigg, R.; Molina, P.; Sridharan, V.; Hughes, M. *Tetrahedron Lett.* **1995**, *36*, 9053–9056; (g) Alcaide, B.; Martin-Cantalejo, Y.; Perez-Castells, J.; Sierra, M. A.; Monge, A. *J. Org. Chem.* **1996**, *61*, 9156–9163; (h) Palomo, C.; Aizpurua, J. M.; Garcia, J. M.; Galarza, R.; Legido, M.; Urchegui, R.; Roman, P.; Luque, A.; Server-Carrio, J.; Linden, A. *J. Org. Chem.* **1997**, *62*, 2070–2079; (i) B. Alcaide, C. Polanco, M. A. Sierra, *Eur. J. Org. Chem.* **1998**, 2913–2921; (j) Alcaide, B.; Rodriguez-Vicente, A. *Tetrahedron Lett.* **1999**, *40*, 2005–2006.
46. Cremonesi, G.; Dalla Croce, P.; Fontana, F.; Forni, A.; La Rosa, C. *Tetrahedron: Asymmetry* **2005**, *16*, 3371–3379.
47. Khasanov, A. B.; Ramirez-Weinhouse, M. M.; Webb, T. R.; Thiruvazhi, M. *J. Org. Chem.* **2004**, *69*, 5766–5769.
48. (a) Borer, B. C.; Balogh, D. W. *Tetrahedron Lett.* **1991**, *32*, 1039–1040; (b) Bhagwat, S. S.; Gude, C.; Chan, K. *Tetrahedron Lett.* **1996**, *37*, 4627–4630; (c) Srirajan, V.; Deshmukh, A. R. A. S.; Bhawal, B. M. *Tetrahedron,* **1996**, *52*, 5585–5590; (d) Joshi, S. N.; Deshmukh, A. R. A. S.; Bhawal, B. M. *Tetrahedron: Asymmetry* **2000**, *11*, 1477–1485.
49. Trabocchi, A.; Lalli, C.; Guarna, F.; Guarna, A. *Eur. J. Org. Chem.* **2007**, 4594–4599.
50. Chincholkar, P. M.; Puranik, V. G.; Deshmukh, A. R. A. S. *Synlett* **2007**, 2242–2246.
51. Niu, C.; Petterson, T.; Miller, M. J. *J. Org. Chem.* **1996**, *61*, 1014–1022.
52. Ojima, I.; Chen, H.-J. C.; Qiu, X. *Tetrahedron* **1988**, *44*, 5307–5318.
53. Palomo, C.; Aizpurua, J. M.; Mielgo, A.; Linden, A. *J. Org. Chem.* **1996**, *61*, 9186–9195.
54. Bandini, E.; Martelli, G.; Spunta, G.; Panunzio, M. *Synlett* **1996**, 1017–1018. See also reference 53.
55. Taggi, A. E.; Hafez, A. M.; Wack, H.; Young, B.; Drury, W. J., III; Lectka, T. *J. Am. Chem. Soc.* **2000**, *122*, 7831–7832.
56. Hodous, B. L.; Fu, G. C. *J. Am. Chem. Soc.* **2002**, *124*, 1578–1579. The catalyst is commercially available from Strem.
57. Lee, E. C.; Hodous, B. L.; Bergin, E.; Shih, C.; Fu, G. C. *J. Am. Chem. Soc.* **2005**, *127*, 11586–11587.
58. [R] (a) Taggi, A. E.; Hafez, A. M.; Lectka, T. *Acc. Chem. Res.* **2003**, *36*, 10–19; [R] (b) Fu, G. C. *Acc. Chem. Res.* **2004**, *37*, 542–547; [R] (c) France, S.; Weatherwax, A.; Taggi, A. E.; Lectka, T. *Acc. Chem. Res.* **2004**, *37*, 592–600.
59. Taggi, A. E.; Hafez, A. M.; Wack, H.; Young, B.; Ferraris, D.; Lectka, T. *J. Am. Chem. Soc.* **2002**, *124*, 6626–6635.
60. Shah, M. H.; France, S.; Lectka, T. *Synlett* **2003,** 1937–1939.
61. Wack, H.; France, S.; Hafez, A. M., Drury, W. J. III; Weatherwax, A.; Lectka, T. *J. Org. Chem.* **2004**, *69*, 4531–4533.
62. France, S.; Shah, M. H.; Weatherwax, A.; Wack, H.; Roth, J. P.; Lectka, T. *J. Am. Chem. Soc.* **2005**, *127*, 1206–1215.
63. Weatherwax, A.; Abraham, C. J.; Lectka, T. *Org. Lett.* **2005**, *7*, 3461–3463.
64. Sereda, O.; Wilhelm, R. *Synlett* **2007**, 3032–3036.
65. Zhang, Y.-R.; He, L.; Wu, X.; Shao, P.-L.; Ye, S. *Org. Lett.* **2008**, *10*, 277–280.
66. Duguet, N.; Campbell, C. D.; Slawin, A. M. Z.; Smith, A. D. *Org. Biomol. Chem.* **2008**, *6*, 1108–1113.
67. [R] Bose, A. K.; Manhas, M. S.; Mathur, A.; Wagle, D. R. In *Recent Progress in the Chemical Synthesis of Antibiotics and Related Microbial Products,* Vol. 2; Lukacs, G., Ed.; Springer-Verlag, Heidelberg, 1993, pp 551.

68. Pathak, V. N.; Gupta, R.; Garg, M. *Heteroat. Chem.* **2004,** *15*, 494–501.
69. Fernandez, R.; Ferrete, A.; Lassaletta, J. M.; Llera, J. M.; Monge, A. *Angew. Chem., Int. Ed. Engl.* **2000**, *39*, 2893–2897.
70. (a) Jarrahpour, A.; Zarei, M. *Molecules* **2007**, *12*, 2364–2379; (b) Jarrahpour, A.; Zarei, M. *Synth. Commun.* **2008**, *38*, 1837–1845.
71. Cremonesi, G.; Dalla Croce, P.; Fontana, F.; La Rosa, C. *Tetrahedron: Asymmetry* **2008**, *19*, 554–561.
72. Berlin, J. M.; Fu, G. C. *Angew. Chem., Int. Ed. Engl.* **2008**, *47*, 7048–7050.
73. (a) Ojima, I.; Habus, I.; Zhao, M.; Zucco, M.; Park, Y. H.; Sun, C. M.; Brigaud, T. *Tetrahedron,* **1992**, *48*, 6985–7012; (b) Holton, R. A.; Liu, J. H. *Bioorg. Med. Chem. Lett.* **1993**, *3*, 2475–2478; (c) Nicolaou, K. C.; Dai, W.-M.; Guy, R. K. *Angew. Chem. Int. Ed. Engl.* **1994**, *33*, 15–44; (d) Srirajan, V.; Deshmukh, A. R. A. S.; Bhawal, B. M. *Tetrahedron* **1996**, *52*, 5585–5590; (e) Kingston, D. G. I. *Chem. Commun.,* **2001**, 867–880.
74. Alajarin, M.; Vidal, A.; Tovar, F. *Tetrahedron* **2005**, 61, 1531–1537;
75. Oshitari, T.; Mandai, T. *Synlett* **2009**, 787–789.
76. Vincent, G.; Williams, R. M. *Angew. Chem., Int. Ed. Engl.* **2007**, *46*, 1517–1520.

Chapter 3. Five-Membered Carbocycles 71

3.1 Danheiser Annulation

Kevin M. Peese

3.1.1 Description

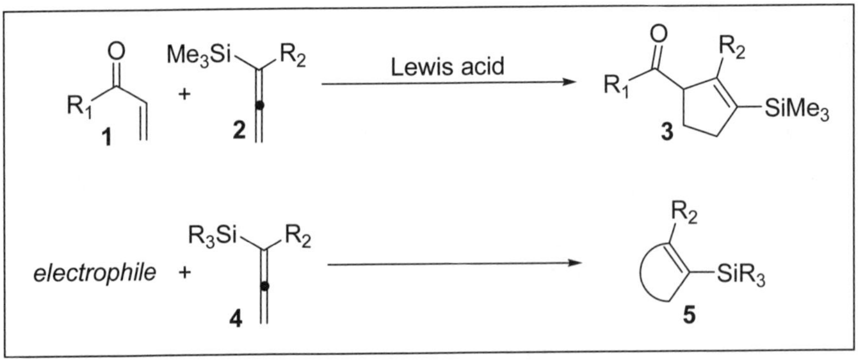

The Danheiser annulation, in its classic form, is the Lewis acid-catalyzed reaction of an *α,β*-unsaturated ketone **1** with a trimethylsilylallene **2** to form a silyl cyclopentene **3**.[1–3] More broadly, the Danheiser annulation ecompasses reactions of silylallenes **4** with electrophilic double bonds to form cyclic products **5**, usually under Lewis acid catalysis. The Danheiser annulation should not be confused with other annulation processes developed by Danheiser, such as Danheiser's aromatic annulation[4] and the Stork–Danheiser alkylation.[5]

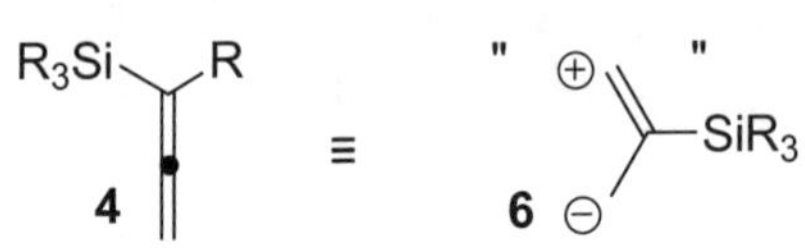

The defining feature of the Danheiser annulation is the use of a silylallene **4** as a three-carbon synthon **6** in a step-wise cycloaddition reaction. The annulation process is always initiated by the attack of the nonsilyl substituted position of the silyl allene onto the electrophilic double bond to generate a silyl-stablized vinyl cation at the central carbon atom. After a 1,2-sp^2-silyl migration, which places the silyl group onto the central carbon of the former allene, ring closure occurs, yielding a 5-membered ring. The electrophile can be activated by complexation to a Lewis acid; however, silylallenes will react with most strongly electrophilic double bonds. A diverse array of electrophiles have been shown to participate in the annulation process, leading to a variety of annulation products beyond cyclopentenes, such as furans, dihydrofurans, dihydro-pyrroles, isoxazoles, and azulenes.

3.1.2 Historical Perspective

The discovery of the Danheiser annulation was serendipitous as it came out of research directed at a different synthetic challenge.[6] In the late 1970s, there were no satisfactory methods available for the addition of propargylic anions, where the anion is centered at the sp^3 carbon, to carbonyl compounds. Whereas addition of allylic organometallics **8** such allylmagnesium halides or allyllithiums to carbonyl compounds **7** efficiently produces homoallylic products **9**, the analogous reaction of propargylic organometallic reagents **10** with carbonyl compounds **7** produces mixtures of homopropargylic alcohols **11** and allenic alcohols **12**. In this context, Danheiser proposed that the use of a silylallene instead of an alkyne-based reagent could deliver the homopropargylic alcohol products selectively.

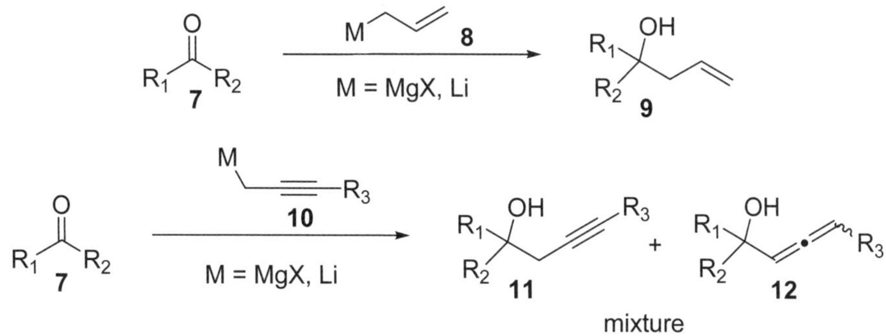

Investigation of the $TiCl_4$-catalyzed electrophilic addition of aldehydes and ketones **7** onto silylallenes **13** produced moderate to high yields of the desired homopropargylic alcohol product **14**.[7] It was surprising that in some cases the reaction produced a mixture of the expected homopropargylic alcohol **14** along with the trimethylsilylvinyl chloride product **16**. This result suggested that the intermediate silylvinyl cation **15** was relatively stable and the desilylation step was not as rapid a process as might have been expected. Preparatively, production of silylchloroalkenes **16** did not prove to be an issue as the crude reaction mixtures could be treated with KF in DMSO to efficiently convert the trimethylsilylvinyl chlorides **16** to the desilylated product alkynes **14**.

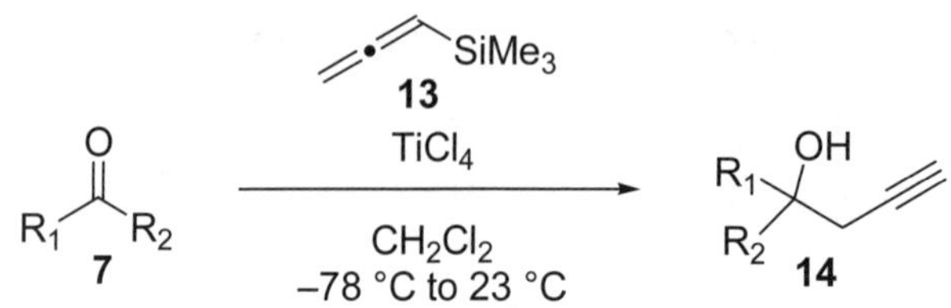

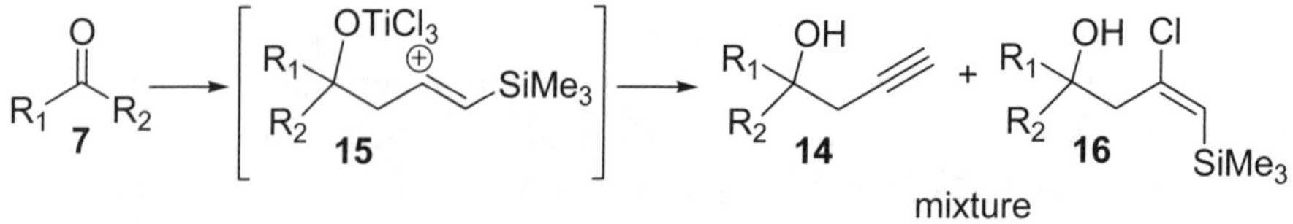

Seeking to expand the scope of the transformation, Danheiser investigated the use of an *α,β*-unsaturated ketone as the electrophilic component in the homopropargylic alcohol synthesis.[6] The first reaction attempted was that of cyclohexenone (**17**) with silylallene **18** using $TiCl_4$ as a Lewis acid catalyst. After treatment of the initial isolated product **19** with fluoride, *α,β*-unsaturated ketone **20** was isolated as the major product. After verification of the structure of *α,β*-unsaturated ketone **20**, the course of the reaction could be deduced.

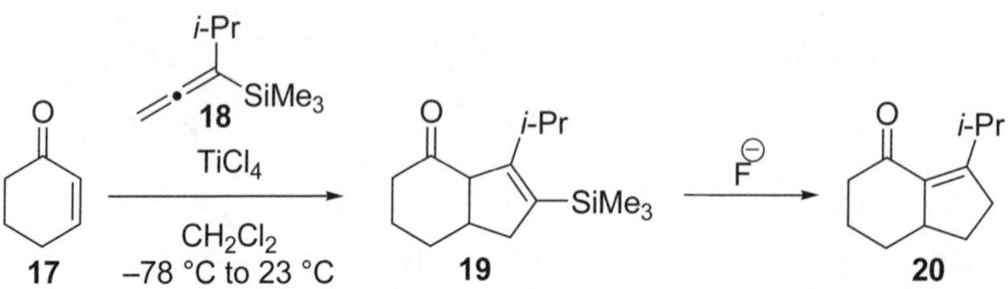

In the expected course of the reaction, the electrophile combines with the silylallene resulting in the intermediate silylvinyl cation **21**, which either undergoes desilylation to produce the expected product alkyne **22** or traps chloride ion to produce vinylchloride **23**. In the case of the annulation reaction, the intermediate silylvinyl cation **21** underoes an apparent 1,2-sp^2-silyl migration process. The resulting isomeric silylvinyl cation **24** is then able to react with the pendent titanium enolate leading to the observed annulation product **25**.

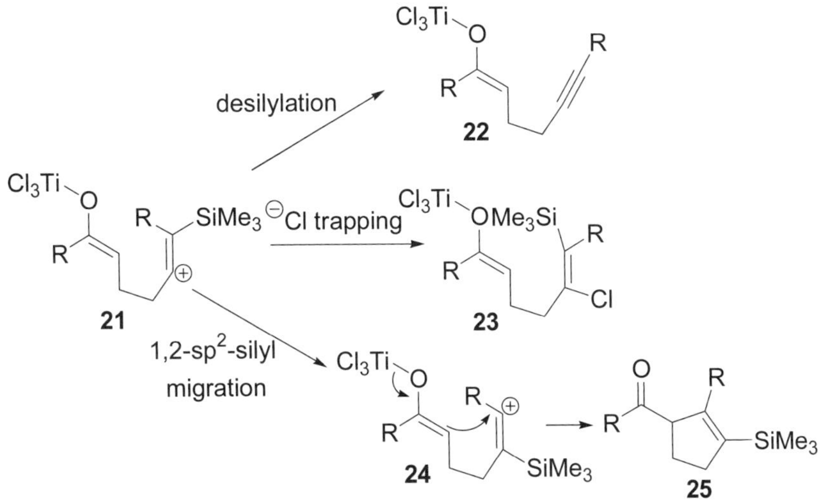

It was fortuitous that isopropy silylallene **18** was the first allene investigated as it was later found that substitution *ipso* to the silyl group is necessary to obtain preparatively useful yields (> 20%).[6] Indeed, in the previous year during investigations of the very similar $TiCl_4$-catalyzed reaction of *α,β*-unsaturated ketones and *α,β*-unsaturated esters with 1-unsubstituted silylallenes, Santelli and Jellal did not report observing the formation of cyclopentenyl annulation products.[8] Further investigation of the annulation process by Danheiser demonstrated it to be quite general and synthetically useful. Results of these initial studies detailing the discovery and scope of the annulation process were then published in 1981.[1] Since the initial publication, a number of interesting synthetically useful variations have been disclosed, greatly expanding the variety of electrophiles that can be used in the annulation reaction (vide infra).

3.1.3 Mechanism

General Mechanism

The generally accepted mechanism of the classic Danheiser annulation involves three basic steps: the Lewis acid-catalyzed electrophilic combination of the *α,β*-unsaturated ketone with the silylallene, a 1,2-sp^2-silyl migration, and a final cyclization step. This mechanism was first proposed by Danheiser in the original publication of the annulation and has been generally accepted but has never been formally investigated.[1] A more detailed account of the reaction pathway is shown below. Treatment of the *α,β*-unsaturated ketone **1** with $TiCl_4$ produces a titanium complex existing as two resonance-stablized cations **26** and **27**. Attack of the 2,3-π-bond of the silylallene **2** onto the strongly electrophilic titanium complex **27** produces the silyl-stablized vinyl cation **21**. Silylvinyl cation **21** exists in equilibrium with the isomeric silylvinyl cation **24** via a 1,2-sp^2-silyl migration. Finally, intramolecular attack of the titanium enolate moiety produces cyclopentene **3**. It is important to note that silylvinyl cation isomer **21** does not undergo annulation, since formation of the cyclobutane product is predicted to be highly disfavored and consequently is not observed.

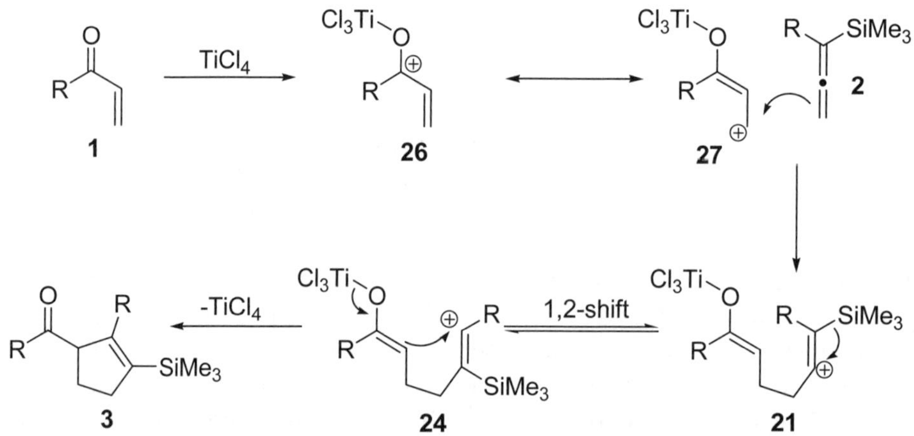

For the reaction to proceed in reasonable yields, substitution on the terminal position of **2** (R ≠ H,) is required. This is likely due to the predicted relative higher energy of the terminal vinyl cation **28** compared to the internal vinyl cation **29**. Thus without substitution, equilibrium of the two vinyl cations **28** and **29** strongly favors the internal cation isomer **29** which does not undergo cyclization but rather undergoes desilylation to afford the terminal alkyne product **30**.[7]

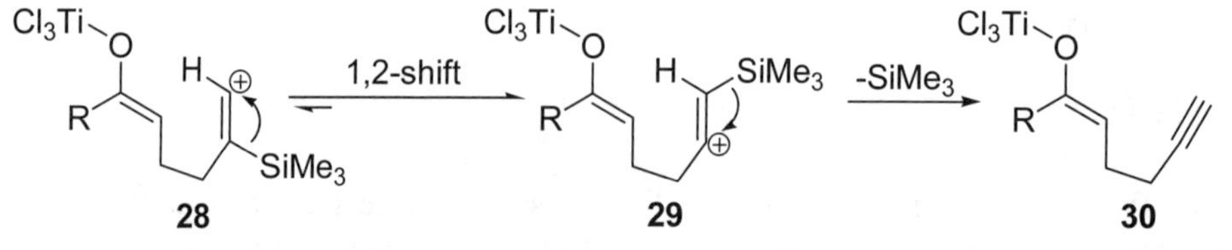

When other electrophiles besides *α,β*-unsaturated ketones are used, the mechanism is analogous, however, care must be taken to ensure that direct desilylation of acyclic intermediate **31**, to form acyclic alkyne **32** does not occur. It is interesting that the use of more sterically encumbered alkylsilanes, for example *tert*-butyldimethylsilyl rather than trimethylsilyl, increases the selectivity of the reaction for the cyclized product versus the acyclic alkyne **32** that is produced by desilylation of silylvinyl cation **31**.[9]

Danheiser has postulated that this is due to the need for chloride to initiate desilylation; sterically large silyl groups should disfavor the approach of chloride to the silyl group. Thus some variations of the Danheiser annultion are practical only with si!anes that are sterically larger than trimethylsilyl in order to avoid desilylation of the vinyl cation intermediate (vide infra).

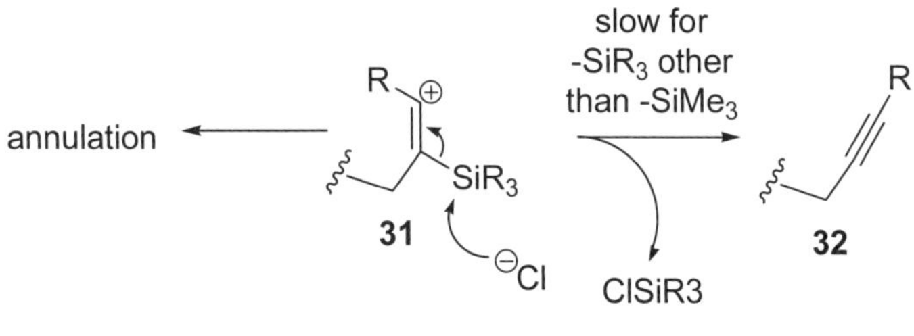

Stereochemical Considerations

Stereochemical investigation of the Danheiser annulation demonstrated that the cyclization usually proceeds with a high degree of stereoselectivity.[2] Annulation of (*E*)-enone **33** affords a single diastereomer, cyclopentene **35**, in which the methyl group and the acetyl group are oriented *anti* to one another, the result of an apparent *suprafacial* attack of the allene on the *α,β*-unsaturated ketone. Likewise, annulation of trisubstituted *α,β*-unsaturated ketone **36** provides a single diastereomer, cyclopentene **37**, once again the result of an apparent *suprafacial* attack of the allene on the *α,β*-unsaturated ketone. For isomeric *α,β*-unsaturated ketone **38**, annulation also proceeds with high stereoselectivity, but in this case the other diastereomer **37**, can be detected in trace amounts. Taken together, these results suggest that the rate of ring closure is significantly faster than the rate of enolate rotation, giving rise to the appearance of *suprafacial* attack of the allene onto the *α,β*-unsaturated ketone. For allenes substituted at the 3-position, the annulation reaction proceeds with moderate to good stereoselectivity. For example, treatment of cyclopentenone (**40**) with allene **41** under standard conditions gives predominantly cyclopentene **42** with the methyl group in an *exo* orientation.

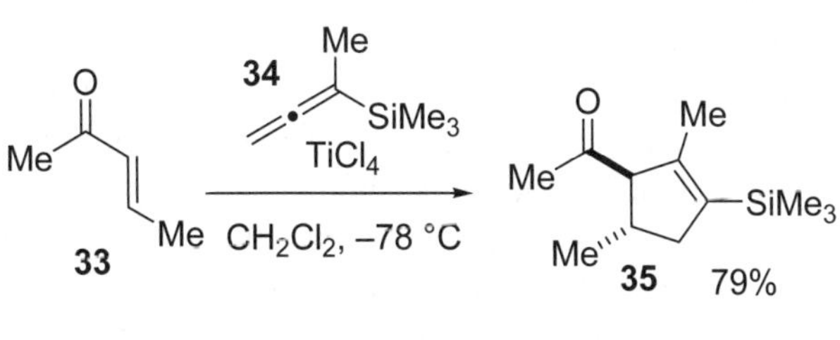

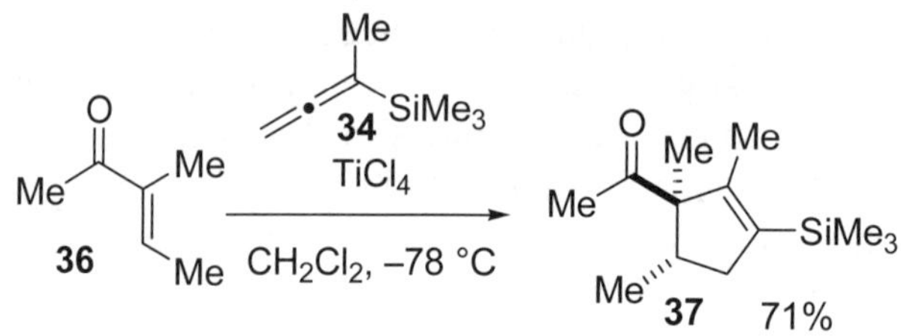

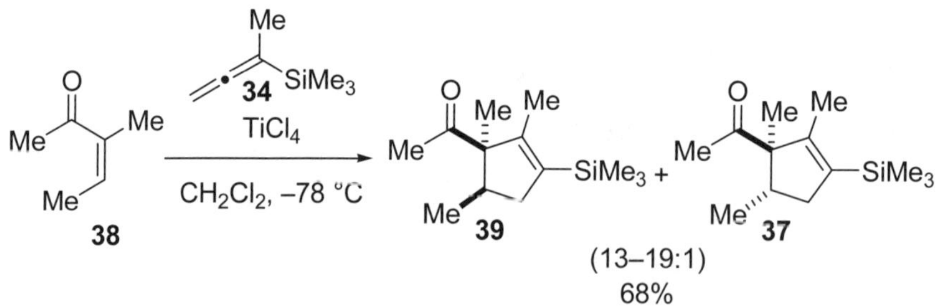

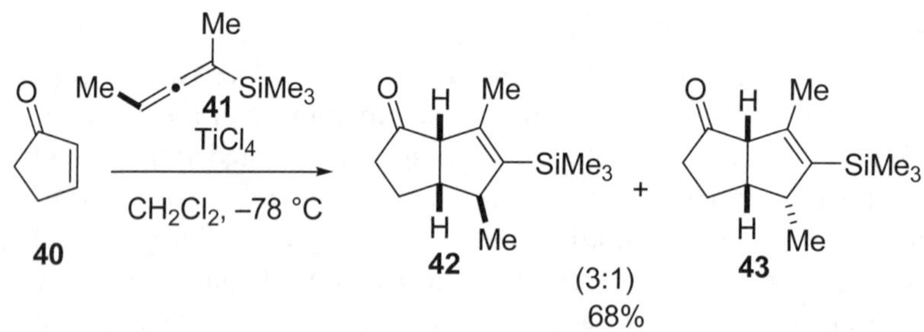

3.1.4 Synthetic Utility

Silyl Allene Scope

The key component of the Danheiser annulation is the silylallene on which there are three points of diversity: the carbon bearing the silyl group; the silyl group itself; and the distal 3-position carbon. Of critical importance is the need for carbon substitution *ipso* to the silyl group in the silyl allene **44** (R ≠ H) to obtain preparatively useful yields (> 20%) of the cyclized product. Without substitution, the 1,2-sp^2-silyl migration becomes disfavoured, leading to predominantly nonannulated products, the acyclic alkyne and the silylvinyl chloride. For example, during early investigations of the annulation process, Danheiser found that reaction of cyclohexenone **17** with the unsubstituted parent allene, trimethylsilylallene **13**, under standard conditions led consistently to only low yields of the annulated cyclopentene product **45** (17–19%).[2] With respect to the substitution on the silyl group, a variety of silyl moieties have been shown to participate in the annulation process including -SiMe$_3$ (-TMS), -SiMe$_2$*t*-Bu (-TBS), and -Si*i*-Pr$_3$ (-TIPS).

For the distal position of the allene, *C*-3 with respect to the silyl group, both hydrogen and alkyl substitution is tolerated. Only at the extreme limit of steric congestion of both the allene and the enone are low yields observed. For example, reaction of *β*-dimethyl *α*,*β*-unsaturated ketone **46** with 3,3-dimethylallene **47** under standard conditions provides only 7% yield of the cyclopentene product **48**.[2]

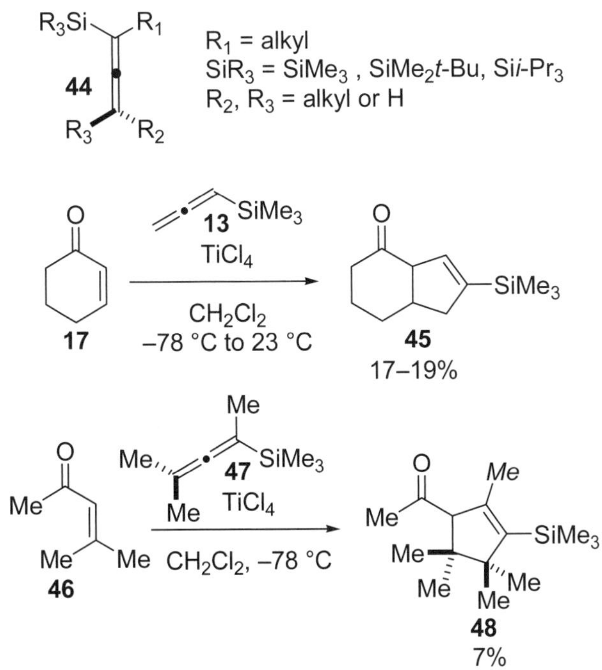

Cyclopentene Synthesis

The original application of the Danheiser cyclization was for the preparation of cyclopentenes employing *α*,*β*-unsaturated ketones as the electophilic component of the annulation process. A variety of *α*, *β*-unsaturated ketones with various alkyl substitution patterns readily participate in the Danheiser annulation.[1,2]

In addition to *α*,*β*-unsaturated ketones, *α*,*β*-unsaturated esters also undergo cyclization, albeit at much slower rates than the corresponding *α*,*β*-unsaturated ketones.[2] Treatment of methyl acrylate (**49**) with silylallene **34** at 25 °C (rather than the standard –78 °C) in the presence of $TiCl_4$ furnished the expected cyclopentene **50** in 49% yield. To access carboxylate derivatives, Danheiser has reported the use of *α*,*β*-unsaturated acylsilanes as the electrophilic component in the annulation process producing the corresponding acylsilane substituted cyclopentenes.[10] The acylsilane can then be readily oxidized to afford the carboxylic acid. For example, treatment of *α,β*-unsaturated acylsilane **51** with silylallene **34** under standard conditions affords cyclopentene **52**, which then undergoes oxidation with H_2O_2/NaOH in aqueous THF at 40 °C to give carboxylic acid **53**. Care must be exercized in maintaining the reaction at –78 °C when using an *α,β*-unsaturated acyltrimethylsilanes as the electrophilic component since at warmer temperatures the acyltrimethylsilanes cyclopentene annulation product undergoes ring expansion under the reaction conditions, providing a six-membered ring product.

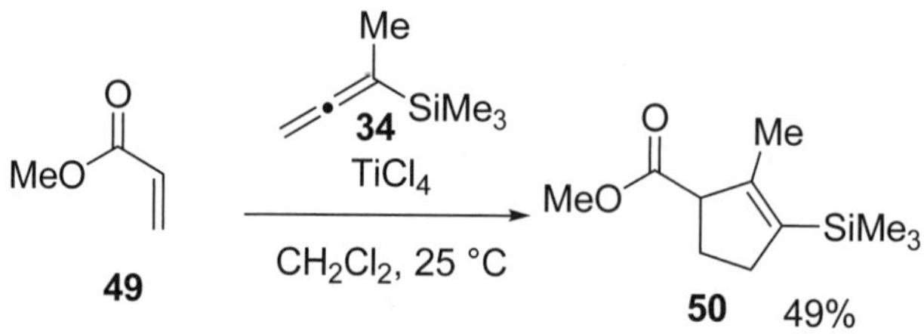

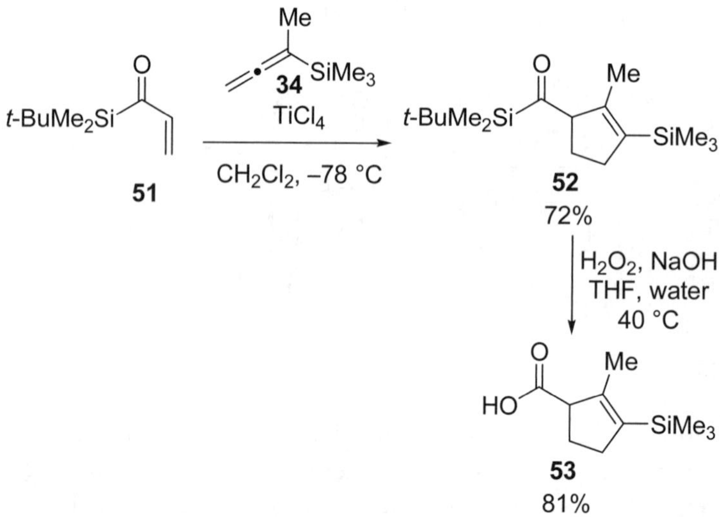

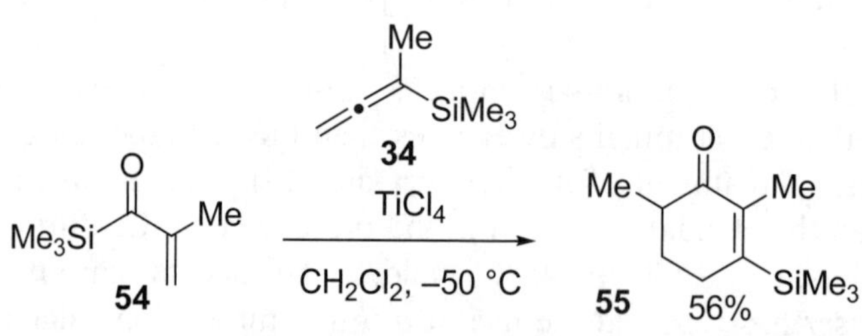

via (proposed):

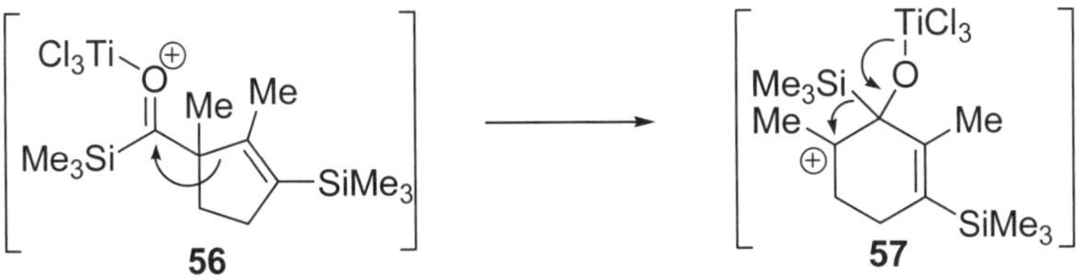

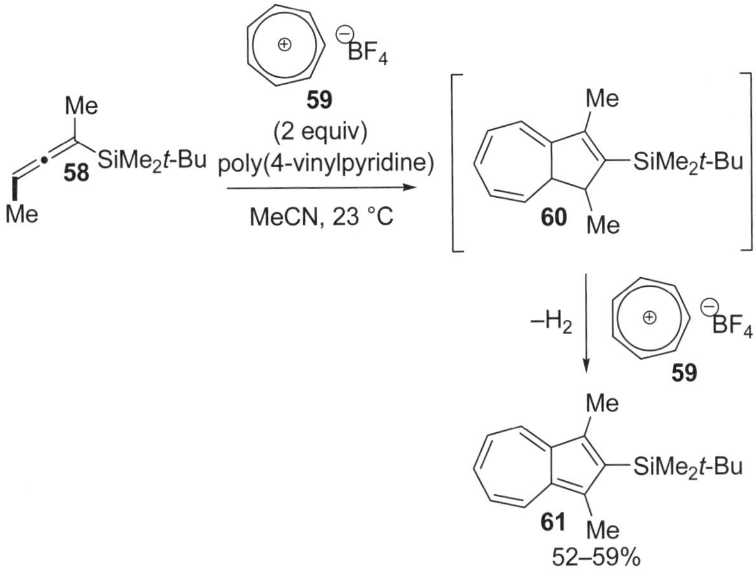

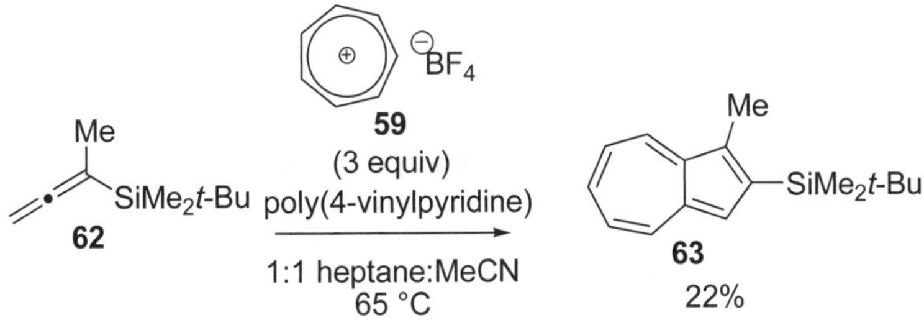

The Danheiser annulation has also been reported to be useful for the preparation of substituted azulene products.[11] For example, treatment of silylallene **58** with tropylium cation **59** at 23 °C produces intermediate cyclopentene dihydroazulene **60**, which is not isolated but rather undergoes in situ dehydrogenation with a second equivalent of tropylium cation **59** to provide azulene **61**. The use of the *tert*-butyldimethylsilyl group as opposed to the trimethylsilyl group is necessary since the trimethylsilyl group tends to undergo premature desilylation rather than cyclization to the azulene. Substitution at the 3-position of the allene was also found to significantly improve the yield of the reaction. For example, the reaction of silylallene **62** (unsubstituted at the 3-position) with tropylium cation **59** afforded the product azulene **63** in only 22% yield compared to the 52–52% yield obtained with the 3-substituted silylallene **61**.

The use of other electrophiles in the Danheiser annulation for the formation of cyclopentenes is relatively unexplored. Ynones have been reported to undergo cycloaddition, although examples are limited.[2] Thus treatment of butynone **64** with silyl allene **47** under the usual conditions delivered cyclopentadiene **65** in 53% yield. Danheiser has reported that nitroalkenes do not provide annulation products furnishing instead acyclic alkynes.[2] For example, treatment of nitroalkene **66** with silylallene **34** under standard conditions gives alkyne **67** rather than the cyclopentene **68**. *α,β*-Unsaturated aldehydes have been reported to be problematic, failing to provide isolable cyclopentene products.[2] Treatment of methacrolein (**69**) with silylallene **34** under the standard conditions produced only a complex mixture of unidentified products.

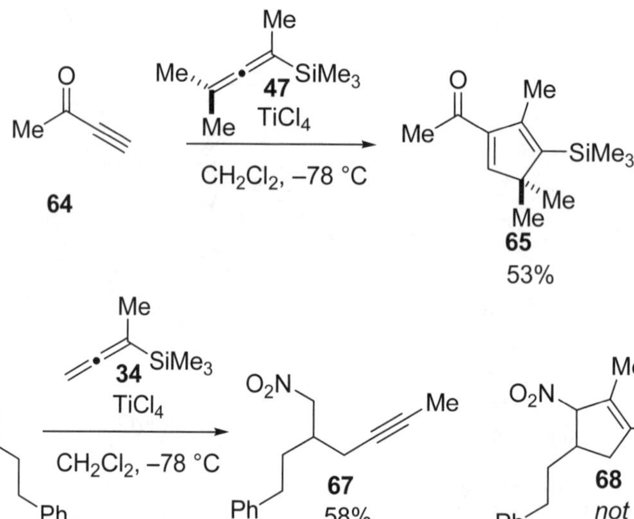

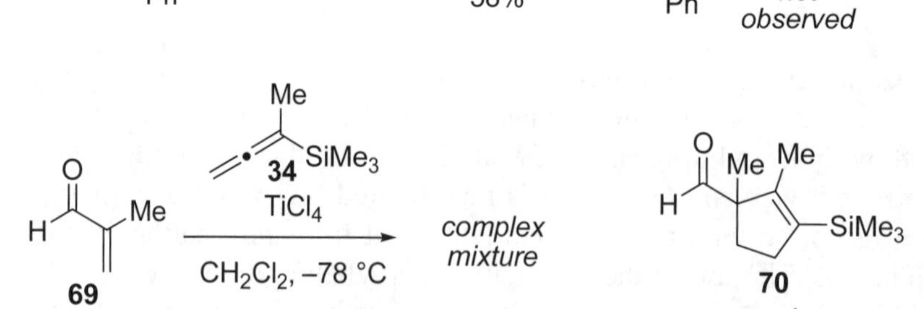

Furan and Dihydrofuran Synthesis

Dihydrofurans[9,12] and furans[13] are accessible through the Danheiser annulation using aldehydes and acid chlorides, respectively, as electrophilic component in the Danheiser annulation. The pivitol advance to allow practical access to dihydrofurans and furans, however, was the recognition that the use of more sterically encumbered alkylsilanes, i.e., *tert*-butyldimethylsilyl rather than trimethylsilyl, is critical to obtaining the annulation product selectively over the undesired noncyclized alkyne product. For instance, treatment of aldehyde **71** with trimethylsilylallene **34**, under standard conditions, results in an 85% yield of acyclic alkyne **72**,[7a] formally the product of propargyl addition to the aldehyde, whereas using the *tert*-butyldimethylsilylallene **62** affords the dihydrofuran **73** in 70% yield.[9]

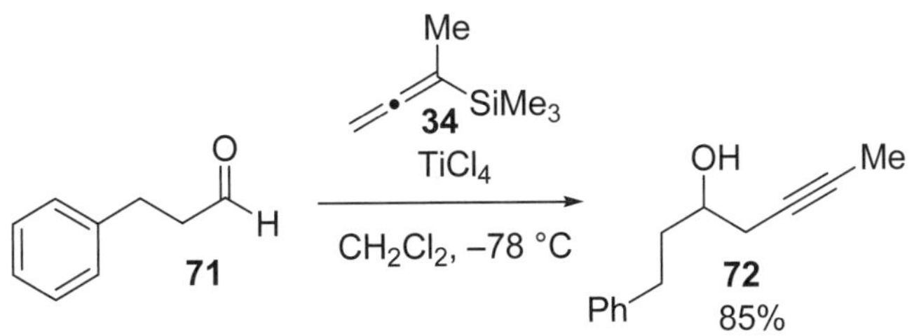

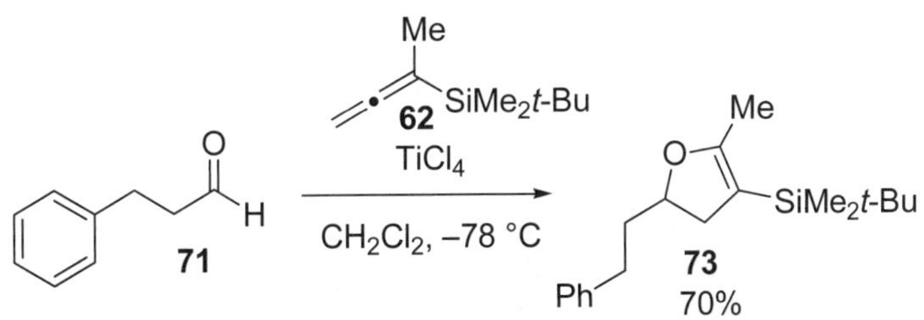

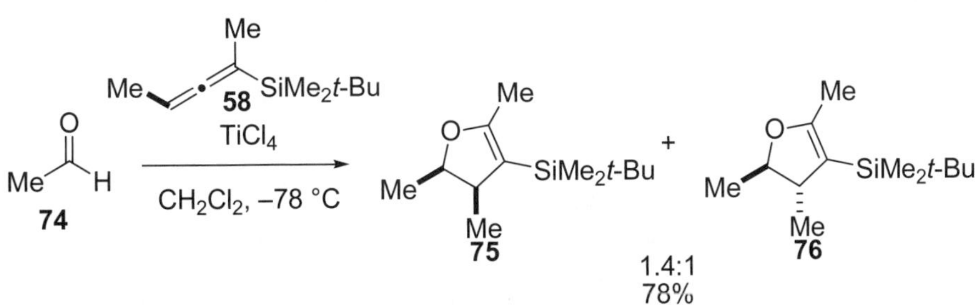

The reaction of silylallenes with aldehydes to form dihydrofurans is applicable to a variety of alkyl aldehydes, including substrates such as acetaldehyde (**74**), cyclohexylaldehyde (**77**), and pivaldehyde (**80**).[9] When the silyllallene is substituted at the 3-position, the reaction diastereoselectively provides the *syn* products, with increasing selectivity as the steric bulk of aldehyde increases.

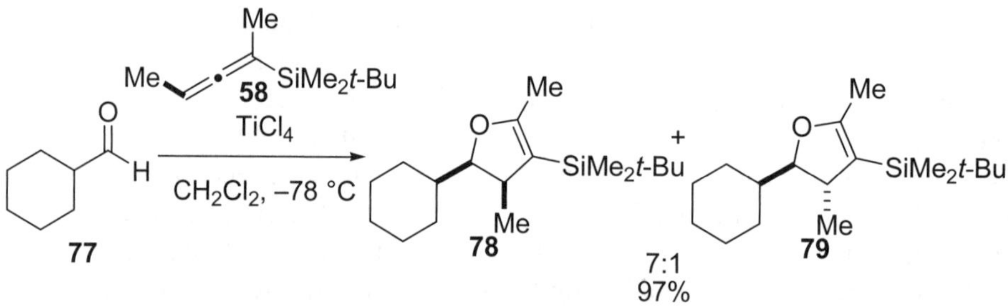

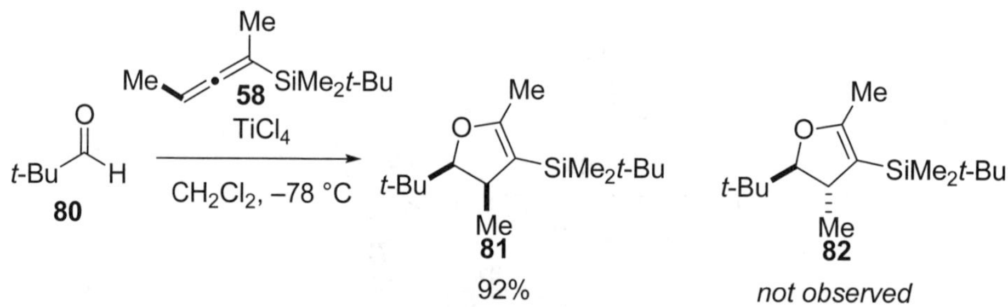

Furans can be prepared using acid chlorides as the electrophilic component in the Danheiser annulation, which allows for the preparation of highly substituted derivatives.[13] Both alkyl and aryl acid chlorides undergo the furan synthesis. For silylallenes lacking substitution at the 3-position, it is necessary to use triisopropylsilyl as the silyl group on the allene. This is due to the propensity of the unsubstituted furan products to undergo further Friedel–Crafts type acylation reactions at the unsubstituted *CH* position of the furan under the reaction conditions. A sufficiently bulky silyl group prevents this from happening since the sterically demanding silyl group effectively shields this site from further reactivity.

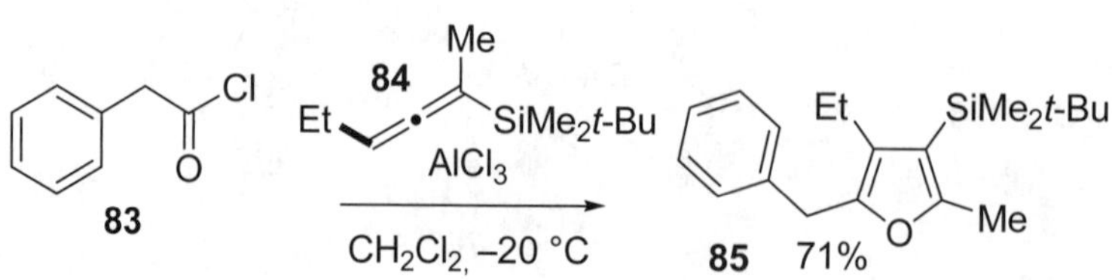

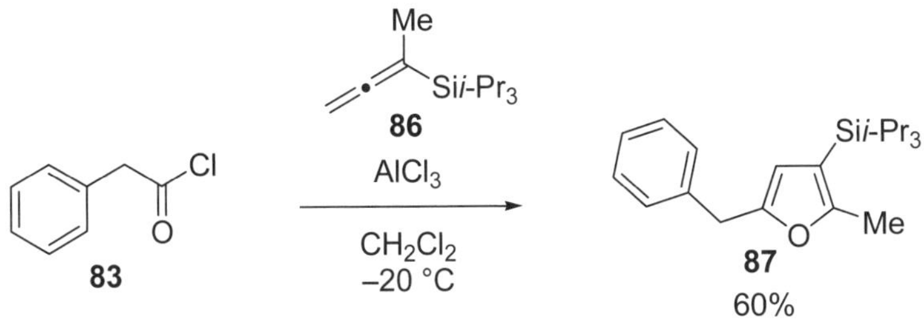

Dihydropyrrole Synthesis

Dihydropyrroles can be prepared through the Danheiser annulation by utilizing *N*-acyliminium ions as the electrophile.[9,14] Similar to the method used to prepare dihydrofurans and furans, sterically encumbered silyl groups are necessary to obtain useful yields of the dihydropyrrole annulation products.[9] For example, treatment of aminal **88** with trimethylsilylallene **34** and $TiCl_4$ provides a mixture of pyrrolizinone **89** (28%) and acyclic alkyne **90** (59%). The analogous reaction using the *tert*-butyldimethylsilylallene analogue **62** affords the annulation product of pyrrolizinone **91** exclusively in 63% yield. A single example involving an acyclic *N*-acyliminium electrophile gave a lower yield suggesting that cyclic *N*-acyliminium substrates are preferred for the annulation reaction.

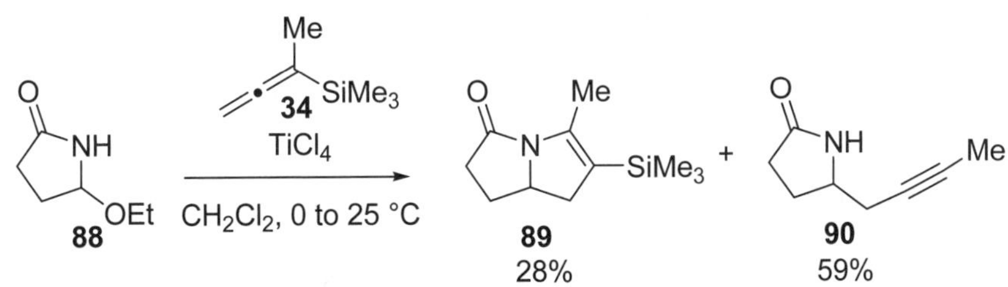

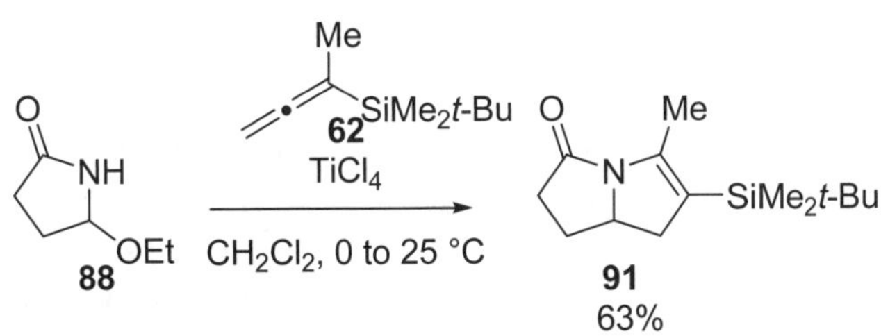

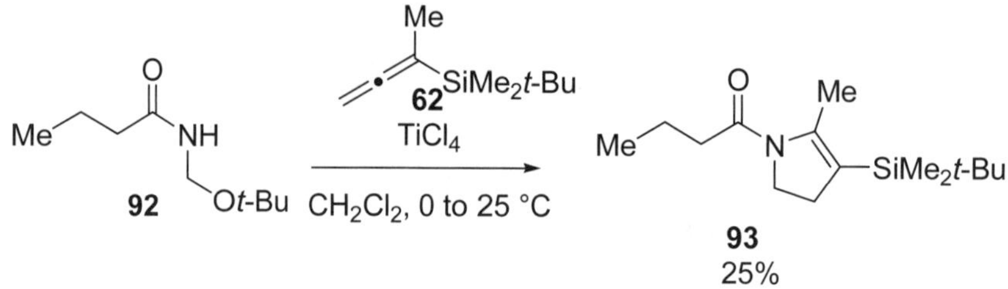

Isoxazole Synthesis

Isoxazoles are accessible through the Danheiser annulation using nitrosonium as the electrophilic component.[15] For the annulation reaction of trimethylsilyl allenes, the corresponding isoxazole annulation product tends to undergo partial protodesilylation during the reaction. The desilylation process can be driven to completion after isoxazole formation is complete by adding water to the reaction and heating until protodesilylation is complete. For *tert*-butyldimethylsilylallenes, no desilylation occurs and the silyl isoxazoles can be isolated directly.

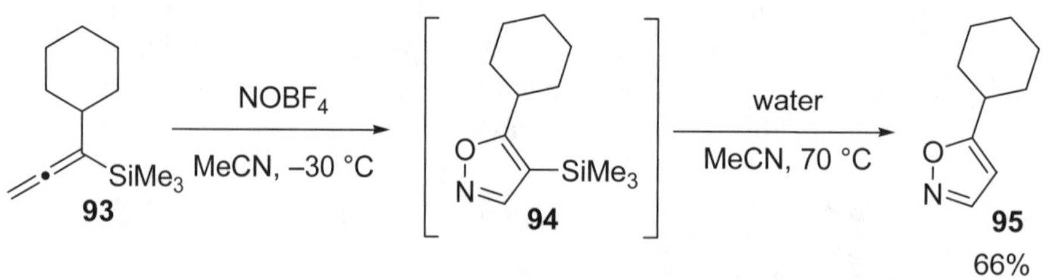

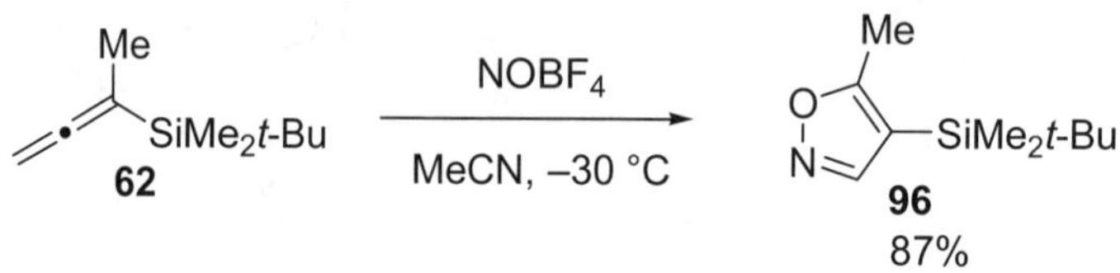

Formal [3 + 3] Processes

Two methods for accomplishing a formal [3 + 3] cycloadditions using the Danheiser annulation have been reported.[16,17] Angle et al. reported that formation of dihydronaphthalene **99** was formed in 77% yield from the reaction of benzyl alcohol **97** with silylallene **98** in the presence of $SnCl_4$.[16] Alternatively, the use of hydroxysubstituted benzyl alcohols provides the spirofused cyclopentene products. Treatment of hydroxybenzyl alcohol **100** with silylallene **98** in the presence of $SnCl_4$ produces spirocyclopentene **101** in 76% yield. Depending on the nature of the substituents on the aryl ring, mixtures of dihydronaphthalenes and spirocyclopentenes may be obtained.

Unfortunately, the method requires an excess of either the benzyl alcohol or the silylallene to obtain moderate yields.

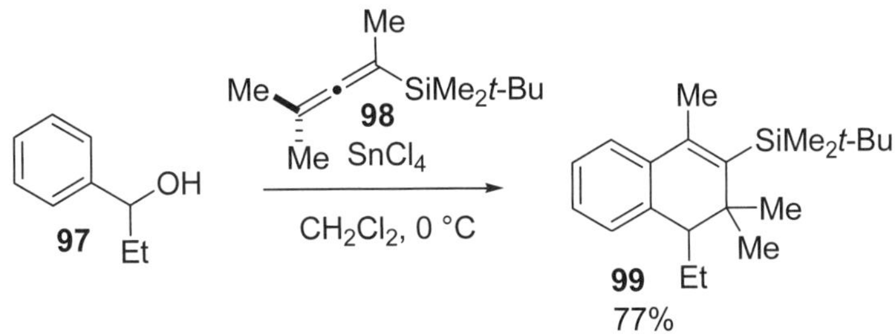

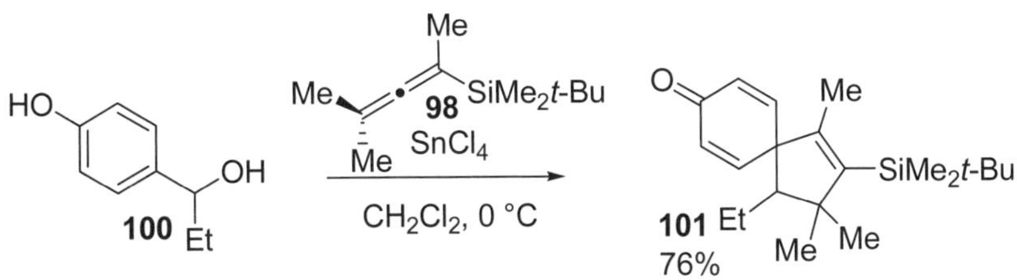

Yadav et al. have reported using cyclopropanes as the electrophilic component in the Danheiser annulation leading to the formation of cyclohexene products.[17] Treatment of cyclopropane **102** with Et_2AlCl at 25 °C resulted in ring opening of the cyclopropane, which then underwent reaction with silylyallene **34** to afford cyclohexene **103**, formally a [3 + 3] annulation product.

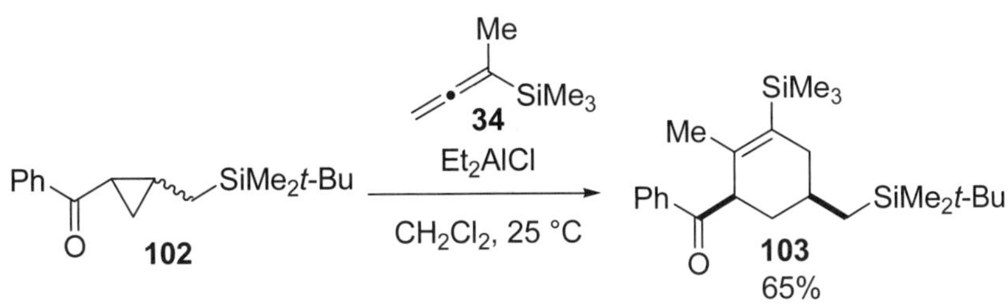

Enantioselectivity

Two reports of enantioselective Danheiser annulations have appeared in the literature.[12,14] Evans et al. have reported the annulation of ethyl glyoxylate (**104**) to form dihydrofuran products in good to excellent enantioselectivity using a chiral scandium catalyst.[12] For example, reaction of ethyl glyoxylate (**104**) with *tert*-butyldiphenylsilylallene **105** in the presence of the chiral scandium catalyst **106** led to dihydrofuran **107** in 91% yield and 92% *ee*. As with Danheiser's previous studies regarding the formation of dihydrofurans,[9] sterically encumbered silanes were found to be necessary to obtain the annulation product as opposed to the acyclic alkyne byproduct.

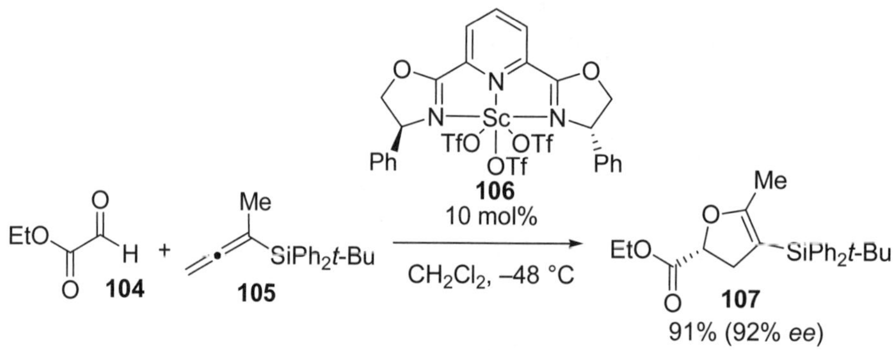

Akiyama et al. have reported the annulation of iminoesters to form dihydropyrrole products with moderate enantioselectivities using a chiral copper catalyst.[14] Reaction of iminoester **108** with triisopropylsilylallene **86** with a chiral copper catalyst afforded dihydropyrrole **110** in 92% yield and 84% *ee.*

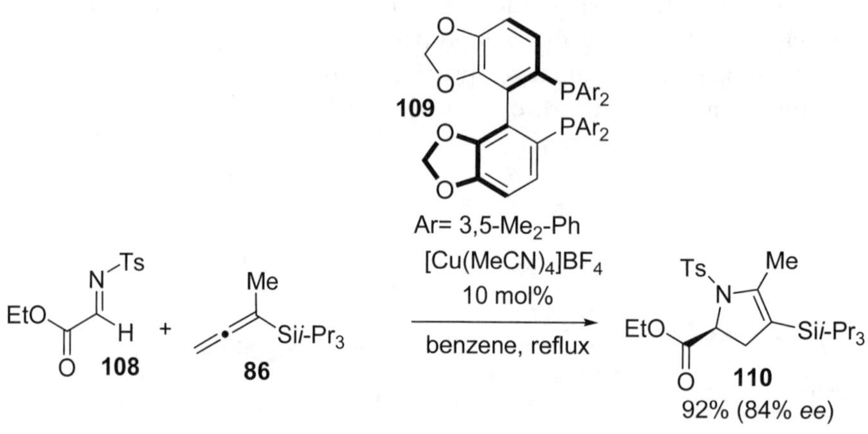

Synthetic Utility of Silylalkene Annulation Products

The silylalkene containing annulation products produced in the Danheiser annulation are themselves useful interemediates in organic synthesis. Danheiser demonstrated the protodesilylation of the silylcyclopentene annulation products in early studies.[2] Treatment of silylalkene **111** with K_2CO_3 in methanol effected isomerisation and protodesilylation to produce isomerized *α,β*-unsaturated ketone **112** in 68% yield. Silyl-substituted aromatic heterocycles readily undergo protodesilylation.[11,15] Silyl-substituted furan **85** underwent smooth desilylation on exposure to HF•pyridine in THF to provide desilylated furan **113** in 86% yield.[11]

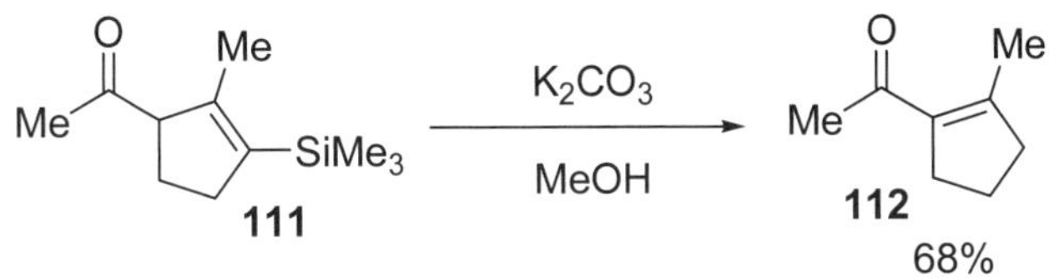

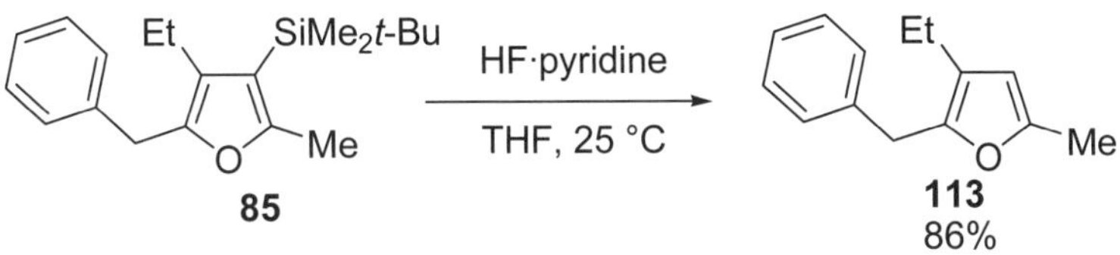

Friedel–Crafts acylation of silylalkene annulation products has been demonstrated. For example, treatment of silylalkene **111** with acetylchloride in the presence of $AlCl_3$ generated *α,β*-unsaturated ketone **114** in 95% yield.[2] More sensitive substrates are also amenable to Friedel–Crafts acylation. Treatment of silylalkene **107**, a chiral dihydrofuran, to similar conditions produced the acyl product **115** in 84% yield.[12] Bromination of silyl substituted isoxazoles has also been demonstrated.[15]

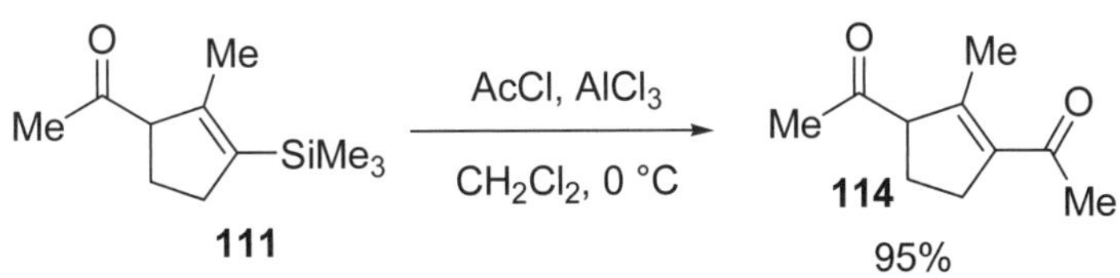

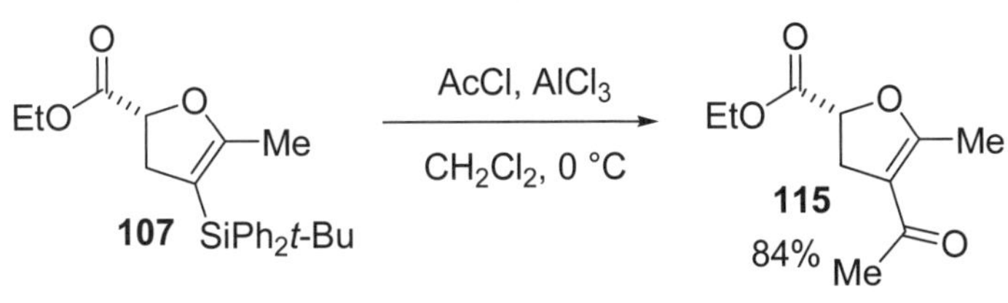

3.1.5 Application in the Total Synthesis of Natural Products

The Danheiser annulation has been featured in the formal total synthesis of the natural products (±)-clavukerin (**116**) and (±)-isoclavukerin (**117**) reported by Schäfer and coworkers.[18] Cycloheptenone **118** was prepared in a three-step sequence from cycloheptadiene. Cycloheptenone **118** was then treated with trimethylsilylallene **34** in the presence of $TiCl_4$ to afford a readily separable mixture of racemic cyclopentenes **119** and **120** in a 1:1 ratio and 91% combined yield. After base-catalyzed isomerization and fluoride-mediated protodesilylation, *α,β*-unsaturated ketone **121** was produced. *α,β*-Unsaturated ketone **121** served as an intermediate in earlier syntheses of (±)-clavukerin (**116**), thus providing a formal total synthesis of **116**.[19] In a similar manner, isomeric cyclopentene **120** was converted to the analogous *α,β*-unsaturated ketone, which was also an intermediate in an earlier synthesis of (±)-isoclavukerin (**117**).[19a]

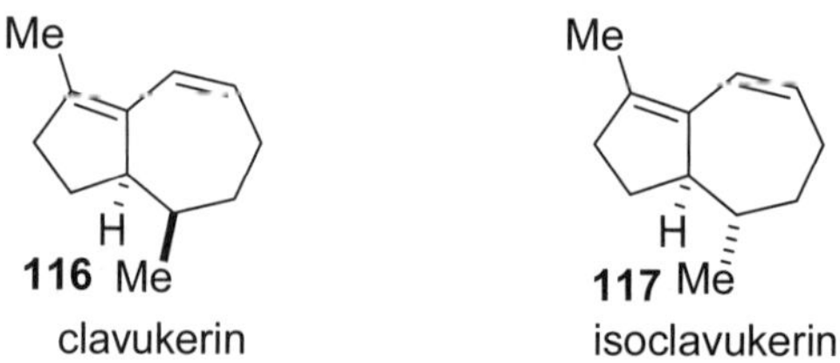

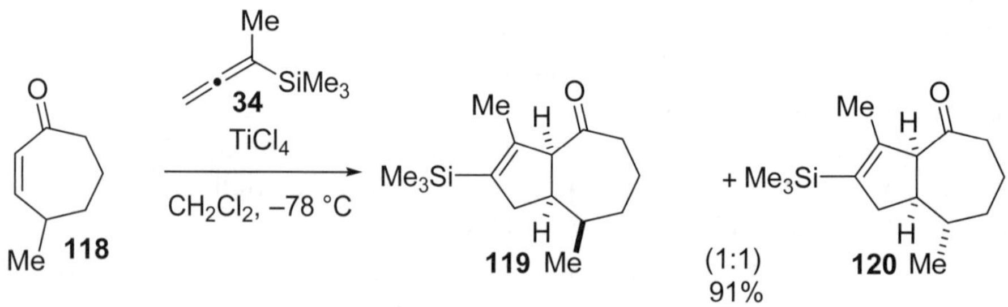

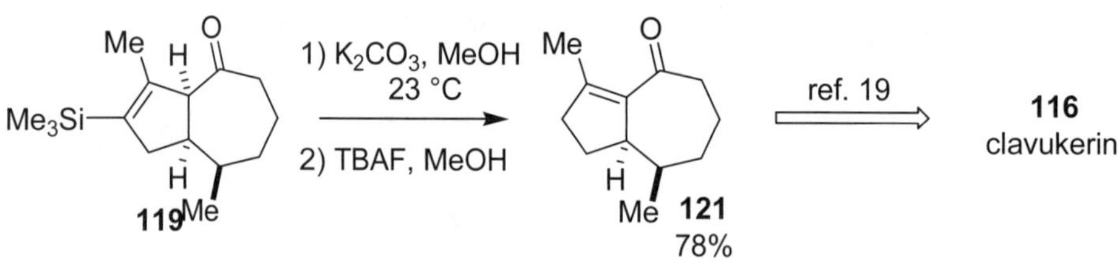

2.1.6 Experimental

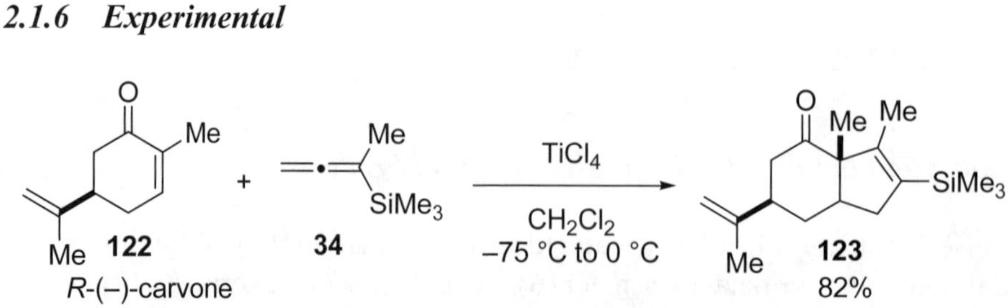

cis-4-exo-Isopropenyl-1,9-dimethyl-8-(trimethylsilyl)bicyclo[4.3.0]non-8-en-2-one (123)[3]

A 500-mL, three-necked, round-bottomed flask is equipped with a 25-mL pressure-equalizing dropping funnel, a mechanical stirrer, and a Claisen adapter fitted with a nitrogen inlet adapter and a low-temperature thermometer. The flask is charged with (*R*)-(–)-carvone **122** (115 g, 0.077 mol), 1-methyl-1-(trimethylsilyl)allene **34** (10.8 g, 0.079 mol), and dry dichloromethane (180 mL), and then cooled below −75 °C with a dry ice–acetone bath while a solution of titanium tetrachloride (17.4 g, 0.092 mol) in dichloromethane (10 mL) is added dropwise over 1 h. After 30 min, the cold bath is removed, and the reaction mixture, which appears as a red suspension, is allowed to warm to 0 °C over approximately 30 min. The resulting dark red solution is poured slowly into a 2-L Erlenmeyer flask containing a magnetically stirred mixture of diethyl ether (400 mL) and water (400 mL). The aqueous phase is separated and extracted with ether (2 × 200 mL). The combined organic phases are washed with water (250 mL) and saturated sodium chloride solution (250 mL), dried over anhydrous magnesium sulfate, and concentrated at reduced pressure using a rotary evaporator. The residual yellow liquid is distilled through a 15-cm Vigreux column at reduced pressure to afford bicyclononenone **123** (17.5 g, 82%) as a very pale yellow liquid.

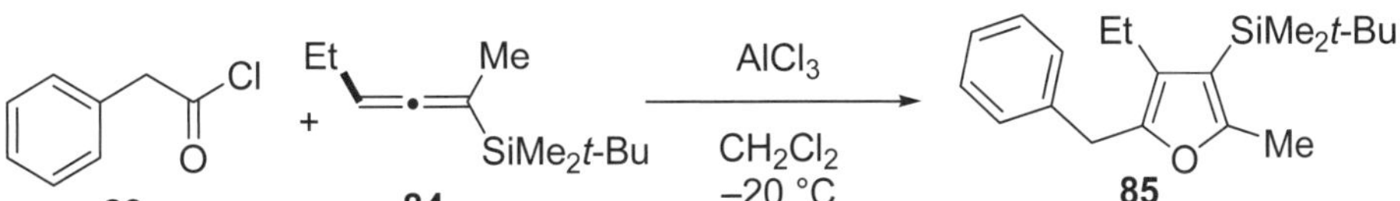

83 **84** **85** 71%

2-Benzyl-4-(*tert*-butyldimethylsilyl)-3-ethyl-5-methylfuran (85)[13]

A 50-mL, one-necked, round-bottomed flask equipped with a three-way argon inlet adapter fitted with a rubber septum was charged with $AlCl_3$, (0.836 g, 6.27 mmol) and of CH_2Cl_2 (12 mL) and then cooled to –20 °C while phenylacetyl chloride (0.97 g, 6.27 mmol) was added rapidly via a syringe over ca. 1 min. After 5 min, a solution of allene **84** (1.294 g, 6.26 mmol) in CH_2Cl_2 (13 mL) was added dropwise via a syringe over the course of 1 min. The resulting orange reaction mixture was stirred at –20 °C for 1 h and then quenched by the addition of triethylamine (0.950 g, 9.39 mmol) in pentane (25 mL). The resulting solution was stirred at room temperature for 10 min, diluted with an additional 25 mL of pentane, and then washed with 10% HCl solution (2 × 50 mL), 3% NaOH solution (50 mL), water (50 mL), and saturated NaCl solution (50 mL). The organic phase was dried over K_2CO_3, filtered, and concentrated to afford a light yellow oil (ca. 1.5 g). Column chromatography on silica gel (elution with 1% triethylamine in petroleum ether) provided (1.39 g, 71%) of silylfuran **85** as a colorless oil.

3.1.7 References

1. Danheiser, R. L.; Carini, D. J.; Basak, A. *J. Am. Chem. Soc.* **1981**, *103*, 1604–1606.
2. Danheiser, R. L.; Carini, D. J.; Fink, D. M.; Basak, A. *Tetrahedron* **1983**, *39*, 935–947.
3. Danheiser, R. L.; Fink, D. M.; Tsai, Y.-M. *Org. Synth.* **1988**, *66*, 8–13.
4. Danheiser, R. L.; Gee, S. K. *J. Org. Chem.* **1984**, *49*, 1672–1674. (b) Smith, A. B. III; Adams, C. M.; Kozmin, S. A.; Paone, D. V. *J. Am. Chem. Soc.* **2001**, *123*, 5925–5937.
5. (a) Stork, G.; Danheiser, R. L. *J. Org. Chem.* **1973**, *38*, 1775 1776. (b) Kende, A. S.; Fludzinski, P. *Org. Synth.* **1985**, *64*, 68–72.
6. Carini, D. J. Personal communication, Jan. 2009.
7. (a) Danheiser, R. L.; Carini, D. J. *J. Org. Chem.* **1980**, *45*, 3925–3927. (b) Danheiser, R. L.; Carini, D. J.; Kwasigroch, C. A. *J. Org. Chem.* **1986**, *51*, 3870–3878.
8. Jellal, A.; Santelli, M. *Tetrahedron Lett.* **1980**, *21*, 4487–4490.
9. Danheiser, R. L.; Kwasigroch, C. A.; Tsai, Y.-M. *J. Am. Chem. Soc.* **1985**, *107*, 7233–7235.
10. Danheiser, R. L.; Fink, D. M. *Tetrahedron Lett.* **1985**, *26*, 2513–2516.
11. Becker, D. A.; Danheiser, R. L. *J. Am. Chem. Soc.* **1989**, *111*, 389–391.
12. Evans, D. A.; Sweeney, Z. K.; Rovis, T.; Tedrow, J. S. *J. Am. Chem. Soc.* **2001**, *123*, 12095–12096.
13. Danheiser, R. L.; Stoner, E. J.; Koyama, H.; Yamashita, D. S.; Klade, C. A. *J. Am. Chem. Soc.* **1989**, *111*, 4407–4413.
14. Daidouji, K.; Fuchibe, K.; Akiyama, T. *Org. Lett.* **2005**, *7*, 1051–1053.
15. Danheiser, R. L.; Becker, D. A. *Heterocycles* **1987**, *25*, 277–281.
16. Angle, S. R.; Arnaiz, D. O. *Tetrahedron Lett.* **1991**, *32*, 2327–2330.
17. Yadav, V. K.; Sriramurthy, V. *Org. Lett.* **2004**, *6*, 4495–4498.
18. Friese, J. C.; Krause, S.; Schäfer, H. J. *Tetrahedron Lett.* **2002**, *43*, 2683–2685.
19. (a) Asaoka, M.; Kosaka, T.; Itahana, H.; Takei, H. *Chem. Lett.* **1991**, 1295–1298. (b) Kim, S. K.; Pak, S. *J. Org. Chem.* **1991**, *56*, 6829–6832. (c) Honda, T.; Ishige, H.; Nagase, H. *J. Chem. Soc., Perkin Trans. 1* **1994**, 3305–3310.

3.2 Dieckmann condensation

Noha S. Maklad

3.2.1 Description

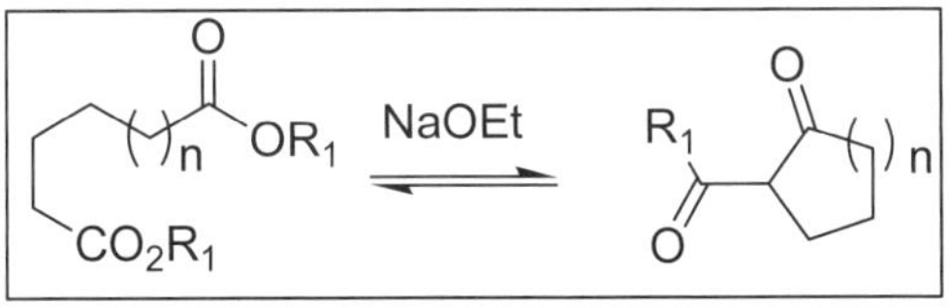

The Dieckmann condensation is an intramolecular reaction by which cyclic β-keto esters are formed using tethered di-esters under basic conditions.[1–4] The name derives from its originator, Walter Dieckmann, who first reported the reaction in 1894 and demonstrated the formation of ethyl β-ketocyclopentanecarboxylate from ethyl adipate using sodium base and, in a similar fashion, the formation of β-ketocyclohexanecarboxylate from ethyl pimelate.[1]

3.2.2 Historical Perspective

Born in Hamburg, Germany on October 8, 1869, Walter Dieckmann studied at the University of Munich and became an assistant to Adolf von Baeyer. He died on January 12, 1925. Dieckmann initiated his work on this condensation reaction at Baeyer's urging. The work intended to test the adaptability of Baeyer's voltage theory in forming 5- and 6-membered rings.[1,4] Although the Dieckmann condensation can be used to produce 4-membered cyclic β-keto esters, the reaction yield is usually low. On the other hand, the Dieckmann condensation is a reliable method to form 5- and 6- membered β-keto esters, and a subsequent de-carboxylation can form the cyclo-keto analogues.[1] Larger rings can be formed under dilute conditions, in which case the yield depends on the size of the ring to be formed.[5] The Dieckmann condensation has been used not only for the synthesis of alicyclic but also heterocyclic β-keto esters.

3.2.3 Mechanism

There are few mechanistic studies for the Dieckmann condensation.[6–10] The reaction can simply be described as an intramolecular Claisen condensation and hence entirely reversible. The summation of the consensus for its mechanism is as follows: the reaction is initiated by anion formation through the abstraction of the most acidic proton on the α-carbon to one of the di-esters (note that there must be at least one α-methylene hydrogen to the di-esters), forming an enolate resonance.[6] This is followed by the C–C bond formation by the attack of the enolate on the carbonyl of the second ester entity and the release of the alkoxylate ion from the tetrahedral transition state (**B**). The C–C bond formation (i.e., the ring formation) step was proven to be the rate-determining step of the reaction, using ^{14}C isotope experiments conducted and published by Carrick and Fry in 1955.[7] The alkoxylate ion then proceeds to remove a proton from the methylene of the keto ester (**C**) to give the enolate (**D**). The final product can be separated after an acidic work-up. The reaction follows first-order kinetic when conducted in alcohol (protic solvent) and when a large excess of the base (e.g., Na ethoxide) is used; this is due to the formation of stable enolate anion of the end product that drives the reaction forward. On the other hand, the reaction is bimolecular in the presence of any solvent of a low di-electric constant (aprotic solvent) and excess base. In the latter case, the reaction rate increases in magnitude with the increase of the strength of the base used, as seen in the following order of bases: $MeO^- < EtO^- \sim Pr^nO^- < Pr^iO^-$.[7] The Dieckmann condensation is similar to the Claisen condensation; the reaction is reversible all the way, and hence the reaction reversal can be conducted to give a new type of reaction, which is now known as retro-Dieckmann or the reversible Dieckmann reaction.[1]

The Thornely and Reed mechanism proposal:

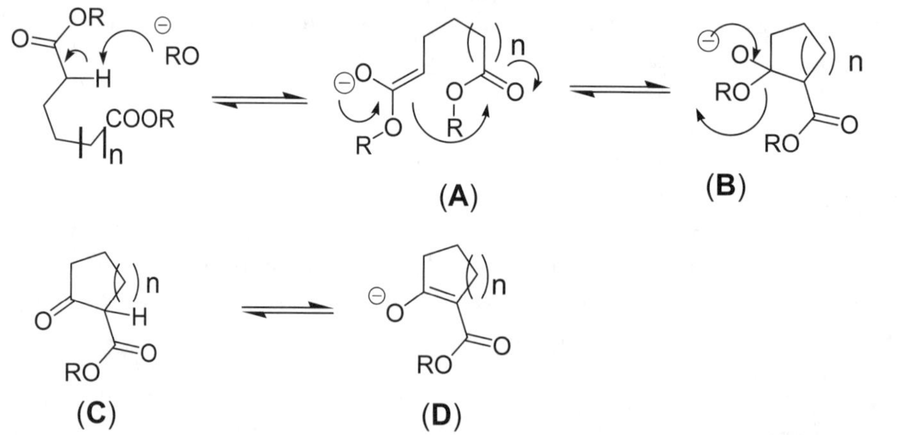

One of the obstacles to designing the ring closure of the Dieckmann condensation is the regioselective control of the reaction. Although the condensation usually occurs in the direction that forms the most stable enolate, there are at least two possible products competing for formation if the starting diester is unsymmetrical. Naturally, if one of the α-carbons to the di-esters is di-substituted, then there is only one way for the ring to close, provided that the other direction would form a stable enolate. Substrates with α-substitution will prefer ring closure where the least hindered enolate is formed, and similarly for β-substituted di-esters, anion formation (and hence the cyclization) would steer away from the substituted methylene group. A simple example for the latter is the synthesis of compound **2** using sodium alkoxide from starting material **1** in 80% yield.[5]

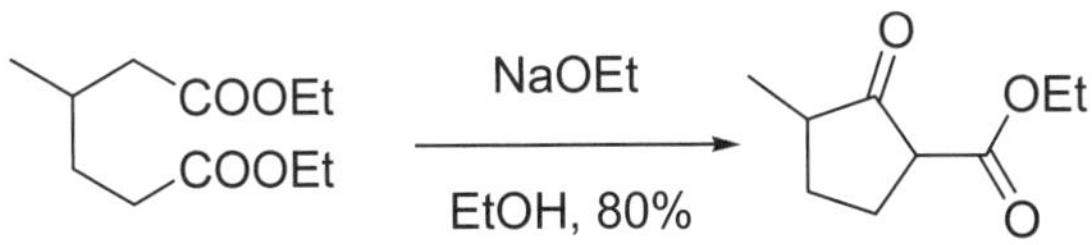

1 **2**

The use of labile auxiliaries to replace one ester moiety was also proven to be a useful strategy for regioselective control of the ring closure as shown by Kondo and coworkers in the case of forming carbapenem **4** or as shown by Déziel and coworkers in the case of forming carbapenem **6**.[11,12] In the first case, the auxiliary dihydrobenzoxazone was used to give the product using TMSCl-promoted cyclization in 82% yield. It is worth noting that the yield of the reaction decreased to 18% in the absence of TMSCl. In the latter reaction, the acyl auxiliary 4,4-dimethyl-1,3-oxazolidine-2-thione was used to form the product **6** using LiHMDS or NaHMDS in THF at –78 °C in 2 min.

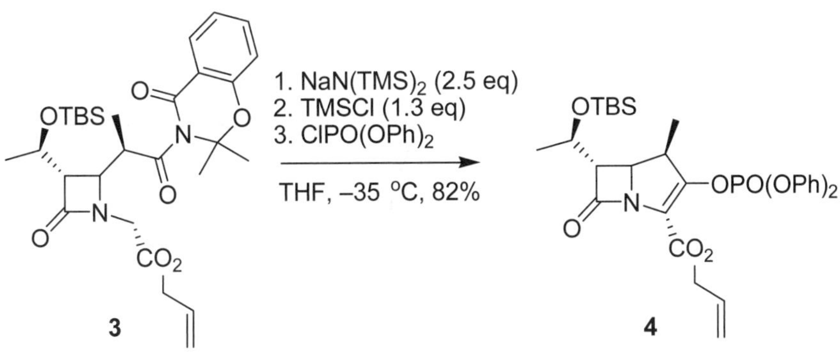

3 **4**

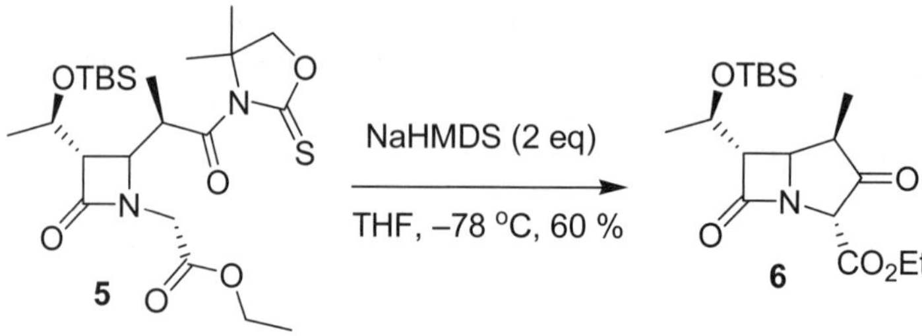

The use of *N*-methoxy-*N*-methylamides auxiliary (Weinreb amide) to replace one of the di-esters has also been used for chemoselective control. The direction of cyclization depends on the kind of base used. The base first removes the proton at the α-position of either the ester side or of the amide side. In the case of KO*t*-Bu, if the deprotonation occurs at α-position to the ester then intermediate **10** is formed. Because potassium's poor chelating ability is combined with the fact that the amide is a poor leaving group, the intermediate reverts back to **7**, and the cyclization goes to form the isomer **8**. On the other hand, the use of LDA or LiHMDS would produce the stable Li-complex **11** and directs the cyclization to form compound **9**.[13]

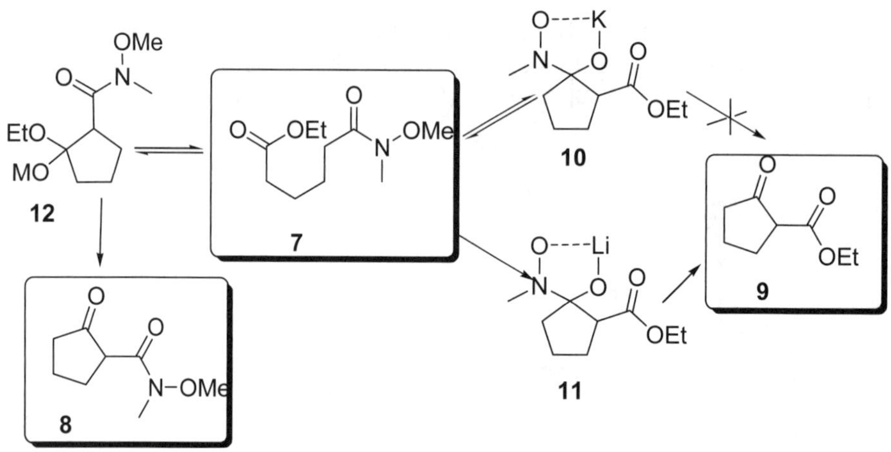

Regio-control for this condensation can also be achieved using sterics. Although six-membered rings are beyond the scope of this chapter, it is important to mention some examples where the size of the esters used can determine the direction of cyclization: for example, the use of a bulky ester such as phenyl esters as exemplified in the synthesis of the carbacephem antibiotic of structure **14**. The ring closure of the β-lactam di-ester **13**, using either potassium *tert*-butoxide or lithium hexamethyldisilazide in tetrahydrofuran at –78 °C, steers the ring closure to only one regioisomer, and the de-protonation occurs α- to the bulky ester.[14]

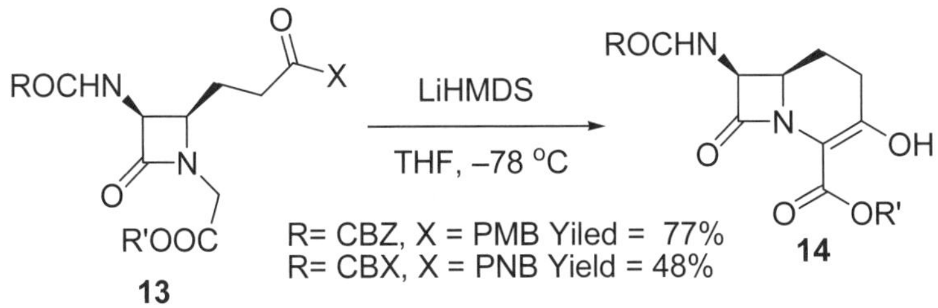

Similarly, if one of the di-esters is replaced by *N,N*-dibenzoyl carbonamide, as in the case of compound **16** (synthesized from compound **15** in 4 steps), the cyclization is regioselective and gives the corresponding carbacephem **17** in the presence of sodium bis(trimethylsilyl)amide as the base.[15]

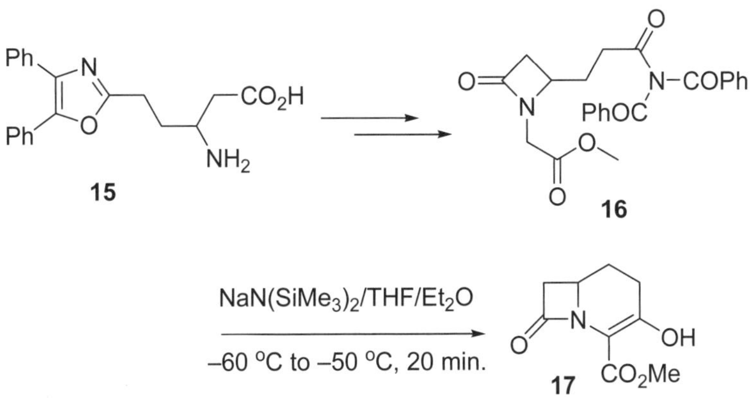

3.2.4 *Standard Method, Variations and Improvements*

To form the 1,3-diketoesters, the starting diester condenses in the presence of a base. The standard procedure for the reaction is to use metal alkoxides of Na, Li, or Mg in the corresponding alcoholic solvent (ethanol, methanol, *t*-butanol etc.) or sodium metal in aprotic solvent such as toluene or xylene. Other bases and solvent combinations are also used such as NaH, or KH in DMF, LiHMDS, KHMDS, or LDA in THF, alkali carbonates, and alkali hydroxides.[5,18,19]

Di-thiol ester version for the Dieckmann-type condensation.[20]

The original Dieckmann condensation conditions require high temperatures and prolonged reaction time, and the decarboxylation that follows to form the corresponding cyclic-keto analogue requires harsh acidic or basic conditions.

Liu et al. in 1979 reported a dithiol version of the Dieckmann condensation. The simple variation offers milder conditions and shorter reaction time. The reaction starts with the appropriate tethered dithiol esters, and under a basic condition the corresponding cyclic β-keto thiol esters are formed in a good yield at room temperature. An example of this variation is the synthesis of S-ethyl-2-cyclopentaneonecarboxythiolate (**19**) in 91% yield from di-*S*-ethyl hexanedithiolate (**18**) using sodium hydride in the presence of a catalytic amount of ethanethiol in dimethoxyethane (DME). The thiol ester group can then be easily removed to give the corresponding cyclopentanone using an excess of Raney nickel under neutral condition. Although this reaction has not received much attention, it was the inspiration for the mono-thiol version of the Dieckamnn condensation, a variation that allowed reasonable chemoselective control of the ring formation.

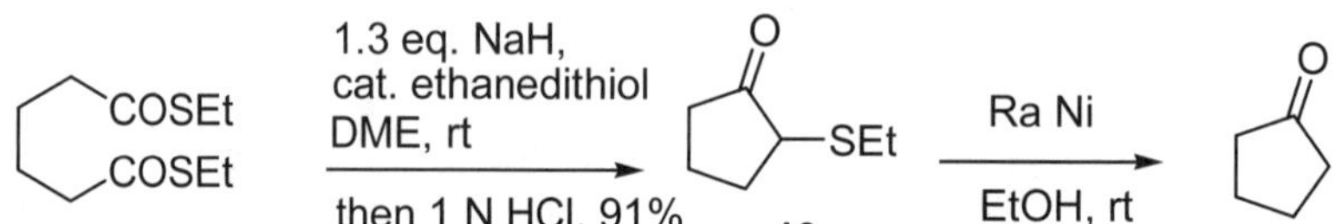

18 **20**

Mono-thiol ester version of the Dieckmann-type condensation[21]

The importance of controlling the direction of ring closure for the unsymmetrical di-ester condensation has prompted researchers to introduce different chemoselective methods for this reaction. The mono-thiol version of the condensation was first developed for that particular purpose by Yamada, Hosaka, and co-workers in 1981. An example is the ring formation of compounds **22**. The ring closure occurs using an appropriate base such as lithium diispropylamide (LDA) in dry THF at –30 °C in the case of X = CH_2 or sodium hydride at room temperature in case of X = N (or S) to form the corresponding carbocyclic or heterocyclic rings **22a–c** in 74–77% yields.

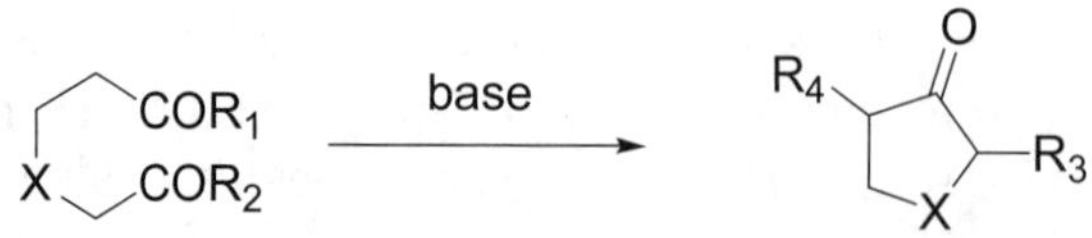

21 a–c **22 a–c**

a. X= CH_2, R_1= SEt, R_2= OEt
b. X= S, R_1= SEt, R_2= OEt
c. X= NCO_2Et, R_1= SEt, R_2= OEt

a. X= CH_2, R_3= CO_2Et, R_4= H
b. X= CH_2, R_3= CO_2Et, R_4= H
c. X= NCO_2Et, R_3= CO_2Et, R_4= H

Ring closure is pushed to the opposite direction to form the corresponding thioesters **23a,b** in 68–70% yields using the Lewis acid ($AlCl_3$) in the presence of an Et_3N base. Another combination of Lewis acid and base is also used to promote this kind of regioselectivity such as $MgCl_2$, $MgBr_2$, and $Sn(OSO_2CF_3)_2$ in the presence of Et_3N or *N*-ethylpiperidine in suitable solvents such as CH_2Cl_2, CH_3CN, or a combination of both. This work was developed by Nagao, Tamai, and co-workers in 1995.[22]

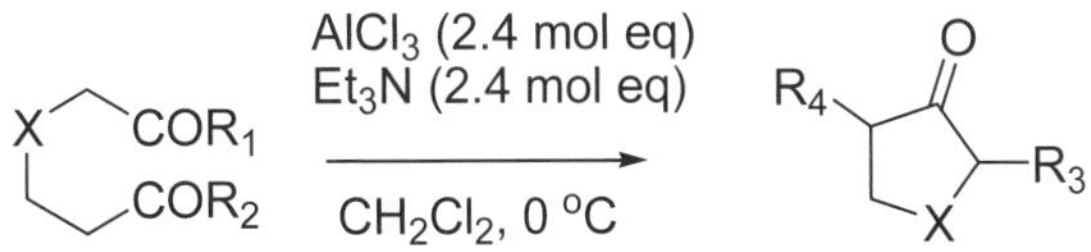

21a,d **23a,d**

a. X= CH2, R1= SEt, R2= OEt a. X= CH2,
d. X= S, R1= SEt, R2= OEt b. X= CH2,

R3= COSEt, R4= H 45 min, 70%
R3= COSEt, R4= H 15 min, 68%

Ti-Dieckmann condensation

In 1989 Tanabe and coworkers published a new variation of the condensation where the Lewis acid dichloro-bis(trifluoromethanesulfonato) titanium(IV) $TiCl_2(OTf)_2$ is used to promote the condensation under mild basic condition (Et_3N for example). In 1997, the group published another paper showing an improvement on the condensation by using $TiCl_4$ and Bu_3N amine (1:1.2) molar ratios with a catalytic amount of $TiCl_2(OTf)_2$ or a catalytic amount of TMSOTf (0.05 equiv). Although this variation can produce the β-ketoesters in reasonable yields, the reaction needs to be at room temperature or heated up to 60 °C to reach completion. In 1999, the group published yet another improvement to perform the reaction at a lower temperature by using TMSCl in catalytic amounts. The use of $TiCl_4/Bu_3N$ with TMSCl (0.05 equiv) allowed the reaction to be run at –78 °C, as seen in the synthesis of **25** from dimethyl adipate (**24**) shown below.

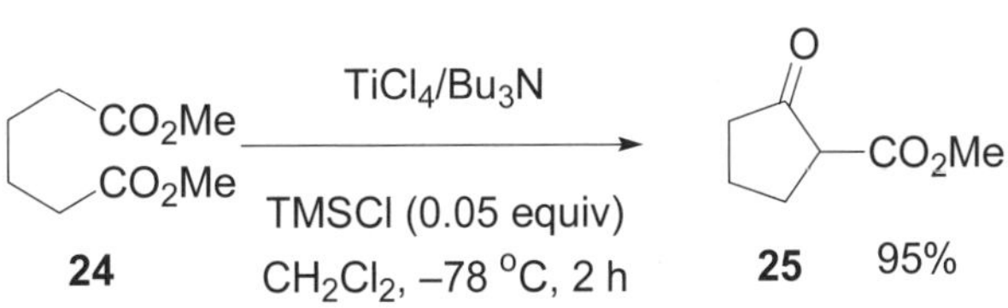

24 **25** 95%

Later, Tanabe and co-workers reported a dehydration version of the Ti-Dieckmann condensation. The reaction can be considered as a combination of the monothiol and Ti-Dieckmann condensations. The cyclization proceeds in the presence of a mild base to produce dehydration adducts that carry the thiol moiety in the end product. Difference from the original Ti-Dieckmann exists in the stoicheometry of the $TiCl_4$ and the base, as well as with the kind of base used. Regioselective synthesis of 4,5-dihydrothiophene-2-carboxylates and methyl thiophene-2-carboxylate could be used as an example of such variation. In this example, there are two possible products, and selectivity can be reached between the formation of the two adducts readily. Subjecting **26** to 2.2 eq. (*sec*-Bu)$_2$NH, and 2.4 equiv of $TiCl_4$ gives the vinyl sulfides adducts **27**. In contrast, the use of 5 equiv $TiCl_4$ and 5.2 equiv Et_3N from –50 to –45 °C gives the totally unsaturated adducts **28** in good 82–87% yields.[23–29]

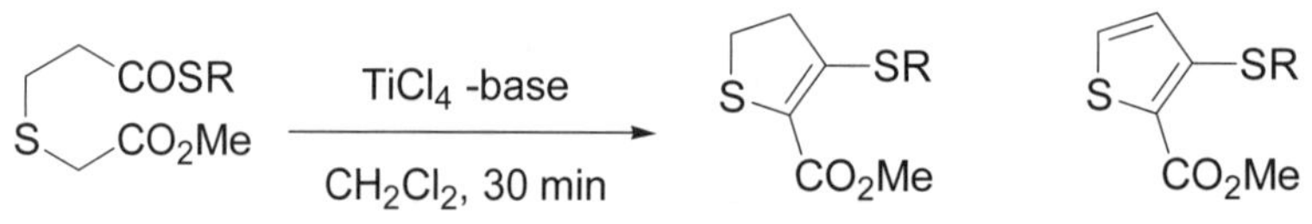

Tanabe et al. proposed mechanism for the dehydration-type Ti-Dieckmann condensation and thiophene formation is shown below.

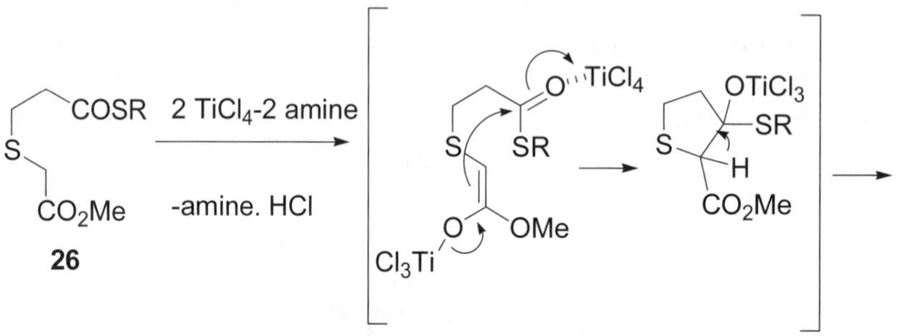

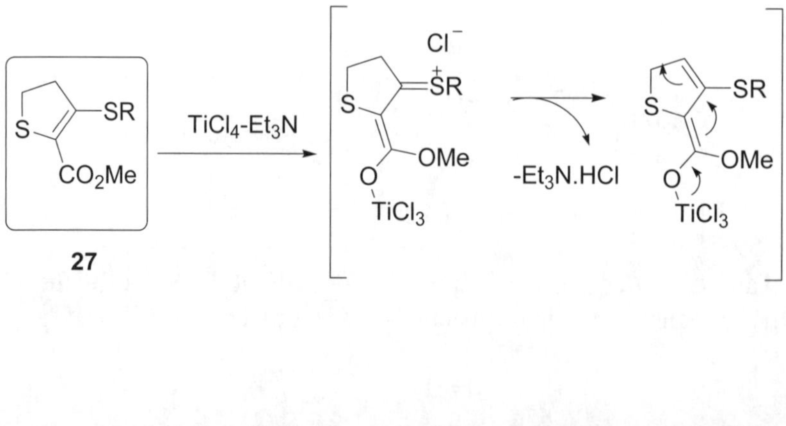

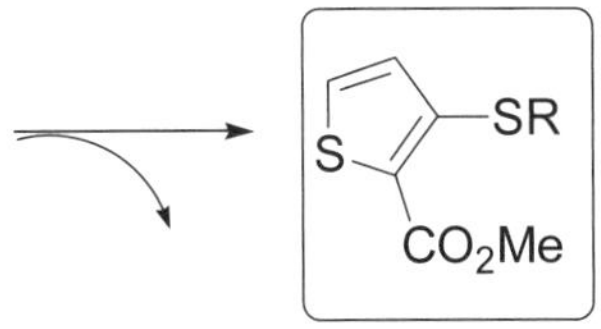

28

The Dieckmann condensation was reported once in the literature to proceed in solvent-free conditions. Toda, Higa and coworkers showed the condensation of adipate (**24**) and pimelate di-esters (**34**) to give the corresponding cyclized products using powdered base such as (BuONa, EtOK, and EtONa). The reaction proceeds with the solid di-esters that have been mixed by using a mortar and a pestle with the powdered base for 10 min, and the mixture was then kept in a desiccator for 60 min and then neutralized with *p*-TSOH•H_2O to give the condensed product in 60–82% yields.[30]

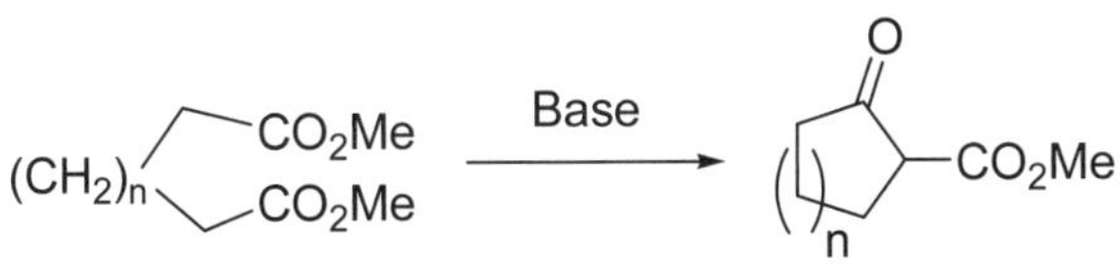

24; n = 1
34; n = 2

25; n = 1
35; n = 2

3.2.5 Synthetic Utility

General Utility

The Dieckmann condensation's main utility is the synthesis of 5- and 6-memebered carbocyclic and heterocyclic β-ketoesters. The cyclization requires tethered di-esters and base to proceed. As discussed before, different variations have resulted in a fine-tuning of the condensation reaction and have allowed it to serve as a wider base for versatile 5- and 6-membered rings.

Regioselective techniques were developed to direct the cyclization of unsymmetrical esters. An interesting example of this aspect of the Dieckmann condensation is the synthesis of 1β-methylcarbapenems by Tanabe and co-workers. 1β-methylcarbapenems is a structure that exists in potent and broad antibacterial compounds, such as meropenem and biapenem, which renders it synthetically interesting for the synthesis of some close analogues **30a–d**. Thioesters **29** was subjected to cyclization under dehydration type Ti-Dieckmann condensation conditions; 3.0 equivalents of the $TiCl_4$ and 3.3 equivalents of the base Bu_3N were used to give the vinyl thiol products where the 4 stereogenic centers were retained, and the cyclization proceeded in good yields (the dehydration reaction is thought to go by a mechanism similar to that of synthesis of compounds **27**).[29]

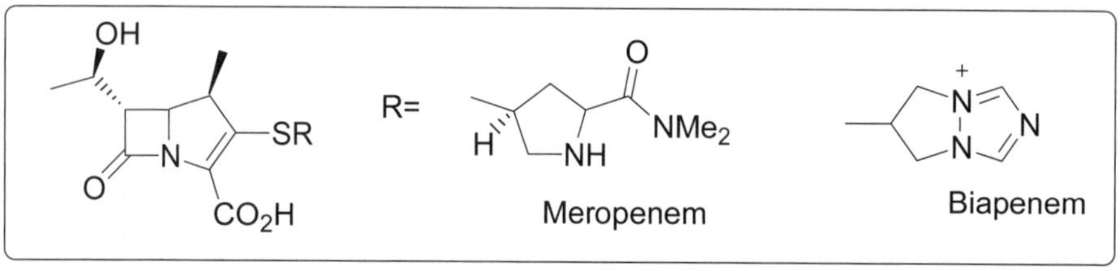

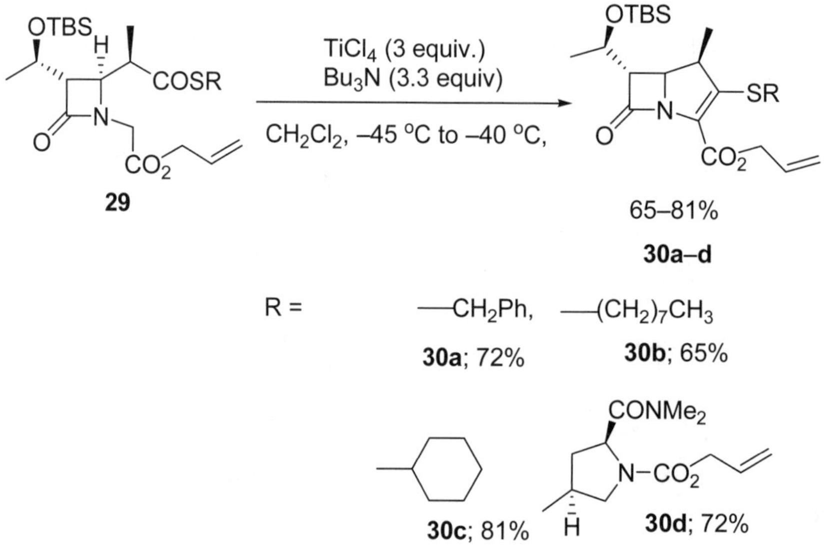

The Dieckmann condensation plays an important role in medicinal chemistry because it provides an easy access to 5-memebered rings. Rapoport asymmetric synthesis of the carbocyclic nucleosides' precursor **33** and **38** is a good example; carbocyclic nucleosides are good nucleosides' isosteres. For the synthesis of **33**, the imidazolyl starting material **27** undergoes Dieckmann condensation using KHMDS to give β-keto ester **28** in a 97% yield and a 3:1 diastereomeric mixture. The cyclized β-keto ester **28** undergoes various transformations to give the lithium carboxylate **29**. The carboxylate is then converted to the *cis*- bicyclic lactam **30** using NaOAc and refluxing in Ac_2O, which is then hydrolyzed with AcOH and 6 M HCl of ratio 2:1. The acid formed is then converted to its methyl ester and reduced to give the intended *cis*-amino alcohol **33**, all in good yields. The amino alcohol is an important precursor for subsequent carbocyclic nucleosides.[31]

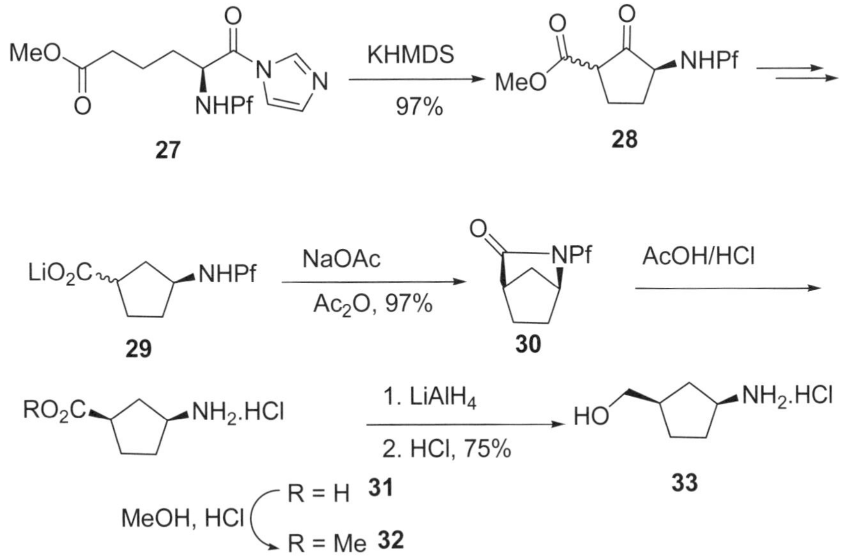

The Rapoport regioselective synthesis of the carbocyclic precursor **38** is another example. The synthesis starts with the reaction of the di-ester **34** with lithium 2,2,6,6-tetramethylpiperidine (LiTMP) to give **35** almost exclusively as *cis* and *trans* isomers. The keto-ester formed is reduced using $NaBH_4$ to give a mixture of 4 diastereomers; the mixture is mesylated and subjected to elimination conditions using potassium *tert*-butoxide to give **36a** in 69% yield; the ester is then reduced using $LiAlH_4$ to give the amino-cyclopentenol **37**, which undergoes a sequence of transformations to give **39**, a carbocyclic analogue of pentostatin (a natural product of microbial origin with potent anticancer activity).[32]

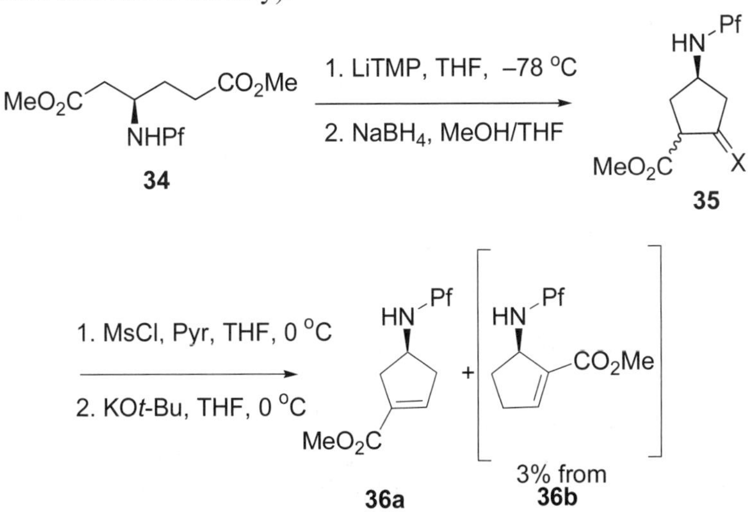

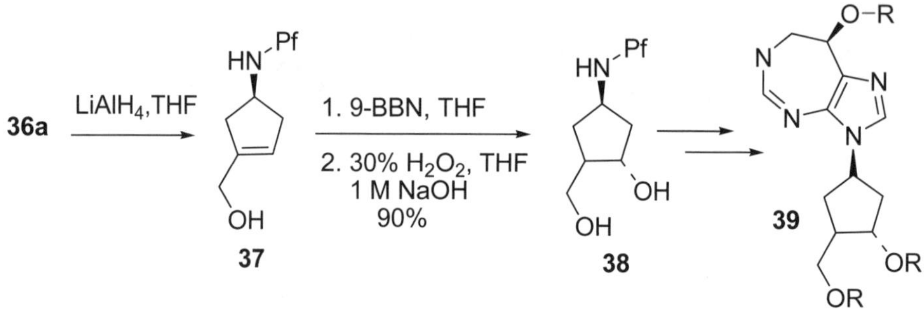

In 1999, Shindo and Sato reported an interesting one-pot tandem [2 + 2]-cycloaddition–Dieckmann condensation. The condensation is a unique approach for the synthesis of functionalized 5- and 6-membered rings. The reaction proceeds by first treating the ynolate anions of structure **41**, which is formed readily in situ from its corresponding α,α-dibromo esters with 4 equiv of *tert*-BuLi at –78 °C. The reaction is warmed up to 0 °C, followed by the addition of γ-keto-esters of structure **40** in THF, after which the temperature is then lowered to –78 °C, and the reaction then proceeds at that temperature. After workup and with no purification, acid-catalyzed decarboxylation in benzene follows to give compounds of structure **42** in good yields.

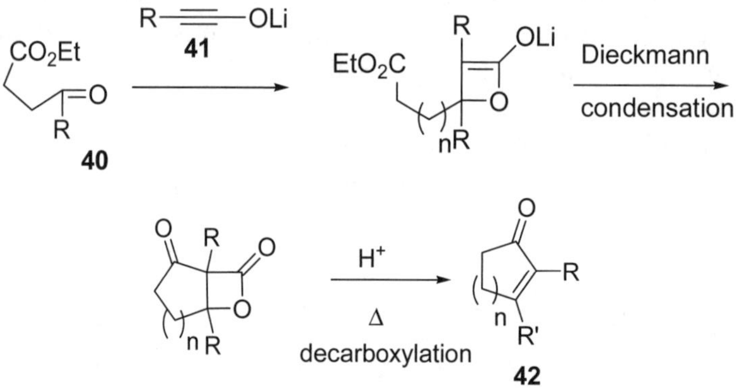

Solid-phase synthesis can provide an attractive direction for the regioselective control of the condensation. In 1970s and 1980s, Rapoport and Crowley pioneered the regioselective Dieckmann condensation for the 6-membered ring. Synthesis of the 5-membered ring using solid-phase Dieckmann condensation conditions (as exemplified by synthesis of tetramic acids) was explored latter by different groups.[36–38]

Tetramic acids are important building blocks in natural products and represent potentially good binding motifs for biological targets such as the asymmetric solid phase synthesis of benzothiadiazine-substituted tetramic acids of structure **46**, which are potentially potent inhibitiors of hepatitis C virus RNA-dependent RNA polymerase. The tetramic acid final product is formed in two steps by first the acylation of the resin bound enantiomerically pure Fmoc-protected L-amino acid (**43**) with the acid **44** to give **45**. The product then undergoes Dieckmann condensation using KO*t*-Bu in THF/*t*-BuOH (2 : 1) or TEA in CH_2Cl_2. The choice of base is important to minimize racemization. The cyclization and the concomitant cleavage from the resin occur at room temperature in 15 min.[39]

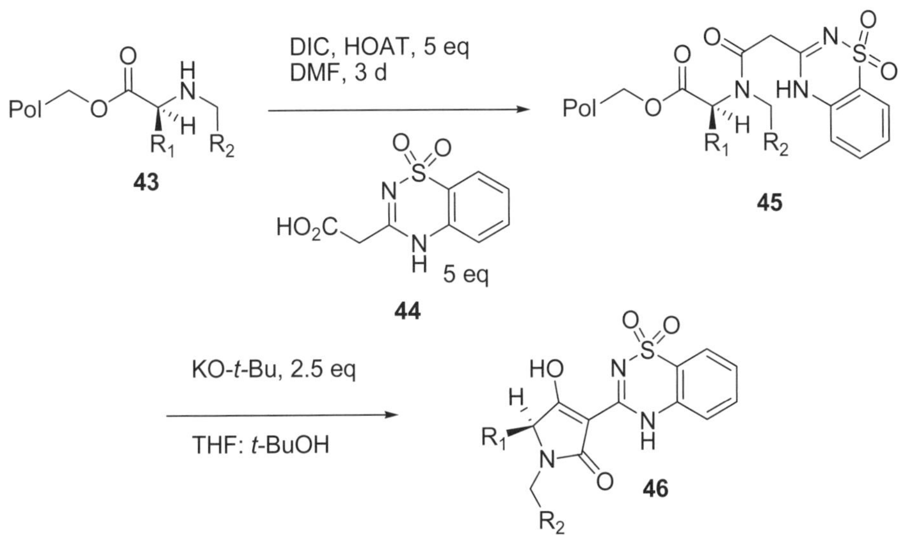

Applications in the total synthesis of natural products

The Dieckmann condensation is implemented in numerous natural product syntheses. Shindo and Shishido use their [2 + 2] tandem cycloadditon–Dieckmann condensation in the construction of the cyclopentenone unit in the total synthesis of cucumin E.[40] The condensation drives the transformation of compound **47** first to the intermediate **49** using lithium ynolate **48** (3 equiv) in THF at –20 °C for 1 h, which is then followed by decarboxylation of the adduct by refluxing in toluene for 13 hin the presence of a catalytic amount of silica gel to give cucumin E in 54% yield in two steps.

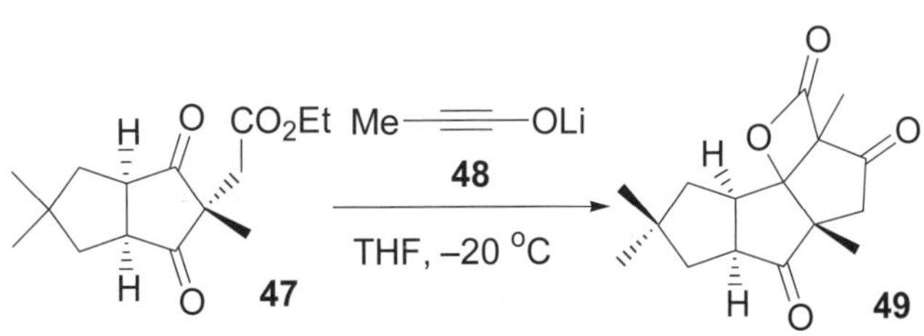

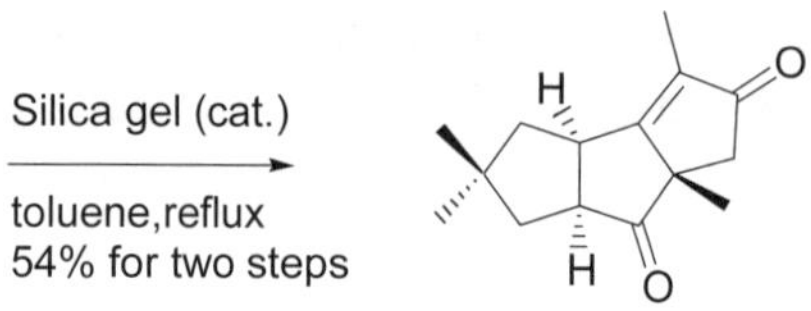

50 cucumin E

Synthesis of physarorubinic acid, which is a polyenoyltetramic acid plasmodial pigment, is another example. The synthesis uses a Lacey–Dieckmann cyclization starting with ester **51**, followed by deprotection to give physarorubinic acid (**52**) in 78% yield for the two steps.[41,42]

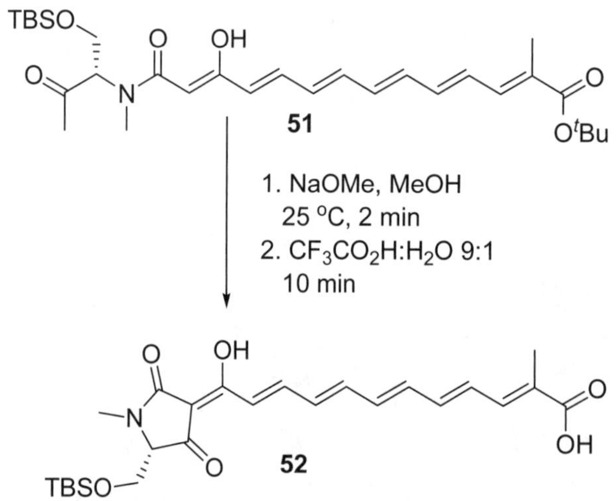

Tetronic acids (4-hydroxy-2(5*H*)-furanones) are also an example of 5-membered rings, and they are compounds of potential activity as antibiotics, antiviral and neoplastic agents. The base-promoted Dieckmann condensation is one of the desired methods to produce 3-acyl-tetronic acids. An example is the intramolecular tetrabutylammonium fluoride-promoted Dieckmann cyclization of **53** to give tetronasin (**54**) in 72% yield.

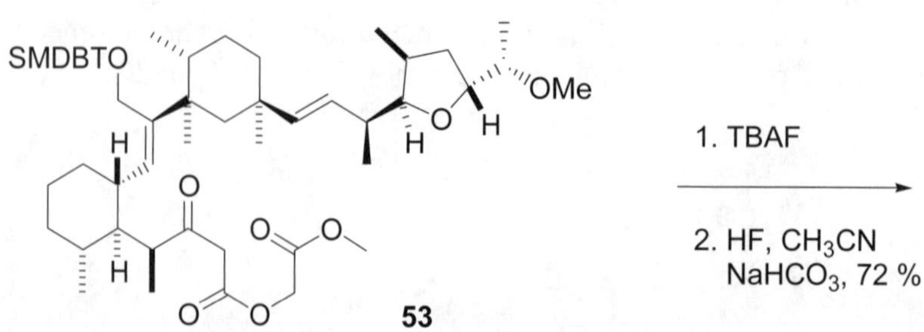

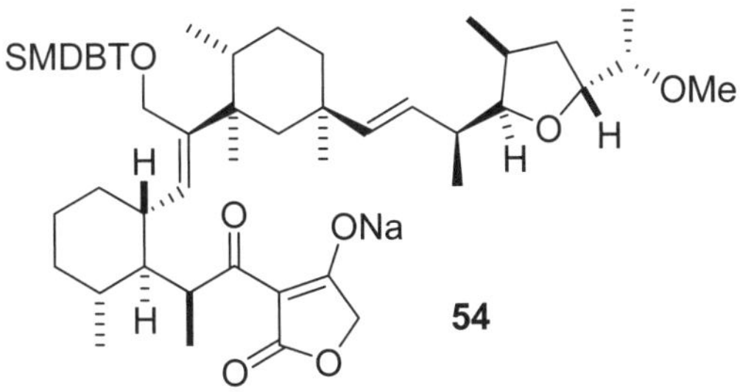

54

3.2.6 Experimental

Dieckmann condensation of ethyl adipate:
2-Carbethoxycyclopentanone (24)

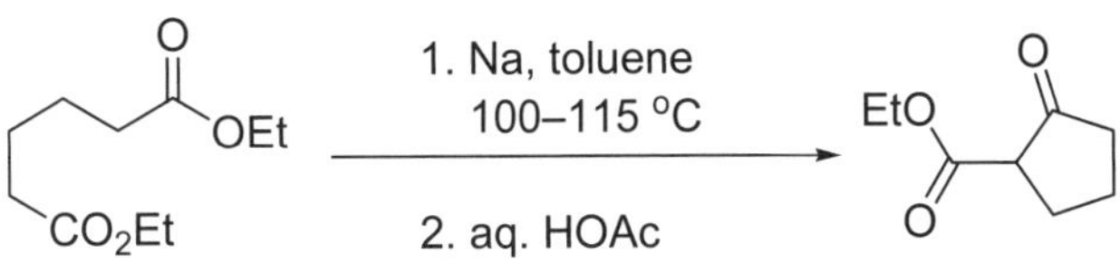

A three-necked, round-bottom flask is fitted with a mercury sealed mechanical stirrer, a 250-mL dropping funnel, and a reflux condenser protected from air by means of calcium chloride tube. In the flask are placed 23 g sodium and 250 mL toluene. The stirrer is started, and 202 g ethyl adipate is added from the dropping funnel at such a rate that the addition is complete in 2 h. The reaction usually starts immediately on addition and for about 5 h longer. Dry toluene is added through condenser from time to time to keep the reaction mixture fluid enough for efficient stirring. Between 750 mL and 1 L of toluene is added in this manner. The reaction mixture is cooled in an ice bath and slowly poured into 1 L 10% acetic acid cooled to 0 °C (ice–salt mixture). The toluene layer is separated, washed once with water, twice with cooled 7% Na_2CO_3 solution, and again with water. The toluene is removed by distillation at ordinary pressure, and the residue is distilled under reduced pressure. The yield is 115–127 g (74–81%).

3.2.7 References

1. Dieckmann, W. *Ber.* **1894**, *27*, 102–103.
2. (a) Dieckmann, W. *Ber.* **1900**, *33*, 2670–2684. (b) Dieckmann, W.; Groeneveld, A. *Ber.* **1900**, *33*, 595–605.
3. (a) Dieckmann, W. *Chem. Zentr.* **1900**, *2*, 892–893. (b) Dieckmann, W. *Chem. Zentr.* **1900**, *1*, 715–716.
4. (a) Dieckmann, W. *Liebigs Ann.* **1901**, 27–109. (b) Dieckmann,W. *Chem. Zentr.* **1901**, *2*, 630–636.
5. [R] Schaefer, J. P.; Bloomfield, J. J. *Org. React.* Wiley & Sons: New York, **1967**, *15*, 1–203.

6. Reed, R. I.; Thornely, M. B. *J. Chem. Soc.* **1954**, *78*, 2148–2150.
7. Carrick, W. L.; Fry, A. *J. Am. Chem. Soc.* **1955**, *77*, 4381–4387.
8. Burinsky, D. J.; Cooks, R. G. *J. Org.Chem.* **1982**, *47*, 4864–4869.
9. Hauser C. R.; Renfrow, W. B. *J. Am. Chem. Soc.* **1937**, *59*, 1823–1826.
10. Nishimura, T.; Sunagawa, M.; Okajima, T.; Fukazawa, Y. *Tetrahedron Lett.* **1997**, *38*, 7063–7066.
11. Déziel, R.; Favreau, D. *Tetrahedron Lett.* **1989**, *30*, 1345–1348.
12. Kondo, K.; Seki, M.; Kuroda, T.; Yamanaka, T.; Iwasaki, T. *J. Org. Chem.* **1997**, *62*, 2877–2884.
13. Sibi, M. P.; Christensen, J. W.; Kim, S.; Eggen, F.; Stessman, C.; Oien, L. *Tetrahedron Lett.* **1995**, *36*, 6209–6212.
14. Jackson, B. G.; Gardner, J. P.; Heath, P. C. *Tetrahedron Lett.* *33*, 6317–6320.
15. Neyer, G.; Ugi. I. *Synthesis* **1991**, *9*, 743–744.
16. Rapoport, H; Crowley, J. *J. Am. Chem. Soc.* **1970**, *92*, 6363–6365.
17. Crowley, J. I.; Rapoport, H. *J. Org. Chem.* **1980**, *45*, 3215–3227.
18. Pátek, M.; Hampl. F. *Collect. Czech. Chem. Commun.* **1989**, *54*, 3267–3277.
19. Svoboda, J.; Nič, M.; Paleček, J. *Collect. Czech. Chem. Commun.* **1992**, *58*, 592–599.
20. Liu, H.; Lai, H. K. *Tetrahedron Lett.* **1979**, *20*, 1193–1196.
21. Yamada, Y.; Ishii, T.; Kimura, M.; Hosaka, K. *Tetrahedron Lett.* **1981**, *22*, 1353–1354.
22. Tamai, S.; Ushirogochi, H.; Sano, S.; Nagao, Y. *Chem. Lett.* **1995**, 295–296.
23. Tanabe, Y. *Bull. Chem. Soc. Jpn.* **1989**, *62*, 1917–1924.
24. Yoshida, Y; Hayashi, R.; Sumibara, H.; Tanabe, Y. *Tetrahedron Lett.* **1997**, *38*, 8727–8730.
25. Yoshida, Y; Matsumoto, N.; Hamasaki, R.; Tanabe, Y. *Tetrahedron Lett.* **1999**, *40*, 4227–4230.
26. Tanabe, Y. *Chem. Commun.* **2001**, 1674–1675.
27. Tanabe, Y.; Makita, A.; Funakoshi, S.; Hamasaki, R.; Kawakusu, T. *Adv. Synth. Catal.* **2002**, *344*, 507–510.
28. Nagase, R.; Gotoh, H.; Katayama, M.; Manta, N.; Tanabe, Y. *Heterocycles* **2007**, *72*, 697–708.
29. (a) Tanabe, Y.; Manta, N.; Nagase, R.; Misaki, T.; Nishii, M.; Sunagawa, M. *Adv. Synth. Catal.* **2003**, *345*, 967–970. (b) Iida, A.; Okazaki, H.; Misaki, M.; Sunagawa, M.; Sasaki, A.; Tanabe, Y. *J. Org. Chem.* **2006**, *71*, 5380–5383.
30. Toda, F.; Suzuki, T.; Higa, S. *J. Chem. Soc., Perkin Trans. 1*, **1998**, 3521–3522.
31. Bergmeier, S. C.; Cobas, A. A.; Rapoport, H. *J. Org. Chem.* **1993**, *58*, 2369–2376.
32. Ho, J. Z.; Mohareb, R. M.; Ahn, J. H.; Sim, T. B.; Rapoport, H. *J. Org. Chem.* **2003**, *68*, 109–114.
33. Shindo, M.; Sato, Y.; Shishido, K. *J. Am. Chem. Soc.* **1999**, *121*, 6507–6508.
34. Rapoport, H.; Crowley, J. I. *J. Am. Chem. Soc.* **1970**, *92*, 6363–6365.
35. Crowley, J. I.; Rapoport, H. *J. Org. Chem.* **1980**, *45*, 3215–3217.
36. Kulkarni, B. A.; Ganesan, A. *Tetrahedron Lett.* **1998**, *39*, 4369–4372.
37. Weber, L.; Iaiza, P.; Biringer, G.; Barbier, P. *Synlett* **1998**, 1156–1158.
38. Romoff, T. T.; Ma, L.; Wang, Y.; Campbell, D. A. *Synlett* **1998**, 1341–1342.
39. Evans, K. A.; Chai, D.; Graybill, T. L.; Burton, G.; Sarisky, R. T.; Lin-Goerke, J.; Johnston, V. K.; Rivero, R. A. *Bioorg. Med. Chem. Lett.* **2006**, *16*, 2205–2208.
40. Shindo, M.; Sato, Y.; Shishido, K. *Tetrahedron Lett.* **2002**, *43*, 5039–5041.
41. Dixon, D. J.; Ley, S.V.; Longbottom, D. A. *J. Chem. Soc., Perkin Trans. 1*, **1999**, 2231–2232.
42. Lacey, R. N. *J. Chem. Soc.* **1954**, 850–854.
43. Tejedor, D.; Garcia-Tellado, F. *Org. Prep. Proced. Int.* **2004**, *36*, 35–59.
44. Pinkney, P. S. *Org. Synth.* **1937**, *17*, 30.

3.3 Favorskii Rearrangement

Brian Goess

3.3.1 Description

When α-haloketones are treated with a nucleophilic base in protic or ethereal solvents, a transformation known as the Favorskii rearrangement occurs to yield carboxylic acid derivatives.[1] Depending on the identity of the incorporated base the final product will be a carboxylic acid, ester, or amide. Cyclic α-haloketones undergo a ring contraction during the course of the rearrangement.

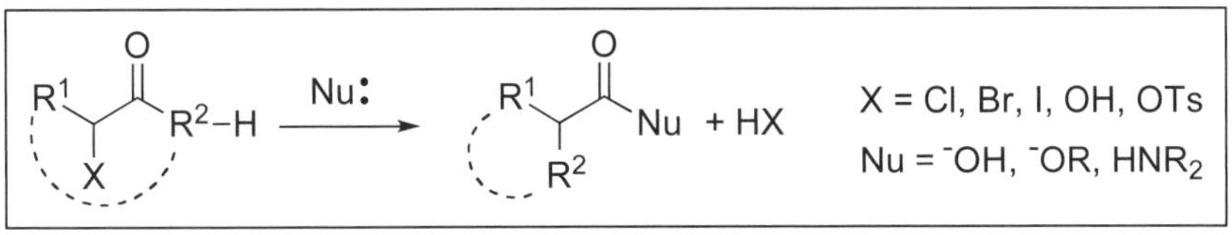

The rearrangement is general for a wide range of α-haloketones and some α-haloketimines; however, yields vary widely due to competitive side reactions that are highly dependent on both solvent and the structures of the substrate and base.[2] When unsymmetrical α-haloketones are used, an unsymmetrical cyclopropanone intermediate is generated that fragments regioselectively to generate one of two possible constitutional isomers. When α-chiral-haloketones are used, the rearrangement occurs with high stereoselectivity if reaction conditions are chosen such that competitive dissociative mechanisms are disfavored.

3.3.2 Historical Perspective

Reactivity between α,α-dihaloketones and bases was initially described in 1890 by Cloëz, who determined that treatment of **1** under saponifying conditions led to formation of α,β-unsaturated acid **2**.[3]

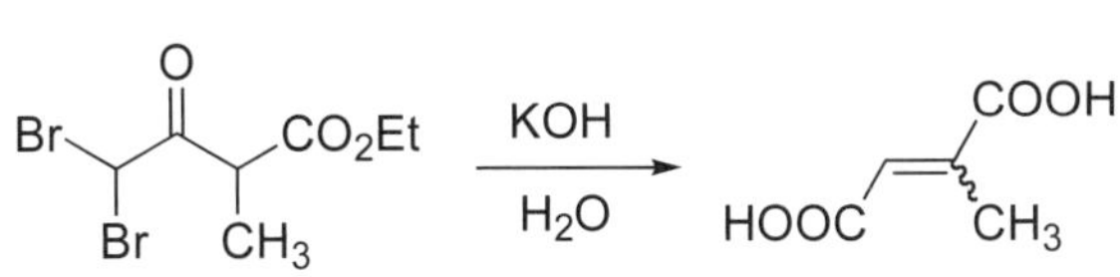

1 **2**

Indeed, 4 years later Favorskii described an analogous transformation on a dichloroketone[4] as part his ongoing investigations into the reactivity of halogenated carbonyls. These reactions are now known as Favorskii rearrangements.

One of the most common early uses of the Favorskii rearrangement was for the preparation of highly branched carboxylic acids and esters. For instance, when 3-bromo-3-methyl-2-butanone is treated with sodium isopropoxide at room temperature, isopropyl trimethylacetate is generated in 64% yield.[5]

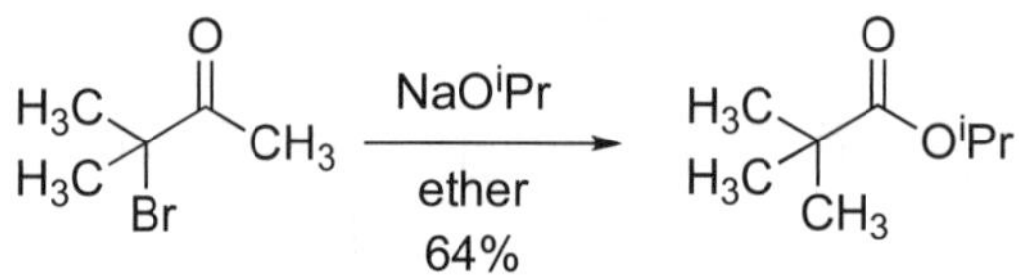

Yields and reaction rates tend to increase with increasing alkyl substitution on the carbon bearing the halide-leaving group due to diminished competition from side reactions such as direct nucleophilic substitution by the base. Other competing reaction products that have been observed include epoxyethers, α-epoxyketals, and α,β-unsaturated ketones.

The Favorskii rearrangement can be highly stereoselective, a finding that has greatly expanded its synthetic utility. When α-haloketone **3** is treated with base in 1,2-dimethoxyethane, ester **4** is formed diastereoselectively.[6] Both this and the preceding reaction exemplify another important feature of the Favorskii rearrangement that has been repeatedly demonstrated over time: In general the reaction is highly regioselective when unsymmetrical ketones are used and favors formation of products derived from the most stable (usually the least substituted) of two possible carbanionic intermediates, although steric factors may also play a role in determining regioselectivity.[2d]

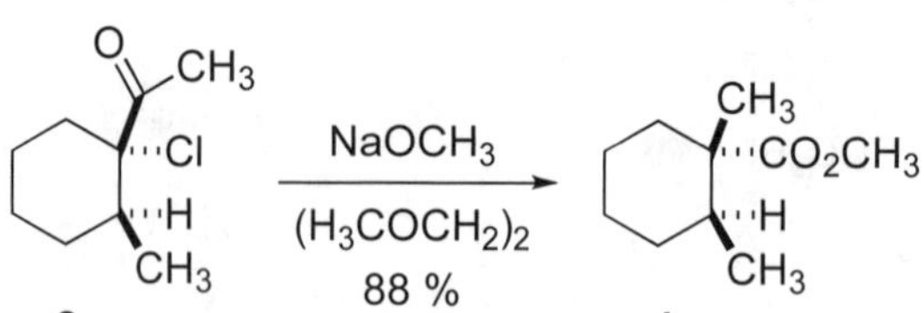

3 **4**

More recent work on the Favorskii rearrangement has focused on testing the ability of increasingly complex substrates to participate in Favorskii and related rearrangements. Efforts have also been directed at determining how reaction conditions affect the precise mechanism of Favorskii rearrangements. Both of these aspects of the Favorskii rearrangement will be discussed in subsequent sections.

3.3.3 Mechanism

The canonical formulation of the mechanism of the Favorskii rearrangement involves initial deprotonation of the α-carbon to generate an enolate, intramolecular displacement of the leaving group on the α'-carbon by the enolate to generate a cyclopropanone, addition of a nucleophile to the cyclopropanone ketone followed by elimination to generate the more stable of two possible carbanions, and protonation to yield the rearranged carboxylic acid derivative.

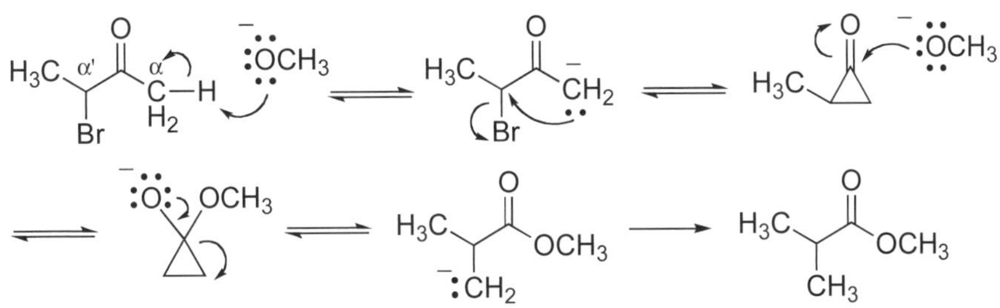

By 1950 five distinct mechanisms had been suggested to account for the formation of the major products of the Favorskii rearrangement. Four involved epoxide, ketene, enol, and carbene intermediates. A fifth mechanism related to the benzylic acid rearrangement was also proposed.[7] Then, in 1951 Loftfield isolated two esters with identical isotope distributions at their α and β carbons from treatment of a radiolabeled, cyclic α-chloroketone with an alkoxide. These two products suggested a symmetrical intermediate, leading Loftfield to postulate the existence of a cyclopropanone along the reaction pathway.[8]

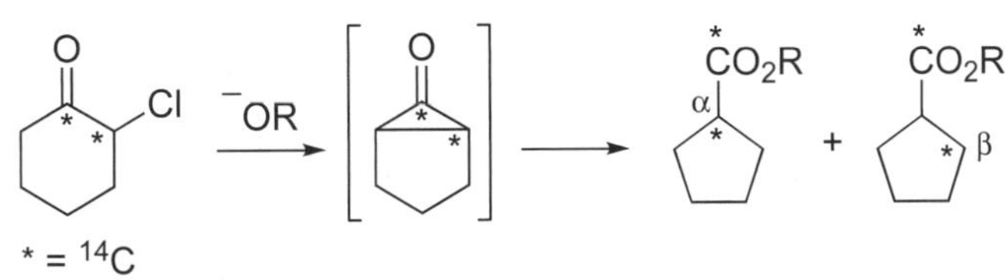

The existence of a cyclopropanone intermediate was supported by subsequent experimental evidence,[9] although in more polar media the presence of dipolar intermediates has also been inferred.[10] For example, when an aromatic α-bromoketone was treated with base in the presence of furan, the major product was a dipolar cycloaddition adduct which was isolated along with the Favorskii rearrangement product.[11]

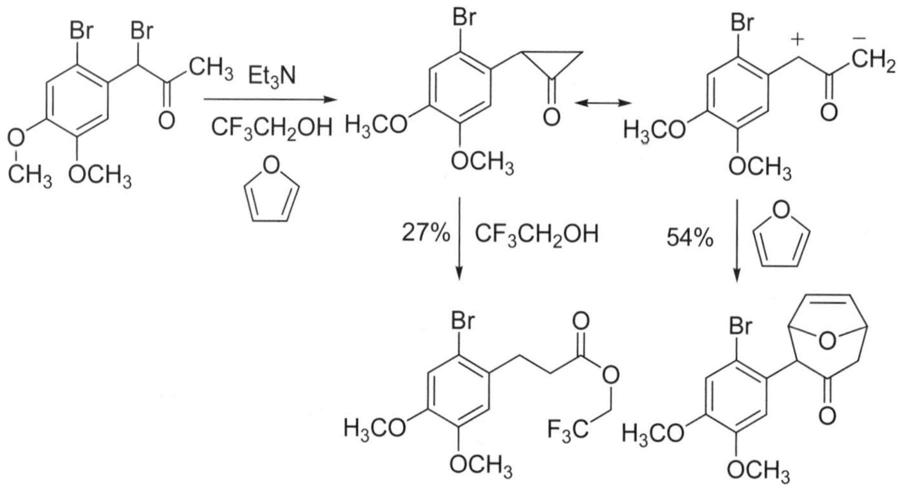

The strained geometry required to achieve good orbital overlap in an intramolecular cyclopropanone formation between an enolate on an α-carbon and an α'-carbon bearing a halide-leaving group has led some to suggest an alternate pathway for cyclopropanone formation.[12] In this pathway, halide departure immediately follows enolate formation to generate a delocalized zwitterion that undergoes an electrocyclization reaction to form the cyclopropanone. The Diels–Alder adduct above can be seen as proceeding through a similar zwitterionic intermediate.

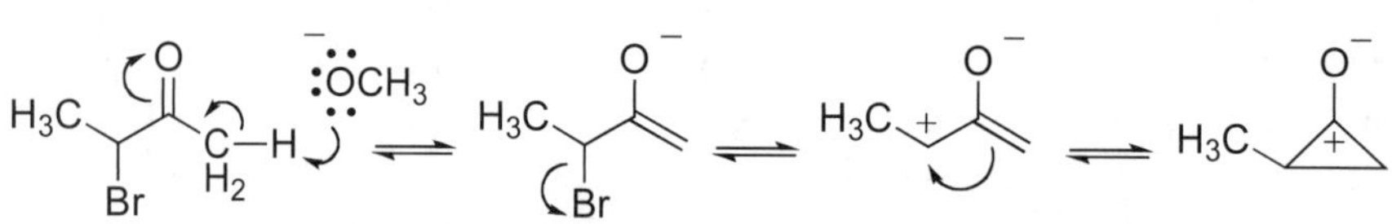

For substrates that lack an enolizable α-proton or for which a cyclopropanone intermediate would be highly strained, Favorskii products can be formed instead through a semibenzylic (also known as quasi-Favorskii) mechanistic pathway. This pathway involves addition of the base to the carbonyl followed by migration of an alkyl or aryl group with concomitant displacement of the halide.

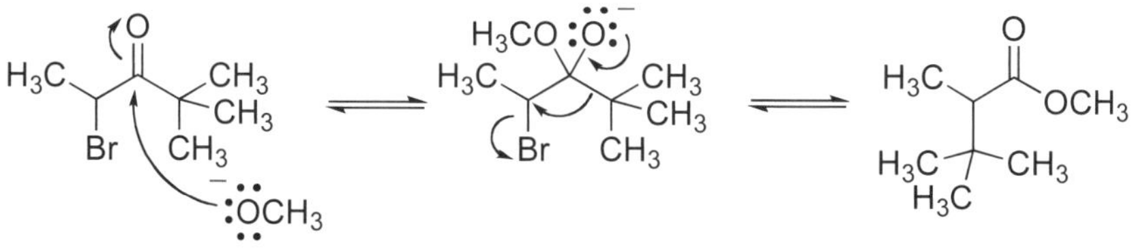

This mechanism is stereospecific and requires inversion at the carbon bearing the halide, a feature that was clearly demonstrated for a pair of conformationally locked substituted cyclohexane diastereomers.[13] Substitution by-products were also observed.

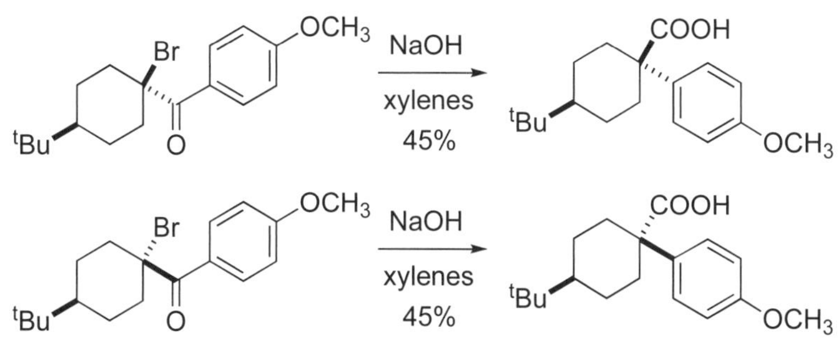

The semibenzylic pathway was also used to prepare an intermediate in the synthesis of the core of tricycloclavulone.[14] The rearrangement took place only on warming and was not successful when vinyllithium was used as the nucleophile. A similar approach was used in a synthesis of sterpurene.[15]

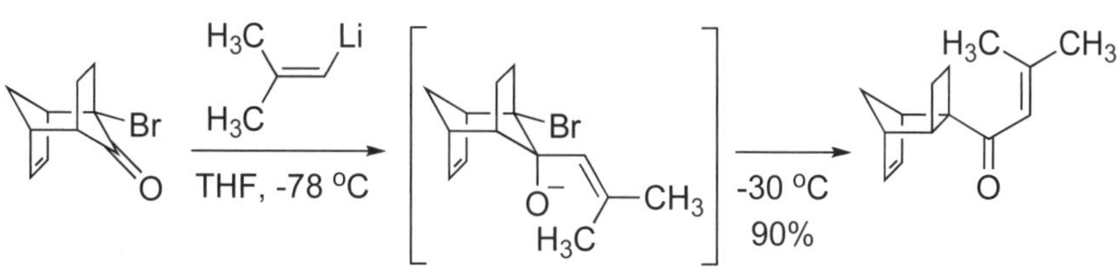

In specific cases, evidence has been obtained that indicates subtle changes in reaction conditions, such as the identity of the base, can alter the preferred mechanism of product formation, leading to the suggestion that both the Favorskii and semibenzylic mechanisms may be simultaneously operative in some Favorskii rearrangements.[2b]

A number of theoretical investigations of Favorskii rearrangement mechanisms have supported the above experimental observations.[16] However, a recent study has elucidated a mechanistic variation whose activation energy for its rate-determining step (RDS; attack of the nucleophile on the cyclopropanone) is lower than that calculated for the canonical Favorskii and semibenzylic mechanisms.[16a] In this newly proposed mechanism, the halide ion generated in the cyclopropanone-forming step reacts first with the cyclopropanone to generate an acyl halide, which is subsequently esterified.

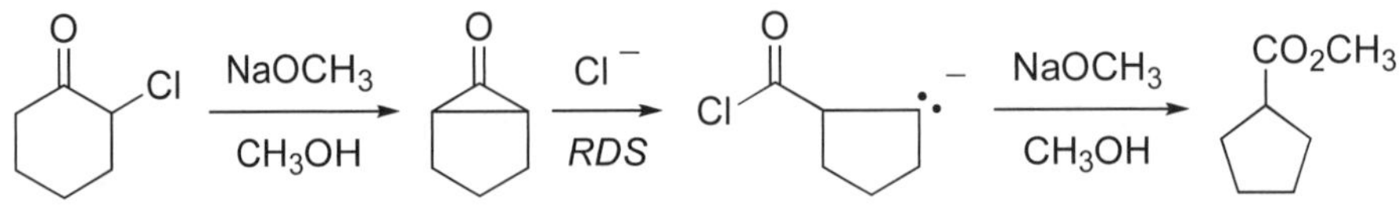

3.3.4 *Variations, Improvements and Modifications*

The canonical Favorskii rearrangement has seen few improvements or modifications, which is not surprising given the simplicity of the reagents used to effect the transformation. However, a number of variations have been developed to expand its scope, though none has received extensive investigation.

The homo-Favorskii rearrangement occurs when enolizable ketones with β-leaving groups are treated with base to yield rearranged products via a cyclobutanone intermediate.[17] Successful examples of this transformation are rare due to competing elimination of the halide to yield an enone and the increased stability of cyclobutanones toward nucleophilic attack. However, in cases where the α' carbon is fully substituted, thus preventing premature elimination of the halide, homo-Favorskii rearrangements are possible. An early example of a base-catalyzed rearrangement of β-dichloroketone **5** is illustrative. Though various mechanisms can be postulated, homo-Favorskii rearrangement via a cyclobutanone intermediate and subsequent elimination(s) leads to two of the observed reaction products obtained in approximately 25% combined yield.

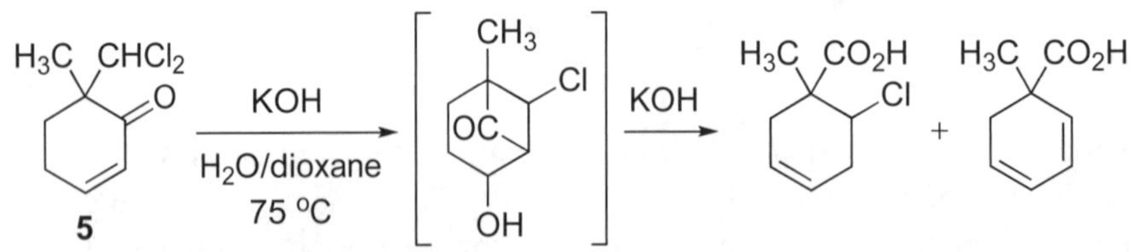

5

Transformations that use the initial steps of the homo-Favorskii rearrangement to prepare cyclobutanones are more common. When treated with base, ketone **6** generates cyclobutanone **8** via oxycyclopropane **7**.[18]

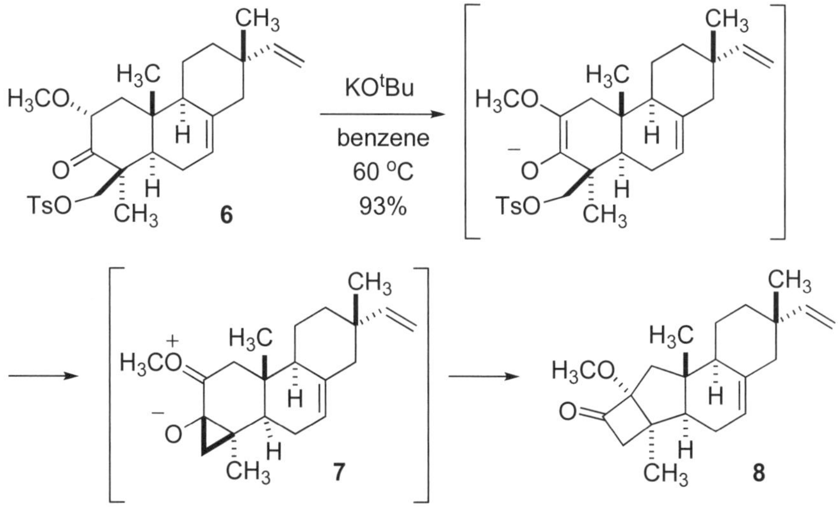

Cyclobutanone precursors to the natural product kelosene were also prepared under homo-Favorskii conditions.[19]

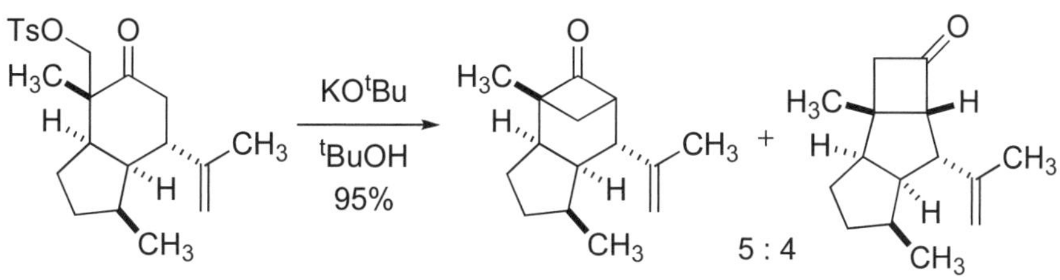

When α-chloroketimines are treated under Favorskii conditions, the corresponding rearranged imidates are generated.[20] When potassium *t*-butoxide is used as the base, the imidates are converted into amides with concomitant elimination of isobutylene.

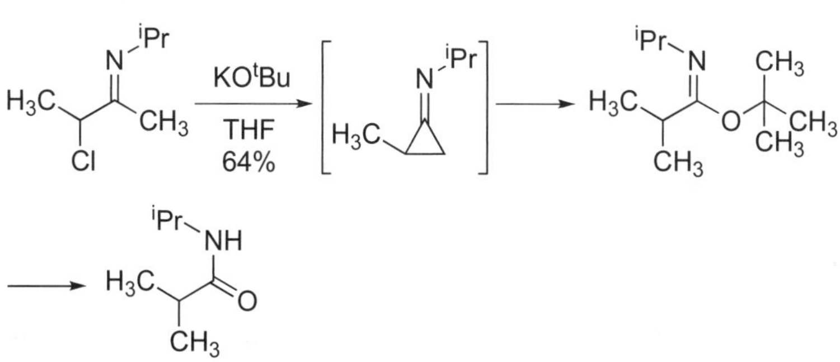

One can generate α-amino cyclopropanecarbonitriles in low yield when α-chloroketimines are treated under Favorskii conditions with KCN as the base.[21] The major products are α-cyanoaziridines which arise from direct attack of cyanide on the imine. Evidence for the intermediacy of a cyclopropylidenamine along the pathway to formation of the minor product was obtained through trapping with an internal nucleophile.[22]

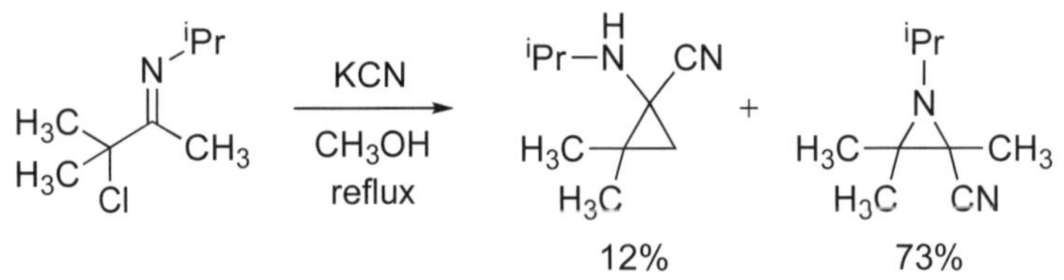

Lewis acids can be used in place of Brønsted bases to promote Favorskii rearrangements.[23]

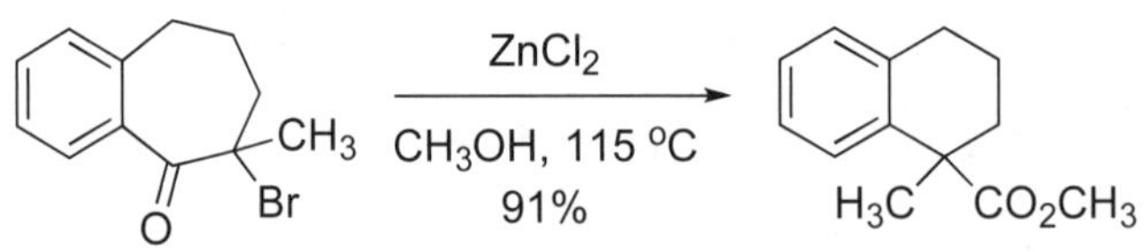

Finally, a Favorskii rearrangement has been proposed in the biosynthesis of the wailupemycins.[24] The enzyme-catalyzed conversion of **9** to **11** can be envisioned to proceed through cyclopropanone intermediate **10**. A Favorskii rearrangement has also been implicated in the biosynthesis of molecules related to okadaic acid.[25]

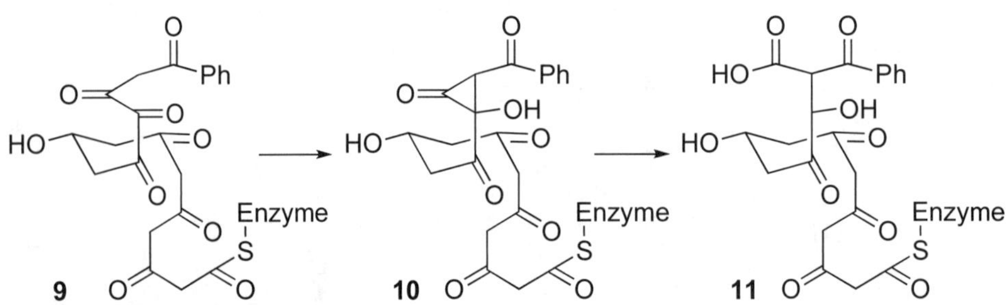

3.3.5 *Synthetic Utility*

Partial Favorskii rearrangements can sometimes be achieved under mild reaction conditions that prevent rupture of the cyclopropane ring. A series of bicyclic nitroxide spin probes were prepared in this manner.[26]

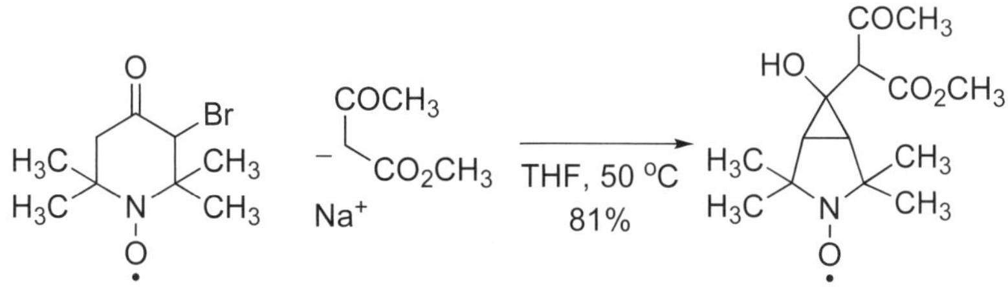

Highly strained structures can be prepared through the Favorskii methodology. For instance, a *cis-t*-butyl ethylene derivative was synthesized from an α,α'-dihaloketone on treatment with a "superbase," KOH in DMSO.[27]

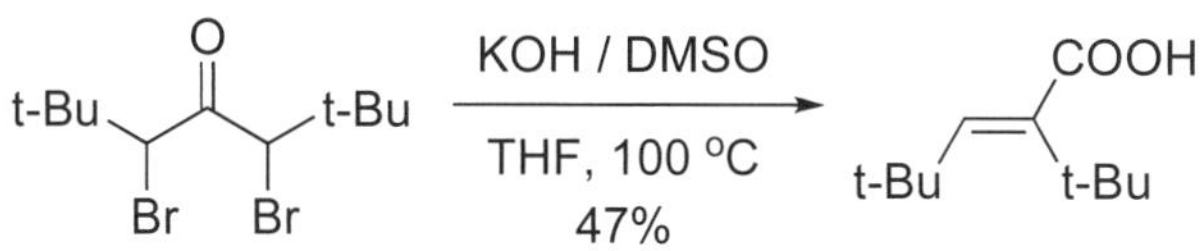

The Favorskii rearrangement has frequently been employed in ring contractions. For example, a Favorskii ring contraction was employed in the synthesis of the cage compound **12** en route to a hexacyclotetradecane,[28] and of cage compound **14** en route to pentaprismane.[29]

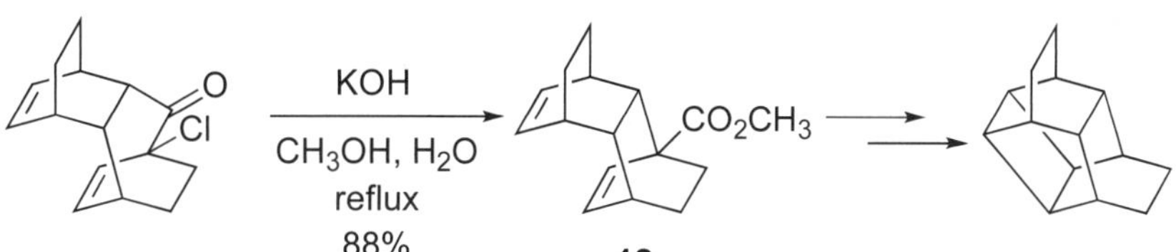

12

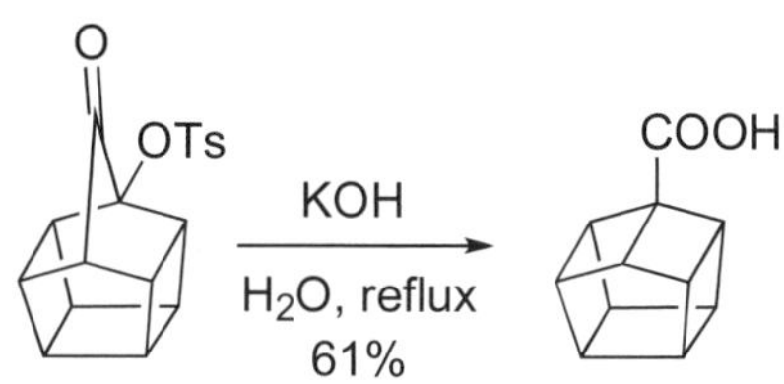

13 **14**

A single cyclopropanone intermediate (**16**) gave rise to both observed rearrangement products obtained from treatment of **15** with DBU.[30] In this case, the nucleophile is an internal enolate derived from deprotonation at C-1 rather than an alkoxide anion. Products of attack of the regioisomeric enolate at C-2 were not observed.

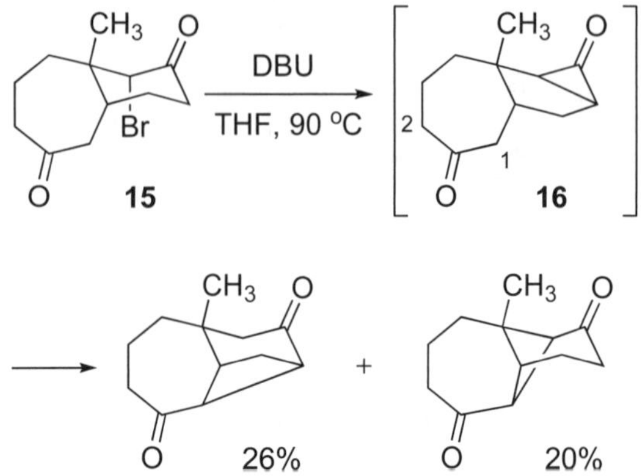

Excellent stereoselectivities can be observed in Favorskii rearrangements. In the ring contraction of cyclohexanone **17**, cyclopentane **18** was observed with greater than 10:1 *dr*, which prompted the authors to study a series of related compounds to determine the factors that control the stereoselectivity. Of compounds **19**, **20**, and **21**, only **19** underwent a Favorskii rearrangement, also with high diastereoselectivity, leading the authors to conclude that a 3-oxy substituent is critical to the success of the rearrangement.[31]

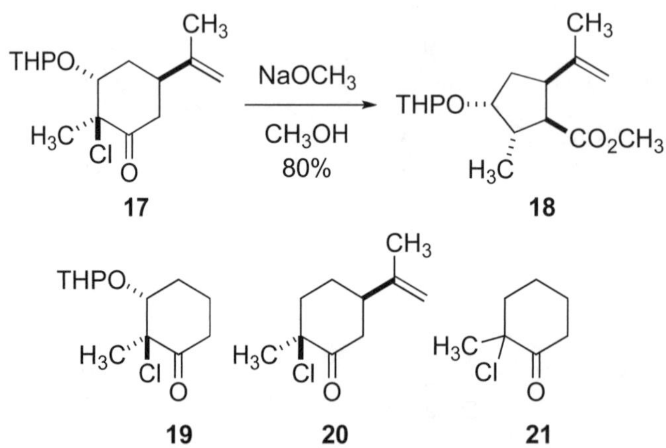

Favorskii ring contraction of ketone **22** followed by diastereoselective protonation led to ester **23**, an intermediate in the synthesis of structures related to the guanacastepene core.[32] In this case, an epoxide served as the leaving group. Favorskii rearrangements of various α,β-epoxyketones have been investigated.[33]

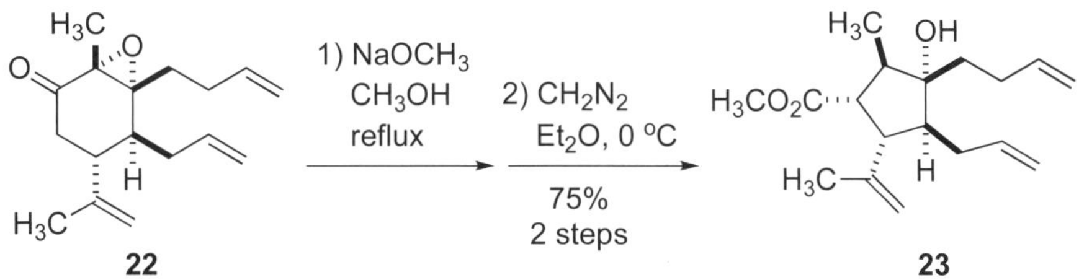

An asymmetric Favorskii rearrangement has been developed that yields optically active α-alkylamides from optically active α-chloroketones. The reaction is presumed to proceed through a chelated intermediate containing a six-membered ring.[34]

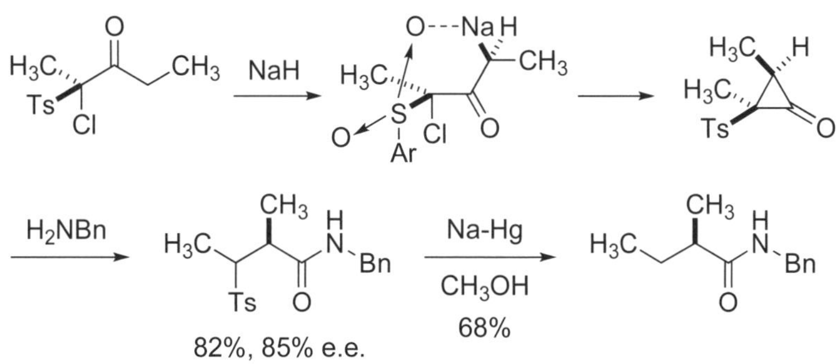

3.3.6 *Experimental*

The examples presented illustrate two of the common ways Favorskii rearrangements are run. The first is a large scale Favorskii ring contraction. The second illustrates the use of the Favorskii rearrangement to form sterically congested carbon frameworks.

Methyl cyclopentanecarboxylate[35]

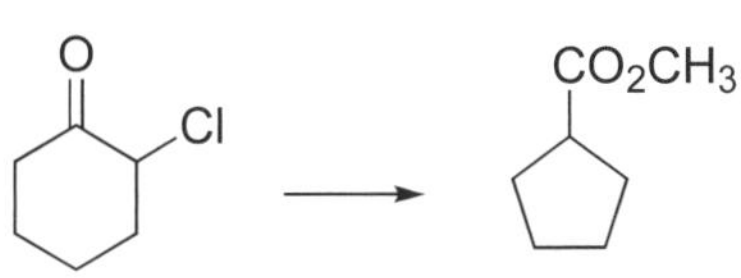

A dry, 1-L, three-necked, round-bottomed flask is equipped with an efficient stirrer, a spiral reflux condenser, and a dropping funnel, and all openings are protected by calcium chloride drying tubes. A suspension of 58 g (1.07 mol) of sodium methoxide in 330 mL anhydrous ether is added, and stirring is begun. To the stirred suspension is added dropwise a solution of 133 g (1 mol) 2-chlorocyclohexanone diluted with 30 mL dry ether. The exothermic reaction is regulated by the rate of addition of the chloroketone; about 40 min. is required for the addition. After the addition of the chloroketone is complete, the mixture is stirred and heated under reflux for 2 h and then cooled. Water is added until the salts are dissolved. The ether layer is separated, and the aqueous layer is saturated with sodium chloride. After extraction of the aqueous layer with two 50 mL portions of ether, the ethereal solutions are combined and washed successively with 100 mL portions of 5% hydrochloric acid, 5% aqueous sodium bicarbonate solution, and saturated sodium chloride solution. The ether solution is dried over magnesium sulfate, and the magnesium sulfate is removed by filtration and washed with ether. Removal of the ether by distillation at atmospheric pressure leaves the crude ester, which is distilled with fractionation at 70–73 °C/48 mm Hg. The yield of methyl cyclopentanecarboxylate is 72–78 g (56–61%).

Compound 14[29]

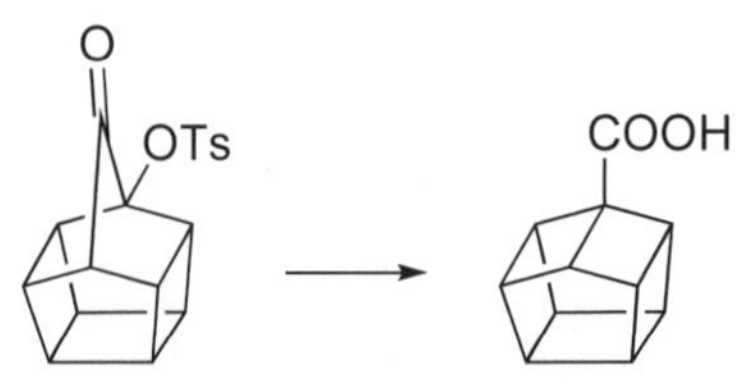

Finely powdered **13** (1.3 g, 4 mmol) was added in portions to a preheated 20% aq. KOH solution (20 mL, 110 °C) and washed in with an additional 20 mL of base solution. The resulting mixture was refluxed rapidly with vigorous stirring for 5 h and became homogenous. Chloroform (50 mL) was added, the mixture was cooled to between 0 and 5 °C and then acidified with ice-cold 50 wt % sulfuric acid. The insoluble material formed was filtered through glass wool and extracted repeatedly with chloroform. The aqueous phase was extracted four times with chloroform (20 mL). The combined organic extracts were washed once with aqueous saturated ammonium sulfate (40 mL) and dried. After removal of solvent in vacuo 470 mg of a pale yellow solid remained. Purification via chromatography on silica gel (dichloromethane) gave 426 mg of **14** (61%).

3.3.7 References

1. [R] (a) Kende, A. S.; *Org. React.* **1960**, *11,* 261–316. (b) Akhrem, A. A.; Ustynyuk, T. K.; Titov, Yu. A. *Russ. Chem. Rev. (Engl. Trans.)* **1970**, *39,* 732–746. (c) Rappe, C. In *The Chemistry of the Carbon-Halogen Bond Part 2* (Patai, S. ed.) Wiley, London, **1973**, pp 1084–1114. (d) Mann, J. *Comp. Org. Syn.* Trost, B. M.; Fleming, I., eds. Pergamon Press, Oxford, **1991**, pp 839–859. (e) Guijarro, D.; Yus, M. *Curr. Org. Chem.* **2005**, *9,* 1713–1735.

2. (a) House, H. O.; Frank, G. A. *J. Org. Chem.* **1965**, *30,* 2948–2956. (b) Warnhoff, E. W.; Wong, C. M.; Tai, W. T. *J. Am. Chem. Soc.* **1968**, 514–515. (c) Loftfield, R. B.; Schaad, L. *J. Am. Chem. Soc.* **1954**, 35–37. (d) Rappe, C.; Knutsson, L.; Turro, N. J.; Gagosian, R. B. *J. Am. Chem. Soc.* **1970**, 2032–2035.
3. Cloëz, M. Ch. *B. Soc. Chim. Fr.* **1890**, 602–605.
4. Favorskii, A. E. *J. Russ. Phys. Chem. Soc.* **1894**, *26*, 559–603.
5. Aston, J. G.; Greenburg, R. B. *J. Am. Chem. Soc.* **1940**, *62*, 2590–2595.
6. House, H. O.; Gilmore, W. F. *J. Am. Chem. Soc.* **1961**, *83,* 3980–3985.
7. These five possibilities are succinctly summarized and referenced to their original sources in Ref. 8.
8. Loftfield, R. B. *J. Am. Chem. Soc.* **1951**, *73,* 4707–4714.
9. Turro, N. J.; Hammond, W. B. *J. Am. Chem. Soc.* **1965**, *87,* 3258–3259.
10. House, H. O.; Frank, G. A. *J. Org. Chem.* **1965**, *30,* 2948–2956.
11. Mann, J.; Wilde, P. D.; Finch, M. W. *Tetrahedron* **1987**, *43*, 5431–5441.
12. Aston, J. G.; Newkirk, J. D. *J. Am. Chem. Soc.* **1951**, *73,* 3900–3902.
13. Baudry, D.; Bégué, J. P.; Charpentier-Morize, M. *B. Soc. Chim. Fr.* **1971**, 1416–1424.
14. Harmata, M.; Wacharasindhu, S. *Org. Lett.* **2005**, *7*, 2563–2565.
15. Harmata, M.; Bohnert, G. J. *Org. Lett.* **2003**, *5*, 59–61.
16. (a) Tsuchida, N.; Yamazaki, S.; Yamabe, S. *J. Org. Biomol. Chem.* **2008**, *6,* 3109–3117. (b) Hamblin, G. D.; Jimenez, R. P.; Sorensen, T. S. *J. Org Chem.* **2007**, *72,* 8033–8045. (c) Moliner, V.; Castillo, R.; Safont, V. S.; Oliva, M.; Bohn, S.; Tuñón, I.; Andrés, J. *J. Am. Chem. Soc.* **1997**, *119*, 1941–1947. (d) Castillo, R; Andrés, J; Moliner, V. *J. Phys. Chem. B.* **2001**, *105*, 2453–2460.
17. Wenkert, E.; Bakuzis, P.; Baumgarten, R. J.; Leicht, C. L.; Schenk, H. P. *J. Am. Chem. Soc.* **1971**, *93,* 3208–3216.
18. Ceccherelli, P.; Curini, M.; Pellicciari, R.; Wenkert, E. *J. Org. Chem.* **1982**, *47*, 3172–3174.
19. Zhang, L.; Koreeda, M. *Org. Lett.* **2002**, *4*, 3755–3758.
20. De Kimpe, N.; Sulmon, P.; Moëns, L.; Schamp, N.; Declercq, J.-P.; Van Meerssche, M. *J. Org. Chem.* **1986**, *51*, 3839–3848.
21. De Kimpe, N.; Sulmon, P.; Verhé, R.; De Buyck, L.; Schamp, N. *J. Org. Chem.* **1983**, *48*, 4320–4326.
22. De Kimpe, N.; Stanoeva, E.; Schamp, N. *Tetrahedron Lett.* **1988**, *29*, 589–592.
23. Maiti, S. B.; Chaudhuri, S. R. R.; Chatterjee, A. *Synthesis* **1987**, 806–809.
24. Xiang, L.; Kalaitzis, J. A.; Moore, B. S. *Proc. Nat. Acad. Sci. USA.* **2004**, *101,* 15609–15614.
25. Wright, J. L. C.; Hu, T.; McLachlan, J. L.; Needham, J.; Walter, J. A. *J. Am. Chem. Soc.* **1996**, *118,* 8757–8758.
26. Babič, A.; Pečar, S. *Synlett* **2008**, 1155–1158.
27. Ionkin, A. S.; Marshall, W. J.; Fish, B. M. *Org. Lett.* **2008**, *10,* 2303–2305.
28. Takeshita, H.; Kawakami, H.; Ikeda, Y.; Mori, A. *J. Org. Chem.* **1994**, *59,* 6490–6492.
29. Eaton, P. E.; Or, Y. S.; Branca, S. J.; Shankar, B. K. R. *Tetrahedron* **1986**, *42,* 1621–1631.
30. Karimi, S.; Christodoulou, J.; Subramaniam, G. *Synth. Commun.* **2006**, *36*, 929–934.
31. (a) Lee, E.; Yoon, C. H. *J. Chem. Soc., Chem. Commun.* **1994**, 479–481. (b) Ley, S. V.; Antonello, A.; Balskus, E. P.; Booth, D. T.; Christensen, S. B. Cleator, E.; Gold, H.; Högenauer, K.; Hünger, U.; Myers, R. M.; Oliver, S. F.; Simic, O.; Smith, M. D.; Søhoel, H. Woolford, A. J. A. *Proc. Nat. Acad. Sci. USA.* **2004**, *101,* 12073–12078.
32. Srikrishna, A.; Dethe, D. H. *Org. Lett.* **2004**, *6,* 165–168.
33. Mouk, R. W.; Patel, K. M.; Reusch, W. *Tetrahedron* **1975**, *31,* 13–19.
34. Satoh, T.; Motohashi, S.; Kimura, S.; Tokutake, N.; Yamakawa, K. *Tetrahedron Lett.* **1993**, *34,* 4823–4826.
35. Goheen, D. W.; Vaughan, W. R. *Org. Synth.* **1959**, *39,* 37–39.

3.4 Nazarov Cyclization

Matthew J. Fuchter

3.4.1 Description

The acid-catalyzed cyclization of divinyl ketones **1** (or their precursors) via pentadienyl cations **2**, is known as the Nazarov cyclization.[1–6] It is a commonly used method for the synthesis of cyclopentenones **3**.

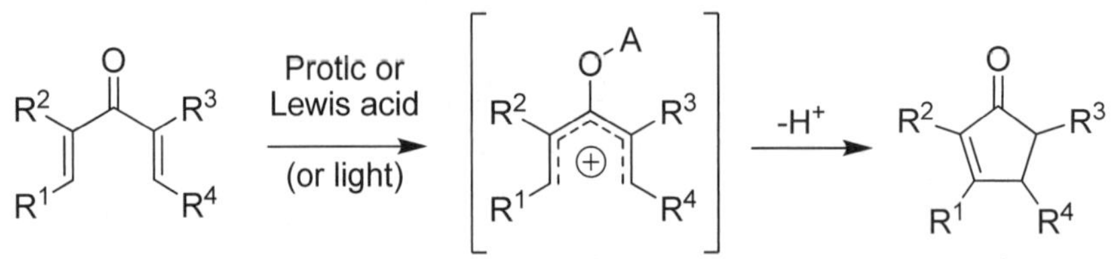

1 2 3

The process can be catalyzed by either protic or Lewis acids; however, it is important to note, that stoichiometric (or even greater) quantities of promoter are often required.[6] The reaction can also be initiated by irradiation, although this has mechanistic consequences (see Section **3.4.3**). In general, any compound that affords a pentadienyl cation **2**, or its equivalent, can be considered to undergo the Nazarov cyclization. Numerous substitution patterns are tolerated, although in some cases, regioisomeric double bond isomers may be obtained. Electron-donating substituents in the α and α' positions (R^2, R^3), accelerate the reaction, whereas they retard the reaction in the β and β' positions (R^1, R^4). Judicious choice of substituents can polarize the conjugated system, accelerating the reaction and improving the regioselectivity of the cyclization. A notable example is the use of trialkylsilyl (or trialkylstannyl) groups in the β and β' positions (R^1, R^4), to ensure the controlled collapse of the cyclopentenylic cation (see Section **3.4.5.1**). The use of chiral substituents can result in transfer of chirality, and limited examples exist for asymmetric catalysis (see Section **3.4.4.4**). The cationic intermediates can be intercepted in alternative reaction pathways and results in 'interrupted' Nazarov processes (see Section **3.4.5.2**).

3.4.2 Historical Perspective

Nazarov's key work on the formation of cyclopentenones was preceded by a number of reports on analogous reaction pathways. For example in 1903, Vorländer and co-workers discovered that the treatment of dibenzylidene-acetone **4** with concentrated sulfuric acid and acetic anhydride, followed by hydrolysis with sodium hydroxide resulted in the formation of a cyclic ketol **5**, the structure of which was unknown at the time.[7] Indeed, there are several examples of reactions in the early literature which presumably follow analogous reaction pathways.[8]

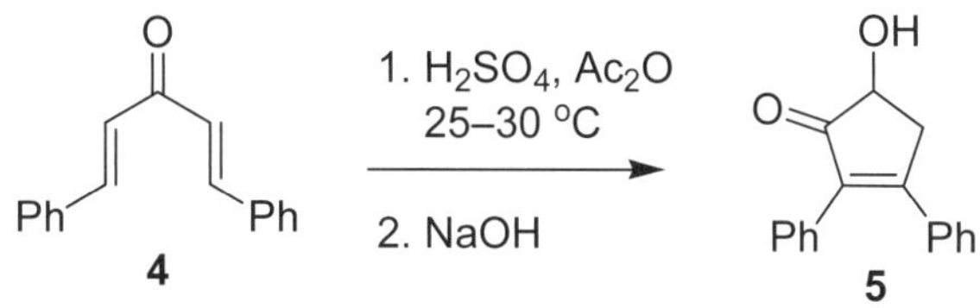

In the 1930s, Marvel and co-workers studied the acid-catalyzed hydration of dienynes,[9,10] and it was this topic that Nazarov revisited in the 1940s and 1950s. He extensively studied this process and demonstrated the cyclization of the intermediate allyl vinyl ketones **7** to 2-cyclopentenones **8** in numerous cases.[11–13] Mechanistic interpretation of the reaction remained unclear, however, until the studies of Braude and Coles in 1952.[14] They demonstrated that the formation of 2-cyclopentenones actually proceeds via divinyl ketones (the allyl vinyl ketones in Nazarov's process isomerize in situ), with the intermediacy of carbocations. Thus the modern interpretation of the Nazarov cyclization was born: The acid-catalyzed closure of divinyl ketones **1** to 2-cyclopentenones **3**.

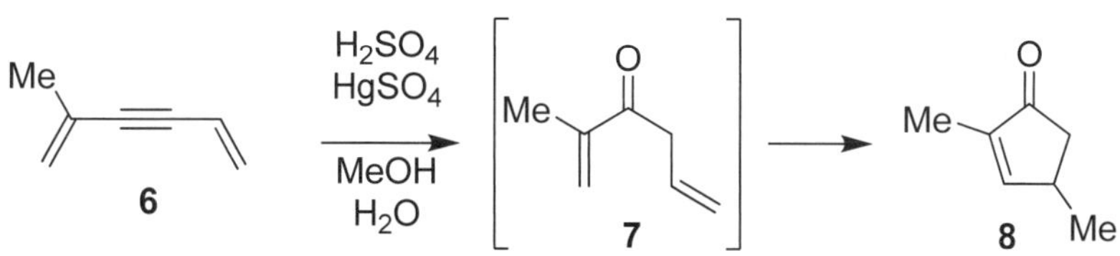

3.4.3 Mechanism

For the protic acid catalyzed reaction, the cyclization commences with protonation of the divinyl ketone **9** and formation of a pentadienyl cation **10**. An analogous process is operational in the case of Lewis acid-catalyzed reactions. The pentadienyl cation **10** then undergoes a 4π electrocyclic closure to give a cyclopentenylic cation **11**. This cyclization is a pericyclic reaction and is governed by the rules for conservation of orbital symmetry. Namely, this means the cyclization occurs stereospecifically in a conrotatory fashion, with predictable relative configurations of the substituents (i.e., the R groups in **11** are *anti*). Elimination of a proton, followed by tautomerization gives product **13**.[4]

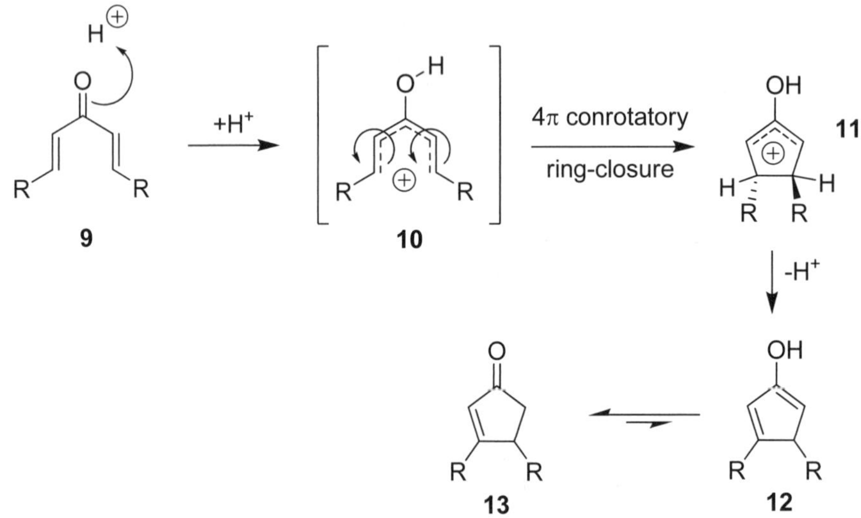

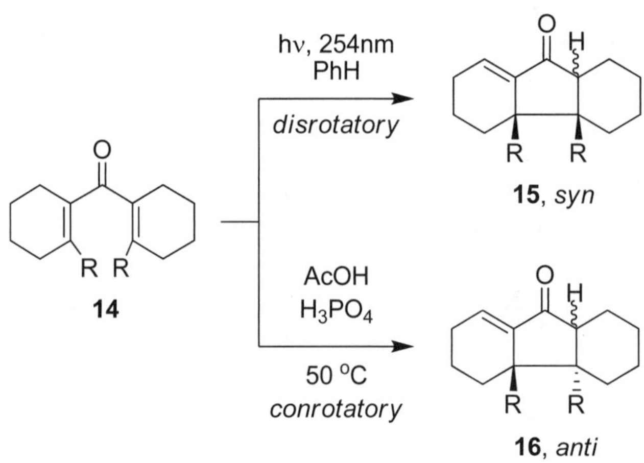

Early attempts to verify the stereochemical predictions of orbital symmetry control were hampered by carbocation rearrangement reactions,[15] such as Wagner–Meerwin shifts, although the very presence of these anomalous pathways is consistent with a cationic pathway. It is now well established that the Nazarov cyclization occurs via a pentadienyl cation **10**, which has been spectroscopically observed,[16] and shown to undergo facile cyclization to a cyclopentenylic cation **11**.[15] The most convincing evidence for the involvement of a pericyclic reaction was the demonstration of complementary rotatory pathways for the cyclization of ketone **14**.[17] Acid-catalyzed thermal cyclization results in conrotatory ring closure and formation of product **16**, whereas under photochemical conditions, disrotatory ring closure occurs, resulting the formation of product **15**, in line with rules for conservation of orbital symmetry. This example also highlights that the Nazarov cyclization can occur on irradiation, although it is important to realize that this results in an alternative rotatory pathway.

The cationic reaction pathway also allows for predictable effects on differential substrate substitution.[4] In the rate-limiting step, the distribution of charge (marked by *) changes from C(1), C(3) and C(5), in pentadienylic cation **18**, to C(2) and C(4), in cyclopentenylic cation **19**. Therefore, substituents that stabilize positive charge (electron-donating groups), accelerate the reaction in the α-position (R^2), or decelerate the reaction in the β-position (R^1). Classically, under rather vigorous conditions, the reactions are under thermodynamic control and result in the formation of product **20**, where the double bond occupies the most substituted position (Saytzeff's Rule).

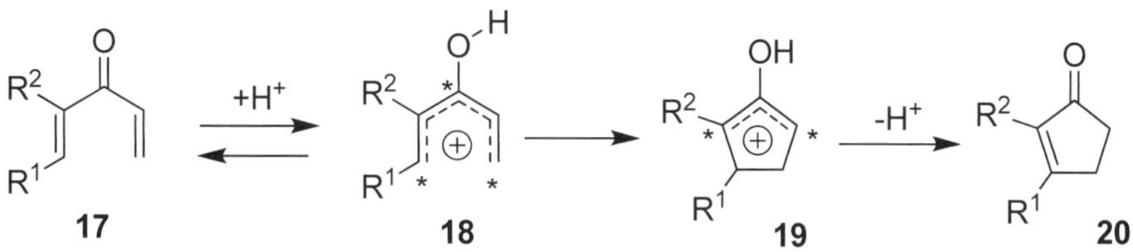

As well as the stereospecific conrotatory cyclization pathway, there is another stereochemical feature of the Nazarov cyclization, which arises from the duality of the allowed electrocyclization reaction.[4] In the presence of a chiral remote substituent, substrate **21** can either undergo clockwise electrocyclization to give product **22** and/or **23**, depending on the regiochemistry of the double bond, or counter-clockwise cyclization to give the corresponding diastereoisomers **24** and/or **25**. The selectivity of this process is governed primarily by steric factors, such as the torsional and nonbonding interactions between substituents in the vicinity of the newly forming bond, and is called *torquoselectivity*.

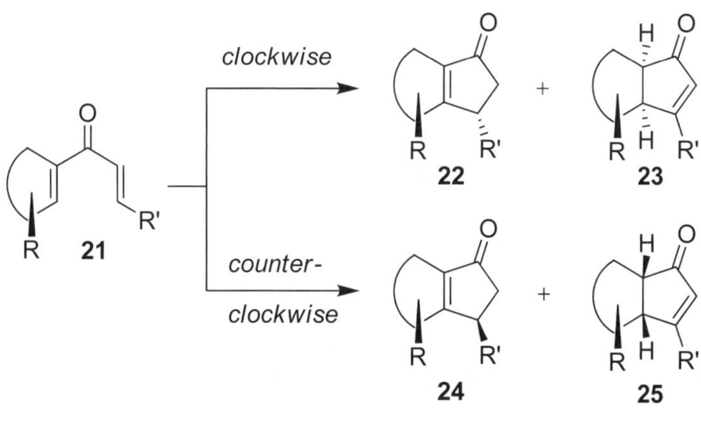

Denmark and co-workers reported a good example of torquoselection in the silicon-directed Nazarov cyclization (see Section **3.4.5.1**). They demonstrated that cyclohexenyl-derived divinyl ketones **26** cyclize to give the relative stereoisomer **27** as the major product (see Section **3.4.5.1** for the mechanism of the silicon-directed reaction).[18,19] The use of bulky alkyl groups (such as *t*-butyl) and/or bulky silicon substituents gave the best selectivity, at the expense of the chemical yield. It is interesting that the corresponding cyclopentenyl-derived systems gave only poor torque-selectivity.

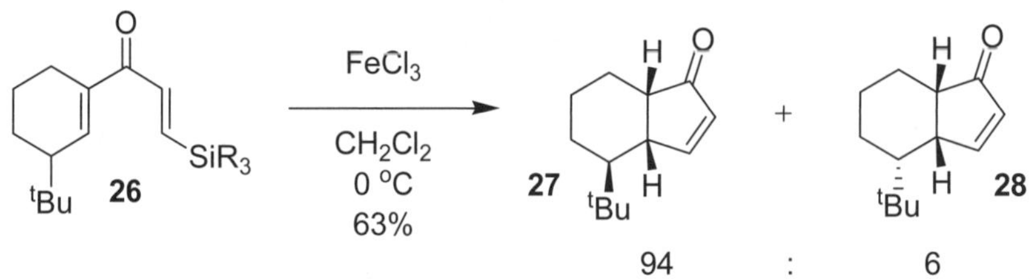

3.4.4 Synthetic Utility

3.4.4.1 Simple Systems

The preparation of simple cyclopentenones can be achieved in modest to good yield using the Nazarov cyclization.[4] Although many substitution patterns can be tolerated, in some cases regioisomeric double-bond isomers may be obtained. When relatively forcing conditions are used (strong acid, heat), the thermodynamically most stable regioisomer is usually produced. Acyclic, monocyclic and bicyclic precursors can all be employed, although the Nazarov cyclization is most frequently used as a cyclopentenone annulation method. In certain cases it is possible to control the torquoselectivity of the reaction. For example, substrate **29** is reported to give cyclopentenone **30** as the sole diastereoisomer in good yield.[20]

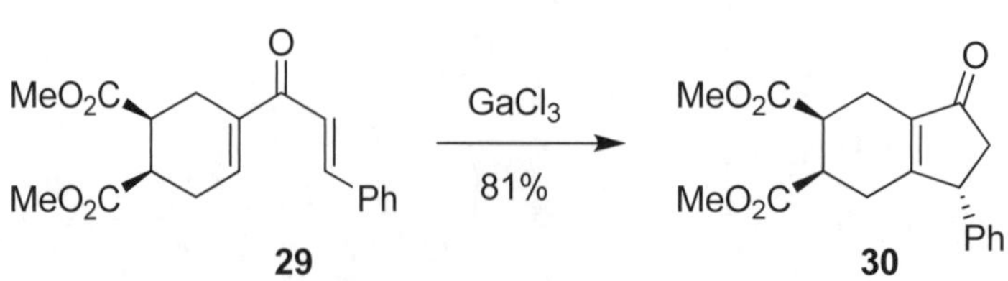

Under sufficiently vigorous conditions aryl vinyl ketones also undergo the cyclization, producing annulated aromatic rings. For example, in 1953 Braude and Forbes reported the cyclization of **31**, principally en route to azulenes.[21] Heating substrate **31** in a mixture of phosphoric and formic acid produced **32** in good yield (80%).

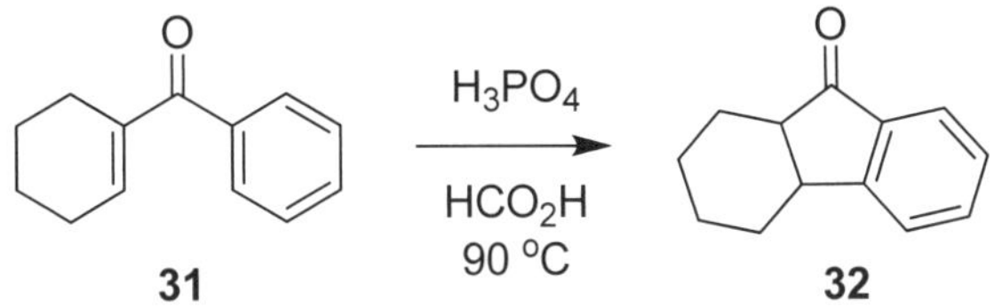

The Nazarov cyclization was the key step in a synthesis of (±)-trichodiene (**35**) by Harding and co-workers.[22] One significant challenge in the preparation of this natural product is the presence of two adjacent quaternary stereocenters, and Harding and co-workers selected the Nazarov cyclization to tackle this problem. Although the reaction was not efficient under protic acid catalysis, the presence of an excess of boron trifluoride etherate enabled the production of **34a/b** in good yield, whereby double bond migration had occurred. These specific reaction conditions were key to the success of this transformation, since under milder conditions or shorter reaction times, the expected α,β-unsaturated ketone was observed.

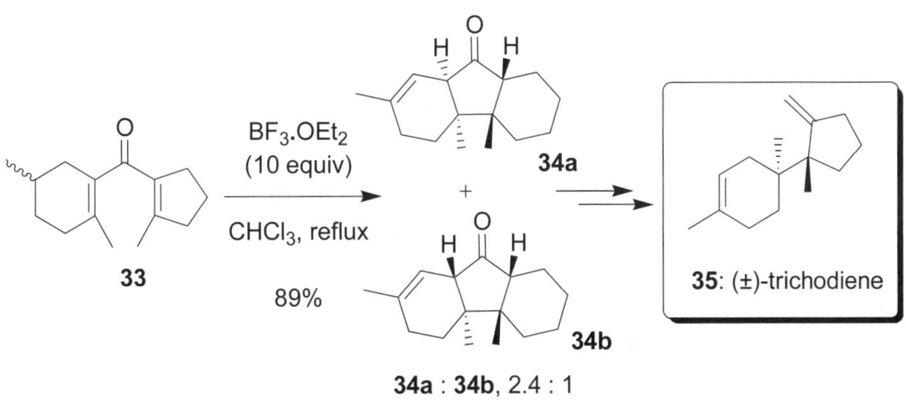

3.4.4.2 α-Substituted Substrates

As a general rule of thumb, α-substitution improves the cyclization efficiency of divinyl ketones. One explanation for this fact is the increased population of the *s-trans/s-trans* conformer **36a** with α-substitution, which is predisposed toward cyclization.[6] The α-substituents experience unfavourable non-bonded interactions in the *s-cis* conformation and, therefore, favour the *s-trans* conformer, whereas substrates containing α-hydrogen favour the *s-cis* conformation.

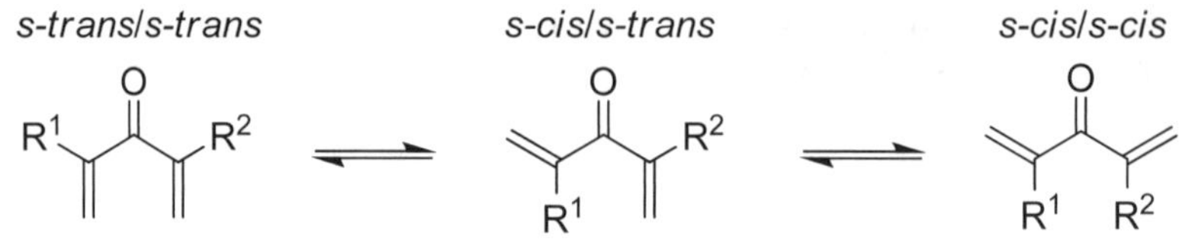

36a **36b** **36c**

As stated previously (see Section **3.4.3**) the cationic reaction pathway of the Nazarov reaction allows for predictable stereoelectronic effects on differential substrate substitution. A theory on the stereoelectronic influence of α- and β-substituents on substrate reactivity was developed by Denmark and co-workers, and stated that electron-donating substituents (cation-stabilizing) in the α-position lower the activation barrier for cyclization by stabilizing the product oxyallyl cation (**19**, see Section **3.4.3**) as opposed to the pentadienyl cation (**18**, see Section **3.4.3**).[23] Electron-donating substituents in the α-position, therefore, facilitate reaction. Coordinating functionalities in the α-position can also aid reactivity by enabling conformation control when the cyclization is mediated by chelating Lewis acids. Coordination of the Lewis acid to both the lone pair of the carbonyl oxygen and the functionality in the α-position enforces the reactive *s-trans* conformation (see **36a**), facilitating cyclization.[6]

The most widely used electron-donating α-substituents are alkoxy groups. Indeed, these substrates are so reactive that they enabled the first truly catalytic examples of Lewis acid-catalyzed Nazarov reactions. Whereas stoichiometric quantities of Lewis acid are often required due to slow protonation of the Lewis acid enolate, α-alkoxy appended systems show efficient catalytic turnover.[6] For example, Trauner and co-workers reported the efficient Nazarov cyclization of **37** in the presence of 10 mol % of aluminium trichloride.[24] Not only do α-alkoxy groups lower the activation barrier to cyclization, but they also localize the resultant positive charge at one α-carbon, ensuring a highly regioselective elimination.

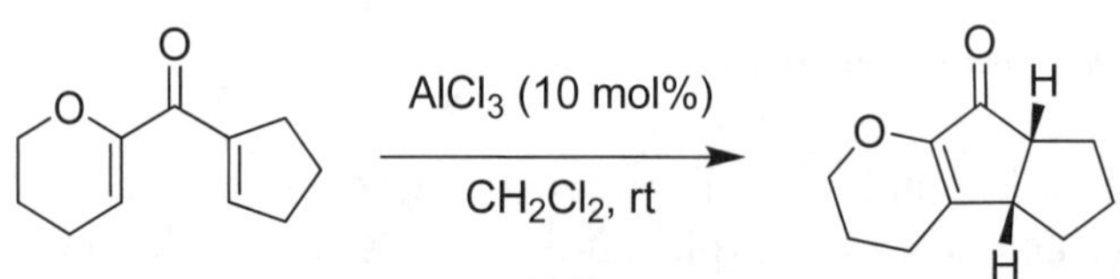

37 **38**

There are many examples of catalytic Nazarov cyclizations using Lewis acid catalysts in combination with α-alkoxy substituents. Tius and co-workers reported one particularly interesting example using palladium catalysts.[25]

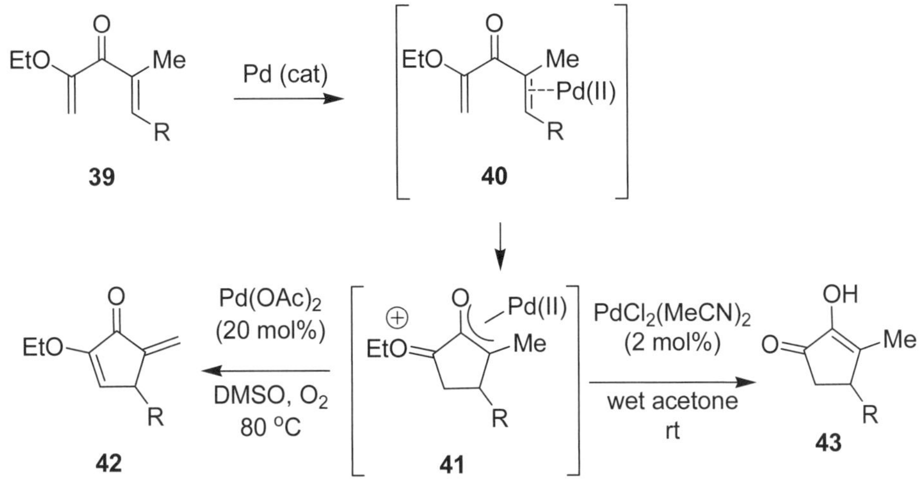

While the product of this reaction somewhat resembles a Nazarov cyclization, it is thought that the reaction proceeds via a different mechanistic pathway and depends on the nature of the palladium catalyst. Activation of the olefin by palladium, followed by cyclization gives palladium enolate **41**. When $Pd(OAc)_2$ is used as the catalyst, β-hydride elimination of **41** produces cyclopentenone **42**, and molecular oxygen reoxidizes the palladium. On the other hand, when $PdCl_2(MeCN)_2$ is used as a catalyst, hydrolysis generates HCl which results in rapid protonation of **41**, giving cyclopentenone **43** as the product.

Several examples of the efficient cyclization of substrates bearing electron-withdrawing groups in the α-position have been report. Indeed, despite the hypothesis that such substituents should retard the rate of Nazarov cyclization,[23] several high-yielding procedures have been reported, and divinyl ketones bearing α-ester and amide groups have proved effective in asymmetric Nazarov reactions (see Section **3.4.4.4**).[26] One important stereochemical consequence of α-electron-withdrawing groups is their predisposition to form *trans*-stereoisomers. This is a consequence of facile equilibration of the α-stereocenter under the reaction conditions, giving the thermodynamically favoured *trans* product.[6]

Flynn and co-workers reported an efficient Nazarov cyclization of β-ketoesters, such as **44**.[27] Under Brønsted acid promotion, a range of substrates gave the *trans*-cyclopentanones after highly regioselective elimination, such as **45**. Although they used predominately *Z*-configured substrates, it is important to note that *E*/*Z* isomerization occurs gradually over time. This can obviously give difficulties in correlating the stereochemical purity of the reagents with that of the products. In this case, it is possible that the predominance of *Z* increases reaction efficiency.

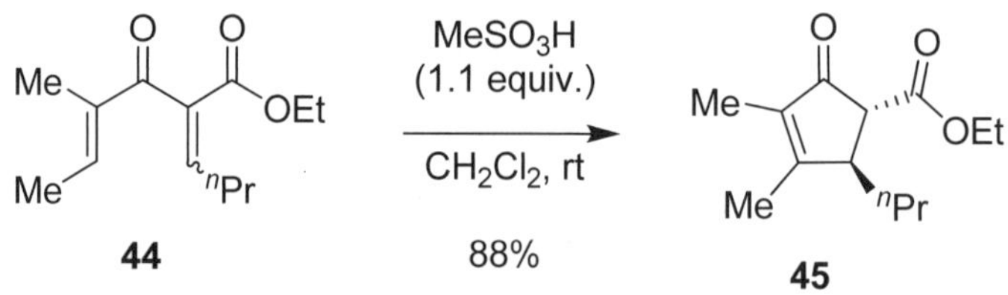

To fully polarize the substrate, Frontier and co-workers have studied substrates bearing both α-electron-donating and α-electron-withdrawing substituents.[28] Compounds such as **46** underwent the cyclization in very high yield, using just 2 mol % of a copper catalyst. Highly regioselective elimination was observed due to the electron-donating substituent, and the *trans*-product was isolated due to the electron-withdrawing substituent. The conclusions of this study regarding α-substituents determined that α-electron-donating substituents increase the reaction rate, whereas the effect of α-electron-withdrawing substituent is complex, possibly affecting the geometry of the substrate, the ability of the catalyst to bind to the substrate, and the facility of catalytic turnover.

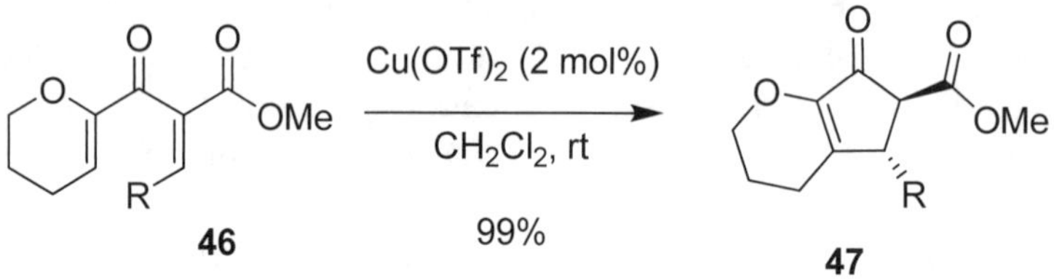

3.4.4.3 β-Substituted Substrates

In general, substitution in the β-position slows the reaction rate of the Nazarov cyclization;[23] however, the nature and geometry of the substituents, with the corresponding steric and stereoelectronic implications has varying effects on the reactivity. For example, substrates with internal β-substitution suffer steric hindrance when the cation adopts the necessary conformation for electrocyclization (see **48**) and thus exhibit poor reactivity.[6] From a stereoelectronic standpoint, electron-donating substituents (cation-stabilizing) in the β-position raise the activation barrier for cyclization by stabilizing the pentadienyl cation (**18**, see Section **3.4.3**) relative to the oxyallyl cation (**19**, see Section **3.4.3**).[23] Therefore the size, nature, and geometry of the substituent is key to the reactivity profile of the substrate.

One of the most important β-substituents in terms of controlling reactivity and selectivity is a trialkylsilyl group, and such systems will be discussed in Section **3.4.5.1**.

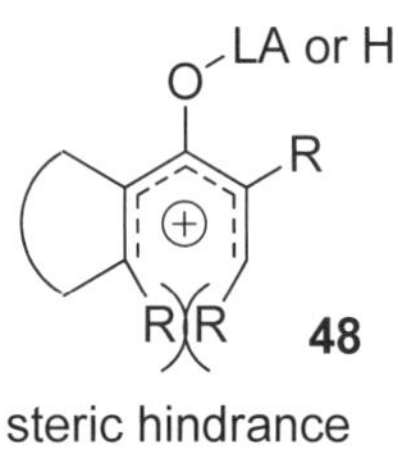

48

Since steric hindrance disfavours cyclization for substrates with internal β-substitution, double-bond isomerization is often a competing pathway. Indeed, West found that reductive Nazarov cyclizations (see Section **3.4.5.2**) of either *trans*- or *cis*-disubstituted enones **49a** or **49b**, both produced a single diastereomeric product **50a**.[29] The stereochemistry of **50a** corresponds to a conrotatory cyclization of *trans*-isomer **49a**, thereby indicating that while the *trans*-isomer **49a** cyclizes, the *cis*-isomer **49b** first isomerizes before cyclization. Recent studies by Frontier and co-workers on polarized Nazarov cyclizations also found that in the case of alkylidene β-ketoester substrates (for example, see **46**), reaction rates depended on the competing rate of isomerization, which depended on the nature of the β-substituent.[28]

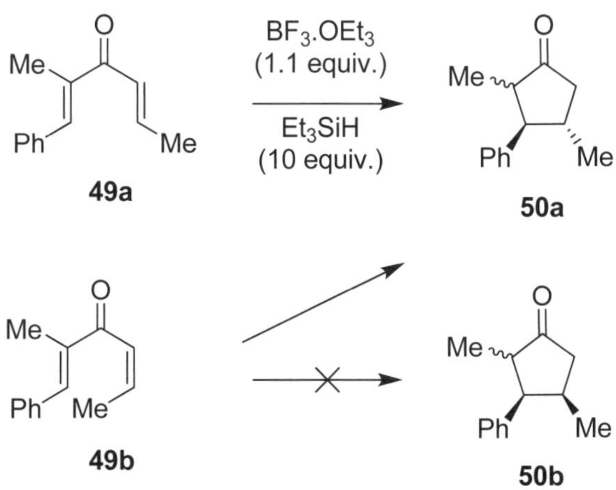

As opposed to these β-sp^2-hybridized systems, allenyl and cumulenyl substrates (β-sp-hybridized) exhibit excellent reactivity in the Nazarov cyclization.[6] This enhanced reactivity is thought to derive from two factors: Minimization of steric interactions at the β-position (increasing the population of the reactive *s-trans* conformer), and relief of allenyl strain in the transition state upon forming the allyl cation. As an example of their facile cyclization, Hashmi and co-workers reported that exposure of ketone **52** to silica gel resulted in conjugated dienone **54**, presumably via the allenyl ketone **53**.[30]

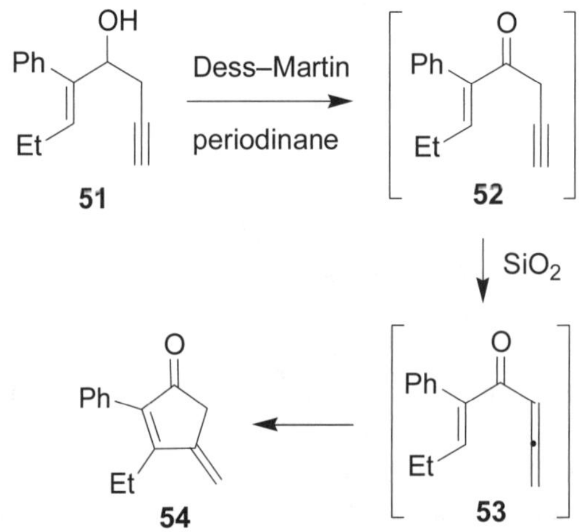

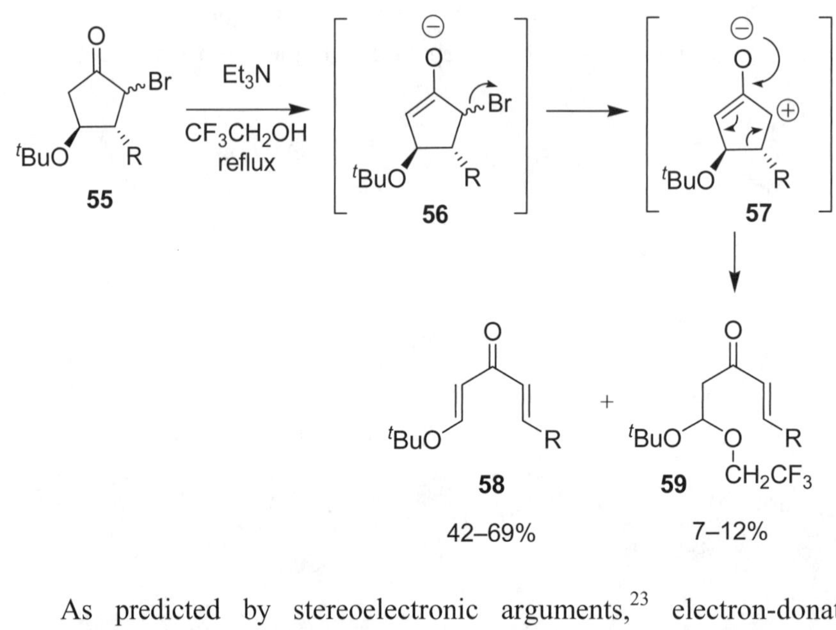

As predicted by stereoelectronic arguments,[23] electron-donating substituents in the β-position raise the activation barrier for cyclization by stabilizing the pentadienyl cation (**18**, see Section **3.4.3**). In 2002, Harmata and Lee reported that β-alkoxy substituents not only stabilize the pentadienyl cation but also promote a retro-Nazarov cyclization processes.[31] Exposure of ketone **55** to triethylamine in trifluoroethanol results in transient formation of an oxyallyl cation **57**, which undergoes retro-electrocyclization to primarily form divinyl ketone **58** (with a small amount of a corresponding conjugate addition product **59**). Efficient retro-cyclization required the presence of an additional pentadienyl cation-stabilizing group such as aryl or alkenyl.

Examples do exist however, of productive Nazarov cyclizations bearing electron-rich heteroatoms in the β-position. For example, cyclization of substrate **60** (albeit with the nitrogen atom contained within an aromatic indole moiety) was reported by Cheng and co-workers in 1996.[32] Exposure of **60** to HCl in refluxing dioxane resulted in the formation of product **61** in moderate yield. This key intermediate was then transformed into inverto-yuehchukene (**62**), a dimer of 2-didehydroprenylindole.

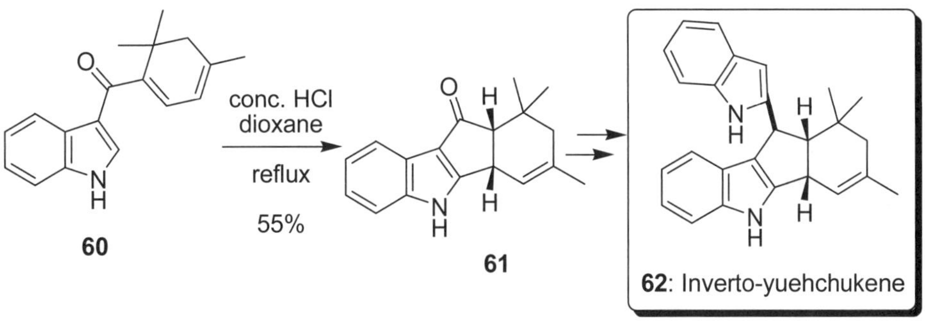

3.4.4.4 Asymmetric Reaction

As with many asymmetric processes, there are three ways to control absolute stereochemistry in the Nazarov cyclization: Asymmetry transfer, the use of chiral auxiliaries, or asymmetric catalysis.[5,6] It is important to realize, however, that there are two distinct processes operating that determine the stereochemistry of the product. To control the absolute stereochemistry of the β-carbon atom(s), it is necessary to control the sense of conrotation, clockwise or counterclockwise (torquoselectivity, see Section **3.4.3**). To control the absolute stereochemistry of the α-carbon atom however, it is necessary to control the facial selectivity for enol protonation.

In terms of asymmetry transfer, several effective means of controlling the absolute asymmetry of the product have emerged. Denmark and co-workers have published extensively on the use of silicon substituents to aid selectivity in Nazarov cyclizations (see Section **3.4.5.1**). In one example of asymmetry transfer, they used a stereogenic trimethylsilyl-bearing carbon atom to control the sense of conrotation.[33] Treatment of ketone **63** with ferric chloride gave product **64** in excellent yield and with complete transfer of asymmetry (see Section **3.4.5.1** for the mechanism of the silicon-directed reaction). The excellent transfer of asymmetry is due to clockwise conrotation maximizing overlap of the C–Si bond with the allylic carbocation all along the reaction coordinate. While the stereochemistry of the β-carbon atoms of **64** is a result of the sense of conrotation, the stereochemistry at the α-carbon atom is thermodynamically favoured and set during proton transfer.

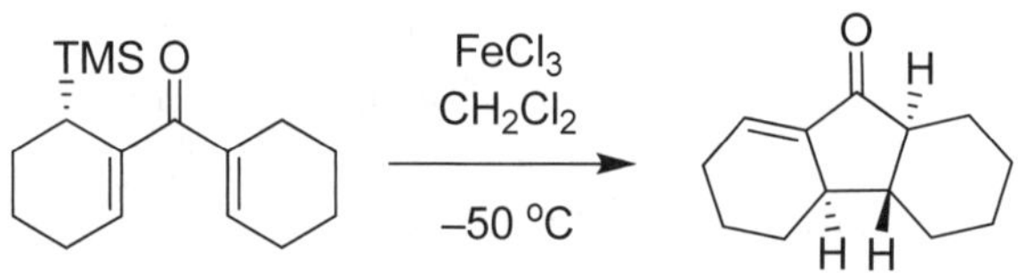

63 **64**
86% *ee* 72%, 86% *ee*

Elegant work by Tius and co-workers has demonstrated that the transfer of asymmetry need not be from an sp^3 hybridized carbon atom. Instead, they have reported examples of the controlled Nazarov cyclization of allenyl vinyl ketones.[5] In one such example, in situ formation of **67** resulted in efficient formation of cyclopentenone **68** with > 95% chirality transfer.[34] The excellent axial to point chirality transfer is a result of the large *tert*-butyl substituent forcing counterclockwise rotation (as viewed by the reader).

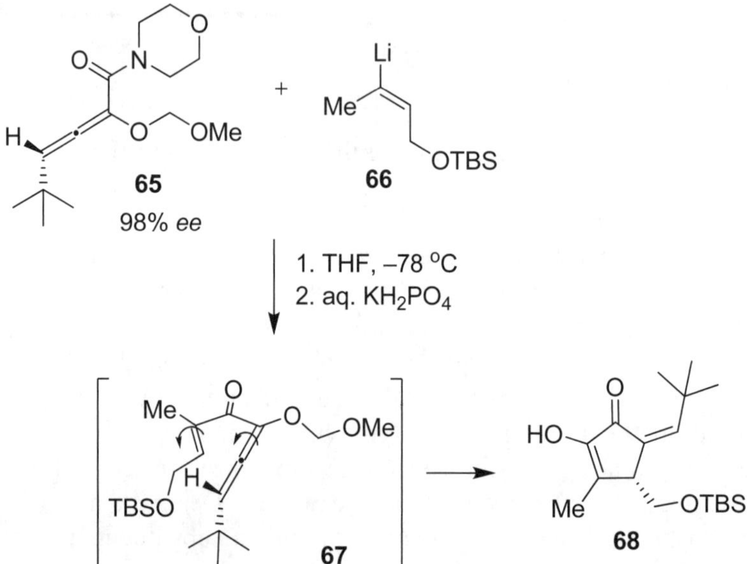

Several groups have reported the use of chiral auxiliaries to control the stereochemical course of the Nazarov cyclization.[5] In general, this strategy has proved effective, with the products isolated in good diastereomeric excess. In one such example, Flynn and co-workers examined the cyclization of **70**, prepared via a palladium-catalyzed hydrostannylation of **69**, followed by acylation.[27] Brønsted acid-mediated cyclization gave the product cyclopentenone **71** in excellent diastereomeric excess and good yield. It is interesting that the kinetically favoured *cis* relative stereochemistry was obtained in this case.

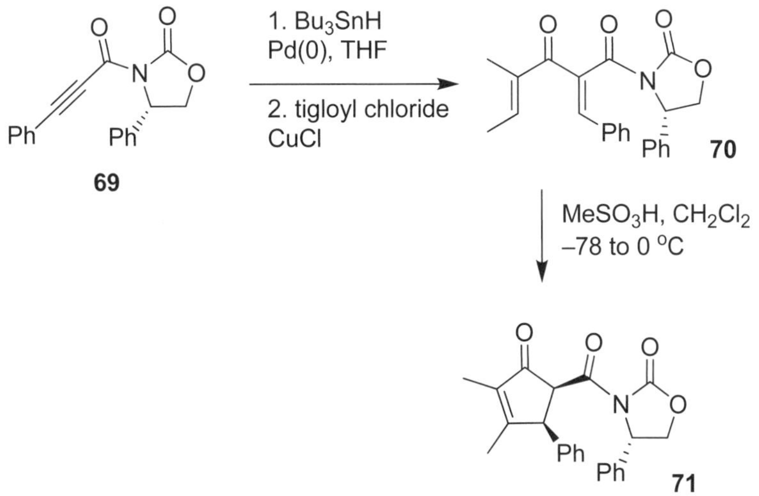

Perhaps the most elegant and attractive method to control absolute stereochemistry, however, is the use of asymmetric catalysis, and several examples of this approach have been applied to the Nazarov cyclization. Trauner and co-workers were the first to report a single example of a successful asymmetric Nazarov cyclization catalyzed by chiral scandium complex in 2003.[24] In the exact same issue of the journal however, Aggarwal and co-workers reported a more in-depth study using copper pyBOX complexes.[26]

Inspired by Evans's work on copper bisoxazoline complexes in asymmetric synthesis, Aggarwal and co-workers hypothesized that these type of complexes offered potential to control the direction of conrotation. Indeed, using stoichiometric amounts of the complex formed from $CuBr_2$, $AgSbF_6$ (which abstracts the halide anions) and **74**, a variety of substrates **72** underwent the cyclization giving the products **73** in up to 88% *ee*, with the thermodynamically favoured *trans*-configuration after proton transfer. Lowering the loading of the complex resulted in lower yields (but not enantioselectivities), and in general a phenyl group in the β-position was required for acceptable reaction rates. Aggarwal and co-workers proposed a stereochemical model for their reactions, whereby in the case of the Cu-pyBOX complexes, the proposed intermediate **75** adopts a square-based pyramid geometry and the alkene substituents are pushed away from the *iso*-propyl groups of the ligand. This places the bonding lobes of the two corresponding orbitals in close proximity, making them predisposed to cyclization in a clockwise manor.

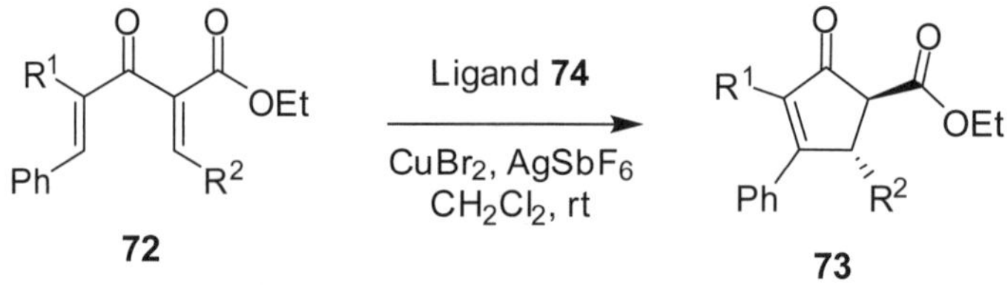

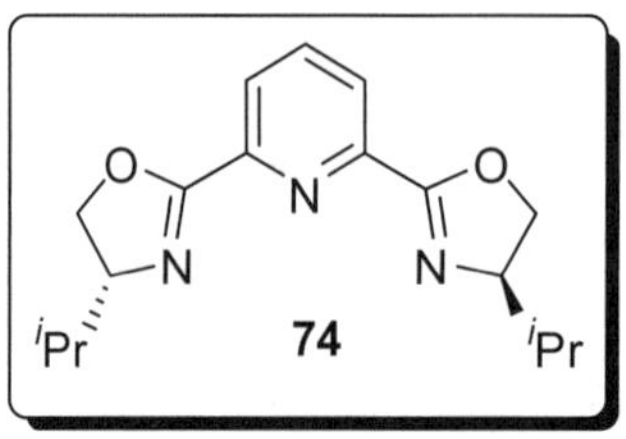

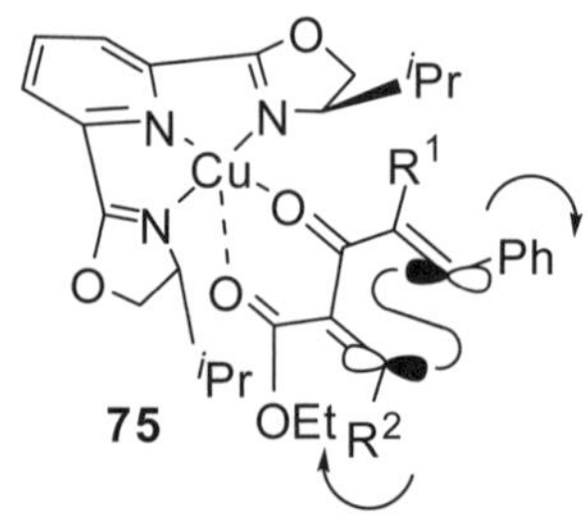

In 2004, Trauner and co-workers published a follow-up communication on their asymmetric catalytic system.[35] Under optimized conditions, they were able to achieve good to excellent levels of enantioselectivity for a variety of substrates using complex **78** with lower catalyst loadings (10 mol %). It is important to note however, that the specific use of an alkoxy dienone substrate lacking a β-substituent on one of the alkenes (such as **76**) was required for high yields and good enantioselectivities. Since the stereocenter formed during electrocyclization is subsequent destroyed on deprotonation of the allylic cation (see Section **3.4.3**), the control of absolute stereochemistry in this case is solely due to facially selective reprotonation of the enolate.

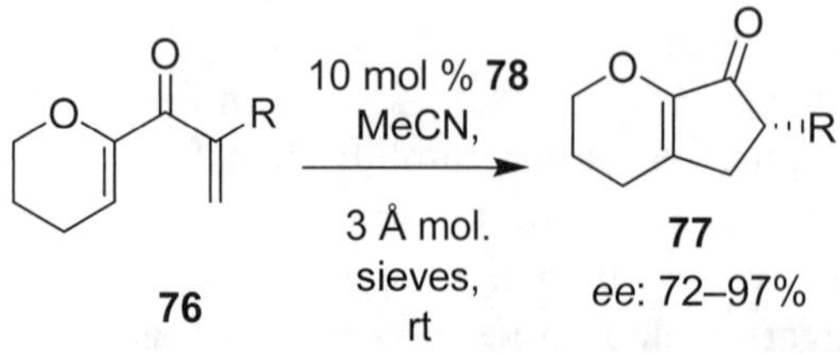

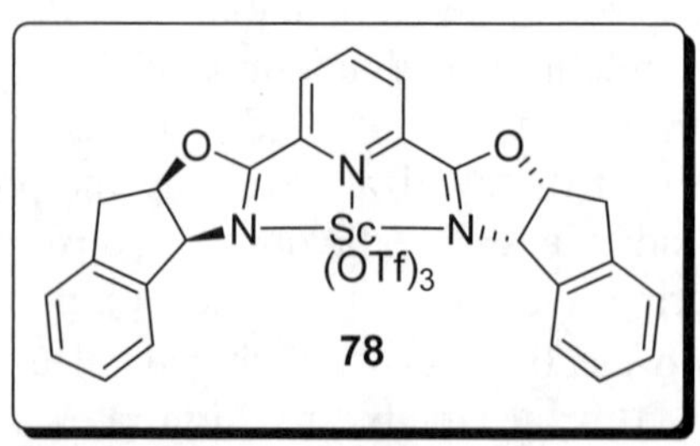

Rueping and co-workers have reported one of the most impressive examples of a Nazarov cyclization under asymmetric catalysis. They have shown that chiral Brønsted-acid catalysis outperforms the chiral Lewis acid catalysts used to date and have demonstrated the efficient cyclization of a variety of substrates **79** with moderate to excellent diastereoselectivity and good to excellent enantioselectivity.[36] Only low catalyst loadings (2 mol %) are required of chiral acid **81** to catalyze the reaction efficiently. Interestingly, the reaction primarily generates the *cis*-cyclopentanones **80a**, as opposed to the Lewis acid-catalyzed reactions that provide the *trans*-product (see **73**).

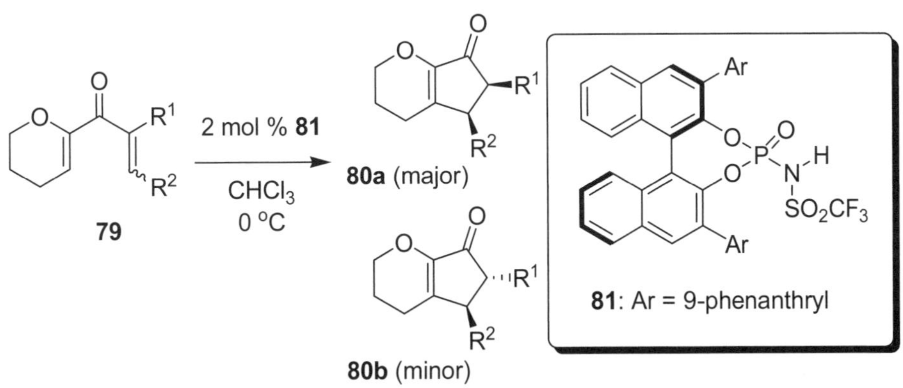

Yield: 45–91%
cis/trans: 1.5:1 to 1:0
ee: 86–98%

3.4.5 Variations and Improvements

3.4.5.1 Silicon-Directed Nazarov Cyclization

Denmark and co-workers have published extensively on the use of β-silyl substituted divinyl ketones (see **82**) in the Nazarov cyclization.[4] Such silyl groups control the collapse of the intermediate cyclopentenylic cations **84**, and thus aid the regioselectivity of elimination, as well as the minimization of side reactions (secondary cationic rearrangements). Such stabilization derives from the known β-cation stabilizing effect of silicon, which through stabilization of **84**, ensures maximum efficiency of the cyclization, with controlled formation of the final double bond. An important consequence of the final elimination step is that the double bond is placed in the thermodynamically less stable position (see **85**). The most common Lewis acid used in the silicon-directed Nazarov cyclization is anhydrous iron(III) chloride, at temperatures below ambient.[4] Alternatively, in cases where the substrate is slow to react, or sensitive to the oxidizing properties of iron(III) chloride, boron trifluoride etherate or zirconium tetrachloride are often employed.

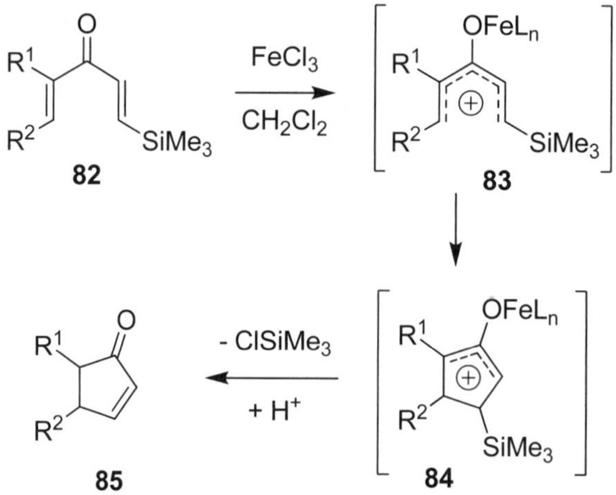

One particularly elegant use of the silicon-directed Nazarov cyclization was in the synthesis of the angular triquinane silphinene (**89**), by Miesch and co-workers.[37] Addition of a large excess of boron trifluoride etherate in refluxing ethylbenzene to **86** ensured annulation of the required third ring. Notably, the benzyloxy group was also eliminated under the reaction conditions, and the product **88** was subsequently converted into the natural product **89**.

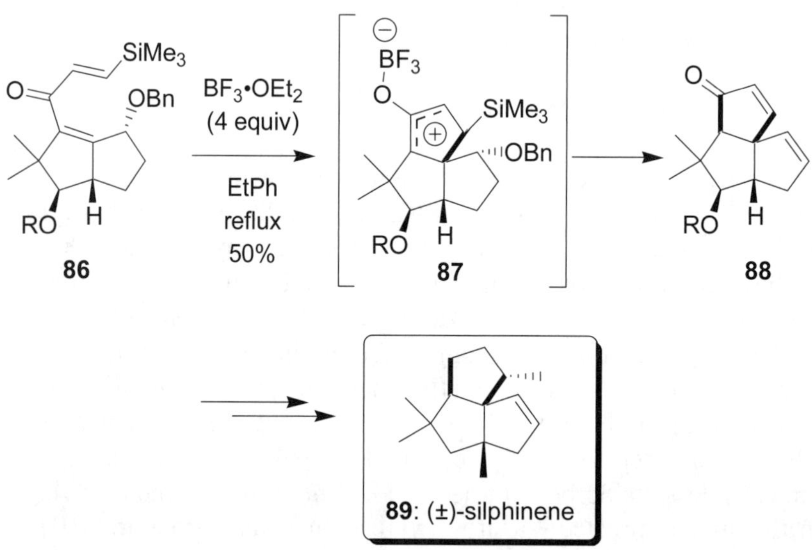

Although the majority of work in this area has used silicon-based directing groups, it is important to note that similar control mechanisms are in operation in β-stannyl appended substrates. Such functionality was harnessed in the synthesis of prostaglandin analogues (see **91**) by Johnson and co-workers.[38] The use of the tributyl stannyl group in substrate **90** ensured the kinetic product was formed, whereby the double bond is located in the least substituted position.

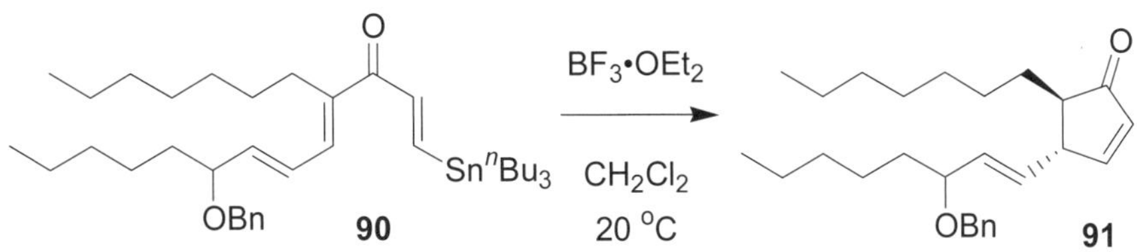

Placement of the directing group need not be at the β-carbon atom destined to become part of the cyclopentenone ring. Indeed, often substrates containing allyl silanes (silyl groups in the α′ position) react at increased rates compared to their β-substituted counterparts.[4] Once again, the positioning of the silyl group determines the regioselectivity of elimination. For example, Kang and co-workers demonstrated that exposure of trimethylsilyl derivatives such as **92** (SiR_3 = $SiMe_3$) to iron(III) chloride results in the formation of the exocyclic double bond (see also **63–64**).[39] On the other hand, Pulido and co-workers exploited the diminished tendency of bulkier silyl groups to undergo protodesilylation and thus isolated product **94** upon treatment of **92** (SiR_3 = Si^tBuPh_2) with TFA.[40]

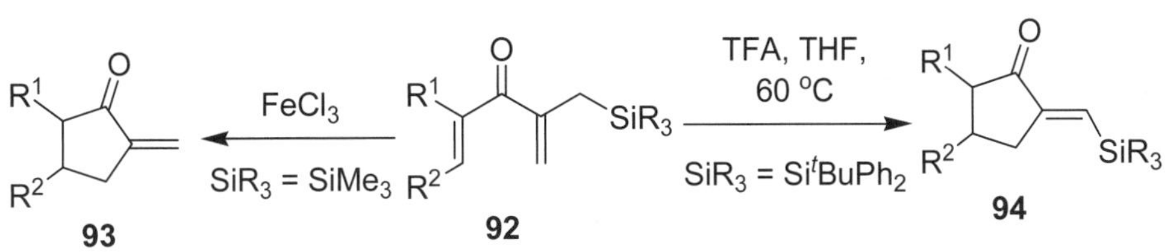

It is important to note that silyl-appended substrates have also demonstrated applicability in the control of torquoselectivity (see Section **3.4.3**), and asymmetric transfer (see Section **3.4.4.4**) in the Nazarov cyclization.

3.4.5.2 "Interrupted" Nazarov Cyclizations

Following electrocyclic ring closure, the resulting cyclopentenylic cation (for example **11**, see **3.4.3**) is stable enough to be intercepted in other reaction pathways. West and co-workers have most extensively explored this reaction manifold and have termed the processes "interrupted" Nazarov cyclization pathways.[41] A variety of examples have emerged such as formal cycloadditions, cationic cascades, and reductive trapping.[6]

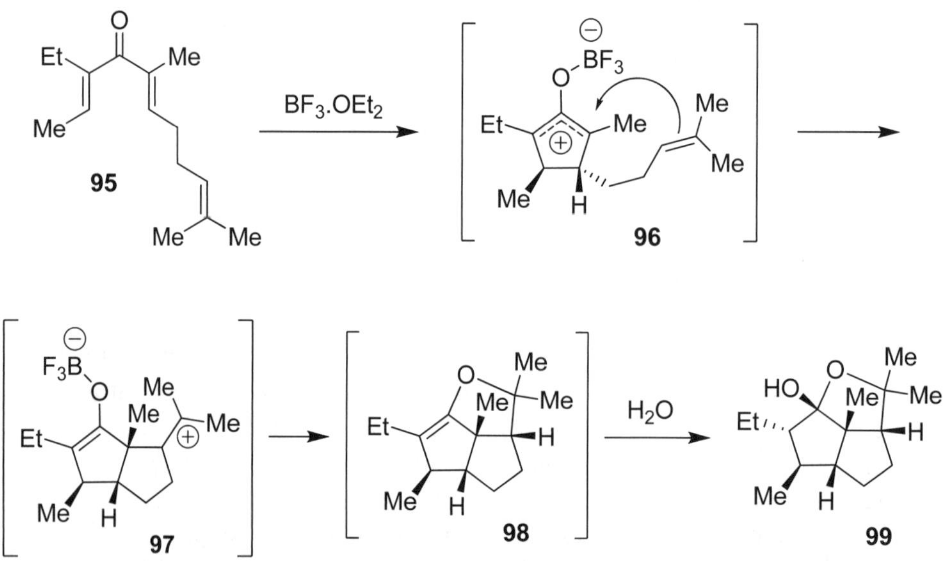

For example, West and co-workers have demonstrated the efficient intramolecular capture of the cyclopentenylic cation by pendent alkenes.[41] Exposure of substrate **95** to boron trifluoride etherate results in electrocyclic ring closure to give **96** under the expected Nazarov pathway, followed by intramolecular trapping by first the alkene and then the enolate oxygen (formal [3 + 2] cycloaddition). Hydration of the enol ether upon workup gives products such as **99** in high stereoselectivity, containing five new stereocenters. This particular pathway is sensitive to substrate structure, whereby the carbon tether between the alkene and dienone must be two carbon atoms long, and there must be substitution in both α-positions. Switching to a terminal alkene results in several alternative pathways, including hydride shifts, chloride anion capture, and proton elimination.[42] Other formal cycloadditions have been demonstrated using alternative tethers.[6]

Capture of the cationic intermediates formed in "interrupted" Nazarov pathways by aromatic groups is equally possible.[6] One particularly impressive example was also reported by West and co-workers.[43] Exposure of substrate **100** to titanium tetrachloride at low temperature resulted in the formation of pentacyclic product **103** in complete diastereoselectivity and excellent yield (98%). This process involves initial electrocyclization to give intermediate **101**, followed by *6-endo* trapping with the pendent olefin and subsequent capture by the phenyl group. During optimization, several intermediate products were observed that could be directly traced back to reactive intermediates **101** and **102**.

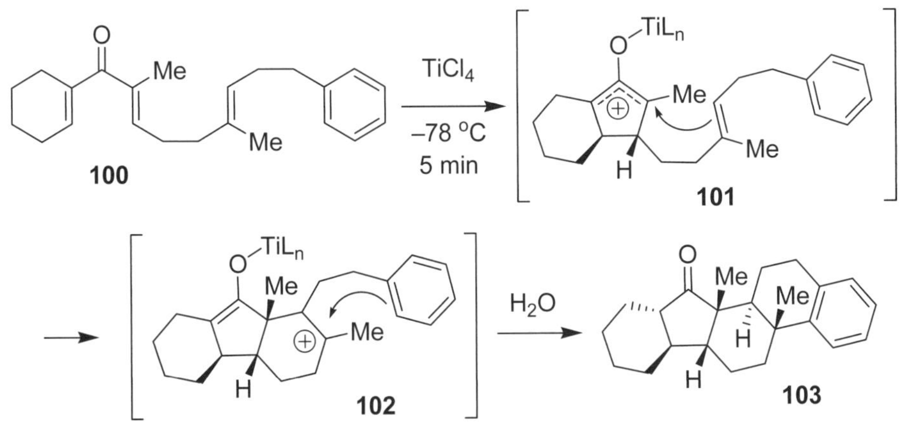

Intermolecular capture of cationic intermediates resulting from Nazarov cyclizations has been demonstrated with added nucleophiles.[6] For example, allyl silanes can be used to either generate α-allyl ketones or [3 + 2] cyclization products.[44]

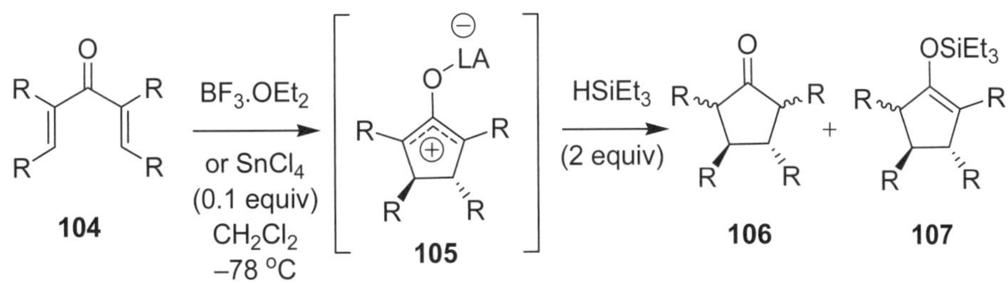

One important development however, was the identification of a reductive trapping pathway, allowing the isolation of saturated cyclopentanones.[29] Lewis acid promoted Nazarov cyclization of substrate **104**, followed by reductive quenching by triethyl silane resulted in the formation of ketones **106** and enol silanes **107**. Such a reaction process requires only 10 mol % of the Lewis acid promoter, with hydride addition occurring at the less-substituted position of the oxyallyl cation **105**. Mixtures of compounds isomeric at the α-positions were isolated in these reactions due to the rapid epimerization of these centers during acidic workup.

3.4.5.3 Related Reactions

It has long been recognized that the construction of cyclopentenones via the Nazarov cyclization can be achieved by using functional equivalents of divinyl ketones or other reaction intermediates. The review published by Denmark and co-workers contains many early examples.[4] For example, exposure of *gem*-dichlorocyclopropylmethanol **108** to acid results in the solvolytic generation of intermediate **109**, followed by a Nazarov-type cyclization and hydrolysis to yield cyclopentenone **110**.[45] Formation of intermediate **109** presumably occurs through cyclopropylcarbinyl cation rearrangement, followed by loss of a proton to give a divinyl dichloride. Ionization of the divinyl dichloride would give intermediate **109**.

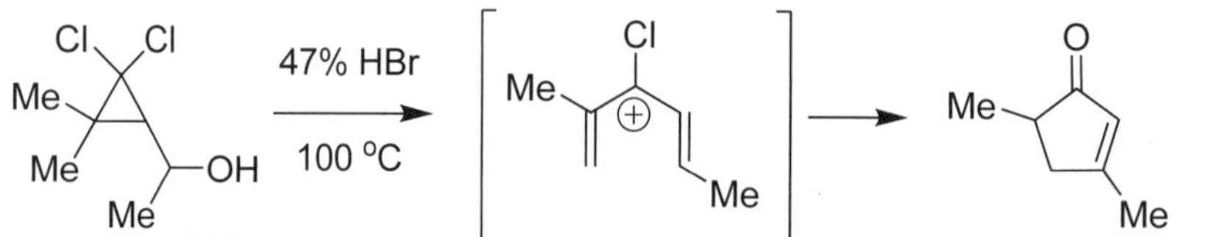

Rautenstrauch reported another mechanistically intriguing example.[46] Treatment of enynol acetate **111** with a palladium(II) catalyst in warm acetonitrile resulted in the formation of cyclopentenone **115**. The proposed mechanism involves generation of divinyl cationic species **113**, followed by electrocyclization, and elimination of the palladium(II) electrofuge in a manner comparable to the silicon-directed Nazarov cyclization (see Section **3.4.5.1**).

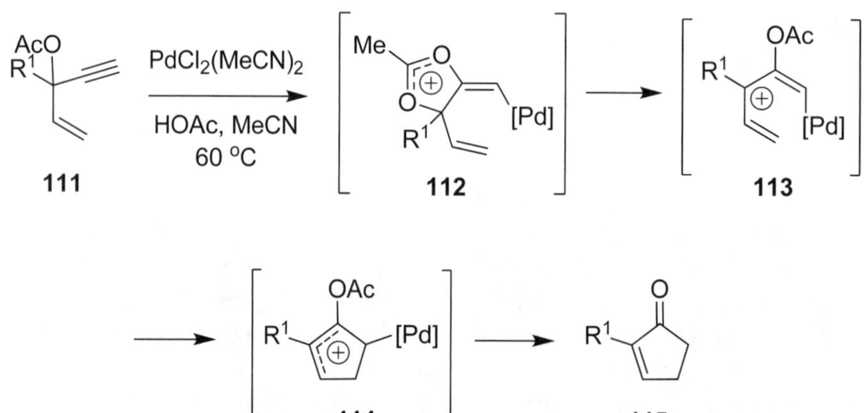

Intermolecular trapping of Nazarov intermediates with oxygen functionalities has been reported by several groups (interrupted Nazarov pathway, see Section **3.4.5.2**).[6] In one report, De Lera and co-workers described the efficient generation of product **119** from acetal **116**.[47] Mechanistically, the reaction most likely proceeds via acid mediated acetal opening to give **117** (pentadienyl cation derived from a conjugated aldehyde equivalent rather than a divinyl ketone), followed by electrocyclization and oxygen trapping to give **119**.

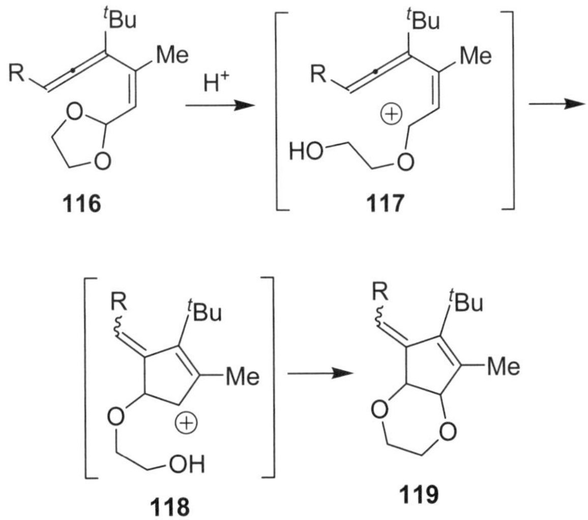

An imaginative approach to the cephalotaxine alkaloids was reported by Li and co-workers and used a functional equivalent of the Nazarov cyclization.[48] Iron-mediated oxidation of substrate **120** gave intermediate **121**. Tautomerization of **121** to **122** followed by electrocyclization gave annulated product **123**. This key intermediate was subsequently converted into cephalotaxine (**124**).

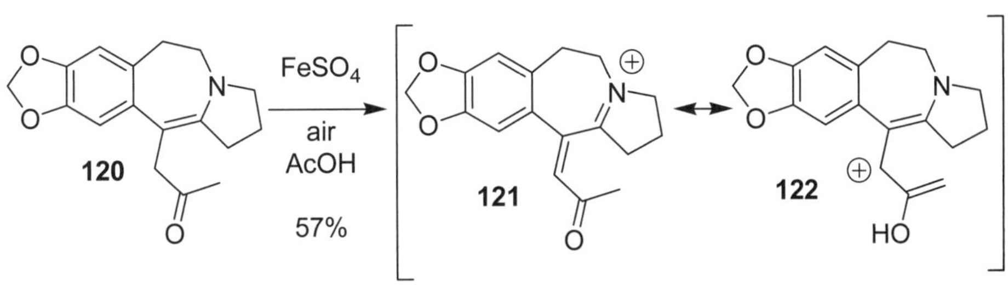

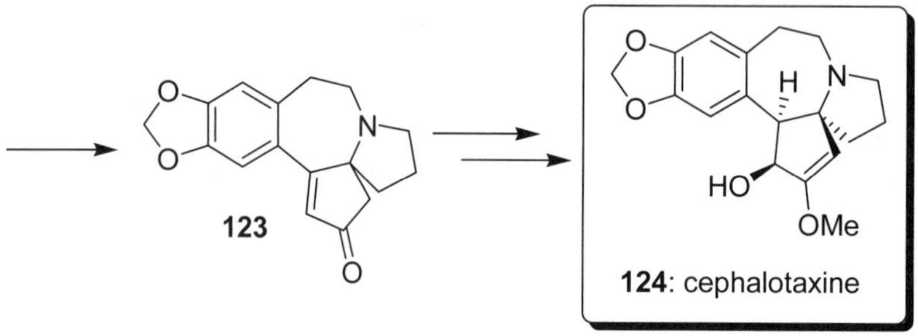

3.4.6 *Experimental*

3.4.6.1 Standard Brønsted acid-mediated conditions

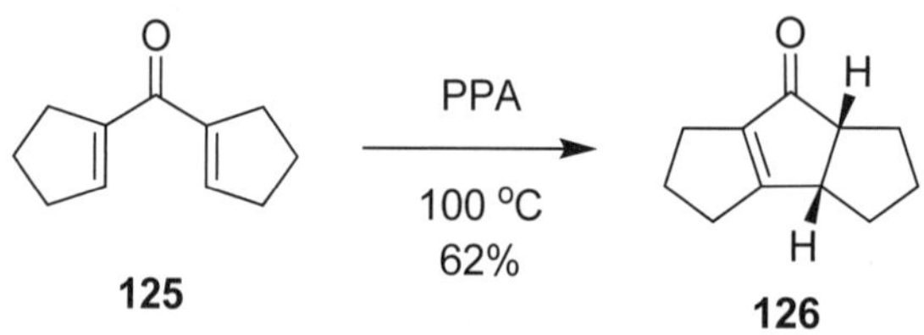

cis-Tricyclo[6.3.0.0^{3,7}]undec-1(8)-en-2-one (126)[49]

1,l'-Dicyclo-pentenyl ketone (19 g) was added with good stirring to hot (100 °C) polyphosphoric acid (100 g) under nitrogen. The colorless PPA immediately turned dark brown. The reaction mixture was stirred for 30 min at 100 °C. After this time, the oil bath was replaced with an ice bath, and ice (100 g) was added immediately to the hot acid. The mixture was stirred for 5 min. A dark precipitate formed during the addition of ice but dissolved on addition of ether. Standard workup (ether) gave a brown oil (19 g), which was distilled carefully to give the tricyclic enone **126**, better than 95% isomerically pure by GLC on OV-225, as a colorless oil (11.9 g, 62%): bp 60–63 °C (0.05 mm Hg).

3.4.6.2 Lewis acid-catalyzed conditions

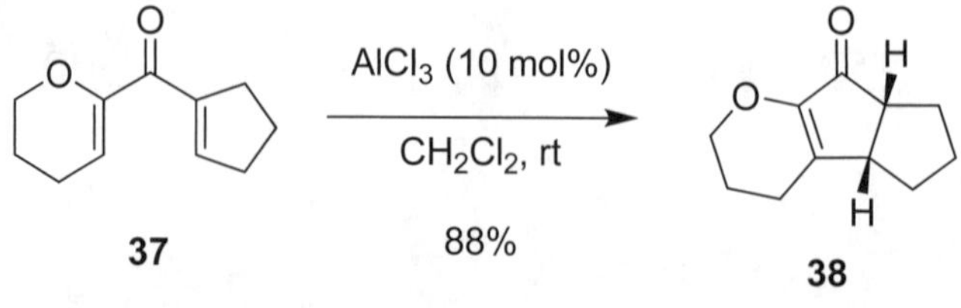

2,3,3a,5,6,8a-Hexahydro-1*H*,4*H*-7-oxa-cyclopenta[α]inden-8-one (38)[24]

To 0.011 g (0.079 mmol) $AlCl_3$ in CH_2Cl_2 (2 mL) was added 0.140 g (0.788 mmol) **37** in CH_2Cl_2 (2 mL). The reaction mixture was stirred for 40 min before it was quenched with water (4 mL). The mixture was further diluted with CH_2Cl_2 (10 mL). The two layers were separated and the aqueous layer was extracted with CH_2Cl_2 (2 × 5 mL). The combined organic layers were washed with brine (10 mL), dried, filtered and concentrated in vacuo. The product was purified by column chromatography (EtOAc:hexanes = 1:4) to afford 0.124 g (88%) **38** as colorless oil. R_f 0.19 (EtOAc:hexanes = 1:4).

3.4.6.3 Silicon-directed Nazarov conditions

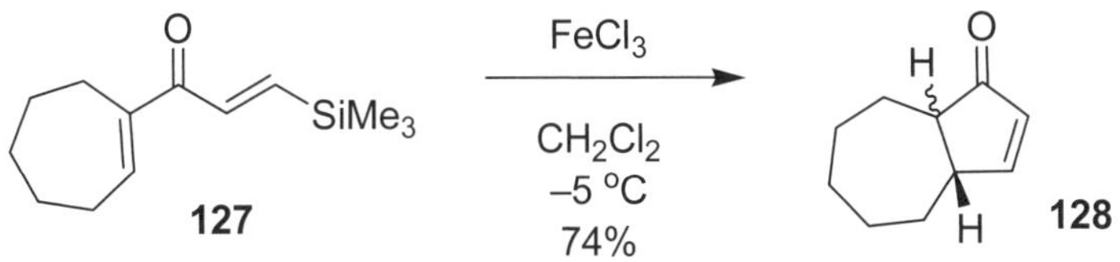

***cis,trans*-1,3,4,5,6,7,8,8a-Octahydroazulen-1-one** (128)[18]

Anhydrous iron trichloride (345 mg, 2.13 mmol) was added in one portion to a cold (–5 °C) solution of (*E*)-1-(1-cycloheptenyl)-3-trimethylsilyl-2-propen-1-one (450 mg, 2.02 mmol) in CH_2Cl_2 (25 mL). The mixture was stirred at –5 °C for 50 min by which time the starting material had been consumed. Water (20 mL) was added, the mixture was diluted with CH_2Cl_2 (10 mL), and the organic layer was removed. The aqueous phase was extracted with CH_2Cl_2 (2 × 30 mL), and the individual organic extracts were washed with saturated aqueous ammonium chloride solution and brine. The combined organic extracts were dried (K_2CO_3) and concentrated. The residue was purified by flash chromatography on silica gel (EtOAc:hexanes = 1:4) followed by distillation, bp 110 °C (0.01 torr) to afford azelenone **128** (225 mg, 74%). GC analysis revealed the product to be an 85/15 mixture of *cis*- and *trans*-isomers.

3.4.7 References

1. [R] Santelli-Rouvier, C.; Santelli, M. *Synthesis* **1983**, 429–442.
2. [R] Denmark, S. E. In *Comprehensive Organic Synthesis*; Trost, B. M., Fleming, I., Eds.; Pergamon Press: Oxford, 1991; Vol. 5, pp 751–784.
3. [R] Krohn, K. *Org. Synth. Highlights* **1991**, 137–144.
4. [R] Habermas, K. L.; Denmark, S. E.; Jones, T. K. *Org. React.* **1994**, *45*, 1–158.
5. [R] Tius, M. A. *Eur. J. Org. Chem.* **2005**, 2193–2206.
6. [R] Frontier, A. J.; Collison, C. *Tetrahedron* **2005**, *61*, 7577–7606.
7. Vorländer, D.; Schroeter, G. *Chem. Ber.* **1903**, *36*, 1490–1497.

8. Francis, F.; Wilson, F. G. *J. Chem. Soc.* **1913**, 2238–2247.
9. Blomquist, A. T.; Marvel, C. S. *J. Am. Chem. Soc.* **1933**, *55*, 1655–1662.
10. Mitchell, D. T.; Marvel, C. S. *J. Am. Chem. Soc.* **1933**, *55*, 4276–4279.
11. Nazarov, I. N.; Zaretskaya, I. I. *Bull. Acad. Sci. U.R.S.S., Classe sci. chim.* **1942**, 200–209.
12. Nazarov, I. N.; Zaretskaya, I. I. *Zh. Obshch. Khim.* **1957**, *27*, 693–713.
13. Nazarov, I. N.; Zaretskaya, I. I. *Zh. Obshch. Khim.* **1960**, *30*, 746–754.
14. Braude, E. A.; Coles, J. A. *J. Chem. Soc.* **1952**, 1430–1433.
15. Sorenson, T. S. *J. Am. Chem. Soc.* **1967**, *89*, 3782–3794.
16. Motoyoshiya, J.; Yazaki, T.; Hayashi, S. *J. Org. Chem.* **1991**, *56*, 735–740.
17. Woodward, R. B. *Chem. Soc.* Special Publication No. 21, **1967**, 217–249.
18. Jones, T. K.; Denmark, S. E. *Helv. Chim. Acta* **1983**, *66*, 2377–2396.
19. Denmark, S. E.; Habermas, K. L.; Hite, G. A.; Jones, T. K. *Tetrahedron* **1986**, *42*, 2821–2829.
20. Wada, E.; Funiuara, I.; Kanemasa, S.; Tsuge, O. *Bull. Chem. Soc. Jpn.* **1987**, *60*, 325–334.
21. Braude, E. A.; Forbes, W. F. *J. Chem. Soc.* **1953**, 2208–2216.
22. Harding, K. E.; Clement, K. S.; Tseng, C. Y. *J. Org. Chem.* **1990**, *55*, 4403–4410.
23. Denmark, S. E.; Habermas, K. L.; Hite, G. A. *Helv. Chim. Acta* **1988**, *71*, 168–194.
24. Liang, G.; Gradl, S. N.; Trauner, D. *Org. Lett.* **2003**, *5*, 4931–4934.
25. Bee, C.; Leclerc, E.; Tius, M. A. *Org. Lett.* **2003**, *5*, 4927–4930.
26. Aggarwal, V. K.; Belfield, A. J. *Org. Lett.* **2003**, *5*, 5075–5078.
27. Kerr, D. J.; Metje, C.; Flynn, B. L. *Chem. Commun.* **2003**, 1380–1381.
28. He, W.; Herrick, I. R.; Atesin, T. A.; Caruana, P. A.; Kellenberger, C. A.; Frontier, A. J. *J. Am. Chem. Soc.* **2008**, *130*, 1003–1011.
29. Giese, S.; West F. G. *Tetrahedron* **2000**, *56*, 10221–10228.
30. Hashmi, A. S. K.; Bats, J. W.; Choi, J.-H.; Schwarz, L. *Tetrahedron Lett.* **1998**, *39*, 7491–7494.
31. Harmata, M.; Lee, D. R. *J. Am. Chem. Soc.* **2002**, *124*, 14328–14329.
32. Cheng, K.-F.; Cheung, M.-K. *J. Chem. Soc., Perkin Trans. 1* **1996**, 1213–1218.
33. Denmark, S. E.; Wallace, M. A.; Walker, Jr., C. B. *J. Org. Chem.* **1990**, *55*, 5543–5545.
34. Hu, H.; Smith, D.; Cramer, R. E.; Tius, M. A. *J. Am. Chem. Soc.* **1999**, *121*, 9895–9896.
35. Liang, G. Trauner, D. *J. Am. Chem. Soc.* **2004**, *126*, 9544–9545.
36. Rueping, M.; Ieawsuwan, W.; Antonchick, A. P.; Nachtshein, B. J. *Angew. Chem., Int. Ed.* **2007**, *46*, 2097–2100.
37. Miesch, M.; Miesch-Gross, L.; Franck-Neumann, M. *Tetrahedron* **1997**, *53*, 2103–2110.
38. Peel, M. R.; Johnson, C. R. *Tetrahedron Lett.* **1986**, *27*, 5947–5950.
39. Kang, K.-T.; Kim, S. S.; Lee, J. C.; U, J. S. *Tetrahedron Lett.* **1992**, *33*, 3495–3498.
40. Barbero, A.; Castreño, P.; García, C.; Pulido, F. J. *J. Org. Chem.* **2001**, *66*, 7723–7728.
41. Bender, J. A.; Blize, A. E.; Browder, C. C.; Giese, S.; West, F. G. *J. Org. Chem.* **1998**, *63*, 2430–2431.
42. Browder, C. C.; West, F. G. *Synlett* **1999**, 1363–1366.
43. Wang, Y.; Arif, A. M.; West, F. G. *J. Am. Chem. Soc.* **1999**, *121*, 876–877.
44. Giese, S.; Kastrup, L.; Stiens, D.; West, F. G. *Angew. Chem., Int. Ed.* **2000**, *39*, 1970–1973.
45. Hiyama, T.; Tsukanaka, M.; Nozaki, H. *J. Am. Chem. Soc.* **1974**, *96*, 3713–3714.
46. Rautenstrauch, V. *J. Org. Chem.* **1984**, *49*, 950–952.
47. de Lera, A. R.; Rey, J. G.; Hrovat, D.; Iglesias, B.; Lopez, S. *Tetrahedron Lett.* **1997**, *38*, 7425–7428.
48. Li, W.-D. Z.; Wang, Y.-Q. *Org. Lett.* **2003**, *5*, 2931–2934.
49. Eaton, P. E.; Giordano, C.; Schloemer, G.; Vogel, U. *J. Org. Chem.* **1976**, *41*, 2238–2241.

3.5 Pauson–Khand Reaction

Louis S. Chupak

3.5.1 Description

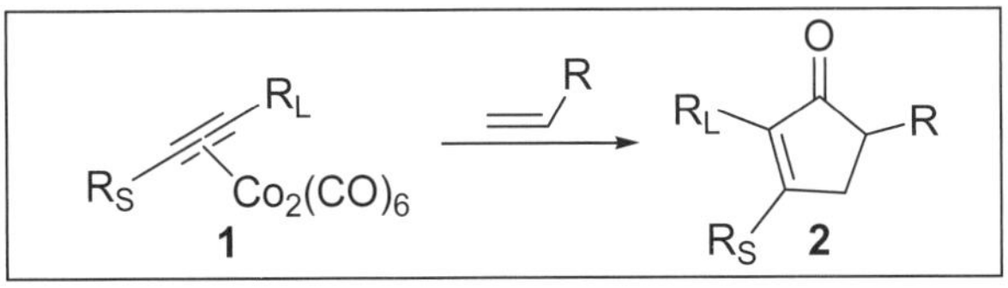

The Pauson–Khand Reaction (PKR) is a formal [2 + 2 + 1] cycloaddition of a cobalt-complexed alkyne **1**, an alkene and carbon monoxide. Three bonds are formed in a single reaction to give a significant increase in structural complexity on going from starting materials to product. With unsymmetrical alkynes, the reaction tends to join the components together regioselectively to place the larger alkyne substituent next to the carbonyl group. Less regioselectivity is seen for the alkene component. The ability to predict regiochemistry, a rapid increase in synthetic complexity and the frequency of 5-membered rings in synthetic targets has made this reaction the focus of numerous mechanistic and synthetic studies as well as multiple reviews.[1–12]

3.5.2 Historical Perspective

In 1971 at the University of Strathclyde in Glasgow, Professor Peter Pauson reported on the retro-Diels–Alder reaction of norbornadiene **3** induced by dicobalthexacarbonyl complexes of acetylene or phenylacetylene **1** to provide dicarbonylcyclopentadienylcobalt complexes **4** in high yield.[13] Almost as an after thought, he mentions "In addition to the above products, the reaction of norbornadiene with complexes **1** yields hydrocarbon and ketonic products derived from norbornadiene, acetylene and carbon monoxide."

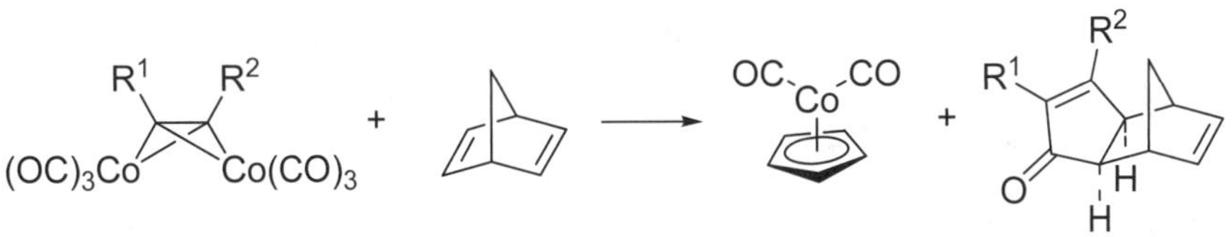

1 **3** **4** **5**

Two years later, in the paper most often cited as the original PKR disclosure, Pauson identifies the "ketonic product" as a cyclpentenone, demonstrates that the larger alkyne substituent prefers to occupy the position adjacent to the carbonyl and that the *endo*-form **5** is the kinetic product when norbornadiene is the alkene.[14] The PKR has tremendous synthetic potential because of the resulting increase in synthetic complexity and the frequency of five-membered rings in synthetic targets. This potential has made the PKR the focus of numerous reviews, mechanistic studies and synthetic strategies. Thus this chapter cannot attempt to provide a complete coverage of the PKR. Instead the aim is to provide readers with enough background to understand new developments and to decide if a PKR can be applied to their work.

3.5.3 Mechanism

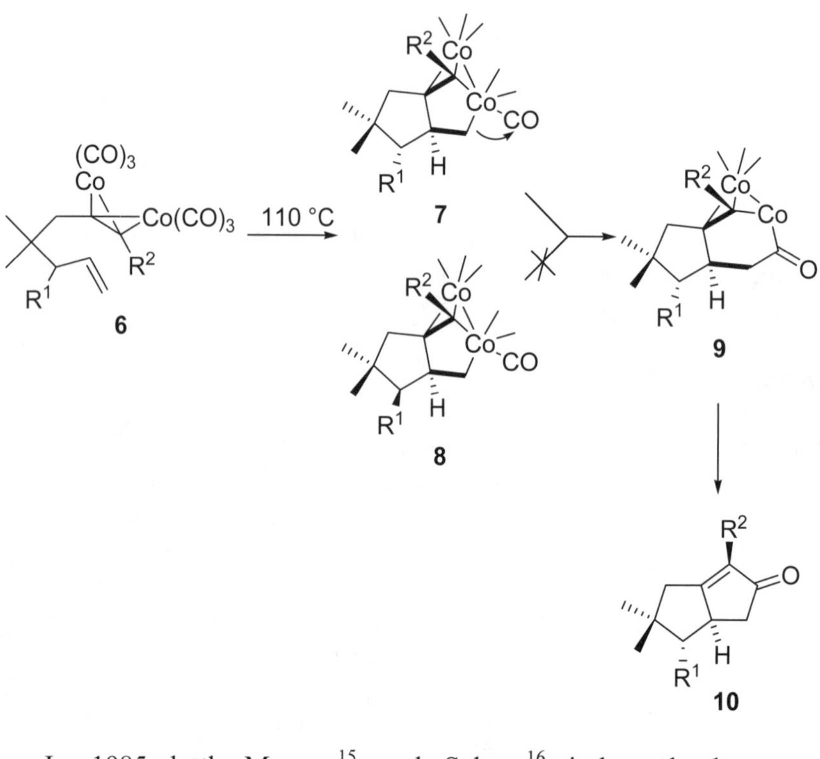

In 1985 both Magnus[15] and Schore[16] independently proposed identical mechanisms based on stereochemical and regiochemical preferences observed in the product. Magnus's hypothesis was based on the stereochemical results for an intramolecular cyclization. He proposed that the product arose from formation of a metallocycle intermediate **7** or **8**, carbon monoxide insertion to give **9**, acyl migration from cobalt to carbon and reductive elimination of cobalt to form **10**. The relative thermodynamic stability of the metallocycles **7** and **8** controlled the final product ratio. In the case of an allylic substituent R^1 as in **6**, the more stable isomer places the R^1 group on the *exo* face of the newly formed bicyclic system as in **7**, *trans* to the R^2 group and *cis* to the adjacent hydrogen at the ring fusion. Thus the stereochemical outcome of the reaction is derived from intermediate **7** and not intermediate **8**. Similarly, when a propargylic group R^1 is present as in **11**, the more stable isomer places the R^1 group on the *exo* face of the newly formed bicyclic system as in **12**, on the opposite face of the R^2 group and on the same face as the hydrogen at the ring fusion. The steric clash between R^1 and R^2 in **13** is avoided in **12**.

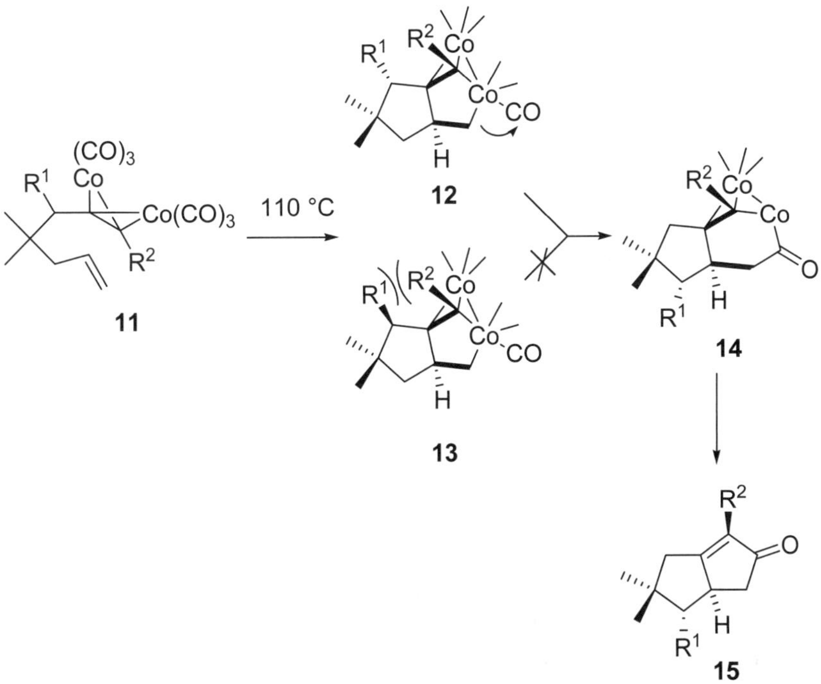

Schore's mechanistic proposal came from observing the preferred regiochemistry of the products derived from the intermolecular reaction of 8-oxabicyclo[3.2.1]oct-6-ene derivitives **16** with alkynes. As in Pauson's norbornene examples, only the *exo*-products resulting from attack on the less-hindered face of the alkene were observed. Schore proposed that the product-determining step was insertion of the less-hindered face of the alkene into the less-hindered alkyne carbon–cobalt bond. Four possible modes of attack, two *syn-* and two *anti-* on the *exo*-face of the alkene are possible. Two of these differ only in the location of the "spectator" cobalt atom and are not shown. In the preffered mode **16-*anti***, the alkyne group R^1 is farthest from the newly forming cobalt–carbon bond. The reacting cobalt is *anti* to the bridgehead group R^1. In **16-*syn*,** a steric clash occurs between the cobalt's carbon monoxide ligands and the allylic R^1 group. Migratory insertion of CO into a metallocycle cobalt–carbon bond, acyl migration from cobalt to carbon, and reductive elimination follow to give the observed products **17**.

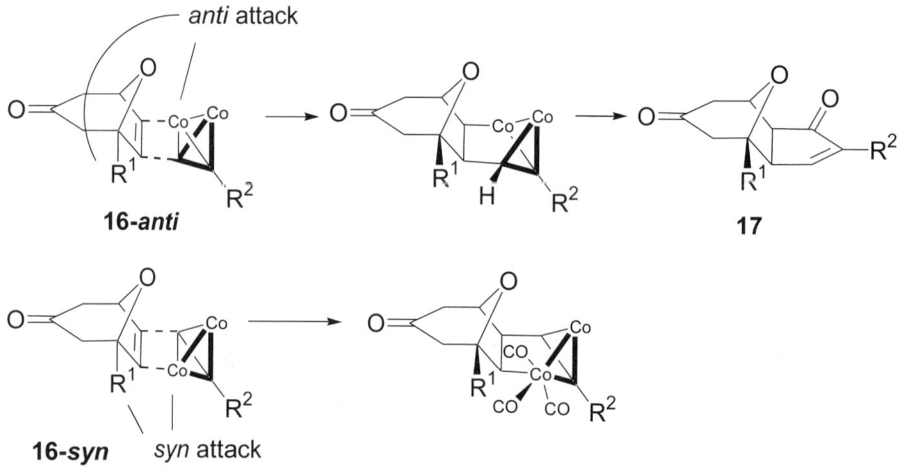

The mechanism resulting from these original studies is shown below. The alkyne **S** is first coordinated by cobaltoctacarbonyl with loss of two equivalents of carbon monoxide to give a well characterized complex **I-1**.

Upon heating, loss of a third equivalent of carbon monoxide opens a coordination site to provide **I-2** and allows cobalt to complex the alkene as in **I-3**. The alkene inserts into the least hindered cobalt–carbon bond to form the cobaltacycle **I-4**. Carbon monoxide inserts into the new cobalt-carbon bond to give **I-5**. Next, acyl migration from cobalt to carbon forms the final carbon–carbon bond as in **I-6**. Finally, a formal reductive elimination of the "spectator" cobalt releases the product cyclopentenone **P** and cobalt–hexacarbonyl. The cobalt is then ready for another catalytic cycle. In practice the cobalt is most often used stoichiometrically. The alkyne cobalt complex is a stable species that can be purified and isolated by silica gel chromatography. Conditions for catalytic turnover of the cobalt have been reported.

The original mechanistic proposal has remained largely unchanged and is supported by subsequent theoretical[17,18] and experimental[19] results. The key findings are (1) Initial loss of carbon monoxide is energetically disfavoured and requires heat, the presence of weak ligands, or activating agents to liberate a CO ligand. (2) The alkene-insertion step determines the regio- and stereochemistry of the final product. This step is also disfavoured energetically and explains the observed importance of reactive alkenes for succesfull reactions. This step, **I-3** to **I-4**, is considered to be rate determining. (3) Acyl migration, **I-5** to **I-6**, is energetically favoured over other possible alternative pathways. (4) The "spectator" cobalt exerts electronic influences on the reacting metal center through the metal–metal bond.

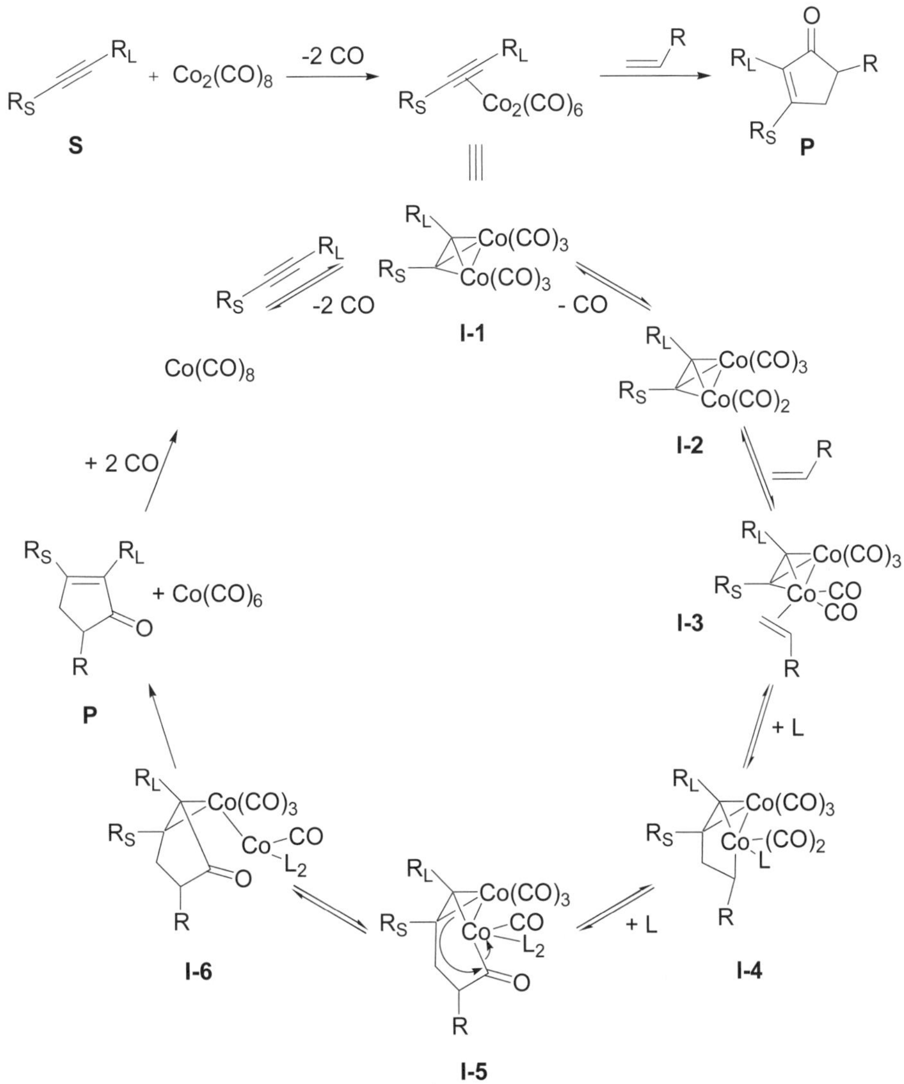

An *X*-ray crystal structure of an alkene cobalt complex has been reported by McGlinchey.[20] A mixture of **18a** and **18b** formed at room temperature and these products were separated by flash chromatography. Although heating did not induce the PKR to occur, **18a** was proposed to be an arrested η^2-alkene–pentacarbonyldicobalt–alkyne complex in the PKR pathway (see intermediate **I-3** above). A comparison of the *X*-ray crystal structures showed the coordinated double bond in **18a** to be slightly longer than the uncoordinated double bond of **18b**, 1.403 Å versus 1.336 Å, respectively. The boat conformation of the seven-membered ring was also more pronounced in **18a** as the cobalt pulled the double bond toward itself. The geometry of the alkyne was similar to that normally observed in hexacarbonyldicobalt–alkyne complexes with the propargylic centers "bent back" from 180° to approximately 143° as expected.

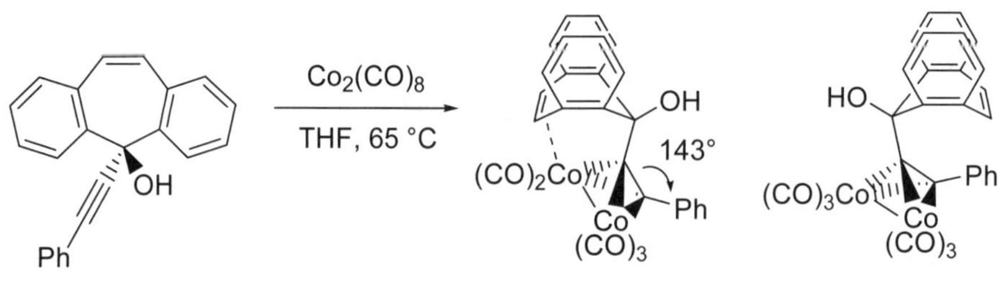

18a **18b**

As will be seen in the following sections, the mechanistic proposal has enabled researchers to improve upon the original reaction conditions and successfully apply the PKR to the synthesis of complex target molecules. In addition, a number of reports describe diverting the reaction to provide alternative products. The subject of alternative products that can arise intentionally and unintentionally is nicely reviewed by Krafft.[21]

3.5.4 Variations and Improvements

Initially, the PKR was limited to reactive alkenes (such as norbornene), required high temperatures (60–110 °C), high CO pressures, and long reaction times. Alkenes were, in general, minimally substitutued or strained. Regioselectivity with nonsymmetric alkenes provided mixtures of regioisomer in contrast to the more predictable final disposition of alkyne substituents. Intramolecular reactions were superior to comparable intermolecular reactions. Early versions of the reaction were performed under an atmosphere of carbon monoxide and gave poor to modest yields. Since its discovery, numerous modifications have been designed to address the issues and expand the scope of the reaction. These modifications include the use of additives, adsorption onto solids before heating, addition of coordinating ligands, inclusion of temporary cleavable tethers, sonication, microwave irradiation, and solvent variation. Finally, great strides have been made toward reactions using substoichiometric cobalt and controlling absolute stereochemistry.

The PKR reaction has always been characterized by good functional group compatibility. Examples abound where amines, amides, sulfonamides, sulfides, alcohols, ketones, esters, and silyl ethers are contained in the reactants. In addition, because substituted alkenes react more slowly than unsubstitued alkenes selectivity can be achieved when multiple alkenes are contained in the reactants. An interesting example of functional group compatibility was demonstrated in which a metal carbene not only survived the PKR but accelerated it.[22] For example, the tungsten carbene **19** formed the cyclopentenone **20** in 80% yield at 0 °C.

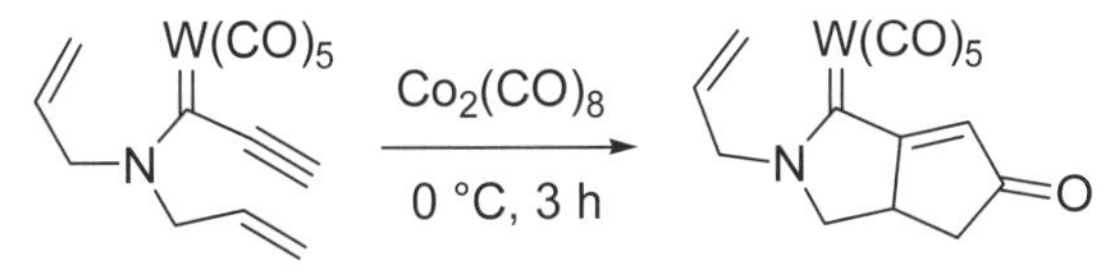

19 **20**

In an attempt to improve on the original PKR, Pauson reported on the effects of various additives.[23] Phosphines, ether as preformed cobalt species or simply added to the PKR, only served to reduce the rate and yield of the reaction. On the other hand, ultrasound and tributylphosphine *oxide* proved equally effective at accelerating the rate of reaction and increasing the overall yield of the PKR. The reaction yields were described as erratic: varying irreproducibly as the reaction atmosphere was changed. The use trimethylamine-*N*-oxide was mentioned but no results were reported.

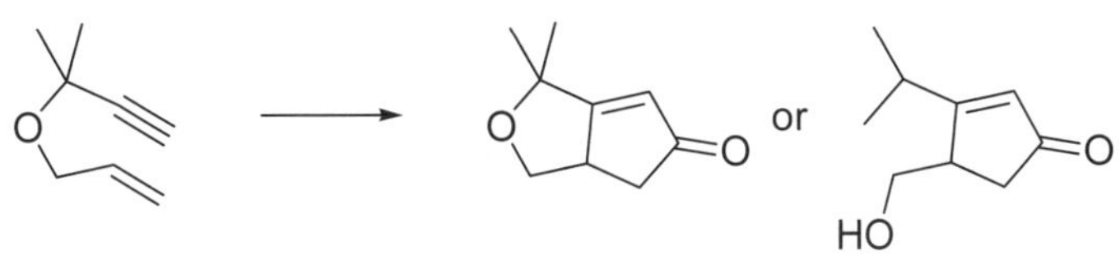

21 **22a** **22b**

Reaction Conditions	% Yield (**22a** : **22b**)
i-octane, 60 °C , 24 h (conventional PKR)	29 : 0
SiO_2, 45 °C, 30 min., O_2	75 : 0
SiO_2, 45 °C, 30 min., Ar	15 : 40

Smit described the rate-accelerating effect of preabsorbing the PKR reactants onto a solid support and removing the solvent before heating. Cyclopentenone **22a** was formed in superior yields at reduced temperatures under these dry state absorption conditions.[24,25] Various solid supports and conditions were examined. Both silica (SiO_2) and alumina (Al_2O_3) worked equally well. The reaction rate was independent of pH. The optimal conditions required 10–20% water content and an oxygen atmosphere. Smit proposed that the rate acceleration was due to a hydrophobic effect that forced the reacting alkene and alkyne together to reduce the entropy barrier for cyclization. Interestingly, when the PKR was performed under an inert atmosphere the dominant product **22b** arose from cleaving the allylic (or perhaps propargyllic) ether bond.

In 1990, a major breakthrough for the utility of the PKR occurred when Schreiber reported the use of *N*-methylmorpholine-*N*-oxide (NMO) as a promoter.[26] As shown, the addition of NMO allowed the reaction to be run at ambient temperature and produced enhanced diastereoselectivity.

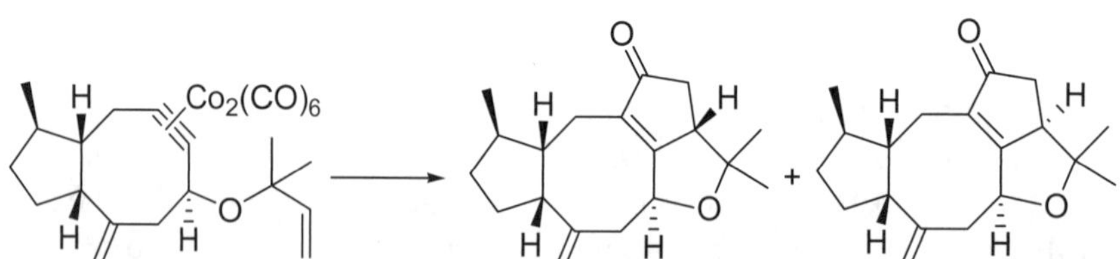

23 24a 24b

Reaction Conditions	% Yield	24a : 24b
NMO, CH_2Cl, room temperature	68	11 : 1
CH_3CN, 82 °C	75	4 : 1
CH_3CN, 45 °C, sonication	45	3 : 1

It was proposed that NMO oxidized one of the carbon monoxide ligands to carbon dioxide to open up a coordination site for the alkene. Alternatively (or additionally), NMO may act to scavange CO to make the dissociation of the CO ligand irreversible. Subsequent to this work, a number of *N*-oxides,[27] including polymer-bound *N*-oxides,[28,29] have been shown to accelerate the PKR. More than one equivalent *N*-oxide is usually required to observe the accelerating effects. Thus polymer-bound *N*-oxide offers the advantage of simplifying the work-up. Trimethylamino-*N*-oxide (TMANO) and NMO are the two most common *N*-oxides used to accelerate the PKR.

Lewis bases[30,31] also accelerate the PKR by supposedly acting as weak ligands to promote the dissociation of CO and to stabilize intermediates. However, density functional theory (DFT) calculations described by Gimbert demonstrate no acceleration for the loss of CO in the presence of a Lewis base.[32] Instead, the calculations suggest that the Lewis base stabilizes the cobaltacycle **I-4**. This stabilization effectively makes the olefin insertion irreversible. Dimethylsufoxide,[33,34] amines[35,36] such as cyclohexyamine,[37] various sulfides,[38] water,[42] alcohols, thioureas,[39–41] and ethers[42] have all shown accelerating effects.

Heterogenous additives such as molecular sieves[43,44] or preabsorbtion of the cobalt catalyst onto charcoal[45] have also been employed to accelerate the PKR. As with other additives, the current view is that these additives promote the dissociation of, or trap, CO to provide the reactive catalyst or increase the concentration of the alkene coordinated intermediate **I-3**.

The improved yields observed with intramolecular versus intermolecular PKRs have encouraged attempts to facilitate the reaction with a cleavable tether. Ether[46,47] and silyl groups have been employed as covalent tethers that can be cleaved after the PKR. Several authors have reported on the use of silicon tethers to promote the PKR.[48–50] However, the success of the PKR varies considerably with the substituent pattern on the silicon, the alkene and the alkyne. Dobbs has explored the use of silyl ether, silyl acetal, and silyl alkyl tethers in the PKR.[51] Under his conditions, only sily ethers produced cyclopentenones. The best results required two isopropyl groups on the silicon. The 6,5-ring system **26** (*n* = 1) was formed in 75% under the best conditions, The 7,5 ring system (*n* = 2) from the homopropargyl silyl ether failed to form under these conditions. This result illustrates one of the current limitations of the PKR. There are very few examples for the formation of medium-size ring systems.[52–54]

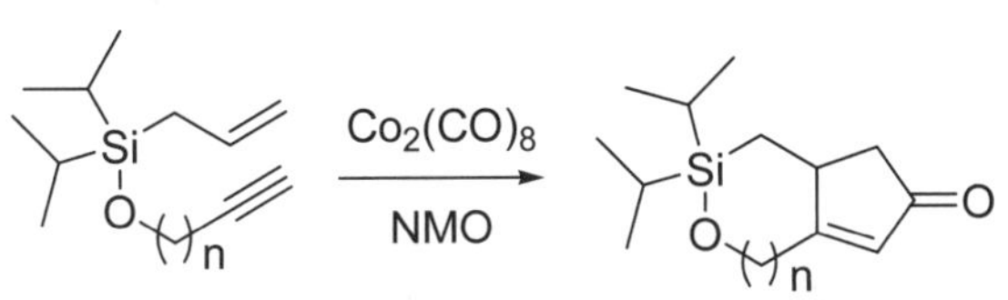

25 **26** (n = 1, 75%; n = 2, 0%)

The PKR can be performed in a wide variety of organic solvents. The reaction is compatible with ionic liquids,[55,56] nonpolar solvents such as hexane, polar protic solvents such as water[57,58] and polar aprotic solvents such as dimethylsulfoxide. Krafft has described the rate-accelerating effects of polar coordinating solvents. Acetonitrile had the greatest accelerating effect: CH_3CN > EtOAc ≈ THF ≈ Acetone > 1 : 1 THF : CH_2Cl_2 > CH_2Cl_2 >> EtOEt > DMSO.[59]

Microwave irradiation (MWI) has been reported as an effective way to promote the PKR.[60–63] Helaja described the use of MWI to promote the PKR for the synthesis of estrone derivatives, **27a** and **27b**.[60] MWI was equally effective at promoting the PKR as was *t*-butylmethylsulfide, but that the regioselectivity was inverted. Reaction yields were improved when the target temperature was no greater than 100 °C and the alkyne was added in portions. It was also observed that reaction yields improved, in select instances, when charcoal was added to these MWI reactions. This intermolecular PKR demonstrates the regioselectivity issues observed with un-symmetrical alkene substitution. The location of the carbonyl in the products with respect to the alkene substituents is essentially random. In contrast, the alkyne substituent is only found α to the carbonyl.

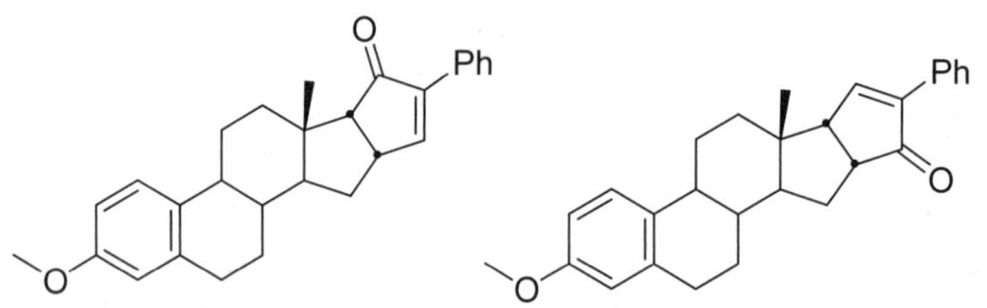

27a **27b**

Conditions	% Yield	**27a : 27b**
MW 100 °C, 6x 0.25 eq. Co-Alkyne	57	1.3 : 1
As Above with charcoal	62	1.3 : 1
*t*BuSCH$_3$	55	1 : 1.6

Billington and Pauson reported on unsuccessful attempts to improve the PKR using ultrasound in 1988.[23] Despite this early failure and mixed results in other reports, Kerr has disclosed improved PKRs using high-intensity ultrasound in combination with TMANO.[64] It was proposed that the standard laboratory ultrasound bath is insufficient to observe the accelerating effect and a high intensity ultrasound source is required. For example, the reaction of phenylacetylene **28** with cycloheptene **29** occurred in 85% yield using high-intensity ultrasound compared with 41% using thermal conditions to provide **30**.

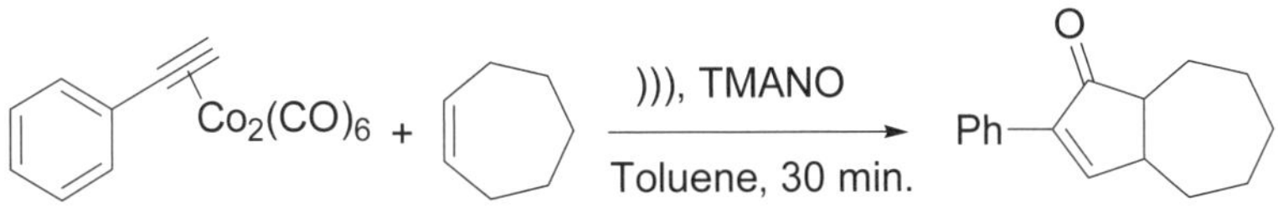

28 **29** **30**

Reactions similar to the PKR that provide cyclopentenones from the [2 + 2 + 1] cycloaddition of an alkyne, an alkene, and carbon monoxide can be achieved with metals other than cobalt. Chromium,[65] iron, iridium,[66,67] molybdenum,[68,69] nickel,[70] palladium,[71] rhodium,[72,73] ruthenium,[74,75] titanium,[76] tungsten[77] and zirconium[78,79] have all been reported to catalyze the cycloaddition. The mechanism, selectivity, and functional group compatibility varies with each metal, making their discussion beyond the scope of this chapter.

Dicobalt octacarbonyl remains a simple and convienent choice for the PKR. The best source and handling of the cobalt catalyst appears to vary, depending on the substrate and the presence of additives. Verdaguer reported no difference in the reaction rate of commercial grade $Co_2(CO)_8$ versus catalyst purified by sublimation.[19] Other precatalyst, such as tetracobalt dodecacarbonyl, have also been used for the PKR.[80,81,19]

From the above discussion it is clear that there are a large number of potential reaction conditions. In practice the most common conditions use a slight excess of cobalt octacarbonyl, in acetonitrile or dichloromethane to form the alkyne complex under an atmosphere of nitrogen or argon. The formation of the alkyne complex is easily monitored by thin layer chromatography on silica gel. The alkene and TMANO or NMO are then added and the reaction was allowed to proceed at room temperature until completion. If the results are not satisfactory, the reaction can be optimized with additives or by changing the temperature.

In her review, Laschat summarizes the strategies that have been employed to control stereochemistry.[1] These strategies include diastereo-selectivity from enantiopure starting materials and enantioselectivity with chiral additives. The use of enantiopure starting materials falls into three catagories: The controlling stereocenter is in the tether for an intramolecular PKR, the controlling stereocenter is in a chiral auxiliary on the alkyne or alkene, or a chiral cobalt complex controls stereochemistry.

As an example of chirality in the tether, Sezer has demonstrated a strategy for controlling quaternary stereocenters using camphor derived enynes.[82] In this system a chair-like transition state, **30**, is invoked to explain the stereochemistry at the ring fusion in **31**.

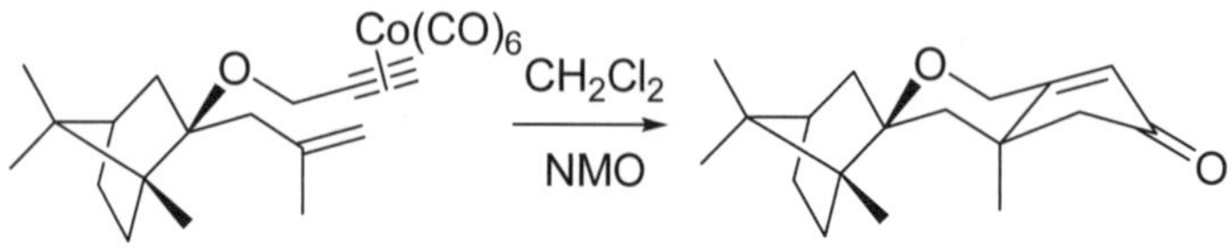

30 31

There are numerous examples of chiral auxiliaries in either the alkyne or alkene producing good stereocontrol. An early example used 2-phenylcyclohexanol to form the enantiopure enol ether **32** in a formal total synthesis of (+)-hirsutene.[83] This enol ether produced **33a** and **33b** with 7:1 diastereoselectivity in 55% overall yield. It is noteable that the enol ether is a suitable PKR substrate even though this example predates the milder reaction conditions enabled by the use of additives.

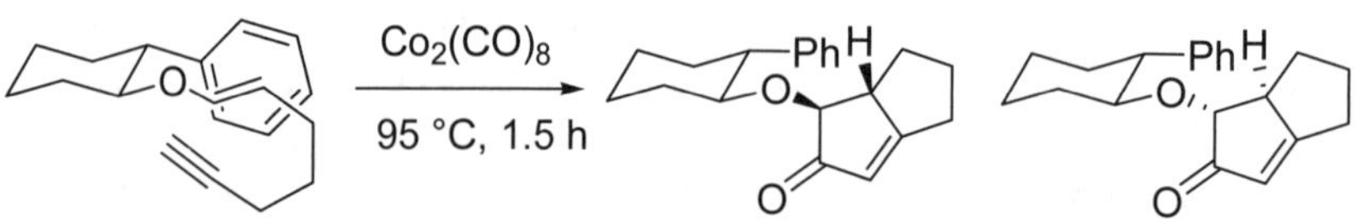

32 33a 33b

7 : 1

Similarly, ynol ether **34** gave cyclopentenone **35** in 82% yield and 24:1 diastereoselectivity when 10-(methylthio)isoborneol was used as the chiral auxiliary. When 2-phenylcyclohexanol was used as the chiral auxiliary the observed diastereoselectivity was only 2.5:1. With 10-(methylthio)isoborneol it is likely that the sulfur's ability to coordinate the cobalt is contributing to the improved selectivity.

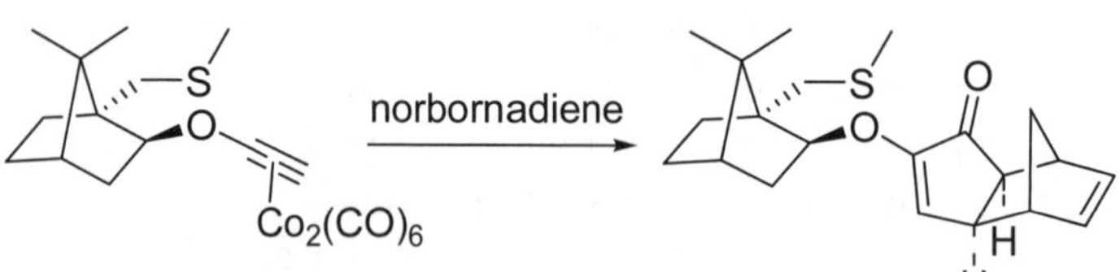

34 35

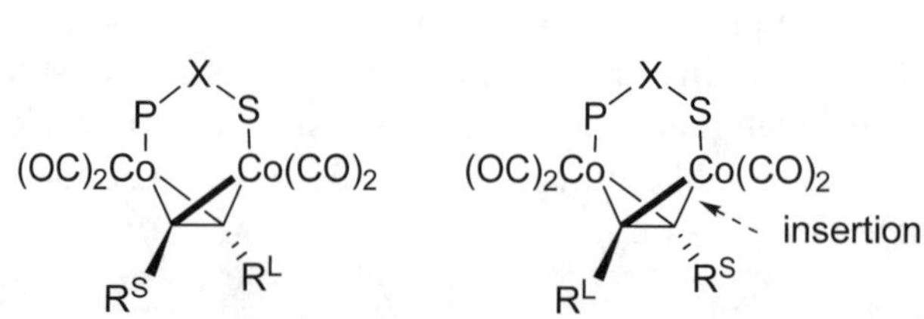

37 38

Chiral cobalt complexes have been used to control stereochemistry. A successful strategy described by Verdaguer uses bidentate ligands containing phosphorous and sulphur.[84–86,133–135] Diastereomeric cobalt complexes **37** and **38** are formed when an unsymmetrical alkyne cobalt complex is reacted with a chiral ligand. The diastereomers can be separated by chromatography or crystallization to provide enantiopure cobalt complexes. In the PKR the alkene inserts into the least hindered and most reactive cobalt–carbon bond. This bond tends to be adjacent to the smaller alkyne substituent and formed from the cobalt coordinated to sulphur. Thus cyclopentenones are formed in high enantiomeric excess from a single diastereomer. The X group can be carbon, nitrogen or absent. Both sulfides and sulfoxides have been employed.

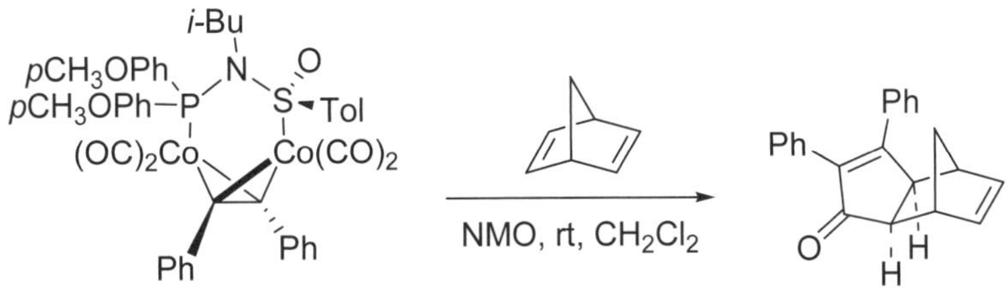

39 **40**

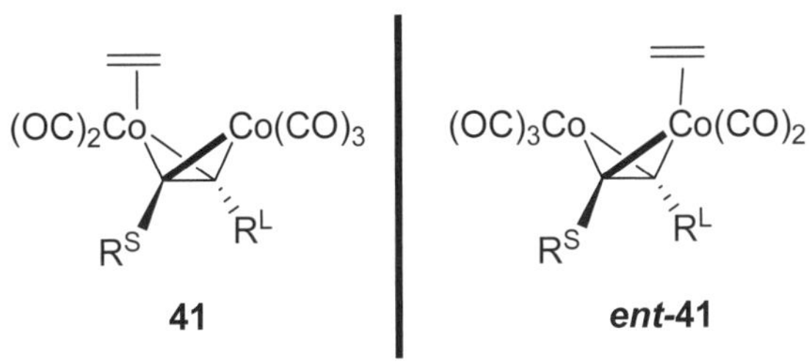

41 ***ent*-41**

In the case of a symmetric cobalt–alkyne complex, a single enantiomeric complex is formed as in **39**. Verdaguer has used enantiopure *N*-phosphino-*p*-tolylsulfinamide (PNSO) ligands to control the absolute stereochemistry to provide **40** in 77% yield and 94% *ee*.[87] There are three critical features of the complex **39**: (1) the greater reactivity of the cobalt coordinated to the sulphur versus the cobalt coordinated to the phosphine, and (2) the chiral sulphur atom is directly bound to the reacting cobalt atom and (3) the strong phosphine–cobalt bond. Thus, sulphur determines which metal center reacts and which metal–carbon bond reacts. Phosphorous glues the complex together and deactivates the metal center to which it is attached. Chiral amine *N*-oxides have been used to enduce stereocontrol in modest *ee* (0 to 78%). The idea is to selectively remove a CO from one of the pro-chiral cobalts. The desymmetrized complex can then coordinate an alkene as in **41** (or ***ent*-41**) and produce enantiopure products. The first and most successful demonstration of this strategy was reported by Kerr.[88] Brucine *N*-oxide promoted the asymmetric PKR of norbornene with tertiary propargyl alcohols. A tertiary alcohol was shown to be required for good *ee*. Neither propargyl alcohol nor trimethylsilyl-protected alcohol gave any enantiomeric excess. In the best example, dimethylpropynol reacted with norbornene in 63% yield and 78% *ee*. The use of dimethoxyethane and low temperatures (–60 °C) were critical for good *ee*.

Catalytic turnover in the PKR is a goal consistent with green chemistry principles. Most applications of the PKR use stoichiometric cobalt. In-roads for the use of catalytic cobalt have been reported.[5] Additives, such as phosphites, alternative sources of cobalt, and other transition metals have been shown to promote the catalytic PKR.[89,90] Most of these procedures require an atmosphere of CO or a way to generate CO during the reaction. The proposed role for CO is reaction with liberated $Co_2(CO)_6$ to prevent this cobalt species from forming oligomers that terminate the catalytic cycle. Pérez-Castells has shown that molecular sieves can be "pre-loaded" with carbon monoxide and used to provide catalytic turnover in the absence of a CO atmosphere.[91] The sieves are heated to 200 °C and allowed to cool under CO. The combination of these pretreated sieves and 0.1 equivalent $Co_2(CO)_8$ gave similar yields to reactions with stoiciometric cobalt. A variety of intermolecular and intramolecular examples were disclosed. Despite these recent advances, a carbon monoxide-free PKR with low or very low cobalt catalyst loading still remains a goal.

3.5.5 Synthetic Utility

3.5.5.1 General Utility

The functional group tolerance of the PKR, often predictable stereochemistry and regiochemistry, and recent modifications for improved yields make it an excellent choice for the synthesis of highly functionalized five-membered carbocycles. In general the alkyne partner is more tolerant of substitution than the alkene. Monosubstituted and disubstituted alkynes function in the PKR and both electron donating and electron withdrawing groups are tolerated. Also alkynes tend to produce predictable regioisomers in useful selectivity. The alkene component is more sensitive to both sterics and electronics. In the intermolecular reaction the alkene is typically limited to reactive alkenes such as norbornene. Mono-substituted alkenes can be good substrates but produce 1:1 mixtures of regioisomeric products. Some of these limitations are overcome in the intramolecular PKR. However, this reaction is, so far, mostly limited to three or four atom tethers to give 5,5 and 6,5 bicyclic products. General application of the PKR to medium-size ring synthesis remains an aspiration.

Alkynes

The chemical literature has examples of both monosubstituted and disubstituted alkynes undergoing the PKR. In general, as explained above, the largest alkyne substituent will be in the α-position of the new cyclopentenone. Exceptions to this "rule" have been observed. For example, Krafft observed that ethyl 2-butynoate reacts with norbornene to place the ester group exclusively in the β-position of **42**.[92] Gimbert and Milet have explained this regioselectivity for monosubstituted and disubstituted alkynes as a balance of triple bond polarization and sterics.[93] They state that in a polarized alkyne–cobalt complex, the alkynal carbon with the greatest electron density will become the β-carbon in the PKR product. Thus the PKR with ethyl 2-butynoate gives **42**. The electron-rich carbon is adjacent to the ester in the product's β-position. In the PKR with propyne,[14] the methyl group polarizes the alkyne such that the unsubstituted alkyne carbon has the highest electron density and becomes the β-carbon of the product **43**. While it is not intuitive, ethylpropiolate is considered essentially unpolarized in the cobalt complex. In unpolarized complexes, sterics dominate and the larger group occupies the α-position as in **44**. When the steric factors are equalized, electronic influences dominate the PKR to produce **45** as the only product. Krafft has used the propensity for the ester of disubstituted alkynes to prefer the β-position in her total synthesis of (±)-asteriscanolide.[94]

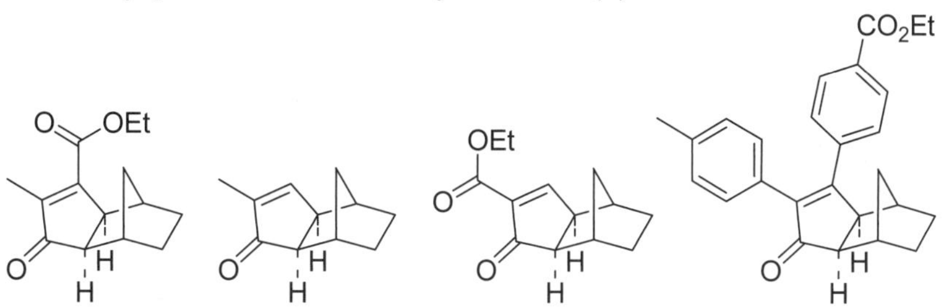

42 **43** **44** **45**

Alkyl and aryl groups are common alkyne substituents in the PKR. The reaction is compatible with a fully substituted propargyllic position, especially in the intramolecular version. Examples of electron-rich alkynes abound in the PKR, but there are fewer examples of electron-deficient alkynes. Riera and Verdaguer reported a study of electron-deficient alkynes in the PKR with norbornadiene.[95] This study clearly demonstrated that both the yield and *exo:endo* ratio were effected by the nature of the alkyne substituent. The *exo*-selectivity decreased as the electron-withdrawing ability of the R goup in **46** increased.

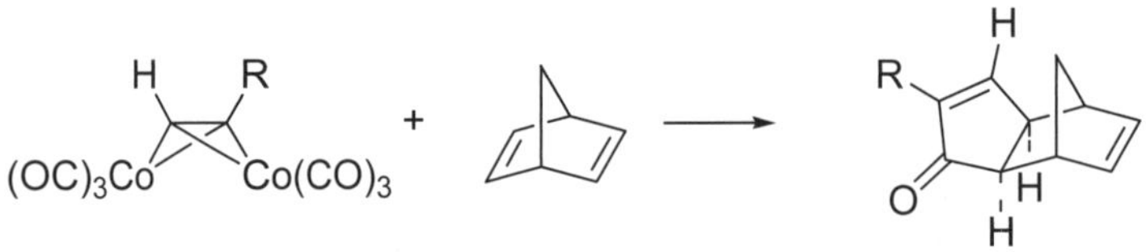

46 **47**

R	% Yield	exo : endo
STol	90	100 : 0
S(O)Tol	50	84 : 16
$S(O)_2Tol$	52	84 : 16
$C(O)N(CH_2CH_3)_2$	85	78 : 22
C(O)NHAr	92	74 : 26

Spicer has reported that dimethylacetylene dicarboxylate reacted with norbornene at room temperature in the presence of TMANO to give the cyclopentenone in 73% yield.[96] However, the cobalt complex is thermally unstable and mild PKR conditions were required for success. The intra-molecular and inter-molecular PKR of alkynones has been reported by Hoye in 30–90% yields.[97] In the intramolecular examples, improved yields were observed when a geminal dimethyl was incorporated into the tether. In the intermolecular examples, the ketone was found in the β-position of the cyclopentenone. The vinylogous **48** ynone cyclised in 62% yield to provide the tricyclic product **49** as a 2.7:1 mixture of *cis/trans* isomers. An interesting intramolecular example was disclosed by de Meijere in which ketone **50** produced the spirocyclopropane **51** in 63% yield.[98]

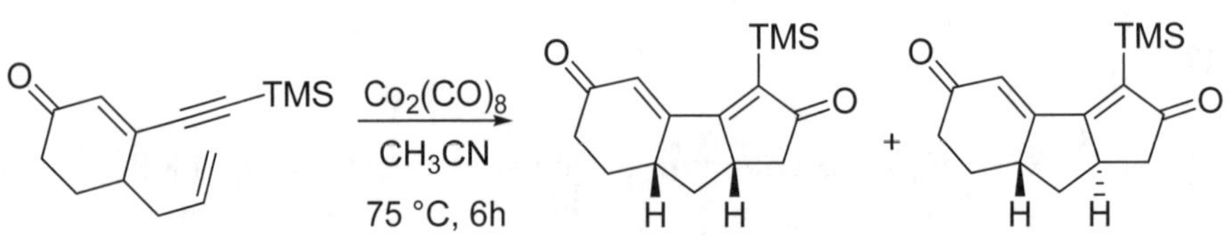

48 ***cis*- 49** ***trans*- 49**

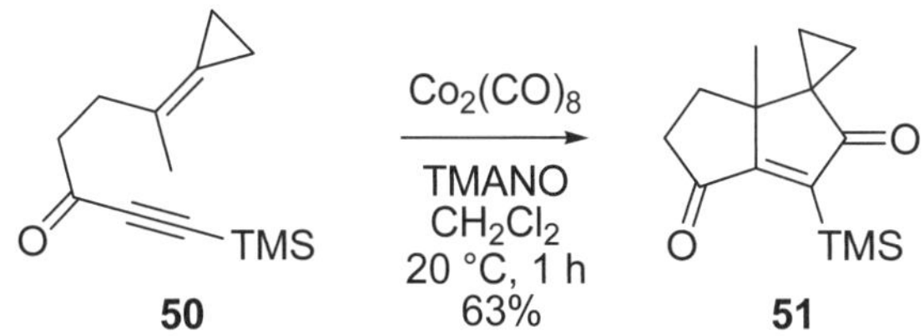

50 **51**

Thus, under the appropriate conditions, the alkyne partner does not appear to be severly limited by electronics or sterics. It is important that the regiochemstry of the alkyne substituents in the products can be predicted with some confidence. The same cannot be said for the alkene.

Alkenes

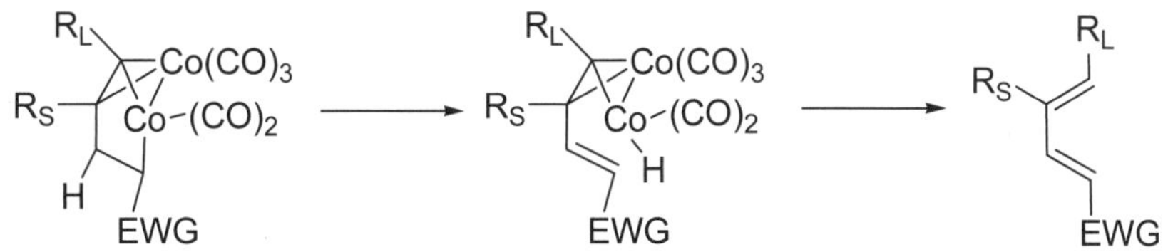

I-4 **52** **53**

The subject of alkene reactivity in the intermolecular PKR has been reviewed by Gibson. She points out several important issues related to reactivity and regiochemistry. With respect to reactivity, it is wellknown that the intermolecular PKR requires a reactive alkene to proceed in useful yields. Gibson compares the reported reactivity to the predicted reactivity calculated from the LUMOs by Milet and Gimbert.[99,100] She concludes that the LUMO predictions are a useful tool for predicting alkene reactivity; a lower lying LUMO is associated with a more reactive alkene. One caveat is that the alkene does not enter a non-PKR reaction manifold. For example, dienes and vinylesters are expected to have low lying LUMOs, but tend to be poor PKR substrates. These alkenes often engage in other reactions under the PKR conditions. The reactivity of electron deficient alkenes has been reviewed by Carretero.[12] In general, α,β-unsaturated nitriles, ketones and esters are poor substrates for the PKR. It is hypothesized that the cobaltacycle intermediate **I-4** undergoes β-hydride elimination to give **52** faster than carbon monoxide insertion. Dienes **53** result from this alternative pathway. Cazes has shown that vinyl esters and sulfones participate in the intermolecur PKR under mild conditions: 0–20 °C and 6 equivalents of NMO.[101] The yields ranged from 0 to 59% for the esters and 49 to 71% for the sulfones. Acrylonitrile gave complex mixture of products under these same conditions. Cyclopent-2-enone, cyclohex-2-enone and methylvinyl-ketone were unreactive.

Chung has taken advantage of the difference in alkene reactivities to perform a one-pot PKR/Diels–Alder reaction.[102] The reaction combined a 2:1:3 mixture of norbornene, 1-ethynylcyclohexene and dimethylfumarate, respectively. A 90% yield of the pentacyclic products **54** and **55** were formed in a 1.1:1 ratio.

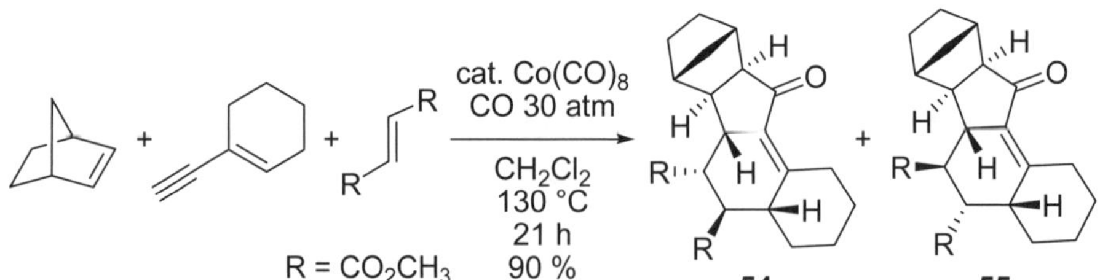

54 **55**

Vinyl sulfones are good substrates for the PKR. Evans has reported the stereoselective PKR of **56** to form **57** in studies directed toward kinabalurine alkaloids.[103]

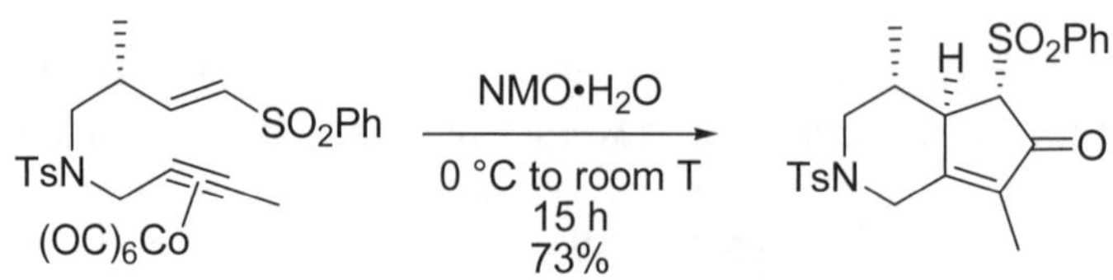

56 **57**

The bond angle of cycloalkenes can be used to estimate the LUMO and predict reactivity. Smaller bond angles correlate with lower LUMOs and greater reactivity. Thus the bond angles and reactivities for cycloalkenes follow the same order: cyclopropene > cyclobutene > cyclopentene = norbornene > cycloheptene = cyclooctene > cyclohexene. Cyclopropene is the most reactive and cyclohexene is the least reactive.

Thus proper choice of substrates and conditions can produce useful products via the intermolecular PKR. For example, Pericàs recently described the preparation of amino acid derivitives using the PKR, **58** to **59**.[104] Unfortunately, the amino acid stereocenter was too removed from the newly formed stereocenters to provide any asymmetric induction. The products were isolated as 1:1 mixtures of diastereomers that required chiral HPLC for separation. Despite the low yields observed in the PKR (18–47%), the method shows promise as a way to assemble peptide-like scaffolds from readily available starting materials.

58

59

Unlike the more predictable regiochemistry for alkynes, monosubstituted or differentially disubstituted alkenes give 1:1 mixtures of regio-isomers as shown for the synthesis of the estrone derivatives **27a** and **27b** above. Krafft addressed the issue of regioselectivity by introducing a group into the alkene to coordinate cobalt.[105–107] Coordination significantly increased the regioselectivity and the reactivity of the alkene. In her examples, a homoallylic sulphur or amine, **61**, gave the best results. Allylic or bishomoallylic heteroatoms gave modest regioselectivity. Alcohols and ethers did not provide any ligand directed regioselectivity. Kerr has shown that diethyl allylphosphonate **63** also provides ligand directed regioselectivity in the PKR when a combination of dichloromethane and acetonitrile is used as the solvent.[108] Tethers containing multiple hetereoatoms[107] and aryl dimethylamino[109,110] groups have been used to take advantage of this directing effect, **64** and **65**, respectively.

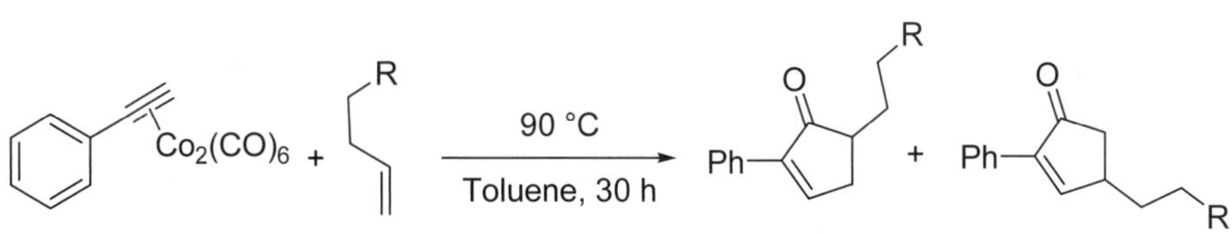

60 **61** **62a** **62b**

R	% Yield	62a : 62b
SCH_3	61 %	18 : 1
$N(CH_3)_2$	72 %	5 : 1

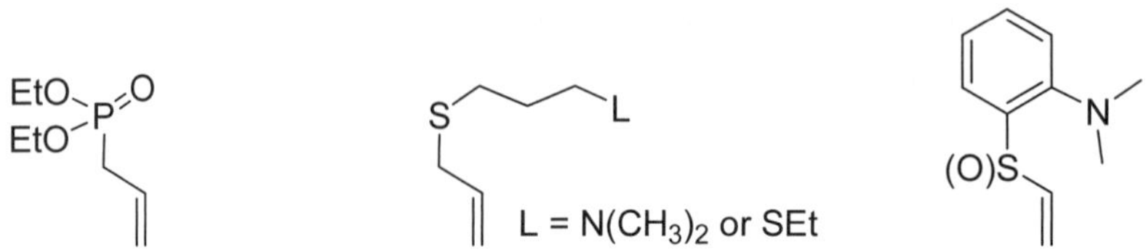

63 **64** **65**

As described above, a temporary tether has also been used to control regiochemistry and enhance reactivity. The improved yields observed with intramolecular versus intermolecular PKRs have encouraged attempts to facilitate the reaction with a cleavable tether that contains ether or silyl groups.[46–51]

The intramolecular PKR has seen the most application for synthesis. There are numerous examples of the formation of carbocyclic and heterocyclic systems. Recent applications show that the PKR is useful for building complex, bridged ring systems. Martin has disclosed 6 examples, **66a–f**, of nitrogen bridged bicyclic heterocycles in modest to excellent yield from the PKR.[111] Several common promoters (NMO, BuSMe and 4 Å molecular sieves) were tried before finding optimum conditions. Stepwise reaction by first forming the cobalt–alkyne complex then adding six equivalents of DMSO as a promoter provided the best conditions. Also, the $Co_2(CO)_8$ catalyst gave the best results when handled and stored under argon. These conditions were applied to a total synthesis of alstonerine (vida infra).

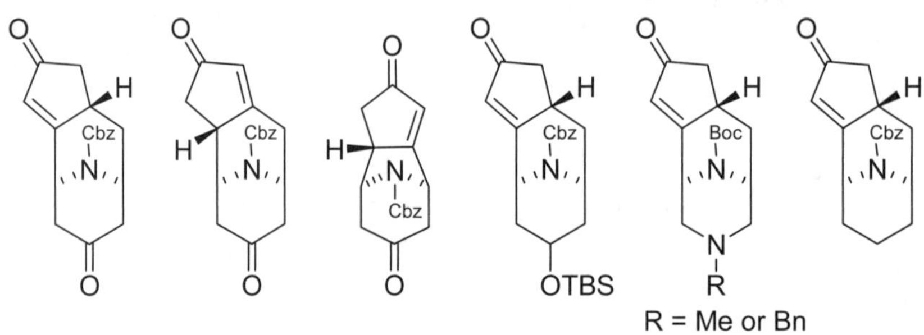

66a (89%) **66b** (91%) **66c** (33%) **66d** (69%) **66e** (>80%) **66f** (74%)

An underused property of cobalt-coordinated alkynes is the stabilization of propargyllic cations. The Nicholas reaction is a propargylic substitution reaction facilitated by the ability of the adjacent cobalt complex to stabilize the propargylic cation, **67** to **68**. Both carbon and heteroatom nucleophiles have been used to effect this transformation.[112–114] This transformation has been been used as a strategy to introduce the alkene component for an intramolecular PKR.[115] Shea has probed the use of an intramolecular-endocyclic-Nicholas PKR sequence for the formation of tricyclic heterocycles, **69** to **71**.[116] The reactions are classified as endocyclic because the cobalt complex is in the ring formed in the Nicolas reaction, **70**. Various ring sizes (m,n) and nucleophiles (X) were tried. When the nucleophile was an alcohol or a sulphonamide the final tricyclic products were obtained. Acids (XH = CO_2H) gave poor yields in the Nicholas reaction, and the ester intermediates failed to undergo the PKR. For the ethers, the [5,7,5] and [5,8,5] ring systems formed in good yield, the [5,6,5] and [5,9,5] systems formed in poor yield and the [5,8,6] system produced no identifiable products. Both the yields and disatereoselectivities varied with the conditions. The sulfonamides were limited to [5,7,5] and [5,8,5] systems. The [5,7,5] ring system **73b**, containing the sulphonamide, formed in exceptional yield and diastereoselectivity.

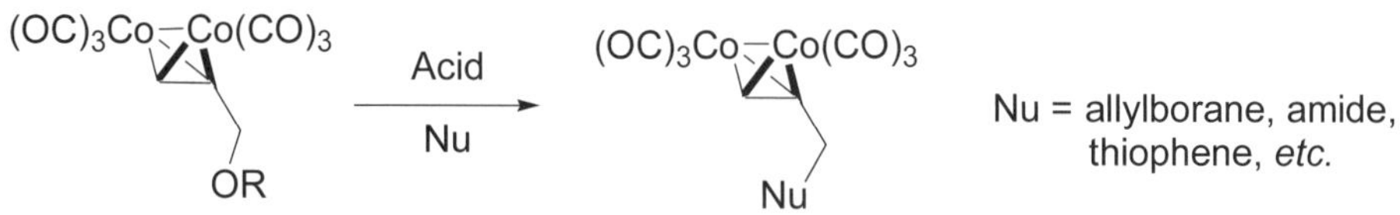

67 **68**

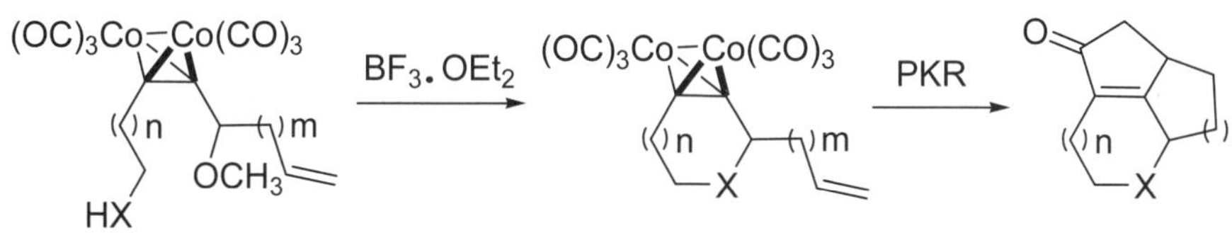

69 **70** **71** (45%)

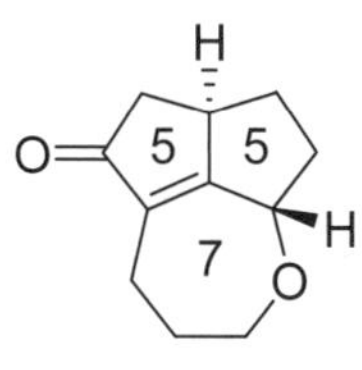

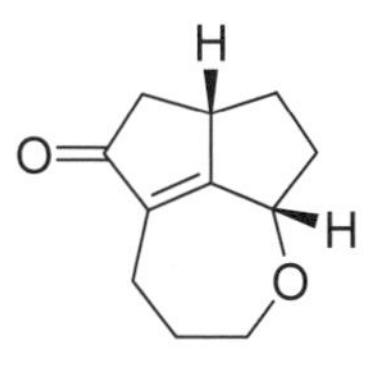

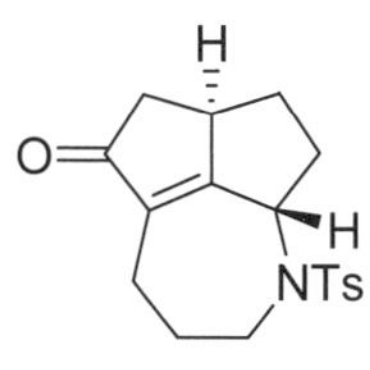

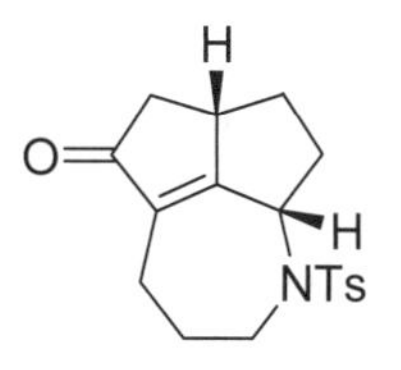

72a **72b** **73a** **73b**

PKR Conditions	Products	% Yield a / b
$CyNH_2$, Heat	**72a / 72b**	38 / 53
NMO	**72a / 72b**	22 / 9
$CyNH_2$, Heat	**73a / 73b**	0 / 98

An alternative method to take advantage of the activating nature of the cobalt alkyne was demonstrated by Smit and Caple.[117] In this strategy, a cobalt coordinated eneyne **74** is acylated with ethyltetrafluoroborate and the intermediate cation is captured with allyl alcohol to provide the starting material for an intramolecular PKR, **75**. Heating in hexane provided the desired tricyclic product **76** in 45% yield. It should be noted that the PKR portion of this sequence predates many of the improved conditions.

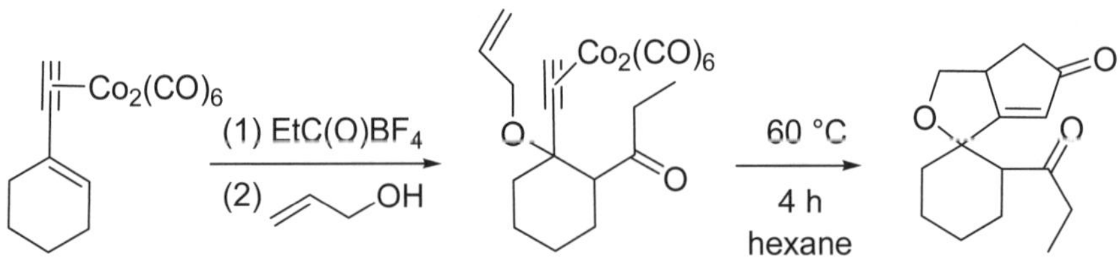

74 **75** (71%) **76** (45%)

3.6.5.2 Applications in total synthesis

The ability to predict regiochemistry, a rapid increase in synthetic complexity, and the frequency of 5-membered rings in synthetic targets has made this reaction the focus of numerous synthetic studies. The PKR has been applied to the synthesis of many natural products: (±)-pentalenene,[118] (+)-hirsutene,[82] (–)-kainic acid,[119] (+)-brefeldin A,[120] (+)-epoxydictymene,[121] (±)-asteriscanolide,[94] (±)-α- and β-cedrene,[122] Japanese Hop Ether,[123,124] (–)-alstonerine[125,111], (–)-incarvilline,[126] paecilomycine,[127] and (+)-α-skytanthine.[128] The Scheme on page 169 shows the PKR disconnections used to assemble these natural products. The wavy blue lines indicate bonds formed from the PKR.

Mukai accomplished the stereosective synthesis of three lycopodium alkaloids (–)-magellanine, (+)-magellaninone, and (+)-paniculatine from a common intermediate.[129] Their strategy employed the use of two intramolecular PKRs to build a tetracyclic intermediate from acyclic starting material; diethyl L-tartrate.

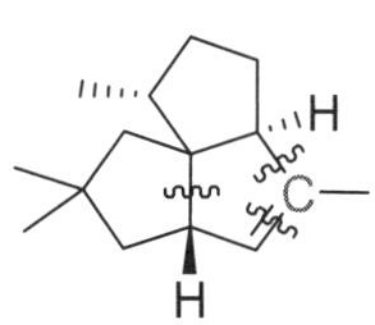
(±)-pentalenene

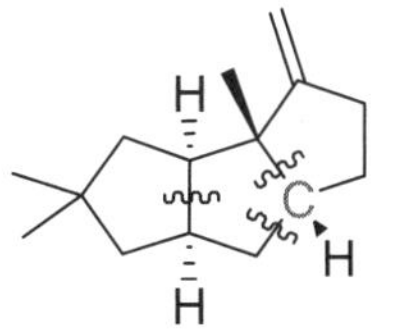
(+)-hirsutene

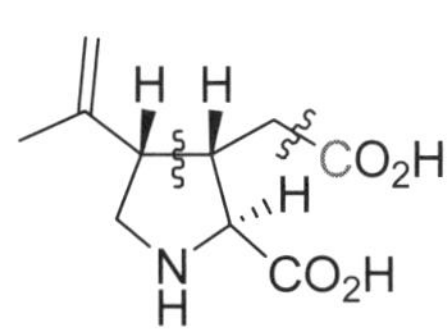
(–)-kainic acid

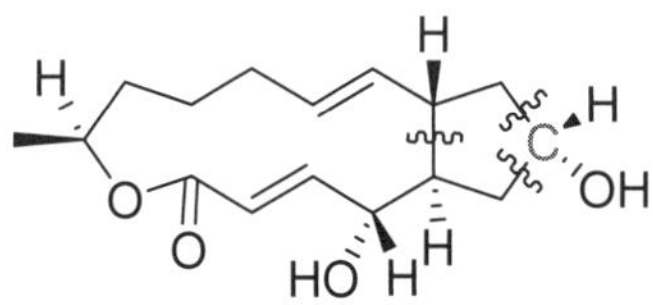
(+)-brefeldin A

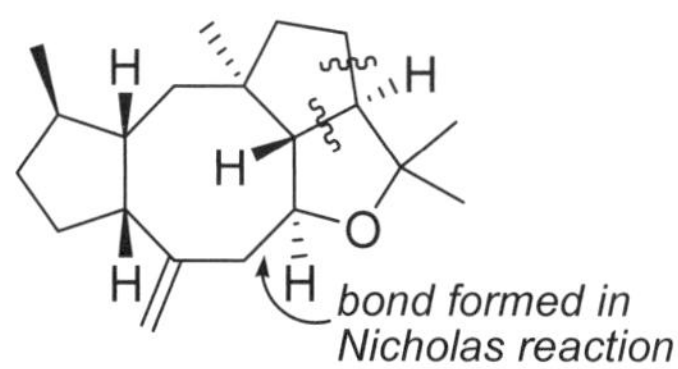

(+)-epoxydictymene

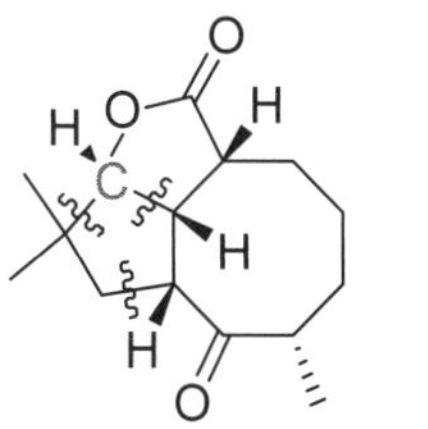
(±)-asteriscanolide

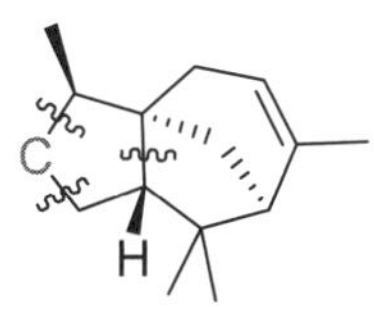
(±)-α-cedrene

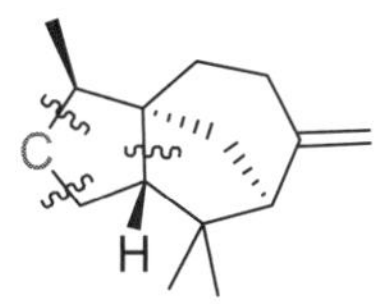
(±)-β-cedrene

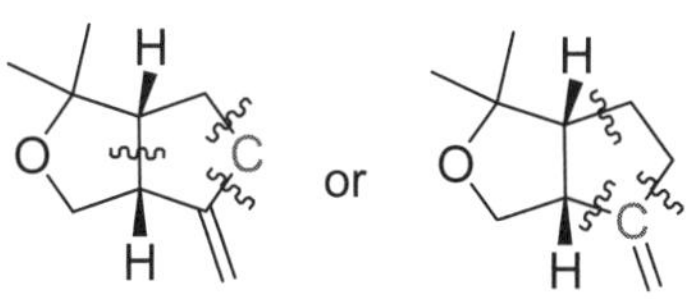
Japanese hop ether

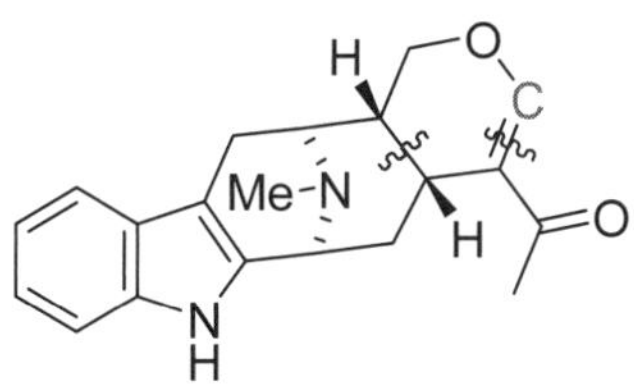
(-)-alstonerine

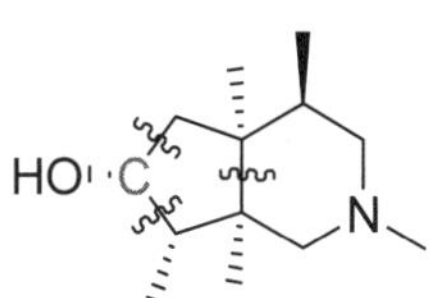
(–)-incarvilline

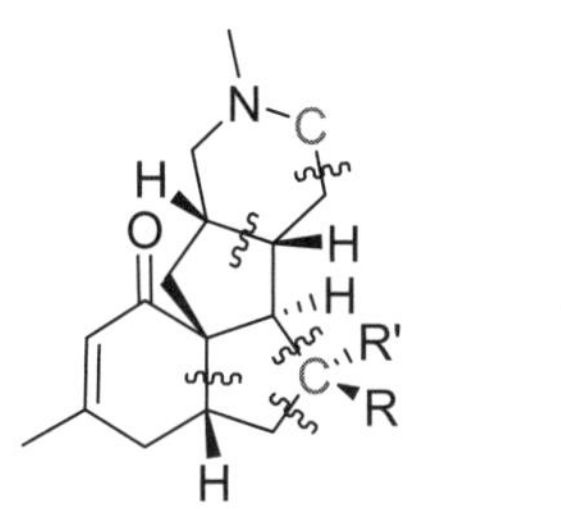
(–)-magellanine R=OH,R'=H
(+)-magellaninone R+R'=O

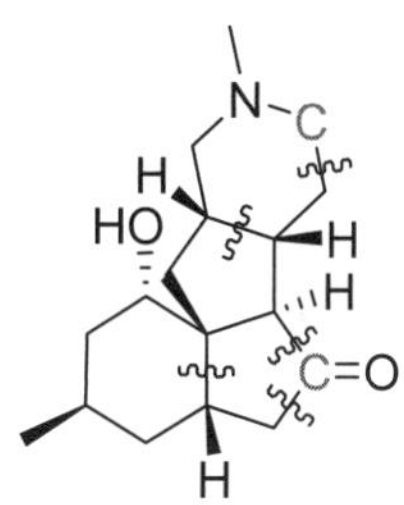
(+)-paniculatine

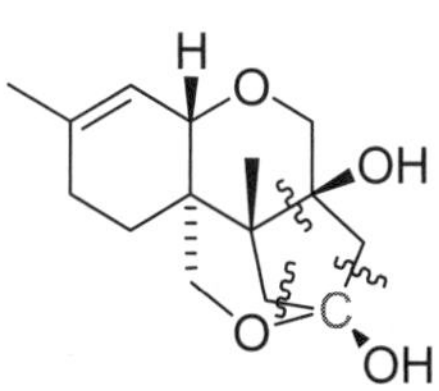
paecilomycine A

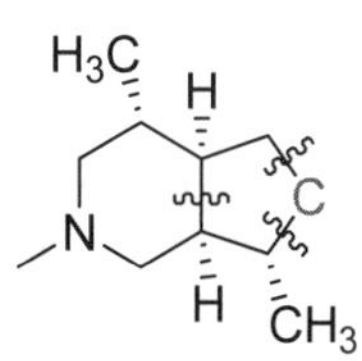
(+)-α-skytanthine

In the synthesis of (±)-pentalenene, the PKR was used to construct the angularly fused triquinane ring system **77** from cyclopentene **76**.

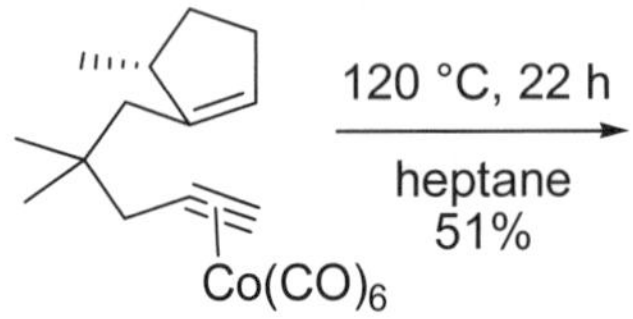

76 **77**

The allylic methyl group in **76** successful directs the bulky cobalt–alkyne complex to the opposite face of the cyclopentene with 8:1 diastereoselectivity. It is notable that, even in the absence of additives, this intramolecular PKR with a trisubstituted alkene provides useful yields of product.

In a more recent application of this allylic directing effect, Dake prepared an advanced intermediate for a potential Fusicoccane synthesis.[130] In his example the allylic position was a quaternary center and excellent, reproducible yields were obtained for the PKR. The observed stereochemistry of the ring fusion was consistent with the Magnus model. The larger of the two allylic groups in **78** prefered the *exo* face of the newly formed ring system, *syn* to the hydrogen at the ring fusion.

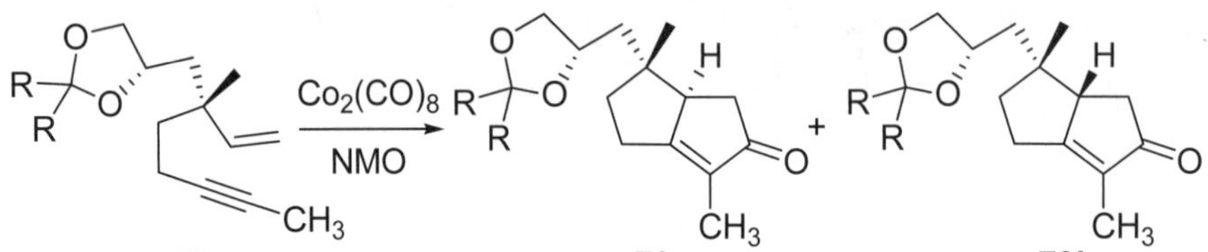

78 **79a** **79b**

Protecting Group	% Yield	79a : 79b
R = CH_3	89	6.1 : 1
R = CH_2CH_3	84	1.3 : 1
R = cyclohexyl (drawn as ring)	79	2.2 : 1

In an attempt to improve the selectivity for the desired isomer **79a**, the protecting group was varied. The expectation was that a larger protecting group would increase the selectivity for the desired isomer. It is surprising that as the size of the acetonide increased the selectivity decreased. No explanation for this observation was offered. It is interesting to constrast this result with Honda's in the synthesis of (+)-α-skytanthine where, again, a simple methyl group in **80** provides the desired diastereomer **81** in 71% isolated yield.

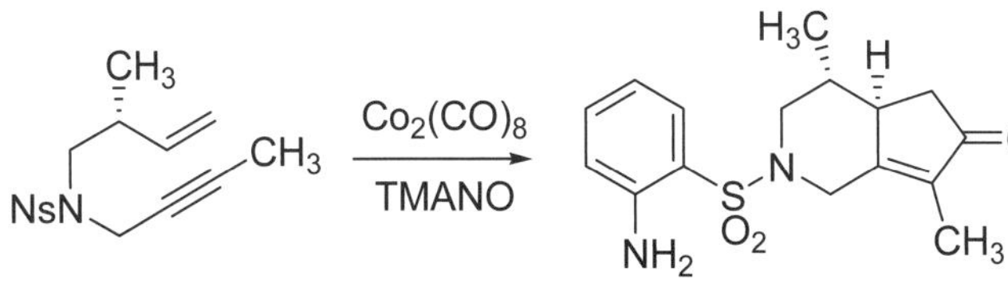

80 **81**

Notice that under the best PKR conditions the nosyl nitro group was reduced to the amine. When the reaction was performed in the absence of a hydrogen source reduction still occurred. The exact mechanism of the reduction remains unclear, but the transition metal-catalyzed reduction of nitro groups with CO is well documented.[131,132]

The disconnections shown for (+)-brefeldin A suggest the use of an intramolecular cyclization to close the macrocycle. In fact, the cyclopentene was formed via an intermolecular PKR/retro-Diels–Alder sequence (**82** goes to **83**) to take advantage of the high reactivity of norbornadiene in the PKR. Furthermore, a new cyclopentenone is revealed by the retro-Diels–Alder reaction so that PKR product **84** is a synthetic equivalent for cyclopentadienone **85**.

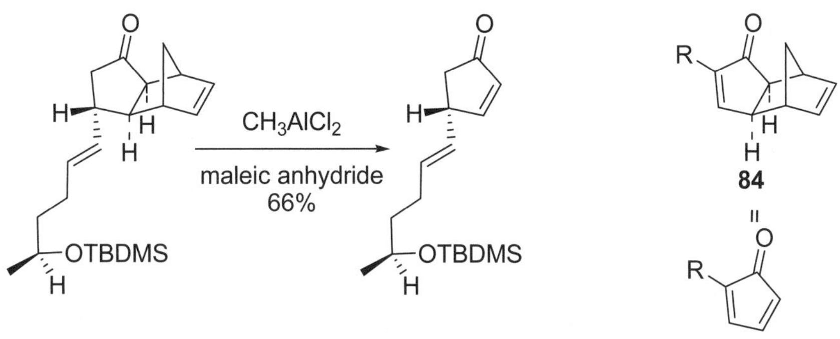

82 **83** **85**

It should be noted for (+)-epoxydictymene the carbon derived from carbon monoxide is not contained in the final product. This carbon is removed after the PKR in a subsequent ring opening. Also of interest is the author's use of a Nicholas reaction to close the 8-member ring. After significant optimization, conditions were found that converted allyl silane **86** to **87** in excellent yield and diastereoselectivity. The diastereomer **87** was favored more than 20:1 over its isomer even though the acetal in **86** existed as a 1:1 mixture.

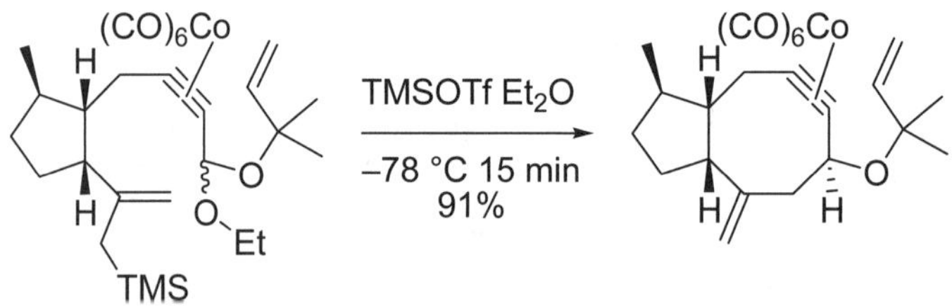

86 **87**

The optimization of the PKR for forming the bridged bicyclic core of (–)-alstonerine was discussed above. The eneyne **88** employed in the successful synthesis was available in enantiopure form from trytophan in four steps. Neither the unprotected indole nitrogen nor the additional sp^2 centers in the piperidine ring adversely affected the PKR. Under the optimized conditions, the desired compound **89** was isolated in 92% yield as a single diastereomer.

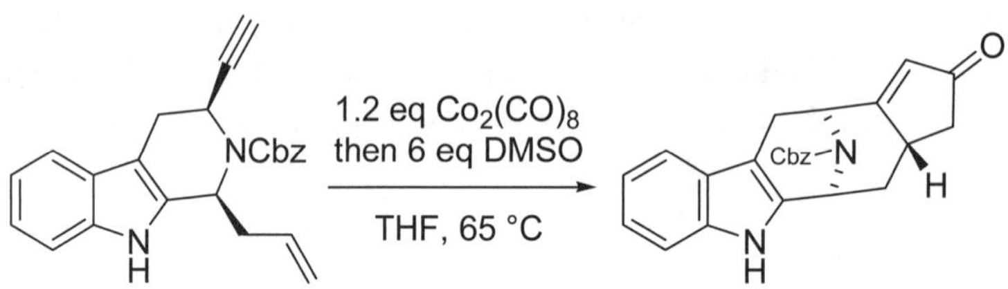

88 **89** (92%)

The syntheses of (–)-alstonerine, (+)-epoxydictymene, and kainic acid all exemplify a strategy of cyclization followed by oxidative ring-opening. For (+)-epoxydictymene, the cyclopentenone is reduced and hydroxylated to **90**. This triol is then opened to provide aldehyde **91**. The carbon originally derived from carbon monoxide is thus lost from the substrate. In (–)-alstonerine the cyclopentenone was converted to the aldehyde **92** in a two-step procedure. First, the sily enol ether was formed with *i*-propylsilyl hydride in the presence of a platinum catalyst. The double bond was then cleaved using catalytic OsO_4 with $NaIO_4$. Takano demonastrates a third ring opening strategy in the synthesis of (–)-kainic acid. A Baeyer–Villiger oxidation is affected with five equivalents of *m*-CPBA to convert ketone **93** to lactone **94**. Thus these syntheses demonstrate methods to expand the use of the PKR beyond five-membered rings.

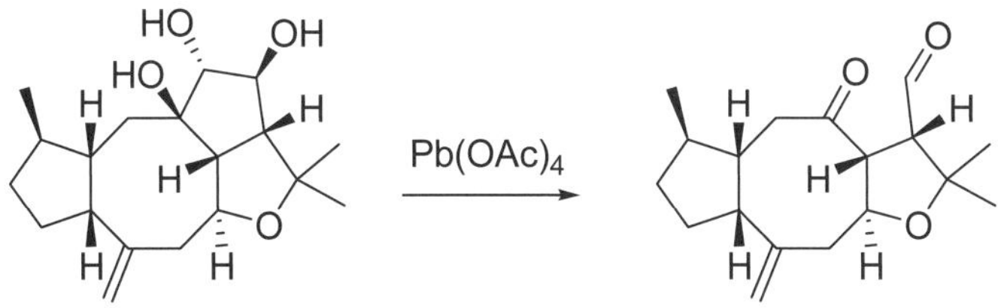

90 **91**

89

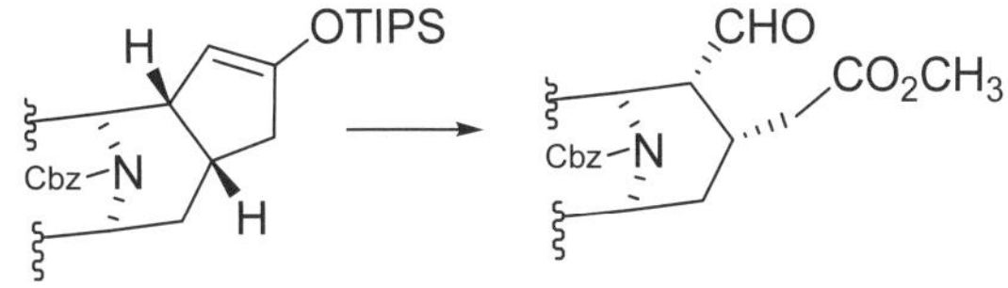

92

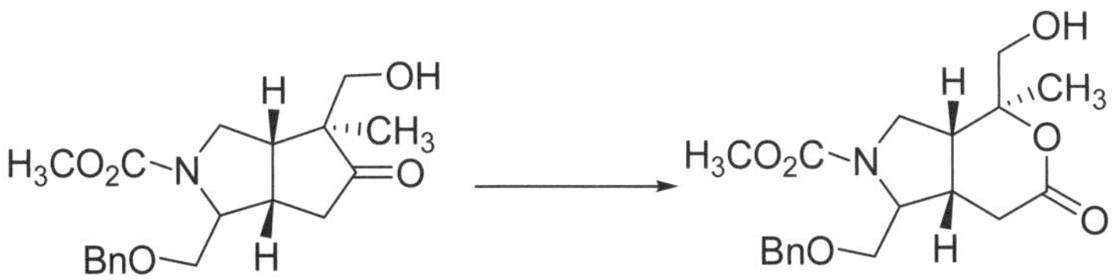

93 **94**

Krafft effectively uses both electronic and steric effects to fully control the PKR regioselectivity in her synthesis of (±)-asteriscanolide. As discussed above, electron-withdrawing groups on the alkyne favor the β-position in the cyclopentenone, while sterically demanding groups favor the α-position. Complete control of regiochemistry was observed in the intermolecular PKR to form intermediate **96** from **95**. The methyl group derived from propene was observed only alpha to the carbonyl. Because asteriscanolide contains a geminal dimethyl group at this alpha position, PKRs with isobutene and allylic thioalkyl groups were attempted. However, no cycloaddition was obtained with these alkenes and the second methyl group was installed by deprotenation and alkylation in 92% yield.

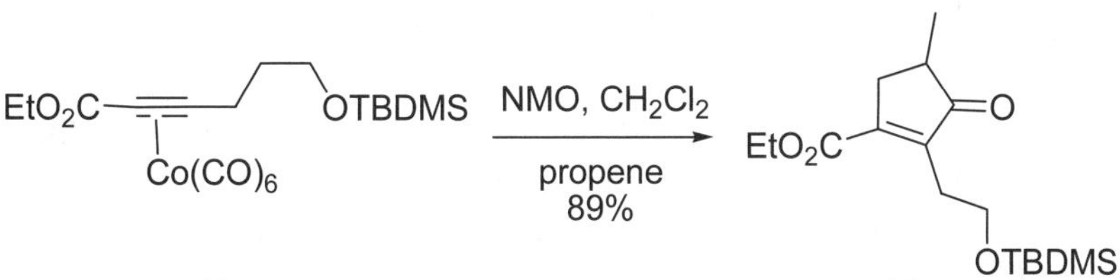

95 **96**

Kerr's syntheses of (±)-α- and β-cedrene are examples of using the PKR to construct compact brigded ring systems from readily available starting materials. This is another example of a trisubstituted alkene reacting well in an intramolecular PKR. As seen previously, the olefin geometry in the starting material **97** is transmitted with high fidelity to the α-keto methyl group in **98**. The ketone **98** was epimerized with lithium hydroxide in a mixture of tetrahydrofuran and water to provide a 9:1 ratio in favour of the desired isomer.

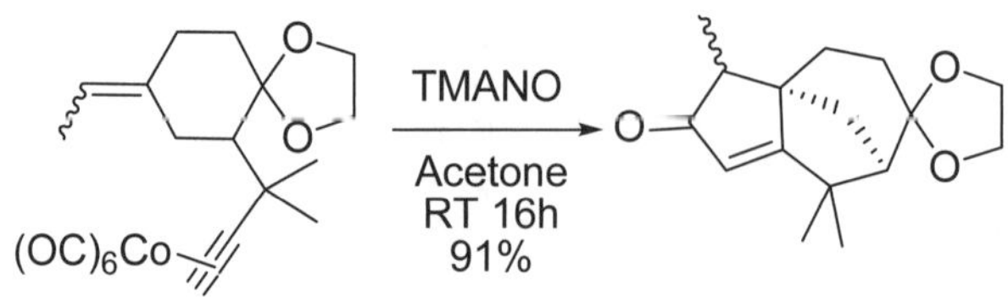

97 2:1 **98** 2:1

Japanese Hop Ether has been prepared by both an intermolecular PKR[123] and an intramolecular PKR.[24] In Kerr's intramolecular version, it was demonstrated that the use of mild conditions (room temperature) allows the double bond stereochemistry to be transmitted to the product with high fidelity: (*E*)-**99** gives exclusively **100** and (*Z*)-**99** gives exclusively **101**. Further manipulation transforms **101** into Japanese Hop Ether.

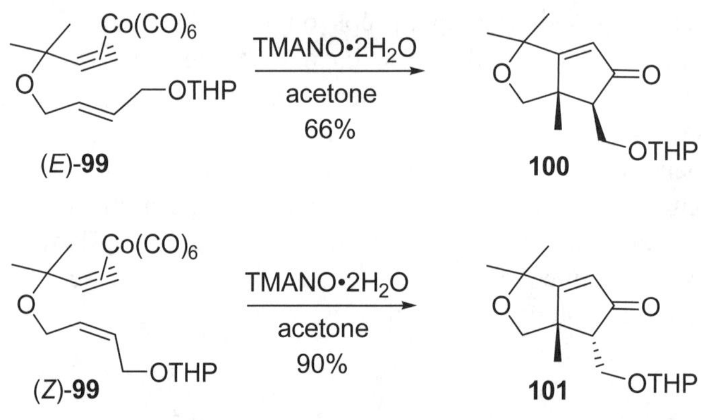

There are few examples where the absolute stereochemistry is determined in the PKR. Verdaguer has demonstrated the use of a bidentate phosphorous–sulphur ligand to provide an enantiopure cyclopentenone in for the synthesis of the HIV drug (–)-abacavir.[133–135] The achiral alkyne cobalt complex is generated first in 1 h at room temperature. Heating with enantiopure *P,S*-ligand **102** provides a separable mixture of diastereomeric alkyne complexes **103a** and **103b**. Reaction of one of these complexes with norbornadiene gave compound **104** in 53% overall yield and greater than 99% enantiomeric excess. This compound was further manipulated to provide the HIV drug (–)-abacavir.

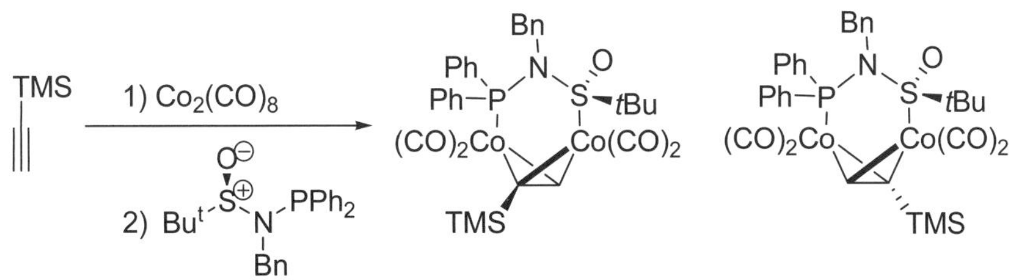

102 **103a** **103b**

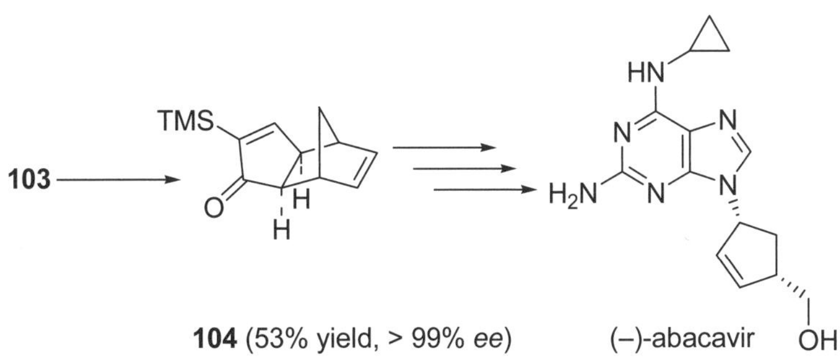

104 (53% yield, > 99% *ee*) (–)-abacavir

As shown above, the PKR has tremendous synthetic potential due to a rapid increase in synthetic complexity from simple starting materials. The frequent presence of five-membered rings in synthetic targets has led to numerous applications. In turn, these applications have inspired improved conditions that allow the PKR to occur under mild conditions with with predictable regiocontrol and stereocontrol. Efforts continue to optimize the reaction for improved alkene regiocontrol, absolute stereocontrol, expanded substrate scope (especially for unreactive alkenes and medium-size ring formation) and catalytic turnover.

From the above reactions, it is clear that there are a large number of potential reaction conditions. In practice the most common conditions use a slight excess of cobalt octacarbonyl, in acetonitrile or dichloromethane to form the alkyne complex under an atmosphere of nitrogen or argon. The formation of the alkyne complex is easily monitored by thin layer chromatography on silica gel. The alkene and TMANO or NMO are then added, and the reaction was allowed proceed at room temperature until complete. If the results are not satisfactory, the reaction can be optimized with additives or by changing the temperature. An experimental example is provided below. If absolute stereocontrol is desired, the reader is encouraged to use the examples sited above as a starting point for the best experimental conditions.

3.5.6 *Experimental*

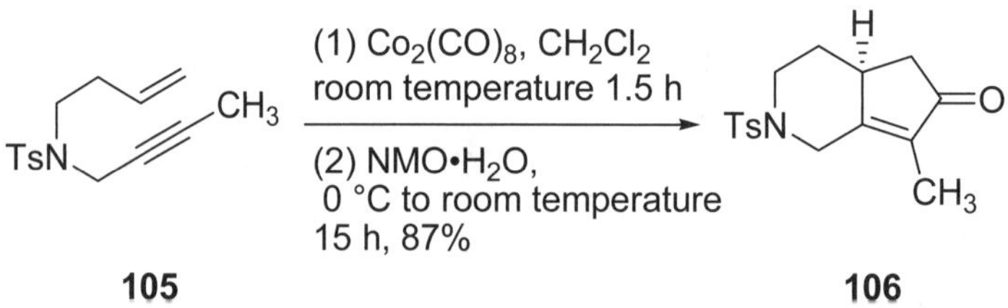

(±)-7-Methyl-2-(toluene-4-sulfonyl)-1,2,3,4,4a,5-hexahydro-[2]pyrindin-6-one 106[103]

At room temperature a solution of the enyne **105** (824 mg, 2.97 mmol, 1 equiv) in dichloromethane (100 mL, 0.029 M) was degassed with a steady stream of N_2 for approximately 0.5 h. $Co_2(CO)_8$ (1.52 g, 4.45 mmol, 1.5 equiv) was added in one portion. Stirring was continued for 1.5 h before TLC analysis indicated the formation of the cobalt complex [brown spot; $R_f \approx$ 0.9 (cyclohexane–EtOAc; 1:1)]. The solution was cooled to 0 °C before NMO·H_2O (2.60 g, 19.26 mmol, 6.5 equiv) was added in two portions. Stirring was continued at 0 °C to room temperature overnight. Silica (ca. 12 g) was added, the solvent was removed under reduced pressure and the residue purified by flash column chromatography (cyclohexane–EtOAc; 1:1) to afford the colourless solid **106** (790 mg, 87%); mp = 116 °C; R_f = 0.35 (cyclohexane–EtOAc; 1 : 1).

3.5.7 *References*

1. [R] Laschat, S.; Becheanu, A.; Bell, T.; Baro, A. *Synlett* **2005,** 2547–2570.
2. [R] Blanco-Urgoiti, J.; Anorbe, L.; Perez-Serrano, L.; Dominguez, G.; Perez-Castells, J. *Chem. Soc. Rev.* **2004,** *33*, 32–42.
3. [R]Struebing, D.; Beller, M. *Top. Organomet. Chem.* **2006,** *18*,165–178.
4. [R] Brummond, K. M.; Kent, J. L. *Tetrahedron* **2000,** *56*, 3263–3283.
5. [R] Gibson, S. E.; Stevenazzi, A. *Chem., Int. Ed.* **2003,** *42*, 1800–1810.
6. [R] Geis, O.; Schmalz, H.-G. *Angew. Chem., Int. Ed.* **1998,** *37*, 911–914.
7. [R] Schore, N.E. *Org. React. (N. Y.)* **1991,** *40*, 1–90.
8. [R] Schore, N. E. *Chem. Rev.* **1988,** *88*, 1081–119.
9. [R] Gibson, S. E.; Mainolfi, N. *Angew. Chem., Int. Ed.*, **2005,** *44*, 3022–3037.
10. [R] Sugihara, T.; Yamaguchi, M.; Nishizawa, M. *Chem. Eur. J.* **2001,** *7*, 1589–1595.
11. [R] Ingate, S. T.; Marco-Contelles, J. *Org. Prep. Proced. Int.* **1998,** *30*, 121–143.
12. [R] Rivero, M. R.; Adrio, J.; Carretero, J. C. *Eur. J. Org. Chem.* **2002**, 2881–2889.
13. Pauson, P. L.; Khand, I. U.; Knox, G. R.; Watts, W. E. *J. Chem. Soc. D* **1971,** 36.
14. Khand, I. U.; Knox, G. R.; Pauson, P. L.; Watts, W. E.; Foreman, M. I. *J. Chem. Soc., Perkin Trans. 1* **1973,** 977–981.
15. Magnus, P.; Principe, L. M. *Tetrahedron Lett.* **1985,** *26*, 4851–4854.

16. La Belle, B. E.; Knudsen, M. J.; Olmstead, M. M.; Hope, H.; Yanuck, M. D.; Schore, N. E. *J. Org. Chem.* **1985,** *50*, 5215–5222.
17. Yamanaka, M.; Nakamura, E. *J. Am. Chem. Soc.* **2001,** *123*, 1703–1708.
18. Pericas, M. A.; Balsells, J.; Castro, J.; Marchueta, I.; Moyano, A.; Riera, A.; Vazquez, J.; Verdaguer, X. *Pure Appl. Chem.* **2002,** *74*, 167–174.
19. Cabot, R.; Lledo, A.; Reves, M.; Riera, A.; Verdaguer, X. *Organometallics* **2007,** *26*, 1134–1142.
20. Banide, E. V.; Mueller-Bunz, H.; Manning, A. R.; Evans, P.; McGlinchey, M. J. *Angew. Chem., Int. Ed.* **2007,** *46*, 2907–2910.
21. [R] Bonaga, L. V. R.; Krafft, M. E. *Tetrahedron* **2004,** *60*, 9795–9833.
22. Camps, F.; Moreto, J. M.; Ricart, S.; Vinas, J. M. *Angew. Chem.* **1991,** *103*, 1540–1542 (See also *Angew Chem., Int. Ed. Engl.* **1991**, *11*, 1470–1472).
23. Billington, D. C.; Helps, I. M.; Pauson, P. L.; Thomson, W.; Willison, D. *J. Organomet. Chem.* **1988,** *354*, 233–242.
24. Simonian, S. O.; Smit, W. A.; Gybin, A. S.; Shashkov, A. S.; Mikaelian, G. S.; Tarasov, V. A.; Ibragimov, I. I.; Caple, R.; Froen, D. E. *Tetrahedron Lett.* **1986,** *27*, 1245–1248.
25. Smit, W. A.; Gybin, A. S.; Shashkov, A. S.; Strychkov, Y. T.; Kyz'mina, L. G.; Mikaelian, G. S.; Caple, R.; Swanson, E. D. *Tetrahedron Lett.* **1986,** *27*, 1241–1244.
26. Shambayati, S.; Crowe, W. E.; Schreiber, S. L. *Tetrahedron Lett.* **1990,** *31*, 5289–5292.
27. Jeong, N.; Chung, Y. K.; Lee, B. Y.; Lee, S. H.; Yoo, S. E. *Synlett* **1991**, 204–206.
28. Kerr, W. J.; Lindsay, D. M.; Kerr, W. J.; Lindsay, D. M. *Chem. Commun.* **1999**, 2551–2552.
29. Brown, D. S.; Campbell, E.; Kerr, W. J.; Lindsay, D. M.; Morrison, A. J.; Pike, K. G.; Watson, S. P. *Synlett* **2000**, 1573–1576.
30. [R] Sugihara, T.; Yamaguchi, M.; Nishizawa, M. *Rev. Heteroat. Chem.* **1999,** *21*, 179–194.
31. Sugihara, T.; Yamaguchi, M.; Nishizawa, M. *Chem.--Eur. J.* **2001,** *7*, 1589–1595.
32. Perez del Valle, C.; Milet, A.; Gimbert, Y.; Greene, A. E. *Angew. Chem., Int. Ed.* **2005,** *44*, 5717–5719.
33. Chung, Y. K.; Lee, B. Y.; Jeong, N.; Hudecek, M.; Pauson, P. L. *Organometallics* **1993,** *12*, 220–223.
34. Whitby, R. J.; Dixon, S.; Maloney, P. R.; Delerive, P.; Goodwin, B. J.; Parks, D. J.; Willson, T. M. *J. Med. Chem.* **2006,** *49*, 6652–6655.
35. Rajesh, T.; Periasamy, M. *Tetrahedron Lett.* **1998,** *39*, 117–118.
36. Sugihara, T.; Yamada, M.; Ban, H.; Yamaguchi, M.; Kaneko, C. *Angew. Chem., Int. Ed. Engl.* **1998,** *36*, 2801–2804.
37. Krafft, M. E.; Bonaga, L. V. R.; Hirosawa, C. *J. Org. Chem.* **2001,** *66*, 3004–3020.
38. Sugihara, T.; Yamada, M.; Yamaguchi, M.; Nishizawa, M. *Synlett* **1999**, 771–773.
39. Tang, Y.; Deng, L.; Zhang, Y.; Dong, G.; Chen, J.; Yang, Z. *Org. Lett.* **2005,** *7*, 593–595.
40. Petrovski, Z.; Romao, C. C.; Afonso, C. A. M. *Synth. Commun.* **2008,** *38*, 2761–2767.
41. Hernandez-Rodriguez, M.; Avila-Ortiz, C. G.; del Campo, J. M.; Hernandez-Romero, D.; Rosales-Hoz, M. J.; Juaristi, E. *Aust. J. Chem.* **2008,** *61*, 364–375.
42. Sugihara, T.; Yamaguchi, M. *Synlett* **1998**, 1384–1386.
43. Perez-Serrano, L.; Blanco-Urgoiti, J.; Casarrubios, L.; Dominguez, G.; Perez-Castells, J. *J. Org. Chem.* **2000,** *65*, 3513–3519.
44. Blanco-Urgoiti, J.; Casarrubios, L.; Dominguez, G.; Perez-Castells, J. *Tetrahedron Lett.* **2002,** *43*, 5763–5765.
45. Son, S. U.; Lee, S. I.; Chung, Y. K. *Angew. Chem., Int. Ed.* **2000,** *39*, 4158–4160.
46. Muto, R.; Ogasawara, K. *Tetrahedron Lett.* **2001,** *42*, 4143–4146.
47. Marco-Contelles, J.; Ruiz, J. *Tetrahedron Lett.* **1998,** *39*, 6393–6394.
48. Reichwein, J. F.; Iacono, S. T.; Patel, M. C.; Pagenkopf, B. L. *Tetrahedron Lett.* **2002,** *43*, 3739–3741.
49. Reichwein, J. F.; Iacono, S. T.; Pagenkopf, B. L. *Tetrahedron* **2002,** *58*, 3813–3822.
50. Ishaq, S.; Porter, M. *Synth. Commun.* **2006,** *36*, 547–557.
51. Dobbs, A. P.; Miller, I. J.; Martinovic, S. *Beilstein J. Org. Chem.* **2007,** *3*.
52. Madu, C. E.; Lovely, C. J. *Synlett* **2007**, 2011–2016.
53. Madu, C. E.; Seshadri, H.; Lovely, C. J. *Tetrahedron* **2007,** *63*, 5019–5029.
54. Lovely, C. J.; Seshadri, H.; Wayland, B. R.; Cordes, A. W. *Org. Lett.* **2001,** *3*, 2607–2610.

55. Mastrorilli, P.; Nobile, C. F.; Paolillo, R.; Suranna, G. P. *J. Mol. Catal. A: Chem.* **2004,** *214*, 103–106.
56. Becheanu, A.; Laschat, S. *Synlett* **2002**, 1865–1867.
57. Krafft, M. E.; Wright, J. A.; Bonaga, L. V. R. *Tetrahedron Lett.* **2003,** *44*, 3417–3422.
58. Krafft, M. E.; Wright, J. A.; Bonaga, L. V. R. *Can. J. Chem.* **2005,** *83*, 1006–1016.
59. Krafft, M. E.; Scott, I. L.; Romero, R. H.; Feibelmann, S.; Van Pelt, C. E. *J. Am. Chem. Soc.* **1993,** *115*, 7199–7207.
60. Fager-Jokela, E.; Kaasalainen, E.; Leppaenen, K.; Tois, J.; Helaja, J. *Tetrahedron* **2008,** *64*, 10381–10387.
61. Iqbal, M.; Vyse, N.; Dauvergne, J.; Evans, P. *Tetrahedron Lett.* **2002,** *43*, 7859–7862.
62. Fischer, S.; Groth, U.; Jung, M.; Schneider, A. *Synlett* **2002**, 2023–2026.
63. Iqbal, M.; Duffy, P.; Evans, P.; Cloughley, G.; Allan, B.; Lledo, A.; Verdaguer, X.; Riera, A. *Org. Biomol. Chem.* **2008,** *6*, 4649–4661.
64. Ford, J. G.; Kerr, W. J.; Kirk, G. G.; Lindsay, D. M.; Middlemiss, D. *Synlett* **2000**, 1415–1418.
65. Jordi, L.; Segundo, A.; Camps, F.; Ricart, S.; Moreto, J. M. *Organometallics* **1993,** *12*, 3795–3797.
66. Shibata, T.; Takagi, K. *J. Am. Chem. Soc.* **2000,** *122*, 9852–9853.
67. Shibata, T.; Toshida, N.; Yamasaki, M.; Maekawa, S.; Takagi, K. *Tetrahedron* **2005,** *61*, 9974–9979.
68. Moradov, D.; Al Quntar, A. A. A.; Youssef, M.; Smoum, R.; Rubinstein, A.; Srebnik, M. *J. Org. Chem.* **2009,** *74*, 1029–1033.
69. Maji, P.; Wang, W.; Greene, A. E.; Gimbert, Y. *J. Organomet. Chem.* **2008,** *693*, 1841–1849.
70. Zhang, M.; Buchwald, S. L. *J. Org. Chem.* **1996,** *61*, 4498–4499.
71. Lan, Y.; Deng, L.; Liu, J.; Wang, C.; Wiest, O.; Yang, Z.; Wu, Y.-D. *J. Org. Chem.* **2009,** *74*, 5049–5058.
72. Kim, D. E.; Lee, B. H.; Rajagopalasarma, M.; Genet, J.-P.; Ratovelomanana-Vidal, V.; Jeong, N. *Adv. Synth. Catal.* **2008,** *350*, 2695–2700.
73. Hayashi, Y.; Miyakoshi, N.; Kitagaki, S.; Mukai, C. *Org. Lett.* **2008,** *10*, 2385–2388.
74. Kondo, T.; Suzuki, N.; Okada, T.; Mitsudo, T. *J. Am. Chem. Soc.* **1997,** *119*, 6187–6188.
75. Itami, K.; Mitsudo, K.; Fujita, K.; Ohashi, Y.; Yoshida, J. *J. Am. Chem. Soc.* **2004,** *126*, 11058–11066.
76. Zhao, Z.; Ding, Y.; Zhao, G. *J. Org. Chem.* **1998,** *63*, 9285–9291.
77. Garcia-Garcia, P.; Fernandez-Rodriguez, M. A.; Rocaboy, C.; Andina, F.; Aguilar, E. *J. Organomet. Chem.* **2008,** *693*, 3092–3096.
78. Takahashi, T.; Xi, Z.; Nishihara, Y.; Huo, S.; Kasai, K.; Aoyagi, K.; Denisov, V.; Negishi, E. *Tetrahedron* **1997,** *53*, 9123–9134.
79. Chen, C.; Liu, Y.; Xi, C. *Tetrahedron Lett.* **2009,** *50*, 5434–5436.
80. Paolillo, R.; Gallo, V.; Mastrorilli, P.; Nobile, C. F.; Rose, J.; Braunstein, P. *Organometallics* **2008,** *27*, 741–746.
81. Arias, J. L.; Cabrera, A.; Sharma, P.; Rosas, N.; Sampere, R. *J. Mol. Catal. A: Chem.* **2006,** *246*, 237–241.
82. Sezer, S.; Guemruekcue, Y.; Sahin, E.; Tanyeli, C. *Tetrahedron: Asymmetry* **2008,** *19*, 2705–2710.
83. Castro, J.; Sorensen, H.; Riera, A.; Morin, C.; Moyano, A.; Pericas, M. A.; Greene, A. E. *J. Am. Chem. Soc.* **1990,** *112*, 9388–9389.
84. Reves, M.; Achard, T.; Sola, J.; Riera, A.; Verdaguer, X. *J. Org. Chem.* **2008,** *73*, 7080–7087.
85. Lledo, A.; Sola, J.; Verdaguer, X.; Riera, A.; Maestro, M. A. *Adv. Synth. Catal.* **2007,** *349*, 2121–2128.
86. Ferrer, C.; Riera, A.; Verdaguer, X. *Organometallics* **2009,** *28*, 4571–4576.
87. Ji, Y.; Riera, A.; Verdaguer, X. *Org. Lett.* **2009,** *11*, 4346–4349.
88. Kerr, W. J.; Lindsay, D. M.; Rankin, E. M.; Scott, J. S.; Watson, S. P. *Tetrahedron Lett.* **2000,** *41*, 3229–3233.
89. Jeong, N.; Hwang, S. H.; Lee, Y.; Chung, Y. K. *J. Am. Chem. Soc.* **1994,** *116*, 3159–3160.
90. [R] Shibata, T., *Adv. Synth. Catal.* **2006,** *348*, 2328–2336.

91. Blanco-Urgoiti, J.; Casarrubios, L.; Dominguez, G.; Perez-Castells, J. *Tetrahedron Lett.* **2002,** *43*, 5763–5765.
92. Krafft, M. E.; Romero, R. H.; Scott, I. L. *Synlett* **1995**, 577–578.
93. de Bruin, T. J. M.; Michel, C.; Vekey, K.; Greene, A. E.; Gimbert, Y.; Milet, A. *J. Organomet. Chem.* **2006,** *691*, 4281–4288.
94. Krafft, M. E.; Cheung, Y.-Y.; Juliano-Capucao, C. A. *Synthesis* **2000**, 1020–1026.
95. Sola, J.; Riera, A.; Pericas, M. A.; Verdaguer, X.; Maestro, M. A. *Tetrahedron Lett.* **2004,** *45*, 5387–5390.
96. Baxter, R. J.; Knox, G. R.; Pauson, P. L.; Spicer, M. D. *J. Organomet. Chem.* **1999,** *579*, 90–96.
97. Hoye, T. R.; Suriano, J. A. *J. Org. Chem.* **1993,** *58*, 1659–1660.
98. de Meijere, A.; Becker, H.; Stolle, A. Kozhushkov, S. I.; Bes, M. T.; Salauen, J.; Noltemeyer, M., *Chem. Eur. J.* **2005,** *11*, 2471–2482.
99. de Bruin, T. J. M.; Milet, A.; Greene, A. E.; Gimbert, Y. *J. Org. Chem.* **2004,** *69*, 1075–1080.
100. de Bruin, T. J. M.; Milet, A.; Robert, F.; Gimbert, Y.; Greene, A. E. *J. Am. Chem. Soc.* **2001,** *123*, 7184–7185.
101. Ahmar, M.; Antras, F.; Cazes, B. *Tetrahedron Lett.* **1999,** *40*, 5503–5506.
102. Kim, D. H.; Chung, Y. K. *Chem. Commun.* **2005**, 1634–1636.
103. Kavanagh, Y.; O'Brien, M.; Evans, P. *Tetrahedron* **2009,** *65*, 8259–8268.
104. Cuevas, F.; Garcia-Granda, C.; Buschmann, H.; Torrens, A.; Yenes, S.; Pericas, M. A. *Synlett* **2007**, 119–122.
105. Krafft, M. E. *J. Am. Chem. Soc.* **1988,** *110*, 968–970.
106. Krafft, M. E. Juliano, C. A.; Scott, I. L.; Wright, C.; McEachin, M. D., *J. Am. Chem. Soc.* **1991,** *113*, 1693–1703.
107. Krafft, M. E.; Juliano, C. A. *J. Org. Chem.* **1992,** *57*, 5106–5115.
108. Brown, J. A.; Janecki, T.; Kerr, W. J. *Synlett* **2005**, 2023–2026.
109. Rivero, M. R.; De la Rosa, J. C.; Carretero, J. C. *J. Am. Chem. Soc.* **2003,** *125*, 14992–14993.
110. Rivero, M. R.; Alonso, I.; Carretero, J. C. *Chem. Eur. J.* **2004,** *10*, 5443–5459.
111. Miller, K. A.; Shanahan, C. S.; Martin, S. F. *Tetrahedron* **2008,** *64*, 6884–6900.
112. [R] Teobald, B. J. *Tetrahedron* **2002,** *58*, 4133–4170.
113. Green, J. R.; Tjeng, A. A. *J. Org. Chem.* **2009,** *74*, 7411–7416.
114. Diaz, D. D.; Betancort, J. M.; Martin, V. S. *Synlett* **2007**, 343–359.
115. [R] Hegedus, L. S. *Transition Metals in the Synthesis of Complex Organic Molecules* University Science Books: Mill Valley, CA, 1994, pp 245–246.
116. Closser, K. D.; Quintal, M. M.; Shea, K. M. *J. Org. Chem.* **2009,** *74*, 3680–3688.
117. Smit, W. A.; Gybin, A. S.; Shashkov, A. S.; Strychkov, Y. T.; Kyz'mina, L. G.; Mikaelian, G. S.; Caple, R.; Swanson, E. D. *Tetrahedron Lett.* **1986,** *27*, 1241–1244.
118. Rowley, E. G.; Schore, N. E. *J. Org. Chem.* **1992,** *57*, 6853–6861.
119. Takano, S.; Inomata, K.; Ogasawara, K. *J. Chem. Soc., Chem. Commun.* **1992**, 169–170.
120. Bernardes, V.; Kann, N.; Riera, A.; Moyano, A.; Pericas, M. A.; Greene, A. E. *J. Org. Chem.* **1995,** *60*, 6670–6671.
121. Jamison, T. F.; Shambayati, S.; Crowe, W. E.; Schreiber, S. L. *J. Am. Chem. Soc.* **1997,** *119*, 4353–4363.
122. Crawford, J. J.; Kerr, W. J.; McLaughlin, M.; Morrison, A. J.; Pauson, P. L.; Thurston, G. J. *Tetrahedron* **2006,** *62*, 11360–11370.
123. Billington, D. C.; Kerr, W. J.; Pauson, P. L. *J. Organomet. Chem.* **1987,** *328*, 223–227.
124. Caldwell, J. J.; Cameron, I. D.; Christie, S. D. R.; Hay, A. M.; Johnstone, C.; Kerr, W. J.; Murray, A. *Synthesis* **2005**, 3293–3296.
125. Miller, K. A.; Shanahan, C. S.; Martin, S. F. *Tetrahedron* **2008,** *64*, 6884–6900.
126. Honda, T.; Kaneda, K. *J. Org. Chem.* **2007,** *72*, 6541–6547.
127. Min, S.-J.; Danishefsky, S. J. *Angew. Chem., Int. Ed.* **2007,** *46*, 2199–2202.
128. Kaneda, K.; Honda, T. *Tetrahedron* **2008,** *64*, 11589–11593.
129. Kozaka, T.; Miyakoshi, N.; Mukai, C. *J. Org. Chem.* **2007,** *72*, 10147–10154.
130. Dake, G. R.; Fenster, E. E.; Patrick, B. O. *J. Org. Chem.* **2008,** *73*, 6711–6715.
131. [R] Tafesh, A. M.; Weiguny, J. *Chem. Rev. (Washington, D. C.)* **1996,** *96*, 2035–2052.
132. Ragaini, F. *Dalton Trans.* **2009**, 6251–6266.

133. Vazquez-Romero, A.; Rodriguez, J.; Lledo, A.; Verdaguer, X.; Riera, A. *Org. Lett.* **2008,** *10*, 4509–4512.
134. Verdaguer, X.; Lledo, A.; Lopez-Mosquera, C.; Maestro, M. A.; Pericas, M. A.; Riera, A. *J. Org. Chem.* **2004,** *69*, 8053–8061.
135. Verdaguer, X.; Moyano, A.; Pericas, M. A.; Riera, A.; Maestro, M. A.; Mahia, J. *J. Am. Chem. Soc.* **2000,** *122*, 10242–10243.

3.6 Weiss–Cook Reaction

Paul Galatsis

3.6.1 Description

The Weiss–Cook reaction[1] entails the formation of *cis*-bicyclo[3.3.0]octane ring systems from the condensation of 1,2-dicarbonyl compounds with 3-oxoglutarate diester derivatives. Decarboxylation of the immediate reaction product affords access to the parent carbon scaffold.

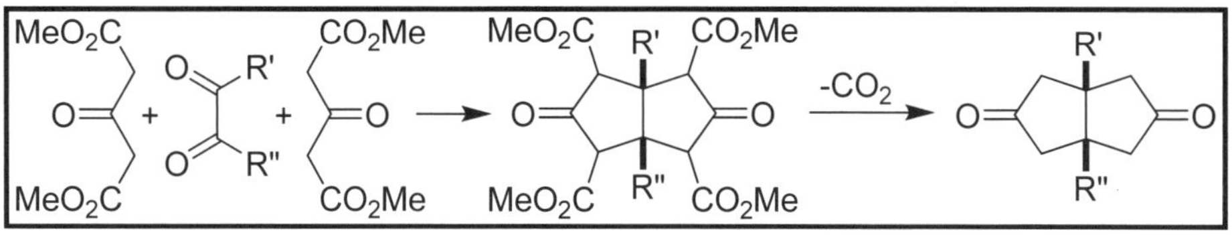

Posner[2] classified this three-component coupling reaction as a 3 + 2 + 3 process, based on the nature of what the reactants contribute to the final product. In addition, the Weiss–Cook reaction has been compared to the Diels–Alder reaction when one appreciates the potential of this reaction to rapidly generate a high level of molecular complexity in a single transformation.[1] This reaction generates four carbon-carbon single bonds and two rings compared to two carbon–carbon single bonds and a single ring for the prototypical Diels–Alder reaction.

3.6.2 Historical Perspective

The bicyclo[3.3.0]octane system had been disclosed in the literature before Weiss. The initial report was by Schroeter in 1922.[3] More definite accounts of this compound were reported later by several groups.[4] These approaches to the bicyclo[3.3.0]octane system suffered from multiple steps with a poor overall yield. Weiss and Edwards, in 1968, published their findings detailing a single-step preparation of this scaffold.[5] Subsequently, Weiss and Cook, working jointly, have elaborated on the mechanism of this reaction and its use in organic synthesis.

3.6.3 Mechanism

From a mechanistic standpoint, the process involved in the bond formation experienced in the Weiss–Cook reaction is a series of Aldol and Michael reactions.[6] This reaction manifold is initiated by an Aldol reaction of oxoglutarate **2** with dicarbonyl **1**. The active species facilitating this bond forming process under acidic conditions is the enol of **2**. Under basic reaction conditions, the corresponding enolate of **2** is the active species. The resultant β-hydroxy carbonyl **3** can undergo an intramolecular Aldol reaction to form the first ring of the bicyclic system found in **6**. Cyclopentenone **5** is produced upon dehydration of **6**. This sets up the system to add the second equivalent of oxoglutarate by a Michael addition to afford **8**. A subsequent dehydration reaction generates **9** that can undergo an additional Michael

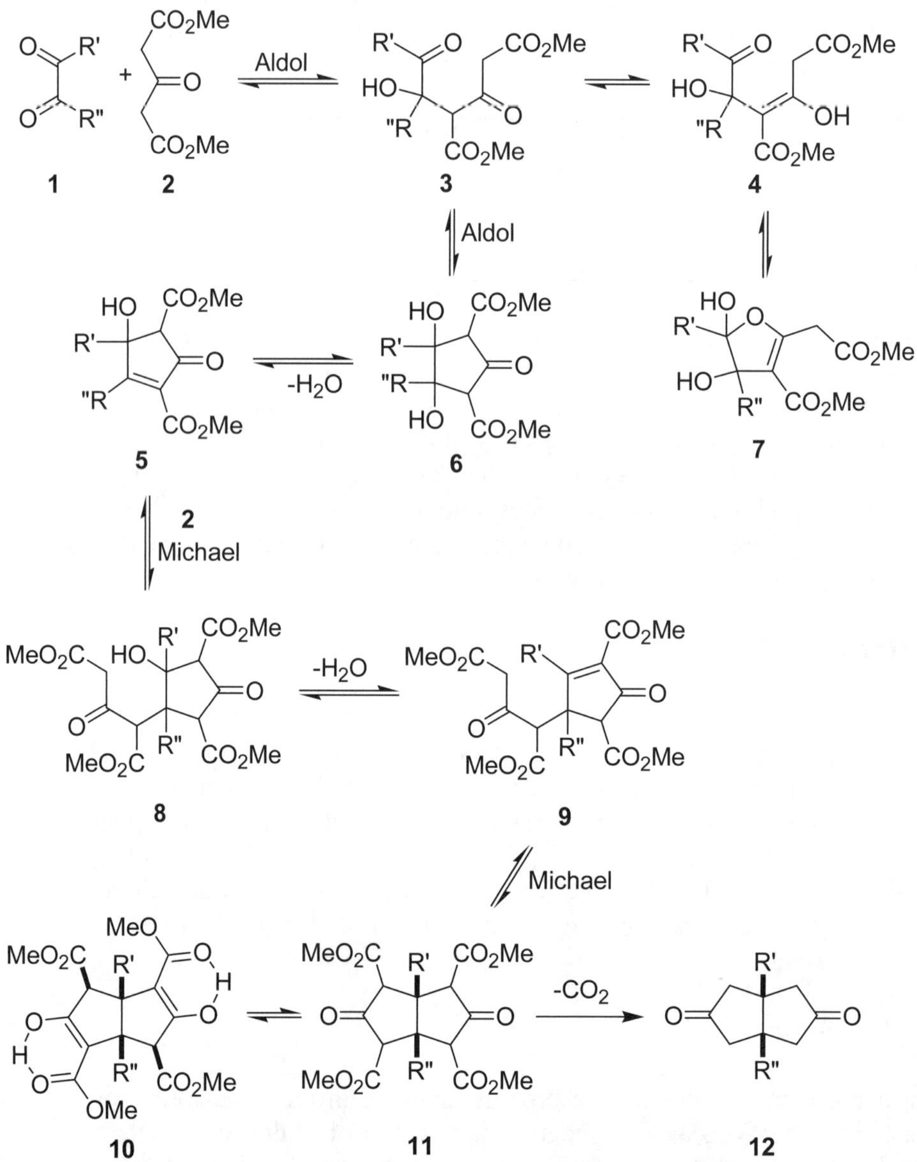

reaction to form **11**, which contains the expected bicyclic scaffold. While most depictions of this product in the literature are as shown in **11**, the compound primarily exists in the *bis*-enol tautomer shown in **10**.[7] Ultimately, decarboxylation of **11** can result in the parent carbon framework **12**. For **11/10** and **12**, *cis* ring fusion is observed as it is the most thermodynamically stable configuration.

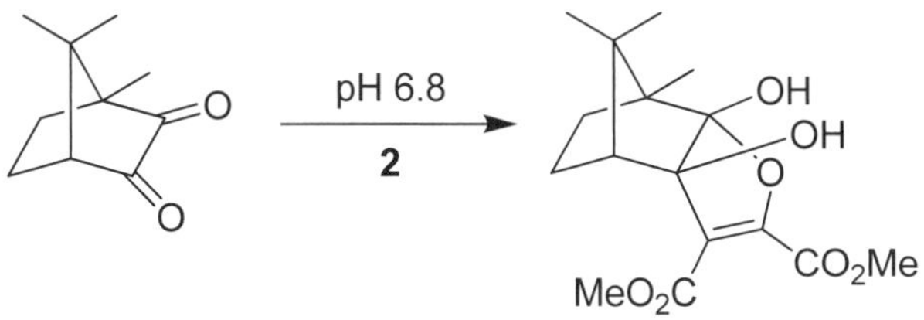

13 **14**

Key to validating this reaction sequence was confirming that cyclopentenone **5** (the 1:1 adduct of **1** and **2**) was an intermediate along the pathway to **11**. An initial branch point for this process was determined when compounds such as camphorquinone **13** was used as component **1**.[6c,8] The steric hindrance associated with diketo **13** resulted in the isolation of **14**. Based on this observation, it was concluded that the initial Aldol adduct **3** could tautomerize to **4**, which could then undergo an intramolecular cyclization to **7**. This intermediate, analogous to **14**, turns out to be the end product for sterically hindered reactants.

The tipping point between steric and electronic effects on the reaction outcome was determined by studying a number of diketo derivatives.[6h,9,10] The product distribution between the 1:1 adduct **5** and the 1:2 adduct **11** was measured for a series of **1** (Table 1). Entries 1–4 show that simple alkyl groups or monosubstituted cyclic derivatives of **1** provide excellent yields of the *bis*-adduct **11**. Similar results were observed for cyclic derivatives of **1** (entries 10–12). The situation changes when derivatives with greater steric requirements are subjected to this reaction (entries 5–9). The cyclopentyl derivative (entry 5) afforded **11** but in poor yield. Homologating by a single carbon to the cyclohexyl deriviative (entry 6) now resulted only in the mono-adduct **5**. Using the phenyl and 2-thienyl derivatives (entries 7 and 8, respectively) also resulted in the formation of **5**. In sharp contrast to these latter results, the 2-furyl derivative (entry 9) afforded only **11**. Thus the sterically more compact 2-furyl moiety appears to provide the appropriate balance of reactivity and steric hindrance to allow reaction with a second equivalent of **2**.

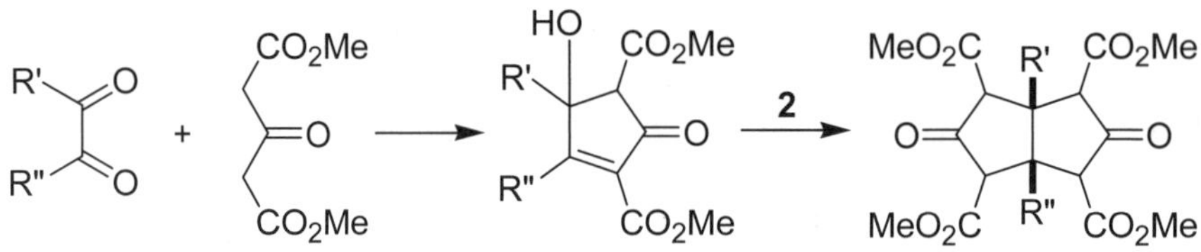

1 2 5 11

Table 1.

Entry	**1** (R′ / R″)	**5** (%)	**11** (%)
1	H / H	0	70
2	Me / H	0	52
3	Ph / H	0	66
4	Ph / Me	0	68
5	cyclopentyl	0	12
6	cyclohexyl	61	0
7	phenyl	70	0
8	2-thienyl	77	0
9	2-furyl	0	60
10	$-(CH_2)_3-$	0	45
11	$-(CH_2)_4-$	0	81
12	$-(CH_2)_6-$	0	80

3.6.4 Variations and Improvements

Turner has described two reaction-type extremes,[11] plateau-type and point-type reactions, based on the reaction yield vs. optimum reaction conditions. While the reaction conditions first disclosed by Weiss were an improvement over the existing methods to assemble the bicyclo[3.3.0] scaffold, the overall yield was low and very sensitive to reaction conditions, i.e., a point-type reaction. To increase the synthetic utility of this chemistry, reaction conditions were investigated to transition it to a plateau-type reaction.

The original preparations[3,4] of the bicyclo[3.3.0]system were multistep processes and not very efficient in terms of overall yield (~ 15%). The initial publication by Weiss and Edwards,[5] while comparable in yield to the previous reports, the efficiency of a multicomponent, single-step process made this variation more attractive. The initial reaction conditions focused on neutral to slightly acidic conditions,[12] which allowed the intermediate tetra-ester **11** to precipitate as the major product from a mixture containing side products **15** and **16**.

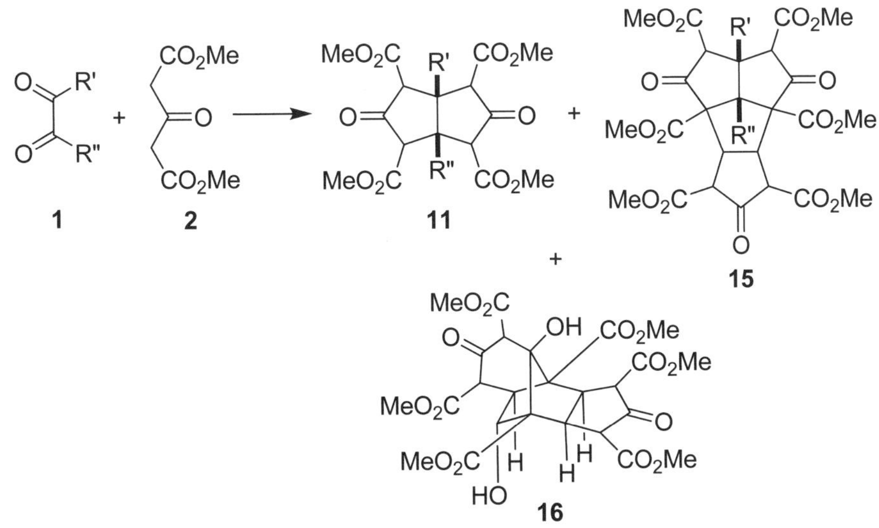

The key to maximize the yield of the desired intermediate **11** was to keep the pH constant. This resulted in a shift of all the equilibria as a consequence of the precipitate formation.

A dramatic improvement in yield was obtained by Bertz and Woodward,[13] when it was determined that basic pH or use of the preformed enolate of **2** in refluxing methanol provided the optimal reaction conditions. A shift from the original 15% yield to more respectable 60–70% yields could be observed and this result could be maintained even upon scale-up of the reaction.

3.6.5 Synthetic Utility

The Weiss–Cook reaction has seen great utility in the total synthesis of non-natural as well as natural products. This has resulted from the ability of the reaction to rapidly assemble a highly functionalized, conformationally constrained carbon scaffold that provides rapid access to advanced intermediates that are readily modified for diverse target synthesis.

During the 1970s and 1980s, the isolation and structure elucidation of a great variety of natural products containing multiple five-membered ring frameworks, the polyquinanes, led to an explosion in approaches to these biologically active targets. Our survey of these approaches begins with the triquinane sesquiterpene isocomene, **22**. Several groups[14] employed the Weiss–Cook reaction for the preparation of **22** and a compilation of the routes began with the reaction of **2** with dicarbonyl compound **17** using slightly acidic conditions of pH 6.8. Hydrolysis of the esters, decarboxylation, and esterification of the pendant acid moiety generated **18**.

Monoprotection of the ketone functionality was accomplished in two steps with *bis*-ketal formation followed by selective monodeprotection to afford **19**. Wolff–Kishner deoxygenation of the free carbonyl, hydrolysis of the ketal, and intramolecular Aldol reaction gave rise to the desired scaffold **20**. Introduction of the remaining methyl groups was accomplished in a stepwise fashion. The second quarternary methyl was added by methylation of the enolate of **20**. Selective mono-ketal formation of the angular cyclopentanone allowed Wittig olefination of the linear cyclopentanone. Hydrolysis of the ketal also resulted in isomerisation of the exocyclic alkene to produce **21**. The ketone was converted to the corresponding enone upon selenoxide elimination. Exposure to dimethyl cuprate resulted in the Michael addition of the final methyl group. The synthesis was completed by Wolff–Kishner deoxygenation to afford **22**.

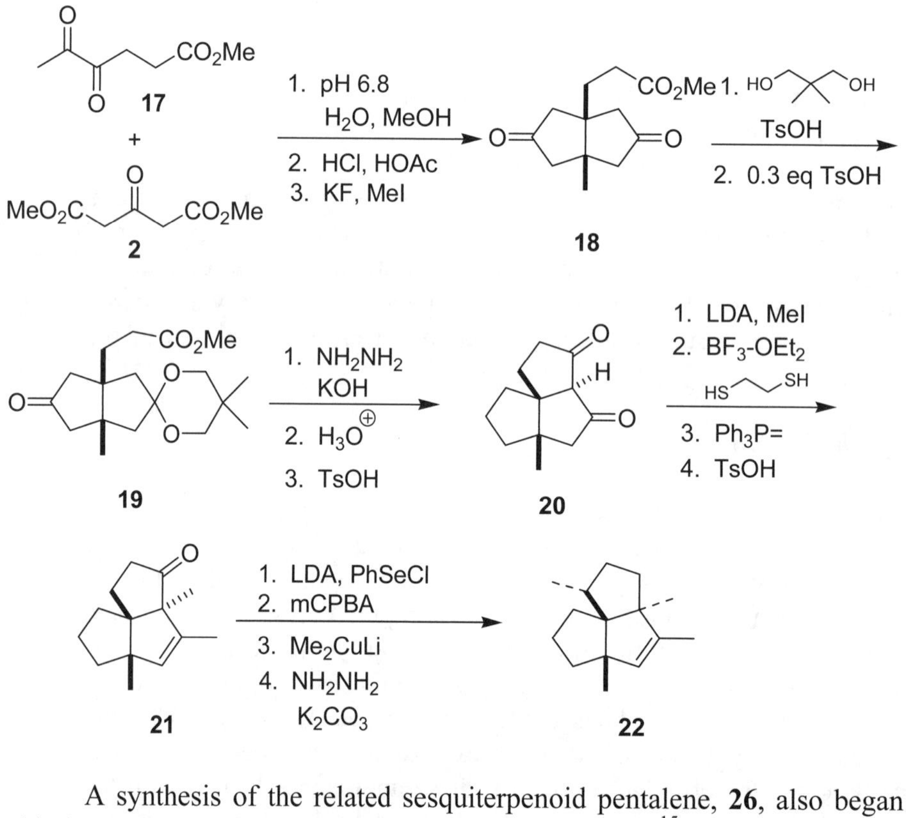

A synthesis of the related sesquiterpenoid pentalene, **26**, also began with the Weiss–Cook reaction product **12** (R′ = R″ = H).[15] Wittig olefination was carried out on the mono-ketal of **12** followed by hydrolysis of the ketal. The resultant ketone was reduced to the corresponding alcohol and the exocyclic alkene underwent cyclopropanation to produce **23**. Oxidation of the alcohol generated a ketone that could be converted to the enone using the conditions of Saegusa. The remaining carbon atoms for the angular cyclopentane ring were now introduced using a cuprate reagent generated for cyclopentane annulation. Based induce ring-closure afforded the advanced intermediate **25** that was eventually transformed into **26**.

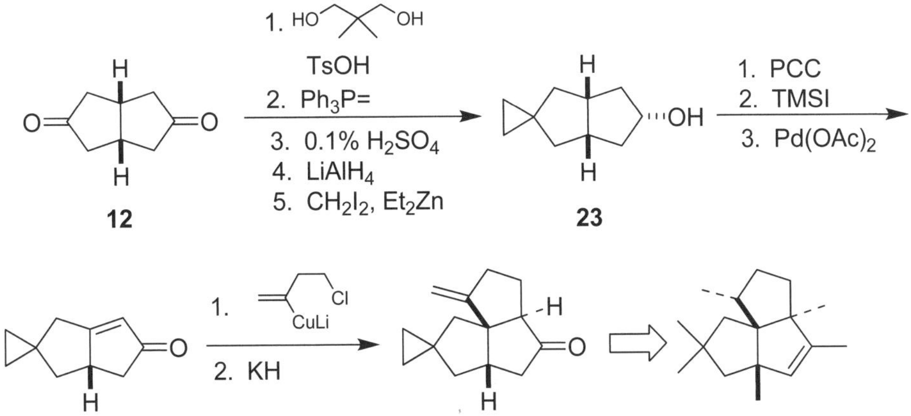

12 **23** **24** **25** **26**

Modhephene, **34**, was the first isolated propellane natural product. As such, the Weiss–Cook reaction was the perfect method for its construction.[16] The process began with the condensation of **2** with diketone **27**. Standard conditions for decarboxylation produced the core scaffold **28**. Hydrogenation of the mono-enol phosphate afforded the monoketone **29**. The cyclopropyl derivative **30** was prepared by copper-catalyzed decomposition of a diazoketone. *gem*-Dimethylation to generate **31** preceded carboxylation and esterification to afford the advanced intermediate **32**. Cuprate-induced cyclopropane ring opening and methylation of the β-ketoester introduced the final carbon atoms giving rise to **33**. Lithium iodide induced decarboxylation preceded reduction of the ketone followed by dehydration with Martin's sulfurane, thus producing **34**.

27

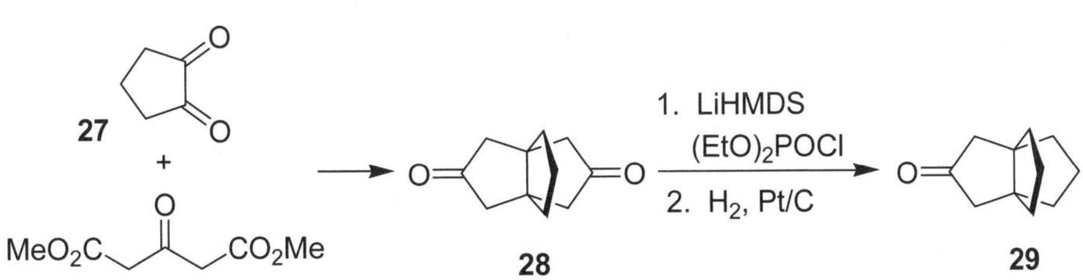

2 **28** **29**

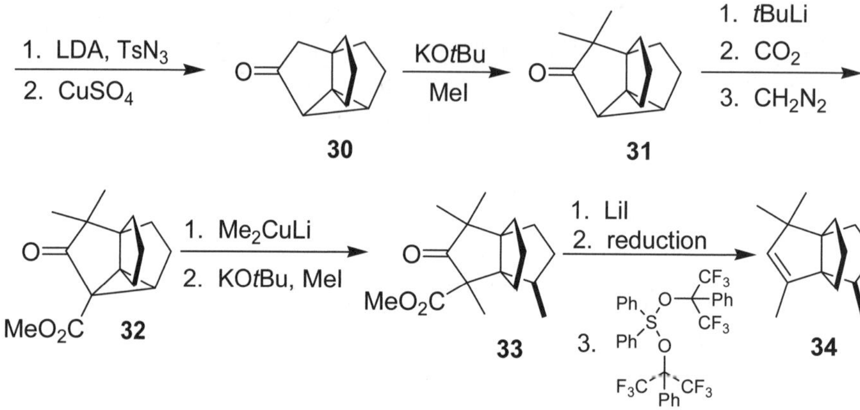

The cytotoxic sesquiterpenoid quadrone, **38**, has an embedded diquinane carbon framework that allowed the Piers's group to leverage the chemistry of the Weiss–Cook reaction.[17] The monoketal **35** is readily available from **12** (R′ = R″ = H). Several transformations converted this compound into the vinyl cyclopropane **36**. Thermal rearrangement of the scaffold employing a Cope process afforded tricyclic **37**. This compound was readily elaborated into **38**.

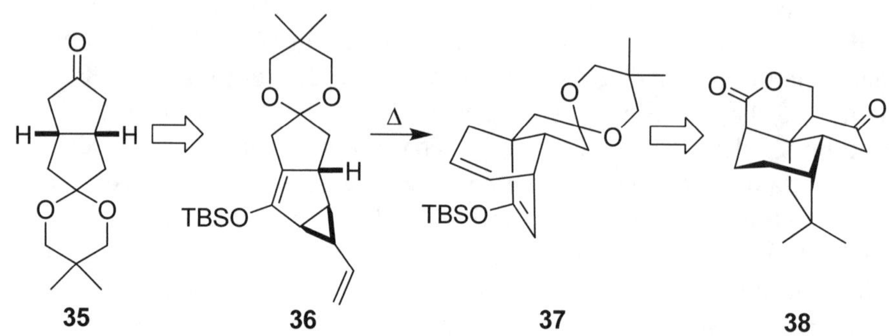

Gymnomitrol, **39**, is another sesquiterpenoid that encompasses an embedded diquinane scaffold. This becomes readily apparent when one examines the alternate representation for the compound. Again the use of the output from the Weiss–Cook reaction was found to be a rapid entry to this system in which two groups employed **40** as their foundation for the synthesis of **39**.[18] As we have seen, **40** is easily obtained from the reaction of **1** (R′ = R″ = Me) and **2**.

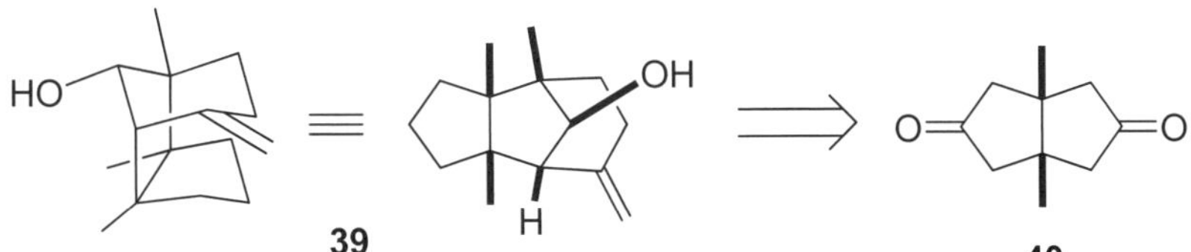

39 40

Carboprostacyclin, **43**, a stable and biologically active analogue of prostacyclin (PGI_2) provides an additional example of the utility of the Weiss–Cook reaction.[19] The Weiss–Cook reaction product **41** was transformed into the advanced intermediate **42**. A Wadsworth–Horner–Emmons reaction provided a method to append the alcohol sidechain to the aldehyde moiety and a Wittig reaction, on the deprotected ketone, introduced the carboxy sidechain for the desired target **43**.

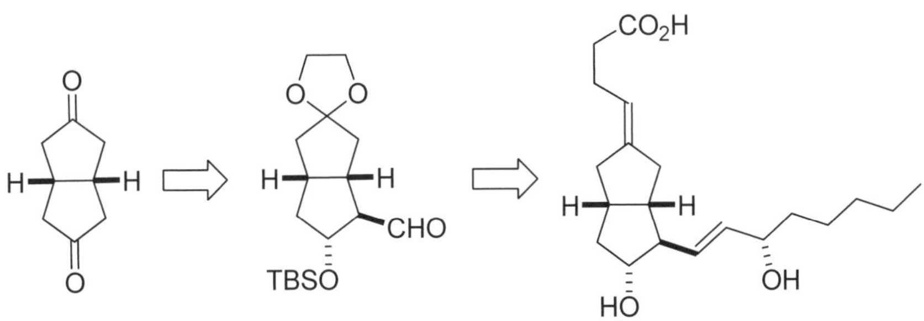

41 42 43

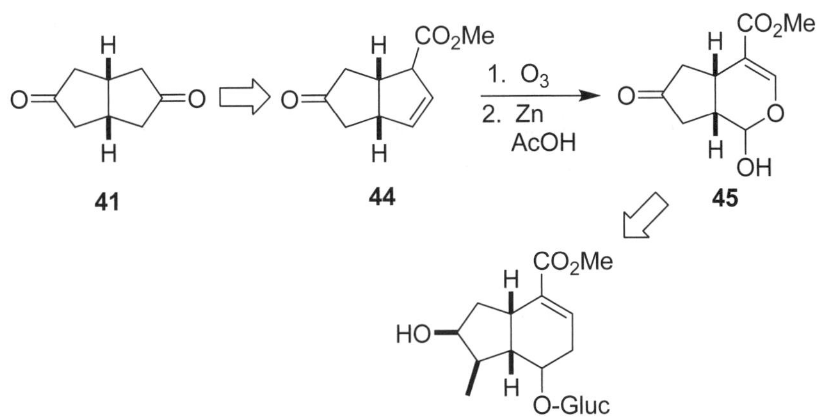

41 44 45

46

The monoterpene glucoside loganin, **46**, is a key intermediate in the biosynthesis of several alkaloid families. The *cis*-fused, bicyclic framework, **41**, derived from the Weiss–Cook reaction was exploited for two of its attributes. One of the rings was fragmented as a precursor to the dihydropyran ring, and the topology of this scaffold was used to control stereochemistry.[20] To this end, **41** was elaborated into **44**, which set the stage for ring fragmentation. Ozonolysis with a reductive work-up generated **45**. This compound embodied a significant amount of the structure of **46** and was easily converted into this desired compound.

The final example, from the natural product category, is that of bifurcarenone **47**. This compound is an inhibitor of mitotic cell division and structurally consists of an unprecedented monocyclic diterpenoid moiety in combination with a hydroquinone scaffold. Retrosynthetically, it was envisioned to arise from the coupling of the three fragments **48**, **49**, and **50**.

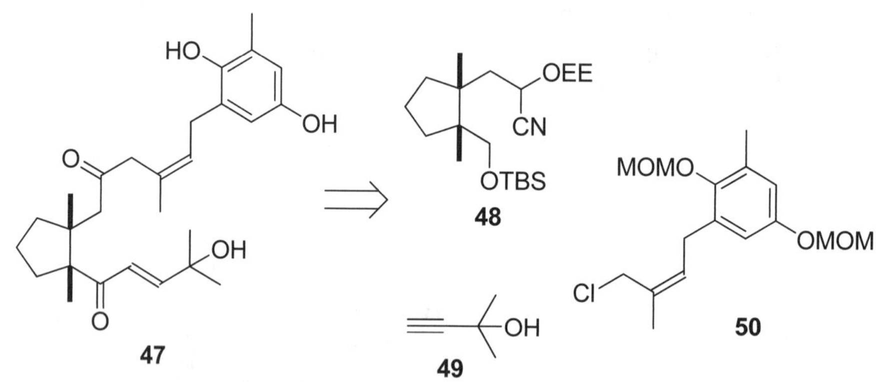

The Weiss–Cook reaction was used to generate the starting material for compound **48**.[21] The advanced intermediate **51** was readily obtained from the reaction of **2** with **1** (R′ = R″ = Me). Baeyer–Villiger reaction followed by reduction afforded lactol **52**. Aldehyde **53** was prepared by ketalization of the latent aldehyde, silylation of the pendant alcohol, and hydrolysis of the ketal back to the aldehyde. The synthesis of **48** was completed by protection of the cyanohydrin of the aldehyde present in **53**.

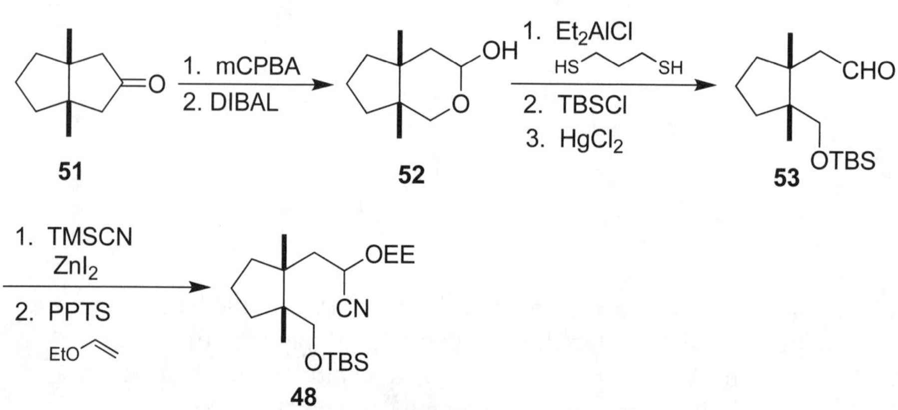

As we have summarized for natural product total synthesis, the Weiss–Cook reaction has enjoyed equal utility in the generation of nonnatural products. This exemplification begins with an overview of the semibullvalene system (Figure 1). The tricyclo[3.3.0.0^{2,8}]octane scaffold is structurally configured to undergo a facile, degenerate Cope rearrangement with a very low energy barrier. The barrier for this interconversion has been calculated to be as low as 3.6 kcal/mol for the parent system.[22]

Figure 1

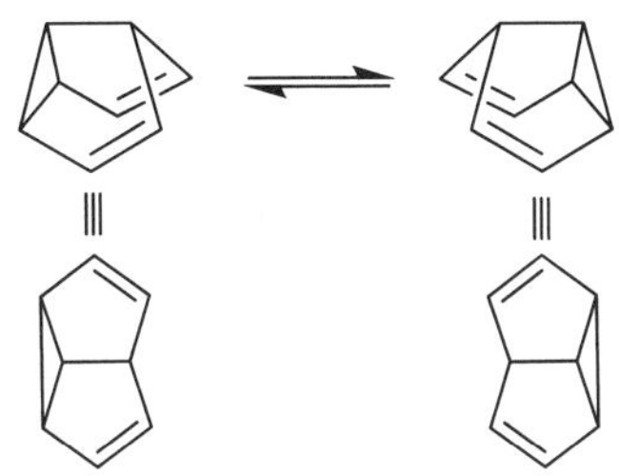

Upon examining the structure of semibullvalene, shown in Figure 1, it becomes obvious how the Weiss–Cook reaction could be implemented in the construction of this framework. The preparation[23] of the substituted semibullvalene **55** began with the Weiss–Cook diketone **11** (R′ = R″ = Me). Bromination followed by cyclizing dehydrobromination gave rise to **54**. Reduction of the ketones and conversion to the corresponding mesylates preceded a conjugate 1,4-elimination reaction to afford the substituted derivative **55**.

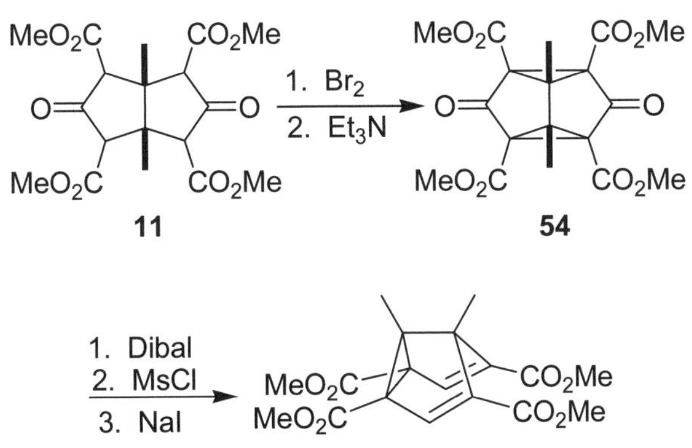

55

A related synthesis[24] converted **12** (R′ = R″ = Me) to the cyanohydrin. This intermediate was dehydrated with phosphorous oxychloride, extensively brominated under radical reaction conditions, and finally zinc–copper couple mediated debromination to afford variant **56**.

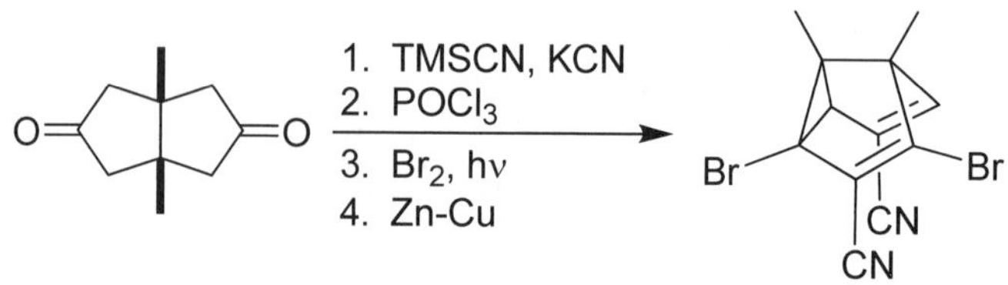

12 56

Triquinacene, **57**, has been a great source of interest for many years due to the potential for it to dimerize to dodecahedrane **58**. This compound was first synthesized by Woodward in 1964 and was the target of several groups.[25] In addition, the potential for the three alkene moieties to participate in conjugation in the form of neutral homoaromaticity led to an entire workstream of several independent laboratories to confirm this phenomenon of theoretical interest.[26]

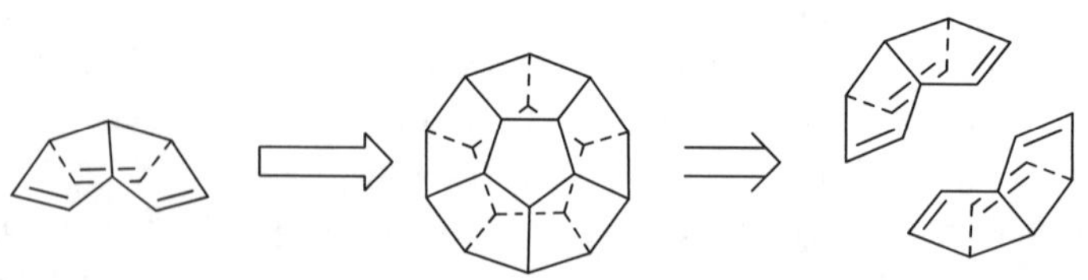

57 58

A more recent approach to **57** began with the Weiss–Cook reaction product **59**.[27] Trapping the enol tautomer of this compound with diazomethane preceded alkylation with allyl iodide and hydrolysis/decarboxylation to generate **60**. The final sequence of reactions involved ozonolysis, aldol cyclization, reduction, and dehydration afforded **57**.

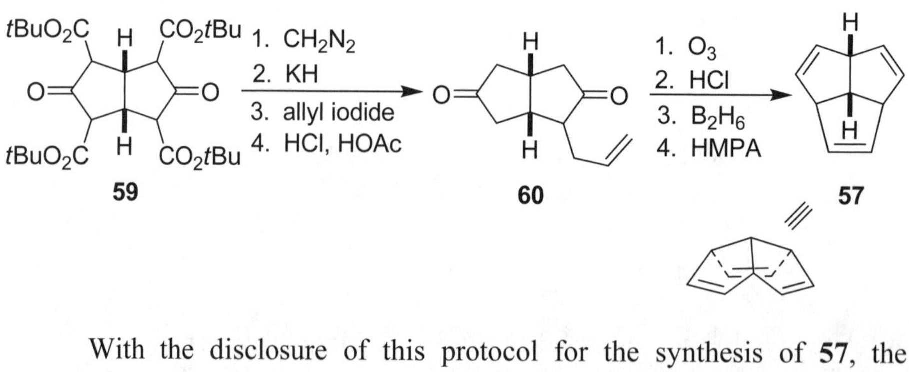

59 60 57

With the disclosure of this protocol for the synthesis of **57**, the opportunity to rapidly generate triquinacene analogs became viable. One of the major side reactions in the attempted dimerization of **57** is that the requisite concave to concave orientation could not be guaranteed. A propellane variant of **57** was prepared to improve the probability of the correct orientation for the photodimerization toward **58**.[28] The approach began with the Weiss–Cook reaction of **61** with **62** with diazomethane enol ether formation produced an intermediate that could intercept the chemistry from the preceeding synthesis. Allylation followed by hydrolysis-decarboxylation afforded **63**. The derivatized triquinacene **64** was produced after ozonolysis, Aldol reaction, reduction, and dehydration.

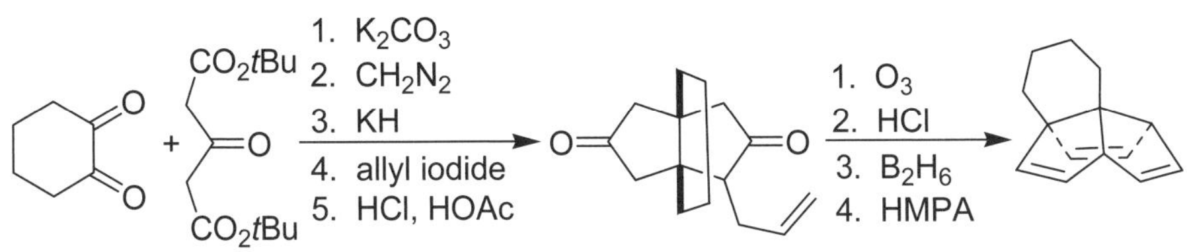

61 **62** **63** **64**

Using a similar strategic approach,[29] ellacene (1,10-decano-triquinacene) **65** has also been prepared. Photochemical and/or high pressure (130 kbar) failed to effect dimerization of this compound to the desired dodecahedrane derivative **66**.

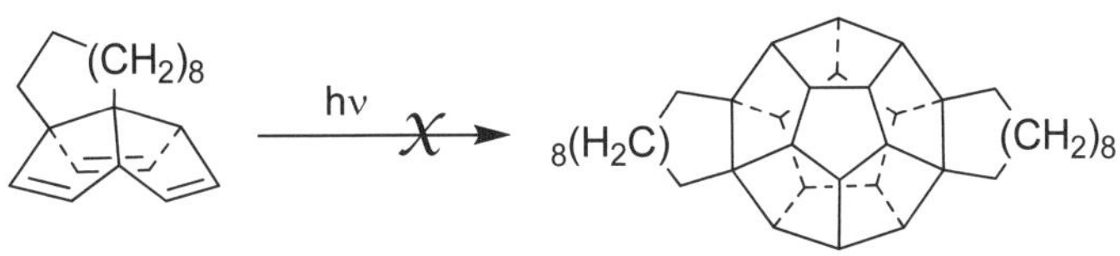

65 **66**

The tetrahedral nature of tetracoordinate carbon is well ingrained into practitioners of organic chemistry. However, Hoffmann has provided a theoretical analysis for stabilizing a tetracoordinate planar carbon.[30] Access to these sorts of compounds was first made possible with the synthesis of stauranes. This name was coined by Cook and Weiss and is derived from the Greek *stauros* meaning "cross".[31] The first member of this family of compounds was **72**. Weiss–Cook reaction of glyoxal **67** with **2** produced **68** after hydrolysis and decarboxylation. Acid-catalyzed Aldol reaction converted **68** into the tetraketo staurane **69**. Diborane reduction and dehydration by heating in HMPA afforded the desired target **72**. Alternatively, diacid **68** could also be obtained from glyoxal **70**.[32] The Weiss–Cook reaction product **71** was elaborated into **68** by ozonolysis of the cyclopentenyl moiety followed by oxidation to the diacid.

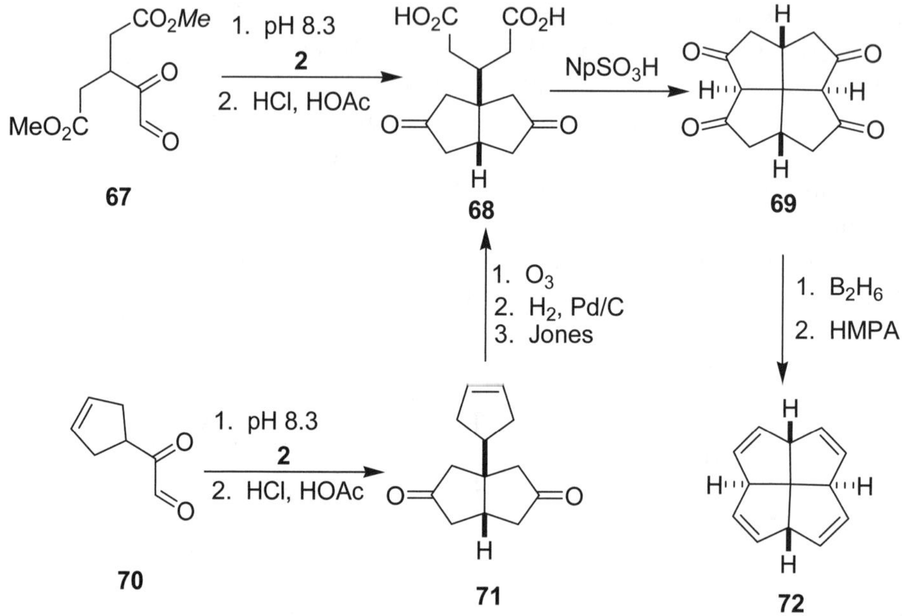

The versatility of the Weiss–Cook reaction was leveraged in the synthesis of an isomer of **72**.[33] To this end, reaction of **73** with **2** produced **74** after hydrolysis/decarboxylation. Ozonolytic cleavage of the cyclooctene moiety resulted in the formation of *bis*-acetal **74** which was able to undergo an acid-catalyzed Aldol reaction to tetracycle **75**. Reduction and dehydration converted **75** into the *bis*-angularly fused polyquinene **76**.

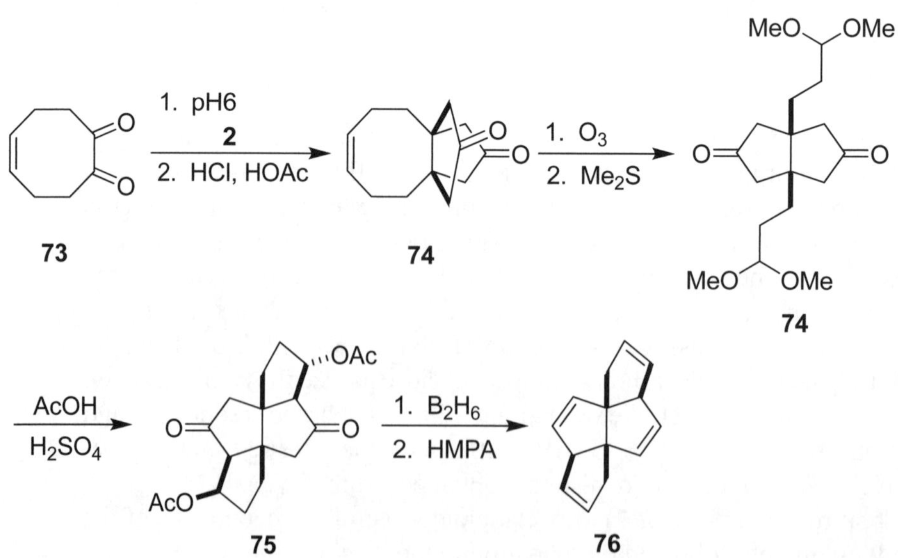

There are many more examples of the utility of the Weiss–Cook reaction in synthesis but this limited survey should provide an indication of what has been disclosed and spark new ideas for additional examples.

3.6.6 Experimental

While the primary literature provides multiple examples of how the Weiss–Cook reaction can be conducted, general reaction conditions that are able to cover a diverse set of glyoxals has been report with yield ranging from 12 to 94%.[34]

cis-1,5-Dimethylbicyclo[3.3.0]octane-3,7-dione

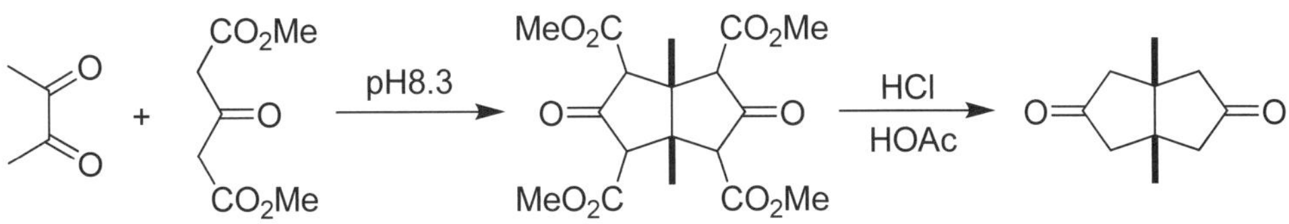

1 **2** **11** **12**

To a solution of $NaHCO_3$ (5.6 g) in 400 mL of water (final pH 8.3) was added dimethyl 1,3-acetonedicarboxylate **2** (70 g, 0.40 mol). Biacetyl **1** (R′ = R″ = Me) (17.2 g, 0.20 mol) was added to the rapidly stirred solution. Over the course of 24 h, a white solid precipitated and was collected by suction filtration. Additional material could be isolated by processing the filtrate. Recrystallization from methanol afforded 58–60 g (73–75%) of tetraester **11** (R′ = R″ = Me).

Tetraester **11** (R′ = R″ = Me, 24 g, 0.060 mol) was refluxed in 200 mL 1 M hydrochloric acid and 40 mL glacial acetic acid for 3–6 h. The reaction mixture was cooled in an ice bath and the solid precipitate collected by suction filtration. Recrystallization from ethanol afforded 7.5–7.7 g (75–77%) diketone **12** (R′ = R″ = Me).

3.6.7 References

1. [R] (a) Mitschka, R.; Oehldrich, J; Takahashi, K.; Cook, J. M.; Weiss, U.; Silverton, J. V. *Tetrahedron* **1981**, *37*, 4521. (b) *Org. Synth. Highlights* **1991**, *121*, 121. (c) Gupta, A. K.; Fu, X.; Snyder, J. P.; Cook, J. M. *Tetrahedron* **1991**, *47*, 3665. (d) Fu, X.; Cook, J. M. *Aldrichimica Acta* **1992**, *25*, 43.
2. [R] Posner, G. H. *Chem. Rev.* **1986**, *86*, 831.
3. Schroeter, G. *Ann.* **1922**, *426*, 1.
4. (a) Paul, H. *Chem. Techn.* **1956**, *8*, 189. (b) Tanaka, S. *J. Am. Chem. Soc.* **1958**, *80*, 5264. (c) Yates, P.; Hand, E. S.; French, G. B. *J. Am. Chem. Soc.* **1960**, *82*, 6347.
5. Weiss, U.; Edwards, J. M. *Tetrahedron Lett.* **1968**, 4885.

6. (a) Edwards, J. M.; Qureshi, I. H.; Weiss, U.; Akiyama, T.; Silverton, J. V. *J Org. Chem.* **1973**, *38*, 2919. (b) Yang, S.; Cook, J. M. *J. Org. Chem.* **1976**, *41*, 1903. (c) Yang-Lan, S. Mueller-Johnson, M.; Oehldrich, J.; Wichman, D.; Cook, J. M. *J. Org. Chem.* **1976**, *41*, 4053. (d) Bertz, S. H.; Adams, W. O.; Silverton, J. V. *J. Org. Chem.* **1981**, *46*, 2830. (e) Kubiak, G.; Cook, J. M.; Weis, U. *J. Org. Chem.* **1984**, *49*, 561. (f) Quast, H.; Roschert, H.; Peters, E-M.; Peters, K.; Schnering, H. G. *Chem. Ber.* **1989**, *122*, 523. (g) Deslongchamps, G.; Mink, D.; Boyle, P.D.; Singh, N. *Can. J. Chem.* **1994**, *72*, 1162. h) Van Ornum, S. G. Li, J.; Kubiak, G. G.; Cook, J. M. *J. Chem. Soc., Perkin Trans. I* **1997**, 3471.
7. Camps, P. *Tetrahedron Lett.* **1974**, 4067.
8. Campos, O.; Cook, J. M. *J. Heterocyclic Chem.* **1977**, *14*, 711.
9. Avasthi, K.; Desphande, M. N.; Han, W-C.; Cook, J. M.; Weiss, U. *Tetrahedron Lett.* **1981**, *22*, 3475.
10. Kubiak, G.; Cook, J. M.; Weiss, U. *Tetrahedron Lett.* **1985**, *26*, 2163.
11. Brewster, D.; Myers, M.; Ormerod, J.; Otter, J.; Smith, A. C. B.; Spinner, M. E.; Turner, S. *J. Chem. Soc., Perkin Trans. I* **1973**, 2796.
12. (a) Rice, K. C.; Sharpless, N. E.; Weiss, U.; Highet, R. J. *Tetrahedron Lett.* **1975**, 3763. (b) Rice, K. C.; Weiss, U. Akiyama, T.; Highet, R. J.; Lee, T.; Silverton, J. V. *Tetrahedron Lett.* **1975**, 3767.
13. (a) Bertz, S. H.; Adams, W. O.; Silverton, J. V. *J. Org. Chem.* **1981**, *46*, 2830. (b) Bertz, S. H.; Rihs, G.; Woodward, R. B. *Tetrahedron* **1982**, *38*, 63. (c) Bertz, S. H. *J. Org. Chem.* **1985**, *50*, 3585.
14. (a) Oehldrich, J.; Cook, J. M.; Weiss, U. *Tetrahedron Lett.* **1976**, 4549. (b) Dauben, W. G.; Walker, D. M. *J. Org. Chem.* **1981**, *46*, 1103. (c) Paquette, L. A.; Han, Y.-K. *J. Am. Chem. Soc.* **1981**, *103*, 1835.
15. a) Piers, E.; Karunaratne, V. *J. Chem. Soc., Chem. Commun.* **1984**, 959. b) Piers, E.; Karunaratne, V. *Can. J. Chem.* **1989**, *67*, 160.
16. Wrobel, J.; Takahashi, K.; Honkan, V.; Lannoye, G.; Cook, J. M.; Bertz, S. H. *J. Org. Chem.* **1983**, *48*, 141.
17. Piers, E.; Moss, N. *Tetrahedron Lett.* **1985**, *26*, 2735.
18. (a) Coates, R. M.; Shah, S. K.; Mason, R. W. *J. Am. Chem. Soc.* **1979**, *101*, 6765. (b) Han, Y-K.; Paquette, L. A. *J. Org. Chem.* **1979**, *44*, 3731. (c) Paquette, L. A.; Han, Y-K. *J. Am. Chem. Soc.* **1981**, *103*, 1831.
19. Nicolaou, K. C.; Sipio, W. J.; Magolda, R. L.; Seitz, S.; Barnette, W. E. *J. Chem. Soc., Chem. Commun.* **1978**, 1067.
20. Caille, J. C.; Bellamy, F.; Guilard, R. *Tetrahedron Lett.* **1984**, *25*, 2345.
21. Mori, K.; Uno, T. *Tetrahedron* **1989**, *45*, 1945.
22. (a) Dewar, M. J. S.; Lo, D. H. *J. Am. Chem. Soc.* **1971**, *93*, 7201. (b) Cheng, A. K.; Anet, F. A. L.; Mioduski, J.; Meinwald, J. *J. Am. Chem. Soc.* **1974**, *96*, 2887.
23. Miller, L. S.; Grohmann, K.; Dannenberg, J. J.; Todaro, L. *J. Am. Chem. Soc.* **1981**, *103*, 6249.
24. (a) Quast, H.; Christ, J.; Gorlach, Y.; von der Saal, W. *Tetrahedron Lett.* **1982**, *23*, 3653. (b) Quast, H.; Gorlach, Y.; Meichsner, G. *Tetrahedron Lett.* **1982**, *23*, 4677.
25. Woodward, R. B.; Fukunaga, T.; Kelly, R. C. *J. Am. Chem. Soc.* **1964**, *86*, 3162.
26. Williams, R. V. *Chem. Rev.* **2001**, *101*, 1185.
27. Bertz, S. H.; Lannoye, G.; Cook, J. M. *Tetrahedron Lett.* **1985**, *26*, 4695.
28. Gupta, A. K.; Cook, J. M.; Weiss, U. *Tetrahedron Lett.* **1988**, *29*, 2535.
29. Fu, X.; Cook, J. M. *J. Org. Chem.* **1992**, *57*, 5121.
30. (a) Hoffmann, R.; Alder, R. W.; Wilcox, C. F., Jr. *J. Am. Chem. Soc.* **1970**, *92*, 4992. (b) [R] Hoffmann, R. *Pure Appl. Chem.* **1971**, *28*, 181.
31. Mitschka, R.; Cook, J. M.; Weiss, U. *J. Am. Chem. Soc.* **1978**, *100*, 3973.
32. Deshpande, M. N.; Jawdosiuk, M.; Kubiak, G.; Venkatachalam, M.; Weiss, U.; Cook, J. M. *J. Am. Chem. Soc.* **1985**, *107*, 4786.
33. Venkatachalam, M.; Jawdosiuk, M.; Deshpande, M. N.; Cook, J. M. *Tetrahedron Lett.* **1985**, *26*, 2275.
34. Bertz, S. H.; Cook, J. M.; Gawish, A.; Weiss, U. *Org. Syn.* **1986**, *64*, 27.

Chapter 4 Six-Membered Carbocycles 197

4.1 Bardhan–Sengupta Phenanthrene Synthesis

Timothy T. Curran

4.1.1 Description

The Bardhan–Sengupta phenanthrene synthesis is a two-step process. The first step is reaction of a hydroxy-substituted phenyl alkane **1** or phenyl-substituted alkene **2** with a phosphorus-containing acid or dehydrating agent, *i.e.*, H_3PO_4, P_2O_5 or polyphosphoric acid (PPA) to generate a carbocationic species, which then undergoes electrophilic aromatic substitution (EAS) and proton elimination to provide the corresponding octahydrophenanthrene **4**. Subsequent dehydrogenation of **4** then provides the phenanthrene ring system **5**.

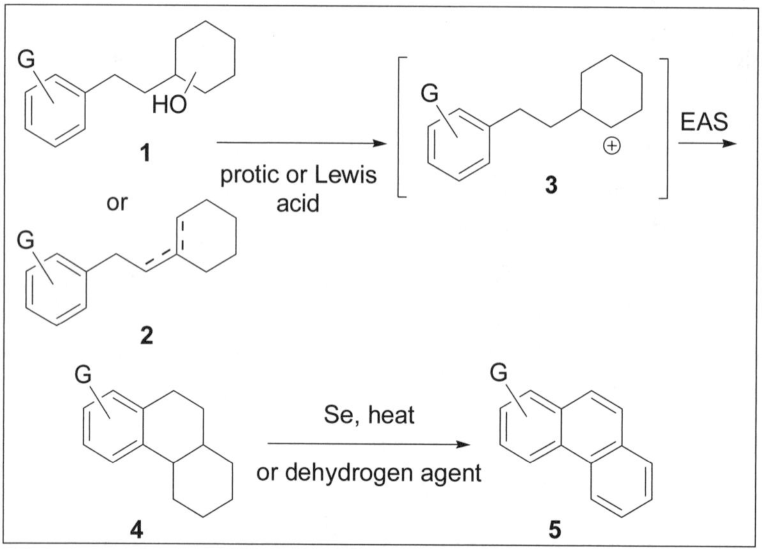

In addition, due to the cationic nature of the cyclization step, there can be an array of byproducts generated depending on the stability of the cation and neighboring groups (which are part of the cyclizing substrate), which may affect the cyclization product. The dehydrogenation was initially done with elemental selenium and heat (~ 300 °C). More mild conditions using a variety of metals have been developed since the initial inception.

4.1.2 Historical Perspective

In 1932, Bardhan and Sengupta[1] reported the use of P_2O_5 to cyclize hydroxy ester **6** that, upon treatment of the crude product with Se metal and heat (300–320 °C), provided **8** presumably through an intermediate like compound **7** (R = H or Et). In an additional publication, the authors provided more experimental detail and examples on the formation of substituted phenanthrenes.[2]

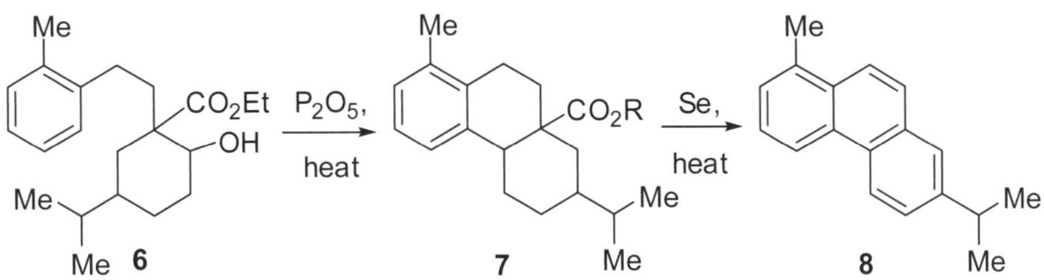

Initial controversy arose when it was suggested that the Se dehydrogenation step provided methyl scrambling (migration) of alkyl groups on the aromatic ring. This notion was put to rest by experimentation in other groups.[3] Like the Bogert–Cook reaction (see Section 4.3), for the cyclization step, the Bardhan–Sengupta cyclization conditions also reportedly formed spirane products. Electron-donating groups (EDG) on the aromatic ring were reported to increase the formation of the spirane.[4] Barnes suggested that intermediates formed during the cyclization containing a neighboring carboxylate group, mitigated spirane formation due to formation of dication **10**.[5–7] Also, for these substrates in which dication formation was suggested, electron-rich aromatics were required for the cyclization to proceed in good yield.

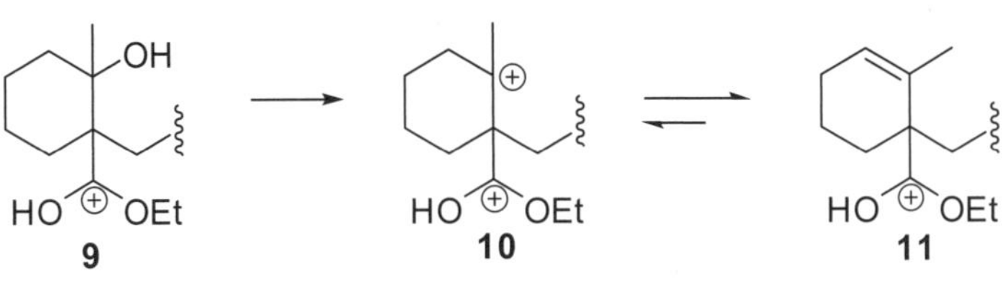

Much like the work conducted by the Bogert and Cook groups, the Bardhan–Sengupta synthesis of phenanthrenes helped lead to the determination of the carbon framework for steroids and was used to prepare terpene-type natural products.

4.1.3 Mechanism

The proposed mechanism of the Bardhan–Sengupta cyclization is similar to that reported for the cyclization by Bogert and Cook.[3,7] Although, there are instances in which the phosphorous containing acid or dehydrating agent appears to be more mild or selective than reported for the H_2SO_4 promoted cyclizations (vida infra), the conformational and nonbonding steric interactions governing the cyclization under acid-promoted conditions remain.

Activation of the alcohol for leaving via protonation or phosphorylation[8] and subsequent leaving would generate carbocation **14**. The carbocation can suffer alkene formation delivering **15**, hydrogen or alkyl 1,2-shift (not shown), or electrophilic aromatic substitution providing the desired cyclized product **16**. It is important to remember that **15** can be protonated by phosphoric acid, getting one back to a carbocationic species, which can subsequently undergo productive cyclization. Aromatization then provides the octahydrophenanthrene **17**.

The subsequent dehydrogenation, which historically employed Se metal and heat, has not been well studied or characterized.[9] The supporting data suggest that dehydrogenation takes place by first complexation of Se-Se with an allyl type system, followed by attachment to the allylic or benzylic carbon atom providing **18**. This bond-forming step could take place by radical formation on Se, subsequent formation of a benzyl radical, followed by addition of the radical to Se_2. At the elevated temperature, the authors propose a radical mechanism breaking the Se–Se bond in compound **18**, generating HSe• radical and the alkyl–selenyl radical **19**. Both **19** and HSe• can then propagate the radical sequence and generate benzyl radical like **20** and alkylselenol **21**. Elimination of H_2Se provides alkene **22**. A continued combination of electrophilic Se addition/elimination and radical dehydrogenation/substitution–elimination sequences may then take place resulting in fully aromatized material **23**. The amount of H_2Se released is not stoichiometric with the dehyrogenations that occur or with the amount of Se spent; therefore, there are multiple reaction pathways leading to ring unsaturation. The mechanism is likely not to be reinvestigated due to the toxicity of Se metal and the development of alternative dehydrogenating conditions, which occur at lower temperature and are less destructive to thermally-sensitive materials.

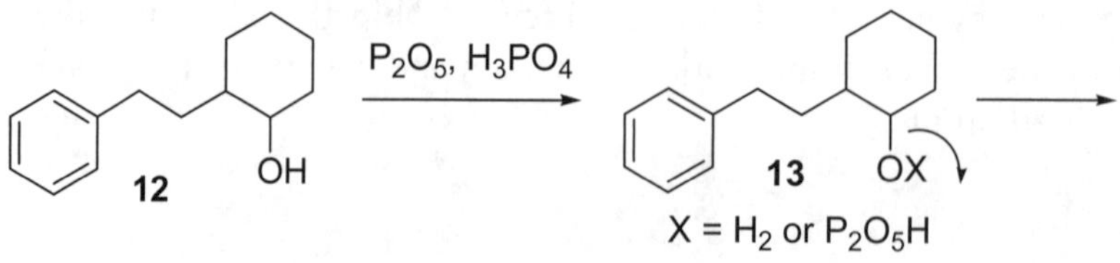

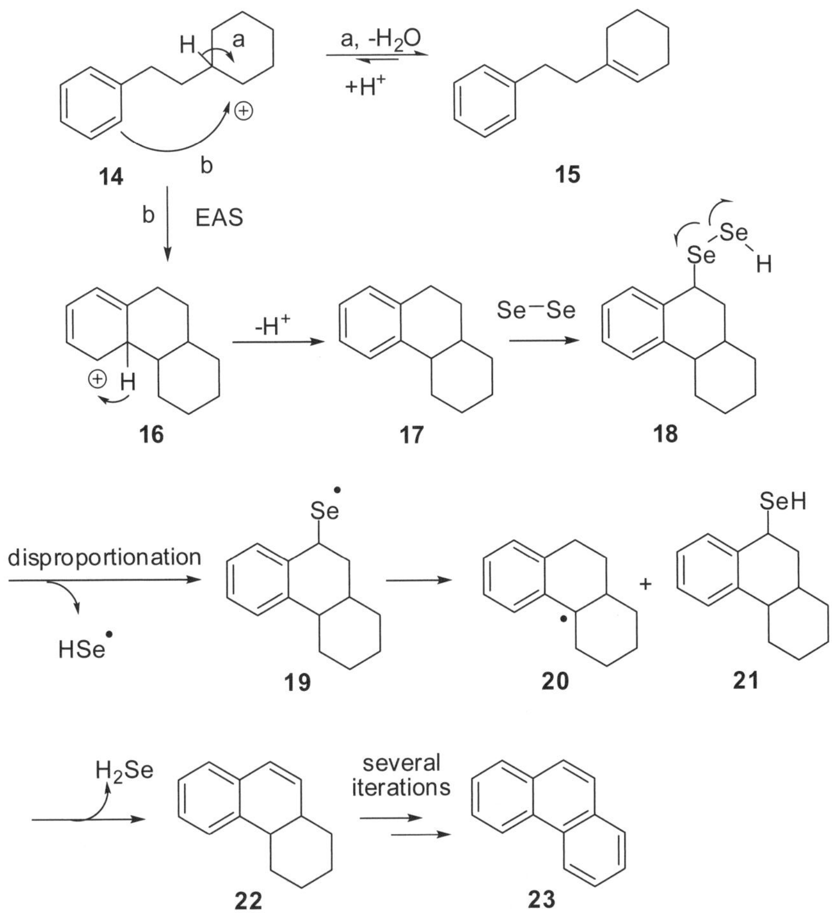

As with the Bogert–Cook cyclization, there have been three possible intermediates proposed as the active ring-forming intermediate in the Bardhan–Sengupta cyclization.[4,7] Specifically, the activated hydroxyl group **24**, the carbocation **25** (which, depending on the substitution, could be a mixture of carbocations with **26** potentially leading to the spirocyclic compound), and the bridged intermediate **27**.

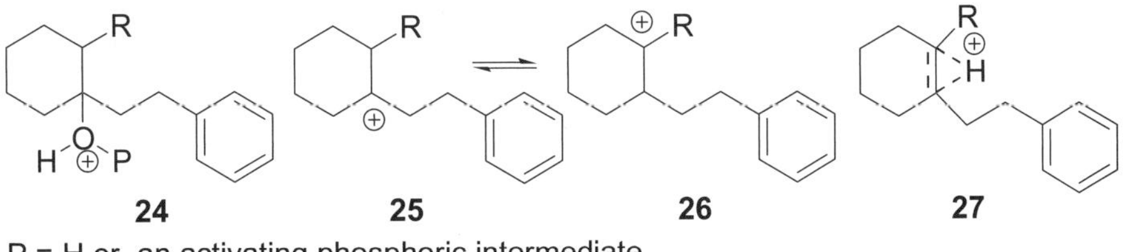

P = H or an activating phosphoric intermediate

Spirane formation was observed when electron-donating groups (EDG) were attached to the aromatic ring and the carbocation formed at the carbon bearing the phenethyl group. For example, cyclizaton of **28** provided **29** in lower yield than **30** was converted to **29** due to the formation of the spirane.[4]

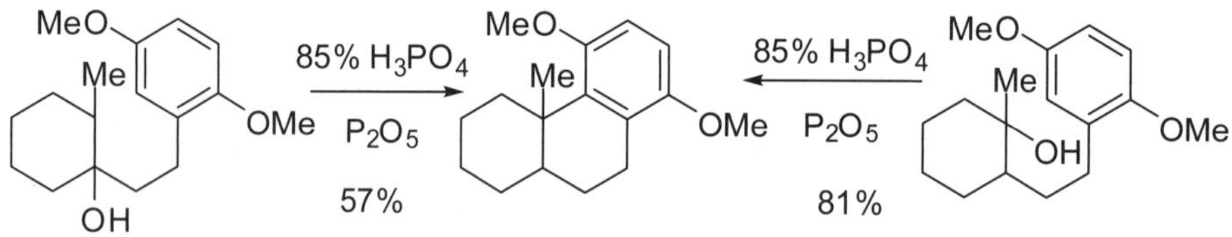

28 **29** **30**

Having EDGs can also work to one's advantage as they provide a reactive aromatic ring poised to attack and not allow carbocation migration as evidenced by the high yield obtained for the conversion of **31** into **32**.

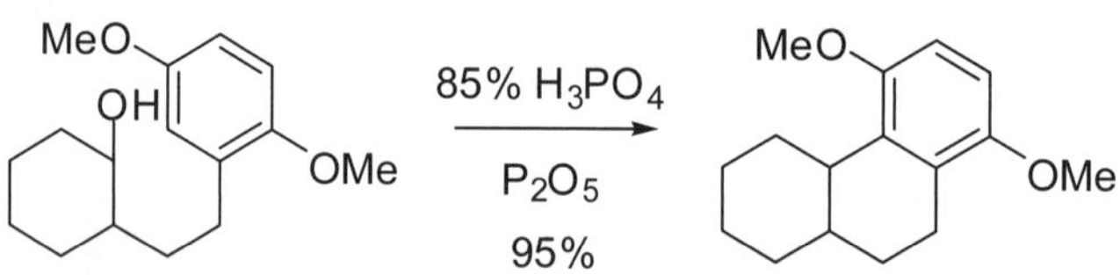

31 **32**

Electronic effects of the ring have been observed in that electron-withdrawing groups (EWG) do slow the rate of cyclization when attached to the nucleophilic aromatic ring. This was nicely demonstrated in a study by Vingiello and Newallis on substrate **33**.[10] While cyclization of **33a** and the chlorosubstituted **33b** gave comparable yields, once the nucleophilic aromatic ring contained a chlorine (**33c,d**), the yield was cut in half under similar conditions.

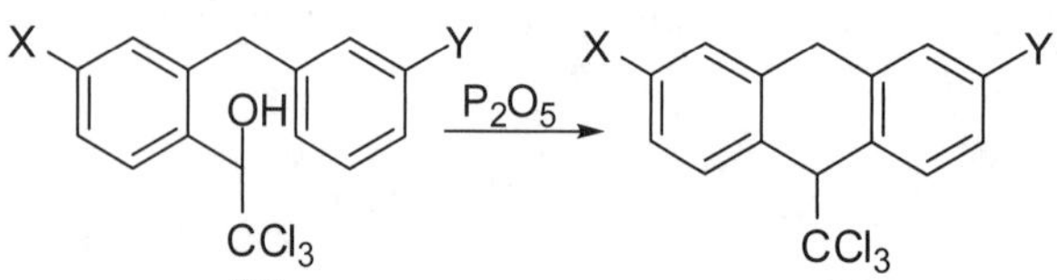

33 **34**

Substitution	yield
34a, X = Y = H	60%
34b, X, = Cl; Y = H	54%
34c, X = H; Y = Cl	26%
34d, X = Y = Cl	27%

4.1.4 *Variation and Improvements*

Cyclization of Keto-Substrates

In addition to the use of hydroxyl or alkenyl substrates, keto substrates have also been cyclized. Polyphosphoric acid (PPA) was used to promote the cyclization of dione **35** into tetracyclic ketone **36** in > 90% yield.[11]

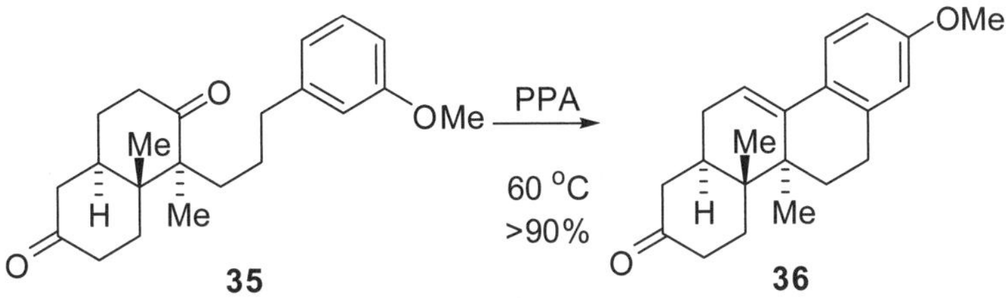

Alternative Dehydrogenating Agents

Several heavy metal-mediated dehydrogenation reactions have been developed in addition to the use of selenium. Pt, Pd, Ni, Ni–Cr, Rh, as and S have all been studied for dehydrogenation of substrates resulting from cyclization.[12] Dehydrogenation of tetrahydrophenanthrene **37** provided phenanthrene **38** in 83% yield using palladium on carbon.[13]

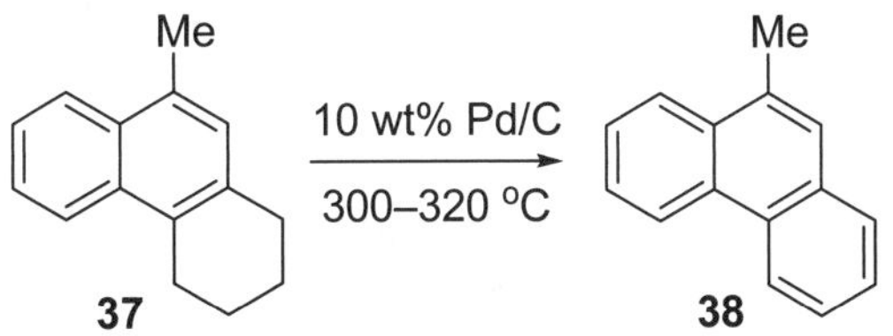

The addition of substrates to enable the transfer of hydrogen, rather than merely allowing the H_2 to be expelled, has improved the safety of dehydrogenation reactions. Additives for such reactions are merely alkenes (cyclohexene, maleic acid, or benzene). While mechanistically use of elemental S in place of Se is proposed to be similar, the dehydrogenation reaction using heavy metals is considered to be merely the reverse of the hydrogenation reaction. A proposal of how the substrate binds to the surface of the catalyst to promote the process has been proposed.[14]

Use of quinones like chloranil and DDQ has also been employed for this dehydrogenation. Halogenation–elimination with bromine or NBS has been a chemical means to promote the dehydrogenation. However, with the bromination/elimination protocol, one must be careful as aromatic halogenation can occur.[12]

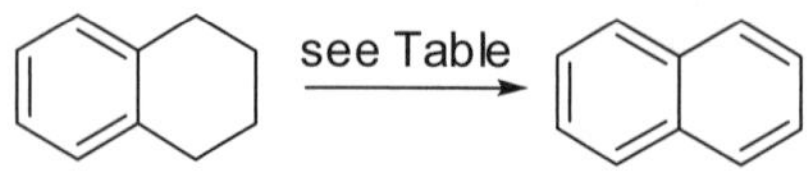

39 40

Conditions	Yield
Pd/C, reflux 22 h	quant
DDQ, PhH, reflux	quant
NBS, CCl_4,	74%

General Selectivity for the Cyclization

In comparison to conditions used for the cyclization of alcohols or alkenes described, use of phosphorous pentoxide (P_2O_5), phosphoric acid (H_3PO_4), or a mixture of the two was shown to enable cyclizations, which did not proceed using classical Bogert–Cook conditions. Notably, the cyclization of primary alcohols like compound **41** failed to provide cyclized material **39** using H_2SO_4 while use of H_3PO_4 provided a 61% yield of **39**.[15]

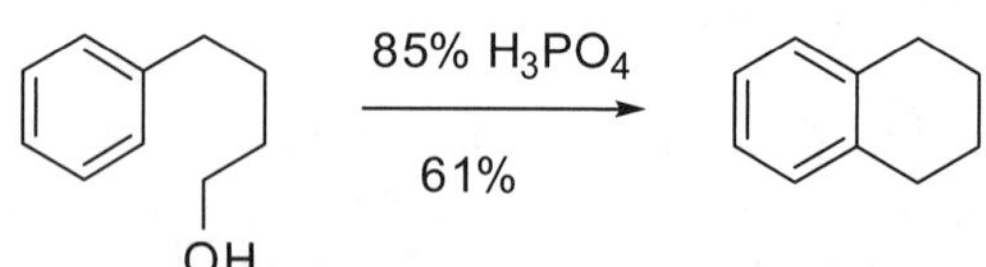

41 **39**

On the other hand, cyclization of primary alcohol **42**, bearing juxtaposed groups for cation stabilization (Me, H), gave predominantly products resulting from rearrangement.

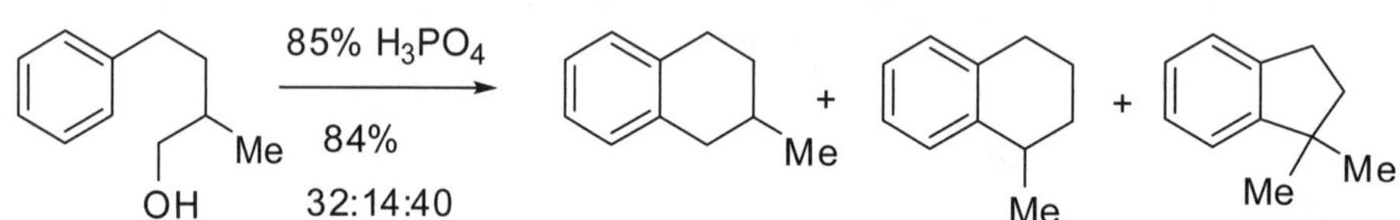

42 **43** **44** **45**

The reaction was also determined to be primarily *cis*-selective. Mossettig and van deKamp reported the cyclization of **46** provided a 70:20 mixture of *cis* and *trans* isomers **47** and **48** using P_2O_5.[16]

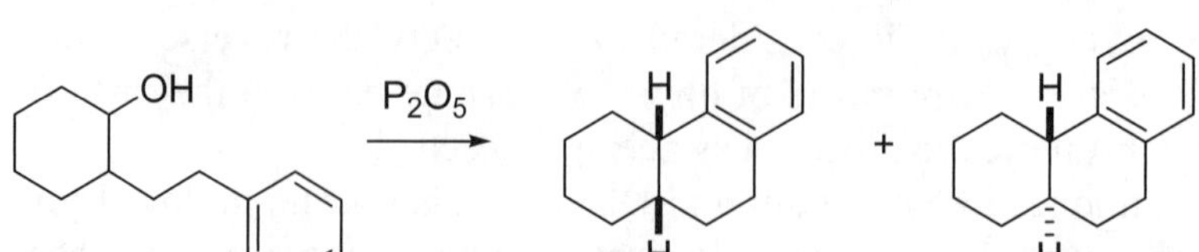

46 **47**, 70% **48**, 20%

Another example in which the phosphoric reaction conditions have proven more selective than other conditions was shown in the conversion of enone **49** into only the *cis*-fused tricyclic compound **50**. Use of other reagents reported, provided a mixture of *cis* and *trans* isomers. Note that this substrate also generates a tetrasubstitued carbon, not necessarily a facile bond to make in excellent yield.[17]

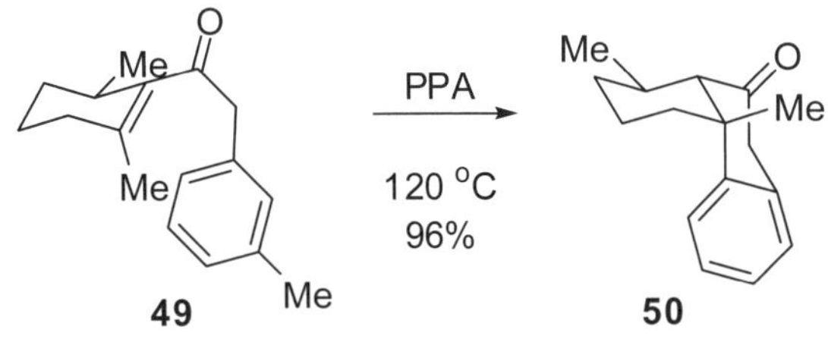

Different Disconnection of the Phenantharene Synthesis

A recent change in the disconnection of putting together the phenanthrene ring system has appeared using conditions similar to the Bardhan–Sengupta method.[18] PPA promoted Freidel–Crafts type acylation followed by electrophilic cationic cyclization of of **51** and **52** gave **53**. Aromatization of **53** with Pd/C then provided the phenanthrenes **54** in overall modest yield. In this work, the authors did not rigorously assign the stereochemistry of the cyclization but suggested that a 1:1 mixture to predominantly the *cis* isomer resulted depending on substitution pattern of the starting material.

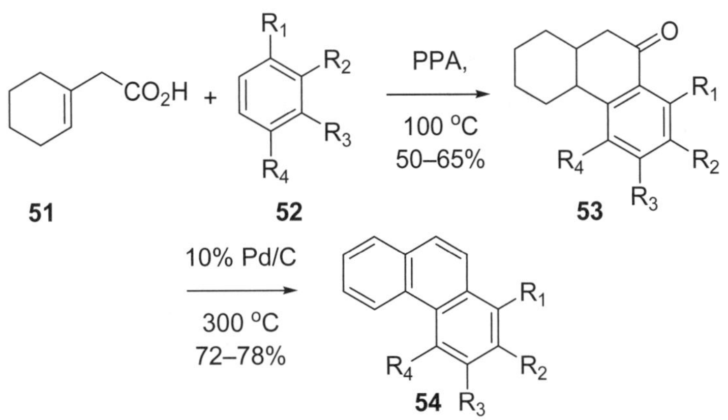

4.1.5 *Synthetic Utility*

General Utility

The cyclization has been used as a key step in Topliss' synthesis of the natural product Ferruginol **57**.[19] P_2O_5 was used to successfully promote the cyclization of alcohol **55** into the octahydrophenanthrene **56** in about 47% yield.

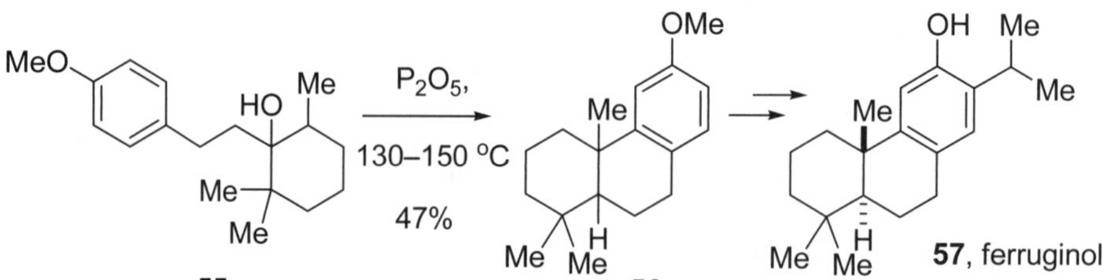

55 **56**

Use of a P_2O_5 to promote cyclization of steroidal backbones was also accomplished by Ireland and co-workers.[11] In this case, the *trans* product **59** proved to dominate the product mixture. The preference was suggested to be due to a combination of A-strain and torsional strain–nonbonding interactions during the cyclization. The stereochemistry of the starting alcohol **58** had no impact on the ratio of the products.

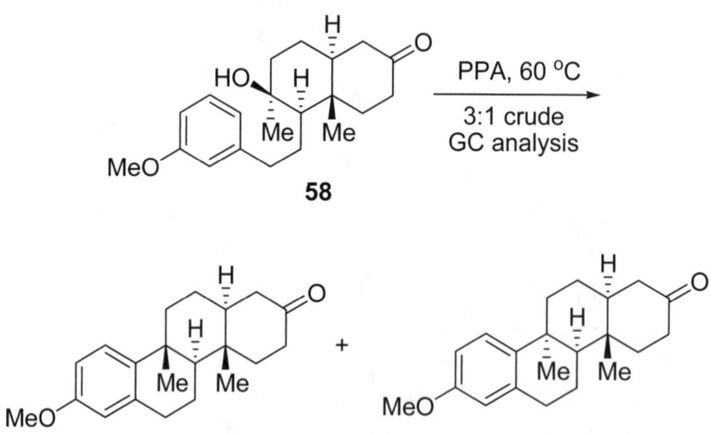

Conditions initially used for the Bardhan–Sengupta cyclization have been applied to a double or cascade cyclization reaction. Thus cyclization of **61** using PPA gave the *cis*-fused ring system **62**.[7]

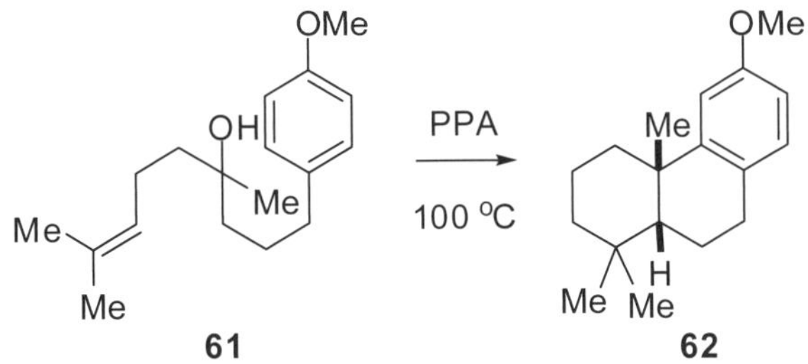

4.1.6 Experimental

Polyphosphoric acid-promoted preparation of 10(9*H*)-1,2,3,4,4a,10aα-hexahydro-1α,4aα-dimethylphenanthrone (50)[16]

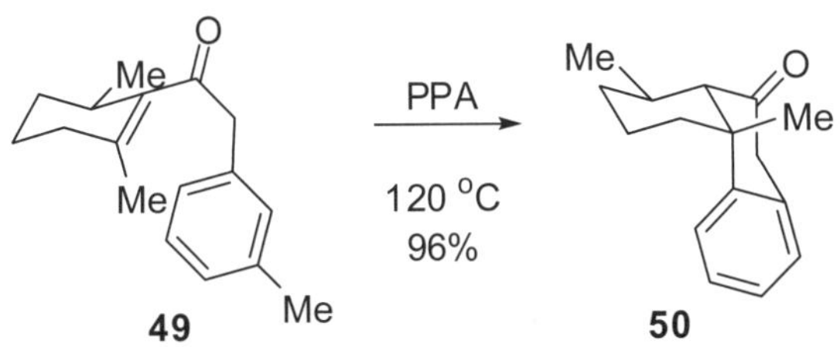

A sample of unsaturated ketone **49** (1.046 g) was heated wth PPA (8 g) at 120 °C for 10 min. The cooled reaction mixture was diluted with water and extracted with ether. The organic layer was washed successively with saline and 10% $NaHCO_3$ solution. The ether solution was dried ($MgSO_4$) then evaporated to dryness to give **50** 1.02 g, 96%. TLC showed one component. The solid was recrystallized from pentane, m.p. 64–65 °C.

Dehydrogenation of 3-methyl octahydrophenanthrene (4)[20]

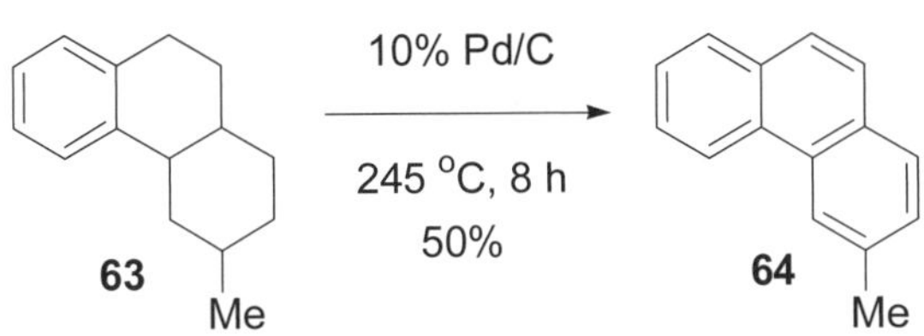

3-Methyl octahydrophenanthrene **63** (20 g, 0.1 mmol, 88:12 mix of 63 and the corresponding spirane) was dehydrogenated with 10% Pd/C (1 g, 5 mol %) at 245 °C for 8 h. The reaction was cooled (solidified), dissolved in PhH, and filtered through dicalite. The filtrate was concentrated in vacuo to provide 19.5 g of a dark product. The product was dissolved in MeOH (200 mL) and was added to a solution of 30 g picric acid in MeOH (200 mL). The resulting picrate solid was collected and recrystallized from hot MeOH to give 19 g of yellow needles. The picrate was decomposed on a column of basic alumina using hexane to give 9 g 3-methyl phenanthrrene (**64**) m.p. 61–62 °C, 50% yield.

4.1.7 References

1. Bardhan, J. C.; Sengupta, S. C. *J. Chem. Soc.* **1932**, 2798–2800.
2. Bardhan, J. C.; Sengupta, S. C. *J. Chem. Soc.* **1932**, 2520–2526.
3. Akin, R. B; Stamatoff, G. S.; Bogert, M. T. *J. Am. Chem. Soc.* **1937**, *59*, 1268–1272.
4. Barnes, R. A. *J. Am. Chem. Soc.* **1953**, *75*, 3004–3008.
5. Papa, D.; Perlman, D.; Bogert, M. T. *J. Am. Chem. Soc.* **1938**, *60*, 319–321.
6. Barnes, R. A.; Hirschler, H. P.; Bluestiein, B. R. *J. Am. Chem. Soc.* **1952**, *74*, 32–34.
7. [R] Barclay, L. R. C. In *Friedel–Crafts and Related Reactions*, Vol. 2, part II, Olah, G. A. Ed., Wiley, New York, NY, 1964, pp 785–977.
8. So, Y.-H., Heeschen, J. P. *J. Org. Chem.* **1997,** *62*, 3552–3561.
9. (a) Fitzpatrick, J. D.; Orchin, M. *J. Am Chem. Soc.* **1957,** *79*, 4765–4771. (b) House, W. T.; Orchin, M. *J. Am Chem. Soc.* **1960**, *82*, 639–642. (c) Silverwood, H. A.; Orchin, M. *J. Org. Chem.* **1962**, *27*, 3401–3404.
10. Vingiello, F. A.; Newallis, P. E. *J. Org. Chem.* **1960**, *25*, 905–907.
11. Ireland, R. E; Baldwin, S. W.; Welch, S. C. *J. Am. Chem. Soc.* **1972**, *94*, 2056–2066.
12. [R] Fu, P. P.; Harvey, R. G. *Chemical Rev.* **1978**, *78*, 317–361.
13. Bachmann, W. E.; Cronyn, M. W.; Struve, W. S. *J. Org. Chem.* **1947**, 596–605.
14. Tsai, M.-C.; Friend, C. M.; Muetterties, E. L. *J. Am. Chem. Soc.* **1982,** *104*, 2539–2543.
15. Khalaf, A. A.; Roberts, R. M. *J. Org. Chem.* **1969,** *34*, 3571–3574.
16. van de Kamp, J.; Mosettig, E. *J. Am. Chem. Soc.* **1936**, *58*, 1062–1063.
17. Schmidt, C.; Thazhuthaveetil, J. *Can. J. Chem.* **1973**, *51*, 3620–3625.
18. Ramana, M. M. V.; Potnis, P. V. *Synthesis* **1996**, 1090–1092.
19. King, F. E.; King, T. J.; Topliss, J. G. *J. Chem Soc.* **1957**, 573–577.
20. Bansal, R. C.; Samad, S. A.; Thomson, J. S.; Eisenbran, E. J. *Org. Prep. Proc. Int.* **1988**, *20*, 305–311.

4.2 Bergman Cyclization

Nessan J. Kerrigan

4.2.1 Description

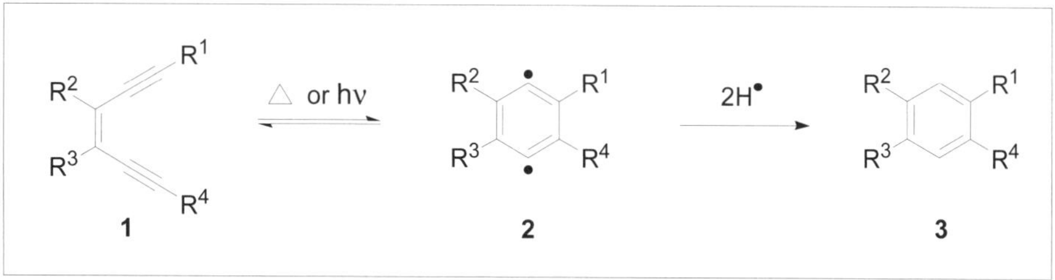

The Bergman cyclization, also known as the Bergman cycloaromatization, is a thermal, photochemical, or metal-mediated cycloaromatization of enediynes **1** that provides access to substituted arenes **3**.[1–5] The cyclization initially forms a 1,4-benzenediyl diradical **2** which, being highly reactive, reacts with a hydrogen donor, such as 1,4-cyclohexadiene, to give an arene **3**.

4.2.2 Historical Perspective

Robert George Bergman (born in 1942) discovered this reaction in 1971 while he was an associate professor at the California Institute of Technology at Pasadena.[1] He later accepted an appointment as professor of chemistry at the University of California, Berkeley. Studies involving the Bergman cyclization were rare until relatively recently due to its limited substrate scope and the availability of alternative methods for construction of substituted arenes.[4] However, since the discovery of naturally occurring enediynes with antitumor activity, interest in the Bergman cyclization has increased greatly, particularly with respect to its role in their mode of action.[4]

4.2.3 Mechanism

Bergman proposed that the reaction mechanism of the cyclization under thermal conditions (200 °C) involved the initial generation of a 1,4-benzenediyl diradical species known as *para*-benzyne (**2**).[1,2] Bergman reported that when the reaction was carried out in a hydrocarbon solvent, such as 2,6,10,14-tetramethylpentadecane, benzene was formed as the final product.[1] This suggests that the hydrocarbon solvent (RH) acts as a hydrogen atom donor to quench the diradical intermediate **2**. This result hints at the radical nature of the mechanism operative in the Bergman cyclization.

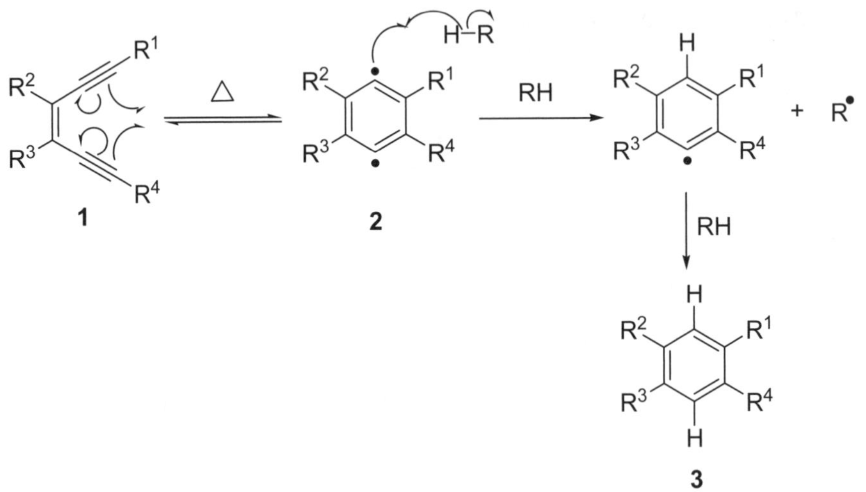

Further support for this mechanism came from the results of studies in other solvents. Carrying out the cyclization in toluene led to the formation of diphenylmethane **4** as the major product, while in carbon tetrachloride, the formation of 1,4-dichlorobenzene **5** was observed. In methanol, benzene and some benzyl alcohol **6** were formed but, notably, no anisole was detected. These results could only be explained by the intermediacy of a free radical species **2a** and clearly ruled out the possibility of a charged (polar) intermediate.[1,2]

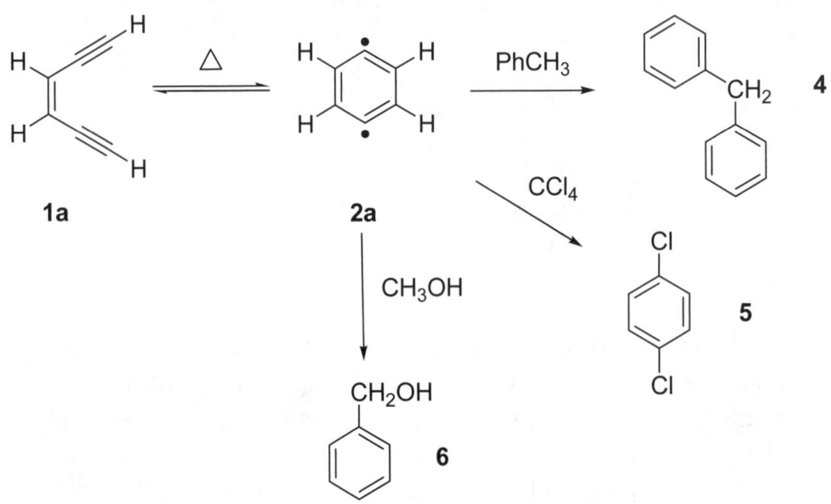

Furthermore, deuterium labeling experiments under gas-phase pyrolysis conditions (200–300 °C) showed that complete scrambling of deuterium between the acetylenic and the vinyl positions of **1b** occurred (to give **1c**). This demonstrated that **1b** is in equilibrium with **1c** and confirmed that **1b** is being converted via an intermediate (**2b**), in which C-1, C-3, C-4, and C-6 are chemically equivalent.[1]

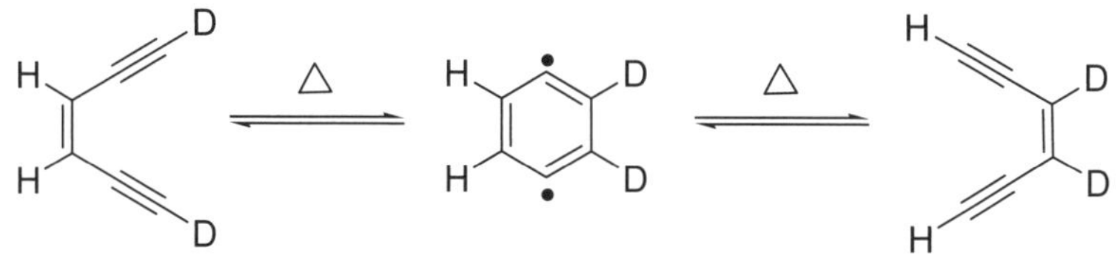

1b **2b** **1c**

Bergman also investigated the spin state of the diradical species **2** produced during the thermolysis of enediynes.[3] On the basis of chemically induced dynamic nuclear polarization (CIDNP) studies, and trapping experiments, which used the spin correlation effect (SCE), it was deduced that most products (including substituted arenes **3**) formed from **2** arise from the singlet state of the diradical species.[3]

4.2.4 Variations and Improvements

While the Bergman cyclization generally requires elevated temperatures (≥ 200 °C) for acyclic enediynes due to their high activation barriers, cyclic enediynes in contrast have lower activation energies and can cyclize at significantly lower temperatures.[6–8] The difference in reactivity has been ascribed to the shortness of the distance (the *cd* distance) between acetylenic groups in cyclic enediynes (3.20–3.31 Å for modest half-life at room temperature) in contrast to the longer distance found in acyclic enediynes (*cd* distance > 3.31 Å). This theory was supported by agreement of molecular mechanics calculations (Macromodel, MM2) concerning *cd* distance with experimental observations for a variety of enediynes.[6] To illustrate this point, 10-membered enediyne **7** underwent cyclization efficiently under moderate conditions (70 °C) to give **8**.[8]

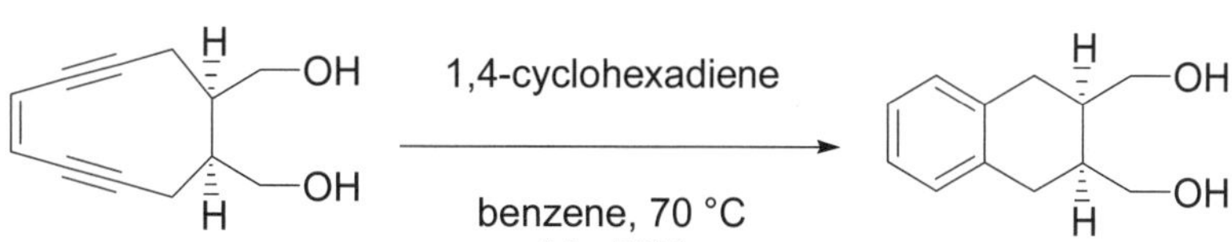

7 **8**

The development of photochemical variants of the Bergman cyclization would contribute to the broadening of its scope and utility. However there are relatively few examples of photoinduced cyclizations, despite the promise that such an approach would have in the photodynamic therapy of cancer.[9–14] Funk and co-workers reported that certain cyclic *ortho*-dialkynylarenes, such as **9**, underwent the photo-Bergman cyclization on irradiation in Pyrex.[10] In 2000 Russell and co-workers showed that 10-membered cyclic pyrimidine enediynes could undergo cyclization efficiently (82–83%) when exposed to light (310 nm) in *i*-PrOH for 2–24 hours.[13]

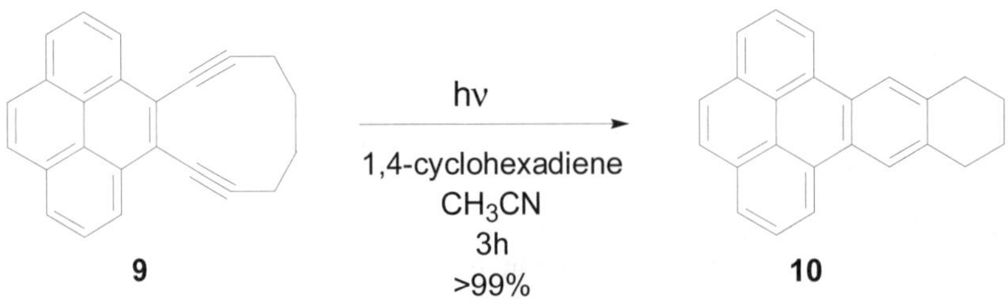

Turro and co-workers showed that simple acyclic aromatic enediynes, such as **11**, could also undergo Bergman cyclization under photochemical conditions. They proposed that the photo-Bergman cyclization of these substrates involved a diradical intermediate **2** identical to that of the thermal reaction.[11]

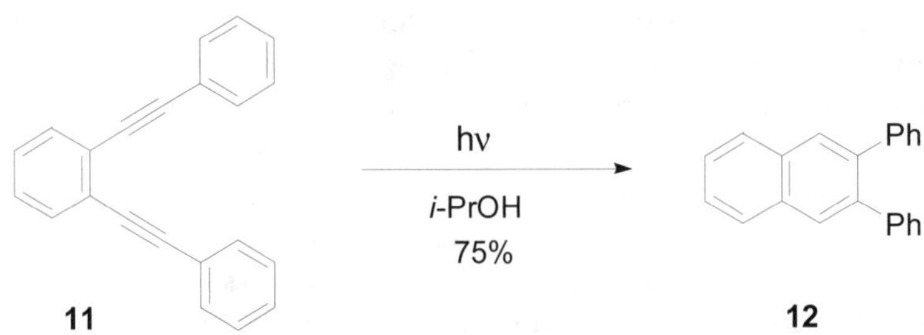

Hirama and co-workers demonstrated that the photo-Bergman cyclization could be extended to both cyclic and acyclic aliphatic enediynes.[12] 1,2-Dipropynylcyclopentene **13** in hexane was converted to indane derivative **14** in 71% yield when irradiated with a low-pressure mercury lamp at room temperature.

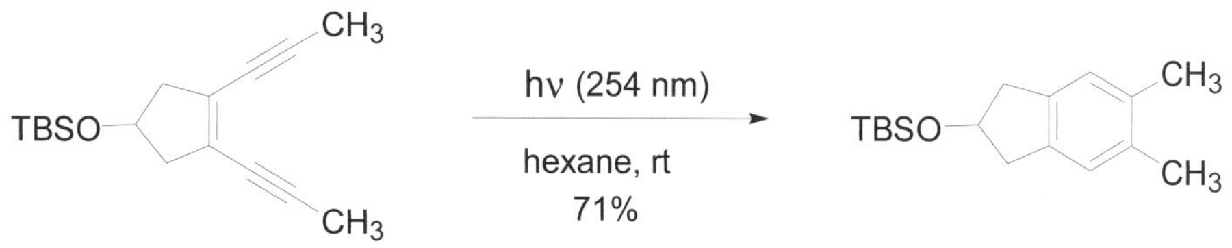

13 **14**

Hirama's group also obtained results to support the intermediacy of a diradical species in the photo-Bergman cyclization. When cyclodeca-1,5-diyn-3-ene **15** in D_8-*i*-PrOH was exposed to photolysis with a low-pressure mercury lamp for 5 min, a significant amount of enediyne **18** (53% based on recovered **15**) was obtained in addition to the expected cyclization product **17** (30% based on recovered **15**). They suggested that **18** must be formed through retro-Bergman reaction of postulated diradical intermediate **16**, and hence provided evidence for **16** acting as an intermediate in the photo-Bergman cyclization.[12]

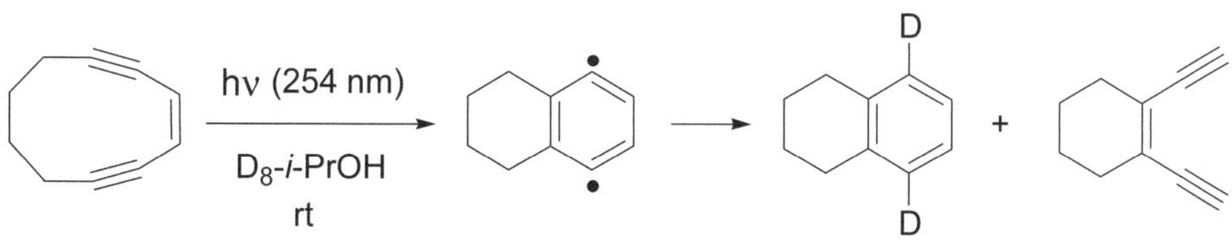

15 **16** **17** (30%) **18** (53%)

A conceptually different approach for Bergman cyclization of acyclic enediynes (and stable cyclic enediynes) relies on the use of metal ion chelation to allow the reaction to proceed at lower temperatures.[15–22] Metal ion chelation requires that the enediyne contain heteroatoms at appropriate positions within the enediyne scaffold. Buchwald et al. reported an elegant use of this approach for the cyclization of a bisphosphane-1,2-diaryl diyne **19**.[17] In this case, Pd^{2+} was found to be the optimal ion and allowed the Bergman cyclization to proceed at 61 °C, instead of 243 °C in the absence of a metal ion. The change in reactivity of the enediyne **19** under metal ion chelation conditions was attributed to both conformational and electronic effects.

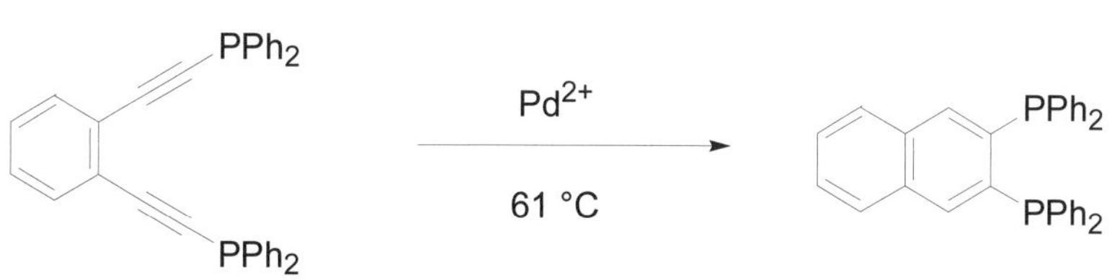

19 **20**

Konig and co-workers also made a significant contribution to this area when they showed that a bipyridyl enediyne **21** could undergo Bergman cyclization at 145 °C in the presence of $Hg(OCOCF_3)_2$.[18] In the absence of the metal salt, a temperature of 237 °C was required to enable cyclization to occur. It was inferred from this result that metal ion coordination brought about a conformational change, facilitating a drop in the cyclization temperature by about 100 °C. Similar observations were made by other groups who carried out studies with amino, sulfonamido, and aldimino enediynes.[19–22]

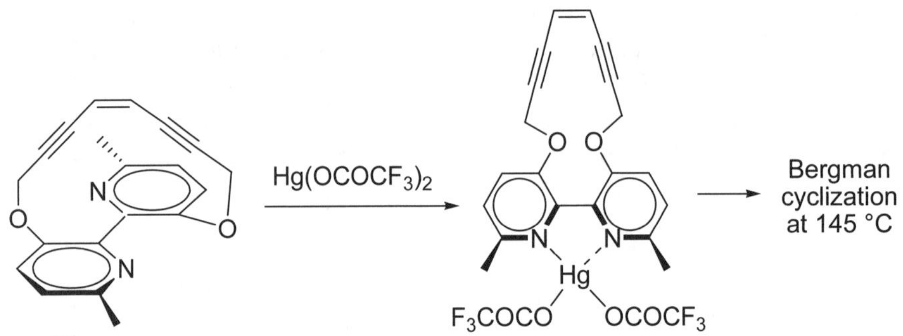

21

A major advance in metal-promoted Bergman cyclizations came about when O'Connor and co-workers revealed that a ruthenium(II) complex **24** could be used to accelerate cyclizations.[23] A major advantage of this approach was that there was no requisite for heteroatoms to be present within the enediyne framework, in contrast to previously reported metal ion chelation approaches. But, rather it was determined that the ruthenium(II) complex activated enediynes to Bergman cyclization through metal-alkyne interactions. In their initial report on the Bergman cyclization, O'Connor and co-workers showed that cyclic enediynes **22** and **23** underwent cyclization in the presence of ruthenium complex **24** under very mild conditions (23 °C in THF) to afford ruthenium complexed products **25** and **26** in good yields of 63% and 71%, respectively. The results of a deuterium labeling experiment implicated 1,4-diradicals as intermediates in the ruthenium-catalyzed Bergman cyclization just as in prototypical thermal Bergman cyclizations.[23]

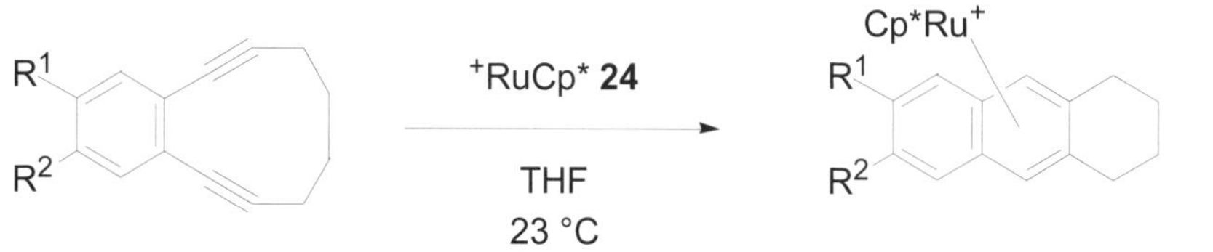

22 R1 = *t*-Bu, R2 = H
23 R1 = R2 = Me

25 R1 = *t*-Bu, R2 = H
26 R1 = R2 = Me

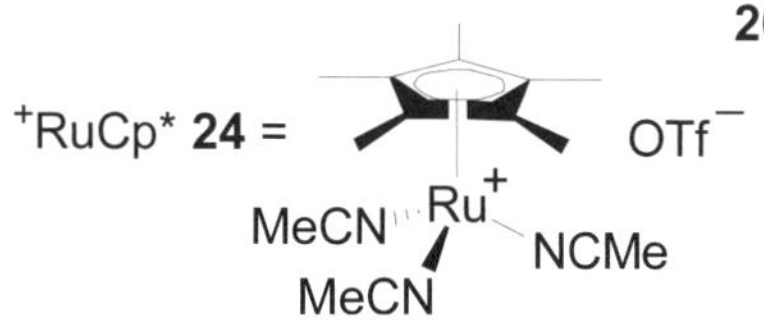

O'Connor's group later reported that **24** could be used to mediate the cycloaromatization of *acyclic* enediynes **27** and **28** in good yields as well.[24] Deuterium-labeling experiments supported the intermediacy of a *p*-benzyne rather than a vinylidene intermediate.[24]

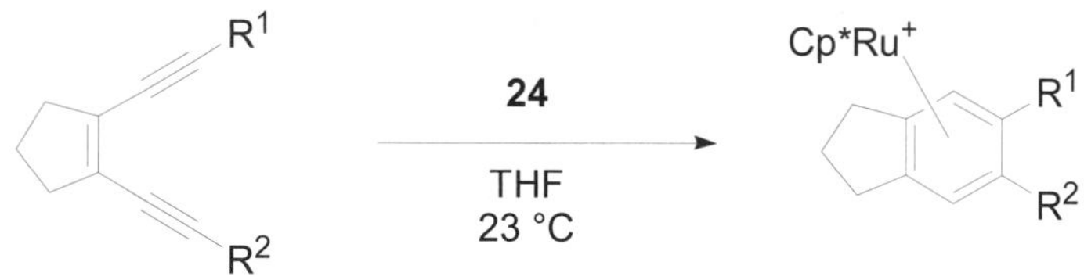

27 R1 = R2 = Me
28 R1 = Me, R2 = TMS

29 R1 = R2 = Me, 64%
30 R1 = Me, R2 = TMS, 77%

More recently, O'Connor and co-workers have reported the first catalytic Bergman cyclization using an iron(II) complex **32** (0.3 equiv) to catalyze cycloaromatization of acyclic enediynes.[25] Mild catalytic activity (3 turnovers) was achievable when photolysis was combined with iron catalysis, enabling decomplexation of the arene product from the iron complex to occur.[25]

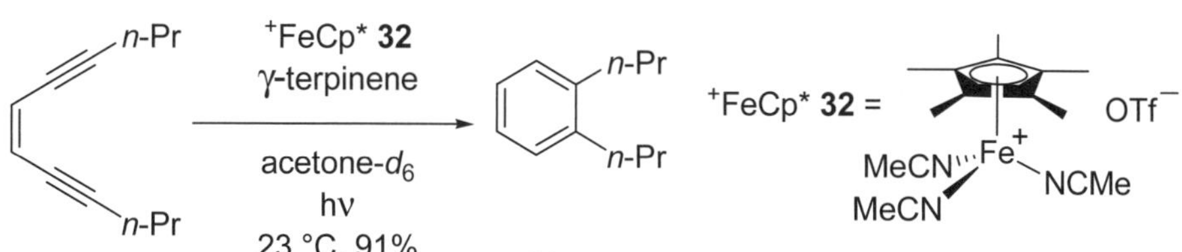

31

33

Related to the Bergman cyclization is the cyclization of an eneyne allene system **34** to provide access to an aromatic system **36**. This variant on the Bergman cyclization is known as the Myers–Saito cyclization.[26–29] Like the Bergman cyclization, it involves a diradical intermediate although in this case the intermediate is a σ,π-diradical **35**.[4,26–29] The enhanced stability of the σ,π-diradical intermediate **35** (and hence the transition state leading to it) over the 1,4-benzenediyl diradical species **2** involved in the Bergman cyclization ensured that the reaction proceeded under much milder conditions (typically 37–100 °C) than are usually required for Bergman cyclizations.[27] Mild thermolysis of **34** in 1,4-cyclohexadiene afforded **36** in 60% yield. It is interesting that methyl substitution at the allene terminus of **34** caused acceleration of the cyclization (approximately six times faster than **34**) and this was presumably due to formation of a more stable diradical intermediate.[27]

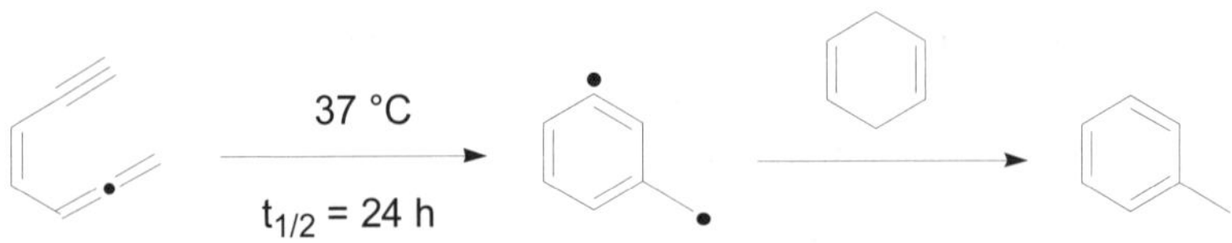

34 **35** **36**

Organometallic reagents may also be employed to facilitate the Myers–Saito cyclization. As the Myers–Saito cyclization proceeds at lower temperatures compared to the Bergman cyclization, strategies have focused on the rearrangement of an enediyne to a more reactive vinylidene eneyne before cyclization. Finn and co-workers reported the preparation of a vinylidene complex **38** through reaction of a benzodiyne **37** with $CpRu(PMe_3)_2Cl$ and NH_4PF_6.[30] They found that vinylidene complex **38** underwent cyclization at a temperature of 100 °C, in contrast to the precursor benzodiyne **37**, which required thermolysis at 190 °C for cyclization to occur.[30] This example clearly demonstrated the superior ability of vinylidene complexes to undergo cyclization in comparison with structurally related enediynes.

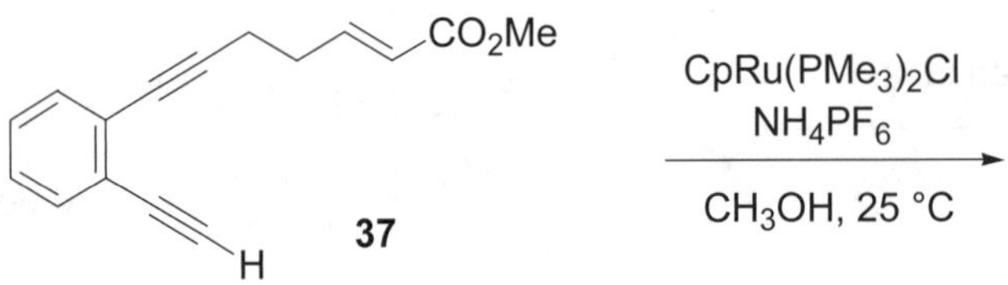

37

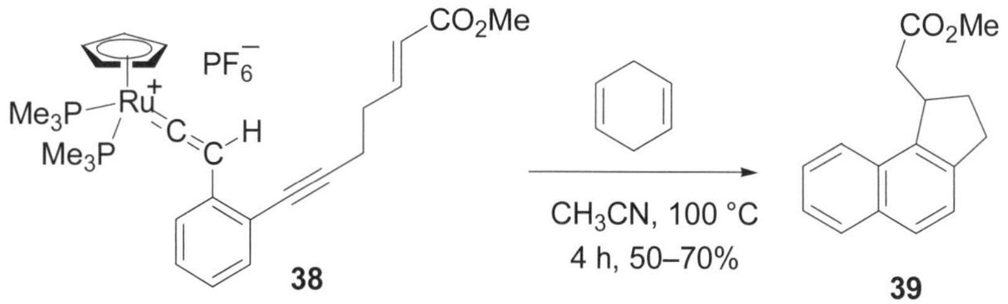

Shortly after Finn's work came to light a catalytic rhodium(I) system was reported.[31] An acyclic enediyne **40** was heated to 50 °C in the presence of just 0.05 equiv of $RhCl(i\text{-}Pr_2P)_2$ and Et_3N in benzene to provide substituted arene **41** in 58% yield. The latter reaction is presumed to involve Myers–Saito cyclization of an in situ formed vinylidene complex.[4,31] A catalytic cycle becomes possible due to steps involving *β*-hydride elimination and reductive elimination.[31]

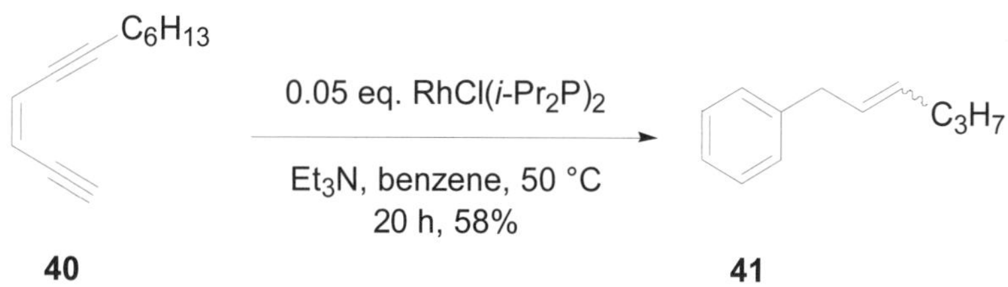

Under thermolysis conditions, cyclization of 1,3-hexadiene-5-ynes occurs to give substituted arenes in a similar fashion to the Bergman cyclization. This variation on the Bergman cyclization is known as the Hopf cyclization.[32] At temperatures up to 550 °C the cyclic intermediate (**43**) in the Hopf cyclization is believed to possess strong diradical character, and so has a lot in common with the Bergman cyclization.[33] Indeed the latter intermediate **43** may be trapped through reaction with various reagents just as in the Bergman cyclization.[34]

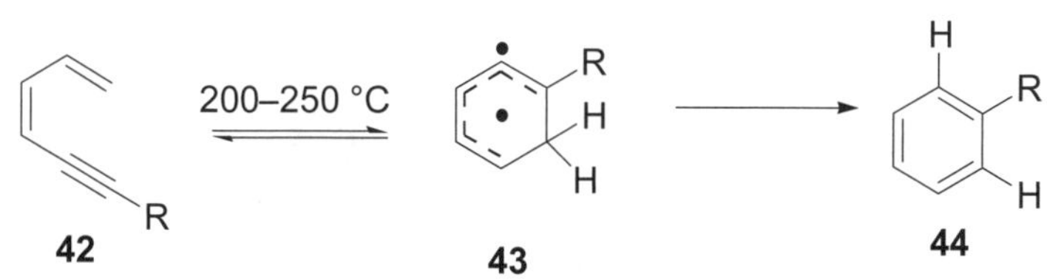

O'Connor and co-workers have shown that the Hopf cyclization can be carried out efficiently under very mild conditions (23 °C) when dienyne **45** is treated with ruthenium complex **24** in THF-d_8 or $CDCl_3$.[24]

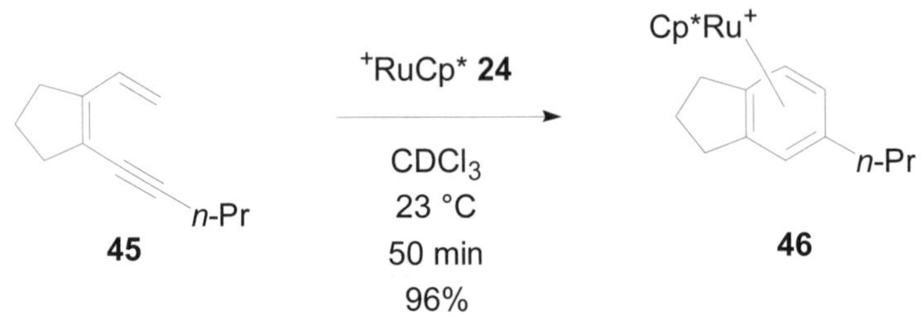

4.2.5 *Synthetic Utility*

Until relatively recently, interest in the Bergman cyclization had been restricted due to its limited substrate scope and the availability of more practical methods for substituted arene construction. Since 1987 with the discovery of natural enediynes that possess cytotoxic activity, there has been a great surge in studies of the Bergman cyclization, specifically with respect to the mode of action of the natural enediynes.[35–38] A number of these natural products were found to undergo Bergman cyclization under much milder conditions (37 °C or lower) than previously thought possible.[37,39]

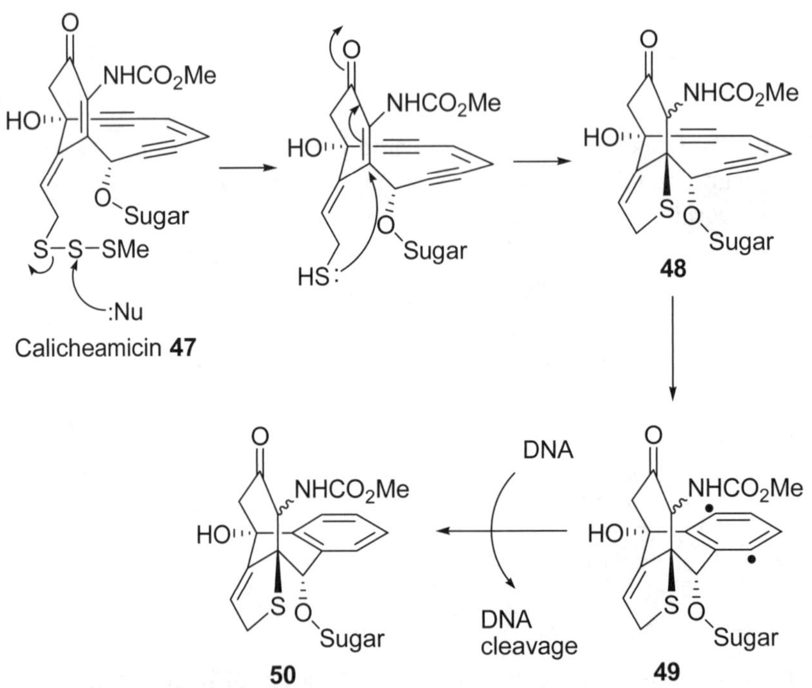

Calicheamicin γ_1^{I} **47** has been proposed to act as an antitumor antibiotic by the mechanism shown.[8,39,40] The enediyne system **47** is triggered to aromatization via Bergman cyclization after nucleophile-mediated cleavage of the methyltrisulfide group of **47** and intramolecular conjugate addition. The change in hybridization from sp^2 to sp^3 at the bridgehead position (on going from **47** to **48**) is critical to the success of the cyclization. This change causes a shortening of the distance between the acetylenic groups (known as the *cd* distance) and hence lowers the activation energy sufficiently for spontaneous cyclization of **48** to occur at physiological temperature.[6,8] Through atom-transfer experiments it has been determined that diradical species **49B** abstracts hydrogen atoms from duplex DNA leading to cleavage of both strands of DNA and, as a result, cell death.[40] Dynemicin also displays significant cytotoxicity through a similar mechanism involving a Bergman cyclization-generated diradical species.[4,38]

Neocarzinostatin's biological activity relies on the involvement of the Myers–Saito cyclization. Its mode of action involves the generation of a diradical species that causes DNA cleavage.[37]

4.2.6 Experimental

Thermal Bergman cyclization: *cis*-2,3-Bis(hydroxymethyl)-1,2,3,4-tetrahydronaphthalene (8)[8]

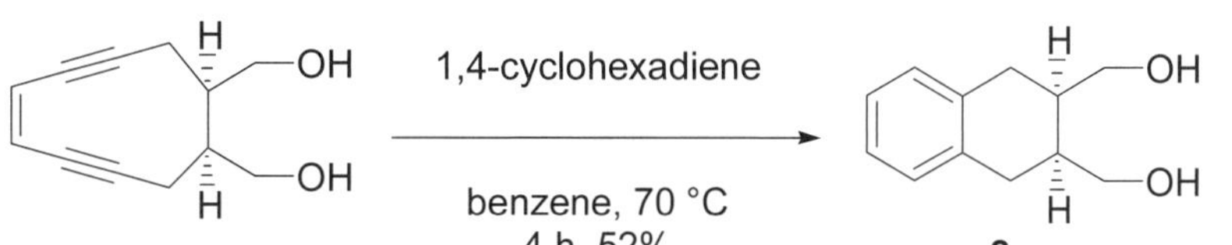

7 **8**

A solution of **7** (38.0 mg, 0.20 mmol) in 1,4-cyclohexadiene (1.90 mL, 20.0 mmol) and degassed benzene (20 mL) was heated at 70 °C for 4 h. The reaction mixture was concentrated, and the residue was purified by preparative thin-layer chromatography (5% methanol in dichloromethane) to afford 20.0 mg (52%) of **8** as a white solid.

Photo-Bergman cyclization: 2,4-Dimethoxy-6,7,8,9-tetrahydro benzo[*g*]quinazolin-6-ol (52)[13]

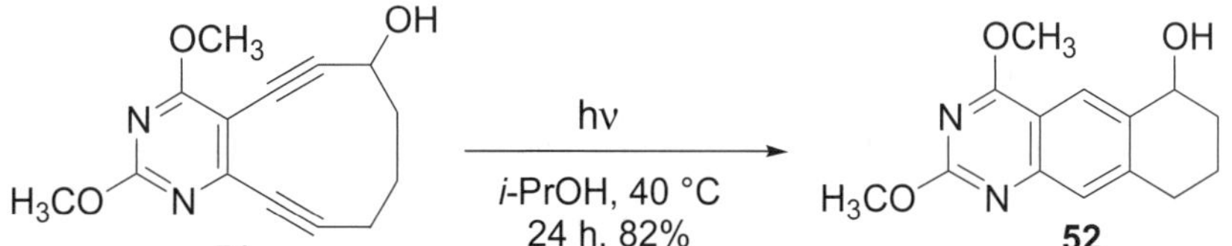

51 **52**

Photolysis of **51** was performed using a Rayonette photochemical reactor equipped with 16 3100 Å lamps. A 0.01 M potassium chromate (K_2CrO_4) solution was used to filter out the 313 nm wavelength. A solution of **51** (2.58 mg, 0.01 mmol) in degassed *i*-PrOH (10 mL) was stirred at 40 °C for 24 h. The solvent was then evaporated under reduced pressure. The resulting residue was purified by column chromatography (SiO_2, hexane/ethyl acetate, 3 : 1), from which pure **52** (82%) was obtained.

Metal-promoted Bergman cyclization: [(1,2,3,4,5-η)-1,2,3,4,5-pentamethyl-2,4-cyclopentadien-1-yl}Ru{(3a,4,5,6,7,7a-η)-2,3-dihydro-6-methyl-1H-inden-5-yl)trimethylsilane}]OTf (30)[24]

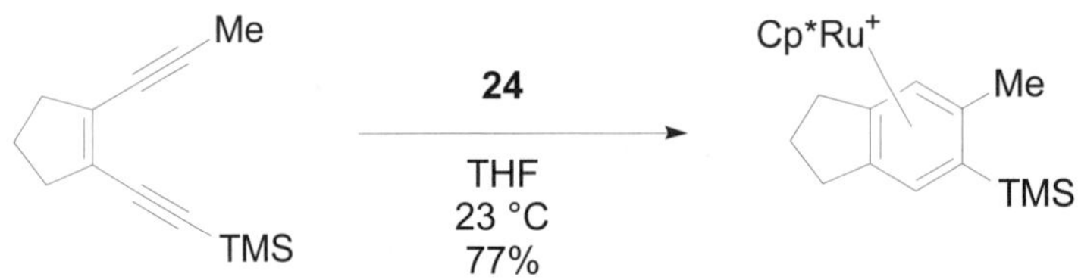

28 30

[Cp*Ru(CH_3CN)$_3$]OTf (**24**, 16 mg, 0.031 mmol) was added to a solution of **28** in THF (ca. 10 mL) at room temperature. An immediate change of solution color from clear to light orange, with progressive darkening, was observed. After 1 h, the volatiles were removed in vacuo to yield a black residue. The residue was then dissolved in a minimum amount of dichloromethane, and ether was added to give 14.2 mg (77%) of **30** as a brown solid.

4.2.7 References

21. Jones, R. R.; Bergman, R. G. *J. Am. Chem. Soc.* **1972**, *94*, 660–661.
22. [R] Bergman, R. G. *Acc. Chem. Res.* **1973**, *6*, 25–31.
23. Lockhart, T. P.; Bergman, R. G. *J. Am. Chem. Soc.* **1981**, *103*, 4091–4096.
24. [R] Basak, A.; Mandal, S.; Bag, S. S. *Chem. Rev.* **2003**, *103*, 4077–4094.
25. Lewis, K. D.; Matzger, A. J. *J. Am. Chem. Soc.* **2005**, *127*, 9968–9969.
26. Nicolaou, K. C.; Zuccarello, G.; Ogawa, Y.; Schweiger, E. J.; Kumazawa, T. *J. Am. Chem. Soc.* **1988**, *110*, 4866–4868.
27. Nicolaou, K. C.; Ogawa, Y.; Zuccarello, G.; Kataoka, H. *J. Am. Chem. Soc.* **1988**, *110*, 7247–7248.
28. Nicolaou, K. C.; Zuccarello, G.; Riemer, C.; Estevez, V. A.; Dai, W.-M. *J. Am. Chem. Soc.* **1992**, *114*, 7360–7371.
29. Turro, N. J.; Evenzahav, A.; Nicolaou, K. C. *Tetrahedron Lett.* **1994**, *35*, 8089–8092.
30. Funk, R. L.; Young, E. R. R.; Williams, R. M.; Flanagan, M. F.; Cecil, T. L. *J. Am. Chem. Soc.* **1996**, *118*, 3291–3292.
31. Evenzahav, A.; Turro, N. J. *J. Am. Chem. Soc.* **1998**, *120*, 1835–1841.
32. Kaneko, T.; Takahashi, M.; Hirama, M. *Angew. Chem., Int. Ed. Engl.* **1999**, *38*, 1267–1268.
33. Choy, N.; Blanco, B.; Wen, J.; Krishan, A.; Russell, K. C. *Org. Lett.* **2000**, 2, 3761–3764.
34. Pandithavidana, D. R.; Poloukhtine, A.; Popik, V. V. *J. Am. Chem. Soc.* **2009**, *131*, 351–356.

35. Konig, B.; Rutters, H. *Tetrahedron Lett.* **1994**, *35*, 3501–3504.
36. Konig, B. Schofield, E.; Bubenitschek, P.; Jones, P. G. *J. Org. Chem.* **1994**, *59*, 7142–7143.
37. Warner, B. P.; Miller, S. P.; Broee, R. D.; Buchwald, S. L. *Science* **1995**, *269*, 814–816.
38. Konig, B.; Hollnagel, H.; Ahrens, B.; Jones, P. G. *Angew. Chem., Int. Ed. Engl.* **1995**, *34*, 2538–2540.
39. Basak, A.; Shain, J. C.; Khamrai, U. K.; Rudra, K. R. *J. Chem. Soc., Perkin Trans. 1* **2000**, 1955-1964.
40. Basak, A.; Shain, J. C. *Tetrahedron Lett.* **1998**, *39*, 1623–1624.
41. Basak, A.; Shain, J. C. *Tetrahedron Lett.* **1998**, *39*, 3029–3030.
42. Rawat, D. S.; Zaleski, M. *J. Am. Chem. Soc.* **2001**, *123*, 9675–9676.
43. O'Connor, J. M.; Lee, L. I.; Gantzel, P. *J. Am. Chem. Soc.* **2000**, *122*, 12057–12058.
44. O'Connor, J. M.; Friese, S. J.; Tichenor, M. *J. Am. Chem. Soc.* **2002**, *124*, 3506–3507.
45. O'Connor, J. M.; Friese, S. J.; Rodgers, B. L. *J. Am. Chem. Soc.* **2005**, *127*, 16342–16343.
46. Nagata, R.; Yamanaka, H.; Okazaki, E.; Saito, I. *Tetrahedron Lett.* **1989**, *30*, 4995–4998.
47. Myers, A. G.; Kuo, E. Y.; Finney, N. S. *J. Am. Chem. Soc.* **1989**, *111*, 8057–8059.
48. Myers, A. G.; Dragovich, P. S. *J. Am. Chem. Soc.* **1989**, *111*, 9130–9132.
49. Nagata, R.; Yamanaka, H.; Murahashi, E.; Saito, I. *Tetrahedron Lett.* **1990**, *31*, 2907–2910.
50. Wang, Y.; Finn, M. G. *J. Am. Chem. Soc.* **1995**, *117*, 8045–8046.
51. Ohe, K.; Kojima, M.; Yonehara, K.; Uemura, S. *Angew. Chem., Int. Ed. Engl.* **1996**, *35*, 1823–1825.
52. Hopf, H.; Musso, H. *Angew. Chem., Int. Ed. Engl.* **1969**, *8*, 680.
53. Zimmermann, G. *Eur. J. Org. Chem.* **2001**, 457–471.
54. Hopf, H.; Berger, H.; Zimmermann, G.; Nüchter, U.; Jones, P. G.; Dix, I. *Angew. Chem., Int. Ed. Engl.* **1997**, *36*, 1187–1190.
55. Lee, M. D.; Dunne, T. S.; Siegel, M. M.; Chang, C. C.; Morton, G. O.; Borders, D. B. *J. Am. Chem. Soc.* **1987**, *109*, 3464–3466.
56. Lee, M. D.; Dunne, T. S.; Chang, C. C.; Ellestad, G. A.; Siegel, M. M.; Morton, G. O.; McGahren, W. J.; Borders, D. B. *J. Am. Chem. Soc.* **1987**, *109*, 3466–3468.
57. Myers, A. G.; Proteau, P. J.; Handel, T. M. *J. Am. Chem. Soc.* **1988**, *110*, 7212–7214.
58. Konishi, M.; Ohkuma, H.; Tsuno, T.; Oki, T. *J. Am. Chem. Soc.* **1990**, *112*, 3715–3716.
59. De Voss, J. J.; Hangeland, J. J.; Townsend, C. A. *J. Am. Chem. Soc.* **1990**, *112*, 4554–4556.
60. De Voss, J. J.; Townsend, C. A.; Ding, W.-D.; Morton, G. O.; Ellestad, G. A.; Zein, N.; Tabor, A. B.; Schreiber, S. L. *J. Am. Chem. Soc.* **1990**, *112*, 9669–9670.

4.3 Bogert–Cook Reaction

Timothy T. Curran

4.3.1 Description

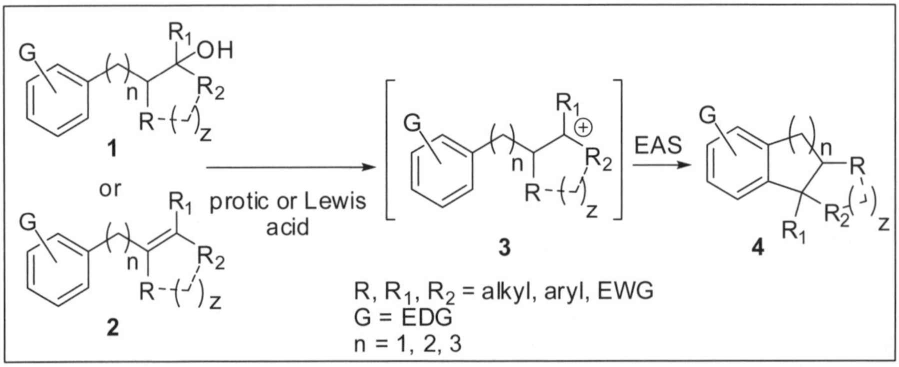

The Bogert–Cook reaction is the reaction of a hydroxy-substituted phenyl alkane **1** or phenyl substituted alkene **2** with an acid, Lewis acid or dehydrating agent (except for phosphorus-containing acids or dehydrating agents, i.e., H_3PO_4 or PPA, which will be reserved for the Bardhan–Sengupta reaction) to generate a carbocationic species which then undergoes electrophilic aromatic substitution (EAS) and proton elimination to provide the corresponding dihydroindene (n = 1), tetrahydronapthalene (n = 2), or tetrahydrobenzo-[7]-annulene (n = 3). While reactions of n = 1 or 3 have been reported, there are limitations to delivering the n = 1 or 3 products in high yield due to cation migration, which generates the preferred tetrahydronapthalene. In addition, due to the cationic nature of the reaction, there can be an array of by-products generated, depending on the stability of the carbocation and neighboring groups.

Due to the mechanistic nature of the reaction, the hydroxyl group or alkene need not be at the site of aryl attack. Cyclization is not the preferred reaction observed when the hydroxyl or alkene is located at the benzylic position of the cyclizing ring.

4.3.2 Historical Perspective

In 1933, Carston Bogert reported the synthesis of phenanthrene from either tertiary or secondary alcohol **5** or **7** via dehydration with dilute H_2SO_4, providing common alkene **6**. Cyclization with conc. H_2SO_4 provided the octahydrophenanthrene **8** and ultimately dehydrogenation with elemental selenium completed the synthesis of phenanthrene **9**.[1] His group followed with a full paper in 1936 describing this work in more detail.[2] During a similar time period, Cook and co-workers reported the analogous cyclization and dehydrogenation as well as application toward preparing the carbon framework of the steroidal backbone.[3] In this work, Lewis and protic acids were used on both alcohols and alkenes to promote the cyclization. Work in the Cook group demonstrated that the cyclization of alkene **10** yielded the incorrect regiochemistry for the angular methyl group in the steroidal backbone, compound **11**, in 68% yield.

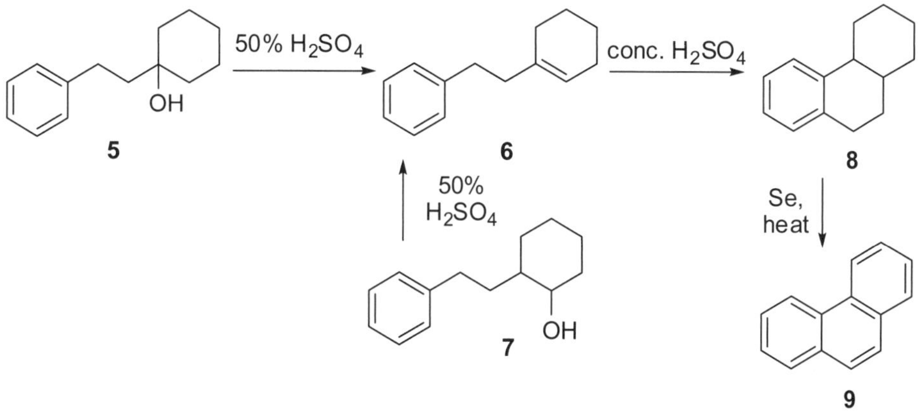

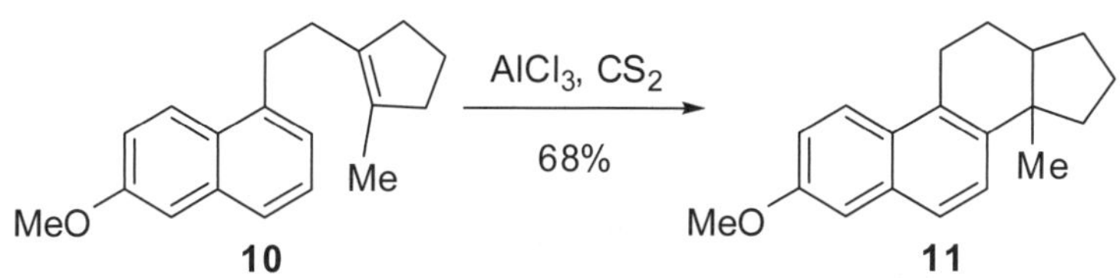

Both the Bogert and Cook groups continued to develop the scope and limitations of these cyclizations and other groups joined the task. During this work, two different spirocycles were reported to be formed. The first at the tertiary alcohol carbon, derived from the point of attachment of the phenylpropanyl moiety. For example, closure to form the six-membered spirocyclic ring for two compounds was reported[4] to occur in about 90% yield from alcohols **12** and **14**.

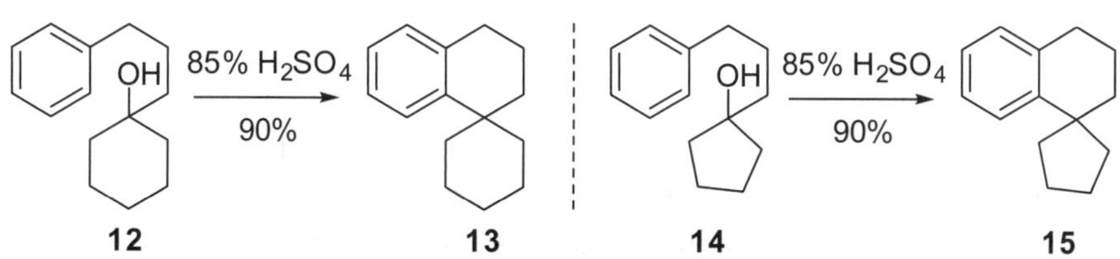

The second type of spirocyclic system resulted from attack of the aromatic ring on the site of attachment of the phenethyl group, which was not necessarily desired. Initially, the regioselectivity of the cyclization conditions were thought to greatly affect the spirocyclic formation but this was discredited by some experiments in the Cook group[3,5] and others, showing that substitution of the alkene with a methyl group or carboxyl group[6] mitigated this type of spirocyclic formation rather than the acid used. For example, cyclization of **16** gave the desired six-membered cyclized adduct in poor yield (27%) presumably due to an appreciable amount of spirocyclic compound **18** formation,[3a] while cyclization of alkene **10** provided compound **11** in 68%. The type of acid used has more recently been thought to alter the *cis*/*trans* product ratio.[7]

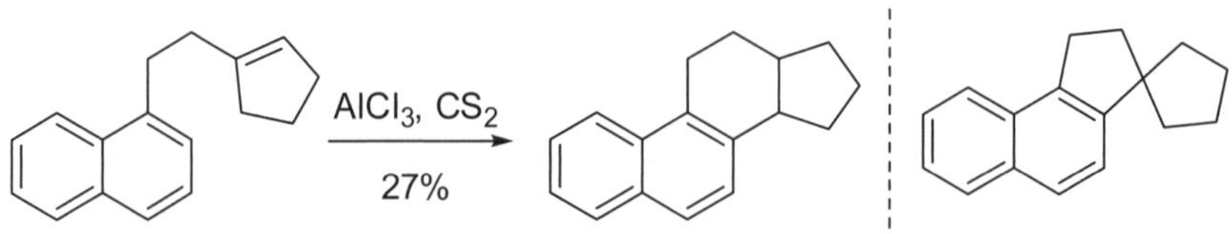

An alkyl carboxyl group (e.g., ester) juxtaposed to the hydroxyl group or the forming carbocation was reported to minimize the amount of spirane formation (resulting from cyclization at the point of attachment of the phenethyl group) due to protonation of this neighboring group, which would require dication formation before cyclization.[6,7] It should be noted that only aromatic rings containing EDG were successful in cyclization with a juxtaposed carboxylate. Six-membered ring formation was shown to be the preferred mode of cyclization. The observation that the reaction was carbocationic in nature was demonstrated by the formation of products, which were best explained by Wagner–Meerwein rearrangement followed by cyclization (vide infra).

Tertiary alcohols reportedly reacted and cyclized faster than secondary alcohols. While some primary alcohols were reported to undergo cyclization at the primary carbon, terminal alkenes have provided the intermediate secondary carbocation that cyclized.[7]

4.3.3 Mechanism

The mechanism of the Bogert–Cook reaction has been shown to be substrate dependent and different transition states have been used to explain different products observed. However, in general, starting with an alcoholic substrate **19a,b**, protonation of the hydroxyl group and loss of water to generate the carbocation **20a,b** (which could either result in alkene formation, cyclization, or carbocation migration) has been proposed. If energy exists in the system to overcome ring and A-strain encountered during the cyclization, then electrophilic aromatic substitution takes place, resulting in a new C–C bond and a phenylcation **22**. Loss of a proton to rearomatize provides product **23**. As can be seen, with the formation of the tertiary carbocation **20b**, cyclization would provide the spirane.

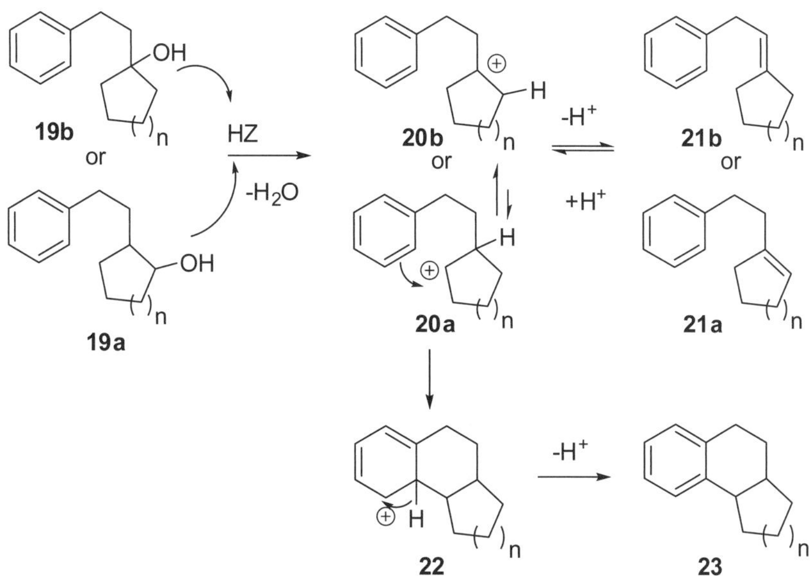

While electron-donating groups are not required on the aromatic ring, as one might expect electron-withdrawing groups might hinder or stop the reaction. When the aromatic ring is electron rich, cyclization is suggested to be less selective providing more spirane.[7]

Initially, Bogert proposed that this reaction always went through an alkene intermediate due to the observation that when the reaction was stopped (quenched before completion) the presence of an olefinic substrate was supported by 30–40 mol % uptake of Br_2,[8] while others have thought alkene or carbocation formation unnecessary. An alternative mechanism for some substrates has been proposed in light of experimental results. The key experiments for this alternative mechanistic view was done by Barnes and Olin, wherein they described the reaction of *trans*-racemic and *trans*-optically-enriched alcohol **24** with 90% H_2SO_4 to provide mainly the *cis*-isolated product **25** which when optically enriched **24** was used, the product **25** rotated plane-polarized light. These results suggested that an alternative mechanism without free cation formation and/or intermediate alkene formation was possible.[9]

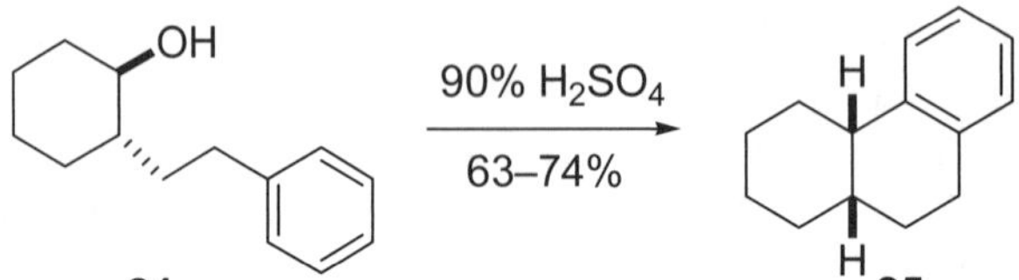

24 25

These authors and others proposed a bridged cationic intermediate **28**. Thus, for the Bogert–Cook reaction, three potential intermediates have been postulated to provide product: activated hydroxyl as a leaving group **26**, free cation formation **27**, and a bridged cationic hydrogen species **28**. The bridged cationic species could be viewed as "partial Wagner–Meerwein" rearrangement or "H stabilization" of the formed cation.[10]

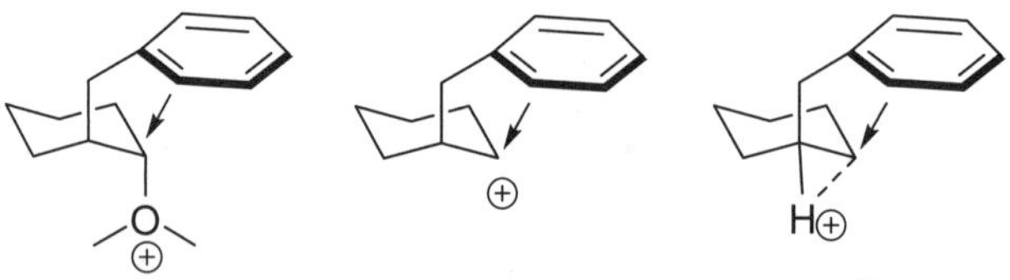

26 27 28

Of these three proposals, structures **27** and **28** are the more likely intermediates for the cyclization. The stability of the formed cation, the reaction conditions, and the stereoelectronic factors for cyclization are all thought to contribute to determining which of the two intermediates best represents the mechanistic path for each reaction.

Evidence for cation formation was partly derived by the observation that for *t*-butyl substituted carbinols, rearrangement occurs to cleanly provide the tetralin or indane. For example, either **29** or **31** reacted with 90% H_2SO_4 to provide the tetralin. Compound **30** resulted from either Me or H 1,2-shift.[11]

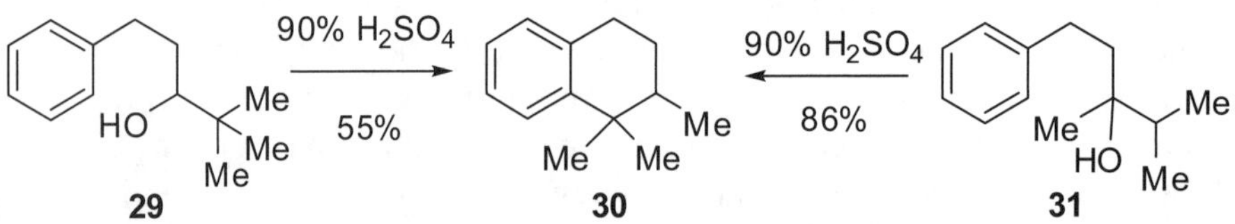

29 30 31

4.3.4 *Selectivity*

Regioselectivity

There is a preference for the formation of the six-membered ring presumably due to increased torsional strain in the transition state to form the 5-

memberred ring in comparison to the formation of the 6-membered ring. It was first noted by Bogert and co-workers[8] that having the hydroxyl group on the 4th carbon from the phenyl ring provided the highest yield of the tetralin.

When there was Me (alkyl) substitution at C3 (with the exception of formation of a tertiary carbocation at the 4th carbon from the cyclizing ring—for example, see conversion of **31** into **30**), then both the indane and tetralin were formed. Many of these experiments were later repeated by Kalif and Roberts who also reported the cyclization of 1-phenyl 3- or 4-Me-substituted pentanols with H_2SO_4. For these cyclizations, the formation of the tetralin predominated over the formation of the indane.[12]

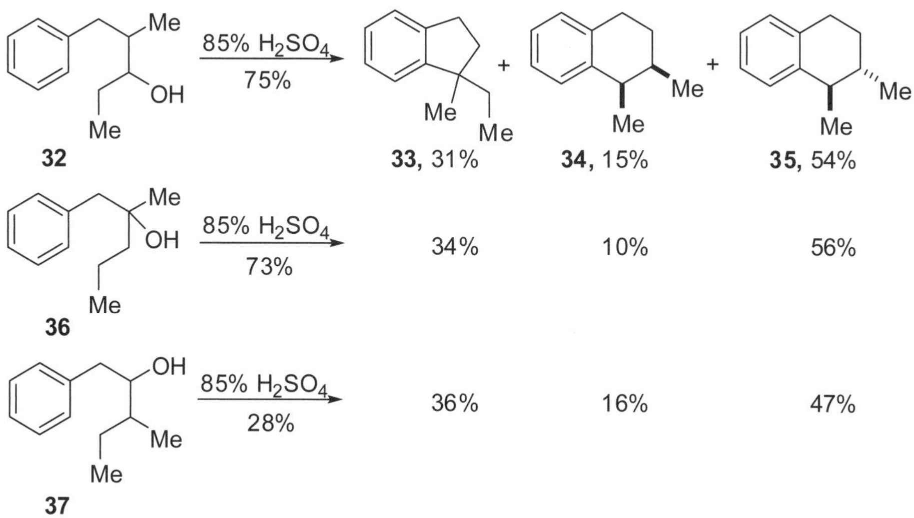

The low yield for the cyclization of **37** was presumably due to polymerization after benzylic cation formation. Additional substrates were prepared to further examine the preference for 6-membered ring formation. Specifically, 5,6-diphenyl-2-hexene **38** was treated with 85% H_2SO_4 and yielded only the 6-membered ring products **39** and **40** in an 87:13 ratio.

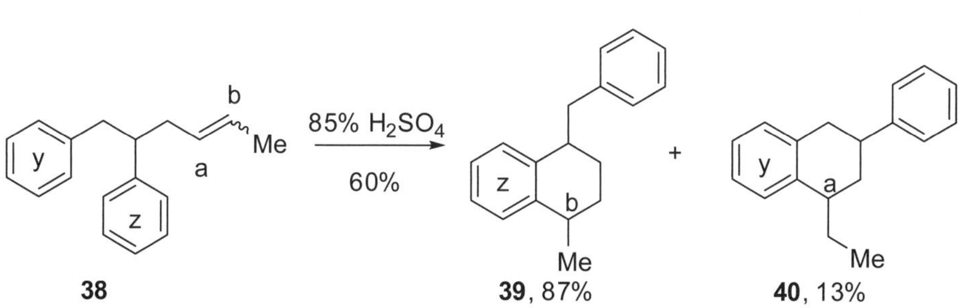

One could obtain both five- and seven-membered ring formation but only the 6-membered ring was observed. Furthermore, formation of the six-membered ring **39** was preferred > 6:1 (cyclization of ring z onto carbon b over cyclization of ring y on carbon a) over **40** presumably due to increased 1,3- steric interactions in forming **40**.

Reaction of a substrate similar to **38**, having the methyl group on carbon 2 of a 4,5-diphenyl-pentene, presumably forming a tertiary carbonium ion, which could cyclize to either the five- or six-membered ring, reportedly gave the six-membered ring product **42**. Again this illustrates the preference for the formation of the 6-membered ring over the five-membered ring.

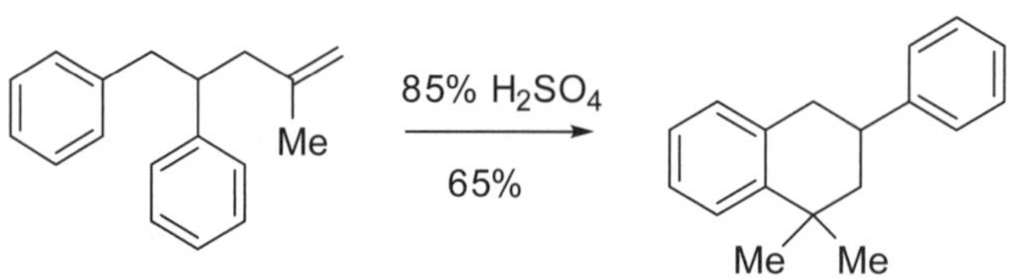

41 **42**

Furthermore a preference of cyclization to a five-membered ring in comparison to the seven-membered ring by reaction of a 4, 6-diphenyl-2-methyl-2-hexanol **43** with H_2SO_4 provided an 85% yield of **44** with no detectable 7-membered ring formed.

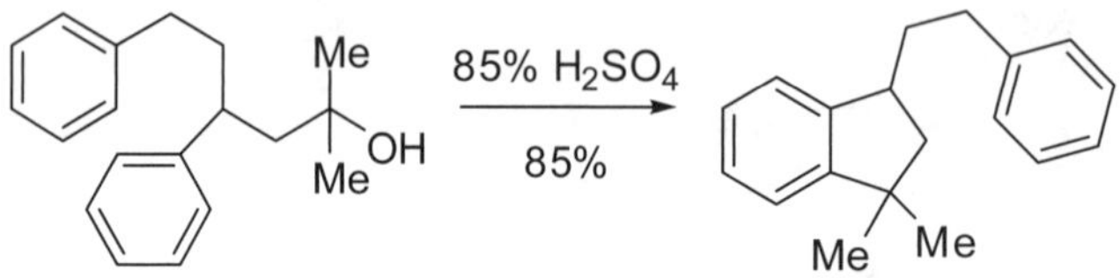

43 **44**

The "forced" formation of a seven-membered ring was accomplished, albeit in poor yield. Treatment of diphenyl substituted pentanol **45** with either 85% H_2SO_4 or $AlCl_3$ in $MeNO_2$ gave 10–15% yield of tetrahydro-benzo-[7]-annulene **46**.

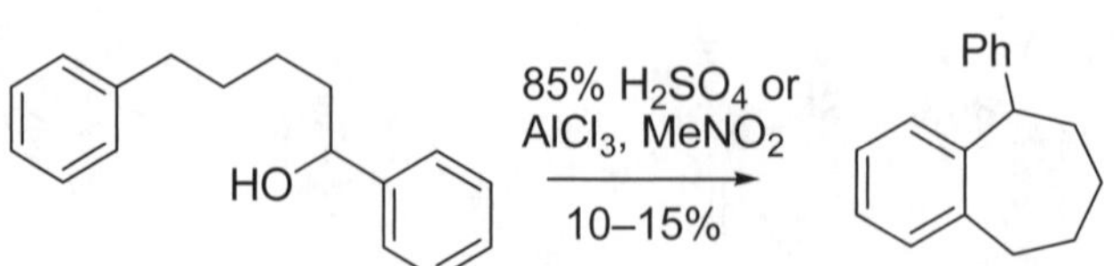

45 **46**

Stereoselectivity

In simple systems, cyclization to the *cis*-fused octahydrophenanthrene has been reported as the major product. Furthermore, it was reported that stronger acidic conditions diminished the selectivity.[13] For example, compound **47**, when treated with 85% H_2SO_4, reportedly gave < 20% *trans* **49** when compared to the amount of *trans* compound formed using 90% H_2SO_4.

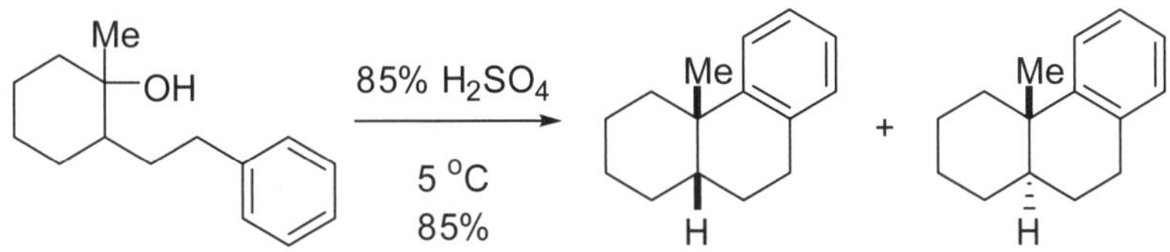

47 **48**, >80% **49**, <20%

The stereochemistry of the carbinol was also shown not to affect the stereochemistry of cyclization. This was reported during the synthesis of estrone by Johnson and co-workers.[14] Whether **50** or **51** was taken into the cyclization, a similar ratio of isomeric products was obtained.

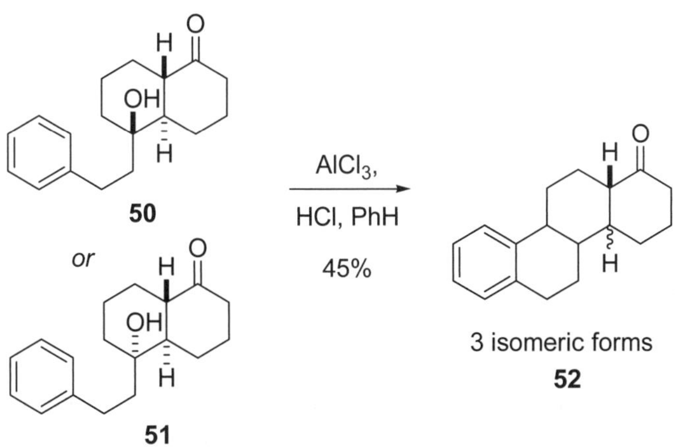

50 **51** **52**

This same notion had previously been reported by Barnes and co-workers in a variety of systems.[7] Ireland and co-workers later applied conformational and approach vector considerations to rationalize the observed stereochemical outcome of reactions.[15]

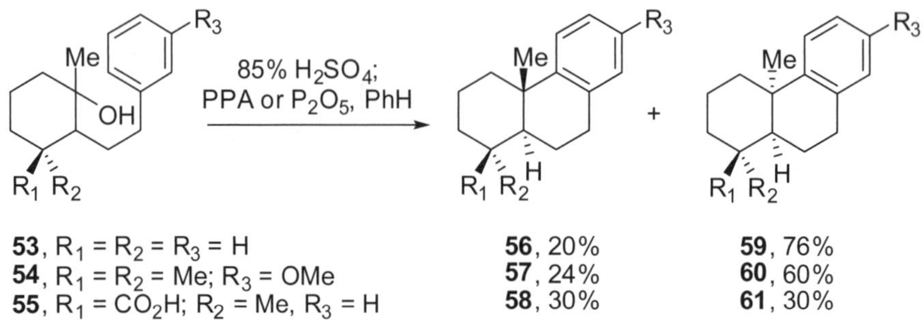

While the *cis*-product predominates, the origin of this preference for **59** and **60** are thought to be due to a combination of $A_{1,2}$ strain, torsional strain, steric approach of the aromatic ring in the transition state, and secondary stabilization effects of the intermediate.

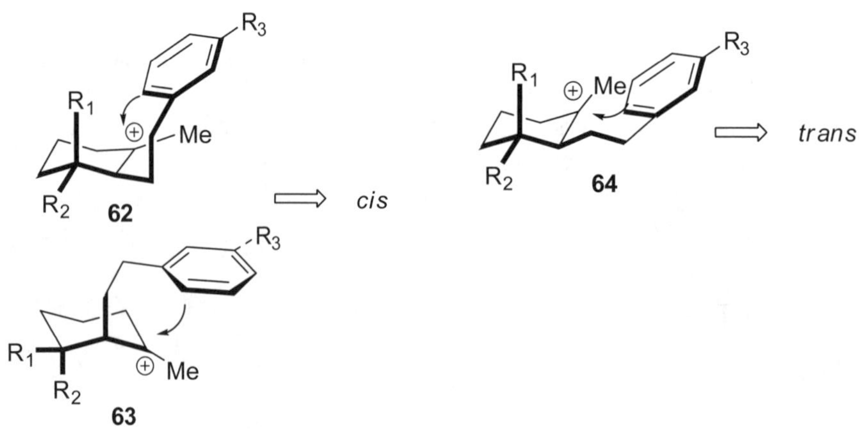

Two possible transition states leading to the *cis* product are depicted in **62** and **63**. In the case of the cyclization of **53**, having the phenethyl group axial should be the preferred or should have a lower energy transition state (TS) depicted as **63**. The torsional strain, which accompanies cyclization via transition state **62** or **64** was presumed to be higher than **63**. Whereas for the cyclization of **54** via TS **63**, the adjacent gem dimethyl substituted carbon should act as a deterrent for *cis* ring formation via **63**, therefore, diminishing the amount of *cis* product. Cyclization of **55** provides a 1:1 mixture of *cis*- and *trans*-isomers. Analogous arguments were used to rationalize the outcome of that reaction with a slight twist introduced due to the participation or stabilization of the adjacent carboxylate. For this interaction to occur, the carboxylate needs to be an axial substituent to stabilize the TS and the ring flip of **62** is likely a contributing conformer. Another important point noted by the authors was the potential for olefin formation and that in the presence of a neighboring asymmetric center (i.e., the asymmetric center in **54**) the stereochemistry of protonation of the alkene could have a significant effect on the product.

It is interesting that there have been reports in which the cyclization reaction did not proceed due to stereochemical considerations. For example, while the model system for the cyclization of **65** to a mixture of **66** and **67** was reported in good yield, the attempted cyclization of **68** did not proceed at all. In this case the authors suggested that the cyclization required a twist boat transition state of the existing cationic ring. The *trans*-fused decalin system renders such transition state inaccessible, and the elimination product was formed. Further proof that this just is not simply due to steric interactions came from the attempted cyclization of **69**, which again gave rise only to the elimination product **71**.[16]

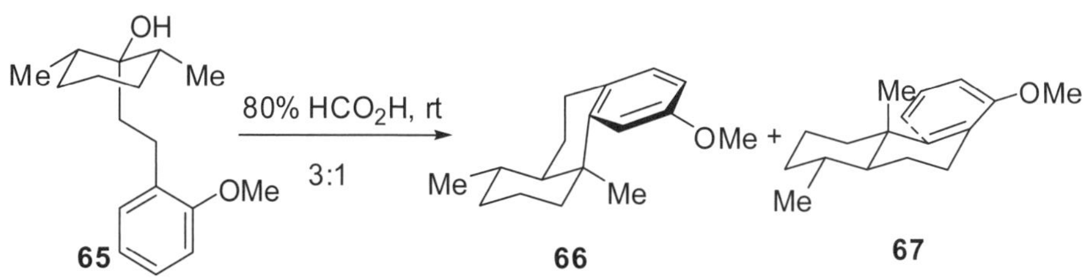

65 **66** **67**

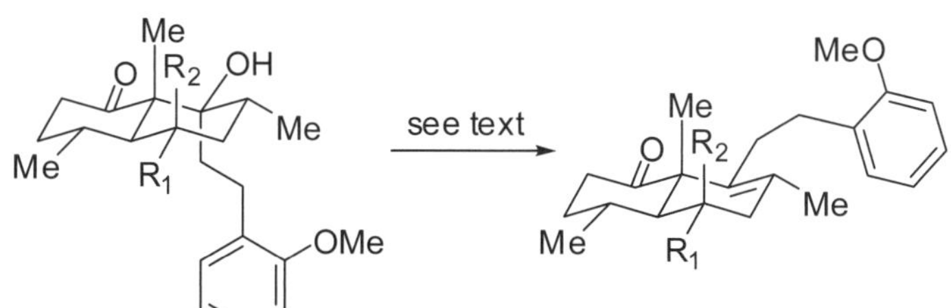

68, R_1 = H, R_2 = OH
69 R_1 = R_2 = O

70, R_1 = H, R_2 = OCOH
71, R_1 = R_2 = O

4.3.6 Variations and Synthetic Utility

General Variations in Conditions and Substrates

A variety of acids and Lewis acids have been described to promote the cyclization from either the alkene or hydroxy compound. While H_2SO_4 was the first reported agent, mixing organic and mineral acids like HCO_2H or AcOH with H_2SO_4 has been utililized. Other mineral acids include HBr, HCl and HF.[17] For example, cyclization of a ketone substrate **72** provided the cyclized material **73** in 87% yield.

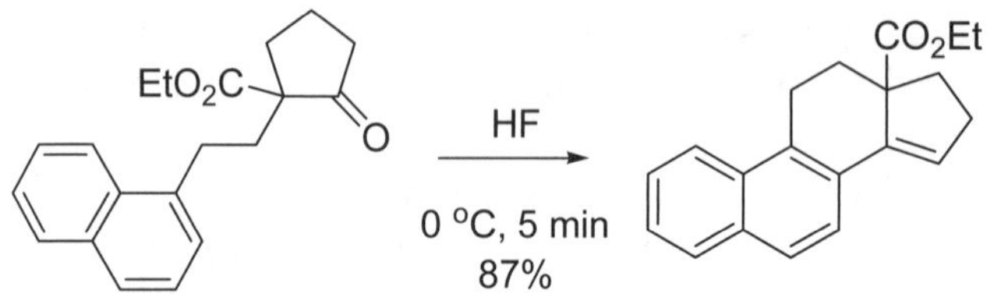

72 **73**

Use of a mineral acid to promote cyclization to an amine-containing substrate yielded the bicyclic alkaloid-skeleton **75** in 62% yield.[18] Reports of cyclization to such bicyclic systems have been limited.

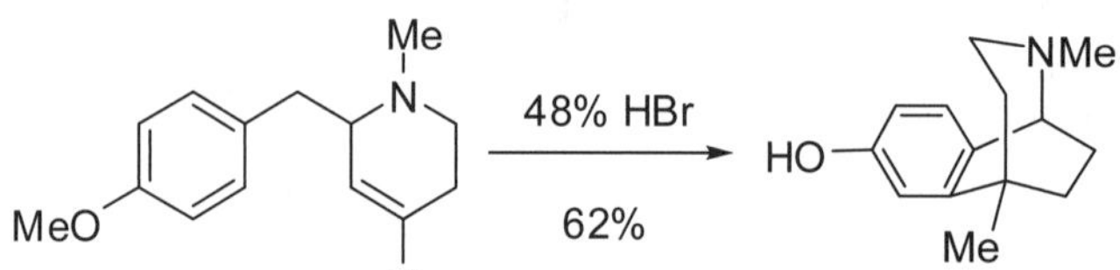

74 **75**

Lewis acids have been limited to just a few, including $AlCl_3$ and $SnCl_4$. For example, cyclization of amide **76** provided a high yield of the cyclized product **77**.[19] Similar conditions have been applied to similar substrates in preparation of experimental therapeutics in the pharmaceutical industry.[20]

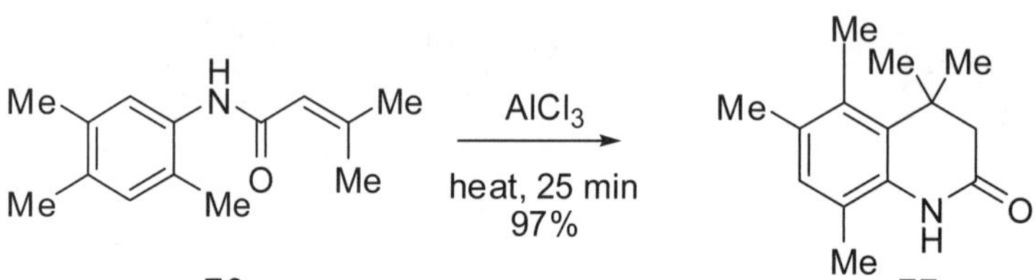

76 **77**

Many alternatives exist, including the lanthanide series of Lewis acids, which should mildly promote the chemistry. The addition of co-solvents has been reported, but due to the harshly acidic conditions, choices are limited to solvents robust to withstand the acidic conditions.[7]

Variation in the Leaving Group

In addition to alkenes or alkanols, ketone **72**, can be dehydrated as easily as an alcohol providing an alkene after cyclization (see compound **73**). Other leaving groups, which can provide a cationic species would also be able to promote cyclization; carboxylic acids could serve this role,[21] epoxides or alkyl halides. For example, cyclization with epoxide opening has been shown to provide good yield of the resulting tetrahydronapthalene. Cyclization to form the six-membered ring was reported as the predominant product for these cyclizations.[22] Thus $SnCl_4$ promoted cyclization of **78** and provided the primary alcohol **79** in 85% yield. The authors suggest an S_N2-like TS, in which the epoxide stabilizes the TS. Cyclizations using epoxide substrates should lend itself to the introduction of asymmetric centers.

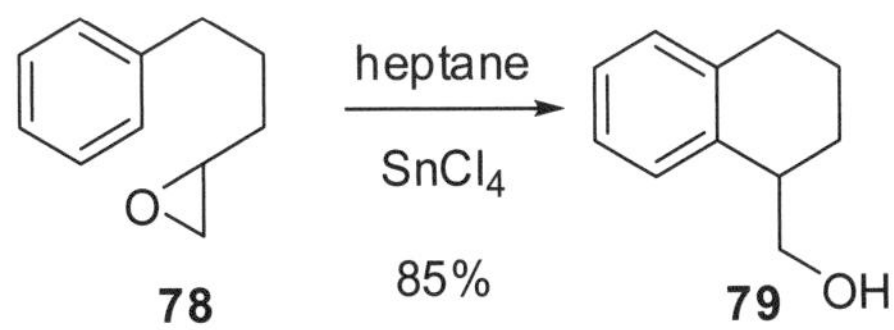

78 **79**

As one might expect, halides serve as good substrates, and Friedel–Crafts type alkylations have been reported.[7,23] In this instance, the cyclization product ratio and yield at similar temperature (25 °C) proved dependent on the solvent. In either case, acceptable yield of the desired products were obtained; the doubly cyclized product was suggested to arise from compound **81** after Lewis acid induced cation formation and further cyclization.

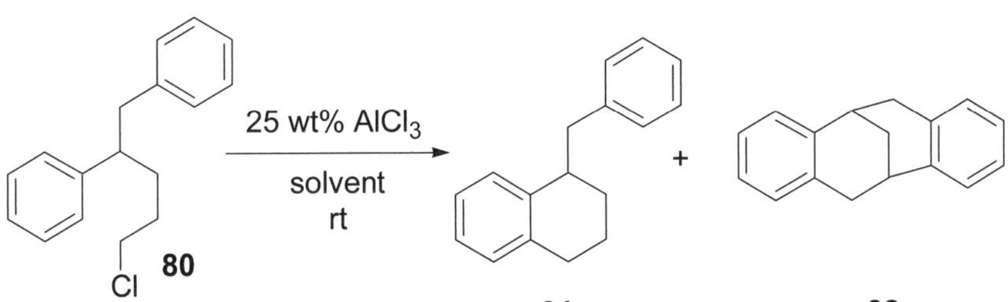

80 **81** **82**

solvent	product ratio **81**:**82**
Pet Ether	47:15
CS_2	76:10

4.3.7 Experimental

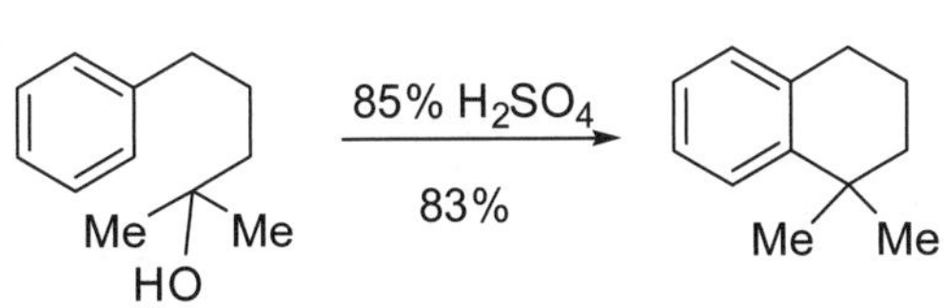

83 **84**

Preparation of 1,1-Dimethyl Tetralin (84)[24]

The procedure described by Bogert and co-workers was essentially followed for tertiary alcohols. One volume of alcohol was dropped into a cooled solution of 85% H_2SO_4 with good stirring while maintaining the temperature ≤ 10 °C. After addition, the reaction appeared complete but was allowed to warm to room temperature and stirred for an additional hour. While Bogert reported diluting with 10–15 volumes of water and distilling, Roberts's modification was to dilute with water and extract with ether. The ether layer was separated, washed with 10% sodium carbonate solution, followed by water, dried over anhydrous sodium sulfate, and finally distilled. For compound **84**, 83% yield, b.p. = 88 °C at 8–10 mm Hg.

Secondary alcohols could be dehydrated using 85% H_2SO_4 but by using 90% H_2SO_4 or by treatment with 85% H_2SO_4 with warming to 35–40 °C, the cyclized product could be realized.

Primary alcohols gave best yield when H_3PO_4 was used. This is covered in Section 4.1.

4.3.8 References

1. Bogert, M. T. *Science* **1933**, *77*, 289.
2. Perlman, D.; Davidson, D.; Bogert, M. T. *J. Org. Chem.* **1936**, 288–299.
3. (a) Cook, J. W.; Hewett, C. L. *J. Chem. Soc* **1933**, 1098–1112. (b) Cohen, A.; Cook, J. W.; Hewett, C. L.; Girard, A. *J. Chem. Soc.* **1934**, 653–658.
4. Perlman, D.; Davidson, D.; Bogert, M. T. *J. Org. Chem.* **1936**, 300–304
5. Barnes, R.; Hirsheler, H. P.; Bernstein, B.R. *J. Am. Chem. Soc.* **1952**, *74*, 32–34.
6. Sterling, E. C.; Bogert, M. T. *J. Org. Chem.* **1939** *4*, 20–28.
7. [R] Barclay, L. R. C. In *Friedel–Crafts and Related Reactions*, Vol. 2, Part II, Olah, G. A., ed., Wiley, New York, NY, 1964, pp 785–977.
8. Roblin, R. O.; Davidson, D.; Bogert, M. T. *J. Am. Chem. Soc.* **1935**, *57*, 151–159.
9. Barnes, R. A.; Olin, A. D. *J. Am. Chem. Soc.* **1956**, *78*, 3830–3833.
10. Barnes, R. A. *J. Am. Chem. Soc.* **1953**, *75*, 3004–3008.
11. Price, D.; Davidson, D.; Bogert, M. T. *J. Org. Chem.* **1938**, 540–545.
12. Khalaf, A. A.; Roberts, R. M. *J. Org. Chem.* **1972**, *37*, 4227–4235.
13. (a) VandeKamp, J.; Mosettig, E. *J. Am. Chem. Soc.* **1936**, *58*, 1062–1063. (b) Barnes, R. A.; Beachem, M. T. *J. Am. Chem. Soc.* **1955**, *77*, 5388–5390. (c) Rarnes, R. A.; Gottesman, R. T. *J. Am. Chem. Soc.* **1952**, *74*, 35–37.
14. Johnson, W. S.; Banerjee, D. K.; Schneider, W. P.; Gutsche, C. D.; Shelberg, W. E.; Chinn, L. J. *J. Am. Chem. Soc.* **1952**, *74*, 2832–2849.
15. Ireland, R. E.; Baldwin, S. W.; Welch, S. C. *J. Am. Chem. Soc.* **1972**, *94*, 2056–2066.
16. Brisse, F.; Lectard, A.; Schmidt, C. *Can. J. Chem.* **1974**, *52*, 1123–1134.
17. Adkins, H.; Hager, G. F. *J. Am. Chem. Soc.* **1949**, *71*, 2965–2968.
18. Eddy, N. B.; Murphy, J. G.; May, E. *J. Org. Chem.* **1957**, *22*, 1370–1372.
19. Smith, L. I.; Prichard, W. W. *J. Am. Chem. Soc.* **1940**, *62*, 778–780.
20. Andreana, T. L.; Cho, S. S. Y.; Graham, J. M.; Gregory, T. F.; Howard, H. R.; Kornberg, B. E.; Nikam, S. S.; Pflum, D. A. WO 2004/026864, April 1, 2004.
21. Smith, L. I.; Spillane, L. J. *J. Am. Chem. Soc.* **1943**, *65*, 202–208.
22. Taylor, S. K.; Hockerman, G. H.; Karrick, G. L.; Lyle, S. B.; Schramm, S. B. *J. Org. Chem.* **1983**, *48*, 2449–2452.

23. Roberts, R. M.; Anderson, G. P.; Khalaf, A. A.; Low, C.-E. *J. Org. Chem.* **1971**, *36*, 3342–3345.
24. Khalaf, A. A.; Roberts, R. M. *J. Org. Chem.* **1969**, *34*, 3571–3574.

4.4 Bradsher Cycloaddition and Bradsher Reaction

Paul Galatsis

4.4.1 Description

The Bradsher cycloaddition[1] is formally a $[4^{+} + 2]$ cycloaddition of quaternary aza-aromatic cations with dienophiles. Specifically, the reaction proceeds by an inverse electron demand Diels–Alder reaction in which the diene conatins a cationic aza-diene moiety.

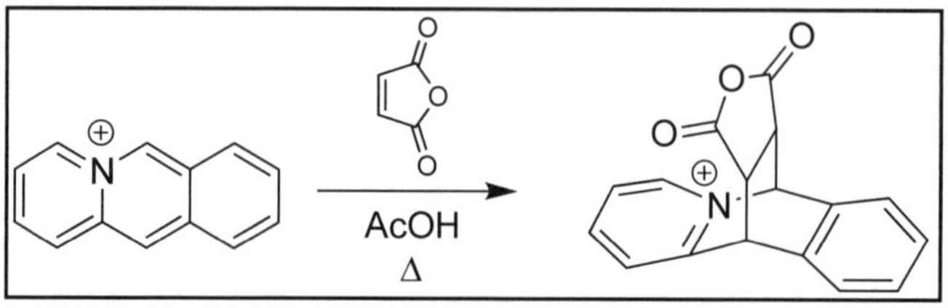

In the literature, the Bradsher name has also been associated with cyclodehydration reactions that give rise to fused aromatic ring systems. The Bradsher reaction[2] uses acidic conditions on diarylmethane carbonyl compounds to facilitate the ring closing process.

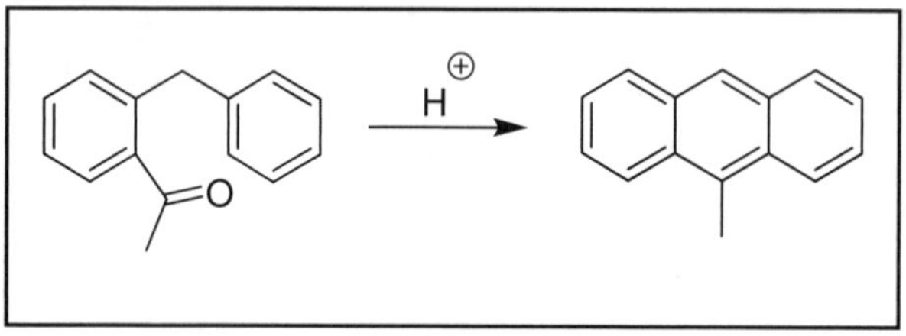

4.4.2 Historical Perspective

When first reported by Bradsher and Solomons in 1957,[3] this reaction represented the first example of a Diels–Alder reaction in which the diene was a quaternary ammonium salt. The reaction of acridizinium bromide with now classical dienophiles, e.g., maleic anhydride, malonate or fumarate esters and acrylonitrile was found to afford the expected Diels–Alder adducts. However, benzoquinone failed to generate any of the predicted products. While the concept of an inverse electron demand Diels–Alder reaction had theoretically been hypothesized much earlier,[4] this was the first practical example of such a reaction.

4.4.3 Mechanism

The Bradsher reaction is formally a $[4^+ + 2]$ Diels–Alder reaction. However, as a consequence of the aza cationic nature of the diene, this reaction proceeds by the inverse electron demand manifold. The classical Diels–Alder reaction employs the partnering of an electron-rich diene and an electron-deficient dienophile to provide the proper interaction of the highest occupied molecular orbital (HOMO) and lowest unoccupied molecular orbital (LUMO) as prescribed by frontier molecular orbital theory (FMO)[5] to generate the observed adducts. Thus FMO theory interprets this reaction proceeding via the HOMO of the diene with the LUMO of the dienophile. In the case of the inverse electron demand reaction, the electronics of the reaction are inverted. Therefore, in the Bradsher reaction, the electron-deficient aza cation diene's LUMO interacts with the HOMO of an electron rich dienophile. This mechanistic pathway provided a rationalization for the lack of reactivity of the electron-deficient tetracyanoethylene (TCNE), while electron-rich styrenes afforded the predicted product from reaction of **1** to generate **2**.[6]

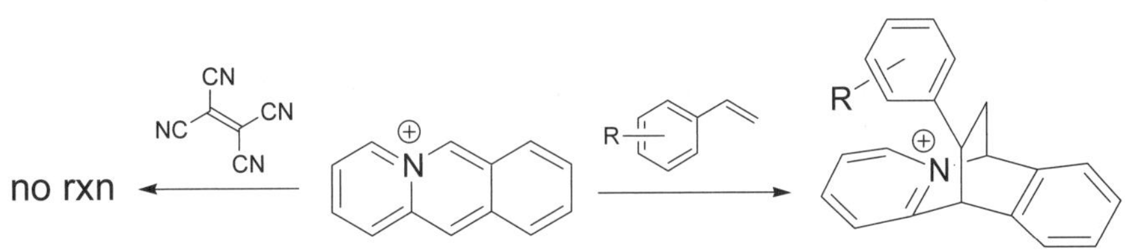

1 **2**

As the substituents on the styrene increase electron density on the alkene moiety the reaction rate increases (see Table 1). This is consistent with FMO theory associated with the inverse electron demand Diels–Alder reaction. As one increases the electron density on the dienophile, the energy of the HOMO also increases and narrows the energy gap with the LUMO of the diene. Thus allowing the reaction to proceed with greater efficiency and a concomitant increase in reaction rate.

Table 1. Effect of Substituents on Rate of Cycloaddition

R	rate
p-NO_2	2.0
H	5.8
p-Me	9.8
p-MeO	25

Additional mechanistics studies have also been published.[7] A Hammett linear free energy relationship was found to exist with the greatest correlation using σ_p substituent constants. While these data tended to imply a two-step mechanistic pathway, with initial bond formation proximal to the cationic nitrogen atom, further multiparametric linear free energy relationships and an analysis using FMO confirmed the polar cycloaddition reaction pathway, albeit in an asynchronous fashion. In addition, the potential for a charge transfer complex preceding the cycloaddition transition state was consistent with the observed data. Most recently,[8] an analysis of the reaction using DFT calculations at the B3LYP/6-31G(d) level add additional support for a reactant-like transition state with asynchronous bond formation.

4.4.4 Variations and Improvements

Most examples of the Bradsher cycloaddition reaction have utilized fused polycyclic aromatics as the cationic aza-diene fragment. Falck and co-workers[9] have reported that one can carry out this reaction using monocyclic quaternary aza-aromatics. The application of this methodology was illustrated using the *N*-(2,4-dinitrophenyl) salt of *N*,*N*-diethylnicotin-amide **3** and ethyl nicotinate **4** in conjunction with enol ethers. The reaction proceeded at room temperature to generate adducts **5**. This was the result of the *exo*-addition at the C2–C5 positions of the pyridyl ring. The resultant iminium ion was then trapped by the methanolic solvent.

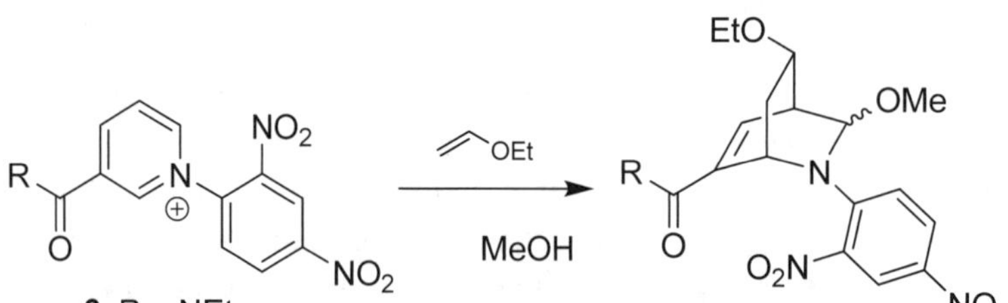

3 R = NEt$_2$
4 R = OEt
5

As alluded (vide supra), some confusion may arise with respect to this named reaction as there is reference in the literature to an alternative reaction with the same name. The Bradsher reaction[2] forms aromatic rings but via an acid-catalyzed Friedel–Crafts-like process. Thus diaryl-methanes having a carbonyl group in the ortho position can undergo a cyclodehydration reaction to generate the corresponding anthracene derivatives. In this respect, the Bradsher reaction is related to the Elbs reaction,[10] which involves the pyrolytic cyclization of diaryl ketones **6** having an *ortho* methyl or methylene substituent for the formation of polycyclic aromatics **7**.

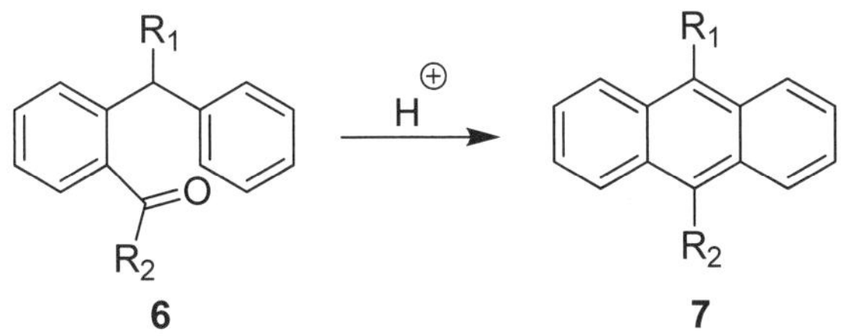

For $R_1 = R_2$ = H, alkyl, the reaction is classified as the proper Bradsher reaction. If R_2 = aryl, the variation has been classified as the Vingiello process.[11] There has been some discussion in the literature as to the mechanistic pathway for these variations and two mechanisms remain to be resolved.[12]

4.4.5 Synthetic Utility

Bradsher cycloaddition reaction

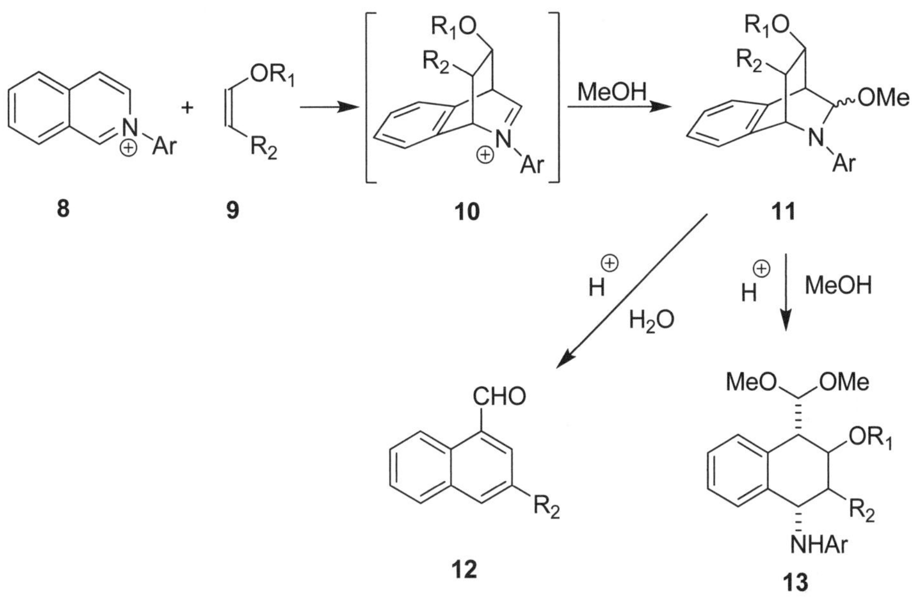

Initially, the Bradsher cycloaddition reaction was used to gain greater understanding of the Diels–Alder reaction. More recently it has found utility in the constuction of carbon frameworks that provided novel entries into the synthesis of natural products. For example, the Bradsher cycloaddition reaction has been reported on the use of isoquinolinium salts for the stereocontrolled synthesis of substituted tetralins.[13] Reaction of isoquinolinium salts **8** (Ar = 2,4-dinitrophenyl) with enol ethers **9**, initially afforded adduct **10**, which was trapped by solvent to generate the azabicyclo[2.2.2] **11**. Exposure to acidic conditions transformed **11**, via aminol hydrolysis, into tetralins **13**. If the process was carried out under acidic aqueous conditions, then naphthalenes **12** were obtained by aromatization of the ring opened species.

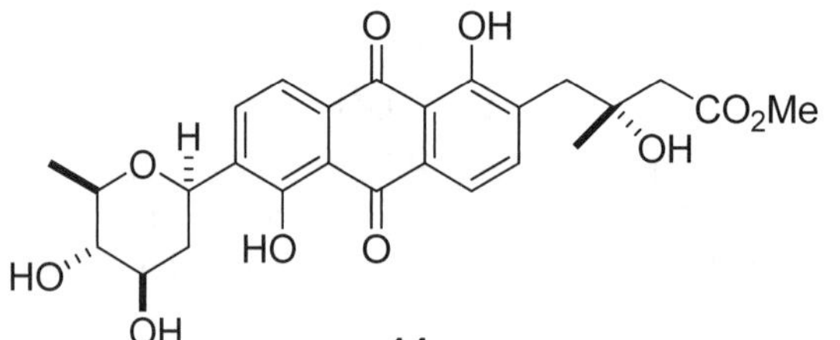

14

The secondary metabolite vineomycinone B_2 methyl ester, **14**, which displayed antitumor/antibiotic activity and has a pharmacophore similar to the anthracyclines, was prepared using a double Bradsher cycloaddition reaction process.[14] This convergent strategy began with the cycloaddition reaction of salt **15** and enol ether **16**. The adduct **17** was produced after intramolecular trapping by the pendant alcohol functionality. Unmasking of the right-hand portion of the molecule affored **18**, which was set up to execute a second Bradsher cycloaddition reaction. Formation of the 2,4-dinitrophenyl salt preceded cycloaddition with enol ether **19**. Acid-catalyzed unmasking of the left-hand portion of the molecule generated **20**, which was elaborated into **14**.

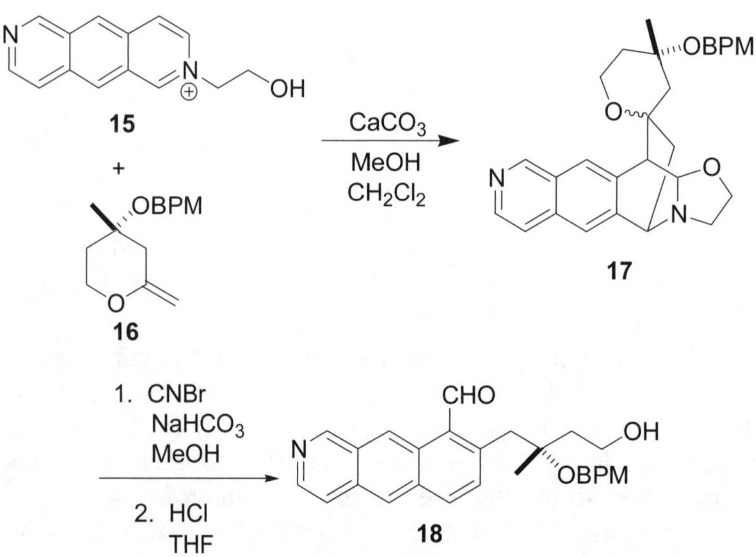

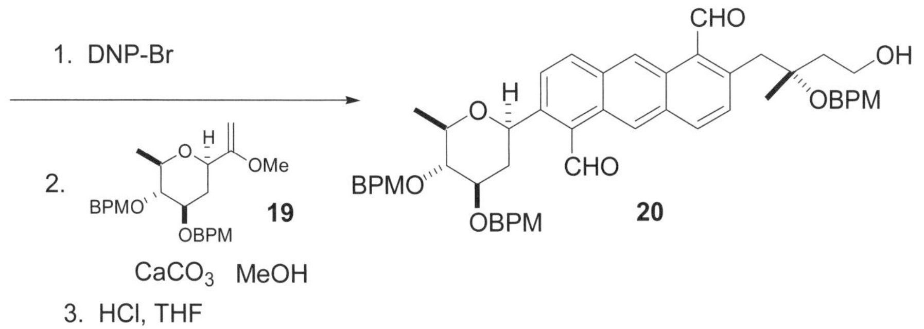

In a convergent approach to the nogalamycin members of the anthracycline family of antibiotics, two cycloaddition reactions were used in the construction of a CDEF-ring analog, **22**, of nogalamycin, **21**.[15] A nitrile oxide furan [3 + 2] cycloaddition was used in the construction of heterocycle **24**. This advanced intermediate was partnered with the isoquinolium salt **23** for participation in a Bradsher cycloaddition reaction. Standard conditions for this reaction did not afford any of the desired product. It was determined that high-pressure conditions were required to drive the reaction manifold toward the requisite intermediate **25** after unmasking of the cycloadduct.

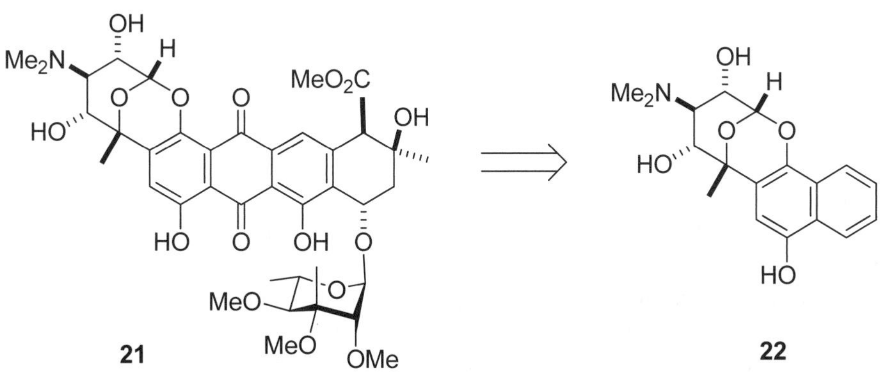

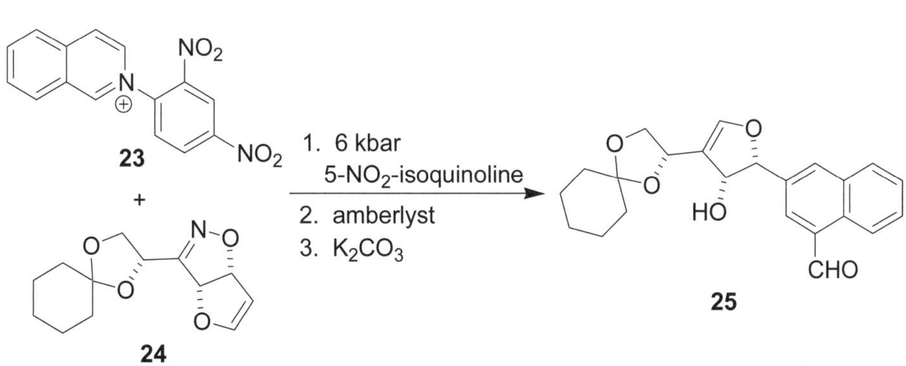

The Franck group[16] has also reported on the use of the Bradsher cycloaddition reaction for the generation of a functionalized B-ring of angucycline antibiotic sakyomicin A, **26**. Their approach began with the cycloaddition reaction of **27** with the TBDMS enol ether of acetaldehyde to generate adduct **28**. A short sequence transformed this compound into **29**, the first synthetic model for the dihydronaphthalene framework of **26**.

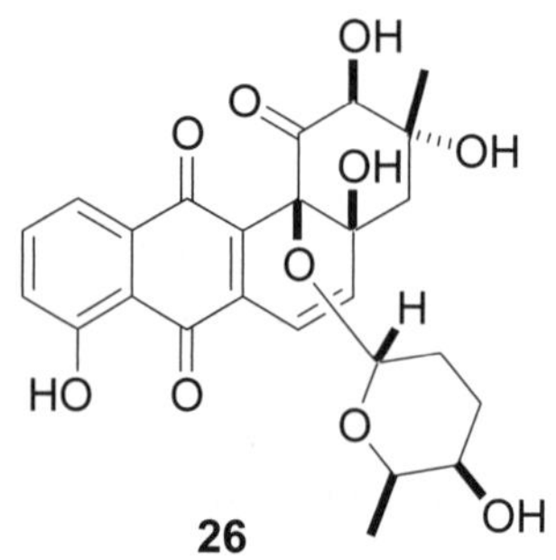

26

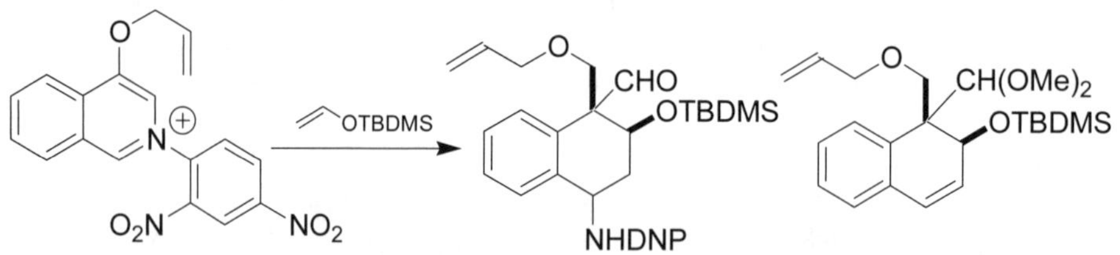

27 **28** **29**

In a series of papers, the Langlois group reported multiple strategies toward the construction of manzamine A, **30**. An approach to the construction of the ABC tricyclic core framework, **31**, employed the Bradsher cycloaddition reaction as the key synthetic step.[17] Thus reaction of isoquinolium salt **32** with ethyl vinyl ether under standard reaction conditions afforded the cycloadduct **33**. The advanced intermediate **31** was then generated in six steps.

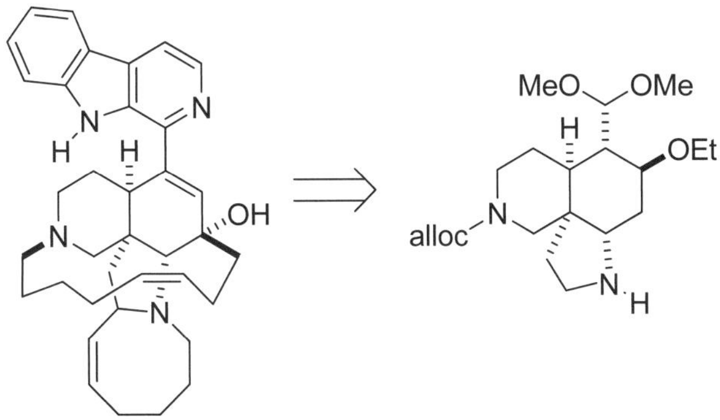

30 **31**

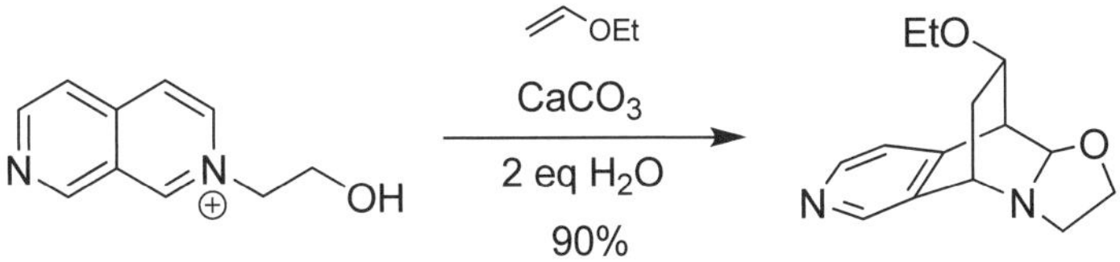

32 **33**

In a related manner the ABE tricyclic core, **34**, of manzamine A, **30**, was reported.[18] Bradsher cycloaddition reaction of naphthyridium salt **35** with an elaborated enol ether afforded cycloadduct **36**. Three steps were required to transform this compound into the advanced intermediate **37**, a synthon for scaffold **34**.

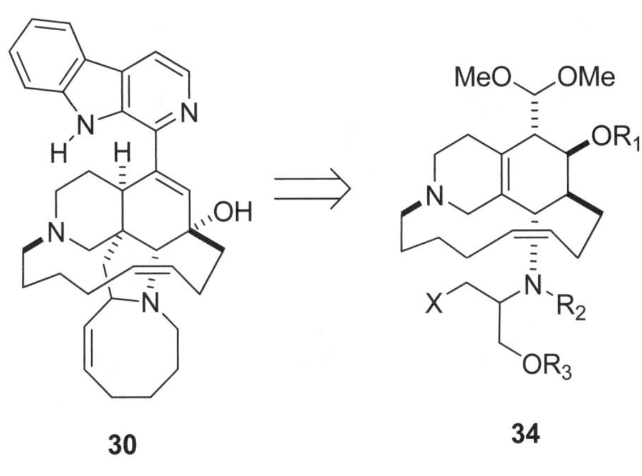

30 **34**

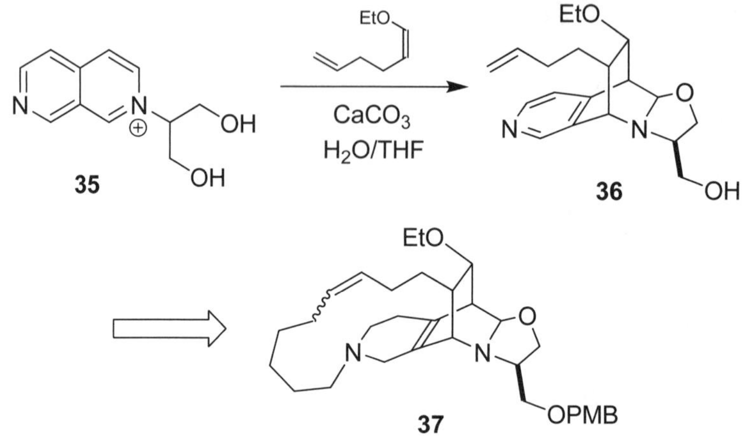

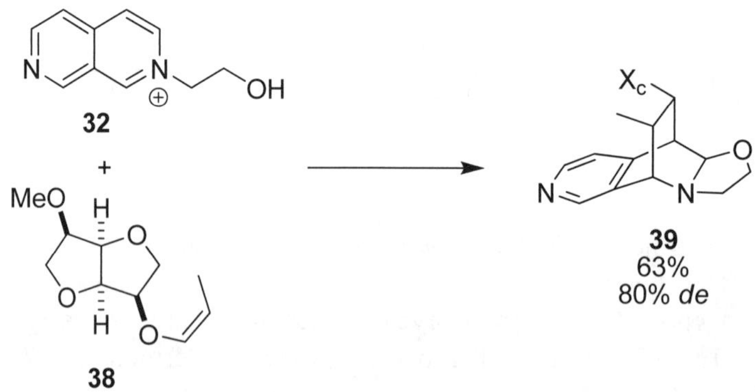

Building on their earlier work and in preparation for an enantioselective approach to manzamine A, **30**, the Langlois group examined asymmetric variations on the Bradsher cycloaddition reaction. An initial report[19] focused on using chiral (*Z*)-enol ether **38** with the previously used naphthyridinium salt **32**. The cycloaddition reaction afforded **39** in good yield with 80% *de*.

An alternative asymmetric Bradsher cycloaddition reaction reversed the sense of chirality, which resulted in the asymmetric center being incorporated into the naphthyridinium salt **40**.[20] Execution of the cycloaddition reaction with the elaborted enol ether **41** afforded cycloadduct **42**. The overall yield and *de* were found to be similar to the previous example.

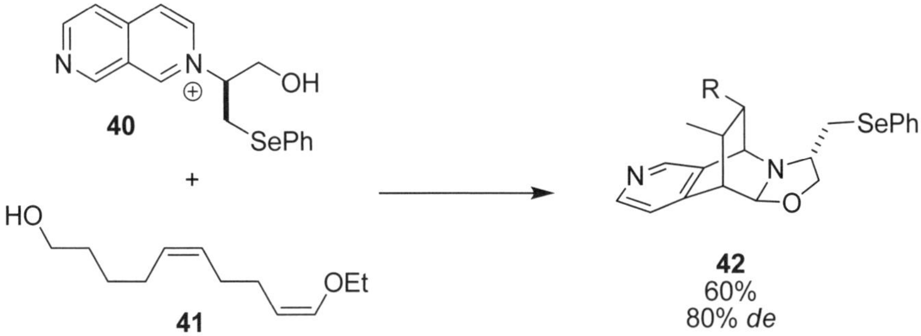

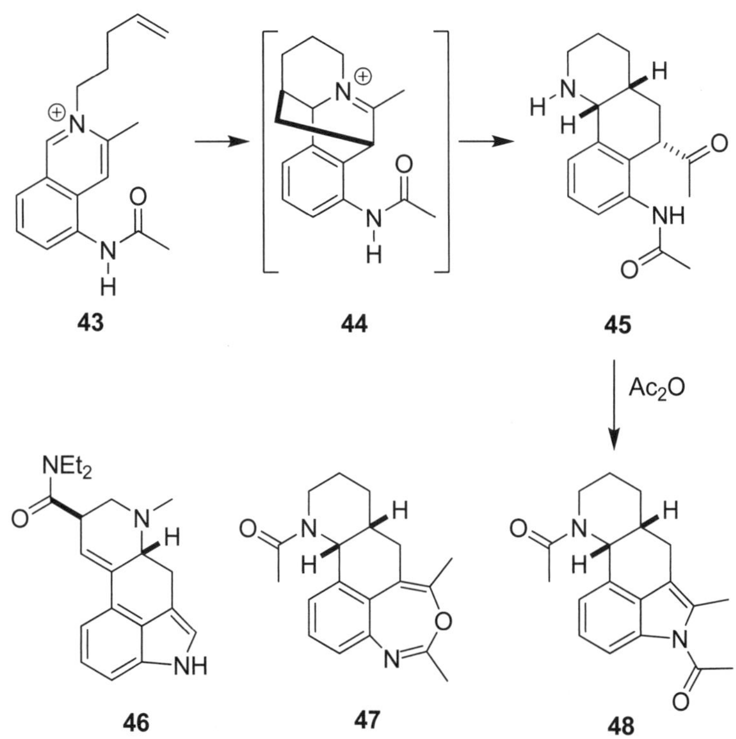

Sammes and co-workers[21] reported studies on an intramolecular variation on the Bradsher cycloaddition reaction. However, the Franck group,[22] publishing on the construction of benz[*c*,*d*]indole frameworks, reported an alternate structure for the product of their reaction. Intramolecular cycloaddition of the isoquinolinium salt **43** initially afforded adduct **44**. This intermediate rapidly converted to **45**, which gave **48** after acetylation. The Sammes group had assigned **47** to the final compound but the Franck group, based on NMR and crystal structure data from a related system, proposed the alternate structure **48**. Thus this chemistry provided entry to the benz[c,d]indole system found in such natural products as lysergic acid diethylamide (LSD), **46**.

Bradsher reaction

In parallel with the previous chemistry, the Bradsher reaction has also found utility in organic synthesis. This reaction has a narrower scope due to the polyaromatic ring framework that results from this cyclodehydration chemistry.

Early reports from the Bradsher group highlighted new routes to anthracene derivatives **7**.[23] In the event, conversion of chloride **49** to ketone **6** was accomplished by treatment with cuprous cyanide followed by 1,2-addition of a Grignard reagent. The cyclized product **7** was obtained by heating with hydrobromic acid. It was reported later that liquid sulfur dioxide as solvent was effective in facilitating the aromatic cyclodehydration.[24]

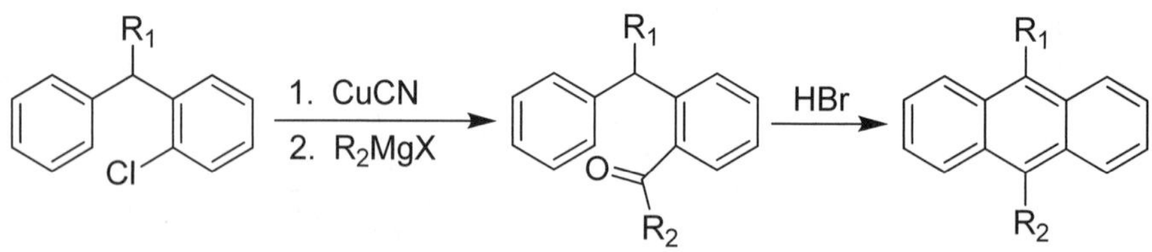

49 **6** **7**

The Bradsher reaction was also found to be effective in the preparation of heterocyclic systems.[25] For example, reaction of indole **50** with bromoacetone afforded indolo[2,3-*a*]acridinium salt **51**.

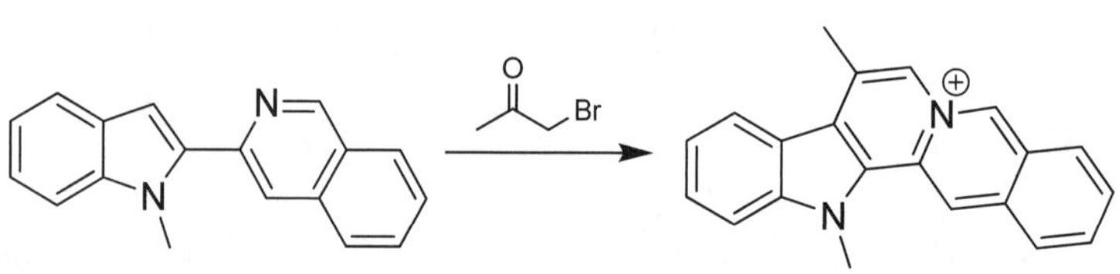

50 **51**

Polycyclic aromatic hydrocarbons can easily be prepared using this methodology.[26] Heating the diketone **52** with hydrobromic acid facilitated the double Bradsher reaction to generate hydrocarbon **53**. This approach was more efficient than the Elbs reaction in that higher yields and no rearrangements were observed.

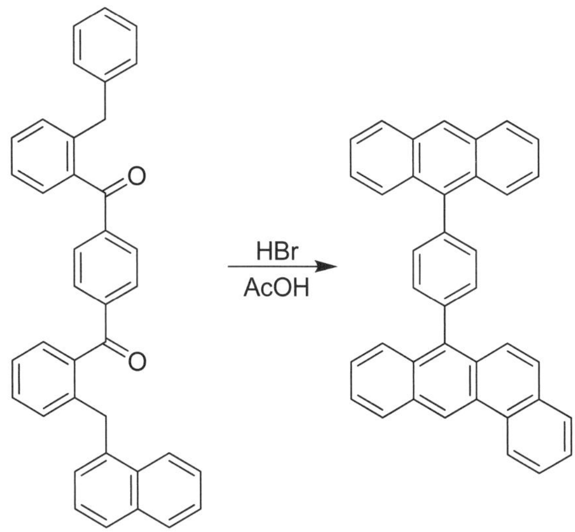

52 53

Application of the Bradsher reaction to the synthesis of natural products has also been reported. The first total synthesis of benzopyrano-benzazepine alkaloid (±)-clavizepine, **54**, was accomplished using a Bradsher reaction as the key step.[27] Thus the tricyclic core of **54** was assembled by stannic chloride induced cyclization of **55** to afford a mixture of **56** and **57**. If methanol-free dichloromethane was used in this cyclization, then **56** could be obtained as the sole product in 66% yield. Alternatively, **56** could be converted in **57** in excellent yield with PPTS.

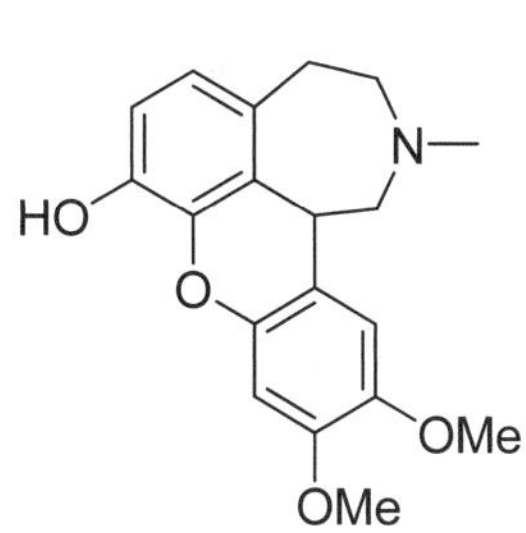

54

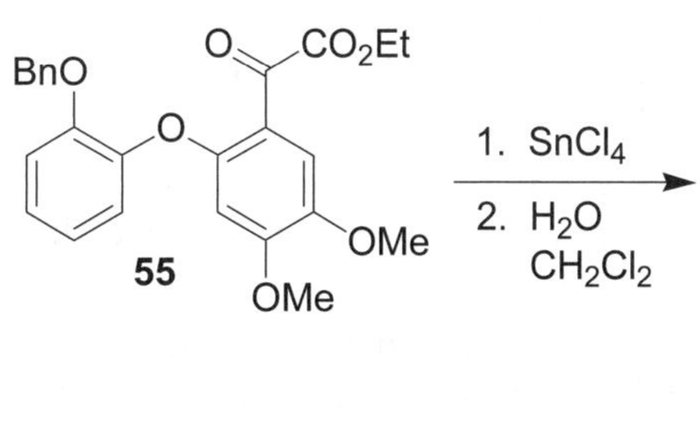

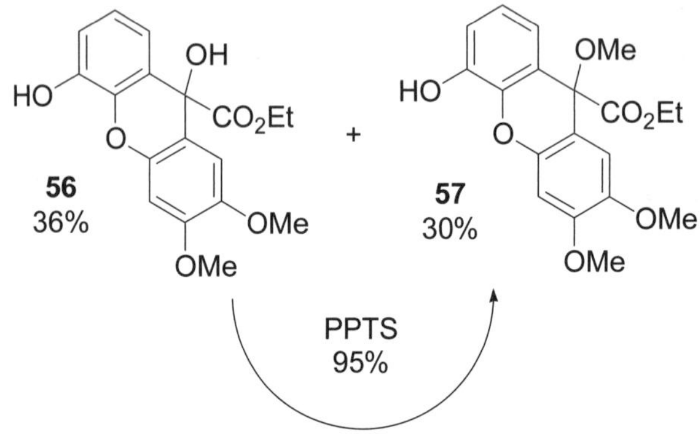

4.4.6 Experimental

Bradsher cycloaddition reaction[13]
(
10-Dimethoxymethyl-3,4,4a,5,10,10a-hexahydro-2*H*-benzo[*g*]chromen-5-yl)-(2,4-dinitro-phenyl)-amine 60

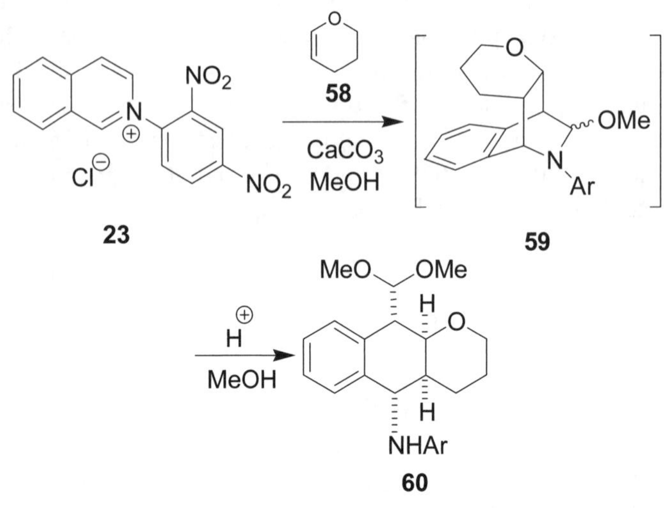

To a solution of isoquinolinium chloride **23** (332 mg, 1 mmol) in 5 mL anhydrous MeOH was added powdered anhydrous calcium carbonate (600 mg, 6 mmol) and vinyl ether **58** (2 mmol). The reaction was stirred under N_2 at 45 °C and judged completed by monitoring for the disappearance of **23** by TLC. The reaction mixture was filtered through Celite and washed with DCM, and the combined organics were concentrated to give the cycloadduct **59**, which was used directly in the next step.

The crude **59** from the previous step was diluted with anhydrous MeOH (10 mL), and Dowex-50×8–400 (250 mg) was added. The mixture was stirred at room temperature for 24 h before filtering the resin. The resin was washed with DCM, the combined filtrates diluted with 150 mL of water, and the aqueous mixture extracted with DCM (3 × 30 mL). The combined extracts were washed once with satd $NaHCO_3$ (20 mL), dried ($MgSO_4$), filtered, and chromatographed to afford **60** as a solid (88%).

Bradsher reaction[28]

12-*m*-Tolyl-benzo[*a*]anthracene 62

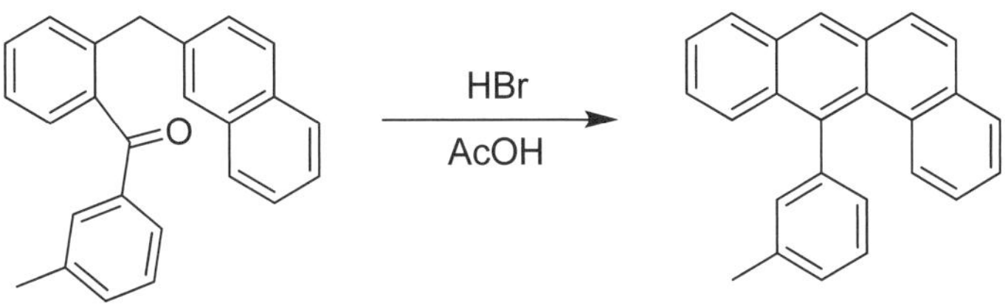

61 62

Ketone **61** (1 g) was dissolved in 15 mL 48% hydrobromic acid and 30 mL glacial acetic acid. The reaction mixture was heated in a sealed tube at 180 °C for 3 h before cooling to room temperature and extracting with benzene. The organic phase was washed with water, dried ($CaCl_2$), filtered, and concentrated. After purification by chromatography, **62** (0.9 g, 95%) was obtained as a colorless solid.

4.4.7 References

1. [R] (a) Bradsher, C. K. *Adv. Heterocyclic Chem.* **1974**, *16*, 289. (b) Katritzky, A. R.; Dennis, N. *Chem. Rev.* **1989**, *89*, 827–861.
2. [R] (a) Bradsher, C. K. *Chem. Rev.* **1946**, *38*, 447. (b) Bradsher, C. K. *Chem. Rev.* **1987**, *87*, 1277.
3. Bradsher, C. K.; Solomons, T. W. G. *J. Am. Chem. Soc.* **1958**, *80*, 933–934.
4. (a) Bachmann, W. E.; Deno, N. O. *J. Am. Chem. Soc.* **1949**, *71*, 3062. (b) Carboni, R. A.; Lindsey, R. V. *J. Am. Chem. Soc.* **1959**, *81*, 4342. (c) Sauer, J.; Wiest, H. *Angew. Chem., Intl. Ed.* **1962**, *1*, 269. (d) Sauer, J. *Angew. Chemie Intl. Ed.* **1967**, *6*, 16. Fields, D. L.; Regan, T. H.; Dignan, J. C. *J. Org. Chem.* **1968**, *33*, 390. [R] (e) Boger, D. L. *Chem. Rev.* **1986**, *86*, 781.
5. For leading references see: [R] (a) Fleming, I. *Frontier Orbitals and Organic Reactions.* Wiley, **1976**, pp 86–181. [R] (b) Anslyn, E. V.; Dougherty, D. A. *Modern Physical Organic Chemistry,* University Science Books **2006**, pp 877–934.
6. Bradsher, C. K.; Stone, J. A. *J. Org. Chem.* **1968**, *33*, 519.
7. (a) Westerman, I. J.; Bradsher, C. K. *J. Org. Chem.* **1971**, *36*, 969. (b) Porter, N. A.; Westerman, I. J.; Wallis, T. G.; Bradsher, C. K. *J. Am. Chem. Soc.* **1974**, *96*, 5104. (c) Bradsher, C. K.; Carlson, G. L. B.; Porter, N. A.; Westerman, I. J.; Wallis, T. G. *J. Org. Chem.* **1978**, *43*, 822. (d) Bradsher, C. K.; Wallis, T. G.; Westerman, I. J. Porter, N. A. *J. Am. Chem. Soc.* **1977**, *99*, 2588. (e) Westerman, I. J.; Bradsher, C. K. *J. Org. Chem.* **1979**, *44*, 727. (f) Gupta, R. B.; Franck, R. C. *J. Am. Chem. Soc.* **1987**, *109*, 5393.
8. Tamilmani, V.; Senthilnathan, D.; Venuvanalingam, P. *J. Chem. Sci.* **2008**, *120*, 225.

9. Falck, J. R.; Wittenberger, S. J.; Rajapaksa, D.; Mioskowski, C.; Boubia, B. *J. Chem. Soc., Perkin Trans. 1* **1990**, 413.
10. [R] Fieser, L. F. *Org. React.* **1942**, *1*, 129.
11. See for example: (a) Bradsher, C. K.; Vingiello, F. A. *J. Org. Chem.* **1948**, *13*, 786. (b) Vingiello, F. A.; Spangler, M. O. L.; Bondurant, J. E. *J. Org. Chem.* **1960**, *25*, 2091. (c) Vingiello, F. A.; Thornton, J. R. *J. Org. Chem.* **1966**, *31*, 659.
12. (a) Brice, L. K.; Katstra, R. D. *J. Am. Chem. Soc.* **1960**, *82*, 2669. (b) Lupes, M. E.; Lupes, S. A. L. *Sci. Bull. Polytech. Instit. Bucharest Chem. Mat. Sci.* **1990**, *52*, 53. (c) Lupes, M.; Lupes, S. *Sci. Bull. Polytech. Instit. Bucharest Chem. Mat. Sci.* **1990**, *52*, 77.
13. a) Gupta, R. B.; Franck, R. W. *J. Am. Chem. Soc.* **1987**, *109*, 5393. b) Gupta, R. B.; Franck, R. W.; Onan, K. D.; Soll, C. E. *J. Org. Chem.* **1989**, *54*, 1097.
14. Bolitt, V.; Mioskowski, C. *J. Am. Chem. Soc.* **1991**, *113*, 6320.
15. Yin, H.; Franck, R. W.; Chen, S.-L.; Quigley, G. J.; Todaro, L. *J. Org. Chem.* **1992**, *57*, 644.
16. Nicolas, T. E.; Franck, R. W. *J. Org. Chem.* **1995**, *60*, 6904.
17. Magnier, E.; Langlois, Y.; Merienne, C. *Tetrahedron Lett.* **1995**, *36*, 9475.
18. Magnier, E.; Langlois, Y. *Tetrahedron Lett.* **1998**, *39*, 837.
19. Sageot, O.; Monteux, D.; Langlois, Y. *Tetrahedron Lett.* **1996**, *37*, 7019.
20. Urban, D.; Duval, E.; Langlois, Y. *Tetrahedron Lett.* **1996**, *41*, 9251.
21. Gisby, G. P.; Sammes, P. G.; Watt, R. A. *J. Chem. Soc., Perkin Trans. 1* **1982**, 249.
22. Soll, C. E.; Franck, R. W. *Heterocycles* **2006**, *70*, 531.
23. (a) Bradsher, C. K. *J. Am. Chem. Soc.* **1940**, *62*, 486. (b) Bradsher, C. K.; Smith, E. S. *J. Am. Chem. Soc.* **1943**, *65*, 451.
24. Bradsher, C. K.; Sinclair, E. F. *J. Org. Chem.* **1957**, *22*, 79.
25. Bradsher, C. K.; Litzinger, E. F., Jr. *J. Org. Chem.* **1964**, *29*, 79.
26. Saraf, S. D.; Vingiello, F. A. *Synthesis* **1970**, 655.
27. Ishibashi, H.; Takagaki, K.; Imada, N.; Ikeda, M. *Tetrahedron* **1994**, *50*, 10215.
28. Vingiello, F. A.; Borkovec, A. *J. Am. Chem. Soc.* **1955**, *77*, 4823.

4.5 Bradsher Reaction

Mukulesh Mondal and Nessan J. Kerrigan

4.5.1 Description

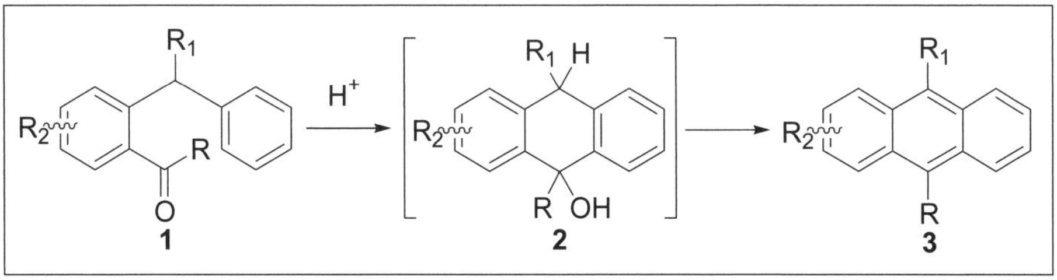

The acid-catalyzed cyclization of an acyl substituted diarylalkane **1** into a cyclic carbinol **2**, followed by a 1,4-dehydration to produce anthracene derivative **3**, is known as the Bradsher Reaction.[1,2]

4.5.2 Historical Perspective

Charles K. Bradsher[2] (born in Petersburg, Virginia, in 1912) introduced this reaction in 1940.[1] After obtaining a Ph.D. in 1933 from Harvard University and carrying out postdoctoral work at the University of Illinois, he joined the faculty at the department of chemistry, Duke University and achieved the rank of James B. Duke Professor in 1965.

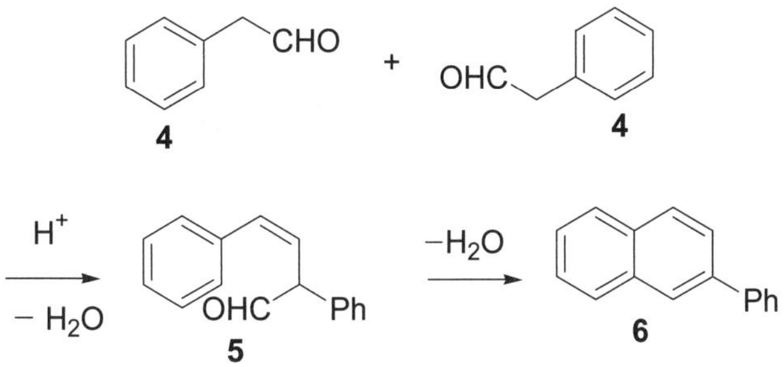

This particular reaction was observed for the first time by Bergmann during the hydrolysis of *o*-benzylbenzaldehyde in boiling hydrochloric acid, when he isolated a small amount of anthracene.[3] A very similar reaction, perhaps the first case of aromatic cyclodehydration was reported by Zincke and Breuer in 1884, when they observed the formation of *β*-phenylnaphthalene from phenylacetaldehyde in boiling sulphuric acid (50%).[4] Zincke postulated that aldol condensation of two molecules of phenylacetaldehyde followed by cyclodehydration furnished the naphthalene derivative **6**.

For many years applications of the Bradsher reaction were restricted due to its limited substrate scope and requirement for harsh reaction conditions.[2,5] However, after the advancement of the arene oxide concept concerning the metabolism of polycyclic aromatic hydrocarbons, synthesis of all the nuclear monohydroxylated derivatives of 7,12-dimethylbenz[*a*]-anthracene (DMBA), diol epoxide metabolites of DMBA, and fluoro derivatives of DMBA was undertaken for carcinogenicity and mutagenicity determination studies.[6–10] Interest in the Bradsher reaction has increased greatly as a consequence of the need to construct these polycyclic aromatic hydrocarbons. Development of fluoroanthracenylmethyl cinchonidine as an efficient phase-transfer catalyst for asymmetric glycine alkylation also expanded the scope of the Bradsher reaction.[11,12]

4.5.3 Mechanism

Bradsher proposed that the mechanism initially involved reversible addition of the proton to the carbonyl group of **1** to form **7**, followed by intramolecular Friedel–Crafts-type electrophilic attack on the aryl ring by the positively charged species in **8**.[2,5,13–15] Restoration of aromaticity through proton loss from **9** followed by acid-catalyzed 1,4-dehydration of carbinol **2** furnished the anthracene derivative **3**.

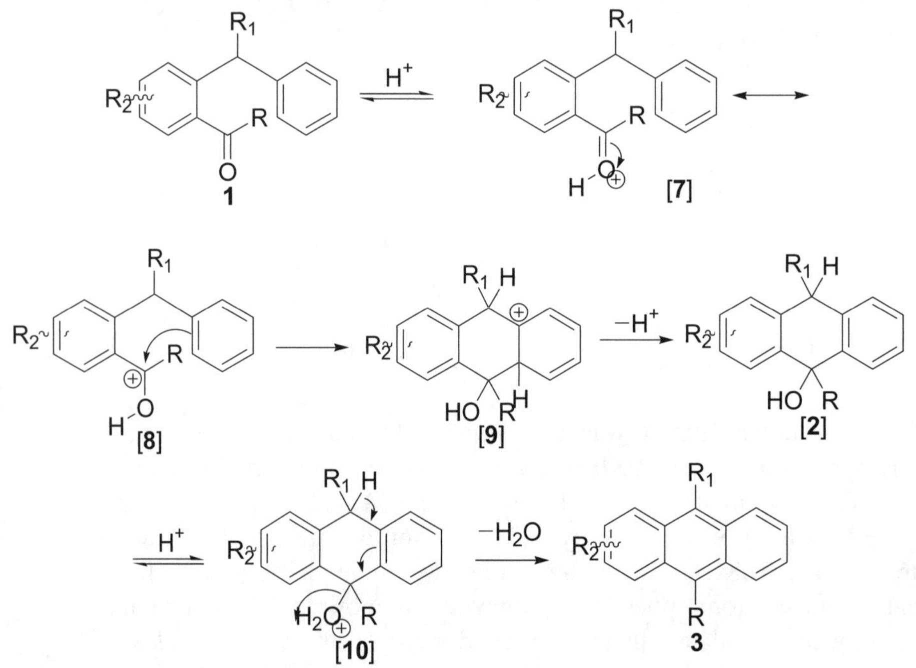

A kinetics study by Brice and Katstra showed that during the cyclization of *o*-benzylbenzophenone **11** to 9-phenylanthracene **12** catalyzed by HBr in an acetic acid-water mixture, the reaction rate retards greatly as the mole fraction of water in the solvent increases.[16] A dramatic increase in cyclization rate can be achieved if the HBr concentration is held constant but the mole ratio of acetic acid and water in the mixture is increased.

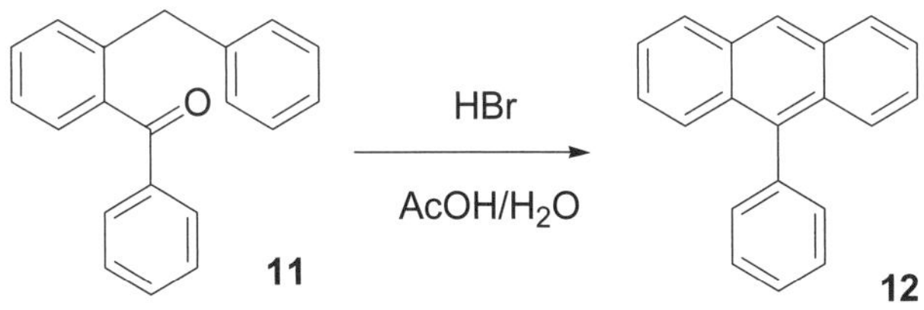

To prove that the cyclization of *o*-benzylbenzophenone (**11**) did not proceed through an enolic intermediate, Bradsher and Smith refluxed the ketimine **13** with hydrobromic acid (48%) and observed that cyclization takes place despite the inability of the expected intermediate **14** to enolize. 9,10-Dihydroanthracene derivative **15** was formed through concomitant reduction under the reaction conditions.[17]

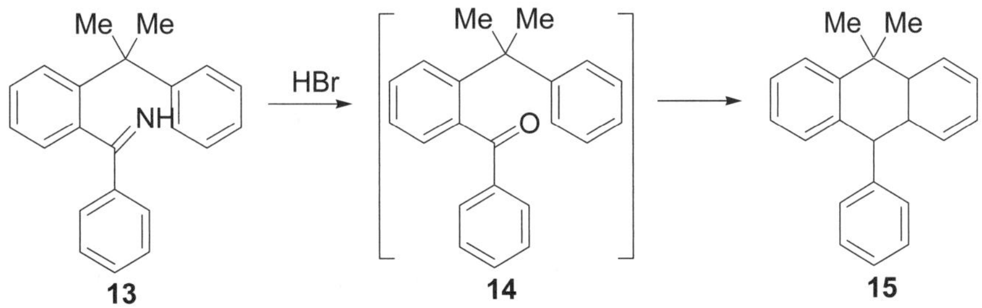

Substituents play an important role in the rate of cyclization of various *o*-benzylphenyl phenylketones. In the case of *o*-benzylphenyl phenyl ketones **1** (R = Ph, $R_1 = R_2 = H$) in which the phenyl group has substituents, it was expected that electron-releasing substituents should stabilize the positive charge of the conjugate acid **8** and hence slow the cyclization to **2**. On the other hand, the same substituents, by increasing the basicity of the ketone **1**, should increase the concentration of **8** at equilibrium.

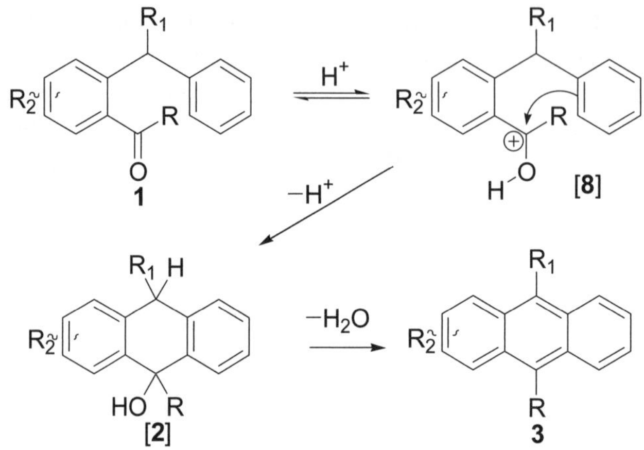

Bradsher and Vingiello observed that when the substituent on the benzoyl group was in the *para* position, the rate of cyclization for methyl, hydrogen, chlorine or bromine substituents was the same, within experimental error of each other.[18] However, in the case of the fluorine substituent, a significant lowering of rate was observed. In a similar experiment, Vingiello's group observed that the rate of cyclization increased slightly for the same halogen substituents when the position is changed from *para* to *meta*.[19] In the case of halogens in the *para* position, the +M (Mesomeric) effect in addition to the –I (Inductive) effect is responsible for a slight lowering of the rate of cyclization compared to the *meta*-substituted case where only –I effect is in operation. However, when the trifluoromethyl group is in the *para* position, with the strong –I effect and electron withdrawing hyperconjugation effect, a higher rate of cyclization than that in the *meta* position, where only –I effect is operative, was observed. Since the +M effect of the methyl group is of hyperconjugative origin and its magnitude is smaller in comparison to that of a halogen, the rate of cyclization for *para*- and *meta*-methyl substituents is almost the same. Hence the electronic effects on the positive nature of the conjugate acid **8** are more important in determining the overall reaction rate than those affecting the acid–base equilibrium.

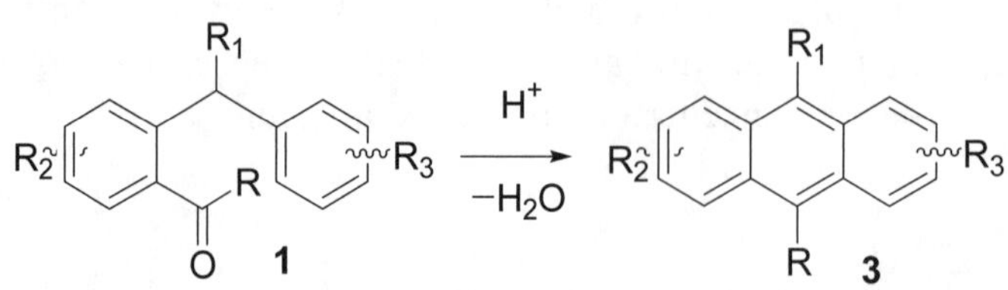

Rates of cyclization of some *o*-benzylphenyl aryl ketones **1**

Entry No.	R	$R_1 = R_2 = R_3$	K, $(h^{-1})\times10^{-2}$
1	*p*-$CH_3C_6H_4$	H	4.2
2	*p*-BrC_6H_4	H	4.2
3	*p*-ClC_6H_4	H	4.1
4	*p*-FC_6H_4	H	2.8
5	*p*-CF_3	H	9.3
6	C_6H_5	H	4.4
7	*m*-$CH_3C_6H_4$	H	4.4
8	*m*-BrC_6H_4	H	5.0
9	*m*-ClC_6H_4	H	5.3
10	*m*-FC_6H_4	H	5.3
11	*m*-CF_3	H	6.4

Analogous *ortho*-substituents on the benzoyl group show comparatively large variations in rates of cyclization.[20] All *ortho*-substituted compounds cyclize more slowly than the unsubstituted 2-benzylbenzophenone. The variation appears to be due to the increasing steric requirements of the substituents in the expected order Br > Cl > F. In this case, the steric bulk of the substituent inhibits the attack of the electropositive carbon atom on the nucleophilic benzene ring.

Rates of cyclization of some *o*-benzylphenylbenzophenone **1** at 150 °C

Entry No.	R	$R_1 = R_2 = R_3$	K, $(h^{-1})\times10^{-2}$
1	C_6H_5	H	55
2	*o*-$CH_3C_6H_4$	H	9.5
3	*o*-BrC_6H_4	H	3.3
4	*o*-ClC_6H_4	H	6.9
5	*o*-FC_6H_4	H	38.6

The rate of cyclization of *o*-(1-phenylethyl)phenyl alkyl ketone **1** (R = alkyl, R_1 = Me, R_2 = H) decreases steadily from methyl to *n*-butyl and then remains constant within the limit of experimental error.[21] The major factor in the large decrease in rate of cyclization with increasing chain length is most likely steric interactions.

In contrast to the small rate change observed by modifying the electronic environment of the carbonyl group of *o*-benzylphenyl aryl ketones, a large change in the rate of cyclization can be made by modifying the availability of electrons at the *ortho* position of the nucleophilic benzene. The introduction of a methyl group at the 3 position of the benzyl ring of **1** (R = Ph, $R_1 = R_2$ = H, R_3 = *m*-CH_3) makes the cyclization 55 times faster compared to the unsubstituted ring, while replacing the methyl group with a trifluoromethyl group (R = Ph, R_1 = R_2 = H, R_3 = *m*-CF_3) makes the cyclization impossible even after 10 days under reflux.[22]

Rates of cyclization of some *o*-(1-phenylethyl)phenyl alkyl ketone **1**

Entry No.	R	R_1	$R_2 = R_3$	K, $(min^{-1})\times10^{-2}$
1	Me	Me	H	4.6
2	Ethyl	Me	H	1.8
3	*n*-Propyl	Me	H	0.99
4	*n*-Butyl	Me	H	0.35
5	*n*-Pentyl	Me	H	0.36
6	*n*-Hexyl	Me	H	0.36

Rates of cyclization of some 2-(substituted benzyl)benzophenones **1**

Entry No.	R	$R_1 = R_2$	R_3	K, $(h^{-1})\times10^{-2}$
1	C_6H_5	H	H	4.4
2	C_6H_5	H	*o*-CH_3	15.4
3	C_6H_5	H	*m*-CH_3	200
4	C_6H_5	H	*m*-CF_3	No reaction
5	C_6H_5	H	*p*-CH_3	13.8

The introduction of either a methyl or phenyl group at R_1 in **1** (R = Ph, R_2 = H) increases the rate of cyclization.[18] In both cases, the enhancement of cyclization rate may be due to an slight increase in electron density at the *ortho* position at which cyclization takes place. Although it should be noted that when R_1 = Ph, there are four such positions available for cyclization with a corresponding increase in probability of the reaction.

Rates of cyclization of some *o*-benzylphenyl ketones **1**

Entry No.	R	R_1	$R_2 = R_3$	K, $(h^{-1})\times10^{-2}$
1	C_6H_5	H	H	4.4
2	C_6H_5	CH_3	H	13
3	C_6H_5	C_6H_5	H	13

In summary, the rate of cyclization of *o*-benzylbenzophenones depends on several factors, the most important of which appear to be (1) the steric nature of R (and perhaps also R_1), (2) the effective positive character of the carbonyl carbon of the conjugate acid, (3) electron density at the *ortho* position of the nucleophilic benzene ring, and (4) the number of such positions available.

4.5.4 Variations and Improvements

The requirement of relatively harsh conditions for the Bradsher cyclodehydration (generally under refluxing hydrobromic acid–acetic acid mixture conditions due to high activation energy barriers) has restricted the use of the Bradsher reaction in the synthesis of complex highly functionalized organic molecules.[23] However there have been considerable efforts devoted to performing the Bradsher cyclodehydration under milder conditions.

Bradsher and Sinclair attempted an aromatic cyclodehydration at lower temperature for heat-sensitive substrates and reported the cyclization of *o*-benzylbenzophenone (**11**) to 9-phenylanthracene (**12**) in the presence of a vigorously stirred fine suspension of phosphorous pentoxide in liquid sulfur dioxide at –10 °C with a moderate yield (53%) of **12** being obtained.[24]

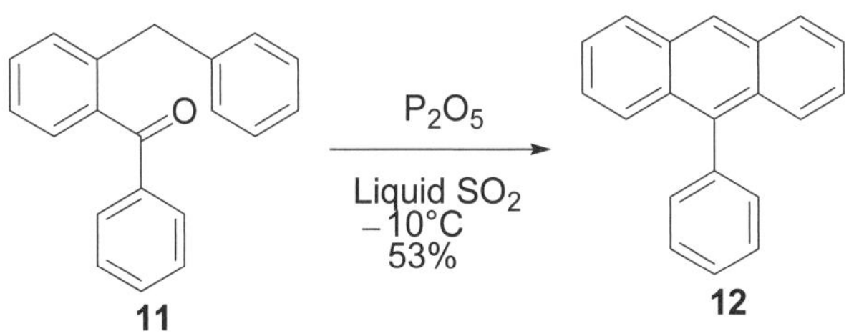

In an innovation, Newman applied polyphosphoric acid (PPA) in place of the usual mixture of hydrobromic acid and acetic acid.[25] PPA had been used earlier in the aromatic cyclodehydration for preparation of phenanthrene derivatives.[26,27] Newman applied PPA under thermal conditions for the aromatic cyclodehydration of an unactivated ketone **16** to afford **17**, an aromatic carcinogen, in moderate yield. In contrast, the use of a mixture of hydrobromic and acetic acids failed to give the cyclized product.

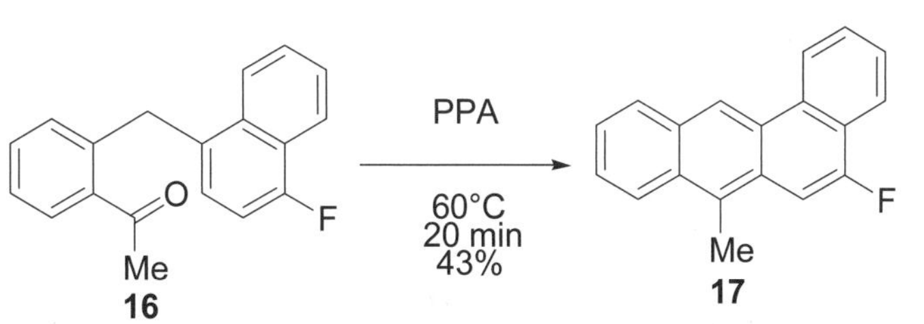

Vingiello and Schlechter described the Bradsher cyclodehydration of *o*-benzylphenyl ketone containing a strongly basic group in the presence of either phenyl acid phosphate or benzenesulfonic acid.[28] 4-Benzylphenyl pyridyl ketone (**18**) underwent cyclization in the presence of phenyl acid phosphate at 190 °C to yield 9-(4-pyridyl)anthracene (**19**) quantitatively, whereas the same cyclization in the presence of benzenesulfonic acid at 150 °C afforded **19** in 65% yield.[8]

Entry No.	Conditions	Yield (%)
1	Phenyl acid phosphate, 190 °C, 5 h	Quant.
2	Benzenesulfonic acid, 150 °C, 5 h	65

Vingiello and Thornton examined the aromatic cyclodehydration of 2-(3-methylbenzyl)phenyl-2-naphthyl ketone (**20**) to 2-methyl-l0-(2-naphthy1)anthracene (**21**) in the presence of either alumina at 250 °C or liquid hydrogen fluoride at room temperature.[29] Results indicate that the use of liquid hydrogen fluoride constitutes very mild conditions and only sterically and electronically favorable substrates give the desired product in good yield, but in those cases liquid hydrogen fluoride was found to be higher yielding than the alumina-catalyzed cyclization.

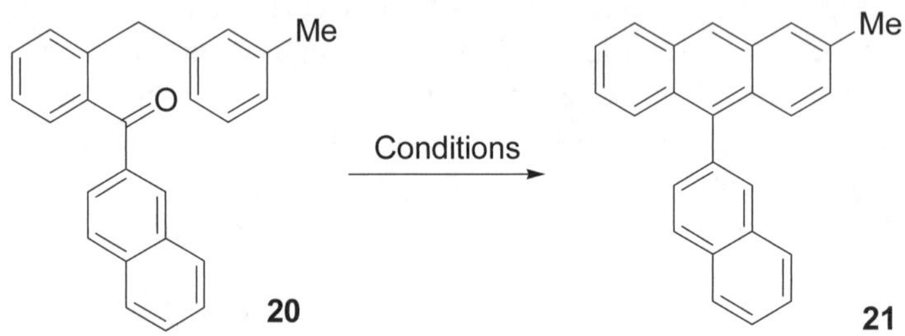

Entry No.	Conditions	Yield (%)
1	Alumina, 250 °C, 2.5 h	50
2	Liquid Hydrogen Fluoride, rt, until acid evaporates	72

The use of zinc chloride in a refluxing mixture of acetic acid and acetic anhydride allow the Bradsher cyclodehydration to be extended to aromatic acids.[8,9,30] Newman and co-workers reported the cyclization of 2-(4-methoxybenzyl)-1-napthoic acid (**22**), on treatment with zinc chloride in boiling acetic acid and acetic anhydride, into 12-acetoxy-10-methoxybenz[a]-anthracene (**23**) in very good yield.[8]

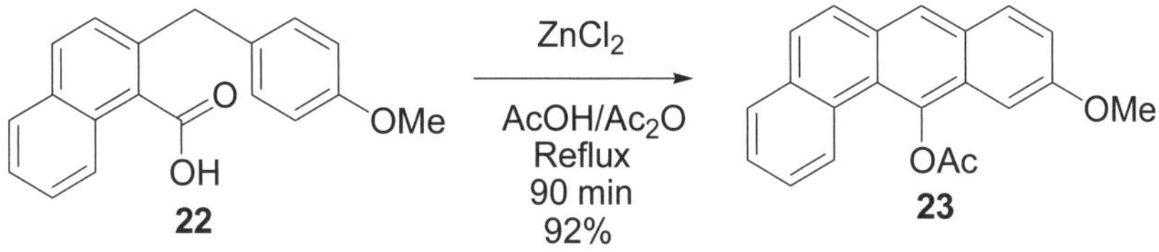

Yamato and co-workers introduced dichloromethyl methyl ether in the presence of titanium tetrachloride as a reagent for one-pot formylation followed by in situ Bradsher cyclization. In this way anthracene regioisomers **25** and **26** were obtained from diarylmethane **24** in good yield.[31] The initially formed anthracene derivative presumably reacts with excess dichloromethyl methyl ether to yield 9- or 10-formylanthracene derivative **25** or **26**.

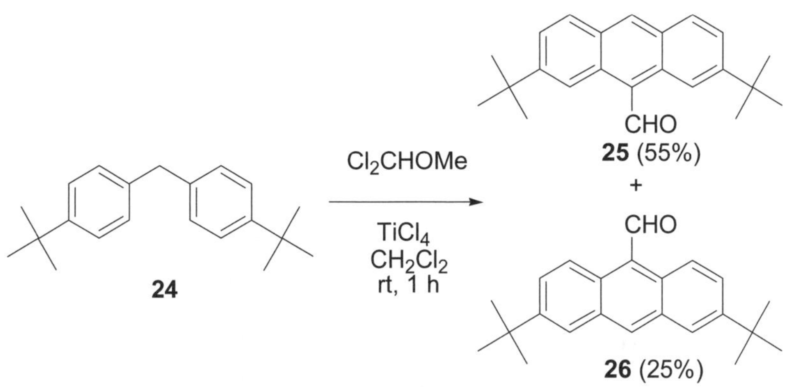

Neckers and co-workers have synthesized two new representatives of pentacyclic [*a*,*d*]-fused ladder-type aromatic sulfur containing materials—namely, thieno[3,2-*f*:4,5-*f*′]bis[1]benzothiophene (**28**-*syn*) and thieno[2,3-*f*:5,4-*f*′]bis[1]benzothiophene (**30**-*anti*)—using Amberlyst-15-mediated Bradsher cyclodehydration under refluxing conditions.[32]

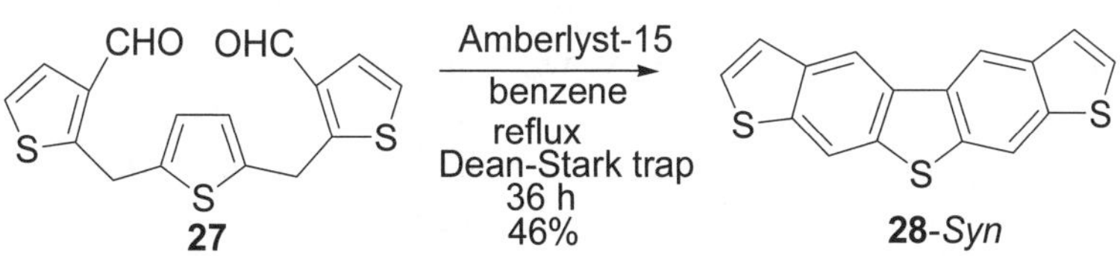

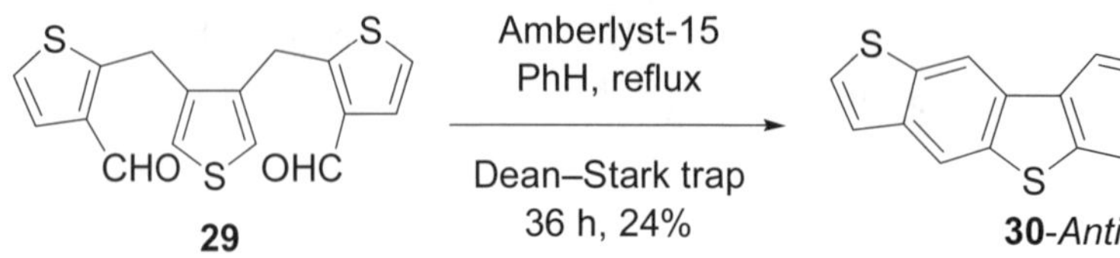

4.5.5 *Synthetic Utility*

Until the mid-1970s, the Bradsher cyclization was not applied to biologically interesting targets due to its substrate limitations and the general requirement of drastic conditions. After the advancement of the arene oxide concept concerning the metabolism of polycyclic aromatic hydrocarbons, Newman and co-workers synthesized 1-, 2-, 3-, 4-, 6-, 9-, and 10-hydroxy-7,12-dimethylbenz[*a*]anthracene (DMBA, **31**–**37**) using the Bradsher reaction, to determine the carcinogenicity and mutagenicity of each compound.[6–8]

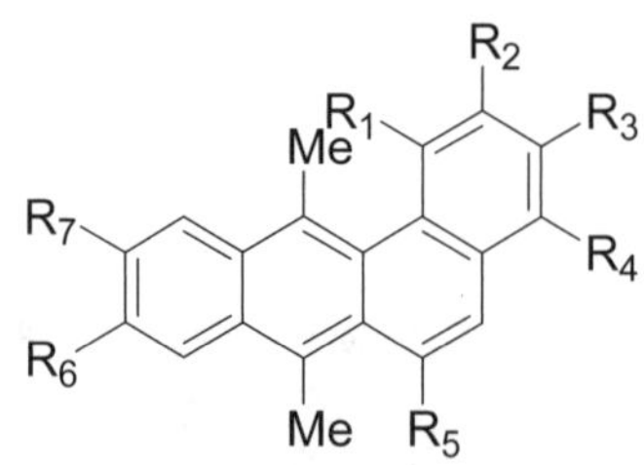

31, R_1 = OH; R_2, R_3, R_4, R_5, R_6 and R_7 = H
32, R_2 = OH; R_1, R_3, R_4, R_5, R_6 and R_7 = H
33, R_3 = OH; R_1, R_2, R_4, R_5, R_6 and R_7 = H
34, R_4 = OH; R_1, R_2, R_3, R_5, R_6 and R_7 = H
35, R_5 = OH; R_1, R_2, R_3, R_4, R_6 and R_7 = H
36, R_6 = OH; R_1, R_2, R_3, R_4, R_5 and R_7 = H
37, R_7 = OH; R_1, R_2, R_3, R_4, R_5, and R_6 = H

Newman and co-workers reported the carcinogenic activity of 7-methylbenz[*a*]anthracene (7-MBA) **38**, 12-methylbenz[*a*]anthracene (12-MBA) **39** and 7,12-dimethylbenz[*a*]anthracene (DMBA) **40** in 1972.[33] The planar hydrocarbon **38** possesses high carcinogenic activity, nonplanar analogue **39** has low carcinogenic activity, while another nonplanar analogue **40** is the most potent carcinogenic polycyclic aromatic hydrocarbon (PAH) commonly employed in carcinogenesis research.[34–36] DMBA-induced rat mammary carcinoma is the standard laboratory animal model in the study of human breast cancer.[35]

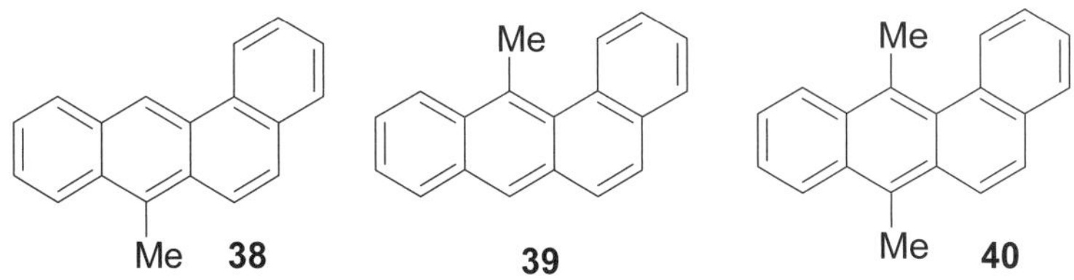

Me 38 39 Me Me 40

Diol epoxide metabolites of DMBA such as *trans*-3,4-dihydroxy-*anti*-1,2-epoxy-1,2,3,4-tetrahydro-DMBA (**41**) or *trans*-3,4-dihydroxy-*syn*-1,2-epoxy-1,2,3,4-tetrahydro-DMBA (**42**) has been implicated as the principle active form of DMBA which binds covalently to DNA in vivo.[36] The intermediacy of **41** and **42** was further supported by the development of methods for their synthesis by using the Bradsher cyclization for the construction of the DMBA moiety, coupled with studies of their mutagenicity, tumorigenicity and DNA binding.[9,37]

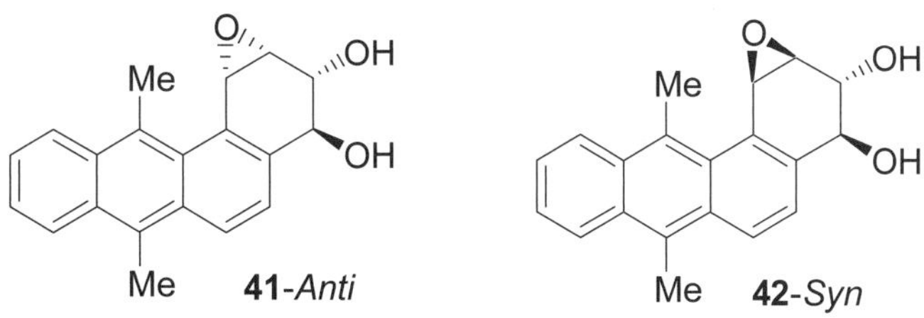

41-*Anti* **42**-*Syn*

The carcinogenic activity of polycyclic aromatic hydrocarbons is often strongly affected by the substitution of fluorine in suitable positions.[38,39] In some cases, the inhibition activity caused by fluorine substitution is due to interference with activation by the P-450 microsomal enzymes to reactive PAH diol epoxide metabolites.[40] Conversely, enhancement of activity by introduction of fluorine into other molecular regions may be considered primarily a consequence of restriction of oxidative metabolism. One of the most thoroughly investigated examples of these effects is the highly potent PAH carcinogen 7,12-dimethylbenz[a]-anthracene (DMBA) (**40**). While the 1-, 2-, 4- and 5-fluoro derivatives of DMBA (**43**–**46**) exhibit markedly reduced activity relative to the parent hydrocarbon **40**, substitution of fluorine in the 8-, 9-, and 11-position (**47**, **48** and **50**) results in no significant loss of activity.[41,42] Tumor-initiating studies on male rats have shown, however, that the carcinogenic activity of 10-fluoro DMBA **49** is greater than that of DMBA.[43] Similar effects were seen for the 7- and 12-monomethyl analogs, 7-MBA **38** and 12-MBA **39**.[44]

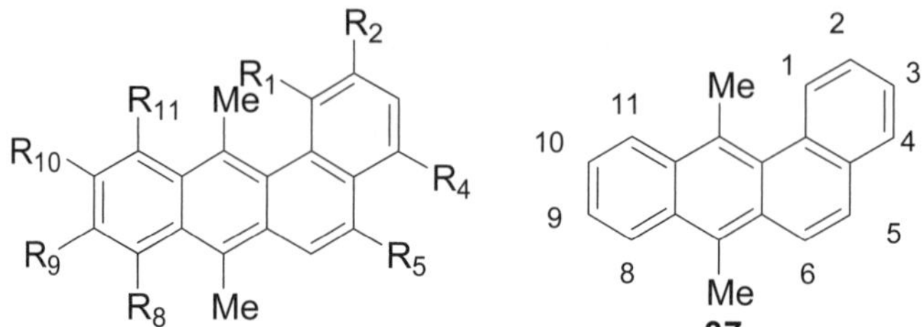

37

43, R_1 = F; R_2, R_4, R_5, R_8, R_9, R_{10} and R_{11} = H
44, R_2 = F; R_1, R_4, R_5, R_8, R_9, R_{10} and R_{11} = H
45, R_4 = F; R_1, R_3, R_5, R_8, R_9, R_{10} and R_{11} = H
46, R_5 = F; R_1, R_2, R_4, R_8, R_9, R_{10} and R_{11} = H
47, R_8 = F; R_1, R_2, R_4, R_5, R_9, R_{10} and R_{11} = H
48, R_9 = F; R_1, R_2, R_4, R_5, R_8, R_{10} and R_{11} = H
49, R_{10} = F; R_1, R_2, R_4, R_5, R_8, R_9 and R_{11} = H
50, R_{11} = F; R_1, R_2, R_4, R_5, R_8, R_9 and R_{10} = H

To verify the possibility that the presence of a fluorine atom in the 9- or 10-positions of the DMBA diol epoxide isomers might remotely influence the reactivity of the epoxide function, Harvey synthesized *trans*-3,4-dihydrodiol of 9- and 10-fluoro-DMBA, 7-MBA and 12-MBA.[10] These fluoro-dihydrodiols are putative proximate carcinogenic metabolites that undergo activation by the P-450 microsomal enzymes to carcinogenic *anti*- and *syn*-diol epoxide metabolites that bind to nucleic acids in vivo. The synthesis of several *anti*-diol epoxide derivatives using the Bradsher cyclization was also reported.

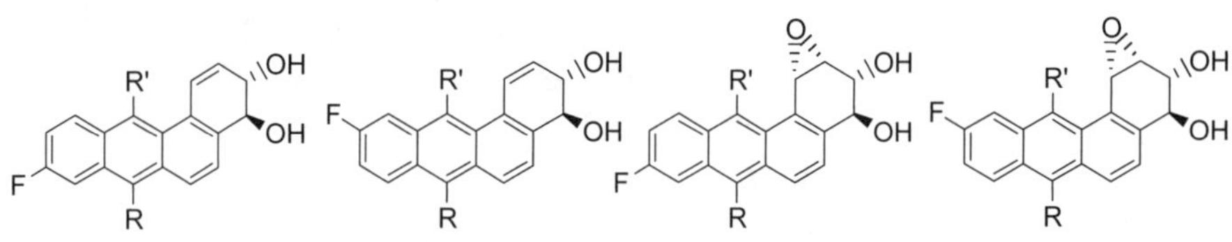

51 **52** **53** **54**

a: R = CH_3, R′ = H; **b**: R = H, R′ = CH_3, **c**: R = R′ = CH_3

Morreal and co-workers reported the synthesis of an antiestrogenic compound, 3,9-diol-7,12-DMBA **55** using the Bradsher reaction.[45] Antiestrogens are known to antagonize the growth of estrogen-dependent human breast cancer.[46] Molecular modeling of **55** showed that its phenolic hydroxyl groups are equivalent to the hydroxyl groups of the natural estrogen 17*β*-estradiol. At a dose of 0.5 mg **55**, a decrease in the percentage of rats in estrus from 78% to 44% was observed. This decrease is identical to that caused by 0.5 mg nafoxidine.

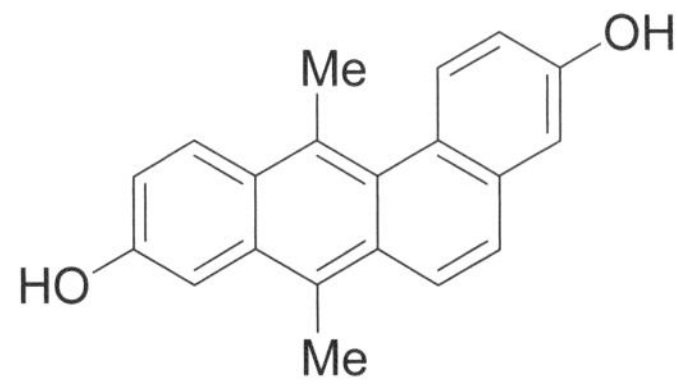

55

Pentacene (**56**), a member of the acene series of linear polycyclic aromatic hydrocarbons, is a fundamental component of organic field-effect transistors (OFET).[47] In an attempt to rectify its shortcomings such as poor solubility, limited stability in solution and unfavourable stacking in the solid state, Neckers and co-workers have synthesized two new representatives of pentacyclic [*a*,*d*]-fused ladder type aromatic sulphur-containing materials.[32] Thieno[3,2-*f*:4,5-*f*′]bis[1]benzothiophene (**28**-*syn*) and thieno[2,3-*f*:5,4-*f*′]bis-[1]benzothiophene (**30**-*anti*) were prepared using an Amberlyst-15-catalyzed Bradsher cyclodehydration under refluxing conditions, and their optical and electrochemical properties were reported.[32]

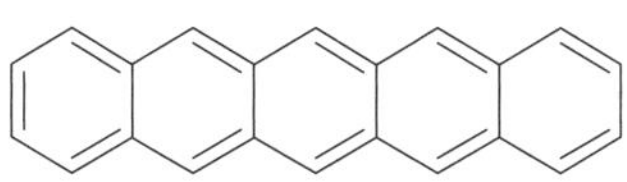

56

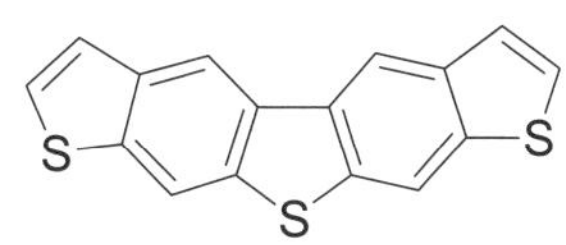

28-*Syn*

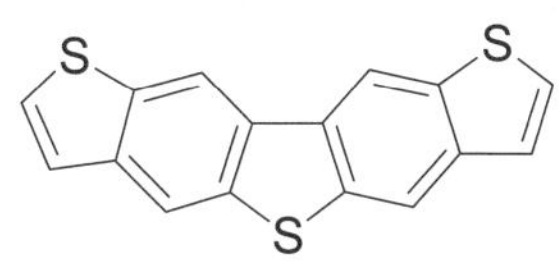

30-*Anti*

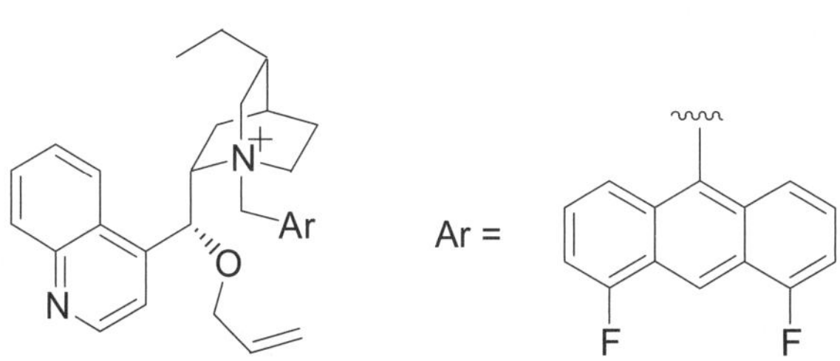

57

Andrus and co-workers developed fluoroanthracenylmethyl cinchonidine **57** as an efficient phase-transfer catalyst, and it was explored for asymmetric glycine alkylation.[11,12] The fluoroanthracenylmethyl precursor was synthesized from an aryloxazolidinone and aldehyde using the Bradsher reaction in the key step. This cinchonidine catalyst promotes highly selective glycine alkylation under mild conditions. The placement of 1,8-difluoro-anthracenyl-10-methyl on the quinuclidine nitrogen accentuates steric interactions and provides a larger region for nonbonded interactions, leading to a significant increase in selectivity. Electronegative fluorine substituents contribute to the overall electron deficiency of the positively charged catalyst, enhancing the degree of ion pairing with the enolate. This effect facilitates both phase transfer and enhances nonbonded interactions leading to improved selectivity.

4.5.6 Experimental

Thermal acid-catalyzed bradsher reaction: 9-phenylanthracene (12)[1]

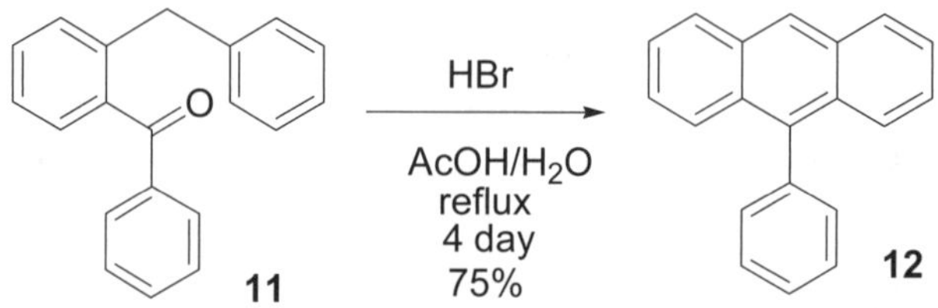

A stirring mixture of **11** (2.06 g, 7.57 mmol) in a mixture of acetic acid (20 mL) and 34% hydrobromic acid (20 mL) was refluxed for 4 days. Upon cooling, the product solidified as fluorescent plates which, after recrystallization from ethanol, yielded 1.44 g (75%) **12** as a crystalline solid.

Phosphorus pentoxide-mediated bradsher reaction at low temperature: 9-phenylanthracene (12)[24]

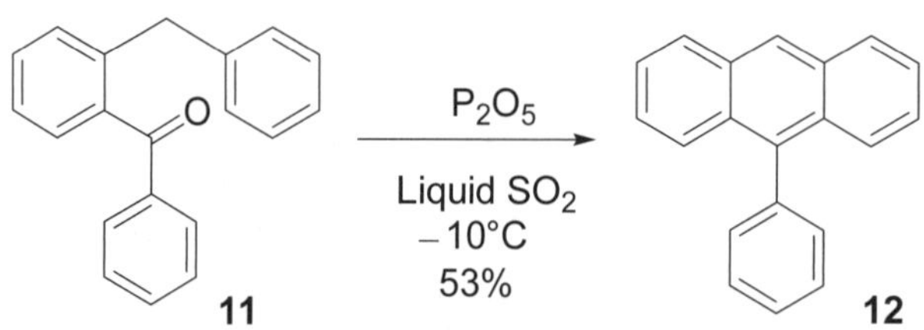

To a stirring mixture of **11** (500 mg, 1.84 mmol) in liquid sulfur dioxide (50 mL), was added phosphorus pentoxide (5.0 g, 17.61 mmol), and the reaction mixure was stirred for several hours at –10 °C. Then carbon tetrachloride (20 mL) was added, and the mixture was allowed to stand until the ice first formed had melted and most of the sulfur dioxide had evaporated. The carbon tetrachloride layer was separated from water, and the aqueous layer was extracted with more carbon tetrachloride. The combined organics were washed with water thrice and dried over calcium chloride. Removal of solvent under reduced pressure followed by recrystallization from ethanol furnished 247 mg (53%) **12** as a crystalline solid.

Amberlyst-catalyzed thermal Bradsher reaction: thieno[2,3-*f*:5,4-*f'*]bis[1]benzothiophene (30-*anti*)[32]

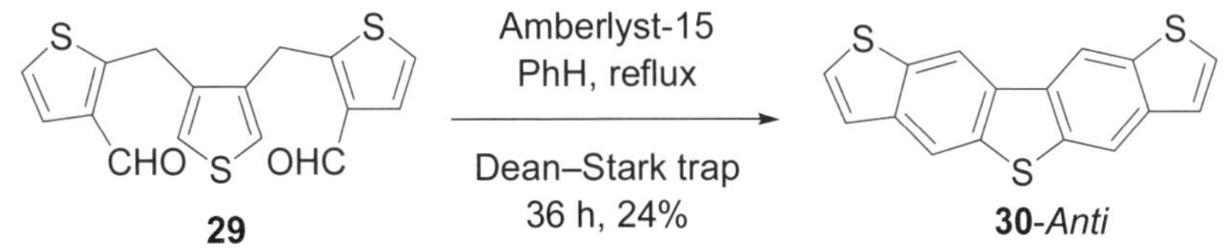

To a stirring solution of **29** (1.63 g, 4.90 mmol) in dry benzene, amberlyst-15 (0.5 g) was added and the reaction mixture was refluxed for 36 h. Water was removed by means of a Dean–Stark trap. After cooling, dichloromethane was added to dissolve the precipitate and the mixture was filtered through a cotton plug. The filtrate was washed with saturated aqueous NH_4Cl solution and dried over anhydrous $MgSO_4$. Evaporation of the solvent followed by silica gel column chromatographic purification using hexane as eluent yielded a pale white solid. Recrystallization from $CHCl_3$ furnished 0.35 g (24%) *anti*-**30** as pale white crystals.

4.5.7 References

1. Bradsher, C. K. *J. Am. Chem. Soc.* **1940**, *62*, 486–488.
2. [R] Bradsher, C. K. *Chem. Rev.* **1987**, *87*, 1277–1297.
3. Bergmann, E. *J. Org. Chem.* **1939**, *4*, 1–13.
4. Zincke, T.; Breuer, A. *Justus Liebigs Ann. Chem.* **1884**, *226*, 23–60.
5. [R] Bradsher, C. K. *Chem. Rev.* **1946**, *38*, 447–499.
6. [R] Jerina, D. M.; Daly, J. W. *Drug Metabolism from Microbe to Man* Taylor & Francis, London, **1976**, 13.
7. Jerina, D. M.; Yagi, H.; Daly, J. W. *Heterocycles* **1973**, *1*, 267–326.
8. Newman, M. S.; Khanna, J. M.; Kanakarajan, K.; Kumar, S. *J. Org. Chem.* **1978**, *43*, 2553–2557.
9. Lee, H.; Harvey, R. G. *J. Org. Chem.* **1986**, *51*, 3502–3507.
10. Harvey, R. G.; Cortez C. *Tetrahedron* **1997**, *53*, 7101–7118.
11. [R] (a) Maruoka, K.; Ooi, T. *Chem. Rev.* **2003**, *103*, 3013–3028. [R] (b) Kacprzak, K.; Gawronski, J. *Synthesis* **2001**, 961–998. [R] (c) O'Donnell, M. J.; *Acc. Chem. Res.* **2004**, *37,* 506–517. [R] (d) Lygo, B.; Andrews, B. I. *Acc. Chem. Res.* **2004**, *37*, 518–526.
12. Andrus, M. B.; Ye, Z.; Zhang, J. *Tetrahedron Lett.* **2005**, *46*, 3839–3842.
13. Berliner, E. *J. Am. Chem. Soc.* **1942**, *64,* 2894–2898.
14. Bradsher, C. K.; Smith, E. S. *J. Am. Chem. Soc.* **1943**, *65*, 854–857.
15. [R] Smith, M. B.; March, J. in *March's Advanced Organic Chemistry,* 5thed., Wiley, New York, **2001**, 720.
16. Brice, L. K.; Katstra, R. D. *J. Am. Chem. Soc.* **1960**, *82*, 2669–2670.
17. Bradsher, C. K.; Smith, E. S. *J. Am. Chem. Soc.* **1943**, *65*, 1643–1645.
18. Bradsher, C. K.; Vingiello, F. A. *J. Am. Chem. Soc.* **1949**, *71*, 1434–1436.
19. Vingiello, F. A.; Van Oot, J. G.; Hannabass, H. H. *J. Am. Chem. Soc.* **1952**, *74*, 4546–4548.
20. Vingiello, F. A.; Spangler, M. O. L.; Bondurant, J. E. *J. Org. Chem.* **1960**, *25*, 2091–2094.

21. Berliner, E. *J. Am. Chem. Soc.* **1944**, *66,* 533–535.
22. Vingiello, F. A.; Van Oot, J. G. *J. Am. Chem. Soc.* **1951**, *73*, 5070–5072.
23. Henson, P. D.; Vingiello, F. A. *J. Org. Chem.* **1967**, *32*, 3205–3207.
24. Bradsher, C. K.; Sinclair, E. F. *J. Org. Chem.* **1957**, *22*, 79–81.
25. Newman, M. S.; MacDowell, D.; Swaminathan, S. *J. Org. Chem.* **1959**, *24*, 509–512.
26. Hauser, C. R.; Murray, J. G. *J. Am. Chem. Soc.* **1955**, *77*, 3858–3860.
27. Bradsher, C. K.; Jackson, W. J., Jr. *J. Am. Chem. Soc.* **1952**, *74*, 4880–4883.
28. Vingiello, F. A.; Schlechter, M. M. *J. Org. Chem.* **1963**, *28*, 2448–2450.
29. Vingiello, F. A.; Thornton, J. R. *J. Org. Chem.* **1966**, *31*, 659–663.
30. Harvey, R. G.; Cortez, C.; Sugiyama, T.; Ito, Y.; Sawyer, T. W.; DiGiovanni, J. *J. Med. Chem.* **1988**, *31*, 154–149.
31. Yamato, T.; Sakaue, N.; Shinoda, N.; Matsuo, K. *J. Chem. Soc., Perkin Trans. 1,* **1997**, 1193–1199.
32. Wex, B.; Kaafarani, B. R.; Kirschbaum, K.; Neckers, D. C. *J. Org. Chem.* **2005**, *70*, 4502–4505.
33. Newman, M. S.; Cunico, R. F. *J. Med. Chem.* **1972**, *15*, 323–325.
34. Jones, D. W.; Sowden, J. M. *Cancer Biochem. Biophys.* **1976**, 281–288.
35. Welsch, C. W. *Cancer Res.* **1985**, *45*, 3415–3443.
36. (a) Moschel, R. C.; Baird, W. M.; Dipple, A. *Biochem. Biophys. Res. Commun.* **1977**, *76*, 1092–1098. (b) Bigger, C. A. H; Sawicki, J. T.; Blake, D, M.; Raymond, L. G.; Dipple, A. *Cancer Res.* **1983**, *43*, 5647–5651. (c) Dipple, A.; Pigott, M.; Moschel, R. C.; Constantino, N. *Cancer Res.* **1983**, *43*, 4132–4135 and references cited therein.
37. (a) Slaga, T. J.; Gleason, G. L.; DiGiovanni, J.;Sukumaran, K. B.; Harvey, R. G. *Cancer Res.* **1979**, *39*, 1934–1936. (b) Sawicki, J. T.; Moschel, R. C.; Dipple, A. *Cancer Res.* **1983**, *43*, 3212–3218.
38. [R] (a) Harvey, R. G. *Polycyclic Aromatic Hydrocarbons; Chemistry and Carcinogenicity*, Cambridge University Press, Cambridge, UK, **1991**. (b) Harvey, R. G. *Polycyclic Aromatic Hydrocarbons*, Wiley-VCH, New York, **1997**.
39. [R] Dipple, A.; Moschel, R. C.; Bigger, C. A. H. In *Chemical Carcinogenesis*, 2nd Edition; C. E. Searle, Ed.; ACS Monograph, 182; American Chemical Society, Washington, D.C., , **1984**, pp. 41–163.
40. [R] Harvey, R. G.; Geacintov, N. E. *Acc. Chem. Res.* **1988**, *21*, 66–73.
41. (a) Harvey, R. G.; Dunne, F. B. *Nature* **1978**, *273*, 566–568. (b) Huberman, E.; Slaga, T. J. *Cancer Res.* **1979**, *39,* 411–414. (c) DiGiovanni, J.; Diamond, L.; Singer, J. M.; Daniel, F. B.; Witiak, D. T.; Slaga, T. J. *Carcinogenesis* **1982**, *3,* 651–655.
42. DiGiovanni, J.; Decina, P. C.; Diamond, L. *Carcinogenesis* **1983**, *4,* 1045–1049.
43. Newman, M. S.; Veeraraghavan, S. *J. Org. Chem.* **1983**, *48*, 32346–32348.
44. Diamond, L.; Cherian, K.; Harvey, R. G.; DiGiovanni, J. *Mutat. Res.* **1984**, *136,* 65–72.
45. Morreal, C. E.; Sinha, D. K.; Schneider, S. L. Bronstein, R. E.; Dawidzik, J. *J. Med. Chem.* **1982**, *25*, 323–326.
46. (a) Dao, T. L.; Suderland, J. *J. Natl. Cancer Inst.* **1959**, *23*, 567-–585. (b) Dao, T. L. *Cancer Res.* **1962**, *22,* 973-981. [R] (c) Dao, T. L. *Prog. Exp. Tumor Res.* **1964**, *5,* 157–216.
47. (a) Sakamoto, Y.; Suzuki, T.; Kobayashi, M.; Gao, Y.; Fukai, Y.; Inoue, Y.; Sato, F.; Tokito, S. *J. Am. Chem. Soc.* **2004**, *126*, 8138–8140. [R] (b) Volkel, A. R.; Street, R. A.; Knipp, D. *Phys. Rev.* **2002**, *B66*, 195336–195343. (c) Kwon, O.; Coropceanu, V.; Gruhn, N. E.; Durivage, J. C.; Laquindanum, J. G.; Katz, H. E.; Cornil, J.; Bredas, J. L. *J. Chem. Phys.* **2004**, *120*, 8186–8194.

4.6 Darzens Synthesis of Tetralin Derivatives

Ewa Krawczyk and Roman Dembinski

4.6.1 Description

Cyclization of 2-benzylpent-4-enoic acid (α-benzyl-α-allylacetic acid) and related compounds **1** to corresponding 1,2,3,4-tetrahydronaphthalene (tetralin) derivatives **2**, catalyzed by a strong acid, has been called Darzens synthesis of tetralin derivatives. The reaction formally falls into the category of a cycloisomerization reaction of alkenyl-substituted arenes.

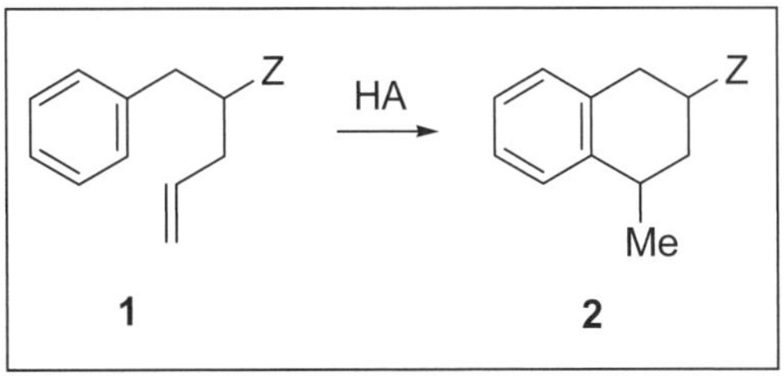

4.6.2 Historical Perspective

Georges Darzens (at the École Polytechnique in Paris, France) in 1926 described the conversion of 2-benzylpent-4-enoic acid **3** into 4-methyl-tetrahydronaphthalene-2-carboxylic acid **4** that has proceeded under the influence of concentrated sulfuric acid.[1] Subsequent dehydrogenation and decarboxylation yielded a naphthalene derivative. This work was followed by a series of extended reports.[2–5] The outcome of Darzens et al. (Lévy, Heinz) sequence of works in this area has been reviewed.[6]

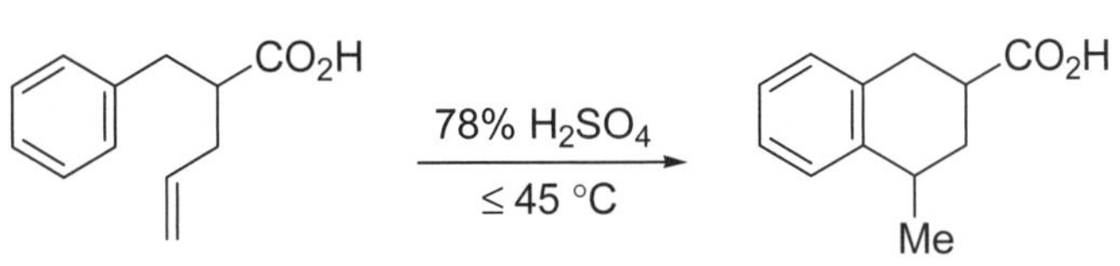

3 **4**

4.6.3 Mechanism

Although in the past a few mechanistic schemes have been drawn,[7] currently the plausible reaction mechanism may be described as the acid-initiated formation of an intermediate carbenium ion followed by S_EAr, and thus falls into the broad category of the Friedel–Crafts alkylation. This may be one of the reasons that the name of the reaction is not often seen in current literature. A plausible mechanistic outline that involves a carbenium ion is illustrated below. Initial protonation of alkene **1** generates a more favorable secondary cationic intermediate **5**, which interacts with an aromatic ring, forming the σ-complex **6**. Deprotonation furnishes the desired product **2**. Migration of the C=C bond via **5** to an internal alkene (not illustrated) may contribute to the mechanistic pathway, also leading predominantly to the product **2**.

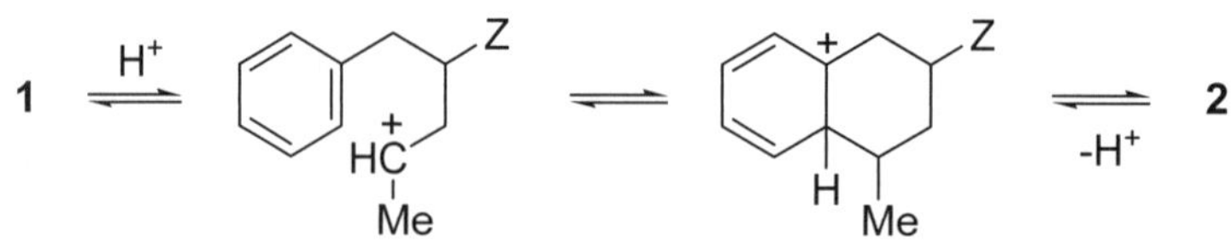

5 **6**

In the case of the Darzens reaction involving 2-benzylpent-4-enoic acid or related compounds, the process is accompanied by the formation of the γ-lactone **7** that was isolated, and later confirmed to also produce tetrahydronaphthalenecarboxylic acid **4**, when treated with 65% sulfuric acid at 120 °C.[8–10] Formation of δ-lactone **8** was also observed.[8]

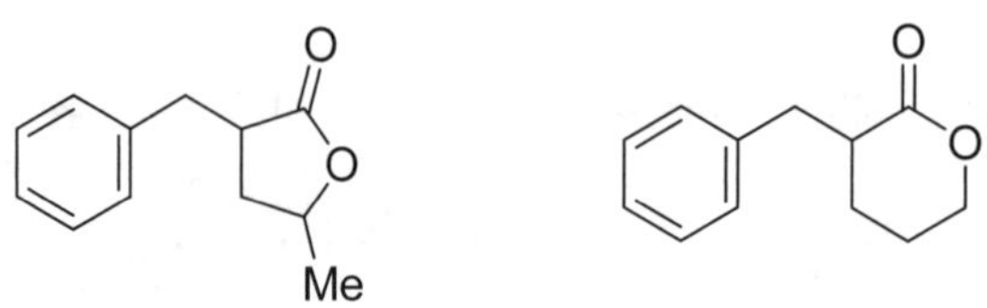

7 **8**

4.6.4 Variations and Improvements

Initially, the cyclization reaction of 2-benzylpent-4-enoic acid **4** was carried out using concentrated sulfuric acid as both the catalyst and the solvent at a temperature below 45 °C.[1] Further, the repetoire of Brønsted acids (used at ambient or higher temperature) has been expanded to include anhydrous hydrogen fluoride (0 °C),[11] trifluoroacetic acid (toluene, 70 °C or acetic acid, 70 °C),[12,13] polyphosphoric acid (xylenes, reflux),[14,15] and anhydrous pyridinium poly(hydrogen fluoride)[14] (room temperature).

Darzens et al. also used readily available malonic acid ester derivatives as substrates in reaction with sulfuric acid.[2–4,16–19] More recent studies and the NMR evidence demonstrated that for the cyclization of 2-methylallyl malonate **9** the use of anhydrous hydrogen fluoride is advantageous and leads to dimethyl tetrahydronaphthalene derivative **10**.[11]

When malonate **9** is treated with polyphosphoric acid at room temperature, formation of lactone **11** is observed, similar to the Darzens procedure with the use of sulfuric acid.[8–10] The structure of lactone is being retained during subsequent hydrolysis and decarboxylation of the exocyclic ester group; followed up treatment with anhydrous hydrogen fluoride gives dimethyl tetrahydronaphthalene carboxylic acid **12**. Hydrolysis and decarboxylation of **10** under standard conditions gives the same product **12**.[11]

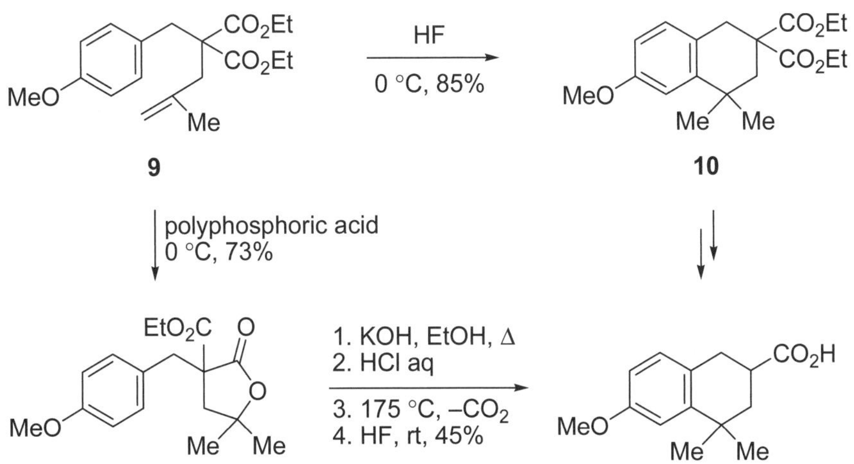

9 **10** **11** **12**

Similar reactions involving an allylic fragment proceed with formation of the S_N2' type products. Indium(III) salts ($InCl_3$ predominantly investigated)[20] have been found to be effective Lewis acids to catalyze conversion of **13** at room temperature in the presence of molecular sieves.[21] The cyclization is accompanied by elimination of hydrogen bromide, leading to the vinyl derivatives of tetrahydronaphthalene **14** and tolerates electron-withdrawing groups.

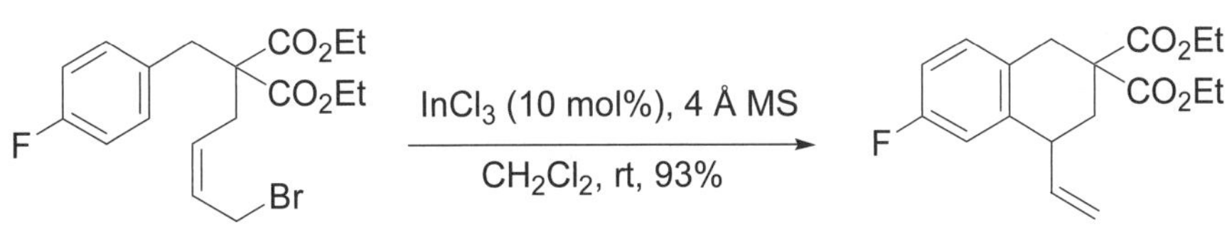

13 **14**

If the aromatic ring is unsymmetrically substituted, the reaction may lead to a mixture of products. Thus *m*-methoxybenzyl amine derivative **15** is transformed into vinyl tetrahydroisoquinolines **16** (major) and **17** (minor). Regioisomers **16** and **17** were formed by *para* and *ortho* cyclization, respectively, with reference to the methoxy group.[21]

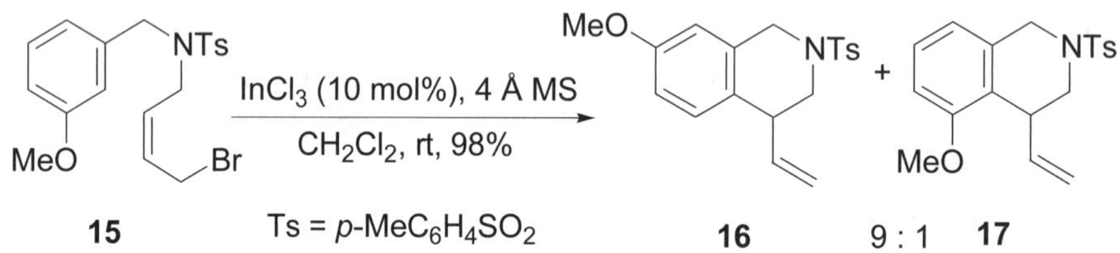

15 16 9 : 1 17

Reaction leading to analogous vinyl tetrahydronaphthalenes with elimination of acetoxy group has also been reported and the role of the substituents at the allylic acetate moiety investigated.[12,13] The presence of alkyl/aryl substituents raise the reaction rate. Best efficiency has been accomplished with the aid of trifluoroacetic acid/acetic acid 3 : 1 ratio. The reaction has been extended to the conversion of cyclohexene ring-containing optically active substrate **18** into optically active tricyclic skeleton **19** in excellent yield and stereoselectivity.

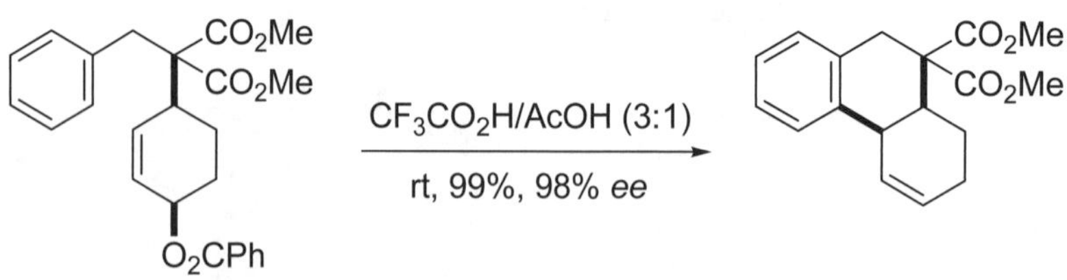

18 19

Vinyl substituted tetrahydronaphthalenes, analogs to **14**, can currently be obtained in high yield by employing a gold catalyst for the cyclization of 4-allenyl arenes.[22] Although the scope of the reaction is limited to electron-rich arenes, acetals are tolerated, and compound **20** leads predominantly to *para*-cyclized product **21** in high yield.

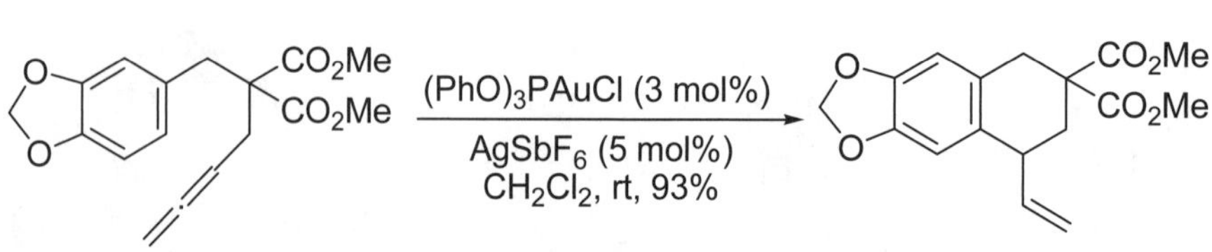

20 21

4.6.5 Synthetic Utility

Influence of the substituents has been investigated. In general, activating substituents at the aryl ring of **4** promote cyclization. If the carboylic group is attached to the benzyl position, the reaction proceeds slower.[4] The presence of isopropyl as a substituent attached to the C=C hampers the cyclization reaction;[17] detailed analysis of scope and limitations is provided in the review.[6]

Darzens reported that compounds of type **4** can be dehydrogenated to 1-methyl-3-naphthoic acid **22** with the use of sulfur or selenium at elevated temperature.[1,2,4] Subsequent decarboxylation using lime yielded 1-methylnaphthalen **23**. This sequence of reaction can be carried out in one pot.[2,4] Through a series of reactions the methyl naphthoic acid **22** can be converted to 4-methyl-2-naphthol **24**, and further, with the aid of Vilsmeyer formylation to alkyl naphthofuran **25**.[23]

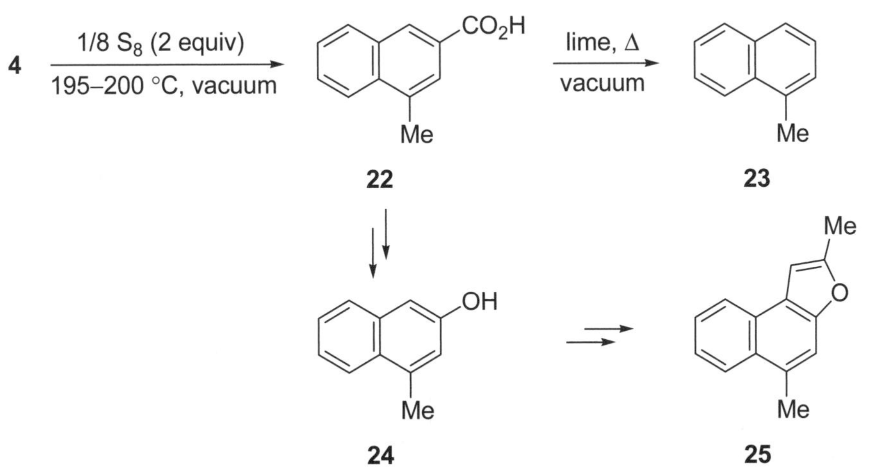

In an analogous way, Darzens and Lévy have synthesized isopropylmethylnaphthalene (eudaline) **26**, which contains a dehydrogenated sesquiterpene "eudesmol" motif.[3] The reaction is applicable for the synthesis of methyl derivatives of phenanthrene such as **27**.[18,19]

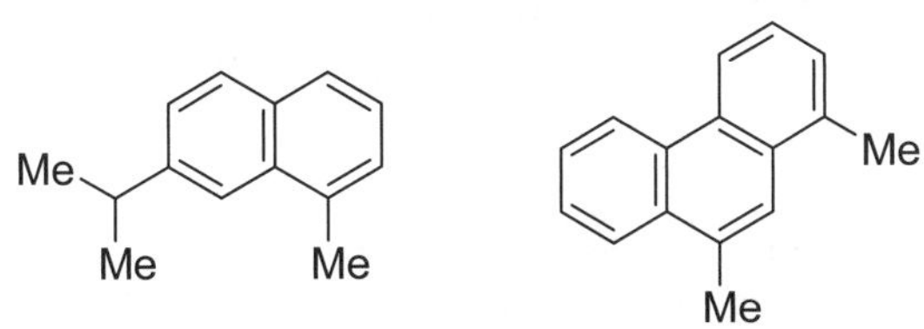

26, eudaline **27**

The reaction tolerates the presence of a phenyl group at the carbon atom that forms the bond with the aryl ring well, and can be stopped before decarboxylation or proceeds further to monocarboxylic acids derivatives of tetralin. As an example, cyclization of malonate **28** at room temperature with the use of anyhydrous hydrogen fluoride gives dicarboxylate **29**.[11] The anhydrous pyridinium poly(hydrogen fluoride) is more efficient in the same reaction.[14] Compound **29** is obtained with almost quantitative yield with purity sufficient to use a crude product for the follow-up synthetic transformations, which lead to derivatives of 1-phenyl-3-aminotetralins **30** that exhibit potential for the treatment of Parkinson disease.[14,15]

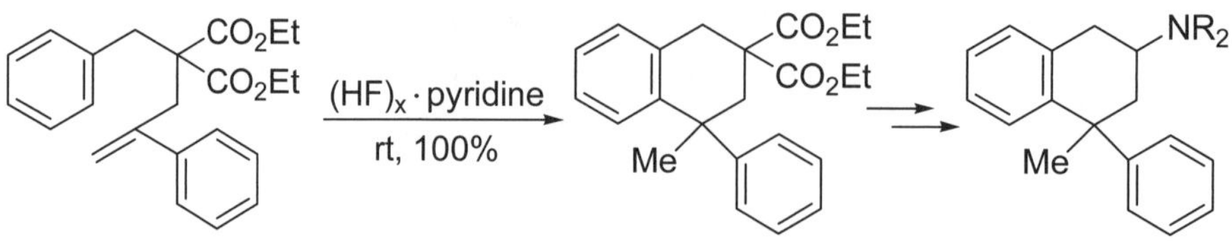

28 **29** **30**

The reaction has been extended to synthesis of tetralone derivatives.[15,24,25] Thus treatement of **31** with polyphosphoric acid leads to the substituted 3-tetralone **32**,[15] which can be also converted to 1-phenyl-3-aminotetraline **33** providing a complementary synthetic route to that one for **30** illustrated above.

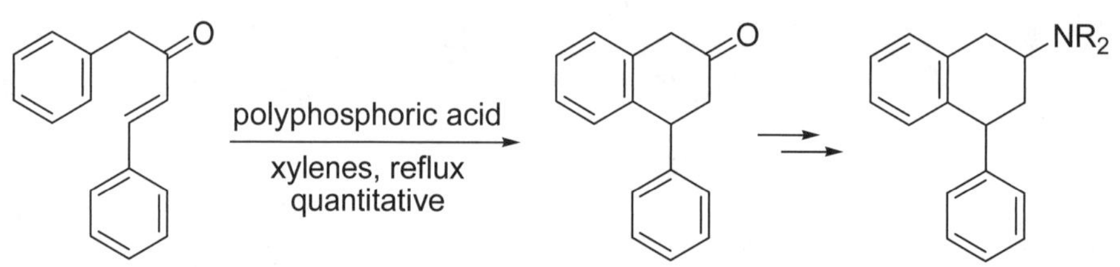

31 **32** **33**

4.6.6 Experimental

Darzens intramolecular cyclization reaction of 2-benzylpent-4-enoic acid (3) with the use of sulfuric acid: 4-methyl-1,2,3,4-tetrahydronaphthalene-2-carboxylic acid (4)[1,26]

Fine powder of compound **3** (1 equiv) and concentrated sulfuric acid (78%, 2.5 mass equiv) were mixed. Precautions were taken to avoid a temperature increase above 45 °C. The mixture was left at ambient temperature for 5 days to allow homogenizing and solidifying. After this time, water was added, and the mixture was neutralized with sodium bicarbonate (12%) until the sodium salt of the acid **4** was obtained. Next, the addition of the excess of hydrochloric acid resulted in the formation of a solid. Crystallization from acetic acid (80%) gave **4** as a white powder (50%, m.p. 121 °C), which can be distilled without decomposition giving a colorless fraction (203–204 °C/20 mm Hg).

Intramolecular cyclization reaction of malonate ester (28) **with the use of anhydrous hydrogen fluoride**[11] **or anhydrous pyridinium poly-(hydrogen fluoride:**[14] **diethyl 1,2,3,4-tetrahydro-4-methyl-4-phenylna-phthalene-2,2-dicarboxylate** (29)

Compound **28** (109.8 g, 0.3 mol) and 400 g anhydrous hydrogen fluoride in a 1-L polyethylene bottle were kept in an ice bath for 12 h. The hydrogen fluoride was then allowed to evaporate spontaneously during 24 h of stirring at room temperature. Water was then added, and the organic material was extracted several times with ether. The ether extracts were combined and washed with a 10% sodium bicarbonate solution until the solution remained slightly basic; then they were washed with a saturated solution of sodium chloride and dried over anhydrous sodium sulfate. The ether was then evaporated, and vacuum distillation (200 °C/0.3 mm Hg) gave a very viscous white oil (98.8 g, 90%). The dicarboxylate crystallized out on standing. Crystallization from alcohol gave **29** as white crystals (100%).

Compound **28** (27.2 g, 0.074 mol) was placed in a 500-mL poly(propylene) bottle, and 75 g of pyridinium poly(hydrogen fluoride) was added. The mixture was then stirred overnight at room temperature. The excess hydrogen fluoride was neutralized by addition of water followed by aqueous NaOH (20%, 200 mL) and the organic material was extracted into ether. The combined ether extracts were dried (Na_2SO_4) and evaporated in vacuo to give **29** as a light orange oil (27.3 g, 100%). Gas chromatography and ^1H NMR analyses indicated essentially pure product, which was used in the next step without further purification.

Intramolecular cyclization reaction of malonate ester (13) **with the use of indium(III) chloride: diethyl 6-fluoro-4-vinyl-3,4-dihydronaph-thalene-2,2(1*H*)-dicarboxylate** (14)[21]

Compound **13** (0.040 g, 0.10 mmol) and dichloromethane (2 mL, 0.05 M) were placed in a small screw-cap scintillation vial equipped with a magnetic stirbar. Powdered 4 Å molecular sieves (0.050 g) and indium(III) chloride (0.0022 g, 0.010 mmol) were added, and the reaction was allowed to stir at room temperature. Upon completion of the reaction (usually 16 h), the mixture was loaded onto a silica gel column directly and eluted using hexane–ethyl acetate (10 : 1) to afford **14** (93%).

4.6.7 References

1. Darzens, G. *Compt. Rend.* **1926**, *183*, 748–750.
2. Darzens, G.; Heinz, A. *Compt. Rend.* **1927**, *184*, 33–35.
3. Darzens, G.; Lévy, A. *Compt. Rend.* **1932**, *194*, 2056–2059.
4. Darzens, G.; Lévy, A. *Compt. Rend.* **1934**, *199*, 1131–1133.
5. Darzens, G.; Lévy, A. *Compt. Rend.* **1934**, *199*, 1426–1428.
6. [R] Linstead, R. P. *Ann. Rep. Prog. Chem.* **1936**, *33*, 312–341.
7. [R] Bergmann, E. *Chem. Rev.* **1941**, *29*, 529–551.
8. Darzens, G.; Lévy, A. *Compt. Rend.* **1935**, *200*, 469–471.
9. Darzens, G. *Compt. Rend.* **1930**, *190*, 1305–1306.
10. Darzens, G. *Compt. Rend.* **1930**, *190*, 1562–1564.
11. Kandeel, E. M.; Anderson, L. J.; Block, J. H.; White, A. I.; Martin, A, R. *J. Pharm. Sci.* **1972**, *61*, 1231–1234.
12. Ma, S.; Zhang, J. *Tetrahedron Lett.* **2002**, *43*, 3435–3438.
13. Ma, S.; Zhang, J. *Tetrahedron* **2003**, *59*, 6273–6283.
14. Wyrick, S. D.; Booth, R. G.; Myers, A. M.; Owens, C. E.; Bucholtz, E. C.; Hooper, P. C.; Kula, N. S.; Baldessarini, R. J.; Mailman, R. B. *J. Med. Chem.* **1995**, *38*, 3857–3864.
15. Wyrick, S. D.; Booth, R. G.; Myers, A. M.; Owens, C. E.; Kula, N. S.; Baldessarini, R. J.; McPhail, A. T.; Mailman, R. B. *J. Med. Chem.* **1993**, *36*, 2542–2551.
16. Darzens, G. *Compt. Rend.* **1926**, *183*, 1110–1112.
17. Darzens, G.; Lévy, A. *Compt. Rend.* **1930**, *191*, 1455–1457.
18. Darzens, G.; Lévy, A. *Compt. Rend.* **1935**, *200*, 2187–2189.
19. Darzens, G.; Lévy, A. *Compt. Rend.* **1936**, *202*, 427–428.
20. Cook, G. R.; Hayashi, R. *Org. Lett.* **2006**, *8*, 1045–1048.
21. Hayashi, R.; Cook, G. R. *Org. Lett.* **2007**, *9*, 1311–1314.
22. Tarselli, M. A.; Gagné, M. R. *J. Org. Chem.* **2008**, *73*, 2439–2441.
23. Chatterjea, J. N.; Lal, S.; Jha, U.; Carnduff, J. *Ind. J. Chem., Sect. B: Org. Chem. Incl. Med. Chem.* **1981**, *20*, 264–267.
24. [R] Silveira, C. C.; Braga, A. L.; Kaufman, T. S.; Lenardão, E. J. *Tetrahedron* **2004**, *60*, 8295–8328.
25. [R] Fieser, L. F.; Fieser, M. In *Current Topics in Organic Chemistry*, Reinhold, New York, 1963, p 548.
26. We are grateful to Edith Chopin, Krzysztof Owsianik, and Maria Zabłocka for the translation of the original Darzens articles.

4.7 Diels–Alder Reaction

Kevin M. Shea

4.7.1 Description

The Diels–Alder reaction is the most important synthetic method for the preparation of six-membered rings. Combination of a conjugated diene (**1**) with a separate alkene (**2**) or alkyne (the dienophile) yields a substituted cyclohexene (**3**). This cycloaddition reaction is often highly regio-selective and stereo-selective, yielding predictable configurations at the four potential new stereocenters formed in the reaction. In the most popular variations, an electron-rich diene is combined with an electron-poor dienophile, and the reaction is promoted by heat or addition of a Lewis acid. Inverse electron demand Diels–Alder reactions involving the combination of an electron-poor diene with an electron-rich dienophile is also possible. The scope of the Diels–Alder reaction is broad with respect to the substituents on the diene and dienophile. In addition, it readily occurs intermolecularly[1] and intramolecularly,[2] and incorporation of heteroatoms in both the diene[3] and dienophile[4] is common (the hetero Diels–Alder reaction). A variety of chiral Lewis acid catalysts enable highly enantioselective reactions for many substrates.[5]

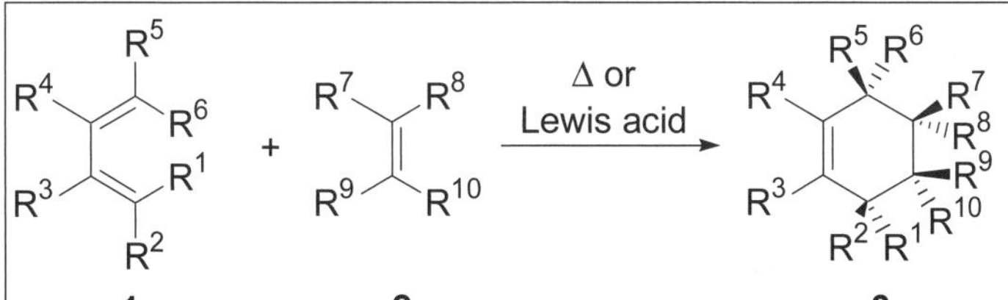

1 2 3

4.7.2 Historical Perspective

Kurt Alder obtained his Ph.D. in 1926 at the University of Kiel, Germany, under the direction of Otto Paul Hermann Diels. After obtaining his degree, Alder continued conducting research in Diels's lab on reactions of unsaturated compounds.[6] Diels and Alder published their seminal paper describing the reaction of 1,3-butadiene (**4**) with acrolein (**5**) to yield substituted cyclohexene **6** in 1928.[7] Although similar reactions had appeared previously in the literature, Diels and Alder's investigations greatly expanded the scope of the "diene reaction" (later renamed in their honor).[8,9]

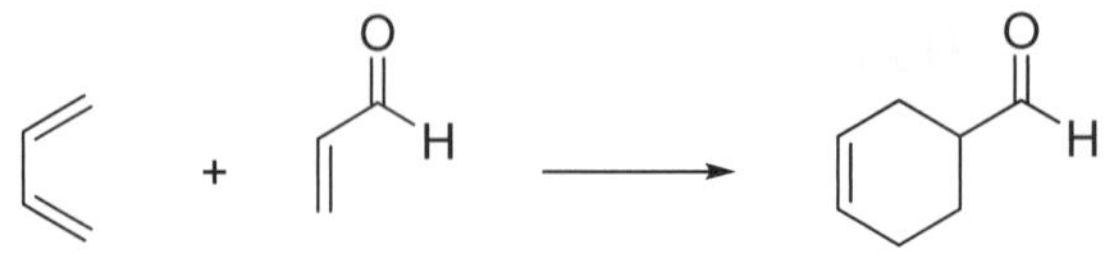

4 5 6

In 1936, Alder left the University of Kiel to pursue research efforts in industry. He continued to study the diene reaction, focusing on the stereochemical course of the transformation. Consequently, the preference for *endo* products in the reaction is generally referred to as the Alder Endo Rule. Diels and Alder's achievements in advancing the utility of this powerful reaction were recognized with the Nobel Prize in chemistry in 1950.[6]

4.7.3 Mechanism

A complete picture of the Diels–Alder mechanism involves explaining the regioselectivity and stereoselectivity of the reaction as well as understanding the molecular orbital description for the transformation. Reactions are impossible for dienes that cannot adopt cisoid conformations (see **4** for an example of a cisoid diene) for steric or other reasons. Cyclic dienes with imposed cisoid geometry are among the most reactive dienes.[5] An arrow-pushing mechanism can be easily drawn using three curved arrows to demonstrate the reorganization of three π bonds into one new π bond and two new σ bonds; however, this simplistic analysis does not help explain why the Diels–Alder reaction occurs thermally but not photochemically.

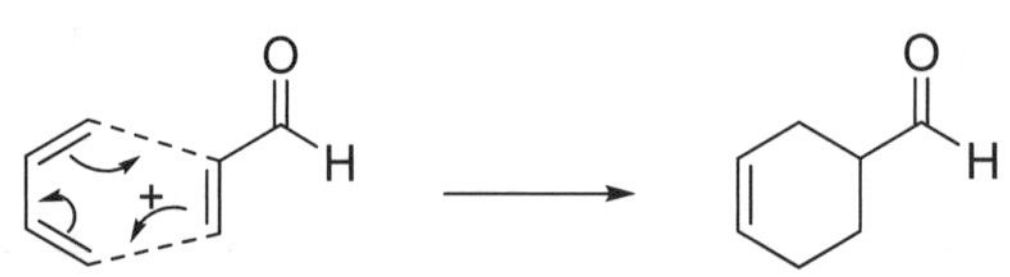

4 5 6

The mechanism of the Diels–Alder reaction was first fully described by Woodward and Hoffmann using their orbital symmetry rules for cycloadditions. The Woodward-Hoffmann rules state that the Diels–Alder reaction is a thermal $[4\pi_s + 2\,\pi_s]$ cycloaddition involving overlap of the diene's highest occupied molecular orbital (HOMO) ψ^2 with the dienophile's lowest unoccupied molecular orbital (LUMO) ψ^2. In the case of an inverse electron demand Diels–Alder reaction, the dienophile's HOMO ψ^1 combines with the diene's LUMO ψ^3.[10]

This is generally a concerted synchronous process with both new bonds forming simultaneously. Combinations of some highly polar substrates undergo concerted asynchronous reactions in which the two new σ bonds are formed at different rates.[5] In some rare cases, the reaction proceeds in a stepwise fashion with the absence of expected stereoselectivity. Orfanopoulos reported an example of a stepwise hetero Diels–Alder reaction in 2009.[11]

Experimental results and theoretical analyses have enabled rational predictions for the regiochemical preferences in Diels–Alder reactions.[12] Known as the ortho, para rule, the strongest electron donating group on the diene ends up either *ortho* (1,2) or *para* (1,4) to the strongest electron-withdrawing group on the dienophile in the final cyclohexene product. For example, combination of isoprene (**7**) with methyl acrylate (**8**) yields mostly *para* product **9**, while reaction of **8** with 1-methylbutadiene **11** furnishes *ortho* cyclohexene **12** as the major product. As shown in these cases, formation of mixtures is common under thermal conditions.[13]

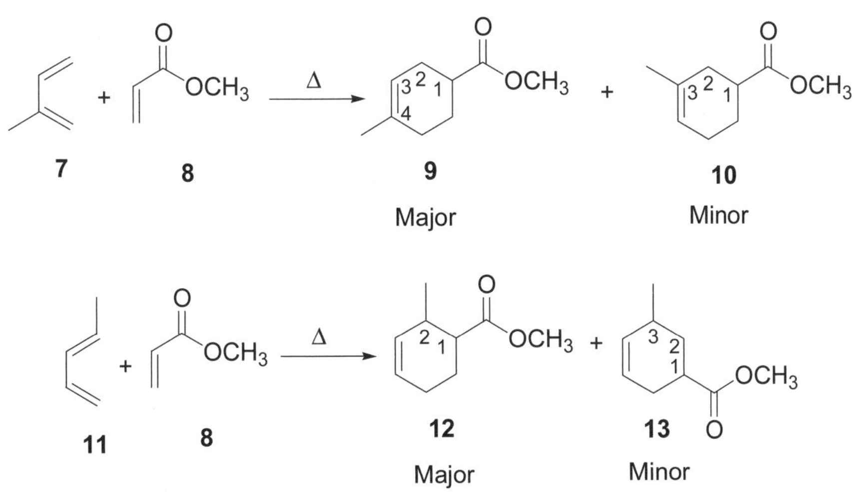

These results are explained by analyzing the frontier molecular orbital (FMO) coefficients for the orbitals involved in the reaction. Combination of the largest coefficient on the diene with the largest coefficient on the dienophile accurately predicts the regiochemical outcome of the reaction.[5] For reaction partners lacking complicated substitution patterns, the *ortho,*

para rule can be derived by drawing resonance structures for the diene and dienophile and connecting the most negative carbon on the diene with the most positive carbon on the dienophile.[14]

The stereochemical outcome of the reaction is also well understood. Due to the concerted nature of the process, it is stereospecific with respect to the diene and dienophile. Substituents that are *cis* on the dienophile are *cis* in the product and likewise for *trans*-substituents. Substitution at carbons 1 and 4 of the diene yield stereocenters in the product. Groups in these positions that pointing "out" on the diene end up on one face of the product, while groups pointing "in" are on the other face of the cyclohexene.[14]

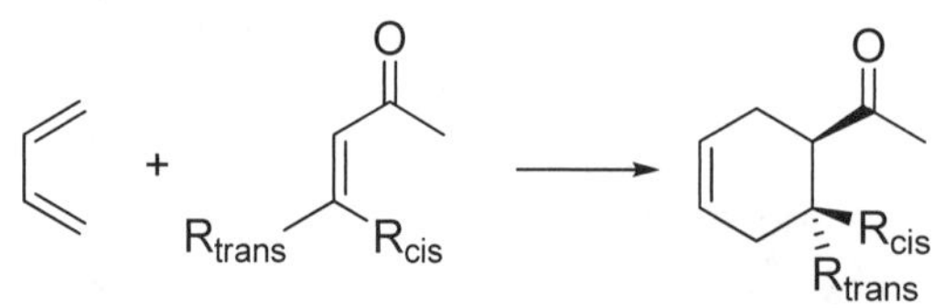

Stereospecific with respect to the dienophile

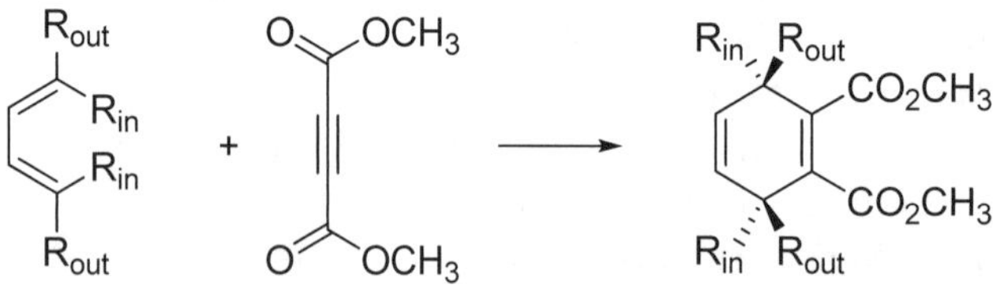

Stereospecific with respect to the diene

The Alder *endo* rule enables predictions of product structures that combine stereocenters generated from both the diene and dienophile. Based on Alder's pioneering studies, he demonstrated that *endo* products are favored over *exo* products.[15] The easiest way to determine the *endo* product is that the withdrawing group on the dienophile ends up *cis* to groups pointing out on the diene in the cyclohexene product.

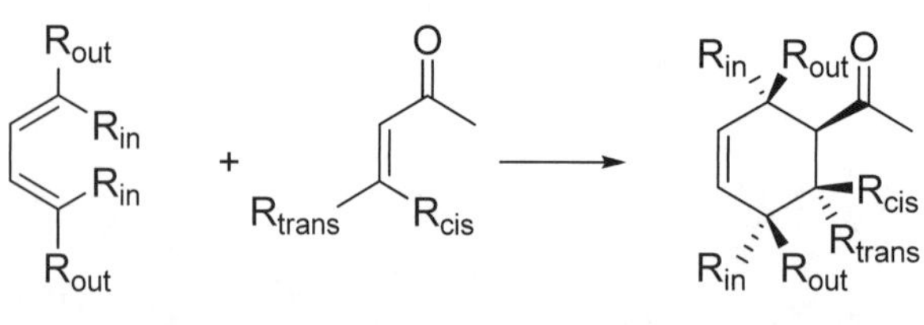

Endo product

The preference for *endo* products is rationalized by interaction of the π orbital of the withdrawing group with the p orbital at C-2 of the diene π system. This secondary orbital interaction overcomes the steric preference for the *exo* product.[14,16]

Endo Reaction 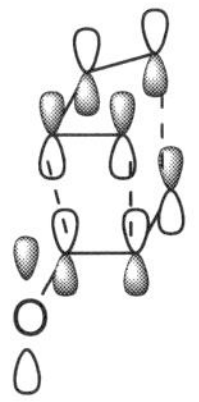

Exo Reaction 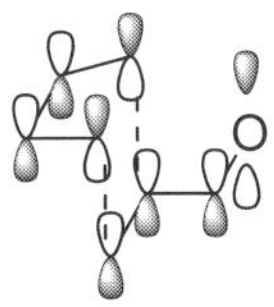

4.7.4 Variations and Improvements

The most important variations and improvements since Diels and Alder's studies include promoting the reaction with something other than heat, incorporating heteroatoms in both the diene and dienophile, reversing the electronics to combine electron rich dienophiles with electron poor dienes, developing asymmetric versions of the reaction, and covalently connecting the diene and dienophile to yield intramolecular reactions. The reverse of the Diels–Alder reaction, the retro Diels–Alder, is also highly useful in organic synthesis.[17]

Lewis Acids and Related Promoters

The most important advance in Diels–Alder methodology was the discovery that reactions could be effectively promoted by addition of a variety of Lewis acids.[18] This eliminated the need to thermally promote the reaction and often afforded the cyclohexene products in higher yield and better regioselectivity and stereoselectivity. As explained by the Woodward–Hoffmann rules for cycloadditions, Lewis acids bind to the electron-withdrawing group on the dienophile and lower the energy of the dienophile LUMO. This decreased diene HOMO-dienophile LUMO gap decreases the activation energy for the reaction. All of the common Lewis acids, $TiCl_4$, $SnCl_2$, $SnCl_4$, $ZnCl_2$, $ZnBr_2$, BF_3, $EtAlCl_2$, Et_2AlCl, and lanthanide complexes, have found successful application in Diels–Alder reactions.[19] Most contemporary studies of Lewis acids are focused on the development of chiral catalysts for asymmetric Diels–Alder reactions (vide infra).

Perlmutter reported a recent example of the superiority of Lewis acid catalysis versus thermal conditions. Thermal reactions between naphthaquinones like **14** and dienes similar to **15** lead to aromatization of the cycloadducts. Perlmutter isolated the target Diels–Alder product **16** in nearly quantitative yield using catalytic BF_3.[20]

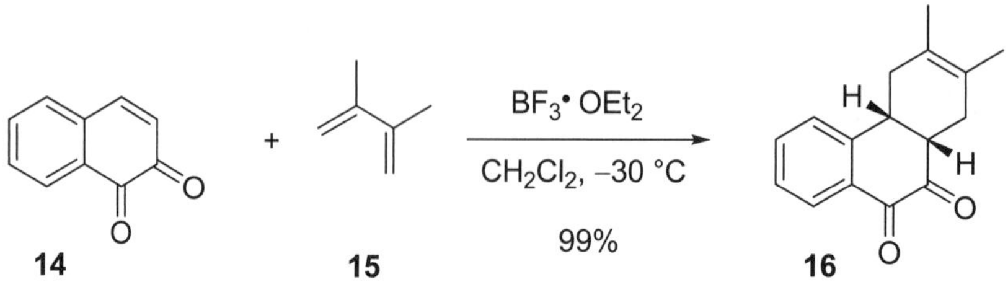

14 15 16

Transition metals are also useful catalysts for the title cycloaddition, and copper catalysts have proven especially effective.[21] In an example from Furstner's lab, unactivated alkyne **17** undergoes an intramolecular Diels–Alder reaction catalyzed by copper thiophene 2-carboxylate (CuTC) to provide bicycle **18** in 95% yield.[22]

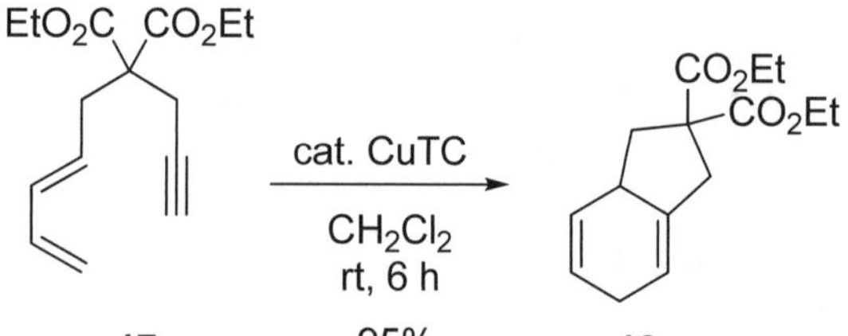

17 18

A variety of other reaction conditions promote the Diels–Alder reaction,[5,14] including high pressure,[23] ultrasound, electron transfer,[24] polar solvents,[25] ionic liquids,[25] supramolecular scaffolds,[26] enzymes,[27] ribozymes,[28] Brønsted acids and bases,[29] and π-basic transition metal complexes.[30]

Hetero Diels–Alder Reactions

Diels–Alder reactions are excellent methods for the production of a plethora of six-membered ring-containing heterocycles. A variety of heterodienes[3] and heterodienophiles[4] readily participate in the reaction with, not surprising, oxygen and nitrogen the most popular heteroatoms for inclusion. Recent reviews highlight the use of imino-containing dienes,[31] imino dienophiles,[32] carbon-phosphorous π bonds,[33] the synthesis of spiroketals,[34] the synthesis of carbohydrate-containing molecular scaffolds,[35] and ring opening of hetero Diels–Alder products.[36] Hetero-Diels–Alder reactions play important roles in inverse electron demand, asymmetric, and intramolecular Diels–Alder reactions and will appear prominently in those sections.

Several recent examples highlight the utility of the hetero-Diels–Alder reaction. Combination of enone **19** with dienophile **20** in the presence of a europium catalyst and high pressure yields cycloadduct **21**, which was ultimately converted into a substituted glycoside.[37]

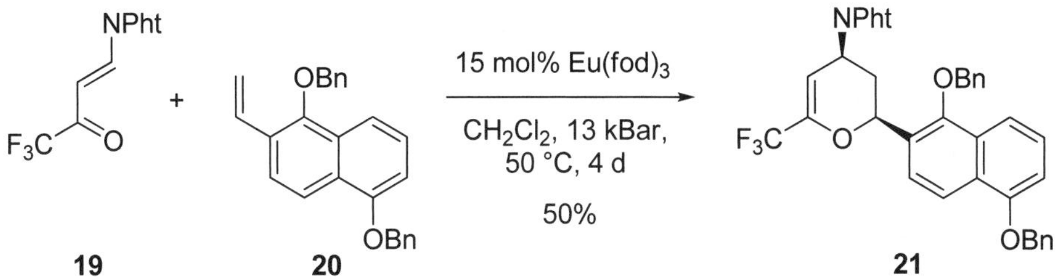

19 **20** **21**

Miller illustrated the use of a nitroso dienophile in his synthesis of Diels–Alder product **24** from ergosterol (**22**) and nitroso **23**. Subsequent cleavage of the N–O bond in **24** furnished an ergosterol derivative with significant anticancer activity.[38]

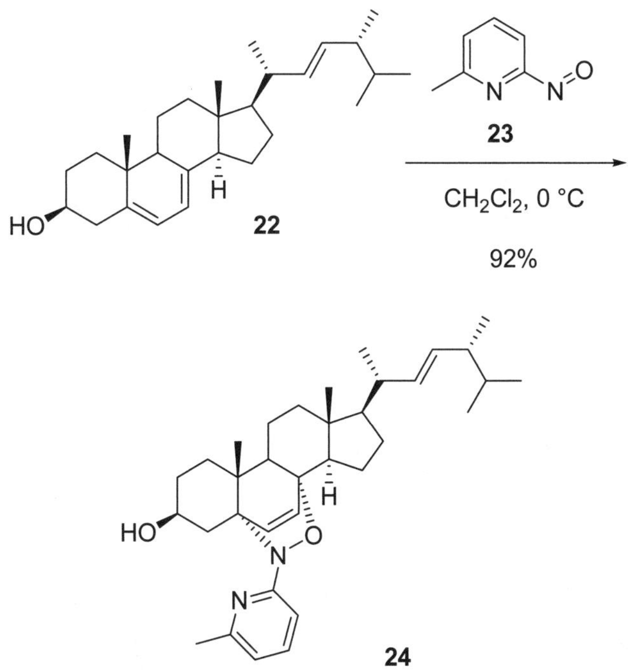

22 **23** **24**

Inverse Electron Demand Diels–Alder Reactions

As already mentioned, electron-poor dienes react with electron-rich dienophiles in what are known as inverse electron demand Diels–Alder reactions.[5,14] The most common versions of this type of transformation are hetero Diels–Alder reactions, though carbocyclic examples do exist.

Yamomoto reported an interesting carbocyclic inverse electron demand Diels–Alder reaction between tropone (**25**) and ketene diethyl acetal (**26**) catalyzed by tris(pentafluoro)phenylborane. Under these conditions, Diels–Alder product **27** is formed almost exclusively, while most other catalysts (including BF_3 and BPh_3) favor sole production of the [8 + 2] cycloadduct **28**. Yamamoto also described the development of an asymmetric version of this Diels–Alder reaction.[39]

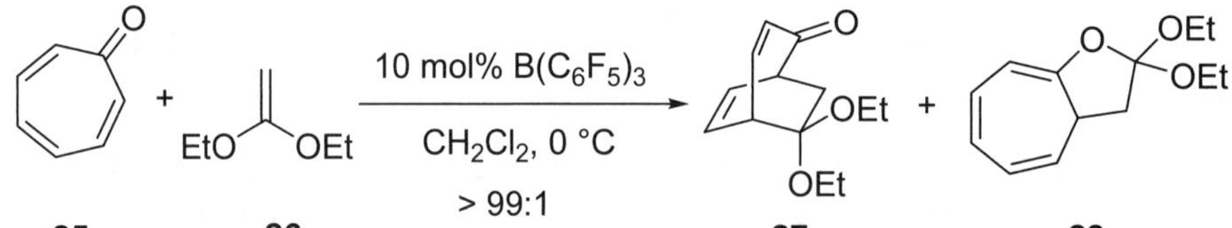

25 **26** **27** **28**

Nelson disclosed a powerful three-component Diels–Alder/alkylation reaction sequence for the synthesis of several nitrogen-containing heterocycles. For example, treatment of enamine **29** with titanium tetrachloride yields electron poor imminium ion diene **30** that reacts with unactivated cyclohexene to furnish imminium ion cycloadduct **31**. Introduction of allyltrimethylsilane into the reaction mixture leads to alkylation of the reactive imminium ion and production of perhydroisoquinoline **32** in 59% overall yield. Due to the lack of secondary orbital interactions, this inverse electron demand Diels–Alder reaction is *exo*-selective to minimize steric interactions.[40]

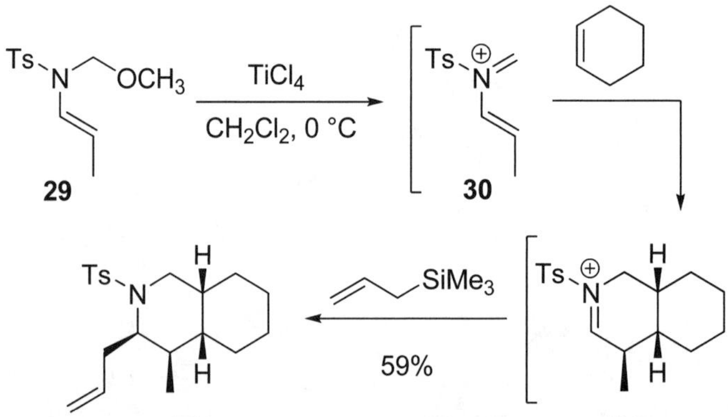

Taylor used an inverse electron demand Diels–Alder/retro-Diels–Alder/aromatization cascade for the synthesis of bicyclic pyridines. Beginning with 1,2,4 triazine **33**, reaction with electron-rich dienophile **34** yields cycloadduct **35**, which eliminates N_2 on a retro Diels–Alder reaction.

Loss of pyrrolidine from **36** provides the target pyridine product **37** in 80% yield.[41] Moody employed a similar Diels–Alder reaction of 1,2,4-triazines as part of a three-step conversion of hydrazides into substituted pyridines.[42]

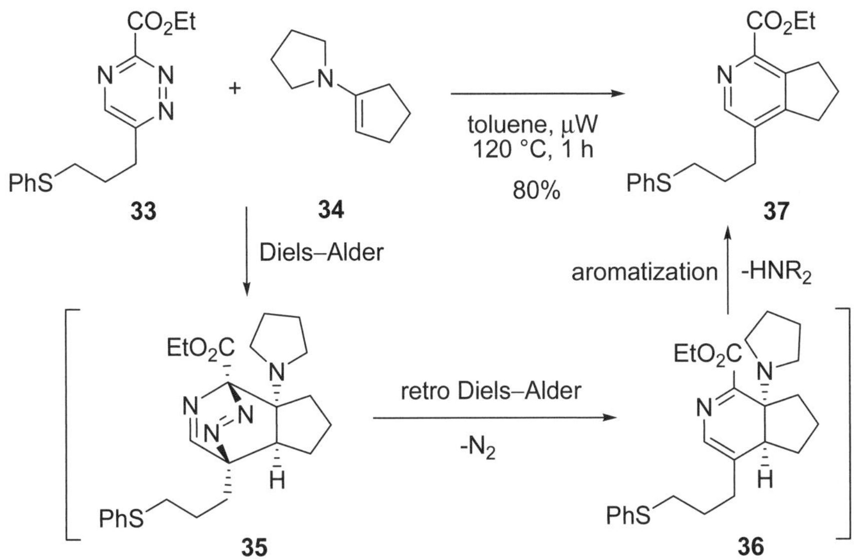

Inverse electron demand Diels–Alder reactions also have applications in biological systems. Fox reported that electron poor tetrazine diene **39** successfully forms a bioconjugate with the protein thioredoxin modified to contain a *trans*-cyclooctene (**38**). In an example of this Diels–Alder reaction in the absence of thioredoxin, tetrazine **39** combines with *trans*-cyclooctene to yield cycloadduct **40** in quantitative yield. Like the synthesis of **37** described above, this reaction proceeds via a Diels–Alder/retro Diels–Alder cascade with elimination of N_2. The reaction works well in organic solvents, water, and cellular media with **41** generated as the final product in protic solvents.[43]

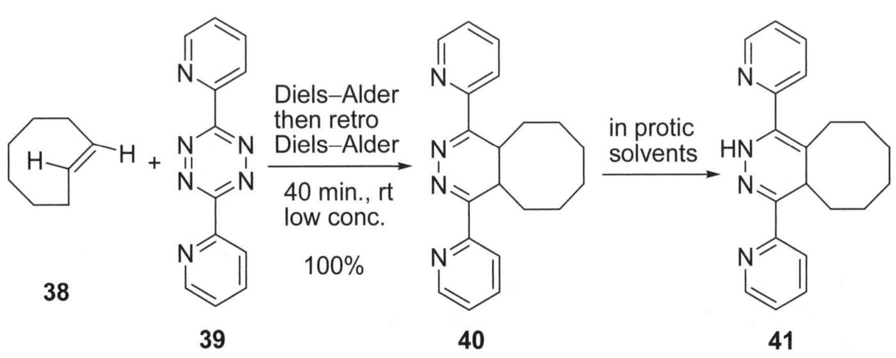

Asymmetric Diels–Alder Reactions

The recent explosion in the development of asymmetric strategies for organic synthesis has fostered investigations into the discovery of methods for enantioselective and diastereoselective Diels–Alder reactions.[44–48] Some early forays into this field focused on the use of chiral auxiliaries[49] covalently attached to one of the reaction partners; however, nearly all recent investigations have centered on developing chiral catalysts. The multitude of new catalysts spans the range of Lewis acids and Brønsted acids and bases[29] as well as metal-based[21] and organic molecules.

One of the most effective classes of asymmetric Diels–Alder catalysts is the family of chiral oxazaborolidines originally developed by Corey. In a recent example from Corey's lab, combining cyclopentadiene (**42**) with quinone **43** in the presence of catalyst **44** furnishes cycloadduct **45** in 99% yield and 99% *ee*.[50] Yamamoto used a similar catalyst for enantioselective Diels–Alder reactions of α,β-unsaturated acetylenic ketones,[51] and Paddon-Row and Houk reported a computational investigation into the reactivity of oxazaborolidine catalysts.[52]

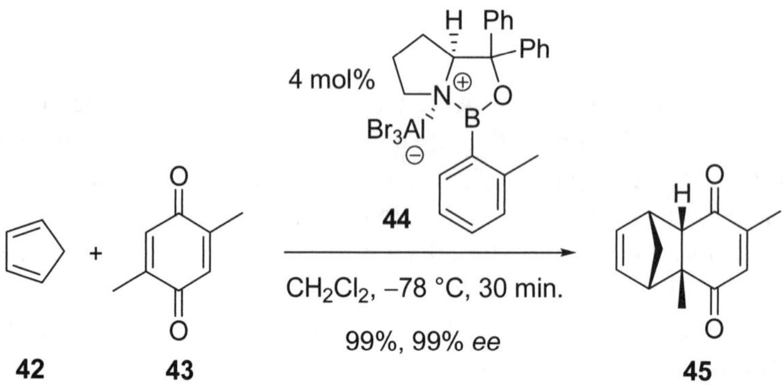

Metal bis(oxazoline) catalysts are also highly efficient at promoting asymmetric Diels–Alder reactions. Ishira developed a highly selective copper bis(oxazoline) catalyst for use in standard intermolecular Diels–Alder reactions.[53] Arrayas and Carretero used a nickel bis(oxazoline) catalyst in inverse electron demand hetero Diels–Alder reactions of 1-azadienes for the production of functionalized piperidines.[54] Sibi studied a variety of metal bis(oxazoline) catalysts in reactions between cyclopentadiene and pyrazolidinone dienophiles and determined that copper and palladium catalysts were most efficient.[55] Sibi applied a copper bis(oxazoline) catalyst in a kinetic resolution experiment in which one enantiomer of dienophile **46** reacts selectively with cyclopentadiene to yield cycloadduct **48** and 98% *ee*

of unreacted **47**. Base-promoted amide cleavage furnished the separate pyrazolidinones **49** and **50**.[56]

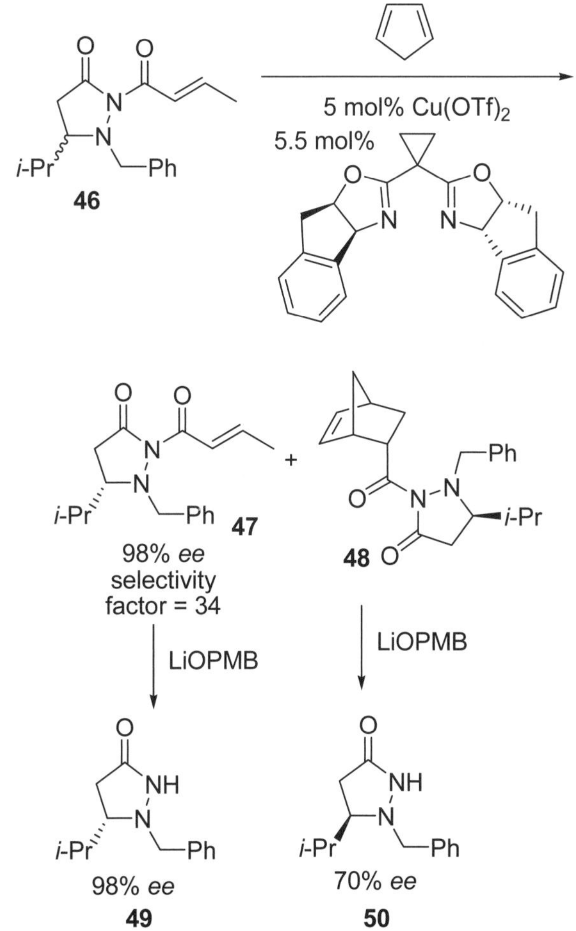

In addition to metal-based catalysts, organocatalysts are also selective promoters of asymmetric Diels–Alder reactions.[57] Several groups reported the use of cinchona alkaloid catalysts in standard Diels–Alder reactions. Deng combined 2-pyrones with α,β-unsaturated ketones,[58] while Bernardi and Ricci focused on the reactions of vinylindoles with quinones and maleimides.[59] Lectka reported enantioselective inverse electron demand hetero Diels–Alder reactions of ketene enolates and *o*-benzoquininone diimides catalyzed by a combination of benzoylquinidine and zinc triflate. For example, subjecting diimide **51** to the standard reaction conditions yields cycloadduct **52** as a single stereoisomer, which can be easily converted to diamine **53**. The proposed mechanism for this Diels–Alder reaction is a stepwise process.[60]

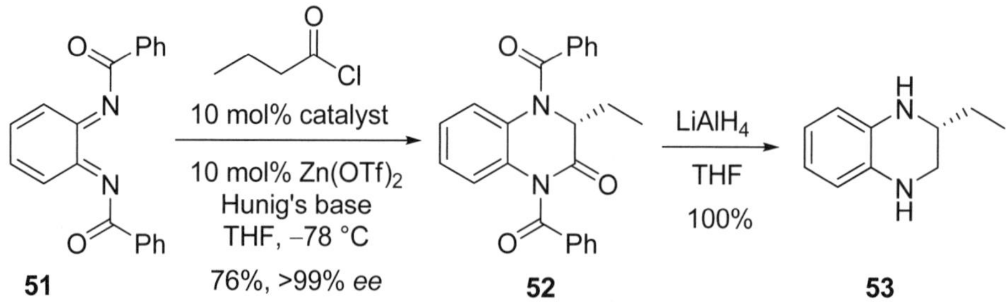

Using a relatively simple chiral secondary amine catalyst (**56**), Chen prepared a series of highly functionalized enantiomerically pure piperidines via inverse electron demand hetero Diels–Alder reactions. For example, amine catalyst **56** reacts with unsaturated aldehyde **55** to furnish a dienamine that functions as the electron-rich dienophile for combination with electron-poor 1-azadiene **54** and provides piperidine **57** in high yield and enantiomeric excess.[61]

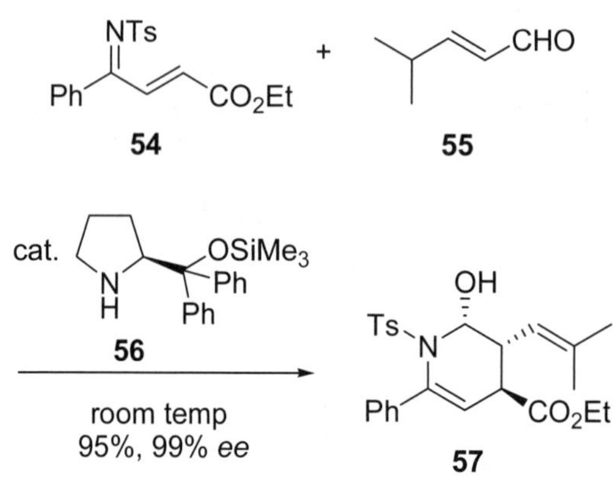

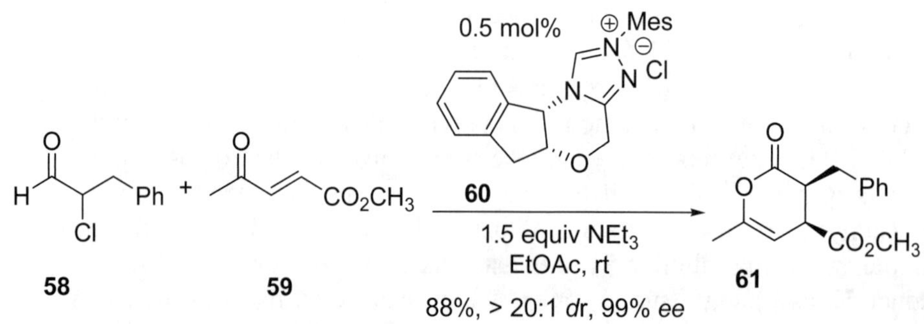

Bode demonstrated the effectiveness of *N*-heterocyclic carbenes as catalysts for inverse electron demand hetero Diels–Alder reactions involving both azadienes and oxodienes.[62] In the oxodiene case, reaction of enones with α-chloroaldehydes affords substituted dihydropyranones. In one example, addition of catalyst **60** to chloroaldehyde **58** followed by elimination of HCl yields the electron-rich dienophile that readily combines with oxodiene **59** to selectively furnish cycloadduct **61**.[62b]

Another important class of enantioselective catalysts for Diels–Alder reactions are BINAP-based catalysts. These are employed either as Brønsted acid or transition-metal complexes. Yamamoto reported that acyclic siloxydienes combine with azopyradines in the presence of a silver BINAP catalyst to yield the target nitrogen-containing heterocycles in high yield and selectivity.[63] Akiyama and Terada each separately studied BINAP-substituted phosphoric acid derivatives as Brønsted acid catalysts for hetero Diels–Alder reactions. Terada demonstrated the ability of these catalysts to promote the stereoselective combination of siloxydienes with glyoxylate to yield substituted dihydropyrans.[64] In another example of an inverse electron demand hetero Diels–Alder reaction, Akiyama selectively generated tetrahydroisoquinolines by combining aldimines and vinyl ethers in the presence of chiral phosphoric acid catalyst **64**. For example, aldimine **62** reacts with ethyl vinyl ether to afford target **65** in high yield as virtually a single stereoisomer.[65]

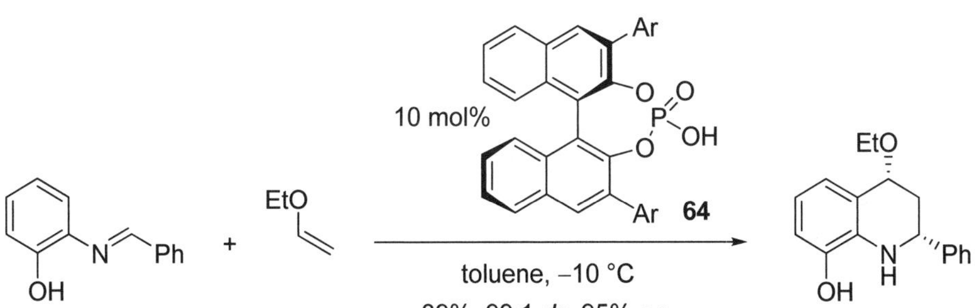

62 63 65

Intramolecular Diels–Alder Reactions

The intramolecular Diels–Alder reaction is most frequently used in natural product total synthesis, and numerous examples will be described in the synthetic utility section. As with the intermolecular variant, intramolecular reactions are highly regioselective and stereoselective and participate in hetero, inverse electron demand, and asymmetric Diels–Alder reactions. One report from 2008 describes the investigation of an intramolecular hetero Diels–Alder reaction in ionic liquids.[66]

Roush has a long-standing interest in the intramolecular Diels–Alder reaction, and he recently reported the use of siloxacyclopentenes as dienophiles in intramolecular Diels–Alder reactions. Prolonged heating of **66** produces intramolecular adduct **67** that can be easily converted to diol **68** via a Fleming–Tamao oxidation. This reaction sequence is notable for the selective production of a racemic mixture of the isomer shown (no other diastereomers observed), and diol **68** is the formal product of an intramolecular Diels–Alder reaction with an enol dienophile.[67]

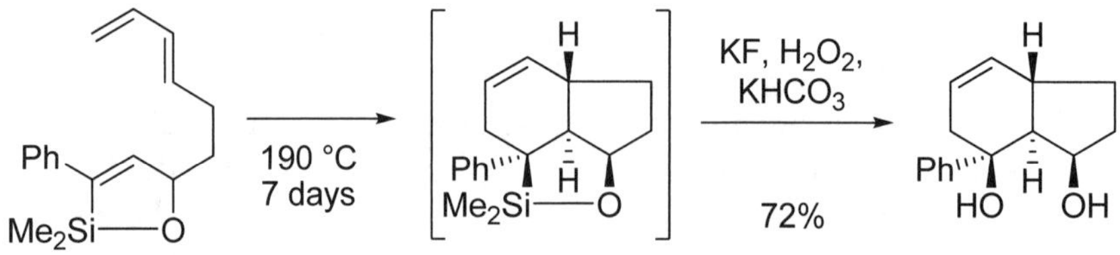

66 **67** **68**

Van de Weghe disclosed an intramolecular hetero Diels–Alder reaction with an imino diene as part of a synthetic strategy for the synthesis of uncialamycin. In the key Diels–Alder reaction, boron trifluoride promotes the cycloaddition followed by DDQ-mediated oxidation to afford aromatic product **70**.[68]

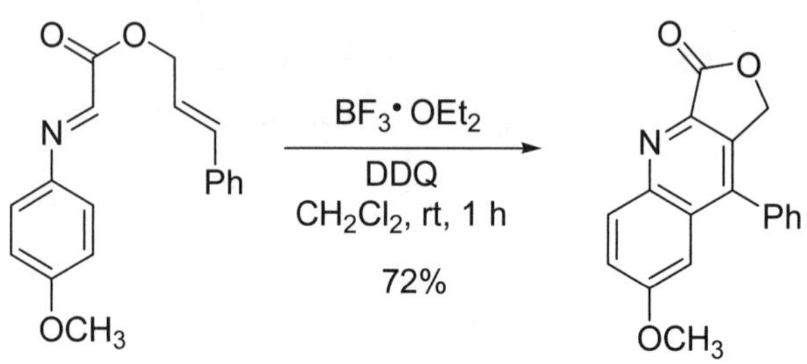

69 **70**

4.7.5 *Synthetic Utility*

The Diels–Alder reaction is a versatile tool for the synthesis of a plethora of target molecules. This section will focus on some of the most common applications, including the use of interesting dienes and dienophiles, tandem/cascade processes, and total syntheses.[5,14]

Interesting Dienophiles

Benzyne is a fascinating dienophile that Lautens recently employed in a diastereoslective synthesis of benzo-fused heterobicycles. Treatment of **72** with fluoride yields benzyne, which readily combines with its diene partner, the functionalized pyrrole **71**, to yield target cycloadduct **73** in high yield and selectivity.[69]

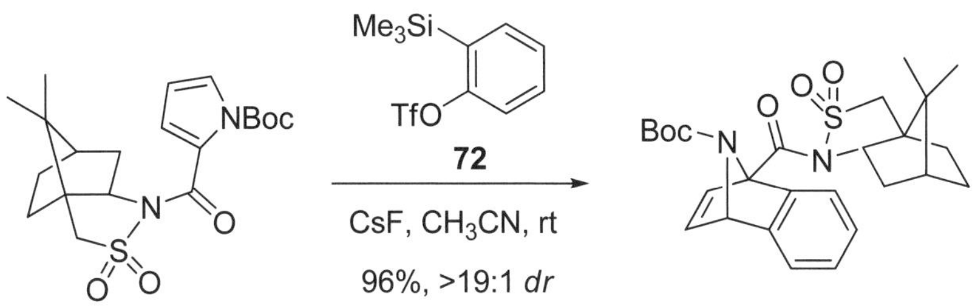

71 **73**

Allenes are another unusual class of synthetically useful dienophiles. Jung demonstrated that intramolecular Diels–Alder reactions of optically active allenic ketones yield substituted oxa-bridged octalones. For example, optically pure allene **74** undergoes a facile cycloaddition to yield a single diastereomer of target **75**.[70]

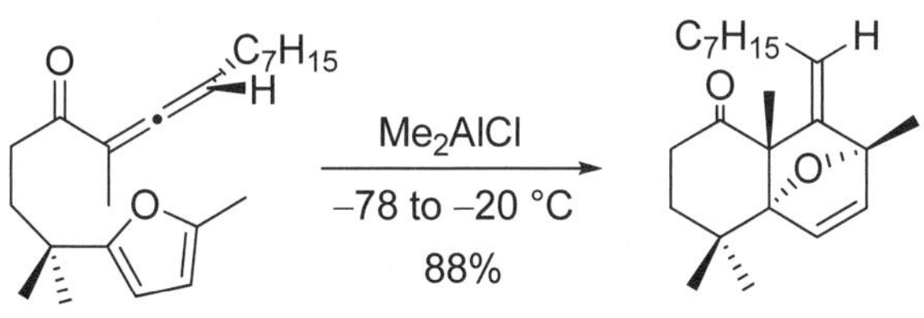

74 **75**

Danishefsky's interest in the development of synthetically useful Diels–Alder reaction methodology has recently extended to two different dienophiles. First, he studied the reactivity of enyne dienophiles to show that ynones react preferentially to enones. Substituted styrene **78** is the exclusive product on intermolecular Diels–Alder reaction of diene **76** with dienophile **77** followed by a retro Diels–Alder reaction (to generate 2-methylpropene) and two desilylations.[71]

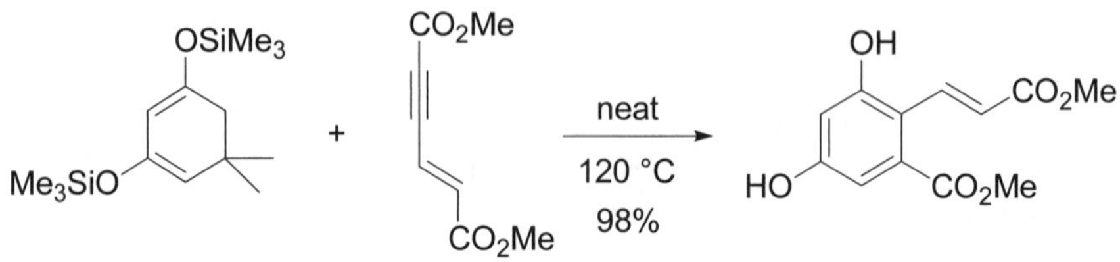

76 **77** **78**

Danishefsky also investigated the development of a *trans* Diels–Alder reaction in seeming violation of the inherent stereoselectivity of the reaction mechanism. The key to this process was the use of 1-nitrocyclohexene (**80**) as the dienophile. After a standard intermolecular Diels–Alder reaction with diene **79**, *cis* cycloadduct **81** could be transformed preferentially into *trans*-fused product **82** upon radical denitration and enol ether hydrolysis.[72]

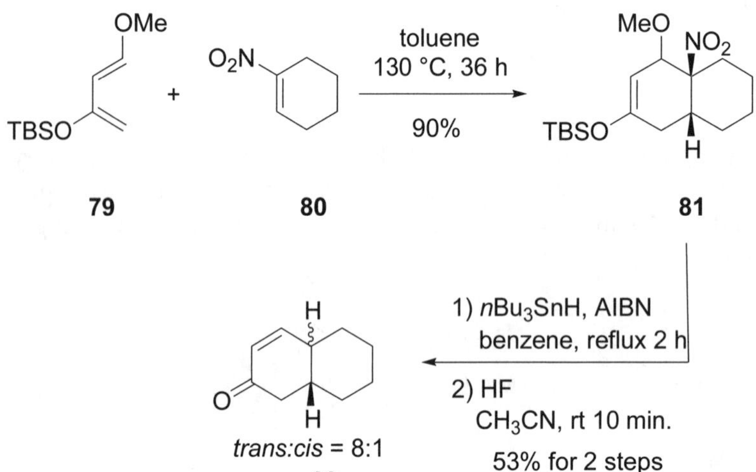

79 **80** **81**

82

Carreno reported a similar strategy for the production of *trans* Diels–Alder products. Instead of nitro-substituted dienophiles, he employed quinones substituted with boronic acids. Reaction of boronic acid dienophile **82** with diene **83** yields the expected cycloadduct as an unstable intermediate that is selectively protonated and then loses boron to yield the ultimate *trans* fused product **84**.[73]

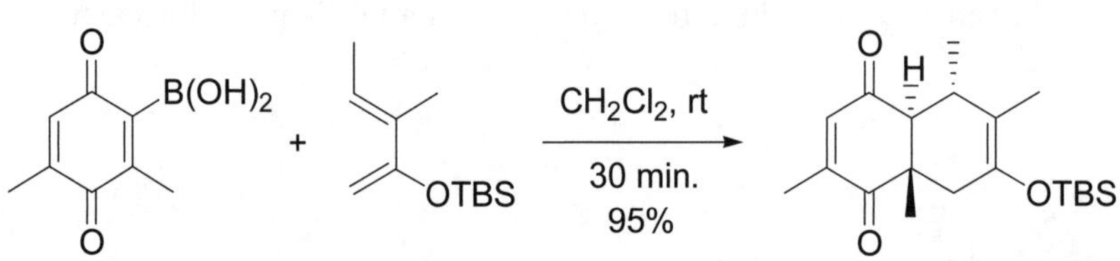

82 **83** **84**

Interesting Dienes

Due to their increased reactivity, cyclic dienes are very useful in the Diels–Alder reaction. Several groups have recently reported interesting applications of functionalized cyclopentadienes. Pearson demonstrated that cyclopentadienones react with aryl alkynes to yield polysubstituted biaryl compounds. Highly functionalized biaryl **87** is available in high yield on reaction of cyclopentadienone **85** and electron-poor aryl alkyne **86**. The mechanism of this reaction includes extrusion of carbon monoxide to yield the pentasubstituted benzene after the initial cycloaddition.[74]

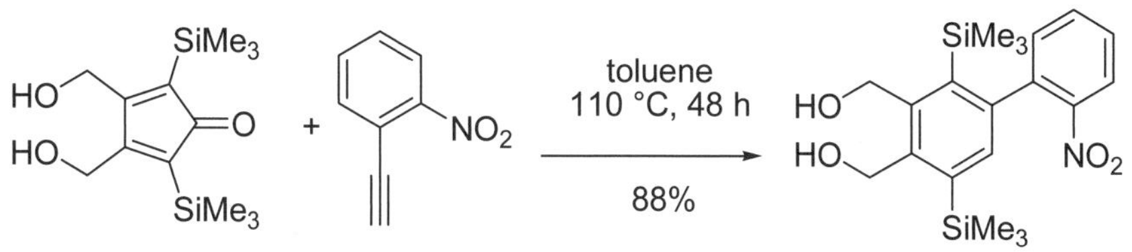

85 **86** **87**

Gleason developed a method to generate stable 5-substituted cyclopentadienes for use as reactive Diels–Alder dienes. Unlike most similarly substituted cyclopendienes, compound **88** does not undergo rapid [1,5]-sigmatropic shifts and reacts quickly enough in Diels–Alder reactions to provide good yields of the desired target compounds. For example, diene **88** combines with dienophile **89** to afford *endo* product **90** in 79% yield. Gleason demonstrated the utility of this diene in a synthetic approach to the E ring of palau'amine.[75]

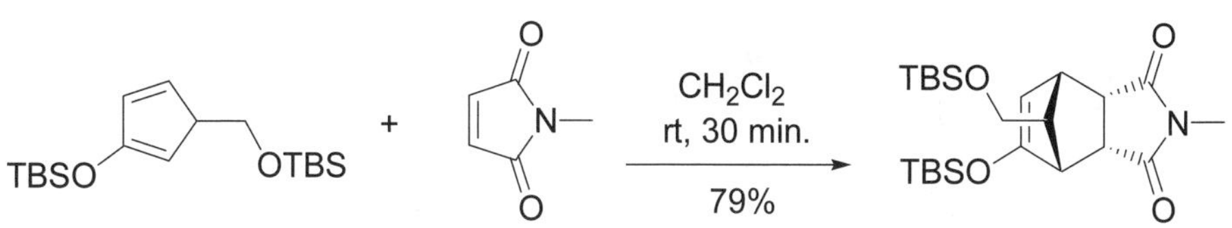

88 **89** **90**

One of the most popular acyclic dienes in the Diels–Alder reaction is Danishefsky's diene (**91**) and numerous research groups have developed chiral catalysts for the reaction of this diene with a variety of dienophiles.[76] For example, Nishida reported that Yb(III)-BINAMIDE complexes catalyze the asymmetric reaction of Danishefsky's diene with electron-deficient alkenes. In one case, diene **91** reacts with dienophile **92** to selectively provide a functionalized cyclohexene that, on hydrolysis, affords the desired chiral cyclohexenone **93** in high yield and high *ee*.[77]

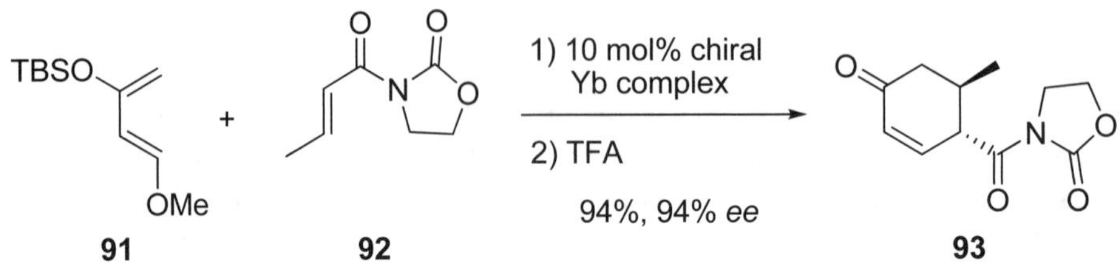

The other recent examples of asymmetric syntheses involving Danishefsky's diene focused on hetero Diels–Alder reactions. Shibasaki and Feng separately reported asymmetric reactions with carbonyl dienophiles. Shibasaki demonstrated successful asymmetric reactions of ketones using a chiral Cu(I)-Walphos catalyst.[78] Feng used a chiral *N,N'*-dioxide/In(OTf)$_3$ catalyst in asymmetric cycloaddition reactions of aldehydes.[79] Imine dienophiles are also amenable to asymmetric Diels–Alder reactions with Danishefsky's diene. Wulff reported enantioselective reactions using a VAPOL-B(OPh)$_3$ catalyst system,[80] while Snapper and Hoveyda disclosed silver-catalyzed enantioselective aza Diels–Alder reactions.[81]

Welker developed a useful synthetic strategy involving silyl substituted dienes.[82] He can easily prepare large quantities of 2-silicon-substituted 1,3-dienes that readily participate in Diels–Alder reactions followed by cross-coupling reactions to yield aryl substituted cyclohexenes. Diene **94** smoothly combines with *N*-phenylmaleimide (**95**) to furnish a silyl cyclohexene that undergoes Hiyama cross-coupling with iodobenzene to provide **96**.[83]

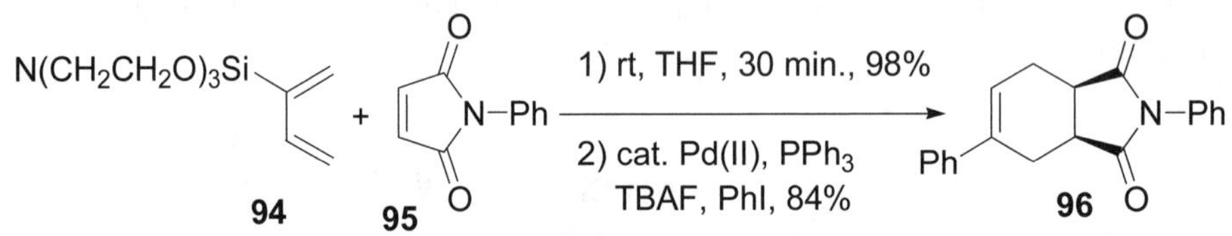

Sherburn reported a robust synthesis of the fascinating diene [4]dendralene (**97**) and its behavior in Diels–Alder reactions with *N*-methylmaleimide (**89**, NMM). Dendralene **97** is available in one step from chloroprene and combines with three equivalents of an *N*-methylmaleimide-methyl aluminium dichloride complex to provide a diastereomeric mixture of **98** after three Diels–Alder reactions.[84]

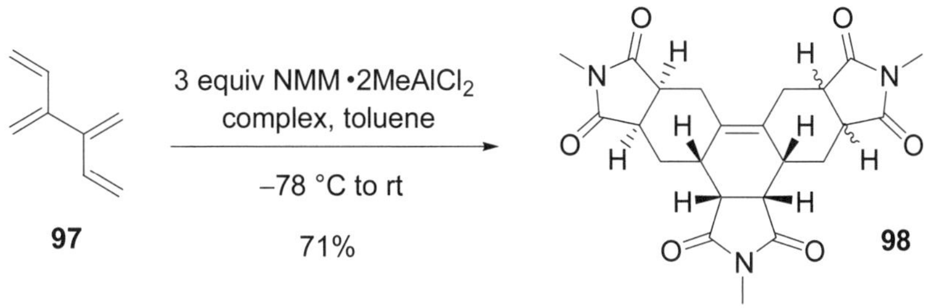

97 **98**

Reactions with conjugated enynes as dienes in Diels–Alder reactions yield cyclohexadiene or benzene products on reaction with alkene or alkyne dienophiles, respectively. These reactions proceed via a stepwise mechanism to avoid formation of a cyclic allene and are referred to as dehydro-Diels–Alder reactions.[85] In 2008, Barluenga and Aguilar demonstrated that gold catalysts promote intermolecular hetero-dehydro-Diels–Alder reactions between dienynes and nitriles. Dienyne **99** combines with phenylnitrile (**100**) to afford substituted pyridine **101**.[86]

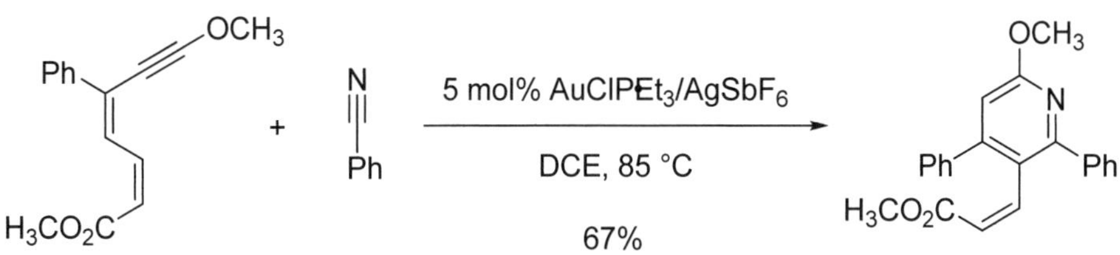

99 **100** **101**

Tandem and Cascade Reactions

The Diels–Alder reaction has a rich history of applications in tandem and cascade processes.[5,14,87,88] Many of the transformations already discussed have a cascade-like nature since processes like retro-Diels–Alder reactions and aromatization reactions often occur after the initial Diels–Alder step. Clearly, the triple Diels–Alder reaction of [4]dendralene described by Sherburn (**97**→**98**) involves a beautiful cascade of cycloaddition reactions.

In 2006, Wood described a tandem aromatic oxidation/Diels–Alder reaction for the synthesis of the carbocyclic core of bacchopetiolone. Oxidation of substituted phenol **102** with bis(trifluoroacetoxy)-iodobenzene (BTIB) provides bicycle **103** that dimerizes via a Diels–Alder reaction to yield polycyclic **104**, which is two decarbonylations away from the target natural product.[89]

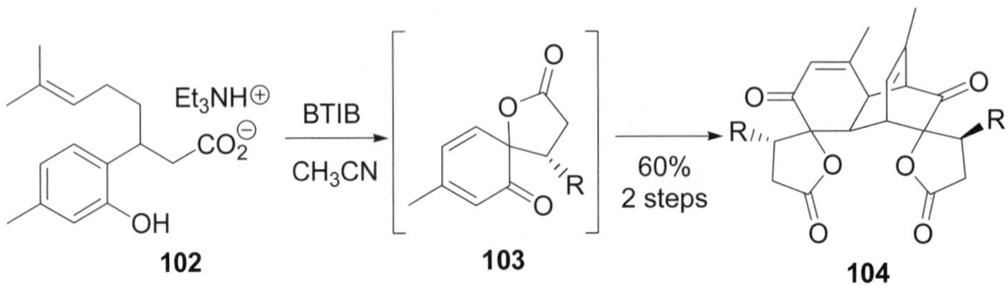

Tietze reported a domino-Knoevenagel–hetero-Diels–Alder reaction involving a three-component reaction between an α-nitroketone, formaldehyde, and an alkyl vinyl ether. In one example, a Knoevenagel condensation between ketone **105** and formaldehyde (**106**) yields electron-poor hetero-diene **108** that undergoes an inverse electron demand Diels–Alder reaction with ethyl vinyl ether to furnish dihydropyran **109**. Tietze subsequently converted **109** into the deoxysugar (+)-forosamin.[90]

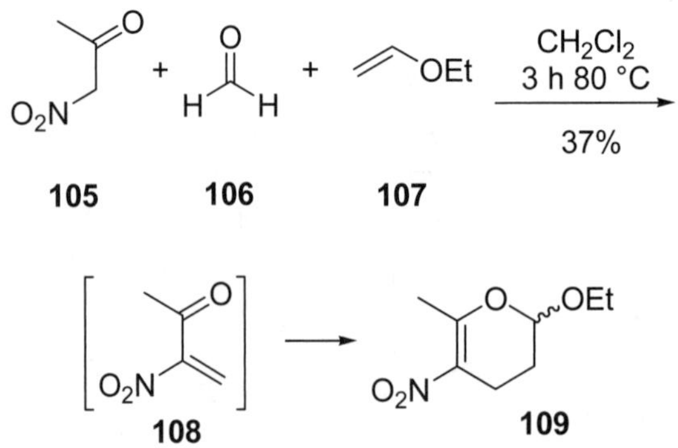

Enders described a fascinating organocatalytic one-pot asymmetric synthesis of tricyclic compounds using a triple-cascade/Diels–Alder reaction sequence. Combination of dieneal **110** with enal **111** and nitro alkene **112** in the presence of a chiral amine catalyst results in a Michael/Michael/aldol condensation sequence to yield cycloaddition precursor **113**. Cooling the reaction mixture and addition of a Lewis acid promotes the desired intramolecular Diels–Alder reaction to selectively afford the highly functionalized tricyclic target **114**.[91]

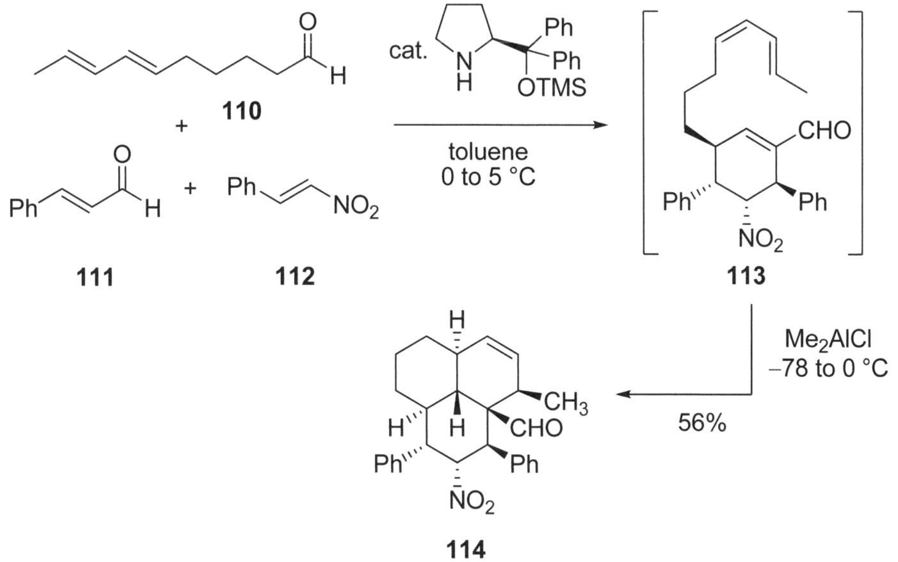

Applications in Total Syntheses

The true scope and limitations of any reaction are discovered when applied in challenging total syntheses. In this arena, the Diels–Alder reaction has proven itself as the most valuable ring-forming method for the construction of six-membered rings at all stages in synthetic execution. Often an intermolecular Diels–Alder reaction is deployed in the first step to generate a cyclohexene- or pyran-containing synthetic intermediate, or in the middle stages an intramolecular Diels–Alder reaction enables formation of a bicyclic decalin system. The title reaction is also frequently used in the ultimate or penultimate step for the generation of a highly complex synthetic target. The versatility of the Diels–Alder reaction in total synthesis is truly remarkable.[92]

Intermolecular Applications

Several groups exploited the power of the intermolecular Diels–Alder reaction early in their syntheses for the formation of substituted cyclohexenes. In his synthesis of platencin, Nicolaou used a Danishefsky-like diene in an asymmetric Diels–Alder reaction for the synthesis of a chiral cylohexenone.[93] Kanai and Shibasaki developed a catalytic asymmetric Diels–Alder reaction promoted by barium isopropoxide for the first step in their synthesis of Tamiflu.[94] Danishefsky constructed the cyclohexene ring in paecilomycine A by employing a highly *endo*-selective Diels–Alder reaction of siloxydiene **115** and enyne dienophile **116** to yield target **117**.[95]

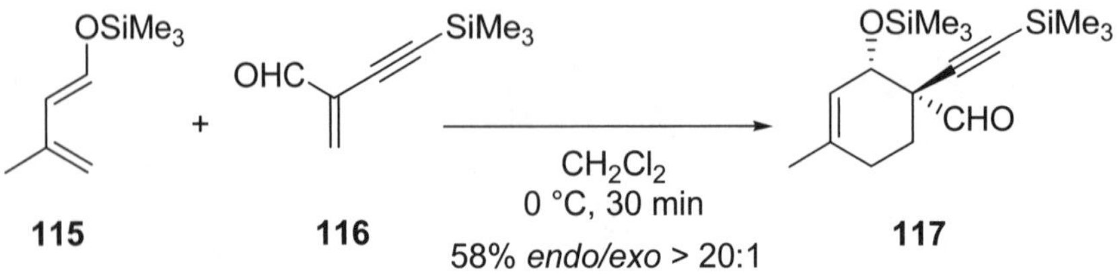

115 **116** **117**

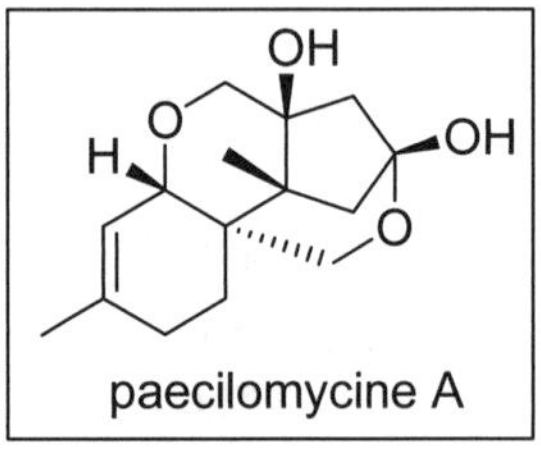

paecilomycine A

Similarly, pyran rings are a common early-stage synthetic intermediate in a variety of syntheses. In Kutay and Gademann's synthesis of anguinomycin C, they prepare the heterocyclic portion of the target by combination an electron-rich diene with an unsaturated aldehyde in the presence of Jacobsen's chromium (III) catalyst.[96] Ghosh used the same asymmetric catalyst to promote the reaction of an aldehyde and an electron-rich diene in his synthesis of brevisamide.[97] Rawal synthesized a pyranone for use in his synthesis of pederin by combination of a chiral dienophile with Danishefsky's diene.[98] In his synthesis of phorboxazole B, Burke treated Brassard diene **119** with chiral aldehyde **118** and a europium catalyst to yield pyranone **120**.[99]

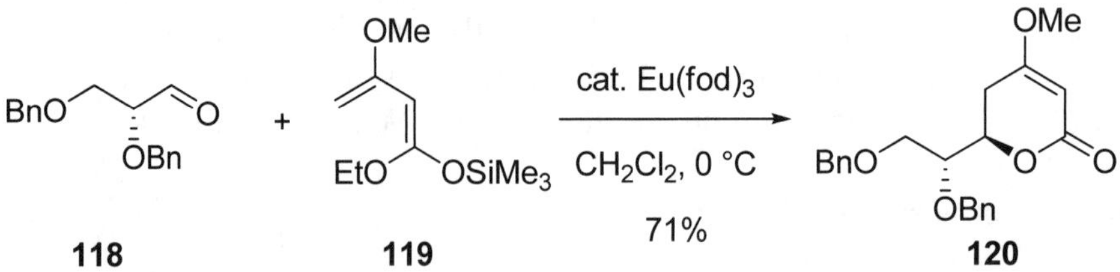

118 **119** **120**

Intermolecular Diels–Alder reactions are also highly useful for the construction of polycyclic carbocycles in natural product total syntheses. Rawal combined a cyclohexadienone with a Danishefsky-type diene to yield a *cis*-decalin in his total synthesis of platencin.[100] Stratakis also generated a *cis*-decalin upon combination of a quinone dienophile and an acyclic diene during the production of acremine G.[101] Nakamura and Hashimoto employed an intermolecular Diels–Alder reaction for the construction of the G ring in pinnatoxin A.[102] Danishefsky reacted a vinylindene diene with a quinoneketal dienophile to form the tetracyclic framework of fluostatin C.[103]

In Nicolaou's synthesis of the bisanthraquinone antibiotic BE-43472B, he initiated a fascinating late-stage cascade by the reaction of a functionalized acyclic diene with a pentacyclic quinone dienophile.[104] In a highly complex example, Nakata performed the penultimate step in the synthesis of sarcophytoate by combining dienophile **121** and diene **122** in a thermal Diels–Alder reaction.[105]

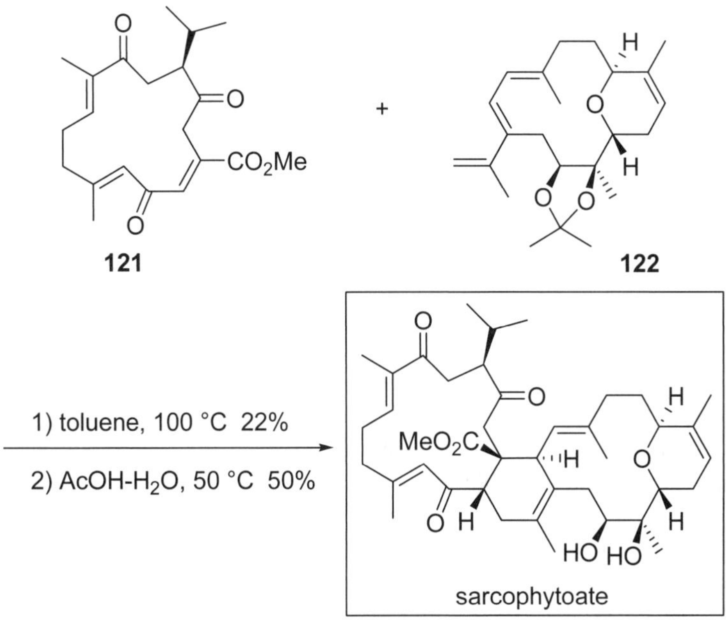

Intramolecular variants are the most common type of Diels–Alder reactions employed in total syntheses with carbocycles the most popular targets. Among the broad variety of structures in this class, bicyclic decalin structures, both *cis* and *trans*, are readily available on intramolecular Diels–Alder reactions. Aubé demonstrated the use of a tandem intramolecular Diels–Alder/Schmidt reaction sequence in his synthesis of three *stemona* alkaloids. The first steps generates *cis*-decalin intermediate **124**, which undergoes a ring expansion in the Schmidt reaction to furnish tricycle **125**.[106]

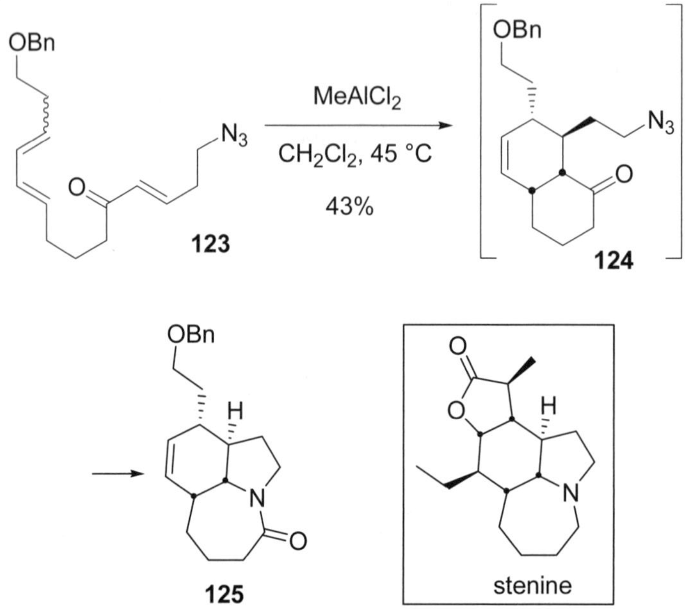

Many research groups generate *trans*-decalins as synthetic intermediates in total syntheses. MacMillan demonstrated the power of his methodology for enantioselective organocatalytic intramolecular Diels–Alder reactions by preparing the *trans*-decalin framework of solanapyrone in a highly selective cycloaddition.[107] Hoye synthesized the *trans*-decalin portion of the heteratriquinane UCS1025A via a thermal Diels–Alder reaction.[108] Evans deployed an intramolecular Diels–Alder reaction for the synthesis of a bicyclic synthetic intermediate during his synthesis of himgaline.[109] Movassaghi prepared the A and B rings of himandrine in an intramolecular Diels–Alder reaction of tetraene **126** to afford *trans*-decalin **127**.[110]

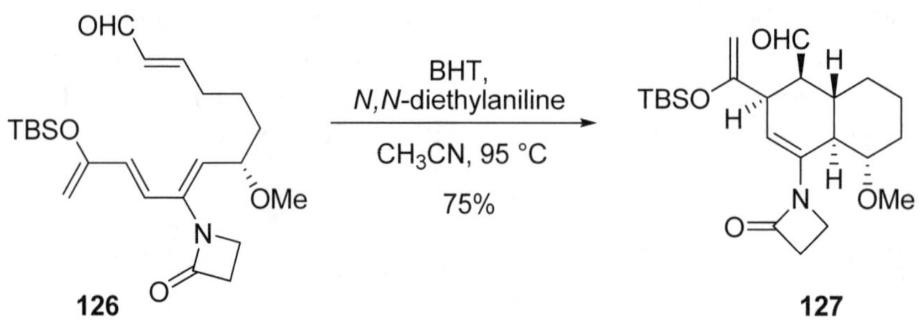

Intramolecular Diels–Alder reactions also enable the construction of cyclohexenes in a variety of molecular frameworks. A late-stage cycloaddition generated the tetracyclic skeleton in Wong's synthesis of pallavicinolide A.[111] Snider and Thomson separately used similar intramolecular Diels–Alder reaction strategies for the synthesis of the key tricyclic framework of deoxysymbioimine and symbioimine, respectively.[112,113] Imagawa and Nishizawa prepared a bicyclic γ-lactone via an intramolecular Diels–Alder reaction in their synthesis of neovibsanin B.[114] Corey deployed an enantioselective Diels–Alder macrobicyclization as a key step in the syntheses of palominol, dolabellatrienone, β-araneosene, and isoedunol.[115] Lebel demonstrated the use of a one-pot copper-catalyzed methylenation-Diels–Alder cyclization as the final step in her synthesis of desoxygaliellalactone.[116] Jacobsen generated the pentacyclic framework of yohimbine in a scandium triflate-catalyzed intramolecular Diels–Alder reaction.[117] Kishi synthesized both the macrocyclic and cyclohexene rings in the pteriatoxins via an intramolecular cycloaddition.[118] Trauner employed a vinyl quinone as a diene in the Diels–Alder reaction that generated the pentacyclic structure of halenaquinone.[119] As demonstrated by Crimmins, the tricyclic core of the eunicellins is readily available from an intramolecular Diels–Alder reaction.[120]

Three specific examples from Gin, Sorenson, and Baran highlight the power of the standard intramolecular Diels–Alder reaction. Gin developed a 1,3-dipolar cycloaddition and dienamine-Diels–Alder reaction sequence for the synthesis of nominine. The dipolar cycloaddition delivered the precursor to dienone **128** which generates dienamine **129** on exposure to hot pyrrolidine. Cycloaddition of **129** provides desired target **130**, which is easily converted into nominine.[121]

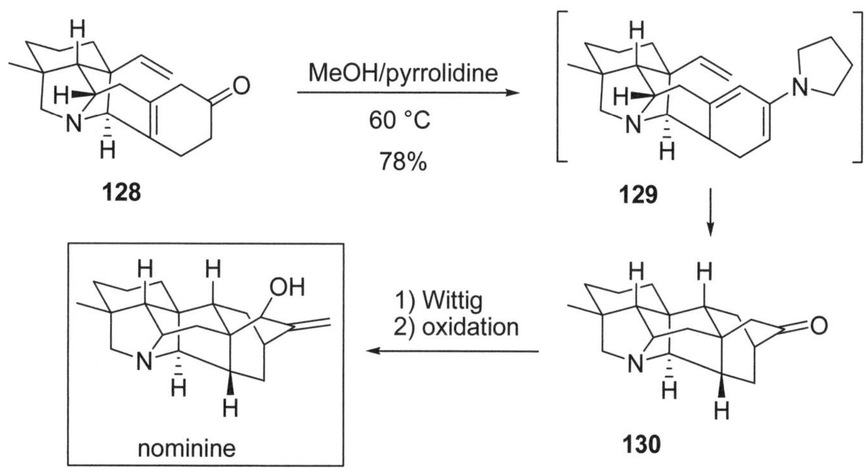

Sorensen's synthesis of abyssomicin C relied on an intramolecular Diels–Alder reaction catalyzed by lanthanum triflate. Heating **131** in the presence of the Lewis acid catalyst promotes both elimination of the silyl ether and the desired cycloaddition to afford tricycle **132**, which is converted into abyssomicin C in three additional steps.[122]

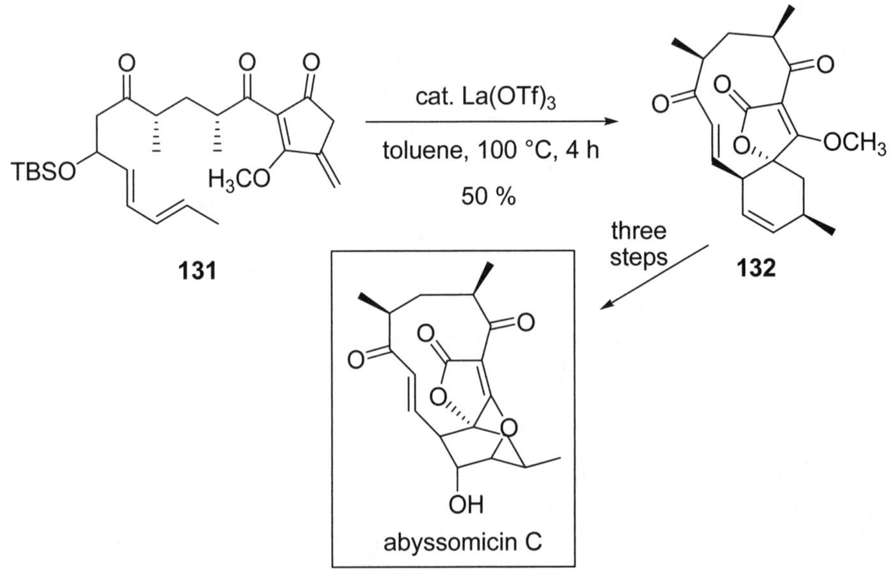

In a spectacular final sequence during his synthesis of haouamine A, Baran employed an intramolecular Diels–Alder reaction with alkyne dienophile **133** to yield cycloadduct **134**. This material then underwent a retro Diels–Alder reaction (producing carbon dioxide) followed by ester hydrolysis to afford haouamine A. This Diels–Alder/retro-Diels–Alder cascade is remarkable since the benzene ring formed in this sequence is not planar.[123]

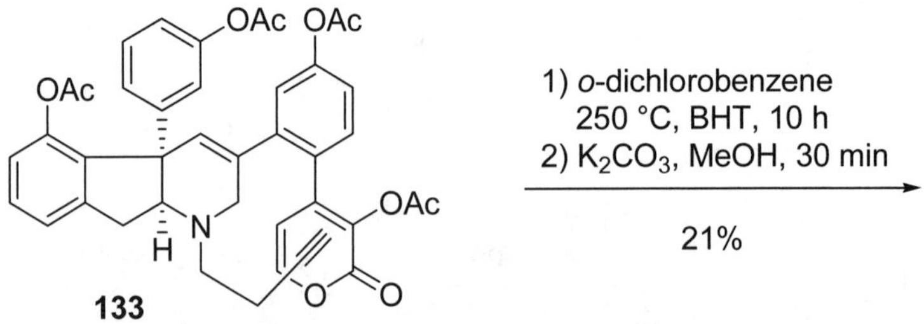

133

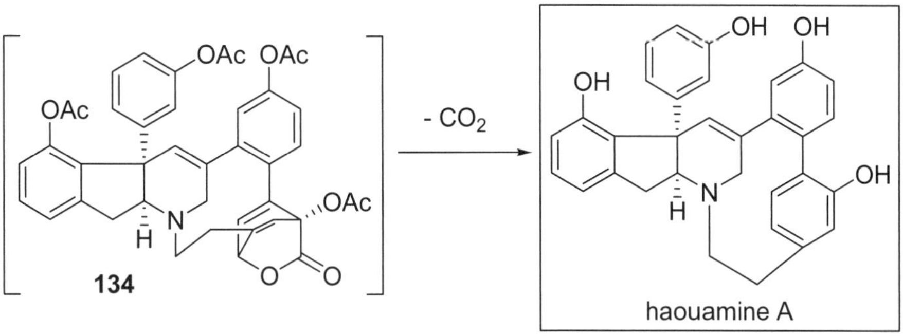

Heterocyclic six-membered rings are often prepared during total synthesis investigations using intramolecular Diels–Alder reactions. Nicolaou employed the reaction of an *o*-quinone diene with a tetrasubstituted alkene to generate the macrocyclic structure of sporolide B,[124] while in Nakada's synthesis of FR182877, he transformed an acyclic compound into an oxygen-containing tetracycle via an intramolecular Diels–Alder/hetero-Diels–Alder cascade.[125] Trauner developed a highly productive cascade sequence to complete his synthesis of rubioncolin B. Treatment of **135** with two equivalents of the fluoride reagent TASF and PhI(OAc)$_2$ led directly to TBS-deprotection and oxidation to the *p*-quinone which readily tautomerized to the cycloaddition precursor **136**. A hetero Diels–Alder reaction followed by demethylation with boron tribromide afforded the target natural product in excellent yield.[126]

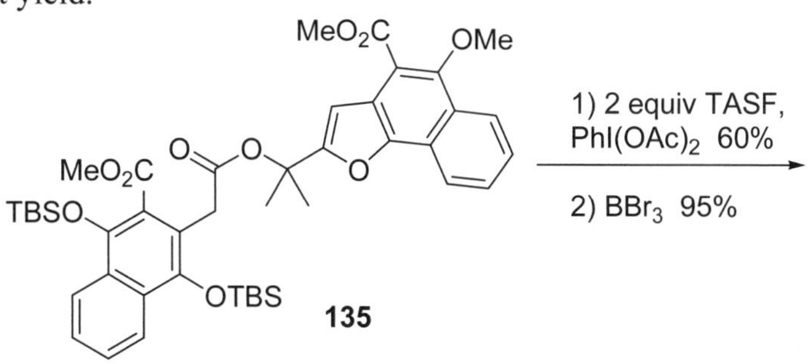

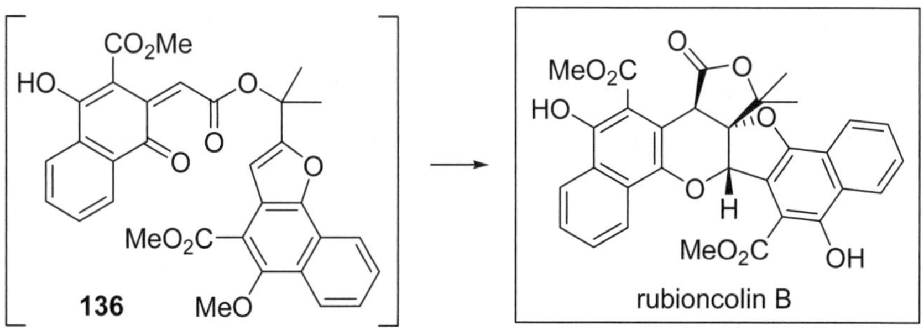

Nitrogen-containing heterocycles are also available via intramolecular hetero Diels–Alder reactions. Williams employed an aza diene to prepare a complex polycyclic synthetic intermediate in his synthesis of versicolamide B.[127] Boger reported a tandem intramolecular hetero Diels–Alder/1,3-dipolar cycloaddition sequence for the synthesis of vindorosine. Cycloaddition precursor **137** undergoes an inverse electron demand Diels–Alder reaction to yield **138**. This compound decomposes via a retro dipolar cycloaddition to generate nitrogen gas and a 1,3-dipole that completes the cascade by reacting with the indole alkene to afford **139**. Seven more steps enable the completion of vindorosine.[128]

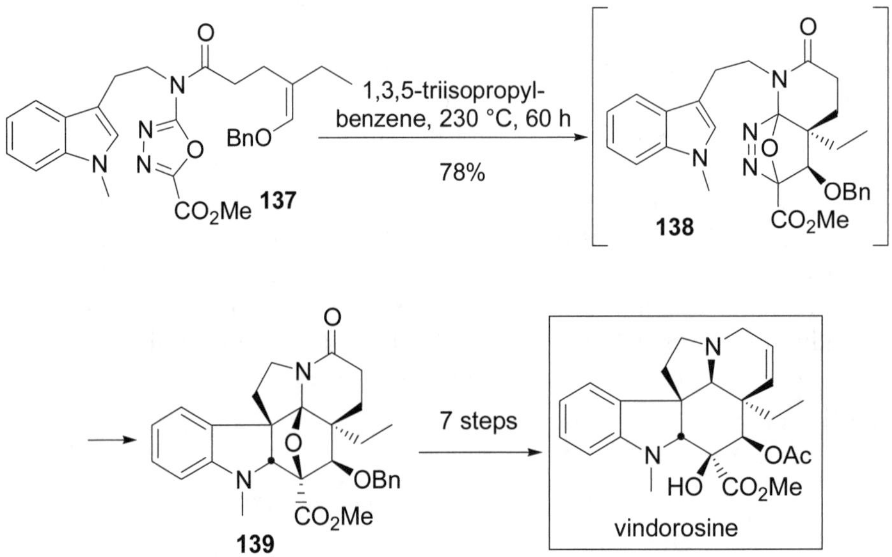

The final type of intramolecular Diels–Alder reaction that finds wide use in natural product total syntheses is the transannular process. Danishefsky exploited the power of this transformation during an oxidative dearomatization/transannular Diels–Alder cascade in his synthesis of 11-*O*-debenzoyltashironin.[129] Deslongchamps produced the tricyclic core of cassaine via a transannular intramolecular Diels–Alder reaction.[130] The tricyclic *cis*-decalin with appended macrocycle framework of superstolide A is also available using this strategy. Roush demonstrated the effectiveness of this approach by heating **140** in toluene to yield cycloadduct **141** that was transformed into superstolide A in four more steps.[131]

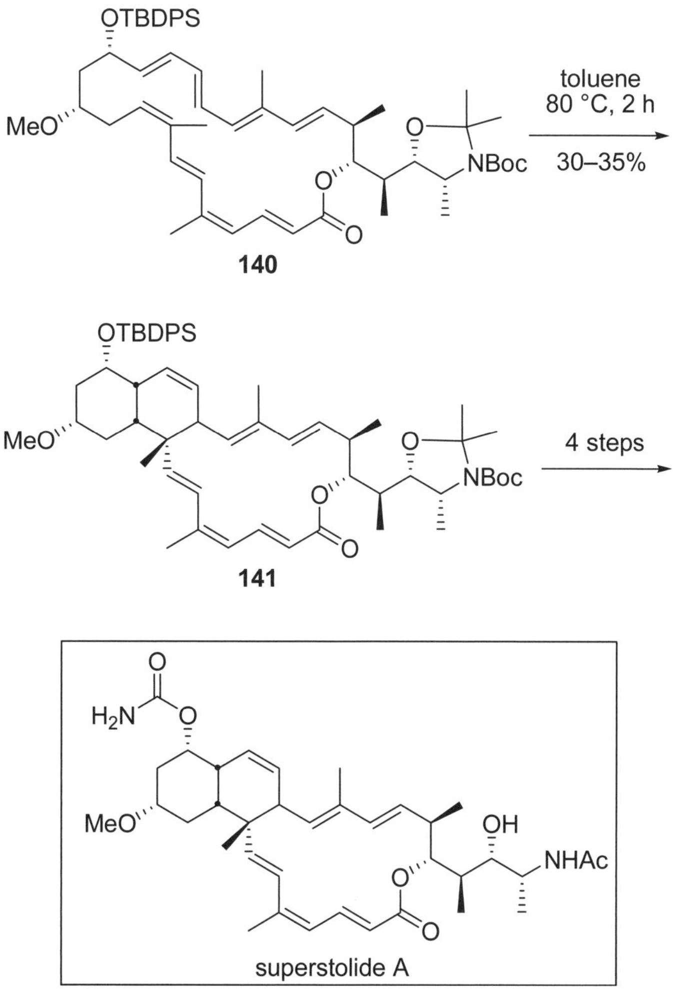

4.7.6 *Experimental*

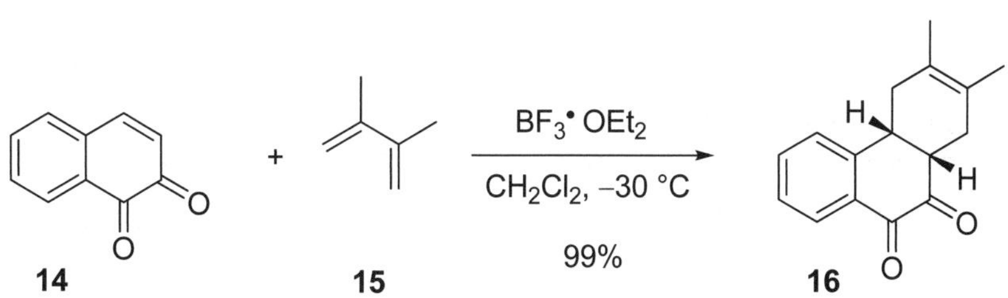

***cis*-2,3-Dimethyl-1,4,4a,10a-tetrahydrophenanthren-9,10-dione** (16):[20]
BF_3•OEt_2 (44μL, 0.35 mmol) was added to a solution of 1,2-naphthoquinone (**14**) (50 mg, 0.32 mmol) in dichloromethane (3 mL) at –78 °C. After 10 min, 2,3-dimethyl-1,3-butadiene (**15**) (72 μL, 0.63 mmol) was added dropwise. The reaction was allowed to slowly warm to –30 °C and maintained for a further 90 min. The reaction was then cooled to –78 °C and brine (3 mL) was added slowly. The reaction was allowed to warm to rt, and the contents were extracted with hexanes (3 × 5 mL). The combined organics were then dried (Na_2SO_4), filtered, and excess solvent removed in vacuo to yield **16** as a yellow powder (75 mg, 99%) without need for further purification.

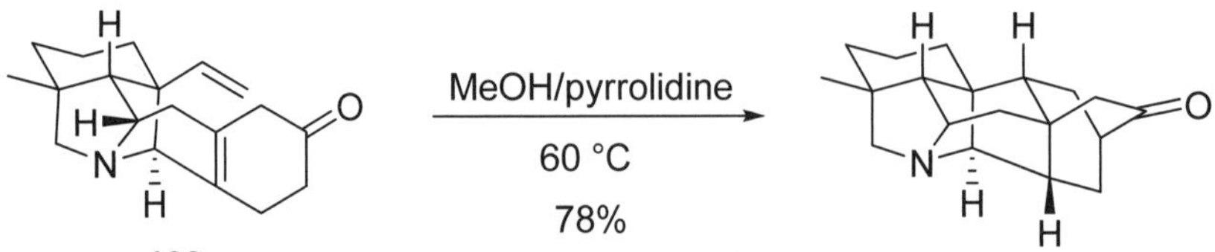

128 **130**

Alkaloid 130:[121]
To a solution of **128** (27 mg, 0.0953 mmol, 1 equiv) in MeOH (4.5 mL) was added pyrrolidine (0.5 mL). The reaction was then heated to 60 °C (oil bath) for 4.5 h, at which time TLC showed no remaining starting material. The reaction was allowed to cool to ambient temperature and then concentrated *in vacuo*. The crude product was purified by flash chromatography (10% methanol in dichloromethane with 1% NH_4OH) to give **130** as an orange solid (21 mg, 78%).

4.7.7 References

1. [R] Oppolzer, W. Intermolecular Diels–Alder Reactions. In *Comprehensive Organic Synthesis*; Trost, B. M., Fleming, I., Eds.; Pergamon: New York, 1991; Vol. 5, pp. 315–399.
2. [R] Roush, W. R. Intramolecular Diels–Alder Reactions. In *Comprehensive Organic Synthesis*; Trost, B. M., Fleming, I., Eds.; Pergamon: New York, 1991; Vol. 5, pp. 513–550.
3. [R] Boger, D. L. Heterodiene Additions. In *Comprehensive Organic Synthesis*; Trost, B. M., Fleming, I., Eds.; Pergamon: New York, 1991; Vol. 5, pp. 451–512.
4. [R] Weinreb, S. M. In *Comprehensive Organic Synthesis*; Trost, B. M., Fleming, I., Eds.; Pergamon: New York, 1991, Vol. 5, pp. 401–449.
5. [R] Fringuelli, F.; Taticchi, A. *The Diels–Alder Reaction: Selected Practical Methods* Wiley, New York, 2002.
6. [R] Walters, L. R. Kurt Alder and Otto Paul Hermann Diels. In *Nobel Laureates in Chemistry 1901–1992*; James, L. K. Ed.; ACS CHF, 1993, pp. 328–337.
7. Diels, O.; Alder, K. *Justus Liebigs Ann. Chem.* **1928**, *460*, 98–122.
8. [R] Kloetzel, M. C. *Org. React.* **1948**, *4*, 1–59.

9. [R] For a detailed look at the discovery of the Diels–Alder reaction, see Berson, J. A. *Chemical Creativity: Ideas from the Works of Woodward, Huckel, Meerwein, and Others* Wiley-VCH, New York, 1999; pp. 9–23.
10. [R] Woodward, R. B.; Hoffmann, R. *Angew. Chem., Int. Ed.* **1969**, *8*, 781–853.
11. Alberti, M. N.; Orfanopoulos, M. *Org. Lett.* **2009**, *11*, 1659–1662.
12. For recent reports exploring regioselectivity in Diels–Alder reactions, see (a) Cui, Y.; Jiang, H.; Li, Z.; Wu, N.; Yang, Z.; Quan, J. *Org. Lett.* **2009**, *11*, 4628–4631. (b) Hayden, A. E.; DeChancie, J.; George, A. H.; Dai, M.; Yu, M.; Danishefsky, S. J.; Houk, K. N. *J. Org. Chem.* **2009**, *74*, 6770–6776. (c) Hilt, G.; Janikowski, J. *Org. Lett.* **2009**, *11*, 773–776.
13. Inukai, T.; Kojima, T. *J. Org. Chem.* **1971**, *36*, 924–928.
14. [R] Sankararaman, S. *Pericyclic Reactions—A Textbook: Reactions, Applications, and Theory*; Wiley-VCH, New York, 2005, pp. 106–168.
15. Alder, K.; Stein, G. *Angew. Chem.* **1937**, *50*, 510.
16. For examples of *exo*-selective Diels–Alder reactions, see Lam, Y.-H.; Cheong, P. H.-Y.; Mata, J. M. B.; Stanway, S. J.; Gouverneur, V.; Houk, K. N. *J. Am. Chem. Soc.* **2009**, *131*, 1947–1957.
17. a) [R] Stajer, G.; Csende, F.; Fulop, F. *Curr. Org. Chem.* **2003**, *7*, 1423–1432. b) [R] Sweger, R. W.; Czarnick, A. W. In *Comprehensive Organic Synthesis*; Trost, B. M., Fleming, I., Eds.; Pergamon: New York, 1991, Vol. 5, pp. 551–592.
18. For the first example (using $AlCl_3$), see Yates, P.; Eaton, P. *J. Am. Chem. Soc.* **1960**, *82*, 4436–4437.
19. (a) [R] Jorgensen, K. A. *Cycloaddition Reactions in Organic Synthesis*; Kobayashi, S., Jorgensen, K. A., Eds.; Wiley-VCH: New York, 2002; pp 301–327. (b) [R] Pindur, U.; Lutz, G.; Otto, C. *Chem. Rev.* **1993**, *93*, 741–761.
20. Gelman, D. M.; Forsyth, C. M.; Perlmutter, P. *Org. Lett.* **2009**, *11*, 4958–4960.
21. [R] Reymond, S.; Cossy, J. *Chem. Rev.* **2008**, *108*, 5359–5406.
22. Furstner, A.; Stimson, C. C. *Angew. Chem., Int. Ed.* **2007**, *46*, 8845–8849.
23. Ballerini, E.; Minuti, L.; Piermatti, O.; Pizzo, F. *J. Org. Chem.* **2008**, *73*, 7909–7915.
24. Sevov, C. S.; Wiest, O. *J. Org. Chem.* **2009**, *74*, 4311–4317.
25. (a) Tiwari, S.; Kumar, A. *Angew. Chem., Int. Ed.* **2006**, *45*, 4824–4825. (b) [R] Kumar, A. *Chem. Rev.* **2001**, *101*, 1–19.
26. Kass, S.; Gregor, T.; Kersting, B. *Angew. Chem., Int. Ed.* **2006**, *45*, 101–104.
27. [R] Kelly, W. L. *Org. Biomol. Chem.* **2008**, *6*, 4483–4493.
28. Zhang, X.; Bruice, T. C. *J. Am. Chem. Soc.* **2007**, *129*, 1001–1007.
29. [R] Shen, J.; Tan, C.-H. *Org. Biomol. Chem.* **2007**, *6*, 3229–3236.
30. Liu, W.; You, F.; Mocella, C. J.; Harman, W. D. *J. Am. Chem. Soc.* **2006**, *128*, 1426–1427.
31. [R] Kouznetsov, V. *Tetrahedron* **2009**, *65*, 2721–2750.
32. [R] Heintzelman, G. R.; Meigh, I. R.; Mahajan, Y. R.; Weinreb, S. M. *Org. React.* **2005**, *65*, 141–599.
33. [R] Bansal, R. K.; Kumawat, S. K. *Tetrahedron* **2008**, *64*, 10945–10976.
34. [R] Rizzacasa, M. A.; Pollex, A. *Org. Biomol. Chem.* **2009**, *7*, 1053–1059.
35. [R] Cacciarini, M.; Menichetti, S.; Nativi, C.; Richichi, B. *Curr. Org. Synth.* **2007**, *4*, 47–57.
36. [R] Hilt, G. *Angew. Chem., Int. Ed.* **2009**, *48*, 6390–6393.
37. Maingot, L.; Leconte, S.; Chataigner, I.; Martel, A.; Dujardin, G. *Org. Lett.* **2009**, *11*, 1619–1622.
38. Yang, B.; Miller, P. A.; Mollmann, U.; Miller, M. J. *Org. Lett.* **2009**, *11*, 2828–2831.
39. Li, P.; Yamamoto, H. *J. Am. Chem. Soc.* **2009**, *131*, 16628–16629.
40. Sarkar, N.; Banerjee, A.; Nelson, S. G. *J. Am. Chem. Soc.* **2008**, *130*, 9222–9223.
41. Catozzi, N.; Edwards, M. G.; Raw, S. A.; Wasnaire, P.; Taylor, R. J. K. *J. Org. Chem.* **2009**, *74*, 8343–8354.
42. Shi, B.; Lewis, W.; Campbell, I. B.; Moody, C. J. *Org. Lett.* **2009**, *11*, 3686–3688.
43. Blackman, M. L.; Royzen, M.; Fox, J. M. *J. Am. Chem. Soc.* **2008**, *130*, 13518–13519.
44. [R] For a review of catalytic asymmetric Diels–Alder reactions, see Hayashi, Y. In *Cycloaddition Reactions in Organic Synthesis*; Kobayashi, S., Jorgensen, K. A., eds.; Wiley-VCH, New York, 2002, pp. 5–56.

45. [R] For a review of catalytic enantioselective Diels–Alder reactions of carbonyl compounds, see Jorgensen, K. A. *Cycloaddition Reactions in Organic Synthesis*; Kobayashi, S., Jorgensen, K. A., Eds.; Wiley-VCH: New York, 2002; pp 151–186.
46. [R] For reviews of stereoselective aza Diels–Alder reactions, see (a) Rowland, G. B.; Rowland, E. B.; Zhang, Q.; Antilla, J. C. *Curr. Org. Chem.* **2006**, *10*, 981–1005. (b) Kobayashi, S. *Cycloaddition Reactions in Organic Synthesis*; Kobayashi, S.; Jorgensen, K. A., eds.; Wiley-VCH, New York, 2002, pp. 187–210.
47. [R] For a review of the use of metal-based catalysts in asymmetric Diels–Alder reactions, see Zhu, H. J.; Jiang, J. X.; Ren, J.; Yan, Y. M.; Pittman, C. U., Jr. *Curr. Org. Synth.* **2005**, *2*, 547–587.
48. [R] For a review on asymmetric nitroso Diels–Alder reactions, see Yamamoto, Y.; Yamamoto, H. *E. J. Org. Chem.* **2006**, 2031–2043.
49. For recent examples of the use of chiral auxiliaries in diastereoselective hetero Diels–Alder reactions, see (a) Gallier, F.; Hussain, H.; Martel, A.; Kirschning, A.; Dujardin, G. *Org. Lett.* **2009**, *11*, 3060–3063. (b) Clark, R. C.; Pfeiffer, S. S.; Boger, D. L. *J. Am. Chem. Soc.* **2006**, *128*, 2587–2593.
50. Liu, D.; Canales, E.; Corey, E. J. *J. Am. Chem. Soc.* **2007**, *129*, 1498–1499.
51. Payette, J. N.; Yamamoto, H. *Angew. Chem., Int. Ed.* **2009**, *48*, 8060–8062.
52. Paddon-Row, M. N.; Anderson, C. D.; Houk, K. N. *J. Org. Chem.* **2009**, *74*, 861–868.
53. Sakakura, A.; Kondo, R.; Matsumura, Y.; Akakura, M.; Ishihara, K. *J. Am. Chem. Soc.* **2009**, *131*, 17762–17764.
54. Esquivias, J.; Arrayas, R. G.; Carretero, J. C. *J. Am. Chem. Soc.* **2007**, *129*, 1480–1481.
55. Sibi, M. P.; Stanley, L. M.; Nie, X.; Venkatraman, L.; Liu, M.; Jasperse, C. P. *J. Am. Chem. Soc.* **2007**, *129*, 395–405.
56. Sibi, M. P.; Kawashima, K.; Stanley, L. M. *Org. Lett.* **2009**, *11*, 3894–3897.
57. [R] For a review of proline-catalyzed enantioselective aza Diels–Alder reactions, see Painter, T. O.; Brummond, K. M. *Chemtracts* **2006**, *19*, 377–384.
58. Singh, R. P.; Bartelson, K.; Wang, Y.; Su, H.; Lu, X.; Deng, L. *J. Am. Chem. Soc.* **2008**, *130*, 2422–2423.
59. Gioia, C.; Hauville, A.; Bernardi, L.; Fini, F.; Ricci, A. *Angew. Chem., Int. Ed.* **2008**, *47*, 9236–9239.
60. Abraham, C. J.; Paull, D. H.; Scerba, M. T.; Grebinski, J. W.; Lectka, T. *J. Am. Chem. Soc.* **2006**, *128*, 13370–13371.
61. Han, B.; He, Z.-Q.; Li, J.-L.; Li, R.; Jiang, K.; Liu, T.-Y.; Chen, Y.-C. *Angew. Chem. Int. Ed.* **2009**, *48*, 5474–5477.
62. a) He, M.; Struble, J. R.; Bode, J. W. *J. Am. Chem. Soc.* **2006**, *128*, 8418–8420. b) He, M.; Uc, G. J.; Bode, J. W. *J. Am. Chem. Soc.* **2006**, *128*, 15088–15089.
63. Kawasaki, M.; Yamamoto, H. *J. Am. Chem. Soc.* **2006**, *128*, 16482–16483.
64. Momiyama, N.; Tabuse, N.; Terada, M. *J. Am. Chem. Soc.* **2009**, *131*, 12882–12883.
65. Akiyama, T.; Morita, H.; Fuchibe, K. *J. Am. Chem. Soc.* **2006**, *128*, 13070–13071.
66. Tiwari, S.; Khupse, N.; Kumar, A. *J. Org. Chem.* **2008**, *73*, 9075–9083.
67. Halvorsen, G. T.; Roush, W. R. *Org. Lett.* **2008**, *10*, 5313–5316.
68. Desrat, S.; van de Weghe, P. *J. Org. Chem.* **2009**, *74*, 6728–6734.
69. Webster, R.; Lautens, M. *Org. Lett.* **2009**, *11*, 4688–4691.
70. Jung, M. E.; Min, S.-J. *J. Am. Chem. Soc.* **2005**, *127*, 10834–10835.
71. Dai, M.; Sarlah, D.; Yu, M.; Danishefsky, S. J.; Jones, G. O.; Houk, K. N. *J. Am. Chem. Soc.* **2007**, *129*, 645–647.
72. Kim, W. H.; Lee, J. H.; Danishefsky, S. J. *J. Am. Chem. Soc.* **2009**, *131*, 12576–12578.
73. Redondo, M. C.; Veguillas, M.; Ribagorda, M.; Carreno, M. C. *Angew. Chem., Int. Ed.* **2009**, *48*, 370–374.
74. Pearson, A. J.; Zhou, Y. *J. Org. Chem.* **2009**, *74*, 4242–4245.
75. Hudon, J.; Cernak, T. A.; Ashenhurst, J. A.; Gleason, J. L. *Angew. Chem., Int. Ed.* **2008**, *47*, 8885–8888.
76. [R] For a review of asymmetric hetero Diels–Alder reactions of Danishefsky's and Brassard's dienes with aldehydes, see Lin, L.; Liu, X.; Feng, X. *Synlett* **2007**, 2147–2157.
77. Sudo, Y.; Shirasaki, D.; Harada, S.; Nishida, A. *J. Am. Chem. Soc.* **2008**, *130*, 12588–12589.
78. Chen, I.-H.; Oisaki, K.; Kanai, M.; Shibasaki, M. *Org. Lett.* **2008**, *10*, 5151–5154.

79. Yu, Z.; Liu, X.; Dong, Z.; Xie, M.; Feng, X. *Angew. Chem., Int. Ed.* **2008**, *47*, 1308–1311.
80. Newman, C. A.; Antilla, J. C.; Chen, P.; Predeus, A. V.; Fielding, L.; Wulff, W. D. *J. Am. Chem. Soc.* **2007**, *129*, 7216–7217.
81. Mandai, H.; Mandai, K.; Snapper, M. L.; Hoveyda, A. H. *J. Am. Chem. Soc.* **2008**, *130*, 17961–17969.
82. For a review of boron- and silicon-substituted dienes, see Welker, M. E. *Tetrahedron* **2008**, *64*, 11529–11539.
83. Pidaparthi, R. R.; Junker, C. S.; Welker, M. E.; Day, C. S.; Wright, M. W. *J. Org. Chem.* **2009**, *74*, 8290–8297.
84. Payne, A. D.; Willis, A. C.; Sherburn, M. S. *J. Am. Chem. Soc.* **2005**, *127*, 12188–12189.
85. [R] Wessig, P.; Muller, G. *Chem. Rev.* **2008**, *108*, 2051–2063.
86. Barluenga, J.; Fernandez-Rodriguez, M. A.; Garcia-Garcia, P.; Aguilar, E. *J. Am. Chem. Soc.* **2008**, *130*, 2764–2765.
87. [R] Vogel, P. *Top. Curr. Chem.* **2008**, *282*, 187–214.
88. [R] Although not focused on true tandem/cascade processes, for a review of the synthesis of polycyclics by a combination of enyne metathesis and the Diels–Alder reaction, see Kotha, S.; Meshram, M.; Tiwari, A. *Chem. Soc. Rev.* **2009**, *38*, 2065–2092.
89. Berube, A.; Drutu, I.; Wood, J. L. *Org. Lett.* **2006**, *8*, 5421–5424.
90. Tietze, L. F.; Bohnke, N.; Dietz, S. *Org. Lett.* **2009**, *11*, 2948–2950.
91. Enders, D.; Huttl, M. R. M.; Runsink, J.; Raabe, G.; Wendt, B. *Angew. Chem., Int. Ed.* **2007**, *46*, 467–469.
92. a) [R] Tadano, K. *Eur. J. Org. Chem.* **2009**, 4381–4394. b) [R] Takao, K.; Munakata, R.; Tadano, K. *Chem. Rev.* **2005**, *105*, 4779–4807. c) [R] Nicolaou, K. C.; Snyder, S. A.; Montagnon, T.; Vassilikogiannakis, G. *Angew. Chem., Int. Ed.* **2002**, *41*, 1668–1698.
93. Nicolaou, K. C.; Tria, G. S.; Edmonds, D. J. *Angew. Chem., Int. Ed.* **2008**, *47*, 1780–1783.
94. Yamatsugu, K.; Yin, L.; Kamijo, S.; Kimura, Y.; Kanai, M.; Shibasaki, M. *Angew. Chem., Int. Ed.* **2009**, *48*, 1070–1076.
95. Min, S.-J.; Danishefsky, S. J. *Angew. Chem., Int. Ed.* **2007**, *46*, 2199–2202.
96. Bonazzi, S.; Guttinger, S.; Zemp, I.; Kutay, U.; Gademann, K. *Angew. Chem., Int. Ed.* **2007**, *46*, 8707–8710.
97. Ghosh, A. K.; Li, J. *Org. Lett.* **2009**, *11*, 4164–4167.
98. Jewett, J. C.; Rawal, V. H. *Angew. Chem., Int. Ed.* **2007**, *46*, 6502–6504.
99. Lucas, B. S.; Gopalsamuthiram, V.; Burke, S. D. *Angew. Chem., Int. Ed.* **2007**, *46*, 769–772.
100. Hayashida, J.; Rawal, V. H. *Angew. Chem., Int. Ed.* **2008**, *47*, 4373–4376.
101. Arkoudis, E.; Lykakis, I. N.; Gryparis, C.; Stratakis, M. *Org. Lett.* **2009**, *11*, 2988–2991.
102. Nakamura, S.; Kikuchi, F.; Hashimoto, S. *Angew. Chem., Int. Ed.* **2008**, *47*, 7091–7094.
103. Yu, M.; Danishefsky, S. J. *J. Am. Chem. Soc.* **2008**, *130*, 2783–2785.
104. Nicolaou, K. C.; Lim, Y. H.; Becker, J. *Angew. Chem., Int. Ed.* **2009**, *48*, 3444–3448.
105. Ichige, T.; Okano, Y.; Kanoh, N.; Nakata, M. *J. Am. Chem. Soc.* **2007**, *129*, 9862–9863.
106. Frankowski, K. J.; Golden, J. E.; Zeng, Y.; Lei, Y.; Aubé, J. *J. Am. Chem. Soc.* **2008**, *130*, 6018–6024.
107. Wilson, R. M.; Jen, W. S.; MacMillan, D. W. C. *J. Am. Chem. Soc.* **2005**, *127*, 11616–11617.
108. Hoye, T. R.; Dvornikovs, V. *J. Am. Chem. Soc.* **2006**, *126*, 2550–2551.
109. Evans, D. A.; Adams, D. J. *J. Am. Chem. Soc.* **2007**, *129*, 1048–1049.
110. Movassaghi, M.; Tjandra, M.; Qi, J. *J. Am. Chem. Soc.* **2009**, *131*, 9648–9650.
111. Dong, J.-Q.; Wong, H. N. C. *Angew. Chem., Int. Ed.* **2009**, *48*, 2351–2354.
112. Snider, B. B.; Che, Q. *Angew. Chem., Int. Ed.* **2006**, *45*, 932–935.
113. Kim, J.; Thomson, R. J. *Angew. Chem., Int. Ed.* **2007**, *46*, 3104–3106.
114. Imagawa, H.; Saijo, H.; Kurisaki, T.; Yamamoto, N.; Kubo, M.; Fukuyama, Y.; Nishizawa, M. *Org. Lett.* **2009**, *11*, 1253–1255.
115. Snyder, S. A.; Corey, E. J. *J. Am. Chem. Soc.* **2006**, *128*, 740–742.
116. Lebel, H.; Parmentier, M. *Org. Lett.* **2007**, *9*, 3563–3566.
117. Mergott, D. J.; Zuend, S. J.; Jacobsen E. N. *Org. Lett.* **2008**, *10*, 745–748.
118. Matsuura, F.; Peteres, R.; Anada, M.; Harried, S. S.; Hao, J.; Kishi, Y. *J. Am. Chem. Soc.* **2006**, *128*, 7463–7465.
119. Kienzler, M. A.; Suseno, S.; Trauner, D. *J. Am. Chem. Soc.* **2008**, *130*, 8604–8605.
120. Crimmins, M. T.; Brown, B. H.; Plake, H. R. *J. Am. Chem. Soc.* **2006**, *128*, 1371–1378.

121. Peese, K. M.; Gin, D. Y. *J. Am. Chem. Soc.* **2006**, *128*, 8734–8735.
122. Zapf, C. W.; Harrison, B. A.; Drahl, C.; Sorensen, E. J. *Angew. Chem., Int. Ed.* **2005**, *44*, 6533–6537.
123. Baran, P. S.; Burns, N. Z. *J. Am. Chem. Soc.* **2006**, *128*, 3908–3909.
124. Nicolaou, K. C.; Tang, Y.; Wang, J. *Angew. Chem., Int. Ed.* **2009**, *48*, 3449–3453.
125. Tanaka, N.; Suzuki, T.; Matsumura, T.; Hosoya, Y.; Nakada, M. *Angew. Chem., Int. Ed.* **2005**, *44*, 2580–2583.
126. Lumb, J.-P.; Choong, K. C.; Trauner, D. *J. Am. Chem. Soc.* **2008**, *130*, 9230–9231.
127. Greshock, T. J.; Grubbs, A. W.; Jiao, P.; Wicklow, D. T.; Gloer, J. B.; Williams, R. M. *Angew. Chem., Int. Ed.* **2008**, *47*, 3573–3577.
128. Elliott, G.; Velcicky, J.; Ishikawa, H.; Li, Y.; Boger, D. L. *Angew. Chem., Int. Ed.* **2006**, *45*, 620–622.
129. Cook, S. P.; Polara, A.; Danishefsky, S. J. *J. Am. Chem. Soc.* **2006**, *128*, 16440–16441.
130. Phoenix, S.; Reddy, M. S.; Deslongchamps, P. *J. Am. Chem. Soc.* **2008**, *130*, 13989–13995.
131. Tortosa, M.; Yakelis, N. A.; Roush, W. R. *J. Org. Chem.* **2008**, *73*, 9657–9667.

4.8 Dötz Benzannulation

Louis Chupak

4.8.1 Description

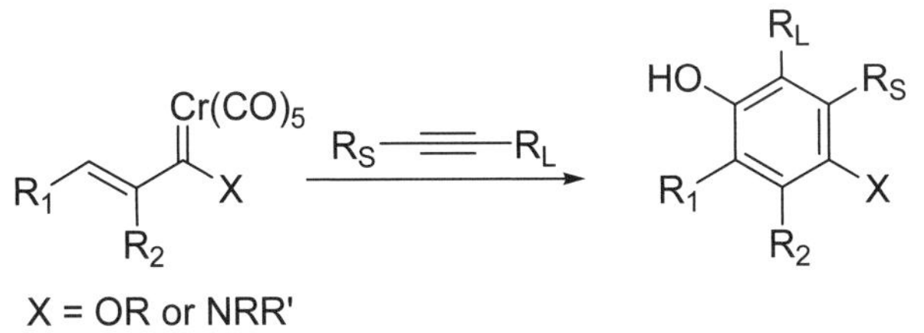

1 2

The Dötz benzannulation reaction (DBR) is the reaction of an α,β-unsaturated Fischer carbene with an alkyne to produce a highly substituted phenol. Alternatively, the DBR can be considered a metal templated 3 + 1 + 2 cycloaddition of an allylic carbine (3 carbon unit), carbon monoxide (1 carbon unit), and an alkyne (2 carbon unit). The initial product of the reaction is the arene chromium tricarbonyl complex of the phenol as in **4**. These complexes are typically unstable in air such that workup and purification of the product lead to the complete loss of the metal. Chromium is the most often used metal for the benzannulation. Molybdinum, tungsten, and manganese have been used but usually give mixtures of products and require harsh reaction conditions.

4.8.2 Historical Perspective

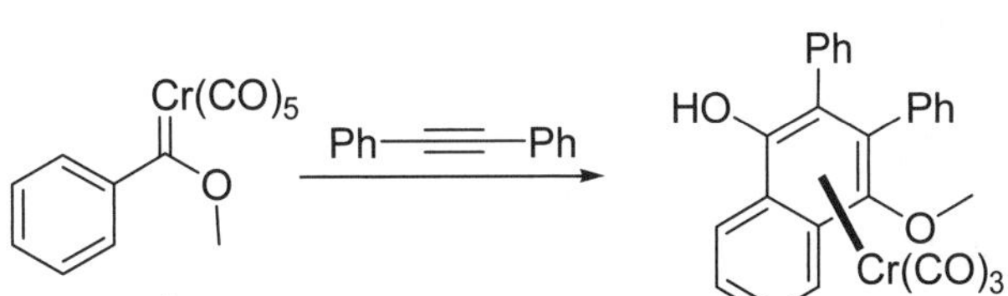

3 4

Karl Heinz Dötz first reported the DBR in 1975 for the reaction of carbene **3** with diphenyl acetylene to give the chromium-coordinated napthol **4**.[1] Since this disclosure, the reaction has found application for the synthesis of numerous heterocyclic and carbocyclic compounds as described in several reviews.[2–6] The DBR has been extensively developed by Dötz and William D. Wulff, and is often reffered to as the Wulff–Dötz reaction.

4.8.3 Mechanism

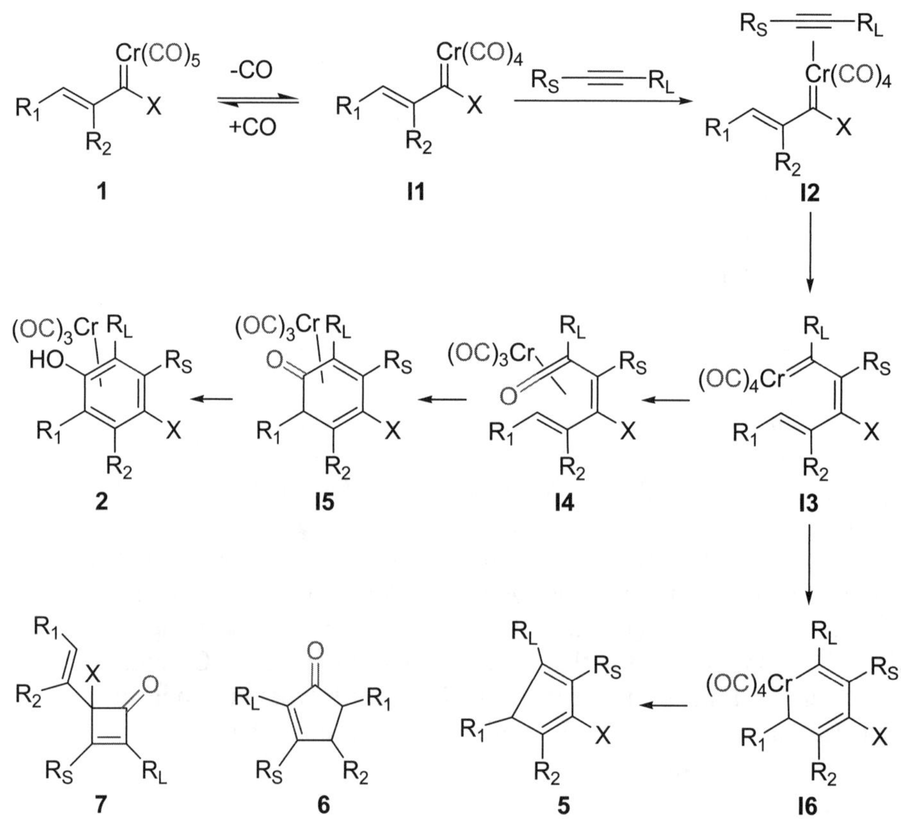

The currently accepted mechanism of the DBR is shown above.[7–8] The rate-determining step is thought to be loss of a carbon monoxide ligand to form a coordinatively unsaturated intermediate **I1**. This process can be facilitated thermally or photolytically. An alkyne can then coordinate to form **I2**. The alkyne inserts into the carbene heteroatom bond to give a new chromium carbene **I3**. At this point there are at least two possible pathways. In the first pathway, carbon monoxide can insert to provide chromium complexed ketene **I4**, which undergoes electrocyclization to give the hexadienone **I5**. Tautomerization completes the reaction to provide the phenol **2**. Alternatively, metallacycle **I6** can form prior to carbon monoxide insertion. Reductive elimination before carbon monoxide insertion leads to pentadiene **5**, a commonly observed by-products of the DBR. Cyclopentanones **6**,[9,10] cyclobutenones **7**, and indenes have also been observed as by-products in the

DBR. In addition, phenolic esters can arise from attack of the DBR products on to a chromium carbon monoxide ligand or the ketene intermediate **I4**.

4.8.4 Variations and Improvements

With unsymmetrical acetylenes ($R_L \neq R_S$) mixtures of products are usually obtained. The ratio of the products depends on the steric difference between the two substituents. The larger group (R_L) is integrated adjacent to the carbon derived from carbon monoxide. Terminal alkynes (R_S = H) are often highly regioselective with little, if any, of the alternative regioisomer observed. The reaction can be used to generate a variety of products: vinyl carbene complexes generate phenols, aryl complexes give naphthols and heteroaryl complexes give benzannulated heterocycles. The DBR is compatible with a wide range of substitutents in the alkyne and the unsaturated carbene side chain allowing for the synthesis of densely substituted phenols and further conversion of these phenols to quinones.

In the alkyne partner a wide range of functionality is tolerated. The benzannulation has been reported with alkynes having selenyl,[11] nitro, boronate, stannyl, aryl, ester, ketone, amide, acetal, ether, enol ether, sulfide, tosyl, and cyano groups. Modest yields have been observed with electron-withdrawing groups. In the extreme, hexafluorobutyne does not participate at all in the DBR. Unprotected alcohols in the reactants can interrupt the DBR by reacting with the ketene intermediate to produce lactones. Alkynes with two large groups often fail to undergo the electrocyclization step to give the DBR product.

Aryl carbene complexes with electron-donating or electron-withdrawing substituents in the *ortho-*, *para-*, or *meta-*positions participate in the DBR. The aryl group can also be naphthyl and heteroaryl such as furan, thiophene, pyrrole, pyrazole, and indole. Simple alkyl substituted vinyl carbene complexes have been extensively examined. The double bond can be in either a cyclic or an acyclic system.

As stated above, cyclopentanones, cyclobutenones, and indenes have been observed as by-products in the DBR. Wulff has studied the effect of solvent, chelation, concentration, and alkyne substitution on the product distribution.[12] He reported that simple α,β-unsaturated chromium carbene complexes typically show excellent selectivity for the benzannulated product. This selectivity is not sensitive to changes in solvent or substituents on the acetylene. However, the reactions of aryl complexes with acetylenes are very sensitive to the nature of both the solvent and the acetylene. For aryl chromium complexes, the highest selectivities and yields for the benzannulated product arise with solvents of low coordinating ability: hexane and benzene. Solvents with intermediate coordinating ability and small size (acetonitrile) give high selectivity for cyclobutenone formation for reactions with disubstituted acetylenes. Solvents with high coordinating ability (dimethyl formamide) give poor selectivity and produce a considerable amount of indene products. The combination of an *o*-methoxy group on the aryl substituent, **8**, of the carbene complex and acetonitrile as solvent alters the product distribution in favor of the cyclobutenone product. Presumably, the formation of the cyclobutenone is favored by coordination of the ether to the metal center. Amide-like complexes [**1**, X = $-N(CH_3)_2$] react with diethylacetylene in THF to give indene products. Similarly, de Meijere has reported that β-amines, as in **9**, completely alter the reaction course so that no benzannulation is observed.[13,14]

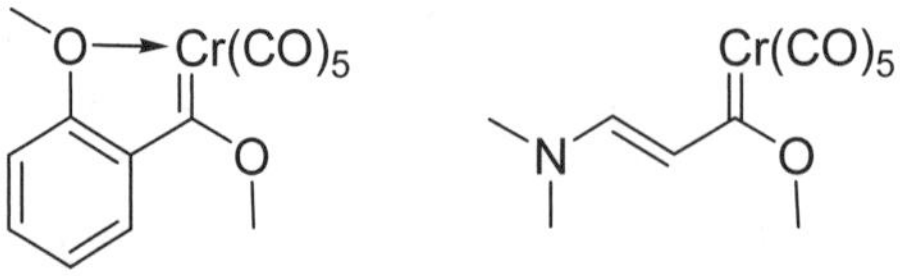

8 **9**

Irradition has been used to initiate the DBR. Ultraviolet irradiation allows reactions to occur at temperatures as low as –78 °C. High selectivity for the benzannulated product is seen with simple aryl complexes, but high selectivity for indene products for complexes having a chelating *o*-methoxy on the aryl ring.

It is interesting that the product distribution from the reaction of the *o*-methoxyphenyl complex with diethylacetylene was found to be dependent on alkyne concentration.[12] One equivalent of alkyne, slow addition of alkyne, or low absolute concentration of alkyne favors the indene product. Reactions performed with excess or high concentrations of alkyne favor benzannulation. The observation, reffered to as the "allochemical effect," was explained as resulting from coordination of reaction intermediates by the alkyne. Thus excess alkyne acts as a ligand to stabilize intermediates along the path to benzannulation. Finally, it was noted that the product distribution was not substantially altered by addition of phosphines, phosphine oxides or sulfides.

Seperately Wulff has described the effect on selectivity and yield of substitution on the aryl ring of chromium carbene complexes.[15] It was shown that electron-withdrawing groups *para-* to the chromium carbene increase the chemoselectivity for benzannulation. Substituents in the *ortho* position are detrimental to phenol formation irrespective of the substituent's electronic nature. Finally, it was shown that large alkoxy groups on the carbene carbon (**1**, X = OR) give increased yields of the desired phenol product.

A vinyl group in the β-position of the carbene, as in **10**, can participate in the reaction to form an eight-membered ring.[16] However, the course of the reaction is sensitive to the size of the alkyne substituent. With smaller groups, R = *n*-butyl or benzyl, the octatrienone **11** is observed. When R is the larger *t*-butyl or trimethylsilyl the cyclohexadienone **12** predominates.

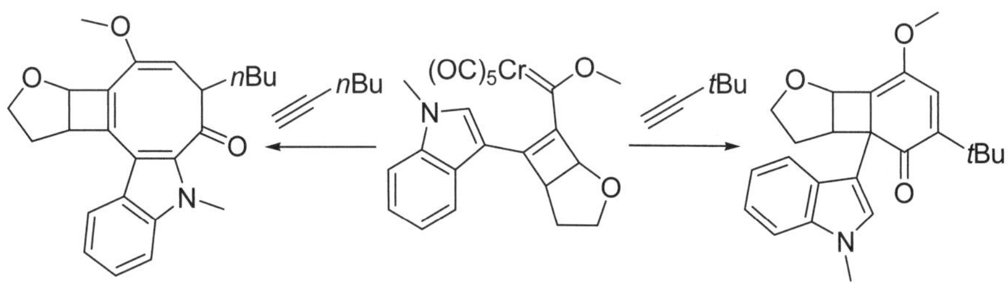

R = *n*-Bu,54%; Bn, 66% **11** **10** R = *t*-Bu, 52%; TMS, 48% **12**

A key driver for the development of the DBR has been the increased availability of the requisite chromium carbene. Fischer carbenes undergo a wide variety of useful reactions and a significant effort has been devoted to their synthesis.[4,5,17] These carbenes undergo many of the same reactions as esters. The α-hydrogens in **13** are quite acidic, with a pKa of approximately 8, that allows for application of the Aldol condensation to form the vinyl-substituted carbene **14**.[18,19] Of course, alkynes insert into these carbenes to form new vinyl substituted carbenes **15**. However, the absence of a heteroatom on the carbene center makes these poor substrates for the DBR. The classical route to Fischer carbenes is the Fischer route: addition of an organolithium to hexacarbonyl chromium and alkylation with a hard electrophile.[20–23] Hoye has also shown that alkyl iodides under phase-transfer conditions can be used to alkylate the lithium alkoxide.[24] Thus reaction of vinyl lithium **16** provides the carbene **17** in 53% over two steps.

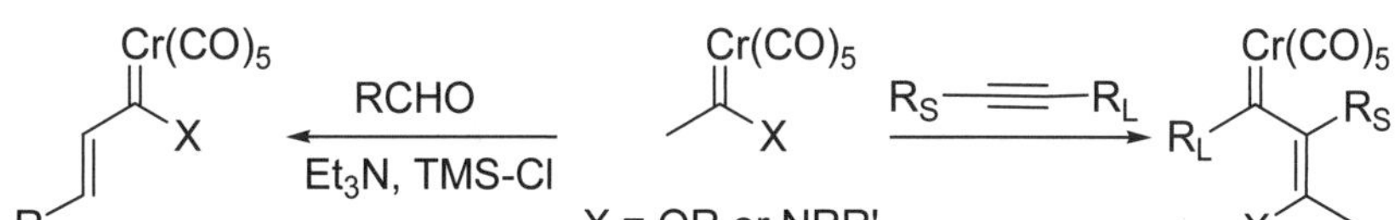

14 **13** **15**

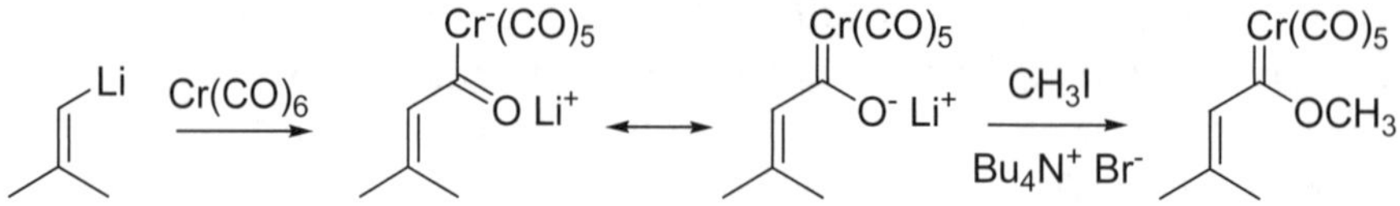

16 **17** 53%

Alternatively, when the counterion of the initial adduct **18** is exchanged with tetramethyl ammonium, the mixed anhydride can be prepared by reaction with pivaloyl chloride. Further reaction with alcohols or amines then yields chromate esters and amides, respectively. Functional groups can be introduced into the heteroatom-linked chains that are not compatible with alkyl and aryl lithium species. The pimary alkyl iodide **19** was prepared by this method.[25]

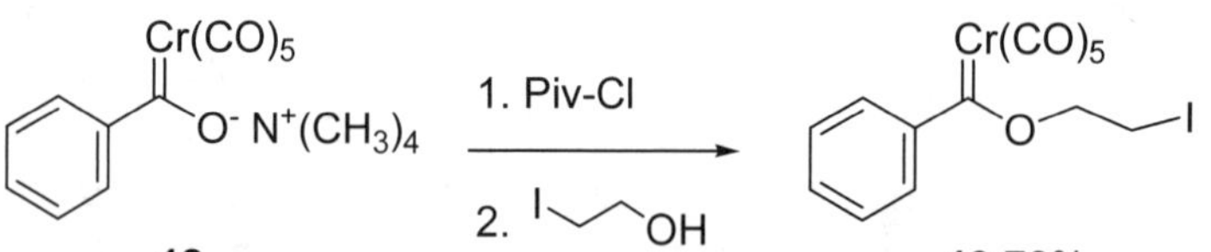

18 **19** 78%

This acylation strategy was used to prepare polymer-supported chromium carbenes. Microwave irradiation on Wang resin shows the same regioselectivity as solution chemistry but fewer side products.[26] The resin-bound phenol **21** is simultaneously released and oxidized to the benzoquinone **22** with ceric ammonium nitrate. Microwave irradiation has been shown to accelerate the DBR and provide high yields of benzannulated products in short reaction times (ca. 5 min).[27]

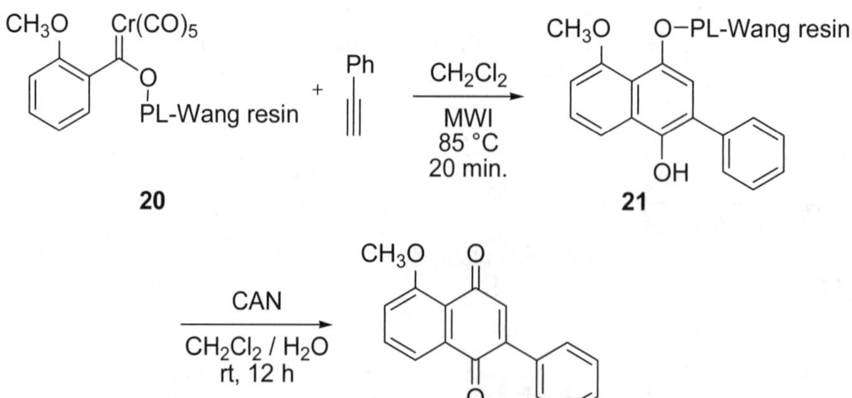

20 **21**

22 80%

Chromium carbenes can also be prepared by the so-called Semmelhack–Hegedus route. Chromium hexacarbonyl is first reduced to a nucleophilic pentacarbonyl dichromate dianion **23** with sodium naphthalenide[28,29] or potassium carbide.[30] Reaction of this dianion species with an acid chloride gives a metal alkoxide that can be quenched with an electrophile to provide the desired chromate ester **24**. Alternatively, the dianion can be added to an amide carbonyl to give a tetrahedral intermediated which collapses to the chromate amide **25** on treatment with trimethylsilyl chloride.

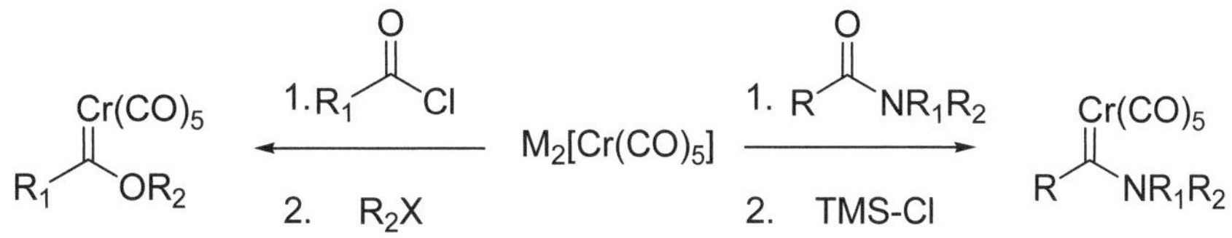

24 **23** **25**

Dry state absorbtion conditions, microwave irradtion, and photolysis have all been used to accelerate the DBR. Recently Kerr has described simple conditions that promote the DBR at near ambient conditions.[31] His optimized procedure used dichloroethane as the solvent with heating at 30 ºC for 18 h with no additives. Dichloroethane and gentle heating were both critical for high yields. It is interesting that ambient laboratory temperature, measured as 14–17 ºC, was insufficient and resulted in sluggish reactions.

4.8.5 Synthetic Utility

General Utility

As stated above, the reaction is highly valuable for both the synthesis of highly substituted phenols and further conversion of these phenols to quinones. Harrity has described optimized conditions for DBR using boronate esters requiring three equivalents of alkyne.[32] Anderson applied these conditions toward sytheses of boronate-substituted quinones. Chromium carbene **26** reacts with boronate **27** in 48% yield over two steps to give the quinone **28**.[33]

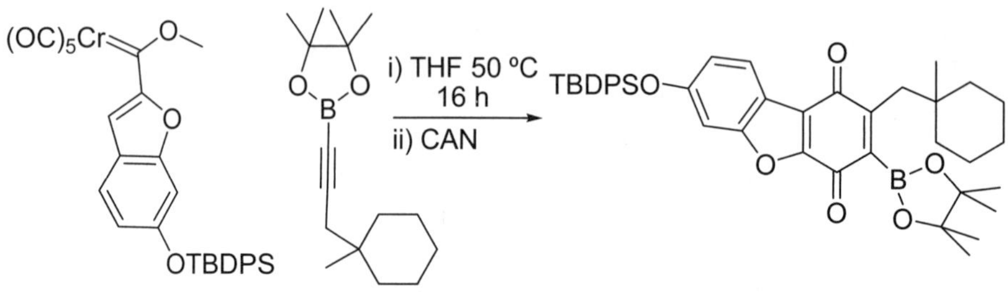

26 **27** **28**

Complex and highly substituted nitrogen- and oxygen-containing heterocycles such as carbazoles, indoles, cyclophanes, naphthoquinones, and furanocoumarins can be assembled from the DBR. Sen, for example, has used the intramolecular DBR to make oxygen-containing heterocycles in good yield under both the standard DBR conditions and under solvent-free conditions.[34] Alkyne **29** was converted to phenol **30** in 56% yield.

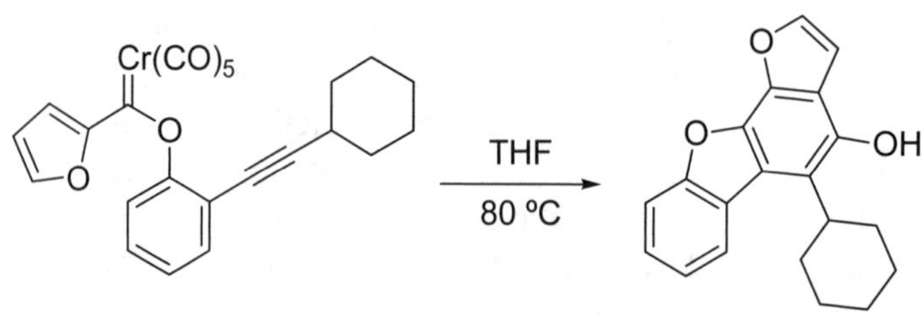

29 **30** 56%

Pulley has reported using the DBR as a strategy to prepare the amino acids isodityrosines[35] and aryl glycines.[36] The styrene-derived carbene **31** reacted with the alkyne **32** to provide the phenol **33** in 87% yield. An interesting observation was that the yields were significantly higher with ultrasonication (59–87%) than with thermal (51–69%) conditions. The targeted amino acid **34** was produced using standard transformations.

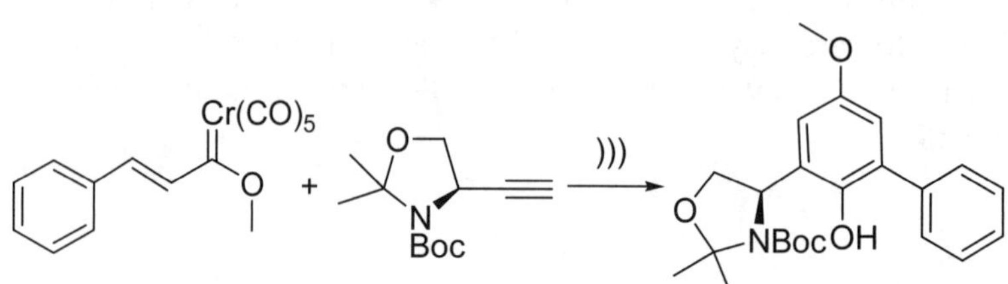

31 **32** **33** 87%

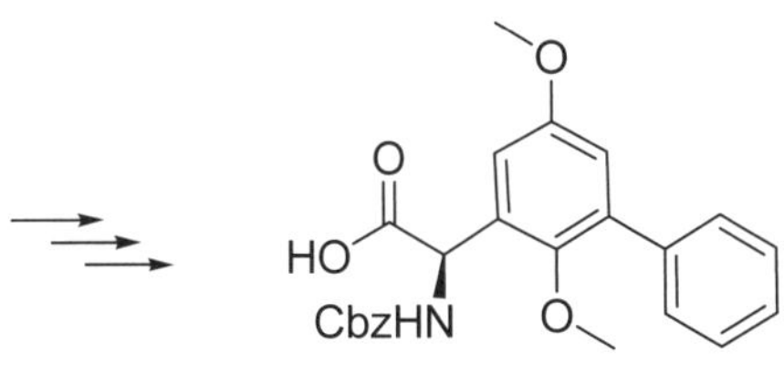

34

The control of stereochemistry with chromium carbine complexes has been reviewed.[37] The DBR can create a new stereocenter in three ways. First, the arene tricarbonyl chromium complex contains a plane of chriality, thus the complexes **35** and ***ent*-35** are enantiomers when RL ≠ R_1 and RS ≠ R_2. Second, when phenyl substituents are included in the reactants the resulting biaryls can posess axial chirality if there is hindered rotation about the new aryl–aryl bond as in **36**.[38] Finally, all DBRs with differentially β-disubstitued alkenes give rise to cyclohexadienones **37** with a new stereocenter adjacent to the carbonyl. When R_1 and R_2 are not hydrogen, tautomerization cannot occur and the final product possesses a chiral center.

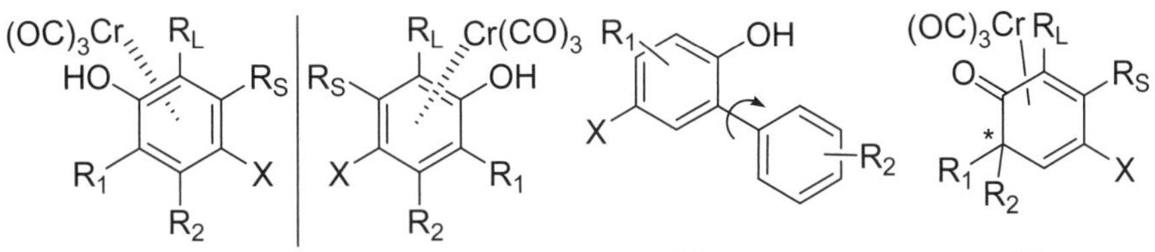

35 ***ent*-35** **36** **37**

enantiomeric chromium arenes

Dötz has reviewed the use of chiral centers in either the alkyne or chromium carbene to control the facial selectivity of the chromium arene complex.[6] Examples of these diastereoselective benzannulations exist with the controlling stereocenter in the alkyne, the chromate ester (or amide), or the unsaturated carbene.[39]

Anderson has studied the transfer of chirality from a substituent *ortho* to the alkyne, as in **39**, to axial chirality in the biaryl **40**.[40] He observed that there is a steric balance where a large *ortho* substituent is required for efficient chirality transfer, but too large a substituent hinders the DBR.

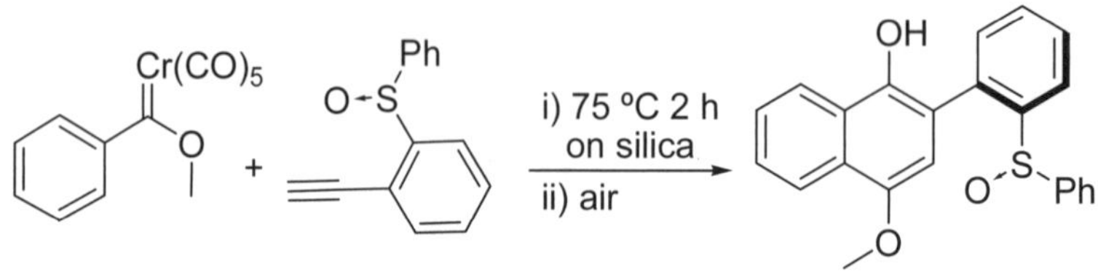

38 **39** **40** 44% 5:1 *dr*

Wulff has studied the use of chiral auxiliaries in the chromate ester (**1**, X = a chiral auxiliary) to control the absolute stereochemistry of the aryl chromium product **35**.[41] In the same disclosure he describes an attempt to control the chiral center adjacent to the ketone formed on electrocyclization. In his example, a pair of chromium carbene atropisomers **41** (only one shown) were prepared. These diastereomers were readily seperable on silica gel. It was shown that each diastereomeric atropisomer reacted stereospecifically with alkynes to produce a different diastereomer **42**. Thus the chiral auxiliary allowed the separation of the atropisomers, but did not directly control the absolute stereochemistry adjacent to the ketone. Instead, the stereochemistry was determined by confining the chromium to one face of the indole in **41** by preventing rotation about the the carbene indole bond.

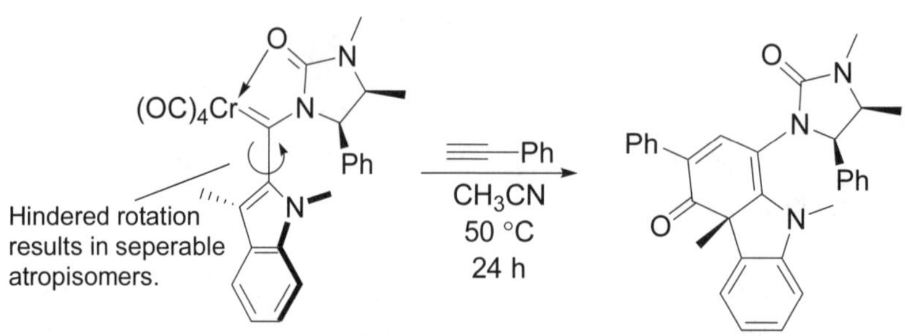

41 **42** 40% 95% *de*

There remain opportunities for further developments in asymmetric induction with the DBR. The generation of axial chirality and control of the stereochemistry α to the ketone are under explored. Finally, to the best of our knowledge, there are no reports of using chiral-coordinating groups to influence the stereochemistry.[37]

Applications in the total synthesis of natural products

Quinones are obvious synthetic targets for the DBR.[42–44] Quinones are numerous, occuring in many biologically active compounds and natural products. Fernandes prepared the quinone (–)-juglomycin A **44** in eight steps in 19% overall yield and > 99% *ee*.[45] The quinone core **43** was rapidly assembled using the DBR and the absolute stereochemistry set with an asymmetric dihydroxylation. A similar strategy was used to prepare the eluetherins.[46]

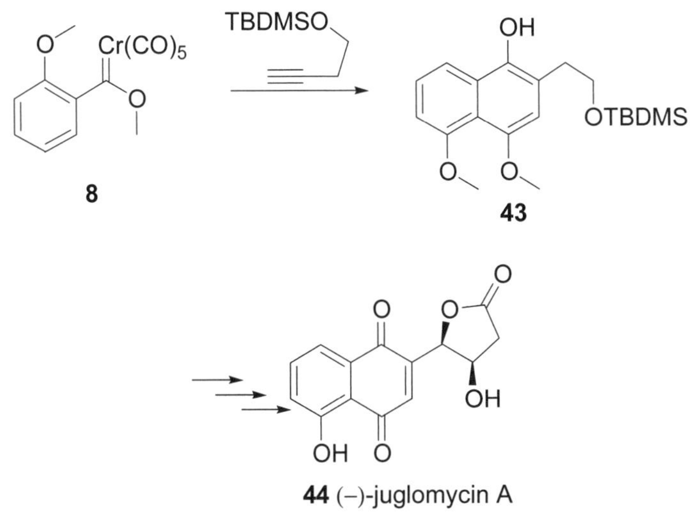

The intermolecular DBR has been used to prepare a variety of interesting synthetic targets. Quayle has reported a formal total synthesis of aflatoxin B2 using the DBR.[47] Reaction of the acetal-containing chromium carbene **45** with an electron rich alkyne occured in 31% yield to give **46**, an advance intermediate for the preparation of aflatoxin.

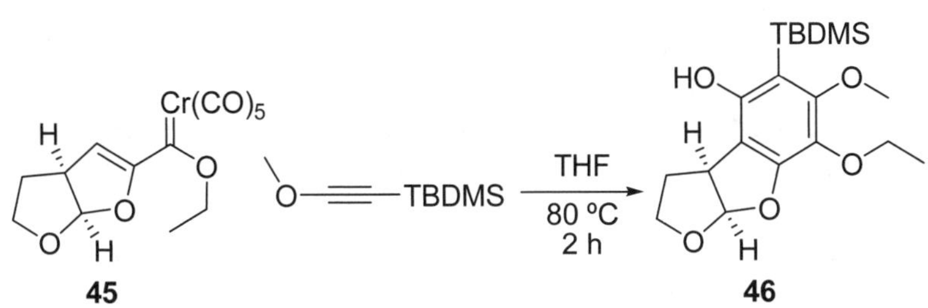

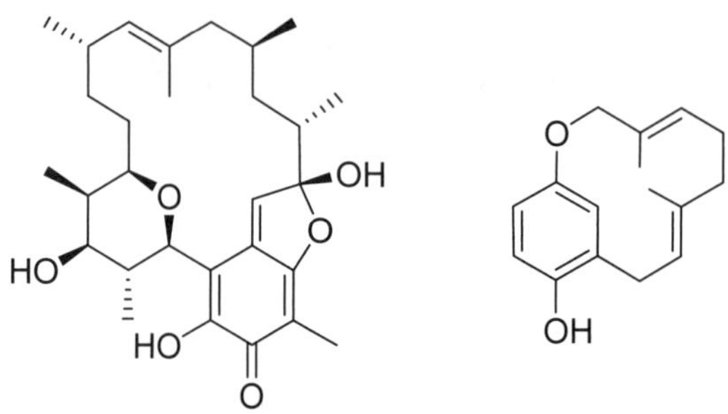

47 (–)-kendomycin **48** arnebinol

The DBR has been used to prepare the densely functionalized aryl ring found in the ansa macrocyles such as kendomycin **47**[48] and arnebinol **48**. Two approaches to this group of targets have been desribed. In the first, the aryl ring is formed in the DBR followed with macrocylization. In the second strategy, the DBR is used to simultaneously control the arene substitution pattern and close the macrocyclic ring.

Saikawa and Nakata have applied the DBR to the synthesis of the *meta*-cyclophane arnebiol **48** using the second strategy.[49] In their approach, the macrocyclic ring is formed via an intramolecular DBR of chromium carbene **49**. In the absence of a tether, sterics would be expected to dominate the regiochemical outcome. Thus the larger alkyne substituent would be incorporated *ortho* to the hydroxyl as in **51**. They show that a short tether provides the complementary regiochemistry observed in **50**. When chain length is increased sterics dominate the regiochemical course of the reaction. When *n* is 2 through to 4 only the *ortho*-cyclophane **50** and the dimer, resulting from intermolecular DBR, are observed. When n is increased to 5 through 13, only the *meta*- cyclophane is isolated. This strategy resulted in DBR to produce arnebinol **48** in 49% yield when the tether contained eight carbons (*n* = 6), including two *E*-alkenes. Wulff applied a similar strategy to form macrocyclic cyclohexadienes as intermediates toward synthesis of phomactin natural products.[50] In one example, a 13-membered ring macrocycle **53** formed in 64% yield from the acyclic precursor **52**. As with the cyclophanes, attempts to produce medium-size rings gave only the dimer from intermolecular reaction.

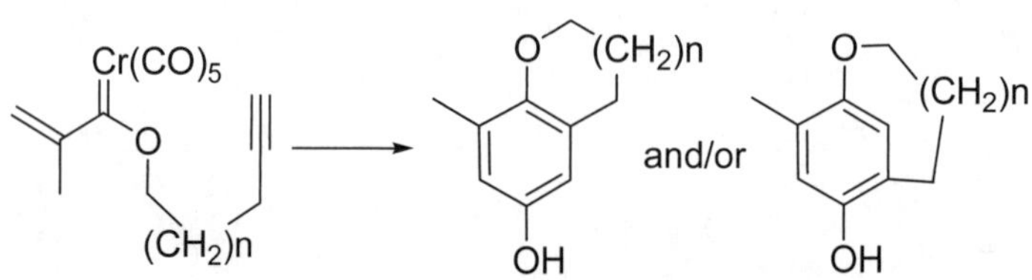

49 **50** **51**

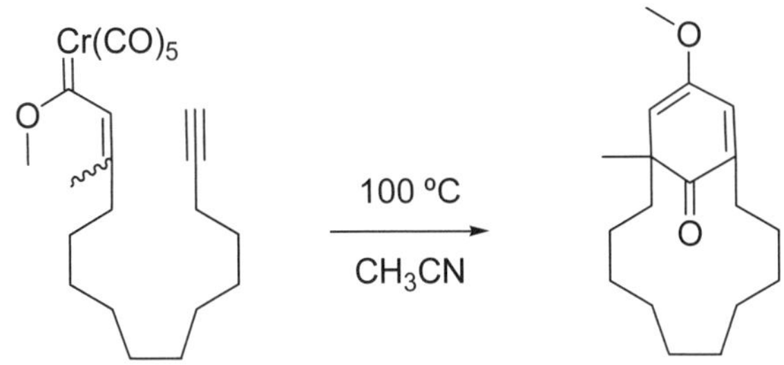

52 53

In summary, the DBR is a powerfull method for the construction of highly substituted phenols, quinones, and heterocycles. Significant effort has gone into exploring the scope of the reaction since its initial discovery. This exploration has enabled the confident prediction of regiochemistry and the application to total synthesis. Further efforts to predict and control stereochemistry are needed to increase the utility of this valuable reaction.

4.8.6 Experimental

4.8.6.1 Preparation of lipoxygenase inhibitor 2-butyl-4-methoxynaphthalen-1-yl acetate 55[51,52]

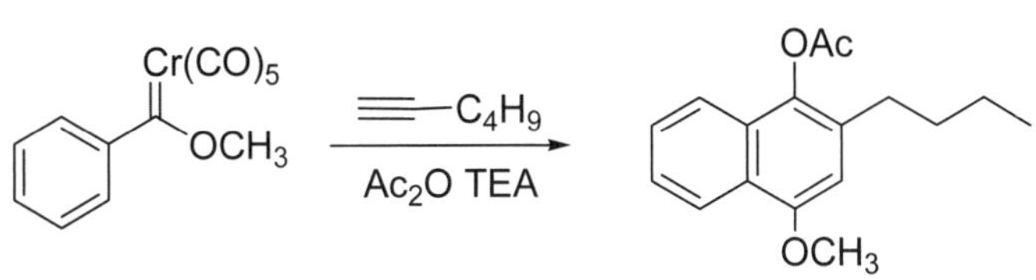

54 55

An oven-dried, 2-L, three-necked, round-bottomed flask, equipped with a nitrogen inlet, magnetic stirring bar, thermometer, and reflux condenser, under an inert nitrogen atmosphere, is charged with 1.22 g (10 mmol) 4-dimethylaminopyridine, 500 mL tetrahydrofuran, 11.0 mL (95.7 mmol) of 1-hexyne, 13.2 mL (140 mmol) acetic anhydride, 9.8 mL (70mmol) of triethylamine), 20.0 g (64.0 mmol) of pentacarbonyl[phenyl(methoxy)-chromium] carbene **54**, and a final 100-mL rinse of tetrahydrofuran. The solution is heated to reflux with an oil bath and heating is maintained until TLC indicates that the chromium complex is totally consumed (45–60 min). The solution is then cooled to ambient temperature, 30 g silica gel is added, and volatile organic material is removed under reduced pressure (rotary evaporator). The green solids are transferred to a filter funnel and washed with hexane until TLC indicates that all products have been removed (5 × 100 mL). The hexane filtrate is then concentrated under reduced pressure to give crude product contaminated with chromium hexacarbonyl. To the mixture is added 20 mL of isopropyl alcohol, and the insoluble chromium hexacarbonyl is removed by filtration. The filtrate is concentrated under reduced pressure to give 14.0 g crude product, which is purified by silica gel chromatography. Appropriate fractions are combined, and the solvent is removed under reduced pressure to give 1-acetoxy-2-butyl-4-methoxynaphthalene (11.8 g, > 95% pure based on HPLC, 68% yield based on the carbene complex as a light yellow oil that crystallizes on standing. If desired, the product can be crystallized from isopropyl alcohol (2.5 mL/g) to give white crystals, m.p. 49–50 °C (> 99% pure based on HPLC).

4.8.7 References

1. Dötz, K. H. *Angew. Chem.* **1975**, *87*, 672–673.
2. [R] Dötz, K. H.; Tomuschat, P. *Chem. Soc. Rev.* **1999**, *28*, 187–198.
3. [R] Dötz, K. H.; Stendel, J. *Chem. Rev.* **2009**, *109*, 3227–3274.
4. [R] Haase, W.-C.; Nieger, M.; Dötz, K. H. *J. Organomet. Chem.* **2003**, *684*, 153–169.
5. [R] De Meijere, A.; Schirmer, H.; Duetsch, M. *Angew. Chem., Int. Ed.* **2000**, *39*, 3964–4002.
6. [R] Minatti, A.; Dötz, K. H. *Top. Organomet. Chem.* **2004**, *13*, 123–156.
7. Barluenga, J.; Aznar, F.; Gutiérrez, I.; Martín, A.; García-Granda, S.; Amparo Llorca-Baragaño, M. *J. Am. Chem. Soc.* **2000**, *122*, 1314–1324.
8. Torrent, M.; Duran, M.; Solá, M. *J. Am. Chem. Soc.* **1999**, *121*, 1309–1316.
9. Eastham, S. A.; Herbert, J.; Ingham, S. P.; Quayle, P.; Wolfendale, M. *Tetrahedron Lett.* **2006**, *47*, 6627–6633.
10. Eastham, S. A.; Ingham, S. P.; Hallett, M. R.; Herbert, J.; Quayle, P.; Raftery, J. *Tetrahedron Lett.* **2006**, *47*, 2299–2304.
11. Caldwell, J. J.; Colman, R.; Kerr, W. J.; Magennis, E. J. *Synlett* **2001**, 1428–1430.
12. Chan, K. S.; Peterson, G. A.; Brandvold, T. A.; Faron, K. L.; Challener, C. A.; Hyldahl, C.; Wulff, W. D. *J. Organomet. Chem.* **1987**, *334*, 9–56.
13. Flynn, B. L.; Schirmer, H.; Duetsch, M.; De Meijere, A. *J. Org. Chem.* **2001**, *66*, 1747–1754.
14. Wu, Y.-T.; Vidovic, D.; Magull, J.; de Meijere, A. *Eur. J. Org. Chem.* **2005**, 1625–1636.
15. Liptak, V. P.; Wulff, W. D. *Tetrahedron* **2000**, *56*, 10229–10247.
16. Barluenga, J.; Fañanás-Mastral, M.; Angel Palomero, M.; Aznar, F.; Valdás, C. *Chem.--Eur. J.* **2007**, *13*, 7682–7700.
17. Barluenga, J.; Martinez, S. *ARKIVOC* **2006**, 129–147.
18. Wang, H.; Hsung, R. P.; Wulff, W. D. *Tetrahedron Lett.* **1998**, *39*, 1849–1852.
19. Lattuada, L.; Licandro, E.; Maiorana, S.; Papagni, A. *J. Chem. Soc., Chem. Commun.* **1991**, 437–438.
20. Aumann, R.; Fischer, E. O. *Angew. Chem., Int. Ed. Engl.* **1967**, *6*, 879–880.
21. Casey, C. P.; Cyr, C. R.; Boggs, R. A. *Syn. Inorg. Metal-Org. Chem.* **1973**, *3*, 249–254.
22. Xu, Y. C.; Wulff, W. D. *J. Org. Chem.* **1987**, *52*, 3263–3275.
23. Hegedus, L. S.; McGuire, M. A.; Schultze, L. M. *Org. Synth.* **1993**, *8*, 216.
24. Hoye, T. R.; Chen, K.; Vyvyan, J. R. *Organometallics* **1993**, *12*, 2806–9.
25. Barluenga, J.; Perez-Sanchez, I.; Suero, M. G.; Rubio, E.; Florez, J. *Chem. Eur. J.* **2006**, *12*, 7225–7235.
26. Shanmugasundaram, M.; Garcia-Martinez, I.; Li, Q.; Estrada, A.; Martinez, N. E.; Martinez, L. E. *Tetrahedron Lett.* **2005**, *46*, 7545–7548.
27. Hutchinson, E. J.; Kerr, W. J.; Magennis, E. J. *Chem. Commun.* **2002**, 2262–2263.
28. Semmelhack, M. F.; Lee, G. R. *Organometallics* **1987**, *6*, 1839–1844.
29. Imwinkelried, R.; Hegedus, L. S. *Organometallics* **1988**, *7*, 702–706.
30. Schwindt, M. A.; Lejon, T.; Hegedus, L. S. *Organometallics* **1990**, *9*, 2814–2819.
31. Irvine, S.; Kerr, W. J.; McPherson, A. R.; Pearson, C. M. *Tetrahedron* **2008**, *64*, 926–935.
32. Davies, M. W.; Johnson, C. N.; Harrity, J. P. A. *J. Org. Chem.* **2001**, *66*, 3525–3532.

33. Anderson, J. C.; Denton, R. M.; Hickin, H. G.; Wilson, C. *Tetrahedron* **2004**, *60*, 2327–2335.
34. Sen, S.; Kulkarni, P.; Borate, K.; Pai, N. R. *Tetrahedron Lett.* **2009**, *50*, 4128–4131.
35. Gupta, A.; Sen, S.; Harmata, M.; Pulley, S. R. *J. Org. Chem.* **2005**, *70*, 7422–7425.
36. Pulley, S. R.; Czakó, B.; Brown, G. D. *Tetrahedron Lett.* **2005**, *46*, 9039–9042.
37. [R] Santamaria, J. *Curr. Org. Chem.* **2009**, *13*, 31–46.
38. Vorogushin, A. V.; Wulff, W. D.; Hansen, H.-J. *Tetrahedron* **2007**, *64*, 949–968.
39. Barluenga, J.; Panday, N.; Santamaria, J.; de Prado, A.; Tomas, M. *ARKIVOC* **2003**, 576–583.
40. Anderson, J. C.; Cran, J. W.; King, N. P. *Tetrahedron Lett.* **2003**, *44*, 7771–7774.
41. Hsung, R. P.; Wulff, W. D.; Chamberlin, S.; Liu, Y.; Liu, R.-Y.; Wang, H.; Quinn, J. F.; Wang, S. L. B.; Rheingold, A. L. *Synthesis* **2001**, 200–220.
42. Caldwell, J. J.; Colman, R.; Kerr, W. J.; Magennis, E. J. *Synlett* **2001**, 1428–1430.
43. Roush, W. R.; Neitz, R. J. *J. Org. Chem.* **2004**, *69*, 4906–4912.
44. Pulley, S. R.; Czakó, B. *Tetrahedron Lett.* **2004**, *45*, 5511–5514.
45. Fernandes, R. A.; Chavan, V. P. *Tetrahedron Lett.* **2008**, *49*, 3899–3901.
46. Fernandes, R. A.; Chavan, V. P.; Ingle, A. B. *Tetrahedron Lett.* **2008**, *49*, 6341–6343.
47. Eastham, S. A.; Ingham, S. P.; Hallett, M. R.; Herbert, J.; Quayle, P.; Raftery, J. *Tetrahedron Lett.* **2006**, *47*, 2299–2304.
48. White, J. D.; Smits, H. *Org. Lett.* **2005**, *7*, 235–238.
49. Watanabe, M.; Tanaka, K.; Saikawa, Y.; Nakata, M. *Tetrahedron Lett.* **2007**, *48*, 203–206.
50. Huang, J.; Wang, H.; Wu, C.; Wulff, W. D. *Org. Lett.* **2007**, *9*, 2799–2802.
51. Yamashita, A.; Timko, J. M.; Watt, W. *Tetrahedron Lett.* **1988**, *29*, 2513–2516.
52. Timko, J. M.; Yamashita, A. *Org. Synth.* **1993**, *71*, 72–76.

4.9 Elbs Reaction

Timothy T. Curran

4.9.1 Description

The Elbs reaction is the cyclic condensation of an *ortho*-methyl- or methylene-substituted diaryl ketone **1** to form the corresponding anthracene adduct **2**. The reaction is typically promoted thermally, occurs at relatively high temperatures (> 300 °C), and provides an equivalent of water.

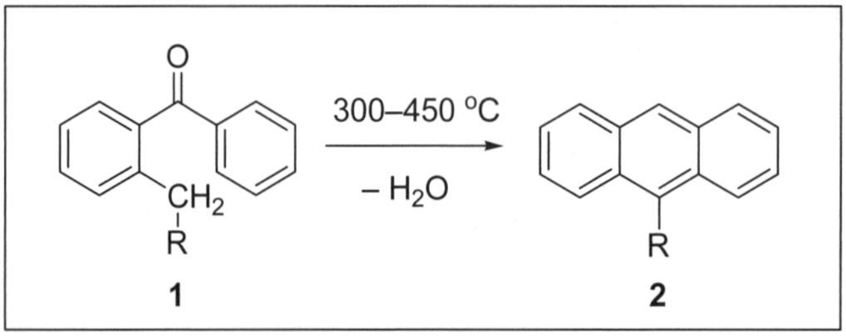

4.9.2 Historical Perspective

In the late 1800s Elbs and co-workers explored the generality and synthetic utility of what is known as the Elbs reaction. While Elbs was not the first to report this pyrolytic, cyclic condensation reaction,[1] the reaction bears his name due to his study of the scope of the reaction.

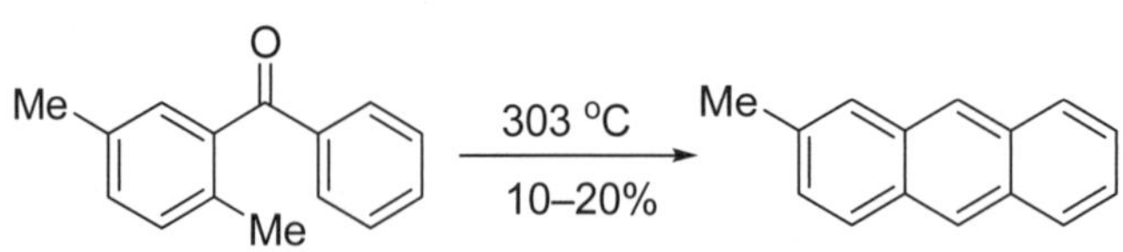

3 **4**

During his study of preparing simple anthracene homologues like **4**, Elbs determined many limitations of the reaction, including low yield and instances in which the Elbs reaction did not give rise to the desired product. Elbs largely discounted the reaction due to these findings along with the fact that other methods were being developed that could prove superior to the Elbs reaction. However, the reaction continues to be used despite its shortcomings due to the ease at which one may assemble the substrate for cyclization and conduct the reaction leading to fairly complicated polyaromatic compounds. Isolation of products has oftentimes been difficult and sometimes requires multiple purification tools in the synthetic chemists' toolbox (chromatography, distillation, crystallization of parents or derivatives, sublimation, *etc.*) to provide pure materials.

4.9.3 Mechanism

The mechanism of the Elbs reaction has not been fully vetted.[1] Cook suggested tautomerization of the ketone **5** into dienol **6**, followed by cyclization forming anthrenol **7**. Subsequent hydride transfer or an additional tautomeric shift of hydrogen to provide dihydroanthrol **8**, followed by water elimination, provides anthracene **9**. Fieser suggested an alternative cyclization step by 1,4-addition of the methyl substituent (without explicit description the methyl species) into the aryl system with the bulk of the intermediates remaining the same as those suggested by Cook. While both of these mechanisms are reasonable, neither author provided sufficient evidence to support their proposal.[1]

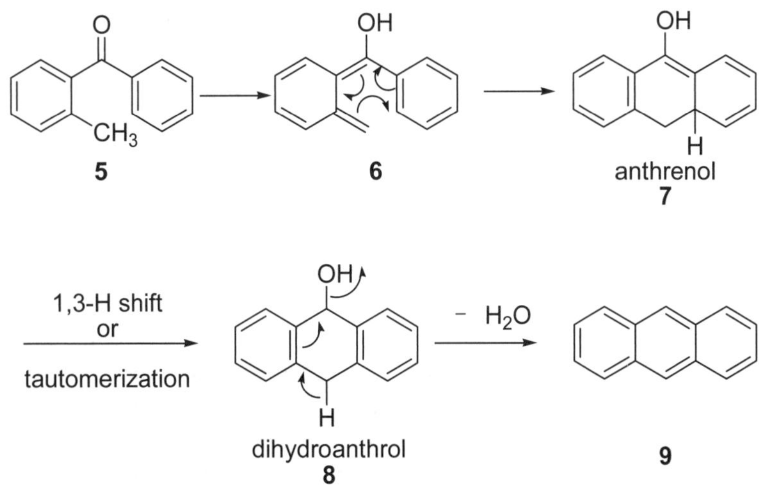

Hurd and Azorlosa[2] performed deuterium labelling experiments of two different substrates **10** and **15**. They subjected these two deuterated *o*-methylated-diaryl ketones to Elbs conditions and monitored where the deuterium went. Subsequent oxidation of the product mixture was used to determine the relative amount of 8*d* vs. 9*d* formation. The fact that more 8*d* isomer was formed than 9*d* isomer is consistent with deuterium isotope effects. The authors found that < 1% of deuterium remained in the water derived from the reaction. The authors suggested that the low percentage of deuterium in the products was due to subsequent by-product formation. These experiments support a transfer of the aromatic H(D) (hydride transfer or tautomeric shift mentioned above—conversion of **7** into **8**) even though such a shift is a forbidden transformation when considering orbital symmetry arguments. It is thought that the extreme conditions used to promote the Elbs reaction can overcome this; perhaps this is a radical or nonconcerted process. In addition, a crossover experiment showed that the *d*-atom of an *o*-deuteroketone is not active enough to exchange at 340 °C.

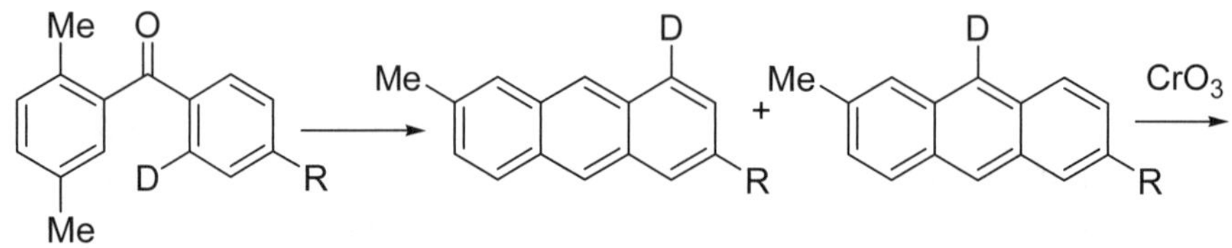

R = H, **10** 325–334 °C, 11% R = H, **11** R = H, **12**

R = Me, **15** 350 °C 26% R = Me, **16** R = Me, **17**

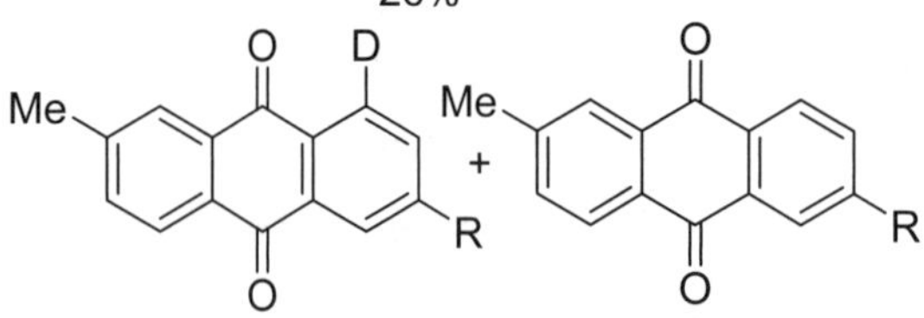

R = H, **13**
R = Me, **18**

R = H, **14**
R = Me, **19**

Compounds	Amt. of D rel. to **10** or **15**
11 and **12**	0.79
16 and **17**	0.67
13 and **14**	0.61
18 and **19**	0.51

Dauben investigated the mechanism by employing a ^{14}C-labeled carbonyl in an Elbs reaction.[3] In this study, he and his co-workers showed that volatile by-products, which were trapped and analyzed accounted for 35% of the radioactivity taken into the reaction. The authors also suggested that optimum yield occurred after about 3 h of reaction time at elevated temperature for the conversion of diarylketone **20** into dibenzanthracene **21**.

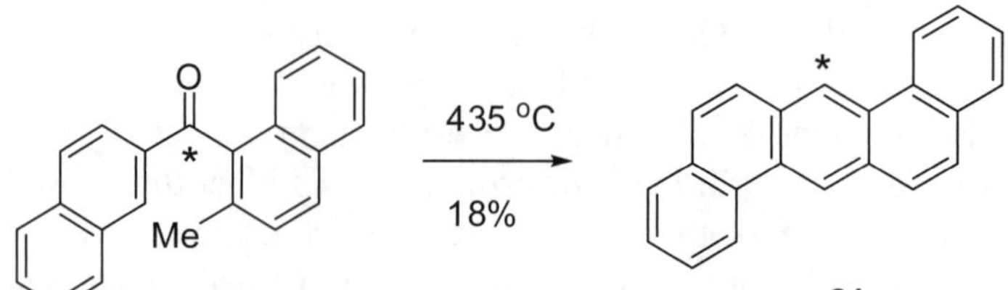

20 **21**

More recently, due to the identification of anthrone and anthroquinone by-products, Badger and Pettit[4] along with Buu-Hoi and co-workers[5] have suggested mechanisms to account for the formation of these by-products. Badger and Pettit suggested a radical mechanism with the first step of the reaction being generation of the benzyl (benzoyl allylogue) radical generating **22**. Cyclization of **22** then provides **23**, which, depending on the preferred electronics of the molecule, may undergo loss of H· to provide anthrone **24** or abstract an H· to generate anthrenol **7**. Oxidation of anthrone to anthraquinone structures are known.[4] Formation of anthrenol **7** intercepts the previously suggested mechanism for the Elbs reaction and requires H· migration forming dihydroanthrol **8**, which required loss of water to provide anthracene **9**.

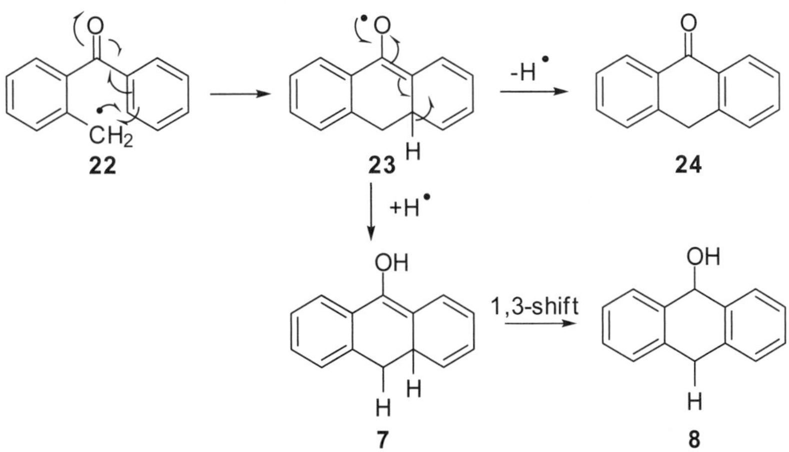

An additional comment from these authors supported the notion that during the pyrolysis, hydrogen species capable of reducing intermediates were present. Evidence for this was provided in the isolation of **26** from **25**.

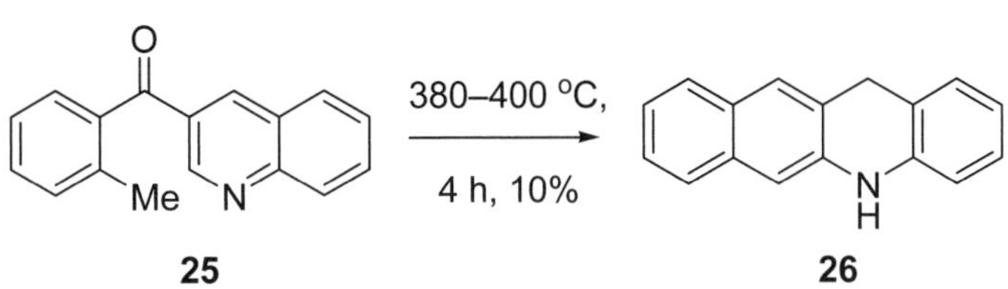

Buu-Hoi and co-workers suggested a more direct formation of the anthrone by dehydrogenation–cyclization of the starting ketone or dehydrogenation of the dihyroanthrol **8** via a hydrogen donor–acceptor mechanism with the involvement of O_2.

4.9.4 Variation and Improvements

Use of Additives

While there has been no general method for the catalysis of the Elbs reaction, various additives have been used. From the mechanistic discussion above, running the reaction in the presence of atmospheric O_2 could have a diminishing impact on the yield of the reaction due to the formation of by-products like anthrone and anthraquinones. While some have claimed additives to improve the yield, more systematic studies by others have been less conclusive. For instance, the addition of Zn was reported to provide superior yield yet after looking into the details of some side-by-side reactions, Zn offered marginal improvements.[1] One positive observation using Zn was reported by Fieser and Hershberg in which the cyclization of **27** with excess Zn occurred more rapidly forming **28** than cyclization without Zn. In additional, an interesting electronic effect was noted when comparing the cyclization of the β-methyl napthyl ketone with the tolyl naphthyl ketone system. The improved yield for the cyclization of **27** over **29** was presumably due to the preferred addition of the methyl group into the more reactive naphthoyl ring system over the benzoyl ring system.

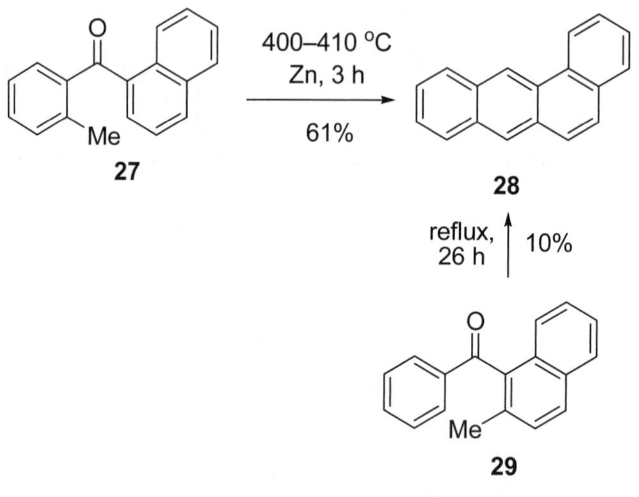

Other additives surveyed and reported to have little effect on improving the Elbs reaction were H_2SO_4, $KHSO_4$, P_2O_5, $ZnCl_2$, and piperidine/Ac_2O. Some surprises in promoting the reaction have come from studies looking at general aryl cyclizations. For example, Vingiello and co-workers[6] reported the Elbs product **31** while studying conditions to promote the formation of the cyclodehydration product **32**. Quite to their surprise, the Elbs product was formed in significant amounts and at relatively low temperature when promoted with AcOH/48% HBr and nearly exclusively using Al_2O_3.

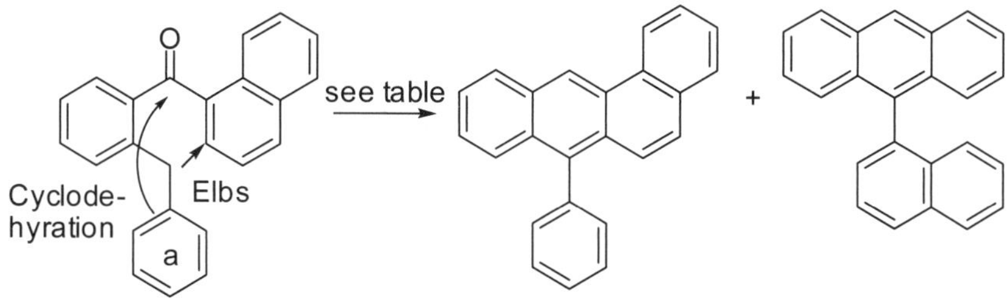

30 **31** **32**

Conditions	Yield **31**	Yield **32**
AcOH, 48% HBr, sealed tube, 180 °C	36%	45%
Al_2O_3, 240-270 °C	49%	4%

In a subsequent report,[7] Vingiello described a more thorough investigation into these conditions in which Elbs conditions (415–430 °C, Zn) were applied to promote the cyclization and gave a 32% yield of **31**; no **32** was observed nor believed to be formed. Vingiello ascribed the general acid or base catalysis in this system to be due to the formation of the Cook type intermediate **6** for the Elbs reaction. In addition a methyl or trifluoromethyl *m*-substituent on ring "a" gave *only* cyclodehydration product or no reaction, respectively. The absence of the Elbs product was ascribed to steric (conformation in which the Me group creates steric issues with the Cook intermediate **6**) and electronic effects (the ring for the cyclodehydration reaction becoming too electron deficient).

Anomalies due to Elimination or Rearrangement

Several anomalies have been observed during studies of the Elbs reaction. There has been the elimination or loss of several groups in a variety of aromatic positions under the high temperature required for the reaction to occur. Loss of both methoxy and chloro groups *para* to the carbonyl of the aromatic ring that is attacked by the methylene group was reported by Fieser.[8] Thus cyclization of **33** gave no detectable amount of the methoxy compound **34** and only a 1% isolated yield of **36** was obtained from **35**.

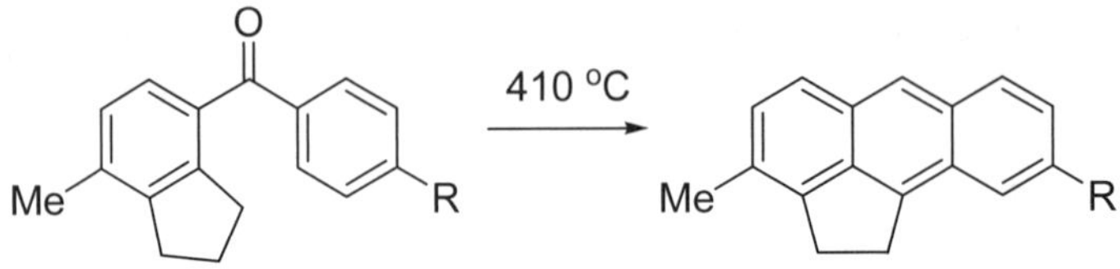

33, R = OMe
35, R = Cl

34, R = OMe, 0%
36, R = Cl, 1%

A report by Bergmann and Blum[9] proposed elimination from the dihydroanthranol derivative **37** after tautomerization to **38** and elimination via push of the enol providing **24** or alternatively via HF elimination to aromatize the system (not shown). In the event, fluoroketone **39** gave a 10% yield of dibenzanthracene **40**; no fluorodibenzanthracene was detected.[10]

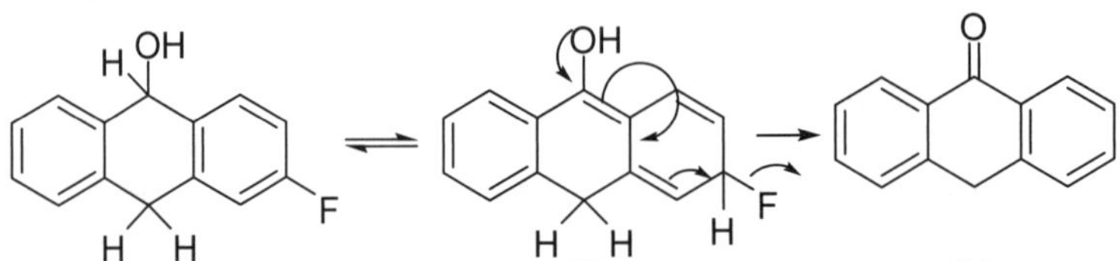

37 **38** **24**

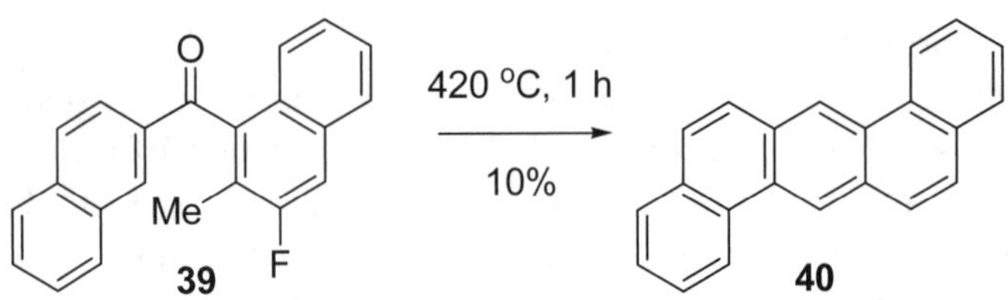

39 **40**

More recently, Newman[11] reported the attempt to synthesize a variety of methylcholanthrene derivatives. The work reported confirmed the above work and determined additional sites in which elimination of certain groups occurred. The authors were able to synthesize the corresponding 8-fluoro- and 10-fluoro-3-methylcholanthrenes **42** and **44** via the Elbs reaction in 21% and 47% yields from the corresponding ketones **41** and **43**, respectively. However, the Elbs reaction starting with appropriately substituted substrates failed to provide the corresponding 10-methoxy, 11-methoxy, 11- fluoro- and 12-fluoro-3-methylcholanthrene derivatives. The 8-methoxy- and 9-methoxycholanthrene derivatives have previously been reported.

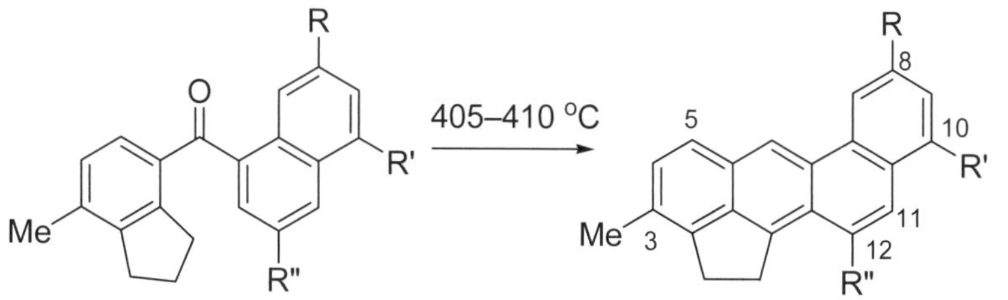

41, R = F, R' = R" = H
43, R' = F, R = R" = H

42, R = F, R' = R" = H; 21%
44, R' = F, R = R" = H; 47%

Another anomalous reaction was shown by Newman[12] in which an optically enriched substrate **45** was taken into the Elbs reaction and yielded completely racemic product **46**.

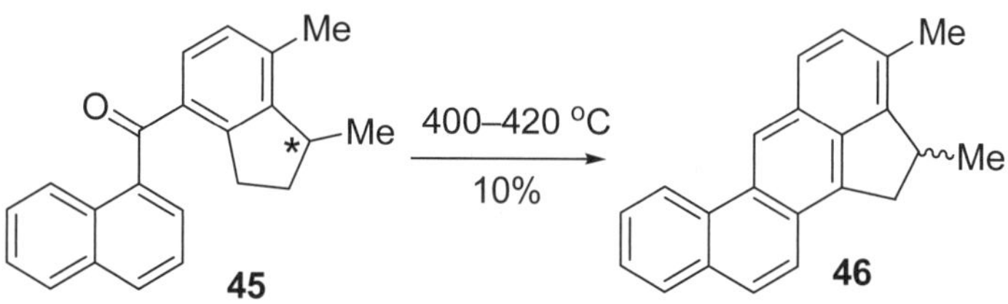

Due to the findings of Dauben cited in the mechanistic section, it should not be surprising to find that rearrangement of the carbonyl does sometimes occur prior to cyclization. An example of such a rearrangement was reported by the Cook group[1] in which two different isomeric naphthyl ketones **47** and **48** gave the same mixture of Elbs products **49** and **50**.

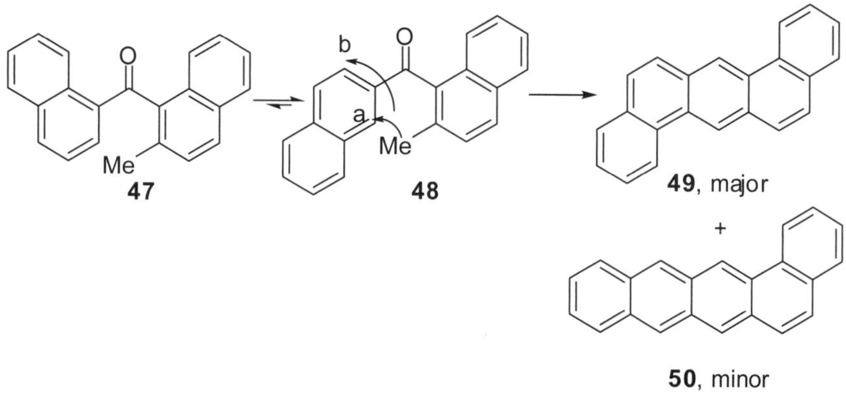

There also have been rearrangements reported after cyclization. This was reported by several authors, most notably by Badger and Christie[13,14] and confirmed in a similar series of more substituted benzothiaphenyl-tolyl substrates by Buu-Hoi and co-workers.[15] The rearrangement proved specific for the benzothiaphenyl-tolylketones like **51** was shown to be the exception rather than the rule. The authors suggested that the initial cyclization occurred followed by a subsequent radical formation by lysis of the S–C bond to form **54** which then recombined to form **55**. Elimination of H_2O then provided the product **52** which was isolated in 37% yield. A by-product, quinone **57**, arising from oxidation of the normal Elbs product, was also reported in 4%.

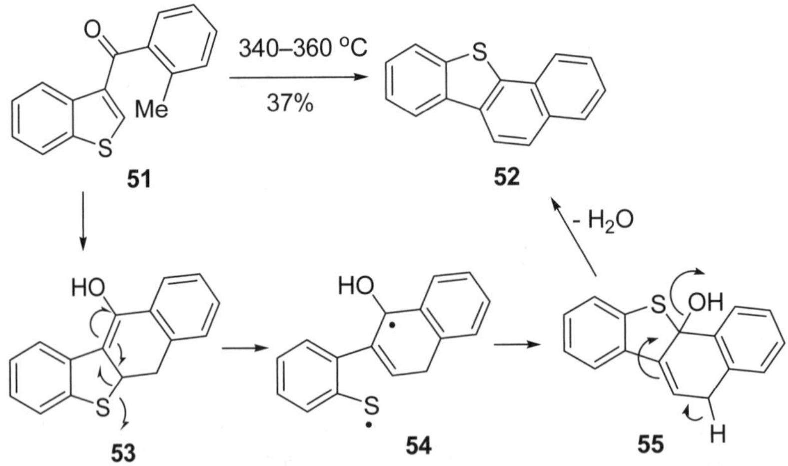

Further proof of the transformation taking place as suggested, or at least demonstration that an intermediate was being intercepted, was given by preparing the expected Elbs product **56** via another route and subjecting that material to pyrolysis (390 °C, 3 h). Over 60% of **56** was recovered, unchanged.

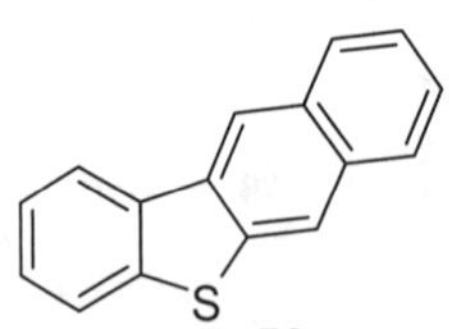

56

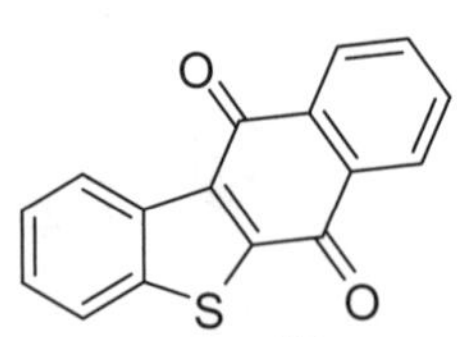

57

Badger and Christie[13] subsequently studied the Elbs reaction of several benzothiaphenylnaphthyl ketones, which provided the normal Elbs product. Ketone **58** was pyrolyzed neat at 450 °C and provided a 54% yield of the thianaphthofluorene **59**.

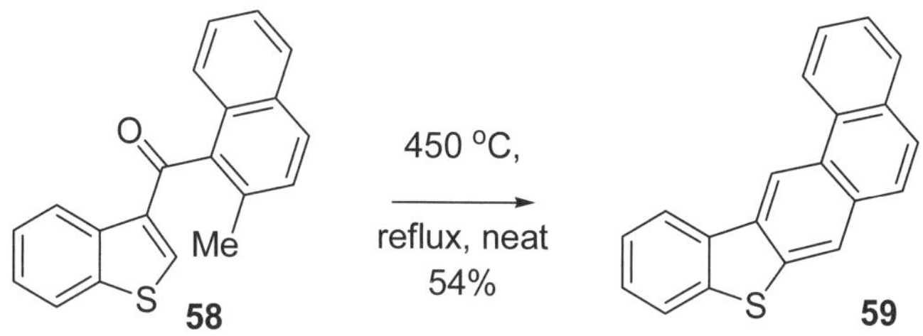

Heterocycles Employed

In addition to the benzothiaphenyl system, indolyl ketones,[16] pyridyl ketones, quinolyl ketones,[4] and dibenzofuryl ketones[17] have all provided Elbs products when subjected to pyrolysis conditions. Typically, yields have been in the 5–30% range.

Photochemical Promoted

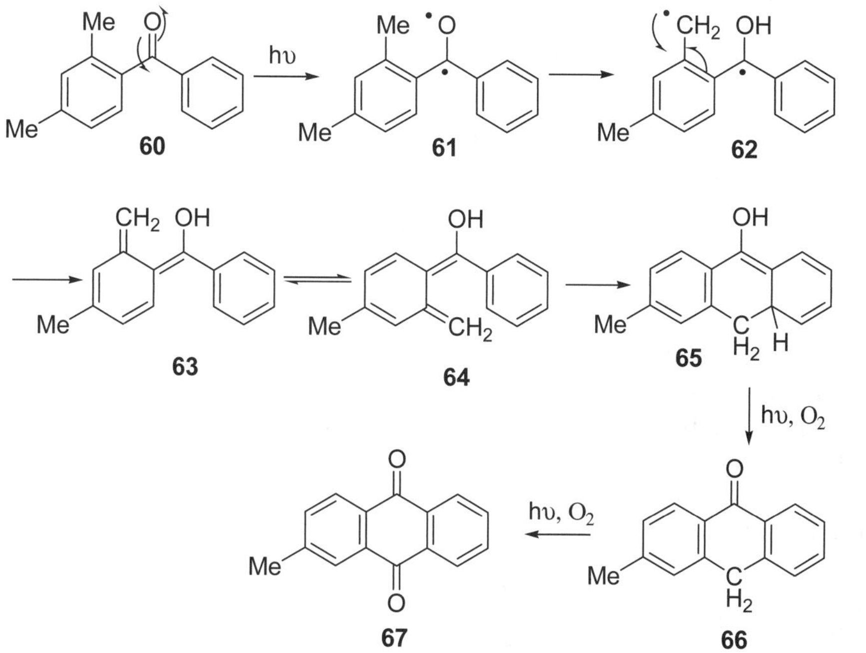

There have been two reports of using light to promote this reaction.[18,19] These studies support the Cook intermediate **6** and also the oxidative by-products previously mentioned. The reaction was proposed to initiate via a type II Norrish reaction forming **62**, which rearranges into a mixture of two dienols. Dienol **64** proved analogous to the Cook intermediate and the believed productive compound to undergo Elbs cyclization to **65**. Further oxidative by-products were observed under the reaction conditions and anthrone **66** and anthraquinone **67** was identified.

4.9.5 Synthetic Utility

The Elbs reaction provides rapid access to polyaromatic systems, particularly due to the ease of preparing many substrates for the reaction. When the Elbs reaction proceeds smoothly, it may provide the best known means to synthesize a certain substrate. For example, 3-methylcholanthrene and derivatives have been shown to be potent carcinogenic agents. The reaction has been used to prepare several methylcholanthrene analogues and other polyaromatic species, the products were isolated and tested in carcinogenic studies to probe characteristics of polyaromatics as potential carcinogenic agents.[1,8,17,20] Despite some of its limitations, the Elbs reaction still offers an easy entry into these complicated polyaromatics.[11]

4.9.6 Experimental

Preparation of 10-fluoro-3-methylcholanthrene (44)[11]

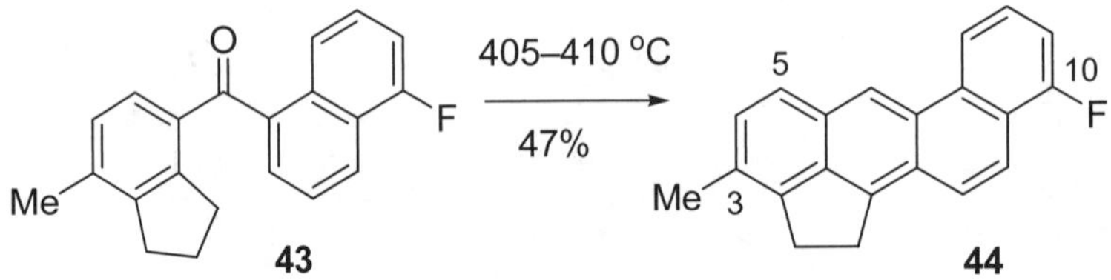

43 **44**

Ketone **43** (2 g) was heated at 405–410 °C for 30 min by means of a sodium nitrate-potassium nitrite salt bath. The product was chromatographed on neutral alumina using benzene-petroleum ether (1:3) to yield 0.9 g (47%) **44** as pale yellow prisms, m.p. 208–209 °C.

4.9.7 References

1. [R] Fieser, L. F. *Org. Reac. I*, **1942**, 129–154.
2. Hurd, C. D.; Azorlosa, J. *J. Am. Chem. Soc.* **1951**, *73*, 37–41.
3. Heidelberger, C.; Brewer, P.; Dauben, W. G. *J. Am. Chem. Soc.* **1947**, 69, 1389–1391.
4. Badger, G. M.; Pettit, R. *J. Chem. Soc.* **1953**, 2774–2778.
5. Buu-Hoi, N. P.; Marie, C.; Jacquignon, P. *Bull. Chem. Soc. Fr.* **1970**, *3*, 1012–1015.
6. Vingiello, F. A.; Borkovec, A.; Zajac, W. *J. Am. Chem. Soc.* **1958**, *80*, 1714–1716.
7. Vingiello, F. A.; Thornton, J. R. *J. Org. Chem.* **1966**, *31*, 659–663.
8. Fieser, L.; Desreux, V. *J. Am. Chem. Soc.* **1938**, *60*, 2255–2262.
9. Bergmann, E. D.; Blum J. *J. Org. Chem.* **1960**, *25*, 474–475.
10. Bergmann, E. D.; Blum, J. *J. Org. Chem.* **1961**, *26*, 3214–3216.
11. Newman, M. S.; Khanna, V. K. *J. Org. Chem.* **1980**, *45*, 4507–4508.
12. Newman, M. S.; Linsk, J. *J. Org. Chem.* **1949**, *14*, 480–483.

13. Badger, G. M.; Christie, B. J. *J. Chem. Soc.* **1958**, 913–915.
14. Badger, G. M.; Christie, B. J. *J. Chem. Soc.* **1956**, 3435–3437.
15. Marie, C.; Buu-Hoi, N. P.; Jacquignon, P. *J. Chem. Soc. (C)* **1971**, 431–434.
16. Buu-Hoi, N. P.; Hoan, N.; Khoi, N. H. *J. Org. Chem.* **1950**, *15*, 131–134.
17. Buu-Hoi, N. P.; Lavit, D. *J. Chem. Soc.* **1959**, 38–41.
18. Porter, G.; Tchir, M. F. *J. Chem. Soc., Chem. Commun.* **1970**, 1372–1373.
19. Heindel, N. D.; Molnar, J. *J. Chem. Soc., Chem. Commun.* **1970**, 1373–1374.
20. Cook, J. W.; deWorms, C. G. M. *J. Chem. Soc.* **1937**, 1825–1828.

4.10 Fujimoto–Belleau Reaction

Nadia M. Ahmad

4.10.1 Description

The Fujimoto–Belleau reaction involves the formation of cyclic α-substituted α,β-unsaturated ketones from enols.

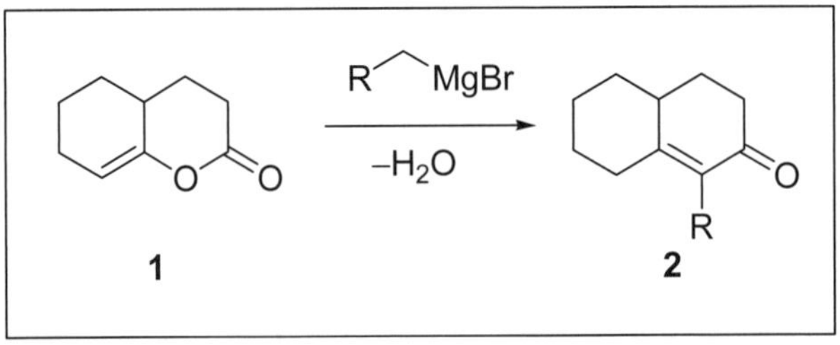

4.10.2 Historical Perspective

The Fujimoto–Belleau reaction was reported independently by both George I. Fujimoto and Bernard Belleau in 1951. Fujimoto was attempting to isotopically label steroids in the A ring using a previously reported synthetic method.[1] However, he found that a modification of said synthesis resulted in a rearrangement which gave a labeled carbon in the 4′-position. His modification consisted of a Grignard reaction followed by treatment with alkali affording the desired product in good yields of 52–60% without isolation of the intermediate. The reaction was applied to testosterone, whereby a methyl Grignard reagent was added to the enol lactone **3**. Cyclisation in the presence of alkali gave testosterone acetate **5** in overall yields of 25–50%.[2]

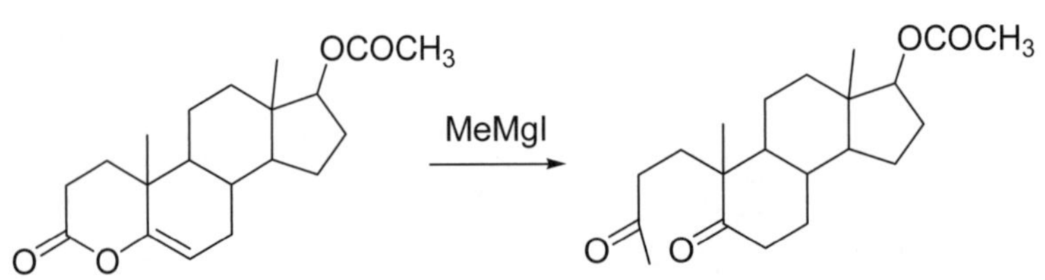

3 **4**

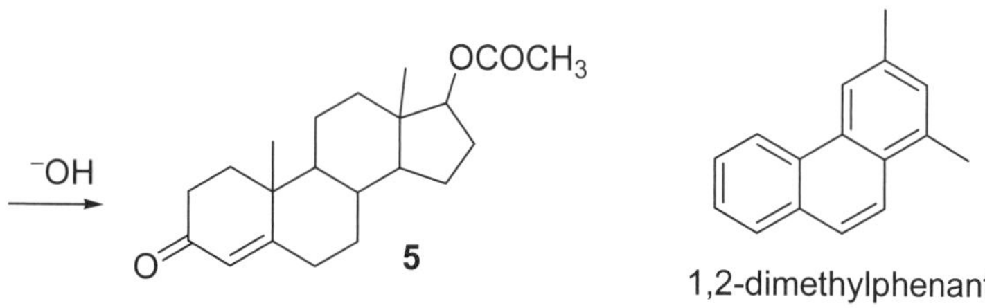

5

1,2-dimethylphenanthrene

Belleau also reported this reaction in his paper titled "The Reaction of Methylmagnesium Iodide with β-(1-Hydroxy-3,4-dihydro-2-naphthyl)-butyric Acid Lactone."[3] Published in the same volume, Belleau described the synthesis of 1,3-dimethylphenanthrene. In this synthesis, enol lactone **6** was converted to ketone **7** by reaction with methylmagnesium iodide and subsequent treatment with hydrochloric acid in 15% yield. The side product of this reaction had an empirical formula of $C_{16}H_{18}$, which when dehydrogenated gave 1,3-dimethylphenanthrene, the actual synthesis that Belleau was aiming for!

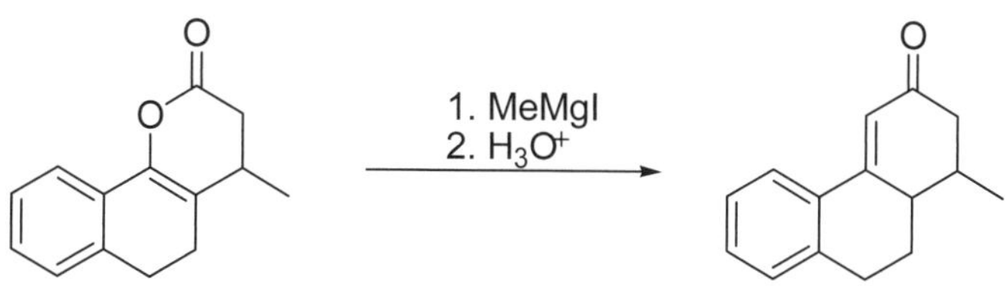

6 7

4.10.3 Mechanism

The Fujimoto–Belleau reaction is a two-step one-pot transformation, which begins with a Grignard reaction (→**1**). This is followed by a hydrogen shift tautomerisation, (**8**→**9**), an aldol reaction, (**9**→**10**), then loss of water through a reverse aldol (**11**→**12**) resulting in the final product.

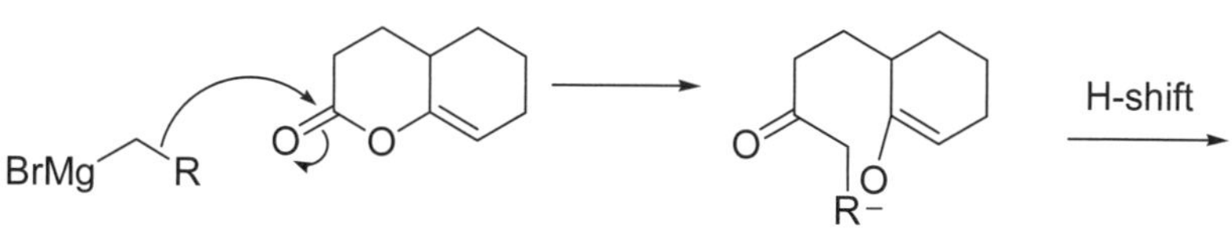

1 8

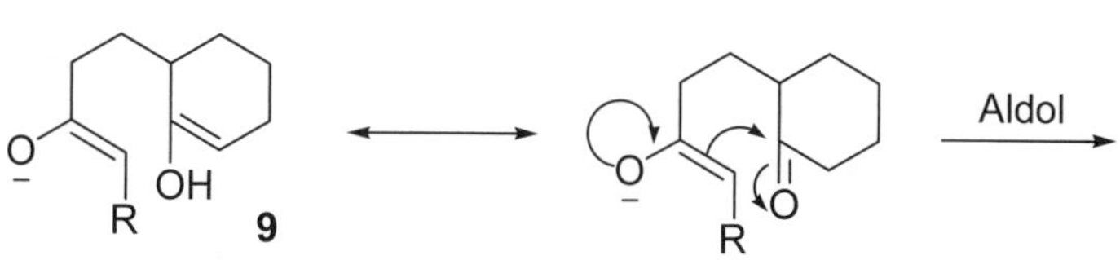

9

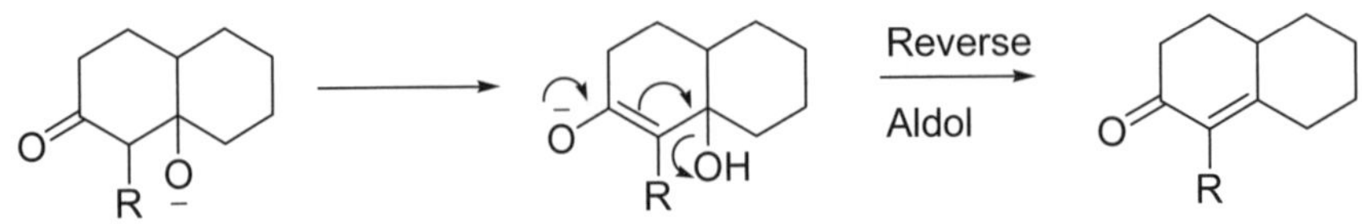

10 11 12

4.10.4 *Variations, Improvements and Modifications*

Organometallic reagents other than Grignards can also be used to effect the Fujimoto–Belleau transformation. In their synthesis of the natural product (–)-nakamurol A, Bonjoch and co-workers used the lithium salt of dimethyl methylphosphonate to convert lactone **13** to the *α,β*-unsaturated cyclic ketone **14**.[4] This procedure gave good yields (75% based on recovered starting material) and was a superior method to using methyllithium as a nucleophile. The use of methyllithium resulted in a methyl ketone, which then had to be subjected to an aldol reaction to obtain the desired product.

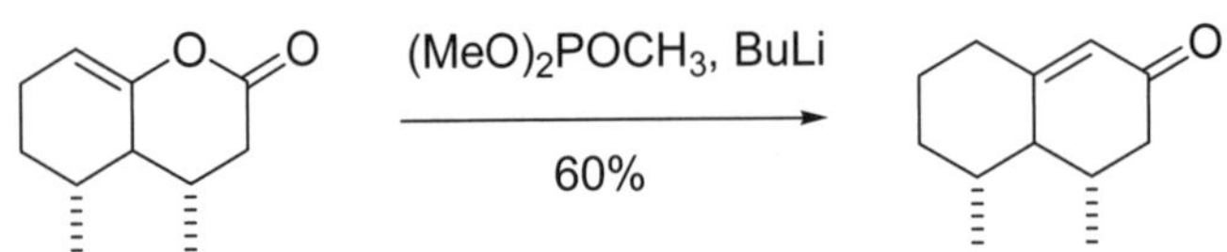

13 14

An example of different substrates amenable to transformation in the Fujimoto–Belleau reaction is shown by the treatment of benzazepine **15** with radioactively labeled methylmagnesium iodide.[5] This was followed by quenching with water to afford diketone **17**. In this two-step procedure the diketone was isolated in 80% yield, then treated with base to give the final *α,β*-unsaturated enone **18**. The synthesis illustrates the utility of the Fujimoto–Belleau reaction in the production of radiolabelled 3-benzazepines using labelled methyl iodide.

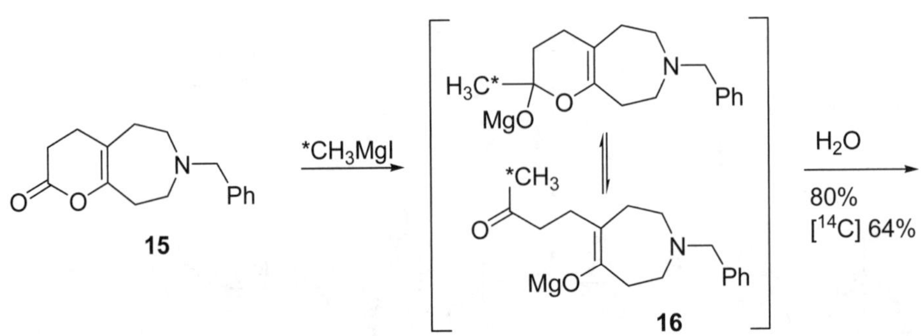

15 16

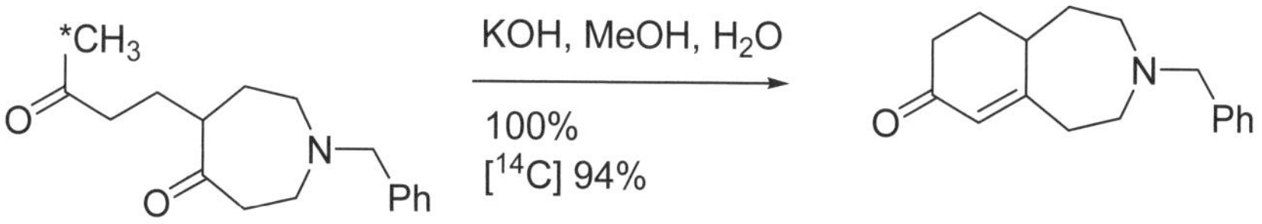

17 18

4.10.5 Synthetic Utility

A relatively early example of the use of the Fujimoto reaction is illustrated by the work of Tournemine and co-workers.[6] In their search for new antiandrogenes, the authors synthesised des-A-steroids **20** by treating lactones **19** with Grignard reagents. These and other analogues were then tested for their relative binding affinities at androgen receptors compared to that of testosterone. In some cases, the analogues were found to have comparable or lower affinity than testosterone itself.

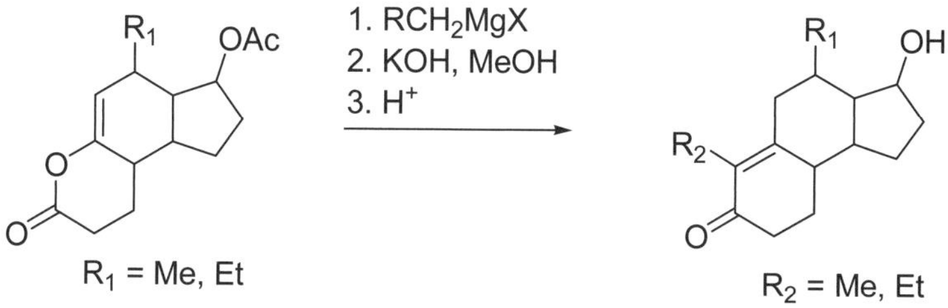

19 20

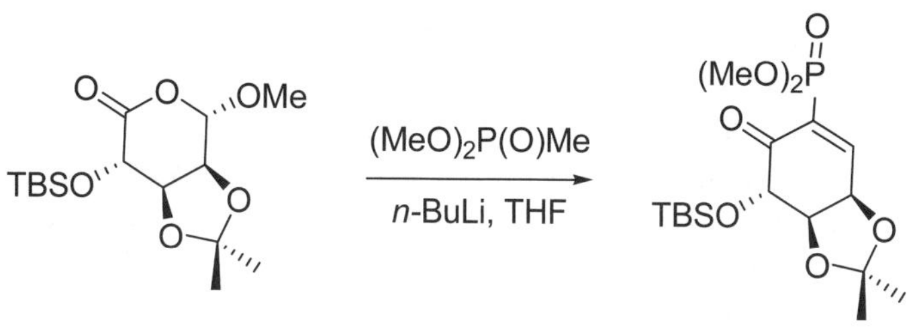

21 22

In their studies on mannose-derived carbocyclic sugars, Trabsa and co-workers applied a Wadsworth–Emmons modification of the Fujimoto–Belleau reaction to give a vinyl phosphonate rather than the expected enone.[7] Originally this procedure was used as an alternative to the Ferrier cyclisation, which requires acidic conditions. However, the authors found that the formation of an enone or vinyl phosphonate depended on the carbohydrate stereochemistry. Lactone **21** gave the vinyl phosphonate **22** when treated with lithium dimethyl methylphosphonate while lactone **23** afforded enone **24** exclusively.

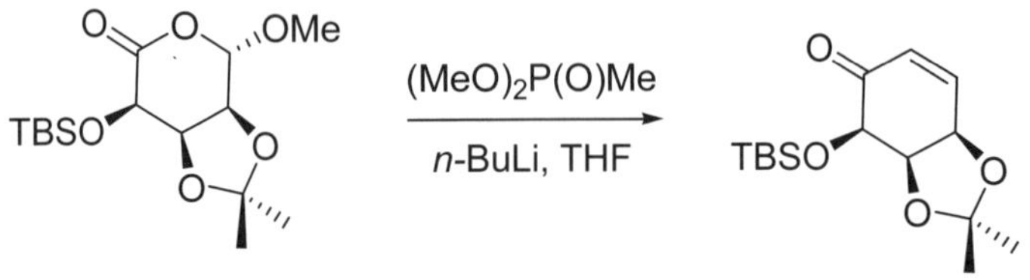

4.11.6 Experimental

3-Hydroxy-4,6-substituted-3,3a,4,5,8,9,9a,9b-octahydro-1*H*-cyclopenta[*a*]naphthalen-7(2*H*)-one 26[6]

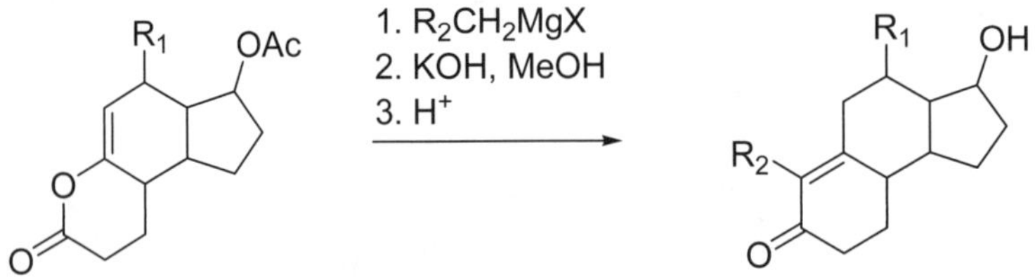

To a solution of the enol lactone **25** (1 mmol) in anhydrous tetrahydrofuran (3 mL) was added the appropriate Grignard reagent (1.5 mmol) in ethereal solution at –60 °C over 30 min. After stirring at –60 °C for 1 h, the reaction mixture was poured onto saturated ammonium chloride solution, and the reaction product extracted with diethyl ether. The organic extract was washed and dried (sodium sulfate). After removal of the solvent, the residue was dissolved in 2 N methanolic potassium hydroxide solution (4 mL), and the resultant mixture heated under reflux for 1 h. The cooled mixture was neutralised with AcOH and evaporated under reduced pressure. The residue was then diluted with water and extracted with diethyl ether or dichloromethane. Chromatography of the crude product on silica gel (eluent 3:7 ethyl acetate/hexanes) afforded pure enones **26** in 20–50% yields.

4.10.7 References

1. Turner, R. B. *J. Am. Chem. Soc.* **1950**, *72*, 579.
2. Fujimoto, G. I. *J Am. Chem. Soc.* **1951**, *73*, 1856.
3. Belleau, B. *J. Am. Chem. Soc.* **1951**, *73*, 5441–5443.
4. Daz, S.; Cuesta, J.; Gozlez, A.; Bonjoch, J. *J. Org. Chem.* **2003**, *68*, 7400–7406.
5. Heys, J. R.; Senderoff, S. G. *J. Org. Chem.* **1989**, *54*, 4702–4706.

6. Morales-Alanis, H.; Brienne, M.-J.; Jacques, J.; Bouton, M.-M.; Nedelec, L.; Torelli, V.; Tournemine, C. *J. Med. Chem.* **1985**, *28*, 1796.
7. Aloui, M.; Lygo, B.; Trabsa, H. *Synlett* **1994**, 115.

4.11 Haworth Reaction

Richard J. Mullins and Everett W. Merling

4.11.1 Description

The Haworth reaction is a classical method for the synthesis of tetralone, beginning with benzene and succinic anhydride. The three-step protocol involves a Friedel–Crafts acylation, followed by reduction of the arylketone, and an intramolecular Friedel–Crafts acylation.[1] The tetralone analog may be further reduced and dehydrogenated to form new aromatic species, in what is known as the Haworth phenanthrene synthesis.

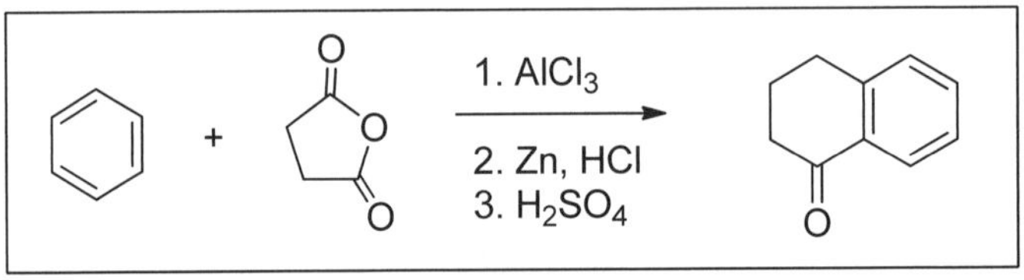

4.11.2 Historical Perspective

Although the modern Haworth reaction is more commonly used to form tetralone analogs, the reaction sequence first targeted the synthesis of phenanthrene and its derivatives. The original protocol used by Robert Downs Haworth in 1932 involved the reaction of naphthalene (**1**) with succinic anhydride (**2**) and aluminum chloride to form nearly equal quantities of naphthoylpropionic acids **3** and **4**.

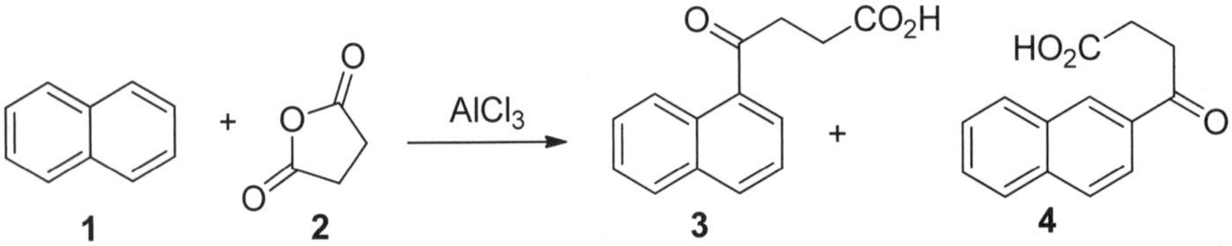

Formation of naphthoylpropionic acids **3** and **4** had been achieved in a similar manner by multiple groups between 1910 and 1920; however, there was a lack of sufficient documentation regarding yields, amounts of each isomer, and methods of separation at the time.[2–6] Haworth was the first to separate the acids by using nitrobenzene instead of either carbon disulfide or benzene as a solvent in the reaction. The isolation of the individual acids **3** and **4** spawned a period of immense productivity for Haworth and co-workers, as these ketoacids were then used to form phenanthrene and multiple derivatives of phenanthrene. Over the course of 2 years, Haworth published at least eight papers on the syntheses of phenanthrene derivatives.[7–14] Certainly, then, it is understandable why this reaction bears his name instead of those of the original discoverers.

4.11.3 Mechanism

The Haworth synthesis consists of a Friedel–Crafts acylation of an aromatic ring with succinic anhydride and aluminum chloride, followed by the Clemmensen or Wolff–Kishner reduction of the resulting ketoacid. An intramolecular Friedel–Crafts acylation then results in the formation of tetralone or tetralone derivatives. Mechanistically, complexation of succinic anhydride (**2**) with $AlCl_3$ yields acylium ion **5**, leading to the resonance-stabilized cation **7** through nucleophilic attack by benzene (**6**). Deprotonation to rearomatize the ring is followed by hydrolysis resulting in the formation of ketoacid **8**. Reduction of the ketone gives **9**, setting the stage for a second Friedel–Crafts acylation via acylium ion **10** to afford tetralone (**11**).

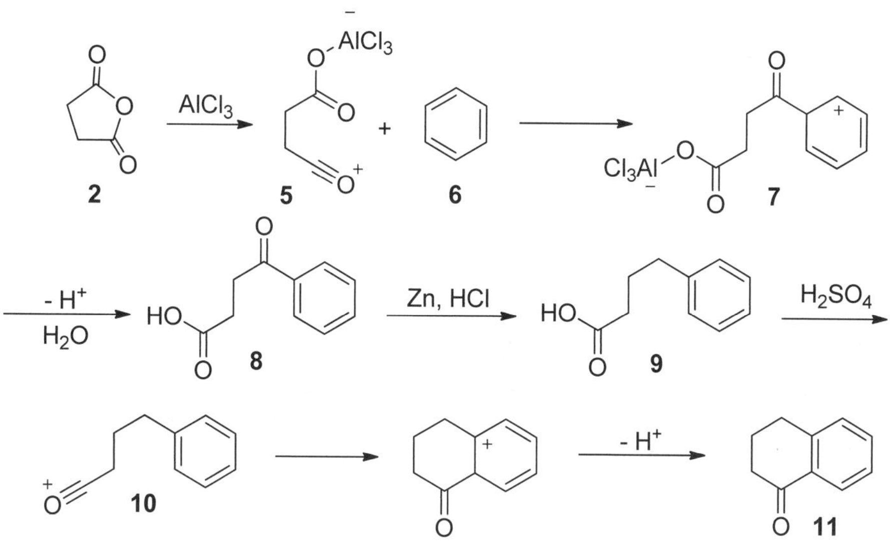

4.11.4 Variations and Improvements

The tetralone derivatives produced by the Haworth reaction are important synthetic intermediates. However, in Haworth's original work, his efforts were directed toward the synthesis of phenanthrene from naphthalene. Thus the synthesis of tetralone derivative **12** by the standard Haworth conditions is followed by a Clemmensen[15–17] or Wolff–Kishner reduction[18,19] of the ketone to give **13**. Ruzicka's[20,21] method, or other methods of dehydrogenation, then results in phenanthrene (**14**). In this review, when the sequence stops at the tetralone product, the reaction will be referred to as the Haworth reaction or Haworth tetralone synthesis; when these additional steps are used the process will be referred to as the Haworth phenanthrene synthesis.

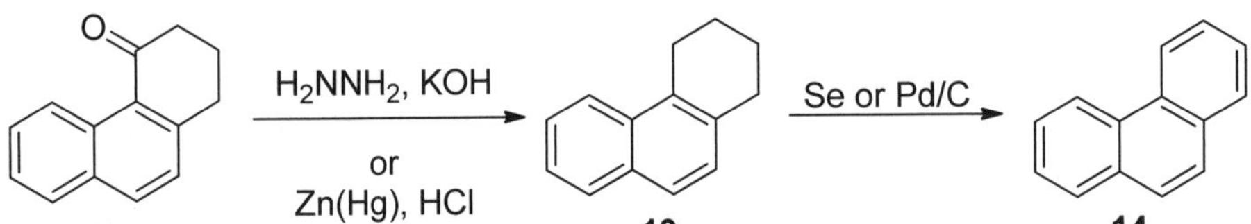

12 **13** **14**

A slight variation on the original Haworth conditions, which promotes the formation of the acylium ion for the intramolecular Friedel–Crafts acylation is the conversion of the carboxylic acid into an acid chloride. The enhanced leaving group ability of the chloride ion allows milder conditions to be used in the final intramolecular acylation step.

4.11.5 Synthetic Utility

Haworth tetralone synthesis

The Haworth synthesis of tetralone derivatives has found substantial utility in natural product and small molecule synthesis.[22–24] Zubaidha and co-workers performed the Haworth reaction on 2-methylanisole (**15**) to yield **18** as an intermediate in the synthesis of (±)-heritol (**19**), a potent itchthyotoxin.[25] Beginning with **15**, acylation to yield **16** is followed by the Clemmensen reduction to produce **17**. Subsequent intramolecular Friedel–Crafts acylation results in the formation of tetralone **18**. Ferraz and coworkers have also synthesized **18** in the same manner as an intermediate for the synthesis of (±)-mutisianthol (**20**), a potential antitumor agent.[26,27] In addition, this substrate **18** has been used for the synthesis of various α-tetralols as substrates for enzymatic resolution studies.[28]

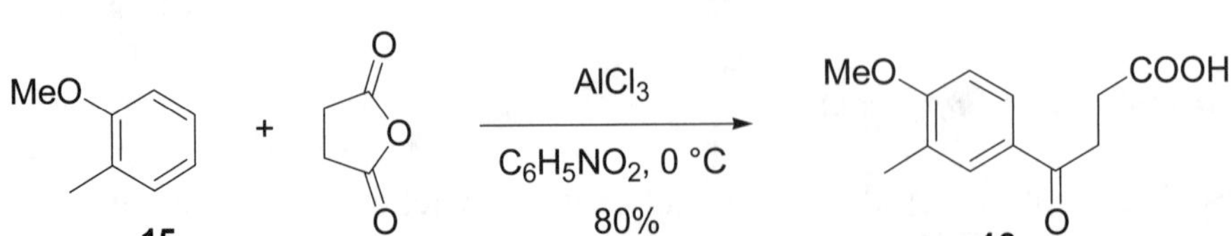

15 **16**

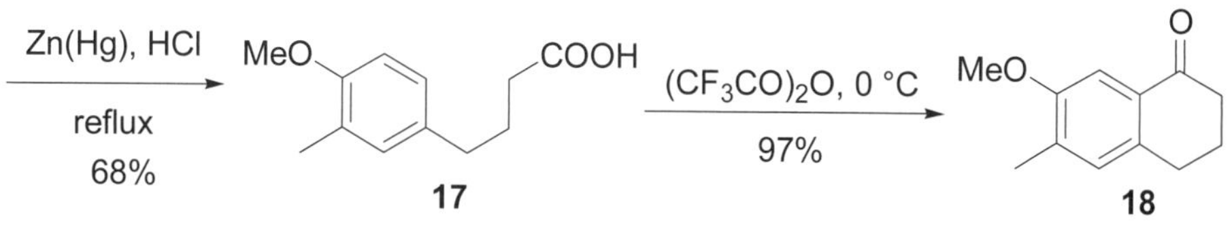

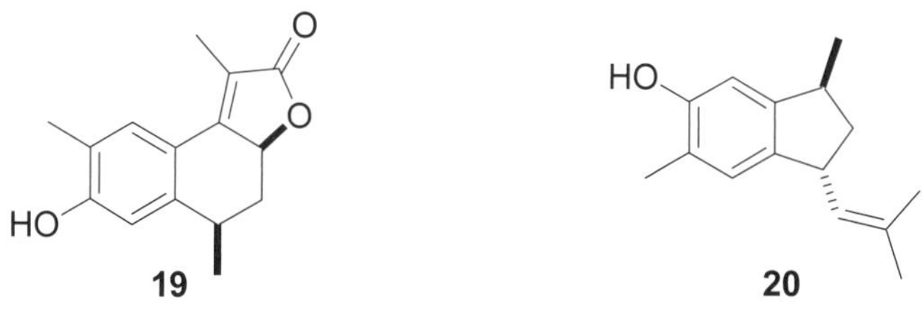

Owing to the facile Friedel–Crafts reaction of electron-rich aromatic rings, methoxy-substituted benzene rings are viable substrates for the Haworth reaction. For example Wipf and Jung[29] used the Haworth reaction with **21** in their formal synthesis of the natural product (+)-diepoxin σ, a fungal metabolite with antitumor properties. Following acylation to yield **22**, Wolff–Kishner reduction resulted in the formation of **23**, which underwent intramolecular acylation, providing **24** in high yield. Green and co-workers have also used the Haworth reaction successfully with polymethoxy-substituted benzyl rings.[30] In addition, Swenton and Chen applied the Haworth reaction to synthesis of the 1,4-dimethoxytetralin ring system as a model for their studies on the effects of allylic substituents on the regiochemistry of bisketal hydrolysis.[31]

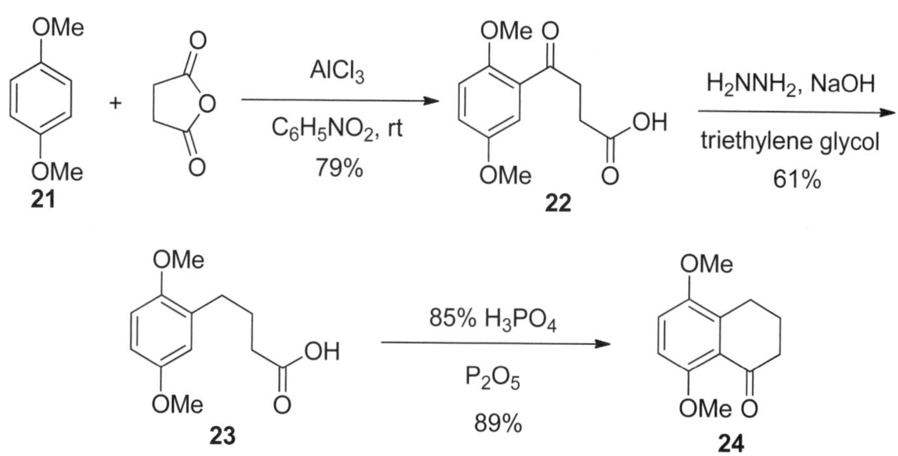

In addition to tolerating substitution on the aromatic ring, the Haworth reaction proceeds with derivatives of succinic anhydride, allowing for broad substrate scope. Norlander and coworkers used an amide-substituted succinic anhydride reagent in the synthesis of 2-amino-6,7-dihydroxy-1,2,3,4-tetrahydronaphthalene (ADTN), a powerful dopamine agonist.[32] In this synthesis, acylation of **25** with substituted succinic anhydride **26** resulted in formation of **27**. Triethylsilane reduction of **27** afforded **28** which was cyclized via a one-pot acid chloride synthesis/Friedel–Crafts acylation sequence to give **29**. Melillo and co-workers similarly used an amide-substituted succinic anhydride derivative for synthesis of *(R,R)*-4-propyl-9-hydroxynaphthoxazine, another potent dopamine agonist.[33]

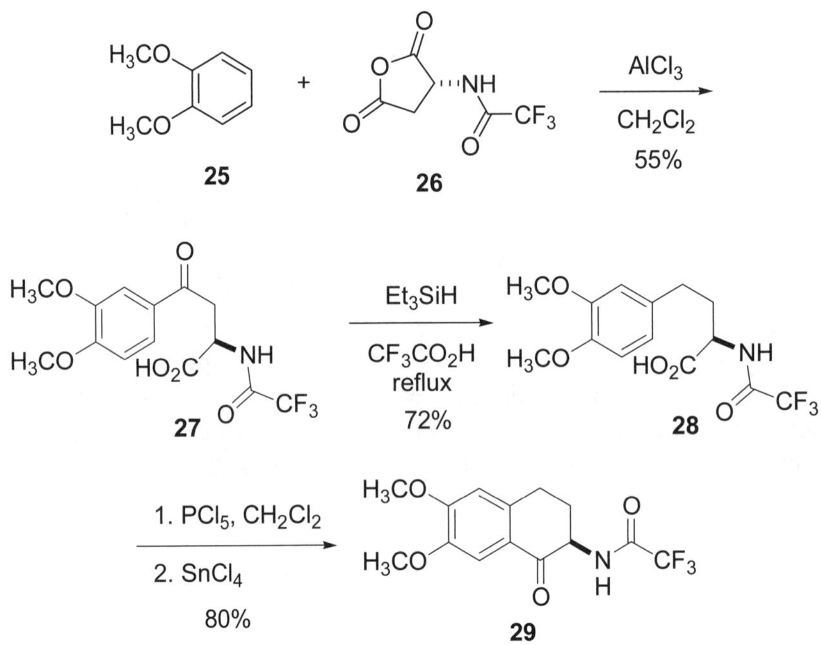

In efforts toward the synthesis of aromatic ring-fused cyclic 1,2-diketones, Ranu and Jana also made use of the Haworth reaction when anisole (**30**) and **31** were treated under Friedel–Crafts conditions to produce **32**. Subsequent reduction and cyclization produced **33** in good yield.[34]

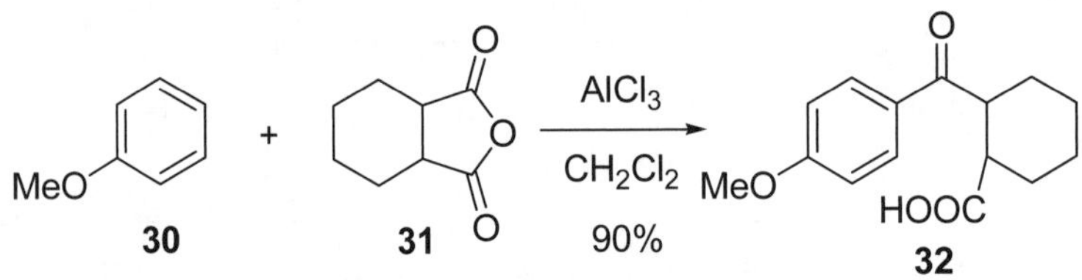

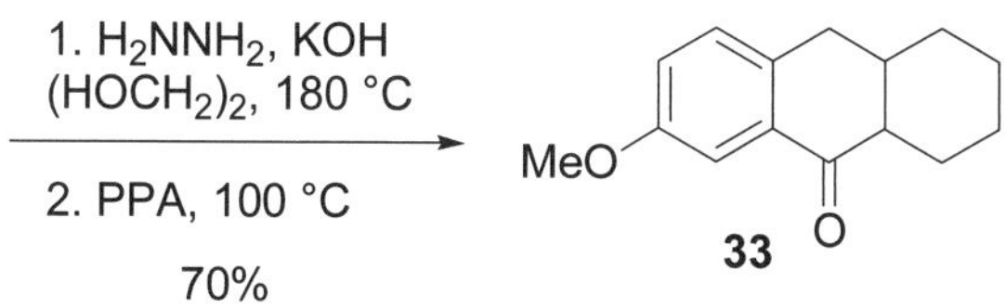

Clive and Wang's synthesis[35] of the marine natural product hamigeran B (**39**), an inhibitor of polio and herpes viruses, uses a slightly modified Haworth reaction to convert *m*-cresol (**34**) into the tetralone derivative **38**. Following acylation and reduction to yield **35**, methylation of the phenolic oxygen produces **36**. Methylation of the α-carbon to form **37** is followed by cyclization in the presence of $POCl_3$, providing **38** in high yield.

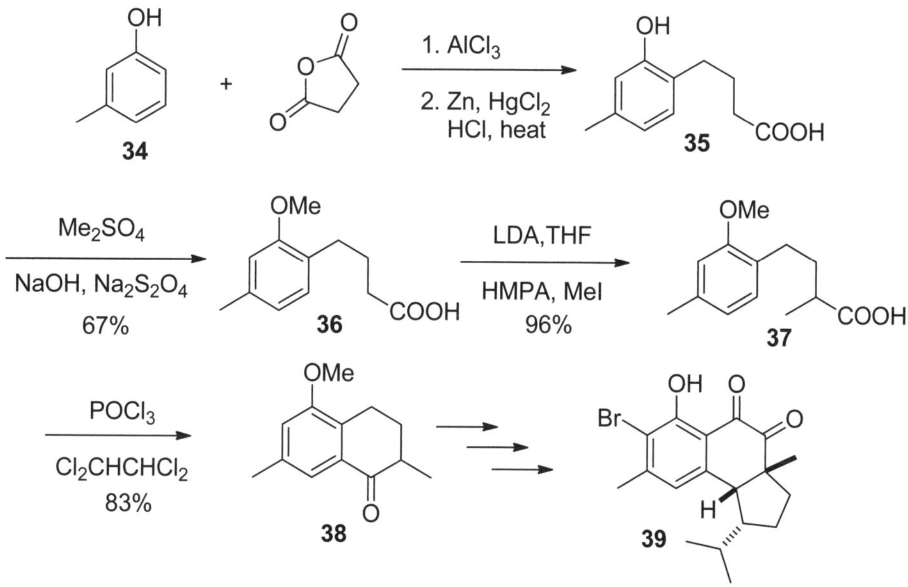

The Haworth synthesis has also found substantial utility in materials synthesis. Within this area, Rheingold and co-workers were able to synthesize rigid spacer–chelators using a monobrominated tetralone prepared under modified Haworth conditions.[36] Following *para*-acylation of bromobenzene (**40**) under vigorous conditions to yield **41**, reduction under modified Wolff–Kishner conditions[37] is followed by cyclization to give **42**.

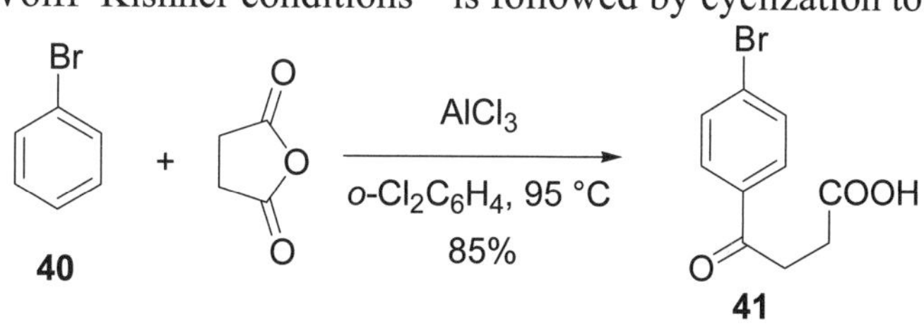

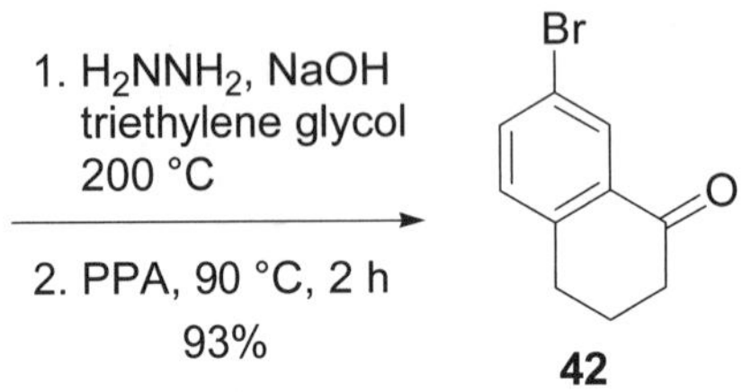

42

Levy and co-workers[38] encountered a problem of selectivity when using the Haworth reaction to synthesize helical transition metal complexes. Although intermediate **51** seems well suited for synthesis via the Haworth procedure on anthracene (**43**), the keto acid obtained was predominately substituted at C-9 (in **44**) rather than C-2 (in **45**). The selectivity issue was creatively remedied using 9,10-dihydroanthracene (**46**) in lieu of anthracene (**43**), which then exclusively formed the acylation product **46**. Following reduction of **47**, the carboxylic acid **48** was necessarily esterified to **49** in order to avoid deactivation of the dehydrogenation catalyst. Dehydrogenation to **50** was then followed by an intramolecular Friedel–Crafts reaction to give **51** as a single regioisomer.

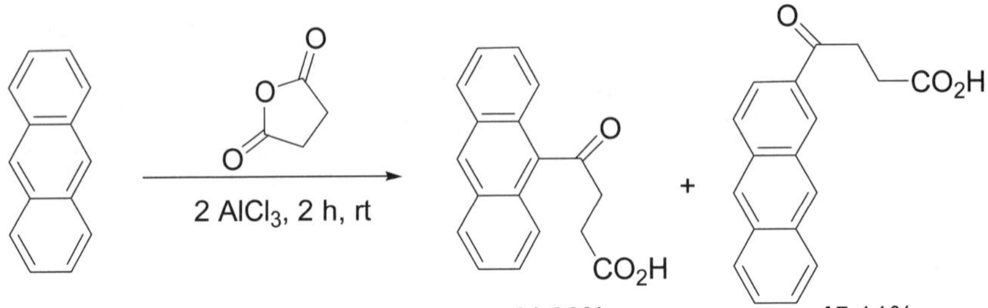

43 **44** 80% **45** 11%

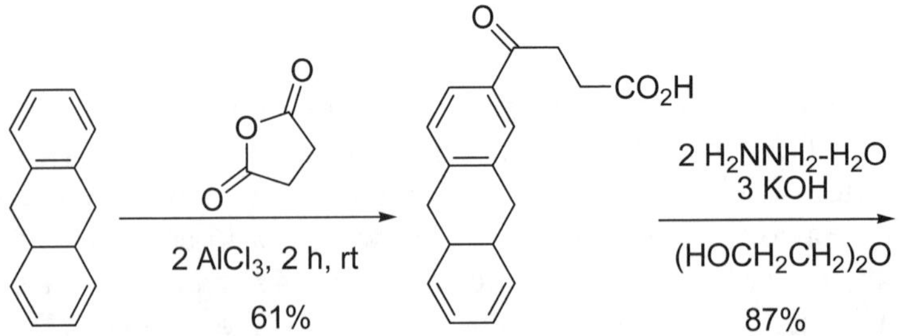

46 **47**

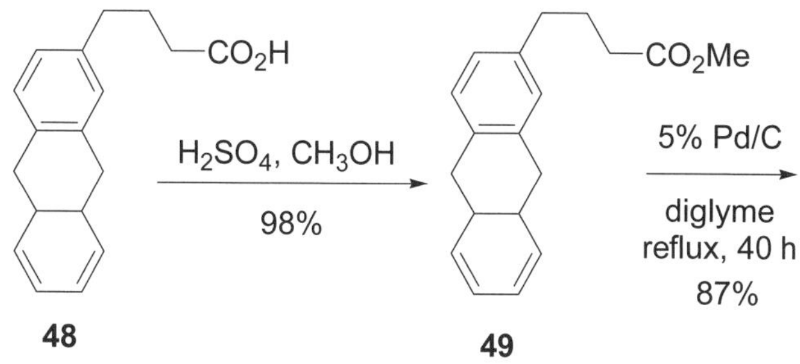

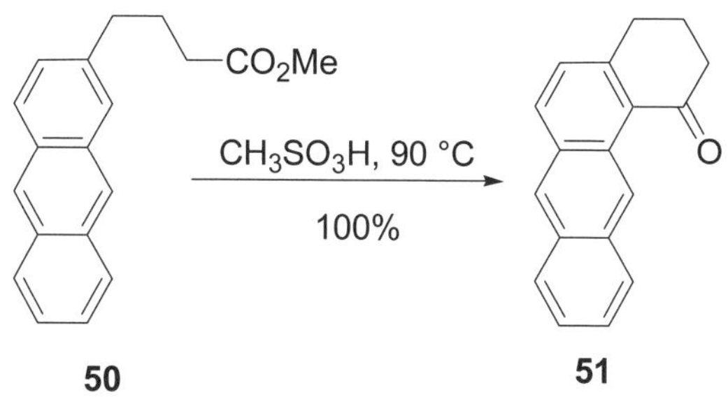

Haworth phenanthrene synthesis

Although not directly resulting in the synthesis of phenanthrene or its analogs, reactions, which can be classified under the Haworth phenanthrene synthesis have been widely exploited in synthetic endeavors,[39] especially in the field of materials chemistry. For example, Ogino and co-workers have demonstrated an interesting way to form substituted pentacenes using the Haworth method.[40] Replacing succinic anhydride with bisanhydride **53** in the Friedel–Crafts acylation of *n*-nonylbenzene (**52**) results in the formation of **54**. Hydrogenation of the arylketones is followed by a second Friedel–Crafts acylation to produce **55**. Finally, reduction with sodium borohydride and acidic workup results in the dehydrogenated product **56**.

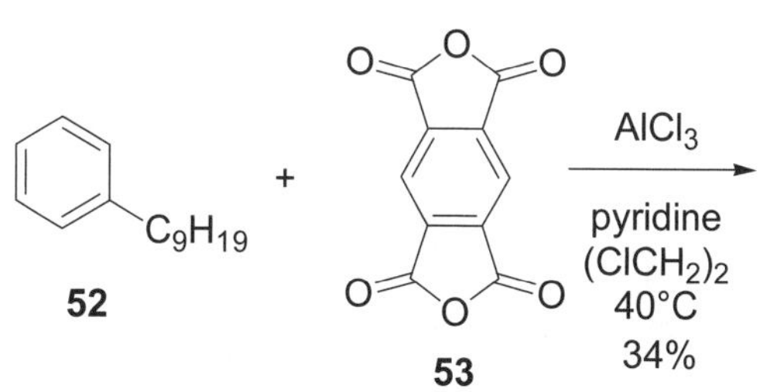

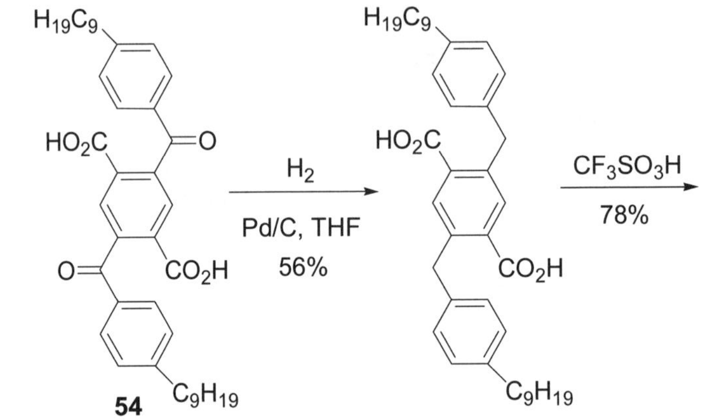

54

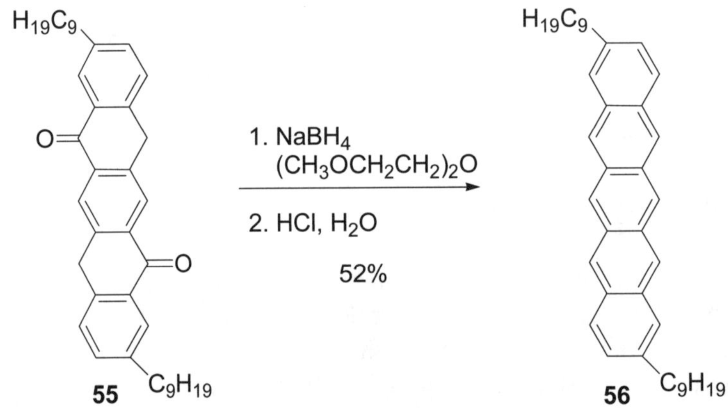

55 56

Rabinovitz and co-workers also used the Haworth phenanthrene synthesis to form complex fused aromatic ring systems.[41] The bowl-like shape of corannulene **57** was extended through the acylation reaction with **58** to produce **59**. Treatment of **59** with red phosphorus under strongly acidic conditions resulted in **60** in good yield.

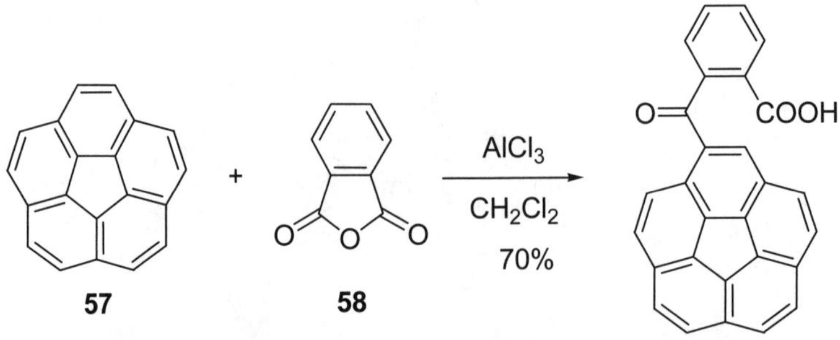

57 58 59

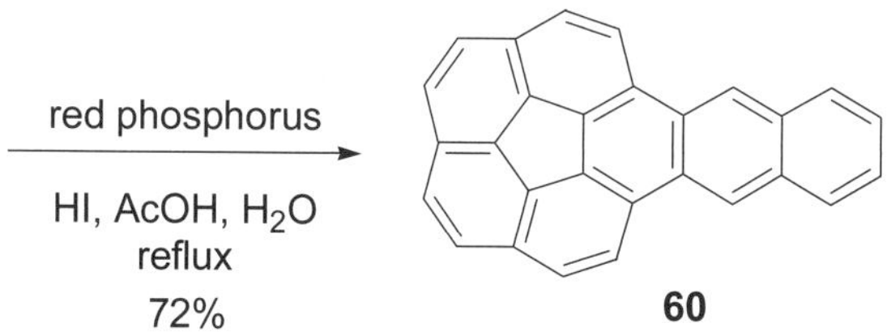

The Haworth phenanthrene synthesis has been extensively used in the synthesis of derivatives of chrysene,[42,43] an environmental pollutant which exhibits tumorigenic and mutagenic properties. For example Harvey and co-workers treated **61** with succinic anhydride under Friedel–Crafts conditions to produce **62**.[44] Reduction of the ketoacid under Wolff–Kishner conditions is followed by esterification to yield **63**. Dehydrogenation of **63** is followed by saponification to yield carboxylic acid **64**. Intramolecular Friedel–Crafts acylation produces tetralone derivative **65**, which undergoes carbonyl reduction and dehydrogenation to produce **66**.

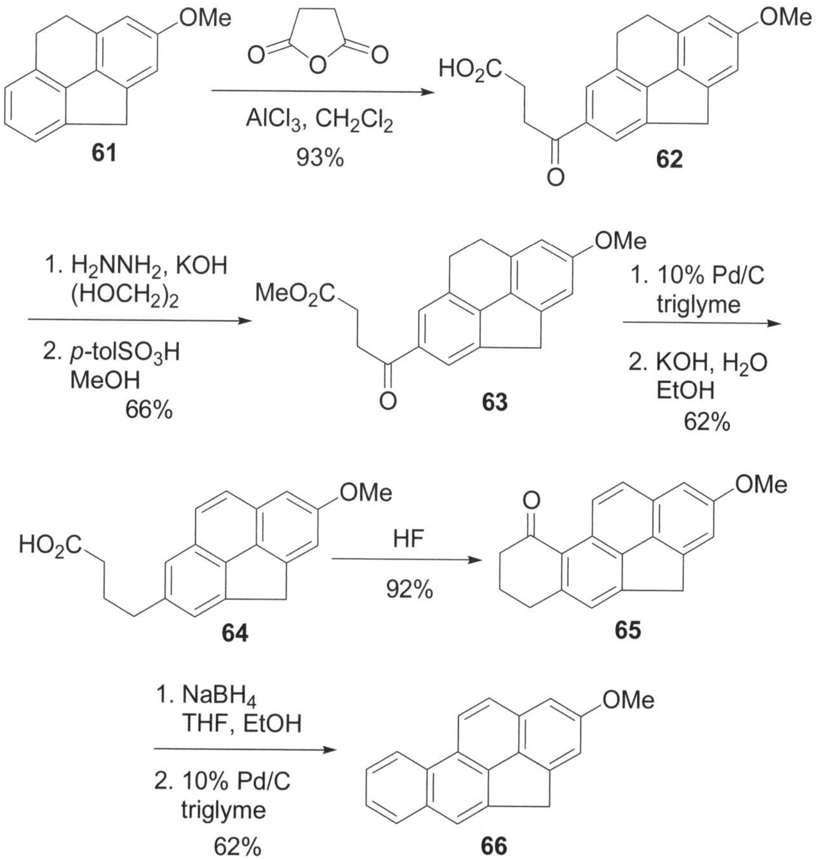

The Haworth phenanthrene synthesis was also employed for the preparation of naphthalene intermediates toward the synthesis of novel HMG-CoA reductase inhibitors.[45] The usual Haworth procedure was followed to secure tetralone **67**. Hydride reduction of the carbonyl produced **68**, which on dehydration to **69**, was subsequently dehydrogenated with DDQ to provide naphthalene **70**. A related procedure was used in the same work to replace the C-7 methyl with a chlorine atom.

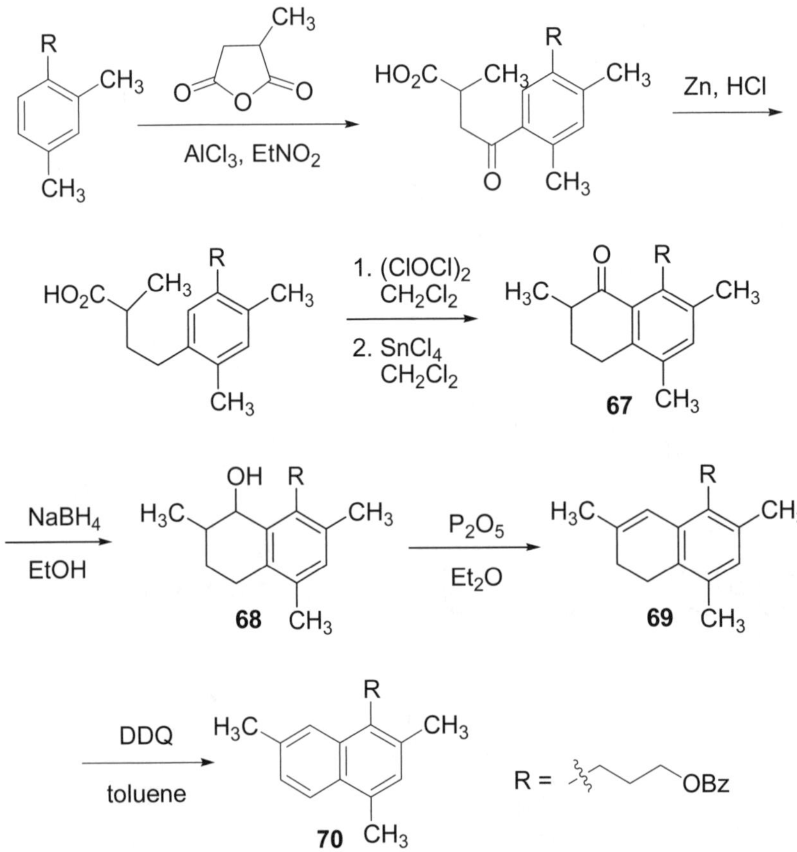

4.11.6 Experimental

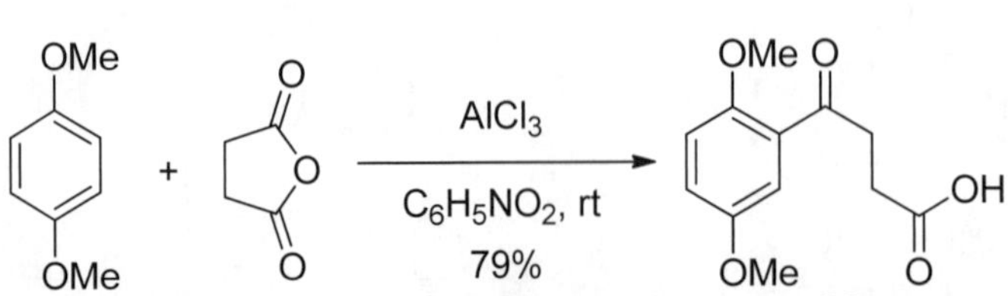

22

4-(2′,5′-Dimethoxyphenyl)-4-oxobutyric acid (22)[29]

To a solution of $AlCl_3$ (80.0 g, 0.60 mol) in nitrophenol (500 mL) were added at 0 °C succinic anhydride (30.02 g, 0.30 mol) and *p*-dimethoxybenzene (37.31 g, 0.27 mol). The mixture was allowed to warm from 5 to 29 °C over a 3.5-h period, and the solution was then promptly poured into ice water. The organic layer was separated and extracted with 10% $NaHCO_3$ solution. The combined aqueous layers were filtered and acidified to pH 1 by concentrated HCl solution in an ice bath. The resulting very pale yellow solid was filtered and dried to afford 51 g (79%) of **22**.

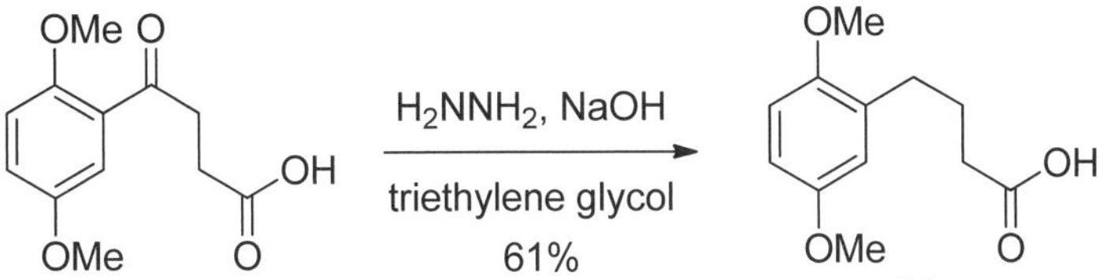

22 **23**

4-(2′,5′-Dimethoxyphenyl)butyric acid (23)

A solution of **22** (50.0 g, 0.21 mol) in triethylene glycol (620 mL) containing sodium hydroxide (31.6 g, 0.79 mol), hydrazine hydrate (26.5 g, 0.53 mol), and water (30 mL) was heated at reflux for 3 h and then heated further without a condenser until the temperature rose to 210 °C. After another hour, sufficient water was added to lower the temperature to 190 °C, and heating was continued for 4 h. The solution was then cooled, poured into a mixture of conc. HCl and ice, and extracted with ether. The combined ether layers were dried (Na_2SO_4) and concentrated in vacuo. Chromatography on SiO_2 (hexanes/EtOAc, 2:1 → 1:1) gave 28.7 g (61%) of **23** as a solid.

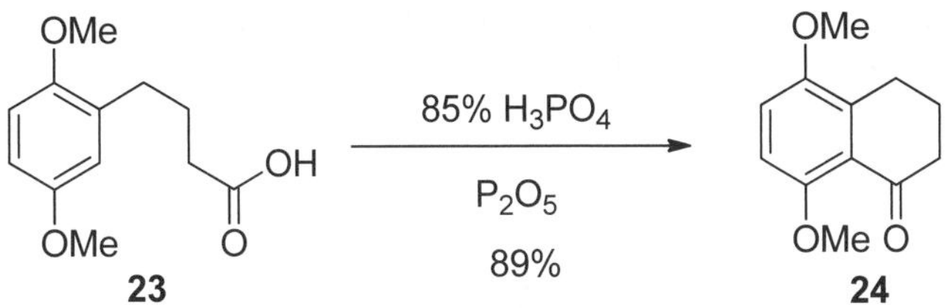

23 **24**

5,8-Dimethoxy-3,4-dihydro-2H-naphthalen-1-one (24)

To polyphosphoric acid prepared from 85% phosphoric acid (305 g) and P_2O_5 (278 g) was added **23** (14.0 g, 62.4 mmol). After the reaction mixture was stirred for 0.5 h at 80 °C, the resulting orange solution was poured into ice water and extracted with ether. The combined ether layers were washed with 1 N NaOH solution, dried (Na_2SO_4), and concentrated in vacuo to afford 11.5 g (89%) of **24** as a pale yellow solid.

4.11.7 References

1. Friedel, C.; Crafts, J. M. *Compt. Rend.* **1877**, *84*, 1392, 1450.
2. Giua, M. *Rend. Soc. Chim. Ital.* **1912**, *9*, 239.
3. Giua, M. *Ber.* **1914**, *47*, 2115–2116.
4. Borsche, W.; Sauernheimer, H. *Ber.* **1914**, *47*, 1645–1648.
5. Krollpfeiffer, F.; Schäfer, W. *Ber.* **1923**, *56*, 620–632.
6. Schroeter, G.; Müller, H.; Huang, J. Y. S. *Ber.* **1929**, *62*, 645–658.
7. Haworth, R. D. *J. Chem. Soc.* **1932**, 1125–1133.
8. Haworth, R. D.; Letsky, B. M.; Mavin, C. R. *J. Chem. Soc.* **1932**, 1784–1792.
9. Haworth, R. D.; Bolam, F. M. *J. Chem. Soc.* **1932**, 2248–2251.
10. Haworth, R. D. *J. Chem. Soc.* **1932**, 2717–2720.
11. Haworth, R. D.; Mavin, C. R. *J. Chem. Soc.* **1932**, 2720–2723.
12. Haworth, R. D.; Mavin, C. R.; Sheldrick, G. *J. Chem. Soc.* **1934**, 454–461.
13. Haworth, R. D.; Sheldrick, G. *J. Chem. Soc.* **1934**, 864–867.
14. Haworth, R. D.; Sheldrick, G. *J. Chem. Soc.* **1934**, 1950–1952.
15. Clemmensen, E. *Ber.* **1913**, *46*, 1837–1843.
16. Clemmensen, E. *Ber.* **1914**, *47*, 51–63.
17. Clemmensen, E. *Ber.* **1914**, *47*, 681–687.
18. Kishner, N. *J. Russ. Chem. Soc.* **1911**, *43*, 582.
19. Wolff, L. *Liebigs Ann. Chem.* **1912**, *394*, 86–108.
20. Hanson, J. RMeyer, J. *Helv. Chim. Acta* **1921**, *4*, 505–510.
22. Yang, T.; Huang, Y.; Cho, B. P. *Chem. Res. Toxicol.* **2006**, *19*, 242–254.
23. Diel, B. N.; Han, M.; Kole, P. L.; Boaz, D. B. *J. Label. Compd. Radiopharm.* **2007**, *50*, 551–553.
24. Beland, F. A.; Churchwell, M. I.; Von Tungeln, L. S.; Chen, S.; Fu, P. P.; Culp, S. J.; Schoket, B.; Gyorffy, E.; Minárovits, J.; Poirier, M. C.; Bowman, E. D.; Weston, A.; Doerge, D. R. *Chem. Res. Toxicol.*, **2005**, *18*, 1306–1315.
25. Zubaidha, P. K.; Chavan, S. P.; Racherla, U. S.; Ayyangar, N. R. *Tetrahedron* **1991**, *47*, 5759–5768.
26. Ferraz, H. M. C.; Aguilar, A. M.; Silva, L. F. *Tetrahedron* **2003**, *59*, 5817–5821.
27. Bianco, G. G.; Ferraz, H. M. C.; Costa, A. M.; Costa-Lotufo, L. V.; Pessoa, C.; de Moraes, M. O.; Schrems, M. G.; Pfaltz, A.; Silva, L. F. *J. Org. Chem.* **2009**, *74*, 2561–2566.
28. Ferraz, H. M. C.; Bianco, G. G.; Teixeira, C. C.; Andrade, L. H.; Porto, A. L. M. *Tetrahedron: Asymmetry* **2007**, *18,* 1070–1076.
29. Wipf, P.; Jung, J.-K. *J. Org. Chem.* **2000**, *65*, 6319–6337.
30. Ameer, F.; Giles, R. G. F.; Green, I. R.; Pearce, R. *Synth. Commun.* **2004**, *34*, 1247–1258.
31. Chen, C.-P.; Swenton, J. S. *J. Org. Chem.* **1985**, *50*, 4569–4576.
32. Nordlander, J. E.; Payne, M. J.; Njoroge, F. G.; Vishwanath, V. M.; Han, G. R.; Laikos, G. D.; Balk, M. A. *J. Org. Chem.* **1985**, *50*, 3619–3622.
33. Melillo, D. G.; Larsen, R. D.; Mathre, D. J.; Shukis, W. F.; Wood, A. W.; Colleluori, J. R. *J. Org. Chem.* **1987**, *52*, 5143–5150.
34. Ranu, B. C.; Jana, U. *J. Org. Chem.* **1999**, *64*, 6380–6386.
35. Clive, D. L. J.; Wang, J. *J. Org. Chem.* **2004**, *69*, 2773–2784.
36. Sommer, R. D.; Rheingold, A. L.; Goshe, A. J.; Bosnich, B. *J. Am. Chem. Soc.* **2001**, *123,* 3940–3952.
37. Minlon, H. *J. Am. Chem. Soc.* **1946**, *68*, 2487–2488.
38. Wiznycia, A. V.; Desper, J.; Levy, C. J. *Dalton Trans.* **2007**, *15*, 1520–1527.
39. Sekiguchi, S.; Hirai, M.; Ota, E.; Hiratsuka, H.; Mori, Y.; Tanaka, S. *J. Org. Chem.* **1985**, *50*, 5105–5108.
40. Okamoto, K.; Kawamura, T.; Sone, M.; Ogino, K. *Liq. Cryst.* **2007**, *34*, 1001–1007.
41. Aprahamian, I.; Preda, D. V.; Bancu, M.; Belanger, A. P.; Sheradsky, T.; Scott, L. T.; Rabinovitz, M. *J. Org. Chem.* **2006**, *71*, 290–298.
42. Silveira, A.; McWhorter, E. J. *J. Org. Chem.* **1972**, *37*, 3687–3691.
43. Mitra, A.; Biswas, T. K.; Ray, R. M. *Tetrahedron* **1992**, *48*, 10353–10362.

44. Dai, W.; Abu-Shqara, E.; Harvey, R. G. *J. Org. Chem.* **1995**, *60*, 4905–4911.
45. Prugh, J. D.; Alberts, A. W.; Deana, A. A.; Gilfillian, J. L.; Huff, J. W.; Smith, R. L.; Wiggins, J. M. *J. Med. Chem.* **1990**, *33*, 758–765.

4.12 Moore Cyclization

Ewa Krawczyk and Roman Dembinski

4.12.1 Description

Moore cyclization is formulated as a cyclization of enyne–ketenes **1** to diradicals **2**. This process usually proceeds by thermal induction, leading to the formation of a benzene ring.

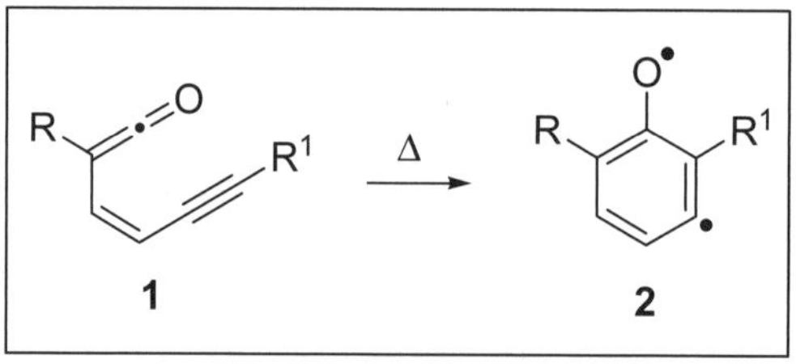

4.12.2 Historical Perspective

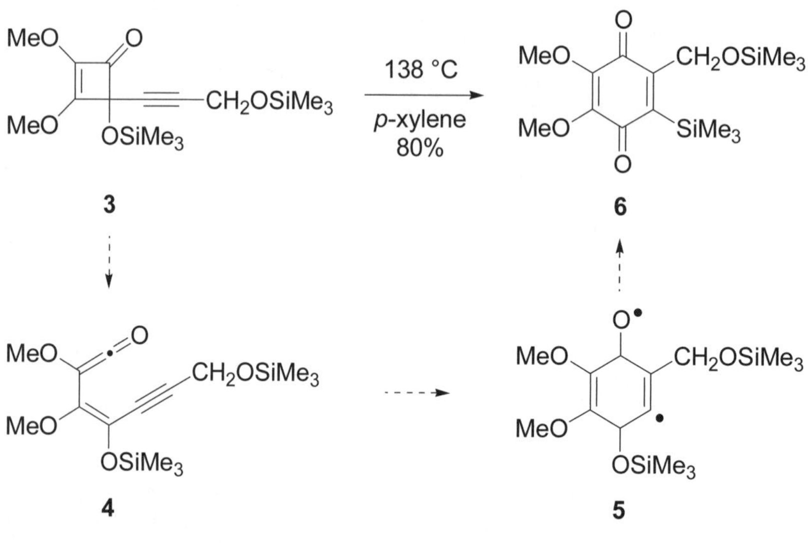

In 1985 Harold W. Moore and co-workers (University of California, Irvine), described the generation and chemistry of 2-alkenylethynyl ketenes, which were accessible from the corresponding alkynylcyclobutenones.[1] For example, 4-alkynyl-4-trimethylsililoxycyclobutenone **3** forms an enyne–ketene **4**. The conjugated ketene **4** undergoes ring closure to produce the diradical **5** (or related zwitterion) which, in turn, proceeds to the substituted

1,4-benzoquinone **6**. The reaction was carried out over 2 h under argon atmosphere in refluxing xylene.

4.12.3 Mechanism

Mechanistic and synthetic studies for the Moore reaction have been reviewed.[2] Equilibrium between alkynylcyclobutenone and (2-alkynylethenyl)ketene **7** at elevated temperature has been assumed in the first mechanism. The enyne–ketene **7**, in turn, undergoes ring closure with the formation of zwitterion intermediate **8**. Finally, transfer of the substituent connected with one oxygen atom, to a negative site in the intermediate **8**, gives 1,4-benzoquinone **9** as the product.[1] Further studies by Moore et al. provided evidence that the quinone-forming rearrangement occurred via intramolecular migration of either the hydrogen atom or other group.[3] The electronic nature of intermediate **8** was discussed and additional evidence for the intermediacy of the diradical was presented. The stereoselective ring opening of alkynylcyclobutenones to **7** takes place under thermal conditions. Therefore, the electrocyclic ring opening affords the desired *Z*-isomer of (2-alkynylethenyl)ketene **7**. During the mechanistic studies on the thermal rearrangement of 4-alkynylcyclobuta-nones to 1,4-benzoquinones, Moore observed a competitive reaction leading, presumably via intermediate **10**, to 2-alkylidene-1,3-cyclopentenediones **11**.[1,3]

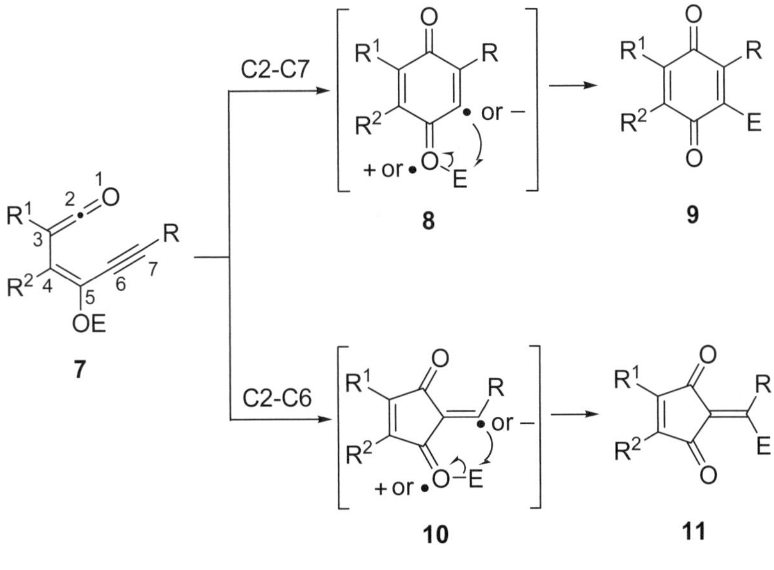

When the alkynyl moiety is substituted with an alkyl group or a proton only, the corresponding benzoquinones were detected as products of thermolysis. But when R = Ph, OR, or Me_3Si, a mixture of both products was formed. Furthermore, the use of 4-alkynylcyclobutanones with electron-withdrawing substituents, like CO_2Et or CH=CHOMe, resulted in the formation of cyclopentediones as a sole product. These results can be explained by the mechanism presented in the scheme above. Intermediate **10** is favored over intermediate **8** when substituent R better stabilizes the adjacent radical site.[3] A ferrocenyl group at the alkyne also provides a stabilization that favors the formation of cyclopentene-1,3-diones.[4]

Computational investigations (ab initio methods for enyne–ketene **1** and substituent effects employing the DFT approach) of the regioselectivity of the cyclization of enyne–ketenes were carried out by Engels et al.[5] For the core system **1** the C2–C7 cyclization proceeds via the most stable diradical intermediate and is kinetically and thermodynamically favored with respect to C2–C6 cyclization. The cyclization modes for the enyne–ketenes are more endothermic but the substituent effects are more pronounced for the enyne–allene (Myers–Saito cyclization). Substituents OMe at the C4 as well as OH at the C5 in the computed molecule core increase the contribution of the C2–C6 cyclization by decreasing the free energy of activation, similar to the phenyl group attached at the alkynyl terminus (C7). A five-membered ring can be formed via the diradical pathway, yielding the σ,π-diradical such as **10**, or via a carbenelike intermediate (not illustrated).[5]

The computational comparison of various pathways of thermal reactions of an analog, in which a heteroatom replaces the central olefinic bond, with possible transformations of parent enyne–ketene **1**, was carried out and showed little relevance.[6]

4.12.4 Variations and Improvements

Enediynyl ethyl ethers like **12** have been applied by Wang's group as precursors of enyne–ketenes **13**, which underwent the Moore cyclization reaction to form diradicals. The intermediate **14** and (after 1,5-hydrogen shift) new diradical **15**, form, upon the cascade transformations, the final products: mainly chromanol **16** and spiro ketone **17**. The latter is a result of subsequent intramolecular reaction of intermediates: *o*-quinones methide and spiro ketone (not illustrated).[7]

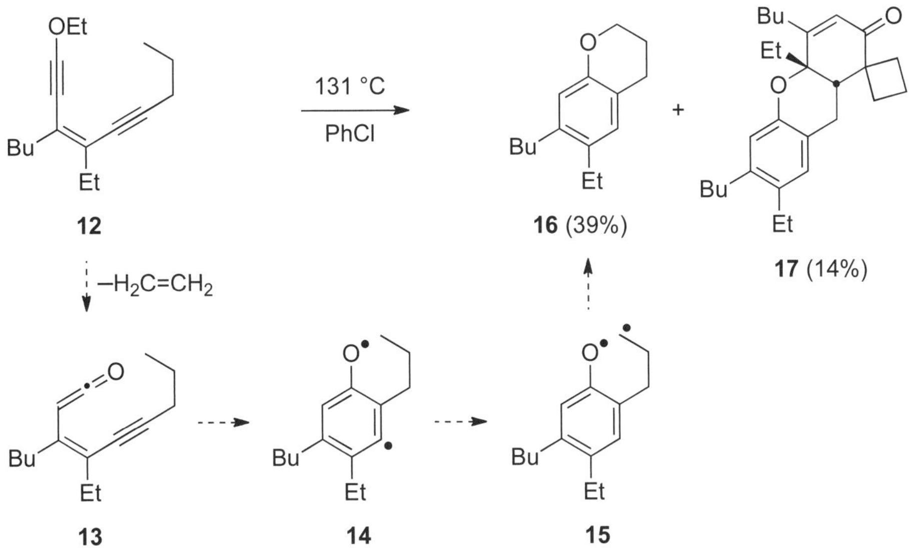

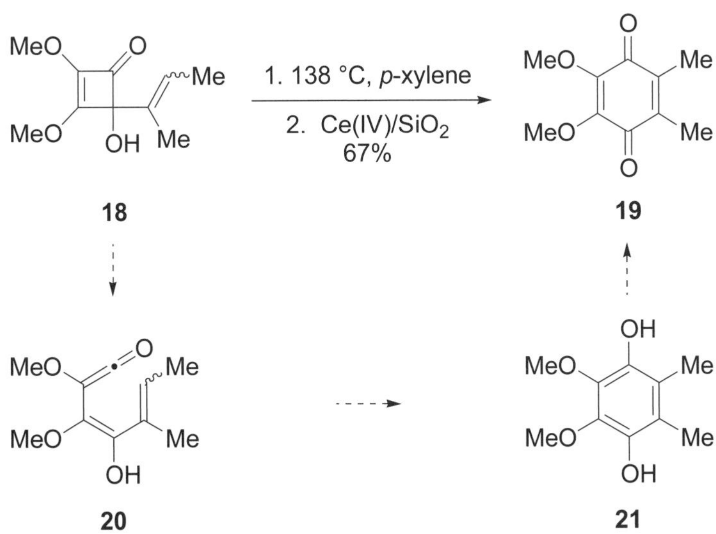

Moore et al. investigated the thermal rearrangements of differently substituted cyclobutenones.[8,9] Reactions of 4-alkenyl-4-hydroxycyclobutenones such as **18**, in which the triple bond is replaced by a double bound, are complementary to the ring expansions of 4-alkynyl-4-hydroxycyclobutenones and provide a route to the differently substituted benzoquinones, such as aurrantiogliocladin **19**. The reaction proceeds via enyne–ketene **20**. Since cyclization produces a derivative of hydroquinone **21**, an additional oxidation step is required that is accomplished with the use of cerium ammonium nitrate on silica.[9] This ring expansion process is independent of the stereochemistry of the 4-alkenyl moiety.[9] 4-Aryl-4-hydroxycyclobute-nones react also in a similar way which is exemplified in the Section 4.12.5.

The thermal rearrangements of 4-substituted-3-methylenecyclobutenes (analogues of 4-alkynylcyclobutanones), leading to phenols, benzylidenecyclohexanones, or acyclic dienones,[10] are discussed in the Chapter 4.13.

Organometallic reagents were used for the synthesis of bicyclic aromatic compounds via Moore-type cyclization. Rahm and Wulff described the new synthesis of 5-hydroxyindolines with the use of a chromium carbene complex bearing alkynyl substituent **22**.[11] The amino-tethered bis-alkynyl carbene complex **22** was transformed into indoline **23** by thermolysis in the presence of a hydrogen source. The low yield of product **23** was improved when the reaction was carried out in the presence of the electrophile, added to protect the phenol function. This process involves the insertion of one carbon monoxide group from the chromium complex into the skeleton of an eneyne compound **24**. The resulting enyne–ketene **25** undergoes a cycloaromatization reaction to afford the 1,4-diradical intermediate **26**. Subsequent demetalation yields product **23**.[11]

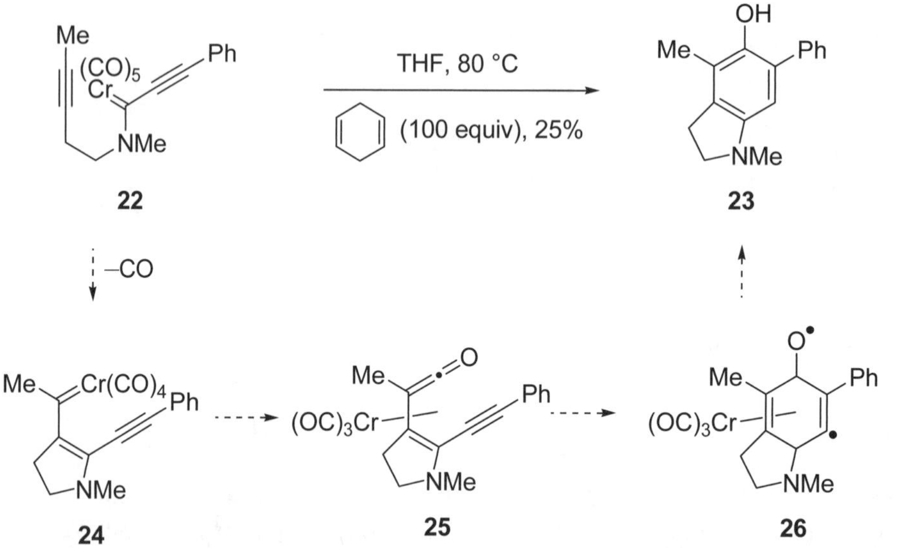

Another example of the application of the chromium carbene complex for the synthesis of benzannulated compounds was described by Herndon and Wang.[12] The coupling of substituted carbene chromium complex **27** with conjugated enediyne **28** results in the formation of intermediate enyne–ketene **29**, which undergoes the Moore cyclization to produce the intermediate chromium-complexed diradical **30**. The decomplexation converts **30** into product **31**, which was identified from the postreaction mixture. In addition, these studies suggest that the chromium complexed diradicals may have alternative reactivity patterns as compared to analogous metal-free diradicals.

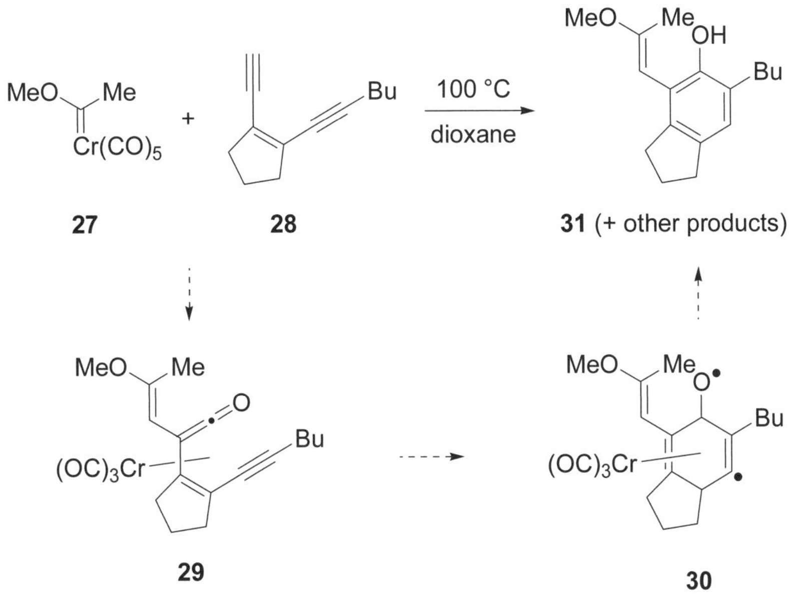

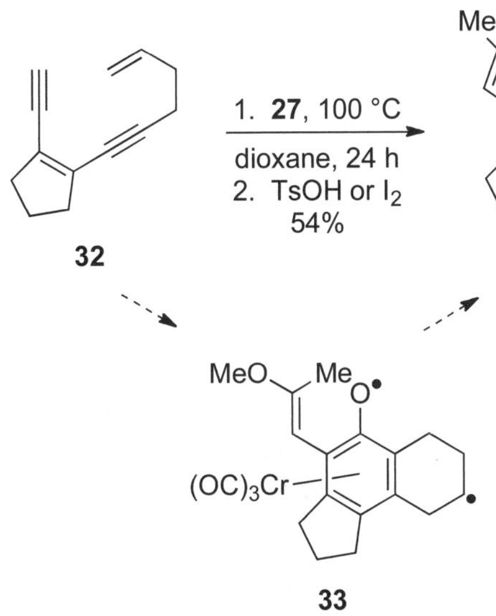

Herndon et al. have investigated the reactions of Fischer carbene chromium complexes with conjugated enediynes that feature a pendant alkene group such as **32**.[13] The experimental results confirm that arene diradicals obtained via Moore cyclization undergo subsequent intramolecular radical cyclization reactions to neighboring alkenes. The latter reaction proceeds predominantly to diradical **33** through the 6-*endo* cyclization mode ultimately giving rise to the highly substituted multicyclic benzofuran derivative **34**.

4.12.5 Synthetic Utility

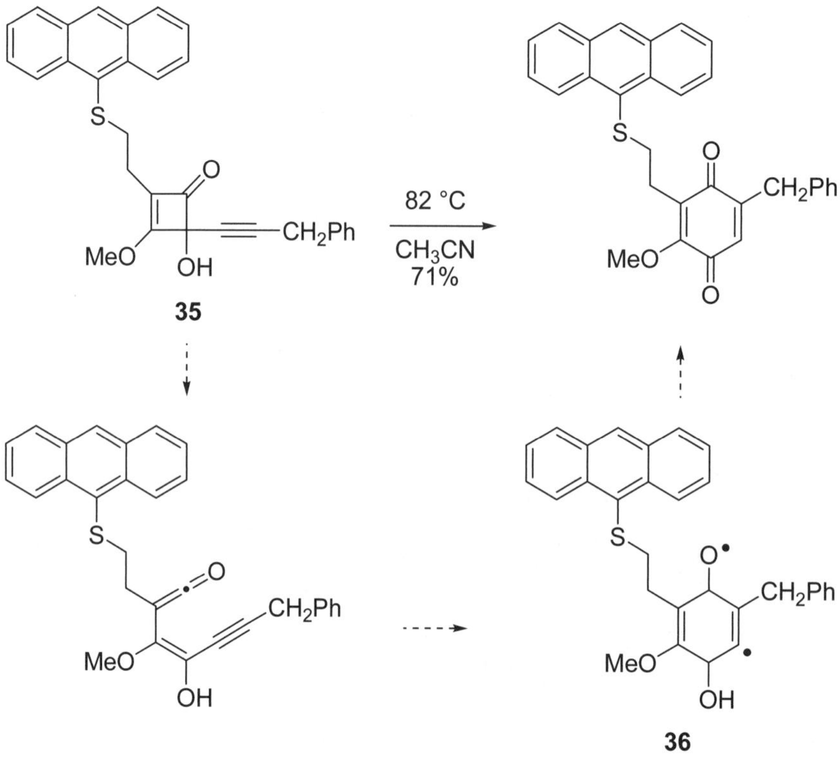

Since 4-alkynylcyclobutanones, precursors of enyne–ketene, are relatively readily available,[1,14,15] subsequent transformation of enyne–ketenes to aromatic diradicals provide interesting synthetic opportunities.

The synthetic utility of the Moore reaction can be illustrated by the efficient synthesis of coenzyme Q_0 (84%, not illustrated) or aurantioglio-cladin **19** (described above, Chapter 4.12.4).[9]

The important application of 4-alkynyl-3-methoxy-4-hydroxycyclo-butenones includes their ability to cleave supercoiled DNA by a mechanism that involves contribution from a diradical intermediate. The 4-alkynyl-4-hydroxycyclobutenones, which bear alkyl group at position 2 can effectively damage DNA, as opposed to the corresponding 2-alkoxy analogs (the latter could rearrange to their epoxides by a facile intramolecular pathway). For example, the 2-[2-(9-antracenylthio)ethyl] derivative **35** and its 2-butyl analog effectively damaged DNA at 49 °C, presumably via diradical intermediate **36**.[16]

Moore's group has also exploited his methodology for the synthesis of a variety of *N*-heterocyclic quinones and hydroquinones such as piperidinoquinones **37**, dihydrophenanthridinediols **38**, benzophenanthridine **39**, and assoanine **40**, a member of a series of biologically active pyrrolophenanthridine alkaloids.[17]

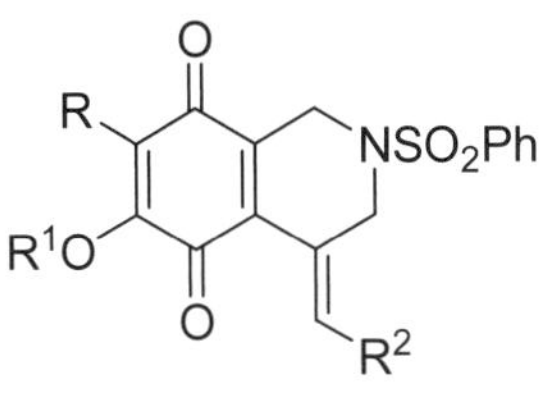

37

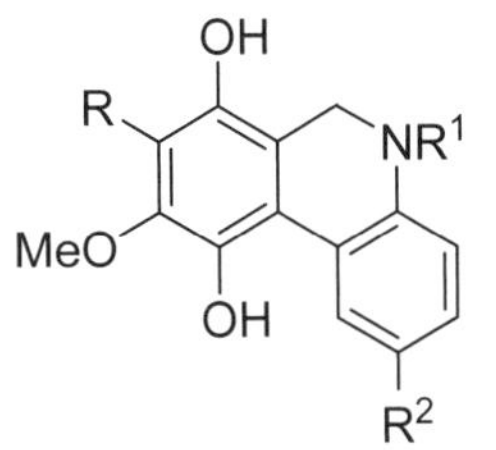

38

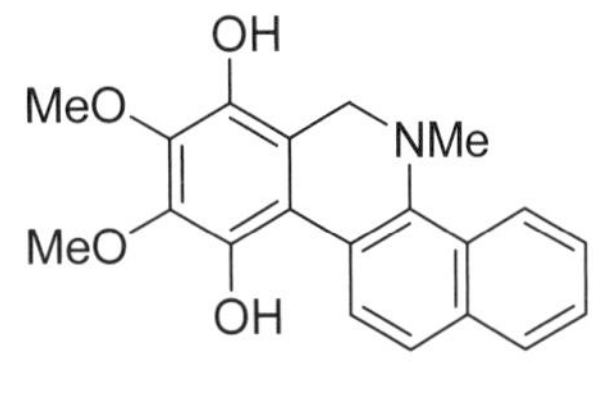

39, benzophenanthridine

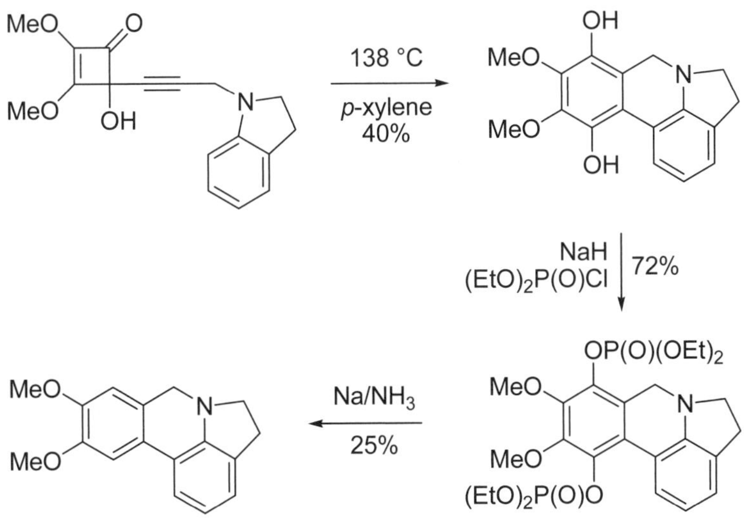

40, assoanine

Thermolysis of 4-hydroxycyclobutenone **41** offers an opportunity for the enantiospecific synthesis of pyranoquinone **42**.[18] The reaction proceeds via rearrangement of initially formed radical **43** to diradical intermediate **44** that cyclizes in a stereopecific fashion.

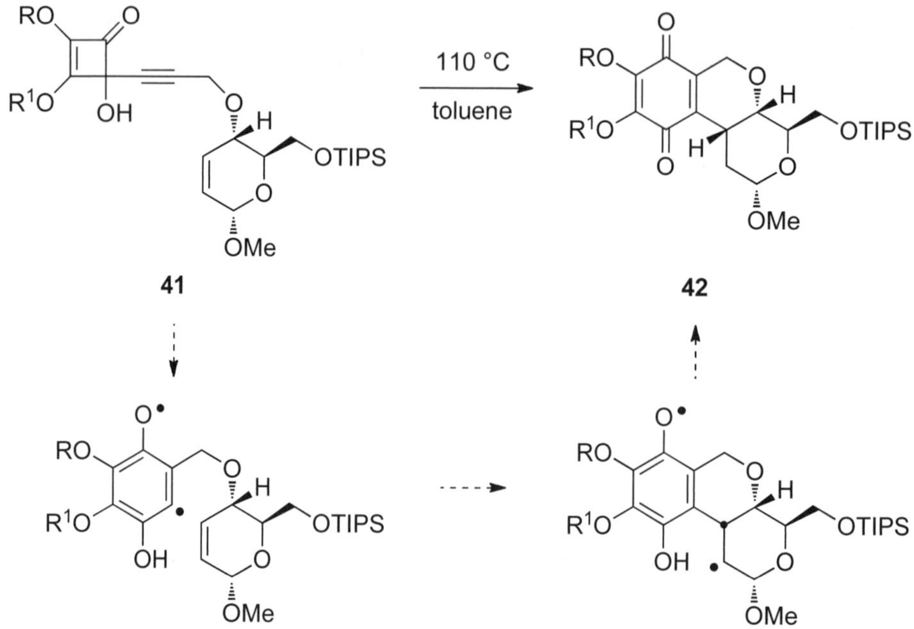

41 42 43 44

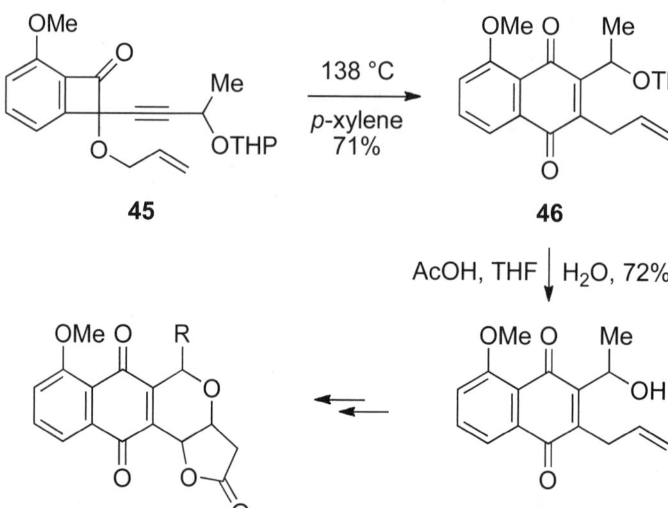

45 46

47 (R = Me)
nanaomycin D

The Moore reaction is also effective for the synthesis of annulated derivatives of quinines. Moore et al. have established that 4-alkynyl cyclobutenones are precursors in the efficient syntheses of some members of the family of naturally occurring isochroman-1,4-naphthoquinones, which show biological activity as antibiotics and antymycotics.[19] Benzocyclobutenedione has been used for the preparation of the corresponding 4-alkynyl benzocyclobutenones such as **45**, which undergoes rearrangement to naphthoquinone **46** via diradical cyclization under thermolytic conditions. Naphthoquinones such as **46** have been employed as key synthetic precursors to biologically important quinones: nanaomycin D **47** and deoxyfrenolicin (analog of **47** in which R = Pr).[19]

Since the C2–C6 cyclization is observed for the phenyl-substituted alkynylcyclobutenones, cyclization of 4-phenylethenyl derivatives provides an effective procedure. The cyclobutenone **48** was prepared from the 4-phenyethynyl derivative by the Lindlar reduction and converted to benzoquinone **49** in a two-step procedure that includes oxidation.[9]

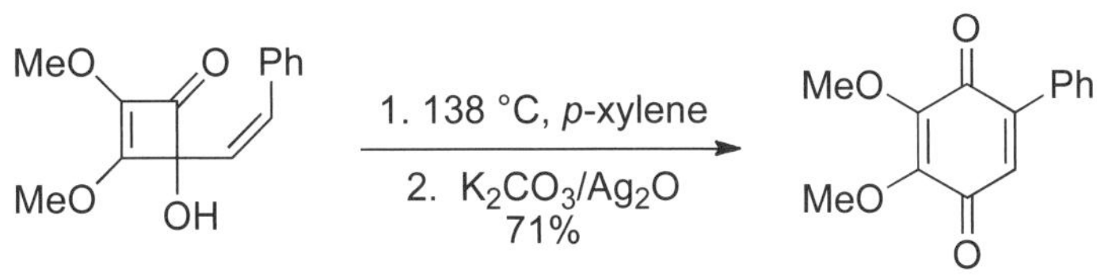

48 **49**

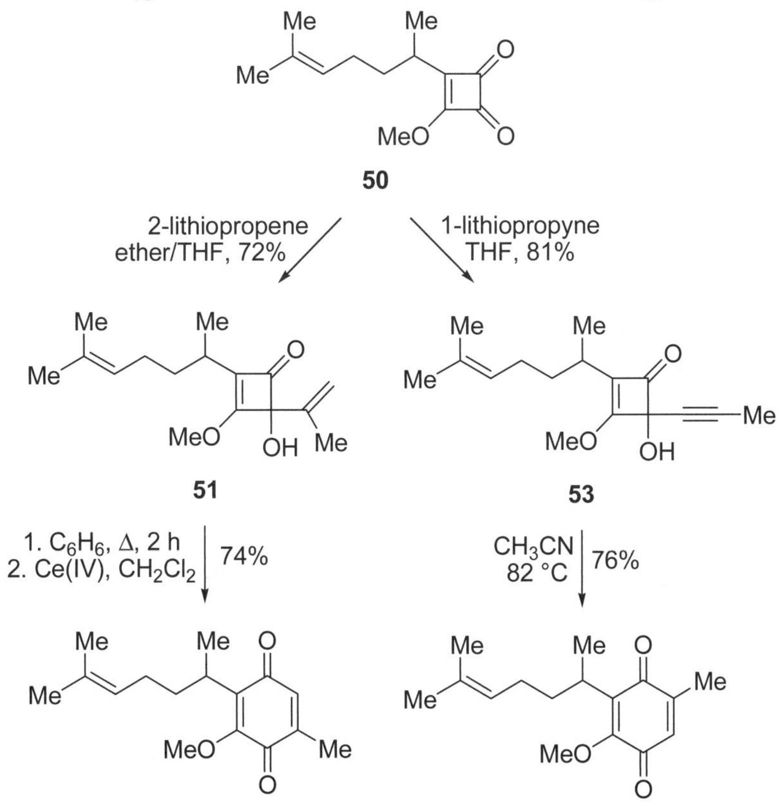

52, *O*-methylperezone **54**, *O*-methylisoperezone

The reaction of cyclobutenediones such as **50** with the appropriate alkenyllithium agent facilitates an alternative route to alkenyl butenones such as **51**.[9] Further synthetic utility of the Moore reaction can be illustrated by the efficient parallel synthesis of (±)-*O*-methylperezone **52** and, proceeding from the alkynyl derivative **53**, its regioisomer (±)-*O*-methylisoperezone **54**.[8,9]

4-Aryl-4-hydroxycyclobutenones react similarly to 4-alkenyl-4-hydroxycyclobutenones and provide a route to the differently substituted benzoquinones. Regiospecific synthesis of a series of hydroxyquinones and annulated derivatives was realized starting from di-*t*-butyl squarate (not illustrated) via 3-*t*-butoxycyclobutane-1,2-dione **55**. Reaction of **55** with phenyllithium leads to aryl-substituted cyclobutanedione **56** that rearranges at elevated temperature to protected quinone **57**. Deprotection of **57** yields natural product-lawsone **58**, a component of henna dyes.[20]

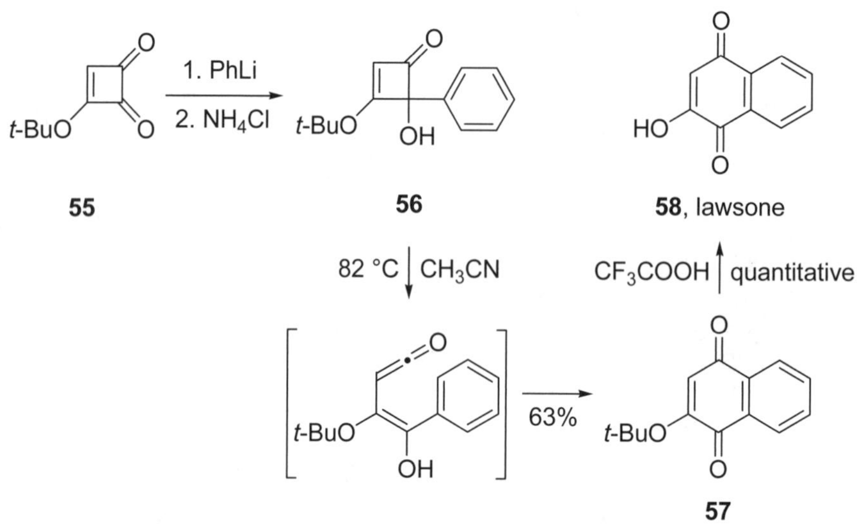

4.12.6 *Experimental*

Moore thermal cyclization of 2,3-dimethoxy-4-(3-trimethylsilyl-1-propynyl)-4-trimethylsiloxycyclobut-2-en-1-one (59): 2,3-dimethoxy-5-trimethylsilyl-6-((trimethylsilyl)-methyl)-1,4-benzoqui-none (60)[21]

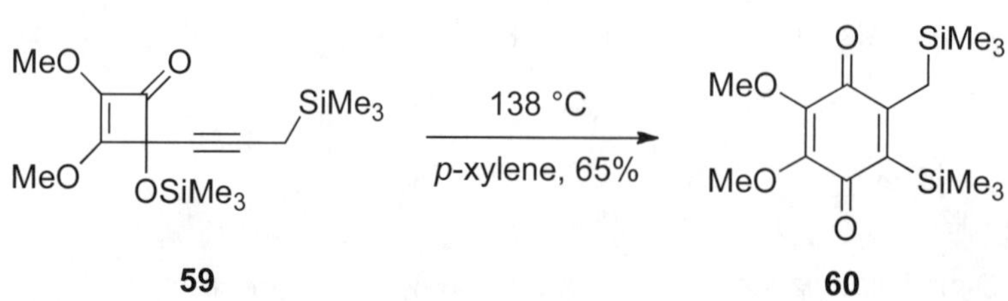

A solution of cyclobutenone **59** (0.297 g, 0.910 mmol) in anhydrous *p*-xylene (60 mL) was heated at reflux for 1 h. Concentration in vacuo followed by flash column chromatography on silica gel (hexanes/ethyl acetate, 10 : 1) gave **60** (0.190 g, 0.582 mmol, 65%) as an orange oil.

Moore thermal cyclization of *N*-benzhydryl-*N*-[3-(1-hydroxy-2,3-diisopropoxy-4-oxocyclobut-2-enyl)prop-2-ynyl]-3-phenylacrylamide (61): 2-benzhydryl-4-benzyl-6,7-diisopropoxy-1,4-dihydro-2*H*-isoquinoline-3,5,8-trione (62).[22]

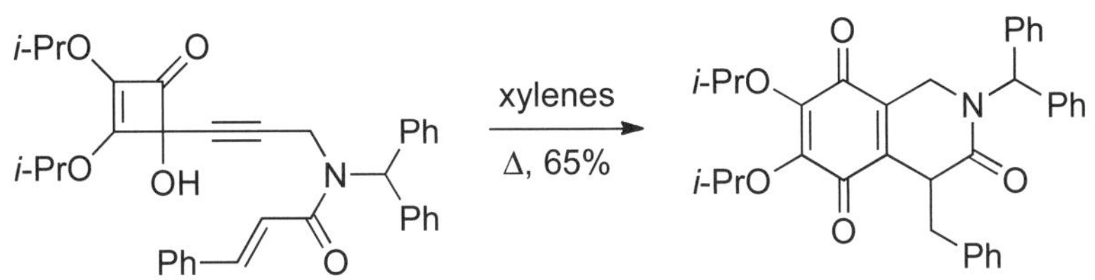

61 **62**

A solution of the cyclobutenone **61** (0.3100 g, 0.5640 mmol) in freshly degassed xylenes (10 mL) was added dropwise over a 20-min period to refluxing, degassed xylenes (40 mL). After the addition was complete, the reaction mixture was cooled to room temperature and concentrated in vacuo. The resulting red oil was purified by column chromatography on SiO_2 (hexanes/ethyl acetate/dichloromethane 6:1:3) to afford isoquinolinone **62** (0.2077 g, 0.3779 mmol, 67%) as a red oil.

4.12.7 References

1. Karlsson, J. O.; Nguyen, N. V.; Foland, L. D.; Moore, H. W. *J. Am. Chem. Soc.* **1985**, *107*, 3392–3393.
2. [R] Wang, K. K. *Chem. Rev.* **1996**, *96*, 207–222.
3. Foland, L. D.; Karlsson, J. O.; Perri, S. T.; Schwabe, R.; Xu, S. L.; Patil, S.; Moore, H. W. *J. Am. Chem. Soc.* **1989**, *111*, 975–989.
4. Zora, M.; Kokturk, M.; Eralp, T. *Tetrahedron* **2006**, *62,* 10344–10351.
5. Musch, P. W.; Remenyi, C.; Helten, H.; Engels, B. *J. Am. Chem. Soc.* **2002**, *124,* 1823–1828.
6. Schreiner, P. R.; Bui, B. H. *Eur. J. Org. Chem.* **2006**, 1162–1165.
7. Tarli, A.; Wang, K. K. *J. Org. Chem.* **1997**, *62,* 8841–8847.
8. Perri, S. T.; Dyke, H. J.; Moore, H. W. *J. Org. Chem.* **1989**, *54,* 2032–2034.
9. Perri, S. T.; Moore, H. W. *J. Am. Chem. Soc.* **1990**, *112,* 1897–1905.
10. Ezcurra, J. E.; Pham, C.; Moore, H. W. *J. Org. Chem.* **1992**, *57,* 4787–4789.
11. Rahm, A.; Wulff, W. D. *J. Am. Chem. Soc.* **1996**, *118*, 1807–1808.
12. Herndon, J. W.; Wang, H. *J. Org. Chem.* **1998**, *63,* 4562–4563.
13. Zhang, Y.; Irshaidat, T.; Wang, H.; Waynant, K. V.; Wang, H.; Herndon, J. W. *J. Organomet. Chem.* **2008**, *693,* 3337–3345.
14. Nakatani, K.; Isoe, S.; Maekawa, S.; Saito, I. *Tetrahedron Lett.* **1994**, *35,* 605–608.

15. Chow, K.; Moore, H. W. *J. Org. Chem.* **1990**, *55,* 370–372.
16. Sullivan, R. W.; Coghlan, V. M.; Munk, S. A.; Reed, M. W.; Moore, H. W. *J. Org. Chem.* **1994**, *59*, 2276–2278.
17. Xiong, Y.; Moore, H. W. *J. Org. Chem.* **1996**, *61,* 9168–9177.
18. Xiong, Y.; Xia, H.; Moore, H. W. *J. Org. Chem.* **1995**, *60,* 6460–6467.
19. Foland, L. D.; Decker, O. H. W.; Moore, H. W. *J. Am. Chem. Soc.* **1989**, *111*, 989–995.
20. Heerding, J. M.; Moore, H. W. *J. Org. Chem.* **1991**, *56*, 4048–4050.
21. Ezcurra, J. E.; Karabelas, K.; Moore, H. W. *Tetrahedron* **2005**, *61,* 275–286.
22. Wipf, P.; Hopkins, C. R. *J. Org. Chem.* **1999**, *64*, 6881–6887.

4.13 Myers–Saito Cyclization

Ewa Krawczyk and Roman Dembinski

4.13.1 Description

The Myers–Saito cyclization is formulated as a cyclization of enyne–allenes **1** to diradicals **2**. It usually proceeds by thermal induction and leads to the formation of a benzene ring.

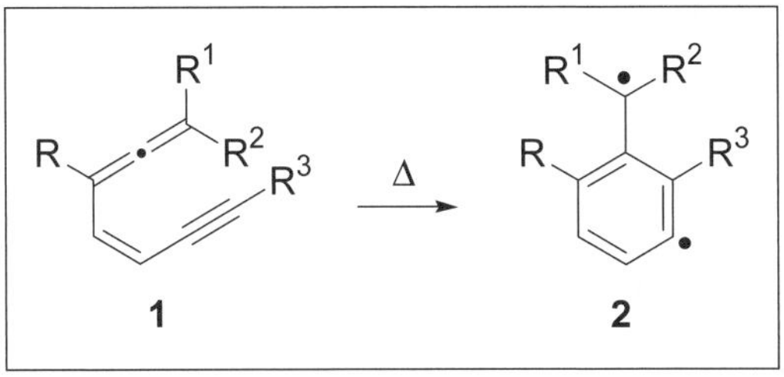

4.13.2 Historical Perspective

The Myers–Saito cyclization was first described independently in 1989 by Isao Saito[1] (Kyoto University, Japan) and Andrew G. Myers[2] (California Institute of Technology, Pasadena) whose findings were submitted for publication on June 7 and June 12, respectively. As a parallel transformation to the Moore cyclization (Chapter 4.12), in which an allenic fragment replaces the ketene moiety in the substrate, the Myers–Saito reaction provides a path to carbon diradicals.[1–5] In its pioneering version, the reaction involved the cyclization of (*Z*)-1,2,4-heptatrien-6-yne (enyne–allene) **3**,[2,4] or its phosphine oxide derivative **5**,[1,3] to substituted α,3-dehydrotoluene diradicals, and subsequently to toluene derivatives **4** and **6**. The reaction proceeds under thermal neutral conditions in 1,4-cyclohexadiene or other organic solvents such as methanol or carbon tetrachloride.

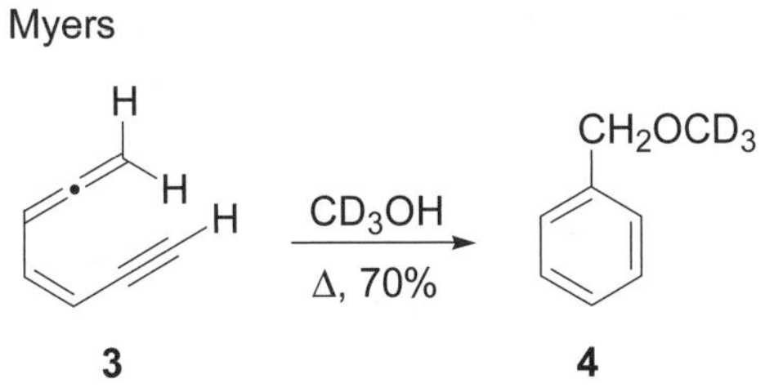

Saito

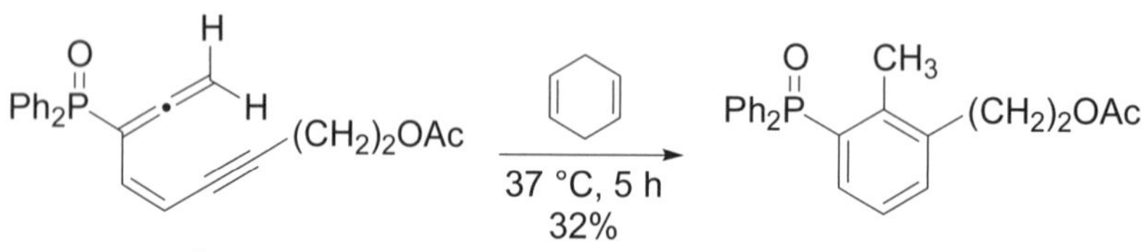

5 6

4.13.3 Mechanism

From a mechanistic point of view, in the initial step, the (*Z*)-1,2,4-heptatrien-6-yne, or compounds containing an equivalently unsaturated core, undergoes a mild, thermal electrocyclization reaction to form an α,3-alkylbenzenediyl, a diradical intermediate with substantial polar character. Dehydroaromatic intermediate **7**, when trapped by the solvent or compounds (e.g., 1,4-cyclohexadiene) present in the reaction medium, forms than aromatic products of type **8**, **9**, and **10**.[6] In methanol, a mixture of hydrogen atom abstraction and polar addition product are obtained.[2]

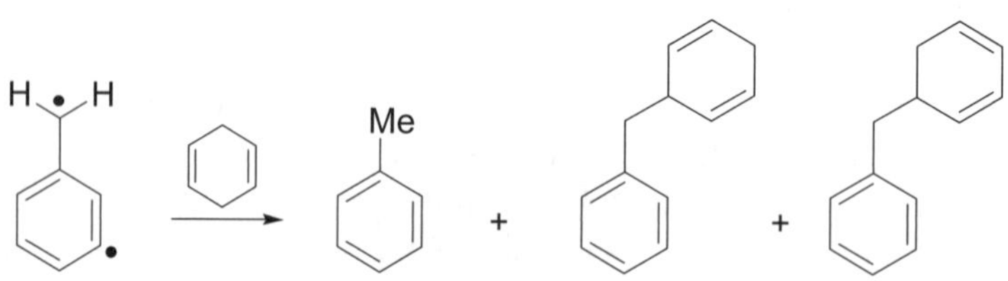

7 8 9 (55%, 1:1) 10

Extensive mechanistic studies of this cyclization reaction were carried out by Myers et al.[6] and extended with theoretical work by Squire's et al.[7,8] It is known that, in contrast to the Bergman cyclization of the ene–diyne (Chapter 4.2), this transformation proceeds as an exothermic process determined by the increased stability of a benzyl radical versus a phenyl radical. The barrier for cyclization from substrate to a diradical product is low and can further be reduced by an appropriate substitution at the allenic terminus of the substrate.[6] The dichotomous (polar and free radical) reactivity is observed on pyrolysis in the presence of polar reactants. Both radical and polar products arise from a common intermediate, which is described as a "polar diradical," a linear combination of limiting structure **7** and zwitterion **11**.[9] According to Squires, "polar diradical" singlet species are involved.[8] Based on computational studies supported by experimental product distribution studies, it has been proposed that both the diradical **7** and zwitterionic **11** forms are produced from the common cyclization transition state by a transition to the zwitterion, which is the excited state of the diradical.[10,11]

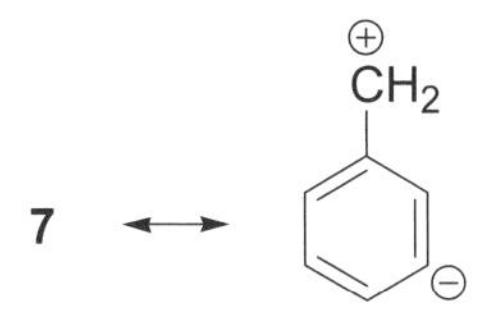

11

The investigation of various parameters of the cyclization of enyne–allene **12**, using the density functional theory (DFT), has been undertaken.[12,13] This includes the thermodynamics for both C2–C7 (leading to **13**) and C2–C6 (leading to **14** and called Schmittel reaction) reaction pathways. Theoretical calculations address the regioselectivity of diradical cyclization in enyne–allene with a different substitution. This study rationalizes the switch between the two radical cyclizations (C2–C7 vs. C2–C6) on the basis of mainly steric (**12**, R^2 = *t*-Bu) or electronic effects (R^2 = Ph) of substituents at the alkyne terminus.[13]

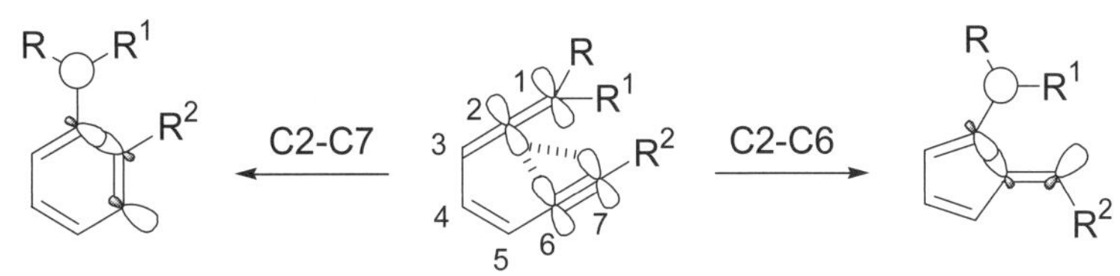

13 **12** **14**

Lipton et al. noticed that the switching of cyclization pathways observed by Schmittel takes place due to a combination of effects. The calculations of the energies of diradicals formed by Myers–Saito and Schmittel cyclization indicate that benzannulation of the eneyne–allene as in **15** plays a significant role in promoting C2–C6 cyclization. The energy of diradicals for the benzannulated system is 10.5 kcal/mol less as compared to those for the parent monocyclic diradicals (**16** vs. **13**, and **17** vs. **14**).[12]

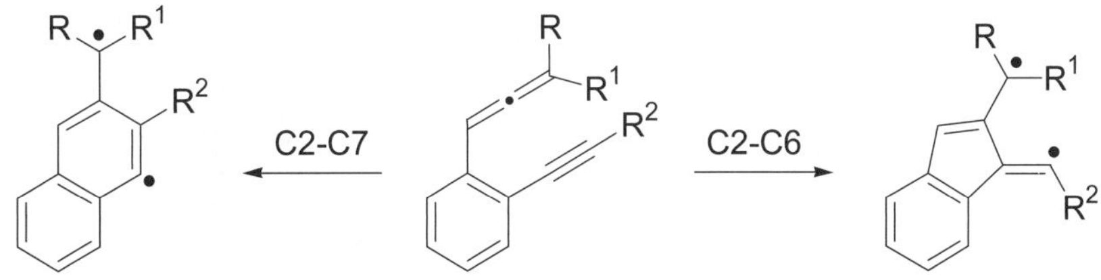

16 **15** **17**

Studies of the diradical (stabilized by a phenyl group) or zwitterionic intermediate have been undertaken for the Schmittel reaction of the enyne–allene with larger substituents ($SiMe_3$, *t*-Bu). It was reported that C2–C6 cyclization occurs via the diradical intermediate and aryl or bulky groups at the alkyne terminus trigger a general thermal reaction pathway for the enyne–allene.[14]

Substitution at the terminus of the allenic moiety, as well as at the terminus of the acetylenic moiety, influences the rate of cyclization. Introduction of one or two methyl groups (**12**, R,R^1 = Me) at the allenic moiety accelerates the reaction rate,[2,5] whereas substitution at the alkyne terminus usually reduces the reaction rate.[15]

Thermal reactions of enyne–allenes initiated by employing different modes such as base,[16] acid,[17] and light[18,19] have also been investigated.

4.13.4 Variations and Improvements

Competition between the Myers–Saito and Schmittel cyclization is also observed for phosphorus-substituted allenes. Terminal alkynes **18** give naphthalene derivatives **19**.

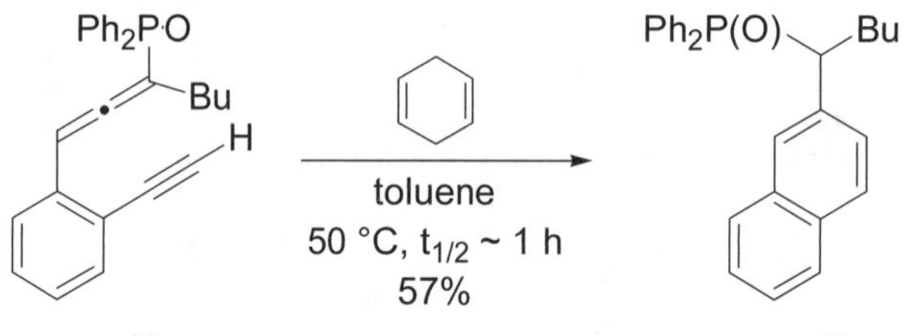

18 **19**

When an aromatic substituent is bound to the alkyne terminus, as illustrated for the terminal alkyne **20**, a competitive, thermally initiated C2–C6 cyclization of an enyne–allene takes place. This process gives indene derivatives such as **21**.[20,21]

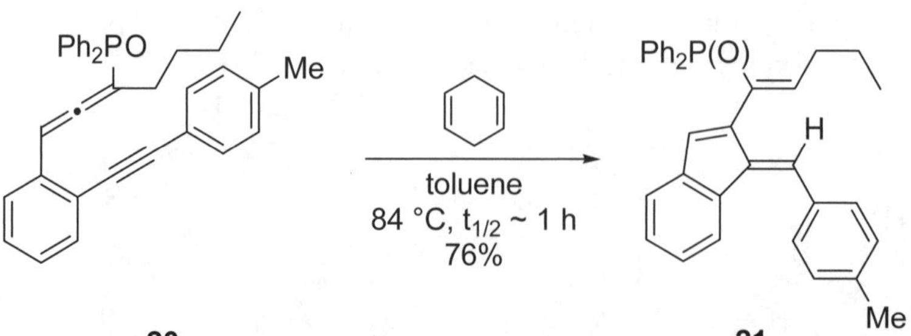

20 **21**

The photochemical cyclization of the Myers–Saito and Schmittel cyclization of enyne was described much later than the Bergman photochemical variant.[18] The theoretical studies indicate that this process is initiated by triplet sensitization.[19] The investigation was carried out using a series of enyne–allenes containing an internal triplet sensitizer unit attached to an allene terminus. The irradiation of enyne–allene containing a biphenyl group **22** at 300 nm in the presence of 1,4-cyclohexadiene resulted in the formation of photocyclization products: Myers–Saito (C2–C7) tetrahydronaphthalene derivative **23** and Schmittel (C2–C6) multicyclic product **24**.

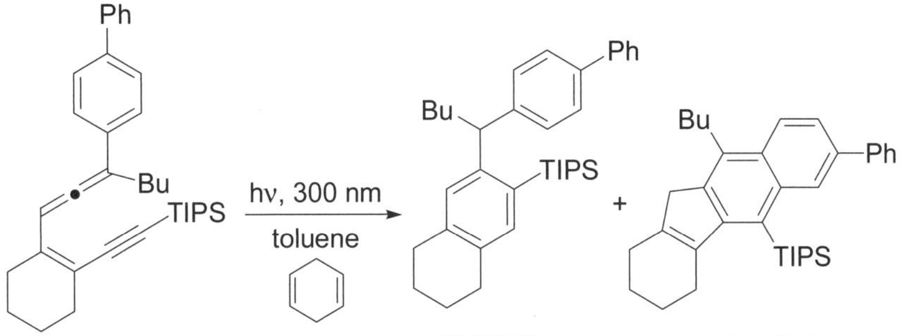

22 **23** (24%) **24** (12%)

For the case of oxyanion, two factors were found to play a role. The size and nature of the ring in which enyne–allene is embedded as well as the steric bulk of the substituents of the allene and alkyne affects the competition between the two cyclizations. The Myers–Saito product is not observed for alkynes with the trimethylsilyl substituent even for cyclohexane–annulated compounds. However, when the alkynyl substituent is changed to phenyl, the resulting relief of steric strain in the C2–C7 transition state permits the cyclization of **25** to occur at low temperature and yields the styrene derivative **26**.[22] The cyclopentane–annulated compounds give the Myers–Saito product, or fail to react when a bulky silyl substituent is present at the alkynyl terminus. In contrary, the benzene-annulated compounds undergo rapid the Schmittel cyclization dominantly or exclusively. The cyclizations of oxyanion-substituted enyne–allenes studied in the cited article occured at far lower temperatures than the analogous cyclizations of neutral enyne–allenes.[22] The presence of the oxyanion presumably stabilizes, by resonance, the diradical in the transition state.[22–24]

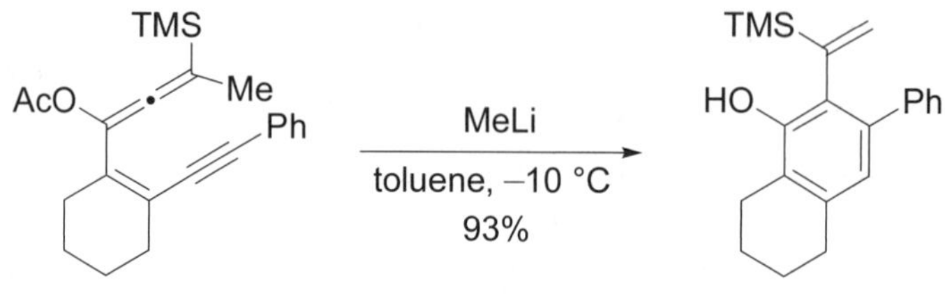

25 26

However, for the cyclization of the oxyanion substituted enyne–allene, substitution causes both the Myers–Saito and Schmittel cyclizations to switch their product formation preferences from diradicals to polar intermediates, as established by DFT.[22,24] The stabilization of the oxyanion-derived Schmittel products is greater than those of the Myers–Saito reaction.[24]

Cyclic 10- and 11-membered ring enediynes, functionalized by a carbonyl group in the β position with respect to the acetylenic terminus, undergo facile cycloaromatization at ambient temperature. Kinetic data and deuterium labeling experiments indicate that cyclization proceeds via a rate-determining tautomerization into a more reactive enyne–allene such as **27**. This tautomer undergoes Myers–Saito cyclization via diradical **28** to give tricyclic product **29**. The process of cycloaromatization exhibits the strong effect of general base catalysis.[16]

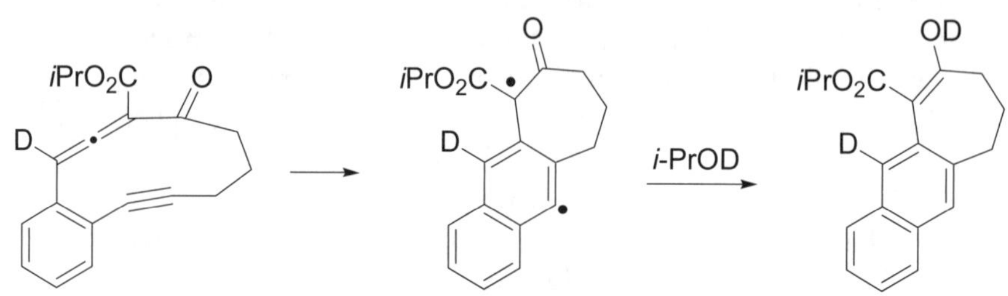

27 28 29

It is interesting that when the enyne–allene moiety is incorporated into a bicyclic system, such as [7.3.2]-eneyne–allene **30**, cyclization does not proceed. In contrast, the Bergmann cycloaromatization has been observed for a related substrate, [7.4.1]-enediyne (not illustrated).[25]

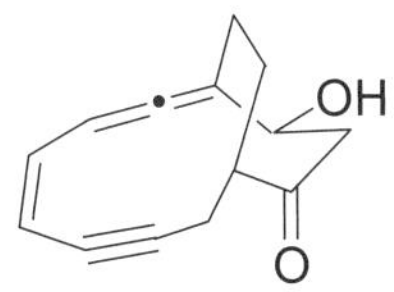

30

Nonconjugated diynes of the type **31** may rearrange to generate enyne–allene **32**, which undergoes Myers–Saito cyclization. When the sulphur-substituted 10-membered ring of 1,6-diyn-3-ene **31** is treated with DBU under aerobic conditions, the compounds **33**–**36** are produced from the diradical intermediate **37**. Anaerobic conditions yield compounds **33** (10%), **34** (19%) and **35** (5%). It has been suggested that the rather stable cyclic diyn–ene **31** under basic conditions undergoes the propargylic rearrangement to **32** to form in turn the intermediate **37**.[26] Selenium(IV) oxide (SeO_2) can also facilitate the cycloaromatization of **31** combined with an oxidation. The reaction proceeds via an enyne–allene seleninic acid radical intermediate which subsequently indergoes cyclization to produce isothiochroman-4-one **36** along with an analog in which R = H.[27]

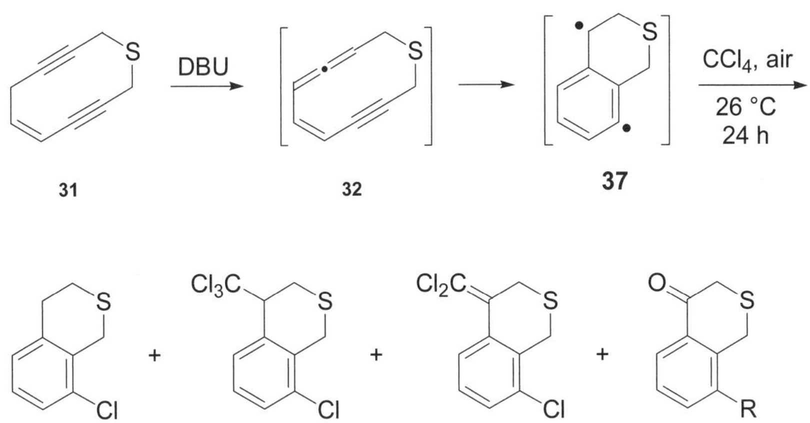

31 32 37

33 (8%) **34** (8%) **35** (4%) **36** (R = Cl, 12%)

Cumulene structures also undergo the Myers–Saito reaction. Cyclization of acyclic enyne[3]cumulenes, on the activation of *Z*-configured dienediyne **38** via acid solvolysis, has been described by Brückner et al.[17] It has been found that **38** dissolved in *t*-BuSH/dichloro-methane and treated with a catalytic amount of triflic acid forms the monocyclic cumulene **39**. Storage of the mixture for 4 days at room temperature gave the corresponding styrene derivatives **40** and **41**; these products form as a result of cycloaromatization via path A (benzoid radical). Independently, after purification of **39**, compound **42**, formed by path B cyclization via a quinoid σ,σ-diradical, was detected.

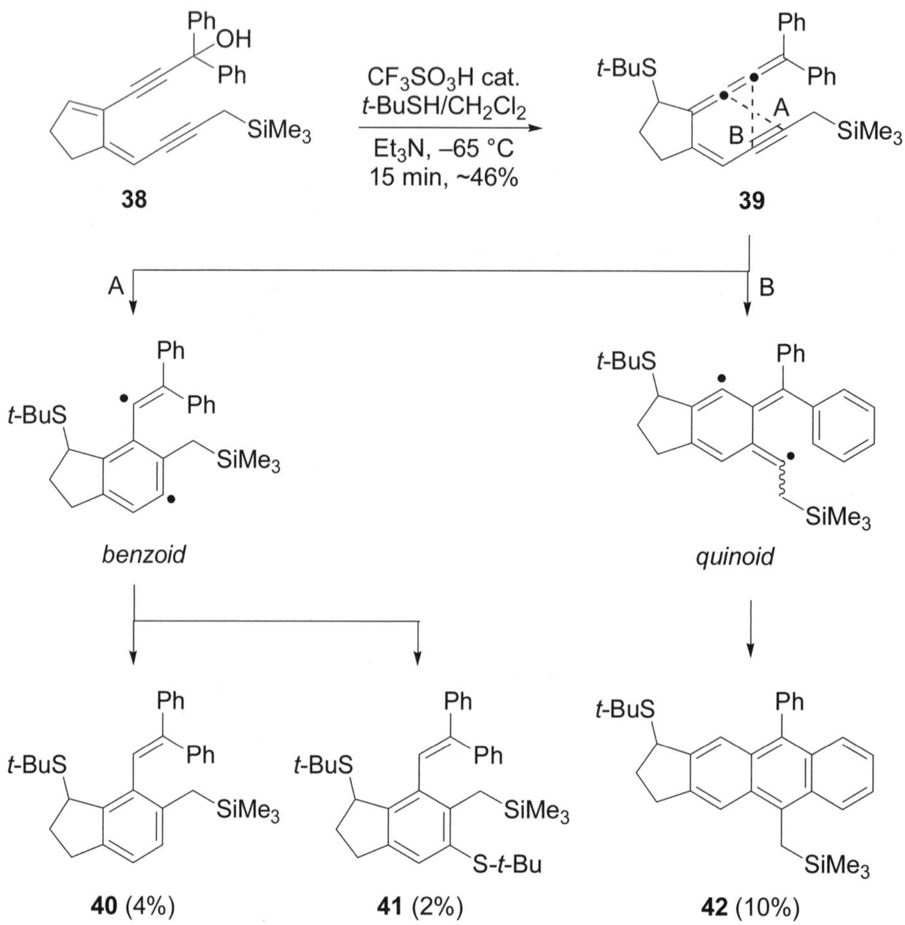

The cyclic cumulene structures were investigated. The 1,6-didehydro[10]annulene **43**, was prepared and characterized spectroscopically at –90 °C by Myers and Finney.[28] This compound undergoes cyclization rapidly at –60 °C to form a localized σ,σ-diradical intermediate **44**, which leads to naphthalene. The kinetics for the cyclization reaction, in CD_2Cl_2 in the presence of 1,4-cyclohexadiene, is found to be first order. Isotope incorporation takes place from a deuterated solvent and yields 1,5-dideuterionaphthalene **45**.

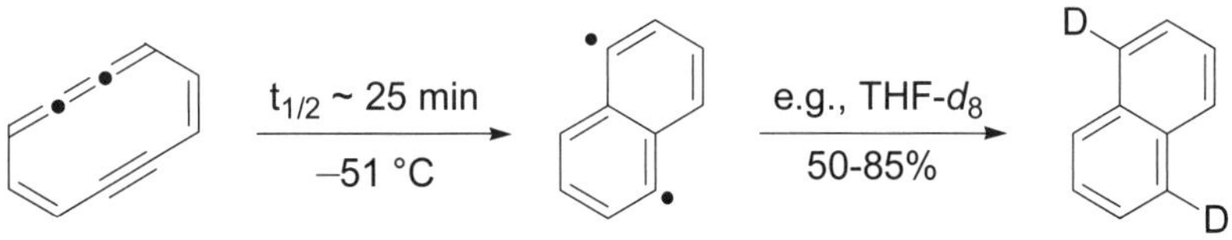

43 **44** **45**

Myers et al. treated the epoxy dienediyne **46** with methyl thioglycolate and triethylamine. The isomeric naphthalene derivatives **47** and **48** were produced as a result of the transformation involving diradical precursors formed from the 1,6-didehydro[10]annulene intermediate **49** via route A and B.[29] Compounds **47** and **48**, with indicated levels of deuterium incorporation, were also formed when the experiment was conducted in a deuterated solvent.

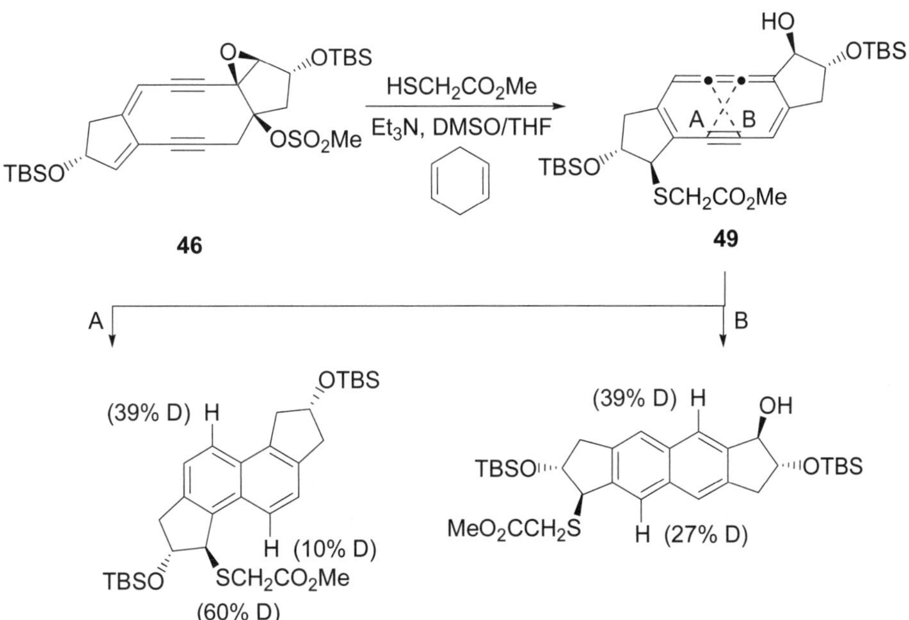

46 **49**

47 (67%) **48** (20%)

An aza-variant of the cycloaromatization of propargyl azaeneynes, such as **50**, via azaenyne–allenes **51**, has been reported by Kerwin et al.[30] The aza-Myers–Saito cyclization provides α,5-didehydro-3-picoline diradical **52**, which affords either polar or radical-based trapping products **53** and **54**, depending on the reaction solvent. The facility of the aza-Myers–Saito cyclization relative to the parent Myers–Saito cyclization was predicted based on DFT calculations; these results also indicate that the corresponding C2–C6 (aza-Schmittel) cyclization, although disfavored in the case of **51**, is more competitive with the aza-Myers–Saito cyclization than the case with enyne–allenes.

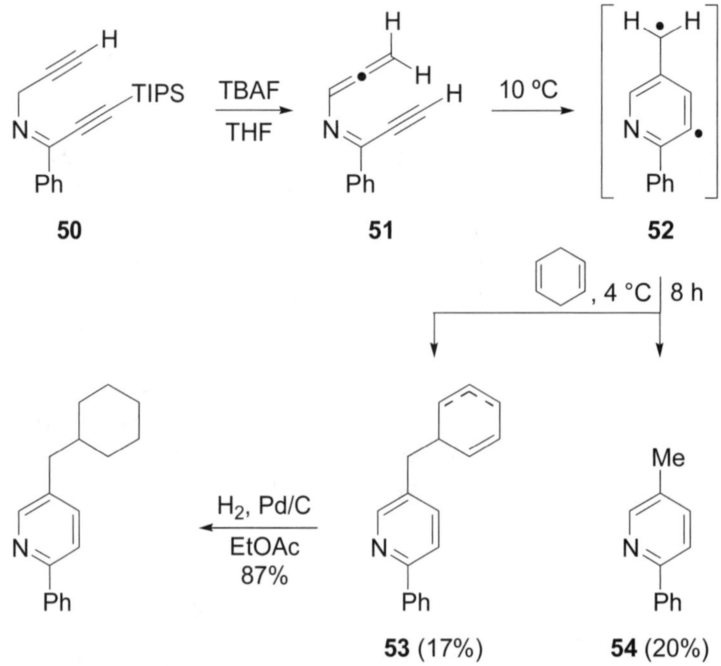

These studies were extended for the synthesis and DNA cleavage chemistry of pyridinium aza-enediynes (2-alkynyl-*N*-propargyl pyridinium salts).[31] The 2-alkynyl-*N*-propropargyl pyridinium triflate **55** cleaves DNA by hydrogen atom abstraction from the deoxyribose backbone, presumably through the intermediacy of diradicals formed by either aza-Myers–Saito or aza-Schmittel cyclization. Attempts to identify trapping products of these intermediates were unsuccessful.[31]

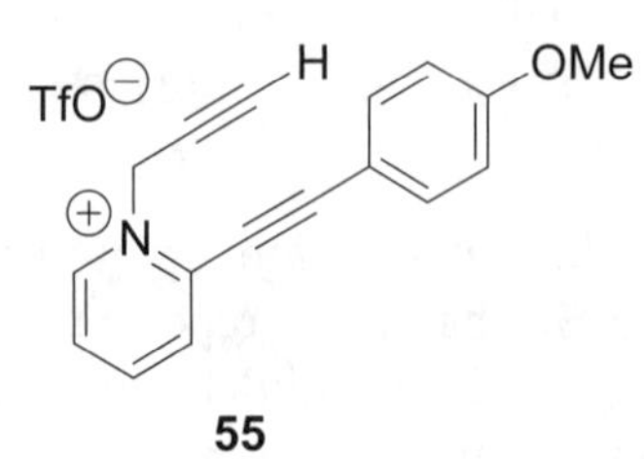

55

4.13.5 Synthetic Utility

The thermal rearrangements of 4-substituted-3-methylenecyclobutenes, which are analogs of 4-substituted-4-hydroxycyclobuten-3-ones (substrates for the Moore reaction, Section 4.12.4), were investigated. The ring expansions of 4-alkynyl-3-methylenecyclobutene **56** provide a route to the differently substituted phenols such as **57** and **58**.[32]

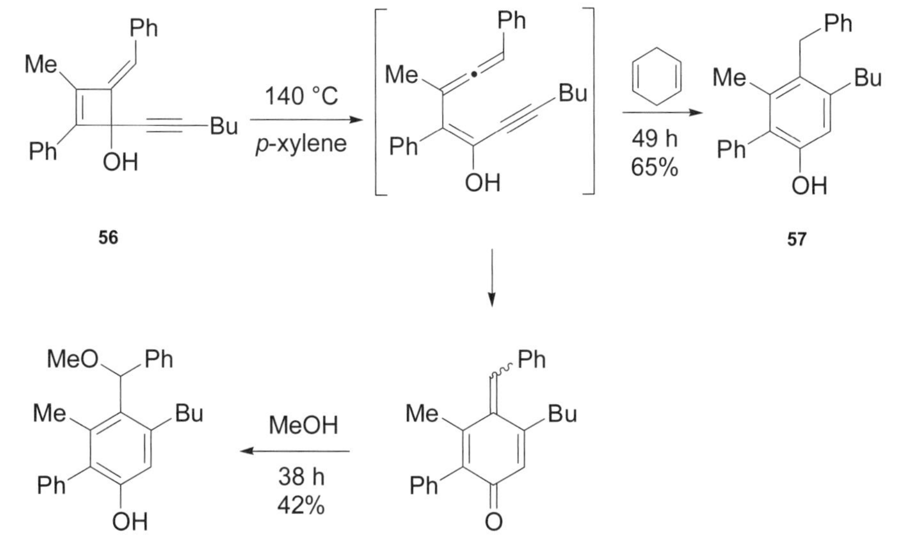

56 **57** **58**

When the analog in which a double bond replaces the triple bond (4-alkenyl-3-methylenecyclobutene **59**) is used, the benzylidenecyclohexanone **60** is obtained. These ring expansion processes represent potentially useful synthetic transformations.[32]

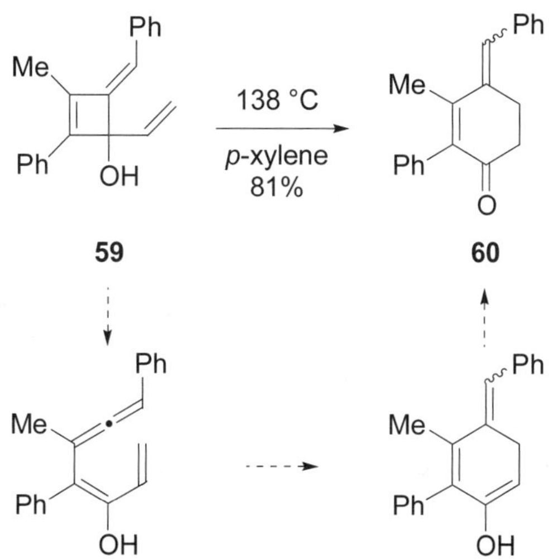

59 **60**

Using the 1,6-addition of lithium dimethylcuprate to dienediynes with an endocyclic double bond **61**, Krause and Hohmann have synthesized the corresponding acceptor-substituted enyne–allenes **62**. These allenes undergo Myers–Saito cyclization to diradicals **63**, which react by intramolecular hydrogen migration to yield the α,β-unsaturated ester **64** (R^1 = *t*-Bu, R^2 = H). It was found that a compound containing two methyl groups at C-2' gives, as the cyclization product, the decaline derivative **65** (R^1 = Pr, R^2 = Me) by intermolecular hydrogen abstraction from the solvent.[33]

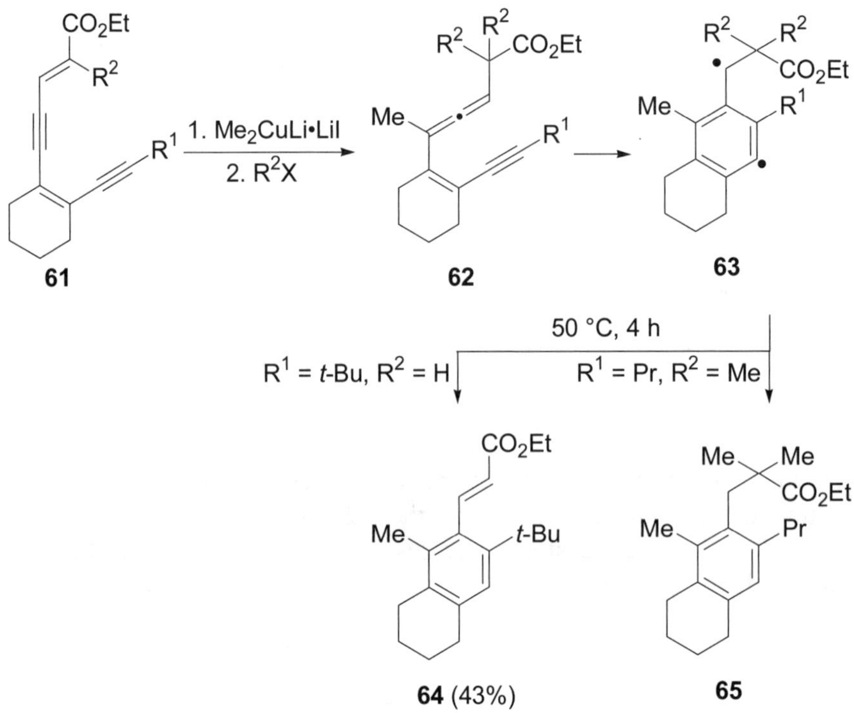

Wang's group investigated the application of the diradicals generated from enyne–allenes for the cascade radical cyclizations allowing for the synthesis of various tetracyclic ring systems.[5,15] Among them, the construction of the tetracyclic steroidal skeleton having an aromatic C-ring was realized.[34] Acyclic enyne–allene **66** bearing 3-butenyl and 7-hexenyl substituents served as a substrate. Its thermal cyclization probably proceeds via intramolecular trapping of a diradical by the carbon–carbon double bond and subsequent 1,5-hydrogen shift to form *o*-quinodimethane **67**. This transformation yields the tetracyclic system having the *trans* ring junction **68**.

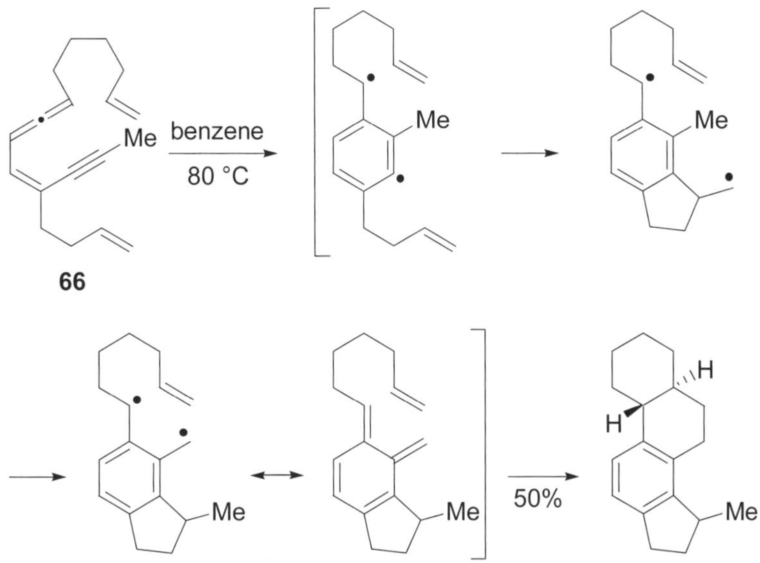

66 67 68

The transition metal catalysis of the Myers–Saito cyclization of enyne–allenes has been investigated. Molybdenum-mediated carbonylation of 1-ethynyl-2-allenylbenzenes occasionally gave minor by products derived from the Myers–Saito rearrangement (not illustrated).[35] Toste's group elaborated an effective synthetic route to aromatic ketones via a transition metal-catalyzed tandem [3,3]-sigmatropic rearrangement/Myers–Saito cyclization of propargyl esters such as **69**.[36] In this method the required enyne–allenes **70** were prepared in situ via a metal-catalyzed sigmatropic rearrangement; following the diradical cyclization at room temperature to afford differently substituted aromatic ketones, such as **71**, in high yield.

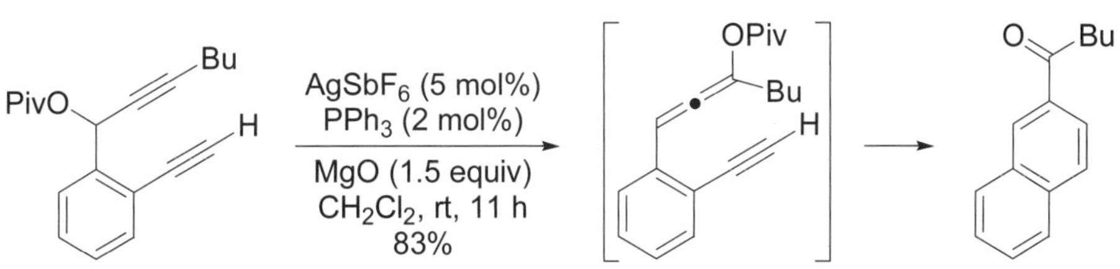

69 70 71

In many cases the silver(I)-catalyzed naphthyl ketone synthesis proceeds as well or better than the analogous gold(I)-catalyzed reaction. However, attempts at silver(I)-catalyzed rearrangement of pyrrole **72** failed to produce the desired aromatic ketone. In this case, the analogous tri-*t*-

butylphosphine-gold(I)-catalyzed reactions delivered indole **73**.[36] A mechanism, in which the metal catalyzes the rearrangement and cyclization process through alkyne activation, was proposed. Both silver and gold procedures are tolerant to air and moisture.

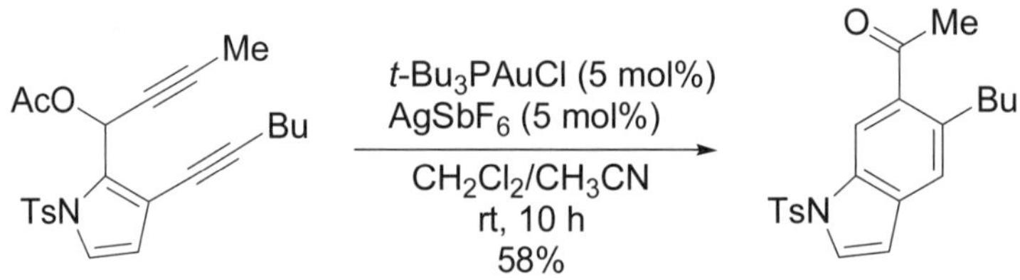

72 **73**

The presence of a cyclic enediyne fragment in antitumor antibiotics initiated a wide interest in application of this reaction for DNA cleaving studies. Neocarzinostatin chromophore, a nonprotein component of the antitumor antibiotic, consists of a diene–diyne structural fragment embeded in a nine-membered ring and forms species capable of cleaving DNA upon activation with thiol. The structure of this species was established as enyne[3]cumulene. The mechanisms of the conversion of analogs of neocarzinostatin chromophore such as **74** involving formation of an enyne–allene intermediate such as **75** were proposed and are exemplified for the ketone analog under anaerobic conditions.[37] Another neocarzinostatin chromophore-related model study is described in Section 4.13.3 (compounds **46**–**49**).

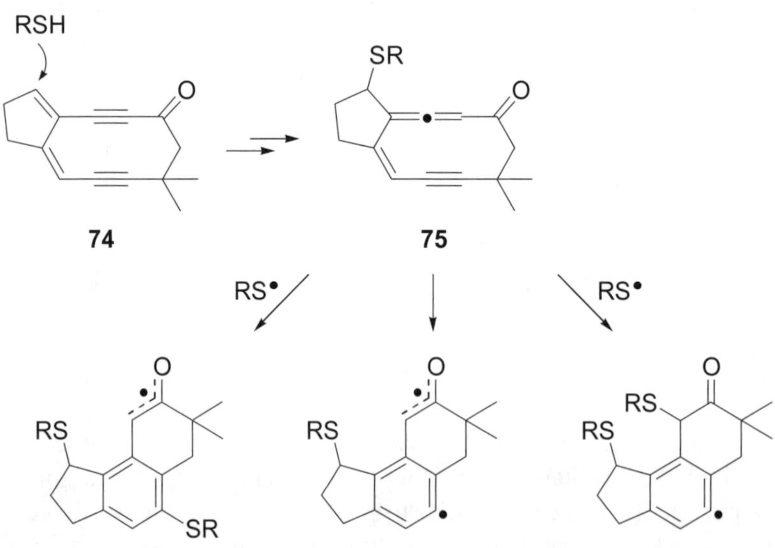

Due to the lower activation energy for the Myers–Saito cyclization versus the Bergman cyclization, even simple, acyclic enyne–allenes can generate diradical intermediates under physiological conditions. Thus the enyne allenes **76** and **77** cleave supercoiled plasmid DNA at 37 °C, pH 8 (phosphorus compounds appeared to be more suitable for practical studies of antitumor activity or DNA-cleaving properties than the corresponding sulfur compounds). It is interesting that products of cleavage activity toward both double-strand and single-strand DNA were observed.[3] However, in general, due to differences in radical character, double-strand DNA cleavage, analogous to that observed for enediynes, would not be expected from enyne–allenes. Another simple, acyclic enyne allene **78**, and its conjugate with the DNA minor groove binding element derived from distamycin A (not illustrated) also displays DNA cleavage activity.[38] The cleavage pattern and kinetics support a mechanism involving hydrogen atom abstraction by the Myers–Saito cyclization-derived diradical, although no details of the DNA cleavage products, indicating the site of hydrogen atom abstraction were reported. Efforts in the area of DNA damage studies via diradical-generating cyclizations was recently reviewed by Kerwin.[39,40]

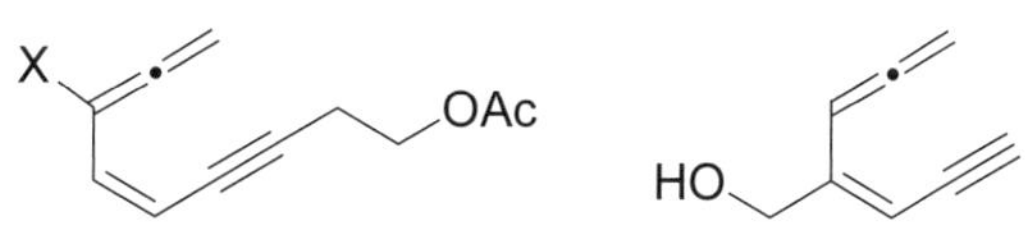

76 R = P(O)Ph_2
77 R = S(O)Ph

78

Outline of an DNA initial interaction with phosphorus-substituted enyne–allenes is exemplified below.[41]

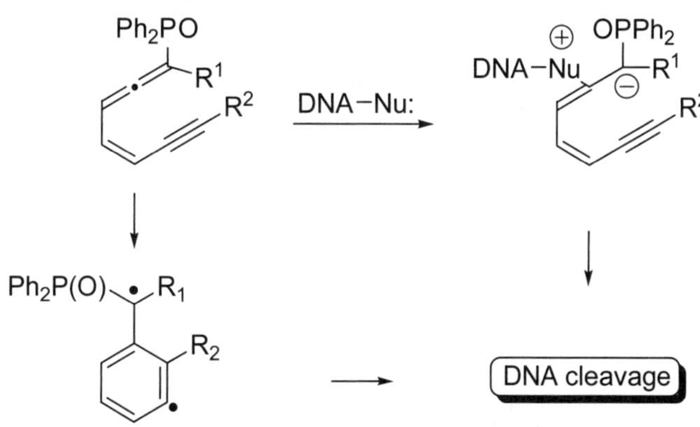

4.13.6 Experimental

Allene–enyne thermal Myers–Saito cyclization of [(4Z)-8-*tert*-butyldimethylsilyloxy-1-(*tert*-butyldimethylsilyloxymethyl)octa-1,2,4-trien-6-yn-1-yl](diphenyl)phosphine oxide (79): **{2- *tert*-butyldimethylsilyloxy-1-[2-(*tert*-butyldimethylsilyloxymethyl)phenyl]ethyl}(diphenyl)-phosphine oxide** (80)[41]

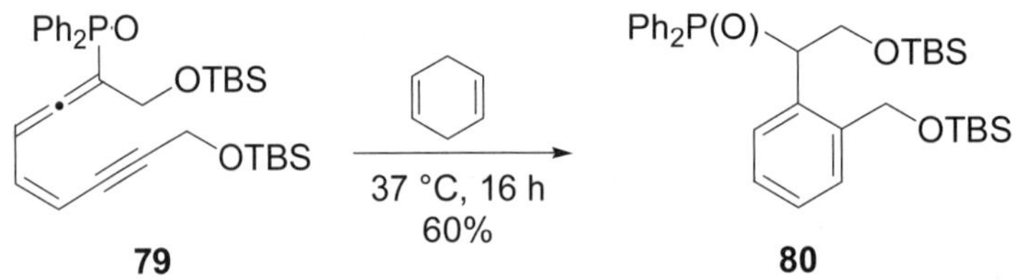

Compound **79** was obtained from the corresponding propargylic derivative (1 equiv) by treatment with diphenylphosphine chloride (1 equiv) in dichloromethane solution at –78 °C for 1 h, in the presence of triethylamine (1 equiv). After [2,3]-sigmatropic rearrangement, the reaction mixture was heated at 37 °C in the presence of cyclohexadiene (0.01 M) for 16 h, ($t_{1/2}$ = 8 h) to give the corresponding phosphine oxide toluene derivative **80** in 60% yield.

Silver(I)–catalyzed [3,3]–Sigmatropic rearrangement and Myers–Saito cyclization of 1-(2-Ethynylphenyl)hept-2-yn-1-yl pivalate (69): **1-(2-Naphthyl)pentan-1-one** (71)[36]

A small screw-cap scintillation vial equipped with a magnetic stir bar was charged with $AgSbF_6$ (5 mol %), triphenylphosphine (2 mol %), MgO (1.5 equiv), and dichloromethane. A solution of propargyl ester **69** (~ 150 mg, 1 equiv) in dichloromethane was added to the cloudy white suspension. The reaction mixture was stirred at room temperature and monitored periodically by TLC. Upon completion, the reaction mixture was loaded directly onto a silica gel column. Flash chromatography (hexanes/ether 19:1) gave the naphthlyl butyl ketone **71** as a light brown solid (83%).

4.13.7 References

1. Nagata, R.; Yamanaka, H.; Okazaki, E.; Saito, I. *Tetrahedron Lett.* **1989**, *30*, 4995–4998.
2. Myers, A. G.; Kuo, E. Y.; Finney, N. S. *J. Am. Chem. Soc.* **1989**, *111*, 8057–8059.
3. Nagata, R.; Yamanaka, H.; Murahashi, E.; Saito, I. *Tetrahedron Lett.* **1990**, *31*, 2907–2910.
4. Myers, A. G.; Dragovich, P. S. *J. Am. Chem. Soc.* **1989**, *111*, 9130–9132.
5. [R] Wang, K. K. *Chem. Rev.* **1996**, *96*, 207–222.
6. Myers, A. G.; Dragovich, P. S.; Kuo, E. Y. *J. Am. Chem. Soc.* **1992**, *114*, 9369–9386.
7. Wenthold, P. G. ; Paulino, J. A.; Squires, R. R. *J. Am. Chem. Soc.* **1991**, *113*, 7414–7415.

8. Wenthold, P. G.; Wierschke, S. G.; Nash, J. J.; Squires, R. R. *J. Am. Chem. Soc.* **1993**, *115*, 12611–12612 and *J. Am. Chem. Soc.* **1994**, *116*, 4529 (erratum).
9. [R] Salem, L.; Rowland, C. *Angew. Chem., Int. Ed.* **1972**, *11*, 92–111.
10. Cremeens, M. E.; Hughes, T. S.; Carpenter, B. K. *J. Am. Chem. Soc.* **2005**, *127*, 6652–6661.
11. [R] Carpenter, B. K. *Chem. Soc. Rev.* **2006**, *35*, 736–747.
12. Wenthold, P. G.; Lipton, M. A. *J. Am. Chem. Soc.* **2000**, *122*, 9265–9270.
13. Engels, B.; Lennartz, Ch.; Hanrath, M.; Schmittel, M.; Strittmatter, M. *Angew. Chem., Int. Ed.* **1998**, *37*, 1960–1962.
14. Schmittel, M.; Kiau, S.; Siebert, T.; Strittmatter, M. *Tetrahedron Lett.* **1996**, *367*, 7691–7694.
15. Andemichael, Y. W.; Gu, Y. G.; Wang, K. K. *J. Org. Chem.* **1992**, *57*, 794–796.
16. Karpov, G.; Kuzmin, A.; Popik, V. V. *J. Am. Chem. Soc.* **2008**, *130*, 11771–11777.
17. Scheuplein, S. W.; Machinek, R.; Suffert, J.; Brückner, R. *Tetrahedron Lett.* **1993**, *34*, 6549–6552.
18. Schmittel, M.; Mahajan, A. A.; Bucher, G. *J. Am. Chem. Soc.* **2005**, *127*, 5324–5325.
19. Spöler, C.; Engels, B. *Chem. Eur. J.* **2003**, *9*, 4670–4677.
20. Schmittel, M.; Wohrle, C. *Tetrahedron Lett.* **1993**, *34*, 8431–8434.
21. Schmittel, M.; Strittmatter, M.; Kiau, S. *Tetrahedron Lett.* **1995**, *36*, 4975–4978.
22. Waddell, M. K.; Bekele, T.; Lipton, M. A. *J. Org. Chem.* **2006**, *71*, 8372–8377.
23. Brunette, S. R.; Lipton, M. A. *J. Org. Chem.* **2000**, *65*, 5114–5119.
24. Cramer, C. J.; Kormos, B. L.; Seierstad, M.; Sherer, E. C.; Winget, P. *Org. Lett.* **2001**, *3*, 1881–1884.
25. Maier, M. E.; Langenbacher, D. *Synlett* **1994**, 713–716.
26. Toshima, K.; Ohta, K.; Ohtake, T.; Tatsuta, K. *Tetrahedron Lett.* **1991**, *32*, 391–394.
27. Toshima, K.; Ohta, K.; Ohtake, T.; Tatsuta, K. *J. Chem. Soc., Chem. Commun.* **1991**, 694–695.
28. Myers, A. G.; Finney, N. S. *J. Am. Chem. Soc.* **1992**, *114*, 10986–10987.
29. Myers, A. G.; Dragovich, P. S. *J. Am. Chem. Soc.* **1993**, *115*, 7021–7022.
30. Feng, L.; Kumar, D.; Birney, D. M.; Kerwin, S. M. *Org. Lett.* **2004**, *6*, 2059–2062.
31. Tuesuwan, B.; Kerwin, S. M. *Biochemistry* **2006**, *45*, 7265–7276.
32. Ezcurra, J. E.; Pham, C.; Moore, H. W. *J. Org. Chem.* **1992**, *57*, 4787–4789.
33. Krause, N.; Hohmann, M. *Synlett* **1996**, 89–91.
34. Andemichael, Y. W.; Huang, Y.; Wang, K. K. *J. Org. Chem.* **1993**, *58*, 1651–1652.
35. Datta, S.; Liu, R.-S. *Tetrahedron Lett.* **2005**, *46*, 7985–7998.
36. Zhao, J.; Hughes, C. O.; Toste, F. D. *J. Am. Chem. Soc.* **2006**, *128*, 7436–7437.
37. Fujiwara, K.; Kurisaki, A.; Hirama, M. *Tetrahedron Lett.* **1990**, *30*, 4329–4332.
38. Myers, A. G.; Parrish, C. A. *Bioconjugate Chem.* **1996**, *7*, 322–331.
39. [R] Kerwin, S. M. In *Radical and Radical Ion Reactivity in Nucleic Acid Chemistry* Greenberg, M., ed.; Wiley, New York, 2009, pp. 389–419.
40. We thank Professor Sean Michael Kerwin for helpful discussions.
41. Nicolaou, K. C.; Maligres, P.; Shin, J.; de Leon, E.; Rideou, D. *J. Am. Chem. Soc.* **1990**, *112*, 7825–7826.

4.14 Robinson annulation

Noha S. Maklad

4.15.1 Description

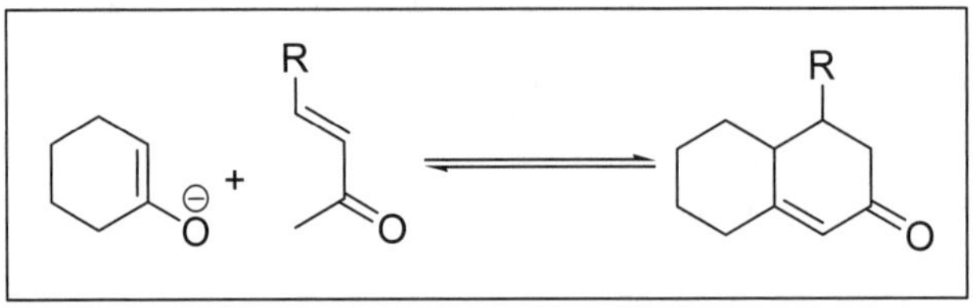

The Robinson annulation was introduced in 1935 by Sir Robert Robinson and William Sage Rapson targeting the synthesis of building blocks of sterols.[1] The condensation is an unprecedented reaction of alkali metal derivative of cyclohexanones and α,β-unsaturated methyl ketones that produces cycloketones and polycycloketones as precursors for sterols synthesis.

4.14.2 Historical Perpective

Robert Robinson was a 1947 Nobel Laureate and a recipient of the order of merit; the tile is the most prestigious honor bestowed on a civilian in the United Kingdom.[2] He was fascinated by developing new methods of steroid synthesis. Steroids are a class of compounds with important and diverse biological activities. In the realm of his fascination with this class of compounds and his diverse interests Robinson and co-workers in the 1930s began a series of 52 publications that have been published over 2 decades under the title "Experiments on the Synthesis of Substances Related to Sterols."[1–10] In 1935 Robinson demonstrated a unique and unprecedented condensation of sodiocyclohexanone and its analogues with α,β-unsaturated methyl ketones. The finding not only has proven an important method for the synthesis of sterols; but, since its discovery, has also been extrapolated to the synthesis of complicated, diverse and important compounds.

4.14.3 Mechanism

The mechanism of the Robinson annulation is explained in three distinct steps. Initially a vinyl ketone undergoes a Michael addition to a cyclic ketone or β-keto ester to give the "3'-oxybutyl" adduct **1**, which in turn is rearranged to give the 1,5-diketo adduct **2**.[5,11] This step is then followed by intramolecular aldol reaction to give the tertiary alcohol **3**, and finally a dehydration step proceeds to give the octalone product **4**.

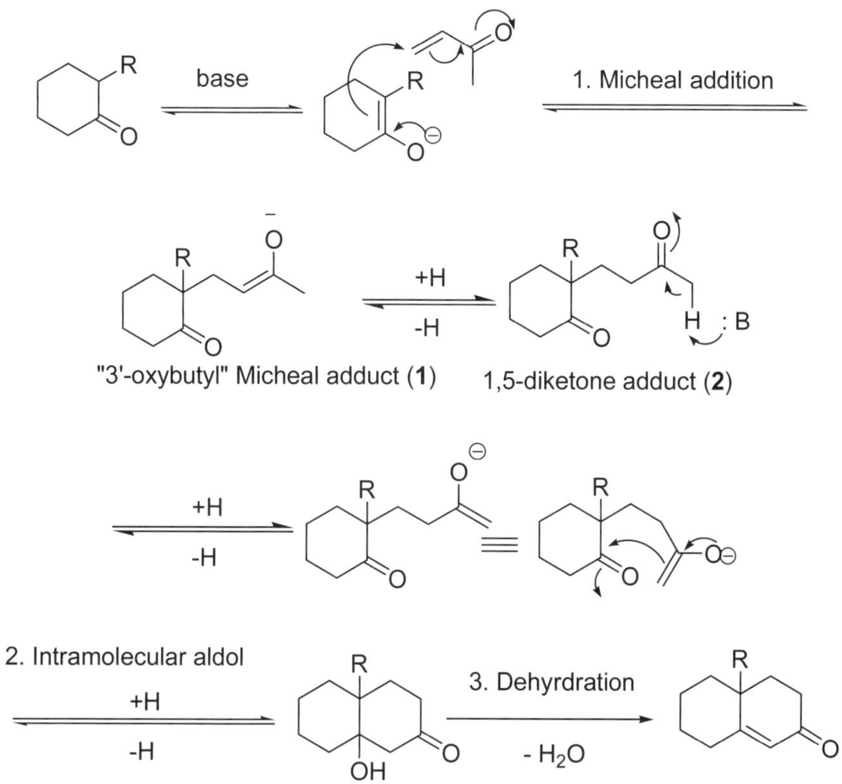

3 **4**

Under basic conditions and shorter reaction time, the ketol **3** and the enone **4** are detected in reaction mixture and can be separated. A higher yield for the enone can be obtained by separation of the Michael adduct **2** or the ketol **3** and either of these can then resubjected to less stringent conditions to form the enone **4** in higher yields.[12,13]

Regiocontrol for the Robinson annulation is an important factor that can influence a synthetic route because there are two α-carbons to the keto group to be considered. Consequently, techniques for regiocontrol have been developed. Under either the basic or acidic standard annulation conditions deprotonation occurs under thermodynamic control at the higher substituted carbon and hence forms an angular substitution at the fused side of the end-product. An exception to this is steric hindrance where alkylation occurs at the least substituted α-carbon, as shown in the synthesis of **9**.[14]

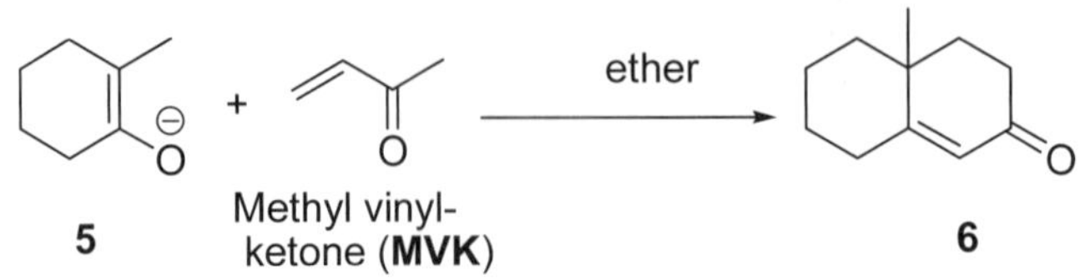

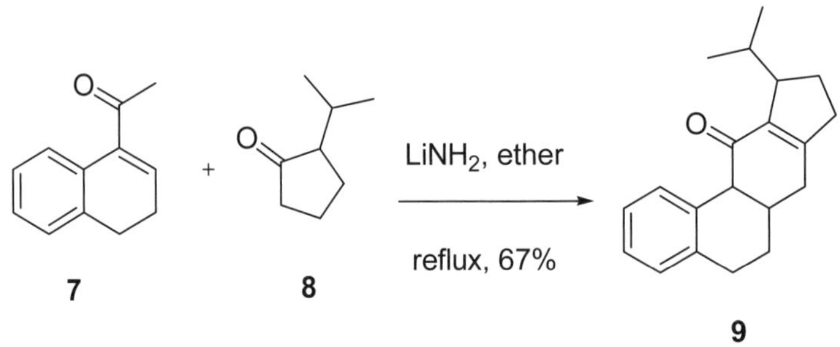

The regiocontrol and stereocontrol aspects of the mechanism of this reaction will be explained under the next section.

4.14.4 Standard Method, Variations, and Improvements

The Robinson annulation is the reaction of alkali metal derivatives of cyclohexanones with α-,β-unsaturated methyl ketones to produce cycloketones and polycycloketones. The standard method for Robinson annulation is exemplified in the mechanism shown above. For the synthesis of the 1,5-diketone side chain, the enolate nucleophile reacts with a Michael acceptor; this Michael acceptor is usually a substituted vinyl ketone or the parent methyl vinyl ketone (MVK), although the latter gives low yield due to its propensity to polymerize under the standard reaction conditions. To overcome the drawbacks for using MVK, Robinson, McQuillin and Du Feu introduced the Robinson–Mannich variation of the annulation reaction.[7,15] This modification uses a quaternized Mannich base formed from the vinyl entity; the Mannich base is made in situ and acts as a methyl vinyl ketone precursor after it is converted to its methiodides. The formed methiodides of the Mannich adduct 4-(trimethylamino-2-butanone) is condensed with sodioderivatives of ketones or with the parent ketone in the presence of sodium ethoxide.

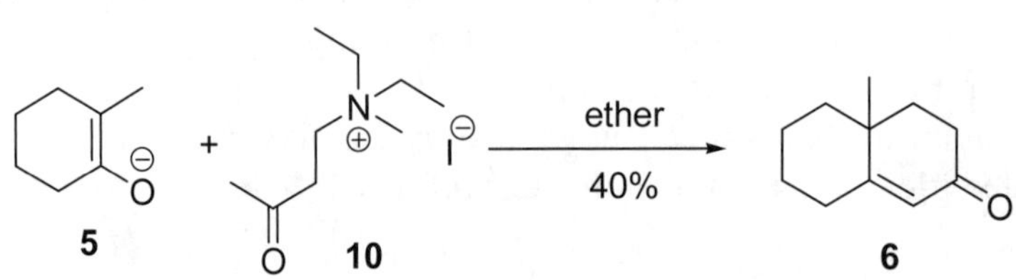

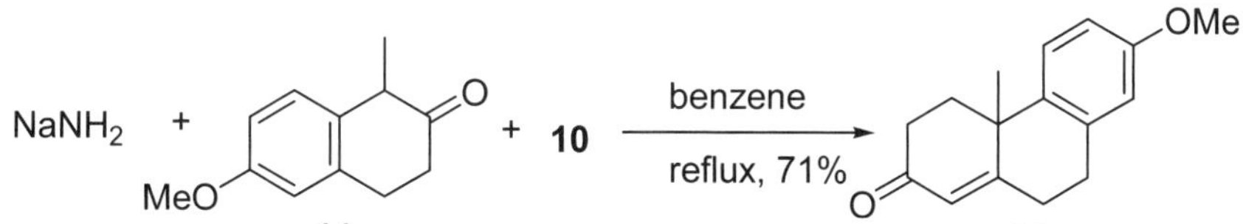

11 12

The generation of the enolate nucleophile can be attempted in different ways. The standard method introduced by Robinson is directed toward a thermodynamically controlled formation of the enolate where the deprotonation occurs at the higher substituted α-carbon to the ketone, thus creating a substitution at the ring juncture. The corresponding alkoxide forms by heating the starting ketone or β-keto ester using sodium or lithium amide in an aprotic solvent such as ether or benzene. Other base/solvent combinations can be used such as lithium diisopropylamide (LDA) in THF at 20 °C.[16] The enolate formation can also be achieved by subjecting the ketone to an alkoxide base in a protic solvent such as sodium methoxide in methanol or by a metal hydroxide base such as potassium hydroxide in methanol or ethanol.[11,17]

The Robinson annulation can also be conducted under acidic conditions. The first example reported was by Ellis and Heathcock in 1971. Acids such as sulfuric acid and *p*-toluenesulfonic acid are used, although Robinson annulations under acidic conditions are not abundant in the literature.[18]

In the past 70 years different variations have immerged for the Robinson annulation reaction to rectify its weaknesses such as polymerization tendencies for the parent vinyl ketone under standard reaction conditions, multiple alkylation products, regiocontrol, and stereocontrol.

Stork–Jung vinylsilanes variant.[19]

In 1974 Stork and Jung introduced vinylsilanes as a carbonyl precursor for the Robinson annulation to be used as the Michael acceptor; the Stork–Jung vinylsilane **14** is shown below. The variation affects the alkylation step under mild conditions, and the release of the di-keto progenitor of cyclohexanones is straightforward. An example is the reaction of the lithium enolate **13** with the vinylsilane **14** in tetrahydrofuran at room temperature to give the silylated intermediate **15** in 91% yield. The silyl group is converted to ketone in the presence of *m*-chloroperbenzoic acid, and the reaction is conducted at 0 °C for a few minutes and then warmed to room temperature to give **16**, which then proceeds to give the octalone **17** using potassium hydroxide in ethanol.

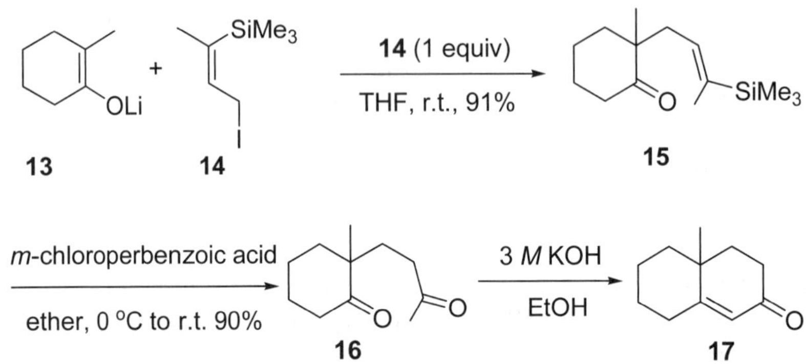

Stork-Enamine ketone variant[17,20–22]

The Stork variation was pioneered by Stork and co-workers in 1963. It entails the reaction of pyrrolidine or morpholine enamine derived from unsymmetrical cyclohexanones with methyl vinyl ketones. The alkylation is directed to the less substituted carbon opposite to the alkylation regioisomer formed by the standard Robinson annulation conditions. Cyclization to the corresponding octalone then occurs. The morpholine enamine is less reactive than the pyrrolidine and hence pyrrolidine enamine has been mostly used for this approach. An example of such annulation is shown in the synthesis of 8-methyl-2-oxo-$\Delta^{1,9}$octalone (**19**). The pyrrolidine enamine of 2-methylcyclohexanone is refluxed in benzene with MVK for 24 h followed by the addition of an acetate buffer and reflux for 4 h, and after the reaction is worked up purification gives **19** in 45% yield.

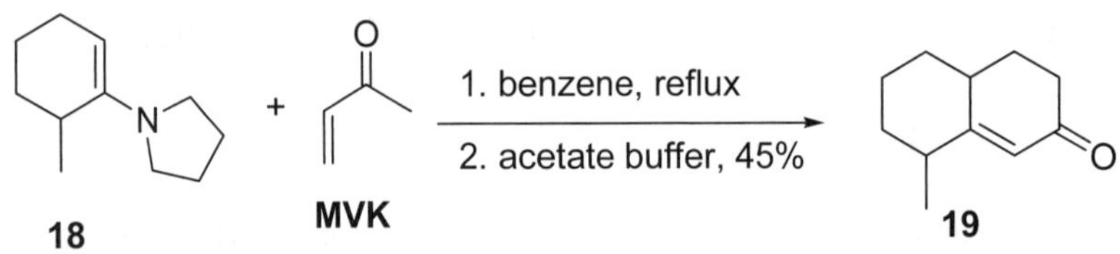

α-Silylated vinyl ketones[23]

The use of α-silylated vinyl ketone is another approach to overcome drawbacks of the standard Robinson annulation conditions such as polymerization of the vinyl ketone. The α-silylated vinyl ketones are stable and can undergo Michael addition in standard aprotic conditions (conditions that induces polymerization for vinyl ketones), as well as protic conditions. Synthesis of the octalone **21** can be used as an example of this variation. The silylated ketone **20** reacts with lithium enolate **13** (generated by methyllithium from its corresponding enol silyl ether in THF) in *t*-butyl alcohol; the reaction mixture is worked up and the crude is refluxed in 10% aqueous sodium hydroxide to give **21** in ~ 60% yield.

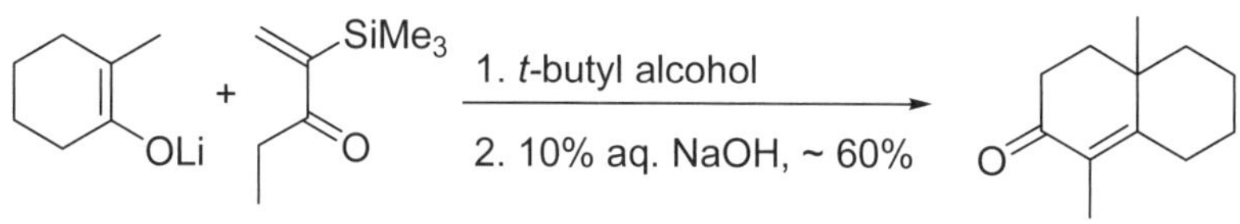

13 **20** **21**

The alkylation can also occur under aprotic conditions such as in the formation of the octalone **23**. The lithium enolate **22** is added to the ketone **20** in tetrahydrofuran at –78 °C, and then the mixture is allowed to reach room temperature. The alkylation process is followed by subjecting the silylated intermediate to 5% sodium methoxide-methanol to give 1-methyl-$\Delta^{1,9}$-octalone (**23**) in 80% overall yield.

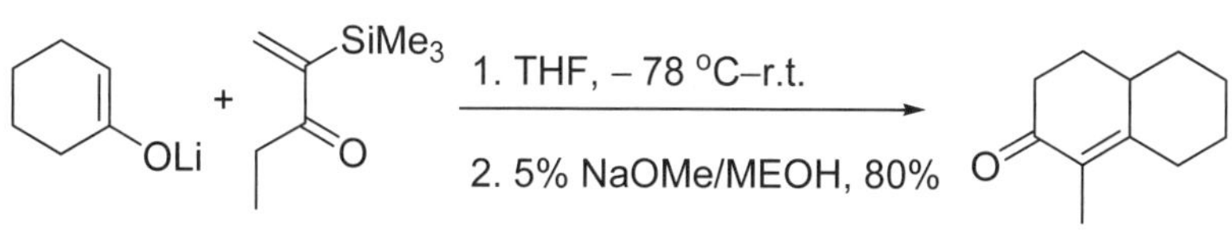

22 **20** **23**

Silyl-enol ether[24–26]

In 1975 Mukayaima's group used Lewis acid for the alkylation of silyl enol ethers with vinyl ketones, although the use of these conditions were only reported once afterward (in 1980) by a Japanese group as an alternative for the Robinson annulation. It was not until 1985 when Huffman, Satish and co-workers introduced a regioselective Lewis acid-catalyzed silyl enol ether variation to the Robinson annulation. The reaction depends on alkylation of silyl enol ethers in the presence of a $TiCl_4$ or Ti(O*i*-Pr)$_4$ catalyst or a combination of both, the latter showing the best results. Huffman and co-workers conducted the alkylation of the enol ethers with various Michael acceptors such as MVK, ethyl vinyl ketone (EVK), or their ketals to avoid polymerization, especially in the MVK case. The optimal alkylation temperatures are in the range of –80 °C to –90 °C. The formed 1,5-diketone can then be cyclized to the final Robinson enone using a base such as potassium hydroxide in ethanol . One of the importances of this approach is the ability to start with the preformed thermodynamic or kinetic enolate of the silyl enol ether and hence control the reaction regioselectively.[27] An example of this is the formation of the 1,5-diketone **26** in 81% yield which is in turn cyclized to the *cis*- and *trans*-analogues **27a** and **b** in 3:1 ratio.

Although the yields of these products are not disclosed in their paper, it provides an interesting regioselective approach for the annulation reaction.

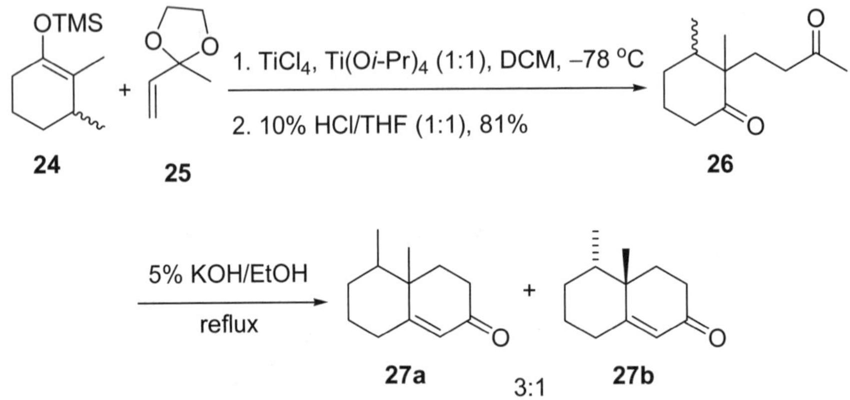

24 25 26 27a 27b

Sato and co-workers in 1991 showed the utilization of organotin triflate as the Lewis acid catalyst for the Michael addition step. Synthesis of octalone **30** is a good example in which silyl enol ether of cyclohexanone and MVK (1.3:1 ratio) reacts in DCM in the presence of 0.05 equivalent of $Bu_2Sn(OTf)_2$ at –78 °C. After alkylation was complete, sodium methoxide in methanol was added, and the mixture was stirred at room temperature to give **30** in an excellent 89% yield.[28]

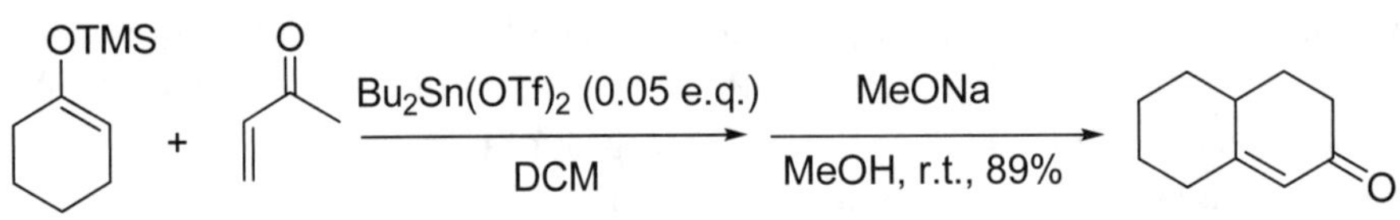

29 MVK 30

Although there are different contributions in the field and the evolution of various Lewis acids used, to this day many chemists refer to the conditions of the reaction of silyl-enol ether with vinyl ketones in the presence of Lewis acid as "Mukayama-like" conditions.

Aza-Robinson annulation[29–31]

There are few examples of this variation in the literature. The variation relies on the alkylation of thiolactams (in replacement of cyclohexanones) by the Michael addition of diazomethyl vinyl ketone (in replacement of methyl vinyl ketones) followed by ring closure using rhodium(II) acetate mediated diazomethane insertion. The diazomethyl entity can also be added after the Michael addition step.[29,30] The reaction affords six-membered rings with nitrogen at the ring juncture. Danishefsky and co-workers have shown a good example of such variation in the synthesis of compound **34a** and **34b**. The synthetic route starts with the alkylation with **32a** to give the intermediate **33a** in excellent 95% yield, followed by ring formation using rhodium(II) acetate and then raney nickel. On the other hand, the alkylation of thiolactam rings of X = H or n > 1 with diazomethyl vinyl ketone shows lower yields. An example is the reaction of the thiolactam **32** (X = H, *n* = 2) with the Michael acceptor **31a**, which shows a low alkylation yield of 28%, by using methyl acrylate **31b** as the alkylating agent followed by the introduction of the diazomethyl entity, which gives **33b**—the Michael adduct was hydrolyzed with NaOH followed by mixed anhydride formation and diazomethane addition—in 70% yield followed by ring formation to give **34b** in 73% yield. This method, although it involves more steps, is more reliable in producing the end product γ-pyridones in good yields.

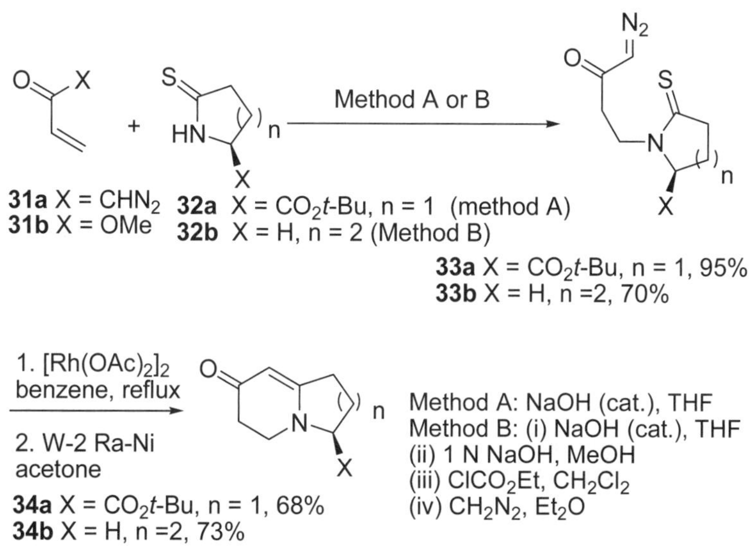

Gurana; Scarpi; et al. in 2000 introduced a modification to this variant which steers away from the use of the diazomethane addition step as shown in the synthesis of **38**. Thiolactam **35** is alkylated with ethyl vinyl ketone in the presence of potassium carbonate and 18-crown-6 in tetrahydrofuran (ethyl vinyl ketone is introduced in excess by slow addition over a 2 h period to prevent *S*-alkylation) to give the *N*-alkylated ketone **36** in 97% yield. A thioiminium ion was formed with Me_2SO_4 followed by cyclization with DBU under reflux to give **38** in a good 67% yield.[31]

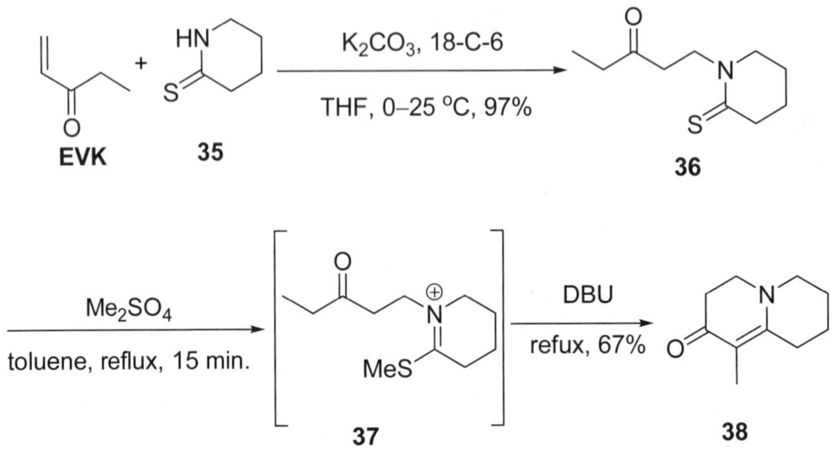

The Hajos–Wiechert reaction[32,33]

Developed in the early 1970s, this reaction, also called the Hajos–Parrish reaction or Hajos–Parrish–Ender–Sauer–Wiechert reaction, is one of the earliest processes for the stereoselective synthesis of Wieland–Miescher ketone, an important building block for steroids and terpenoid synthesis.[34] This reaction is a proline mediated asymmetric variation to the Robinson annulation. Hajos and Parrish of Hoffmann–La Roche Inc. in 1971 and 1974 published an asymmetric aldol cyclization of triketones such as that of structure **39**, which affords optically active annulation products in the presence of catalytic amounts of (*S*)-proline (*L*-proline).[32] One of the early examples is the synthesis of **41** from the triketone **39** (a product of the Michael addition of MVK to the corresponding 2-methylcyclopentane-1,3-dione), the reaction is performed in two steps: first by ring formation in the presence of 3 mol % of (*S*)-proline in DMF to afford the ketol **40** in 100% yield after crystallization with 93% *ee* and then by reaction with toluenesulfonic acid to give the dehydrated adduct **41**. The formation of the Wieland–Miescher Ketone **44** follows the same synthetic route, starting from the tri-ketone **42** to give the end product in 75% optical purity and 99.8% of optical yield.

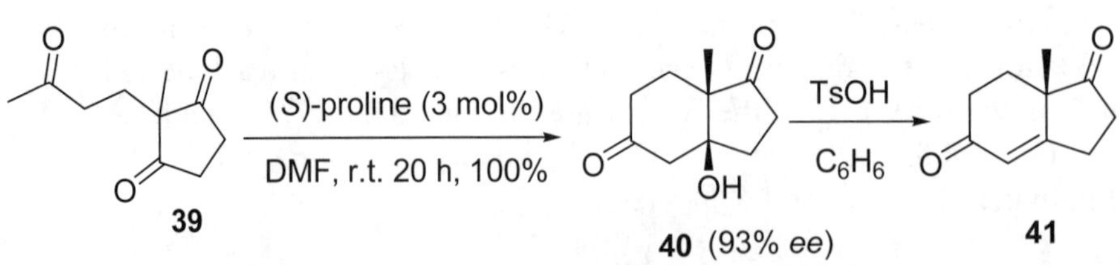

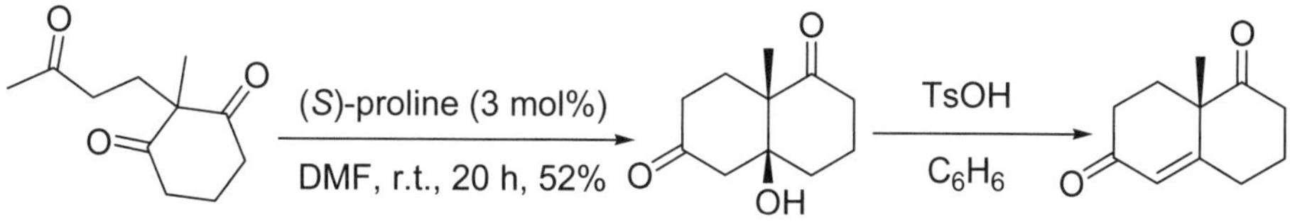

42 43 (*S*)-**44**

At the same time Eder, Sauer and Wiechert of Schering A.G. in 1971 published a one-pot reaction for the asymmetric synthesis of chiral bicyclics. The group affected this transformation from prochiral triketones in the presence of chrial amines or amino acids such as proline (similar to Hajos and Parrish). The (*S*)-amine or (*S*)-amino acid induces (*S*)-configuration bicyclics, while the (*R*)-configuration outcomes varies. The best results for this annulation are shown in the synthesis of **41** and **44**. The reaction is performed in the presence of (*S*)-proline and perchloric acid. The reaction mixture is heated under reflux to give the products in good 87% and 83% yields, respectively, and 84% and 71% *ee*, respectively.

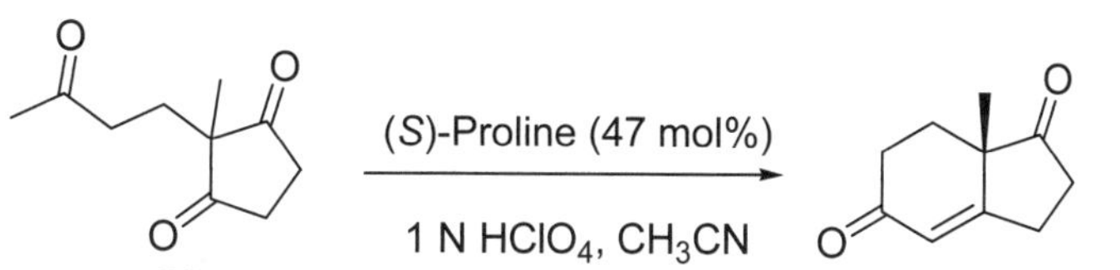

39 **41** (84% *ee*)

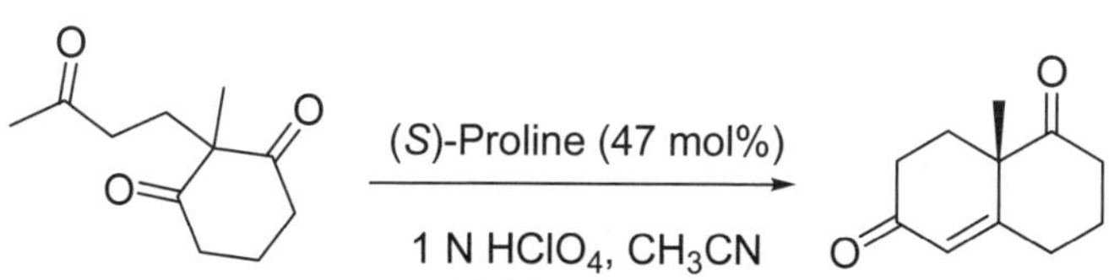

42 (*S*)-**44** (71% *ee*)

Bui and Barbas in the year 2000 introduced a single-step enantioselective synthesis of the Wieland–Miescher ketone (**44**). The keto-enone **44** is formed by the reaction MVK, 2-methyl-1,3-hexadione, in the presence of (*S*)-proline in 49% and 76% *ee*. By screening the different catalysts, **46**–**48** show synthesis of the (*S*)-(+)-**44** in 60–75% *ee*.[34]

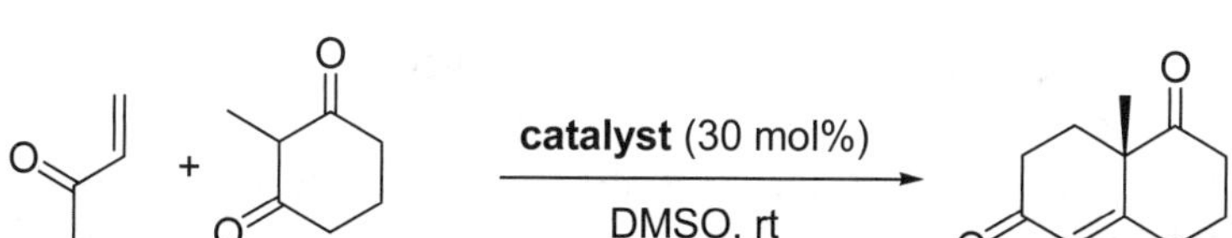

MVK **45** (*S*)-**44** (60–75% *ee*)

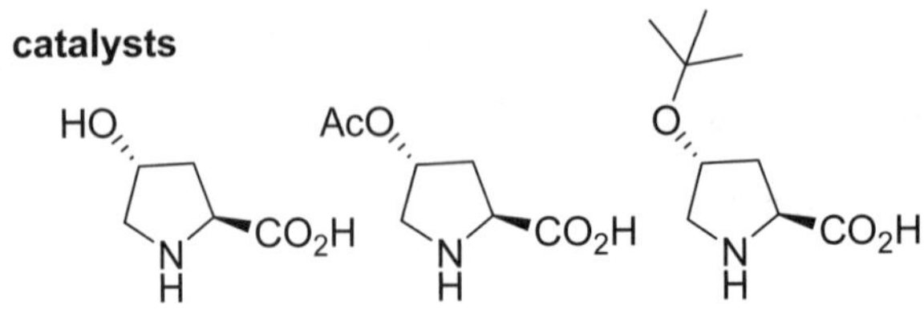

46 47 48

Asymmetric aspects for the Robinson annulation

In the Robinson annulation, cyclization to give the ketol intermediate can produce possibly five stereocenters. The dehydration process that follows minimizes the number of possible chiral centers to three or less.[11b]

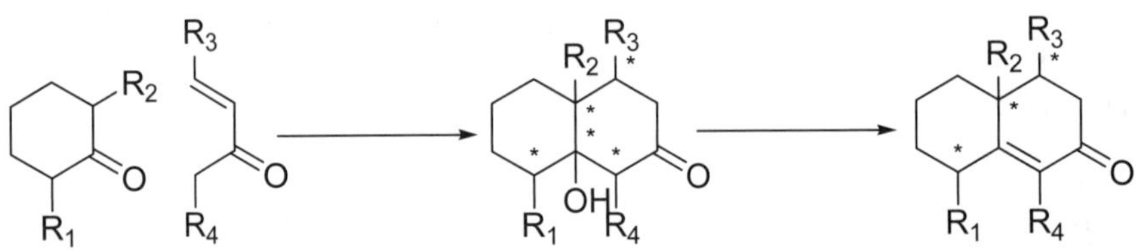

There are different approaches for stereocontrol for the Robinson annulation; the control can either arise from the inherent nature of the starting ketone and/or the vinyl ketones substituents in combination with the reaction conditions, or by the use of a chiral catalyst. In the first case, an example is the stereoselective aldol cyclization to give the ketol intermediate **50**.[12c] In this case the cyclization is kinetically controlled under protic basic conditions of sodium ethoxide and ethanol as it gives the *cis*-fused adduct rather than the more stable *trans*-fused ketol, which is not detected at any time during the reaction.

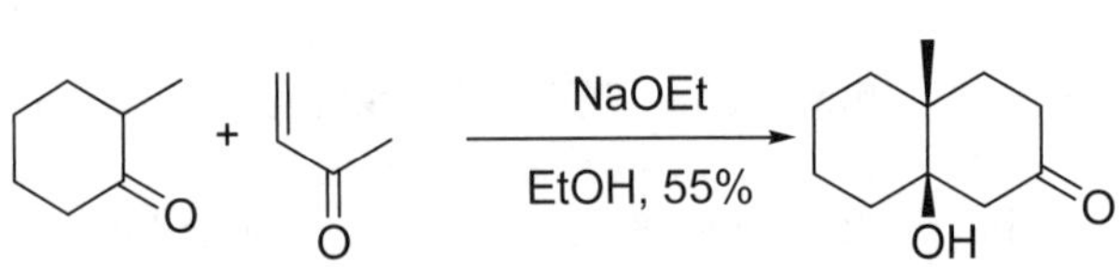

49 **MVK** **50**

Sterics can influence the stereochemical outcome of the aldol cyclization, as in the case of the synthesis of the ketol **52** from the triketone **51** where R_2 = OAc.[35] In this case sterics stabilize the transition state that brings about the aldol cyclization to give the *cis*-fused ketol **52**, and a similar result occurs when $R_2 = CH_3$. Contrary results occurs when R = H or CN, and under kinetically controlled conditions the outcome relies on the stability of the end-product and so gives the more stable *trans*-fused ketol.[11]

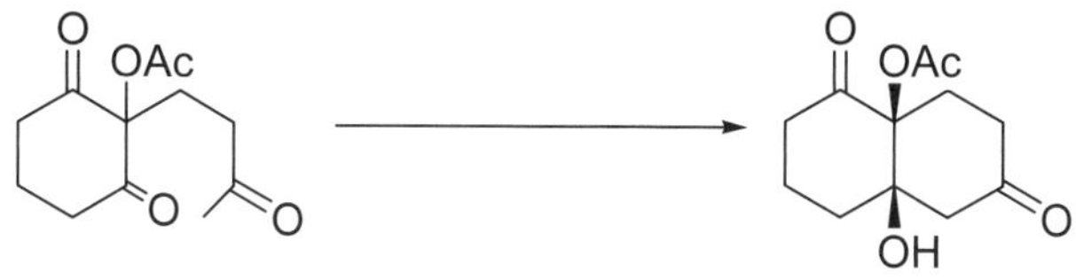

51 **52**

Solvents can influence the stereochemical outcome of the Robinson annulation. An interesting example is the one-step annulation using the sodium enolate of **49** to give the octalone **54** and **55**. The reaction of the sodium enolate of the 2-methyl-cyclohexanone with trans-3-penten-2-one (**53**) in the presence of dioxane at room temperature for 100 h gives the *cis*-4,10-dimethyl-$\Delta^{1,9}$octal-2-one (**54**) in 65% yield, while the reaction gives the *trans*-isomer **55** in 72% yield in 3 h under same conditions.[36]

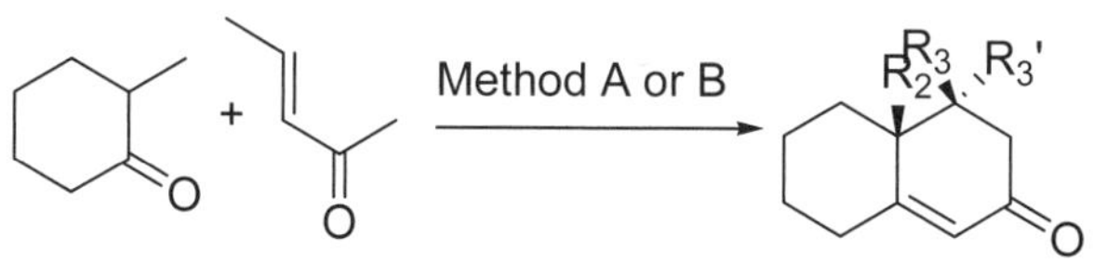

49 **53**

A = Dioxane, rt, 100 h, 65%
B = DMSO, rt, 3 h, 72%

54, $R_2 = R_3 = CH_3$, $R_3' = H$
55, $R_2 = R_3' = CH_3$, $R_3 = H$

A more efficient approach to control the stereochemical outcome for the Robinson annulation can be through the use of chiral catalysts such as in the case of the enantioselective Hajos–Wiechert variation introduced earlier. There are other chiral agents other than the popular (*S*)-proline-mediated annulation reaction that are used for these transformations—for example the use of (*S*)-2-(pyrolidinylmethyl)pyrrolidine in the presence of Brønsted acid such as trifluoroacetic (TFA).[37] This new catalyst for the Robinson annulation was reported in 2007 by Endo et. al., where the Brønsted acid, contrary to Hajos–Wiechert reaction, gives the (*R*)-isomer of the Wieland–Miescher ketone **44** in a moderate yield of 47% and 75% *ee*.

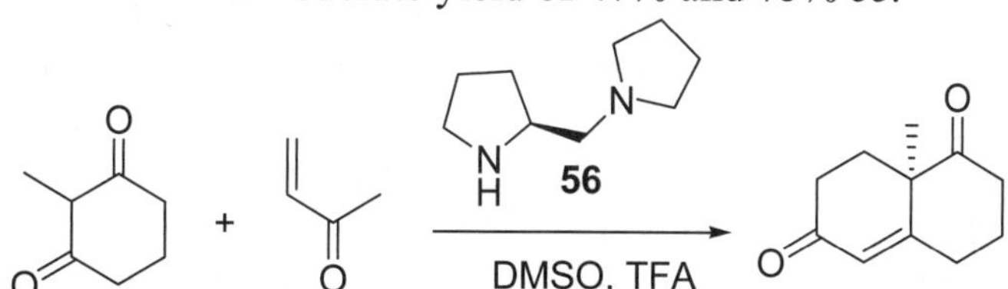

43 **MVK** (*R*)-**44** (75% *ee*)

Amines such as (*S*)-1-phenylethylamine and (*R*)-1-phenylethylamine can be used as chrial auxiliaries for the Robinson annulation. The secondary enamine of the ketone **57** with (*S*)-1-phenylethylamine reacts with MVK in tetrahydrofuran for 30 h. This is then followed by the addition of 20% of acetic acid for 30 min. The intermediate formed is then cyclized with sodium methoxide in methanol to form (*R*)-**59** in 50% overall yield and 92% *ee*.[38]

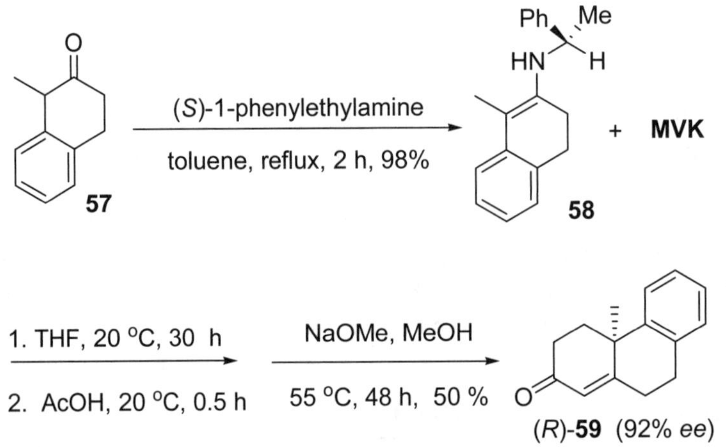

In 1997 Barbas, Danishefsky, Zohng, and co-workers reported an antibody-catalyzed enantioselective Robinson annulation. The antibody used (Ab38C2) catalyzes the cyclodehydration step of the Robinson process for the prochiral starting triketone **60** to give (*S*)-**44** in > 95% *ee* and with 96% optical purity. The reaction is carried out at room temperature for 10 days to give the product in a 94% yield.[39]

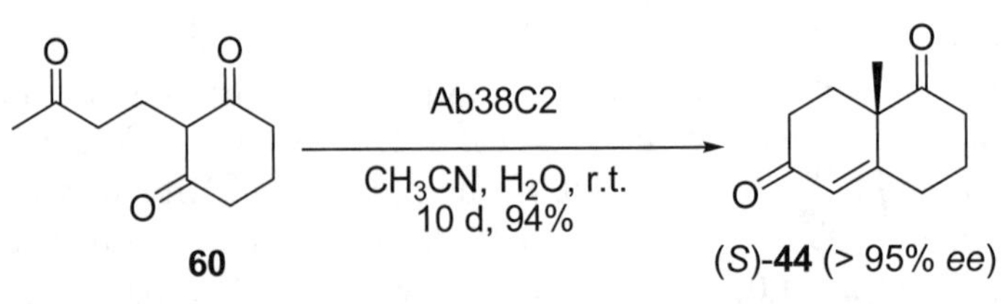

In 2009 Miro et al. reported the use of phosphoric acids as a chiral catalyst for enantioselective transformation of the Robinson annulation.[40] Chrial phosphoric acids **61** and **62** are used in sequence first for the Michael reaction step and are then followed by the cyclization step. Synthesis of the annulation adduct **64** is shown as an example in the group's report. The cyclized adduct is formed from the reaction of the β-keto ester **63** in the presence of the phosphoric acid **61** at 40 °C for 24 h and is followed by refluxing in toluene for 48 h in the presence of the second phosphoric acid **62** to give the product in 64% yield and 96% *ee*.

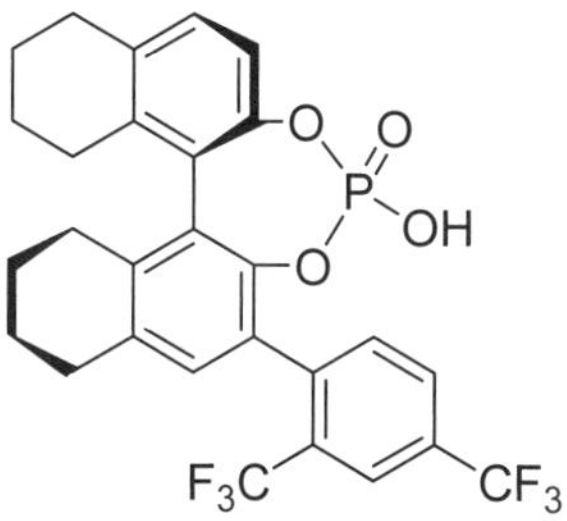

61

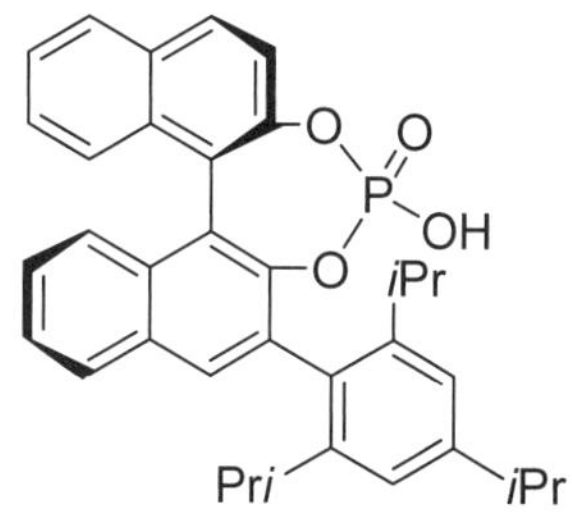

62

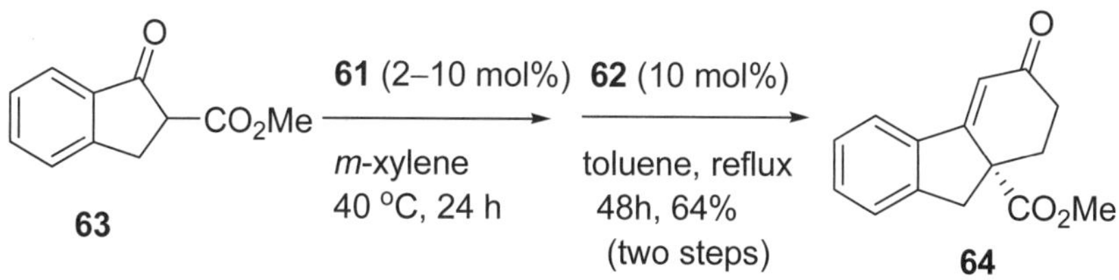

Use of directing group

Hydroxymethylene group is used as a directing group for the annulation reaction; the group is added to the α-carbon of the starting ketone using ethylformate. The formation of the hydroxymethylene group directs alkylation to the less substituted position.[11,41] A good example can be see in Robert Woodward's synthesis of intermediate **68** in his total synthesis of cholesterol.[42] The bicyclic ketone **65** reacts with ethylformate in the presence of sodium methoxide in benzene, to give the formylated intermediate in 94% yield. The crude is then reacted with EVK and with a catalytic amount of potassium hydroxide in *t*-butanol to give the alkylated adduct **67** in 54% yield. Without further purification the crude was cyclized and the formyl group was removed in one step by potassium hydroxide in a mixture of water and dioxane to give **67** in 88% yield.

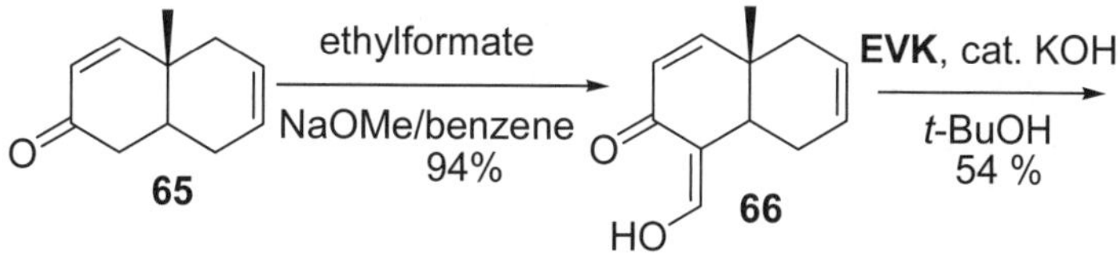

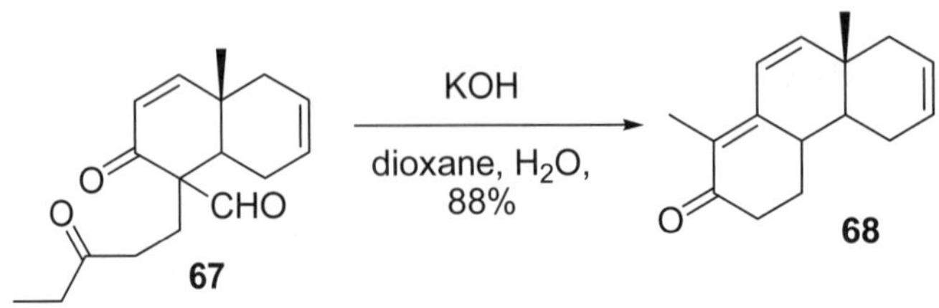

4.14.5 *Synthetic utility*

General utility

Robinson annulation has played a pivotal role in the synthesis cycloketones and polycycloketones for decades, and since its discovery it has been the keystone of many impressive syntheses.

The decalin system is an abundant structure motif in terpenoid natural products, the *trans*-isomers more so than the *cis*-isomers. Robinson annulation is a robust strategy for the synthesis of such molecules or their unsaturated congeners. Its importance lies especially in its ability to synthesize steroselectively the *trans*-decalin skeleton. In the synthesis of compound **71** Robinson annulation assists in the formation of the compound's ABC ring motif.[43] The ketone **69** is first alkylated with methyl formate to give the β-keto carbonyl adduct **70** in 89% yield after purification. The product reacts with MVK and triethylamine at 0 °C, and the crude is subjected to concomitant cyclization and removal of the formyl group after workup using sodium methoxide in methanol which gives **71** as a single isomer in excellent 81% yield.

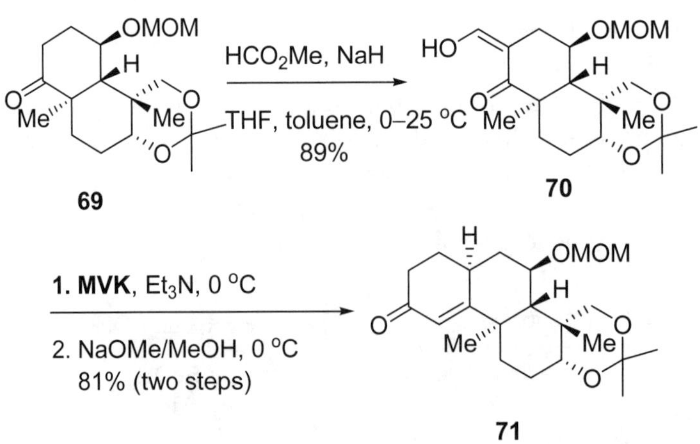

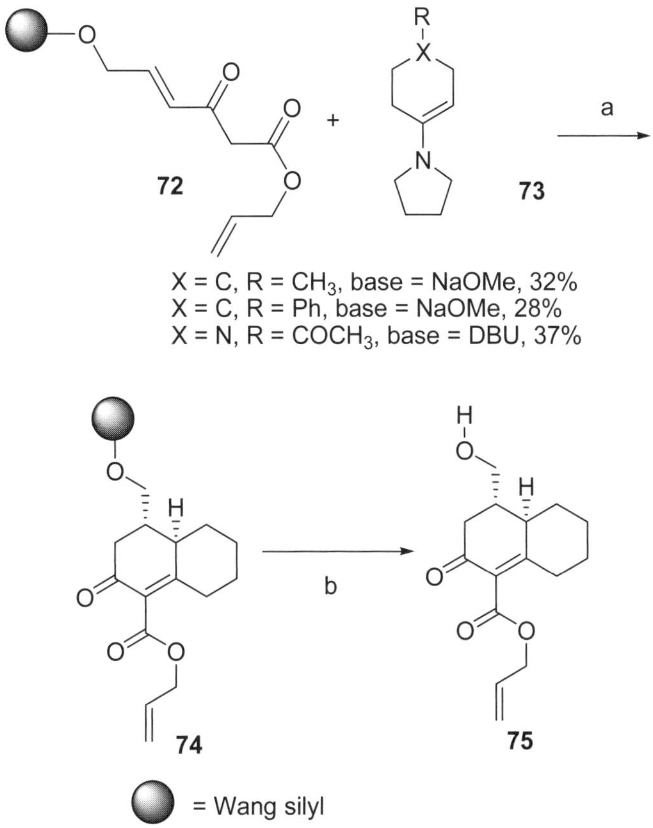

(a) MeOH, H_2O (3:1) then base
(b) TBAF, THF, r.t., overnight

Solid-phase techniques are also used for synthesis of compounds with *trans*-decalin motif where the Robinson annulation is used for synthesis of some natural product inspired structures such as that of **75**.[44] The protocol uses an immobilized solid phase bound Nazarov reagent that reacts with the enamine of the starting ketone under basic conditions. After cyclization the product is released from the solid phase by using TBAF at room temperature overnight. For the products **75a**–**c** only one stereoisomer was formed (*de* > 98%) with modest 22–38% yields.

Robinson annulation is abundantly used in medicinal chemistry as a means for producing natural-product like structures such as Amgen's (AMG 067). The compound is a potent melanin-concentrating hormone receptor 1 (MCHr1) antagonist for the treatment of obesity.[45] The benzylpiperidone (**76**) starting material is converted to the enamine **77** by gradual heating with pyrrolidine in toluene until reflux; the crude is then subjected to Michael addition of 3-penten-2-one (EVK) followed by cyclization to give a racemic mixture of *syn*- and *anti*-isomers (methyl related to the bridgehead substituent) and a mixture of α,β- and β,γ-enones **78a–d**. The mixture is treated with di-*p*-toluoyl-L-tartaric acid (L-DTTA) in 80% ethanol/water which crystallizes the desired diastereomeric salt as an off-white solid. The process gives **76**•L-DTTA of the (8*S*,8*R*) enantiomer, and so the optical purity revealed that the enantiomer is obtained as 97% *ee* and is produced in 28% yield from the starting benzylpiperidone.

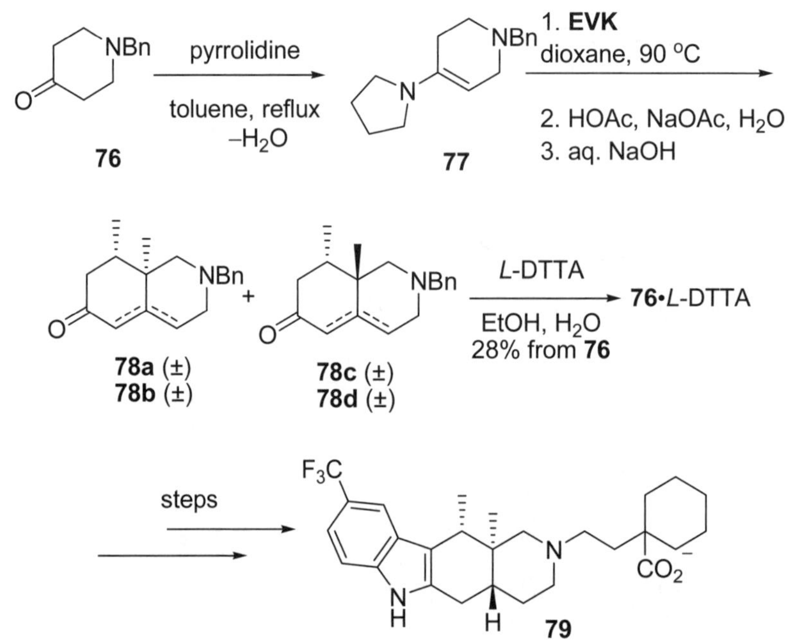

Robinson annulation although not extensively has been used in synthetic routes toward complex carbohydrate structures such as in the synthesis of **82**.[46] The starting ketone **80** reacts with (trimethylsilyl)but-3-en-2-one at –78 °C in the presence of lithium 2,2,6,6-tetramethylpiperidide (LTMP) base to give the alcohol **81**. The α-methyl inverts to the axial position. The alcohol then produces the Robinson annulation adduct **82** in the presence of catalytic amount of methanolic potassium hydroxide in 58% yield.

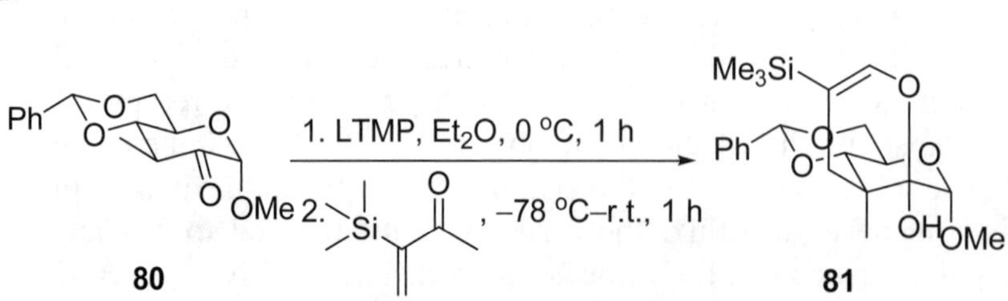

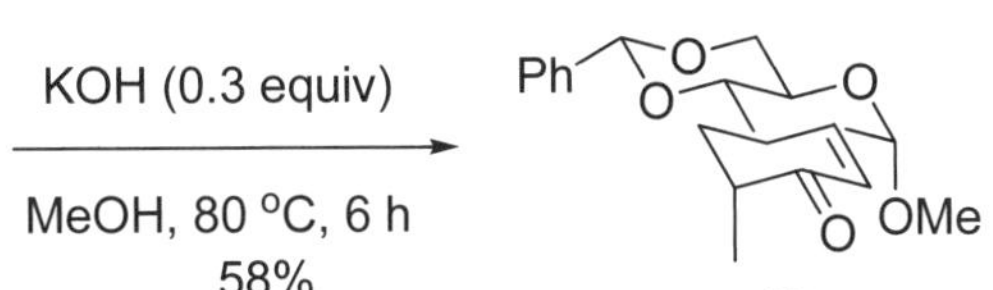

82

The so-called homo-Robinson annulation, as pioneered and reported by Danishefsky in 2005, is an innovative Robinson annulation that can be used for the synthesis 7,5-fused rings such as bicyclo[5.3.0]decane ring (hydroazulene core). The annulation protocol uses the Robinson annulation techniques to form the starting ketone which undergoes ring expansion to give a 7,5-fused ring. In 2008 Danishefsky et al. reported a revised protocol to eliminate unwanted side reaction. On the route toward the synthesis of a natural product core, compound **88** was synthesized using this transformation. The Robinson annulation intermediate is synthesized through Mukayama-like conditions (BF_3•Et_2O, acetic acid, and MVK) to give the Michael adduct **84**. Cyclization was brought about by the use of sodium methoxide as a base. Ring expansion steps proceed with the cyclopropanation of a silyl enol ether (derived from the Robinson annulation adduct) with halocarbene that is followed by silver-promoted ring expansion to give the bromodienone in 59% yield (three steps from **85**), which is reduced to give the hydroazulene adduct **88** in 41% yield.

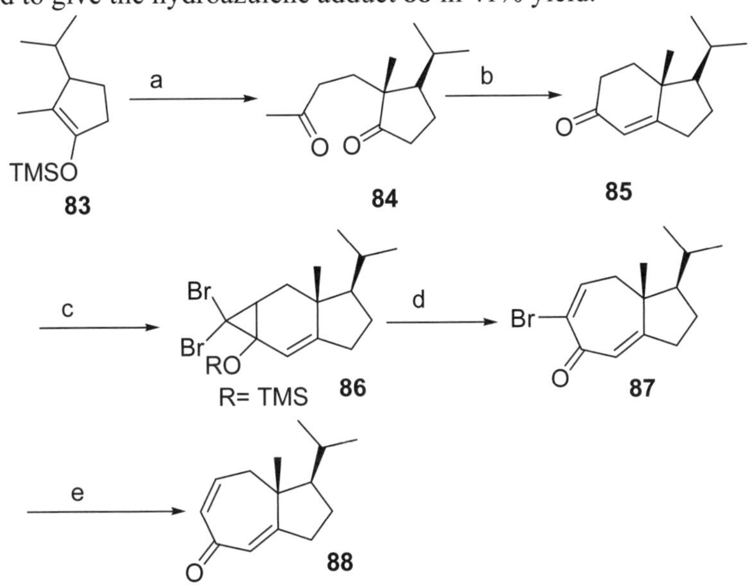

(a) MVK, AcOH, BF_3•Et_2O, –20 °C, 97%; (b) NaOMe, 98%;
(c) (i) LDA, TMSCl, THF, –78 °C, (ii) $CHBr_3$, KO*t*-Bu, pentane 0–23 °C;
(d) $AgNO_3$, pyr, EtOH, 23 °C, 1.5 h, 59% (three steps from 3);
(e) BF_3•OEt_2, NaI, CH_3CN, 0 °C, 41%

Applications in the total synthesis of natural products

Robinson and Rapson first discovered this important transformation during there synthetic trials toward the synthesis of compounds with steroids structures such as sex hormones, e.g. oestrone and androgen, as well as their grand synthesis of cholesterol and many more syntheses.[1,48a–d] In the 1950s cholesterol synthesis was considered to be one of the most important accomplishment that this annulation was used for during the 16 years since its initial discovery. Robinson reported the total synthesis of cholesterol almost at the same time as Robert Woodward of Harvard University. The synthetic route was published in its entirety in 1951, although parts of the synthetic routes were being worked on earlier, such as the pivotal step of the transformation of 1,6-dimethoxynaphthalene (**84**) to 5-methyl-β-tetralone (**86**) as reported by Robinson and Cornforth in 1942. In the final Robinson–Cornforth route, alkylation of **86** with methyl iodide in the presence of sodium methoxide gives **11** in 62% yield. The Robinson annulation then follows by refluxing the product with the Mannich adduct **10** (methyl vinyl ketone precursor) to give the intermediate **12** in 71% yield. The synthesis proceeds to a total of 38 steps as one of the first completed synthetic routes for cholesterol.

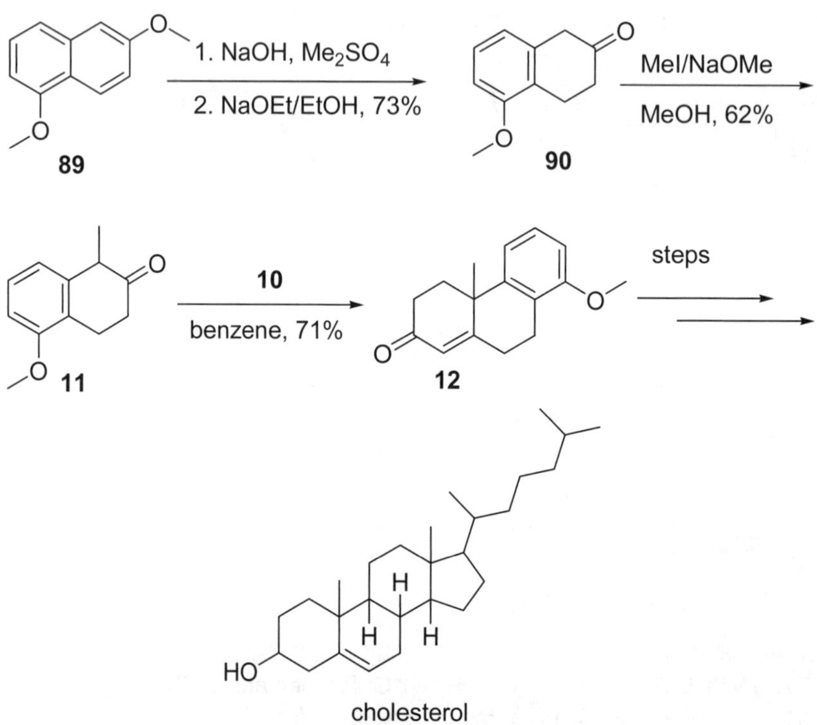

Corey's enantiospecific synthesis of pseudopteroxazole shows a use of a modified Robinson annulation for the synthesis of the octalone ring **93**.[49] Silyl enol ether of the ketone **9** was synthesized under kinetic conditions of LDA and chlorotrimethylsilane. The intermediate is subjected to the Mukayama–Michael reaction conditions by reacting with 1-benzyloxy-3-methyl-but-3-en-2-one and $SnCl_4$, which gives the alkylation adduct **92** in 61% yield (as a mixture of diastereomers). The product is then cyclized through standard aldol cyclization conditions of 0.01 M of ethanolic solution of potassium hydroxide, which give a ketol intermediate. The hydroxy group of the tertiary alcohol is then eliminated to give the diastereomeric α,β-enone in 83% yield. The cyclized enone **93** then undergoes a different transformation to give the pseudopteroxazole (**94**).

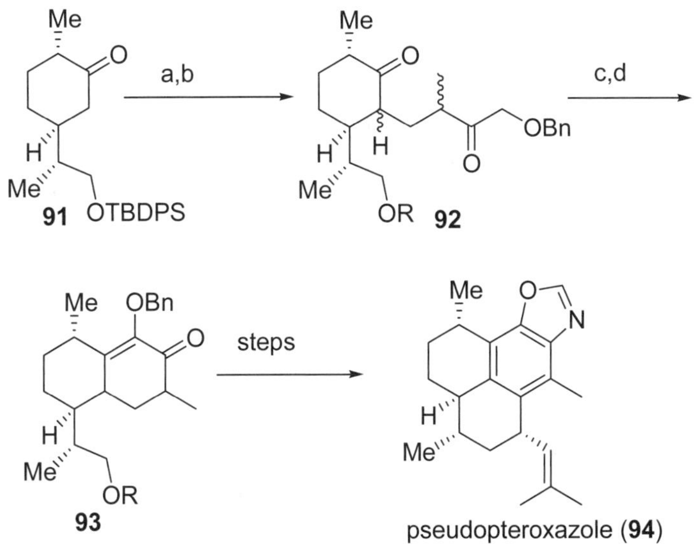

(a) LDA, TMS-Cl, –78 °C, 100%. (b) 1-Benzyloxy-3-methyl-but-3-en-2-one, $SnCl_4$, CH_2Cl_2, –78 °C, 61%.
(c) KOH, EtOH, –10 °C, 83%. (d) $SOCl_2$, pyridine, 83%.

4.14.6 Experimental

8-Methoxy-4a-methyl-4,4a,9,10-tetrahydrophenanthren-2(3H)-one (12)[8]

Diethylaminobutanone (15.05 g, prepared according Ref. 50) was swirled gently in a 1-L flask and cooled in ice during the addition of methyl iodide (15.0 g) in portions over 0.5 h. The swirling was regulated so as to obtain the crystalline methiodide as an even coating on the walls of the flask for 0.5 h and then under the tap for 45 min. A solution of the ketone (**11**) (20.0 g) in dry, thiophen-free benzene (100 cc) was added, air was expelled from the flask by dry nitrogen, and a solution of potassium (6.5 g) in dry ethanol (100 cc) added with ice cooling over 5 minutes. Swirling was continued until the methiodide had all dissolved (about 30 min) and was replaced by a precipitate of potassium iodide. After it had been kept in ice for another hour, the mixture was boiled gently for 25 minutes. An excess of 2 N-sulphuric acid was then added and the nitrogen stream stopped. After addition of enough water to dissolve the potassium sulphate the benzene layer was separated and the aqueous layer extracted twice with ether. The united extracts were washed with water, clarified with a little magnesium sulphate, and evaporated. The residue was distilled and 23.2 g were collected up to 180 °C/0.1 mm. The distillate was warmed until fluid, and ether added gradually until the total weight was 40 g. Crystallisation set in at once and was allowed to proceed at 0 °C overnight; the ketone (**12**) (17.0 g; m.p. 115–117 °C) was then collected and washed with chilled ether. The mother liquors after fractional distillation afforded an additional 1 g; the total yield was thus 71%.

1,6-Dioxo-8a-Methyl-1,2,3,4,6,7,8,8a-Octahydronaphthalene (44)[21]

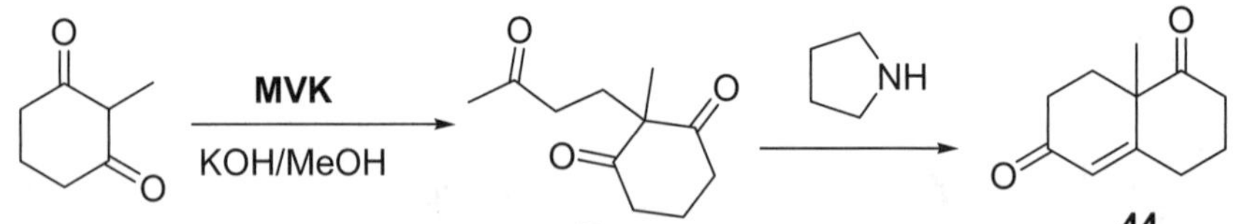

A mixture of 63.1 g (0.5 mol) 2-methyl-1,3-cyclohexanedione (**45**), 52.6 g (0.75 mol) methyl vinyl ketone, about 0.25 g (3 pellets) potassium hydroxide, and 250 mL absolute methanol is placed in a 500-mL round-bottomed flask fitted with a reflux condenser and a drying tube. The mixture was heated under reflux for 3 h, and the dione gradually went into solution. At the end of this period, methanol and the excess methyl vinyl ketone were removed by distillation under reduced pressure. The residual liquid was dissolved in 250 mL benzene, a Dean–Stark phase-separating head was attached, and 20 mL solvent was removed by distillation at atmospheric pressure to remove traces of water and methanol. The solution was cooled well below the boiling point, 3 mL of pyrrolidine is added and the mixture held at reflux for about 30 min, during which time about 9 mL of water collects in the trap. Refluxing was continued for an additional 15 min after the separation of water ceases. The water collected was removed, and then 50 mL of solvent was distilled. The reddish reaction mixture is cooled to room temperature and diluted with 150 mL ether. This solution was washed with 100 mL distilled water containing 15 mL 10% hydrochloric acid and 100 mL water. The aqueous extracts were extracted with 50 mL ether, and the combined organic layers were washed with three 100-mL portions water, then with saturated salt solution and dried over magnesium sulfate. The solvents were then removed, and on distillation of the residue (82–85 g) at 0.5–1.0 mm the material, b.p. 117–145 °C, was collected and diluted with 5 mL of ether. The distillate was placed overnight in a refrigerator; the resulting crystals were then collected by rapid filtration and washed with about 25 mL cold ether. The first crop of diketone weighed 50–53 g and was colorless. The combined mother liquors were redistilled to obtain a further 4–6 g crystalline product. A yield of 56–58 g (63–65% based on dione) 1,6-dioxo-8a-methyl-1,2,3,4,6,7,8,8a-octahydronaphthalene (**44**), m.p. 47–50 °C, suitable for most other purposes, was obtained.

4.14.7 References

1. Rapson, W. S.; Robinson, R. *J. Chem. Soc.* **1935**, 1285–1288.
2. [R] Laylin, K. *Nobel Laureates in Chemistry*; American Chemical Society & Chemical Heritage Foundation, Washington, D.C., **1993**.
3. Lewis, H. J.; Ramage, G. R.; Robinson, Robert *J. Chem. Soc.* **1935**, 1412–1414.
4. Robinson, R.; Schlittler, E. *J. Chem. Soc.* **1935**, 1288–1291.
5. Crowfoot, D. M.; Rapson, W. S.; Robinson, R. *J. Chem. Soc.* **1936**, 757–759.
6. Rapson, W. S. *J. Chem. Soc.* **1936**, 1626–1628.
7. Du Feu, E. C.; McQuillin, F. J.; Robinson, R. *J. Chem. Soc.* **1937**, 53–60.
8. McQuillin, F. J.; Robinson, R. *J. Chem. Soc.* **1938**, 1097–1099.
9. Robinson, R.; Rydon, H. N. *J. Chem. Soc.* **1939**, 1394–1405.
10. Nunn, J. R.; Rapson, W. S. *J. Chem. Soc.* **1949**, 825–831.
11. [R] (a) Yung, M. E. *Tetrahedron* **1976**, *32*, 3–31, [R] (b) Gawley, R. E. *Synthesis* **1976**, 777–794.
12. (a) Chen, E. Y. *Synth. Commun.* **1983**, *11*, 927–31. (b) Jansen, B. J. M.; Kreuger, J. A.; De Groot, A. *Tetrahedron.* **1989**, *5*, 1447–1452. (c) Marshall, J. A.; Fanta, W. I. *J. Org. Chem.* **1964**, *29*, 2501–2505.
13. Naussbaumer, C. *Helv. Chim. Acta.* **1990**, *73*, 1621–1636.
14. Birch, A. J.; Robinson, R. *J. Chem. Soc.* **1944**, 503–506.
15. Cook, J. G.; Robinson, R. *J. Chem. Soc.* **1941**, 391–393.
16. Ziegler, F. E.; Hwang, K.-J. *J. Org. Chem.* **1983**, *48*, 3349–3351.
17. Stork, G.; Brizzolarha, A.; Landesman, H.; Szmuszkovicz, J. *J. Am. Chem. Soc.* **1963**, 207–222.
18. (a) Heathcock, C. H.; Ellis, J. E.; *Tetrahedron Lett.* **1971**, *52*, 4995–4996 (b) Paquette, L. A.; Belmont, D. T.; Hsu, Y. *J. Org. Chem.* **1985**, *50*, **4667**–4672.
19. Stork, G.; Jung, M. E. *J. Am. Chem. Soc.* **1974**, *96*, 3682–3684.
20. Telschow, J. E.; Reusch, W. *J. Org. Chem.* **1975**, *40*, 862–865.
21. Ramachandran, S.; Newman, M. S. *Org. Synth.* **1973**, *Coll. Vol. 5*, 486.
22. Coates, R. M.; Shaw, J. E. *J. Am. Chem. Soc.* **1970**, *92*, 5657–5664.
23. Stork, G.; Ganem, B. *J. Am. Chem. Soc.* **1973**, *95*, 6152–6153.
24. Narasaka, K.; Soai, K.; Aikawa, Y.; Mukaiyama, T. *Bull. Chem. Soc. Japan.* **1976**, *49*, 779–783.
25. Yanami, T.; Miyashita, M.; Yoshikoshi. A. *J. Org. Chem.* **1980**, *45*, 607–612.
26. Huffman, J. W.; Potnis, S. M.; Satish, A. V. *J. Org. Chem.* **1985**, *50*, 4266–4270.
27. Stork, G.; Hudrlik, P.F. *J. Am. Chem. Soc.* **1968**, *90*, 4462–4464.
28. Sato, T.; Wakahara, Y.; Nozaki, H. *Tetrahedron.* **1991**, 9773–9782.
29. [R] Bergmann, E. D.; Ginsburg, D.; Pappo, R. *Org. React.* **1959**, *10*, 179–555.
30. Fang, G. F.; Prato, M.; Kim, G.; Danishefsky, S. J. *Tetrahedron Lett.* **1989**, *30*, 3625–3628.
31. Guarna, A.; Lombardi, E.; Machetti, F.; Occhiato, E. G.; Scarpi, D. *J. Org. Chem.* **2000**, *65*, 8093–8095.

32. (a) Hajos, Z. G.; Parrish, D. R. German patent DE 71–2102623 (issued to Hoffman-LaRoche). (b) Hajos, Z. G.; Parrish, D. R. *J. Org. Chem.* **1974**, *10*, 1615–1621.
33. (a) Eder, U.; Sauer, G.; Wiechert, R. *Angew. Chem., Int. Ed. Engl.* **1971**, *10*, 496–497; (b) Eder, U.; Sauer, G.; Wiechert, R. German patent DE 70–214757 (issued to Schering A.G.)
34. Bui, T.; Barbas, C. F., III *Tetrahedron Lett.* **2000**, *41*, 6951–6954.
35. Spencer, T. A.; Schmiegel, K. K.; Williamson, K. L. *J. Am. Chem. Soc.* **1963**, *85*, 3785–3793.
36. Scanio, C. J. V.; Starrett, R. M. *J. Am. Chem. Soc.* **1971**, *93*, 1539–1540.
37. Akahane. Y; Inage, N.; Nagamine, T.; Inomata, K.; Endo, Y. *Heterocycles* **2007**, *74*, 637–648.
38. (a) D'Angelo, J.; Desmaele, D.; Dumas, F.; Guingant, A. *Tetrahedron: Asymmetry* **1992**, *3*, 459–505. (b) Volpe, T.; Revial, G.; Pfau, M.; d'Angelo, J. *Tetrahedron Lett.* **1987**, *28*, 2367–2370.
39. Zhong, G.; Hoffmann, T.; Lerner, R. A.; Danishefsky, S.; Barbas, C. F., III. *J. Am. Chem. Soc.* **1997**, *119*, 8131–8132.
40. Akiyama. T.; Katoh, T.; Mori, K. *Angew. Chem., Int. Ed.* **2009**, *48*, 4226–4228.
41. Corey, E. J.; Nozoe, S. *J. Am. Chem. Soc.* **1963**, *85*, 3527–3528.
42. Woodard, R. B.; Sondheimer, F.; Taub, D.; Heusler, k. *J. Am. Chem. Soc.* **1952**, *74*, 4223–4251.
43. Ghosh, S.; Rivas, F.; Fischer, D.; Gonzalez, M. A.; Theodorakis, E. A. *Org. Lett.* **2004**, 6, 941–944.
44. Röttger, S.; Waldmann, R. *Eur. J. Org. Chem.* **2006**, 2093–2099.
45. Andersen, D.; Storz, T.; Liu, P.; Wang, W.; Li, L.; Fan, P.; Chen, X.; Allgeier, A.; Burgos, A.; Tedrow, J.; Baum, J.; Chen, Y.; Crockett, R.; Huang, L.; Syed, R.; Larsen, R. D.; Martinelli, M.; *J. Org. Chem.* **2007**, *72*, 9648–9655.
46. (a) Bonnert, R. V.; Jenkins, P. R. *J. Chem. Soc., Chem. Commun.* **1987**, *1*, 6–7. (b) Bonnert, R. V.; Howarth, J.; Jenkins, P. R.; Lawrence, N. J. *J. Chem. Soc., Perkin. Trans. 1.* **1991**, *5*, 1225–1229.
47. (a) Yun, H.; Danishefsky, S. J. *Tetrahedron. Lett.* **2005**, *46*, 3879–3882. (b) Singh, M. P.; Janso, J. E.; Luckman, S. W.; Brady, S. F.; Clardy, J.; Greenstein, M.; Maiese, W. M. *J. Antibiot.* **2000**, *53*, 256–261.
48. (a) Cornforth, J. W.; Cornforth, R. H.; Robinson, R. *J. Chem. Soc.* **1942**, 689–691. (b) Cornforth, J. W.; Robinson, Robert. *Nature* **1947**, *160*, 737–739. (c) Cardwell, H. M. E.; Cornforth, J. W.; Duff, S. R.; Holtermann, H.; Robinson, R. *Chem. Ind.* **1951**, 389–390. (d) Cardwell, H. M. E.; Cornforth, J. W.; Duff, S. R.; Holtermann, H.; Robinson, R. *J. Chem. Soc.* **1953**, 361–384.
49. Davidson, J. P.; Corey, E. J. *J. Amer. Chem. Soc.* **2003**, *125*, 3486–3489.
50. Wilds, A. L.; Shurk, C. H. *J. Am. Chem. Soc.* **1943**, *65*, 471–475.

4.15 Scholl Reaction

Richard J. Mullins and Michael T. Corbett

4.15.1 Description

The Scholl reaction, formally representing a Friedel–Crafts arylation reaction, is the aryl–aryl coupling of two aromatic species **1** via dehydrogenation under Lewis acid and protic acid conditions.

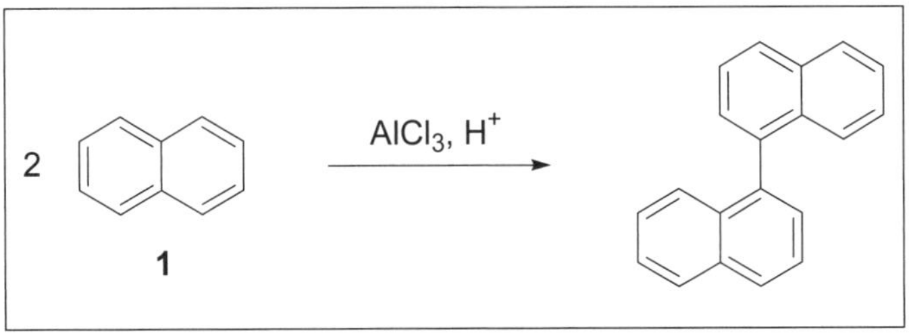

4.15.2 Historical Perspective

One of the more important and useful organic reactions, known as the Friedel–Crafts reaction, was discovered in 1877 by Charles Friedel and James Crafts.[1] This class of reactions is generally thought to include all sets of electrophilic alkylation and acylation reactions on aromatic rings promoted by Lewis acids (traditionally $AlCl_3$ or $FeCl_3$) typically under anhydrous reaction conditions. Due to their synthetic utility, Friedel–Crafts reactions have been extensively studied and utilized across a broad and diverse area of chemical research.

The aryl–aryl coupling reaction, currently known as the Scholl reaction, was first crudely observed by Friedel and Crafts in 1885 when it was determined that naphthalene in the presence of $AlCl_3$ formed an appreciable amount of β,β-dinapthyl at high temperatures.[2] Similar observations were made by Homer in 1907;[3] however, it was not until the early 1910s that Roland Scholl and co-workers began to investigate further and generalize this class of aryl–aryl coupling reactions under Friedel–Crafts conditions beginning with the syntheses of *meso*-naphthodianthrone[4] and 3-hydroxy-1,2-benzfluorenone.[5]

Neither the synthetic scope nor mechanism was significantly investigated until 1935, when Baddeley and co-workers provided new insights into the Scholl reaction. Their work provided the first mechanistic interpretation, suggesting that the homocoupling reaction of naphthalene proceeds via an arenium cation intermediate.[6] Based on experiments in which HCl (a decomposition product of $AlCl_3$ in the presence of water) was excluded from the reaction, Baddeley determined that the presence of HCl was essential for carrying out the Scholl reaction.[7] It was also shown that the addition of NaCl results in the formation of $NaAlCl_4$, which does not promote the Scholl reaction.

The Scholl reaction is traditionally run in the presence of a Lewis acid, such as $FeCl_3$ or $AlCl_3$, under typical Friedel–Crafts conditions. Recently, however, various other oxidants have also been shown to effectively promote the reaction including $CuCl_2$, $Cu(OTf)_2$, $MoCl_5$, $SbCl_5$, etc., allowing for milder reaction conditions.[8]

4.15.3 Mechanism

There are two plausible mechanistic pathways under debate for the Scholl reaction. The mechanistic uncertainty revolves around whether the reaction proceeds through cationic or radical cation intermediates. Both of these mechanisms are outlined below.

The accepted mechanism, which proceeds through the arenium cation intermediate, occurs as follows.[9,10] First, protonation of the aryl species **2** occurs to afford the electrophilic σ-complex **3**. What follows is a two-step electrophilic aromatic substitution process whereby the electrophilic intermediate **3** is attacked by the adjacent aromatic ring to form a new carbon-carbon bond in intermediate **4**. Deprotonation of **4** regenerates the aromatic species in intermediate **5**. Finally, oxidation/aromatization of the product occurs with the formal expulsion of H_2, resulting in the formation of **6**.

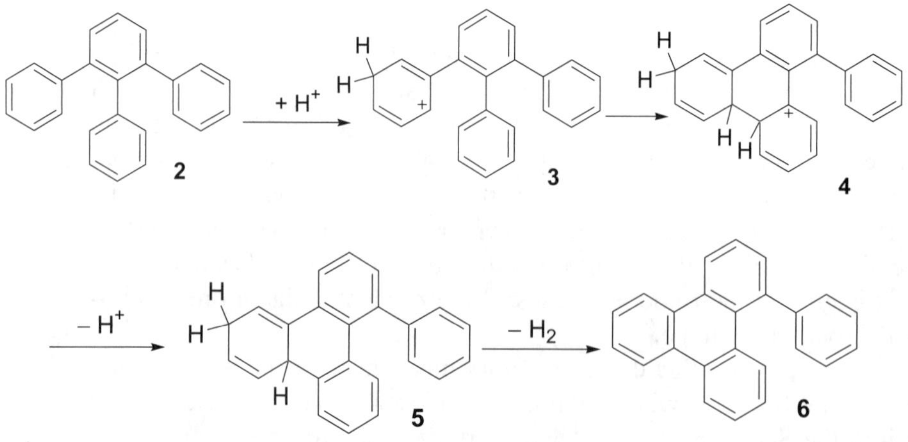

The radical cation mechanistic pathway utilizing an oxidizing agent such as $CuCl_2$ is presumed to occur via a stepwise process characterized by isolable intermediates.[9,11] In the mechanism, the aromatic species **7** undergoes a one-electron oxidation to provide the radical cation species **8**. This radical cation **8** then undergoes an electrophilic carbon–carbon bond forming reaction with an adjacent aromatic ring to provide the ring-closed intermediate **9**. Subsequent deprotonation of **9** regenerates the aromatic species in intermediate **10**. Finally, the formal loss of an H-atom via the intermediate **10** results in aromatization and complete formation of the coupling adduct **11**.

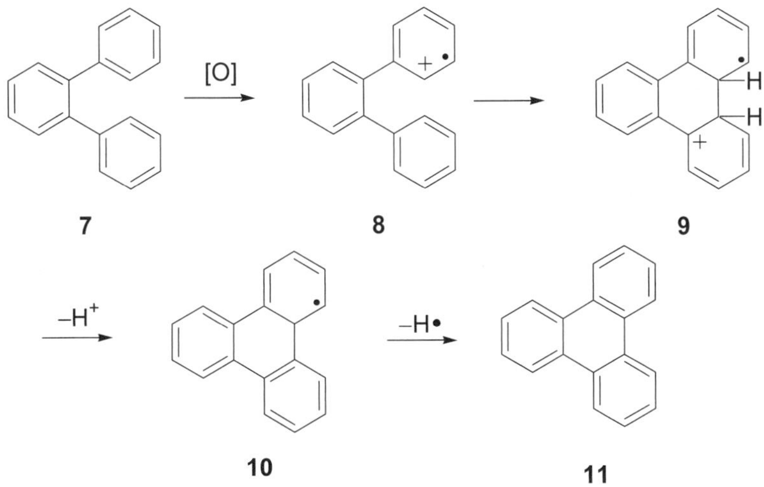

The arenium cation mechanistic pathway was first proposed in 1935 by Baddeley and was later reinforced by Balaban and Nenitzescu.[6,12,13] Computational calculations performed by King and co-workers have produced several important conclusions.[9,14] First, it was determined that the arenium cation mechanistic pathway was thermodynamically favored in studies under both vacuum and solvated conditions due to lower energy transition states than those found in the radical cation mechanistic pathway. Second, due to the increasingly exergonic nature of the reaction and the observed nonaccumulation of intermediates, C–C bond formation, in the case of hexaphenylbenzene, was found to occur slowest for the first bond and fastest for the last bond. The arenium cation mechanistic pathway is further supported by evidence that the Scholl reaction can proceed in acidic solutions that do not promote radical formation, such as anhydrous HF.[15] The cationic mechanism has been shown to predominate in strongly protic conditions. ESR[16] and EPR[17] studies have concluded that the radical cations previously observed during the Scholl reaction are not part of the actual reaction, but rather are formed by the interaction of the polycyclic products with $AlCl_3$ in solution.

The radical cation mechanistic pathway was first proposed in 1961 by Rooney and Pink[18] and is also consistent with mechanisms proposed based on related studies performed on analogous classes of reactions.[19,20] ESR studies have been performed to investigate the aromatic cation radicals present during the Scholl and have determined that radical species are present in reactions with $AlCl_3$.[21,22] Calculations by Di Stefano and Negri also support the radical cation mechanistic pathway, which is consistent with observations made by Müllen and co-workers.[11] The radical cation mechanistic pathway, which is stepwise, is supported by experimental evidence that intermediates can be isolated during the reaction, which is inconsistent with the arenium cation mechanism.[23] Further research by Kovacic into the role of oxidizing agents in the formation of radical cation species in the Scholl reaction has also provided some experimental evidence to support this pathway.[8]

The limited knowledge of the mechanism to date has caused a schism to form due to insufficient physical evidence to support either mechanistic pathway definitively. Thus the reader is directed to the aforementioned papers for a complete understanding of the mechanism.

4.15.4 Variations and Improvements

While the original Scholl conditions called for both a Lewis and protic acid, research by Kovacic and Kyriakis into the role of oxidizing agents (that facilitate the formation of radical cations) in the Scholl reaction led to the observation that benzene (**12**) when reacted in the presence of a heterogeneous mixture of anhydrous aluminum chloride and copper chloride afforded poly(*para*-phenylene) (**13**).[8,24,25] It is important that this reaction was conducted under mild conditions (25–35 °C) and was complete in only 2 h. As discoverers of one of the first Scholl reactions conducted under mild conditions, Kovacic and Kyriakis have greatly increased the synthetic utility of this reaction.

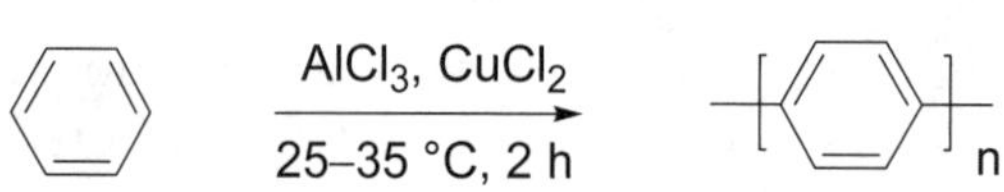

12 **13**

4.15.5 Synthetic Utility

Although the Scholl reaction has been known for almost a century, due to the harsh reaction conditions, its synthetic utility had, until recently, not been extensively explored. With the discovery by Kovacic and co-workers, its synthetic scope has been significantly broadened through the adoption of efficient and mild reaction conditions.[25] For an extensive survey of the Scholl reaction before 1970, the reader is led to the review by Balaban and Nenitzescu.[10]

One of the earliest examples of the Scholl reaction was performed during the ring-closing synthesis of 3-hydroxy-1,2-benzfluorenone (**15**).[5] Using 4-hydroxy-1-benzoylnaphthalene (**14**) under Friedel–Crafts conditions, the desired product **15** was obtained in high yield along with only trace amounts of **16**, suggesting important substituent effects in the reaction. Substituted 1,2-benzfluorenes have also been accessed using this approach.[26]

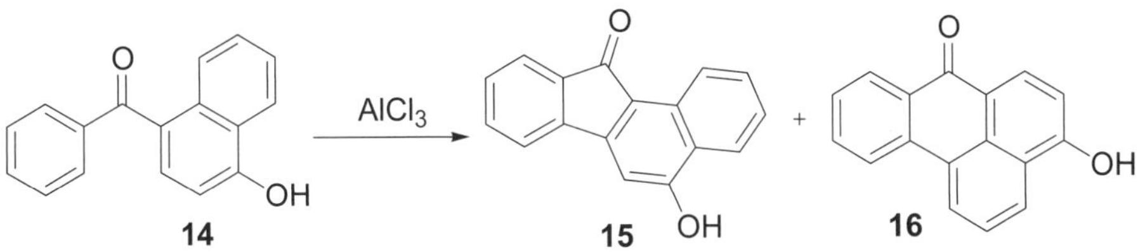

14 **15** **16**

Musgrave and Buchan developed a procedure for the preparation of triphenylene-1,4-quinones from 2,3-diaryl-1,4-benzoquinones using an acid-catalyzed intramolecular Scholl reaction.[27] Using the 2,3-diaryl-1,4-benzoquinone **17**, the $AlCl_3$-mediated cyclization occurred to afford a quinol intermediate that was then oxidized in the presence of $FeCl_3$ to afford the triphenylenequinone **18** in moderate yield under mild reaction conditions. Inclusion of 2,3-dichloro-5,6-dicyano-1,4-benzoquinone (DDQ) as an oxidant has also been shown to increase the yield of the reaction up to 50%.

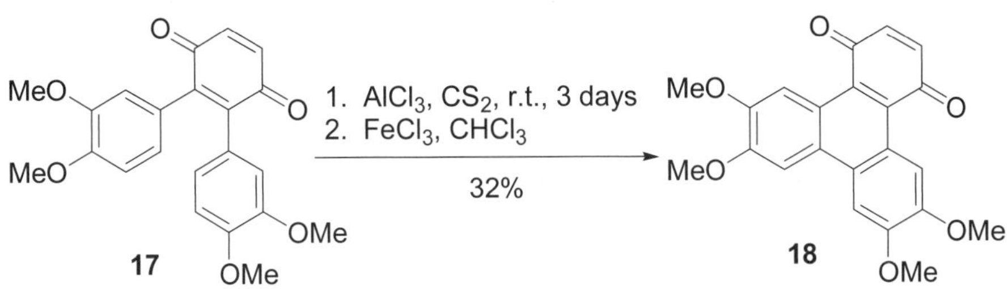

17 **18**

The intermolecular version of the Scholl reaction is much less prevalent in the literature due to the difficulty in controlling the formation of the desired products. Bushby and co-workers have used an intermolecular version of the Scholl reaction in efforts toward the synthesis of chirally discotic liquid crystals featuring the triphenylene nucleus.[28] The fluorobenzene derivative **20** was coupled with the symmetric biphenyl **19** to afford the desired product **21**, the structure of which was determined by NOE analysis.

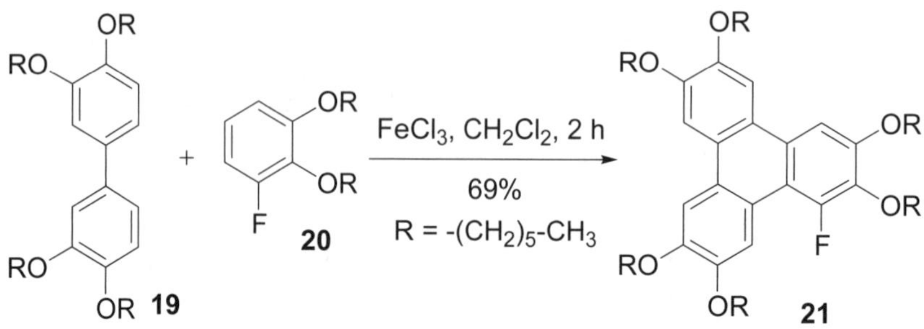

In the early 1990s, an increased interest in molecular electronics has led to the desire to obtain functionally and geometrically interesting planar organic species via facile synthetic routes.[29,30] As such, the Scholl reaction has become increasingly useful due to its ability to effect intramolecular aryl–aryl coupling reactions in a controlled manner. To meet the needs in this area, milder reaction conditions have been developed, making the Scholl reaction more applicable to the synthesis of compounds with sensitive functionality.

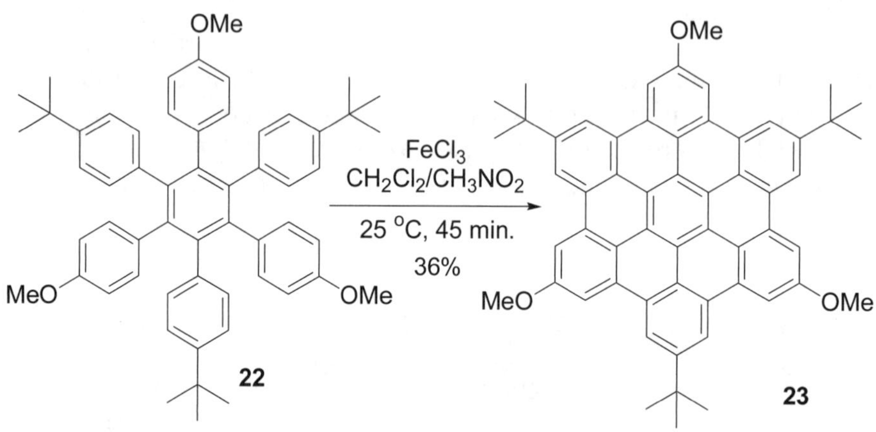

The Scholl reaction has been extensively used to obtain a wide range of functionalized and fully cyclized hexa-*peri*-hexabenzocoronenes (HBCs). These compounds are useful because of their potential application as organic semiconductors for use in a wide range of electronic and optoelectronic devices due to their robustness and ability to effectively π-stack. An example is the structurally interesting C_3 symmetric *meta*-trimethoxy substituted HBC with alternating methoxy and *tert*-butyl substituents synthesized by Müllen and co-workers.[31] The Scholl reaction precursor **22** was synthesized via a

$Co_2(CO)_8$-catalyzed cyclotrimerization. The subsequent $FeCl_3$-mediated planarization of **22** under mild reaction conditions afforded the HBC **23** in moderate yield.

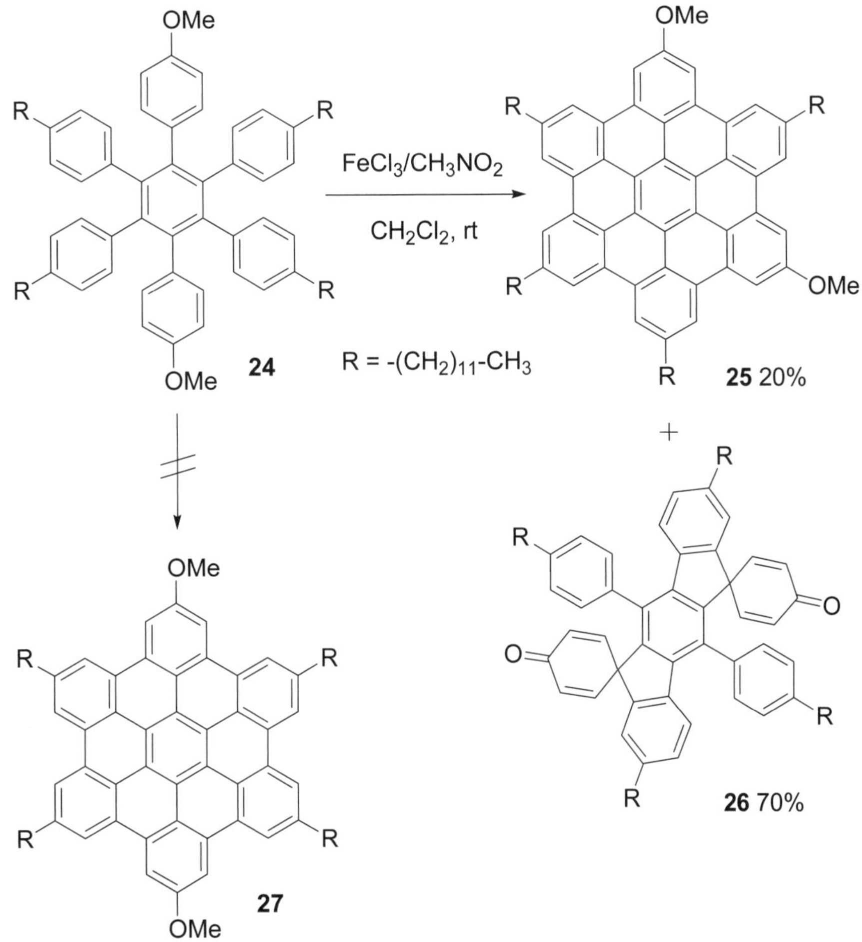

24, **25** 20%, **26** 70%, **27**

The arrangement of substituents in the hexaphenylbenzene precursor for most HBCs has been known to play a significant role in the outcome of the Scholl reaction. Therefore, acquiring some HBCs is not as synthetically straightforward as it may be perceived. Rearrangements during the Scholl reaction were first observed by Müllen and co-workers during the proposed synthesis of dimethoxy-substituted HBCs.[32] Subjection of the *para*-dimethoxy HBC **24** to Scholl conditions provided a mixture of the only observed HBC product **25** (20%) and bis-spirocyclic dienone **26** (70%) while none of the predicted product **27** was observed. The observed rearrangements are consistent with both arenium cationic and radical cation mechanisms.

The implications of substituent arrangements on the course of the Scholl reaction along with synthetic strategies to bypass these roadblocks have been extensively studied by King and co-workers.[33,34] An example of this is work by Rathore and co-workers that sought to avoid the spirocycle formation first observed by Müllen by employing new precursor geometries.[35] Using the alkoxy-precursor **28**, the hexaalkoxy HBC **29** can be obtained in nearly quantitative yields with no observed spirocycle formation. Long alkyl groups are necessary for the reaction to proceed, however, since no desired product is observed when R = CH_3.

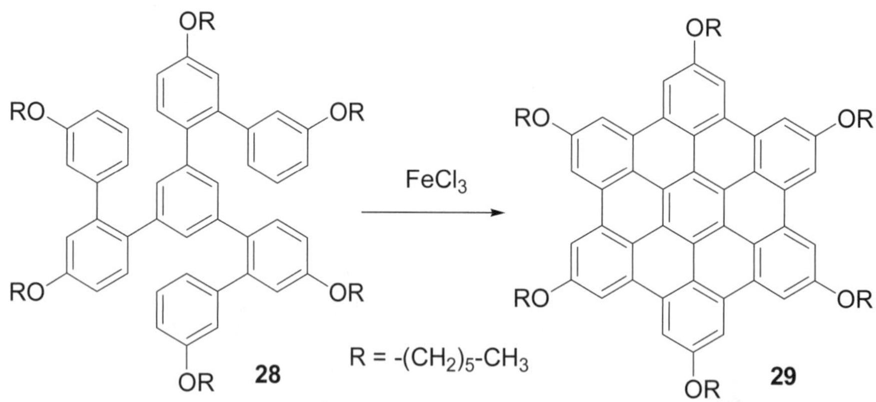

28 **29**

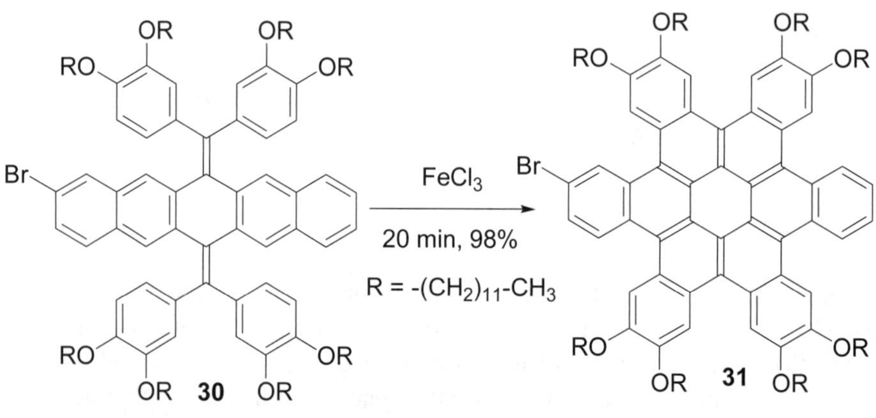

30 **31**

Nontraditional HBC architectures have also gained synthetic interest recently. Contorted HBCs were expeditiously synthesized in three facile steps by Nuckolls and co-workers using a $FeCl_3$-mediated Scholl reaction.[36] The Scholl reaction precursor **30** was obtained via a two-step synthesis from a soluble pentacene quinone derivative which could then undergo an intramolecular cyclization to afford the contorted HBC **31** in high yield.

Although some substrates do not undergo complete cyclization, the Scholl reaction has been shown to be broadly applicable as an effective route towards these types of HBCs.

Research into the preparation of heterocyclic HBCs has also come to the foreground in the literature. The development of pyrimidyl-*penta*-phenylbenzene systems was sought by Gourdon and co-workers to access novel optoelectronic materials from hetero-oligophenylenes.[37] Substitution on the pyrimidyl moiety greatly altered the degree of cyclization observed in the products. When precursor **32** was treated under Scholl conditions, the fully-cyclized diaza-hexa-*peri*-benzocoronene **33** was obtained in moderate yield. It is interesting that without the *t*-butyl substituent at the 2 position of the pyrimidine ring, incomplete cycloaddition was observed, only occurring *ortho* to the nitrogens in the pyrimidine moiety. This partial cyclization[38] has been exploited to the advantage of Moore and co-workers for the preparation of semifused HBCs, with interesting fluorophoric properties.[39] The isolation of these semi-fused HBCs has been cited as evidence for the stepwise radical cation mechanism proposed by Müllen.[23]

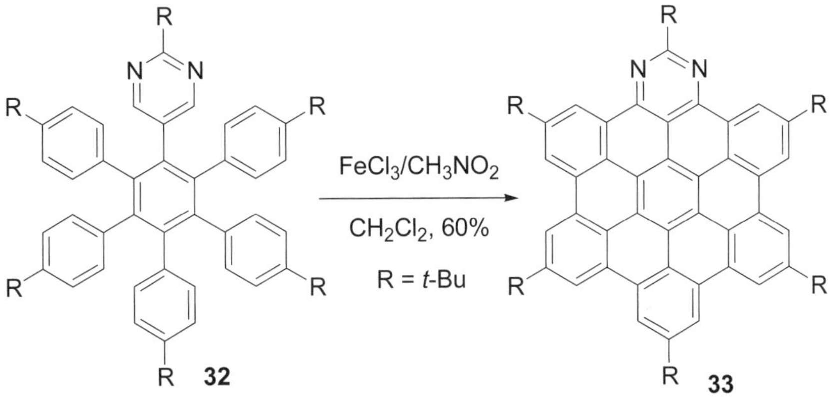

32 33

In addition to the ongoing research into the preparation of functionalized hexa-*peri*-benzocoronenes, there has been significant interest in the preparation of extended aromatic networks. Vingiello and co-workers used the Scholl reaction to obtain polycyclic aromatic compounds from 1-arylbenz[*a*]anthracenes.[40] 1-Phenylbenz[*a*]anthracene (**34**) when reacted under Scholl conditions in the presence of stannic chloride afforded dibenzo[*a*,*l*]pyrene (**35**) in moderate yield after just 5 min.

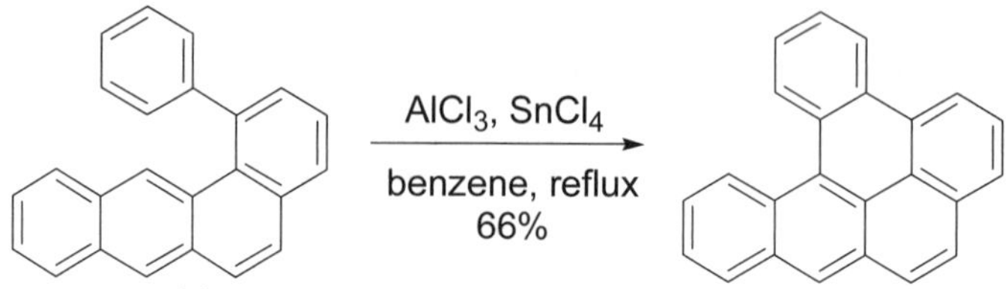

34 35

Müllen and co-workers have developed a procedure for converting stilbenoids to extended aromatic systems using a combination of [2 + 2]-cycloadditions followed by a Scholl cyclodehydrogenation.[41] Subsequent intermolecular and intramolecular cycloaddition reactions provide the polybenzenoid arene **36**, which exists as a mixture of conformations **36a** and **36b** in equilibrium at room temperature. As result of this conformational equilibrium, the AlCl$_3$-mediated cyclodehydrogenation afforded the isomeric mixture of **37** and **38**.

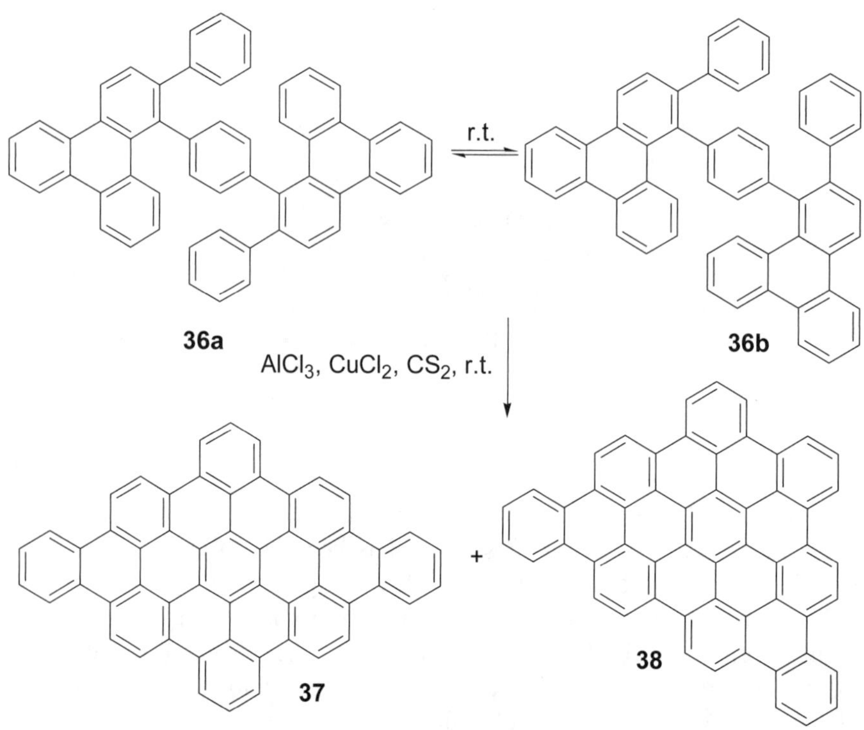

36a 36b 37 38

Research by Müllen and co-workers has continued to expand into the field of large polycyclic aromatic networks. Using similar approaches, molecular propellers have been synthesized using simple cyclodehydrogenation reactions of dendrimeric HBC-precursor materials.[42] Further work has led to the formation of up to 12-nm-long two-dimensional graphene nanoribbons from a nonplanar precursor.[43] The sterically hindered polyphenylene **39** was obtained through a Suzuki–Miyaura coupling polymerization. The $FeCl_3$-mediated oxidative cyclodehydrogenation of **39** afforded the aromatic ribbon **40** in good yield.

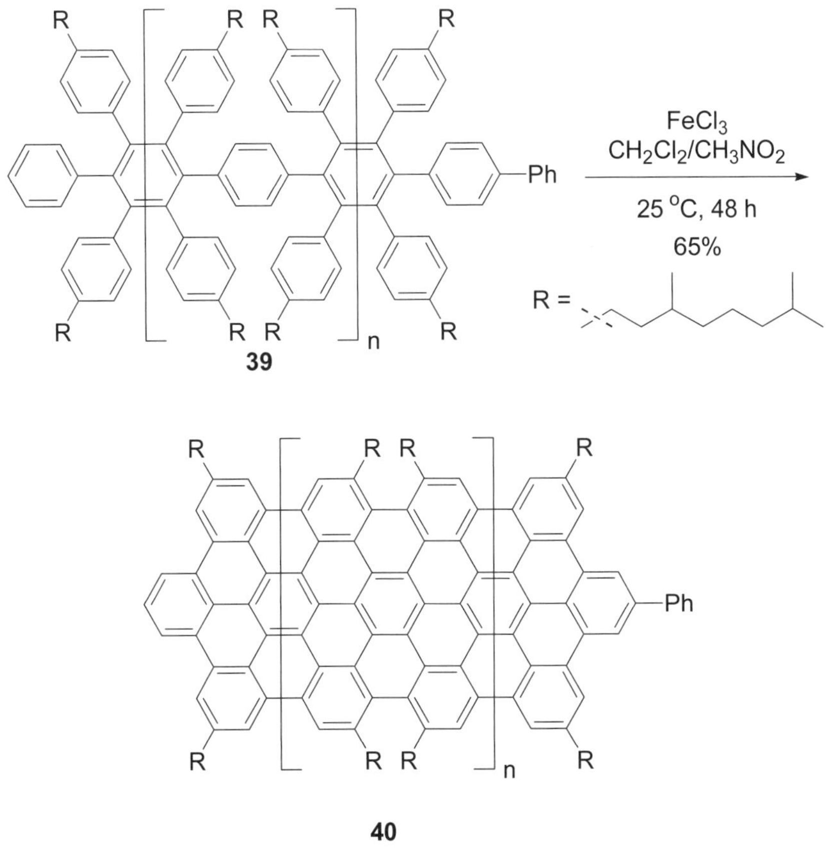

Use of the Scholl reaction for polymerization of substituted aromatic rings has precedence in the literature as well due to the facile and often predictable aryl–aryl coupling using these conditions. Geerts and co-workers used the Scholl reaction to access poly(*para*-phenylene) (**13**), which is used in blue light-emitting diodes, and other soluble poly(*para*-phenylene)s such as **41**.[44] The use of $MoCl_5$ as the oxidant afforded the desired polymers **13** and **41** in one-step with moderate molecular weights, low PDIs, and high *para* regioselectivity. For a more thorough review of the Scholl reaction's synthetic utility for the synthesis of poly(phenylene)s, the reader is directed to a review by Kovacic and Jones.[8]

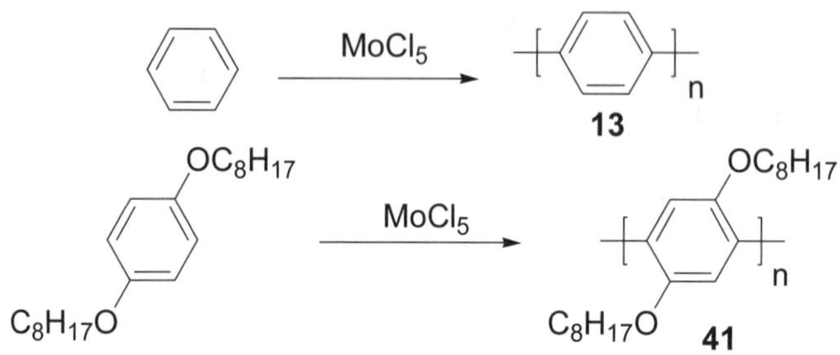

Percec and co-workers have also employed the Scholl reaction in a variety of polymerization reactions, especially for the synthesis of aromatic polyethers.[45] The 1-naphthol derivative **42** underwent the cation–radical polymerization under Scholl conditions to afford the polyether **43** in high conversion. As a major contributor to this area, Percec has used the Scholl reaction in a similar manner in various other polymerization studies of aromatic polyethers.[46–55] The Scholl reaction also been employed in polymerizations of binapthyl,[56–58] 4-methyltriphenylamine,[59] and 1,3-di-*n*-butoxybenzene[60]-based systems.

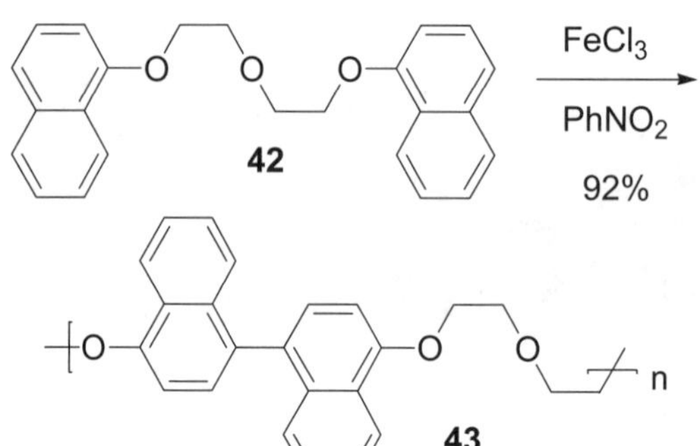

4.15.6 Experimental

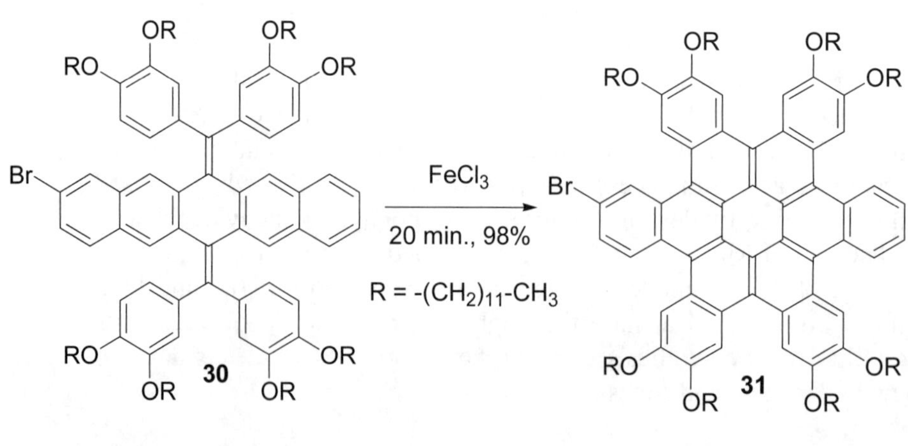

10-Bromo-2,3,6,7,14,15,18,19-octakis(dodecyloxy)trinaphtho[1,2,3,4-*fgh*:1',2',3',4'-*pqr*:1'',2'',3'',4''-za_1b_1]trinaphthylene (31)[36]

In a 1000-mL round-bottomed flask were added **30** (700 mg, 0.32 mmol) and 200 mL CH_2Cl_2. This solution was closed with a rubber septum and bubbled with argon while adding a solution of $FeCl_3$ (840 mg, 5.2 mmol) in 2.0 mL nitromethane. The solution was bubbled for 5 min and then the solution was stirred under an argon atmosphere for 30 min. Methanol (300 mL) was then added and the solution was stirred for 10 min. The CH_2Cl_2 was removed by rotary evaporation and lead to a yellow precipitate. This solid was filtered with a Millipore vacuum filtration apparatus to give 682 mg of a yellow solid (**31**, 98%).

4.15.7 References

1. Friedel, C.; Crafts, J. M. *Compt. Rend.* **1877**, *84*, 1450–1454.
2. Friedel, C.; Crafts, J. M. *Compt. Rend.* **1885**, *100*, 692–698.
3. Homer, A. *J. Chem. Soc., Trans.* **1907**, *91*, 1103–1114.
4. Scholl. R.; Mansfeld, J. *Ber.* **1910**, *43*, 1734–1746.
5. Scholl, R.; Seer, C. *Ann.* **1912**, *394*, 111–123.
6. Baddeley, G.; Kenner, J. *J. Chem. Soc.* **1935**, 303–309.
7. Baddeley, G. *J. Chem. Soc.* **1950**, 994–997.
8. [R] Kovacic, P.; Jones, M. B. *Chem. Rev.* **1987**, *87*, 357–379.
9. Rempala, P.; Kroulík, J.; King, B. T. *J. Org. Chem.* **2006**, *71*, 5067–5081.
10. [R] Balaban, A. T.; Nenitzescu, C. D. In *Friedel–Crafts and Related Reactions*; G. A. Olah, Ed.; Interscience, New York, **1964**, Ch. 23.
11. Di Stefano, M.; Negri, F.; Carbone, P.; Müllen, K. *Chem. Phys.* **2005**, *314*, 85–99.
12. Nenitzescu, C. D.; Balaban, A. T. *Chem. Ber.* **1958**, *91*, 2109–2116.
13. Balaban, A. T.; Nenitzescu, C. D. *Acad. Repub. Pop. Rom., Fil. Cluj. Stud. Cercet. Chim.* **1959**, *7*, 521–529.
14. Rempala, P.; Kroulík, J.; King, B. T. *J. Am. Chem. Soc.* **2004**, *126*, 15002–15003.
15. Simons, J. H.; McArthur, R. E. *J. Ind. Eng. Chem.* **1947**, *39*, 364–367.
16. Rang, H.; Kispert, L. D.; Sang, H. *J. Chem. Soc. Perkin Trans. II* **1989**, 1463–1469.
17. Wang, T.; Wu, A.-A.; Gao, L.-G.; Wang, H.-Q. *Chin. J. Chem. Phys.* **2009**, *22*, 51–56.
18. Rooney, J. J.; Pink, R. C. *Proc. Chem. Soc.* **1961**, 142–143.
19. Pummerer, R.; Luther, F. *Ber. Dtsch. Chem. Ges.* **1928**, *61*, 1102–1107.
20. Rooney, J. J.; Pink, R. C. *Trans. Faraday Soc.* **1962**, *58*, 1632–1641.
21. Lehnig, M.; Reiche, T.; Reiss, S. *Tetrahedron Lett.* **1992**, *33*, 4149–4152.
22. Bakker, M. G.; Claridge, R. F. C.; Kirk, C. M. *J. Chem. Soc., Perkin Trans. II*, **1986**, 1735–1741.
23. Kübel, C.; Eckhardt, K.; Enkelmann, V.; Wegner, G.; Müllen, K. *J. Mater. Chem.* **2000**, *10*, 879–886.
24. Kovacic, P.; Kyriakis, A. *Tetrahedron Lett.* **1962**, 467–469.
25. Kovacic, P.; Kyriakis, A. *J. Am. Chem. Soc.* **1963**, *85*, 454–458.
26. Gross, M. E.; Lankelma, H. P. *J. Am. Chem. Soc.* **1951**, *73*, 3439–3442.
27. Buchan, R.; Musgrave, O. C. *J. Chem. Soc., Perkins Trans. I* **1975**, 568–572.
28. Boden, N.; Bushby, R. J.; Cammidge, A. N.; Duckworth, S.; Headdock, G. *J. Mater. Chem.* **1997**, *7*, 601–605.
29. [R] Berresheim, A. J.; Müller, M.; Müllen, K. *Chem. Rev.* **1999**, *99*, 1747–1785.
30. [R] Watson, M. D.; Fechtenkötter, A.; Müllen, K. *Chem. Rev.* **2001**, *101*, 1267–1300.
31. Feng, X.; Pisula, W.; Takase, M.; Dou, X.; Enkelmann, V.; Wagner, M.; Ding, N.; Müllen, K. *Chem. Mater.* **2008**, *20*, 2872–2874.

32. Dou, X.; Yang, X.; Bodwell, G. J.; Wagner, M.; Enkelmann, V.; Müllen, K. *Org. Lett.* **2007**, *9*, 2485–2488.
33. King, B. T.; Kroulík, J.; Robertson, C. R.; Rempala, P.; Hilton, C. L.; Korinek, J. D.; Gortari, L. M. *J. Org. Chem.* **2007**, *72*, 2279–2288.
34. Ormsby, J. L.; Black, T. D.; Hilton, C. L.; Bharat; King, B. T. *Tetrahedron* **2008**, *64*, 11370–11378.
35. Wadumethrige, S. H.; Rathore, R. *Org. Lett.* **2008**, *10*, 5139–5142.
36. Plunkett, K. N.; Godula, K.; Nuckolls, C.; Tremblay, N.; Whalley, A. C.; Xiao, S. *Org. Lett.* **2009**, *11*, 2225–2228.
37. Nagarajan, S.; Barthes, C.; Gourdon, A. *Tetrahedron* **2009**, *65*, 3767–3772.
38. Iyer, V. S.; Wehmeier, M.; Brand, J. D.; Keegstra, M. A.; Müllen, K. *Angew. Chem., Int. Ed.* **1997**, *36*, 1604–1607.
39. Lu, Y.; Moore, J. S. *Tetrahedron Lett.* **2009**, *50*, 4071–4077.
40. Vingiello, F. A.; Yanez, J.; Campbell, J. A. *J. Org. Chem.* **1971**, *36*, 2053–2056.
41. Müller, M.; Mauermann-Düll, H.; Wagner, M.; Enkelmann, V.; Müllen, K. *Angew. Chem., Int. Ed.* **1995**, *34*, 1583–1586.
42. Simpson, C. D.; Mattersteig, G.; Martin, K.; Gherghel, L.; Bauer, R. E.; Räder, H. J.; Müllen, K. *J. Am. Chem. Soc.* **2004**, *126*, 3139–3147.
43. Yang, X.; Dou, X.; Rouhanipour, A.; Zhi, L.; Räder, H. J.; Müllen, K. *J. Am. Chem. Soc.* **2008**, *130*, 4216–4217.
44. de Halleux, V. M.; Geerts, Y. H. *React. Funct. Polym.* **2000**, *43*, 145–151.
45. Percec, V.; Wang, J. H.; Yu, L. *Polym. Bull.* **1992**, *27*, 503–510.
46. Percec, V.; Nava, H. *J. Polym. Sci., Part A: Polym. Chem.* **1988**, *26*, 783–805.
47. Percec, V.; Wang, J. H.; Oishi, Y. *J. Polym. Sci., Part A: Polym. Chem.* **1991**, *29*, 949–964.
48. Percec, V.; Wang, J. H. *Polym. Bull.* **1991**, *25*, 9–16.
49. Percec, V.; Wang, J. H.; Oishi, Y.; Feiring, A. E. *J. Polym. Sci., Part A: Polym. Chem.* **1991**, *29*, 965–976.
50. Percec, V.; Wang, J. H.; Okita, S. *J. Polym. Sci., Part A: Polym. Chem.* **1991**, *29*, 1789–1800.
51. Percec, V.; Okita, S.; Wang, J. H. *Macromolecules*, **1992**, *25*, 64–74.
52. Percec, V.; Wang, J. H.; Okita, S. *J. Polym. Sci., Part A: Polym. Chem.* **1992**, *30*, 429–438.
53. Percec, V.; Wang, J. H.; Oishi, Y. *J. Polym. Sci., Part A: Polym. Chem.* **1992**, *30*, 439–448.
54. Percec, V.; Wang, J. H. *J. Mater. Chem.* **1991**, *1*, 1051–1056.
55. Percec, V.; Wang, J. H.; Yu, L. *Polym. Bull.* **1992**, *27*, 503–510.
56. Tanaka, M.; Nakashima, H.; Fujiwara, M.; Ando, H.; Souma, Y. *J. Org. Chem.* **1996**, *61*, 788–792.
57. Abd-El-Aziz, A. S.; de Denus, C. R.; Todd, E. K.; Bernardin, S. A. *Macromolecules* **2000**, *33*, 5000–5005.
58. Okada, T.; Ueda, M. *React. Funct. Polym.* **1996**, *30*, 157–163.
59. Strzelec, K.; Fugino, N.; Ha, J.; Ogino, K.; Sato, H. *Macromol. Chem. Phys.* **2002**, *203*, 2488–2494.
60. Okada, T.; Fujiwara, N.; Ogata, T.; Haba, O.; Ueda, M. *J. Polym. Sci., Part A: Polym. Chem.* **1997**, *35*, 2259–2266.

5.1 Buchner Reaction

Yong-Jin Wu

5.1.1 Description

The Buchner reaction describes cyclopropanation of an aromatic double bond with the α-ketocarbene derived from an α-diazocarbonyl compound **1** to produce an unstable norcaradiene intermediate **2**, which is in thermal equilibrium with the more stable cycloheptatriene tautomer **3**. This tautomer undergoes thermally or photochemically induced electrocyclic ring opening to give other cycloheptatriene isomers **4**–**6**.[1]

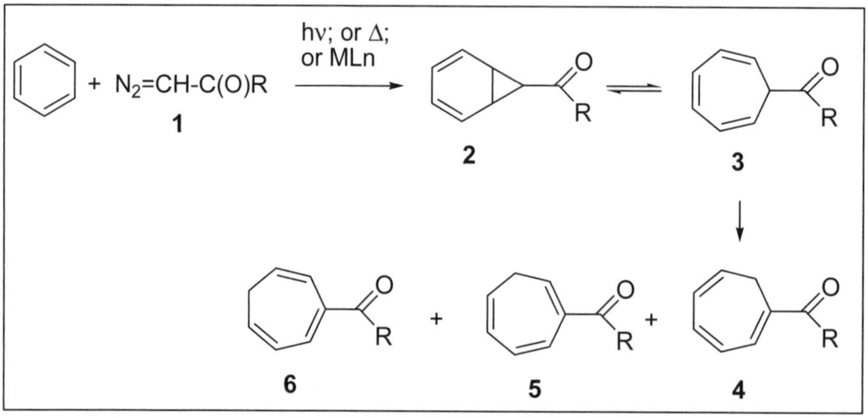

5.1.2 Historical Perspective[2]

Since Theodor Curtius reported the synthesis of ethyl diazoacetate in 1883,[3] Buchner had investigated its reactions with carbonyl compounds, alkenes, alkynes, and aromatic compounds for more than 30 years.[1b] His extensive contributions in this area resulted in two reactions named in his honor: the Buchner–Curtius–Schlotterbeck reaction (formation of ketones from aldehydes and aliphatic diazo compounds) and the Buchner reaction. The prototypical example of the latter involves the thermal or photochemical reaction of ethyl diazoacetate with benzene to give (via norcaradiene **7**) a mixture of four isomeric cycloheptatrienes **8**–**11**. Initially, Buchner believed that a single norcaradiene product **7** was generated from this reaction, but later, he realized that the hydrolysis of the product afforded a mixture of four isomeric carboxylix acids. The norcaradiene formulation persisted until 1956 when Doering reinvestigated this reaction.[4]

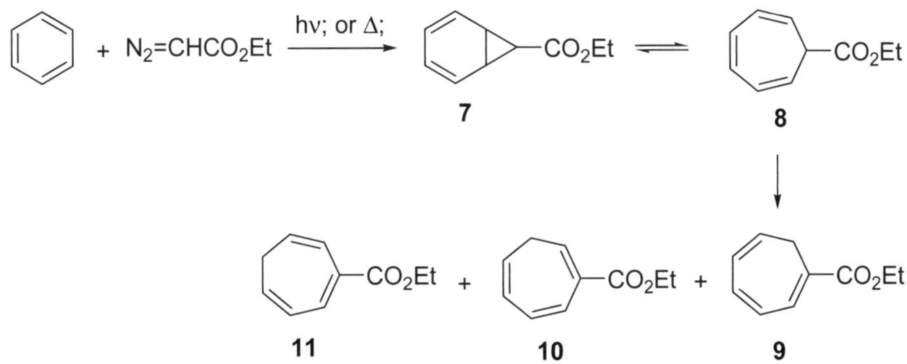

The thermal or photochemical Buchner reactions produce complex mixtures of cycloheptatrienyl esters, and the daunting complexity of the product mixtures was reduced or even eliminated with the advent of transition-metal catalysts, at first copper-based, then in the early 1980s rhodium(II) catalysts, which were developed by the Belgium group led by Noels and Hubert.[5] The rhodium(II)-catalyzed cyclopropanations of aromatics, especially intramolecular cyclopropanations, have enjoyed a certain popularity due to their high regioselectivity and stereoselectivity. Since the intramolecular Buchner reaction is much more widely used in organic synthesis than the intermolecular version, the former is the focus of this review.

The Buchner reaction resulted from his pioneering contribution to the reactions of arenes with α-diazoketones. Eduard Buchner (1860–1917) obtained his Ph.D. in Munich in 1888 under the supervision of Theodor Curtius (1857–1928). His graduate studies resulted in "A New Synthesis of Trimethylene Derivetives," in which he demonstrated the existence of a cyclopropanol ring. As a postdoctoral researcher, Buchner synthesized pyrazole for the first time in 1891. Since then, Buchner had taken over and further developed the chemistry of doazoalkanes from his mentor Theodor Curtius.

Eduard Buchner made substantial contributions not only to organic chemistry but also to biochemistry. About half of his 120 scientific publications are dedicated to his research in biochemistry, and in fact, he is regarded as one of the fathers of modern biochemistry. In 1897, Eduard Buchner (together with his brother Hans Buchner) discovered quite by accident that fermentation could occur outside living cells, thus disproving a long held belief, asserted by Louis Pasteur in 1860, that fermentation is inextricably tied to living cells. His chance discovery, which opened the door to modern biochemistry, led him to the award of the Nobel Prize for chemistry in 1907.

Buchner's life ended tragically. He was killed in Romania in 1917 while serving as a major in World War I.

5.1.3 Mechanism[6]

The transition-metal-catalyzed decomposition of a diazo carbonyl compound **12** generates a metal-carbene complex **13**, which is electrophilic. This carbene intermediate undergoes [2 + 2] cycloaddition to give a four-membered ring intermediate **16** that incorporates the metal. Reductive elimination of **16** leads to the cyclopropanation product **17**. Alternatively, neucleophilic addition of the benzene aromatic double bond with the carbene complex **13** gives a polar species **15**, which undergoes 1,3-bond formation to give **17**. The additions of metal–carbene complexes to aromatic double bonds are stereospecific, suggesting that if an open-chain is involved, it must collapse to the product more rapidly than single-bond rotations to avoid any loss of stereoselectivity.

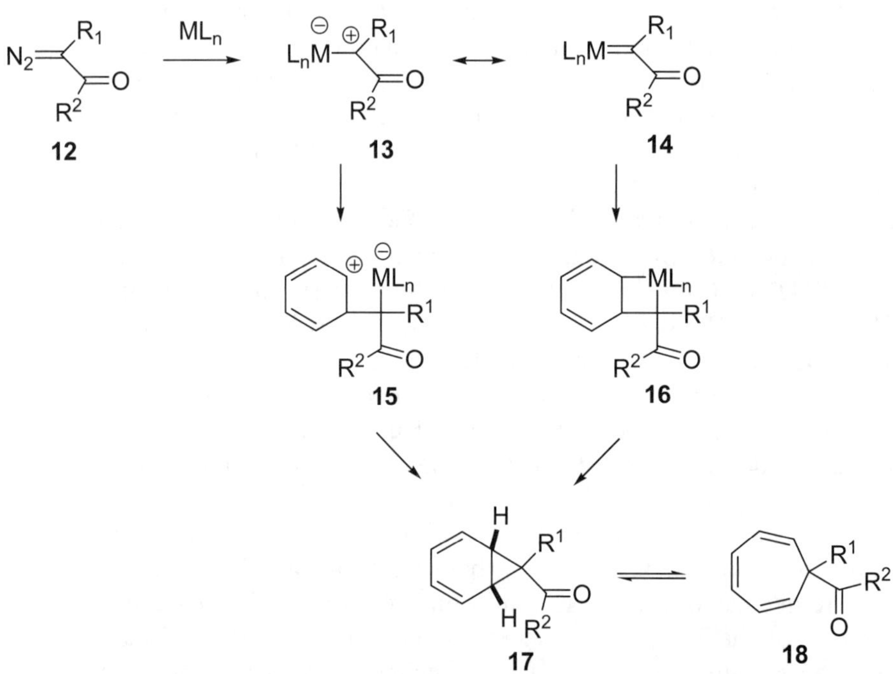

The catalytic cycle of the transition-metal-catalyzed Buchner reactions can be generally described as below:

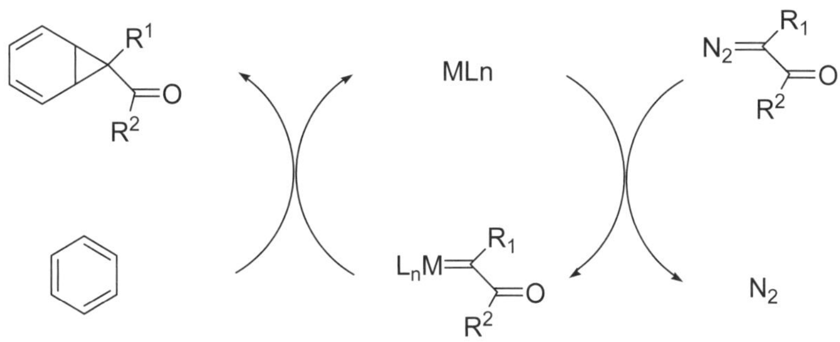

5.1.4 *Transition Metal-Catalyzed Buchner Reactions*

The problems endemic to the thermal and photochemical Buchner reactions were solved comprehensively in 1980 when rhodium(II) catalysts were introduced.[5] The measurement of improvement using Rh(II) catalysts can be appreciated by comparing the thermal reaction of ethyl diazoacetate with anisole (35% yield, seven products) with its rhodium trifluoroacetate-catalyzed counterpart (83% yield, three products **19**–**21**). The methoxy substituent clearly exerts a directive effect in favor of the 4-methoxy isomer **19**, and all the products are kinetically controlled unconjugated esters. In general, the rhodium(II)-catalyzed decomposition of alkyl diazoacetates in the presence of a large excess of aromatic substrates at room temperature affords kinetically controlled cycloheptatrienyl esters in excellent yield.

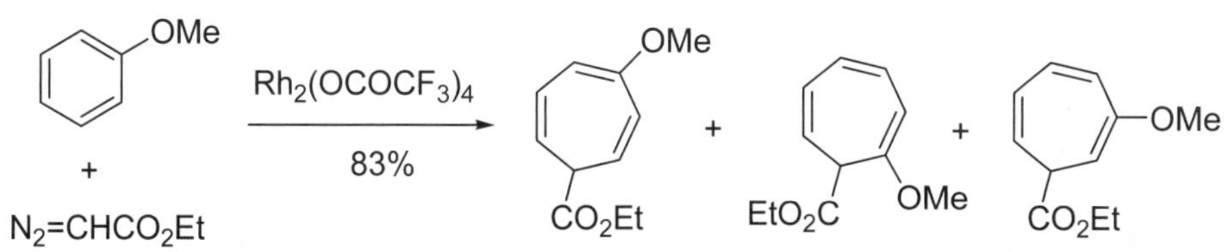

19 (56%) **20** (29%) **21** (8%)

Double cyclopropanation of benzene occurs in the Rh(II)-catalyzed reaction of dimethyl diazomalonate **22**.[7] Heating a benzene solution of this diazo compound and rhodium(II) acetate (1 mol %) under reflux gives a mixture of [**23** + **24**] (19%), **25** (8% yield) and bis-cyclopropanation product **26** (58% yield). When the same reaction is carried out using rhodium(II) trifluoroacetate instead of rhodium(II) acetate as the catalyst, a vastly different product distribution is obtained: [**23** + **24**], 64%; **25**, 32%; **26**, 4%. The low yield of double cyclopropanation product **26** obtained with rhodium(II) trifluoroacetate is comparable to other carbenoid reactions with aromatic substrates, where double cyclopropanation is rare.

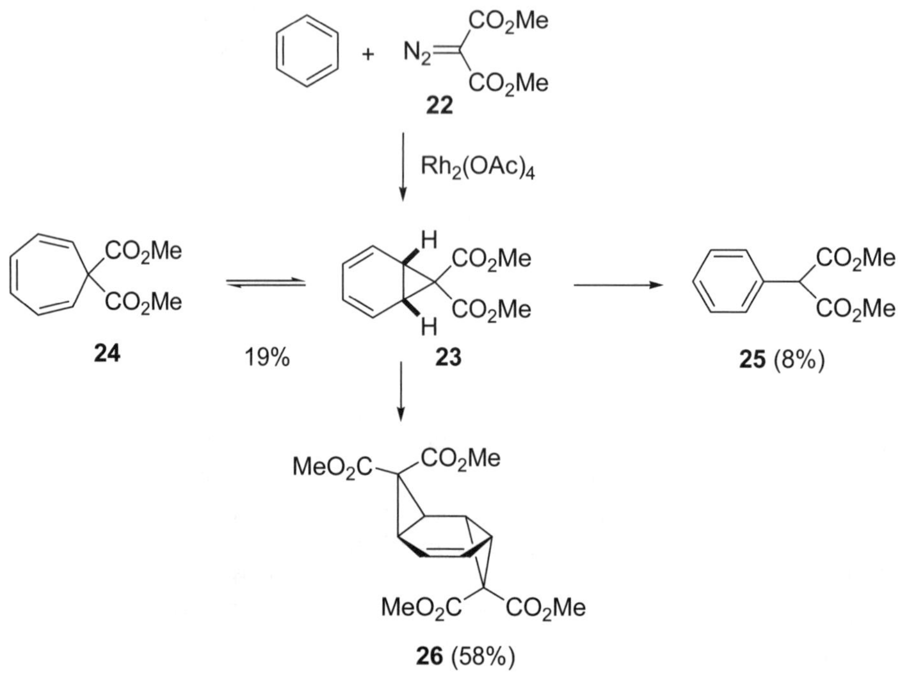

The intramolecular Buchner reaction of aryl diazoketones has been carried out using both copper(I) and rhodium(II) catalysts. For example, 1-diazo-4-phenylbutan-2-one **27a** cyclizes in bromobenzene with copper(I) chloride catalysis, furnishing 3,4-dihydroazulen-1(2*H*)-one **30** in 50% yield after purification by chromatography over alumina.[8] Trienone **30** is not the primary cyclization product, and the less conjugated isomeric trienone **29a** is first produced, but contact with alumina causes isomerization to **30**. The yield of this cyclization is further improved when rhodium(II) acetate is used as the catalyst instead of copper(I) chloride. Thus a catalytic amount of rhodium(II) acetate brings about the nearly quantitative conversion of **27a** to **29a** within minutes in hot dichloromethane.[9] Compound **29a** isomerizes to **30** on treatment with triethylamine, and rearranges to 2-tetralone **31a** when exposed to silica gel or acid.

The rhodium(II) trifluoroacetate-catalyzed cyclization of the disubstituted diazoketone **27b** also proceeds smoothly, but the product is a equilibrium mixture of cycloheptatrienone **29b** and norcaradiene-like tricyclic tautomer **28b**, with the tricyclic form **28b** the dominant partner at room temperature.[9]

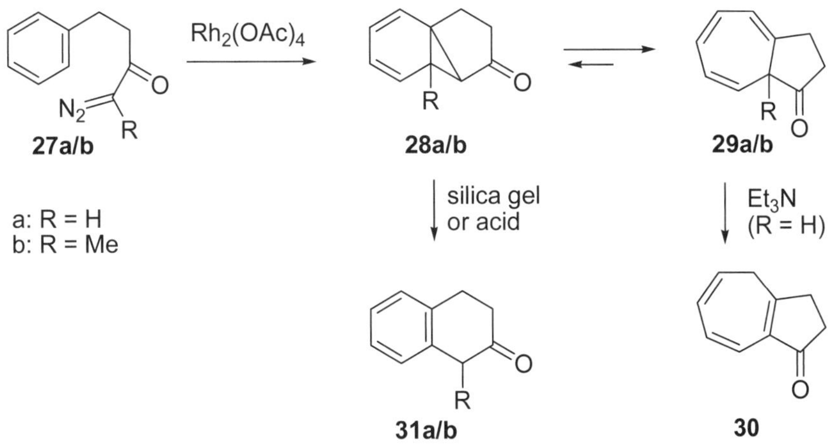

The effect of *ortho*- and *meta*-substitution in the above-mentioned intramolecular Buchner reactions has been examined. When the 2-methoxy-substituted diazoketone **32** is subjected to rhodium(II) acetate catalysis, a single cycloheptatrienone **34** is obtained in 94% yield.[10] This result is consistent with the outcome of the rhodium(II) trifluoroacetate-catalyzed intermolecular reaction of ethyl diazoacetate with anisole, which yields no product arising from addition of the ketocarbenoid on the most hindered site of the anisole. Dihydroazulenone **34** rearranges to tetralone **36** under acidic conditions, and isomerizes to the conjugated ketone **35** under basic conditions. It is interesting that the catalyzed decomposition of the *para*-methoxy derivative **37** provides exclusively 6-methoxy-2-tetralone **40** with no trace of the putative trienone **39**.[9]

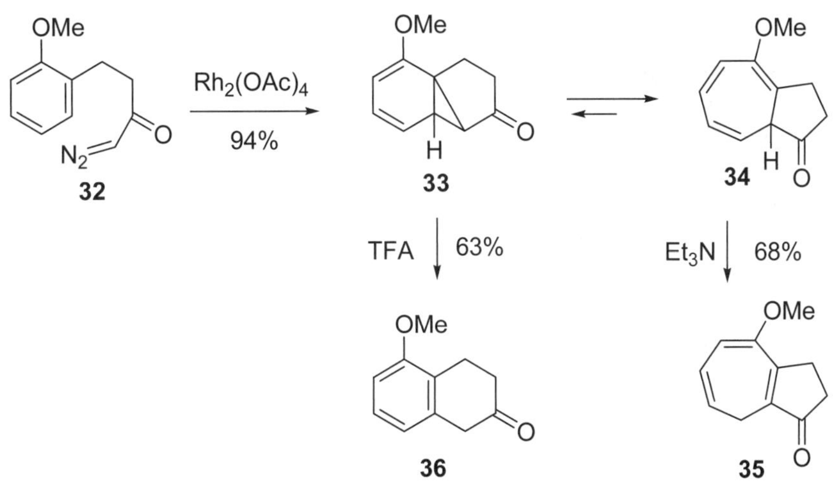

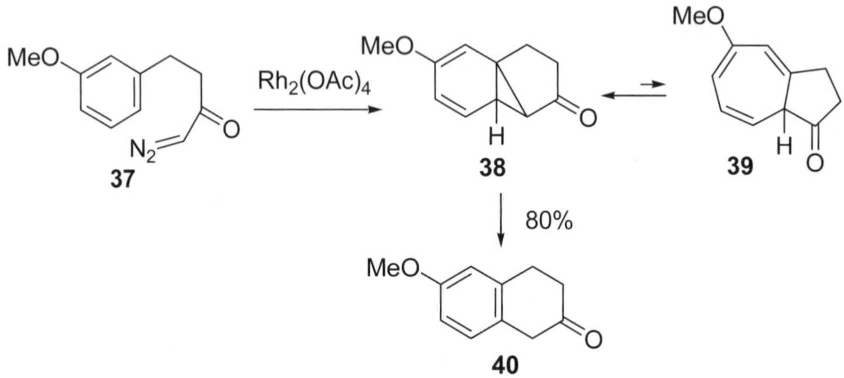

More highly substituted aromatics have also been studied in the course of natural product synthesis. For example, rhodium(II) mandelate-catalyzed cyclization of diazoketone **41** produces the ring expanded product **42**, which on hydrogenations furnishes the tricyclic lactone **43**.[11]

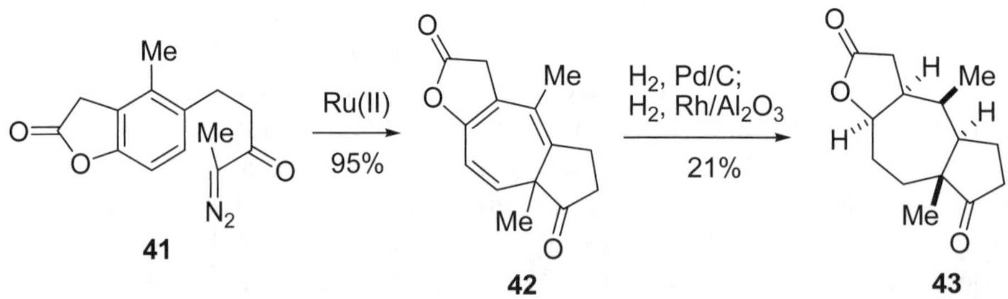

The first direct chemical evidence for the formation of the norcaradiene system in the intramolecular Buchner reaction was obtained in the rhodium(II)-catalyzed decomposition of 1-diazo-4-(2-naphthyl)butan-2-one **44**.[12] This reaction provides the tetracyclic norcaradiene **45** and tricyclic ketone **52** in 71% and 8% yield, respectively. When a catalytic amount of trifluoroacetic acid is added to **45**, tricyclic ketone **51** is formed. It is surprising that compound **45** is recovered quantitatively after treatment with triethylamine in dichloromethane under reflux. The formation of **52** is explained in terms of an attack of the carbenoid carbon of **44** on the 2,3-double bond of the naphthalene nucleus followed by double bond migration in the tricyclic nonconjugated ketone **49**.

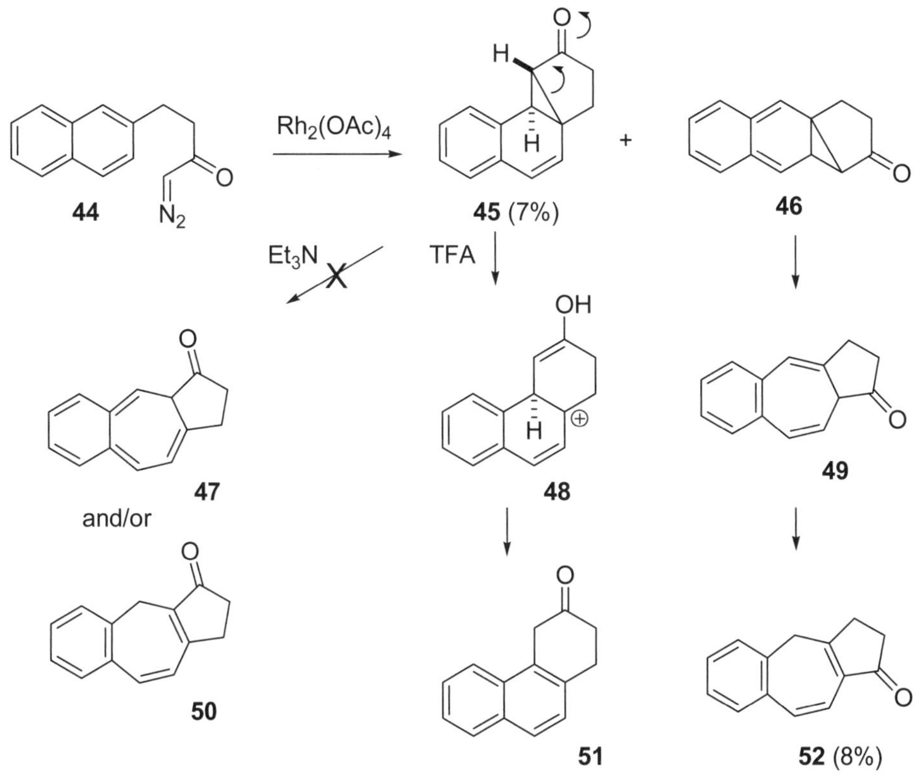

The intramolecular cyclopropanation reactions of the aromatic ring in tetrahydronaphthyl diazomethyl ketones initiated by transition metal catalysts also afford stable norcaradiene products.[13] These compounds are expected to be thermodynamically more stable than the tautomeric cyclohepatrienes because of the geometric constraints imposed by the bridging ring system. Indeed, decomposition of diazomethyl ketones **53** with rhodium(II) acetate affords 71% yield of the norcaradiene **54** along with 14% yield of the benzylic C–H insertion product, cyclopentanone **56** (vide infra). Several copper-based catalysts have also been evaluated, and a comparable result is obtained with Cu(II)(acac)$_2$. These norcaradiene derivatives are valuable intermediates in natural product synthesis as they rearrange to polycyclic ring systems under various acidic conditions. For example, diketone **58** is obtained from norcaradiene **54** by treatment with aqueous hydrochloric acid, whereas on exposure to Lewis acids under anhydrous conditions, the initial fragmentation of the three-membered ring is followed by rearrangement to give tricyclic enone **61**.

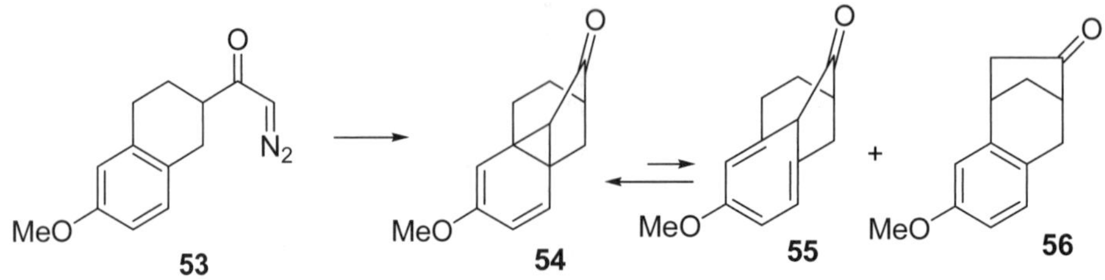

$Rh_2(OAc)_4$: **54** (71%), **56** (14%)
$Rh_2(TPA)_4$: **54** (51%), **56** (38%)
$Cu(acac)_2$: **54** (61%), **56** (7%)

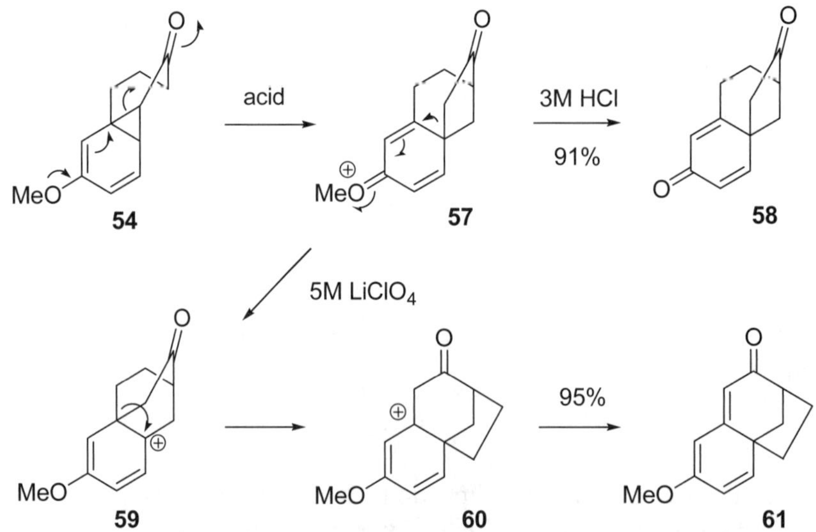

A novel carbene insertion reaction in the Grubbs's second-generation catalyst **62a** promoted by carbon monoxide has been described.[14] The reaction pathway represents a novel carbene insertion into the mesityl group on the *N*-heterocyclic carbene supporting ligand, a Buchner reaction of a ruthenium carbene. Treatrment of complex **62a** with carbon monoxide immediately results in the formation of **63a** in excellent yield. Similarly, the methylidene **62b** gives a cycloheptatrienyl insertion product **63b** in 63% yield. The Buchner reaction pathway can also be triggered by adding isocyanide ligands.

The carbene insertion into the mesityl group is promoted by carbon monoxide binding. This binding causes the carbene to cyclopropanate the closest "double bond" of the mesityl group and electrocyclic ring opening of the resulting cyclopropane provides the cycloheptatriene. The high regioselectivity of the carbene insertion suggests that the carbene is still encumbered to the ruthenium centre and is not reacting as a free carbene.

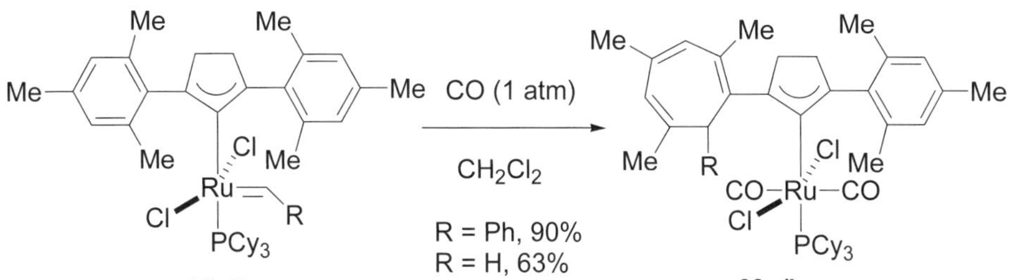

62a/b
63a/b

There are several examples of catalyzed aromatic cycloadditions leading to heterocyclic systems. The rhodium(II) acetate-catalyzed intramolecular Buchner reactions of *N*-benzyldiazoacetamides **64a/b** afford azabicyclo[5.3.0]decatrienes **66a/b** in excellent yields.[15] In contrast, the *N*-methyl derivative **64c** gives **66c** in moderate yield. Use of rhodium(II) perfluorobutyrate ($Rh_2(pfb)_4$) in place of rhodium(II) acetate increases the yield to 54%. Unlike its carbon counterpart, dihydroazulenone **29a** (vide supra), **66a** is insensitive to either trifluoroacetic acid or boron trifluoride etherate, even in excess, and the unrearranged reactant is recovered intact even after prolonged treatment at room temperature.

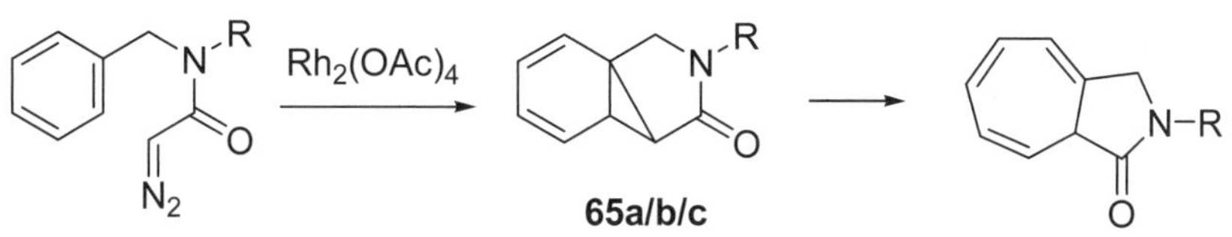

64a/b/c
65a/b/c
66a/b/c

a: R = *t*-Bu, 100%; b: R = Bn, 93%; c: R = Me, 37%

Other heterocyclic systems formed via Buchner reactions include cycloheptafuranones **69**. Thus diazoketones **67** with α-phenoxy substituents undergo cyclization catalyzed by copper(II) bis(hexafluoroacetonate) to furnish mixtures of cycloheptafuranones **69** and chromanones **70**.[16] The product compositions depend on substituents in the precursors. These cycloheptafuranones rearomatize readily to chromanones **70** upon contact with silica gel.

In addition to copper and rhodium, silver-catalyzed Buchner reactions have also been explored.[17]

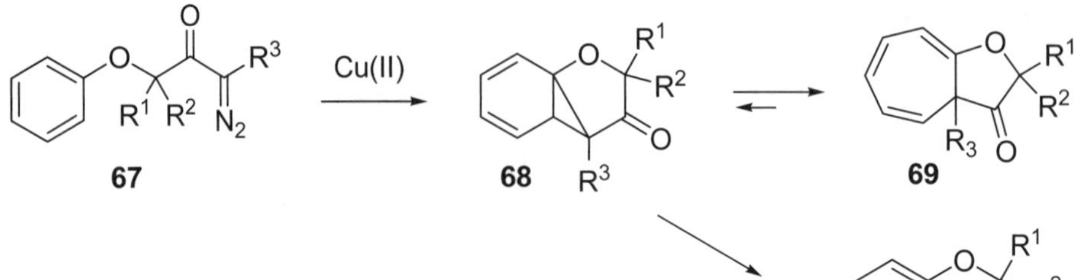

$R^1 = R^2 = H$, $R^3 = Me$ 65% yield (**69/70** = 2/3)
$R^1 = R^2 = R^3 = Me$, 95% yield (**69/70** = 9/1)
$R^1 = R^3 = Me$, $R^2 = H$, 95% yield (**70** only)
$R^1 = Ph$, $R^2 = H$, $R^3 = Me$, 88% yield (**70** only)

5.1.5 Buchner Reaction vs. C–H Insertion Reaction

Transition metal-catalyzed Buchner reactions of arene substrates proceed via electrophilic carbenoids. In addition to cyclopropanation of the arene double bond, these α-diazoketones possessing an aromatic ring can also participate in C–H insertion reactions.[18] As shown in the decomposition of diazomethyl ketone **53**, the benzylic C–H insertion product **56** is obtained as a minor product (vide supra). The rhodium(II) acetate-catalyzed reaction of diazoketone **71** also affords cycloheptatriene derivative **73** along with the benzylic C–H insertion product, γ-lactam **72**, in a ratio of 1:2.[19] Treatment of **71** with the more electron-rich rhodium(II) caprolactamate [$Rh_2(Cap)_4$] favors more C–H insertion, but the cycloaddition pathway is still significant; the ratio of **73** to **72** is 1:3.5.

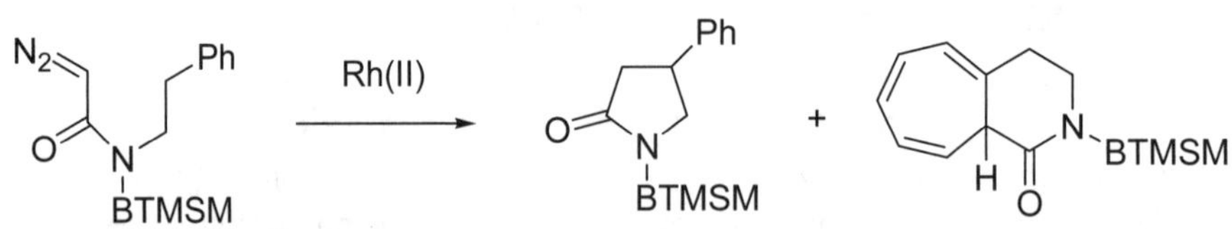

$Rh_2(OAc)_4$, 99% yield **72/73** = 2/1
$Rh_2(Cap)_4$, 97% yield **72/73** =3.5/1

Decomposition of **74** in the presence of rhodium(II) acetate gives stable norcaradiene **75**, indicating that the electron-rich aromatic ring is more receptive than the enone moiety to attack by the ketocarbenoid.[20] Introduction of a bromine substituent at C8 diverts carbene attack from the aryl ring, and insertion into the allylic C–H bond is now the preferred outcome. The suppression of the carbene addition to the benzenoid ring is expected since on both steric and electronic grounds the aryl ring becomes less reactive when it bears a bulky halogen.

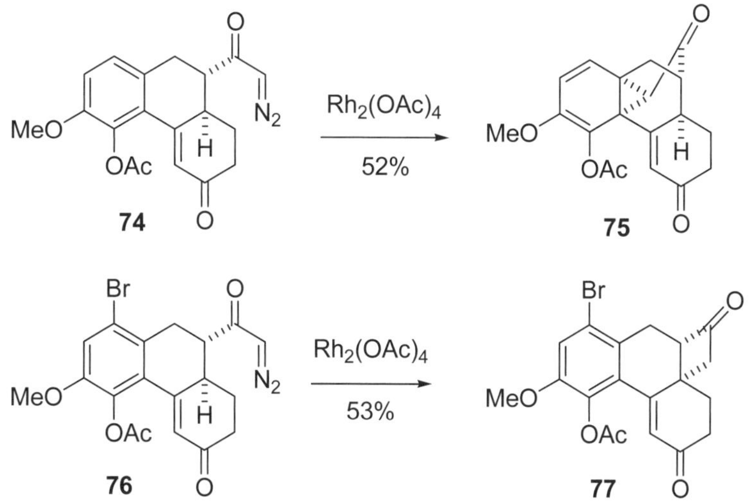

The choice of transition-metal catalysts can play an important role in determining reaction pathways as shown in the model studies toward the synthesis of harringtonolide. The epimeric C–H insertion products **80** are obtained in 75% and 40–50% yield, respectively, with $Rh_2(tpa)_4$ and rhodium mandelate. In contrast, bis(*N*-*t*-butylsalicyl-aldiminato) copper(II) generates the very labile cycloheptatriene **82** (50% yield), which is converted to the more stable isomer **83** upon treatment with 1,8-diazabicyclo[5.4.0]undec-7-ene (DBU).[21]

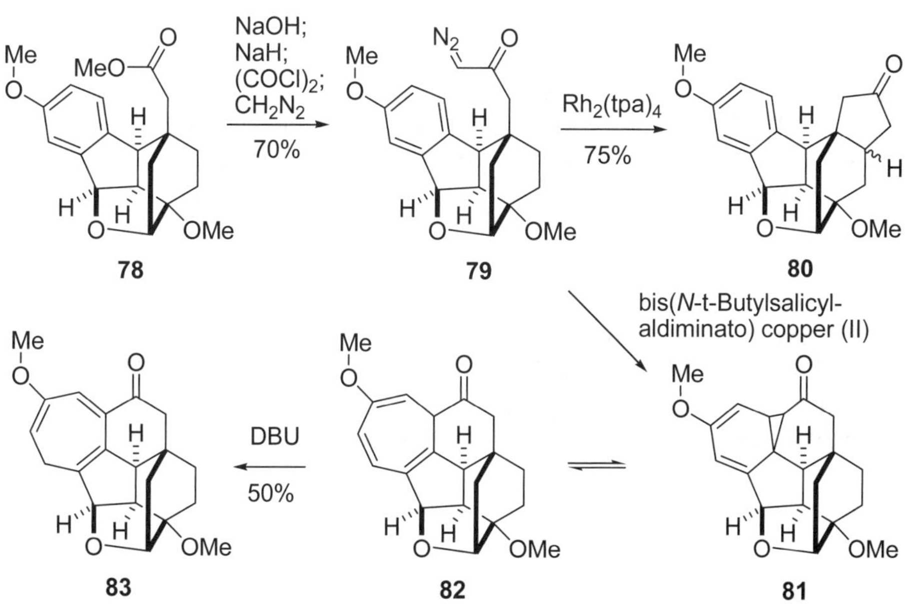

The α-(phenylsulfonyl)- and α-(ethoxyphosphoryl)-diazoacetamides **84d/e** are exclusively converted to formal aromatic C–H insertion products **86d/e** upon rhodium(II) perfluorobutyramide ($Rh_2(pfb)_2$) catalysis.[22] The unsubstituted diazoacetamide **84a** affords exclusively the Buchner ring expansion product **85a**, and the Buchner reaction remains the favorable pathway with diazo substrates **84b/c**, which bear relatively small α-substituents. The predominant formation of the Buchner products in these cases can be rationalized on the basis of steric effects. Various isoquinolinones are synthesized intramolecularly via six-membered ring formation with high regioelectivity and diastereoselectivity, while averting the common Buchner reaction.

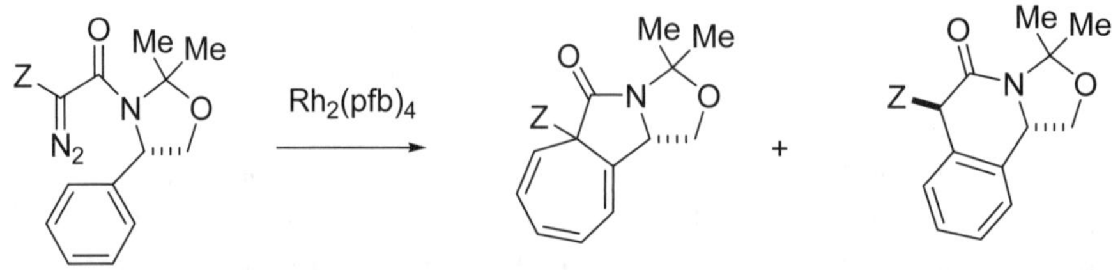

84a–e 85a–e 86a–e

	85a–e	86a–e
a: Z = H	85%	
b: Z = Ac	65%	12%
c: Z = CO_2Me	66%	15%
d: Z = SO_2Ph		93%
e: Z = $P(O)(OEt)_2$		86%

The rhodium(II) acetate-catalyzed decomposition of diazoamide **87** gives the benzylic C–H insertion product, the *trans* β-lactam **90** (27% yield), together with the Buchner ring expansion product, the cycloheptapyrrolone **89** (5% yield).[23] When rhodium(II) perfluorobutyramide is used as the catalyst, the aromatic C–H insertion proceeds exclusively to give indole **88** in excellent yield after silylation. In the case of the *N*-benzyl diazoamide **91**, a higher yield of the cycloheptatriene product is expected since the oxindole formation pathway is precluded. Indeed, the rhodium(II) acetate-catalyzed reaction of **91** gives the expected *trans*-β-lactam **93** (39%), together with its *cis*-isomer **94** (28%), and the cycloheptapyrrolone **92** (22%). Application of the prefluorobutyramide ligand favors attack on the double bond of the aromatic ring and gives the cycloheptapyrrolone **92** in good yield.

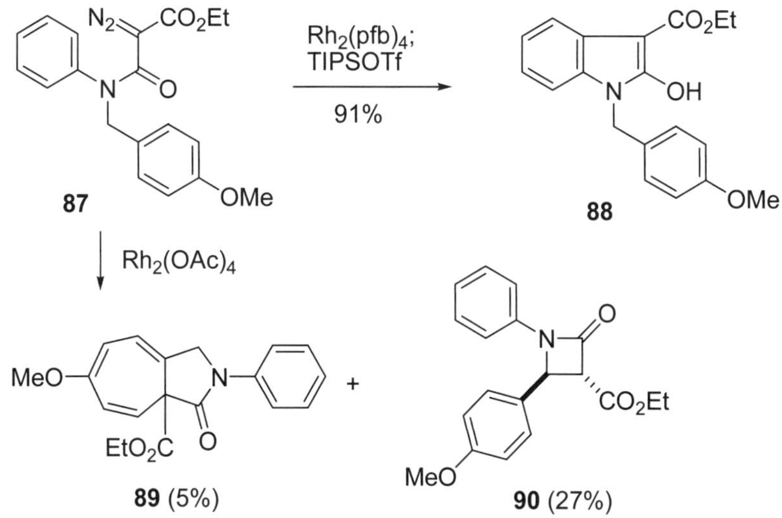

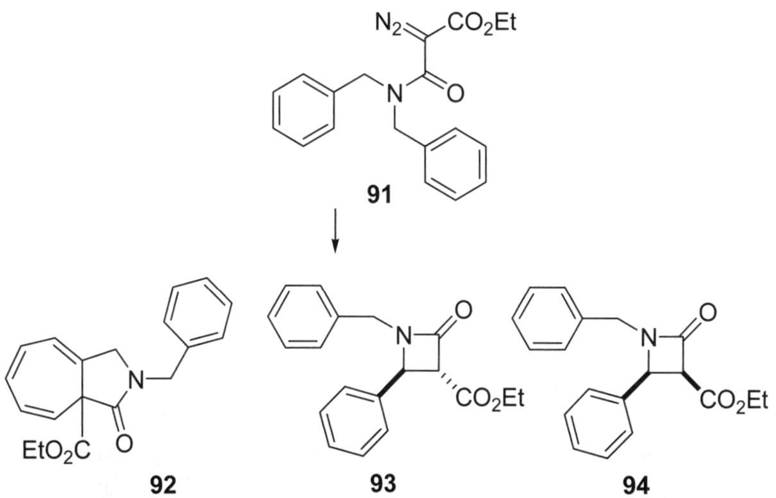

	92	93	94
$Rh_2(OAc)_4$	22%	39%	28%
$Rh_2(pfb)_4$	70%	none	12%

The Buchner reaction can be shut down by arene chromium tricarbonyl complexation.[24] Thus benezenechromium tricarbonyl **95** and even electron-rich *para*-dimethoxybenzenechromium tricarbonyl (structure not shown) fail to react with ethyl diazoacetate and rhodium(II) trifluoroacetate. In contrast, the same reaction with benzene provides a single isomer of the cycloheptatriene in 98% yield. The Buchner reaction of *pseudo*-C_2 symmetric substrate **97** clearly demonstrates the effect of chromium tricarbonyl complexation on arene cyclopropanation. Decomposition of diazoacetamide **97** with rhodium(II) acetate brings about exclusive addition to the noncomplexed ring to give lactam **98**, with no addition to the complexed ring detected. Thus in both intermolecular and intramolecular Buchner reactions, chromium complexation protects arenes from cyclopropanation. Conceivably, the lack of activity of arene complexes toward carbenes arises from the electron-withdrawing nature of the chromium tricarbonyl moiety which is comparable to that of a nitro group. Carbene additions to electron-poor arene substrates are known to afford minimal, if any, the desired cyclopropanation products, with a propensity for benzylic C–H insertion reactions instead.

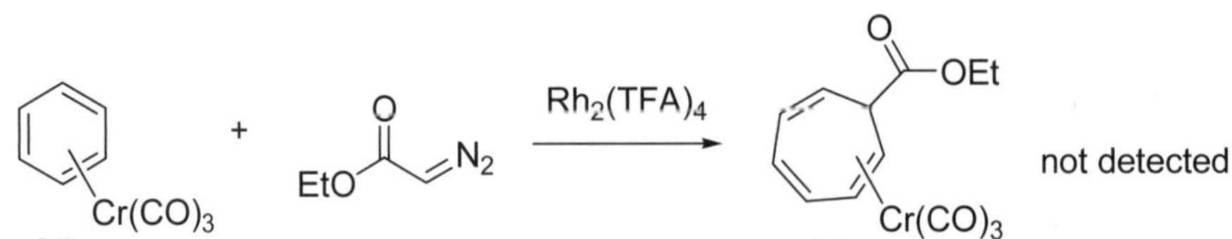

95 **96**

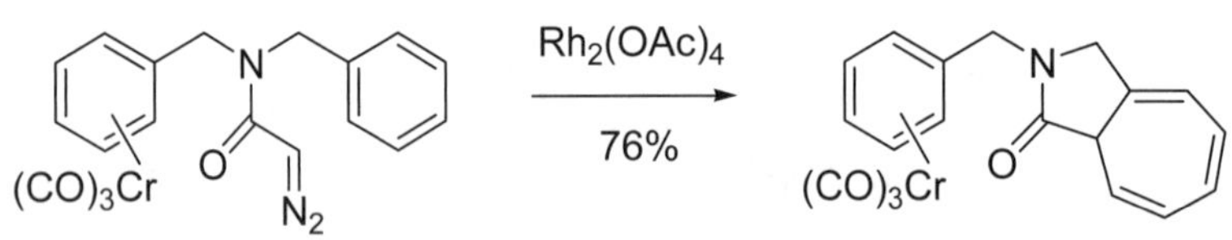

97 **98**

5.1.6 Macrocyclic Buchner Reactions

Decomposition of diazoacetate **99a** using rhodium(II) perfluorobutyrate in refluxing dichloromethane results in the formation of three aromatic cycloaddition products from reaction with the remote benzyl group at the 1,2-, 2,3-, and 3,4-positions in 75:12:13 ratio in 68% yield.[25] This reaction is remarkabley free of byproducts including those from addition/substitution to the original benzenedimethanol unit or C–H insertion into the oxygen activated benzylic position. The *para*-methoxybenzyl analog **99b** also undergoes rhodium(II) perfluorobutyrate-catalyzed diazo decomposition to give two aromatic cycloaddition products, the major isomer of which (**101b**) results from addition to the 3,4-position of the benzylic group. This regiochemical preference is parallel to that of intermolecular aromatic cycloaddition to *para*-disubstituted benzene derivatives.

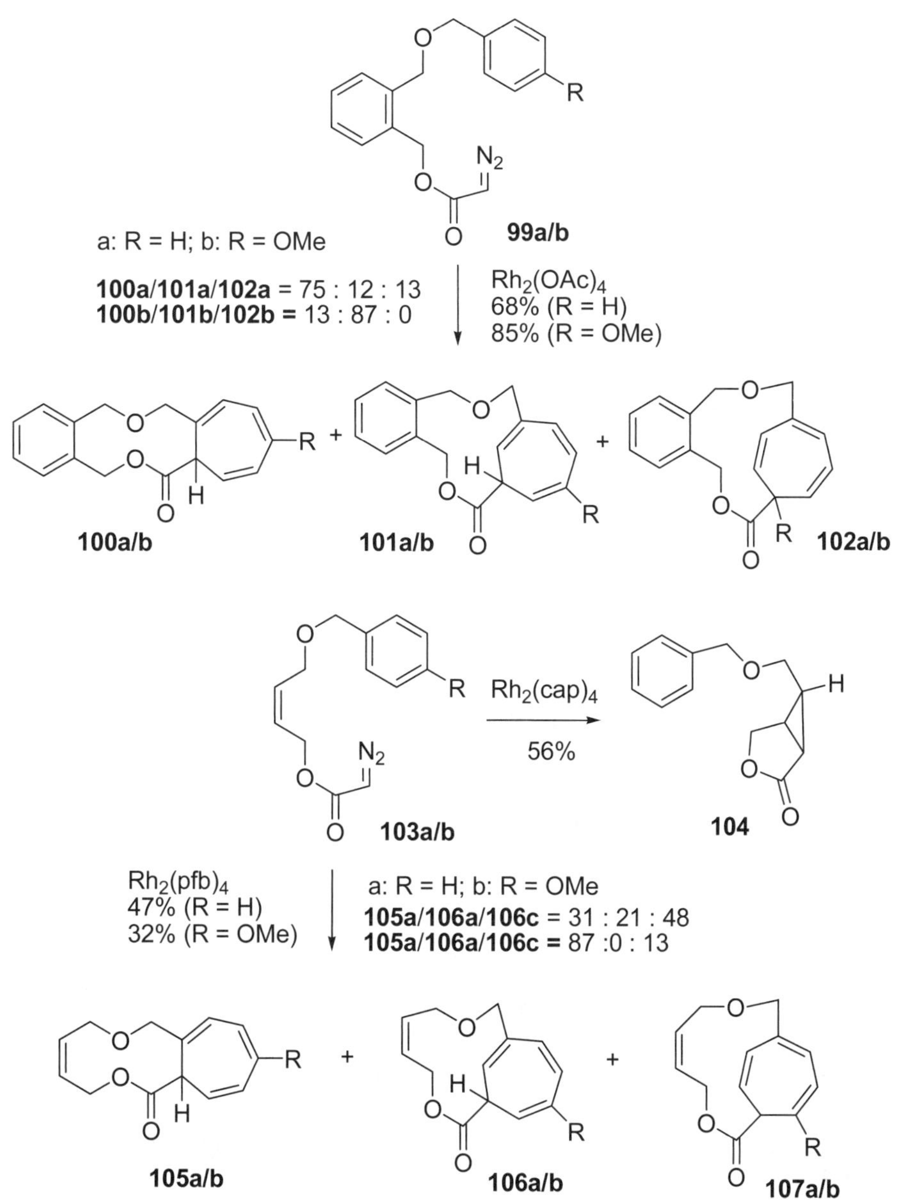

The analogous *cis*-2-buten-1,4-diyl derivatives **103a** undergoes rhodium(II) perfluorobutyrate-catalyzed diazo decomposition to produce a moderate yield of three cycloheptatriene products resulting from addition to the 1,2-, 2,3-, and 3,4-positions of the benzene ring. The product ratio (**105a**:**106a**:**107a** = 31:21:48) is comparable to that obtained from intermolecular aromatic cycloaddition of ethyl diazoacetate to toluene catalyzed by rhodium(II) trifluoroacetate (18:24:58). It is interesting that cyclopropanation of the allylic double bond is not observed in reactions catalyzed by rhodium(II) perfluorobutyrate, but this occurs exclusively with rhodium(II) caprolactamate. With the *para*-methoxybenzyl analog **103b**, both the product from aromatic cycloaddition to the 3,4-position and that from cycloaddition to the 1,2-position are obtained in 87:13 ratio in the rhodium(II) perfluorobutyrate-catalyzed reaction (32% yield).

Diazoketones also undergo macrocyclic aromatic cycloaddition reactions. Decomposition of **108** with rhodium(II) prefluorobutyrate yields aromatic cycloaddition products **109** and **110** in 30% and 9% yield, respectively. When cycloheptatriene **109** is exposed to neutral alumina, isomerization to **111** occurs. It is interesting that the 1,4-isomer **110** is inert to rearrangement on both silica and alumina. The ability to formally connect a carbene to a remote aromatic ring provides new opportunities for the construction of macrocyclic compounds.

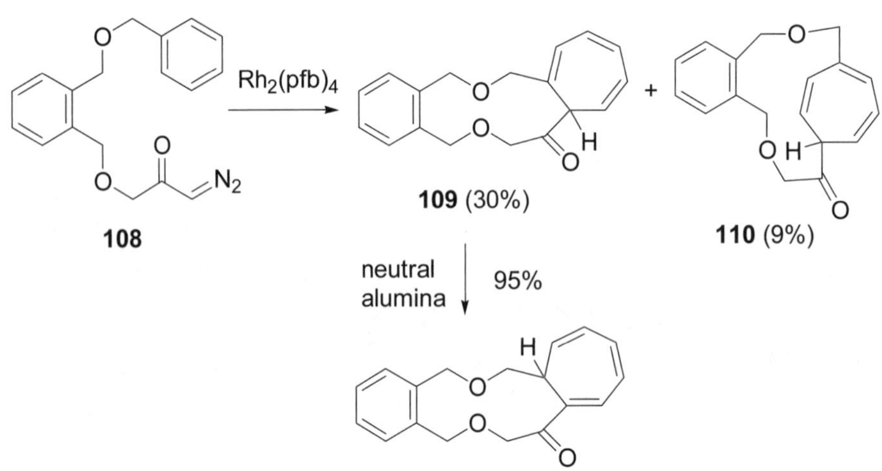

108 **109** (30%) **110** (9%) **111**

5.1.7 *Tandem Alkyne Insertion/Buchner Reactions*

The dirhodium tetra(triphenylacetate) (Rh_2TPA_4)-catalyzed decomposition of ethyl 2-diazo-3-phenylpropanoate in the presence of aryl alkynes yields the angularly substituted dihydroazulenes **115**.[26] The formation of **115** presumably takes place by a tandem alkyne insertion/Buchner ring expansion pathway via intermediates **112–114**. The reactivity of aryl alkynes with 2-diazo-3-phenylpropanoate is altered dramatically by changing the catalyst from dirhodium tetra(triphenylacetate) to dirhodium tetrapivalate (Rh_2Piv_4): cyclopropenes **116** are formed in 40–75% yields.

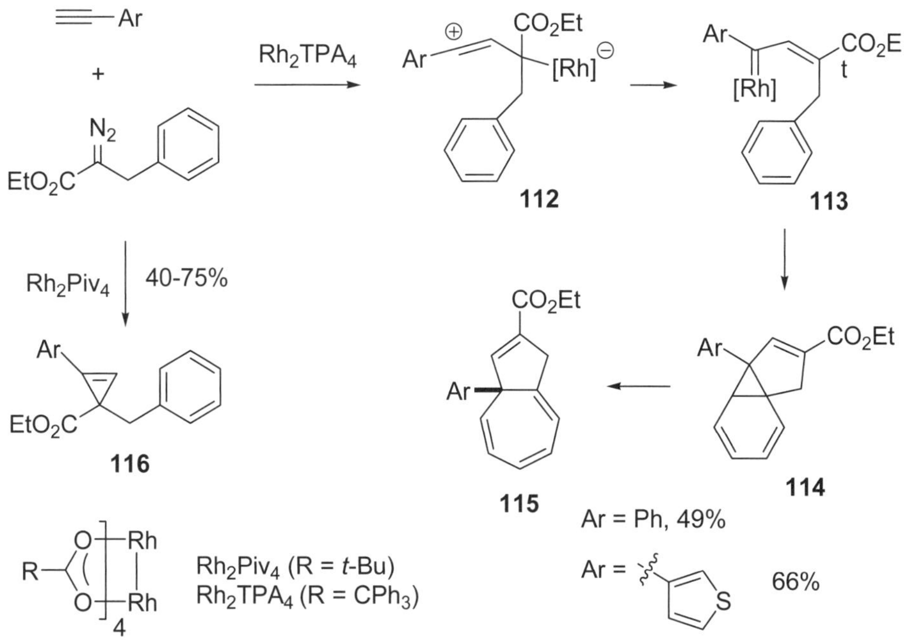

5.1.8 *Asymmetric Buchner Reactions*

Treatment of **117** with rhodium(II) acetate produces macrocyclic cyclopropene **119** in 69% yield. However, application of chiral catalyst $Rh_2(4R\text{-MEOX})_4$ to this system results in the cyclopropanation product **118** in 66% yield and 73% *ee*.[27]

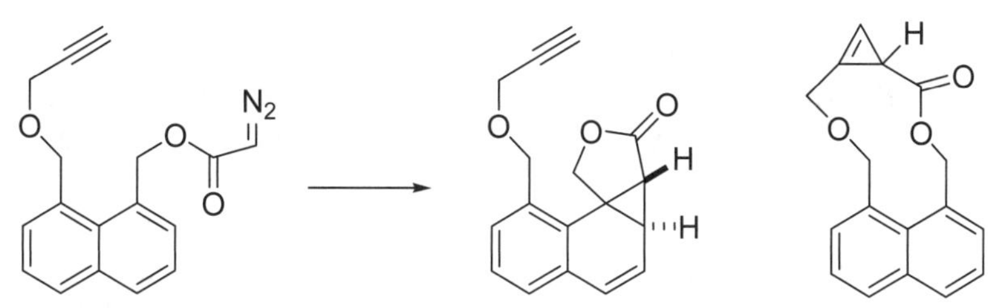

	117	**118**	**119**
$Rh_2(OAc)_4$		none	69%
$Rh_2(4S\text{-MEOX})_4$		66% (73% *ee*)	none

The competition between ylide formation and aromatic cycloaddition has also evaluated. Decomposition of diazo acetate **120** in the presence of $Rh_2(4S\text{-MEOX})_4$ leads to the sole production of the aromatic cycloaddition product **121** in 55% yield and 84% *ee*.

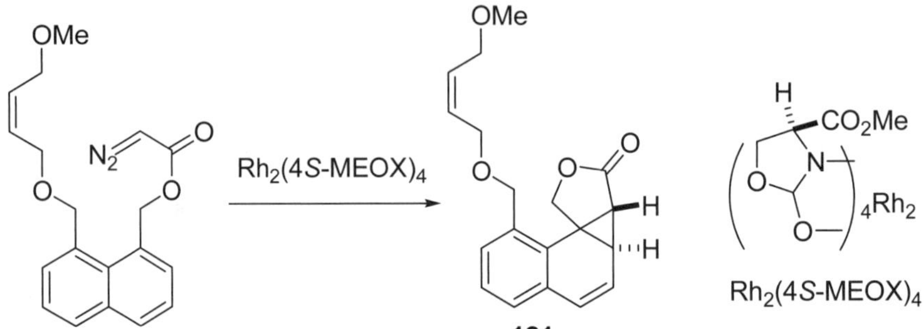

120 **121**

55% yield; 84% *ee*

A number of catalysts have been investigated for aromatic cycloaddition on the basic naphthalene system **122**. In this case, Rh_2(4*S*-IBAZ)$_4$ is superior even to Rh_2(4*R*-MEOX)$_4$ catalyst for highly enantioselective aromatic cycloaddition.

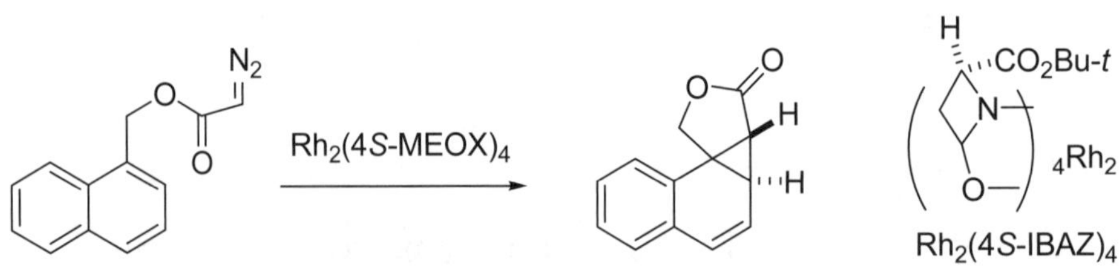

122 **123**

Rh_2(4*R*-MEOX)$_4$: 76% yield; 56% ee
Rh_2(4*S*-IBAZ)$_4$: 87% yield, 81% ee

Asymmetric Buchner reactions using chiral auxiliary have also been undertaken.[28] The diazoketo substrate **126** for the chiral tethered Buchner reaction is prepared from optically pure (2*R*,4*R*)-2,4-pentanediol in three steps: the Mitsunobu reaction with 3,5-dimethylphenol, esterification with diketene, and diazo formation/deacetylation. Treatment of **126** with rhodium(II) acetate results in a quantitative yield of **127** with more than 99% *ee*. This compound is reduced with lithium aluminium hydride, and the resulting diol **128** undergoes epoxidation and concurrent acetal formation to give **129** as a single diastereomer. Hydrogenation of **129** with Raney nickel proceeds stereoselectively to yield saturated diol **130**, which is converted to aldehyde **132** via acid hydrolysis followed by oxidation. Compound **132** is a versatile intermediate for natural product synthesis.

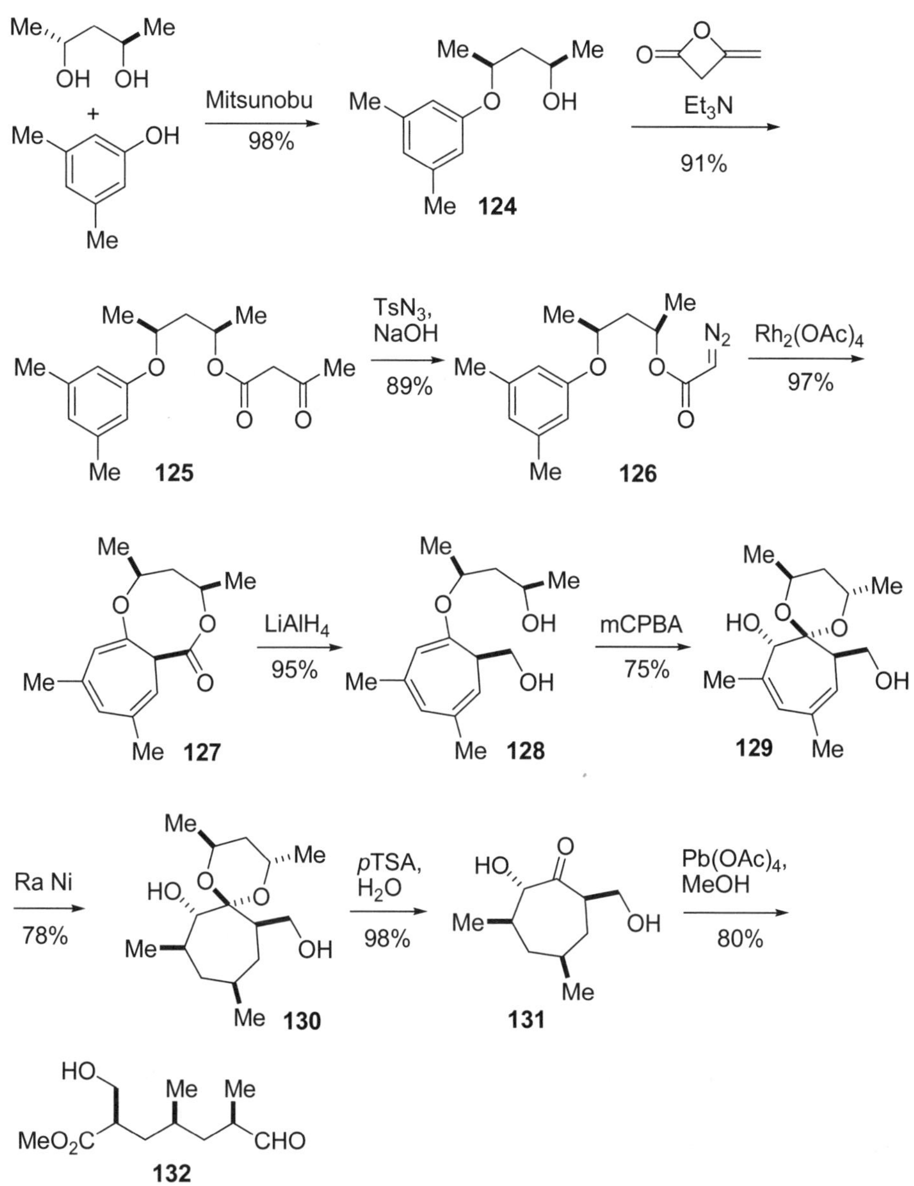

5.1.9 Synthetic Utility

The power of the Buchner reaction, especially the intramolecular version, has been demonstrated in the syntheses of numerous natural products, and only a few representatives are described as below.

Azulene(Pfaz–Plattner)[29] and Substituted Azulenes (Danheiser)[30]

Azulene is an isomer of naphthalene, but their colors are different: azulene is dark blue, whereas naphthalene is a colorless. The name of azulene is derived from the Spanish word *azul*, meaning "blue." Derived from the German chamomile flowers (*Matricaria recutita*), azulene is known for its superior skin-soothing properties. Azulene has a long history, dating back to the 15th century as the azureblue chromophore obtained by steam distillation of German chamomile flowers. The chromophore was discovered in yarrow and wormwood and named in 1863 by Septimus Piesse. Azulene has been shown to be a highly effective anti-inflammatory and soothing agent, and it has been used professionally in topical applications for sensitive skin and in sun care and burn products, as well as calming face and body creams.

The first chemical synthesis of azulene was reported by Pfau and Plattner in 1937. Their synthesis takes advantage of the ring enlargement of indane **133** on addition of diazoacetic ester to give cycloheptatriene **135**. This compound is converted into azulene via a three-step sequence: hydrolysis, dehydrogenation, and decarboxylation of the resulting acid.

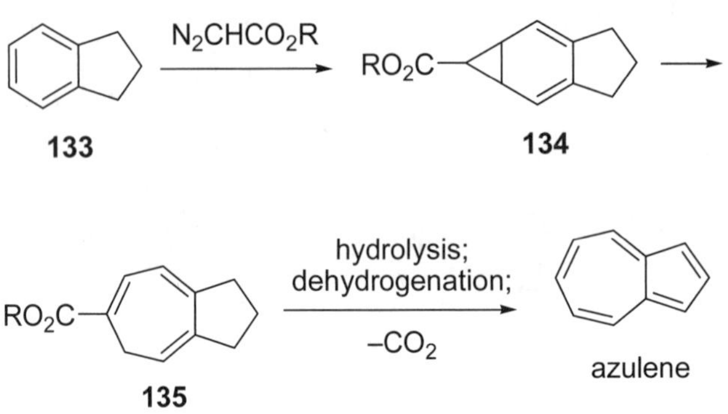

Early approaches to the synthesis of azulenes, including the Pfau-Plattner approach involve low-yield dehydrogenation steps and are limited to the preparation of relatively simple azulene analogs. To this end, a ring expansion–annulation strategy for the synthesis of substituted azulenes has been developed, and this methodology is based on the reaction of β-bromo-α-diazo ketones with rhodium carboxylates. The key transformation involves the intramolecular Buchner reaction of diazoketone **136** followed by β-elimination of the bromide **138**, tautomerization and in situ trapping of the resulting 1-hydroxyazulene as a carboxylate **139** or triflate **140**. This triflate undergoes various coupling reactions to provide the substituted azulenes **141**.

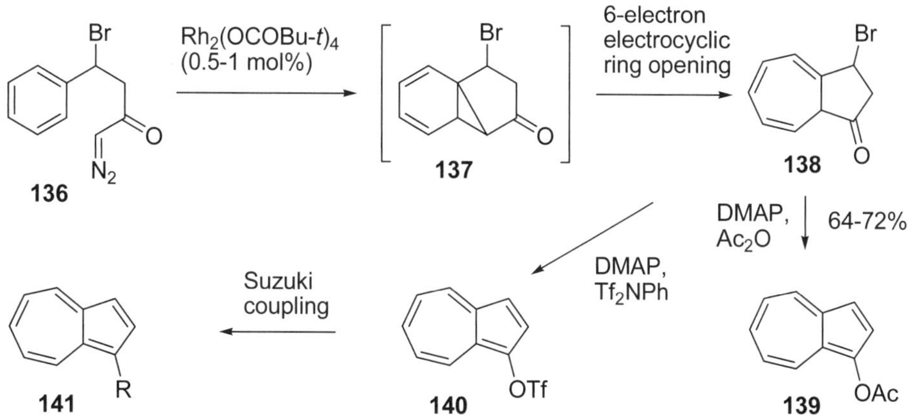

(±)-Confertin (McKervey, 1991)[31]

A formal total synthesis of confertin, a member of the pseudoguaianolide sesquiterpenoid family, employs the Buchner reaction to construct the seven-membered ring moiety. Thus decomposition of diazoketone **143** with rhodium(II) mandelate in hot dichloromethane furnishes a single ring-expanded product consisting of an equilibrium mixture of bicyclic cycloheptatriene **145** and the tricyclic norcaradiene **143**. This equilibrium mixture is reduced with lithium tri-*tert*-butoxyaluminium hydride to give a mixture of the epimeric alcohols **146** where norcaradiene moiety is absent. This alcohol is converted to confertin via a multistep sequence.

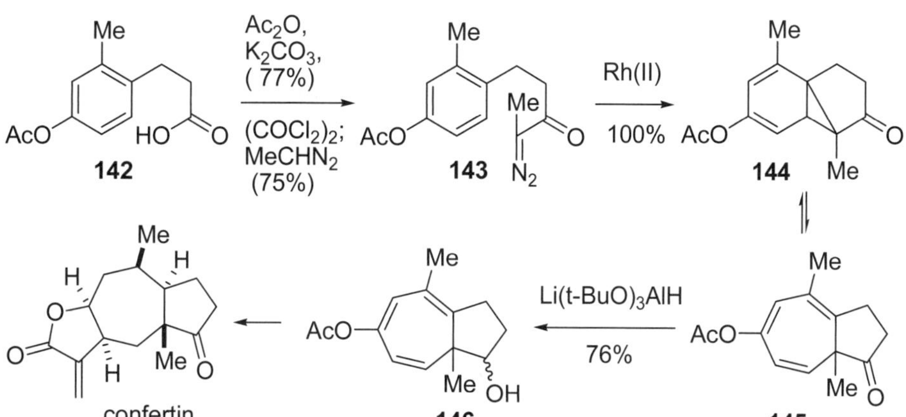

Hainanolidol and Harringtonolide (Mander, 1998)[32]

The diterpenoid tropone, harringtonolide, first isolated in North America from seeds of *Cephalotaxus harringtonia* (Taxaceae) and independently

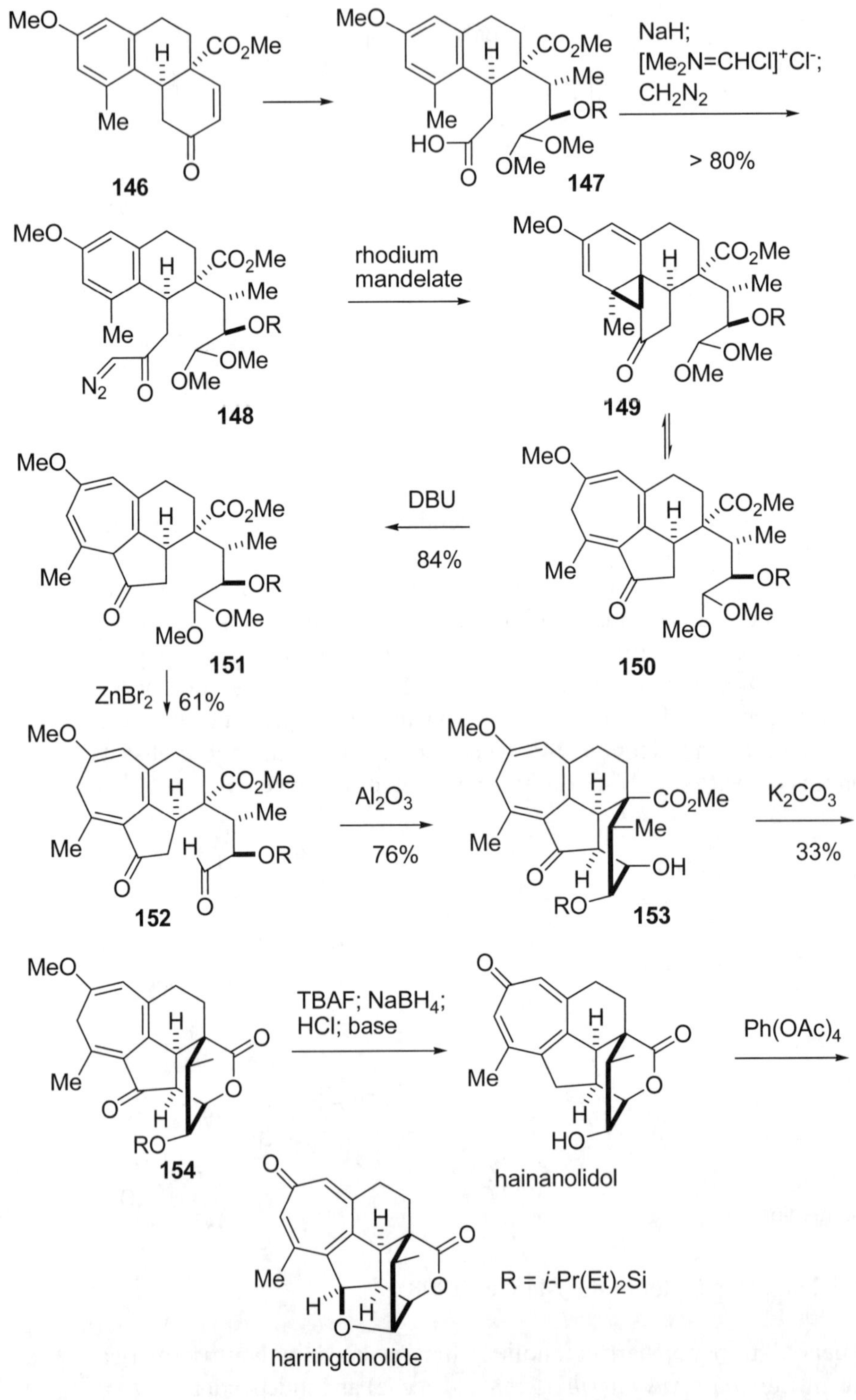
146
147
NaH;
[Me$_2$N=CHCl]$^+$Cl$^-$;
CH$_2$N$_2$
> 80%
148
rhodium mandelate
149
150
DBU
84%
151
ZnBr$_2$ 61%
152
Al$_2$O$_3$
76%
153
K$_2$CO$_3$
33%
154
TBAF; NaBH$_4$;
HCl; base
hainanolidol
Ph(OAc)$_4$
harringtonolide
R = *i*-Pr(Et)$_2$Si

from the bark of the related Chinese species *Cephalotaxus hainanensis*, has been shown to have promising antineoplastic and antiviral peoperties. In *C. hainanensis*, harringtonolide is accompanied by the closely related, but biologically inactive carbinol, Hainanolidol. Both diterpenoids were synthesized by Mander who constructed the tropone moiety via the intramolecular Buchner reaction of the diazo ketone **148**. This diazo substrate is prepared by treating the sodium carboxylate of **147** with Vilsmeier reagent and adding the reaction mixture directly to an excess of diazomethane, affording **148** in 80% overall yield from **147**. Cyclopropanation catalyzed by rhodium(II) mandelate furnishes an unstable adduct **150** that is immediately treated with 1,8-diazabicyclo[5.4.0]undec-7-ene (DBU) to give the less labile cycloheptatriene **151** in 84% overall yield. Liberation of the aldehyde function from dimethyl acetal is carried out with zinc bromide, and the resulting aldehyde **152** undergoes intramolecular aldol reaction with basic alumina to give the desired aldol **153** in 76% yield. Treatment of **153** with potassium carbonate in aqueous methanol furnishes lactone **154** in moderate yield. Desilylation is effected smoothly with tetrbutylammonium fluoride (TBAF), and the ketone is reduced to diol, which, when briefly exposed to acid, affords hainanolidol in > 50% overall yield. Transannular oxidation of hainanolidol with lead tetraacetate generates harringtonolide.

A drawback of the above-mentioned approach is the need to carry out multiple operations in the presence of the highly reactive cycloheptatriene moiety. An alternative route involves the assembly of the cycloheptatriene array to a much later stage.[33] To this end, diazo ketone substrates **155** and **158** were prepared, but unfortunately, decomposition of both compounds in

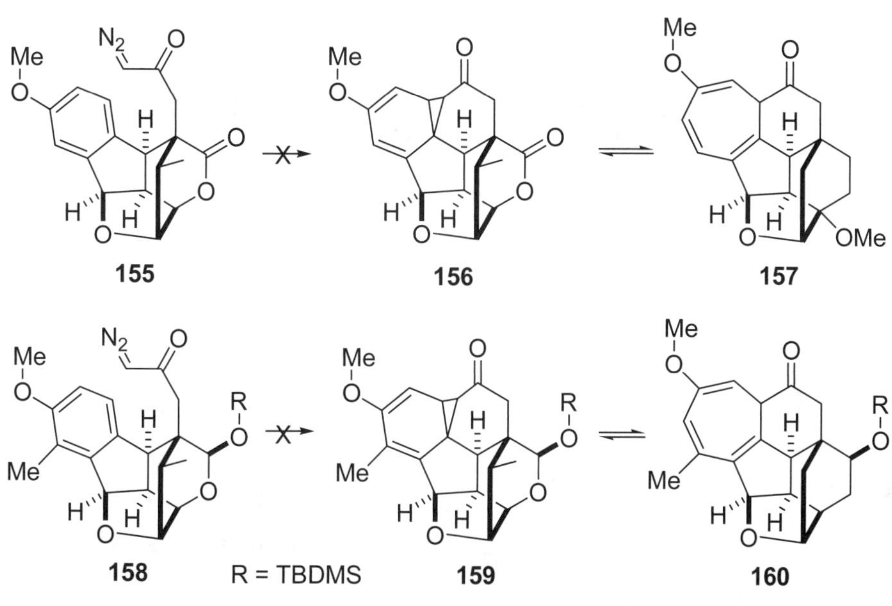

the presence of various Rh(II) and Cu(II) catalysts failed to deliver any desired Buchner reaction products

Gibberellin (±)-GA_{103} (Mander, 1997)[34]

Norcaradienes from intramolecular cyclopropanation reactions of diazomethyl ketones are valuable intermediates for the synthesis of polycyclic diterpenoids such as gibberellin (±)-GA_{103}, a representative of a family of hexacyclic gibberllins isolated in trace amounts from developing apple seeds. Decomposition of **161** with $Cu(acac)_2$ affords norcaradiene **162**, which undergoes Diels–Alder reaction with 3-methylfuran-2,5-dione **163** to give adduct **164** in 75% yield over two steps. Solvolysis of the anhydride function, in situ reduction of the ketone with sodium borohydride, and then hydrolysis of the enol ether during the acidic workup afford ketoacid **165**. This acid is converted to the diazoketone **166** in three operations: iodolactonization, reduction of the iodide, and diazo-transfer reaction with trisyl azide. Wolff rearrangement of **166** proceeds smoothly, furnishing ester **167** in excellent yield. Gibberellin (±)-GA_{103} is made from **167** in six steps.

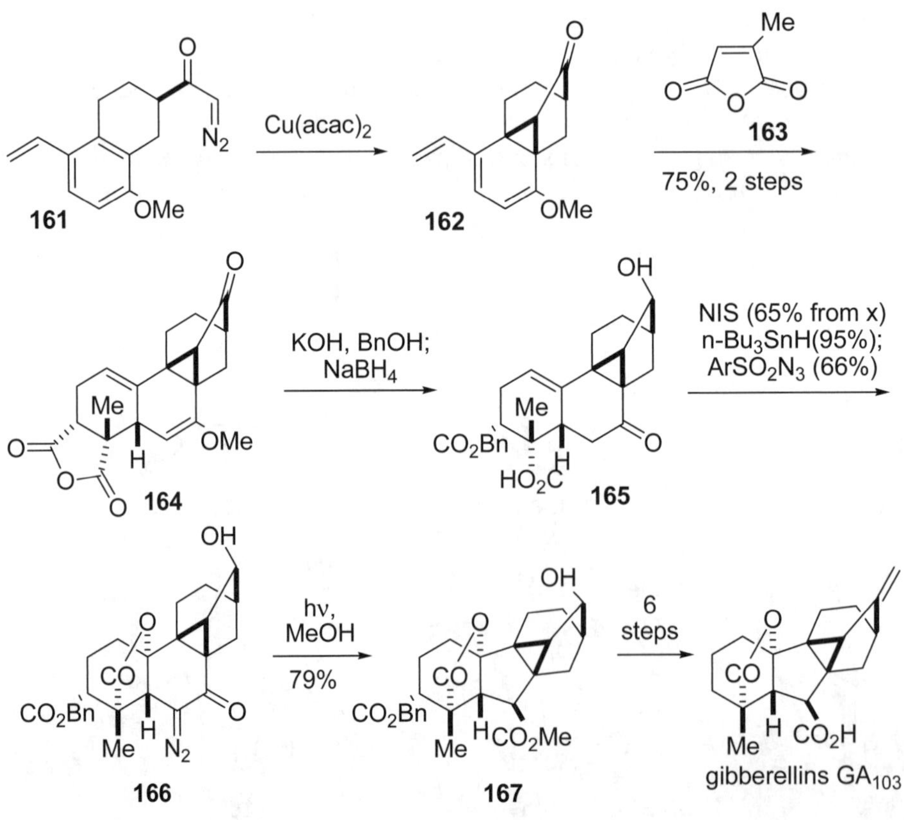

5.1.10 Experimental

2,3-Dihydroazulen-1(4H)-one (30a)[35]

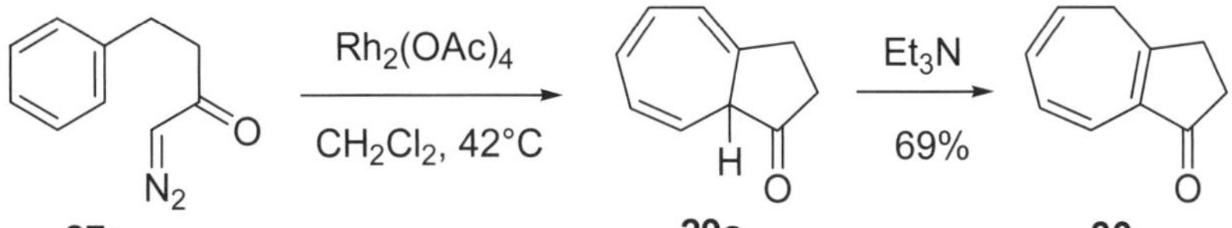

In a 750-mL, three-necked flask, equipped with an efficient condenser, thermometer, and nitrogen inlet, rhodium(II) acetate (52 mg, 0.12 mmol) was dissolved in dichloromethane (430 mL). The blue-green solution was heated under reflux (42 °C) under nitrogen, and a solution of 1-diazo-4-phenylbutan-2-one (**27a**, 37.8 g, mmol) in dichloromethane (10 mL) was added within 15 h via a syringe driven by a dosage pump, whereby the tip of the elongated hollow needle was placed in the refluxing solvent stream at the condenser. As a result of the very slow addition of 1-diazo-4-phenylbutan-2-one and the additional dilution by the refluxing solvent, an optimally high dilution effect for the catalyzed intramolecular Buchner reaction could be attained. After the addition of 1-diazo-4-phenylbutan-2-one was complete, the resulting solution was heated under reflux for another hour and then cooled to room temperature. A small amount of triethylamine (0.20 mL) was added, which caused a transitory darkening from yellow to brown and spontaneous warming of the solution to 40 °C. Finally, the initial yellow color returned. After 30 min, the solution was filtered over silica gel, and the solvent was removed by distillation, and the yellow residue was purified by silica gel chromatography eluting with hexanes/*t*-BuOMe (5:1) to give a greenish yellow oil, which was recrystallized from hexanes/*t*-BuOMe (20:1) to give (5*Z*,7Z)-2,3-dihydroazulen-1(4*H*)-one (**30a**) as colorless needles (21.87 g, 69%).

5.1.11 References

1. [R] (a) Marchard, A. P.; Brockway, N. W. *Chem. Rev.* **1974**, *74*, 431. (b) Ye, T.; McKervey, M. A. *Chem. Rev.* **1994**, *94*, 1091. (c) Padwa, A.; Krumpe, K. E. *Tetrahedron*, **1992**, *48*, 5385; (d) Doyle, M. P.; McKervey, M. A.; Ye, T. *Modern Catalytic Methods for Organic Synthesis with Diazo Compounds*, Wiley, New York, **1998**; Ch. 6. (d) Davies, H. M. *Comprehensive Organic Synthesis*, Trost, B. M., ed.; Pergamon, Oxford, **1991**, Vol. 4, Ch.4.8.
2. [R] (a) Jaenicke, L. *Angew. Chem., Int. Ed.* **2007**, *46*, 6776. (b) Stryer, L. Biochemistry, 4th ed. Freeman, New York, **1999**, pp. 484–484.
3. Curtius, T. *Ber. Dtsch. Chem. Ges.* **1883**, *16*, 2230.
4. von E. Doering, W.; Knox, L. H. . *J. Am. Chem. Soc.* **1957**, *79*, 352.
5. Anciaux, A. J.; Demonceau, A.; Noels, A. F.; Hubert, A. J.; Warin, R.; Teyssie, P. *J. Org. Chem.* **1981**, *46,* 873.

6. [R] Carey, F. A.; Sundberg, R. J. *Advanced Organic Chemistry*, Part B, 3rd ed. Prenum Press, **1990**, pp. 522–528.
7. Yang, M.; Webb, T. R.; Livant, P. *J. Org. Chem.* **2001**, *66*, 4945.
8. (a) Scott, L. T. J. *J. Chem. Soc., Chem. Commun.* **1973**, 882. (b) Scott, L. T.; Minton, M. A.; Kirms, M. A. *J. Am. Chem. Soc.* **1980**, *102*, 6311.
9. Kennedy, M.; McKervey, M. A.; Maguire, A. R.; Tuladhar, S. M.; Twohig, M. F. *J. Chem. Soc., Perkin Trans. I* **1990**, 1047.
10. Cordi, A. A.; Lacoste, J.; Hennig, P. *J. Chem. Soc., Perkin Trans. I* **1993**, 3.
11. Duddeck, H.; Ferguson, G.; Kaitner, B.; Kennedy, M.; McKervey, M. A.; Maguire, A. R. *J. Chem. Soc., Perkin Trans. I* **1990**, 1055.
12. Manitto, P.; Monti, D.; Speranza, G. *J. Org. Chem.* **1995**, *60,* 484.
13. Morris, J. C.; Mander, L. N.; Hockless, D. C. R. *Synthesis* **1998**, 455.
14. Galan, B. R.; Gembicky, M.; Dominiak, P. M.; Keister, J. B.; Diver, S. T. *J. Am. Chem. Soc.* **2005**, *127*, 15702.
15. Doyle, M. P.; Shanklin, M. S.; Pho, H. Q. *Tetrahedron Lett.* **1988**, *29*, 2639.
16. Pusino, A.; Saba, A.; Rosnati, V. *Tetrahedron Lett.* **1986**, *42*, 4319.
17. Lovely, C. J.; Browning, R. G.; Badarinarayana, V.; Dias, H. V. R. *Tetrahedron Lett.* **2005**, *46*, 2453.
18. For a review on rhodium-mediated intramolecular C–H insertion, see Taber, D. F.; Stiriba, S. *Chem. Eur. J.* **1998**, *4*, 990.
19. Wee, A. G. H.; Duncan, S. C. *Tetrahedron Lett.* **2002**, *43*, 6173.
20. White, J. D.; Hrnciar, P.; Stappenbeck, F. *J. Org. Chem.* **1999**, *64,* 7871.
21. Zhang, H.; Appels, D. C.; Hockless, D. C. R.; Mander, L. N. *Tetrahedron Lett.* **1998**, *39*, 6577.
22. Park, C. P.; Nagle, A.; Yoon, C. H.; Chen, C.; Jung, K. W. *J. Org. Chem.* **2009**, *74,* 6555.
23. Miah, S.; Slawin, A. M. Z.; Moody, C. J.; Sheehan, S. M.; Marino, J. P. Jr.; Semones, M. A.; Padwa, A.; Richards, I. C. *Tetrahedron* **1996**, *52*, 2489.
24. Merlic, C. A.; Zechman, A. L.; Miller, M. M. *J. Am. Chem. Soc.* **2001**, *123*, 11101.
25. Doyle, M. P.; Protopopova, M. N.; Peterson, C. S.; Vitale, J. P. *J. Am. Chem. Soc.* **1996**, *118*, 7865.
26. Panne, J. M.; Fox, J. M. *J. Am. Chem. Soc.* **2007**, *129*, 22.
27. Doyle, M. P.; Ene, D. G.; Forbes, D. C.; Pillow, T. H. *J. Chem. Soc., Chem. Commun.* **1999**, 1691.
28. Sugimura, T.; Im, C. Y.; Sato, Y.; Okuyama, T. *Tetrahedron* **2007**, *63*, 4027.
29. Pfau, A. S.; Plattner, P. A. *Helv. Chim. Acta.* **1939**, *22*, 202.
30. Crombie, A. L.; Kane, J. L. Jr.; Shea, K. M.; Danheiser, R. L. *J. Org. Chem.* **2004**, *69*, 8652.
31. Kennedy, M.; McKervey, M. A.; Maguire, A. R. *J. Chem. Soc., Perkin Trans. I* **1991**, 2565.
32. Frey, B.; Wells, A. P.; Rogers, D. H.; Mander, L. N. *J. Am. Chem. Soc.* **1998**, *120*, 1914.
33. O'Sullivan, T. P.; Zhang, H.; Mander, L. N. *Org. Biomol. Chem.* **2007**, *5*, 2627.
34. King, G. R.; Mander, L. N.; Monck, N. J. T.; Morris, J. C.; Zhang, H. *J. Am. Chem. Soc.* **1997**, *119*, 3828.
35. Rüedi, G.; Hansen, H. *Helv. Chim. Acta.* **2001**, *84*, 1017.

5.2 de Mayo Reaction

Yong-Jin Wu

5.2.1 Description

The de Mayo reaction is a sequence of reactions involving the photocycloaddition of an olefin with an enol or enol derivative of a β-dicarbonyl compound, followed by a *retro*-aldol fragmentation reaction to give a 1,5-diketone.

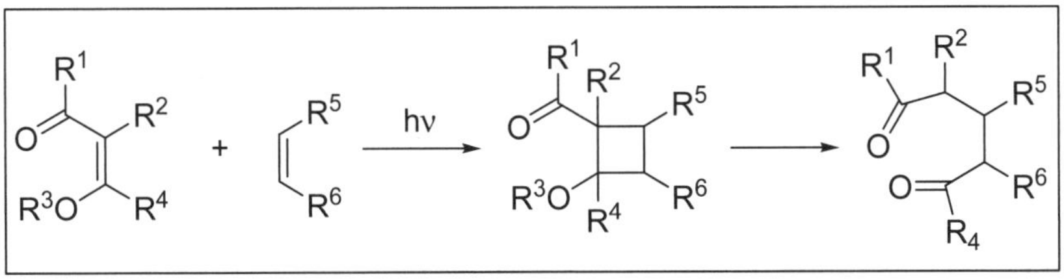

5.2.2 Historical Perspective

The photochemical [2 + 2] cycloaddition of two alkenes to generate a four-membered ring is a powerful and general method in organic synthesis. The discovery of this reaction started with the work of Ciamician/Silber[1] and Stobbe[2] more than a century ago when they investigated the photochemistry of stibene and styrene derivatives. The most well-known example of their work is the photodimerization of cinnamic acid to the truxinic and truxillic acids. In 1908, Cimician also observed that exposure of carvone to Italian sunlight for 1 year led to the formation of carvonecamphor.[3] This conversion represents the first example of an intramolecular enone–olefin photocycloaddition. It is surprising that very little attention was paid to this type of reaction until Büchi confirmed the carvone photoisomerizationin in 1957.[4] One year later, Cookson reported that irradiation of the *endo* Diels–Alder adduct **1** derived from cyclopentadiene and *para*-benzoquinone, generates the cage structure **2** via an intramolecular [2 + 2] photocycloaddition.[5] In 1964, Eaton completed the synthesis of the platonic solid, cubane, using the intramolecular enone–olefin photocycloaddition of intermediate **3** to give compound **4**.[6] The intramolecular version of this reaction was first applied to the total synthesis of a natural product in 1968 by Wiesner who prepared 12-epilycopodine.[7]

In the meantime, Corey, Eaton, de Mayo, and others examined intermolecular enone–olefin photocycloadditions. In 1962, Eaton reported the photocycloaddition of 2-cyclopentenone to cyclopentene to give adduct **5**.[8] In 1964, Corey disclosed the [2 + 2] photocycloadditions of 2-cyclohexenone to a variety of alkenes and established many of the characteristic features of this reaction.[9] The potential of this type of enone-olefin reactions was first shown by Corey in his landmark syntheses of *dl*-caryophyllene and *dl*-isocaryophyllene in 1963, and α-caryophyllene alcohol in 1964.[10,11]

In 1962, de Mayo reported an ingenious application of the intermolecular enone–olefin photocycloadditions by irradiating β-diketones in the presence of olefins to generate 1,5-diketones.[12] This reaction, which is now known as the de Mayo reaction,[13] proceeds through the enol of a 1,3-diketone (e.g., pentane-2,4-dione), which exists rigidly in a six-membered ring by an intramolecular hydrogen bond (e.g., **7**). Photoaddition of an olefin (e.g., cyclohexene) to this enol leads to a β-hydroxy ketone (e.g., **8**), which undergoes spontaneous retroaldolization to give a 1,5-diketone product (e.g., **9**). The synthetic power of this reaction was not fully realized until the late 1970s when Oppolzer, Pattenden, and others, reported the facility with which complex macrocyclic structures can be constructed via an intramolecular photoaddition sequence.[13c,13g] Over the years, the de Mayo reaction has been extended to vinylogous esters and amides and dioxolenones as β-keto ester equivalents. These developments have culminated in the total syntheses of complex natural products, including saudin, ingenol, vindorosine, and manzamine A by Winkler.[13h] In fact, Winkler's approach to manzamine using intramolecular photocycloaddition of vinylogous amides has been recognized as a classic piece of total synthesis.[14]

The de Mayo reaction resulted from his extensive contribution to photochemistry during the 1960s and 1970s. Paul Jose de Mayo (1924–1994)[15] completed his Ph.D. in organic chemistry at the Birkbeck College in 1954 under the supervision of Nobel Laureate D. H. R. Barton. In 1955 and 1957, he moved with Barton to the University of Glasgow and Imperial College, respectively, and served as a lecturer. The introduction of photochemistry in his research during this time would prove to be pivotal in his career. de Mayo conducted postdoctoral studies with Nobel Laureate R. B. Woodward at Harvard (1958–1959). He started his independent research career at the University of Western Ontario in 1959, and 3 years later, he discovered the de Mayo reaction. In his university, he founded a photochemistry department unit that contributed over 500 papers to the field.

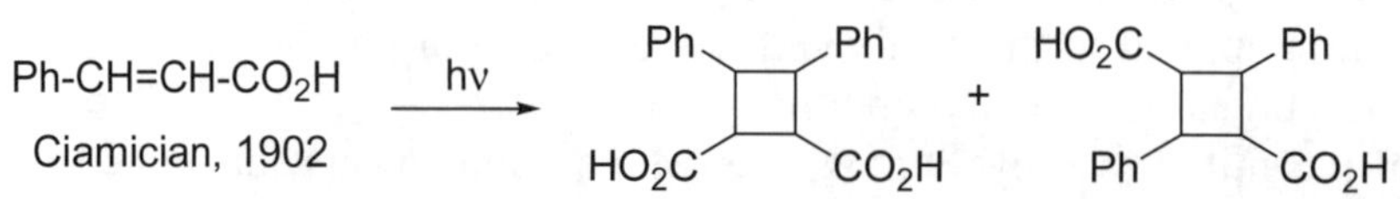

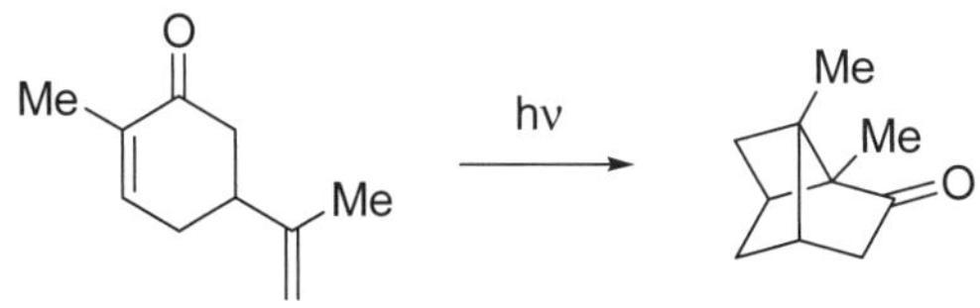

carvone carvonecamphor

Ciamician, 1908

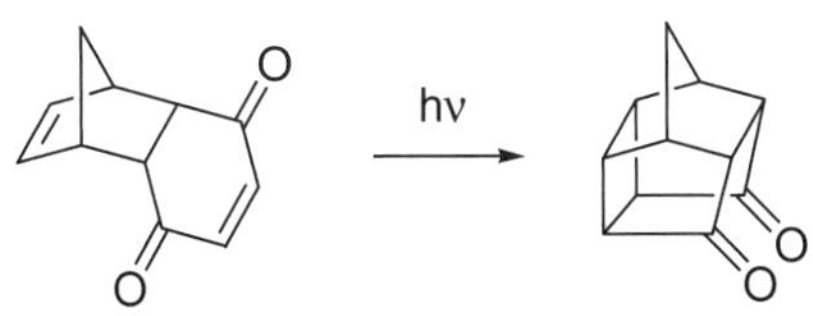

1 2

Cookson, 1958

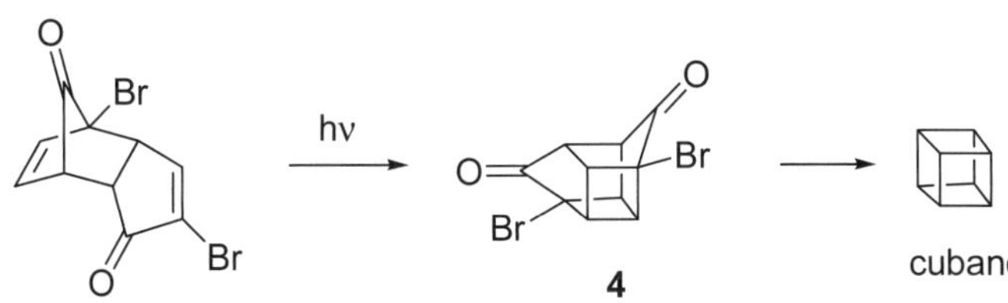

3 4 cubane

Eaton, 1964

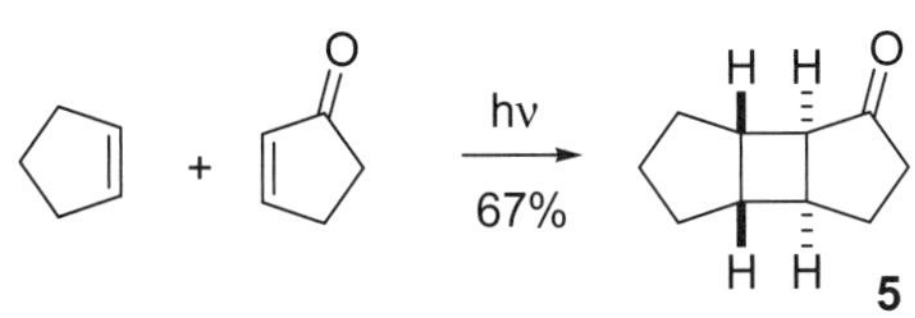

5

Eaton, 1962

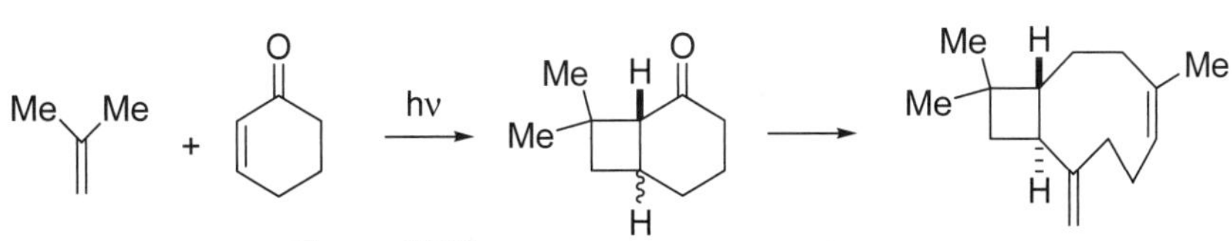

Corey, 1963 6 isocaryophyllene

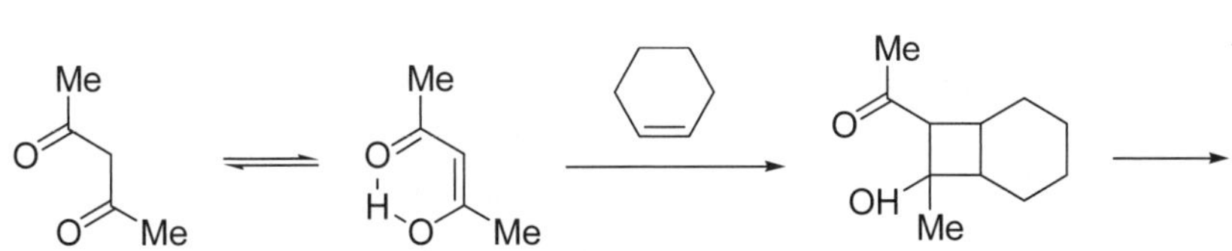

7 8

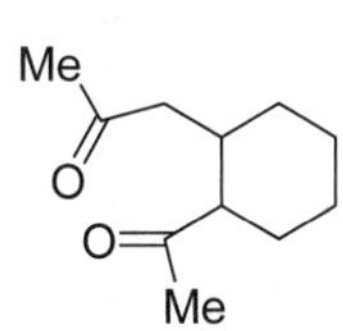

9

de Mayo, 1962

5.2.3 *Mechanism of Photoadditions*[13f,13g,16]

The photocycloaddition process is still not well understood, and a proposed mechanism is briefly discussed here. Excitation of the ground state enone probably via n $\rightarrow$ π^* produces the excited singlet, which undergoes intersystem crossing to either an n $\rightarrow$ π^* or $\pi \rightarrow \pi^*$ excited triplet. The next step is the complexation of the triplet state with the olefin to form an exciplex. Even though this exciplex has not been yet directly observed, it is consistent with the regiochemistry of some intermolecular photocycloaddition reactions and the observation that photocycloaddition reactions are much faster than those of normal radical additions to olefins. The exciplex is collapsed to a 1,4-diradical, and this process may involve initial bond formation at either C_α or C_β of the enone. Finally, the triplet 1,4-diradical must undergo spin inversion to the singlet diradical before ring closure to form the cyclobutane. Stereospecificity is lost if the intermediate 1,4-diradical undergoes bond rotation faster than ring closure. As a result, photocycloadditions are not always stereospecific.

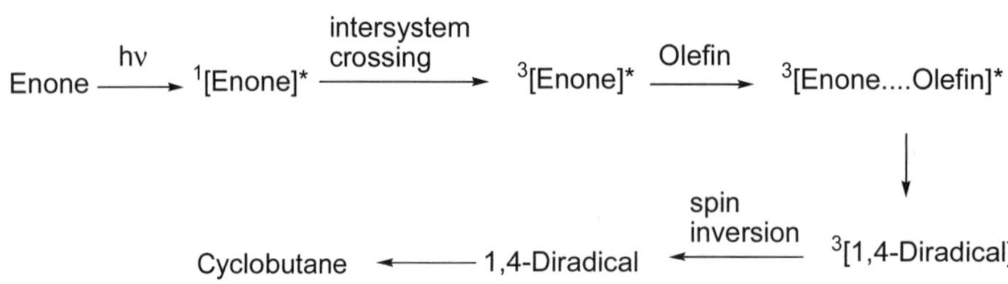

5.2.4 *Regioselectivity of Photocycloadditions*[13f,13g]

In the intermolecular photocycloadditions of enones to unsymmetrical alkenes, the regiochemistry of the product may result from head-to-head or head-to-tail addition. In many cases, however, a mixture of both types of these regioisomers is formed. Unfortunately, regiospecificity in [2 + 2] intermolecular photocycloadditions does not follow a simple rule.[17,18] In contrast, β-dicarbonyl compounds or their enol derivatives bearing appropriate olefin moieties undergo highly regioselective intramolecular photocycloadditions. The regioselectivity of the intramolecular photocycloaddition is generally high in systems where the two double bonds are separated by two, three, or four atoms. Formation of five-membered rings is preferred, and if a five-membered ring cannot be formed, then formation of six-membered systems is favored. This observation is termed "rule of fives" by Hammond and Srinivasan,[19] and is similar to the observation by Beckwith that 5-hexenyl radical undergoes cyclization to the cyclopentylmethyl radical

75 times faster than to the cyclohexyl radical.[20] Since the intramolecular de Mayo reaction is much more widely utilized in organic synthesis than the intermolecular version, the former is the focus of this review.

5.2.5 *β-Diketones and Their Derivatives*

Classic de Mayo Reactions

The classic de Mayo reaction involves the [2 + 2] photocycloaddition of an alkene to the hydrogen-bonded enol tautomer of a β-dicarbonyl compound as exemplified by the formation of 1,5-diketone **9** from pentane-2,4-dione and cyclohexene (vide supra). In addition to alkenes, allenes are also used as the olefinic component. For example, irradiation of a mixture of dimedone and allene results in the formation of 3,3-dimethyl-7-methylenecycloocta-1,5-dione **12** via the cyclobutane intermediate **11**, together with the corresponding head-to-tail product **13**, which spontaneously dimerizes to the hetero Diels–Alder adduct **14**.[21] Diketone **12** is a versatile building block for the preparation of substituted cyclooctadienones and δ-valerolactones.

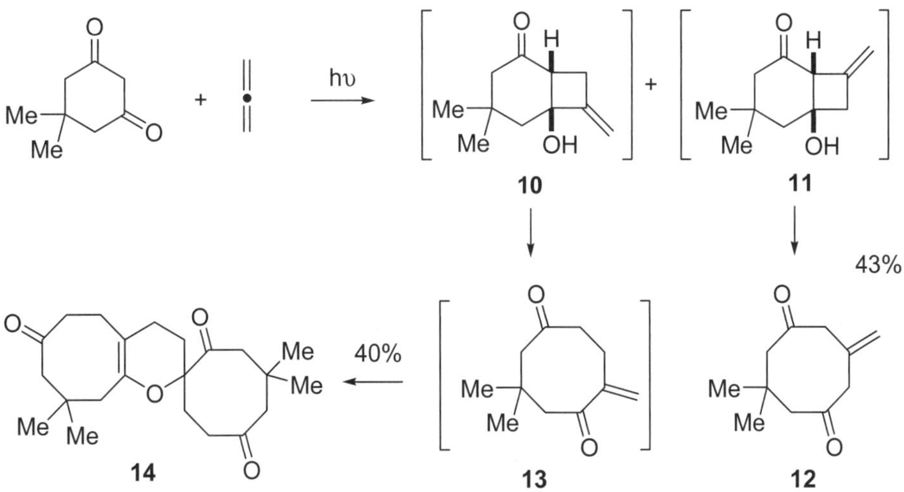

The intramolecular de Mayo reaction using an isocarbostyril substrate with a functionalized tether on nitrogen is employed in the synthesis of the galanthan ring system. Irradiation of **15** in acetonitrile gives a single product **18** in good yield.[22] Ring closure to the galanthan skeleton is carried out under basic conditions with piperidine to generate ketone **19** in 78% yield.

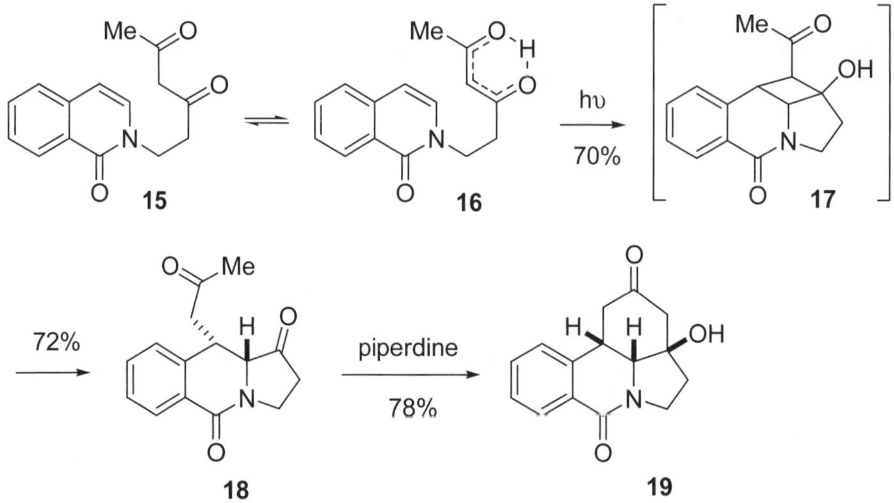

Enol Esters of β-Diketones

Both enol esters and enol ethers (vinylogous esters) undergo similar intramolecular photocycloaddtions, but enol esters have advantage over enol ethers in terms of fragmentation process. Fragmentation of the cyclobutane photoadducts from enol esters is conveniently carried out under basic conditions, whereas that from enol ethers often requires additional operation to convert ether to ester or lactone (vide infra). For this reason, enol esters are more widely used than enol ethers.

Photoaddition of enol acetate **20** produces the straight adduct **21** in high yield.[23,24] This regiochemistry is consistent with the general preference for the formation of five-membered rings when possible. Adduct **21** is fragmented under basic conditions to generate diketone **22**.

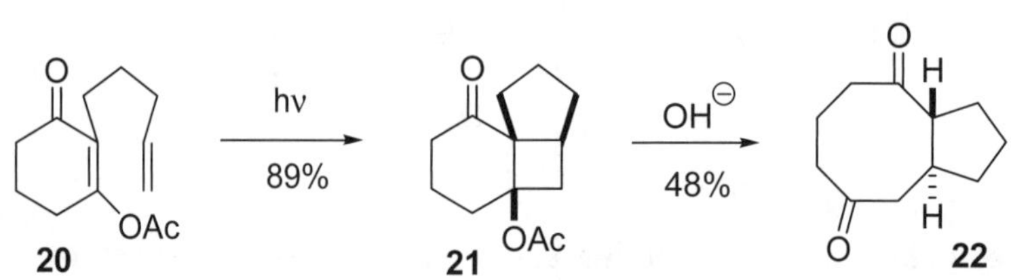

Photocycloaddtion of the cyclopentyl derivative **23** also proceeds regioselectively to give the *cis*- and *trans*-cyclobutane **24** and **25** in 70% and 12% yield, respectively.[23] Base-induced fragmentation of both isomers furnishes a single diketone **26** in excellent yield.

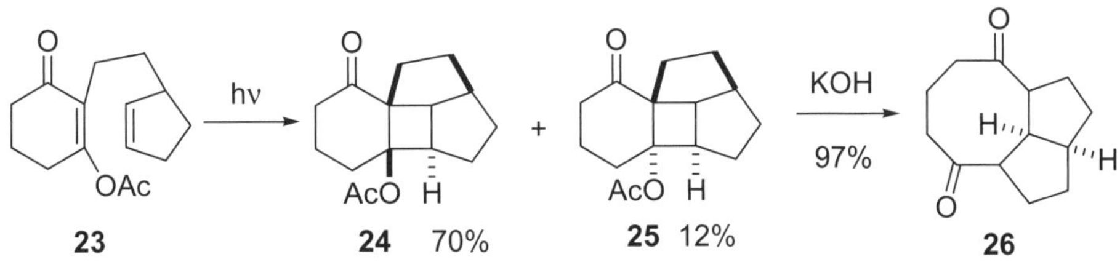

An exception to the rule of fives is observed in the case of enol acetate **27**.[23] In contrast to **23**, irradiation of the cyclohexenyl substrate **27** gives the tricyclic ketoacetate **31** as the major product and the expected tetracyclic photoadduct **30** as the minor product. The major product probably arises from hydrogen atom abstraction pathway via the 1,4-diradical intermediate **29**, whereas the pathway to the minor adduct **30** may involve the radical intermediate **28**. Presumably, there is significant geometric strain in the transition state for the initial cyclization to **28**, therefore the alternative initial cyclization to **29** is favored. Exposure of adduct **30** to basic conditions brings about saponification followed by retro-aldol fragmentation to give tricyclic diketone **32**.

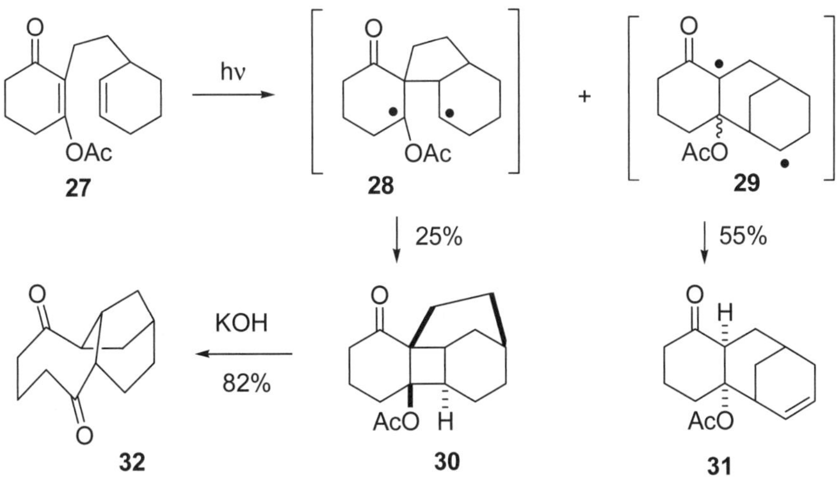

Irradiation of enol acetate **33** produces a 2:3 ratio of **36** to **37**.[25] The poor regioselectivity observed in this case is not entirely unexpected, since there is little stability difference between the anticipated seven-membered and eight-membered biradical intermediates, **34** and **35**, respectively. Both adducts **36** and **37** are fragmented under basic conditions to the bicyclic compounds, **38** and **39**, respectively.

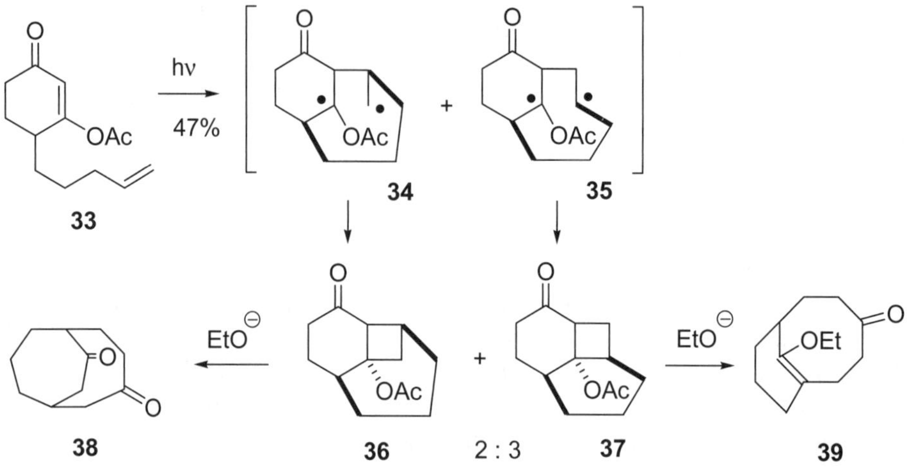

The regioselectivity of photoaddition of the enol acetate of **40** depends on reaction temperature, and the ratios of **41** to **42** are 11:89, 2:3, and 51:49 at –70 °C, 25 °C, and 65 °C, respectively.[25,26] Of note is that the acetylation of 1,3-diketone **40** is not regiospecific, but the two enol acetates interconvert via a photo-Fries process. However, only the enol acetate leading to **41** and **42** participates in the cycloaddition. Fragmentation of adducts **41** and **42** gives diketone **44** and the aldol product **45** via diketone **43**.

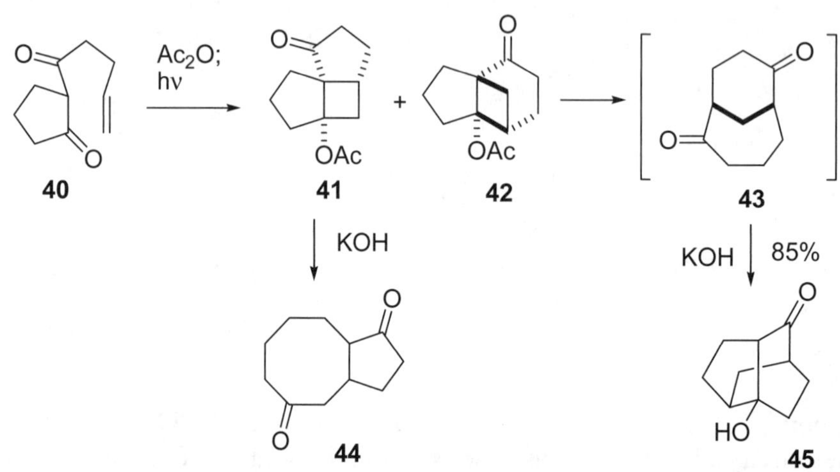

In contrast to the enol acetate of diketone **40**, the methyl-substituted enol acetate **46**, undergoes photocycloaddition to give exclusively the straight adduct **49**.[27] This has been explained on the basis of a steric interaction between the methyl group and the cyclopentane methylene hydrogens in the exciplex **48**. This interaction is much reduced in exciplex **47**.

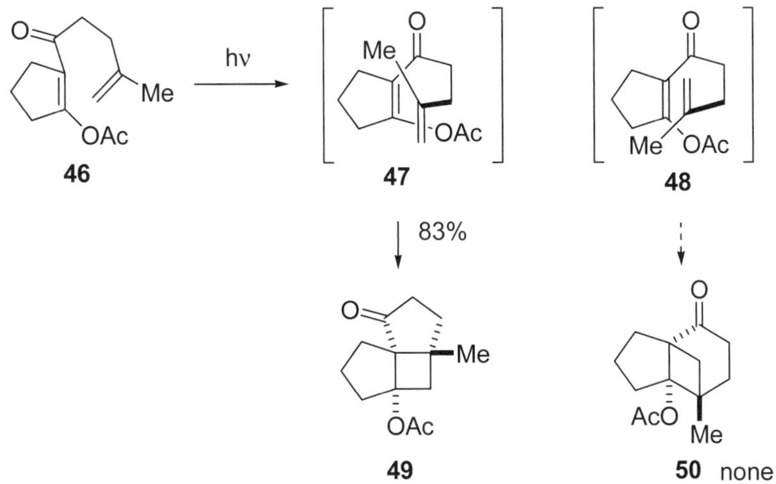

Irradiation of enol acetate **51** affords the exclusive straight adduct **52** as a 3:1 mixture of methyl epimers.[27] Base-induced fragmentation and subsequent spontaneous retroaldolization–realdolization sequence generates **54**, which has been converted to daucene.

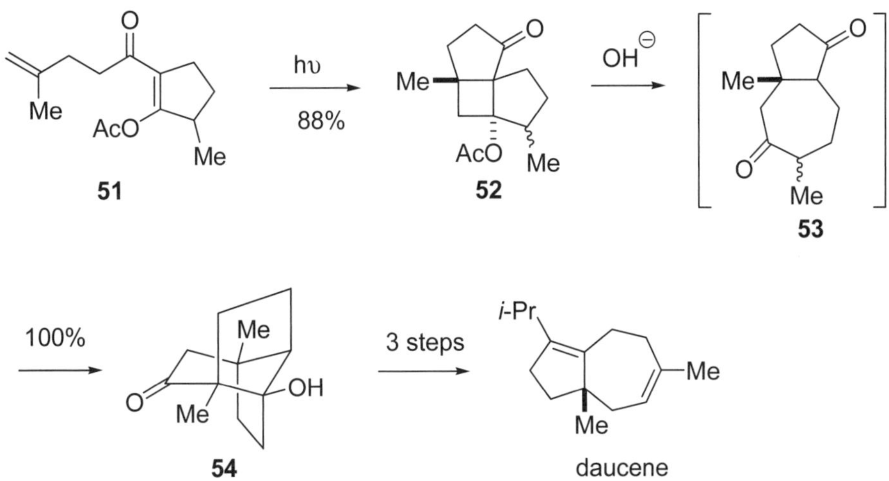

The 1,7-dienes **55** undergo highly regioselective cycloaddition to give **56** whether R is H or Me.[28] Both substrates afford the straight adducts **56** exclusively, the general tendency for 1,7-dienes, since the formation of a six-membered ring is preferred over seven-membered ring. Treatment of **56** with aqueous potassium hydroxide gives the tricyclic alcohol **58**, the result of a tendem *retro*-aldol/intramolecular aldol reaction sequence.

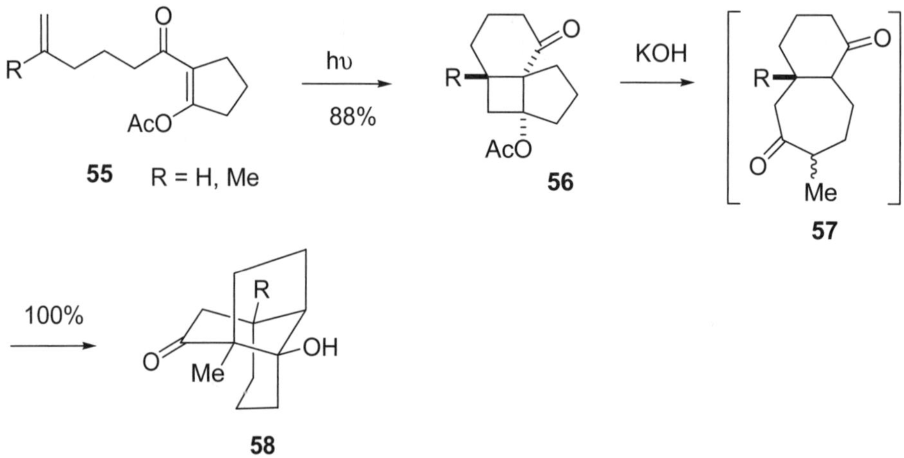

Regioselective photocycloaddition of enol carbonate **59** with allene results in the formation of adduct **60** in 83% yield.[29] The protecting group is removed under palladium-catalyzed conditions, and the resulting alcohol undergoes *retro*-aldol reaction and olefin isomerization to give the 1,5-diketone **62** in excellent yield. This compound contains the AB ring system of the taxane skeleton.

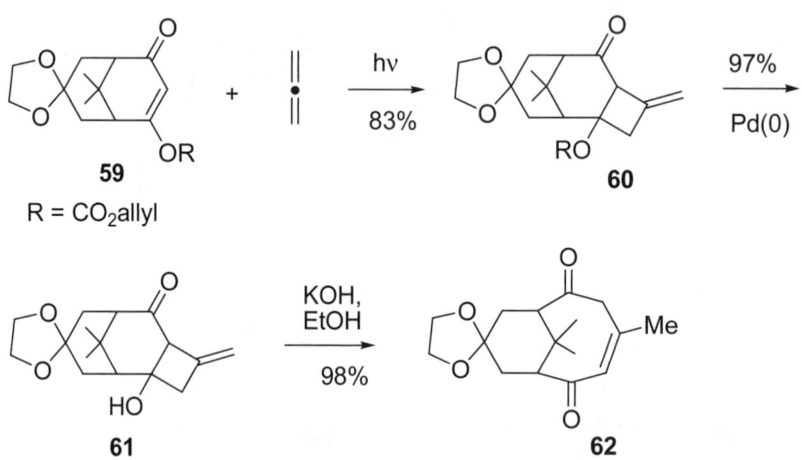

Vinylogous Esters

Intramolecular de Mayo reactions have also been carried out on vinylogous esters of cyclic 1,3-diketones, and the adducts can be fragmented under appropriate conditions. Thus irradiation of the vinylogous ester **63** furnishes the crossed product **64**,[30] which, upon gentle warming in water, fragments to ketoalcohol **65** in 53% yield.[31] It is interesting that when the geminal dimethyl substituted analog **66** is irradiated under similar conditions, the expected photoadduct is not formed, and instead the bicyclic ketone **67** is obtained.[32] The formation of **67** may be considered as a photochemical version of the intramolecular "ene" reaction. The 1,6-diene **68a** undergoes photocycloaddition in accord with the rule of fives to give the straight adduct **69a**. When this adduct is treated with $BF_3{\cdot}Et_2O$ followed by aqueous workup, cyclooctanedione **70** is generated. The photocycloaddtion of the homologous 1,7-diene **68b** also gives the straight adduct **69b**.[33]

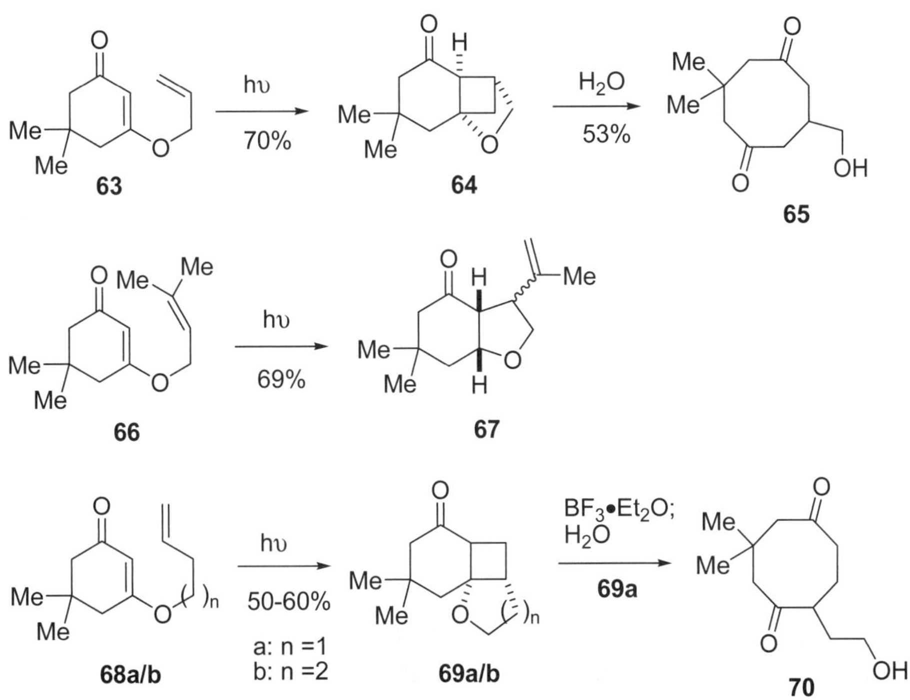

The photocycloaddtion/fragmentation approach of vinylogous esters has been applied to the synthesis of the BC ring system of the taxane skeleton. Photocycloaddition of **71** generates a 47% yield of the single diastereomer **72**, which is converted to the *n*-propyl-substituted cyclobutane derivative **73** using three operations.[34] This cyclobutane is cleaved with trimethylsilyl triflate to give the silylenol ether **74**, which is further hydrolyzed to the tricyclic ketone **75**. Treatment of **73** with acid leads directly to **75** in a slightly lower yield. Alternatively, the tetrahydrofuran moiety of **73** is oxidized with RuO_4 to afford the lactone **76**, which is saponified with concomitant cyclobutane cleavage to give the bicyclic 1,5-diketone **77**. This diketone resembles the BC ring system of the taxane skeleton, and the whole skeleton has also been constructed using a similar strategy (vide infra).

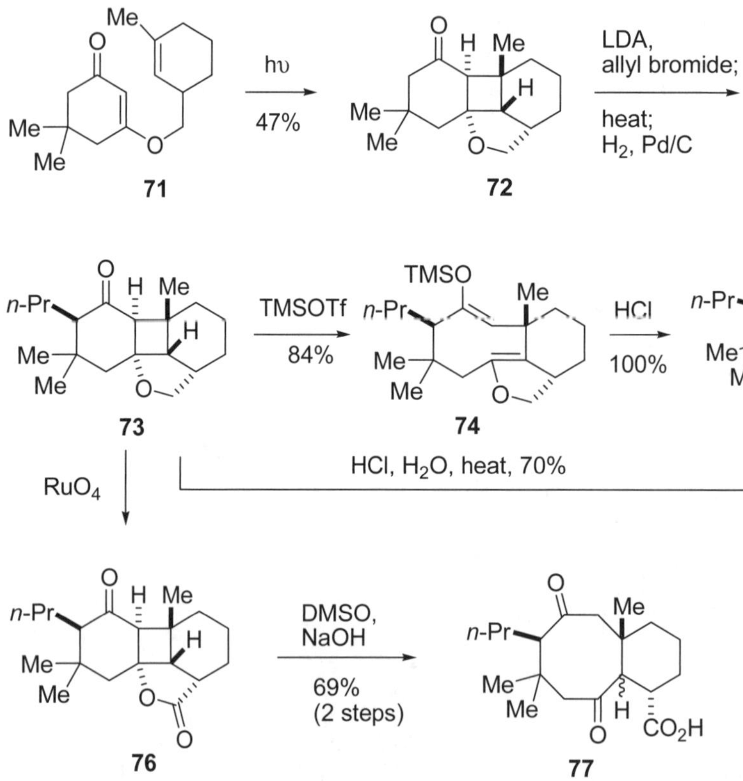

In contrast to cyclopentenones and cyclohexenones, medium-ring vinylogous esters are not suitable for photocycloaddition reaction unless the olfin coupling partner is substituted. For example, when (*E*)-3-(but-3-enyloxy)cyclooct-2-enone **78a** is irradiated under a variety of conditions, no intramolecular cycloadduction occurs.[35] However, upon substitution of a vinyl or phenyl on the olefin, the cycloaddition proceeds efficiently to give diastereomeric mixtures **79b/c**, and **80b/c**, respectively. The dramatically enhanced yields and rates of the photoaddition reactions upon olefin substitution result from the stabilization of the 1,4-biradical by a vinyl or phenyl. A mixture of diastereomers is formed presumable because the rotational relaxation of the intermediate 1,4-diradical is faster than [2 + 2] ring closure.

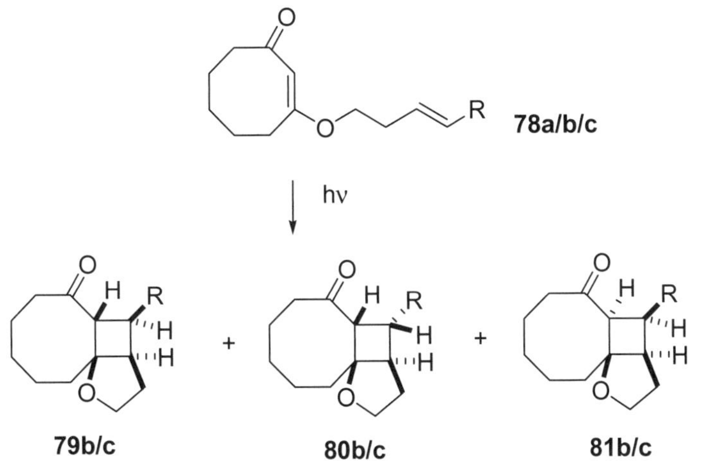

a: R = H	No Reaction		
b: R = vinyl	21%	25%	17%
c: R = Ph	6%	21%	17%

The intramolecular photoaddition of vinylogous esters with allenes has also been explored.[36] Thus photocycloadition of **82** leads to the formation of the bicyclic furan **84** in moderate yield. This furan derivative is prepared from the cyclobutane intermediate **83**, which results from the parallel addition to the terminal olefin of the allene. However, this approach does not work for the acyclic vinylogous ester **85**. Irradiation of this ester provides none of the expected furan product **87** and results only in isomerization of the vinylogous ester.

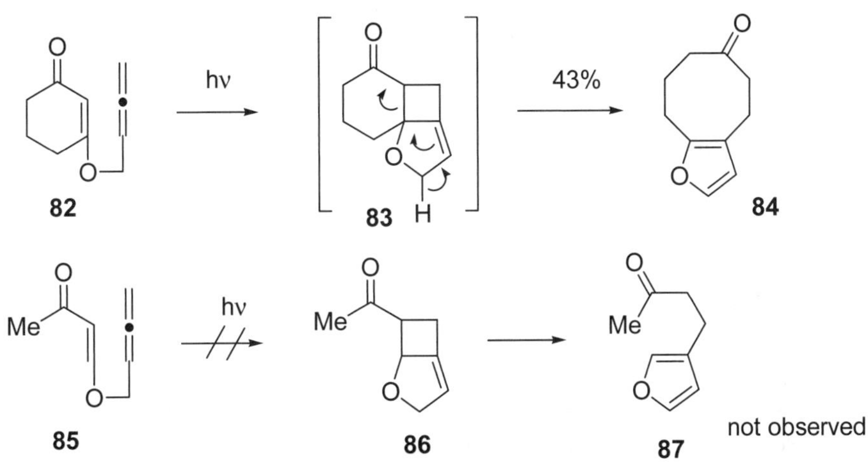

Vinylogous Amides (Aza-de Mayo Reaction)[37]

Vinylogous amides undergo a similar sequence to the classic de Mayo reaction with the exception that the intermediate cyclobutanes cleave via *retro*-Mannich pathway rather than a *retro*-aldol process. The photocycloaddition/*retro*-Mannich reaction of vinylogous amides is sometimes referred to as the *aza*-de Mayo reaction in the literature. For example, photolysis of **88** leads to the exclusive formation of the crossed photoadduct **89** in 50–60% yield.[32,38] *Retro*-Mannich fragmentation of **89** proceeds readily in refluxing water to give a 94% yield of an equilibrium mixture of tautomers: ketoaminal **91**, diketoamine **90**, and keto imminium ion **92**. When the geminal dimethyl substituted substrate **93** is irradiated, none of the desired *aza*-de Mayo product is obtained, and instead the unusual "photochemical ene product" **94** is isolated as the sole product in 66% yield.

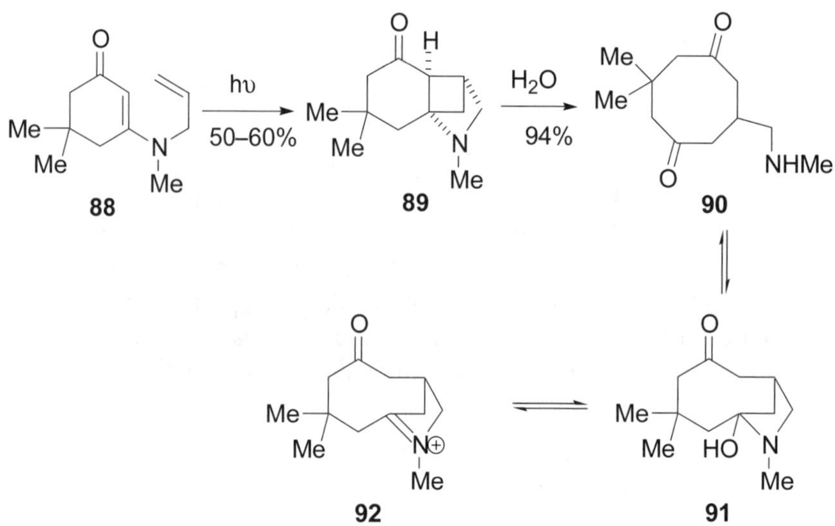

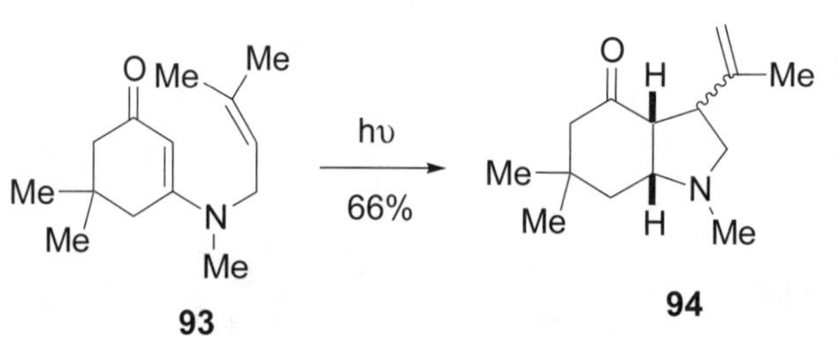

Irradiation of **95** gives the *retro*-Mannich fragmentation product 1,5-ketoimine **97**, presumably through the intermediacy of the straight the photoaddition adduct cyclobutane **96**.[39] When the linking unit is shortened, photo-*aza*-Claisen rearrangement competes with photocycloaddition.[40] Thus photocycloaddition of **98** provides a mixture of two compounds, the photo-*aza*-Claisen rearrangement product **101**, and the 1,5-ketoimine **100**. This imine derives from the straight photoaddition intermediate **99**; it is surprising that no keto imine **103** (resulting from the crossed adduct **102**) is formed. This appears to be an exception to the rule of fives. Of note is that the structural assignment of ketoimine **100** (based on NMR and IR data) has not yet been confirmed by *X*-ray studies. It is interesting that the *N*-methyl derivative **104** leads to a mixture of the photo-*aza*-Claisen rearrangement product **106**, and the photoadduct **105** via cross addition, according to the rule of fives.

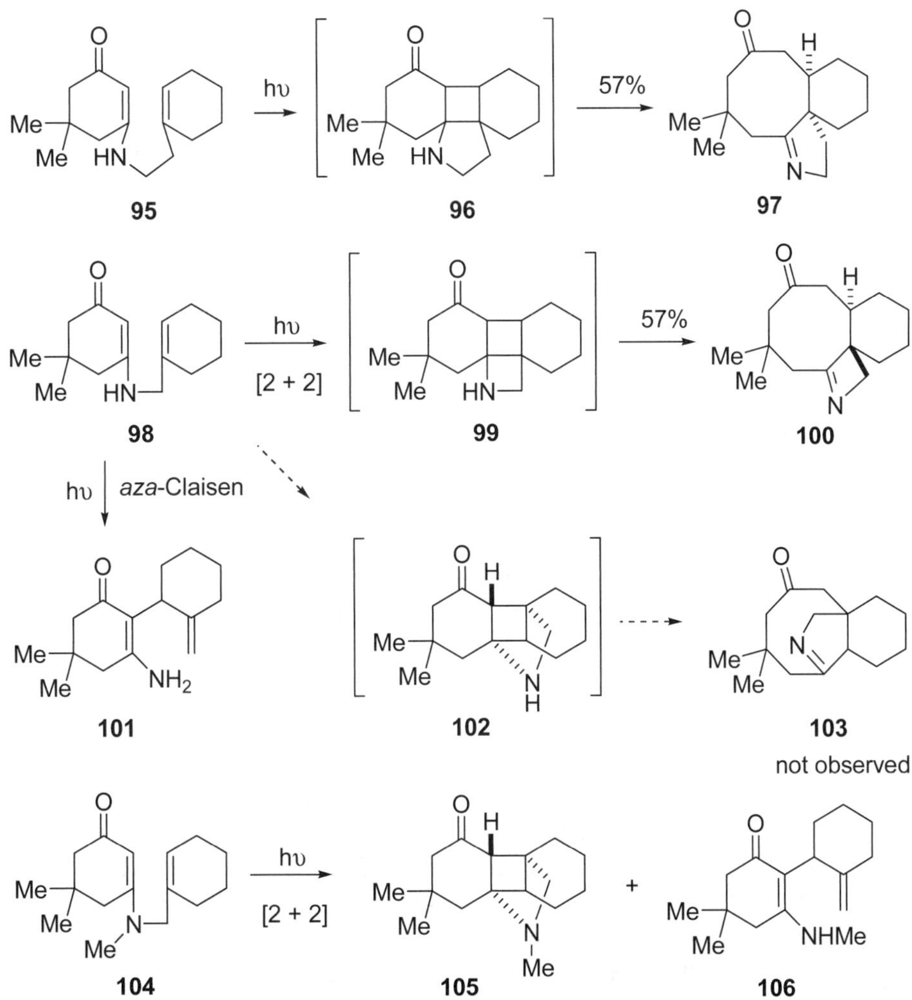

The highly substituted *N*-alkenoyl β-enaminone **107** undergoes photocycloaddition to give the tetracyclic adduct **108** as a single diastereomer.[41] Fragmentation of the photoadduct **108** using hydrochloric acid in aqueous dioxane affords the fused bicyclic ring system **109** in 60% yield.

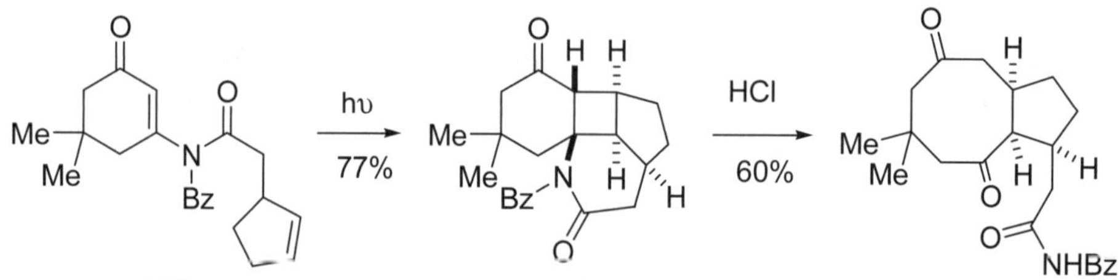

107 **108** **109**

The appropriately substituted vinylogous amides can undergo an intramolecular photocycloaddition-*retro*-Mannich–Mannich sequence. This sequence is analogous to the photocycloaddition-*retro*-aldol-aldol sequence shown in the formation of **19** from **15** (vide supra). Thus, irradiation of **110** leads to the formation of ketoimine **112**, the product of photoaddition followed by *retro*-Mannich fragmentation. Reaction of **112** with 1 equiv of trimethyloxonium tetrafluoroborate, followed by treatment of the resulting iminium ketone with aqueous hydrochloric acid, provides the photocycloaddition-*retro*-Mannich-Mannich product **113** in 50% yield from the acyclic photosubstrate **110**.[42]

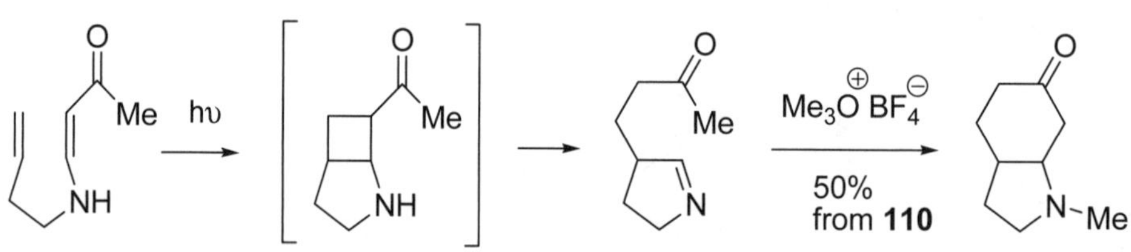

110 **111** **112** **113**

The photocycloaddition-retro-Mannich–Mannich methodology is featured in a concise synthesis of mesembrine.[42] Irradiation of vinylogous amide **114** effects photocycloaddition-*retro*-Mannich sequence to give product **116** via the cyclobutane intermediate **115**. Methylation with trimethyloxonium tetrafluoroborate followed by treatment with DMAP produces mesembrine in 84% yield. Other applications include construction of the bicyclic core of peduncularine[43] and synthetic approaches to hetisine alkaloids[44a] and 8-substituted 6-azabicyclo[3.2.1]octan-3-ones.[44b]

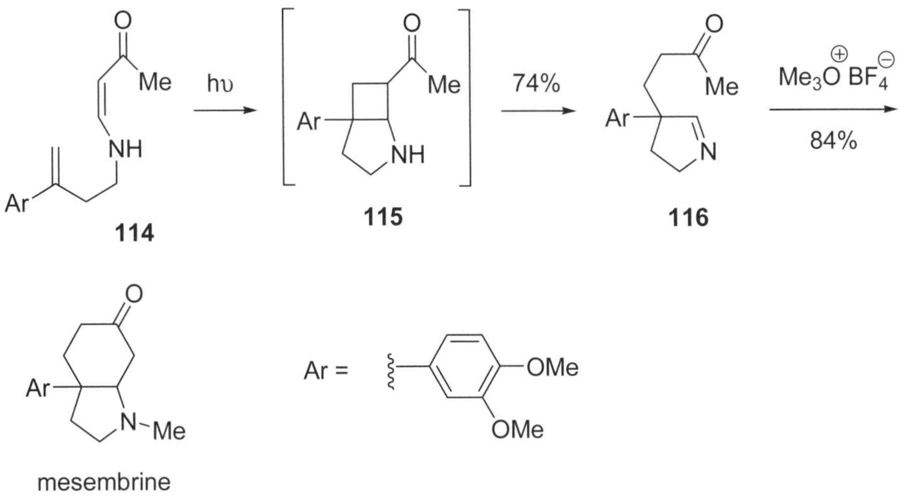

Intramolecular photoaddition of vinylogous imide **117a** results in the formation of the bridged bicyclic compound **120a** in 52% yield.[36] This compound is generated via the crossed photoaddition of **117a** to generate intermediate **118a**. This intermediate then undergoes *retro*-Mannich fragmentation to afford zwitterionic intermediate **119a**, cyclization of which provides the observed product **120a**. The *N*-Boc protected photosubstrate **117b** undergoes the same transformation, but the corresponding *N*-Boc product **120b** is unstable to purification. However, when vinylogous amides **121** are irradiated, only pyrroles **124** are obtained. These pyrroles are formed via the intermediacy of cyclobutanes **122** resulting from the parallel addition to the terminal olefin of the allene. This represents the first example of parallel intramolecular photocycloaddition to the terminal olefin of an allene. Similarly, irradiation of cyclohexane-1,3-dione-derived vinylogous amides **125** affords pyrrole products **126** in excellent yields, but it is surprising that only moderate yield of pyrrole **128** is obtained from the corresponding cyclopentenone **127**.

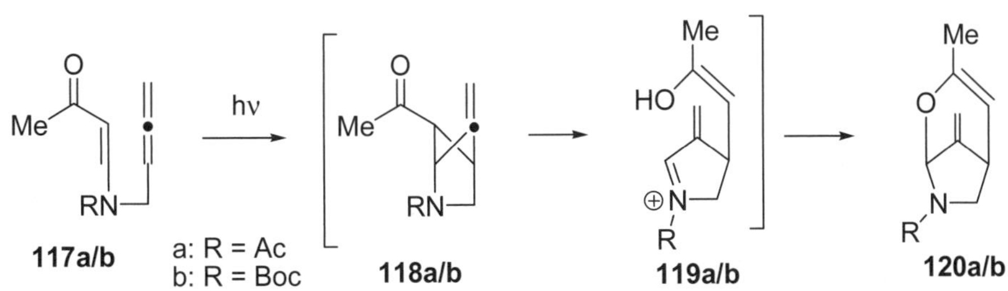

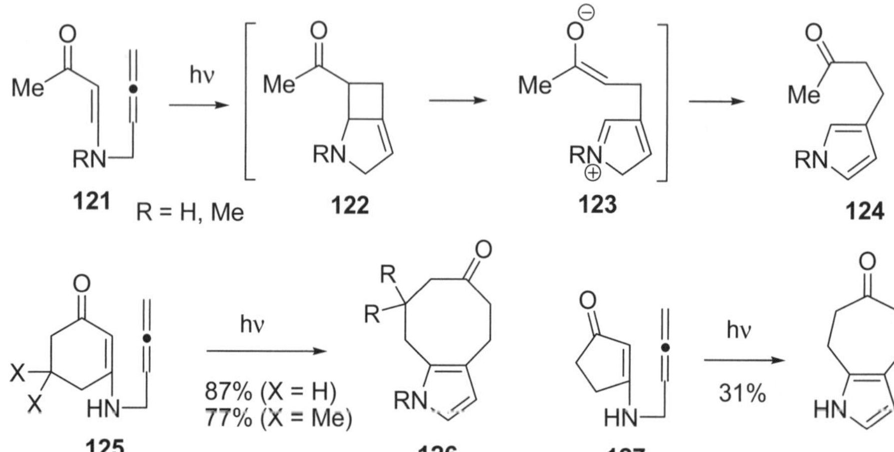

121 122 123 124

125 126 127 128

5.2.6 *β-Keto Ester Derivatives*[13h]

The photochemical reactivity of β-ketoesters is different form that of β-diketones.[13h] Irradiation of a β-ketoester in the presence of an alkene produces oxetane via the ketone carbonyl instead of the desired cyclobutane ring system. Therefore, it is necessary to covalently "lock" the ketoesters as the enol tautomers. To this end, silyl enol ethers, **129** and **132a**, and enol acetates, **130** and **132b**, were prepared, but these substrates still fail to undergo the desired intramolecular [2 + 2] photocycloaddition with olefins. The only new products observed in these reactions result from the photo-Fries rearrangement of the cyclic enol acetate (**130** to **131**) and *cis-trans* isomerization of both acyclic substrates **132a/b**. However, tetronates are appropriate substrates for both intermolecular and intramolecular photocycloadditions with olefins. In addition, enol acetates and silyl enol ethers of β-keto esters are known to undergo [2 + 2] photoaddition with cyclic enones (vide infra).

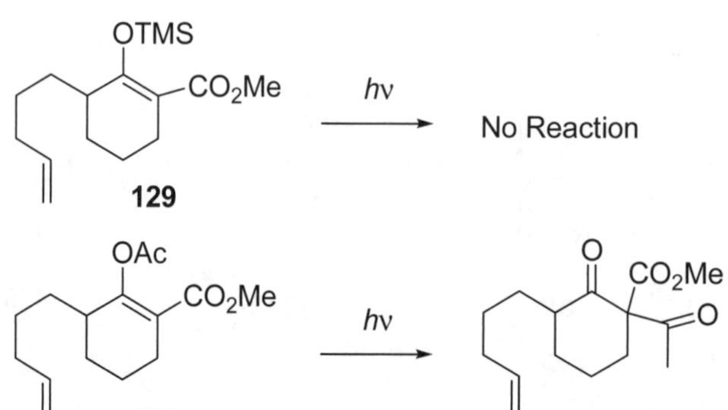

129

130 131

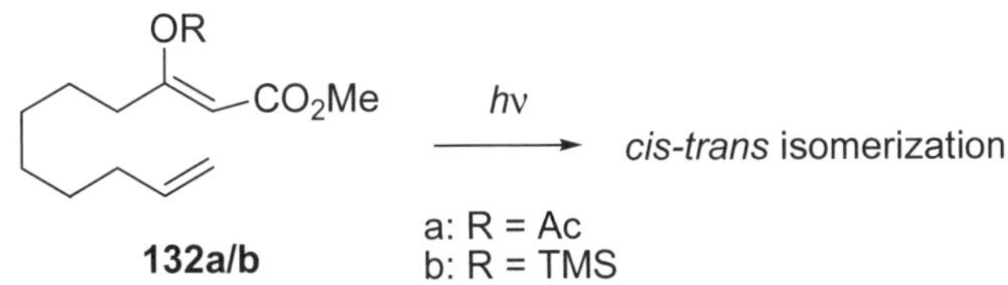

132a/b

Dioxolenones as β-Keto Ester Equivalents

In 1980, Baldwin developed a modification of the de Mayo reaction using dioxinone heterocycles as covalently locked enol tautomers of β-keto esters.[45] Thus intermolecular cycloaddition of 2,2,6-trimethyl-1,3-dioxolenone **133** occurs in good yield using stoichiometric quantities of a variety of alkenes. For example, irradiation of **133** with tetramethylethylene yields the cyclobutane adduct **135** in 90% yield. This adduct is converted to cyclohexenone **138** in two steps. Controlled reduction of **135** with diisobutylaluminum hydride (DIBAL) gives keto aldehyde **137** (after spontaneous loss of acetone from hemiacetal **136** and *retro*-aldol cyclobutanol fragmentation), which on exposure to acidic conditions affords cyclohexenone **138** in 76% yield.

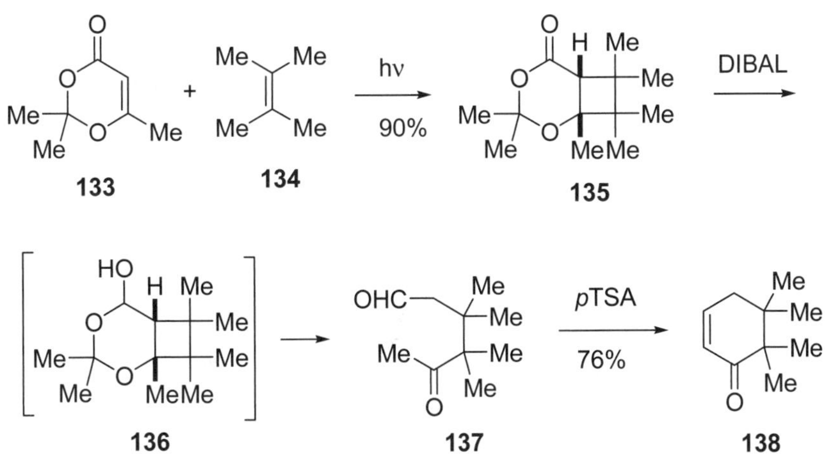

2,2-Dimethyl-4*H*-1,3-dioxin-4-one **139** has been used as the photochemical equivalent of formylacetate, which is inactive as the enone partner in the photoaddition with alkenes. Irradiation of dioxinone **139** in the presence of symmetrical cyclopentene **140** leads to the cyclobutane adduct **141**. This intermediate upon subjection to hot water undergoes *retro*-aldol fragmentation and spontaneous lactonization to give lactone **143**, a useful prostaglandin intermediate.[46] Enantiomerically pure lactone **143** is prepared by using a chiral version of dioxinone **139**.[47]

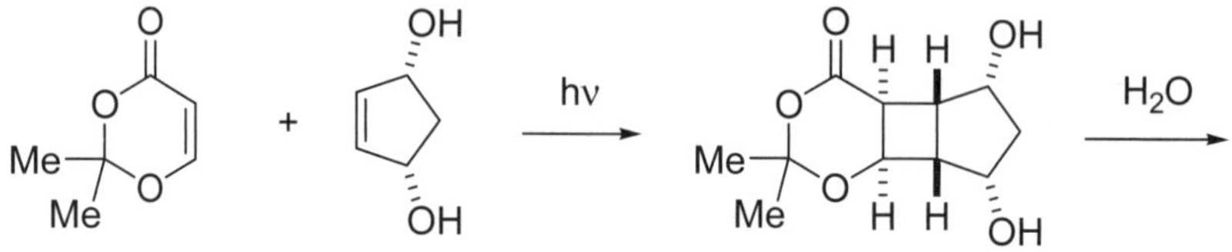

139 140 141

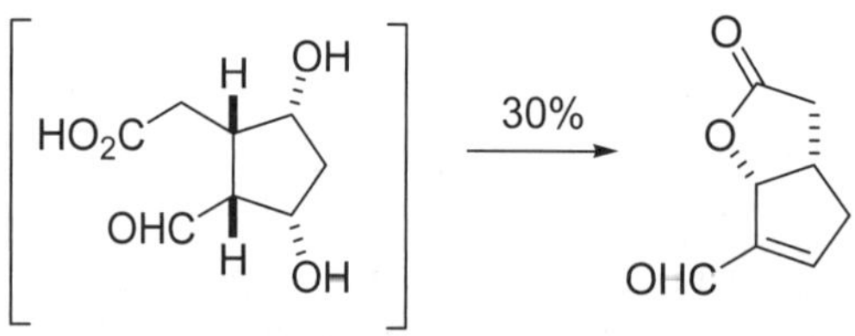

142 143

Like enol derivatives of β-diketones, the regiochemical outcome of the intermolecular cycloaddition of dioxolenone substrates with unsymmetrical alkenes can be difficult to predict on the basis of existing enone photocycloaddition prediction models. The intramolecular version of the dioxinone photocycloaddition reaction, however, provides greater regiochemical control. For example, photocycloaddition of **143** occurs regioslectively to afford the desired crossed adduct **144** as a single diastereomer in excellent yield.[48] Treatment of this adduct with aqueous potassium hydroxide brings about selective saponification of the six-membered lactone with concomitant *retro*-aldol fragmentation to give a 78% yield of the ester **145** (from photosubstrate **143**) upon methylation with diazomethane.

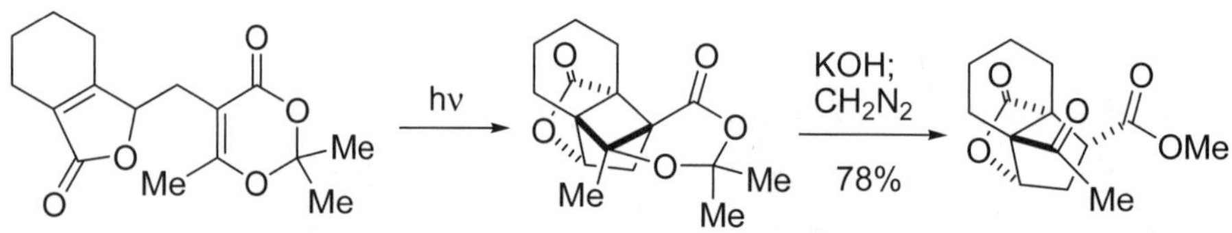

143 144 145

An intramolecular dioxinone photocycloaddition is used in the synthesis of (+)-valeranone.[49] Irradiation of dioxinone **146** gives a high yield of cycloadduct **147**, which undergoes reduction followed by spontaneous *retro*-aldol fragmentation. The resulting diketoaldehyde intermediate **149** is cyclized to enone **150** via aldol condensation. This enone is converted to (+)-valeranone in a few steps.

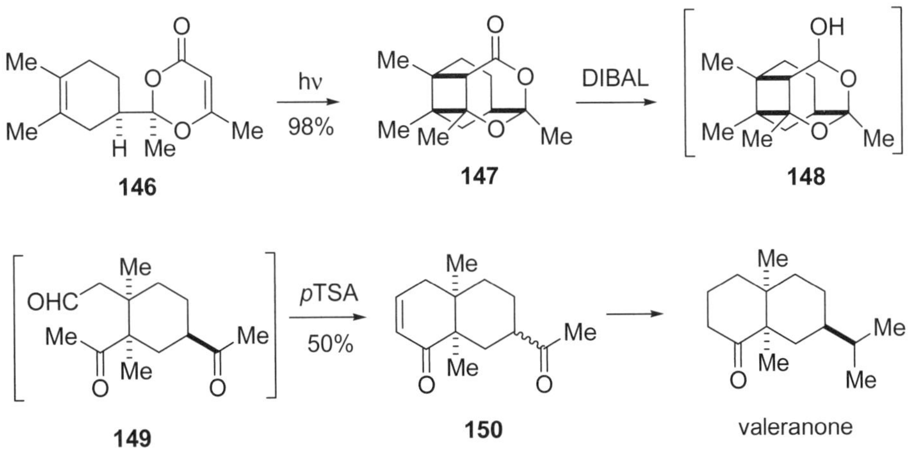

The intramolecular dioxolenone photocycloaddition/fragmentation approach provides easy access to six-, seven-, eight-membered ring keto esters **153** in excellent yields with high (> 50:1) levels of regiochemical control.[50]

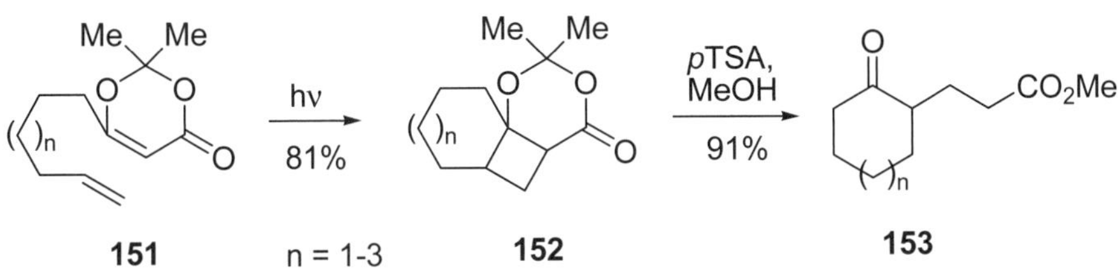

The intramolecular photocycloaddition of dioxenones with alkynes is also reported.[51]

Tetronates[52]

The photocycloaddition of tetronates is analogous to that of vinylgous esters as β-diketone equivalents (vide supra). The MEM-protected tetronate **155** is irradiated in the presence of cyclopentene, cyclohexene and *cis*-cyclooctene, and the resulting adducts **156** are immediately deprotected with titanium tetrachloride to give the hydroxyl lactones **157** as mixture of diastereomers in about 50% yield. These adducts are subjected to cesium carbonate in THF under microwave irradiation to give the ring-opened oxepanediones **158** in 74–84% yield.

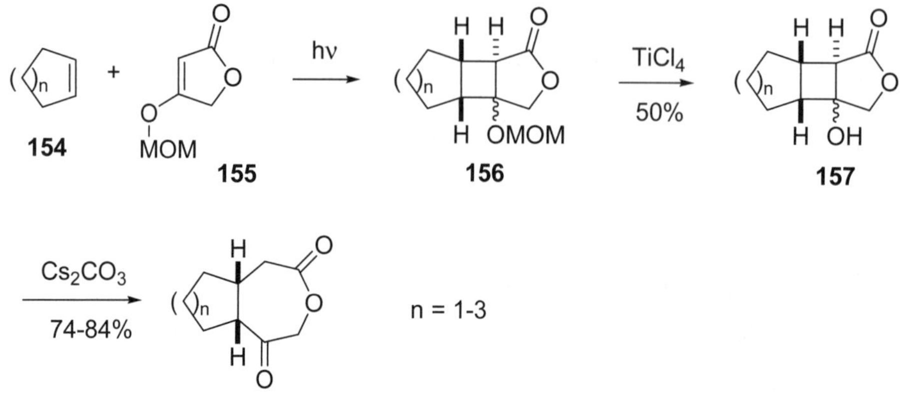

154 155 156 157 158

The C-4 substituted tetronates **159** also undergo intramolecular photocycloaddition to give the tricyclic adducts **160** with the heterocyclic ring directly anellated to the cyclobutane moiety. In consistency with the rule of fives, these photocycloadditions proceed exclusively to furnish the straight adducts.

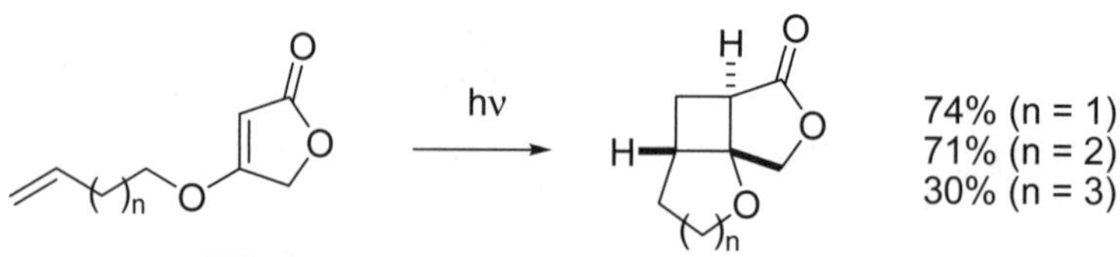

159 160

5.2.6.3 Enol Esters and Silyl Enol Ethers of β-Ketoesters

As described previously, no photocycloaddition occurs between alkenes and silyl enol ethers, **129** and **132a**, and enol acetates **130** and **132b** (vide supra). However, this type of enol derivatives may be suitable for photoaddition with cyclic enones. For example, irradiation of **162** in the presence of enone **161** proceeds with high regioselectivity to give adducts **163** (40%) and **164** (40%) along with a dimer of enone **161**.[53] As expected, irradiation of **161** with an excess of either **167** or its ethyl enol ether **168** gives in less than 10% yield the desired head-to-head adducts **169**. Reduction of **164** with sodium borohydride gives lactone **165**, and treatment of this lactone with dilute sodium methoxide results in transesterification of the acetoxy group followed by a *retro*-aldol type reaction to give the fragmentation product **166**.

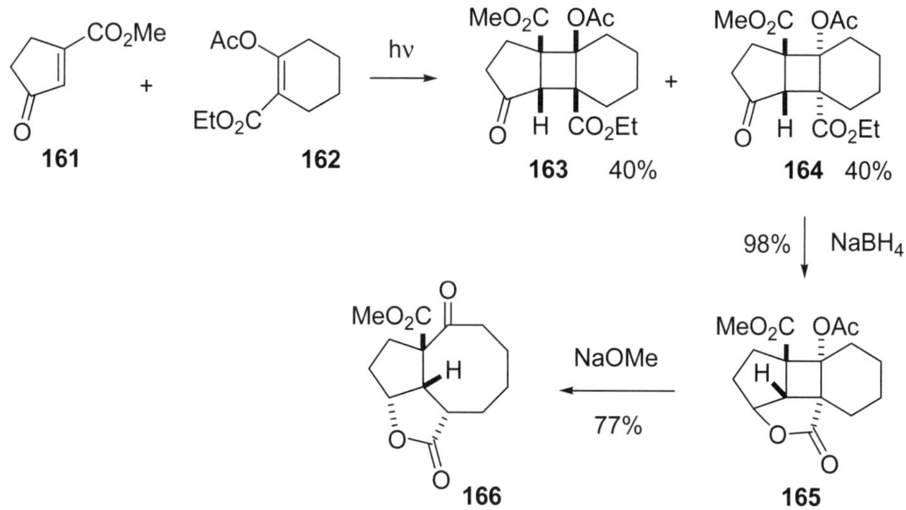

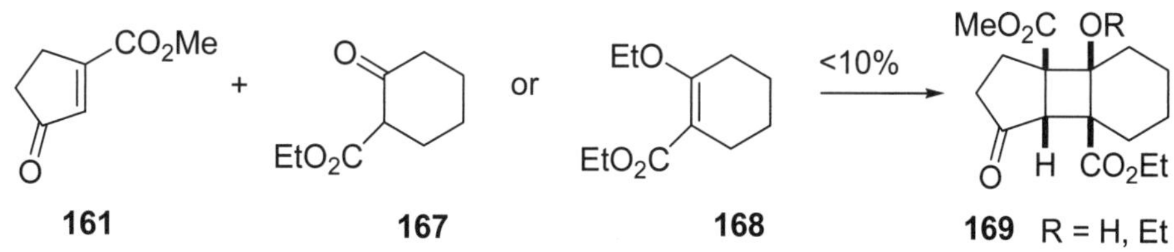

The trimethylsilyl ether **171** also undergoes photoaddition with 2-cyclopentenone to give the head-to-head *cis-anti-cis* adduct **172** albeit in a modest 35% yield along with a considerable amount of enone dimer.[53] Reduction of **173** with sodium borohydride gives lactone **174**, which, upon

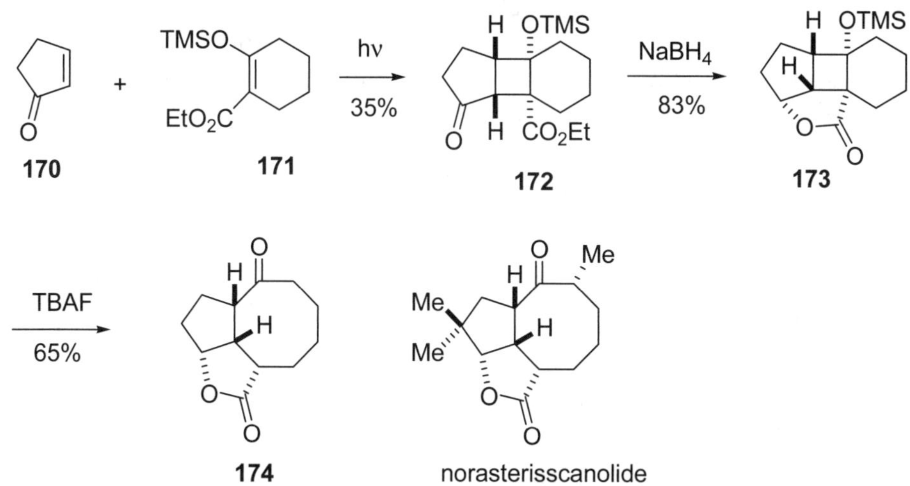

treatment with fluoride ion, leads to the 6/7 fused ketone **174** via *retro*-aldol fragmentation. This compound has all the structural features of asteriscanolide, but lacks the three methyl groups.

Irradiation of isophorone and a 15-fold excess of enol ether **177** results in the formation of a diastereomeric mixture of adducts **178** in quantitative yield.[54] On brief treatment with tetrabutylammonium fluoride, the mixture of photoadducts **178** undergo cyclobutane ring-opening to give two epimeric diketo esters **179** in 70% yield. The Claisen condensation of these diketo esters proceeds on treatment with sodium hydride in refluxing THF, and a single enolized trione **180** is isolated in good yield. This photochemical approach provides easy access to a series of bicyclic 1,3-cyclohexanediones.

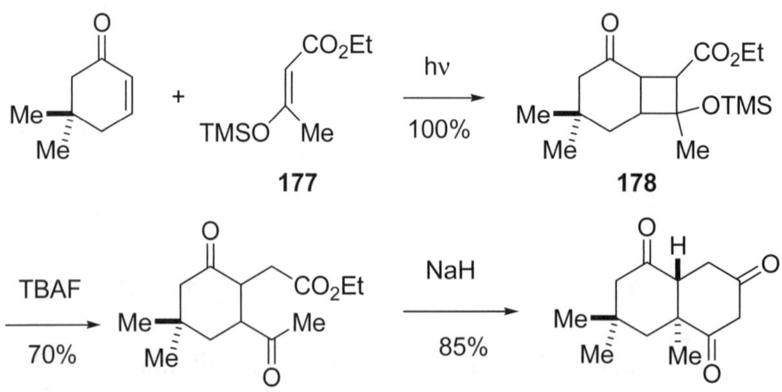

Photocycloaddition of β,β-diethoxyacrylate **182** with cyclopent-2-enone proceeds with excellent regioselectivity to give **183** as a mixture of diastereomers.[55] Reduction of this mixture with sodium borohydride leads to alcohol **184**, exposure of which results in fragmentation and lactonization to give lactone **185** in 59% yield from enone **181**.

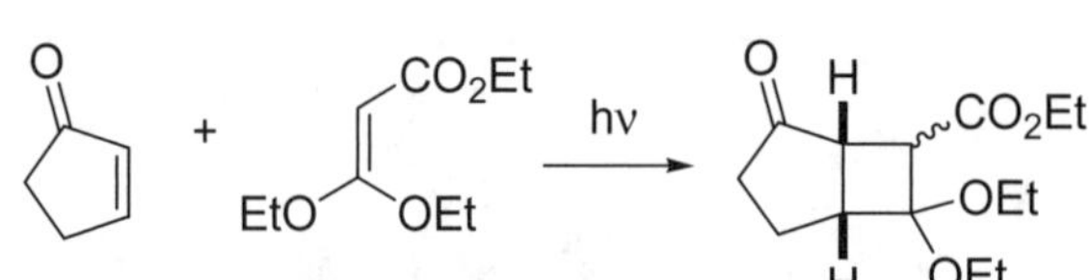

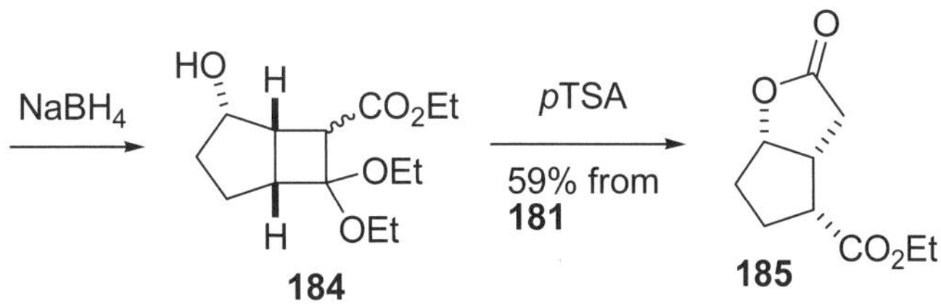

5.2.7 *Synthetic Utility*

The power of the de Mayo reaction, especially the intramolecular version, has been demonstrated in the syntheses of numerous natural products, and only a few representatives are described as below.

12-epi-Lycopodine (Wiesner, 1968)[7]

Wiesner carried out pioneering studies on the photocycloaddtion of vinylogous imides in the 1960s, and he first applied the intramolecular photochemical [2 + 2] enone-olefin cycloaddition reaction in natural product synthesis. In his landmark synthesis of 12-*epi*-lycopodine, compound **186** is irradiated to give a 70% yield of photoadduct **187**. Protection of the ketone, epoxidation of the exocyclic double bond, and reduction of the epoxide give ketal alcohol **188**. Deprotection of the ketal and spontaneous *retro*-aldol fragmentation lead to the diketone **190**. Of note is that the bond cleaved in this *retro*-aldol pathway (i.e., **189** to **190**) is different from that of typical de Mayo reactions. Diketone **190** transforms to the tetracyclic alcohol **191** via aldol reaction, and this alcohol is converted to 12-*epi*-lycopodine in four steps.

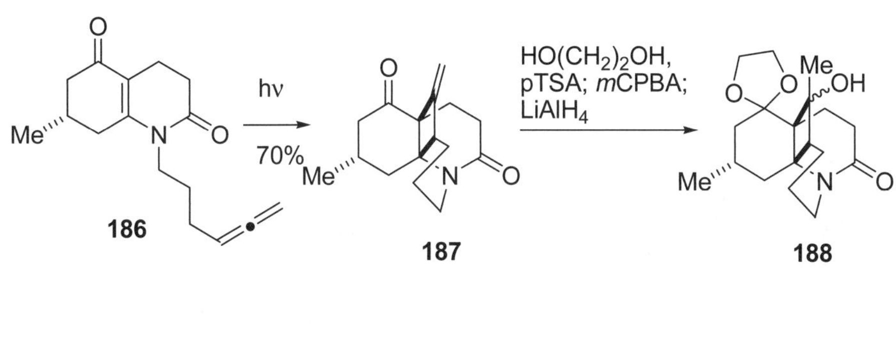

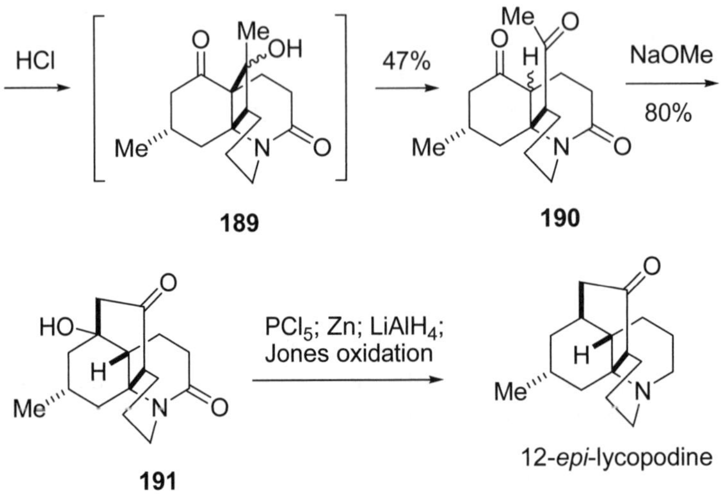

(±)-β-Himachalene (de Mayo, 1969)[56]

Photocycloaddition of enol acetate **193** and ketal **192** proceeds with high regioselectivity to give adduct **194** in excellent yield. Reduction of the ketone and mesylation of the resulting alcohol give compound **195**, which undergoes hydrolysis and spontaneous fragmentation under basic conditions to give ketone **196** in 35% yield from enol acetate **193**. This ketone is converted to β-himachalene in a few steps.

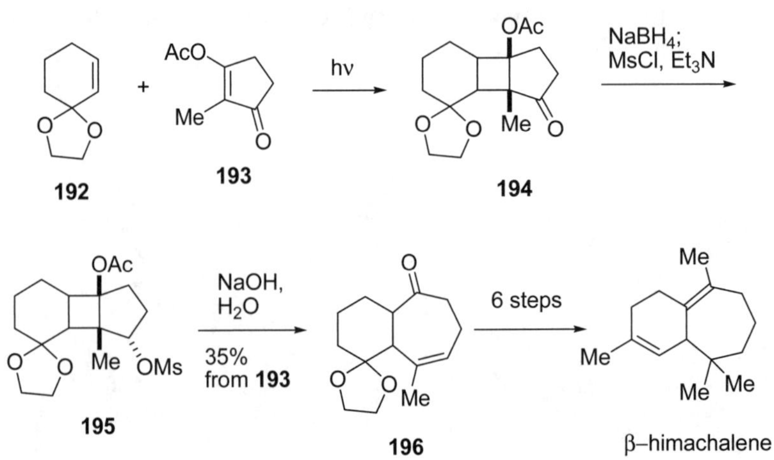

Longifolene (Oppolzer, 1978);[57] Sativene (Oppolzer, 1984)[58]

Longifolene is a sesquiterpene isolated from a variety of plant sources, and like many other smaller terpenoids, it has found use in the perfume industry due to its rich woody odor. This compound is relatively inexpensive and has limited commercial use, but the longifolene skeleton has served as a subject for synthetic planning and strategy. This compound was first synthesized by Corey in 1961,[59] and it was among the first molecules on which Corey demonstrated his new retrosynthetic analysis theory.[60] A number of impressive total syntheses have been accomplished since then, including the photochemical approach by Oppolzer in 1980. Thus irradiation of the crude enol acetate **198a** and **199a** (derived from diketone **197**) through Pyrex affords regioselectively adducts **200a** as a 1:3 mixture of stereoisomers in 78% yield. As described previously, the acetylation of 1,3-diketone **107** is not regiospecific, but the two enol acetates interconvert via a photo-Fries process. However, only the enol acetate **199a** participates in the cycloaddition. Hydrolytic cleavage of the acetoxy group requires heating of **200a** with 4% potassium hydroxide in dioxane/water at 100 °C, and under these harsh conditions the resulting *retro*-aldol product **201** undergoes further

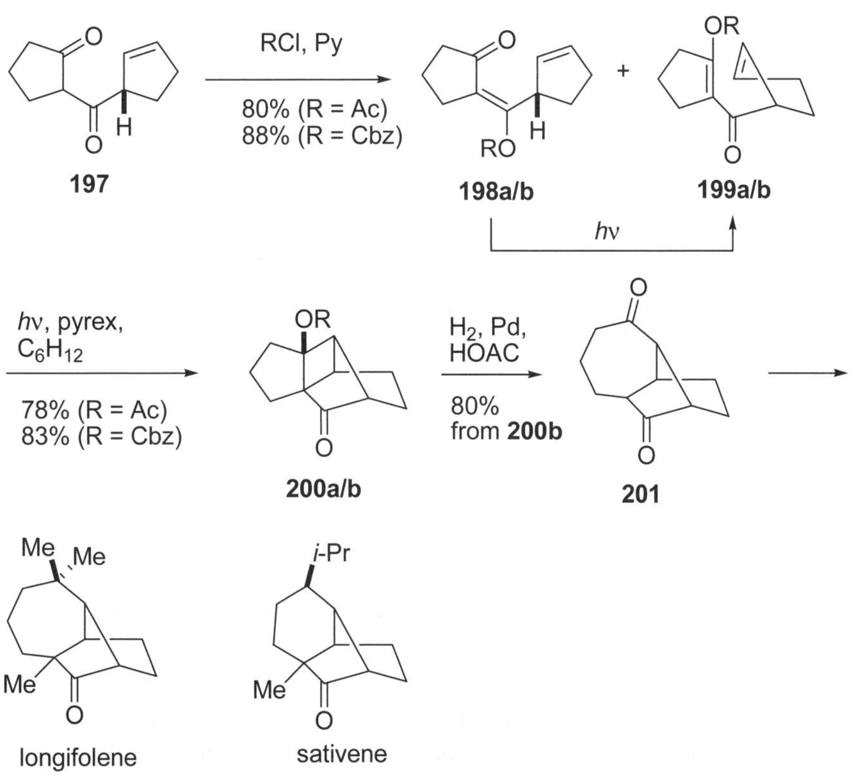

intramolecular aldol reaction. To this end, the mixture of the crude enol carbonates **198b** and **199b** is irradiated to give adducts **200b** as a 2:3 mixture of stereoisomers in 83% yield. Hydrogenolysis of **200b** results in clean *retro*-aldol cleavage to give the 1,5-diketone in 83% yield. The exclusive formation of the crossed adduct is typical of systems with two atoms between the two double bonds. Diketone **201** has been converted to longifolene and sativene.

(±)-$\Delta^{8,9}$-Capnellene (Pattenden, 1980)[61]

The enol benzoate **202** undergoes regioselective and stereoselective photocycloaddition to give adduct **203** in 98% yield. Treatment of the keto-benzoate **203** with lithium hexamethyldisilazide (LHMDS), followed by methyl iodide, leads to the geminal dimethyl substituted adduct, which on fragmentation under basic conditions, gives the trimethyl substituted cyclooctane-1,5-dione **204**. This dione is transformed into *epi*-precapnelladiene in a six-step sequence. Treatment of *epi*-precapnelladiene with boron trifluoride etherate results in a clean transannular cyclization to give $\Delta^{8,9}$-capnellene in more than 50% yield accompanied by two minor isomeric capnellenes.

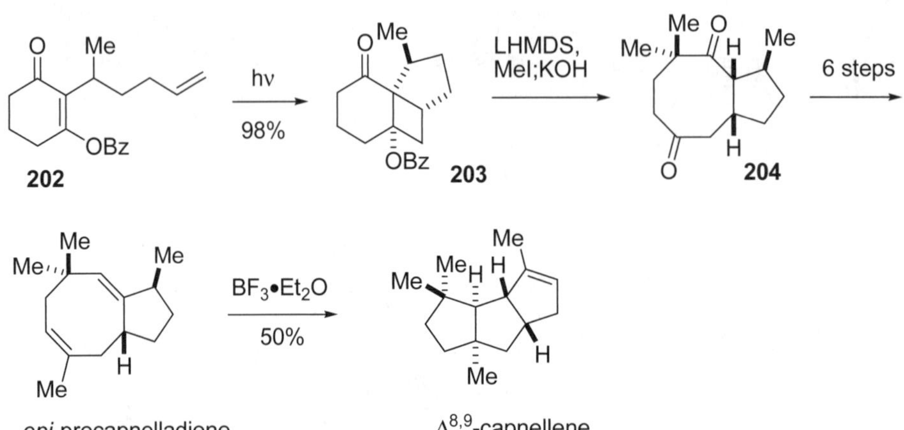

(±)-Zizaene (Pattenden, 1981)[62]

Irradiation of the 6:4 mixture of enol acetates **205** and **206** produces a 7:3 mixture of the photo adducts **208** and **207**. Reduction of the major photoadduct **208** followed by mesylation of the resulting alcohol leads to a mixture of isomers of the mesylate **209** in 70% yield. Treatment of the mesylate with sodium hydroxide effects simultaneous saponification and the Grab fragmentation, with the formation of a 1:2 mixture of α- and β-methyl epimers of the tricyclic alkene **210**. Hydrogenation of the alkene gives the tricyclic ketone **211**, accompanied by the epimer. Tricyclic ketone is a precursor to (±)-zizaene.

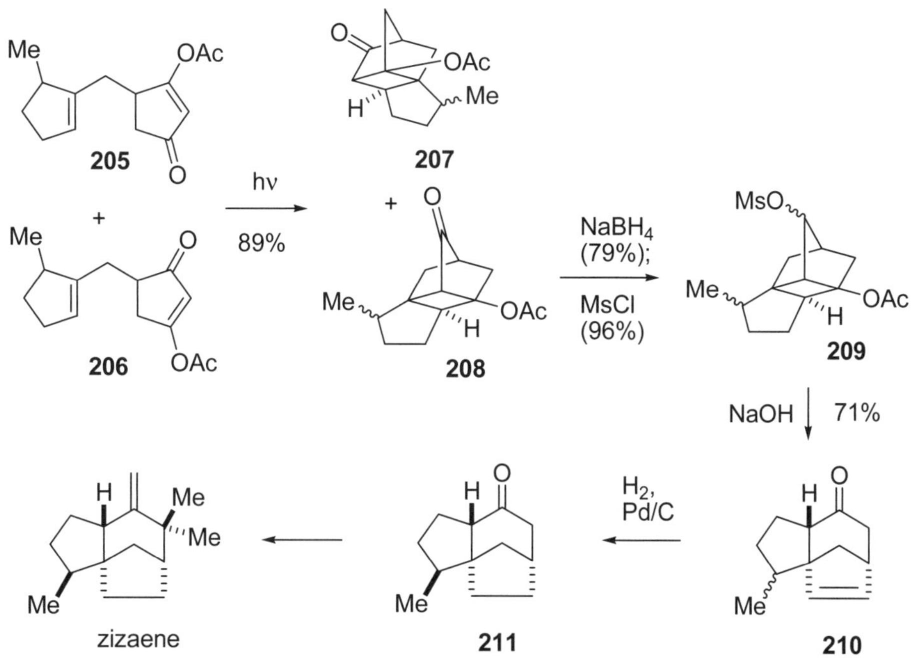

(±)-Pentalenene (Pattenden, 1984)[63]

Irradiation of the silyl enol ether **212** results in regioselective intramolecular [2 + 2] photocycloaddition producing only the tricyclic ketone **213** in 81% yield from the β-diketone. The silyl enol ether of the β-diketone **212** is chosen as a photosubstrate instead of the enol ester (acetate, benzoate) because of the comparative neutrality of silyl ethers toward organometallic reagents and because of the known affinity of silicon for fluoride ion to trigger the Grob fragmentation step. Addition of the tricyclic ketone **213** with Me_3CuLi_2 brings about stereoselective formation of the tertiary alcohol, which undergoes the smooth Grab fragmentation upon treatment with hydrofluoric acid to give the enone **215** in 73% yield. This enone is converted to the cycloocta-1,5-diene **216** following the Witting reaction with methylenetripehylphosphorane and isomerisation of the resulting olefin with rhodium trichloride. Treatment of the cycloocta-1,5-diene **216** with boron trifluoride etherate results in stereoselective transannular cyclization, producing pentalenene in 38% yield, along with *iso*-pentalenene as a minor isomer.

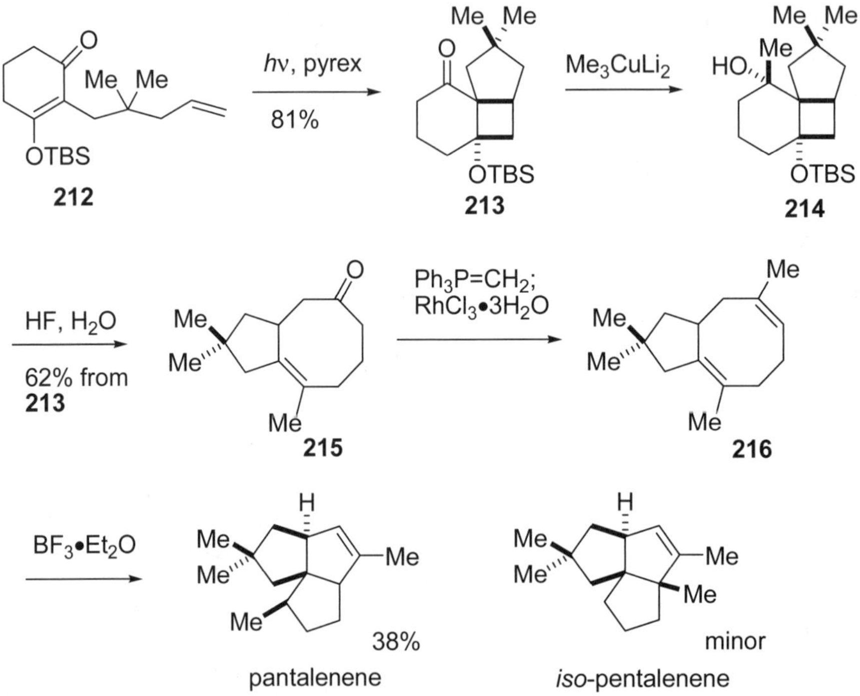

(±)-Hirsutene (Weedon, 1985)[64]

Photocycloaddition of 5,5-dimethylcyclohexane-1,3-dione to 2-methyl-2-cyclopentenol followed by in situ silylation of the crude photoadducts, **219a** and **220a**, with *tert*-butyldimethylsilyl chloride to give an 89% yield of the 1:1 mixture of the isomeric silylated, hydroxyl-substituted cyclooctane-1,5-diones **219b** and **220b**. The protection of the hydroxyl group in the photoadducts **219a** and **220a** proves necessary due to their instability. Treatment of the mixture of silyl ethers with a low-valence titanium reagent (McMurry coupling reaction) results in intramolecular reductive coupling of the dione function to give the hirsutene carbon skeleton **221** with the desired regiochemistry in 17% yield. Desilylation with fluoride ion, followed by sequential catalytic hydrogenation and the Jones oxidation, completes the formal synthesis of hirsutene. This route suffers from poor regioselectivity in the photocycloaddition step, typical of the intermolecular de Mayo reaction.

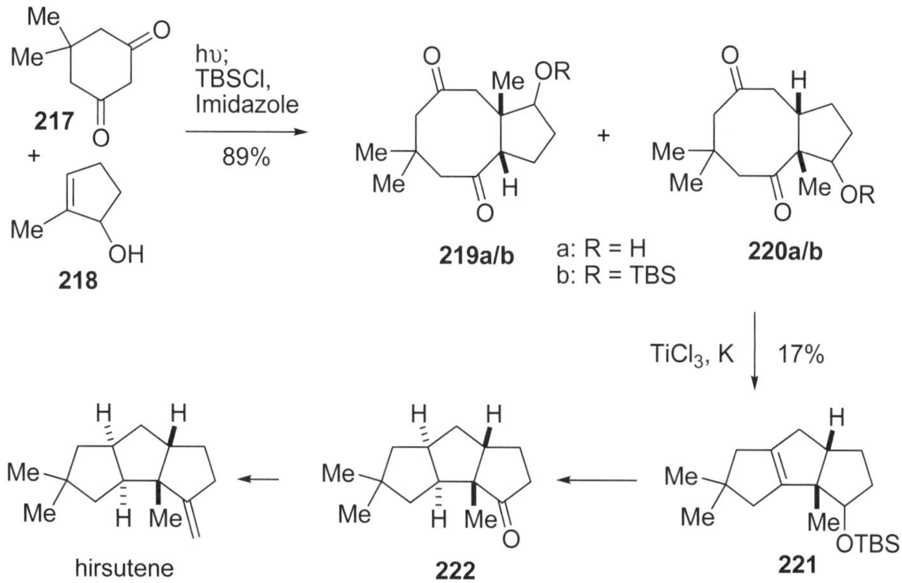

Taxane Skeleton (Inouye, 1985;[65] Winkler, 1989[66])

A system analogous to the taxanes has been prepared by using the intramolecular de Mayo reaction of vinylogous esters. Irradiation of the mixture of diastereomers **223** results in cycloaddition of one diastereomer to give adduct **225** and recovery of the other isomer **224**. The tetrahydrofuran is oxidized with RuO_4 to the lactone, which is saponified with concomitant cyclobutane cleavage to the diketo acid **227**.

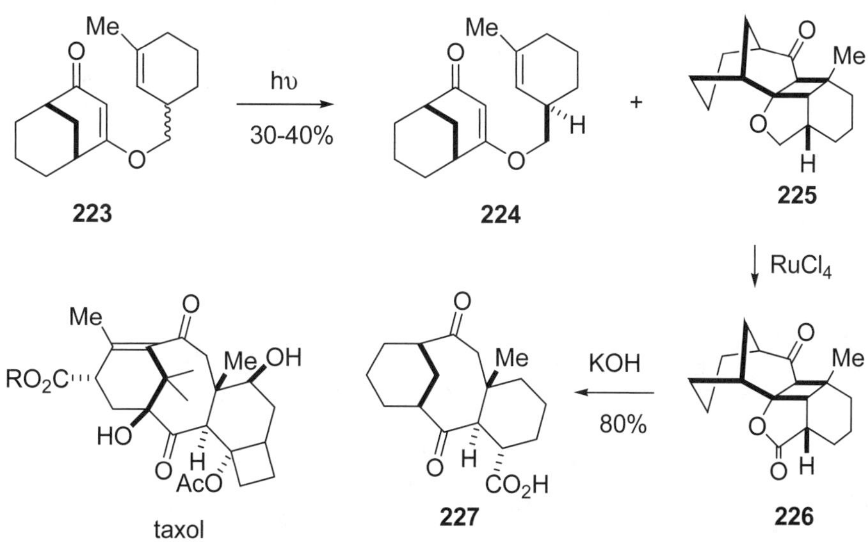

Photocycloaddition of dioxinone **228**, with the correct C-1/C-3 relative stereochemical relationship for taxane construction, leads to the formation of a single diastereomer **229**. Fragmentation under basic conditions and treatment of the resulting keto acid with diazomethane generates the keto ester **230** and its C-15 epimer (3:1 ratio) in quantitative yield.

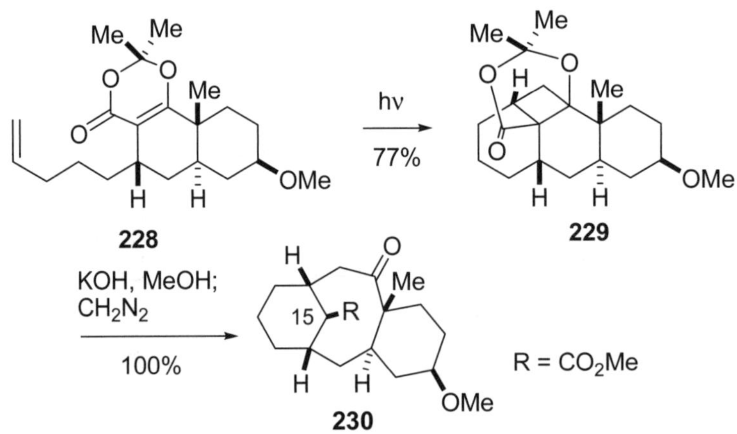

Vindorosine (Winkler, 1990)[67]

Irradiation of **231** gives, after the *retro*-Mannich fragmentation of the photoadduct **232**, a 91% yield of **233**. The single stereogenic center in the photosubstrate leads to complete stereochemical control in the formation of the cyclobutane intermediate **232**, which contains two new strereogenic centers. Treatment of **233** with lithium diisopropylamide, followed by an excess of *tert*-butyldimethylsilyl triflate and reaction of the crude product with tetrabutylammonium fluoride, results in the formation of the desired tetracyclic product **234** in 51% yield. This compound is converted to tetracyclic ketone **235**, which is an advanced intermediate in Büchi's synthesis of vindorosine.

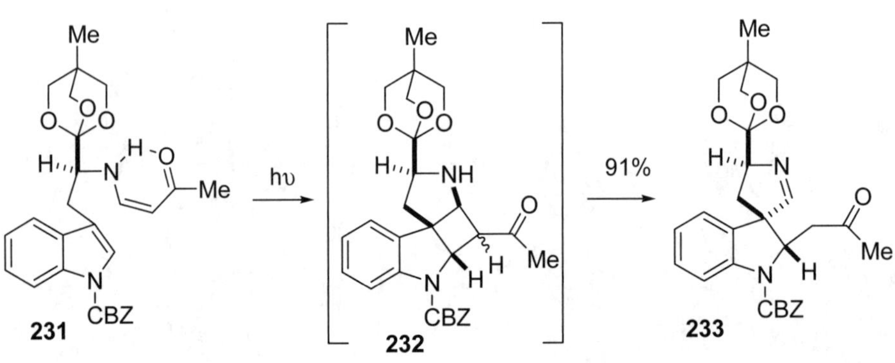

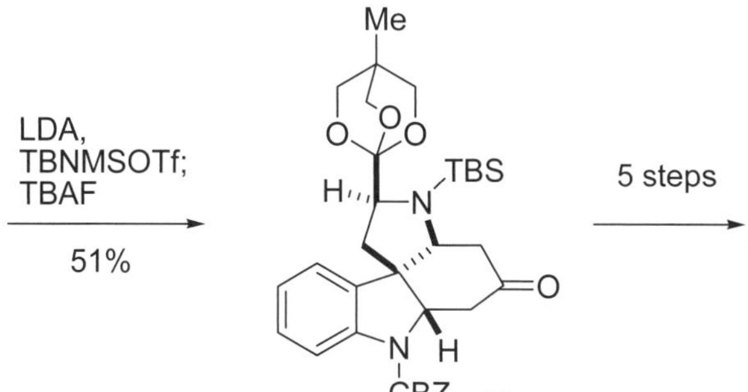

234

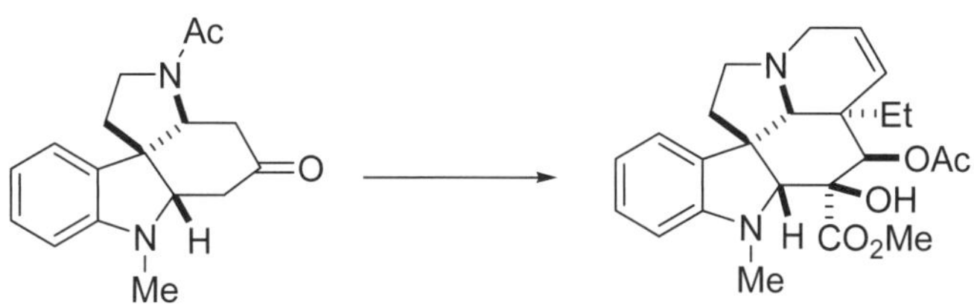

235

vindorosine

Manzamine A (Winkler, 1998)[68]

Manzamine A contains an unprecedented pentacyclic core consisting of 13- and 8-membered rings attached to a central pyrrolo[2,3-*i*]isoquinoline system and an array of five stereocenters. It would require the identification of a novel synthetic strategy bolstered by a supply of synthetic transformations to conquer such a complex molecule. To this end, an elegant photochemical approach to manzamine A has been developed by Winkler. Thus photoaddition and *retro*-Mannich fragmentation of the tertiary vinylogous amide **236** leads, via *O*-closure of the ketoiminium intermediate **238**, to aminal **239**. The isomerisation of **239** to the manzamine tetracycle **241** proceeds upon exposure to pyridinium acetate to give **241** in 20% overall yield from photosubstrate **236** (an average of 60% yield/step for photoaddition, fragmentation, and Mannich closure). As expected, **241** is produced as a single stereoisomer, presumably because of the overwhelming influence of the lone C-34 stereocenter in **236** controlling the formation of the resulting stereogenic centers during the cascade sequence. Of note is that the C-12 stereochemistry is not assigned at this stage of the synthesis and thus not defined in the scheme, but a single stereochemical arrangement occurs at this site. In fact, the stereochemistry at C-12 does not matter for the synthesis since it would be destroyed during subsequent transformations to install the functional requirements of manzamine A. This tetracyclic compound is converted to manzamine A in a 12-step sequence.

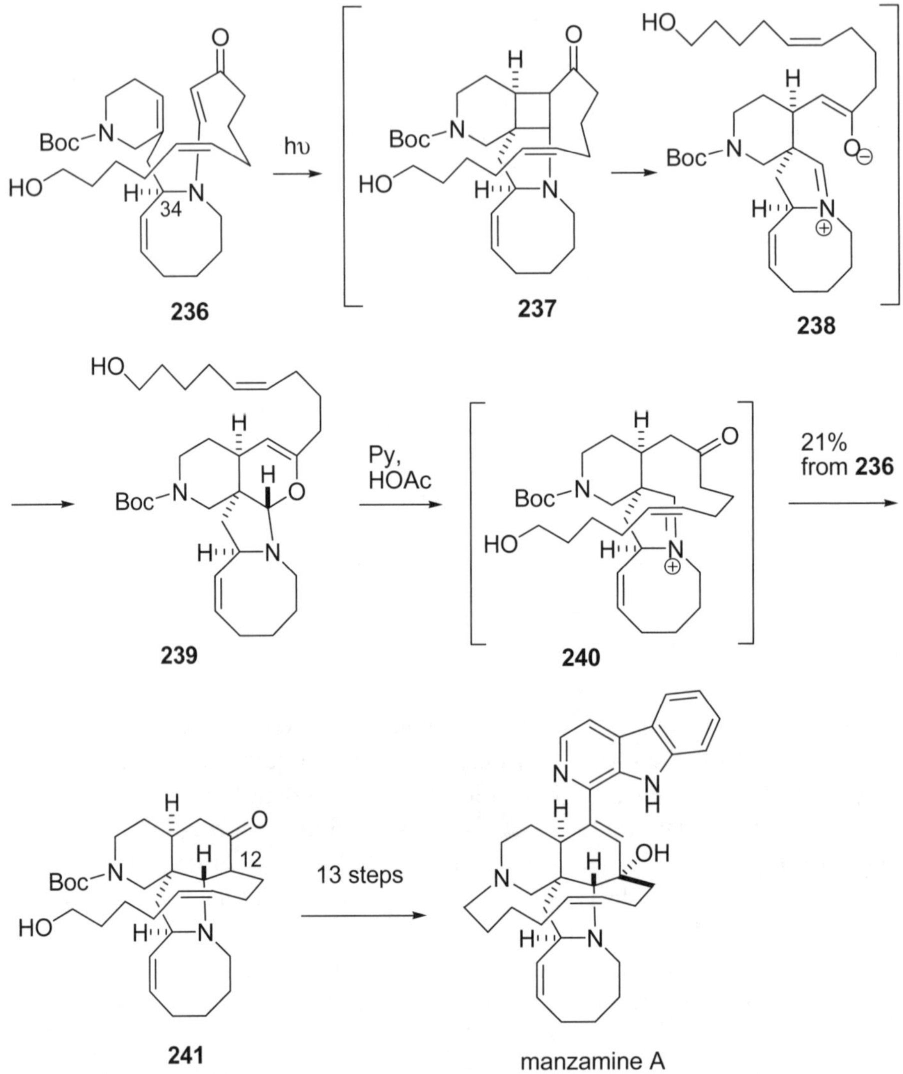

(±)-Saudin (Winkler, 1999)[69]

Irradiation of **242** leads to the formation of **243** as a single diastereomer in 80% yield. Treatment of **243** with *n*-BuLi and Tf_2O in the presence of TMEDA gives the enol triflate, which undergoes Stille coupling with 3-furyltributylstannane to give the furyl enol ether **244** in 95% yield. Exposure of **244** to lithium hydroxide and cyclization of the resulting hemiketal keto-acid **245** furnishes (±)-saudin in 52% yield.

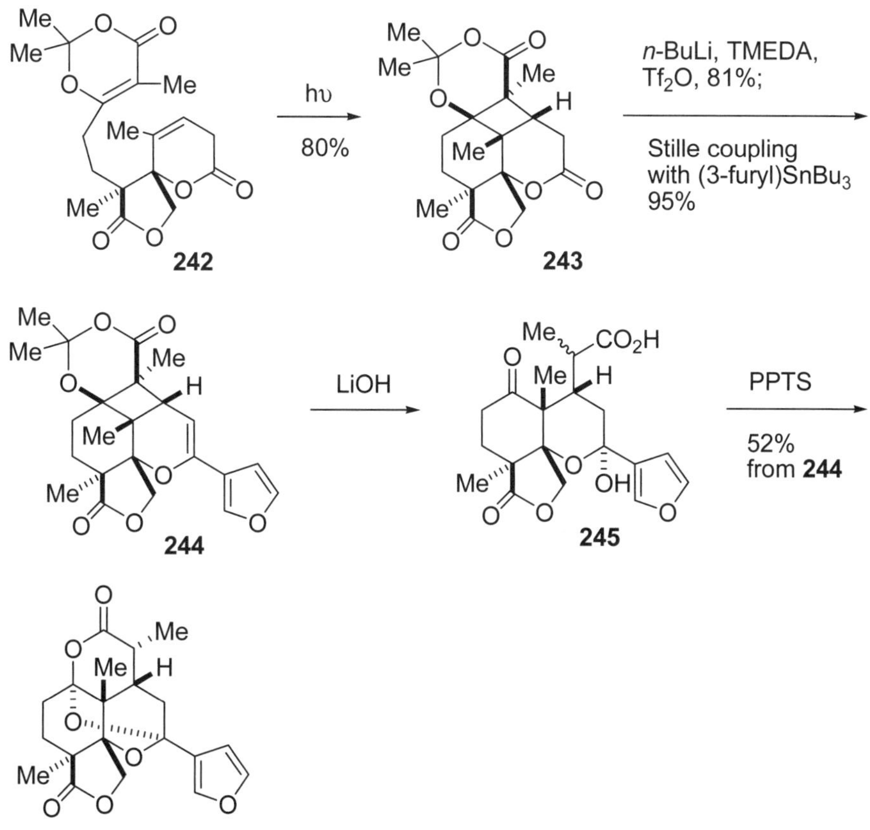

(±)-Ingenol (Winkler, 2002)[70]

Photocycloaddition of the allylic chloride **246** proceeds in 60% yield to give the desired photoadduct **247**. Fragmentation of **247** with methanolic potassium carbonate generates ester **248**. Reduction of the ester with lithium aluminium hydride, elimination of the chloride with DBU, and silylation of the primiary alcohol with *tert*-butyldimethylsilyl chloride (TBSCl), give **249** as a 7:1 mixture of the C-6 α:β epimers in 35% yield from **247**. Ingenol is derived from **247** in a 24-step sequence.

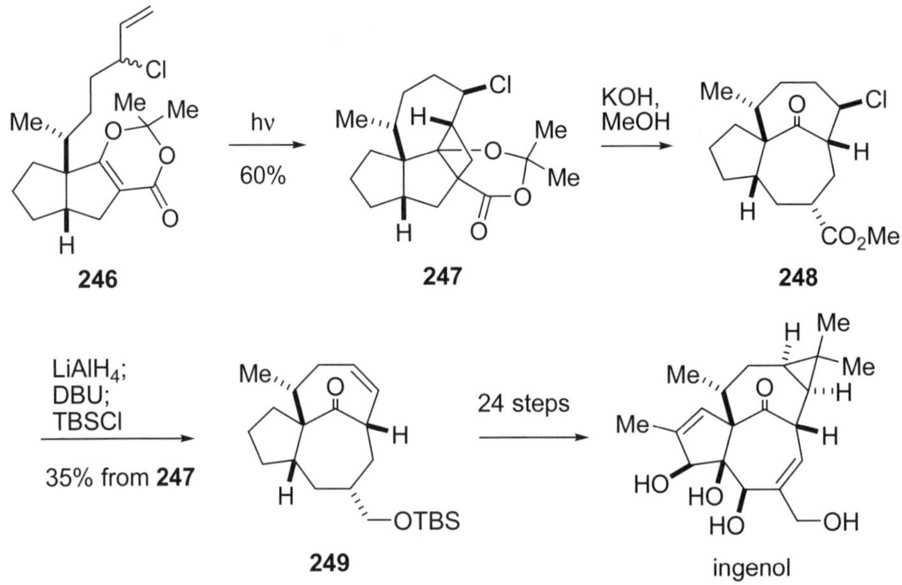

5.2.8 *Experimental*

(10*S*,10a*R*)-10-(2-Oxopropyl)-2,3,10,10a-tetrahydropyrrolo[1,2-*b*]isoquinoline-1,5-dione[22]

A solution of 6-(1-oxoisoquinolin-2(1*H*)-yl)hexane-2,4-dione (101 mg, 0.39 mmol) in acetonitrile (10 mL) was purged with nitrogen for 15 min. The reaction mixture was irradiated through a Pyrex filter at room temperature for 1.5 h, and then evaporated in vacuo. The residue was purified by silica gel flash chromatography eluting with 2:1 ethyl acetate/hexane to give the title compound as a white crystalline solid (72 mg, 72% yield).

5.2.9 *References*

1. Ciamician, G.; Silber, P. *Ber.* **1902**, *35*, 4129.
2. Stobbe, H. *Ber.* **1912**, *45*, 3396.
3. Ciamician, G.; Silber, P. *Ber.* **1908**, *41*, 1928.
4. Buchi, G.; Goldman, I. M. *J. Am. Chem. Soc.* **1957**, *79*, 4741.
5. Cookson, R. C.; Crundwell, E.; Hudec, *J. Chem. Ind.* **1958**, 1003.
6. Eaton, P. E.; Cole, T. W. Jr. J. *J. Am. Chem. Soc.* **1964**, *86*, 962; and *J. Am. Chem. Soc.* **1964**, *86*, 3157.
7. Wiesner, K.; Musil, Wiesner, K. J. *Tetrahedron Lett.* **1968**, 5643.
8. Eaton, P. E. *J. Am. Chem. Soc.* **1962**, *84*, 2454.
9. Corey, E. J.; Bass, J. D.; Lemahieu, R.; Mitra, R. B. *J. Am. Chem. Soc.* **1964**, *86*, 5570.
10. (a) Corey, E. J.; Mitra, R. B.; Uda, H. *J. Am. Chem. Soc.* **1963**, *85*, 362. (b) Corey, E. J.; Mitra, R. B.; Uda, H. *J. Am. Chem. Soc.* **1964**, *86*, 485. (c) Corey, Nozoe, S. *J. Am. Chem. Soc.* **1964**, *86*, 1652. (d) Corey, Nozoe, S. *J. Am. Chem. Soc.* **1965**, *87*, 5733.

11. [R] Corey, E. J.; Cheng, X. -M. *The Logic of Chemical Synthesis*, Wiley, New York, **1989**, pp. 153–155.
12. de Mayo, P.; Takeshita, H.; Sattar, A. M. B. A. *Proc. Chem. Soc. London*, **1962**, *119*; de Mayo, P.; Takeshita, H. *Can. J. Chem.* **1963**, *41*, 440
13. [R] For reviews on enone-olefin cycloadditions including de Mayo reactions, see: (a) Eaton, P. E. *Acc. Chem. Res.* **1968**, *1*, 50. (b) de Mayo, P. *Acc. Chem. Res.* **1971**, *4*, 41. (c) Oppolzer, W. *Pure Appl. Chem.* **1981**, *53*, 1181. (d) Oppolzer, W. *Acc. Chem. Res.* **1982**, *15*, 135. (e) Schuster, D. I.; Lem, G.; Kaprimidis, N. A. *Chem. Rev.* **1993**, *93*, 3. (f) Schuster, D. I.; Lem, G.; Kaprimidis, N. A. *Chem. Rev.* **1993**, *93*, 3. (g) Crimmins, M. T. *Chem. Rev.* **1988**, *88*, 1453. (h) Winkler, J. D.; Bowen, C. M.; Liotta, F. *Chem. Rev.* **1995**, *95*, 2003.
14. [R] Nicolaou, K. C.; Snyder, S. A. *Classics in Total Synthesis II*, Wiley-VCH, **2003**, pp. 211–236.
15. www.uwo.ca/sci/publications/history/deMayo.html.
16. (a) Loutfy, R. O.; de Mayo, P. *J. Am. Chem. Soc.* **1977**, *99*, 3559; Froese, R. D.; Lange, G. L.; Goddard, J. D. *J. Org. Chem.* **1996**, *61*, 952.
17. Lange, G. L.; Organ, M. G.; Lee, M. *Tetrahedron Lett.* **1990**, *31*, 4689.
18. Quevillon, T. M.; Weedon, A. C. *Tetrahedron Lett.* **1996**, *37*, 3939.
19. (a) Srinivasan, R.; Carlough, K. H. *J. Am. Chem. Soc.* **1967**, *89*, 4932; (b) Liu, R. S. H.; Hammond, G. S. *J. Am. Chem. Soc.* **1967**, *89*, 4930.
20. Beckwith, A. L. *Tetrahedron* **1981**, *37*, 3063.
21. Bajgrowicz, J. A.; Petrzilka, M. *Tetrahedron* **1994**, *50*, 7461.
22. Minter, D. E.; Winslow, C. D. *J. Org. Chem.* **2004**, *69*, 1603.
23. Oppolzer, W.; Bird, T. G. C. *Helv. Chim. Acta.* **1979**, *62*, 1199.
24. Bagley, M. J.; Mellor, M.; Pattenden, G. *J. Chem. Soc., Chem. Commun.* **1979**, 235.
25. Bagley, M. J.; Mellor, M.; Pattenden, G. *J. Chem. Soc., Perkin Trans. I* **1983**, 1905.
26. Seto, H.; Hirokawa, S.; Fujimoto, Y.; Tatsuno, T. *Chem. Lett.* **1983**, 989.
27. Seto, H.; Fujimoto, Y.; Tatsuno, T.; Yoshika, H. *Synth. Commun.* **1985**, *15*, 1217.
28. Seto, H.; Tsunoda, S.; Ikeda, H.; Fujimoto, Y.; Tatsuno, T.; Yoshika, H. *Chem. Pharm. Bull.* **1985**, *33*, 2594.
29. Kaczmarek, R.; Blechert, S. *Tetrahedron Lett.* **1986**, *27*, 2845.
30. Tamura, Y.; Kata, Y.; Ishibashi, H.; Ikeda, M. *J. Chem. Soc., Chem. Commun.* **1971**, 1167.
31. Tamura, Y.; Y.; Kata, Y.; Ishibashi, H.; Ikeda, M. *Tetrahedron Lett.* **1972**, 1977.
32. Tamura, Y.; Ishibashi, H.; Ikeda, M. *J. Org. Chem.* **1976**, *41*, 1277.
33. Ikeda, M.; Ohno, K.; Homma, K.; Ishibashi, H.; Tamura, Y. *Chem. Pharm. Bull.* **1981**, *29*, 2062.
34. Kojima, T.; Inouye, Y.; Kakisawa, H. *Bull. Chem. Soc. Jpn.* **1985**, *58*, 1738.
35. Pirrung, M. C.; Webste, N. J. G. *J. Org. Chem.* **1987**, *52*, 3603.
36. Winkler, J. D.; Ragains, J. R. *Org. Lett.* **2006**, *8*, 4031.
37. For pioneering studies on the photocycloaddtion of vinylogous imides, see (a) Bohme, E. H.; Valenta, Z.; Wiesner, K. *Tetrahedron Lett.* **1965**, 2441. (b) Wiesner, K.; Jirkovsky, I.; Fishman, M.; Williams, C. A. *Tetrahedron Lett.* **1967**, 1523.
38. Tamura, Y.; Y.; Kata, Y.; Ishibashi, H.; Ikeda, M. *Tetrahedron Lett.* **1972**, 1977.
39. Schell, F. M.; Cook, P. M. *J. Org. Chem.* **1984**, *49*, 4067.
40. Vogler, B.; Bayer, R.; Meller, M.; Kraus, W. *J. Org. Chem.* **1989**, *54*, 4165.
41. Amougay, A.; Pete, J.-P.; Piva, O. *Tetrahedron Lett.* **1992**, *33*, 7347.
42. Winkler, J. D.; Muller, C. L.; Scott, R. D. . *J. Am. Chem. Soc.* **1998**, *110*, 4831.
43. Ragains, J. R.; Winkler, J. D. *Org. Lett.* **2006**, *8*, 4437.
44. (a) Kwak, Y.; Winkler, J. D. *J. Am. Chem. Soc.* **2001**, *123*, 7429. (b) Winkler, J. D.; Fitzgerald, M. E. *Synlett* **2009**, 562.
45. Baldwin, S. W.; Wilkinson, J. M. *J. Am. Chem. Soc.* **1980**, *102*, 3634.
46. Sato, M.; Sekiguchi, K.; Kaneko, C. *Chem. Lett.* **1985**, 1057.
47. Sato, M.; Takayama, K. Abe, Y.; Furuya, T.; Inukai, N.; Kaneko, C. *Chem. Pharm. Bull.* **1990**, *38*, 336.
48. Blaauw, R. H.; Briere, J.; Jong, R. de; Benningshof, J. C. C.; Ginkel, A. E. van; Fraanje, J.; Goubitz, K.; Schenk, H.; Rutjes, F. P. J. T.; Hiemstra, H. *J. Org. Chem.* **2001**, *66*, 233.
49. Takeshita, H.; Cui, Y.; Kato, N.; Mori, A. *Bull. Chem. Soc. Jpn.* **1993**, *66*, 2694.
50. Winkler, J. D.; Hey, J. P.; Hannon, F. J. *Heterocycles* **1987**, *25*, 55.

51. Winkler, J. D.; McLaughlin, E. *Org. Lett.* **2005**, *7*, 227.
52. Kemmler, M.; Herdweck, E.; Bach, T. *Eur. J. Org. Chem.* **2004**, 4582.
53. Lange, G. L.; Organ, M. G. *J. Org. Chem.* **1996**, *61*, 5358.
54. Liu, H.; Dieck-Abularach, T. *Tetrahedron Lett.* **1982**, *23*, 295.
55. Audenhove, M. V.; Termont, D.; Keukeleire, D. D.; Vandewalle, M.; Claeys, M. *Tetrahedron Lett.* **1978**, *31*, 2057.
56. Challand, B. D.; Kornis, H.; Lange, K. G.; de Mayo, P. *J. Org. Chem.* **1969**, *34*, 794.
57. Oppolzer, W.; Godel, T. *J. Am. Chem. Soc.* **1978**, *100*, 2583.
58. Oppolzer, W.; Godel, T. *Helv. Chim. Acta.* **1984**, *67*, 1154.
59. Corey, E. J.; Ohno, M.; Vatakencherry, P. A.; Mitra, R. B. *J. Am. Chem. Soc.* **1961**, *83*, 1251.
60. [R] Nicolaou, K. C.; Montagnon. *Molecules That Changed the World*, Wiley-VCH, **2007**, pp. 107–112.
61. (a) Birch, A. M.; Pattenden, G. *J. Chem. Soc., Chem. Commun.* **1980**, 1195. (b) *J. Chem. Soc. Perkin Trans. I* **1983**, 1913.

(a) Parker, A. J.; Pattensen, G. *Tetrahedron Lett.* **1981**, *22*, 2599. (b) Parker, A. J.; Pattensen, G. *J. Chem. Soc., Perkin Trans. I* **1983**, 1901.

(a) Pattenden, G.; Teague, S. J. *Tetrahedron Lett.* **1984**, *25*, 3021. (b) Pattenden, G.; Teague, S. J. *Tetrahedron* **1987**, *43*, 5637.

(a) Disanayaka, B. W.; Weedon, A. C. *J. Chem. Soc., Chem. Commun.* **1985**, 1282; (b) Disanayaka, B. W.; Weedon, A. C. *J. Org. Chem.* **1987**, *52*, 2905;

62. Kojima, T.; Inouye, Y.; Kakisawa, H. *Chem. Lett.* **1985**, 323.
63. Winkler, J. D.; Lee, C.; Rubo, L.; Muller, C. L. *J. Org. Chem.* **1989**, *54*, 4491.
64. Winkler, J. D.; Scott, R. D.; Williard, P. G. *J. Am. Chem. Soc.* **1990**, *112*, 8971.
65. (a) Winkler, J. D.; Axten, J. M. *J. Am. Chem. Soc.* **1998**, *120*, 6425. (b) Winkler, J. D.; Axten, A.; Hammach, A. H.; Kwak, Y.; Lengweiler, U.; Lucero, M. J.; Houk, K. N. *Tetrahedron* **1998**, *54*, 7045.

(a) Winkler, J. D.; Doherty, E. M. *J. Am. Chem. Soc.* **1999**, *121*, 7425. (b) Winkler, J. D.; Doherty, E. M. *Tetrahedron Lett.* **1998**, *39*, 2253.

66. Winkler, J. D.; Rouse, M. B.; Greaney, M. F.; Harrison, S. J.; Jeon, Y. T. *J. Am. Chem. Soc.* **2002**, *124*, 9726.

5.3 Ring-closing Metathesis

Nicole L. Snyder and Kevin W. Graepel

5.3.1 Description

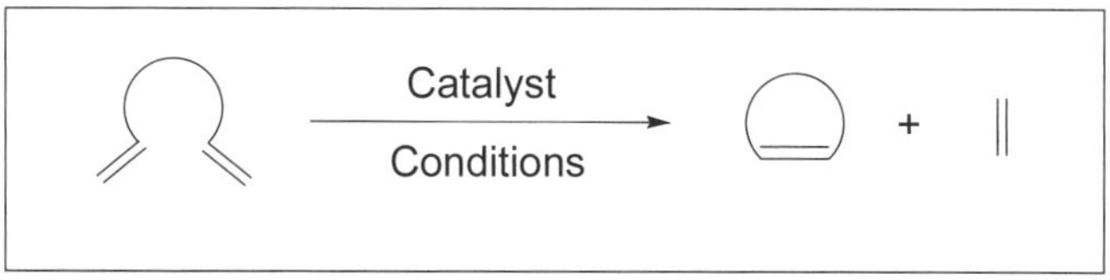

Catalysts:

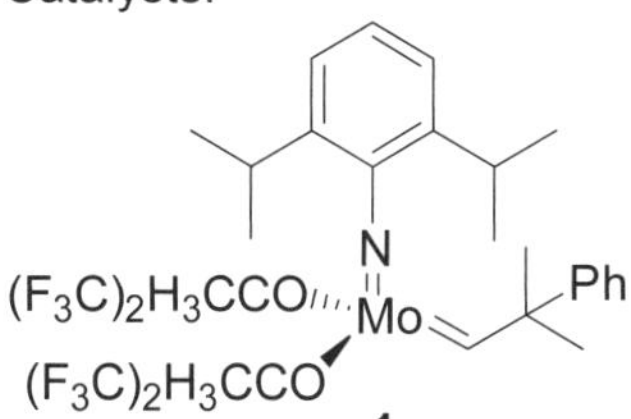

1

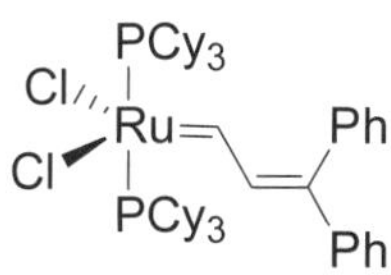

2

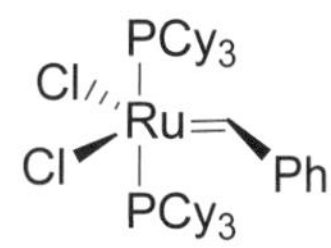

3

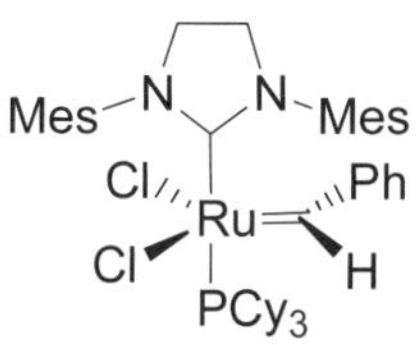

4

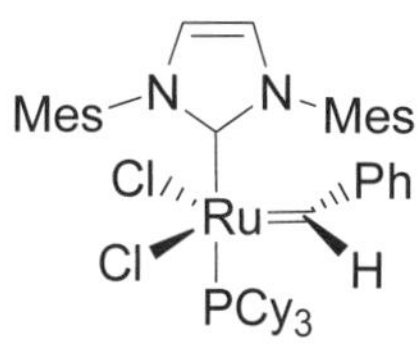

5

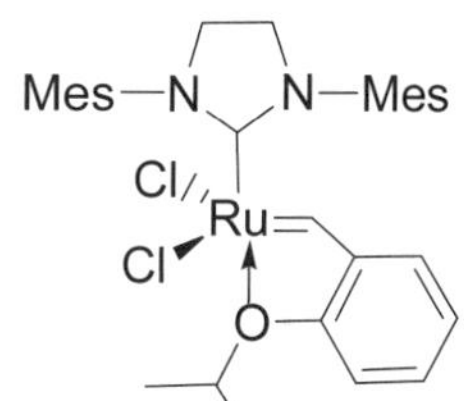

6

Ring-closing or olefin metathesis is the intramolecular redistribution of alkylidene moieties between two alkenes in the presence of a catalytic amount of a metal carbene to provide for a new product olefin and a byproduct olefin that is usually volatile in nature.

5.3.2 Historical Perspective

Ring-closing metathesis reactions were first used in 1980 by Tsuji[1] and Villemin.[2] However, it was not until the early 1990s that well-defined, single-component catalytic systems were developed independently in the research laboratories of Richard R. Schrock[3] and Robert H. Grubbs.[4] In 2005, Schrock and Grubbs shared the Nobel Prize in chemistry with Yves Chauvin for their work in this area. Schrock's molybdenum catalyst (**1**), Grubbs first- and second-generation catalysts (**2–5**), and the Grubbs–Hoyveda catalyst (**6**) are the most common catalysts used in ring-closing metathesis today.

Over the past 20 years ring-closing metathesis has become a powerful tool for the synthesis of a wide range of carbocyclic compounds. Carbocycles containing as few as 5 and as large as 18 carbon atoms have been prepared. Traditionally, five-, six-, and seven-membered ring carbocycles have been the most common targets for ring-closing metathesis. Recent developments in catalyst design, and a better understanding of substrate and reaction condition requirements for ring-closing metathesis have led to the increased use of this reaction in the synthesis of large carbocycles and macrocycles.

5.3.3 Mechanism

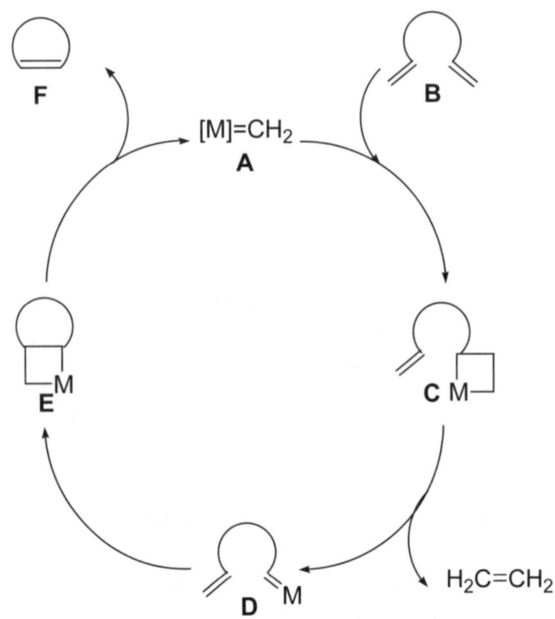

The mechanism for ring-closing metathesis using ruthenium complexes **3** and **4** has been well established both experimentally[5] and theoretically.[6] Entry into the catalytic cycle begins with the initial dissociation of the phosphine ligand to form the active 14-electron metalacarbene complex (**A**). This complex then coordinates with the an α,ω-diene (**B**) to form a 16-electron system. Migratory insertion, presumably via a [2 + 2] cycloaddition, gives rise to the corresponding metallacyclobutane intermediate (**C**), which has been characterized by NMR spectroscopy.[7] Metallacyclobutane (**C**) then undergoes a *retro*-[2 + 2] cycloaddtion to form a new 16-electron carbene intermediate (**D**) with the concomitant release of an alkene, usually ethylene. Intermediate **D** then undergoes an intramolecular [2 + 2] cycloaddition reaction to form a new metallocyclobutane **E**, which subsequently undergoes a *retro*-[2 + 2] cycloaddition to produce the desired cyclic olefin (**F**), while regenerating the active 14-electron metallacarbene complex **A**. The reaction is entropically driven and there is still some debate as to the rate-limiting step of the reaction.

Schrock's catalyst **1**, a molybdenum imido complex, has been shown to be highly active; however, its sensitivity to oxygen, moisture, and some polar functional groups has limited the utility of this catalyst in the synthesis of complex and highly functionalized compounds. On the other hand, Grubbs's first- and second-generation ruthenium carbene catalysts **2**–**5** and the Grubbs–Hoyveda catalyst **6** are less sensitive to oxygen and moisture. In addition, the functional group tolerance exhibited by **2**–**6** has made them the catalysts of choice in the synthesis of many complex synthetic targets.[8] Despite the sensitivity and functional group tolerance of catalysts **2**–**6**, Schrock's catalyst **1** is still the best catalyst in many cases for forming rings with high steric and electronic demands.

In general, ring-closing is fastest for smaller rings (5–7-membered rings) for which enthalpic and entropic factors are favorable.[9] The formation of eight-, nine-, and ten-membered rings are particularly problematic due to conformational constraints. Conformationally directed ring-closing metathesis using cyclic precursors has been widely used to overcome many of the problems associated with the synthesis of larger rings.[10] *gem*-Dialkyl[11] and Thorpe–Ingold effects[12] have also been exploited in an effort to produce larger rings via ring-closing metathesis. In addition, Hoye and co-workers recently demonstrated that substrates containing allylic alcohols are activated towards ring-closing metathesis, suggesting that the installation of such groups might be useful when employing ring-closing metathesis for the synthesis of larger rings.[13]

The reaction conditions for ring-closing metathesis have also been studied.[14] In general, low concentrations (0.1 M–0.1 mM) of catalyst favor ring formation, while higher catalyst concentrations favor cross-metathesis and polymerization reactions. Ring-closing metathesis reactions have also been shown to strongly depend on the solvent and temperature of the reaction. Experimental evidence has shown that polar solvents lead to higher initiation rates, as they are able to better stabilize the active 14-electron metalacarbene complex.[15] A quick survey of the literature reveals the polar solvent dichloromethane is one of the most frequently used solvents for ring-closing metathesis, supporting this observation. Benzene and toluene are also commonly employed, as catalyst degradation tends to be slower with these solvents. It is interesting that a recent study by Adjiman, Taylor, and co-workers suggests that cyclohexane and acetic acid may, in fact, be the best solvents for ring-closing metathesis.[16] Finally, the use of higher temperatures over long reaction periods has been shown to lead to catalyst degradation.

5.3.4 Variations and Improvements

The most notable improvements for ring-closing metathesis have been in the development of new catalysts.[17] In particular, the "phosphine mimic's" **4–6**, as well as several variants of these complexes have proven to be more reactive with electronically deactivated and sterically hindered alkenes, both of which have presented significant challenges for ring-closing metathesis. For example, the Blechert catalyst **7**[18] and Grela catalyst **8**[19] have found application in the ring-closing metathesis of substrates that showed little or no reactivity when catalyzed with **1–6**. Grubbs–Hoyveda derivative **9** has also recently been employed in the synthesis of tetra substituted olefins.[20]

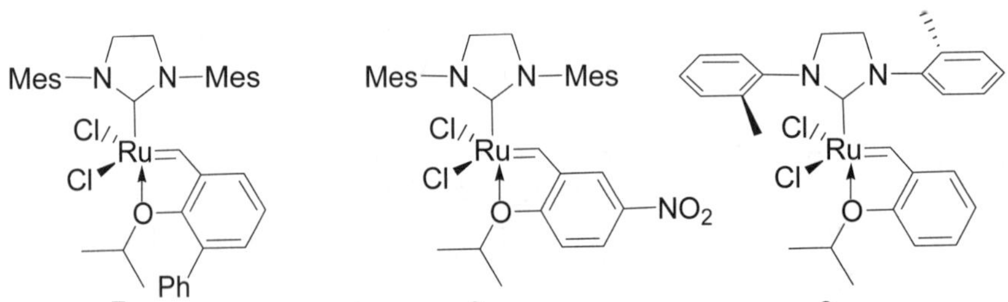

7 **8** **9**

In addition to sterics, a major limitation with the ruthenium-based catalysts has been the formation of stable Fischer carbenes, which tend to be electronically deactivated toward metathesis. Previously, most groups were able to circumvent this issue by building sterics into olefins prone to Fischer carbene formation, thus driving the reaction forward by forcing the catalyst to coordinate with the less reactive alkene. The advent of new catalysts, such as the commercially available second generation catalysts **4–5** and the Grubbs–Hoyveda catalyst **6**, as well as catalysts **7–9** has opened new opportunities in this area. For systems that remain problematic, relay ring-closing metathesis has proven useful in directing catalyst coordination.[21]

The development of several water-soluble catalysts, including several derivatives of the Grubbs's first-generation (**10a,b**,[22] **11**[23]), and second-generation (**12**[24]) and the Grubbs–Hoyveda catalysts (**13a**,[25]**b**,[26]**c**[27]) have expanded the scope of these catalysts to substrates that require more polar reactions conditions. These catalysts have also found wide applications in green chemistry processes[28] and for olefin metathesis in chemical biology.[29] The recent development of ruthenium olefin metathesis catalysts bearing carbohydrate ligands (**14–15**) is also of promise in this area.[30]

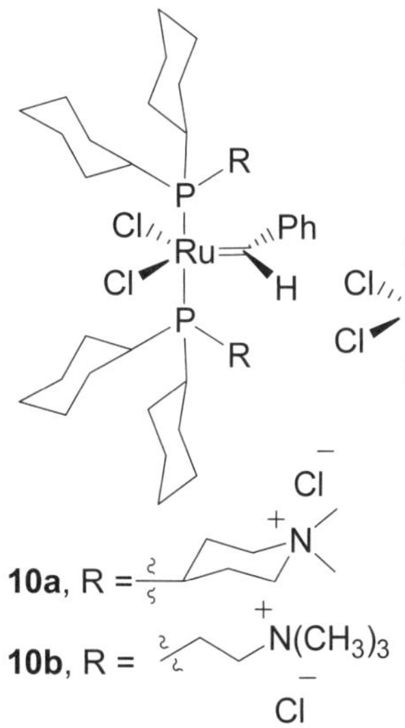

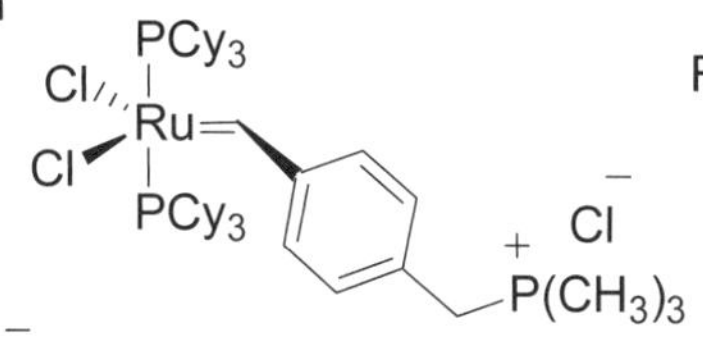

11

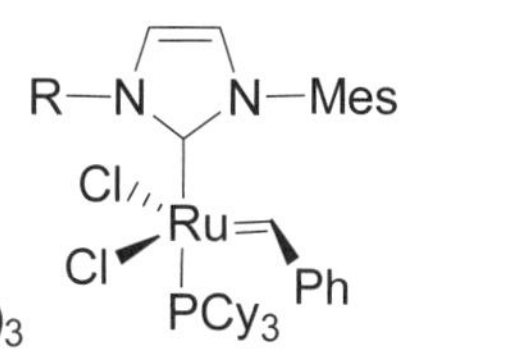

R = $CONH(CH_2)_2OPEG\text{-}CH_3$

12

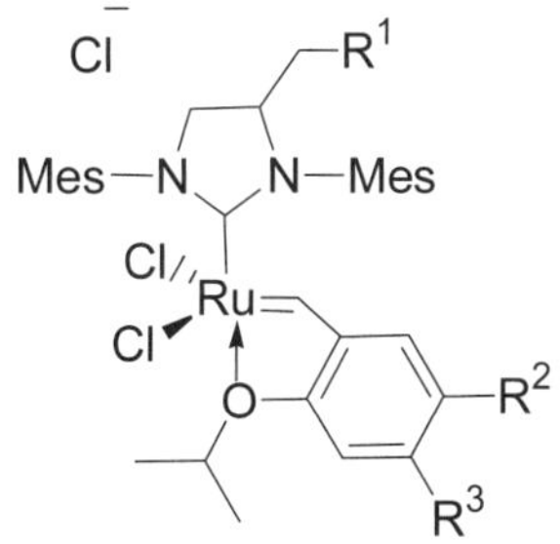

13a, $R^1 = CH_2OPEG\text{-}CH_3$
$R^2, R^3 = H$
13b, $R^1 = CH_2NH_3^+Cl^-$
$R^2 = CH_2N(CH_3)_3^+Cl^-$
$R^3 = H$
13c, $R^1, R^2 = H$
$R^3 = NEt_2CH_3^+I^-$

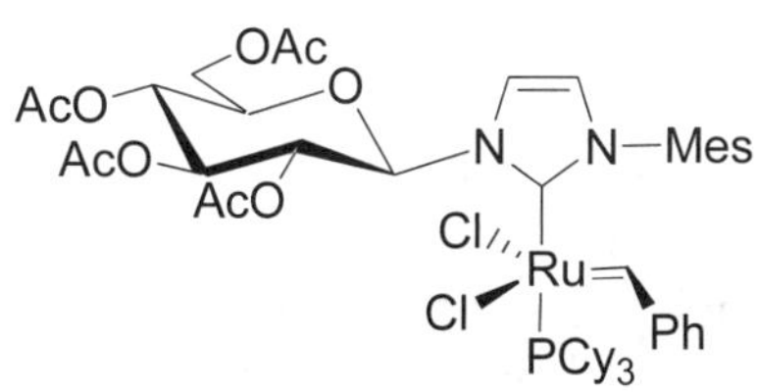

14

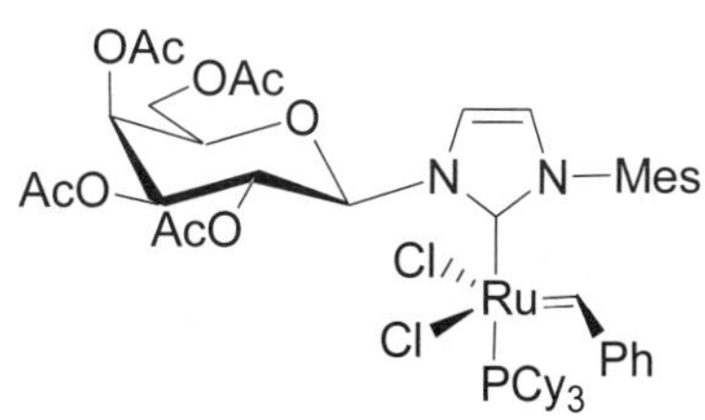

15

Two additional limitations with ruthenium-based catalysis are catalyst lifetime and the efficient removal (and subsequent recovery) of residual catalyst and decomposition products from the desired reaction products.[31] Currently, only the Grubbs–Hoyveda catalyst **6** can be readily recovered by flash chromatography. Numerous special workup and purification methods have been developed to solve the later problem.[32] However, the recent immobilization of ring-closing metathesis catalysts via the *N*-heterocyclic carbene to a solid support is gaining increasing attention as solution to both of these problems. In most cases, these immobilized catalysts provide both greater efficiency and easy recovery.[33] A recent report by Grubbs and co-workers on a well-defined silica-supported olefin metathesis catalyst for catalysis offers new promise in this area.[34]

5.3.5 Synthetic Utility

General Utility

In 1993, Grubbs and co-workers reported the first examples of ring-closing metathesis using **1**[35] and **2**[36] to prepare carbocyclic compounds. Reaction of **16** with 2–4 mol % of either **1** or **2** in benzene at room temperature furnished the corresponding five-membered ring carbocycle in 91% (using **1**) and 85% (using **2**).

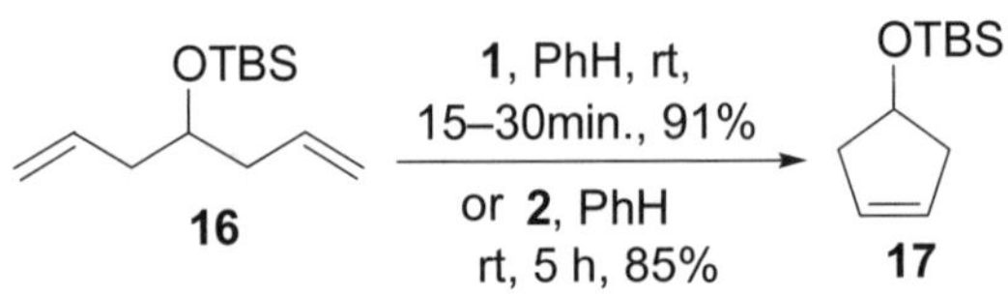

Since their initial report, this methodology has been extended to the formation of six-, seven-, eight- and larger ring carbocyclic systems. The following sections highlight significant application of the ring-closing metathesis strategy in the synthesis of these systems.

Five-Membered Rings

Shibasaki and co-workers used a ring-closing metathesis approach to prepare a number of five-, six-, and seven-membered rings from electron-deficient olefins.[37] Treatment of acyclic enol ether **18** with 7 mol % of **3** in refluxing benzene provided the corresponding cyclic enol ether **19** in 94% yield. Deprotection of the silyl ether **19** (not shown) resulted in the corresponding cyclic ketone, a valuable synthetic intermediate in natural products synthesis and a number of industrial processes. The authors reported additional examples of the synthesis of five-membered ring carbocycles as part of this study.

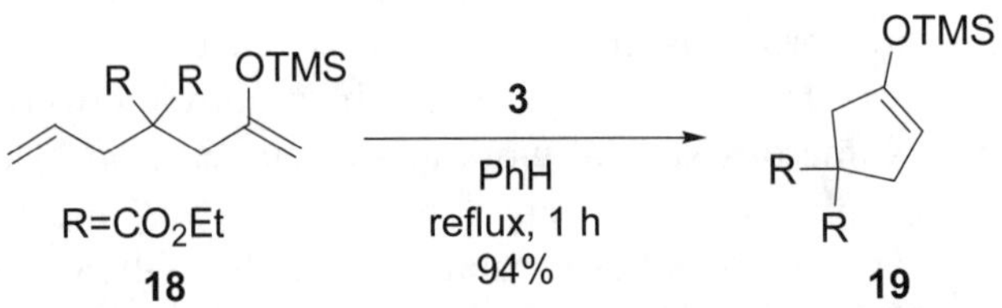

Independently, Aggarawa and co-workers prepared five-, six-, and seven-membered carbocyclic methyl and silyl enol ethers bearing phenyl and cyclohexyl substituents.[38] Dienes **20** and **22** required upward of 20 mol % of **4** in refluxing benzene to achieve the desired products **21** and **23** in 60 and 65% yeild, respectively. The authors attributed the modest yields to a lack of *gem*-diester substituents, a pattern, which has been known to dramatically increase the rates of cyclization.[39]

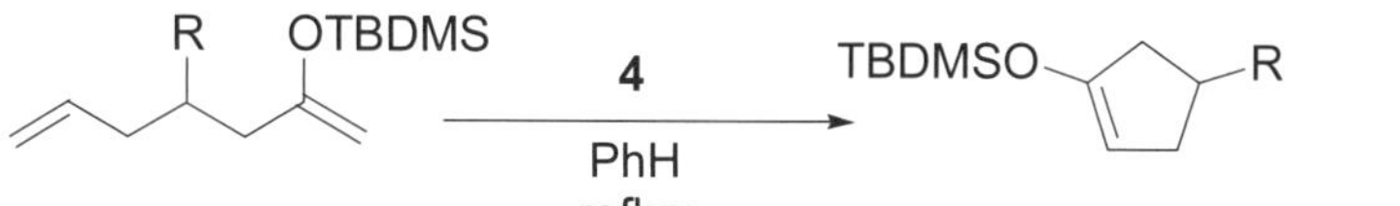

20, R = Ph
22, R = Cy

21, R = Ph 60%
23, R = Cy 65%

Roush and co-workers used olefin metathesis as a key step in their enantioselective and diastereoselective preparation of several cyclic β-hydroxy allylsilanes.[40] Reaction of β-hydroxy allylsilane **24** with 10 mol % of **4** in toluene gave the corresponding carbocycle **25** in 73% yield. The authors also applied this strategy to the formation of six-membered ring carbocycles bearing cyclic β-hydroxy allylsilanes.

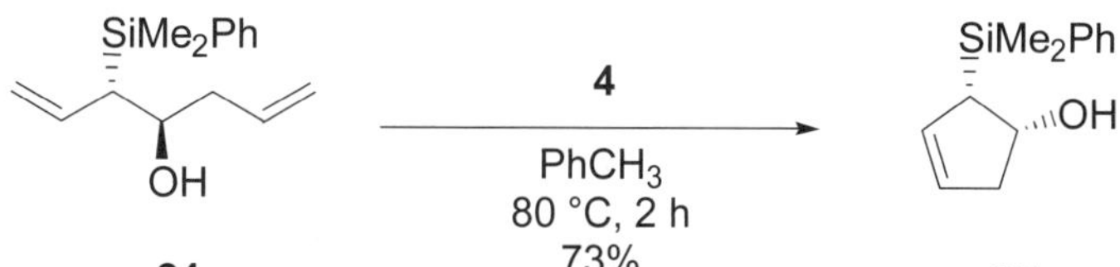

24 **25**

6,6-Bicyclic malonamides are important intermediates in the synthesis of natural products and have recently gained attention as potential donor atoms for ligand-metal complexes. Hutchinson and co-workers employed ring-closing metathesis as a key step in the synthesis of several malonamides.[41] Treatment of diethyl malonate diene **26** with a catalytic amount of **3** in dichloromethane provided the corresponding carbocycle **27** in quantitative yeild in a half an hour.

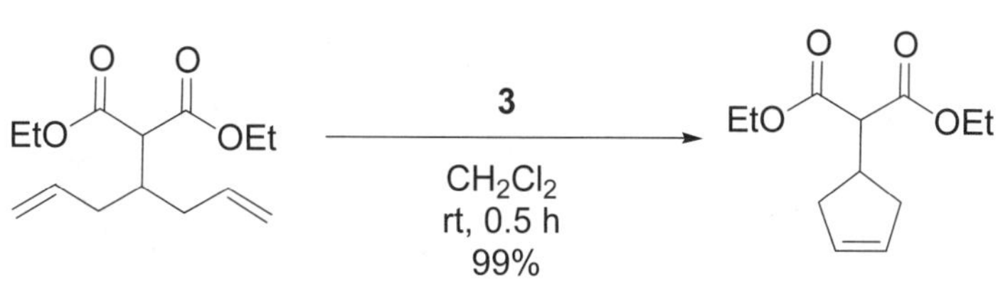

26 **27**

Weinreb and co-workers reported one of the first examples of a ring-closing metathesis strategy employing vinyl chlorides.[42] Vinyl chloride **28**, when reacted in the presence of 10 mol % of **4** in refluxing benzene, provided the corresponding carbocyclic derivative **29** in near quantitative yield. Unfortunately, reactions with the analogous vinyl bromides under similar reaction conditions gave none of the desired carbocylic products. The authors speculated that this was due to the formation of a stable, unreactive Fischer-type carbene.

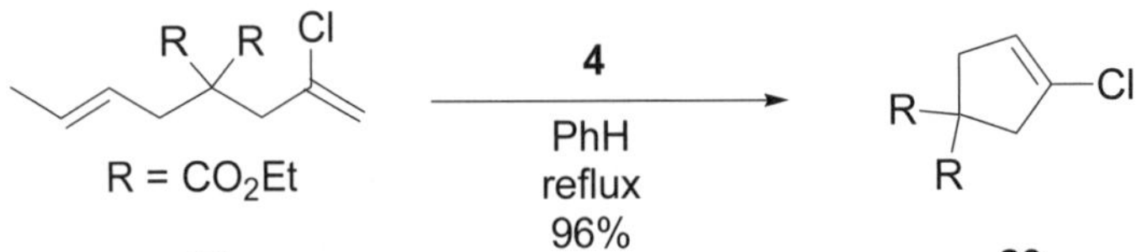

28 **29**

A number of research groups have used ring-closing metathesis to prepare conformationally constrained α- and β-amino acids. The corresponding peptides that incorporate these unusual amino acid residues often exhibit interesting biological properties. Several examples of constrained amino acid residues incorporating five-membered ring carbocycles are illustrated below.

Kotha and co-workers used a ring-closing metathesis strategy in their synthesis of α,α-dialkylated amino acids for the preparation of novel conformationally constrained peptide therapeutics.[43] Treatment of diene precursors bearing different amino acid substitutions with 10 mol % of **3** in refluxing toluene, provided the corresponding α,α-amino acid derivatives in yields ranging from 49–90%. Undheim and co-workers employed a similar ring-closing metathesis strategy in their synthesis of α,α-dialkylated constrained amino acids,[44] and Ple and co-workers also used olefin metathesis to prepare several constrained examples of α-alkoxy and α-amino esters (not shown).[45]

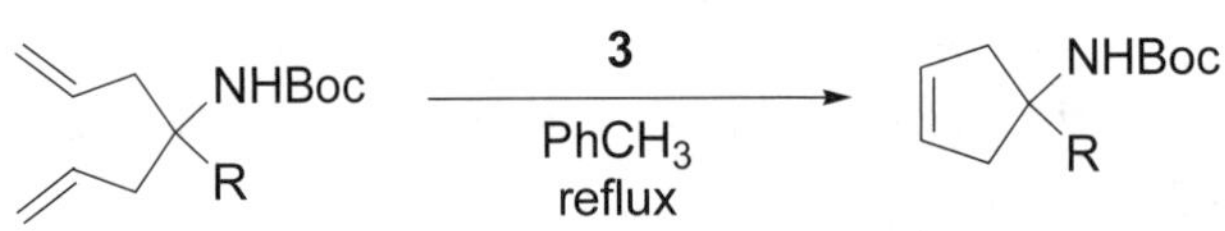

30, R = CONH(L)PheOCH$_3$ **31**, R = CONH(L)PheOCH$_3$ 75%
32, R = CONH(L)ValOCH$_3$ **33**, R = CONH(L)ValOCH$_3$ 90%
34, R = CONH(L)Ala(L)LeuOCH$_3$ **35**, R = CONH(L)Ala(L)LeuOCH$_3$ 50%
36, R = CONH(L)Leu(L)AlaOCH$_3$ **37**, R = CONH(L)Leu(L)AlaOCH$_3$ 49%
38, R = CONH(D)Val(L)ValOCH$_3$ **39**, R = CONH(D)Val(L)ValOCH$_3$ 53%
40, R = CONH(D)Val(L)LeuOCH$_3$ **41**, R = CONH(D)Val(L)LeuOCH$_3$ 75%

Abell and co-workers used ring-closing metathesis to synthesize a number of five-, six-, and seven-membered ring β-amino acids for incorporation into β-peptide mimetics.[46] β-Peptide mimetics have been shown to selectively disrupt bacterial cell membranes over mammalian cell membranes, and as such are key targets for the synthesis of new antibiotics. Treatment of dienes **42** and **44** with a 5 mol % of **4** in benzene furnished the corresponding β-amino acid derivatives **43** and **45** in 92 and 93% yield, respectively. Attempts to increase yields by refluxing in toluene led to a reduction in the amount of product produced, presumably due to catalyst decomposition.

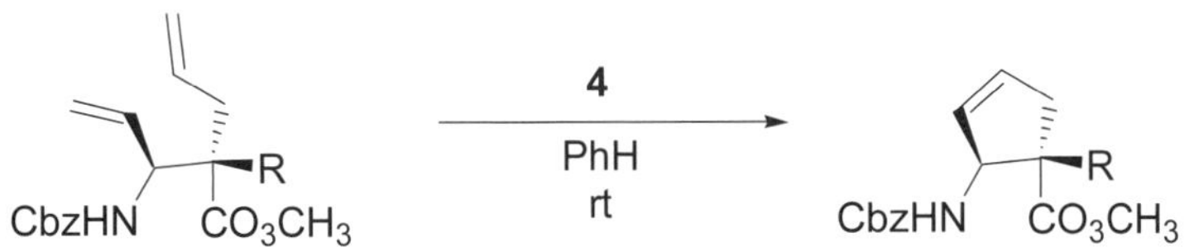

42, R = H
44, R = CH_3

43, R = H 92%
45, R = CH_3 93%

Davies and co-workers also applied a ring-closing metathesis strategy in their preparation of constrained β-amino acid derivatives.[47] Their work employed a chiral auxiliary, *S*-(1-phenyl-ethyl)-carbamic acid benzyl ester, to achieve the desired β-amino acid derivatives in a stereoselective fashion. Treatment of diene precursors **46** and **48** with 4 mol % of **3** in refluxing dichloromethane provided the corresponding carbocyclic β-amino acid derivatives **47** and **49** in 85 and 35% yield, respectively, with greater than 95% diastereoselectivity in both cases. The diminished yield in the case of **49** is presumably due to sterics effects.

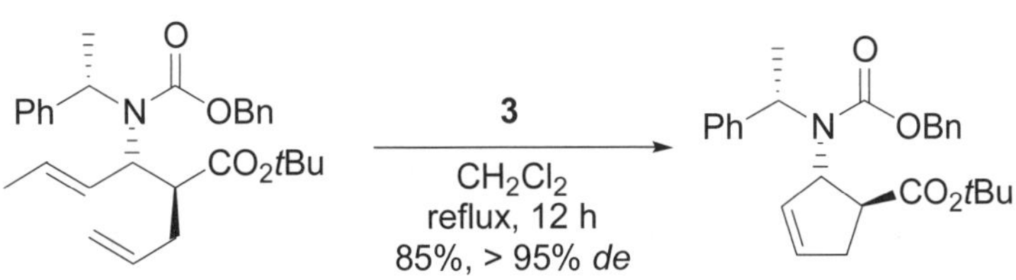

46 **47**

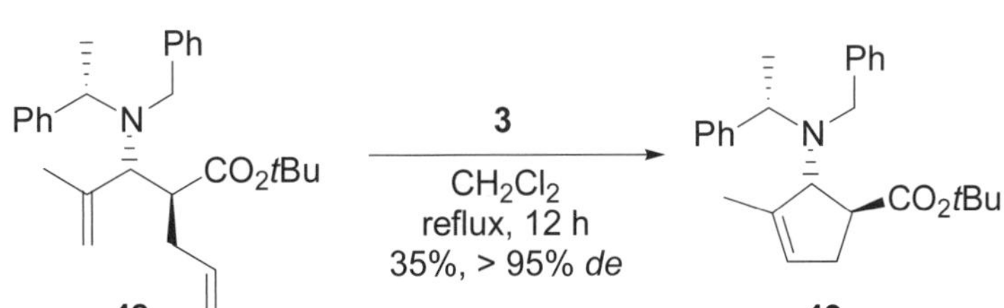

48 **49**

Carbasugars are structurally similar to natural sugars except that a carbon atom replaces the oxygen atom in the ring. A number of five-membered ring carbacycles have been synthesized to date, most of these mimicking the reverse transcriptase inhibitor carbanucleoside (–)-carbovir. The advantage of carbaucleosides over other nucleosides (such as AZT) is that they are more resistant to phosphorylation and subsequent degradation.

A number of reviews have been published highlighting advancements in the synthesis of carbanucleosides using ring-closing metathesis.[48] As such, only some of the more recent and synthetically challenging syntheses of these compounds are reported here.

In an effort to prepare new anti-HIV compounds with low cytotoxicity, Park and co-workers synthesized L-*N*-MCd4T, a carbocyclic nucleoside containing a fused cyclopropane functionality.[49] Reaction of cyclopropyl dienes **50** and **52** as a mixture of diastereomers with a catalytic amount of **4** in dichloromethane furnished the corresponding fused carbacycles **51** and **53** in 85% and 84% yield, respectively. Compound **53** was readily converted to the desired nucleoside analog in three steps.

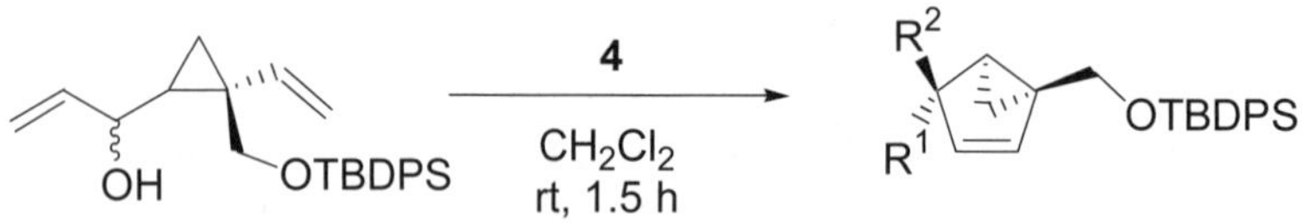

50, (2*S*): R^1 = OH, R^2 = H
52, (2*R*): R^1 = H, R^2 = OH

51, (2*S*): R^1 = OH, R^2 = H 85%
53, (2*R*): R^1 = H, R^2 = OH 84%

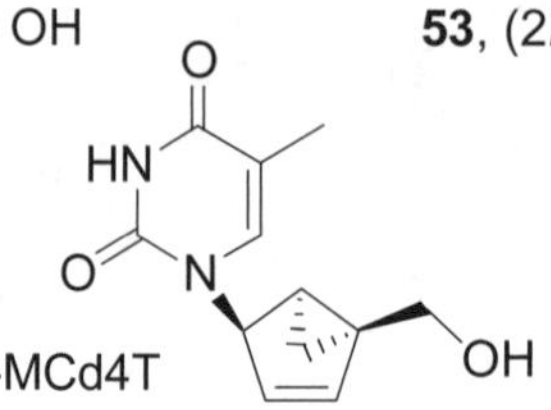

L-*N*-MCd4T

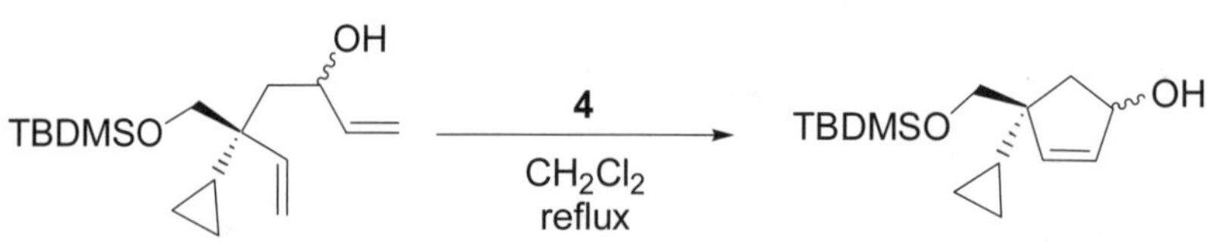

54α
56β

55α 41%
57β 42%

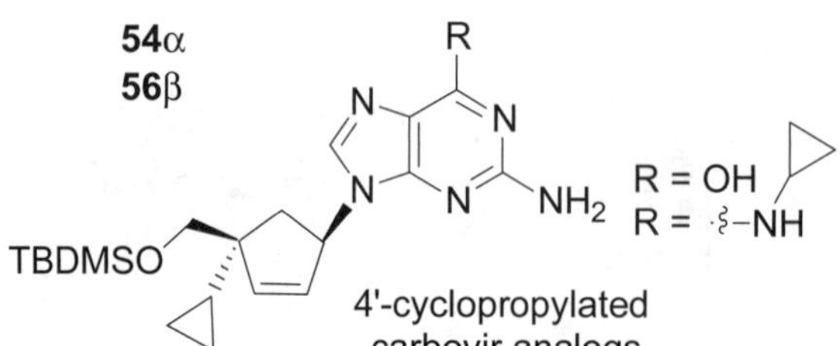

4'-cyclopropylated carbovir analogs

In a separate study, Liu and co-workers synthesized a nucleoside incorporating a cyclopropyl group at the 4′-position.[50] Reaction of cyclopropyl dienes **54** and **56** with 3 mol % of **4** in refluxing dichloromethane gave a mixture of the corresponding carbasides **55** and **57** in 41 and 42% yeild, respectively. Three additional steps were required to convert compound **57** to the desired nucleoside analog.

Recently, Li and co-workers reported on the synthesis of a 4′-branched exomethylene carbocyclic nucleoside analog as potential mimic of olefinic carbocyclic nucleosides.[51] Treatment of acyclic triene **58** with 10 mol % of **4** in refluxing dichloromethane gave the corresponding carbocycle in 74% yield. It is surprising that isomerisation of the exocyclic double bond, which has been known to occur ruthenium-based metathesis catalysts, did not occur under the reaction conditions reported.[52] Carbocycle **59** was then readily converted to the corresponding exomethylene carbocyclic nucleoside analog in just a few additional steps.

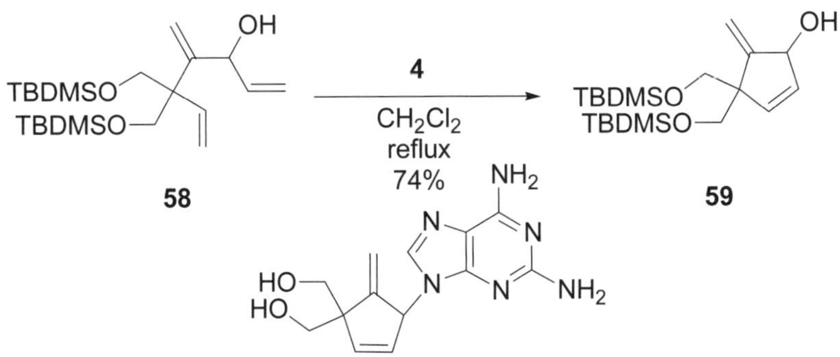

exomethylene carbocyclic
nucleoside analog

Terpenes are a class of compounds composed of one or more isoprene units. Nearly all organisms produce terpenes, and many of these compounds have interesting biological activities. Many terpenes also exhibit complex architectures that pose a significant challenge for synthetic organic chemists. For these reasons, a number of scientists have set out to synthesize terpenes and terpenoid derivatives, and many recent attempts have employed a ring-closing metathesis strategy.

Srikrishna and co-workers used ring-closing metathesis to synthesize HM-3 and HM-4, two aromatic sesquiterpenes with antioxidant and antibiotic activity.[53] Treatment of a diastereomeric mixture of allyl alcohols **60** and **62** with 5 mol % of **3** in dichloromethane, followed by oxidation with PCC gave the corresponding enones **61** and **63** in 93 and 96% yield, respectively. These compounds were readily converted to the corresponding aromatic sesquiterpenes in just a few additional steps. Srikrishna and co-workers applied a similar strategy to the construction of the five-membered ring moieties of (±)-12-methoxyherbertenediol dimethyl ether,[54] (±)-laurokamurene B,[55] and (±)-herbertenediol (not shown).[56]

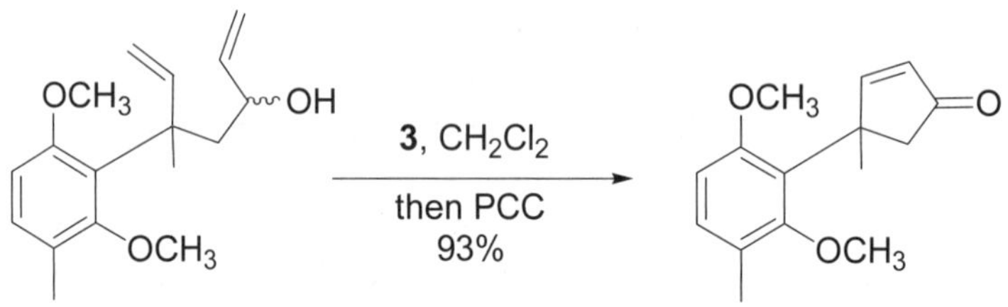

60 61

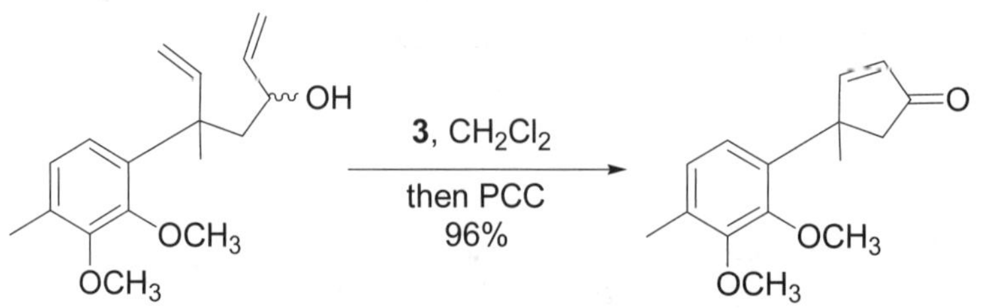

62 63

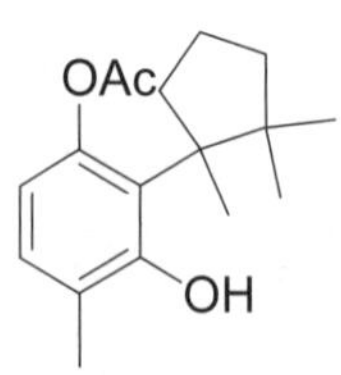

HM-3

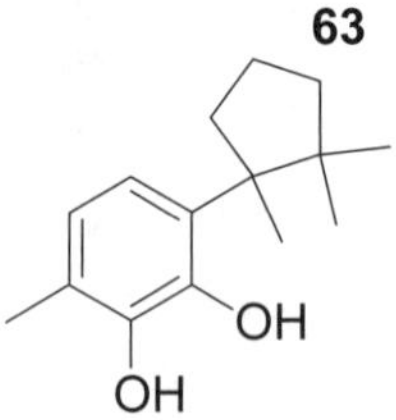

HM-4

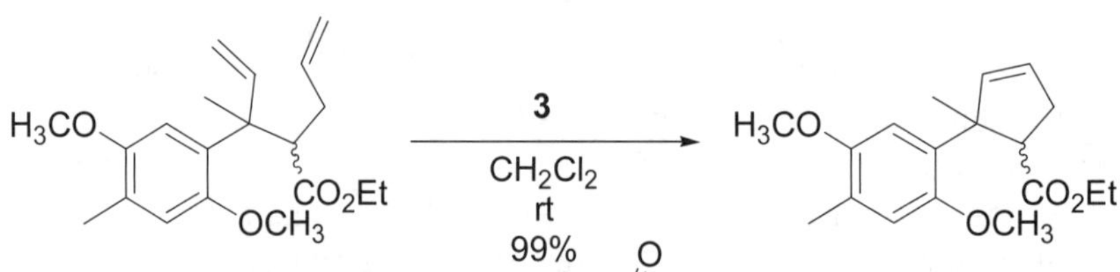

64 65

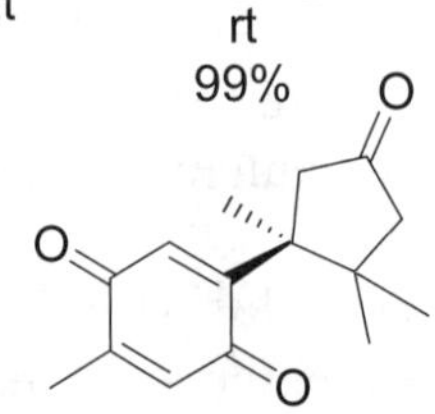

(±)-lagopodin A

A ring-closing metathesis strategy was also employed by Srikrishna and workers in the first total synthesis of (±)-lagopodin A, a fungal sesquiterpene.[57] The authors found the sterically congested 1-aryl-1,2,2-trimethylcyclopentane component especially challenging to construct. Olefin metathesis of heptadiene **64** in the presence of 5 mol % of **3** in dichloromethane furnished the corresponding cyclopentene ester **65** in quantitative yield. A similar strategy was used by Kulkarni and co-workers in their 2006 report on the synthesis of (±)-β-cuparenone, a related compound (not shown).[58]

Another sesquiterpene, fomannosin, has caused considerable concern in the southeastern United States due to its toxicity, especially toward certain species of pine and select symbiotic bacteria. Paquette and co-workers recently synthesized both (+)- and (–)-fomannosin from *D*-glucose in an effort to further investigate the properties of these compounds.[59] Ring-closing metathesis of cyclobutyl diene precursors **66** and **68** in the presence of 5 mol% of **4** in refluxing benzene furnished the corresponding carbocyclic products **67** and **69** in 77 and 91% yield, respectively.

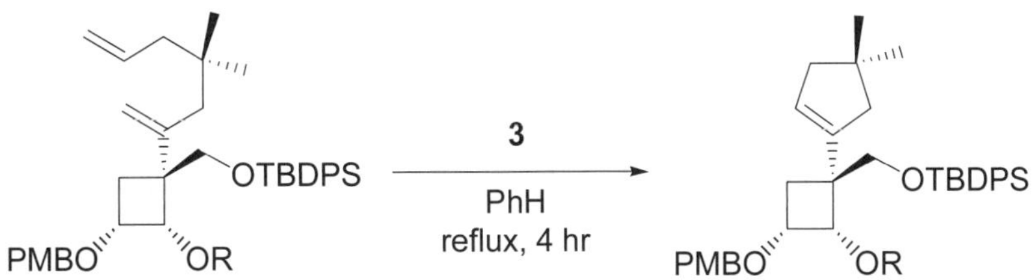

66, R = MEM
68, R = TBS

67, R = MEM 77%
69, R = TBS 91%

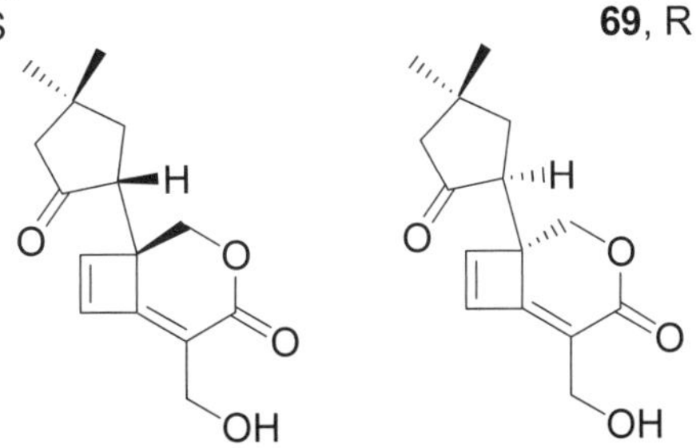

(+)-fomannosin (–)-fomannosin

Srikrishna and Dethe employed a ring-closing metathesis strategy in their enantiospecific synthesis of the pacifigorgiane sesquiterpene carbocyclic core which contains a fused 6,5 system.[60] Treatment of **70** as a mixture of diastereomers with 10 mol % of **3** in refluxing dichloromethane provided the corresponding pacifigorgia-2,7-dien-4-one **71** in quantitative yield. Srikrishna and co-workers later applied this strategy in their synthesis of functionalized bicyclo[4.3.1]decanes, which are key intermediates in the construction of vibsane diterpenoids (not shown).[61]

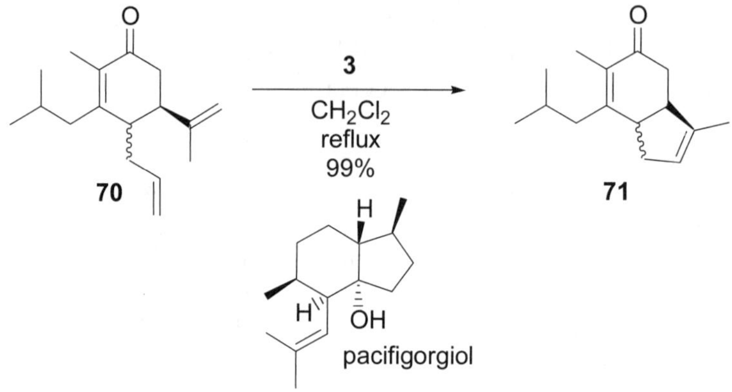

Spirocycles are found in a number of natural products with therapeutic and industrial applications. Until the development of Schrock's and Grubbs's catalysts, the formation of these desirable, rigid compounds presented a significant challenge for synthetic organic chemists. In recent years, many carbocyclic spirocycles have been prepared in high yields using ring-closing metathesis as a key step. Several examples of five-membered ring carbocyclic spirocycles are illustrated in the parargraphs below.

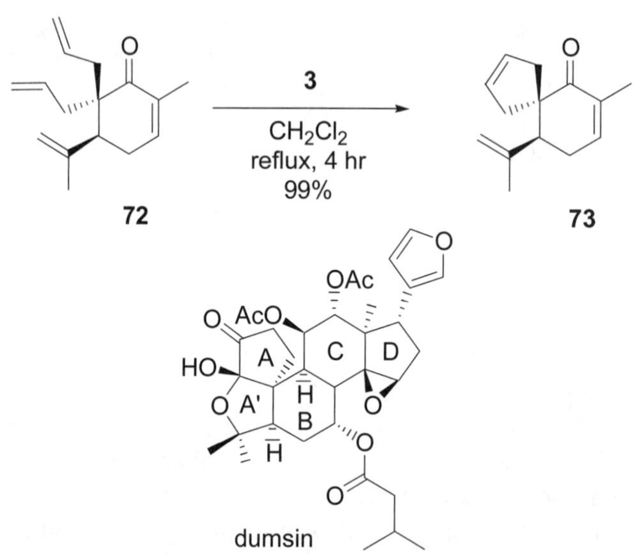

Dumsin, a highly complex tetranortriterpene containing 18 stereogenic centers and a spirocyclic A, A′, B system, has recently gained attention due to its potential as a selective and potent pesticide. Srikrishna and Babu employed olefin metathesis as a key step in their rapid and enantiospecific synthesis of the ABC ring system of dumsin.[62] Treatment of 6,6-bis-allyl carvone **72** with 5 mol % of **3** in dichloromethane provided the corresponding spirocyclic A ring **73** in quantitative yield. No competeing side reactions were observed with the more sterically hindered olefin in the ring-closing step, highlighting the selectivity of **3** for less sterically demanding olefins. Srikrishna and Babu also used a similar approach to construction the C ring of dumsin, yielding the correspoinding tricyclic ABC system (not shown).

Marquez and Hobson, in an effort to produce spirocycles with functional handles for diversification, synthesized several highly functionalized spirocyclic pyrans using ring-closing metathesis.[63] Reaction of bis-alkenyl compound **74** with 5 mol % of **3** in refluxing dichloromethane gave the corresponding spirocycle **75** in 85% yield. Six-, seven-, and eight-membered ring spirocycles were also prepared via this method, and the authors noted that yields decreased correspondingly with an increase in the size of the spirocycle, presumably due to the lack of conformational constraints.

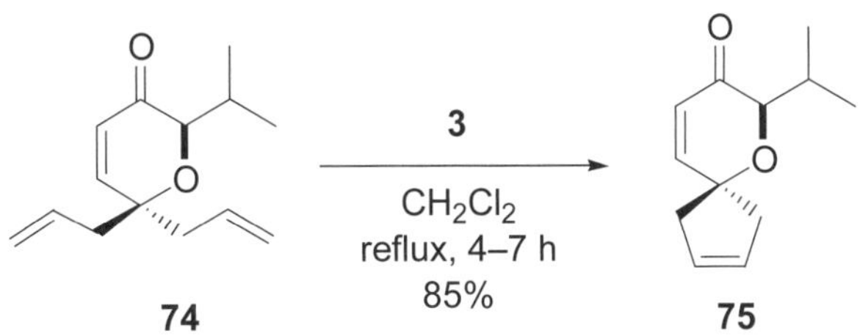

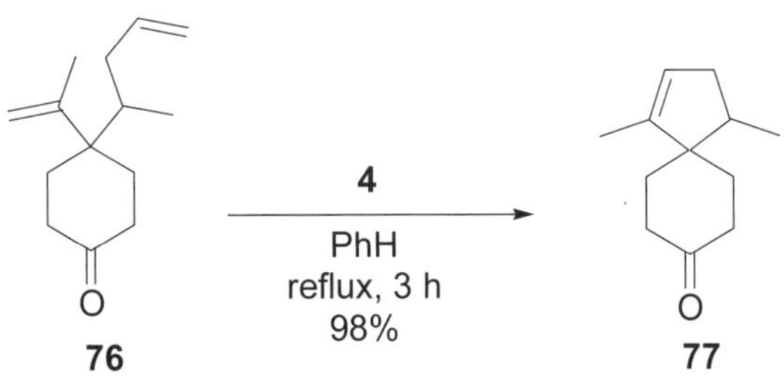

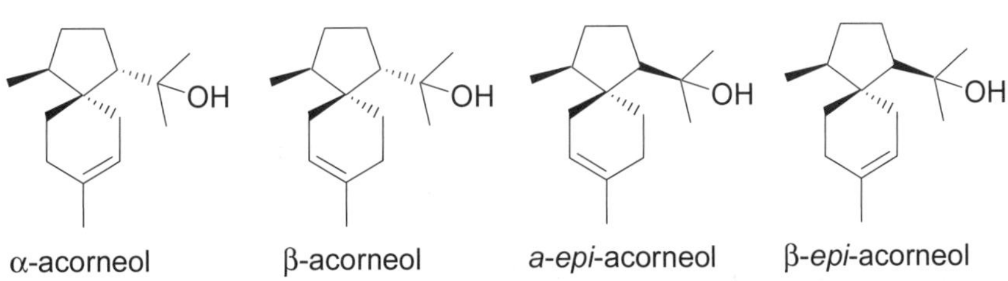

Srikrishna and co-workers employed a similar strategy for the general synthesis of a spirocyclic core inherent to several acorneols.[64] Treatment of diene **76** with 3 mol % catalyst loading of **4** in refluxing benzene provided the corresponding spirocyclic carbocycle **77** in near quantitative yield. Five additional steps were required to produce the desired acorneols.

Gurjar and co-workers prepared several novel carbohydrate-based spirocycles and a spirocyclic proline derivative for applications in peptide, nucleoside, and carbohydrate synthesis.[65] Ring-closing metathesis of carbohydrate diene precursor **78** furnished the corresponding spirocycle **80** in 88% yield using a catalytic amount of **3** in dichloromethane. Reaction of proline derivative **79** under similar conditions gave the corresponding spirocyclic peptide **81** in 96% yield.

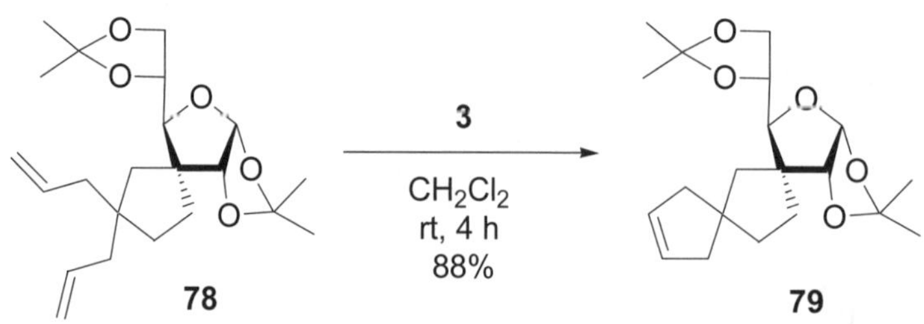

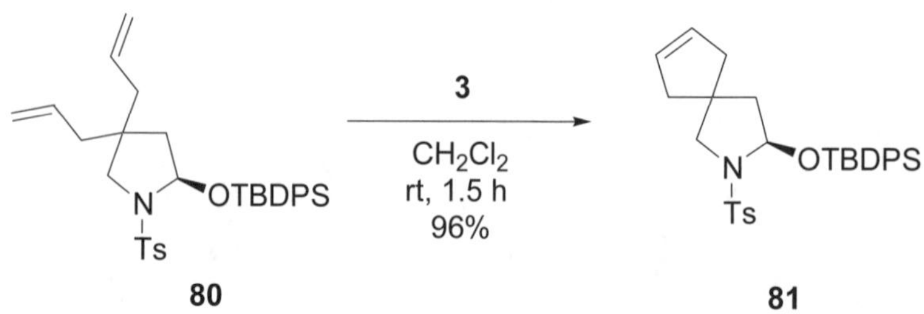

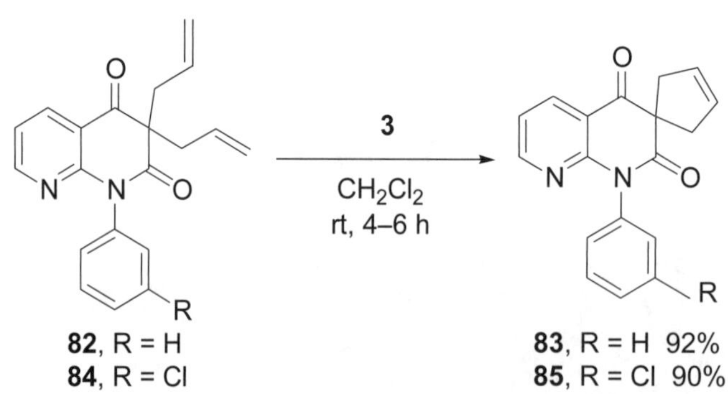

Majumdar and co-workers employed an olefin metathesis strategy in their synthesis of several spironaphthyridinone derivatives.[66] Spironaphthyridinones are a class of compounds that have shown promise in the treatment of autoimmune and other immune disorders. Ring-closing metathesis of **82** or **84** using 10 mol % of **3** in dichloromethane gave the corresponding spirocycles **83** and **85** in 92% and 90% yield, respectively.

Hughes and co-workers employed a ring-closing metathesis strategy as a general route for the synthesis of enantiopure five-, six-, and seven-membered ring spirocarbocycles using zizaene as a chiral auxiliary.[67] Diene precursors **86** and **88** required 15 mol % catalyst loading of **3** (added in three equal portions every 4–6 hours) in refluxing dichloromethane to achieve the desired five-membered ring spirocycles in 91 and 89% yield, respectively. The authors noted that as the size of the spirocycle increased, the yields obtained from ring-closing metathesis decreased due to conformational constraints.

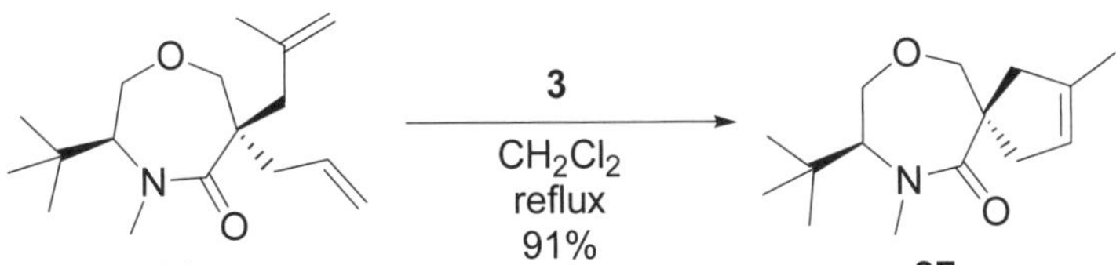

86 **87**

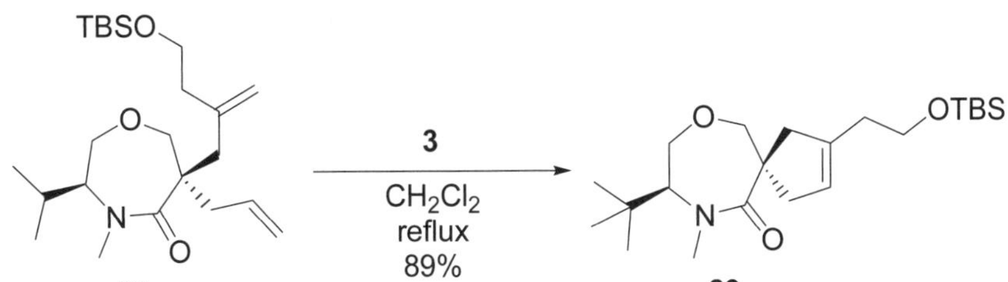

88 **89**

Undheim and co-workers used a similar strategy to prepare an interesting class of five-, six-, and seven-membered ring spirocyclic carbocycles for use as templates in natural product synthesis.[44] Reaction of **90** with 2 mol % of **3** furnished the corresponding spirocycle **91** in 53% yield. The authors were able to produce six- and seven-membered ring spirocycles of this class, but were unable to access the analogous eight-membered ring derivatives, presumably due to conformational constraints.

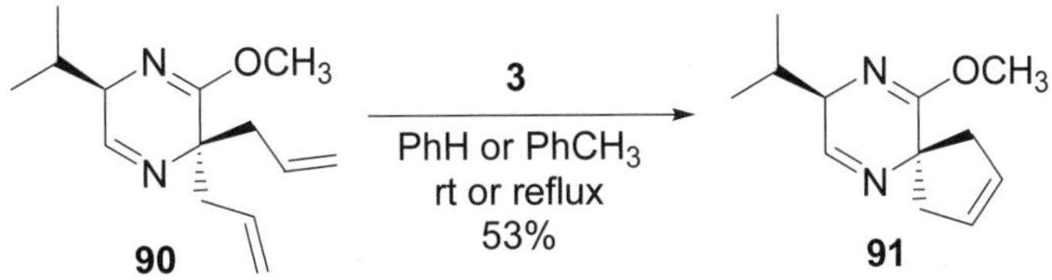

90 **91**

Cyclopentanone prostaglandins have recently garnered attention as potential cancer therapeutics. Unfortunately, native prostaglandins are quickly metabolized and exhibit limited water solubility. Florent and co-workers, in an effort to prepare metabolically stable and water-soluble prostaglandins, subjected acyclic allyl alcohols **92** and **94** (as a mixture of diastereomers) to 1 mol % of **3** in the presence of dichloromethane to produce the corresponding cyclic allyl alcohol derivatives (not shown).[68] The carbocyclic allyl alcohols were subsequently oxidized to the corresponding enones using NMO and TPAP to give **93** and **95** in 89% and 90% yield, respectively over two steps. Enones **93** and **95** were further functionalized to give the desired prostaglandin derivatives which showed good activity against L1210 leukemia cells.

92, R = $(CH_2)_3OCH_2$-*p*FPh **93**, R = $(CH_2)_3OCH_2$-*p*FPh 89%
94, R = $(CH_2)_7CH_3$ **95**, R = $(CH_2)_7CH_3$ 90%

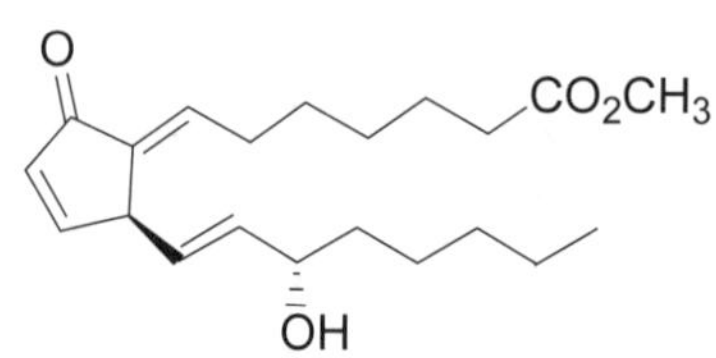

Δ^7-PGA$_1$

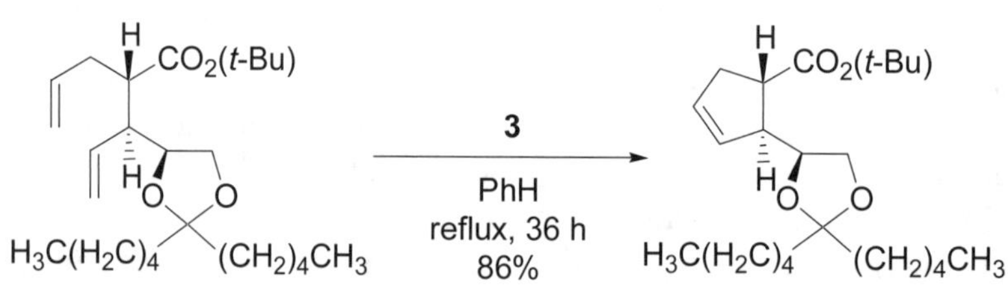

96 **97**

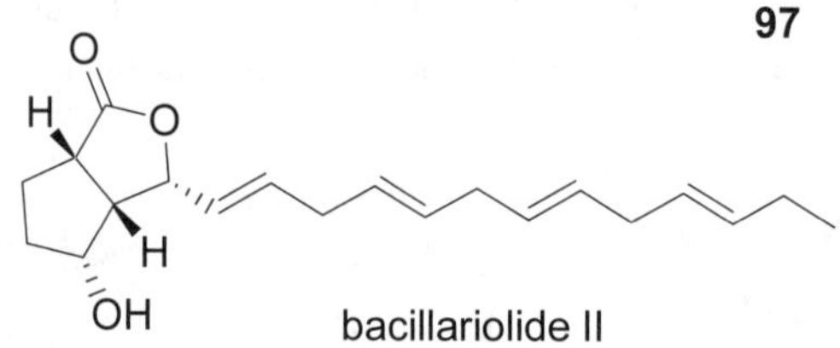

bacillariolide II

Ghosh and co-workers used olefin metathesis as a key step in their synthesis of the carbocyclic core of *ent*-bacillariolide II, an oxylipin.[69] Oxylipins are densely functionalized naturally occurring bicyclic systems with four contiguous stereocenters. In addition to the synthetic challenges these compounds present, several members of the oxylipin family have shown significant inhibitory activity against phopholipase A_2, an enzyme that plays a key role in regulating the inflammatory response. Reaction of diene **96** with 6 mol % of **3** in benzene led to the desired cyclopentene **97** in 86% yield.

Covey and co-workers used an abnormal Beckmann fragmentation/ring-closing metathesis strategy in their synthesis of 18-nor-$\Delta^{13(17)}$-androgens, derivatives of 3α-hydroxysteroids.[70] These compounds have been shown to modulate ion channels within the central nervous systems of animals. Treatment of diene **98** in the presence of a catalytic amount **4** in dichloromethane gave the corresponding steroid **99** in 98% yield. Reaction of **99** with *m*-CPBA furnished the desired 18-nor-$\Delta^{13(17)}$-androgen as a mixture of diasteromers (not shown).

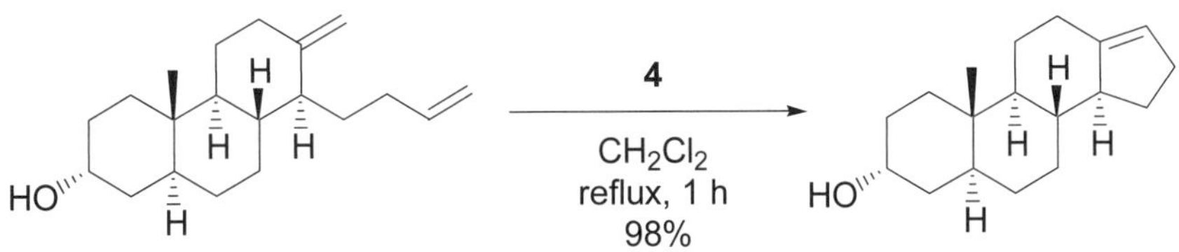

98 99

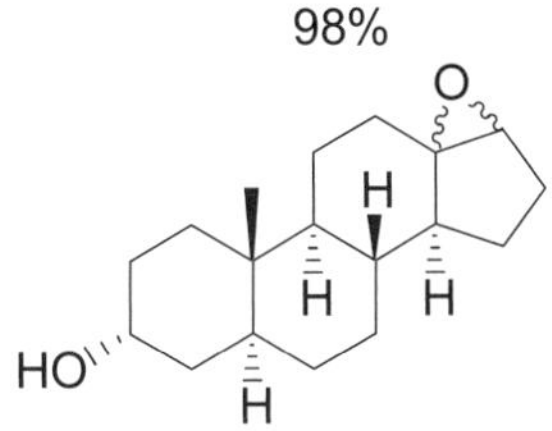

18-nor-$\Delta^{13(17)}$-androgens

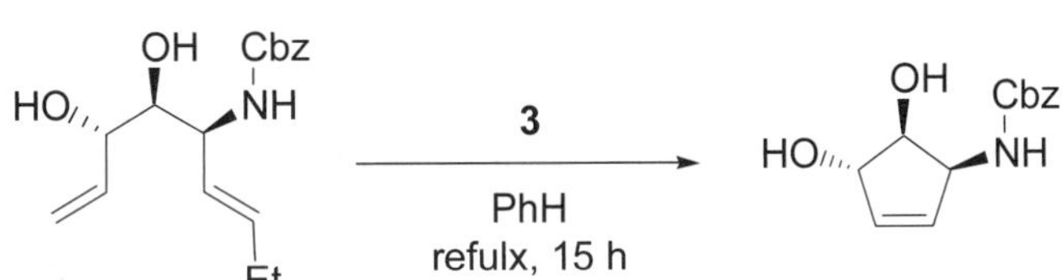

100 101

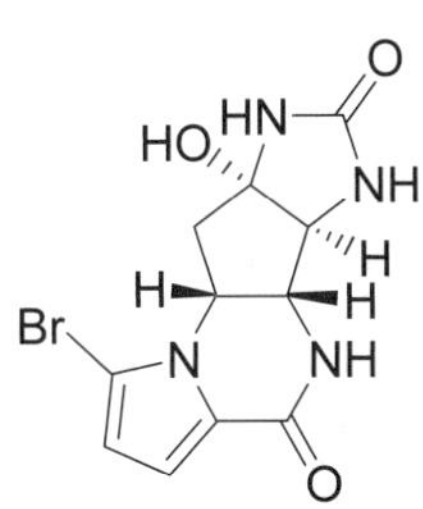

(−)-agelastatin

(–)-Agelastatin A is an architecturally unusual tetracyclic compound that exhibits significant antitumor activity. Ichikawa and co-workers recently improved on an earlier syntheses of (–)-agelastatin A by using ring-closing metathesis as a key step in the construction of the five-membered carbocycle core.[71] Reaction of diene **100** with 5 mol % of **3** in benzene produced the corresponding highly functionalized cyclopentene ring **101** in 83% yield. This compound was converted via several steps to provide (–)-agelastatin A.

The sesquiterpene merrilactone A is an important neurotrophic factor of considerable therapeutic interest for the treatment of several neurodegenerative disorders. It remains a considerable challenge synthetically due to its densely oxygenated pentacyclic architecture containing seven stereogenic centers, two γ-lactone moieties, and four quaternary carbon atoms. Mehta and co-workers recently reported on the use of a ring-closing metathesis strategy to synthesize the fused tricyclic core of merrilactone A.[72] Treatment of lactone **102** with 10 mol % of **3** in dichloromethane furnished the desired tricyclic system **103** in 95% yield.

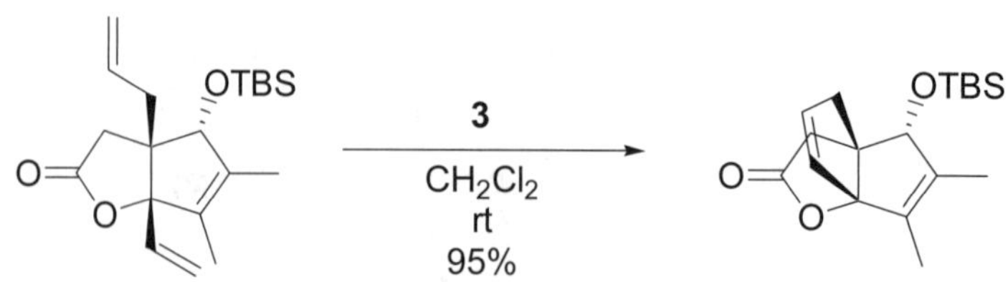

102 **103**

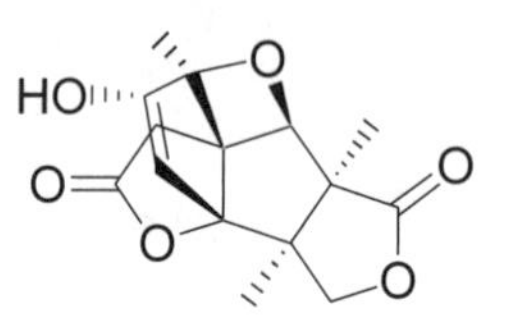

merrilactone A

Triquinane natural products have recently gained attention due to their complex architectures, and cytotoxic and antibacterial properties. Recently, Srikrishna and Beeraiah used a ring-closing metathesis strategy to synthesize both the *cis*, *syn*, *cis*- and *cis*, *anti*, *cis*-linear triquinanes.[73] Treatment of diene **104** with 5 mol% of **3** in dichloromethane gave the corresponding *cis*, *syn*, *cis*-triquinane **105** in quantitative yield. A similar strategy was employed in the synthesis of the *cis*, *anti*, *cis*-triquinane, with the ring-closing metathesis proceeding smoothly in 97% yeild (not shown).

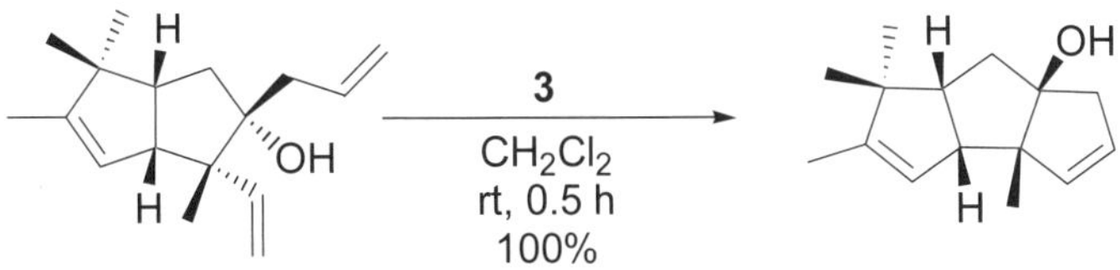

104 105

Six-Membered Rings

A general example of the versatility of ring-closing metathesis in the formation of six-membered ring carbocycles is demonstrated by Shibasaki and co-workers, who employed their ring-closing metathesis strategy for five-membered ring carbocyclic enol ethers to six-membered ring carbocyclic enol ethers. Carbocyclic enol ether **107** was readily prepared from the corresponding electron deficient olefin **106** in 92% yield using only 7 mol % of catalyst **3** in refluxing benzene.

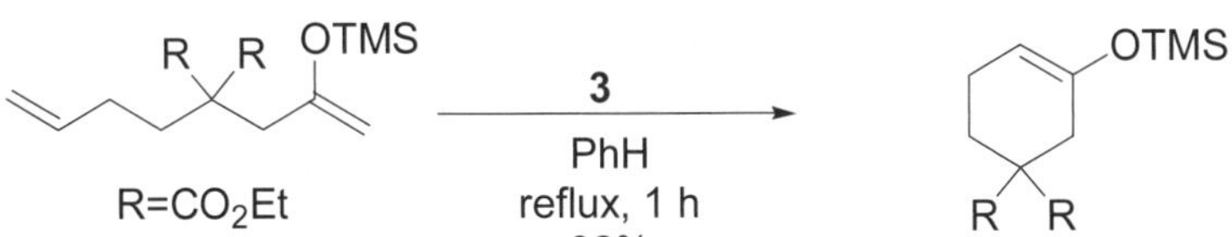

106 107

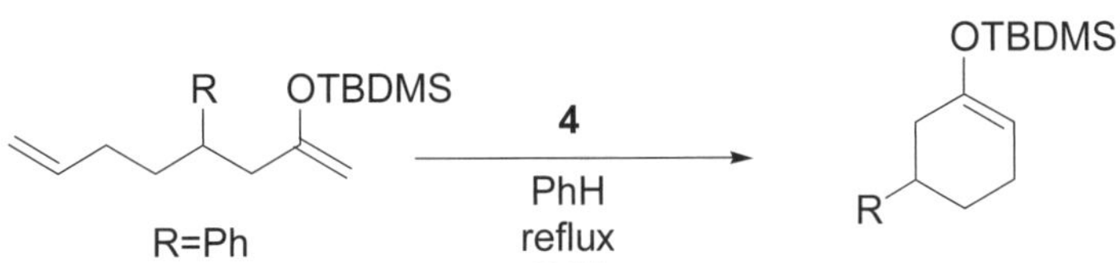

108 109

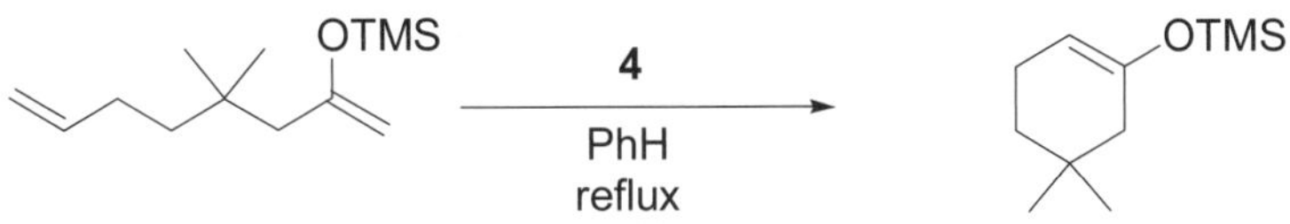

110 111

Aggarawa and co-workers also used their ring-closing metathesis strategy for the preparation of five-membered ring carbocyclic silyl enol ethers to prepare a six-membered ring carbocyclic silyl enol ether bearing a phenyl substituent.[38] In their study, treatment of **108** with upward of 20 mol % of **4** in refluxing benzene led to a 65% yield of the desired product **109**. The authors attributed the modest yield to the lack of *gem*-substituents, and tested this theory by using olefin metathesis to synthesize a cyclic silyl enol ether bearing a simple *gem* dialkyl group (**110**). This substrate required only 10 mol % of **4** to provide the corresponding carbocycle **111** in 89% yield.

Weinreb and co-workers used their olefin metathesis strategy for the preparation of chloro-cyclopentene derivatives from vinyl chlorides for the synthesis of chloro-cyclohexene derivatives.[42] Vinyl chloride **112**, when treated with 10 mol % of **4** in benzene, gave the corresponding cyclohexene derivative **113** in quantitative yields.

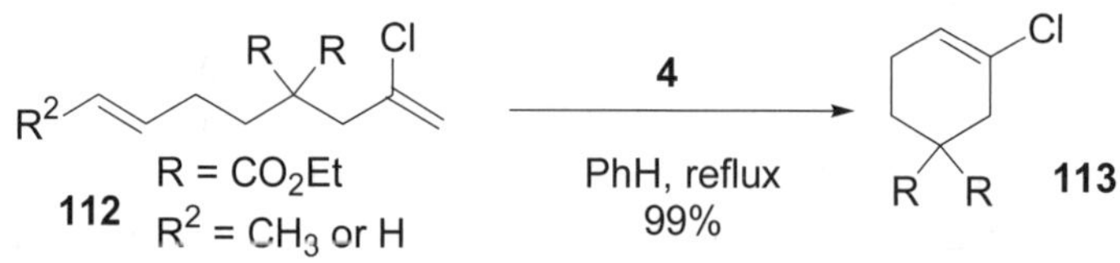

Piscopio and co-workers pioneered a method for synthesizing functionalized carbocycles via an ester enolate Claisen/ring-closing metathesis strategy.[74] Substrates **114**, **116**, **118**, and **120** were subjected to olefin metathesis using 2.5 mol % of **3** in the presence of dichloromethane to furnish the corresponding carbocycles, including fused bicyclic (**119**) and spirocyclic (**121**) systems, in upward of 86% yield. Haudrechy and co-workers applied a similar approach to synthesize a number of highly functionalized carbocycles,[75] and recently Sutherland and co-workers used an analogous aza-Claisen/ring-closing metathesis route in the preparation of functionalized carbocyclic amides (not shown).[76]

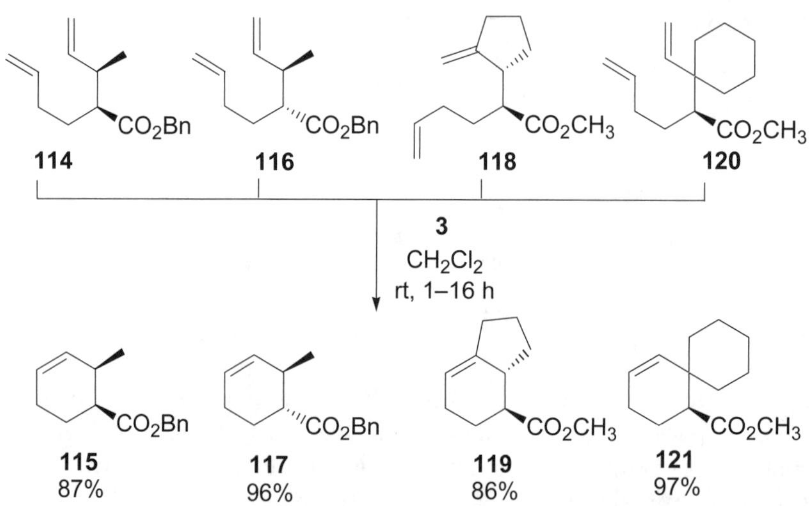

Martin and co-workers applied a ring-closing metathesis strategy in the synthesis of medium-size carbocycles fused to butyrolactones.[77] Fused butyrolactones are ubiquitous in nature and play an important role in the structural integrity and biological activity of many natural products. Butyrolactones **122** and **124** were subjected to ring-closing metathesis using 10 mol% of **4** in refluxing dichloromethane to furnish the corresponding α,β fused γ-lactones **123** and **125**, both in 85% yield.

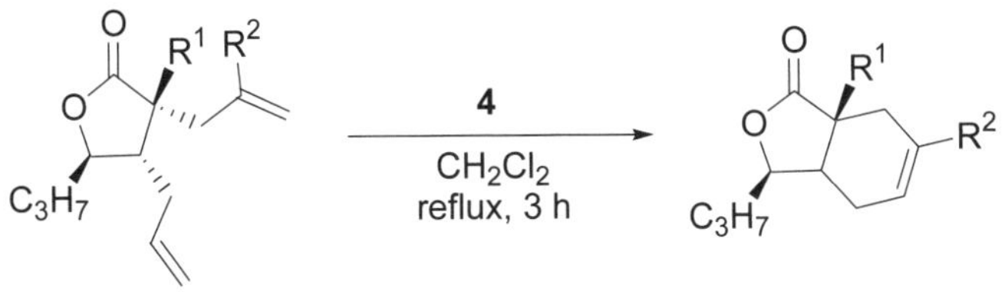

122, $R^1 = SO_2Ph$, $R^2 = H$
124, $R^1 = SO_2Ph$, $R^2 = CH_3$

123, $R^1 = SO_2Ph$, $R^2 = H$ 85%
125, $R^1 = SO_2Ph$, $R^2 = CH_3$ 85%

Spino and co-workers employed a relay ring-closing metathesis strategy using a chiral auxiliary in their enantioselective synthesis of several amino carbocycles.[78] Treatment of diene **126** with a catalytic amount of **3** in refluxing dichloromethane gave the corresponding aromatic allyl amine **127** in quantitative yields as a single enantiomer with concomitant loss of the chiral auxiliary.

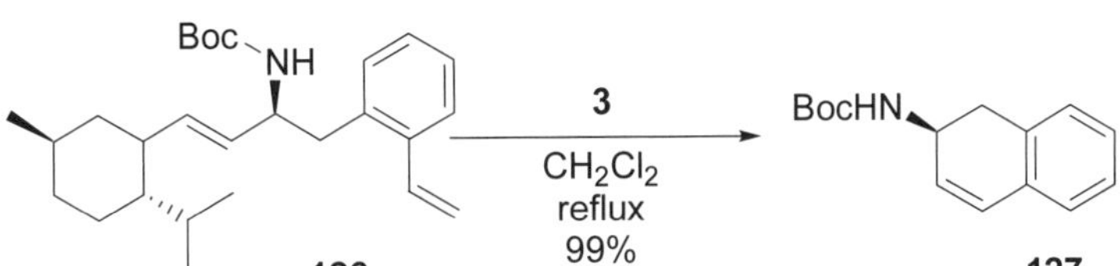

126 **127**

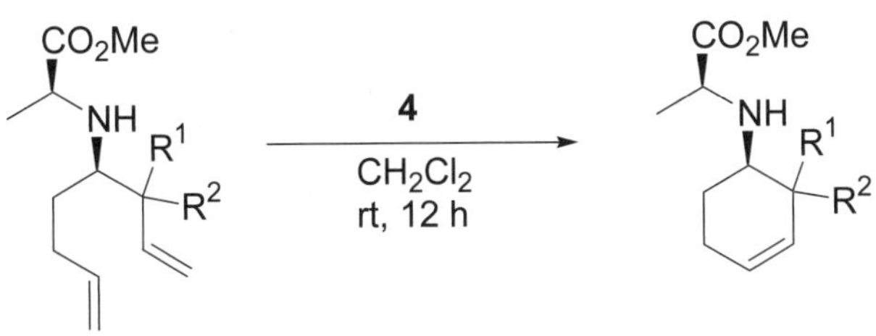

128, $R^1 = CH_3$, $R^2 = H$
130, $R^1 = CH_3$, $R^2 = CH_3$
132, $R^1 = Ph$, $R^2 = H$

129, $R^1 = CH_3$, $R^2 = H$ 84%
131, $R^1 = CH_3$, $R^2 = CH_3$ 92%
133, $R^1 = Ph$, $R^2 = H$ 86%

Loh and co-workers used olefin metathesis to synthesize a number of homoallylic amines, which serve as important intermediates in the synthesis of alkaloid natural products and nitrogen heterocycles.[79] Reaction of substituted dienes **128, 130**, and **132** with 10 mol % of **4** in dichloromethane provided the corresponding amines in high yield (84–92%).

A number of research groups have used ring-closing metathesis to prepare conformationally constrained α- and β-amino acids containing six-membered ring carbocycles. Several key examples are discussed in the paragraphs below.

Abell and co-workers applied their ring-closing metathesis strategy for the synthesis of β-amino acid residues constrained by five-membered carbocycles to produce a host of β-amino acid residues constrained by six-membered ring carbocylcles.[46] Treatment of the dienes **134**, **136**, and **138** with a catalytic amount of **3** in refluxing benzene provided the corresponding constrained β-amino acids in upwards of 90% yield. In a later report, they were able to verify the yields of these compounds using **4** under similar reaction conditions, and expanded their work to synthesize a host of five-, six- and seven-membered ring cyclic β-amino acid analogs (not shown).[80]

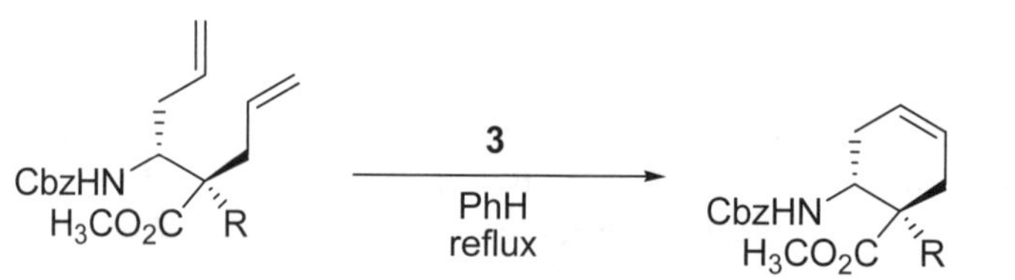

134, R = H
136, R = CH_3
138, R = Et

135, R = H 96%
137, R = CH_3 96%
139, R = Et 94%

Ring-closing metathesis has also been applied to the synthesis of a number of carbasugars containing six-membered rings. Many of these carbacycles are natural products; however, several unnatural carbasugars have also been prepared and evaluated for their potential to serve as biological mimics of natural pyranose sugars. Many carbasugars have gained attention as potent glycosidase inhibitors, while others have been used as precursors in the synthesis of higher order natural products. Several examples highlighting the diversity of six-membered ring carbasugars are illustrated below.

A number of carbocyclic mimetics of pyranose carbohydrates have been prepared using ring-closing metathesis as a key step. One of the first examples by Madsen and co-workers, employed a novel zinc-mediated domino reaction to convert 5-iodo-ribofuranosides to the corresponding diene precursors **140** and **143** (not shown).[81] These diene precursors were then subjected to ring-closing metathesis using up to 10 mol% of **3** in the presence of dichloromethane to provide the corresponding carbasugar derivatives **141** and **143** in near quantitative yields. Dihydroxylation of compounds **141** and **143** using OsO_4 provided the corresponding carbasugars (+)-*talo*-quercitol and (–)-*gala*-quercitol (not shown) in high yeilds. Their report signified the shortest enantioselective synthesis of both these compounds, and the authors were able to extend this approach to synthesize a number of other derivatives.

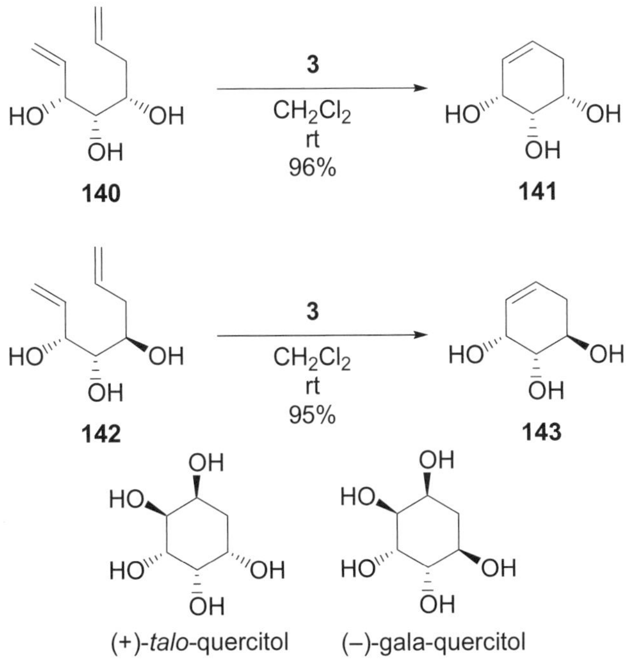

Since Madsen's initial report, several additional pyranose carbacycles have been synthesized. Vankar and co-workers synthesized *muco*-quercitol, (+)-*gala*-quercitol, and 5-amino-5-deoxy-D-*vibo*-quercitol, from D-mannitol using olefin metathesis as a key step.[82] The later compound, 5-amino-5-deoxy-*D-vibo*-quercitol is especially noteworthy as it an important component of a number of aminoglycoside antibiotics. Van Boom and co-workers used olefin metathesis in the synthesis of 2-deoxystreptamine, a important carbohydrate component of neomycin B and kanamycin B, two aminoglycoside antibiotics.[83] Gallos and co-workers also employed a ring-closing metathesis approach in their synthesis of 6a-carba-β-D-fructopyranose, a potential sweetener.[84] Kumaraswamy and co-workers used a similar approach in their preparation of glucose and galactose based carbacycles as intermediates in drug development.[85] And finally, Cumpstey and co-workers recently published the synthesis of two β-hexosaminidase inhibitors, valienamine and 1-*epi*-valienamine from either D-glucose or L-sorbose,[86] or D-mannose[87] using ring-closing metathesis as the key step in their syntheses.

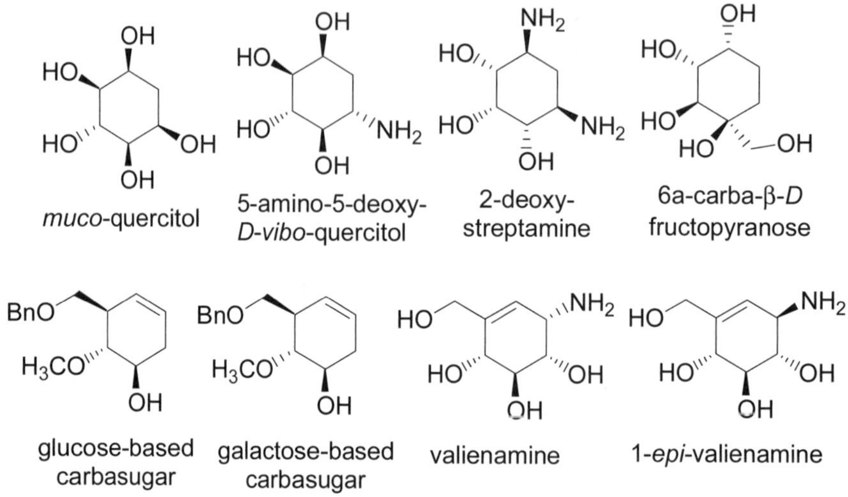

In 2004, Halcomb and co-workers reported the first successful synthesis of an isostere of an *O*-linked glycopeptide in an effort to study how sugar modifications affect protein folding in inflammatory diseases and cancer.[88] Treatment of diene **144** with 10 mol % of **4** in refluxing toluene produced the desired product **145** in 76% yeild as a mixture of diastereomers.

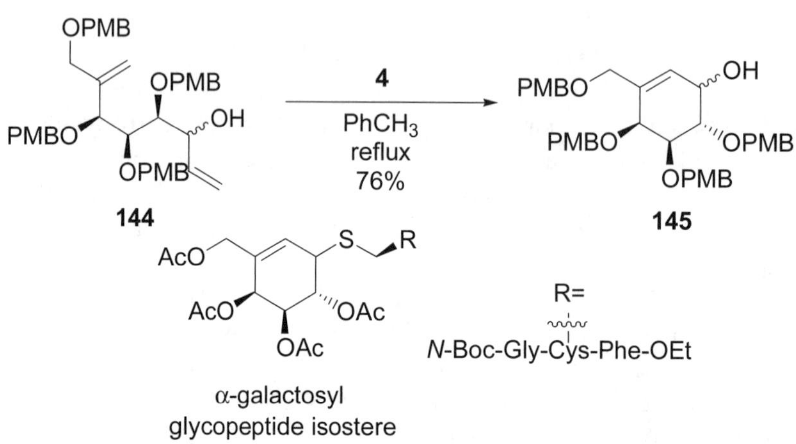

Holt and co-workers used olefin metathesis to synthesize a number of enantiomerically pure annulated carbohydrate systems containing carbocycles, as precursors for the synthesis of taxoids and other natural products.[89] Treatment of dienes **146** and **148** with a catalytic amount of **3** in benzene gave the corresponding 6,6,6-carbocycles, *cis*-**147** and *trans*-**149**, in 89 and 80% yield, respectively. The authors attempted a similar strategy to prepare 6,6,5-annulated systems; however, they achieved little to no yield of the desired products, presumably due to ring strain. Holt and co-workers followed up on their initial report, using a similar strategy to prepare a host of 6,6,6-, 6,6,7-, and 6,6,8-, and 6,6,9-membered ring carbocyclic systems as well as several oxygen containing spirocycles (not shown).[90]

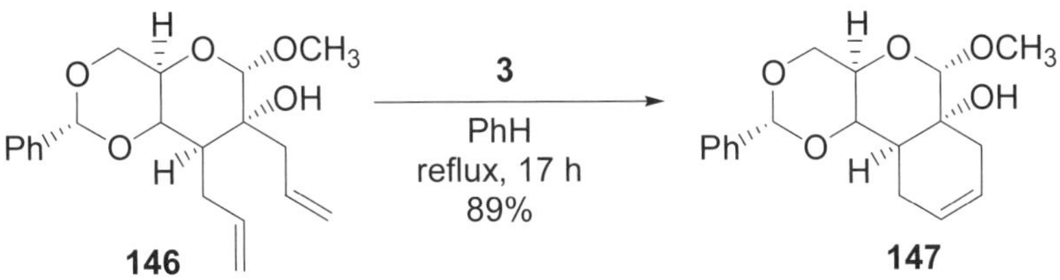

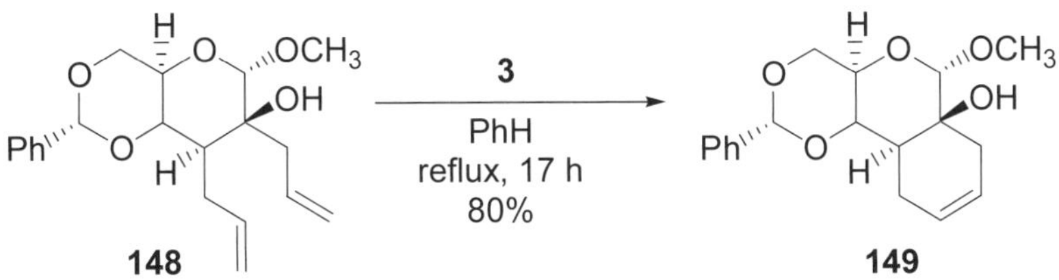

Korinenko and co-workers used ring-closing metathesis as a key step in their synthesis of several analogs of pancratistatin, a potent anticancer natural product.[91] Dienes with varying aryl substituents (**150**) were prepared and subjected to ring-closing metathesis using 3 mol % of **3** in dichloromethane to furnish the corresponding carbocycles in 80–95% yields.

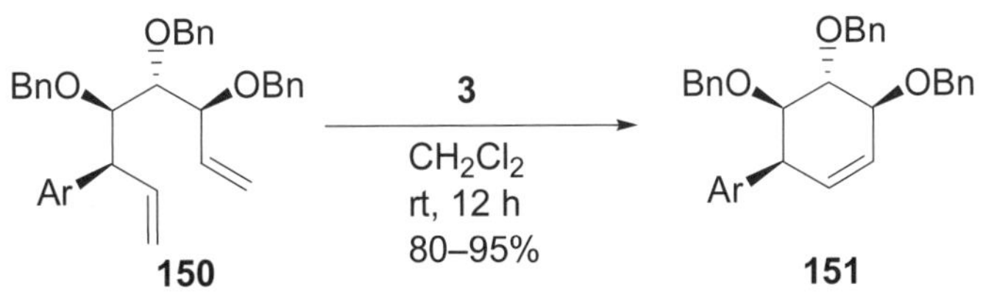

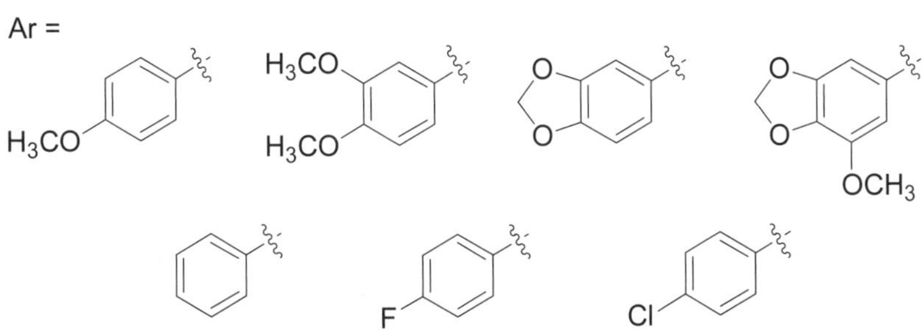

Ring-closing metathesis has also been applied to the synthesis of a number of terpene derivatives containing six-membered rings. The terpene derivatives prepared by this approach have been as simple as cuparene and as complex as tricycloillicinone. Several key examples are illustrated below.

Prasad and co-workers employed ring-closing metathesis as a key step in their synthesis of (±)-cuparene.[92] Diene precursor **152** underwent quantitative conversion to the corresponding carbocycle in refluxing benzene using only 3 mol % catalyst loading of **4**. Dehydration and aromatization of the corresponding tertiary alcohol (not shown) gave the desired product in high yield.

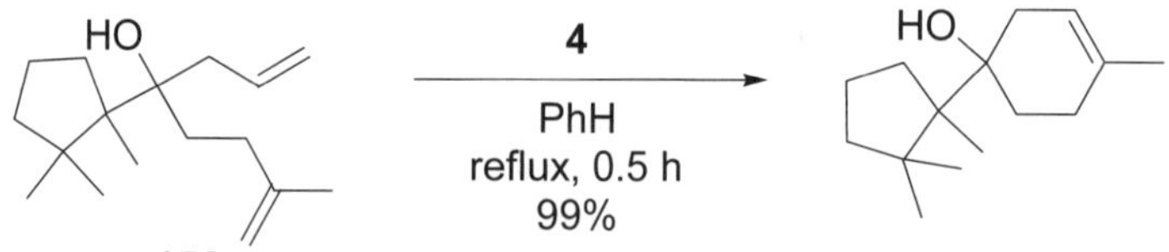

152 **153**

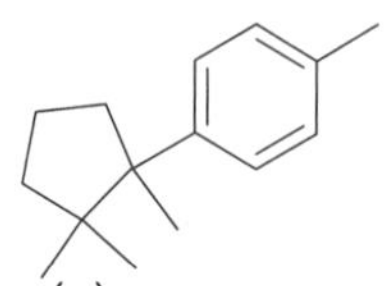

(±)-cuparene

Ring-closing metathesis was also used as a key step in the enantioselective synthesis of (+)-carissone, a eudesmane sesqueterpenoid, by Stoltz and co-workers.[93] Reaction of enone **154** with 3 mol % of **4** in benzene gave the corresponding carbocycle **155** in quantitative yields. Seven additional steps were required to access (+)-carissone (not shown).

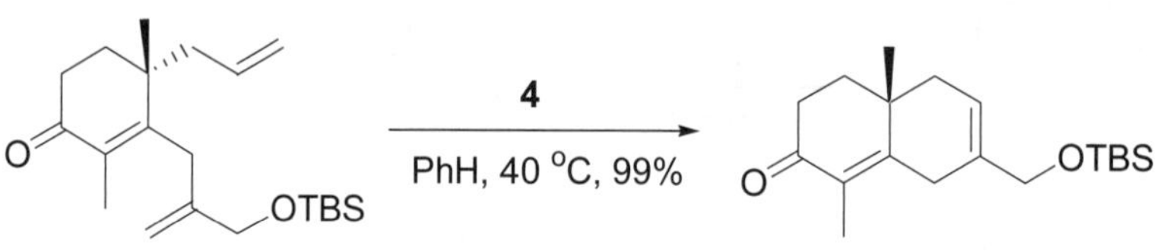

154 **155**

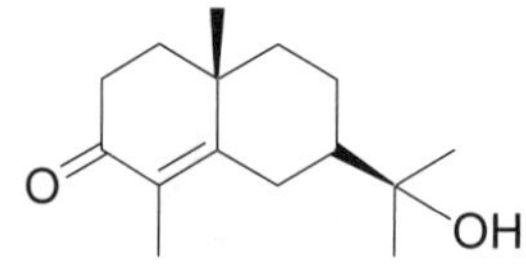

(+)-carissone

Saicic and co-workers used olefin metathesis as one of the final steps in their synthesis of (±)-periplanone C, a C-macrolide pheromone.[94] α-Hydroxy ketoester **156**, when treated with 3 mol% of **3** in dichloromethane at room temperature over the course of 23 h, gave the desired carbocyle **157** in 81% yield. The authors applied this methodology to access a number of other intermediates in the periplanone family.

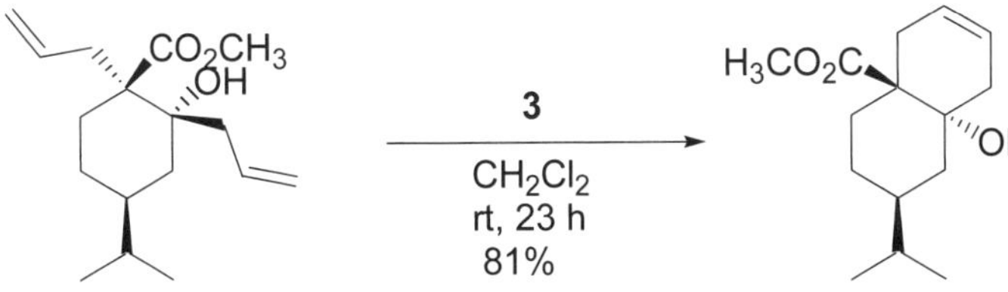

156 157

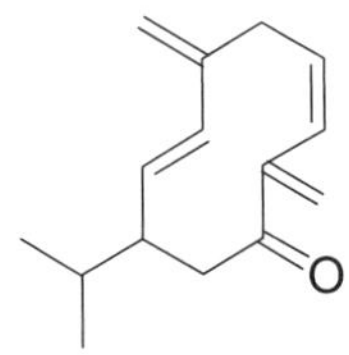

periplanone C

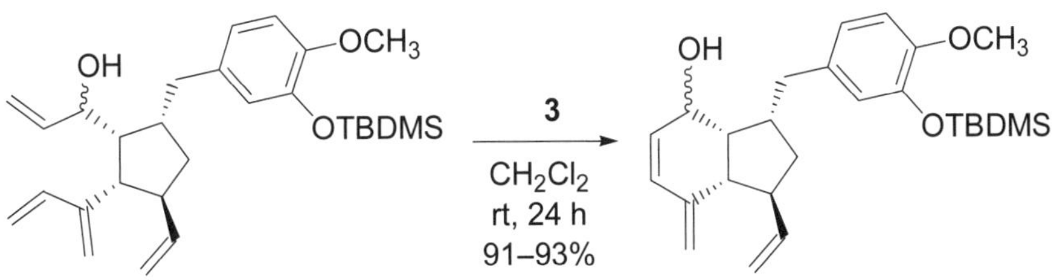

158 159

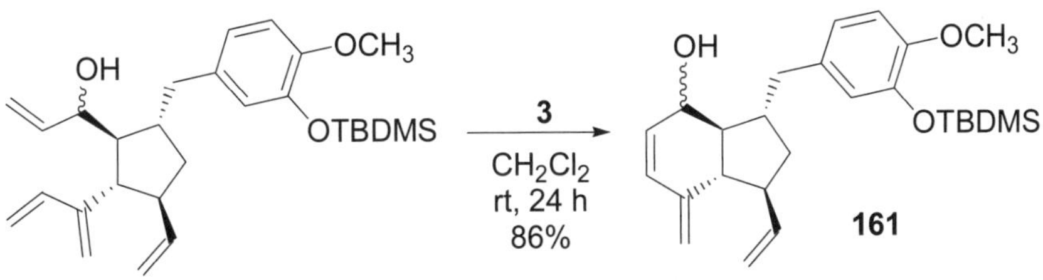

160 161

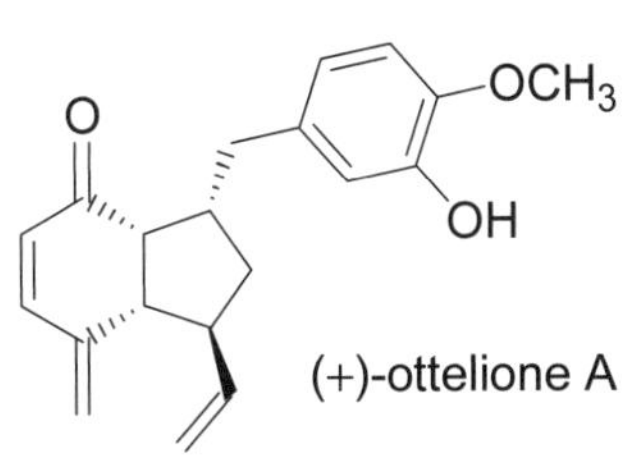

(+)-ottelione A

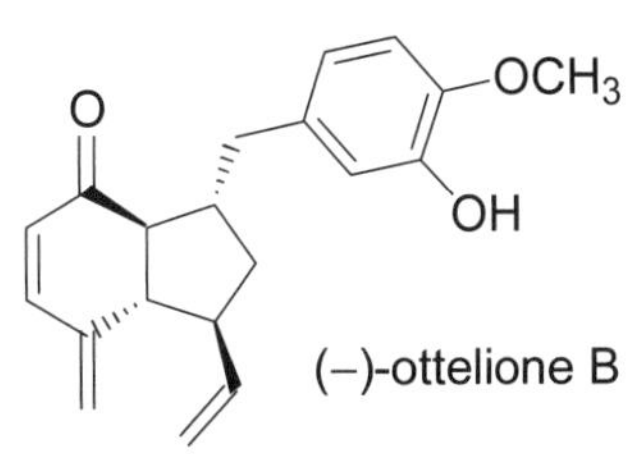

(–)-ottelione B

Clive and Liu used a ring-closing metathesis strategy to access the six-member ring carbocyclic components of the anticancer agent's ottelione A and B.[95] Olefin metathesis of **158** or **160** with catalytic amount of **3** in the presence of dichloromethane provided the corresponding carbocycles **159** and **161** in 93 and 86% yields, respectively. The authors also attempted this reaction using similar reaction conditions with **4**; however, they found no improvement in the overall reaction yields (not shown).

Srikrishna and co-workers employed ring-closing metathesis as a general strategy for the preparation of the 6,6,5-tricyclic core of the elisabethane diterpenes, a family of compounds which exhibit a wide range of biological properties.[96] It is interesting that ring-closing metathesis of 3,4,4-trisallylcarvone **162** in the presence of 5 mol % of **3** in refluxing dichloromethane did not proceed as expected to provide the desired tricycle **163**. However, olefin metathesis of trisallylcarveol **164** proceeded smoothly under similar conditions to give the targeted tricyclic system **165** in 40% yield, highlighting the importance of the allyl alcohol in substrate activation.

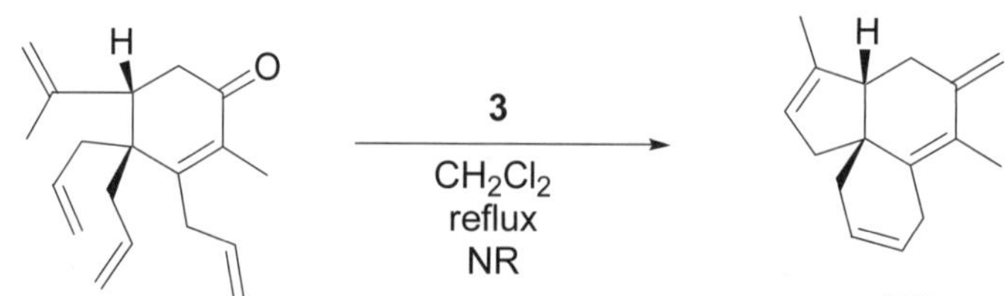

162 **163**

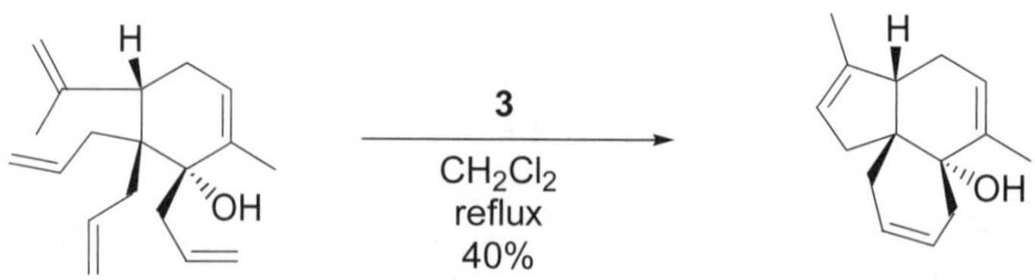

164 **165**

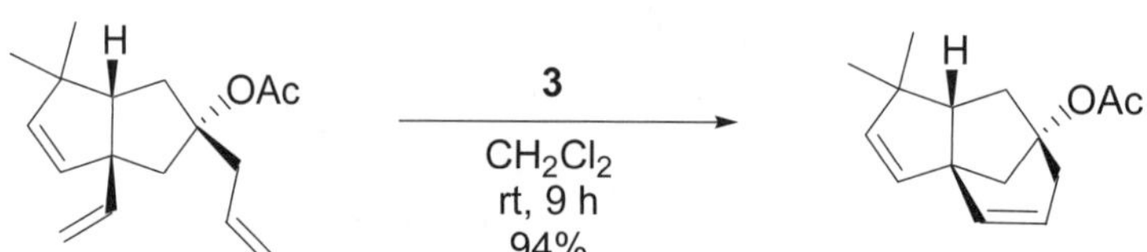

166 **167**

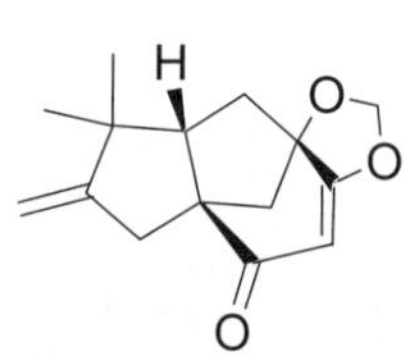

tricycloillicinone

Tricycloillicinone is a novel C6–C3 prenylated compound with an interesting 3,4,4-trimethyltricyclo[5.3.1.0^{1,5}]undecane ring. This compound has recently attracted attention for its ability to increase choline acetyltranferase activity and as such may hold promise as a therapeutic for the treatment of neurological disorders. Recently, Srikrishna and co-workers employed a ring-closing metathesis approach to synthesize the tricyclic carbocyclic core of tricycloillicinone.[97] Reaction of acetate **166** with 10 mol % **3** in dichloromethane provided the corresponding tricyclic acetate **167** in 94% yield. Ongoing efforts by Srikrishna and co-workers are focusing on the transformation of this compound to the desired tricycloillicinone product.

Kim and co-workers applied an olefin metathesis strategy in the total synthesis of (–)-perrottentinene, a biogeneic precursor to of (–)-Δ^1-*trans*-tetrahydrocannabinol.[98] Treatment of diene precursor **168** with a catalytic amount of **4** in the presence of refluxing dichloromethane gave the desired product **169** in 90% yield. Their report signified the first successful example of olefin metathesis using a proximal *o*-phenol group.

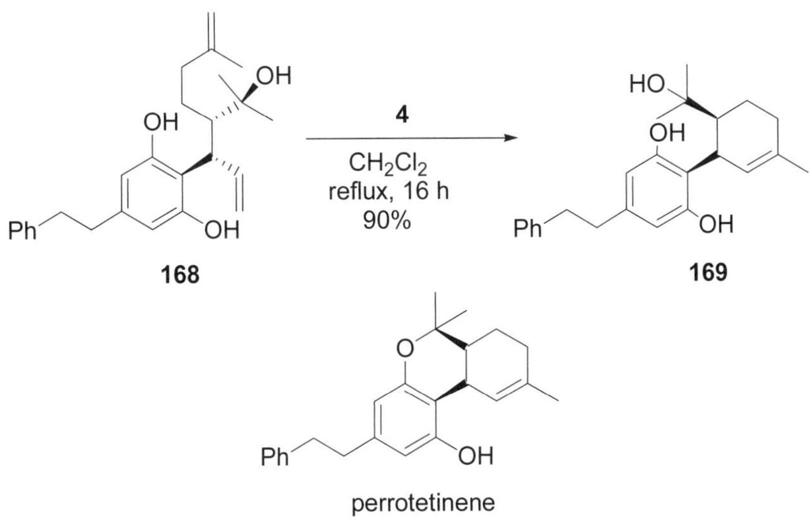

Olefin metathesis was also employed by Trost and Dogra in the synthesis of (–)-Δ^9-*trans*-tetrahydrocannabinol, the psychomimetic component of marijuana.[99] Treatment of **170** as a mixture of diastereomers with a catalytic amount of **4** gave a mixture of the corresponding *anti*-**171** and *syn*-**172** cyclohexenes. Equilibration in sodium methoxide and methanol over the course of 3 days gave almost exclusively the *anti* product. Four additional steps were required to achieve (–)-Δ^9-*trans*-tetrahydrocannabinol.

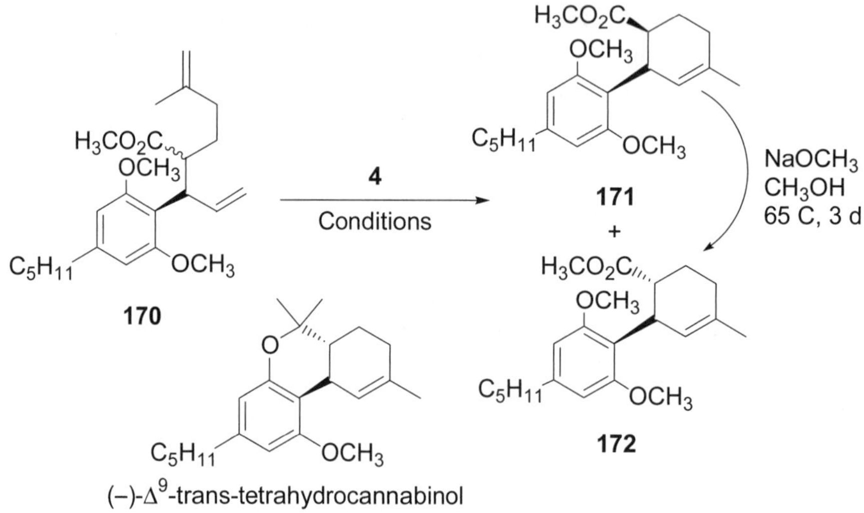

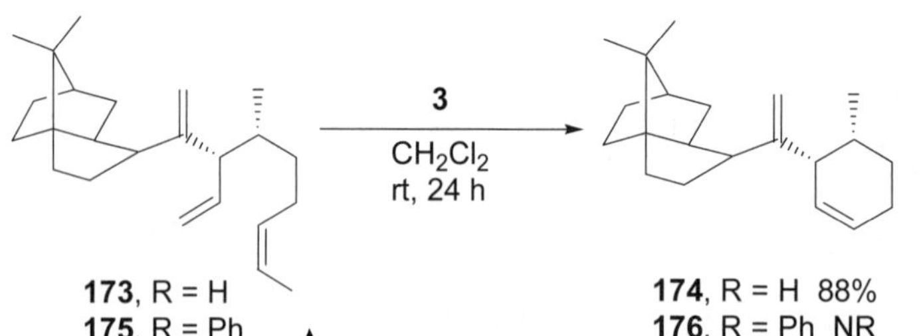

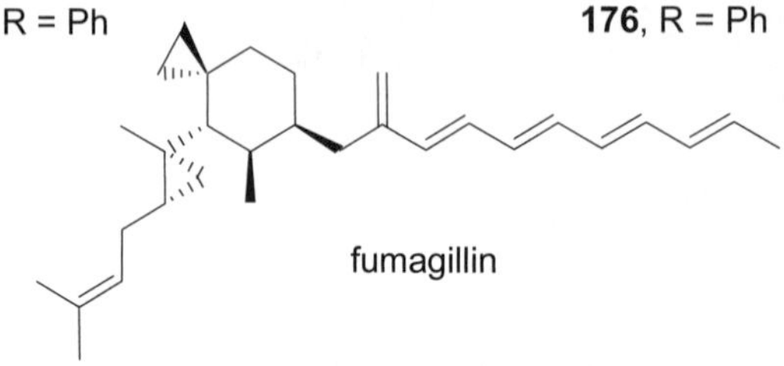

Fumagillin and several related analogs have been shown to inhibit angiogenesis, and as such are of interest as potential treatments for cancer. Watson and co-workers recently employed a ring-closing metathesis strategy using a chiral auxiliary to enantioselectively construct the cyclohexene core of fumagillin.[100] Diene **173**, where R = H, when subjected to ring-closing metathesis using 5 mol % of **3** in dichloromethane, provided the corresponding carbocyclic ring **174** in 88% yield as a single enantiomer. Interestingly, diene **175**, where R = Ph, failed to undergo ring-closing metathesis. The authors did not speculate as to why the reaction failed; however, it is likely that sterics may have played a role.

Spirocycles incorporating six-membered rings are important synthons in organic synthesis. These structural motifs have also been found in a number of natural products. In general, six-membered ring carbocycles can be readily prepared in high yields using olefin metathesis. Several examples of the synthesis of six-membered ring carbocycles are illustrated below.

In 2008, Gals and co-workers reported a general method for the synthesis of asymmetric spiroketals, which are important intermediates in the synthesis of a number of biologically active compounds.[101] α,α-Dienyl dihydropyrans **177** and **179** underwent ring-closing metathesis using a catalytic amount of **4** in dichloromethane to provide the corresponding spirocyclic carbocycles **178** and **180** in high yields. The authors observed that the Lewis basic sulfoximine group had little effect on the ring-closing metathesis reaction.

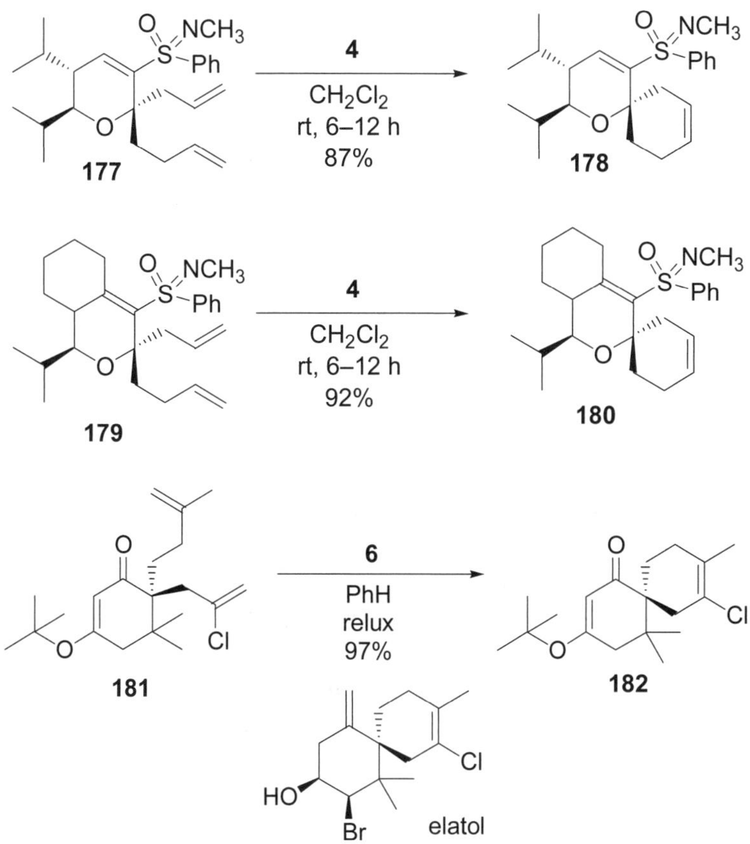

Grubbs, Stoltz, and co-workers used olefin metathesis as a key step in the first catalytic asymmetric total synthesis of elatol.[102] Elatol, a chamigrene sesquiterpene with antibiofouling, antibacterial, antifungal, and cytotoxic properties, is a particular challenge synthetically due to its sterically congested spirocyclic system, exocyclic olefin, vinyl halide and *cis*-halohydrin functionality. Treatment of α,ω-diene **181** with 5 mol % of **6** in benzene gave the corresponding chloroalkene **182** in 97% yield. Four additional steps were required to access the desired natural product.

Kim and and co-workers used ring-closing metathesis to construct the spirocyclic core of (–)-lepadiformine.[103] (–)-Lepadiformine is a tricyclic perhydropyrrolo[2,1-*j*]quinolone that has recently gained attention due to its moderate in vitro tumor cell cytotoxicity and positive cardiovascular effects. Treatment of diene **183** with a catalytic amount of **4** in the presence of refluxing dichloromethane furnished the corresponding azaspiro cyclohexene **184** in near quantitative yield.

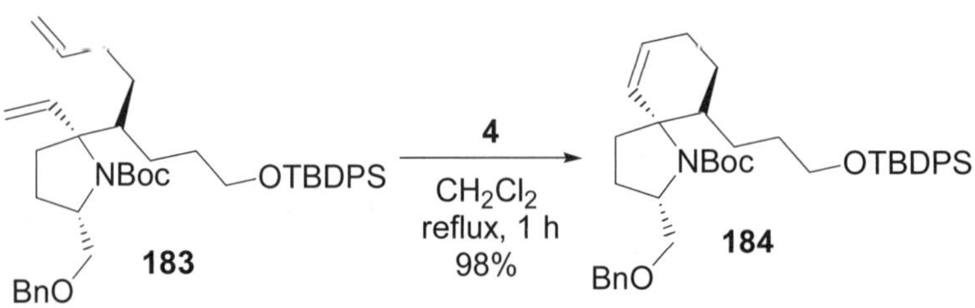

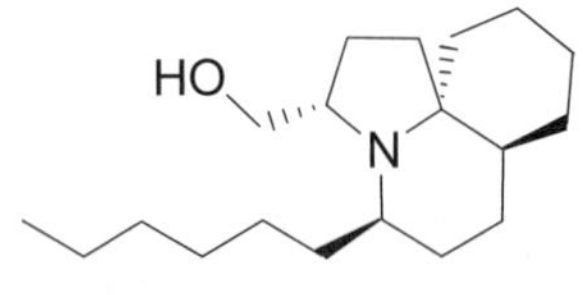

(–)-lepidformine

Martin and co-workers employed a ring-closing metathesis strategy in their total synthesis of the azatricyclic skeleton of FR01483.[104] (–)-FR01483 is an immunosuppressant that has been shown to prolong graft survival time by inhibiting purine nucleotide synthesis. Reaction of **185** as a mixture of dienes with 10 mol % of **4** in dichloromethane gave a separable mixture of the corresponding azaspiro cyclohexene derivatives **186** and **187** in near quantitative yield. Unfortunately, the desired diastereomer, **186**, was produced as the minor product; however, their report constituted an improvement over their previous work on this compound.[105]

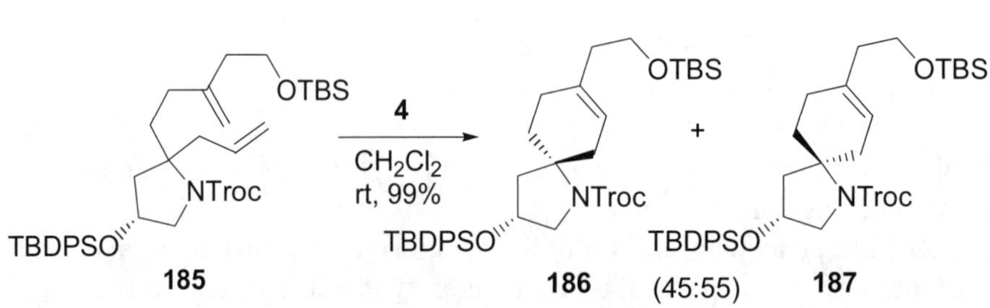

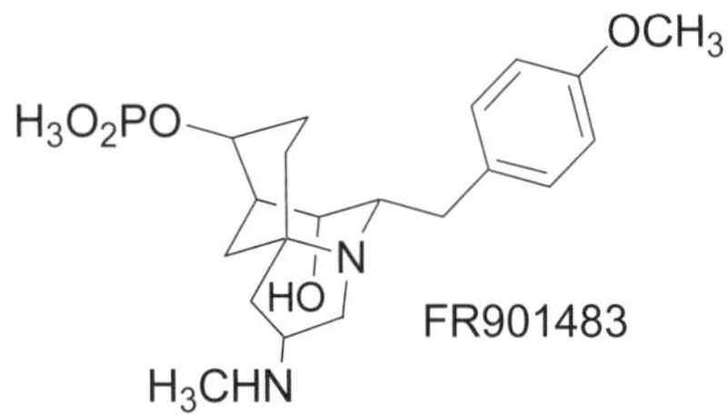

FR901483

Brimble and Trzoss used a double-alkylation/ring-closing metathesis approach to synthesize several spiroimines.[106] Spiroimines are important functional moieties in a number of shellfish toxins suchs as gymnodimine. Lactams **188**, **190**, and **192** were subjected to ring-closing metathesis using 5 mol % of **3** in dichloromethane to furnish the corresponding 5,6-(**189**), 6,6-(**191**), and 7,6-(**193**) spirolactams in upwards of 85% yield.

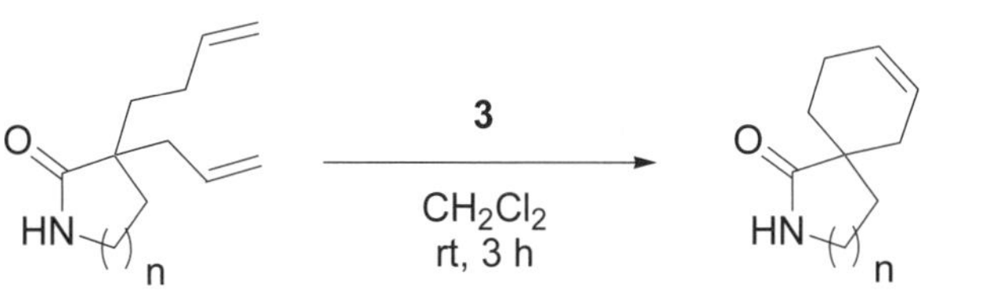

188, n = 1
190, n = 2
192, n = 3

189, n = 1 86%
191, n = 2 90%
193, n = 3 88%

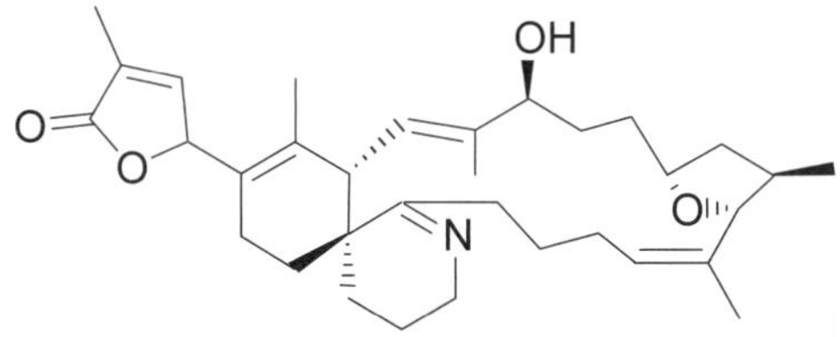

gymnodimine

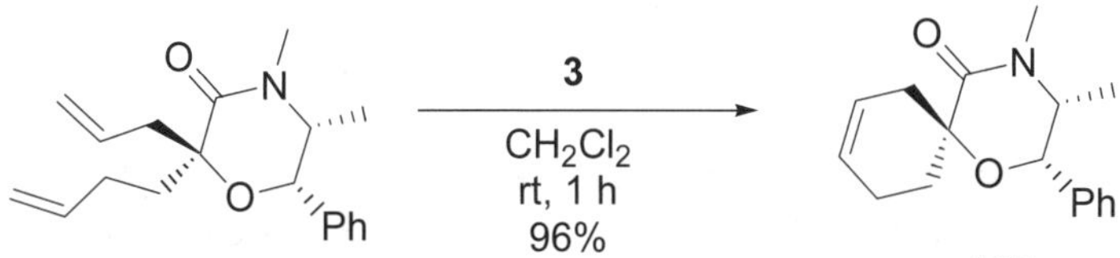

194

195

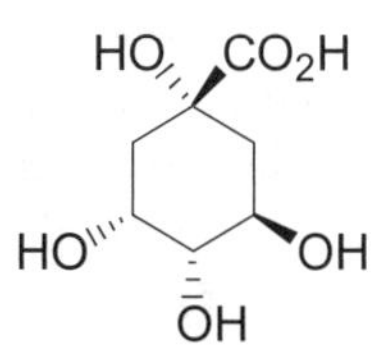

(–)-quinic acid

Pansare and Adsool employed a ring-closing metathesis strategy in their synthesis of (–)-quinic acid.[107] Quinic acid is an important regulator in the biosynthesis of aromatic compounds via the shikimic pathway, and may serve as a potential antifungal, antibacterial and antiparasitic. Ring-closing metathesis of diene **194** using 7 mol % of **3** in dichloromethane gave the corresponding spiro-morpholinone **195** in 96% yield.

Hughes and co-workers extended their ring-closing metathesis strategy for the synthesis of enantiopure five-membered ring spirocylcles using zizaene as a chiral auxiliary to the synthesis of enantiopure six-membered ring spirocarbocycles.[67] Diene precursor **196** required 15 mol % catalyst loading of **3** in refluxing dichloromethane to achieve a near quantitative yield of the desired spirocycle **197**.

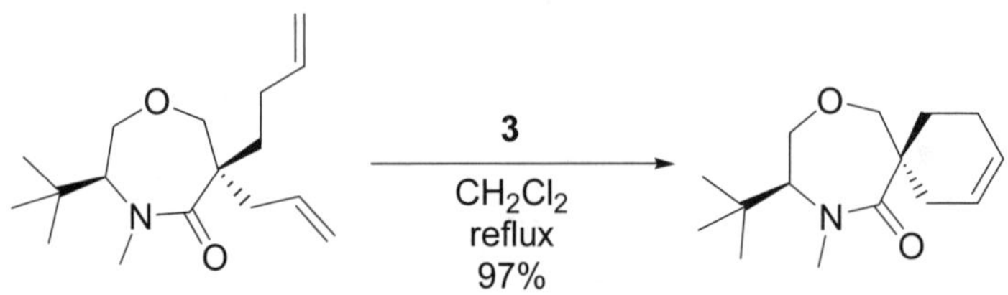

196 **197**

Undheim and co-workers applied their olefin metathesis strategy for the synthesis of five-membered ring spirocyclic carbocycles to synthesize six-membered ring spirocyclic carbocycles for use as templates in natural product synthesis.[68] Reaction of **198** with 2 mol % of **3** furnished the corresponding spirocycle **199** in greater than 95% yield.

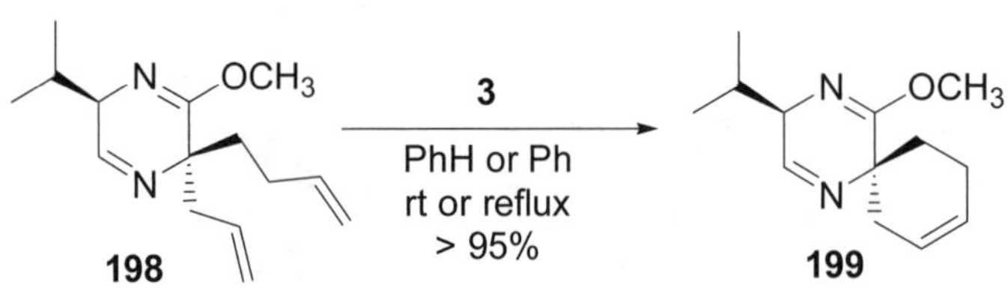

198 **199**

Frejd and co-workers used a ring-closing metathesis strategy as the key transformation in their synthesis of several novel spiro-cyclohexene bicyclo[2.2.2]octane derivatives.[108] Spiro-cyclohexene bicyclo[2.2.2]octane derivatives are rare molecular frameworks, with only a few examples reported to date. Olefin metathesis of diastereomeric homoallylic alcohols **200** and **202** using 10 mol % of **3** in refluxing dichloromethane provided the corresponding spirocyclic carbocycles **201** and **203** in 84 and 85% yields, respectively.

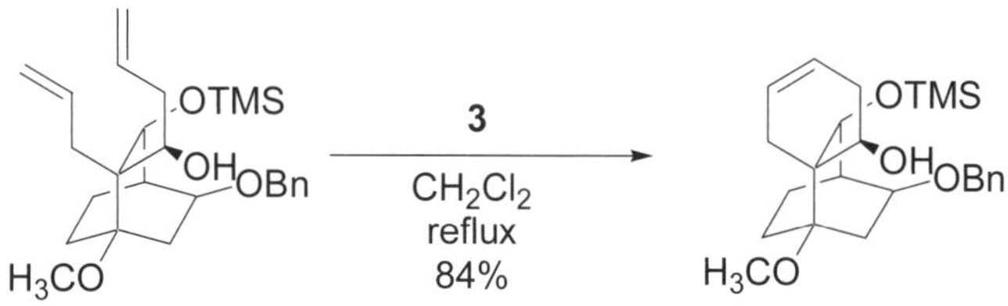

200 201

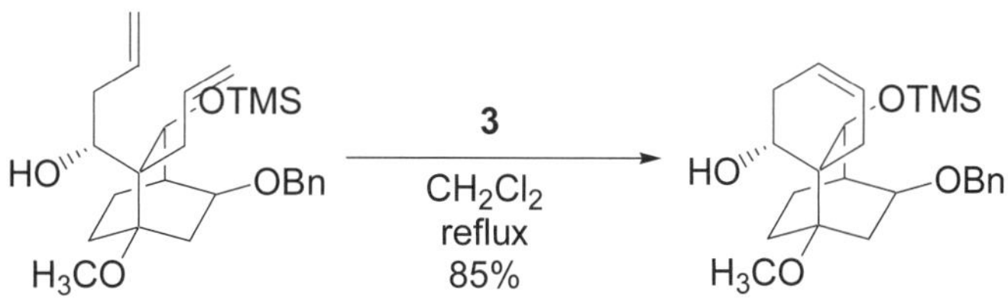

202 203

Singh and co-workers recently reported a general method for the synthesis of embellished spiro-fused bicyclo[2.2.2]octane systems using a Diels–Alder cycloaddition/ring-closing metathesis route.[109] These systems, which are key intermediates in the synthesis of atisane diterpenoides such as serofenic acid A, have been shown to possess neuroprotective activity. Treatment of allyl alcohol **204** with 10 mol % of **3** in dichloromethane provided the corresponding spirocyclic carbocycle **205** in 86% yield. The authors also prepared seven- and eight-membered ring analogs of **205**.

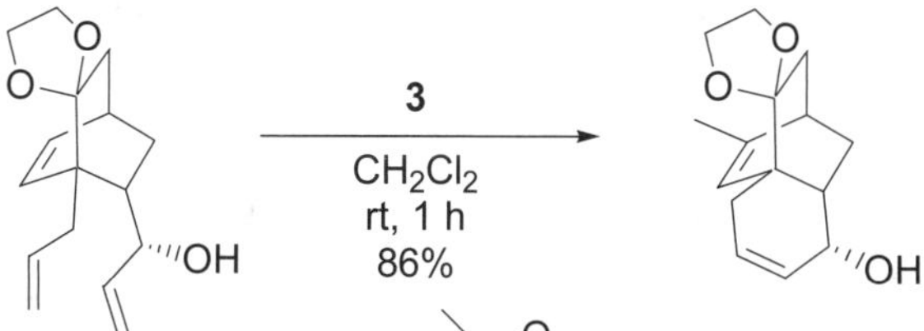

204 205

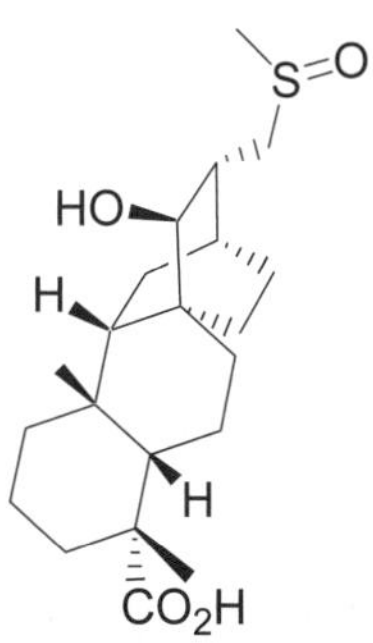

serofenic acid A

Dysinosin A is a highly oxygenated novel inhibitor of thrombin and factor VIIa. Hanessian and co-workers employed a ring-closing metathesis strategy in their preparation of the 6,5-fused core of dysinosin A.[110] Olefin metathesis of **206** and **208** using only 1 mol % of **4** gave quantitative yields of the desired carbocycles **207** and **209**. Several additional steps were required for the construction of the final target.

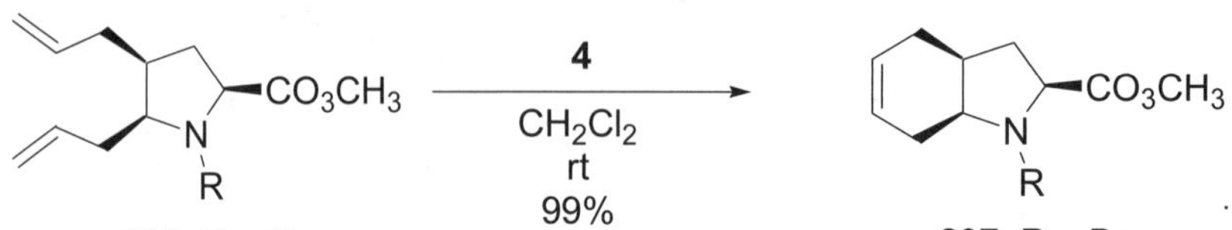

206, R = Boc
208, R = Cbz

207, R = Boc
209, R = Cbz

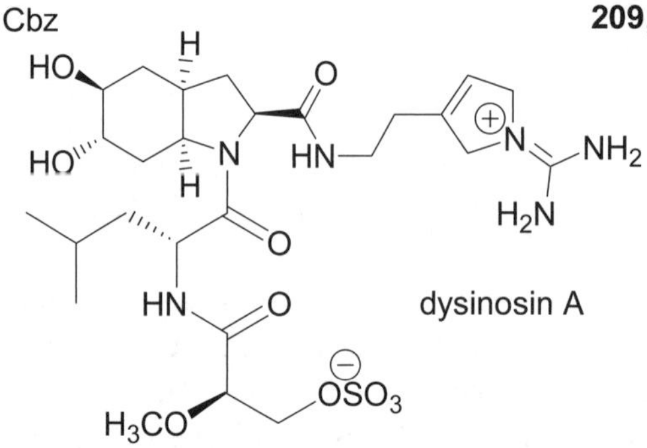

dysinosin A

A similar strategy was employed by Robichaud and Termblay in the enantioselective synthesis of (+)-compactin, a hydroxymethylglutaryl coenzyme A (HMG-CoA) reductase inhibitor.[111] Olefin metathesis of trienes **210** and **212** in the presence of a catalytic amount of **4**, provided the corresponding conjugated dienes **211** and **213** in 86% and 71% yield, respectively.

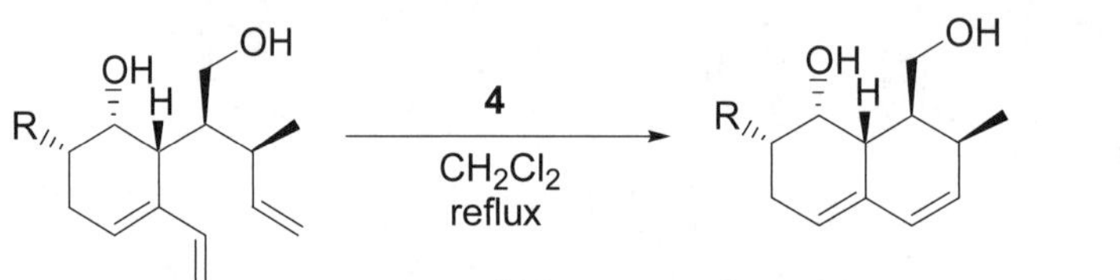

210, R = TMS
212, R = BDMS

211, R = TMS 86%
213, R = BDMS 71%

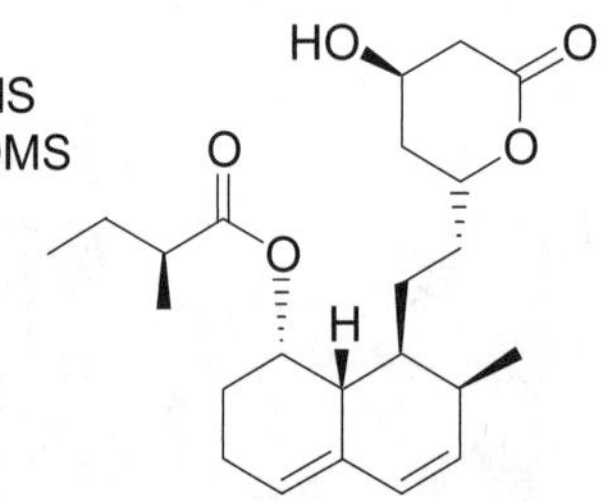

(+)-compactin

Enev and co-workers employed olefin methathesis in their approach to the *cis*-decalin core of branimycin, an antibiotic.[112] Until their report in 2008, preparations of branimycin almost exclusively employed the use of

Diels–Alder reactions to construct the *cis*-decalin system. Metathesis of triene **214** in the presence of 5 mol % of **4** in toluene gave the corresponding *cis*-decalin system **215** in quantitative yields. The authors reported on the ring-closing metathesis of several additional trienes differing only in the stereochemistry of the groups with similar success.

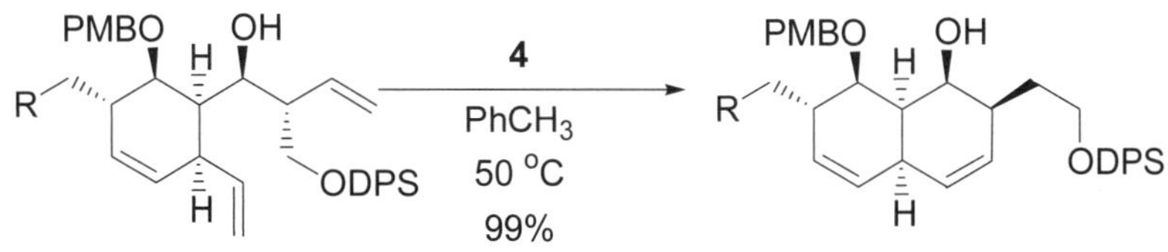

$R=Si(CH_3)_2Ph$

214 **215**

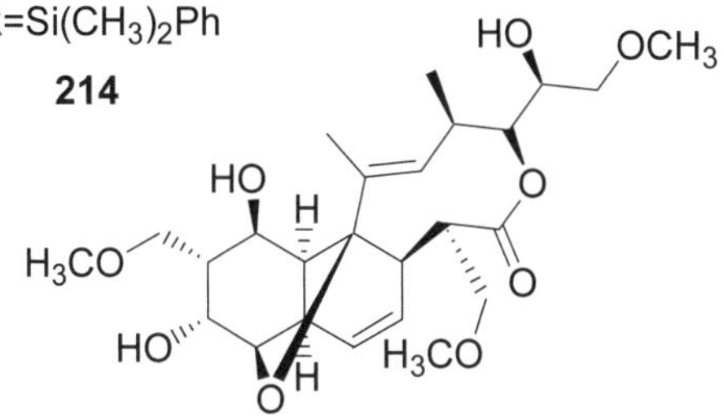

branimycin

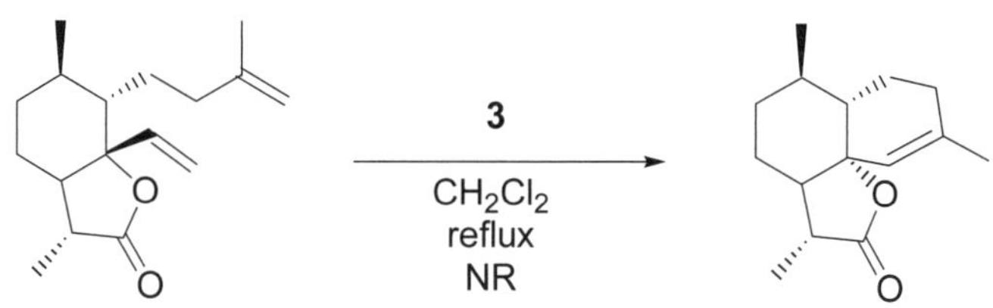

216 **217**

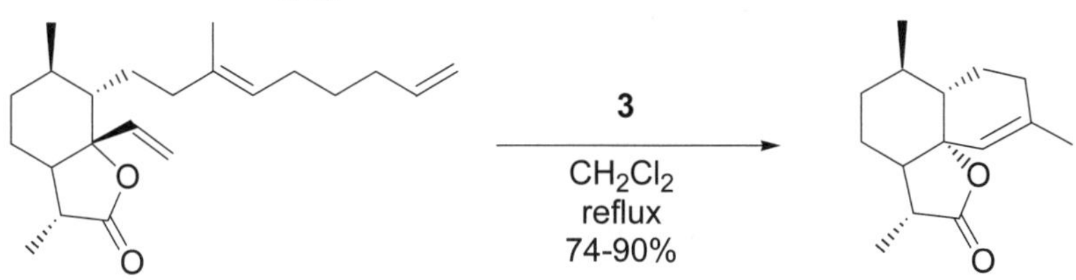

218 **219**

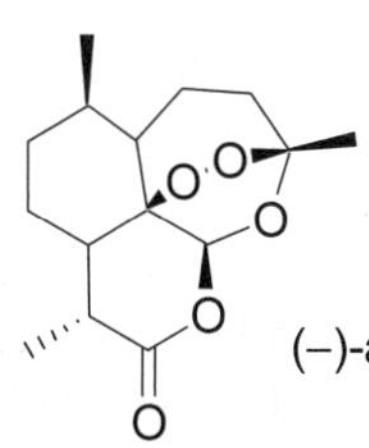

(–)-artemisinin

Dudley and co-workers used a relay ring-closing metathesis strategy in the synthesis of (+)-dihydro-*epi*-deoxyarteannuin B, a key biogenic precursor in the synthesis of artemisinin.[113] Initially, they attempted olefin metathesis of diene **216** using a catalytic amount of **3** in refluxing dichloromethane; however, the expected carbocylic product **217** was not obtained. The authors speculated that steric congestion might be the cause of the failed cyclization. Exentsion of **216** to provide **218** followed by relay ring-closing metathesis using a catalytic amount of **3** under similar conditions gave the desired tricyclic system **219** in 74–90% yield (multiple attempts).

Castle and co-workers used a ring-closing metathesis strategy in their synthesis of the carbocyclic core of (±)-hasubanonine, a member of the hasubanan alkaloid family.[114] Hasubanan alkaloids are of interest due to their structural similarity to morphine alkaloids. Treatment of diene **220** with a catalytic amount of **4** in refluxing dichloromethane furnished the desired phenanthrene core **221** in quantitative yield. Six additional steps were required to obtain the desired natural product.

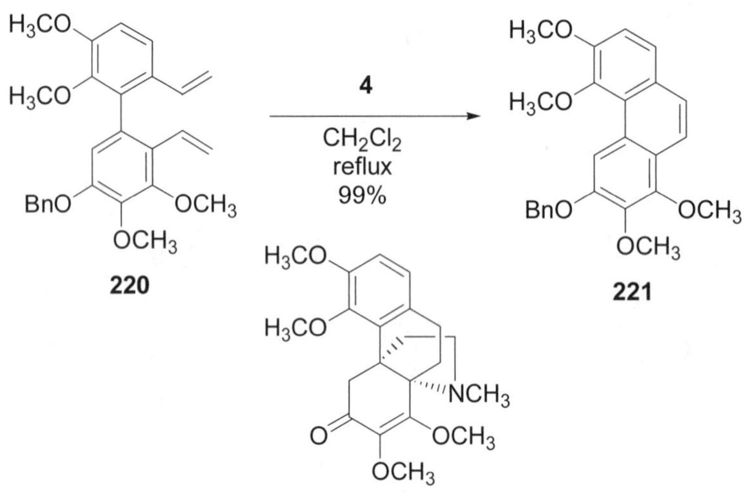

220 **221**

(±)-hasubanonine

Wipf and co-workers employed olefin metathesis in their synthesis of (±)-thiohalenaquinone.[115] (±)-Thiohalenaquinone is of particular interest due to its complex carbocyclic core and wide range of biological activities. In particular, it has been shown to inhibit protein tyrosine kinase, phosphatidylinositol 3 kinase, and Cdc25B dual specificity phosphatase. Ruthenium catalyzed bond migration, followed by ring-closing metathesis of diene **222** using 20 mol% of **6** in refluxing dichloromethane, provided the corresponding pentacyclic core **223** in 56% yield over two steps. Oxidation of **223** provided (±)-thiohalenaquinone (not shown).

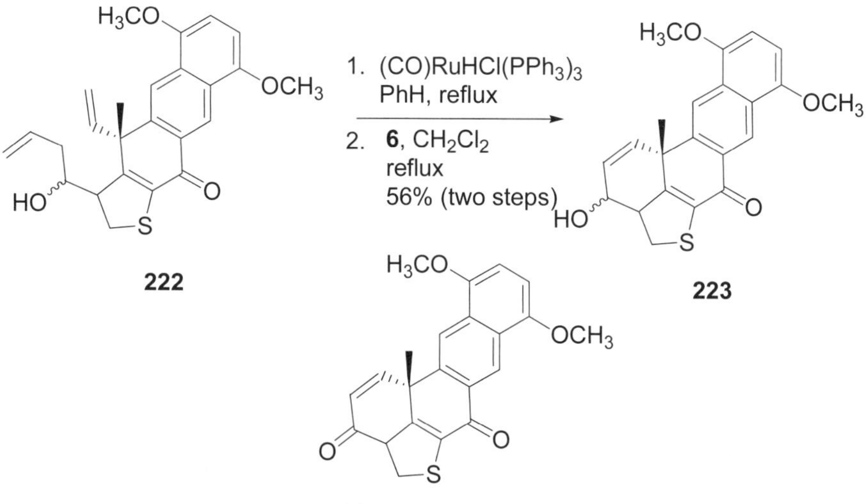

222 223 (±)-thiohalenquinone

Smith and co-workers employed ring-closing metathesis to synthesize a host of benzoporphyrins.[116] Benzoporphyrins are particularly attractive in medicine as agents for photodynamic therapy, and in industry for use as electro-optic materials. Treatment of diene **224** with a catalytic amount of **4** in dichloromethane provided the corresponding porphyrin analog **225** in good yield. Oxidation of **225** with DDQ (not shown) furnished the desired benzoporphyrin in nearly quantitative yield. The authors reported on the synthesis of several additional monosubstituted and trisubstituted derivatives as part of their study.

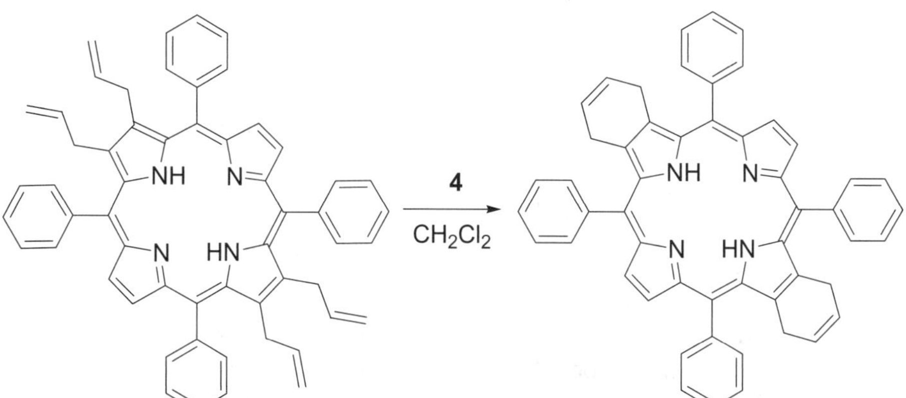

224 225

Seven-Membered Rings

Ring-closing metathesis has been most useful in the synthesis of larger carbocycles, such as seven-membered rings. Seven-membered ring carbocycles are found in a number of biologically active natural products, but have proven difficult to prepare in high yields using standard ring forming strategies. The development of ring-closing metathesis as a tool for the construction of these systems has provided for the synthesis of a host of structurally diverse compounds and natural products incorporating seven-membered ring systems, often in high yield. The following section highlights a number of these examples.

Shibasaki and co-workers used a ring-closing metathesis approach to prepare seven-membered rings from electron-deficient olefins.[37] Reaction of acyclic enol ethers **226** and **228** with 7 mol % of **3** in benzene provided cyclic enol ethers **227** and **229** in 88 and 93% yield, respectively.

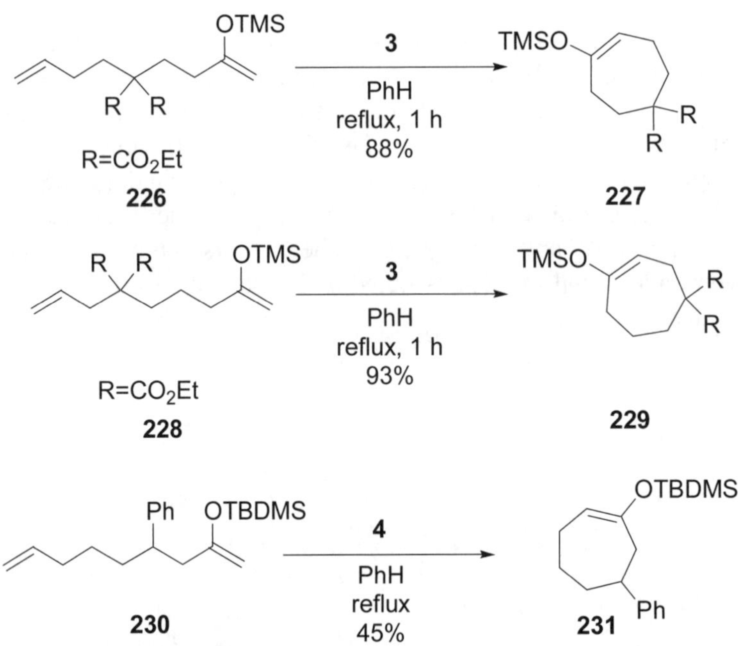

Aggarawa and co-workers applied their strategy for the synthesis of five- and six- membered ring carbocyclic silyl enol ethers to the the synthesis of seven-membered ring carbocyclic silyl enol ether bearing a phenyl substituent.[38] Substrate **230** required upward of 20 mol % of **4** in refluxing benzene to achieve modest yields of the desired product **231**.

Paley and co-workers used enantiopure η^4-(1-sulfinyldiene)iron(0) tricarbonyl complexes as templates for the enantioselective construction of carbocyles via a ring-closing metathesis strategy.[117] Enantiopure homoallyl alcohol adduct **232**, when treated with 8 mol % of **3** in toluene, gave the corresponding seven-membered ring carbocyle **233** as a single enantiomer in 90% yield. This strategy was also applied to form eight- and nine-membered ring carbocycles with similar success.

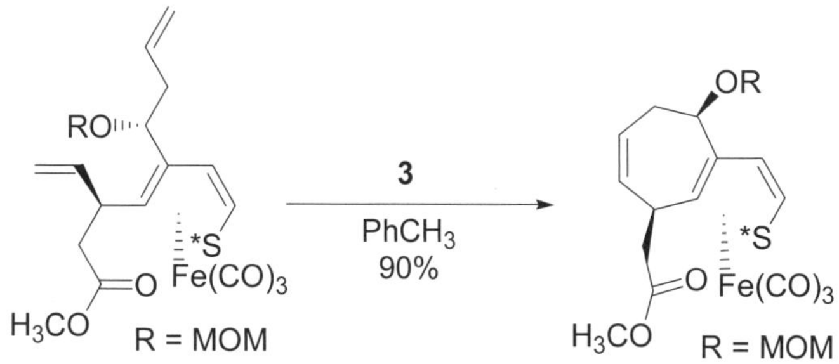

232 **233**

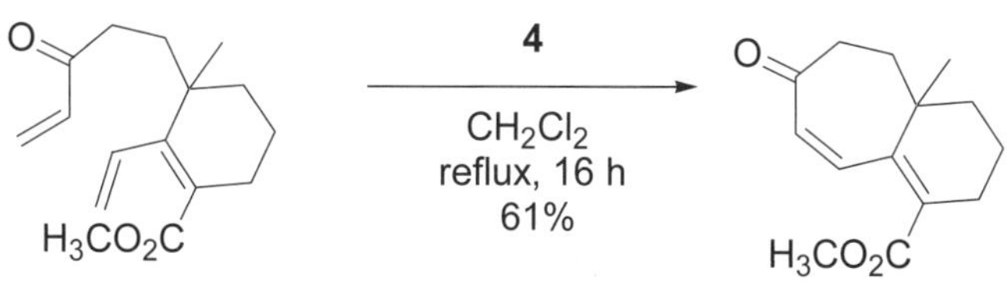

234 **235**

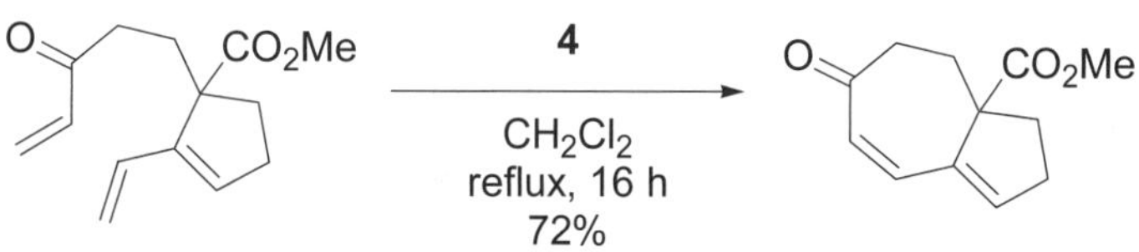

236 **237**

Funk and co-workers employed a ring-closing metathesis strategy in their synthesis of fused haloethyl vinyl ketones.[118] These compounds are important synthetic intermediates in the production of a number of natural and unnatural products. Treatment of the five (**234**)- or six (**236**)-membered enone with a catalytic amount of **4** in the presence of refluxing dichloromethane gave the corresponding vinyl ketones **235** and **237** in 61% and 72% yield, respectively. The reported yields were based on a two step process with the first step, a retrocycloaddition of the dioxin precursor catalyzed by $ZnCl_2$, giving rise to the enone (not shown).

Martin and co-workers applied their ring-closing metathesis strategy for the preparation of 5,6-fused-γ-butyrolactones to 5,7-fused-γ-butyrolactones.[77] Diene precursors **238** and **240** were subjected to ring-closing metathesis using 10 mol % **4** in refluxing dichloromethane to produce the corresponding α,β-fused γ-lactones **239** and **241** in 85 and 89% yield, respectively.

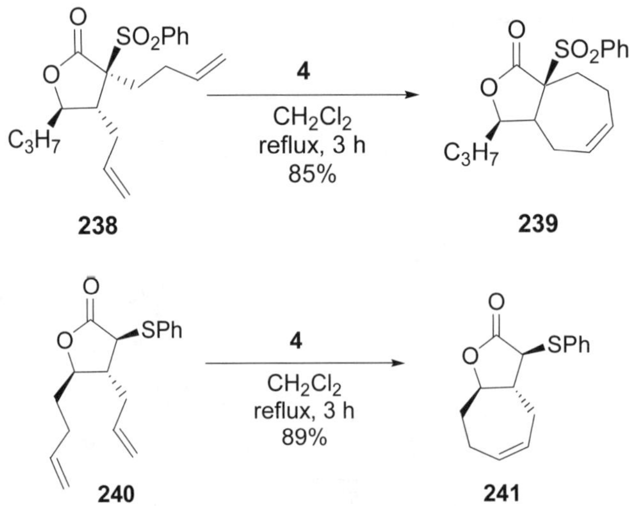

A number of research groups have used ring-closing metathesis to prepare α- and β-amino acids constrained by seven-membered ring carbocycles. Several examples of the synthesis of these interesting molecules are provided in the following paragraphs.

Kotha and co-workers used an olefin metathesis strategy in their synthesis of a conformationally constrained seven-membered ring carbocyclic α-amino acid derivative.[43] Treatment of **242** with 10 mol % of **3** in refluxing toluene provided the corresponding carbocycle **243** in 69% yield.

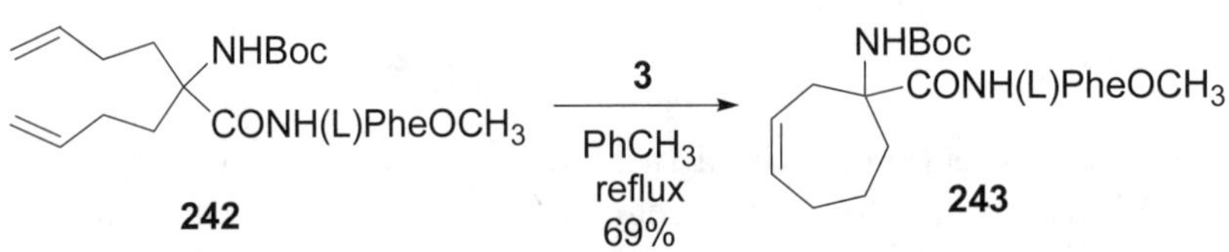

Abell and co-workers applied their ring-closing metathesis strategy for the synthesis of β-amino acids constrained by five- and six-membered rings to synthesize a seven-membered ring β-amino acid for incorporation into β-peptide mimetics.[46] Treatment of diene **244** with a 5 mol % of **4** in benzene provided the corresponding β-amino acid derivative **245** in 96% yield.

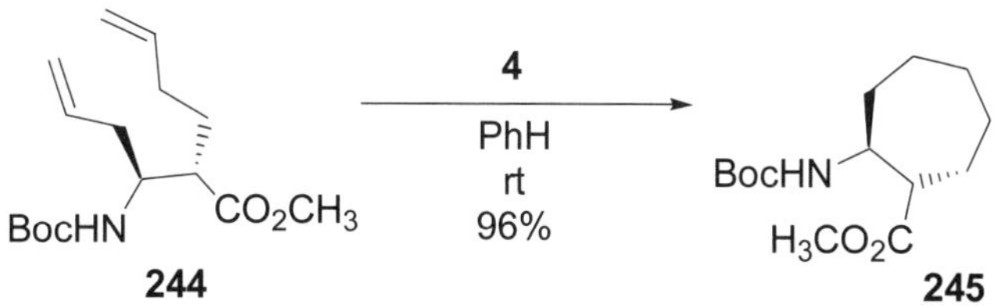

Davies and co-workers also employed a ring-closing metathesis strategy in their preparation of a constrained seven-membered ring β-amino acid derivative.[119] Treatment of lactam precursor **246** with 4 mol % of **3** in refluxing dichloromethane gave the corresponding carbocyclic β-amino acid derivative **247** in 93% yield with greater than 95% diastereoselectivity.

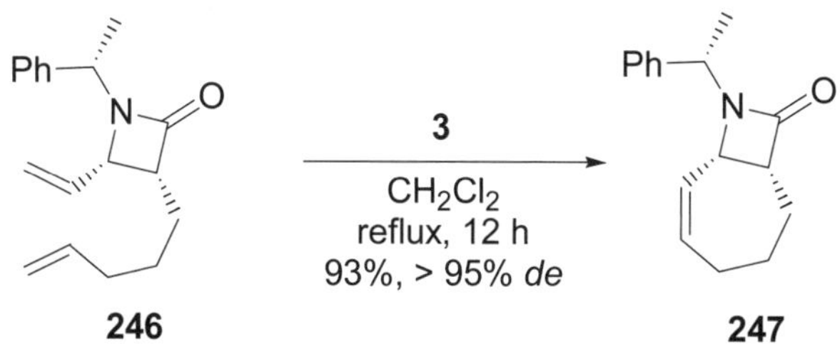

A number of seven-membered ring carbacyles have also been prepared using ring-closing metathesis as a key strategy. Many of these carbacycles are natural products or important synthons for the production of natural products. Several key examples are highlighted below.

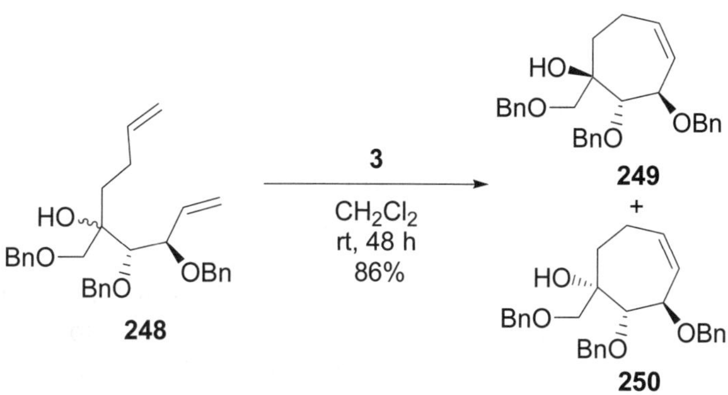

Zhang and co-workers used an olefin metathesis stragtegy in their preparation of 8-oxa-bicyclo[3.2.1]octane derivatives from D-arabinose.[120] Treatment of diene **248**, prepared in three steps from D-arabinose, with a catalytic amount of **3** in dichloromethane provided the corresponding cycloheptene products **249** and **250** in 86% yield as a mixture of diastereomers.

In 2000, Hanna and co-workers employed a ring-closing metathesis strategy for the preparation of seven-membered ring carbasugars from methyl 6-deoxy-6-iodo-3,4-isopropylidene-2-*O*-(*tert*-butyldimethyl-silyl)-D-galactopyranoside and methyl 6-deoxy-6-iodo-1,2:3,4-di-*O*-isopropylidene-D-galactopyranoside in high yields (not shown).[121] Later, in 2001, Hanna and Boyer used this strategy to synthesize (+)-calystegine B_2, a member of a family of compounds known for their nutritional mediation properties in the plant rhizoshpere.[122] Ring-closing metathesis of **251** in the presence of 8 mol % of **3** in dichloromethane furnished the corresponding carbasugar in near quantitative yield. Oxidation, followed by hydrogenolysis and deprotection (not shown) led to the production of the desired product. Madsen and Skaanderup applied a similar strategy in their 2001 synthesis of (+)-calystegine B_2 (not shown).[123]

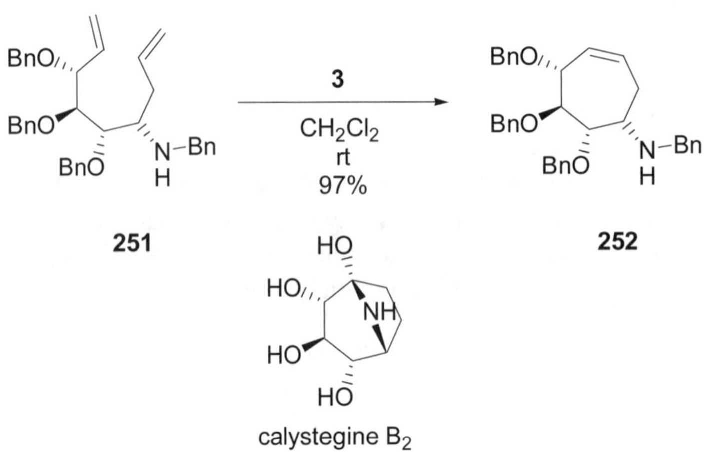

In 2002, Marco-Contelles and Opazo[124] used an alternative ring-closing metathesis strategy in an effort to improve on the 2001 syntheses of (+)-calystegine B_2, with limited success. Acetate **253** was subjected to ring-closing metathesis using 10 mol % of **3** in dichloromethane to furnish the corresponding carbacycle **254** in only 8% yield. Reaction of ketone **255** under similar conditions using $Ti(Oi\text{-}Pr)_4$ as a promoter, led to the production of the desired carbacycle **256** in only 4% yield. The authors hypothesized that the low yields were due to steric interactions between the catalyst and the functional groups on the carbacycle.

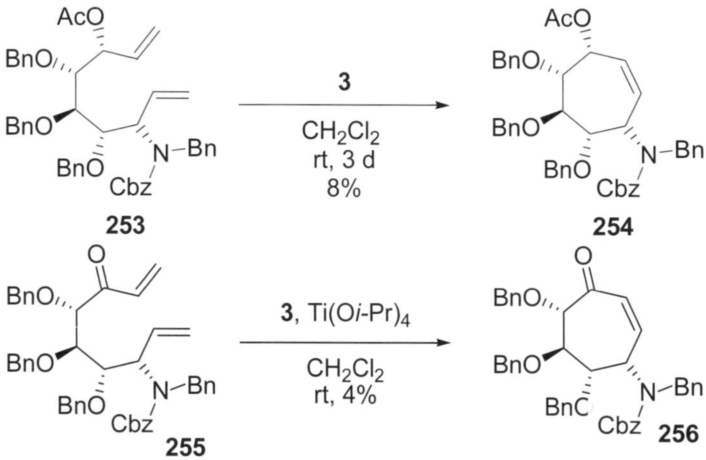

Csuk and co-workers employed a ring-closing metathesis strategy in their total synthesis of calystegine A_7, another member of the calystegine family.[125] Olefin metathesis of diene **257** using 10 mol % of **3** in dichloromethane gave the desired cycloheptene derivative **258** in 80% yield.

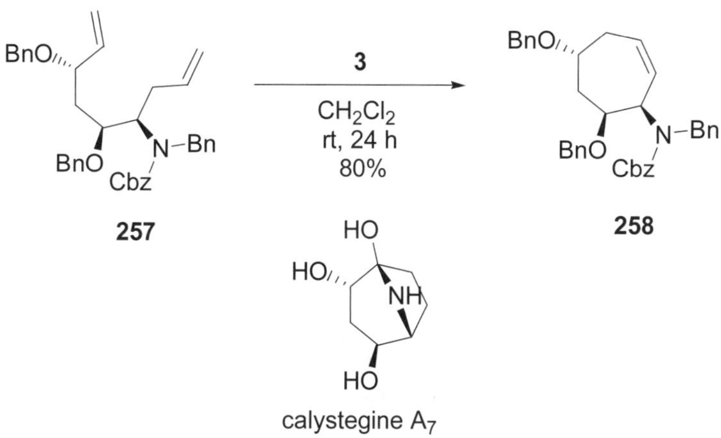

Holt and co-workers applied their ring-closing metathesis strategy for the synthesis of annulated carbohydrate systems contining six-membered rings to the synthesis of enantiomerically pure annulated carbohydrate systems containing seven-membered ring carbocycles.[90] Treatment of dienes **259** and **261** with a catalytic amount of **3** in refluxing benzene gave the corresponding 6,6,7-carbocycles, *cis*-**260** and *trans*-**262**, in 89% and 80% yield, respectively.

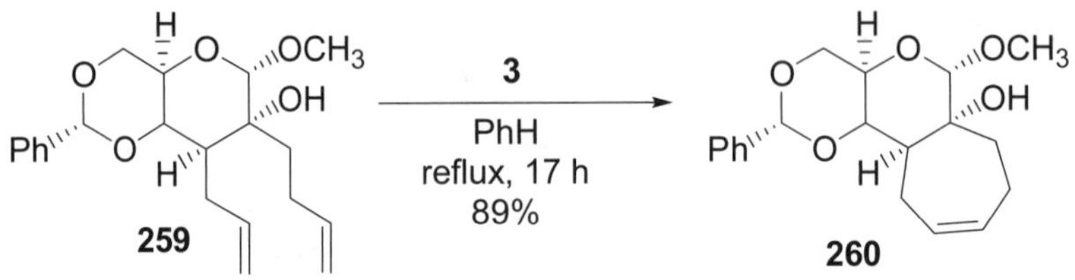

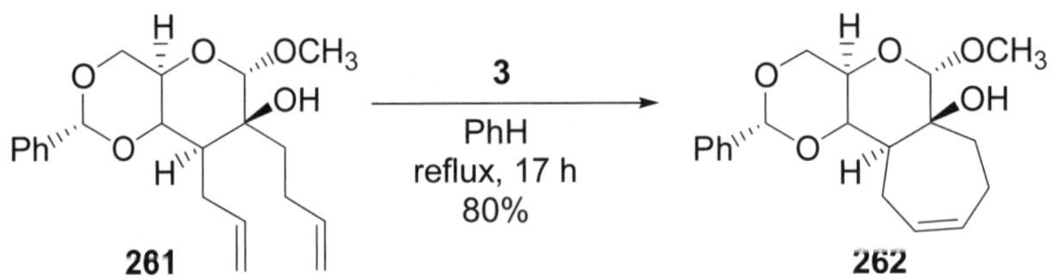

A number of terpenes and terpenoids bearing seven-membered rings have also been synthesized using an olefin metathesis strategy. Several examples are highlighted below.

Ring-closing metathesis was employed by Shishedo and co-workers in their total synthesis of (+)-sundiversifolide, a herbicide.[126] Treatment of enone **263** with 5 mol % of **4** in refluxing dichloromethane furnished the corresponding α/β unsaturated ketone **264** in 95% yield.

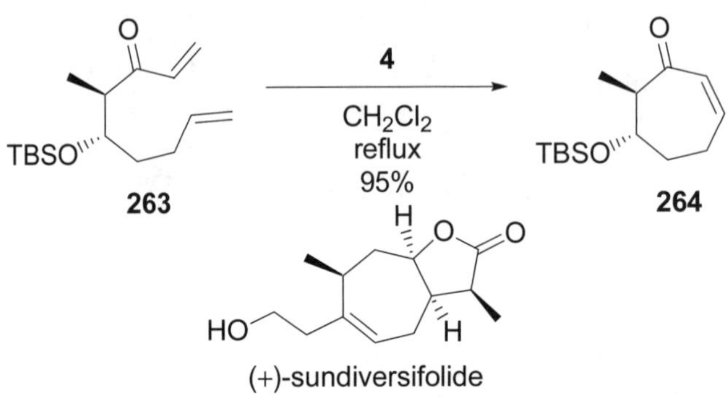

Nosse and co-workers used ring-closing metathesis to synthesize a 5,7,5-fused lactone.[127] These tricyclic frameworks are the core components of the thapsigargin family, a family of sesquiterpene lactones with the ability to restore apoptotic function in cancer cell lines. Thapsigargins are currently under investigation as potential therapeutics for the treatment of prostate cancer. Reaction of lactone **265** with 10 mol % of **4** and a catalytic amount of TBAF in toluene while irradiating at 300 W gave the corresponding tricyclic system **266** in near quantitative. The authors noted that nitrogen sparging to remove ethylene was required to force the equilibrium of the reaction towards the product. Nosse's work presented a slight improvement over Ley's 2003 report using a similar approach to construct the same tricyclic core (not shown).[128]

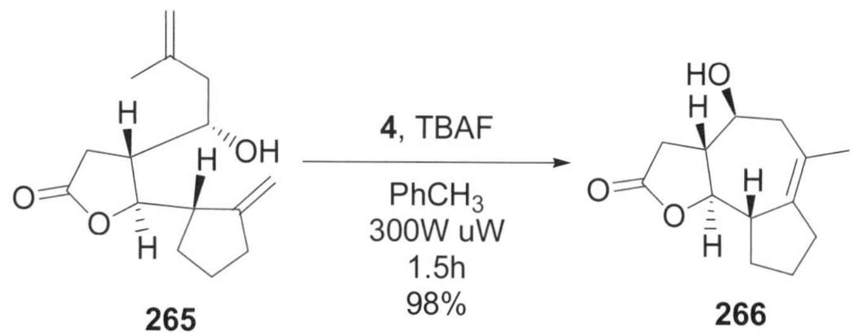

265 266

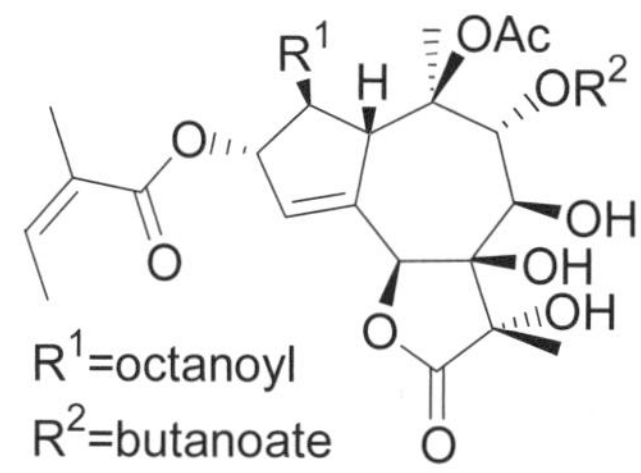

thapsigargin

Krafft and co-workers used olefin metathesis to prepare several inside-outside medium-size rings as scaffolds for natural products synthesis.[129] Reaction of bicyclic lactone **267** with 10 mol % of **4** in refluxing dicholormethane gave the corresponding tricyclic lactone **269** in 88% yield after only two hours.

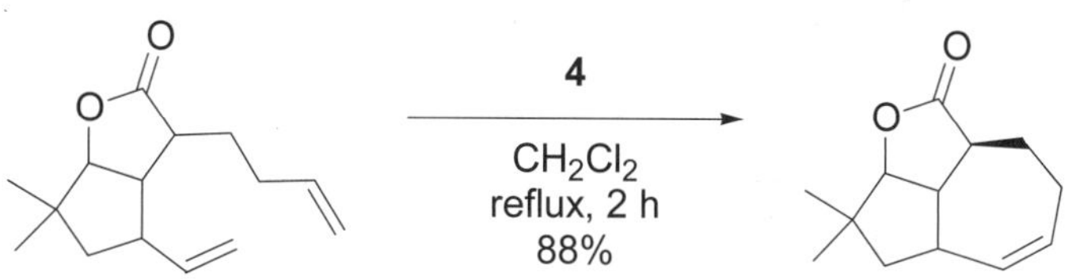

267 268

Mehta and Lakshminath employed a ring-closing metathesis strategy to generate the seven-membered ring carbocycle of the tricyclic core of rameswaralide as part of their ongoing efforts to synthesize this compound.[130] In addition to having a complex and highly functionalized 5,7,6-fused tricarbocyclic core incorporating a stable enol functionality and six stereogenic centers, ramewaralide is a potential anti-inflammatory compound with activity against TNF-α, IL-15, IL-5, and COX_2. Treatment of enone **269** in the presence of 10 mol % of **4** in dichloromethane gave the corresponding tricyclcic system **270** in 90% yield. Srikrishna and Dethe employed a similar ring-closing metathesis strategy for the BC and AB rings of rameswaralide (not shown).[131]

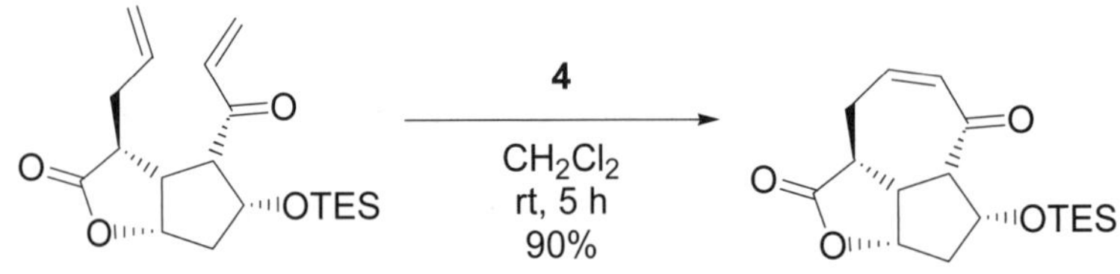

269 **270**

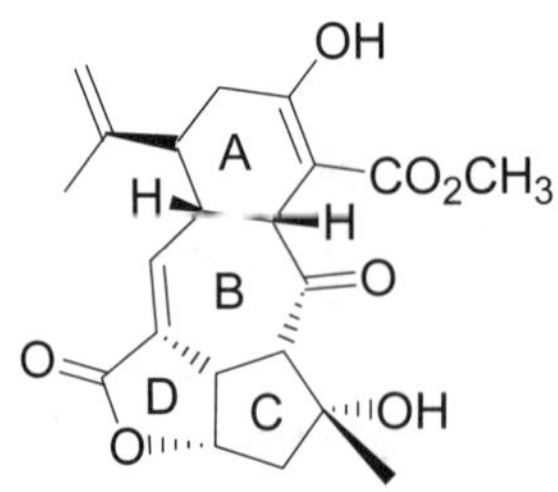

rameswaralide

Ring-closing metathesis has also been applied to the synthesis of a number of terpene derivatives containing seven-membered rings. Several key examples are illustrated below.

Tori and co-workers used ring-closing metathesis as a key step in their synthesis of several sphenolobane-type diterpenoids in an effort to determine the absolute configuration of these compounds.[132] Higher yields were observed when **271**, **273**, and **275** were reacted with **1** in the presence of toluene, in comparison to **3** in refluxing dichoromethane. However, twice the molar percentage of **1** was required.

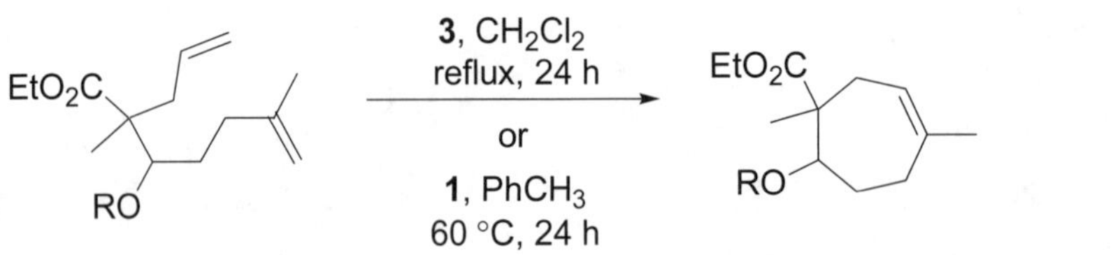

271, R = TES
273, R = OAc
275, R = MOM

272, R = TES **3**-94%, **1**-99%
274, R = OAc **3**-61%, **1**-84%
276, R = MOM **3**-21%, **1**-37%

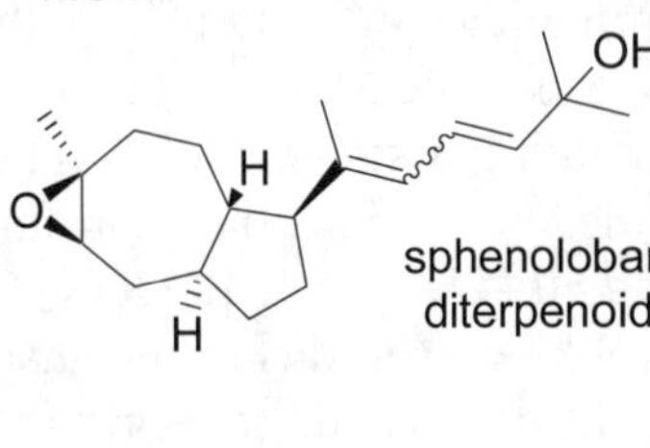

sphenolobane diterpenoids

Olefin metathesis was also used as a key step in Wicha and coworkers synthesis of the carbocyclic core of several di- and sesquiterterpenes.[133] Reaction of **277** with 5 mol % of **4** in refluxing benzene gave the corresponding bicyclic system **278** in 60% overall yield.

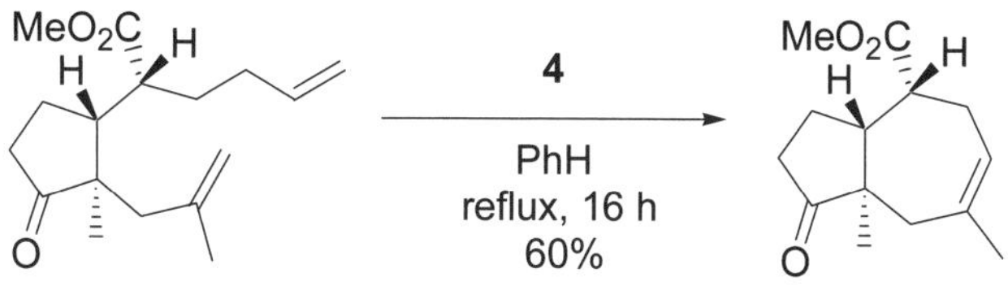

Mehta and Likhite employed an olefin metathesis approach in their synthesis of (±)-frondosins A and B.[134] These novel meroterpenoids have shown promise as therapeutics for the treatment of inflammatory diseases. Ring-closing metathesis of **279** and **281** in the presence of a catalytic amount of **3** in refluxing benzene gave the corresponding bicyclic tertiary alcohols **280** and **282** in 75 and 85% yields, respectively.

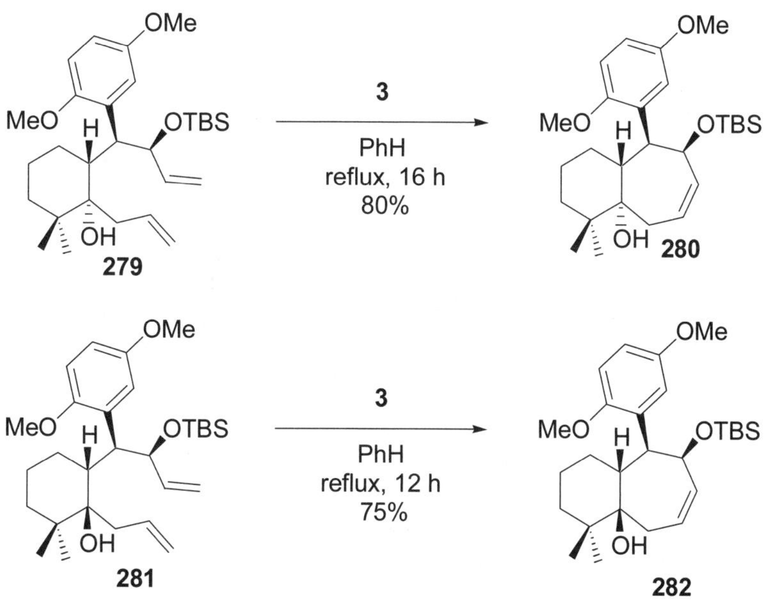

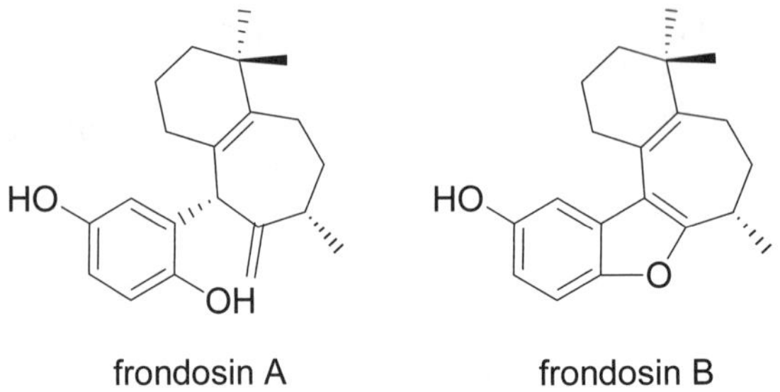

Cha and co-workers[136] used cyclopropanols[135] in an effort to determine the conformational constraints of ring-closing metathesis. Treatment of diene **283** with 10 mol % of **4** in dichloromethane provided the corresponding tricyclic system **284** in 71% yield as a single diastereomer. The authors noted that only one isomer underwent ring-closing metathesis, presumably due to conformational constraints.

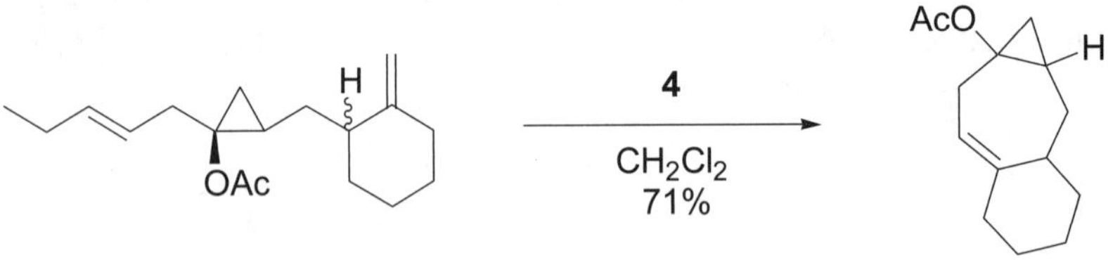

283 **284**

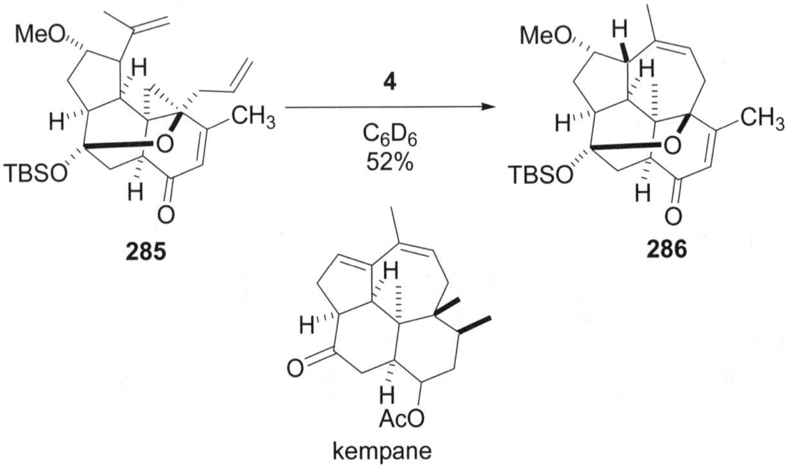

285 **286**

A ring-closing metathesis strategy was employed by Burnell and Zhao in their synthesis of the tetracyclic core of several kempane derivatives.[137] Kempanes are a group of complex tetracyclic diterpenes that serve as key defense molecules for many species of termites. The tetracyclic core, incorporating seven contiguous stereocenters, is especially challenging from a synthetic stand point. Most notably, the generation of the cycloheptene ring proved to be a considerable challenge in a previous synthesis of this compound.[138] Ring-closing metathesis of diene **285** with 3 mol % of **4** in deuterated benzene was moderately successful in generating the cycloheptene ring, and subsequently the corresponding tetracycle **286** in 52% yield.

Several spirocyclic compounds incorporating seven-membered rings have also been prepared by ring-closing metathesis. These molecules are important structural motifs that are found in a number of natural and unnatural products. A number of examples are illustrated in the paragraphs below.

Brimble and Trzoss used a double-alkylation/ring-closing metathesis approach in their synthesis of spiroimines containing seven-membered ring carbocycles.[106] Lactams **287**, **289**, and **291** were subjected to ring-closing metathesis using 5 mol% of **3** in dichloromethane to provide the corresponding 5,7-(**288**), 6,7-(**290**) and 7,7-(**292**) spirolactams in upward of 90% yield. It is interesting that the yields of seven-membered ring spirolactams are higher than the analogous six-membered ring spirolactams, presumably due to the greater flexibility of the diene precursors for the seven-membered rings.

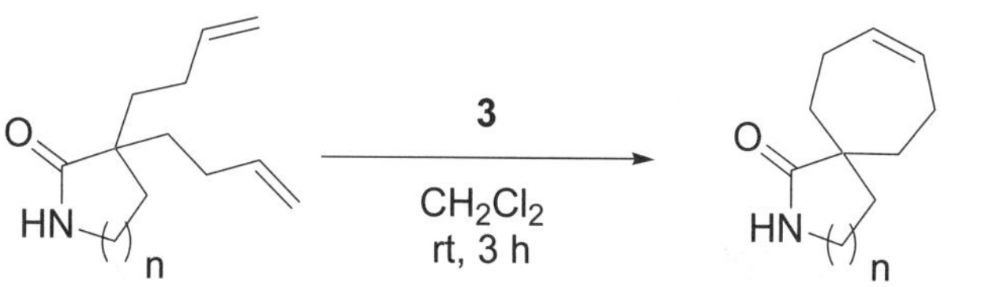

287, n = 1
289, n = 2
291, n = 3

288, n = 1 93%
290, n = 2 95%
292, n = 3 92%

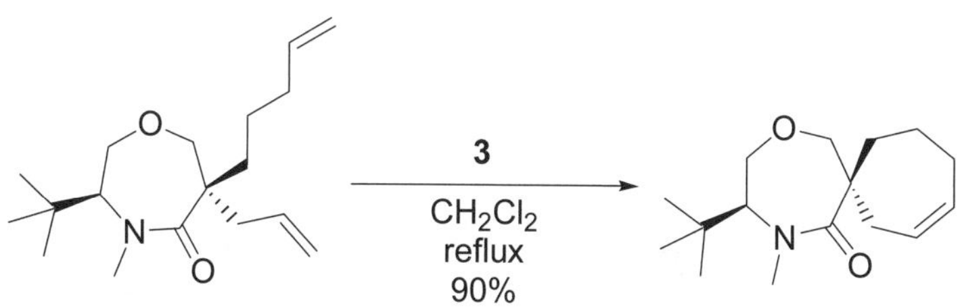

293 **294**

Hughes and co-workers applied their ring-closing metathesis strategy using zizane as a chiral auxiliary to the synthesis of enantiopure seven-membered ring spirocarbocycles.[67] Diene precursor **293** required 5 mol % catalyst loading of **3** in refluxing dichloromethane to achieve a 90% yield of the desired spirocycle **294**. The authors reported on the synthesis of several additional seven-membered ring spirocycles as part of this effort.

Undheim and co-workers employed their olefin metathesis strategy for the synthesis of five- and six-membered ring spirocyclic carbocycles to prepare seven-membered ring spirocyclic carbocycles for use as templates in natural product synthesis.[68] Reaction of diene **295** with 2 mol % of **3** furnished the corresponding spirocycle **296** in only 60% yield. However when diene **297** was reacted under similar conditions, the desired spirocycle **298** was obtained in 90% yield. Although no specific reason is given for the lower yield of **296**, it is possible that conformational constraints and/or sterics may have played a role.

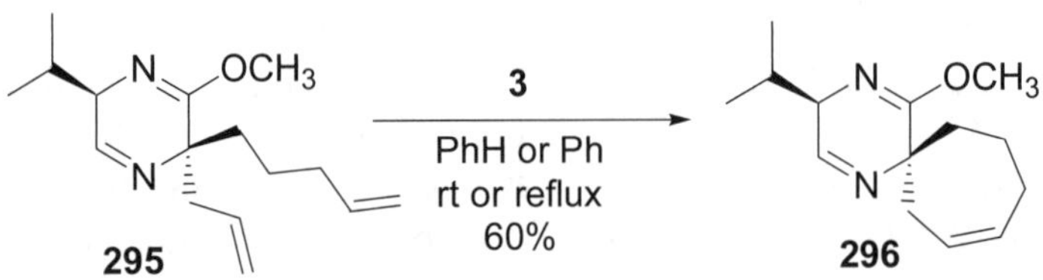

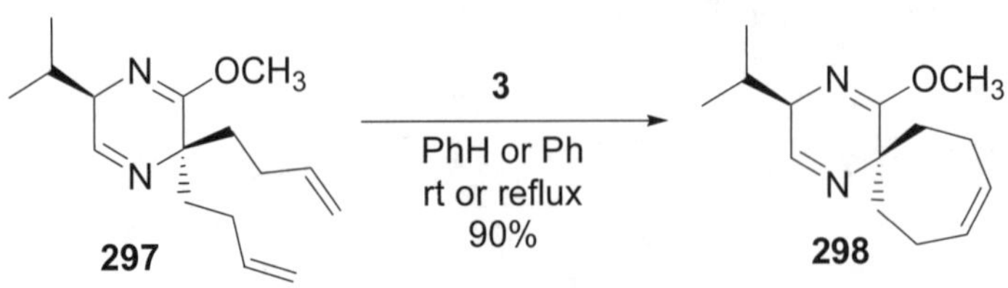

Singh and co-workers applied their methodology for the synthesis of embellished spiro-fused bicyclo[2.2.2]octane systems to the synthesis of seven-membered ring carbocyclic derivatives.[139] Treatment of diene **299** with 10 mol % of **3** in dichloromethane gave the corresponding spirocyclic carbocycle **300** in 86% yield.

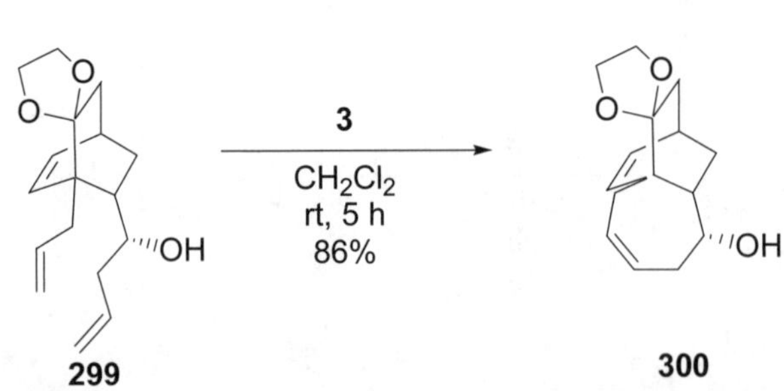

Bennasar and co-workers used ring-closing metathesis to prepare 2,3-fused indole derivatives, which are prominent heterocyclic components of a number of natural products such as ervitsine.[140] Treatment of dienes and **301** and **303** with 10 mol % **3** in refluxing dichloromethane gave the corresponding cyclohepat[*b*]indoles **302** and **304** in 65 and 60% yields, respectively. The modest yields in these examples were likely do to conformational constraints.

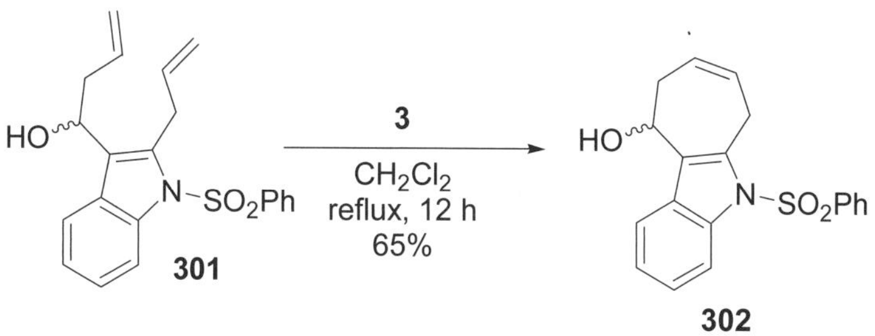

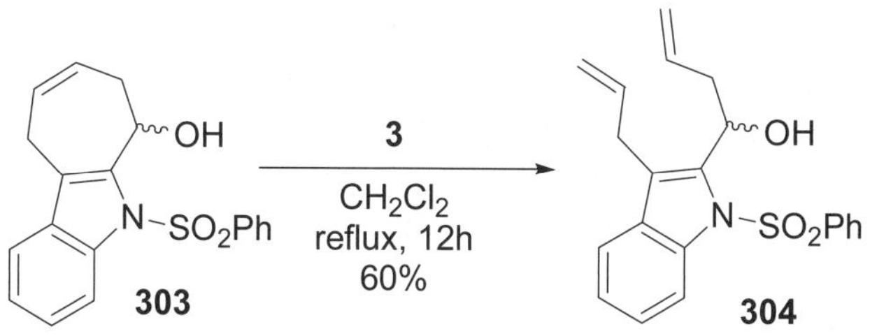

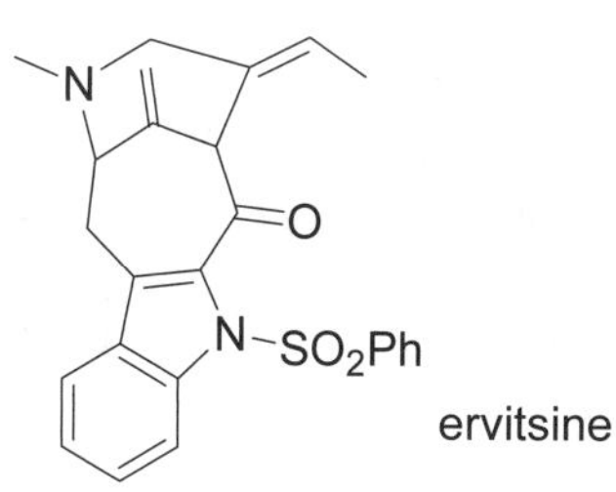

Chattopadhyay and co-workers applied an olefin metathesis strategy in the preparation of a carbocyclic naphthalene derivative containing a seven-membered ring carbocycle with the goal of generating cycloheptanaphthalene, an important enzyme inhibitor.[141] Reaction of diene **305**, prepared in two steps from commercially available 2,7-dihydronaphthalene using a double Claisen rearrangement/acetylation sequence (not shown), with 5 mol % **3** in dichloromethane produced the corresponding tricyclic system **306** in 79% yield.

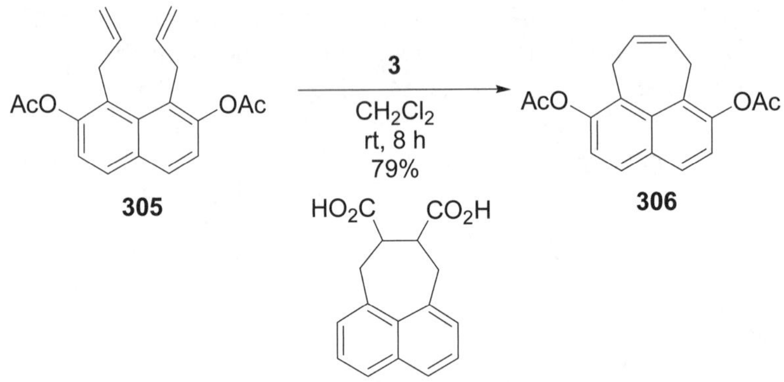

305 306

cyclohepta
naphthalene inhibitor

Eight-Membered Rings

The synthesis of cycloctonoids using ring-closing metathesis was first addressed by Grubbs and co-workers in 1995.[142] Their initial studies revealed that intermolecular cross-metathesis was favored over the intramolecular ring-closing process. This problem was eventually solved by building conformational constraints into the diene precursors such that preorganization favors ring-closing metathesis. For example, 1,2-*trans*-disubstituted cyclohexane derivative **307**, when treated with 5 mol % of **2** in benzene, gave the corresponding fused bicyco[6.4.0]dodecane derivative **308** in 75% yield after only 4 hours. In contrast, treatment of the *cis*-cyclohexane precursor **309** under similar reaction conditions gave only 33% yield of the desired product **310**, and a number of side products, highlighting the importance of conformational constrains in ring-closing metathesis.

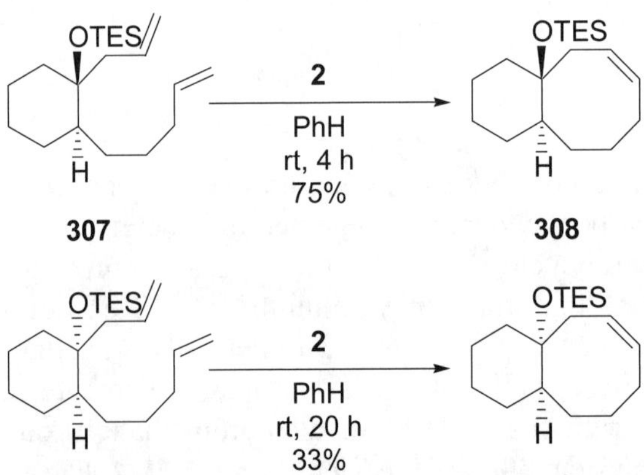

307 308

309 310

More recently, Percy and co-workers used a ring-closing metathesis approach to synthesize highly functionalized difluorinated cylooctenones.[143] These compounds have been used in the design and synthesis of protease inhibitors, as they have been shown to be effective transition-state mimetics. In addition, the electrophilic nature of the ring system makes them attractive targets for adduct formation with active site nucleophiles such as serine. Reaction of allyl alcohol **311** with or without (not shown) the presence of $Ti(Oi\text{-}Pr)_4$ as a precatalyst and 5 mol % of **3** failed to provide the corresponding cyclooctenol **312**. However, when β-hydroxy ketone **313** was used instead, the corresponding difluorinated cyclooctenone **314** was produced in 78% yield. The authors later reported a shorter route to a similar difluorinated system using the same ring-closing metathesis approach (not shown). Unfortunately, the substrate required longer reaction times (166 h) and gave lower yields.

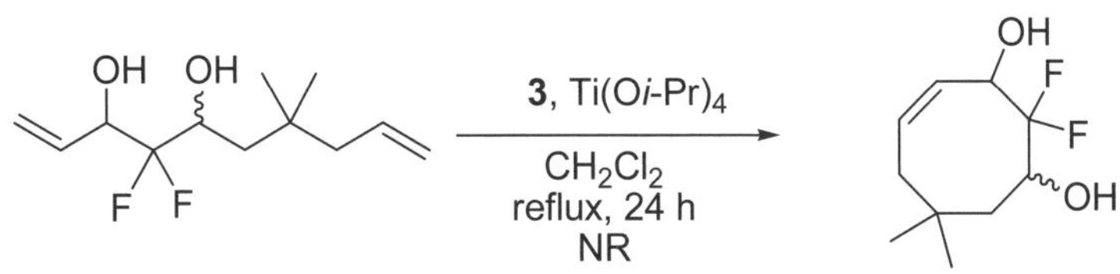

311 **312**

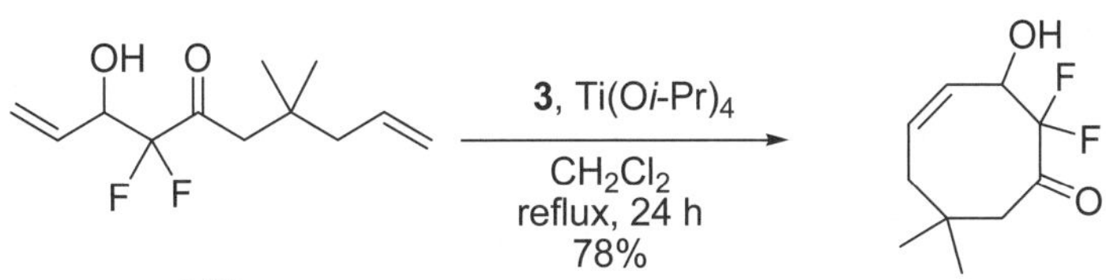

313 **314**

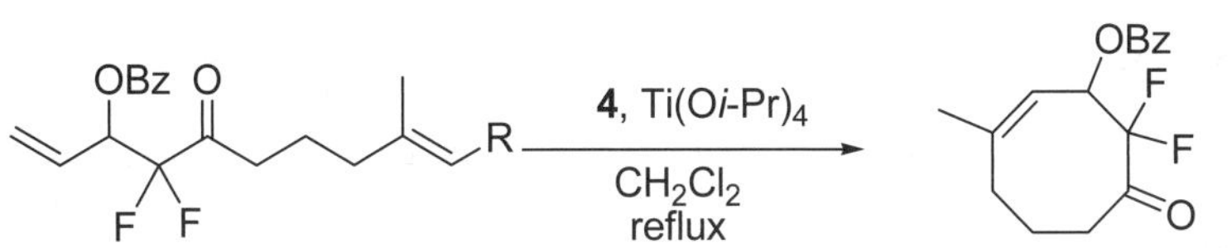

315, R = H
317, R =

316, R = H 24%
318, R = 57%

Percy and co-workers applied a similar strategy in the synthesis of a trisubstituted cyclooctene derivative in an effort to determine the limits of relay ring-closing metathesis.[144] Reaction of diene **315** (R = H) in the presence of **4** (three additions: 10 mol %, then 5 mol % then 5 mol % over a 12-day period) and $Ti(Oi\text{-}Pr)_4$ in refluxing dichloromethane furnished the corresponding octocycle **316** in only 24% yield. However, when a relay approach was employed, diene **317** (R = $(CH_2)_3CHCH_3$) gave the corresponding cyclooctene **318** in 57% yield, a marked improvement over the previous synthesis.

Paley and co-workers used enantiopure η^4-(1-sulfinyldiene)iron(0) tricarbonyl complexes as templates for the enantioselective construction of eight-membered ring carbocycles.[117] Enantiopure homoallyl alcohol adducts **319** and **321**, when treated with 8 mol % of **3** in toluene, gave the corresponding octocycles **320** and **322** as single diasteromers in 87 and 89% yield, respectively.

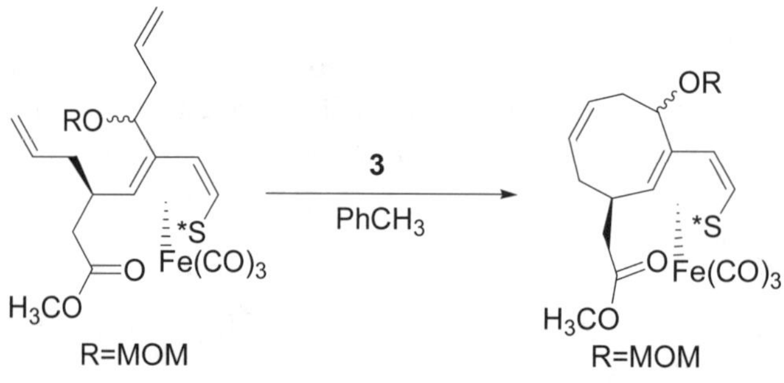

319, *(7S)*
321, *(7R)*

320, *(7S)* 87%
322, *(7R)* 89%

Porco and co-workers used a diversity oriented/ring-closing metathesis approach to construct several complex systems as small molecule protein modulators.[145] Microwave irradiation of Michael adduct **323** in the presence of a catalytic amount of **4** in refluxing dichloromethane gave the corresponding cyclooctene derivative **324** in modest yield. The authors included several additional examples as part of their report.

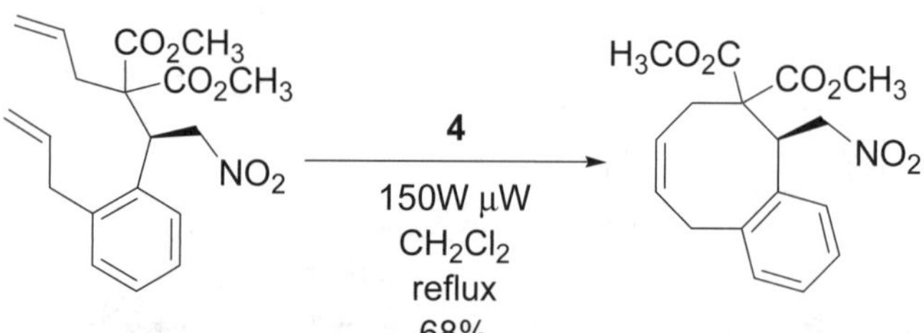

323

324

Hanna and Ricard prepared several eight-membered ring carbocyclic rings from methyl 6-deoxy-6-iodo-3,4-isopropylidene-2-*O*-(*tert*-butyldimethyl-silyl)-D-galactopyranoside and methyl 6-deoxy-6-iodo-1,2:3,4-di-*O*-isopropylidene-D-galactopyranoside using a ring-closing metathesis strategy.[121] Diene **325**, prepared in two steps from methyl 6-deoxy-6-iodo-1,2:3,4-di-*O*-isopropylidene-D-galactopyranoside, gave the corresponding cyclooctene derivative **326** in 94% yield when treated with 5 mol % **3** in refluxing dichloromethane. Several derivatives with various protecting groups were also reported, and the authors applied this strategy to construct a number of carbohydrate-based cyclooctanoids in a later report (not shown).[146]

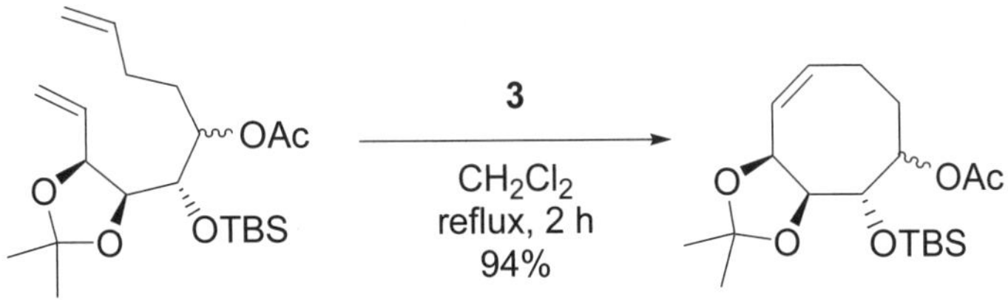

325 **326**

Le Merrer and co-workers employed a ring-closing metathesis strategy to access a number of polyfunctionalized cyclooctane carbasugars as part of their ongoing effort to prepare new glycosidase inhibitors and non-insulino-based compounds for the treatment of diabetes.[147] Olefin metathesis of **327** using up to 13 mol % of **3** in dichloromethane gave the corresponding *cis*-cyclooctene product **328** in 87% yield.

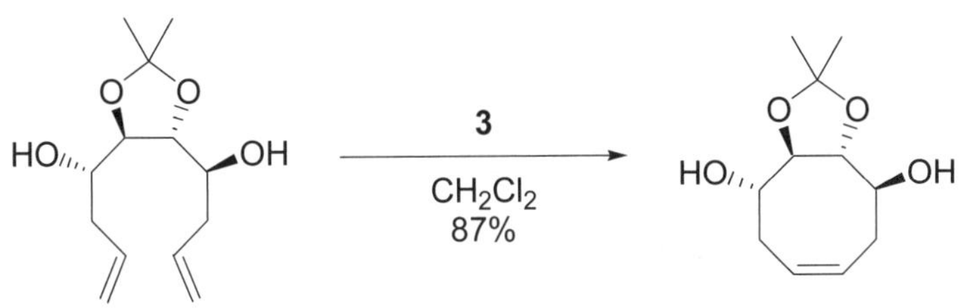

327 **328**

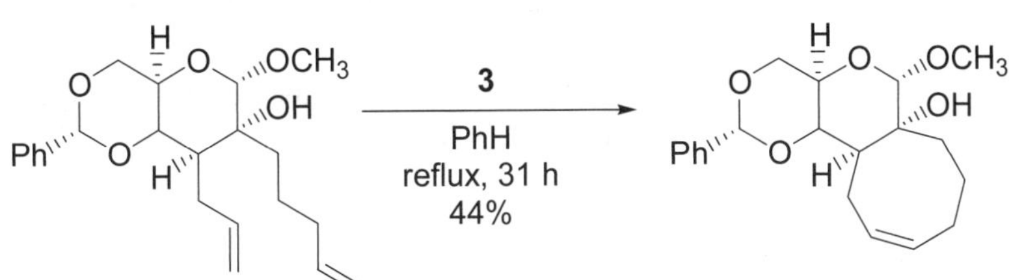

329 **330**

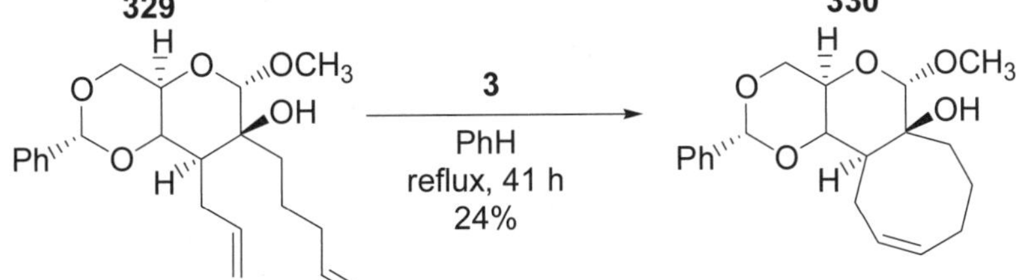

331 **332**

Holt and co-workers extended their ring-closing metathesis strategy for the synthesis of enantiomerically pure annulated six- and seven-membered ring carbohydrate systems to systems containing containing eight-membered ring carbocycles.[90] Treatment of 5-hydroxy-1,9-diene precursors **329** and **331** with a 9 mol % **3** in refluxing benzene provided the corresponding 6,6,8-*cis*-**330** and *trans*-**332** annulated systems in 44 and 24% yield, respectively. The authors noted that conformational constraints and the presence of the tertiary alcohol likely played a role in the successful transformation.

In their continuing efforts to synthesize the carbocyclic core of sesquiterterpenes such as ophiobolin A, Wicha and co-workers constructed a 5,8-fused carbocyclic system.[148] Reaction of **333** with 5 mol % of **4** in refluxing dichloromethane furnished the desired bicyclic system **334** in 95% overall yield.

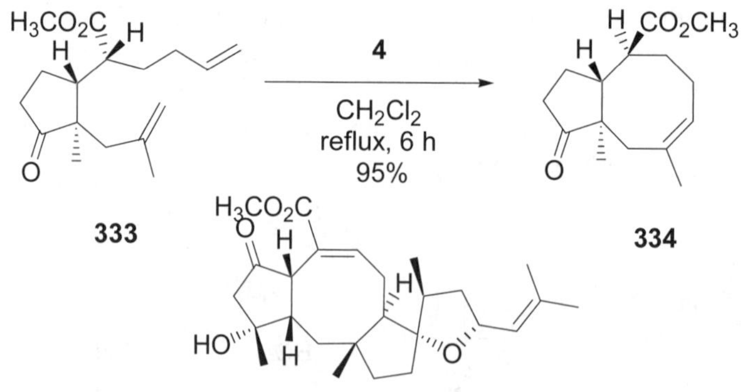

333 334

ophiobolin A

Martin and co-workers applied their ring-closing metathesis strategy to the synthesis of 5,6- and 5,7-fused butyrolactones to 5,8-fused butyrolactone systems.[77] Diene precursor **335** was subjected to ring-closing metathesis using 10 mol % of **4** in refluxing dichloromethane to produce the corresponding α,β-fused γ-lactone **336** in 45% yield.

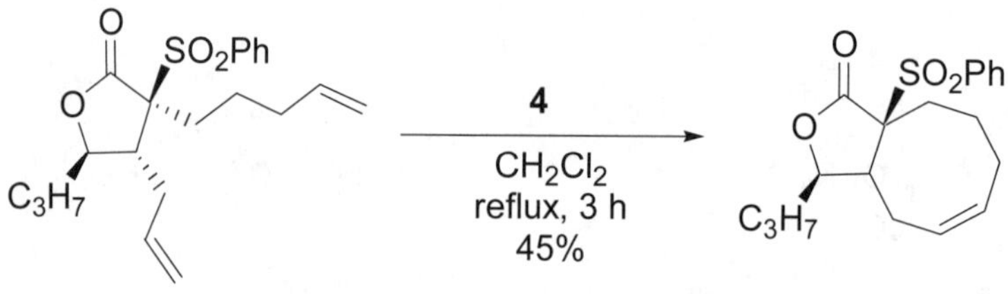

335 336

Krafft and co-workers used ring-closing metathesis to prepare several 'inside-outside' medium-size rings, including eight-membered carbocycles, as scaffolds for natural products synthesis.[129] Reaction of bicyclic lactone **337** with 10 mol % of **4** in refluxing dichloromethane gave the corresponding tricyclic lactone **338** in 85% yield after only 2 hours. Additional eight-membered ring carbocycles incorporating functional handles were also reported, though significantly higher catalyst loadings (upward of 50 mol %) were required to achieve high yields. Attempts to apply this strategy to the synthesis of nine-membered ring carbocycles led solely to the production of dimeric products (not shown).

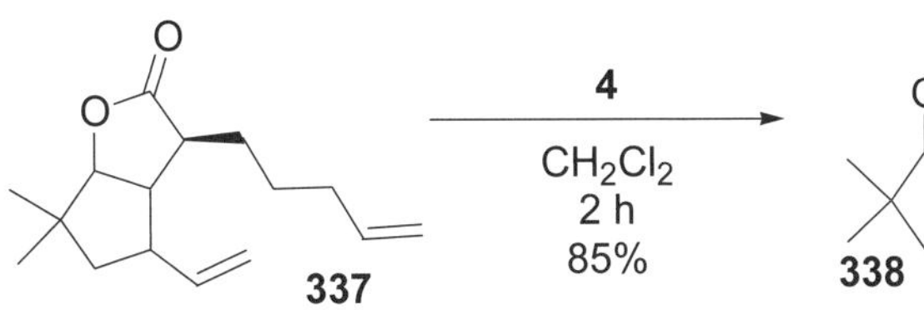

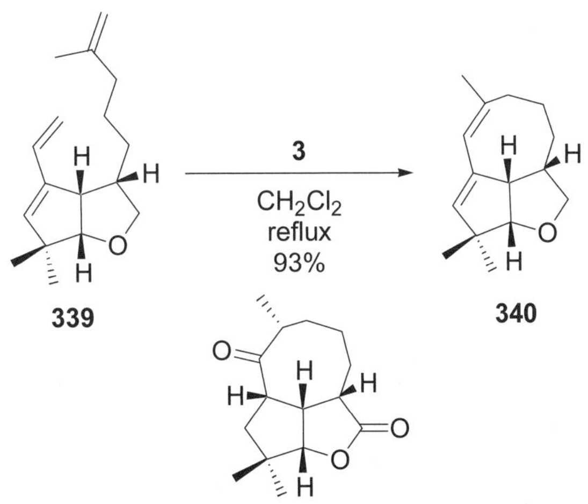

Paquette and co-workers employed a ring-closing metathesis strategy in their total synthesis of (+)-asteriscanolide.[149] The sesquiterpenoid framework of (+)-asteriscanolide consists of a rather uncommon bicyclo-[6.3.0]undecane ring system bridged by a butyrolactone fragment. Ring-closing metathesis of 5,5-diene precursor **339** using 10 mol % of **3** in refluxing dichloromethane proceeded smoothly to give the corresponding carbocyclic core (**340**) of (+)-asteriscanolide in 93% yield. The authors speculated that limited conformational flexibility of the diene substrate was critical to the high yields achieved in this transformation. Four additional steps were required to access (+)-asteriscanolide. Krafft and co-workers applied a similar ring-closing metathesis strategy in their 2001 synthesis of (+)-asteriscanolide (not shown).[150]

Tadoano and co-workers used olefin metathesis as a key step in their synthesis of (±)-mycoepoxydiene, a novel octacycle with an oxygen-bridged-[4.2.1]nona-2,4-diene core.[151] Reaction of diene **341** with 20 mol % of **3** in refluxing benzene furnished the corresponding oxygen-bridged cyclooctene derivative **342** in 83% yield. The authors noted that a highly dilute solution was essential in obtaining high yields of the desired product.

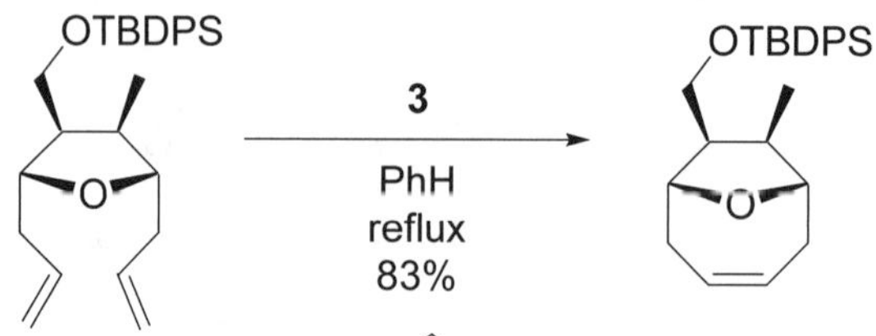

341 **342**

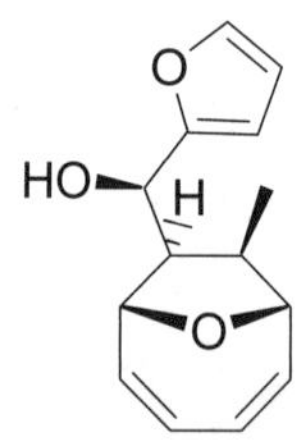

mycoepoxydiene

Mascarenas and co-workers employed a similar strategy in their synthesis of eight-membered ring carbacycles using a ring-closing metathesis/ring fragmentation strategy.[152] Treatment of **343** as a mixture of diasteromers with 5 mol % of **3** in refluxing dichloromethane provided the desired bridged bicyclic compound **344** as the only product in 15% yield. The unreactive *trans*-isomer was easily separated from the bicyclic system after deprotection of the silyl ether. Further studies showed that enantiopure alcohol **345** underwent ring-closing metathesis using 5 mol % **3** in dichloromethane at room temperature in only nine hours to give the corresponding bicylic system **346** in 95% yield. Treatment with lead acetate led to the production of the corresponding cycloctanoid in near quantitative yields (not shown).

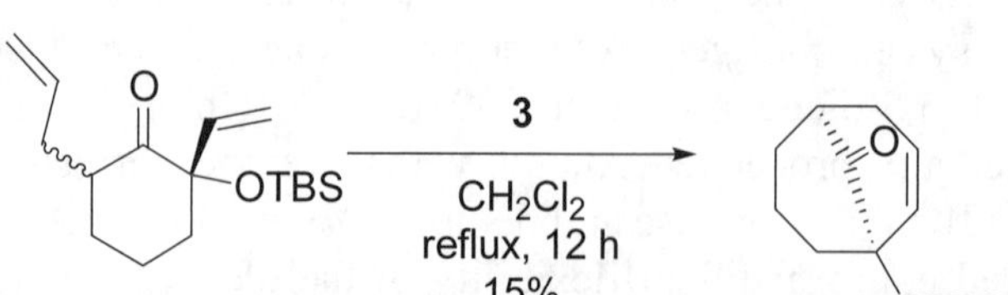

343 **344**

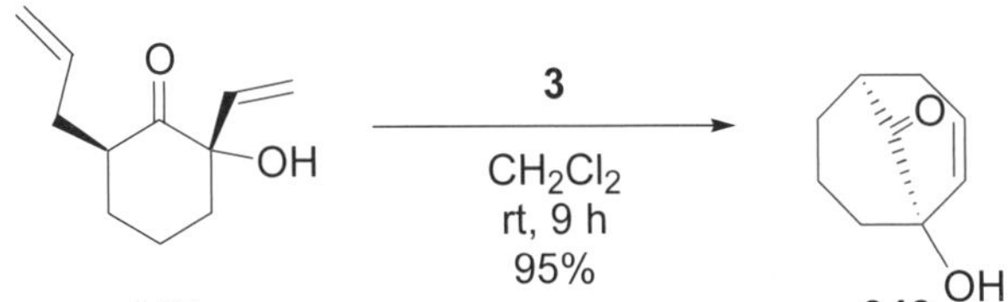

345 **346**

Rodriguez and co-workers applied a similar strategy in the synthesis bicyclo[4.2.1]nonane derivatives as precursors to functionalized cyclooctanes.[153] Reaction of ketones **347**, **349** and **351** in refluxing dichloromethane with 2 mol % of **4** gave the corresponding cyclooctene derivatives in 68–74% yield. Unfortunately, these transformations required long reactions times and the modest yields observed with substrates **350** and **352** were presumably due to sterics and catalyst decomposition. On the other hand, treatment of alcohols **353** and **355** using only 1 mol % of **4** in refluxing dichloromethane furnished the corresponding carbocycles in high yields (92–98%) in only two to three hours.

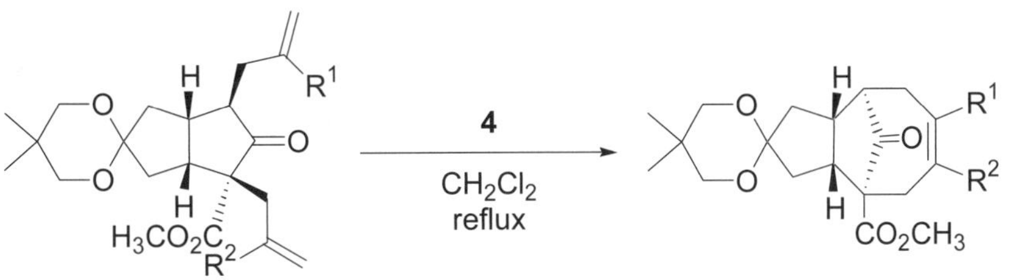

347, $R^1 = H$, $R^2 = H$
349, $R^1 = CH_3$, $R^2 = H$
351, $R^1 = H$, $R^2 = CH_3$

348, $R^1 = H$, $R^2 = H$ 84%
350, $R^1 = CH_3$, $R^2 = H$ 70%
352, $R^1 = H$, $R^2 = CH_3$ 68%

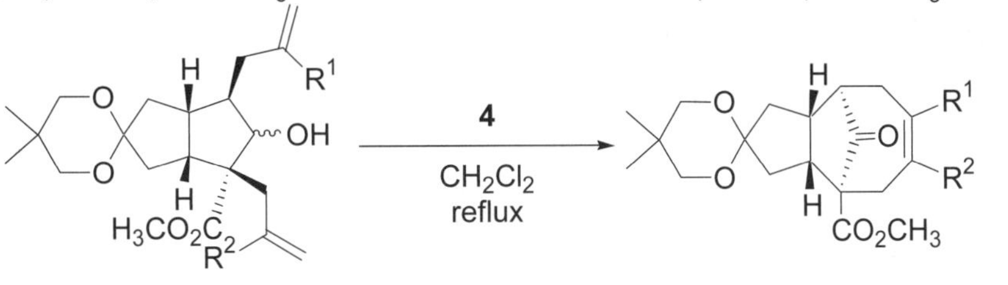

353, $R^1 = H$, $R^2 = H$
355, $R^1 = H$, $R^2 = CH_3$

354, $R^1 = H$, $R^2 = H$ 92%
356, $R^1 = H$, $R^2 = CH_3$ 98%

Singh and co-workers used Rodriquez's ring-closing metathesis strategy in the stereoselective synthesis of several expanded homologs of the cucumin family.[154] These compounds exhibit cytotoxic and antibacterial properties. Diquinane derivative **357** underwent ring-closing metathesis in the presence of **4** in dichloromethane in under an hour to provide the corresponding tricyclic carbocycle **358** in modest yield.

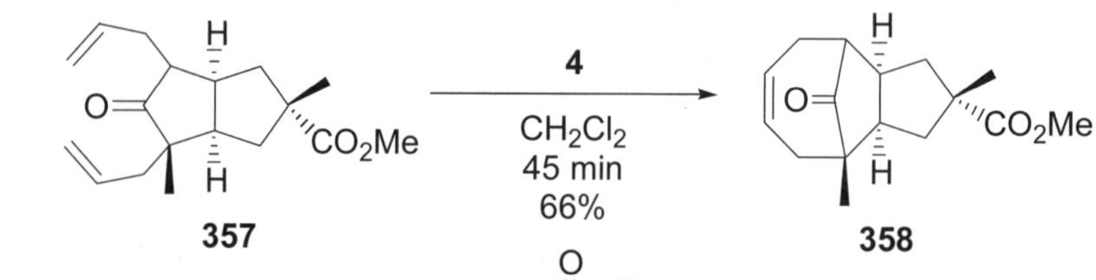

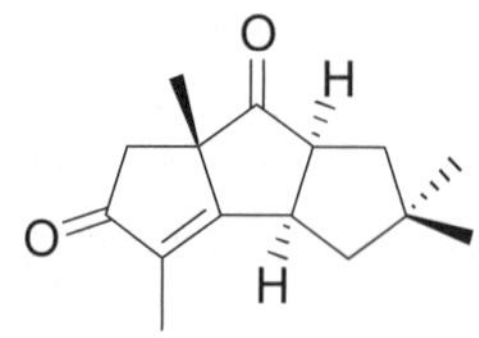
cucumin E

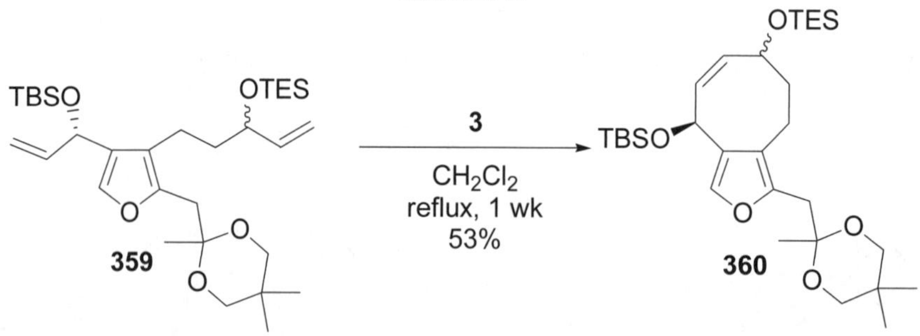

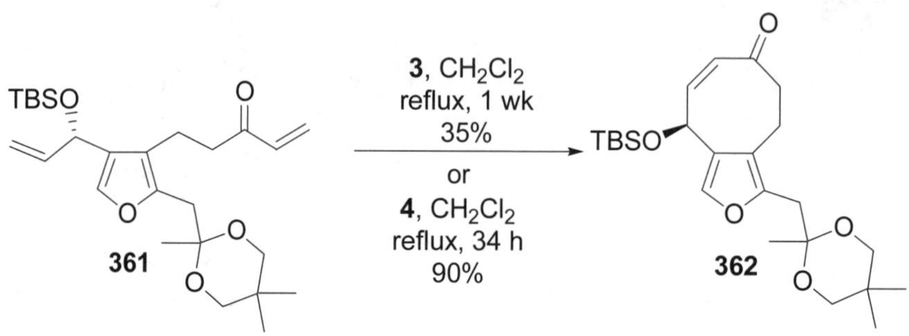

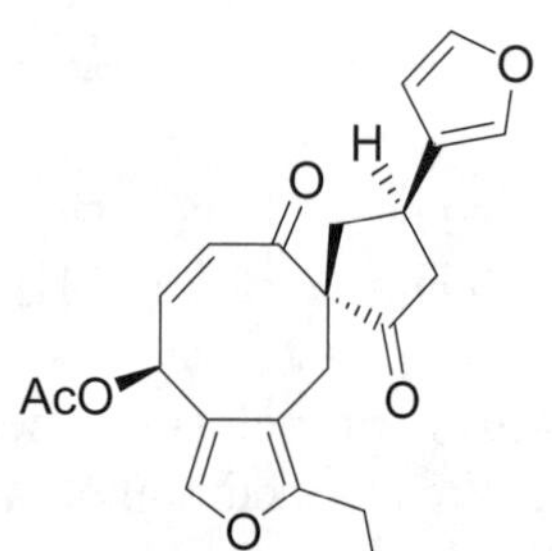
teubrevin G

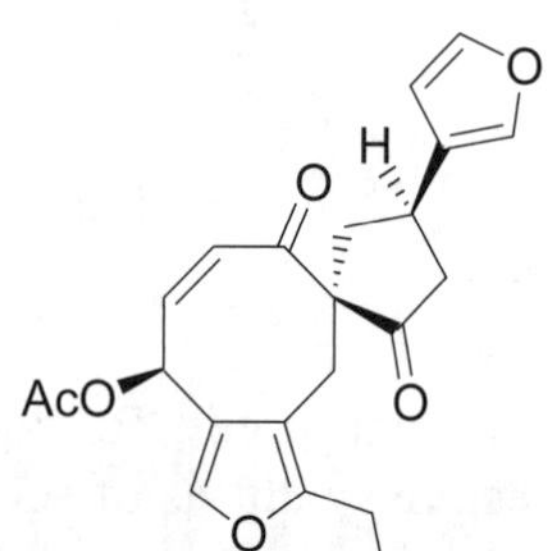
teubrevin H

Paquette and Efremov used a ring-closing metathesis strategy in their first total synthesis of the rearranged neo-clerodanes, teubrevin G and teubrevin H.[155] These compounds feature a cycloctanene core fused and spiroannulated to smaller oxygen containing rings. Treatment of triethylsilyl ether **359** with 30–35 mol % of **3** in refluxing dichloromethane for one week furnished the corresponding fused octocycle **360** in 53% yield. Olefin metathesis with enone **361** under similar reaction conditions gave only 35% yield of the desired product **362**. However, reaction of enone **361** with 10 mol% of **4** in refluxing dichlromethane gave the desired product in 90% yield after 34 hours.

Prunet and co-workers employed ring-closing metathesis as a key step in their synthesis of the BC bicycle of Taxol, a potent treatment for breast and ovarian cancer. Previous research showed that dienes **363** and **365**, when treated with 10 mol % of **3** in refluxing benzene or 5 mol % of **5** in refluxing 1,2-dichloroethane, gave the corresponding C9–C10 cyclooctene derivatives in good yield.[156] However, in an effort to prepare the C10–C11 cyclooctene system using the same strategy, the authors observed that only diene **367** underwent ring-closing metathesis suggesting a higher energy barrier to preorganzation for these substrates.[157] The authors went on to demonstrate that catalysts **3** (30 mol %), **4** (10 mol%), and **5** (5 mol %) could be successfully employed to produce the desired C10–C11 cyclooctene system (not shown) in modest yields (65%, 69%, and 72% yield, respectively).

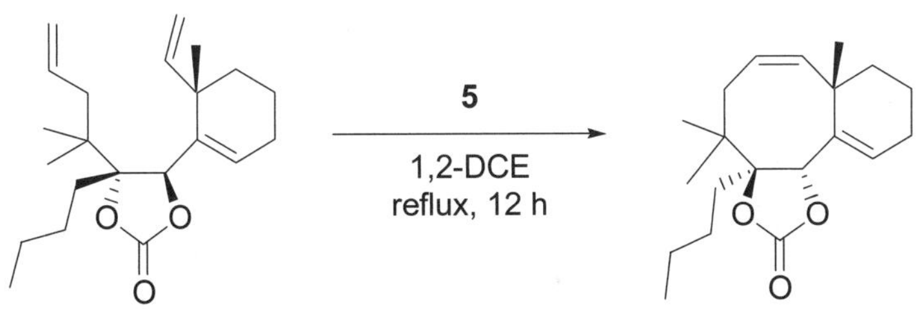

363 **364**

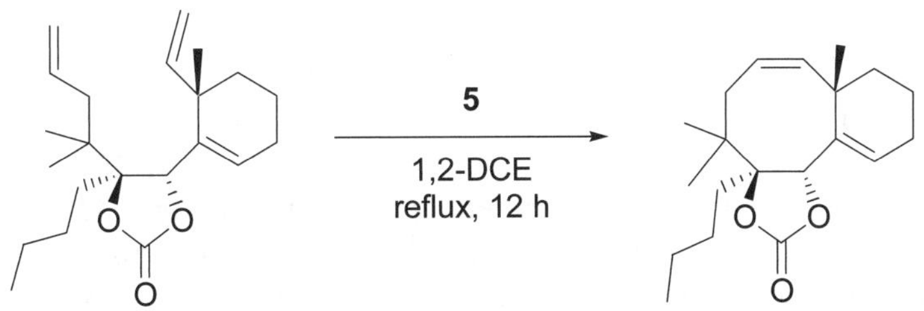

365 **366**

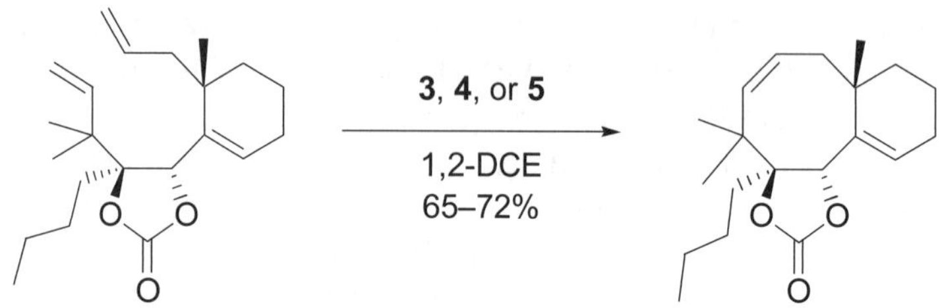

367 368

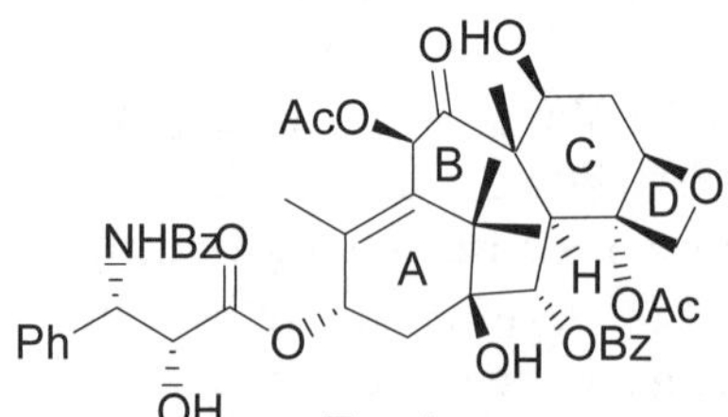

Taxol

Granja and co-workers employed a ring-closing metathesis strategy in their synthesis of a novel steroid-like polycycle incorporating a 6,8,6-fused carbocyclic system.[158] These molecules purportedly mimic the putative transition state structure of the isomerization reaction of previtamin D_3 to vitamin D_3. Treatment of diene **369** with near quantitative amounts of **3** in refluxing dichloromethane gave the desired tricyclic system **370** as a mixture of diastereomers in 95% yield after 6 days. In a later report, several additional examples bearing various substitutions patterns were reported (not shown).[159]

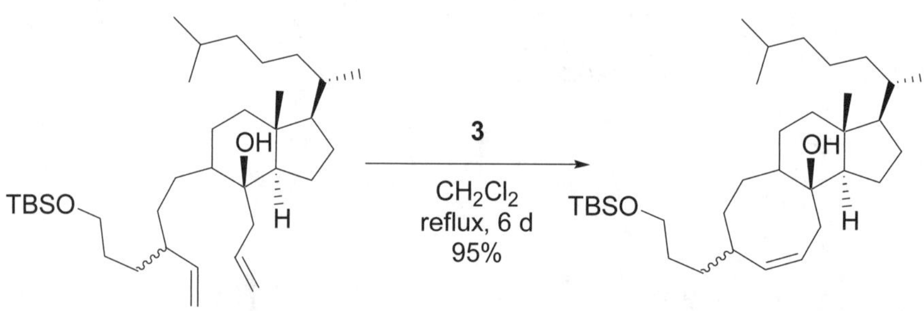

369 370

Singh and co-workers also applied their methodology for the synthesis of embellished spiro-fused bicyclo[2.2.2]octane systems to the synthesis of eight-membered ring carbocyclic derivatives.[160] Treatment of diene **371** with 5 mol % of **4** in dichloromethane gave the corresponding spirocyclic carbocycle **372** in 89% yield.

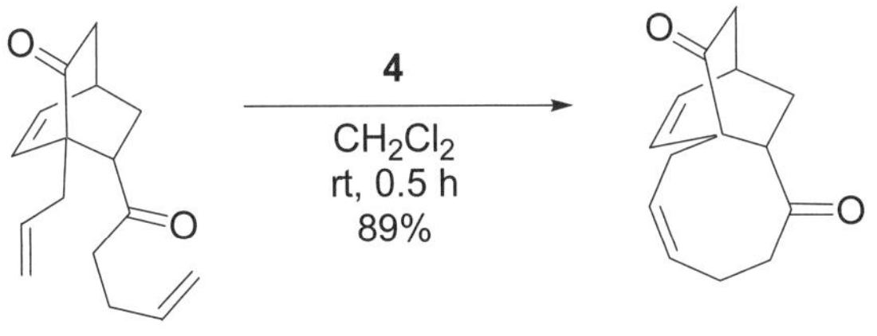

371 **372**

Bennasar and co-workers used their ring-closing metathesis strategy for the formation of 2,3-fused indole derivatives containing seven-membered ring cabocycles to prepare 2,3-fused indole derivatives containing eight-membered ring carbocycles.[161] Treatment of diene **373** with 10 mol % **3** in refluxing dichloromethane furnished the corresponding cyclooctindole derivative **374** in 85% yield.

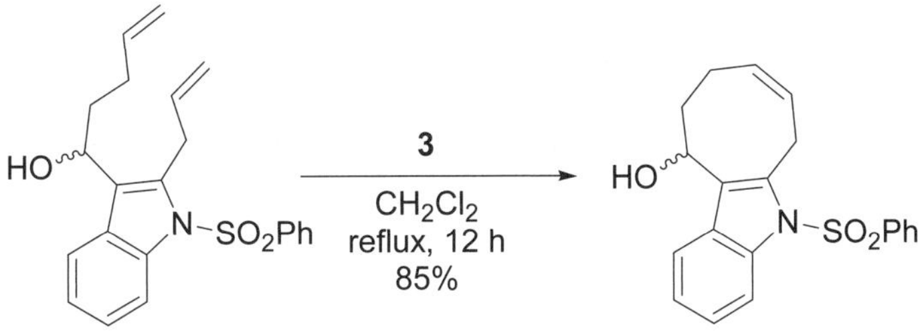

373 **374**

Large-Membered Rings

Several large-membered ring systems have been synthesized using ring-closing metathesis. This section is organized by ring size and highlights a number of examples including nine-, ten-, eleven-, twelve-, thirteen-, fifteen-, and sixteen-membered ring carbocycles.

Paley and co-workers used enantiopure η^4-(1-sulfinyldiene)iron(0) tricarbonyl complexes as templates for the enantioselective construction of nine-membered ring carbocycles, in addition to seven- and eight- membered ring carbocycles.[117] Enantiopure homoallyl alcohol adduct **375**, when treated with 8 mol % of **3** in toluene, gave the corresponding nine-membered carbocyclic ring in 88% yield with a *cis*/*trans* ratio of 35:1.

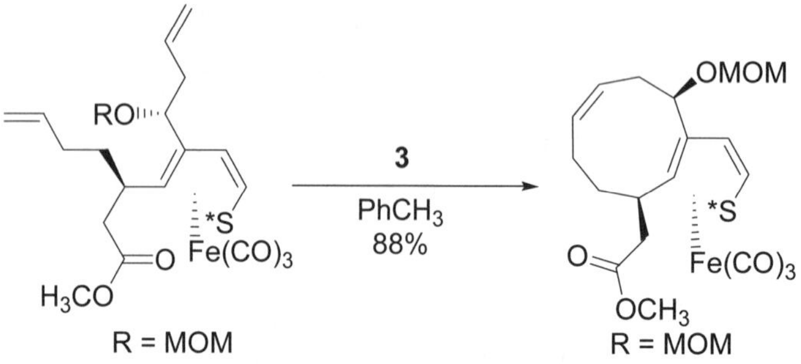

375 **376**

Holt and co-workers applied their ring-closing metathesis strategy for the synthesis of six-, seven-, and eight-membered ring annulated carbohydrate systems to synthesize a number of enantiomerically pure annulated carbohydrate systems containing nine-membered ring carbocycles.[90] Treatment of *cis*-diene **377** with a catalytic amount of **3** in refluxing benzene gave the corresponding 6,6,9-carbocycle **378** in 52% yield. However, treatment of the *trans*-diene **379** under similar conditions gave none of the desired product, highlighting the importance of conformational constraints on nine-membered ring systems.

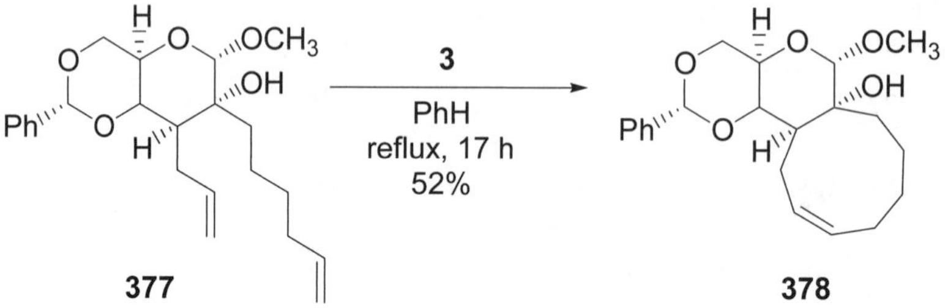

377 **378**

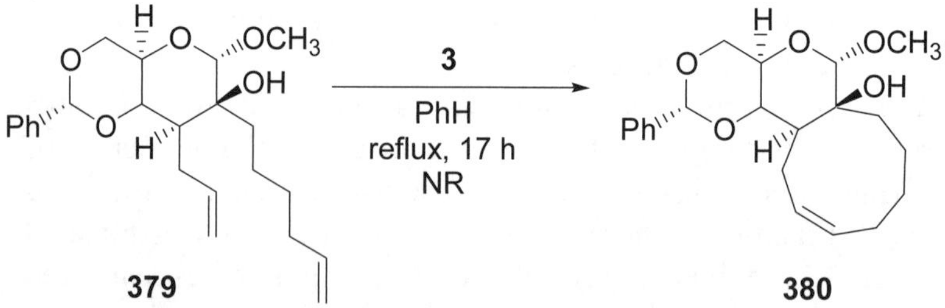

379 **380**

Clark and co-workers[162] used the olefin metathesis strategy pioneered by Rodriquez and co-workers for the synthesis of eight and nine-membered ring carbocycles,[163] in their synthesis of the nine-membered carbocyclic core of a class of herbicidial nonadrides known as the cornextins. Treatment of diene **381** with either catalytic **3** in refluxing dichloromethane, or **4** in toluene, produced the desired nine-membered ring carbocycle **382** in 70% and 61% yield, respectively as mixtures of diastereomers. In a later report, Clark and co-workers employed a ring-closing fragmentation strategy complete the synthesis of (±)-5-*epi*-hydroxycornexistin (not shown).[164]

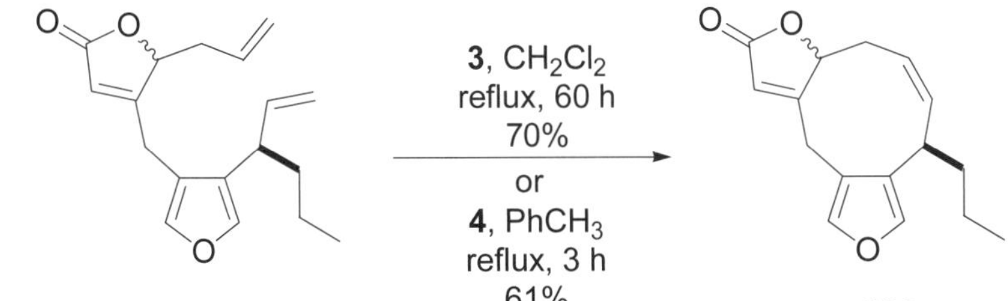

381 **382**

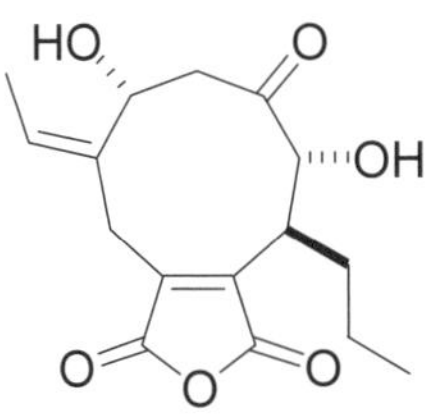

cornexistin

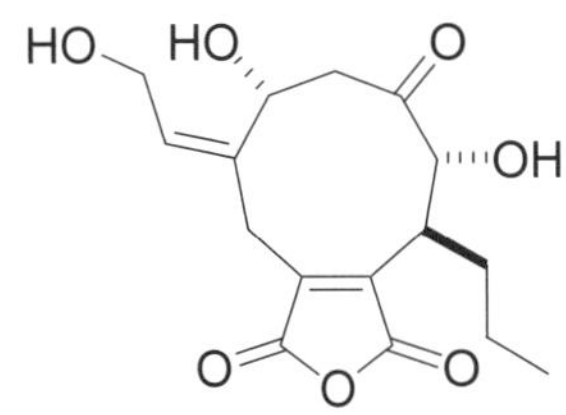

hydroxycornexistin

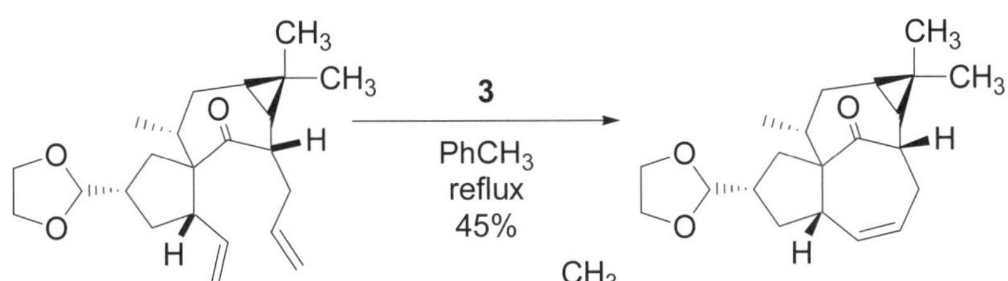

383 **384**

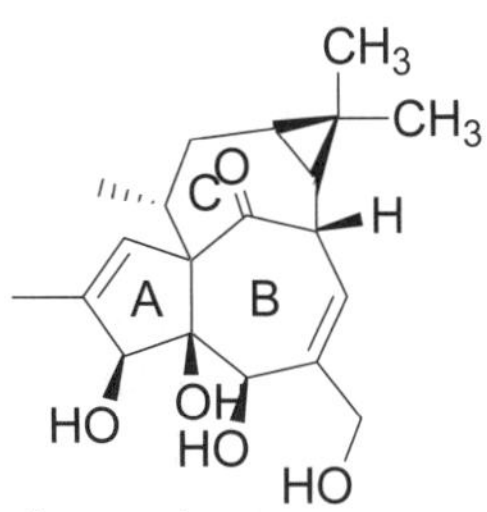

ingenol

Wood and co-workers employed a ring-closing metathesis strategy in their construction of the carbocyclic core of ingenol.[165] Ingenol esters have been shown to mimic diacylglycerol and function as PKC activators. These compounds are particularly challenging from a synthetic stand point due to their high degree of oxygenation, including a *cis*-triol and highly strained inside-outside BC ring system. Treatment of diene **383** in refluxing toluene with four additions of 20 mol % of **3** every 45 minutes led to the construction of the desired inside-outside ring system **384** in 45% yield. The research groups of Winkler[166] and and Kigoshi[167] have independently applied similar strategies in their syntheses of ingenol (not shown).

ent-Clavilactone B, a unique compound with antifungal and antibacterial properties, was synthesized by Barrett and co-workers using olefin metathesis as a key step in their strategy.[168] Extensive optimization led to the slow addition of 40 mol % of **4** to diene **385** in the presence of 80 mol % tetrafluorobenzoquinone in toluene to afford the desired 10-membered ring **386** in 65% yield. The authors noted that the reaction proceeded smoothly without affecting the strained epoxide ring. Treatment of **386** with CAN (not shown) furnished the desired target.

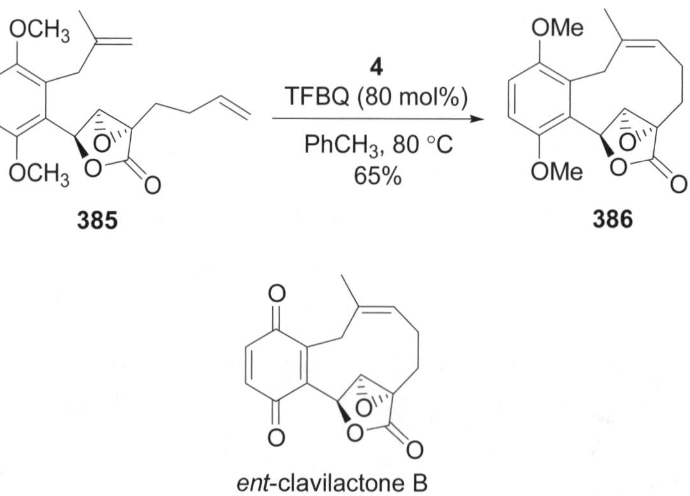

Gennari and co-workers employed a ring-closing metathesis strategy in their synthesis of C-7 substituted eleuthesides.[169] These molecules are analogs of the sarcodictyin family, which have shown significant microtubule stabilizing activity in tumor cell lines, in addition to the ability to inhibit Taxol resistant tumor cell lines. Dienes **387** and **389** were subjected to ring-closing metathesis using 6 mol % of **4** in dichloromethane, first at room temperature and then at reflux, to provide the corresponding *Z*-alkenes **388** and **390** in 78% and 71% yields, respectively. A number of additional alkenes bearing various functional handles were also synthesized.

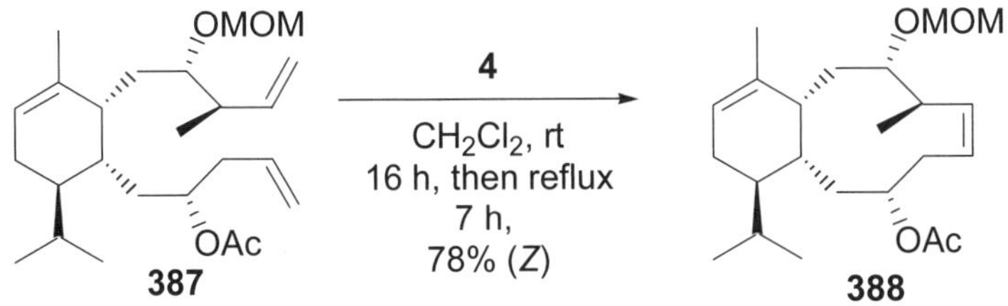

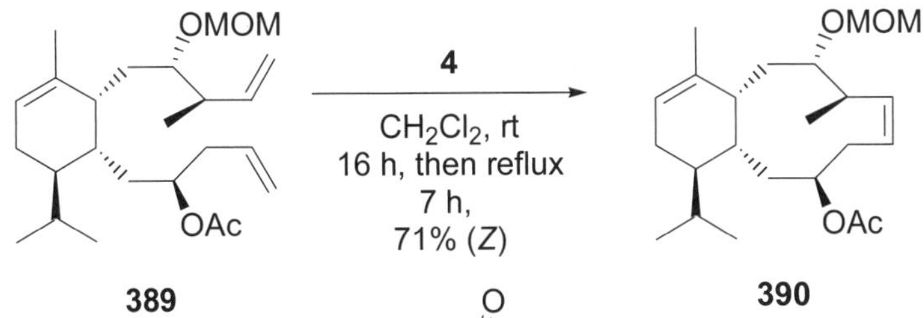

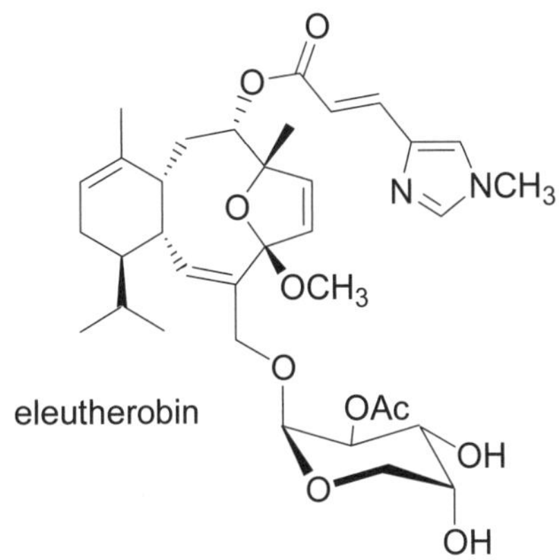

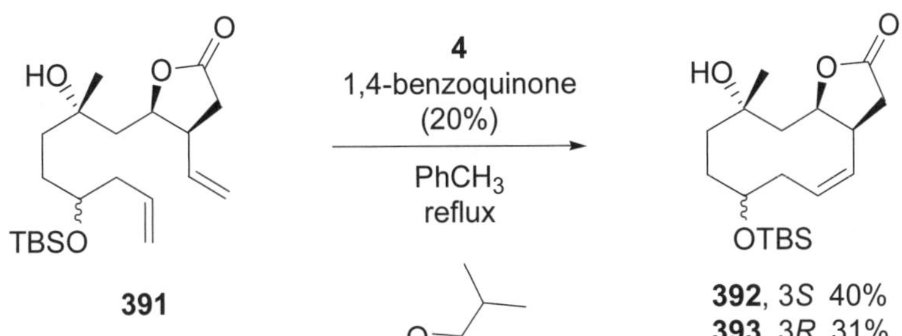

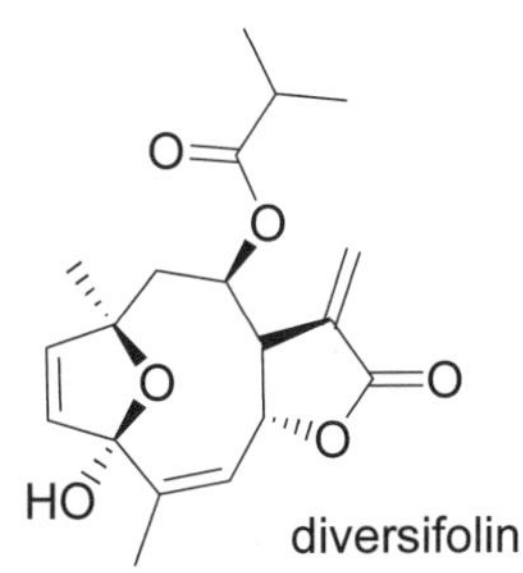

Kobayashi and co-workers employed an olefin metathesis strategy in their synthesis of the 11-oxabicyclo[6.2.1]undec-3-ene core of diversifolin, a densely oxygenated germacrane-type sesquiterpene with the ability to inhibit transcription factor NF-kB.[170] Reaction of **391** as a mixture of diasteromers with 20 mol% of **4** in the presence of 20 mol % of 1,4-benzoquinone in refluxing toluene gave the desired bicyclic lactones 3*S*-(**392**) and 3*R*-(**393**) in 71% overall yield. It is interesting that when only 10 mol % of **4** was used without 1,4-benzoquinone, only the 3*R* isomer **393** was obtained (27% yield) in addition to a trace amount of the 3*S* isomer, recovered starting material, and an isomerisation product (not shown). The reactions were also attempted using **3** with no yield of the desired products.

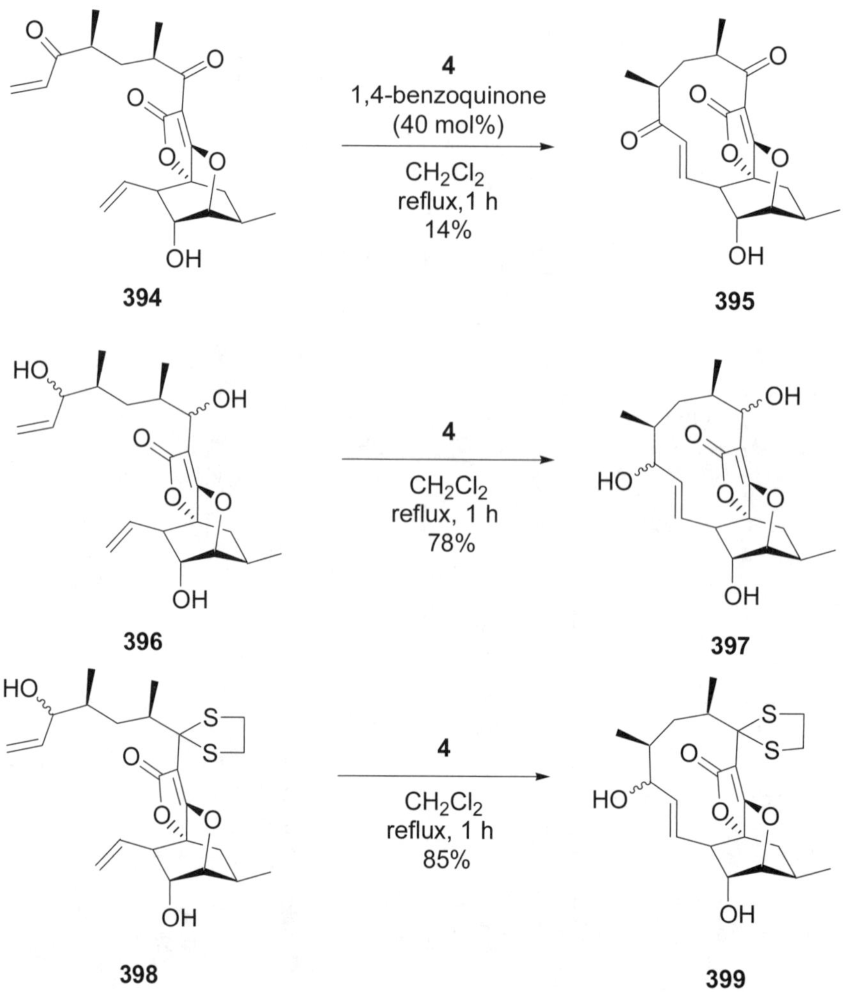

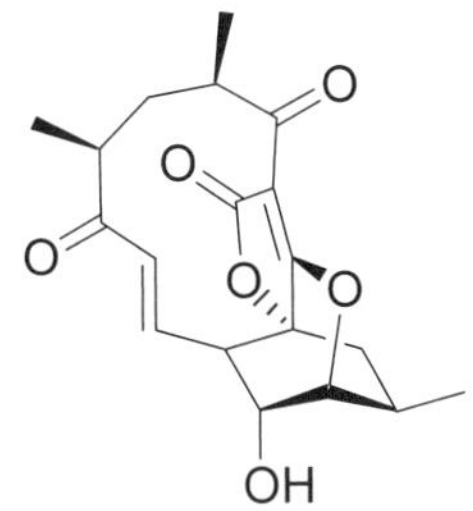

abyssomicin C

Nicolaou and Harrison used a ring-closing metathesis strategy to construct the carbocyclic core of abyssomicin C, one of the only compounds to exhibit antibiotic activity via inhibition of the *p*-aminobenzoic acid biosynthetic pathway.[171] The larger, strained 11-membered ring containing four stereogenic centers presented some specific challenges. Treatment of vinyl ketone **394** with 10 mol % of **4** and 20 mol % of 1,4-benzoquinone in refluxing dichloromethane gave the desired product, abyssomicin C **395** in only 14% yield. However, when a diastereomeric mixture of vinylic triol **396** was treated with 5 mol % of **4** in refluxing dichloromethane, the corresponding carbocyclic core **397** was generated in 78% yield. Unfortunately, the authors were unable to find a suitable oxidation strategy to produce abyssomicin C. Speculation that the two additional sp^2 centers of **394** were responsible for the modest conversion of **394** to **395**, in addition to difficulties in oxidizing **397** led to the design of **398**, which was devoid of the intramolecular hemiketalization problem. This substrate smoothly underwent ring-closing metathesis using 5 mol % of **4** in refluxing dichoromethane to provide the desired carbocycle **399** in 85% yield. **399** could be readily oxidized and deprotected to afford the desired product (not shown).

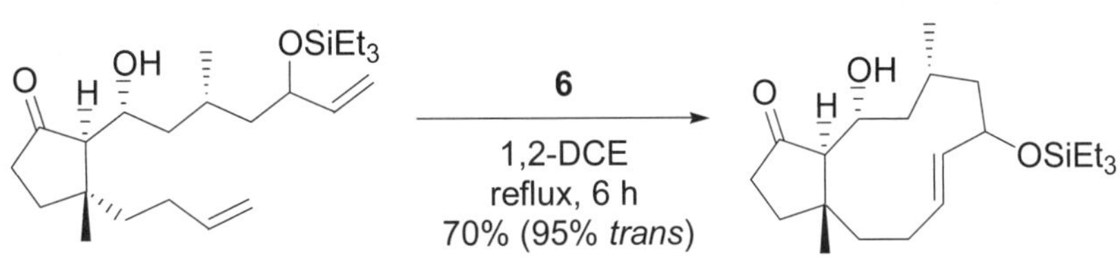

400 **401**

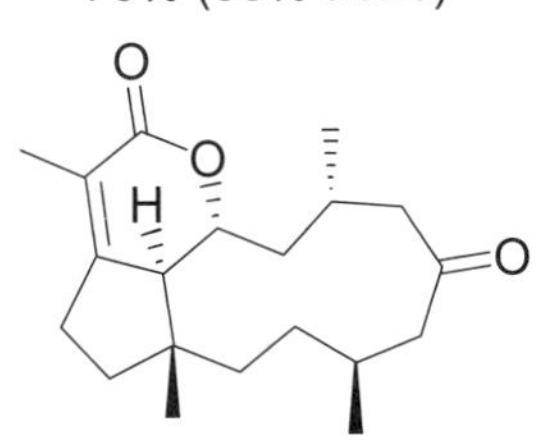

clavirolide C

Hoyveda and Brown used olefin metathesis to synthesize clavirolide C, a member of the dolabellane family of diterpenes.[172] The *trans*-bicyclo[9.3.0]tetradecane architecture of clavirolide C presents an interesting synthetic challenge. Reaction of diene **400** in the presence of 10 mol % of **6** in refluxing dichloroethane gave the corresponding carbocyclic macrocycle **401** in 70% yield with greater than 95% *trans* selectivity. The authors noted that their attempts to employ catalyst **4** led to less than 10% conversion to the desired macrocycle. In addition, ring-closing metathesis using either the free allylic alcohol or corresponding ketone led to complex mixtures of products (not shown).

Botta and co-workers employed a ring-closing metathesis strategy in the stereoselective synthesis of advanced intermediates in route to the total synthesis of taxuspines U and X, the biogenic precursors for Taxol.[173] In addition to an interesting and synthetically challenging architecture, taxuspines U and X are believed to have similar microtubule stabilizing properties to Taxol, and as such are of current interest for their medicinal properties. Ring-closing metathesis with **402** in the presence of either **3** or **4** using a number of reaction conditions failed to provide the desired 3,8-secotaxane diterpenoid **403**. The authors speculated that the failed ring-closing metathesis was the result the catalyst complexing with the more electron-rich alkyne in the presence of the electron rich OTBDPS group. Reduction of the alkyne to the alkene (not shown) and subsequent treatment of the corresponding ω-ω′-diolefin **404** with 10 mol % of **4** gave the corresponding 3,8-secotaxane diterpenoid **405** in 20% yield. In a later report, Botta and co-workers were finally able to achieve cyclization with a number of alkynyl substrates using 20 mol % **1** in toluene in low yeild (20% yeild).[174] This led the authors to speculate that the molecular constraints for cyclization when the alkyne is present may require too high of energy barrier for ring-closing metathesis to occur.

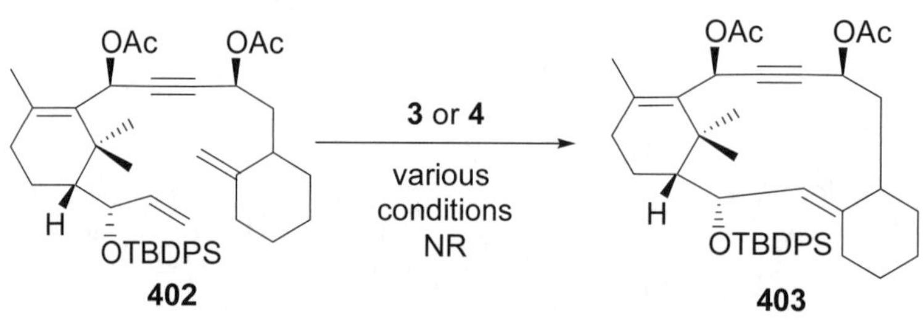

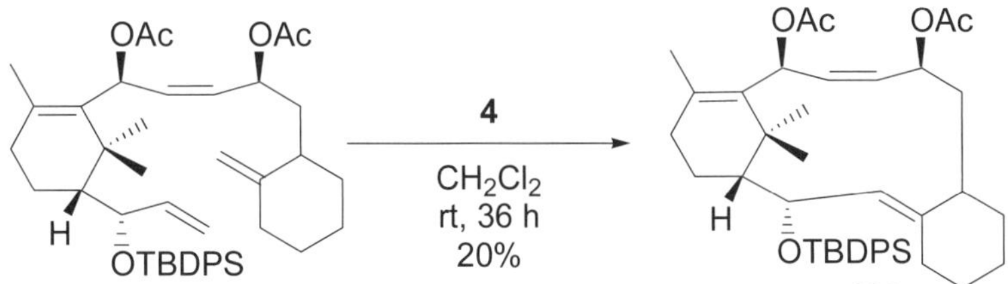

404 **405**

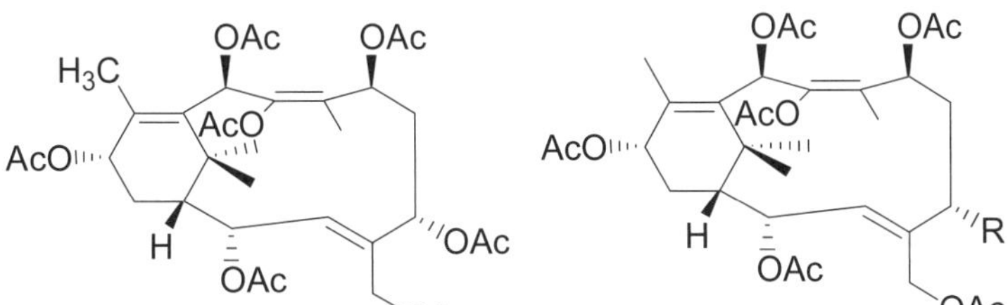

taxuspine U

taxuspine X
R = *trans*-cinnamoyl

Blechert and co-workers used ring-closing metathesis to synthesize the central bridged bicyclo[5.3.1]undecane moiety of Taxol.[175] Diene precursor **406**, prepared in nine steps from commercially available (–)-β-pinene, was subjected to ring-closing metathesis using 10 mol % of **3** in refluxing dichloromethane to give 58% of the desired product (**407**) in four hours. The authors noted that only one vinyl acetate cyclized to form the desired macrocycle, presumably due to the sterics of the bridgehead.

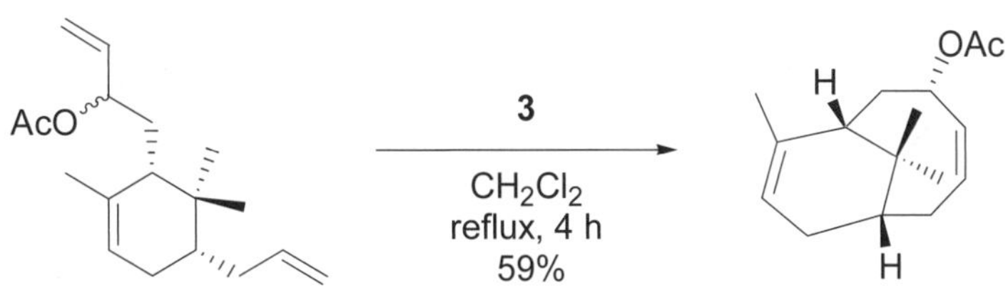

406 **407**

Deslongchamps and co-workers approach to the synthesis of Kempane diterpenes employed a ring-closing metathesis strategy in the construction of the 13-membered ring carbocycle, which was subsequently subjected to a transannular Diels–Alder reaction to produce the tricyclic core.[176] Treatment of tetraene ester **408** with **6** in refluxing toluene gave the corresponding triene **409** with a *trans-cis-cis* geometry as the only product in 74% yield.

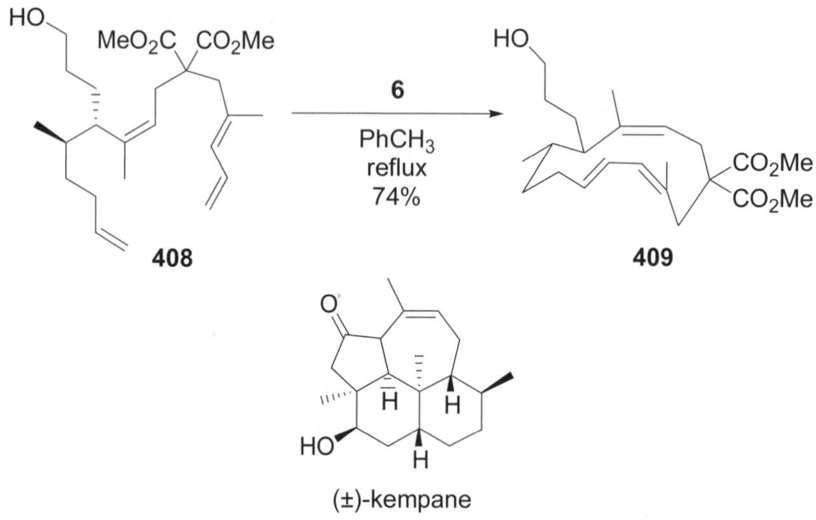

A number of ring-closing metathesis strategies have been employed in the the 13-membered core of roseophilin, a novel antitumor antibiotic with a unique pentacyclic skeleton. The first synthesis by Fuchs and co-workers employed diene **410** as a mixture of diasteromers.[177] Treatment of **410** with 30 mol % of **3** in refluxing dichloromethane gave the corresponding ansa-bridged silylether **411** as a single diastereomer in 60% yield. The modest yield was attributed the conformationally biased diene precursor. The research groups of Fürstner,[178] Hiemstra,[179] and Boger[180] used a similar approach in their syntheses of roseophilin in 1999, 2000, and 2001, respectively, although notably Boger achieved an 88% yield of the *ansa*-bridged macrocycle as a 1:1 mixture of the the *E* and *Z* isomers using a triene analog of **410** (not shown).

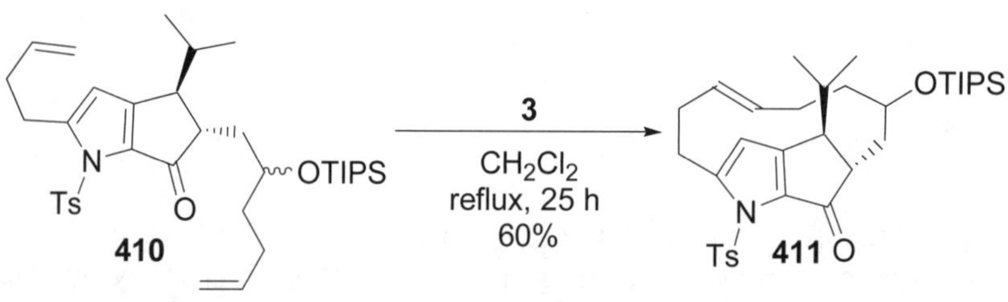

Tius and co-workers also used a ring-closing methathesis strategy in their construction of the macrocyclic ring of roseophilin.[181] Treatement of the more conformationally flexible olefin **412** with 30 mol % of **3** in refluxing dicholormethane gave the corresponding 13-membered ring carbocycle **413** in 90% yield.

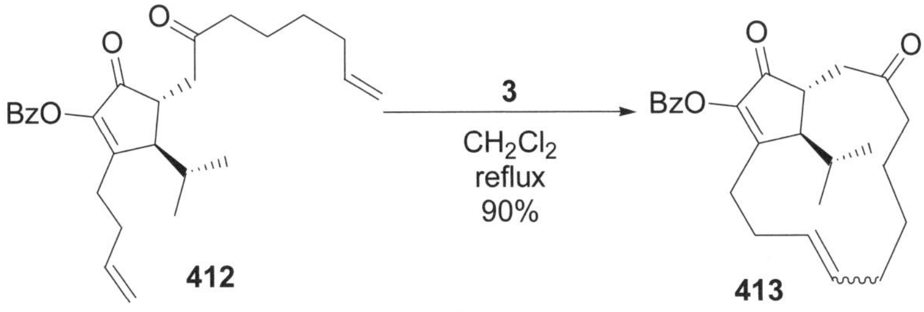

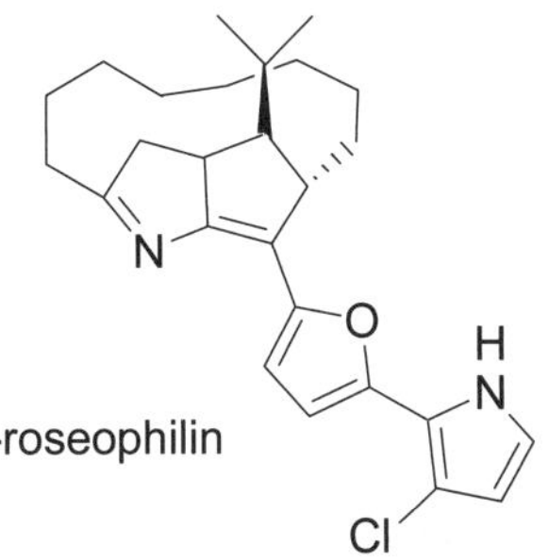

ent-roseophilin

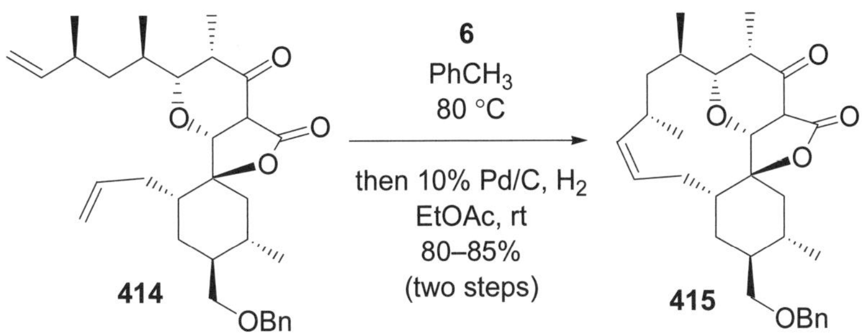

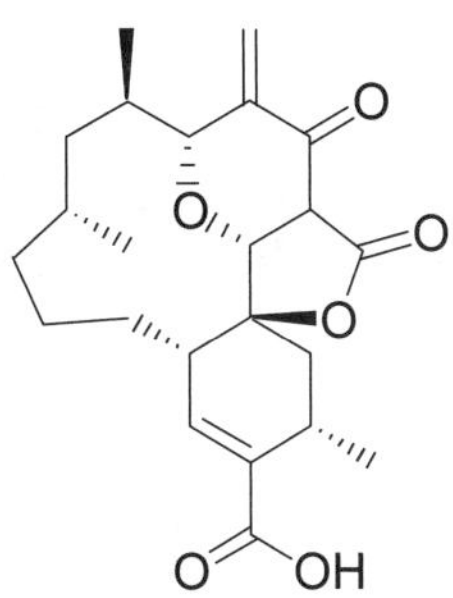

(–)-okilactomycin

Smith and co-workers used a ring-closing metathesis strategy in their synthesis of the 13-membered ring of (–)-okilactomycin, a novel polyketide antitumor antibiotic.[182] (–)-Okilactomycin is a considerable challenge synthetically due to its highly functionalized cyclohexene ring complete with a spirocenter and a 2,6-*cis*-tetrahdyropyronone moiety. Ring-closing metathesis of lactone **414** in the presence of 30 mol % of **6** in toluene followed by catalytic hydrogenolysis, gave the corresponding *cis*-alkene **415** in 80–85% yield over two steps. The authors noted that to achieve optimal yields and prevent dimerization, **6** was decomposed by air before concentration. A similar strategy was employed in their most recent report on (–)-okilactomycin (not shown).[183]

Nakata and co-workers recently employed a ring-closing metathesis strategy to construct the 14-membered carbocyclic core of methyl sarcophytoate, a biscembranoid.[184] Olefin metathesis of diene **416** using a stoichiometric amount of **4** in refluxing benzene gave the corresponding macrocycle **417** in 43 % yield.

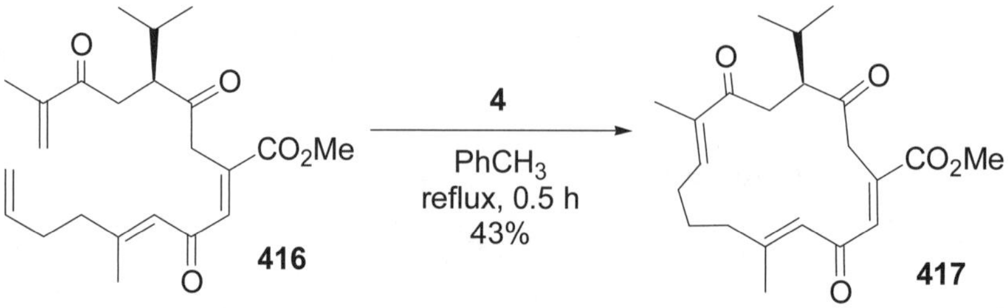

416 **417**

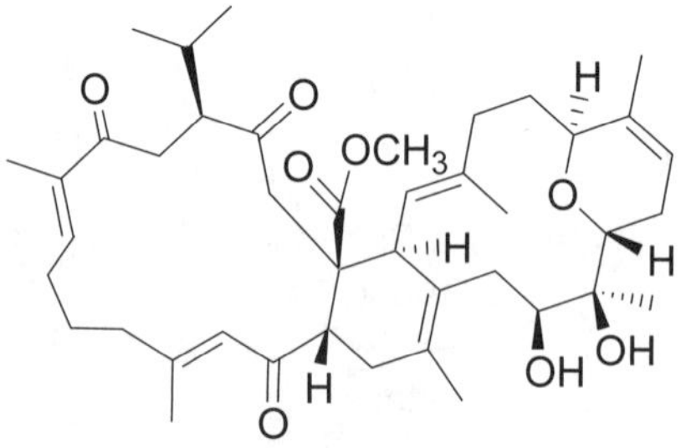

methyl sarcophytoate

A ring-closing metathesis strategy was first employed by Furstner and co-workers in the synthesis of civetone, a macrocyclic musk.[185] Treatment of diene **418** using 5 mol % of **2** in refluxing dichloromethane led to the corresponding 15-member ring macrocycle **419** in 72% yield with an *E*:*Z* selectivity of 4.6 to 1.

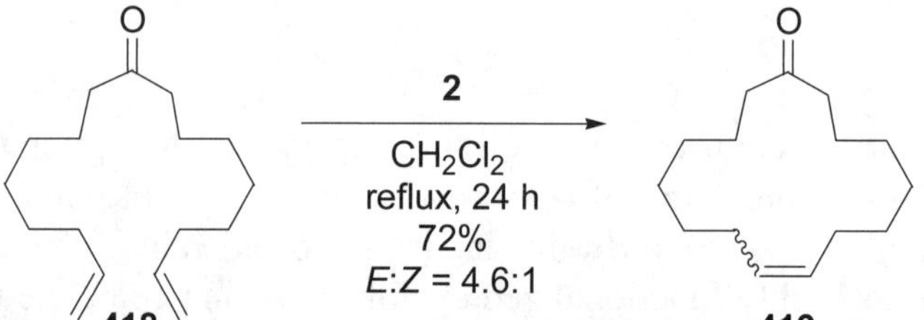

418 **419**

civetone

Hagiwara and co-workers employed a similar metathesis strategy in their synthesis of (*R*)-(–)-muscone from (+)-citronellal.[186] Treatment of ketone **420**, generated in seven steps from commercially available (+)-citronellal, with 5 mol % of **3** furnished the corresponding 15-membered ring carbocycle **421** in 78% yield as a mixture of *E* and *Z* isomers. Hydrogenation of **421** led to the final desired product (*R*)-(–)-muscone (not shown). A similar protocol was employed in a recent synthesis by the same group.[187]

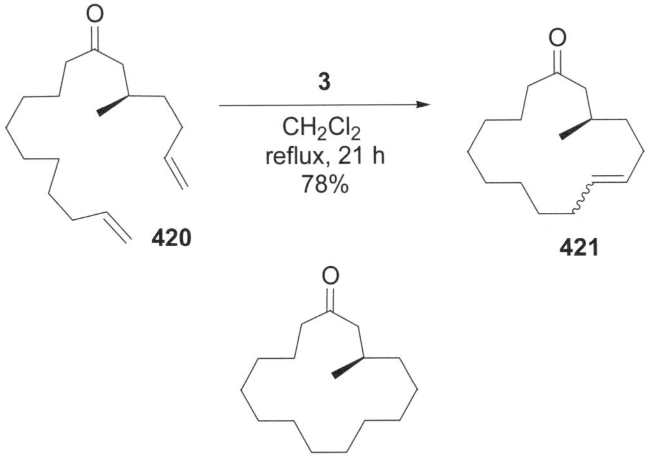

(*R*)-(–)-muscone

Trost and co-workers used ring-closing metathesis in their total synthesis of (–)-terpestacin.[188] Terpetacin is a known inhibitor of syncytia, produced by HIV infected cells, and has also been shown to inhibit angiogenesis. Treatment of **422** with 10 mol % of **4** in benzene gave the corresponding 15-membered ring carbocycle **423** in 35–45% yields. The authors noted that temperature and the presence of the allylic alcohol were critical in the chemoselectivity of the ring-closing metathesis.

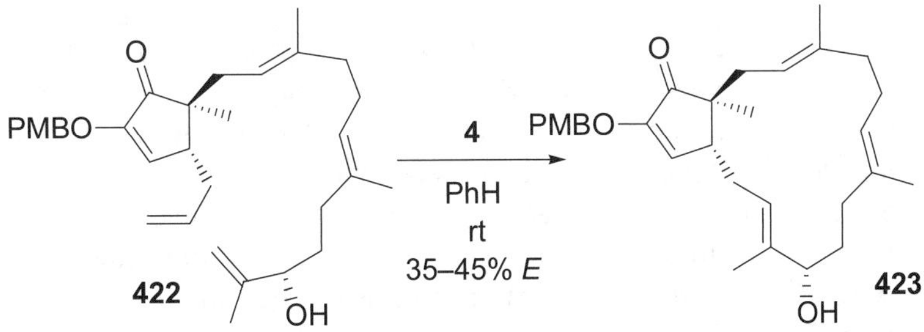

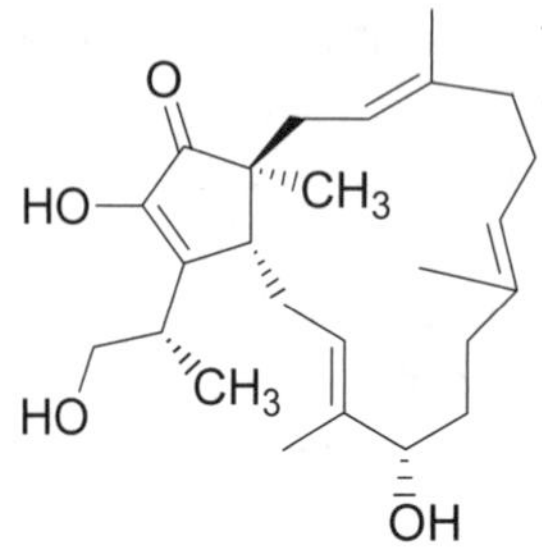

terpestacin

Collins and co-workers used relay ring-closing metathesis to synthesize the macrocyclic component of longithorone C, a farnesylated quinone with a macrocyclic [12]paracyclophane skeleton.[189] Reaction of ester **424** with 10 mol % of a **4** provided the corresponding macrocycle **425** in 68% yield, based on recovery of starting material. Collins recently expanded on this work using catalyst **8** to provide comparable yields (not shown).[190] Several other groups including Smith,[191] Kotha[192] and Suzuki[193] have also employed ring-closing metathesis strategies in the synthesis of related cyclophane derivatives (not shown).

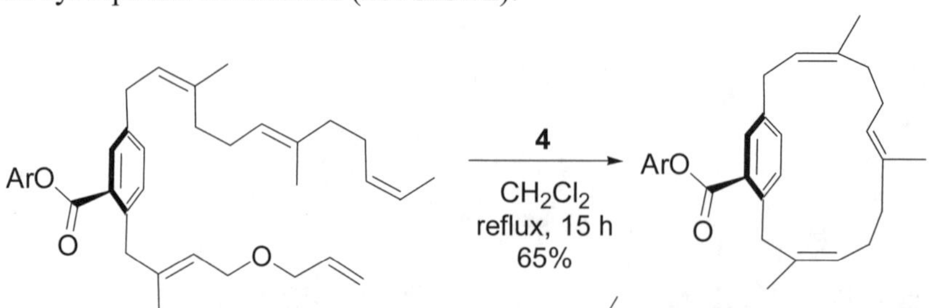

424, Ar=3,5-bis(CF_3)Ph

425, Ar=3,5-bis(CF_3)Ph

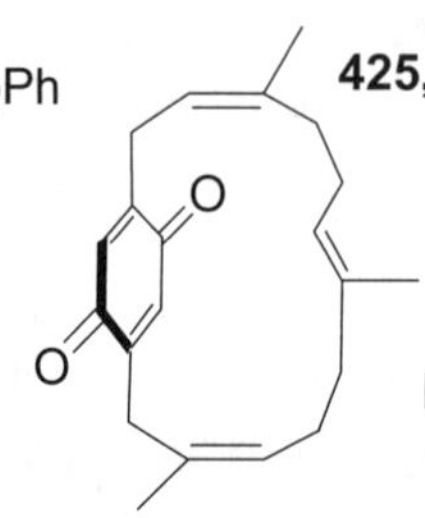

longithorne C

Smith and co-workers recently used a Petasis–Ferrier rearrangement/ ring-closing metathesis to construct the 18-membered ring macrocyclic core of (–)-kendomycin, a polyketide macrocyclic endothelin receptor antagonist and anti-osteoporotic.[194] Reaction of diene **426**, with 10 mol % of **4** in refluxing dichloromethane provided macrocycle **427** in 57% yield as the *Z* isomer, rather than the desired *E* isomer. Despite obtaining the undesired stereioisomer, Smith's report is the first synthesis of an α-branched trisubstituted olefin in a a large macrocylic structure. Compound **427** was cleverly converted to the *E* isomer in four steps, and three additional steps were required to reach the final product, (–)-kendomycin.

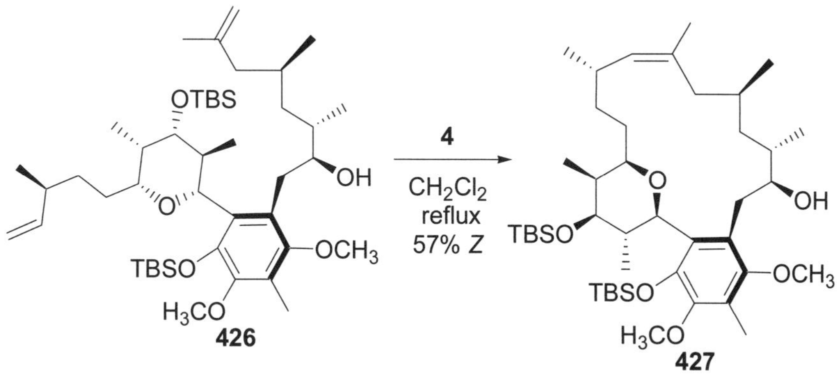

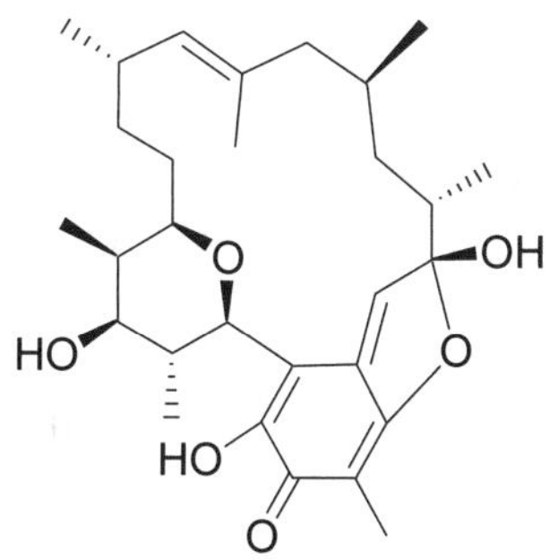

(–)-kendomycin

5.3.6 *Experimental*

The following example, adapted from Grubbs original report on ring-closing metathesis is still considered the standard for most ring-closing metathesis reactions.

tert-Butyl-(cyclopent-3-enyloxy)-dimethyl-silane[36]

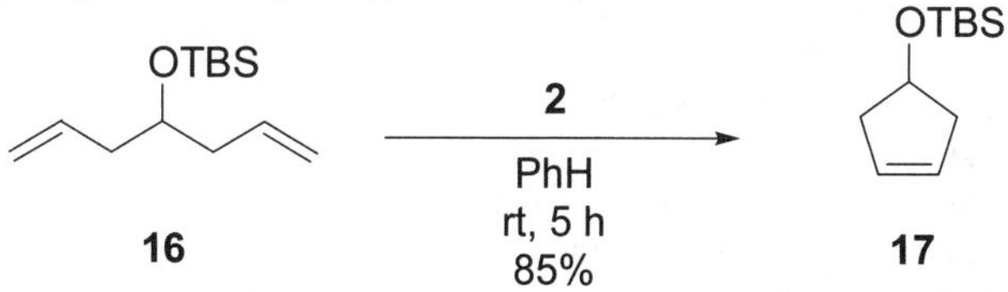

Typical experimental procedure: The diene **16** (0.5 mmol) was added to a homogenous orange red solution of **2** (0.01 mmol) in dry PhH (15 mL, 0.001 M) under argon. The resulting mixture was stirred at 20 °C for 5 h, at which time TLC showed the reaction to be complete. The reaction mixture was quenched by exposure to air, concentrated, and purified by flash chromatography (0 to 6% diethyl ether/hexanes) to give a colorless oil.

The following procedure, adapted from Percy and co-workers synthesis of difluorinated cyclooctenoids, is a useful method when $Ti(Oi\text{-}Pr)_4$ is to be used as a precatalyst.

2,2-Difluoro-3-hydroxy-7,7-dimethyl-cyclooct-4-enone[143]

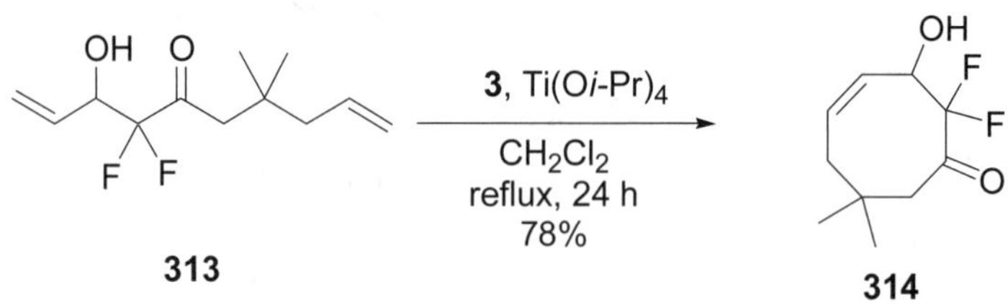

Diene (1.28 mmol), $Ti(Oi\text{-}Pr)_4$ (0.422 mmol), and catalyst (0.064 mmol) were dissolved in dried, degassed dichloromethane (512 mL). The solution was allowed to reflux under inert atmospheric conditions for 24 h or until complete as monitored by ^{19}F NMR. The solvent was removed under reduced pressure and the residue was taken up in diethyl ether (5 mL), filtered and concentrated under reduced pressure. The residue was taken up in methanol (1 mL) then eluted through a Stratospheres DPE tube, eluting with methanol (5 × 2 mL). The solution was concentrated under reduced pressure to afford a brown oil, which was purified by flash chromatography (silica gel, 20% diethyl ether/hexane) to give the cycloctenone as colorless solid.

5.3.7 References

1. Tsuji, J.; Hashiguch, S. *Tetrahedron Lett.* **1980**, *21*, 2955–2958.
2. Villemin, D. *Tetrahedron Lett.* **1980**, *21*, 1715–1718.
3. Schrock, R. R.; Murdzek, J. S.; Bazan, G. C.; Robbins, J.; DiMare, M.; O'Regan, M. *J. Am Chem. Soc.* **1990**, *112*, 3875–3886.
4. (a) Fu, G. C.; Grubbs, R. H. *J. Am. Chem. Soc.***1992**, *114*, 5426–5427. (b) Fu, G. C.; Grubbs, R. H. *J. Am. Chem. Soc.* **1992**, *114*, 7324–7325. (c) Fu, G. C.; Nguyen, S. T. Grubbs, R. H. *J. Am. Chem. Soc.***1993**, *115*, 9856–9857. (d) Nguyen, S. T. Grubbs, R. H.; Ziller, J. W. *J. Am. Chem. Soc.***1993**, *115*, 9858–9859
5. a) Adlhart, C.; Hinderling, C.; Baumann, H.; Chen, P. *J. Am. Chem. Soc.* **2000**, *122*, 8204–8214. (b) Sanford, M. S.; J. A. Love, J. A.; Grubbs, R. H. *J. Am. Chem. Soc.* **2001**, *123*, 654–6554.
6. (a) Cavallo, L. *J. Am. Chem. Soc.***2002**, *124*, 8965–8973. (b) Adlhart, C.; Chen, P. *J. Am. Chem. Soc* **2004**, *126*, 3496–3510. (c) Straub, B. F. *Angew. Chem.* **2005**, *117*, 6129–6132.; (d) Straub, B. F. *Angew.Chem., Int. Ed.* **2005**, *44*, 5974–5978
7. Romero, P. E.; Piers, W. E. *J. Am. Chem. Soc.* **2005**, *127*, 5032–5033.
8. Armstrong, S. K. *J. Chem. Soc., Perkin Trans. 1* **1998**, 371–388.

9. [R] Illuminati, G.; Mandolin L. *Acc. Chem. Res.* **1981**, *14*, 95–102.
10. Maier, M. E. *Angew. Chem., Int. Ed.***2000**, *39*, 2073–2077.
11. (a) Bruice, T. C.; Pandit, U. K. *J. Am. Chem. Soc.***1960**, *82*, 5958–5865. (b) Schleyer, P. V. R. *J. Am. Chem. Soc.* **1961**, *83*, 1368–1373.
12. [R] Jung, M. E.; Puzzi, G. *Chem. Rev.* **2005**, *105*, 1735–1766, and references therein.
13. Hoye, T. R.; Zhao, H. *Org. Lett.* **1999**, *1*, 1123–1125.
14. [R] For a recent review on the advances and limitation in ring-closing metathesis, see Conrad, K.; Fogg, D. E. *Curr. Org. Chem.* **2006**, *10*, 185–202.
15. Sanford, M. S.; Love, J. A.; Grubbs, R. H. *J. Am. Chem. Soc.* **2001**, *123*, 6543–6554.
16. Adjiman, C. S.; Clarke, A. J.; Cooper, G.; Taylor, P. C. *Chem. Comm.* **2008**, 2806–2808.
17. [R] For a recent review on new developments in olefin metathesis catalysts see: Michrawska, A.; Greala, K. *Pure Appl. Chem.* **2008**, *80*, 34–43.
18. (a) Dunne, A. M.; Mix, S.; Blechert, S. *Tetrahedron Lett.* **2003**, *44*, 2733–2736. (b) Buschmann, N.; Wakamatsu, H.; Blechert, S. *Synlett* **2004**, *4*, 667–670.
19. Bieniek, M.; Michrowska, A.; Gulajski, L.; Grela, K. *Organometallics* **2007**, *26*, 1096–1099.
20. Stewart, I. C.; Ung, T.; Petnev, A. A.; Berlin, J. M.; Grubbs, R. H.; Schrodi, Y. *Org. Lett.* **2007**, *9*, 1589–1592.
21. Hoye, T. R.; Jeffrey, C. S.; Tennakoon, M. A.; Wang, J.; Zhao, H. *J. Am. Chem. Soc.* **2004**, *126*, 10210–10211.
22. (a) Mohr, B.; Lynn, D. M.; Grubbs, R. H. *Orgnaometallics* **1995**, *15*, 4317–4325. (b) Krikland, T. A.; Lynn, D. M.; Mohr, B.; Grubbs, R. H. *J. Org. Chem.* **1998**, *63*, 9904–9909. (c) Lynn, D. M.; Mohr, B.; Grubbs, R. H.; Henling, L. M.; Day, M. W. *J. Am. Chem. Soc.* **2000**, *122*, 6601–6609. (d) Lynn, D. M.; Grubbs, R. H. *J. Am. Chem. Soc.* **2001**, *123*, 3187–3193.
23. Roberts, A. N.; Cochran, A. C.; Rankin, D. A.; Lowe, A. B.; Schanz, H-J. *Organometallics* **2007**, *26*, 6515–6518.
24. Gallivan, J. P.; Jorda, J. P.; Grubbs, R. H. *Tetrahedron Lett.* **2005**, *46*, 2577–2580.
25. Hong, S. H.; Grubbs, R. H. *J. Am. Chem. Soc.* **2006**, *128*, 3508–3509.
26. Jordan, J. P.; Grubbs, R. H. *Angew. Chem., Int. Ed.* **2007**, *46*, 5152–5155.
27. (a) Michrowska, A.; Gulajski, L.; Kaczmarska, Z.; Mennecke, k.; Krischning, A.; Grela, K. *Green Chem.* **2006**, *8*, 685–688. (b) Rix, D.; Clavier, H.; Gulajski, L.; Grela, K.; Mauduit, M. *Chem. Comm.* **2007**, 3771–3773.
28. Clavier, H.; Grela, K.; Kirschning, A. Mauduit, M.: Nolan, S. P. *Angew. Chem., Int. Ed.* **2007**, *46*, 6786–6801.
29. Binder, J. B.; Raines, R. T. *Curr. Opin. Chem. Biol.* **2008**, *12*, 767–773.
30. Keitz, B. K.; Grubbs, R. H. *Organometallics* ASAP.
31. [R] For a recent review on sustainable concept in olefin metathesis, see Clavier, H.; Grela, K.; Krisching, A.; Mauduit, M.; and Nolan, S. P. *Angew. Chem., Int. Ed.* **2007**, *46*, 6786–6801.
32. For representative examples, see (a) Maynard, H. D.; Grubbs, R. H. *Tetrahedron Lett.* **1999**, *40*, 4137–4140. (b) Ahn, Y. M.; Yang, K.-L.; Georg, G. I. *Org. Lett.* **2001**, *3*, 1411–1413. (c) Cho, J. H.; Kim, B. M. *Org. Lett.* **2003**, *5*, 531–533. (d) Nicola, T.; Brenner, M.; Donsbach, K.; Kreye, P. *Org. Process Res. Dev.* **2005**, *9*, 513–515. (e) Yee, N. K.; Farina, V.; Houpis, I. N.; Haddad, N.; Frutos, R. P.; Gallou, F.; Wang, X.-J.; Wei, X.; Simpson, R. D.; Feng, X.; Fuchs, V.; Xu, Y.; Tan, J.; Zhang, L.; Xu, J.; Smith-Keenan, L. L.; Vitous, J.; Ridges, M. D.; Spinelli, E. M.; Johnson, M.; Donsbach, K.; Nicola, T.; Brenner, M.; Winter, E.; Kreye, P.;

Samstag, W. *J. Org. Chem.* **2006**, *71*, 7133–7145. (f) Goldup, S. M.; Pilkington, C. J.; White, A. J. P.; Burton, A.; Barrett, A. J. M. *J. Org. Chem.* **2006**, *71*, 6185–6191. (g) Semeril, D.; Olivier-Bourbigou, H.; Bruneau, C.; Dixneuf, P. H. *Chem. Commun.* **2002**, 164–167.
33. For representative examples see: (a) Schurer, S. C.; Gessler, S.; Buschmann, N.; Blechert, S. *Angew. Chem.* **2000**, *112*, 4062–4065. (b) Schurer, S. C.; Gessler, S.; Buschmann, N.; Blechert, S. *Angew. Chem., Int. Ed.* **2000**, *39*, 3898–3901. (c) Mayr, M.; B. Mayr, B.; Buchmeiser, M. R. *Angew. Chem.* **2001**, *113*, 3957–3960. (d) Mayr, B.; Buchmeiser, M. R. *Angew. Chem., Int. Ed.* **2001**, *40*, 3839–3842. (e) Mayr, M.; Buchmeiser, M. R.; Wurst, K. *Adv. Synth. Catal.* **2002**, *344*, 712–719. (f) Pruhs, S.; Lehmann, C. W.; Furstner, A. *Organometallics* **2004**, *23*, 280–287. (g) Mayr, M.; Wang, D.; Kroll, R.; Schuler, N.; Pruhs, S.; Furstner, A.; Buchmeiser, M. R. *Adv. Synth. Catal.* **2005**, *347*, 484–492. (h) Gallivan, J. P.; Jordan, J. P.; Grubbs, R. H. *Tetrahedron Lett.* **2005**, *46*, 257–2580. (i) Hong, S. H.; Grubbs, R. H. *J. Am. Chem. Soc.* **2006**, *128*, 3508–3509. (j) Sußner, M.; Plenio, H. *Angew. Chem.* **2005**, *117*, 704–7048. (k) Sußner, M.; Plenio, H. *Angew. Chem., Int. Ed.* **2005**, *44*, 6885–6888.
34. Allen, D. P.; Van Wingerden, M. M.; Grubbs, R. H. *Org. Lett.* **2009**, *11*, 1261–1264.
35. Fu, G. C.; Grubbs, R. H. *J. Am. Chem. Soc.* **1993**, *115*, 3800–3801.
36. Fu, G. C.; Nguyen, S.T.; Grubbs, R. H. *J. Am. Chem. Soc.* **1993**, *115*, 9856–9857.
37. Okada, A.; Ohshima, T.; Shibasaki, M. *Tetrahedron Lett.* **2001**, *42*, 8023–8027.
38. Aggarawa, V.; Daly, A. M. *Chem. Comm.* **2002**, 2490–2491.
39. Jung, M. E. *Synlett* **1999**, 843–846.
40. Heo, J.; Micalizio, G. C.; Roush, W. R. *Org. Lett.* **2003**, *5*, 1693–1696.
41. Parks, B. W.; Gilbertson, R. D.; Domaille, D. W.; Hutchison, J. E. *J. Org. Chem.* **2006**, *71*, 9622–9627.
42. Chao, W.; Weinreb, S. M. *Org. Lett.* **2003**, *5*, 2505–2507.
43. Kotha, S.; Sreenivasachary, N.; Mohanraja, K.; Durani, S. *Bioorg. Med. Chem. Lett.* **2001**, *11*, 1421–1423.
44. Undheim, K.; Efskind, J. *Tetrahedron* **2000**, *56*, 4847–4857.
45. Probst, N. P.; Haudrechy, A.; Ple, K. *J. Org. Chem.* **2008**, *73*, 4339–4341.
46. Gardiner, J.; Anderson, K. H.; Downard, A.; Abell, A. *J. Org. Chem.* **2004**, *69*, 3375–3382.
47. Chippindale, A. M.; Davies, S. G.; Iwamoto, K.; Parkin, R. M.; Smethurst, C. A. P.; Smith, A. D.; Rodriquez–Solla, H. *Tetrahedron* **2003**, *59*, 3253–3265.
48. [R] For recent reviews see: (a) Madsen, R. *Eur. J. Org. Chem.* **2007**, 399–415. (b) Plumet, J.; Gomez, A. M.; Lopez, J. C. *Mini-Rev. Org. Chem.* **2007**, *4*, 201–216. (c) Argrofoglio, L. A.; Nolan, S. P. *Curr. Top. Med. Chem.* **2005**, *5*, 1541–1558.
49. Park, A.; Moon, H. R.; Kim, K. R.; Chun, M. W.; Jeong, L. S. *Org. Biomol. Chem.* **2006**, *4*, 4065–4067.
50. Liu, L. J.; Yoo, J. C.; Hong, J. H. *Nucleos. Nucleot. Nucl. Acids.* **2008**, *27*, 1186–1196.
51. Li, H.; Yoo, J. C.; Hong, J. H. *Nucleos. Nucleot. Nucl. Acids.* **2008**, *27*, 1238–1249.
52. [R] For a recent review see: Donohoe, T. J.; O'Riordan, T. J. C.; Rosa, C. P. *Angew. Chem., Int. Ed.* **2009**, *48*, 1014–1017.
53. Srikrishna, A.; Ravikumar, P. C. *Tetrahedron* **2006**, *62*, 9393–9402.
54. Srikrishna, A.; Kumar, S. R.; Ravikumar, P. C. *Synth. Comm.* **2007**, *37*, 4123–4140.
55. Srikrishna, A.; Khan, I. A.; Babu, R. R.; Sajjanshetty, A. *Tetrahedron* **2007**, *63*, 12616–12620.
56. Srikrishna, A.; Satyanarayana, G. *Tetrahedron.* **2006**, *62*, 2892–2900.

57. Srikrishna, A.; Lakshmi, B. V.; Ravikumar, P. C. *Tetrahedron Lett.* **2006**, *47*, 1277–1281.
58. Kulkarni, M. G.; Davawala, S. I; Shinde, M. P; Dhondge, A. P.; Borhade, A. S.; Chavan, S. W.; Gaikwad, D. D. *Tetrahedron Lett.* **2006**, *47*, 3027–3029.
59. Paquette, L. A.; Peng, X.; Yang, J.; Kang, H. *J. Org. Chem.* **2008**, *73*, 4548–4558.
60. Srikrishna, A.; Dethe, D.H. *Tetrahedron Lett.* **2003**, *44*, 7871–7820.
61. Srikrishna, A.; Pardeshi, V. H.; Satyanarayan, G. *Tetrahedron: Asymm.* **2008**, *19*, 1984–1991.
62. Srikrishna, A.; Babu, R. R. *Tetrahedron Lett.* **2007**, *48*, 6916–6919.
63. Majumdar, K. C.; Islam, R.; Rahaman, H.; Roy, B. *Org. Biomol. Chem.* **2006**, *4*, 2393–2398.
64. Srikrishna, A.; Babu, R. R. *Tetrahedron Lett.* **2007**, *48*, 6916–6919.
65. Gurjar, M. K.; Ravindranadh, S. V.; Sankar, K.; Karmakar, S.; Cherian, J.; Chorghade, S. *Org. Biomol. Chem.* **2003**, *1*, 1366–1373.
66. Hobson, S. J.; Marquez, R. *Org. Biomol. Chem.* **2006**, *4*, 3808–3814.
67. Hughes, R. C.; Dvorak, C. A.; Meyers, A. I. *J. Org. Chem.* **2001**, *66*, 5545–5551.
68. Roulland, E.; Monneret, C.; Florent, J. *J. Org. Chem.* **2002**, *67*, 4399–4406.
69. Ghosh, S.; Sinha, S.; Drew, M. G. B. *Org. Let.* **2006**, *8*, 3781–3784.
70. Wang, C.; Rath, N. P.; Covey, D. F. *Tetrahedron Lett.* **2006**, *47*, 7837–7839.
71. Ichikawa, Y.; Yamaoka, T.; Nakano, K.; Kotsuki, H. *Org. Lett.* **2007**, *9*, 2989–2992.
72. Mehta, G.; Singh, S. R. *Tetrahedron Lett.* **2005**, *46*, 2079–2082.
73. Srikrishna, A.; Beeraiah, B. *Tetrahedron: Asymm.* **2008**, *19*, 884–890.
74. Miller, J. F.; Termin, M.; Koch, K.; Piscopio, A. D. *J. Org. Chem.* **1998**, *63*, 3158–3159.
75. Francias, A.; Bedel, O.; Picoul, W.; Meddour, A.; Courteiu, J.; Haudrechy, A. *Tetrahedron: Asymm.* **2005**, *15*, 1141–1155.
76. Swift, M. D.; Donaldson, A.; Sutehrland, A. *Tetrahedron Lett.* 2**009**, *50*, 3241–3244.
77. (a) Rodriquez, C. M.; Ravelo, J. L.; Rodriquez, C. M. *Org. Lett.* **2004**, *6*, 25; (b) Ravelo, J. L.; Rodriguez, C. M.; Martin, B. S. *J. Organomet. Chem.* **2006**, *691*, 5326–5335.
78. Gagnon, D.; Lauzon, S.; Godbout, C.; Spino, C. *Org. Lett.* **2005**, *7*, 4769–4771.
79. Lee, C.-L. K.; Ling, H. Y.; Loh, T.-P. *J. Org. Chem.* **2004**, *69*, 7787–7789.
80. Gardiner, J.; Anderson, K. H.; Downard, A.; Abell, A. *J. Org. Chem.* **2004**, *69*, 3375–3382.
81. Hyldtoft, L.; Madsen, R. *J. Am. Chem. Soc.* **2000**, *122*, 8444–8452.
82. Doddi, V. R.; Kumar, A.; Vankar, Y. D. *Tetrahedron* **2008**, *64*, 9117–9122.
83. Verhelst, S. H. L.; Wennekes, T.; van der Marel, G. A.; Overkleeft, H. S.; van Boeckel, C. A. A.; van Boom, J. H. *Tetrahedron* **2004**, *60*, 2813–2822.
84. Totokotsopouos, S. M.; Koubis, A. E.; Gallos, J. K. *Tetrahedron* **2008**, *64*, 3998–4003.
85. Kumaraswamy, G.; Sadaiah, K.; Ramakrishna, D. S.; Police, N.; Sridhar, B.; Bharatam, J. *Chem. Comm.* **2008**, 5324–5326.
86. Cumpstey, I.; Gehrke, S.; Erfan, S.; Cribiu, R. *Carbohydr. Res.* **2008**, *343*, 1675–1692.
87. Ramstadius, C.; Hekmat, O.; Eriksson, L.; Stalbrand, H.; Cumpstey, I. *Tetrahedron: Asymm.* **2009**, *20*, 795–807.
88. Whalen, L. J.; Halcomb, R. L. *Org. Lett.* **2004**, *6*, 3221–3224.
89. Holt, D. J.; Barker, W. D.; Jenkins, P. R.; Davies, D. L.; Garratt, S.; Fawcett, J.; Russell, D. R.; Ghosh, S. *Angew. Chem., Int. Ed.* **1998**, *37*, 3298–3300.

90. Holt, D. J.; Barker, W. D.; Jenkins, P. R. *J. Org. Chem.* **2000**, *65*, 482–493.
91. Kireev, A. S.; Nadein, O. N.; Agustin, V. J.; Bush, N. E.; Evidente, A.; Manpadi, M.; Ogasawara, M. A.; Rastogi, S. K.; Rogelj. S.; Shors, S. T.; Kornienko, A. *J. Org. Chem.* **2006**, *71*, 5694–5707.
92. Srikrishna, A.; Satayanarayana, G.; Prasad, K. R. *Syn. Commun.* **2007**, *37*, 1511–1516.
93. Levine, S. R.; Krout, M. R.; Stoltz, B. M. *Org. Lett.* **2009**, *11*, 289–292.
94. Matovic, R.; Ivkovic, A.; Manojlovic, M.; Tokic-Vujosevic, Z.; Saicic, R. N. *J. Org. Chem.* **2006**, 71, 9411–9419.
95. Clive, D. J.; Liu, D. *J. Org. Chem.* **2002**, *41*, 2396–2398.
96. Srikrishna, A.; Pardeshi, V. H.; Satyanarayana, G. *Tetrahedron Lett.* **2007**, *48*, 4087–4090.
97. Srikrishna, A.; Beeraiah, B.; Gowri, B. V. *Tetrahedron* **2009**, *65*, 2649–2654.
98. Song, Y.; Hwang, S.; Gong, P.; Kim, D.; Kim, S. *Org. Lett.* **2008**, *10*, 269–271.
99. Trost, B. M.; Dogra, K. *Org. Lett.* **2007**, *9*, 861–863.
100. Ciampini, M.; Perlmutter, P.; Watson, K. *Tetrahedron: Asymm.* **2007**, *18,* 243–250.
101. Lejkowski, M.; Banerjee, P.; Runskink, J.; Gals, H.-J. *Org. Lett.* **2008**, *10*, 2713–2716,
102. White, D. E.; Stewart, I. C.; Grubbs, R. H.; Stoltz, B. M. *J. Am. Chem. Soc.* **2008**, *130*, 810–811.
103. Lee, M.; lee, T.; Kim, E.-Y.; Ko, H.; Kim, D.; Kim, S. *Org. Lett.* **2006**, *8*, 745–748.
104. Simila, S. T. M.; Martin, S. F. *J. Org. Chem.* **2007**, *72*, 5342–5349
105. Simila, S. T. M.; Reichelt, A.; Martin, S. F. *Tetrahedron Lett.* **2006**, *47,* 2933–2936.
106. Brimble, M. A.; Trzoss, M. *Tetrahedron* **2004**, *60*, 5613–5622.
107. Pansare, S. V.; Adsool, V. A. *Org. Lett.* **2006**, *8*, 2035–2037,
108. Thornqvist, V.; Manner, S.; Wendt, O. F.; Frejd, T. *Tetrahedron* **2006**, *62*, 11793–11800.
109. Singh, B.; Sahu, P. K.; Sahu, B. C.; Mobin, S. M. *J. Org. Chem.* **2009**, *74*, 6092–6104.
110. Hanessian, S.; Margarita, R.; Hall, A.; Johnstone, S.; Termblya, M.; Parlanti, L. *J. Am. Chem. Soc.* **2002**, *124*, 13342–13343.
111. Robichaud, J.; Tremblay, F. *Org. Lett.* **2006**, *8*, 597–600.
112. Enev, V. S.; Drescher, M.; Mulzer, J. *Org. Lett.* **2008**, *10*, 413–416.
113. Dudley, G. B.; Engel, D. A.; Ghiviriga, I.; Lam, H.; Poon, K. W. C.; Singletary, J. A. *Org. Lett.* **2007**, *9*, 2839–2842.
114. Jones, S. B.; He, L.; Castle, S. L. *Org. Lett.* **2006**, *8*, 3757–3760.
115. Wakefield, B.; Halter, R. J.; Wipf, P. *Org. Lett.* **2007**, *9*, 3121–3124.
116. Jiao, L.; Hao, E.; Fronczek, F. R.; Graca, M.; Cicente, H.; Smith, K, M. *Chem. Commun.* **2006**, 3900–3902.
117. Paley, R. S.; Estroff, L. A.; Gauget, J-M.; Hunt, D. K.; Newlin, R. C. *Org. Lett.* **2000**, *2*, 365–368.
118. Greshock, T. J.; Funk, R. L. *Tetrahedron Lett.* **2006**, *47*, 5437–5439.
119. Chippindale, A. M.; Davies, S. G.; Iwamoto, K.; Parkin, R. M.; Smethurst, C. A. P.; Smith, A. D.; Rodriquez-Solla, H. *Tetrahedron* **2003**, *59*, 3253–3265.
120. Liu, Y.; Han, T.-X.; Yang, Z.-J.; Zhang, L.-R.; Zhang, L.-H. *Tetrahedron Asymm.* **2007**, *18*, 2326–2331.
121. Hanna, I.; Ricard, L. *Org. Lett.* **2000**, *2*, 2651–2654.
122. Boyer, F.-D.; Hanna, I. *Tetrahedron Lett.* **2001**, *42*, 1275–1277.
123. Skaanderup, P. R.; Madsen. R. *Chem. Comm.* **2001**, 1106–1107.
124. Marco-Contelles, J.; de Opazo, E. *J. Org. Chem.* **2002**, *67*, 3705–3717.
125. Czuk, R.; Prell, E.; Reissmann, S. *Tetrahedron* **2008**, *64*, 9417–9422.

126. Yokoe, H.; Sasaki, H.; Yoshimura, T.; Shindo, M.; Yoshida, M.; Shishido, K. *Org. Lett.* **2007**, *9*, 969–971.
127. Nosse, B.; Schall, A.; Jeong, W. B.; Resiser, O. *Adv. Synth. Catal.* **2005**, *347*, 1869–1874.
128. Oliver, S. F.; Hogenauer, K.; Simic, O.; Antonello, A.; Smith, M. D.; Ley. S. B. *Angew. Chem., Int. Ed.* 2**003**, *42*, 5996–6000.
129. Krafft, M. E.; Cheung, Y. Y.; Kerrigan, S. A.; Abboud, K. A. *Tetrahedron Lett.* **2003**, *44*, 839–843.
130. Mehta, G.; Lakshminath, S. *Tetrahedron Lett.* **2006**, *47*, 327–330.
131. Srikrishna, A.; Dethe, D. H. *Org. Lett.* **2004**, *6*, 165–168.
132. Nakashima, K.; Inoue, K.; Sono, M.; Tori, M. *J. Org. Chem.* **2002**, *67*, 6034–6040.
133. Michalak, K.; Michalak, M.; Wicha, J. *Tetrahedron Lett.* **2005**, *46*, 1149–1153.
134. Mehta, G.; Likhite, N. S. *Tetrahedron Lett.* **2008**, *49*, 7113–7116.
135. Llyod-Jones, G. C.; Murray, M.; Stentiford, R. A.; Worthington, P. A. *Eur. J. Org. Chem.* **2000**, 975–985.
136. Lysenko, I. L.; Lee, H. G.; Cha, J. K. *Org. Lett.* **2006**, *8*, 2671–2673.
137. Zhao, L.; Burnell, D. J. *Org. Lett.* **2006**, *8*, 155–157
138. Paquette, L. A.; Sauer, D. R.; Cleary, D. G.; Kinsella, M. A.; Blackwell, C. M.; Anderson, L. G. *J. Am. Chem. Soc.* **1992**, *114*, 7375–7387.
139. Singh, B.; Sahu, P. K.; Sahu, B. C.; Mobin, S. M. *J. Org. Chem.* **2009**, *74*, 6092–6104.
140. Bennasar, M.-L.; Zulaica, E.; Tummers, S. *Tetrahedron Lett.* **2004**, *45*, 6283–6285.
141. Chattopadhyay, S. K. Ghosh, D.; Neogi, K. *Synth. Comm.* **2007**, *37*, 1535–1543.
142. Miller, S, J.; Kim, S.-H.; Chen, Z.-R.; Grubbs, R. H. *J. Am. Chem. Soc.* **1995**, *117*, 2108–2109.
143. Kariuki, B. M.; Owton, W. M; Percy, J. M.; Pinant, S.; Smith, C. A.; Spencer, N.; Thomas, A. C.; Watson, M. *Chem. Comm.* **2002**, 228–229.
144. Ashworth, I. W.; Miles, J. A. L.; Nelson, D. J.; Percy, J. M. Singh, K. *Tetrahedron* **2009**, *65*, 9637–9646.
145. Comer, E.; Rohan, E.; Deng, L.; Porco, J. A. *Org. Lett.* **2007**, *9*, 2123–2126.
146. Boyer, F.-D.; Hanna, I.; Nolan, S. P. *J. Org. Chem.* **2001**, *66*, 4094–4096.
147. Gravier-Pelletier, C.; Andriuzzi, O.; Le Merrer, Y. *Tetrahedron Lett.* **2002**, *43*, 245–248.
148. Michalak, K.; Michalak, M.; Wicha, J. *Tetrahedron Lett.* **2005**, *46*, 1149–1153.
149. Paquette, L. A.; Tae, J.; Arrington, M. P.; Sadoun, A. H. *J. Am. Chem. Soc.* **2000**, *122*, 2742–2748.
150. Krafft, M. E.; Cheung, Y. Y.; Abboud, K. A. *J. Org. Chem.* **2001**, *66*, 7443–7448.
151. Takao, K.; Watanabe, G.; Yasui, H.; Tadano, K. *Org. Lett.* **2002**, *4*, 2941–2943.
152. Rodriguez, J. R.; Castedo, L.; Mascarenas, J. L. *Chem. Eur. J.* **2002**, *8*, 2923–2930.
153. Michaut, A.; Miranda-Garcia, S.; Menendez, J. C.; Rodriguez, J. *Org. Lett.* **2004**, *6*, 3075–3078.
154. Singh, V.; Pal, S.; Tosh, D. K.; Mobin, S. M. *Tetrahedron* **2007**, *63*, 2446–2454.
155. Paquette, L. A.; Efremov, I. *J. Am. Chem. Soc.* **2001**, *123*, 4492–4501.
156. Bourgeois, D.; Pancrazi, A.; Ricard, JL.; Prunet, J. *Angew. Chem., Int. Ed. Engl.* **2000**, *39*, 725–728. (b) Bourgeois, D.; Mahuteau, J.; Pancrazi, A.; Nolan, S. P.; Prunet, J. *Synthesis* **2000**, 869–882.
157. Schiltz, S.; Ma, C.; Ricard, L.; Prunet, J. *J. Organomet. Chem.* **2006**, *691*, 5438–5443.
158. Codesido, E. M.; Rodriquez, J. R.; Castedo, L.; Granja, J. R. *Org. Lett.* **2002**, *4*, 1651–1654.

159. Garcia-Fandino, R.; Aldegunde, M. J.; Codesido, E. M.; Castedo, L.; Granja, J. R. *J. Org. Chem.* **2005**, *70*, 8281–8290.
160. Singh, B.; Sahu, P. K.; Sahu, B. C.; Mobin, S. M. *J. Org. Chem.* **2009**, *74*, 6092–6104.
161. Bennasar, M.-L.; Zulaica, E.; Tummers, S. *Tetrahedron Lett.* **2004**, *45*, 6283–6285.
162. Clark, J. S.; Marlin, F.; Nay, B.; Wilson, C. *Org. Lett.* **2003**, *5*, 89–92.
163. Rodriquez, J. R.; Costedo, L.; Mascarenas, J. L. *Org. Lett.* **2000**, *2*, 3209–3212.
164. Clark, J. S.; Northall, J. M.; Marlin, F.; Nay, B.; Wilson, C.; Blake, A. J.; Waring, M. J. *Org. Biomol. Chem.* **2008**, *6*, 4012–4025.
165. Tang, H.; Yusuff, N.; Wood, J. L. *Org. Lett.* **2001**, *3*, 1563–1566.
166. Winkler, J. D.; Rouse, M. B.; Greany, M. F.; Harrison, S. J.; Jeon, Y. T. *J. Am. Chem. Soc.* **2002**, *124*, 9726–9728.
167. Hayakawa, I.; Asuma, Y.; Ohyoshi, T.; Aoki, K.; Kigoshi, H. *Tetrahedron Lett.* **2007**, *48*, 6221–6224.
168. Larrosa, I.; Da Silva, M. I.; Gomez, P. M.; Hannen, P.; Ko, E.; Lenger, S. R.; Linke, S. R.; White, A, J. P.; Wilton, D.; Barrett, A. G. M. *J. Am. Chem. Soc.* **2006**, *128*, 14042–14043.
169. (a) Caggiano, L.; Castoldi, D.; Beumer, R.; Bayon, P.; Tesler, J.; Gennari, C. *Tetrahedron Lett.* **2003**, *44*, 7913–7919. (b) Castoldi, D.; Caggiano, L.; Bayon, P.; Costa, A. M.; Capella, P.; Sharon, O.; Gennari, C. *Tetrahedron* **2005**, *61*, 2123–2139.
170. Nakamura, T.; Oshida, M.; Nomura, T.; Nakazaki, A.; Kobayashi, S. *Org. Lett.* **2007**, *9*, 5533–5536.
171. Nicolaou, K. C.; Harrison, S. T. *J. Am. Chem. Soc.* **2007**, *129*, 429–440.
172. Brown, M. K.; Hoyveda, A. H. *J. Am. Chem. Soc.* **2008**, *130*, 12904–12906.
173. Renxulli, M. L.; Rocheblave, L.; Avramova, S. I.; Galletti, E.; Castagnolo, D.; Tafi, A.; Corelli, F.; Botta, M. *Tetrahedron* **2007**, *63*, 497–509.
174. Galletti, E.; Avramova, S. I. Renzulli, M. L.; Corelli, F.; Botta, M. *Tetrahedron Lett.* **2007**, *48*, 751–754.
175. Wenz, M.; Grossbach, D.; Beitzel, M.; Blechert, S. *Synthesis* **1999**, 607–614.
176. Caussanel, F.; Wang, K.; Ramachandran, S. A.; Deslongchamps, P. *J. Org. Chem.* **2006**, 71, 7370–7377.
177. Kim, S. H.; Figueroa, I.; Fuchs, P. L. *Tetrahedron Lett.* **2004**, *38*, 2601–2604.
178. Furstner, A.; Gastner, T.; Weintritt, H. *J. Org. Chem.* **1999**, *64*, 2361–2366.
179. Bamford, S. J.; Lucker, T.; Speckamp, W. N.; Hiemstra, H. *Org. Lett.* **2000**, *2*, 1157–1160.
180. Boger, D. L.; Hong, J. *J. Am. Chem. Soc.* **2001**, *123*, 8515v8519.
181. Harrington, P. S.; Tius, M. A. *Org. Lett.* **1999**, *1*, 649–651.
182. Smith, A. B.; Basu, K.; Bosanac, T. *J. Am. Chem. Soc.* **2007**, *129*, 14872–14874.
183. Smith, A. B.; Bosanac, T.; Basu, K. *J. Am. Chem. Soc.* **2009**, *131*, 2348–2358.
184. Ichige, T.; Okano, Y.; Kanoh, N.; Nakata, M. *J. Org. Chem.* **2009**, *74*, 230–243.
185. Furstner, A.; Langemann, J. *Synthesis* **1997**, 792–798.
186. Kamat, V. P.; Hagiwara, H.; Katsumi, T.; Suzuki, T.; Ando, M. *J. Chem. Soc., Perkin Tranas. 1,* **1998**, 2253–2254.
187. Kamat, V. P.; Hagiwara, H.; Katsumi, T.; Hoshi, T.; Suzuki, T.; Ando, M. *Tetrahedron* **2000**, *56*, 4397–4403.
188. Trost, B. M.; Dong, G.; Vance, J. A. *J. Am. Chem. Soc.* **2007**, 129, 450–4541.
189. El-Azizi, Y.; Schmitzer, A.; Collins, S. K. *Angew. Chem., Int. Ed.* **2006**, *45*, 968–973.
190. Zakarian, J. E.; El-Azizi, Y.; Collins, S. K. *Org. Lett.* **2008**, *10*, 2927–2930.

191. Smith, A. B.; Adams, C. M.; Kozmin, S. A.; Paone, D. B. *J. Am. Chem. Soc.* **2001**, *123*, 5925–5937.
192. Kotha, S.; Mandal, K. *Eur. J. Org. Chem.* **2006**, 5387–5393.
193. Mori, K.; Ohmori, k.; Suzuki, K. *Angw. Chem., Int. Ed.* **2009**, *48*, 5658–5641.
194. Smith, A. B.; Mesaros, E. F.; Meyer, E. A. *J. Am. Chem. Soc.* **2006**, *128*, 5292–5299

5.4 Thorpe–Ziegler Cyclization

Richard J. Mullins and Michael W. Danneman

5.4.1 Description

The Thorpe–Ziegler cyclization is the intramolecular condensation of dinitriles to yield imines which ultimately tautomerize to the corresponding enamine.[1] The enamine can be hydrolyzed to yield the β-ketonitrile; under more harsh conditions, hydrolysis of the nitrile results in decarboxylation to yield the ketone.

5.4.2 Historical Perspective

In 1904, Jocelyn Field Thorpe and co-workers conducted the first studies of the intermolecular dimerization of nitriles.[2] A short time later, the first example of a dinitrile cyclization was described by Moore and Thorpe.[3] Initially, all products of dinitrile cyclization reactions were characterized as imines. However, in 1955, it was determined that the product of the cyclization of adiponitrile was in actuality an enamine.[4] The enamine structure for the product of dinitrile cyclizations has since been confirmed through numerous other studies, and makes sense chemically. The conjugated enamine would be expected to be more stable than the corresponding nonconjugated β-iminonitrile.

In 1933, Ziegler and co-workers demonstrated that Thorpe's reaction could be applied toward the synthesis of cyclic ketones ranging from seven to 33 carbons in size.[5,6] In a series of studies, it was demonstrated that these cyclic ketones were optimally produced by conducting the reaction under highly diluted conditions to avoid bimolecular reactions. Diethyl ether was found to be the ideal solvent, and soluble amide bases were found to give the most efficient cyclization.[6–11] The contributions made by Ziegler toward the development and generalization of this reaction have resulted in its being known as the Thorpe–Ziegler cyclization. For a more thorough discussion of the history of the Thorpe–Ziegler cyclization, the reader is directed here.[12]

5.4.3 *Mechanism*

The Thorpe–Ziegler cyclization is characterized by a relatively straightforward mechanism. As demonstrated below, deprotonation of dinitrile **1** results in the formation of the anionic species **2**. Intramolecular cyclization in a manner similar to the well-known Dieckmann condensation yields **3**. Workup under aqueous conditions then produces imine **4**, which immediately tautomerizes to the conjugated enamine **5**.

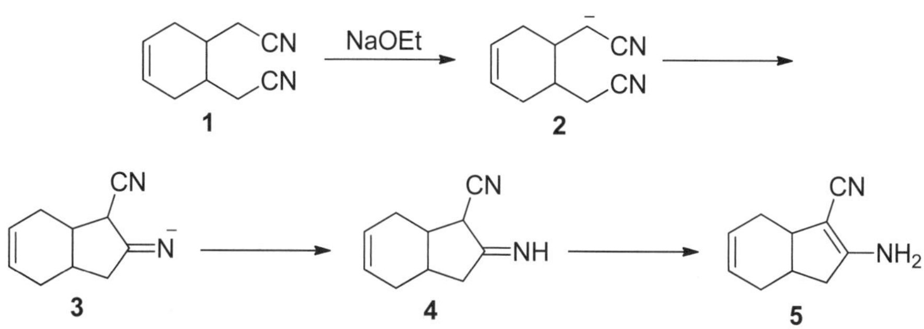

5.6.4. *Variations and Improvements*

Although the Thorpe–Ziegler name is typically limited to the self-condensation of nitriles, many other intramolecular condensation reactions are often referred to in the same manner when a nitrile is the electrophile.[13] While this review focuses specifically on the dinitrile variant of the reaction, the reader's attention is directed to a wealth of literature on these related reactions, commonly used in heterocycle synthesis.[14–26] A few examples of these reactions are shown below, as used for the preparation of **6**,[27] **7**,[15] and **8**.[28]

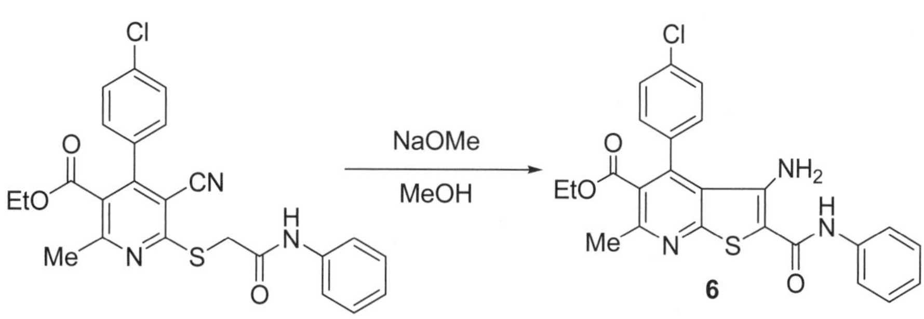

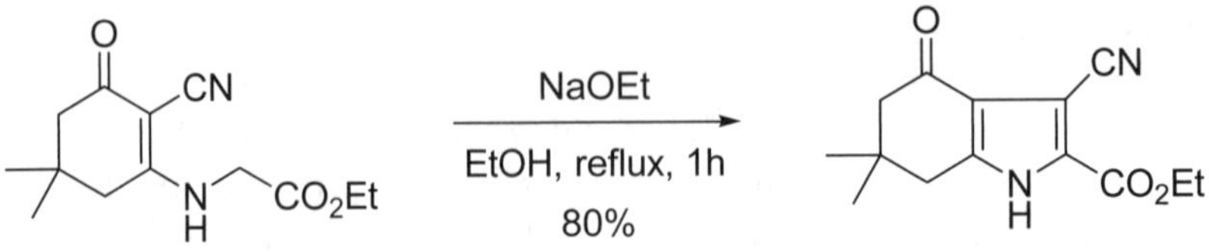

7

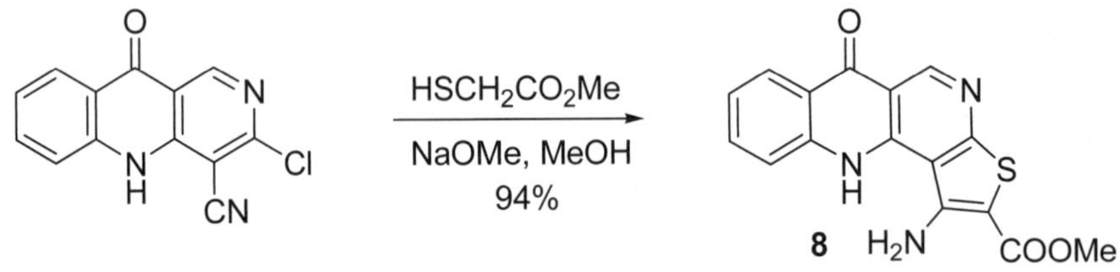

5.4.5 *Synthetic Utility*

The Thorpe–Ziegler cyclization has found substantial utility in synthetic endeavors. Specifically, it has found broad scope for a variety of heterocyclic compounds. As demonstrated in the work of Salaheldin and co-workers, a one-pot alkylation/Thorpe–Ziegler sequence resulted in the synthesis of 3-aminopyrroles.[29] Treatment of β,β-enaminonitrile **9** with chloroacetonitrile under basic conditions results in alkylation to produce intermediate **10**, which undergoes a spontaneous Thorpe–Ziegler cyclization to create the 3-aminopyrrole **11**. Notably, the use of Et_3N as base resulted in a substantial improvment compared to the use of K_2CO_3 in DMF.

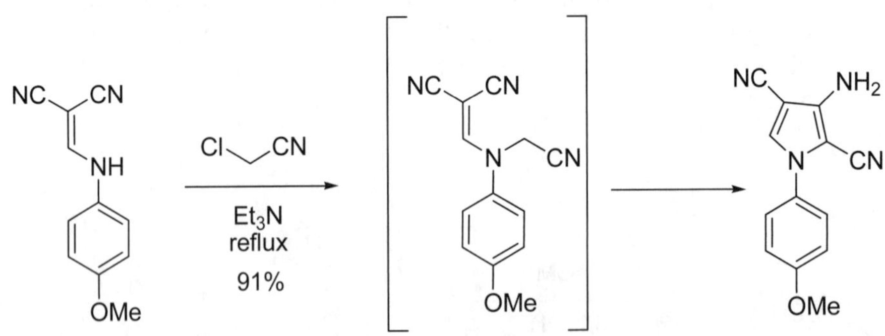

9 **10** **11**

A similar procedure, demonstrated by Tsolomitis and co-workers, has found utility in the synthesis of pentasubstituted pyrroles.[30] Treatment of amide **12** with $POCl_3$ results in intermediate **13**, which, upon reaction with malononitrile is presumed to yield intermediate **14**. Before isolation of **14**, the Thorpe–Ziegler cyclization occurs to give pyrrole **15**.

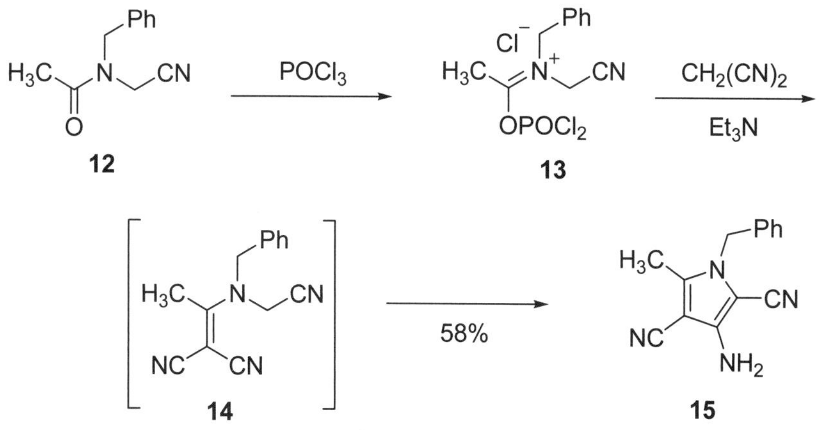

The Thorpe–Ziegler reaction has also been used for the preparation of functionalized thienopyridines. Alkylation of **16** by chloroacetonitrile first results in the formation of **17**. Treatment of **17** with sodium ethoxide results in a Thorpe–Ziegler reaction to produce **18** in excellent yield. Thienopyridine **18** can also be secured directly by treatment of **16** with chloroacetonitrile in the presence of ethoxide.[20]

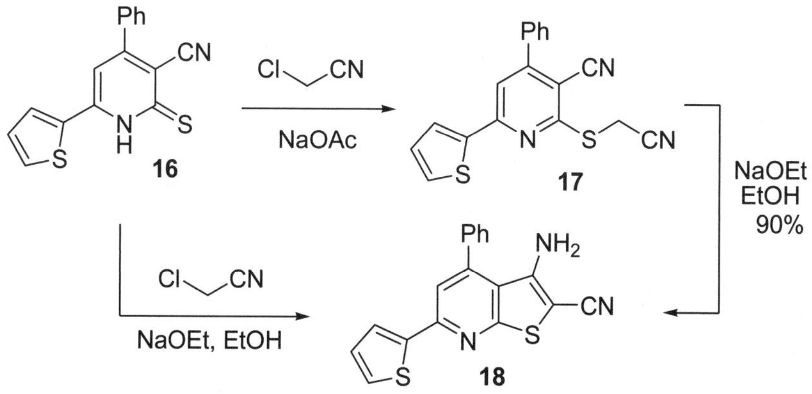

An interesting rearrangement was observed in the Thorpe–Ziegler reaction of **19**.[31] Aimed at the synthesis of a benzothiazepine, treatment of **19** with sodium hydride resulted in the formation of intermediate **20**. While protonation of **20** would have provided the desired benzothiazepine, instead, intramolecular attack by the enolate resulted in the formation of **21**. Subsequent expulsion of sulfur and aromatization provided the 2-morpholinoquinoline **22** in good yield.

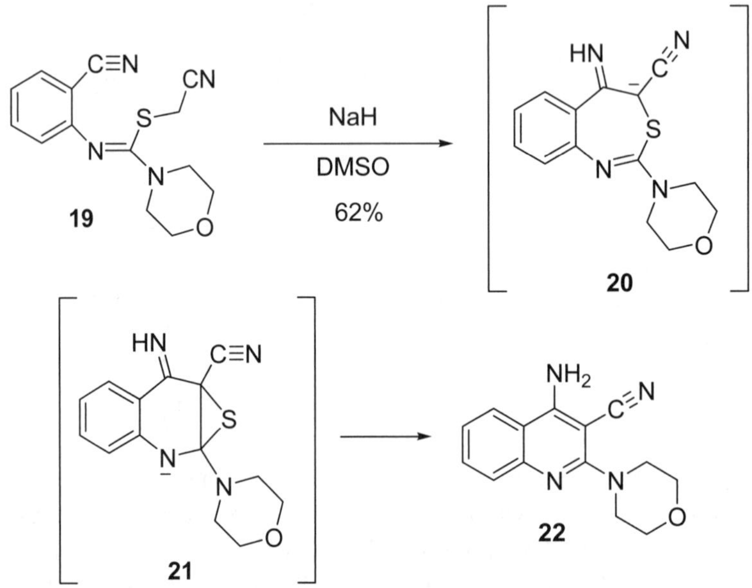

The Thorpe–Ziegler reaction has been commonly used for the synthesis of spirocyclic ketones.[32] An example of this comes from the work of Chande and coworkers in their efforts toward the synthesis of novel thiobarbituric acid derivatives.[33] In this synthesis, 1,3-ditolylthiobarbituric acid **23** is reacted with acrylonitrile in the presence of sodamide in DMF at ambient temperature. Following successive Michael additions to produce **24**, a spontaneous Thorpe–Ziegler cyclization occurs to provide enamine **25**. Exhaustive hydrolysis and subsequent decarboxylation of **25** result in the desired spirocyclohexanone derivative **26**.

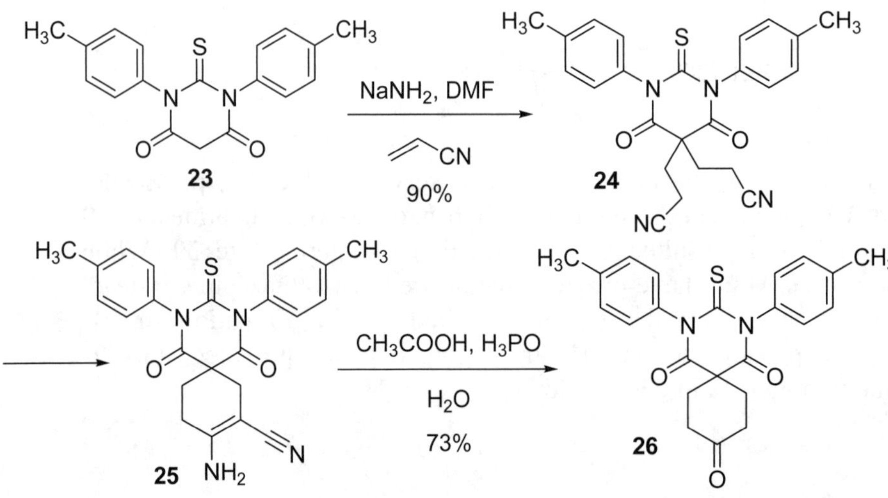

The Chande group has similarly used the Thorpe–Ziegler annulation for synthesis of spiroketones containing antibacterial, antitubercular and anticancer properties.[34] For example, the Michael addition of **27** with acrylonitrile in the presence of sodamide yields **28**. When diadduct **28** is exposed to base, the Thorpe–Ziegler intramolecular ring closure occurs to yield the desired derivative **29**. Hydrolysis and decarboxylation then occurs to give compound **30**, possessing antitubercular activity. Ketone **31**, which possesses anticancer and antibacterial activity, has been produced in a similar manner.

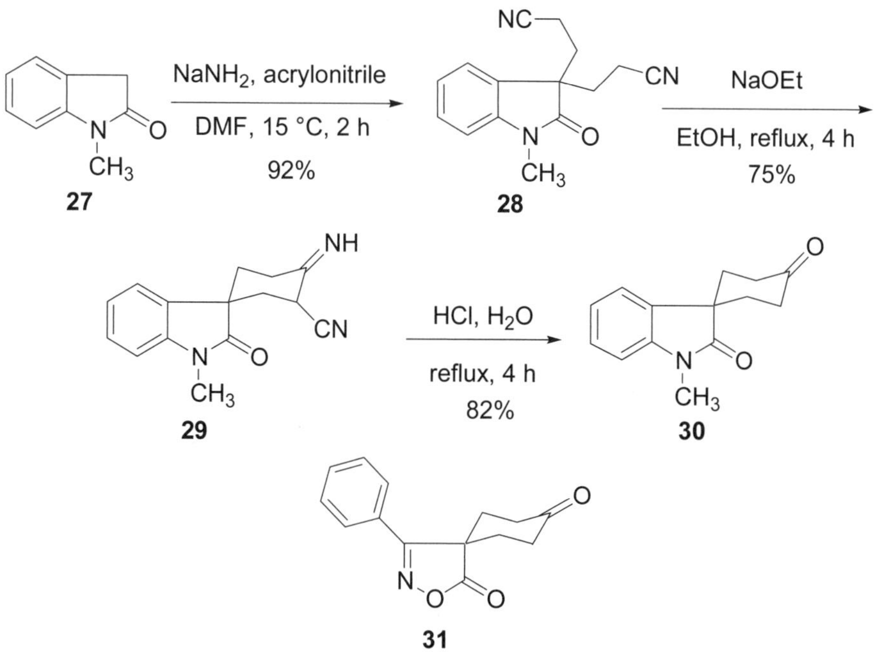

The Thorpe–Ziegler reaction is well-known for its ability to construct medium sized rings and large macrocycles.[12,35–40] Studies done by Anderson and Breazeale focused on the formation of a 1,3-bridged azulene structure utilizing an intramolecular ring closure via a Thorpe–Ziegler reaction.[41] Exposure of dinitrile **32** to base under high-dilution conditions results in **33** upon hydrolysis. Although the yield was modest, these efforts resulted in the first synthesis of a 1,3-bridged azulene.

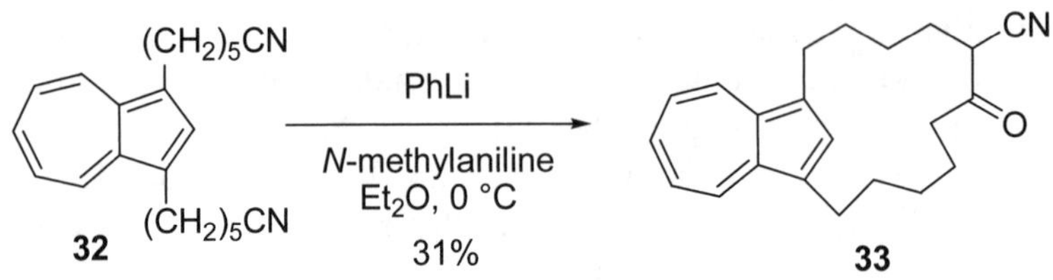

The Thorpe–Ziegler reaction has been widely utilized in the synthesis of natural products.[42–48] Deslongchamps and co-workers used the cyclization for construction of the D-ring of the stemodane skeleton in the natural product (+)-maritimol. (+)-Maritimol (**36**), isolated from *Stemodia maritime*[43] was used as a Caribbean folk medicine for the treatment of venereal diseases. Cyclization of dinitrile **34** was followed by acidic hydrolysis to yield tetracycle **35**. The synthesis of **35** represented a formal synthesis of the natural product, as this intermediate could be elaborated to **36** using conditions developed by Piers and co-workers.[47]

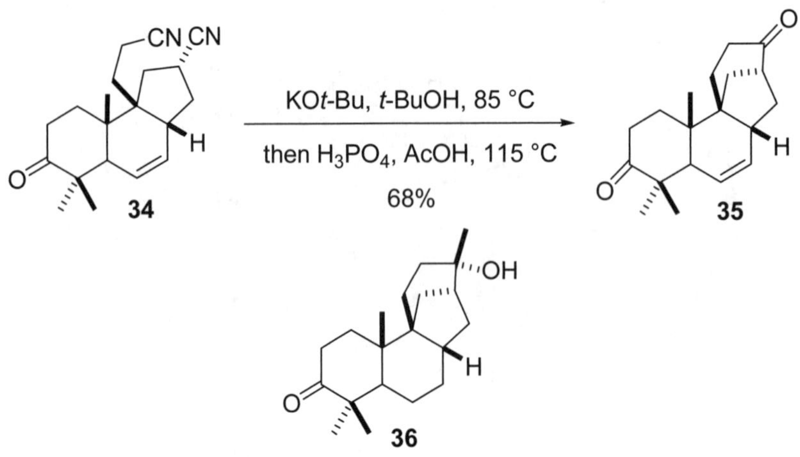

Malassene and co-workers recently examined the application of the Thorpe–Ziegler cyclization for synthesis of perhydrohistrionicotoxin, a skin extract of the Colombian poison arrow frog, *Dendrobates histrionicus*.[42,49] Under standard conditions, dinitrile **37** was directly converted into spiropiperidine **38** through the Thorpe–Ziegler cyclization, completing construction of the carbon framework of perhydrohistrionicotoxin.

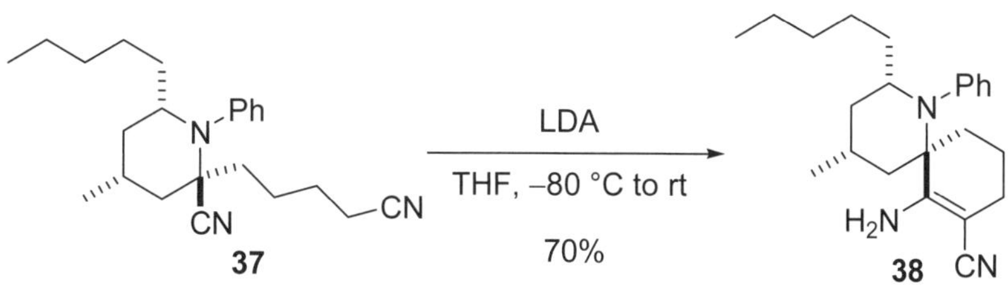

Satoh and Wakasugi have developed a creative method for the synthesis of cyclopentenones involving a Thorpe–Ziegler cyclization.[50,51] Beginning with α-chlorovinylsulfoxide **39**, conjugate addition of cyanomethyllithium results in the preparation of **40**. Notably, this reaction, which results in a quaternary center, proceeds in very high yield for a variety of vinylsulfoxide substrates. Treatment of **40** with LDA results in the formation of an α-sulfinyl carbenoid which is trapped by 2-lithiopropionitrile. Subsequent Thorpe–Ziegler cyclization and elimination of the toluenesulfenyl anion then occurs to produce enaminonitrile **41**. Hydrolysis and decarboxylation finally results in the isolation of cyclopentenone **42**. Of note, this reaction has proven general for a large number of substrates. Several different vinylsulfoxides and 2-lithionitriles have been utilized to expand the scope of this reaction. As demonstrated in the preparation of **43**, when a chiral sulfoxide is utilized, the reaction proceeds with a high degree of enantioselectivity for formation of the quaternary chiral center at C-4 of the cyclopentenone.[50]

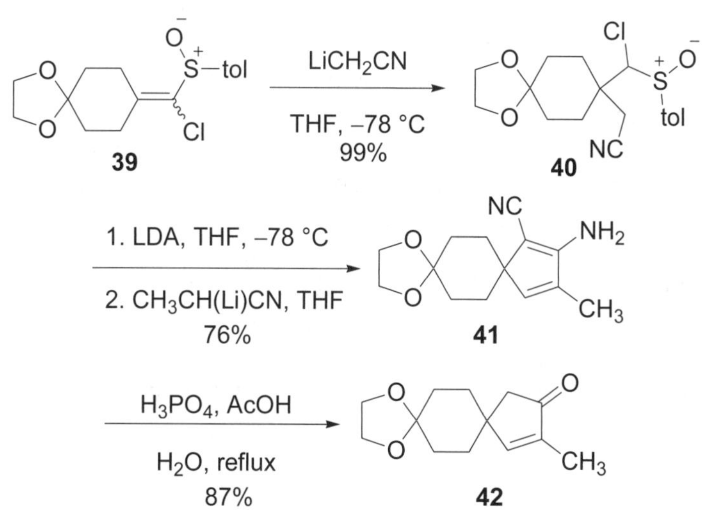

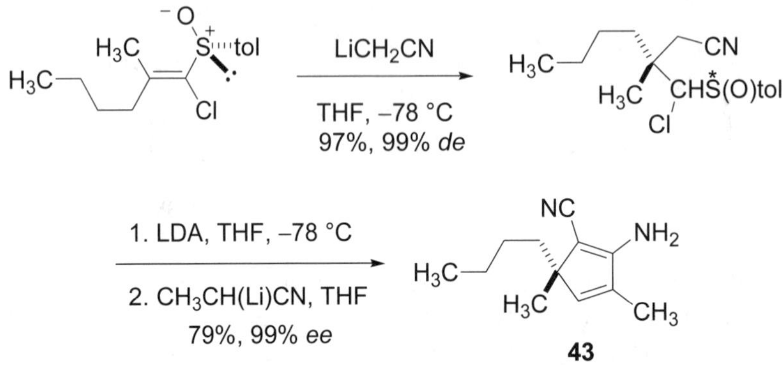

5.4.6 *Experimental*

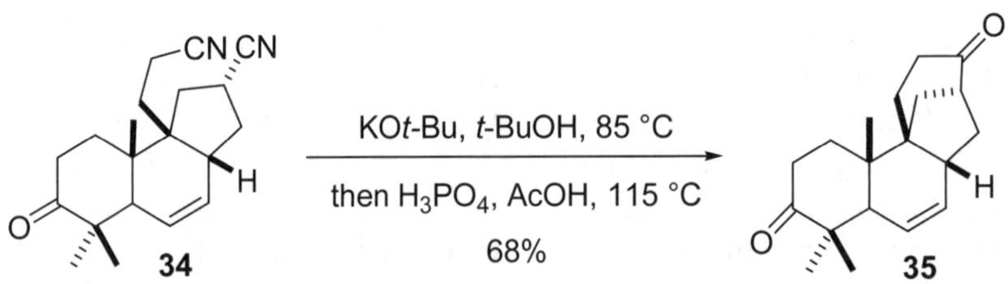

Enedione (35)[43]

Dinitrile **34** (85 mg, 274 μmol) in a deoxygenated *tert*-BuOK solution (2 mL, 1% in *tert*-BuOH) was heated in an oil bath for 2.5 h at 85 °C. Upon cooling, the mixture was diluted with CH_2Cl_2 and hexane, acetic acid (10 μL) was added and the mixture was evaporated several times from hexane. The residue was dissolved in CH_2Cl_2, washed with water, dried over $MgSO_4$, filtered through a cm long silica pad, which was eluted with ether. Evaporation of the solvent afforded the enaminonitrile (85 mg, 100%) as an off-white solid. It was carried over to the hydrolysis step without delay.

The enaminonitrile from the previous experiment was heated for 37 h in a deoxygenated acid mixture (2 mL, made by admixture of 4 mL 85% H_3PO_4, 10 mL AcOH and 1 mL H_2O) in an oil bath at 110 °C. Upon cooling, the mixture was diluted with CH_2Cl_2, washed with water, dried over $MgSO_4$ and evaporated. Flash chromatography (hexane/ether/CH_2Cl_2 3:3:1) of this crude material afforded enedione **35** (53 mg, 68%) as white plates.

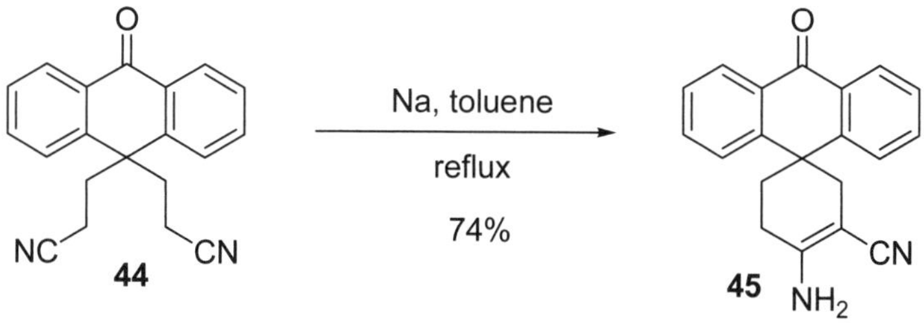

4′-Imino-10-oxospiro[anthracene-9,1′-cyclohexane]-3-carbonitrile (45).[32] Compound **44** (1.5 g, 5 mmol) was dissolved in 25 ml of toluene. To this, pulverized sodium (0.12 g, 5 mmol) was then added. The reaction mixture was refluxed for 8 h. To this reaction mixture, a little methanol was added to react with the unreacted sodium metal. From the reaction mixture, solvent was removed by vacuum distillation and the remaining solid contents were poured onto crushed ice and acidified using aqueous 2 N HCl to pH 2–3. The dark colored solid obtained was then filtered, washed with water, vacuum dried, and recrystallized from benzene/pet ether to afford pure compound **45** (1.11 g, 74%).

5.4.7 References

1. Li, J. J. *Name Reactions: A Collection of Detailed Reaction Mechanisms*, 3rd ed. Springer-Berlin-Heidelberg, 2006, pp. 592–593.
2. Baron, H.; Remfry, F. G. P.; Thorpe, Y. F. *J. Chem. Soc.* **1904**, *85*, 1726–1761.
3. Moore, C. W.; Thorpe, J. F. *J. Chem. Soc.* **1908**, *93*, 165–187.
4. Hammer, C. F.; Hines, R. A. *J. Am. Chem. Soc.* **1955**, *77*, 3649–3650.
5. [R] Taylor, E. C.; McKillop, A. *The Chemistry of Cyclic Enaminonitriles and o-Amino Nitriles*, Interscience, New York, 1970; Ch. 1.
6. Ziegler, K.; Eberle, H.; Ohlinger, H. *Liebigs Ann.* **1933**, *504*, 94–130.
7. Ziegler, K.; Lüttringhaus, A. *Liebigs Ann.* **1934**, *511*, 1–12.
8. Ziegler, K.; Weber, K. *Liebigs Ann.* **1934**, *512*, 164–171.
9. Ziegler, K.; Aurnhammer, R. *Liebigs Ann.* **1934**, *513*, 43–64.
10. Ziegler, K.; Hechelhammer, W. *Liebigs Ann.* **1937**, *528*, 114–142.
11. Ziegler, K.; Holl, H. *Liebigs Ann.* **1937**, *528*, 143–154.
12. [R] Schaefer, J. P.; Bloomfield, J. J. *Org React.* **1967**, *15*, 1–44.
13. [R] Granik, V. G.; Kadushkin, A. V. *Adv. Heterocycl. Chem.* **1998**, *72*, 79–125.
14. Ivanov, A. S.; Tugusheva, N. Z.; Solov'eva, N. P.; Granik, V. G. *Russ. Chem. Bull., Int. Ed.* **2002**, *51*, 2121–2128.
15. Ivanyuk, T. V.; Kadushkin, A. V.; Solov'eva, N. P.; Granik, V. V. *Mendeleev Commun.* **1993**, *4*, 160–161.
16. Fedorov, A. E.; Shestopalov, A. M.; Belyakov, P. A. *Russ. Chem. Bull., Int. Ed.* **2003**, *52*, 2063–2069.
17. Dotsenko, V. V.; Krivokolysko, S. G.; Litvinov, B. P.; Chernega, A. N. *Chem. Heterocycl. Compd.* **2007**, *43*, 599–607.
18. Koditz, J.; Rudorf, W. D.; Hartung, H.; Heinemann, F. *Liebigs Ann. Chem.* **1993**, 1003–1007.
19. Ryndina, S. A.; Kadushkin, A. V.; Solov'eva, N. P.; Granik, V. G. *Russ. Chem. Bull., Int. Ed.* **2002**, *51*, 854–859.

20. Abdel-Rahman, A. E.; Bakhite, E. A.; Moamed, O. S.; Thabet, E. A. *Phosphorus, Sulfur, and Silicon* **2003**, *178*, 89–106.
21. Abdel-Rahman, A. E.; Bakhite, E. A.; Al-Taifi, E. A. *J. Chem. Res.* **2005**, *7*, 461–468.
22. Kaigorodova, E. A.; Konyushkin, L. D.; Kambulov, Yu.; Krapivin, G. D. *Chem. Heterocycl. Compd.* **1997**, *33*, 752–753.
23. Semioshkin, A. A.; Artemov, V. A.; Ivanov, V. L., Ptashits, G. M; Petrovskii, P. V.; Shestopalov, A. M.; Bregadze, V. I.; Litvinov, V. P. *Chem. Heterocycl. Compd.* **1998**, *34*, 688–691.
24. Wei, H.; Cai, J.; Sun, M.; Ji, M. *Heteroat. Chem.* **2007**, *18*, 236–238.
25. Yakovlev, M. Y.; Kadushkin, A. V.; Solov'eva, N. P.; Granik, V. G. *Heterocycl. Commun.* **1998**, *4*, 245–252.
26. Botsi, S.; Tsolomiti, G.; Hamilakis, S.; Tsolomitis, A. *Heterocycl. Commun.* **2008**, *14*, 233–236.
27. Mohamed, O. S.; Al-Taifi, E. A.; El-Emary, T. I.; Bakhite, E. A.-G. *Phosphorus, Sulfur, and Silicon* **2007**, *182*, 1061–1082.
28. Ivanov, A. S.; Tugusheva, N. Z.; Alekseeva, L. M.; Granik, V. G. *Russ. Chem. Bull., Int. Ed.* **2003**, *52*, 1182–1189.
29. Salaheldin, A. M.; Oliveira-Campos, A. M. F.; Rodrigues, L. M. *ARKIVOC* **2008**, *14*, 180–190.
30. Tsolomiti, G.; Tsolomiti, K.; Tsolomitis, A. *Heterocycl. Commun.***2007**, *13*, 235–238.
31. Fathalla, W.; Marek, J.; Pazdera, P. *Heterocycl. Commun.* **2002**, *8*, 79–82.
32. Chande, M. S.; Khanwelkar, R. R.; Barve, P. A. *J. Chem. Res.* **2007**, *8*, 468–471.
33. Chande, M. S.; Suryanarayan, V. *Heterocycl. Commun.* **2006**, *12*, 415–418.
34. Chande, M. S.; Verma, R. S.; Barve, P. A.; Khanwelkar, R. R.; Vaidya, R.B., Ajaikumar, K. B. *Eur. J. Med. Chem.* **2005**, *40*, 1143–1148
35. Hurd, R. N.; Shah, D. H. *J. Org. Chem.* **1973**, *38*, 390–394.
36. Fry, E. M.; Fieser, L. F. *J. Amer. Chem. Soc.* **1940**, *62*, 3489–3494.
37. Allinger, N. L.; Nakazaki, M.; Zalkow, V. *J. Am. Chem. Soc.* **1959**, *81*, 4074–4080.
38. Allinger, N. L.; Greenberg, S. *J. Amer. Chem. Soc.* **1959**, *81*, 5733–5736.
39. Allinger, N. L.; Greenberg, S. *J. Am. Chem. Soc.* **1962**, *84*, 2394–2402.
40. Rodriguez-Hahn, L.; Parra M., M.; Martinez, M. *Synth. Commun.* **1984**, *14*, 967–972.
41. Anderson, A. G.; Breazeale, R. D. *J. Org. Chem.* **1969**, *34*, 2375–2384.
42. Malassene, R.; Vanquelef, E.; Toupet, L.; Hurvois, J.; Moinet, C. *Org. Biomol. Chem.* **2003**, *1*, 547–551.
43. Toro, A.; Nowak, P.; Deslongchamps, P. *J. Am. Chem. Soc.* **2000**, *122*, 4526–4527.
44. Dansou, B.; Pichon, C.; Dhal, R.; Brown, E.; Mille, S. *Eur. J. Org. Chem.* **2000**, *8*, 1527–1533.
45. Linders, J. T. M.; Flippen-Anderson, J. L.; George, C. F.; Rice, K. C. *Tetrahedron Lett.* **1999**, *40*, 3905–3908.
46. Luyten, M.; Keese, R. *Tetrahedron* **1986**, *42*, 1687–1691.
47. Piers, E.; Abeysekara, B. F.; Herbert, D. J.; Suckling, I. D. *J. Chem. Soc., Chem. Commun.* **1982**, *7*, 404-406.
48. [R] Fleming, F. F.; Shook, B. C. *Tetrahedron* **2002**, *58*, 1–23.
49. Malassene, R.; Toupet, L.; Hurvois, J.-P.; Moinet, C. *Synlett* **2002**, *6*, 895–898.
50. Satoh, T.; Wakasugi, D. *Tetrahedron Lett.* **2003**, *44*, 7517–7520.
51. Wakasugi, D.; Satoh, T. *Tetrahedron* **2005**, *61*, 1245–1256.

6.1 Blanc Chloromethylation

Richard J. Mullins and Reid M. Faylor

6.1.1 Description

The Blanc reaction is the Lewis acid-promoted installation of a chloromethyl group onto an aromatic or heteroaromatic ring, using formaldehyde, or a synthetic equivalent, in combination with hydrochloric acid.[1]

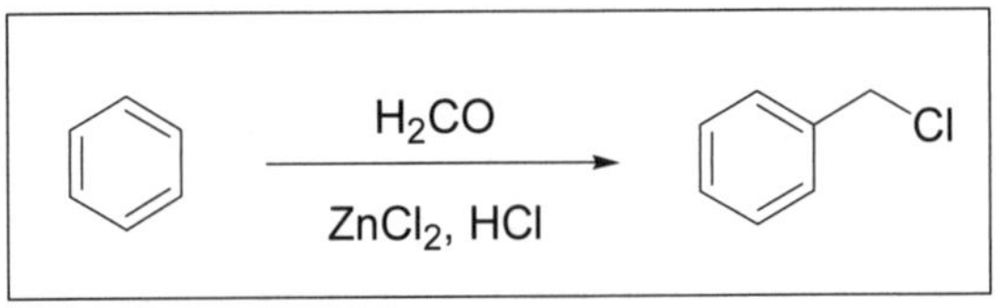

6.1.2 Historical Perspective

Discovered in 1877, the Friedel–Crafts reaction is one of the more important and useful organic transformations. This class of reactions is generally considered to include all Lewis acid-promoted electrophilic alkylations and acylations of aromatic rings. Due to their synthetic utility, Friedel–Crafts reactions have been extensively studied and used across a broad and diverse area of chemical research.

In 1898, the first chloromethylation of benzene was performed by Grassi and Maselli.[2] Twenty-five years later, the reaction was extensively redeveloped by Gustave Louis Blanc[3] while he was director at the Intendance militaire aux Invalides.[1] Although his reaction conditions do not differ from the original conditions of Grassi and Maselli, his efforts earned him the honor of having the reaction bear his name. That the reaction seems closely related to the Friedel–Crafts reaction should not be surprising, as Blanc studied under Charles Friedel in Paris.

In 1932, Quelet[4] used the Blanc procedure, replacing formaldehyde with aliphatic aldehydes in the reaction with phenolic ethers. The resulting reaction mixtures were found to contain α-chloroalkyl derivatives. Although the conditions are virtually identical and the reaction proceeds via the same basic mechanism, the Blanc chloromethylation is often referred to as the Quelet reaction.

Over time, the reaction has become a rather important synthetic method in organic chemistry, as the chloromethyl group can be easily converted into a variety of functional groups. A number of different procedures have been used to effect this transformation. In addition to formaldehyde, several formaldehyde equivalents have been used, including paraformaldehyde, α-trioxymethylene, and diethyl- or dimethyl-formal, among others.[5] Several different catalysts have also been employed, with zinc chloride, sulfuric acid, and acetic acid being especially useful among these.[5] While the reaction was first applied to aromatic hydrocarbons, a variety of aromatic rings containing activating and deactivating substituents have been employed with considerable success.

For a more thorough discussion of the history of the Blanc chloromethylation, its scope and limitations as well as its use in synthesis prior to 1963, the reader is directed to excellent reviews by Fuson and McKeever[5] and Olah and Tolgyesi.[6]

6.1.3 Mechanism

There are two basic mechanistic pathways, which have been proposed for the Blanc reaction. Both involve electrophilic aromatic substitution, but differ with regards to the identity of the electrophilic species and the point at which the halogenation occurs. The subtle differences between the two mechanisms may depend on the conditions involved.

Studies by Olah and Yu[7] have suggested the following mechanism is operable when a Lewis acid such as $ZnCl_2$ is employed. Following protonation of formaldehyde and addition of chloride ion, a second protonation occurs giving rise to complex **1**, which acts as the electrophile in the reaction with benzene (**2**). Attack of **1** by benzene results in the formation of the resonance-stabilized cation **3**, with displacement of a water molecule. Finally, removal of a proton from **3** results in rearomatization of the benzene ring to provide **4**.

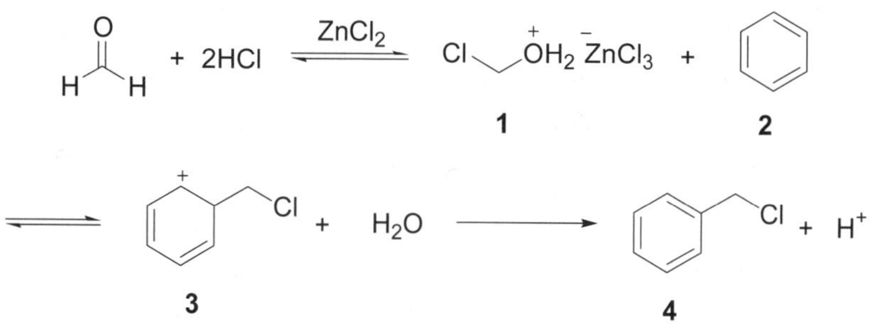

An alternate mechanism, suggested by Ogata and Okano,[8] has precedence in the absence of a Lewis acid. Protonation of formaldehyde activates it for nucleophilic attack by formation of cation **5**. Attack of **5** by benzene once more results in the formation of resonance stabilized cation **6**. Loss of a proton regenerates the aromatic ring, forming benzyl alcohol (**7**), which then undergoes subsequent displacement by HCl to produce benzyl chloride (**4**).

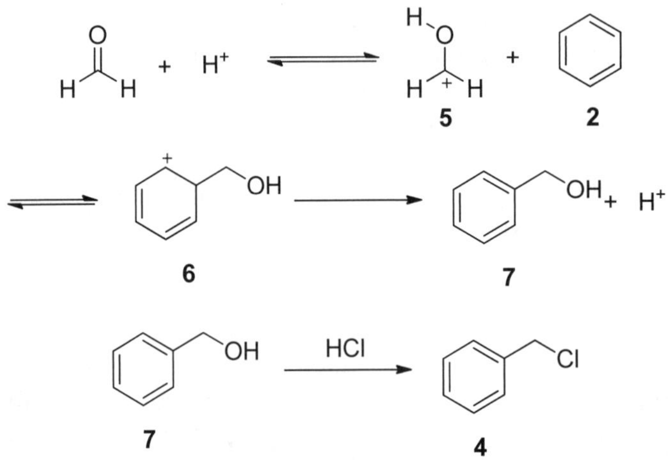

Although the mechanism above is considered generally correct, there have been multiple studies,[7–12] which have attempted to delineate the more subtle aspects of the mechanism. For an extensive review on some of the earlier mechanistic studies, the reader is directed here.[13]

6.1.4 Variations and Improvements

In lieu of the catalysts $ZnCl_2$, $SnCl_4$ and BF_3 traditionally employed in the Blanc chloromethylation reaction, Sugi and co-workers have pioneered the use of Group 3 and 4 metal triflates to effect the transformation.[14] Three of these triflates, $Sc(OTf)_3$, $Yb(OTf)_3$ and $Sm(OTf)_3$, were shown to provide yields similar to those obtained under standard conditions. It is important that these Lewis acids can be used catalytically with loadings as low as 10 mol %. In addition, recovery and reuse of the catalyst is possible, without significant loss in activity. This work is significant, in that it avoids several problems associated with the use of traditional Lewis acids in the Blanc reaction. Primarily, there must be a stoichiometric amount of catalyst in regard to the substrate, which makes work up procedures tedious. The high corrosivity, susceptibility to water and low recoverability can cause environmental problems.[14] Furthermore, these catalysts can form carcinogenic chloromethylethers and promote chlorination of the aromatic ring.[15]

One drawback of the above procedure is the expense of the triflates, making them unsuitable for industrial purposes. Thus the search for other environmentally conscious and effective Lewis acid catalysis systems has resulted in the use of ionic liquids. Desired for their low vapor pressure, stability in a wide temperature range, and recyclability, ionic liquids are popular green alternatives to traditional solvent systems. Wang and co-workers have used one such ionic liquid, 1-ethyl-3-methylimidazolium tetrafluorborate ([emim]BF_4) as a promoter for the chloromethylation of a number of aromatic hydrocarbons.[16] In the absence of [emim]BF_4, the chloromethylation of *m*-xylene (**8**) underwent slow conversion, requiring 48 h to achieve 63% conversion. In contrast, the use of 0.3 equiv [emim]BF_4 resulted in 92% conversion in just 7 h. Similar improvements including shorter reaction times, higher conversion, milder conditions and easy recycling were realized using the ionic liquid [C_{12}mim]Br as promoter.[17]

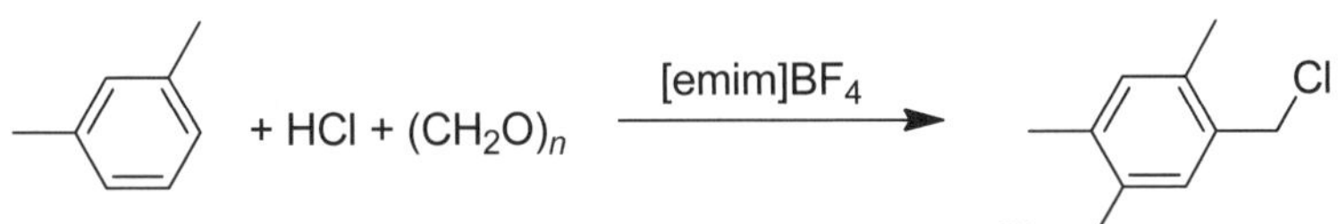

8 **9**

Issues of regioselectivity in the Blanc chloromethylation and related Friedel–Crafts reactions have been studied extensively. As is common with a majority of electrophilic aromatic substitution reactions, substitution typically occurs *ortho* or *para* to electron-donating substituents, with issues of steric strain playing a role in the relative ratio of *ortho* and *para* products. The Blanc reaction is typically somewhat regioselective, favoring the *para*-isomer but accompanied by lesser amounts of the *ortho* product.[5]

An additional difficulty in the Blanc reaction is the tendency for activated aromatic rings to undergo polychloromethylation under the typically harsh reaction conditions. For example, in the chloromethylation of benzene, the product benzylchloride (**4**) is often accompanied by small amounts of *p*-xylylene dichloride **10**, as well as a small amount of diphenylmethane, the product resulting from Friedel–Crafts alkylation of benzene with benzylchloride (**4**).[3,5] With more activated ring systems, such as phenols, the reaction is increasingly difficult to control, resulting often in the formation of polymeric materials.[5]

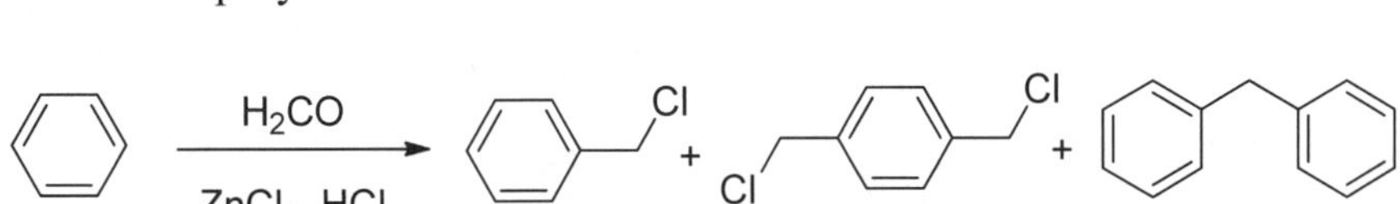

2 **4** **10** **11**

79% 5%

One solution to the aforementioned problem, realized by Tundo and co-workers, involved the addition of quaternary ammonium salts to the typical Blanc reaction mixture.[18] For example, addition of hexadecyltrimethylammonium bromide to normal Blanc reaction conditions with cumene results in a two-phase system, where the reaction proceeds with high conversion (~ 89%) and high selectivity for formation of the *mono*-chloromethyl derivative (99%). Under these conditions, virtually none of the diphenylmethane product is observed.

6.1.5 Synthetic Utility

The Blanc chloromethylation has found use across a wide spectrum of aromatic and heteroaromatic compounds. For a more thorough discussion of the use of the Blanc chloromethylation in synthesis before 1963, especially involving aromatic hydrocarbons, the reader is directed to excellent reviews by Fuson and McKeever[5] and Olah and Tolgyesi.[6] To ensure that this review has relatively broad scope, a few examples from these reviews are highlighted below.

The use of methylchloromethyl ether in conjunction with $SnCl_4$, as an alternate procedure for chloromethylation, is demonstrated in the reaction of 1,3,5-trisopropylbenzene (**12**), yielding benzyl chloride **13**.[5]

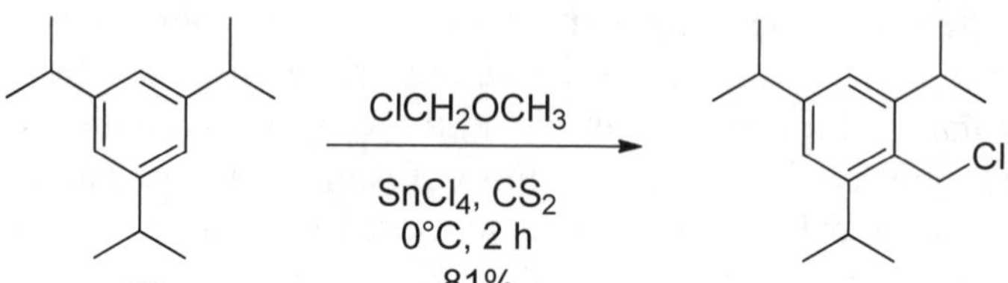

12 **13**

While acetophenones are typically resistant to chloromethylation due to the electron-withdrawing ability of the carbonyl, some highly substituted acetophenones have been known to undergo this reaction. In the case of **14**, treatment with HCl and paraformaldehyde results in the production of **15** in high yield.[19]

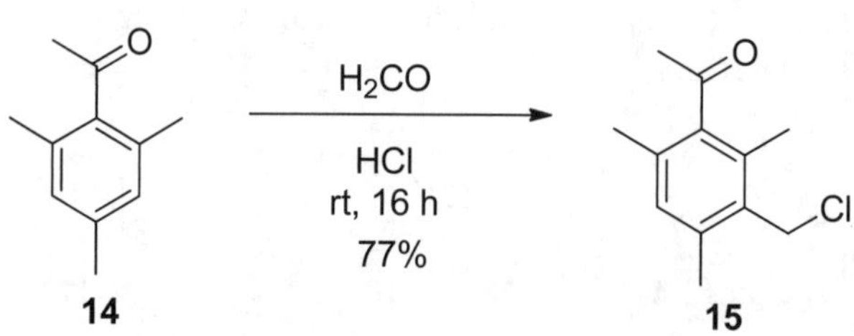

14 **15**

Although phenols are known to react very rapidly under Blanc conditions and often result in polymeric materials, when substituted with an electron-withdrawing substituent they become suitable substrates for this reaction. As shown below, *p*-nitrophenol (**16**) is chloromethylated using the dimethylacetal of formaldehyde giving **17** in relatively high yield.[20]

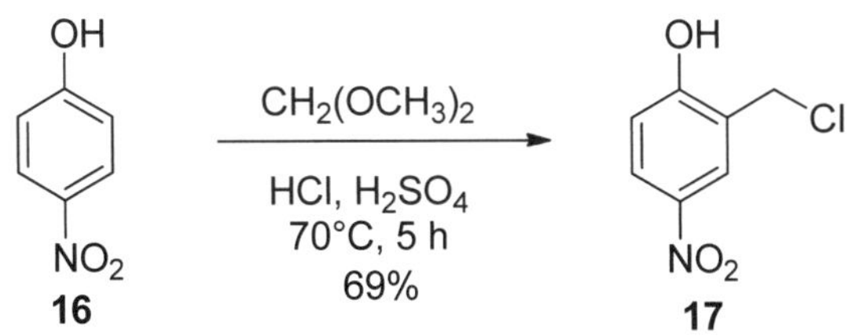

In a similar manner, work done by Quelet also focused on the chloromethylation reaction of substituted aromatic rings such as *p*-methylanisole.[4] Mallory and co-workers used this reaction as a key step in their synthesis of 2-methylphenanthrene (**21**).[21] In the initial stages of synthesis, *p*-methylanisole (**18**) was converted to 2-(chloromethyl)-4-methylanisole (**19**). Subsequent ylide formation and Wittig reaction with benzaldehyde to give **20** was followed by a photocyclization to produce **21**. A similar chloromethylation reaction has been used in efforts toward the synthesis of macrocyclic ligands.[22]

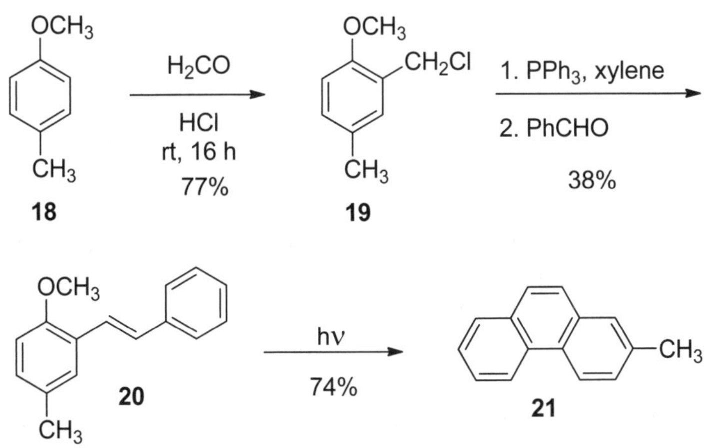

Owing to their high electron density, furans have been especially useful in the Blanc chloromethylation.[23–25] In particular, Pevzner and co-workers have recently published numerous papers using this combination of reactants.[26–30] Directed toward the synthesis of dialkoxyphosphorylmethyl derivatives of furans, introduction of a chloromethyl side chain was envisioned to proceed through the Blanc reaction. Thus treatment of **22**

under standard conditions resulted in the formation of **23** in high yield.[31] Notably, in spite of the steric hindrance imposed by the *t*-butyl group at C-5 of the furan ring, substitution occurs exclusively at C-4. This regioselectivity is likely a result of the electron-withdrawing ester at *C*-2, which prevents substitution at *C*-3.

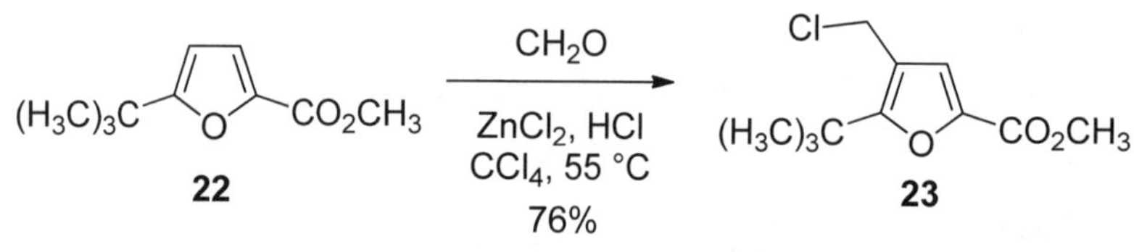

Alternatively, the direct chloromethylation of diethoxyphosphinoyl-methyl compounds **24** and **26** was carried out to give **25** and **27**, respectively, both of which were obtained in high yield.[32] The fact that these reactions proceed at low temperatures demonstrates the high reactivity of furans when not substituted with an electron-withdrawing group, as in **22** above.

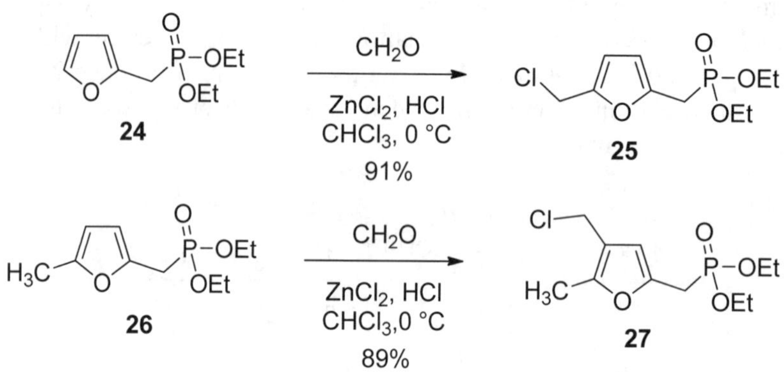

Hara and co-workers used the Blanc reaction for the synthesis of a key intermediate toward a series of benzodiazepines.[33] Since their goal was ultimately the installation of an ethyl substituent at C-4, in lieu of a formaldehyde precursor like trioxane or paraformaldehyde, acetaldehyde was used, in a manner similar to the Quelet reaction. As commonly observed under these conditions, alkene **30** was isolated along with product **29**. Given that the alkene **30** was also a desired product, the reaction mixture was simply heated with pyridine, resulting in complete conversion to **30**. To complete the installation of the ethyl substituent, the alkene of **30** was subsequently reduced.

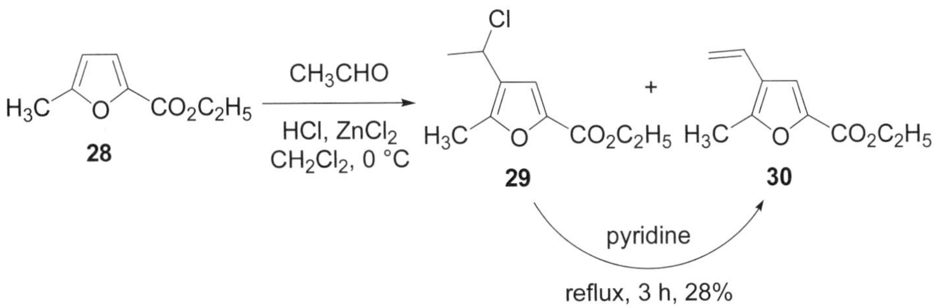

Thiophene, another π-excessive heterocycle, was originally used in the Blanc reaction in 1942 by Blicke and Burckhalter.[34] Following their procedure, a large scale synthesis of **32** has been developed. Notably, in reaction with un-substituted thiophene (**31**), the chloromethyl group is selectively installed at the 2-position.[35] This particular chloromethylation has been frequently used as a method for incorporation of the thiophene core into other molecules.[36,37]

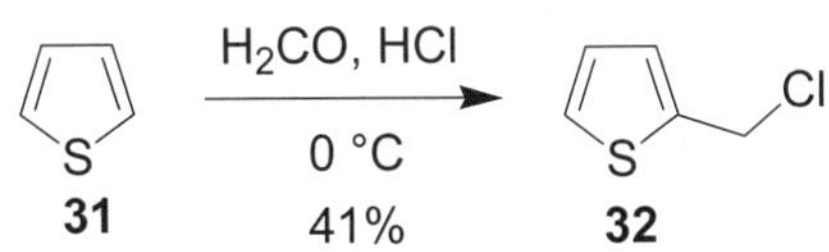

Benzothiophenes have also found utility in the Blanc chloromethylation, especially in the area of medicinal chemistry.[38–40] As demonstrated by Cross and co-workers during their efforts toward the synthesis of selective thromboxane synthetase inhibitors, treatment of **33** under the standard conditions resulted in **34**.[41] While benzothiophene typically substitute preferentially at C-3, in this case the methyl at C-3 forces substitution to occur at *C*-2.

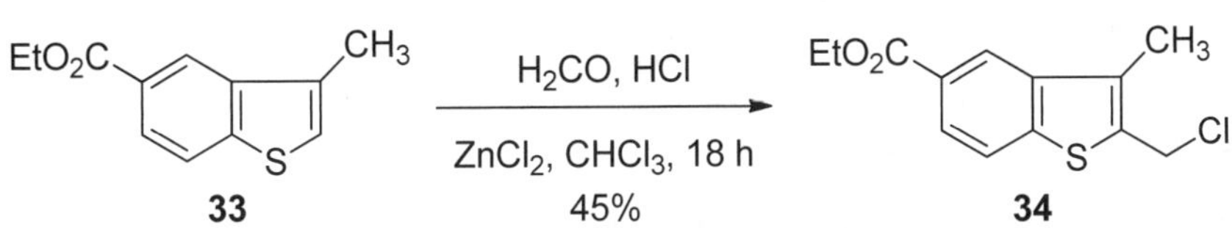

6.1.6 Experimental

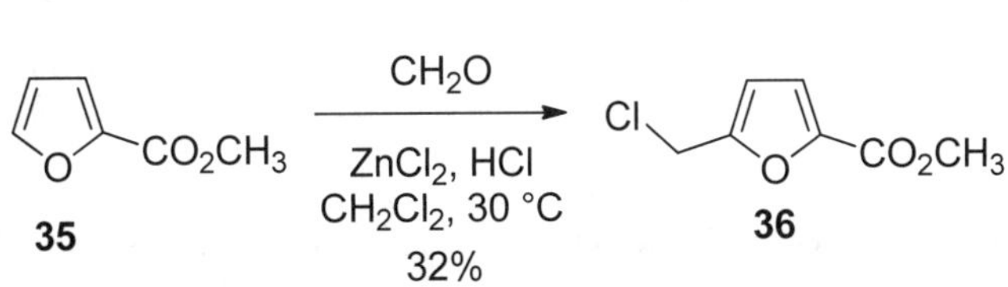

Methyl 5-(chloromethyl)furan-2-carboxylate (36)[25]

Dry HCl gas was introduced into the stirred mixture of methyl furan-2-carboxylate (**35**, 25.2 g, 0.2 mol), paraformaldehyde (6.0 g, 0.2 mol), and $ZnCl_2$ (30.0 g, 0.22 mol) in CH_2Cl_2 (100 mL) for 3 h at 30–35 °C. The reaction mixture was poured into 80 mL water. The mixture was extracted with CH_2Cl_2 (3 × 30 mL). The organic phase was combined and dried with anhydrous Na_2SO_4. The solution was evaporated to dryness. The remaining residue was distilled and the distillate collected as a colorless oil in 32.3% yield (11.3 g).

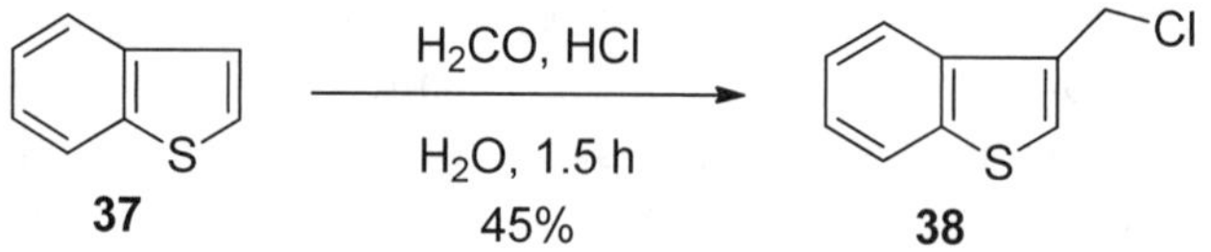

3-(Chloromethyl)benzo[*b*]thiophene (38)[38]

HCl(g) was bubbled vigorously through a mixture of thianaphthene (**37**, 17.0 g, 126.68 mmol), 37% aqueous formaldehyde (15 mL), and concentrated HCl (15 mL) until the reaction temperature rose to 65 °C. At this time, the flow of HCl gas was reduced to a slow stream which was maintained for 1.5 h. The reaction mixture was diluted with H_2O (50 mL) and subsequently extracted with ether (2 × 50 mL). The combined ethereal extracts were dried over Na_2SO_4 and concentrated under reduced pressure to yield a straw-colored liquid: 21.0 g (90.7%).

6.1.7 References

1. [R] Li, J. J. *Name Reactions: A Collection of Detailed Reaction Mechanisms*, 3rd ed. Springer, New York, **2006**, p. 61.
2. Grassi, G.; Maselli, C. *Gazz. Chim. Ital.* **1898**, *28*, 477.
3. Blanc, G. L. *Bull. Soc. Chim. Fr.* **1923**, *33*, 313.
4. Quelet, R. *Compt. Rend.* **1932**, *195*, 155.
5. [R] Fuson, R. C.; McKeever, C. H. *Org. React.* **1942**, *1*, 63–90.
6. [R] Olah, G.; Tolgyesi, W. S. In *Friedel–Crafts and Related Reactions,* vol. II, part 2, Olah G., ed. Interscience, New York, **1963**, pp. 659–784.
7. Olah, G. A.; Yu, S. H. *J. Am. Chem. Soc.* **1975**, *97*, 2293–2295.
8. Ogata, Y.; Okano, M. *J. Am. Chem. Soc.* **1956**, *78*, 5423–5425.
9. Szmant, H. H.; Dudek, J. *J. Am. Chem. Soc.* **1949**, *71*, 3763–3765.
10. Brown, H. C.; Nelson, K. L. *J. Am. Chem. Soc.* **1953**, *75*, 6292–6299.
11. Tyrlik, S. K.; Radziwonka, Z.; Piasecka-Maciejewska, K. *Bull. Pol. Acad. Sci., Chem.* **1986**, *34*, 351–358.
12. Uribe, M. I. O.; Salvador, A. R.; Gulías, A. I. *Ind. Eng. Chem. Res.* **1987**, *26*, 1725–1735.
13. [R] Belen'kii, L. I.; Vol'kenshtein, Yu. B.; Karmanova, I. B. *Russ. Chem. Rev.* **1977**, *46*, 891–903.
14. Kishida, T.; Yamauchi, T.; Kubota, Y.; Sugi, Y. *Green Chem.* **2004**, *6*, 57–62.
15. Kishida, T.; Yamauchi, T.; Komura, K.; Kubota, Y.; Sugi, Y. *J. Mol. Catal. A: Chem.* **2006**, *246*, 268–275.

16. Wang, Y.; Shang, Z.-C.; Wu, T.-X. *Synth. Commun.* **2006**, *36*, 3053–3059.
17. Fang, Y.; Deng, Y.; Qinggang, R.; Huang, J.; Zhang, S.; Huang, B.; Zhang, K. *Chin. J. Chem. Eng.* **2008**, *16*, 357–360.
18. Selva, M.; Trotta, F.; Tundo, P. *Synthesis* **1991**, *11*, 1003–1004.
19. Fuson, R. C.; McKeever, C. H. *J. Am. Chem. Soc.* **1940**, *62*, 784–785.
20. Buehler, C. A.; Kirchner, F. K.; Diebel, G. F. *Org. Synth.* **1940**, *20*, 59.
21. Mallory, F. B.; Rudolph, M. J.; Oh, S. M. *J. Org. Chem.* **1989**, *54*, 4619–4626.
22. Dijkstra, P. J.; Skowronska-Ptasinska, M.; Reinhoudt, D. N.; den Hertog, H. J.; van Eerden, J.; Harkema, S.; de Zeeuw, D. *J. Org. Chem.* **1987**, *52*, 4913–4921.
23. Schmuck, C.; Machon, U. *Eur. J. Org. Chem.* **2006**, 4385–4392.
24. Elliott, M.; Janes, N. F.; Pearson, B. C. *J. Chem. Soc. (C)* **1971**, *17*, 2551–2554.
25. Su, H.; Nebbioso, A.; Carafa, V.; Chen, Y.; Yang, B.; Altucci, L.; You, Q. *Bioorg. Med. Chem.* **2008**, *16*, 7992–8002.
26. Pevzner, L. M. *Russ. J. Gen. Chem.* **2005**, *75*, 774–781.
27. Pevzner, L. M. *Russ. J. Gen. Chem.* **2003**, *73*, 1877–1880.
28. Pevzner, L. M. *Russ. J. Gen. Chem.* **2002**, *72*, 1085–1089.
29. Pevzner, L. M. *Russ. J. Gen. Chem.* **2003**, *73*, 260–266.
30. Pevzner, L. M. *Russ. J. Gen. Chem.* **2004**, *74*, 1182–1184.
31. Pevzner, L. M. *Russ. J. Gen. Chem.* **2003**, *73*, 1715–1724.
32. Pevzner, L. M. *Russ. J. Gen. Chem.* **2008**, *78*, 206–209.
33. Hara, T.; Kayama, Y.; Mori, T.; Itoh, K.; Fujimori, H.; Sunami, T.; Hashimoto, Y.; Ishimoto, S. *J. Med. Chem.* **1978**, *21*, 263–268.
34. Blicke, F. F.; Burckhalter, J. H. *J. Am. Chem. Soc.* **1942**, *64*, 477–480.
35. Wiberg, K. B.; McShane, H. F. *Org. Synth.* **1949**, 31–33.
36. Lee, W.-H.; Kong, H.; Oh, S.-Y.; Shim, H.-K.; Kang, I.-N. *J. Polym. Sci., Part A: Polym. Chem.* **2009**, *47*, 111–120.
37. Hou, J.; Tan, Z. A.; Yan, Y.; He, Y.; Yang, C.; Li, Y. *J. Am. Chem. Soc.* **2006**, *128*, 4911–4916.
38. Efange, S. M. N.; Mash, D. C.; Khare, A. B.; Ouyang, Q. *J. Med. Chem.* **1998**, *41*, 4486–4491.
39. Abadi, A. H. *Arch. Pharm. Med. Chem.* **2004**, *337*, 383–390.
40. Podea, P. V.; Irimie, F. D.; Toşa, M. I.; Paizs, C.; Irimie, F. D. *Tetrahedron: Asymmetry* **2008**, *19*, 500–511.
41. Cross, P. E.; Dickinson, R. P.; Parry, M. J.; Randall M. J. *J. Med. Chem.* **1986**, *29*, 1643–1650.

6.2 Asymmetric Friedel–Crafts Reactions: Past to Present

Jeffrey A. Campbell

6.2.1 Description

This chapter will review basic concepts of Friedel–Crafts (F–C) reactions and then will present a chronological breakdown of key highlights of F–C asymmetric bond homologation methodology applied to aromatic ring systems developed over the years. The F–C reaction is of broad scope and for the purpose of this chapter will be limited to two categories, the alkylation or acylation reactions of aromatic or heteroaromatic compounds to afford the corresponding alkyl and acyl arenes, respectively. Both F–C alkylation and acylation reactions involve replacement of a hydrogen atom of an aryl moiety swith an alkyl or acyl group by the reaction of an alkylating or acylating agent in the presence of a catalyst (Lewis acid with and without a cocatalyst, Brønsted acid, etc.).

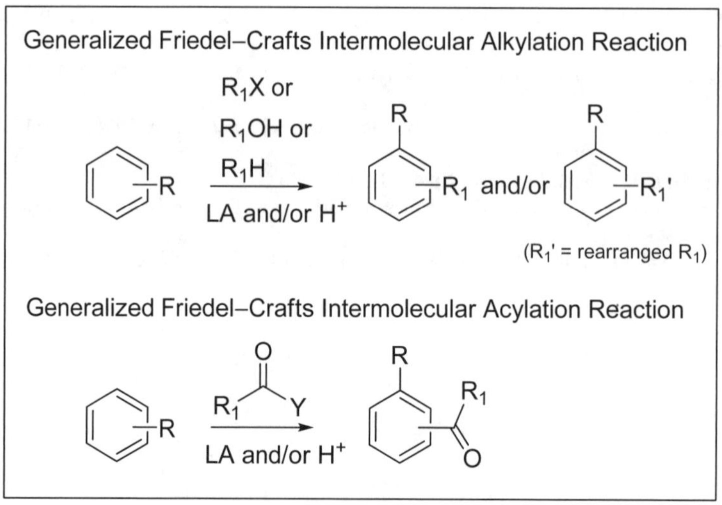

In the F–C alkylation reaction, carbocation intermediates, generated from the reaction of a wide variety of catalysts with alkyl halides, alkenes, and alcohols, are used to alkylate an aromatic ring. Other alkylating agents that can be used include alkynes, esters, aldehydes, ketones, amines (via diazotization), epoxides, thiols, thiocyanates, ethers, sulfides, sulfates, nitro groups and acyclic/cyclic alkanes. Depending on substrate and reaction conditions, skeletal rearrangements, elimination, polyalkylation, isomerization, cracking and polymerization reactions also fall within the scope of this reversible reaction. Monoalkylation of the arene moiety without skeletal rearrangement (R_1 to R_1') of the electrophile or the pendant alkyl side chain of the alkylated arene product, polyalkylation, and complexation of the Lewis Acid with the reagents/product(s), are the greatest synthetic limitations of this reaction. In addition, incorporation of substituents onto a monoalkyl arene scaffold under kinetic versus thermodynamic control, can result in different dialkylarene product orientations (*ortho/para* versus *meta* dialkyl substitution, respectively).

The F–C acylation reaction, industrially useful for the production of aromatic ketone and aldehydes, involves generation of an acylium containing electrophile which is used to acylate an arene moiety. It is generated from the reaction of Lewis acids with carboxylic acids, carboxylic acid halides or carboxylic active esters. In addition, ketenes and nitriles may be used as the source of the electrophile. Overall this reaction is generally not reversible and skeleletal rearrangements of the reagent or products seldom occur. Polyacylation does not usually occur primarily for two reasons: (1) the acylium ion is more stable than the corresponding carbocation so the R_1 group cannot easily rearrange, and (2) the acyl moiety of the product of this reaction is usually complexed more strongly with the Lewis acid than the starting arene which deactivates the ring toward further acylation. The inherent limitation of F–C alkylation reaction in preparation of primary alkyl arenes due to skeletal rearrangement can be overcome using a two-step sequence involving F–C arene acylation followed by reduction of the resulting ketone to a methylene moiety.[1–5]

6.2.2 Historical Perspective

The Friedel–Crafts (F–C) reaction was discovered at the Sorbonne (Paris, France) in 1877 by Charles Friedel (1832–1899) and James Mason Crafts (1839–1917) who met while studying under C. A. Wurtz in 1861. Due to common research interests involving organosilicon compounds, they began a collaboration during the period of 1863–1865. This partnership continued for 17 years, involved many areas of chemistry, including the F–C reaction, and resulted in close to 100 publications.[3,5,6] Their first paper describing a general method for the synthesis of aromatic hydrocarbons and ketones, such as amylbenzene and benzophenone, proved revolutionary and is now referred to as F–C alkylation and acylation reactions, respectively.[7]

Over the course of their collaboration they extended the scope of the use of catalytic aluminium chloride to a wide variety of organic reactions: (1) reaction of alkyl, acyl chloride, and unsaturated compounds with aliphatic and aromatic hydrocarbons; (2) reactions of acid anhydrides, oxygen, sulfur, sulfur dioxide, carbon dioxide, and phosgene with aromatic hydrocarbons; (3) cracking of aliphatic and aromatic hydrocarbons, and (4) polymerization of unsaturated hydrocarbons. The breadth and chemical diversity introduced by these chemical pioneers was brilliant.[8]

6.2.3 General Friedel–Crafts Alkylation Mechanism

The traditional Friedel–Crafts alkylation reaction involves the reaction of an aromatic substrate with alkyl halide catalyzed by a Lewis acid to form an alkylated arene moiety. Alternative reactants other than alkyl halides, such as alcohol, alkenes and cyclic/acyclic alkanes, can be also be used in reaction with the aromatic substrate. The mechanism of other alkylating agents that provide sources of carbocations such as alkynes, esters, aldehydes, ketones, amines (via diazotization), epoxides, thiols, thiocyanates, ethers, sulfides, nitro groups, and sulfates will be inferred from the general mechanism and not discussed in detail.[1–5,8] Many comprehensive F–C alkyl halide alkylation reviews, including books on this subject, have been reported by Price (1946)[9], Olah (1964, 1973),[3,5] Roberts and Khalaf (1984),[4] and Olah (1991 and 2005).[1,8] In all cases, a carbocation intermediate is generated which is capable of reacting as the electrophile in a reversible electrophilic aromatic substitution (EAS) reaction.

Mechanism of Alkyl Halides and Alcohols as Alkylating Agents

Alkyl halides and alcohols can coordinate with typical Lewis Acids (LA) such as $AlCl_3$, $TiCl_4$, SbF_5, BF_3, $ZnCl_2$ or $FeCl_3$, to form LA complex-**I**, which can act as the electrophile. It should be noted that alcohols may be activated also by the use of Brønsted acids to form similar complexes forming water as a byproduct. Alternatively, tight ion pair complex-**I** where the R_1 moiety can form a stable carbocation-**II**, can serve as the initial electrophile or rearrange to complexes containing a carbocation of similar or greater stability (**III**, $R_1'^{+}$). It is likely that primary alkyl groups proceed through a tight ion pair **I** (polar donor-acceptor complex), rather than via a free carbocation intermediate (**II**) which would be rapidly trapped by the arene moiety. The extent of polarization of the R_1–X/R_1–OMX_3/R_1–OH_2^+ bond resulting in the formation of **I**, **II** or **III** depends on the structure of R_1, the nature of the acid catalyst used, and the solvent of the reaction. In general, the order of reactivity of alkyl halides with Lewis acids is C–F > C–Cl > C–Br > C–I.[1–5,8,10]

Generalized Friedel-Crafts Intermolecular RX/ROH Alkylation Mechanism

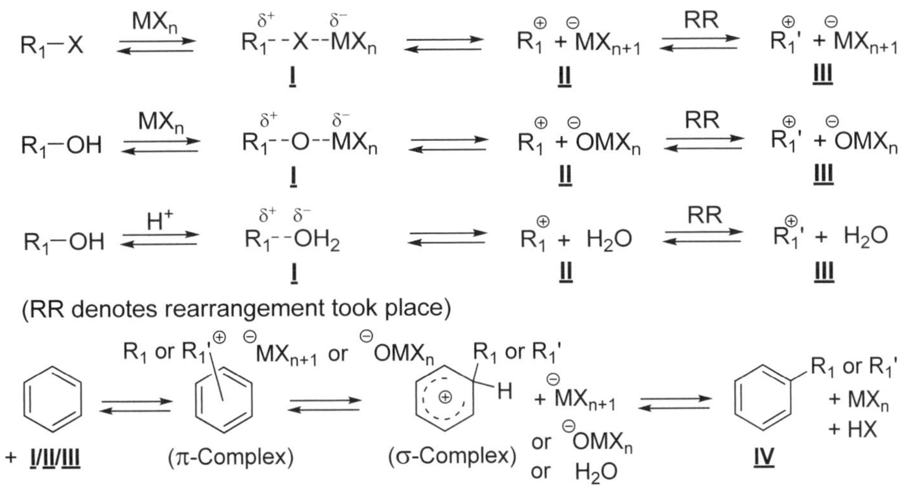

The arene moiety can form a rapid π-complex with the formed complex ion **I/II/III** which can rearrange slowly to a σ-complex, a cyclohexadienyl cation known as a Wheland intermediate. This step is usually the rate-determining step (RDS) whose existence was first postulated by Brown in the very first kinetics study of this reaction.[11] In further support of the mechanistic intermediate, the detection of several σ-complexes stable at low temperatures have been reported.[12,13] The aromatic ring moiety is generally regenerated by fast base abstraction of a proton with regeneration of the Lewis acid and HX formed. The actual RDS and reversibility of the overall mechanism, however, depends on a complex interplay of substrate and reaction conditions.

Some evidence that this reaction may not proceed completely through true carbocation intermediates such as tight-ion pairs complexes of **I**, is that alkylation of arenes in the presence of Lewis catalysts with optically active oxiranes can give up to 100% inversion (vide supra).[14,15] The majority of cases, however, show partial or total racemization indicating the significant involvement of carbocation intermediates. More compelling evidence is that a competitive ethylation study of toluene using methyl bromide and methyl iodide, gave different ortho/para/meta ratios.[14,16] A duality (carbocation versus tight-ion pair) of mechanism can exist of which the actual operating mechanism depends on many factors: (1) nucleophilicity of arene, (2) nature and reactivity of the alkylating agent, (3) solvent, (4) catalyst, and (5) reaction temperature.[17]

Olah and Olah reported a competitive alkylation mechanism study of naphthalene in 1976 and postulated that the positional and substrate selectivities of napthalene F–C alkylation shed insight on the kinetically versus thermodynamic controlled product composition of the reaction. It was suggested that a π-complex such as **I/II** was involved when highly electrophilic/highly basic arene moieties, resulting in the formation of the kinetically controlled product (1-substituted naphthalene analogue). In contrast, when weakly electrophilic Lewis acids or less basic arenes were used, the thermodynamic alkylation product (2-substituted naphthalene analogue) was produced via the intermediacy of the σ-complex (**II**, Wheland intermediate). These results were later supported by related observations of Nakane and co-workers in 1978.[1,18,19]

Mechanism of Alkenes, Alkynes and Alkane as Alkylating Agents (π- and σ-donors)

Reaction of Alkenes proceed through a similar intermediate **V**:

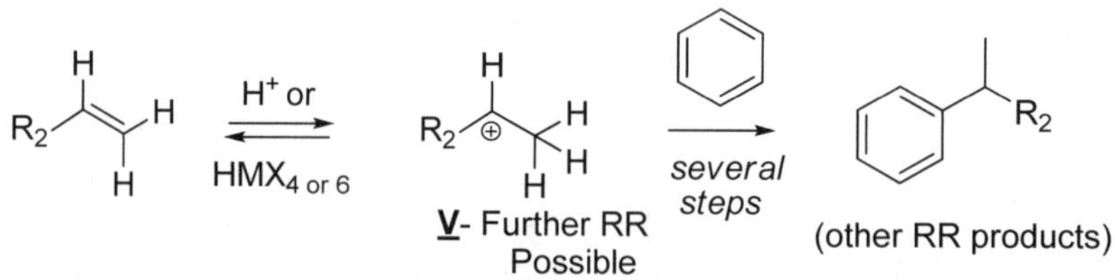

Reaction of alkanes via proton abstraction produce carbocation **VI**:

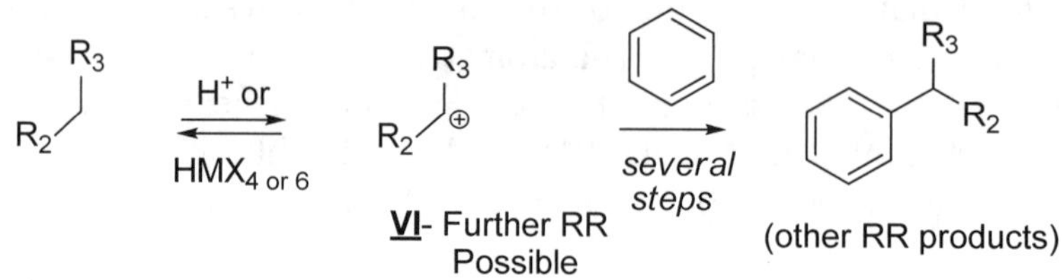

Both alkenes and alkynes (π-donars) react with Brønsted acids as well as conventional Lewis acids (MX_3, MX_5, etc.) in the presence of a proton source such as water or acids of the formula HMX_4 or HMX_6 give similar carbocation intermediates (**V**) following Markovnikov's rule and can be trapped by arenes via the same EAS mechanism.[5,8] Neat Lewis acids or even super Lewis acids (stronger than $AlCl_3$) are unsuccessful in these reactions without the addition of a proton donor cocatalyst. The addition of the cocatalyst is essential in allowing formation of a strong conjugate acid or carbocation to be formed.

Alkanes and cycloalkanes (σ-donors) can undergo hydride abstraction reactions to produce reactive carbocation intermediates of type **VI** which can be also captured by arenes in similar fashion.[8] Usually, strong acid donors on the order of super acids (equal or stronger than sulfuric acid) are required to generate the carbocation intermediate provided the C–H bond being abstracted has an appropriately low enough bond dissociation energy.

6.2.4 Key Friedel–Crafts Alkylation Reaction Variables

Lewis acid Catalysts: Use and Limitations

Many kinds of aprotic and protic acid catalysts used as single agents and in combination have been reported: (1) Lewis acids (metal halides, metal alkyl/alkoxides), (2) Acidic oxides-sulfides (acidic chalcogenides such as alumina, silica as single agent/combination, clays and zeolites), (3) acidic-cation exchange resins (Dowex, Amberlite), (4) Brønsted acids (regular and super proton acids), (5) Brønsted–Lewis superacid combination, (6) solid superacids that include acidic and shape selectivity (ZSM–zeolites), and (7) metathetic cation forming agents such as silver salts (non-catalytic/stoichiometric use).[1]

Some frequently used Lewis acids include AlX_3, $BeCl_2$, $CdCl_2$, $ZnCl_2$, BF_3, BX_3, FeX_3, GaX_3, TiX_4, SnX_4, and various antimony halides (where X = Br, Cl). Comparative studies ranking Lewis acid activities have been reported.[8,20a] Actual activity, however, depends heavily on reaction conditions (reagent substrates, solvent, and temperature).[20] Since most Lewis acid catalysts are destroyed in the aqueous workup step, this serves as a limitation for the recycling of moisture sensitive catalysts. Special measures can be taken, depending on the nature of the catalyst and appropriate reaction workup conditions employed, to permit effective recycling. The catalyst BF_3 can be recovered from reaction mixtures and reused as it is a low boiling gas (b.p. –101 °C). Fujiwara (1986) has shown the use of lanthanide trihalide-based Lewis acids to promote F–C alkylations with both recovery and reuse of the catalyst without loss of catalytic activity.[1,8,21]

Effect of Cocatalyst on F–C Alkylation Reaction

It has been determined that impurities such as moisture can accelerate this reaction, as $AlCl_3$-promoted reactions conducted under strictly anhydrous conditions can occur at a significantly lower rate of reaction. Due to this observation, it was subsequently discovered that the addition of cocatalysts can accelerate the rate of this reaction, such as oxygen, proton-releasing substances (ROH), Brønsted acids, and cation/carbocation producing substances.[8,22] For example, the F–C alkylation of arenes with alkenes and alkynes promoted by $AlCl_3$ was found to be cocatalyzed by trace amounts of moisture effecting an increase in both the rate and yield of the reaction; whereas the opposite was observed with use of $FeCl_3$ as the cocatalyst.[1,8,22]

Reactivity and Orientation of Monosubstituted Arenes

Monosubstituted arenes show good reactivity with moderately activated substituents such as alkyl but do not proceed under normal conditions in the presence of other strongly deactivating electron-withdrawing groups, such as carboxyl, nitro, and sulfonyl functional group containing moieties. The reactivity limit appears to be arenes with a moderately deactivating group such as halogen and trihalomethyl. Substrates with more electron-releasing functional groups that are basic or containing easy exchangeable protons such as amino and hydroxyl, lead to loss of catalytic activity. In these cases, the reaction may proceed if the Lewis acid is used in greater 1:1 stoichiometry. One exception due to lower overall basicity, is the preparation of alkylated arylamines from arylamines and alkenes using aluminium anilides as catalyst.[17]

The orientation of an incoming alkylating agent to a monoalkylated arene under conditions favoring kinetic control is *ortho* or *para*, with the latter usually predominating due to steric reasons. There are many experimental parameters favoring kinetic control of the reaction. They include use of a reactive electrophile in a preferential noncoordinating solvent with basic arene as nucleophiles, as well as use of low reaction temperature, short reaction times, and minimal amount of *weaker* Lewis acid catalysts. Under conditions favoring thermodynamic control (elevated temperature, long reactions times, large amount of catalyst, stronger Lewis acid, less reactive arenes, less reactive electrophiles and absence of solvent), *meta* substitution usually predominates.[4,8] Allen and Yats (1961) concluded that analysis of various *t*-butylations of toluene produce either 7/93 or 67/33 *meta*:*para*-toluene with no *ortho* isomer formed under conditions of kinetic and thermodynamic control, respectively.[23] Other monosubstituted arene having highly active electron-releasing groups (OR, NRCOR, NR_2) and moderately deactivating groups such as halo due to arene resonance contribution are usually *ortho*/*para* directing under either kinetic or thermodynamic conditions.

Alkyl Group Rearrangements and Reactions Involving Dealkylation and/or Transalkylation (Disproportionation).

Since the alkylation reaction is reversible, rearrangement of the alkylating agent,[24] as well as rearrangement of the pendant alkyl moieties of the formed alkylarene **III** can occur. In addition, the gain and loss of alkyl moieties of the formed alkylarene can occur through dealkylation–transalkylation disproportionation reactions. These reactions have some limitations, but may also be utilized to synthetic advantage. The likelihood of rearrangement is related to the stability of the carbocation formed (benzyl, $3^{o} > 2^{o} > 1^{o}$);

whereas, the tendency of disproportionation of the R_1 moiety is observed in similar order: *t*-butyl > *i*-butyl > *n*-butyl > Et > Me.[8,25]

General Disproportionation Reaction Scheme:

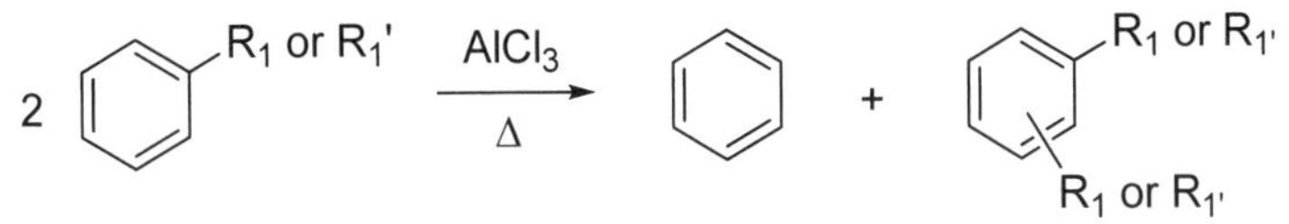

The likelihood of rearrangement of the alkyl group is clearly affected by temperature. One example is the alkylation of neat benzene by Ipatieff (1940) with *n*-chloropropane in the presence of $AlCl_3$ catalyst, where a 1.5:1 selectivity is observed in favor of *n*-propyl (**1**) over *i*-propylbenzene (**2**) at –6 °C; whereas, the reverse is true at 35 °C.[26] Other variables such as nature of the alkylating moiety, choice of solvent, and catalyst are also important in determining product composition.[4]

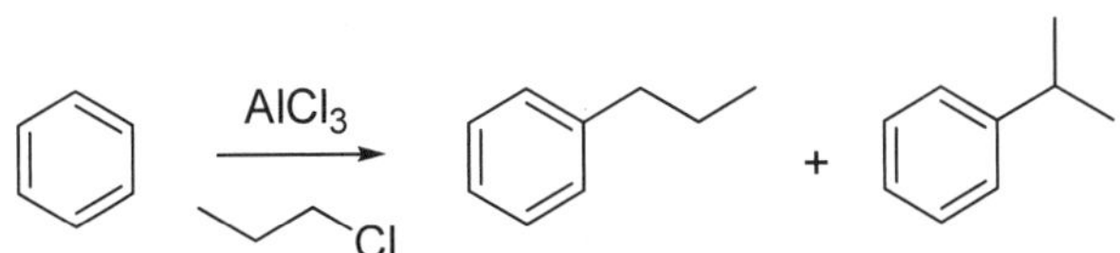

	1	**2**
Temp -6 °C	(60%)	(40%)
Temp 35 °C	(40%)	(60%)

Alkylation–Dealkylation Reactions Sequences. Introduction of Removable Blocking Groups (tert-Butyl Moiety)

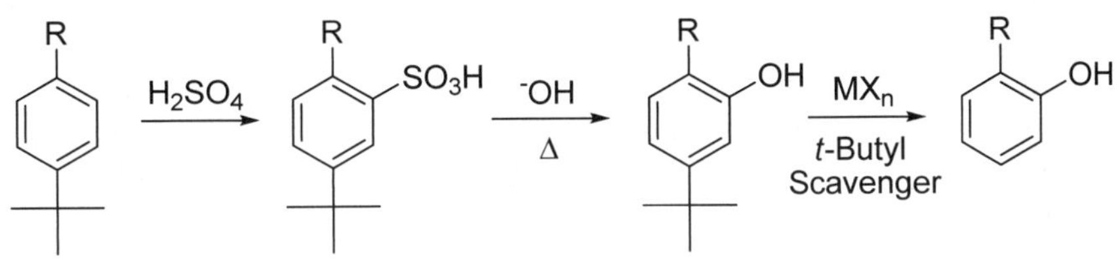

3 **4** **5a** **6a**

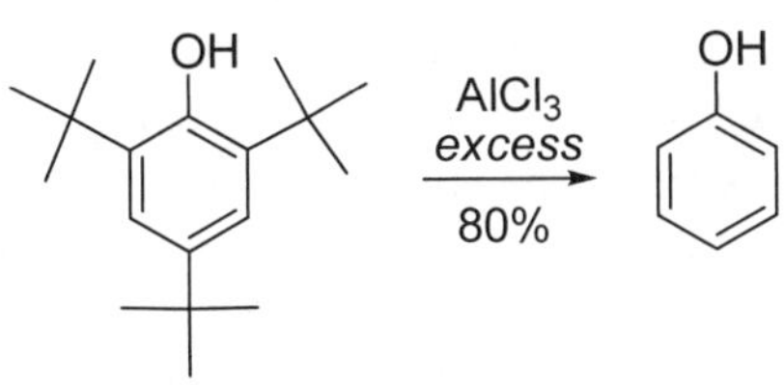

5b **6b**

The reversibility of the reaction allows alkyl blocking groups such as the *tert*-butyl moiety to be incorporated early into a reaction sequence and later removed after suitable functional group manipulation.[27,10] An interesting general example is the preparation of an *ortho* substituted alkyl phenol **6a** by selective *ortho*-sulfation of a *para*-alkyl substituted *tert*-butylbenzene **3**, saponification of the resulting sulfonic acid **4** and *tert*-butyl removal of the corresponding phenol **5a** via a retro-F–C alkylation sequence (Lewis acid usually $AlCl_3$ and arene based *tert*-butyl scavenger). Recently, use of TFA/$Na_2S_2O_4$[28] or excess Lewis acid such as $AlCl_3$[29] has shown to eliminate the need for arene based scavengers such as toluene, anisole or dimethylaniline in removal of the *tert*-butyl moiety. For example, removal of as many as three *tert*-butyl moieties of **5b** using excess $AlCl_3$ afforded **6b** in 80% yield.[29] In addition, selective removal of an *ortho*- versus *para*-substituted *tert*-butyl moiety has also been shown to be possible.[29]

Polyalkylation, Cracking, Isomerization, and Polymerization Reactions.

Polysubstitution can be an issue and is primarily due to preferentially partitioning of the alkylated product with the catalyst because of heterogeneity in the reaction mixture. Minimization of this side product may be accomplished using the arene as solvent and with high speed stirring. This may seem contrary to the popular notion that electron-releasing groups overactivate the ring to induce polyalkylation. Kinetic studies clearly show that electron-releasing alkyl groups show only a modest rate enhancement (only 1.5–3.0× as fast as benzene) and do not fully account for polysubstitution. In the industrial preparation of ethylbenzene from ethylene and benzene in the presence of $AlCl_3$ and cocatalyst water, the polyethylbenzene side product can be effectively recycled back to ethylbenzene.[8,17,30a]

The F–C alkylation reaction has played a vital role in the historical evolution of many core chemical industries, including the petroleum industry. In fact, $AlCl_3$ was the first catalyst to demonstrate catalytic cracking, the isomerization and conversion of larger hydrocarbon chain petroleum products to smaller more branched hydrocarbon products necessary to the production of high-performance (high-octane) gasoline. This process, mediated through formation of both radical and carbocationic intermediates, however, was limited industrially by excessive coke (nonvolatile carbon) and other polymeric by product formation. It was subsequently discovered (late 1920s) that the use of natural clays or zeolites, was far more effective promoting catalytic cracking over $AlCl_3$. This process was adopted widespread industrially and use of higher performance aviation fuel that resulted was believed to have been responsible for the air superiority of the allies over Germany during WWII.[30b]

Zeolites are complex bridging tetrahedral arrangements of silicon and alumina, each having four oxygen atoms as ligands. The silicon atom is neutral, whereas each aluminium atom has a negative charge that has associated with it a cation (such as natural clays that contain a sodium ion, Na^{+1}). These resulting networks create pores of regular dimensions (order of multiple Å size) that can be used to trap small organic molecules or the sodium cation exchanged with other cations to modulate its use as a powerful Lewis acid or Brønsted acid. For example, using a cation exchange process, the sodium may be exchanged for ammonia (NH_4^+) which can be heated with the extrusion of ammonia. The result is that the ammonium cation is replaced by a proton producing a powerful Brønsted acid. Zeolites may be used or modified from natural sources or produced artificially to modulate shape and size selectivity of the channels as well as the polarity.[30b,c]

Relative Reactivity of Dihaloalkanes

Boron Halide Catalyzed Haloalkylation with fluorohaloalkanes:

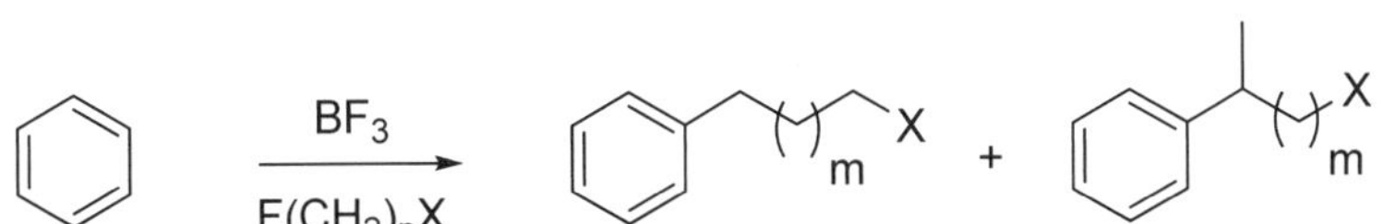

7 8

(n=2, m=0, X=I,Br,Cl)	(63%, 92%, 94%)	(N/A)
(n=3, m=1, X=Br,Cl)	(0%)	(89, 90%)
(n=4, m=2, X=Br,Cl)	(0%)	(80, 84%)

The reactivity order of fluorohaloalkanes carbon halogen bond (type R_1–X) with Lewis acid such as boron trihalides (BX_3) was C–F > C–Cl > C–Br > C–I whereas the reactivity order of the boron trihalide catalyst in the same study was BI_3 > BBr_3 > BCl_3 > BF_3. Complete monohalide alkylation selectivity of various primary alkyl monofluorohalides (n = 3,4) with regioselective displacement of the fluoro moiety in the presence of BF_3 catalyst was achieved by Olah (1964), albeit with rearrangement of **7** to **8** via capture of the more stable carbocation that was formed.[31] A similar trend has been observed with other more reactive Lewis acids, such as $AlCl_3$, though dialkylation in this case may be competitive.[4] It should be noted that alkylation of dihaloalkanes containing the same halogen atom leads to either intramolecular or intermolecular dialkylation, depending on the size of the methylene linker between the halogens of the reacting electrophile, the catalyst, and choice of reaction solvent.[4]

6.2.5 *General Friedel–Crafts Acylation Mechanism*

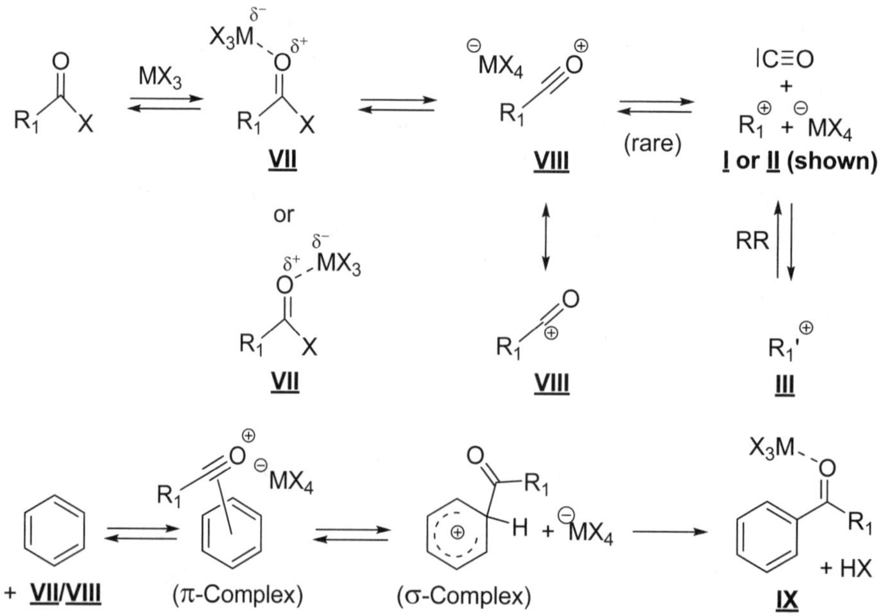

The F–C acylation is very similar to F–C alkylation and can also display a duality of mechanism proceeding through a tight ion and acylium π- and σ-complex (**VII** and **VIII**, respectively) of the arene moiety. The actual operating mechanism depends on a complex interplay of substrate and reaction conditions as previously discussed with the alkylation mechanism.[17] Although the EAS mechanism is the same, there is a significantly important difference. The arene σ-complex is often the rate-determining step (RDS) of the reaction; however, the reaction is not generally reversible as the final acylarene product of the reaction forms a tight complex (**IX**) with the catalyst. The formation of this complex can, but not always, prevent effective recycling of the catalyst from the catalyst-acylated product complex via ineffective halogen exchange between the catalyst and acylating agent.[8]

The acylium ion is particularly stable and generally does not easily undergo rearrangement reactions unlike F–C alkylations. If the acylium ion **VIII** can produce an especially stable carbocation and reacted with an arene of modest reactivity, side products can form via a competing F–C alkylation pathway. An example of such is the competing F–C acylation–alkylation pathways in the attempted pivaloylation of neat benzene versus anisole. The reaction of benzene with pivaloyl chloride in the presence of $AlCl_3$ catalyst produces *tert*-butylbenzene as the major product, whereas the reaction of anisole produces primarily *p*-methoxy pivaloylphenone. Choice of substrate and reaction conditions can be important criteria in which pathway predominates.[8,17]

6.2.6 *Key Friedel–Crafts Acylation Reaction Variables*

Acyl Halide Reactivity and Catalyst Selectivity

The majority of the catalysts discussed in the F–C alkylation section (vide supra) can also be used for the acylation reaction. A few commonly used Lewis acids are $AlCl_3$, SbF_5, and BF_3, but Brønsted acids, in general, are not effective in catalyzing acylation reactions with acyl halides. Some common electrophiles include carboxylic acids, halides, anhydrides, mixed active esters, ketenes, *O*-activated amidates derived from amides, nitriles, and carbon monoxide. The order of reactivity of acyl halides is usually but not exclusively I > Br > Cl > F. Due to complexation of Lewis acid catalyst with the carbonyl moiety, at least one and two equivalents of catalyst are usually required with acid chlorides and anhydrides, respectively. Cases where the LA catalyst may be conducted with high turnover involve use of arene substrates of high nucleophilicity with catalysts of moderate coordinating activity ($FeCl_3$, I_2, $ZnCl_2$, and Fe). In similar fashion, use of the catalyst may be eliminated all together when highly active acylating agents such as mixed carboxylic sulfonic anhydrides are combined with highly nucleophilic arene substrates.[14]

Synthetic Advantages of Friedel–Crafts Acylation Versus Alkylation.

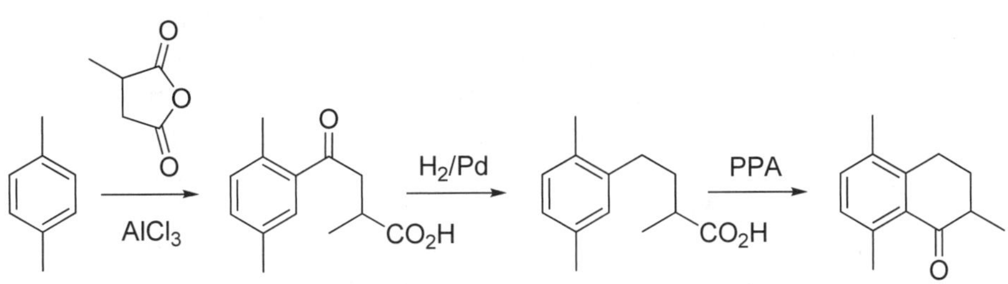

9 10 11 12

Commercially inexpensive acyl chlorides, anhydrides and nitriles are frequently used in industry as acylating agents; however, carboxylic acids using strong Brønsted acids (H_2SO_4, H_3PO_4, PPA), can also be particularly effective in promoting intramolecular F–C acylation.[14] A particularly powerful synthetic example involves a three-step Haworth-type reaction sequence for introducing a fused ring onto an arene scaffold involving initial intermolecular acylation of *p*-xylenes **9** with 2-methyl succinic anhydride at the less hindered carbonyl moiety, reduction on the intermediate arylketone **10** with H_2/Pd, and intramolecular acylation of the terminal carboxylic acid of **11**with PPA (polyphosphoric acid) to afford **12**. Note that the intermediate reduction step allows complete regioselective incorporation of a non-branched methylene moiety next to the aryl ring, a limitation of the F–C alkylation reaction.[32,33a,b]

An interesting one-pot tandem Lewis Acid Friedel–Crafts inter-molecular acylation method with concomittant in situ reduction of the aryl ketone product has been reported by Jaxa-Chamiec.[33c] Treatment of toluene (**13**) with 4-chlorobutanoyl chloride and $AlCl_3$ gave an intermediate ketone **15** which was reduced in situ by use of either of Et_3SiH of PMSH (polymethylhydroxysilane) to afford **16** in an impressive 85% yield from **13**.

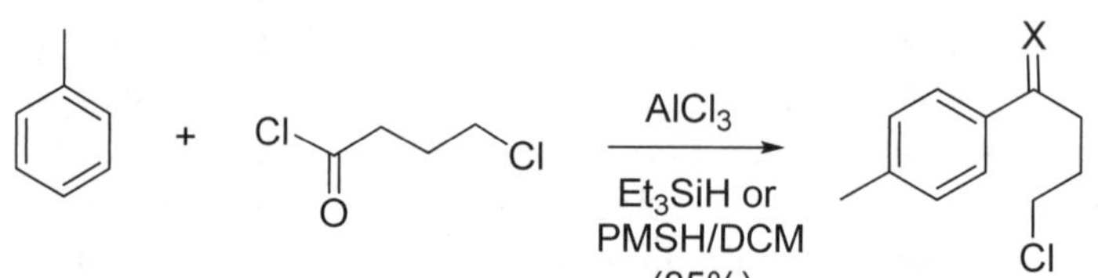

13 **14** **15** (X = O) **16** (X = H,H)

Substrate Selectivity-Orientation in Synthesis

The F–C acylation reaction does not generally undergo polyacylation due to a deactivating complexation of the product with Lewis acid catalyst. For this reason, arenes containing *m*-directing deactivating electron-withdrawing groups (CX_3, sulfonyl, nitro, and various carbonyl-containing functional groups) usually prove unreactive. Known *ortho*/*para*-directing groups such as alkyl, halogens, alkoxy, acetamido moieties work well, but aromatic amines and phenols suffer from competing *N*- versus *O*-alkylation side reactions. Protection of amine and alcohol functional groups is often required to ensure selective acylation. Many other complex regiochemical issues can be encountered in the F–C acylations of electron-rich arenes or heterocyclic arenes such as a complex interplay of competing electronic and steric directional control elements, which can work in synergy or against one another. Thus careful control of reaction conditions can be essential for success in these types of reactions.[5]

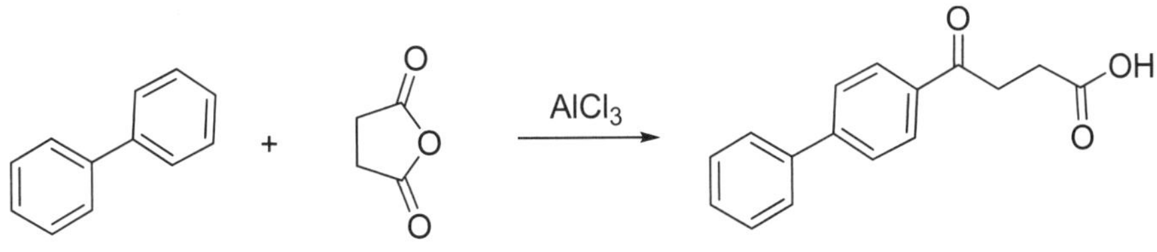

17 **18** **19** Fenbufen (80%)

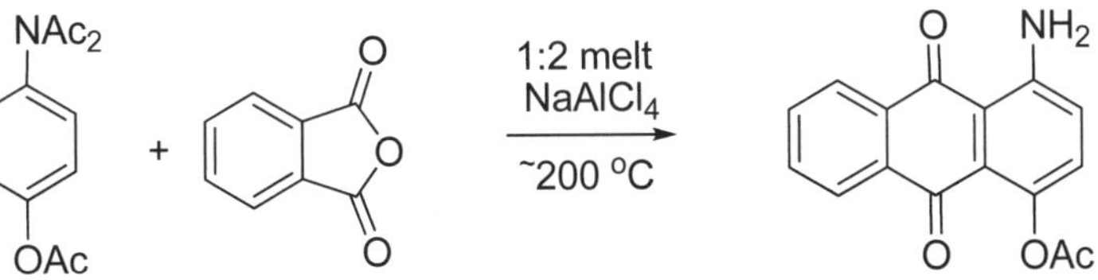

20 **21** **22** (45%)

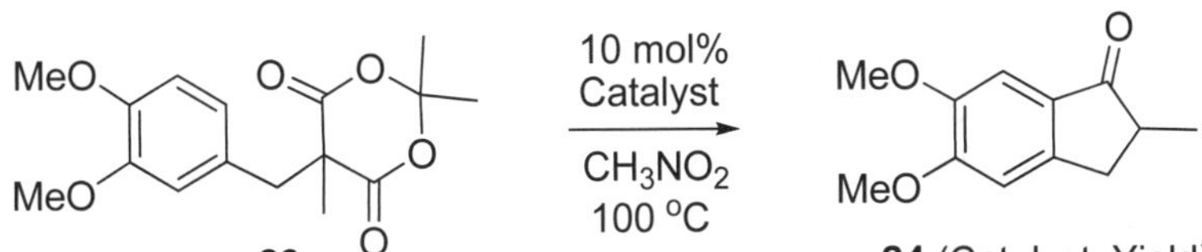

23

24 (Catalyst, Yield)

BF_3•OEt_2	90%
$Sc(OTf)_3$	85%
TfOH	86%

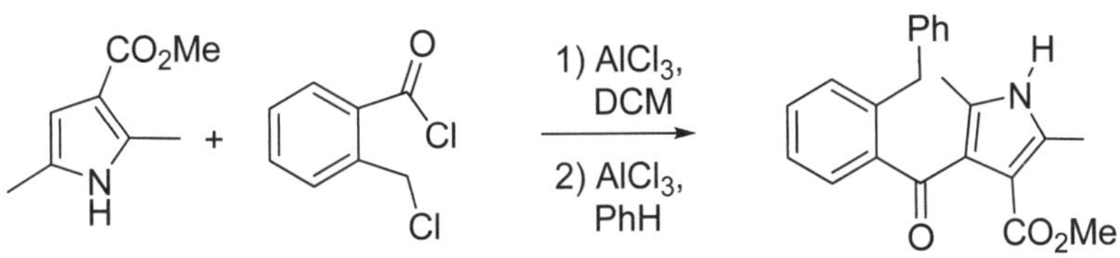

26 **27** **25** FPL64176 (60%)

Carboxylic anhydrides can react once or twice with arenes, depending on conditions. The regioselective monoacylation of biphenyl (**17**) with succinic anhydride (**18**) in the presence of $AlCl_3$ results in a 80% yield of fenbufen (**19**), a nonsteroidal antiinflammatory (NSAID) analgesic drug.[33d] Triacetyl protected aminophenol **20**, however, can be bisacylated under harsher conditions by reacting pthalic anhydride **21** in the melt with $NaAlCl_4$ at ~ 200 °C with concomitant cleavage of both the amino *N*-acyl protecting groups to afford **22** in 45% yield.[33e]

Regioselective F–C acylations of carboxylic anhydrides can be accompanied by other Lewis acid mediated reactions in tandem. Recently, a general method for the Lewis acid mediated intramolecular cyclization of β-keto ester alkyl substituted arenes to functionalized 1-indanones was reported by Fillion. Thus treatment of Meldrum's acid analogue **23** under a variety of

Lewis acid conditions gave an intermediate F–C acylation adduct regioselectively which underwent decarboxylation at 100 °C to afford 1-indanone **24** in 85–90% yield.[33f–h] An impressive example of reaction regioselectivity involving a tandem Friedel–Crafts acylation–alkylation sequence is the elegant synthesis of benzopyrrole calcium channel blocker **25** (FPL64716).[33i] Selective acylation of substituted pyrrole **26** with 2-chloromethyl benzoyl chloride **27** using $AlCl_3$ in DCM, followed by phenyl alkylation of the intermediate benzylic chloride with benzene and $AlCl_3$ gave adduct **25** in an impressive 60% overall yield for the two-step sequence.

Orientation of the incoming acyl group with an electron-releasing monosubstituted benzene, however, is predominantly *para* over *ortho* but selectivity can depend also on reaction conditions.[14] For acylation or benzoylation of simple benzene monoalkylbenzene derivatives, the *para*-selectivity can approach 20:1 or greater.[34] The reason for this was initially attributed to steric factors but a systematic analysis has shown that the relative reactivity of the electrophilic acylating agents can also affect the selectivity. For example, acylation of more reactive less stable acylium ion derivatives such as formyl and dinitrobenzoyl ion, the *ortho*:*para* selectivity can diminish greatly (~ 1:1).[32,35a]

Regioselectivity is also critical in the acylation of electron-rich heterocycles other than substituted benzenes. High regioselectivity (94:6 *C*-4 versus *C*-2 acylation) observed in the $AlCl_3$ promoted intramolecular acylation of **28** a 3-(indol-3-yl)propionyl chloride analogue, to adduct **29**, a precursor in the synthesis of Uhle's ketone **30**.[35b] A donor–acceptor complex between $AlCl_3$ and chloroacetyl chloride was proposed to drive the acylation reaction. A sodium methoxide mediated cleavage of the *N*-BOC moiety of **29** completed the synthesis of **30** in 95% yield.

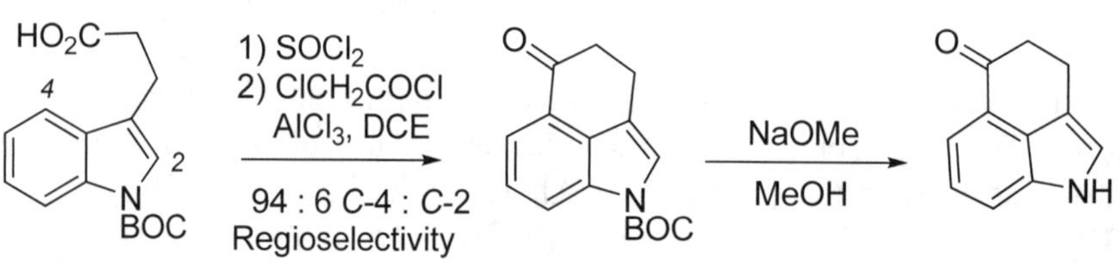

28 **29** (83%) **30** (95%)

Solvent Effects

Highly coordinating solvents such as nitrobenzene, nitromethane, and carbon disulfide not only reduce the activity of the Lewis acid but also complex the acyl cation intermediate. This is also an issue with the use of more contemporary ionic liquids and supercritical CO_2 as solvents used to potentiate chemical reactivity.[35c] This complex modulates the overall reactivity of the electrophile with the resulting increase of steric bulk of this intermediate further increasing selectivity of the reaction with the arene moiety (increasing *para-* over *ortho*-selectivity) for the less encumbered position. For example, acylation of chrysene and phenanthrene in nitrobenzene or carbon disulfide occurs at the sterically less encumbered outer ring; whereas, reaction with naphthalene under similar conditions leads to reaction primarily at the less sterically encumbered and less reactive 2-position.[8] Differences in the coordinating activity of these solvents can be exploited as well. For example, acetylation of naphthalene occurs primarily at the more reactive 1-position in carbon disulfide (kinetic control), but in nitrobenzene reaction occurs at the 2-position (thermodynamic control).[17]

6.2.7 Key Asymmetric Friedel–Crafts Developments, Improvements and Utility

F–C Alkylation Reactions involving an Asymmetric Center[8]

F–C alkylations, involving manipulation or creation of an asymmetric center, require a delicate fine tuning of substrate and reaction conditions to achieve an adequate balance of yield (regioselectivity) and optical purity (enantioselectivity). Regioselectivity is complicated by selectivity of the electrophile (alkylating agent) with the nucleophile (arene component) and the coupling selectivity of the arene itself (i.e., *ortho*/*para*-selectivity or positional selectivity of a heterocycle). For the enantioselective creation of a new asymmetric center using a prochiral electrophile as substrate, one can couple typically an electron-rich arene with a chiral Lewis acid catalyst or by performing a diastereoeselective alkylation reaction using a removable chiral auxillary and catalyst.

The chronological evolution of selected enantioselective F–C alkylations of arenes and related topics will be presented in the following sections: (1) Section A-chiral and achiral alkylating agents resulting in inversion or retention of configuration at the newly formed stereocenter, (2) Section B in 2 Parts-creation of new asymmetric centers by alkylation of carbonyl compounds (Section B, Part 1) and imines (Section B, Part 2) through use of chiral ligands coordinated to Lewis acids or use of chiral Brønsted acid catalysts, (3) Section C-indole polyalkylation issues related to Parts 1-2 of Section B, (4) Section D-asymmetric Pictet–Spengler reactions and related *N*-acyliminium cyclizations, and 5) Section E-asymmetric F–C Michael-type alkylations using chiral Lewis acids and chiral organocatalysts.

A Reaction with Chiral and Achiral Alkylating Agents

Historically, the majority of cases involving F–C alkylation of an asymmetric center derived from an acyclic substrate, result in almost complete racemization. The alkylation of cyclic substrates, however, exhibited better level of stereocontrol, perhaps by the enforced proximity of the leaving group. The stereospecific F–C alkylation of (*R*)-1,2-epoxypropane (1969) and (*R*)-1,2-epoxybutane (1975) with benzene and CS_2 in the presence of either $AlCl_3$ or $SnCl_4$, first reported by Nakajima and Suga, were landmark cases which shed some insight into the mechanistic complexicity of achieving both regio- and stereoselectivity in the reaction.[15]

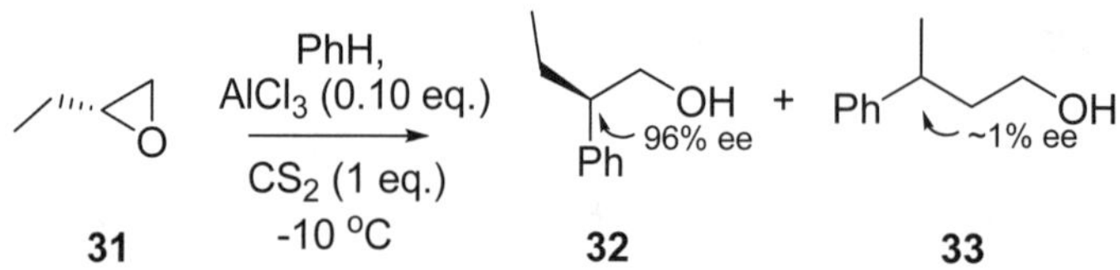

31 **32** **33**

(40% **32:33**, 51:49)

In the latter study, treatment of (*R*)-1,2-epoxybutane (**31**) with 0.10 equiv of $AlCl_3$ and 1 equiv of CS_2 at –10 °C gave **32** with almost complete inversion of configuration (96% *ee*), presumably via an intimate ion pair intermediate (vide supra). Poor regiochemical control was observed, however, due to competing chloride capture reactions (products not shown) and hydride transfer/phenyl capture isomerization reactions. Note that migration of the phenyl from the 2- to 3-position occurred in equal amount (51 : 49 of **32** and **33**, respectively) in 40% yield from **31**. With the weaker coordinating $SnCl_4$ as catalyst at the same temperature (–10 °C), the ratio of **32**:**33** changed to 98:2; however, with a reduced 83% *ee* of the 2-phenyl alkylated product **32**.

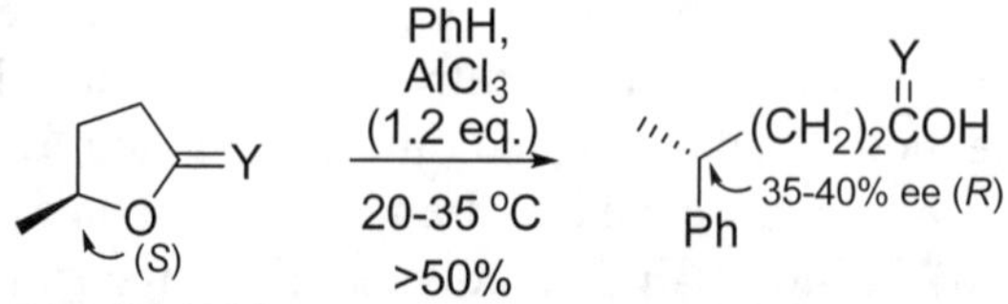

34 (Y=H,H)
35 (Y=O)

36 (Y=H,H)
37 (Y=O)

Some earlier studies by Brauman (1968–1969) in the alkylation of with benzene and $AlCl_3$ (> 1 equiv) with less reactive 2-methyl substituted cyclic ether[36] and 4-methyl lactone[37] substrates (**34** and **35**, respectively) at 20–35 °C gave much reduced stereochemical control (35:40% *ee*, **36**:**37**), with a reasonable yield (> 50%) of the phenyl alkylated products **36** and **37**, respectively. Thus chiral 5-membered cyclic ether and lactones are less reactive electrophiles than oxiranes, requiring greater than stoichiometric amounts of $AlCl_3$ (1.25 equiv), possibly the cause of the higher level of racemization observed. Tuning of the substrate with the reaction conditions appears critical in maximizing the utility of this reaction.

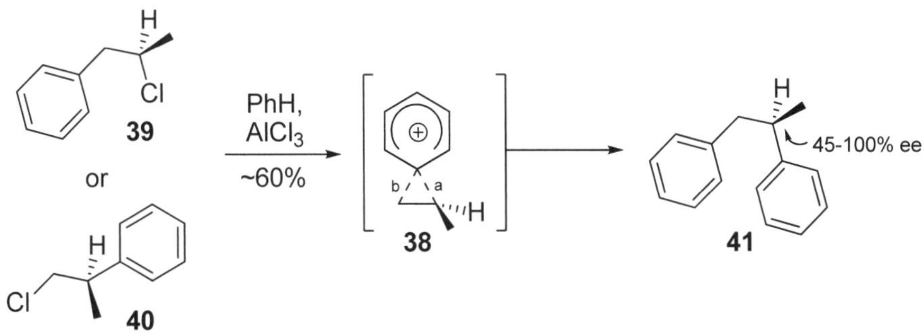

A particularly interesting case, reported by Masuda in 1983, involved induction of asymmetry via a phenyl π-assisted cation nonsymmetrically bridged intermediate (**38**) with stereospecific cleavage of bond a over b. Thus reaction of either (–)-2-chloropropane (**39**) and (+)-1-chloro-2-phenylpropane (**40**) with benzene in the presence of $AlCl_3$ catalyst, gave the identical product, (–)-1,2-diphenylpropane (**41**), in yields as high as 60% and optical purities of 45–100% *ee*, depending on reaction conditions.[38] The best results were obtained involved use of 0.1 equiv of $AlCl_3$ at 0 °C in benzene as solvent (30–40 equiv) to provide **41** in 100% optical purity from either **39** or **40** via a net retention and inversion of stereochemistry, respectively.

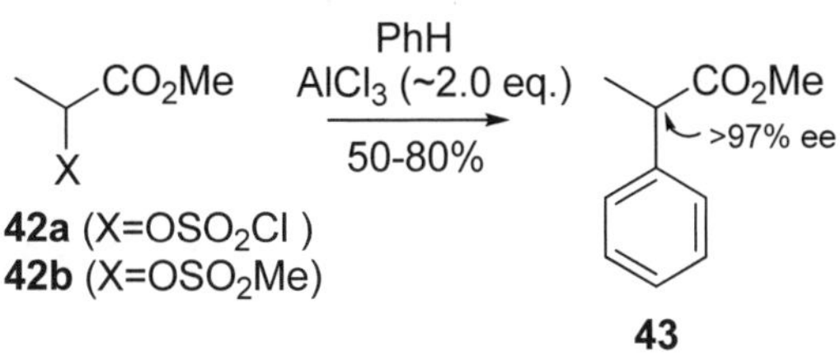

A landmark enantioselective F–C alkylation of benzene was reported by Piccolo in 1985 involving the use of acyclic substrates (sulfonates of alkyl lactates) with essentially complete transfer of chirality.[15] Alkylation of benzene with either the (*R*)- or (*S*)-enantiomer of chiral chlorosulfonates or mesylates (**42a** and **42b**, respectively) derived from methyl lactate, in the presence of $AlCl_3$ as catalyst, affords 50–80% yields of methyl-2-phenylpropionate (**43**) with essentially complete inversion of configuration

(> 97% *ee*). It is likely that use of a moderately nucleophilic arene neat as a low polarity solvent and coordination of the pendant amide and sulfonate leaving group to the Lewis acid contributed to the high degree of stereochemical control. Monosubstituted benzenes also appear to undergo alkylation with similar level of stereocontrol but suffered from very poor *ortho*:*para*-arene regioselectivity typical with most F–C alkylations.

Kronenthal (1990) extended this reaction to cycloalkyl sulfates by demonstrating that the 4(*R*) and 4(S) diastereoisomers of *N*-benzoyl-4-mesyloxy-*L*-proline (**44** and **45**) also alkylate benzene with complete inversion of configuration to afford the corresponding *trans*- and *cis*-4-phenyl analogues **46** and **47** in 91% (75% isolated) and 76.3% yields, respectively.[39] The isolation of optically pure **46** from **44** was reported to be accomplished on several hundred gram scale. A large excess (3.6 equiv) of $AlCl_3$ was required due to complexation with the carbonyl moieties. In either reaction, less than 0.2% of the other 4-phenyl epimer was detected. In addition, isolation of 8.5% and 14.5% of the corresponding 4-Cl epimer adducts (not shown) were isolated from **46** and **47**, respectively. It was postulated that a Lewis acid complex with the benzoyl carbonyl moiety (electron-withdrawing group) inductively destabilizes formation of a carbocation intermediate, leading to the high inversion stereospecificity observed occurring through an exclusive S_N2 pathway.

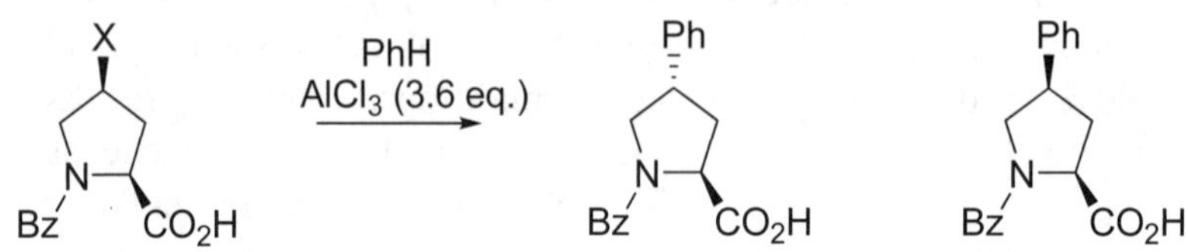

		46	47
44: X= OMs,	Temp 10 °C,	91% (75%)	----
48: X= F	Temp 25 °C,	70%	30%
49; X= Cl	Temp 80 °C,	75%	16.5%

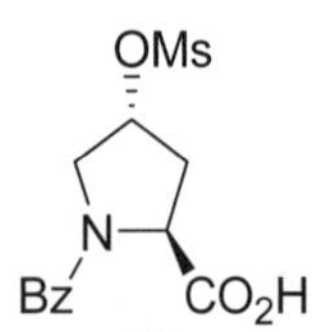

45	Temp 10 °C,	0.2%	76.3%

Replacement of the 4(*R*)-mesyloxy group (**44**) with fluoride (**48**) and chloride moieties (**49**), respectively, gave 70% and 75% of the inverted 4-phenyl adduct (**46**). These reactions required progressively higher reaction temperatures, consistent with the reactivity order of halides with Lewis acids (vide supra). The lower reactivity of the 4(*R*)-chloride (**49**) corresponded with isolation of 2% and 16.5% of the 4(*R*)-chloro (not shown) and phenyl epimers (**47**), respectively.

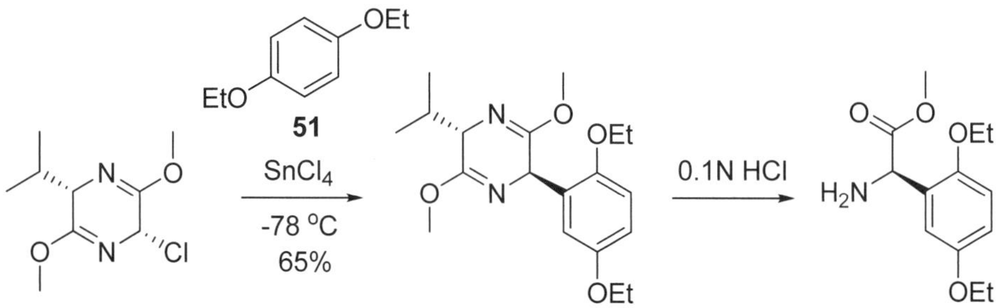

50 **52** (94% dr) **53** (>94% ee)

The use of a chiral auxillary directed F–C asymmetric alkylation was reported in 1987 by Schollkopf. Optically pure 2-arylglycine esters, important antibiotic pharmacophores, were prepared via halide inversion of diastereomerically pure alkyl chiral bislactim ester ethers. Treatment of **50**, the chiral bislactim ester ether of cyclo-(L-Val-Gly), with an activated electron-rich arene such as 1,4-diethoxybenzene (**51**) and $SnCl_4$ in DCM at –78 °C afforded predominantly bislactim adduct **52** (94% *dr*) in 65% yield. A hydrolysis operation then provided the 2-arylglycine ester **53** in > 94% *ee*. Treatment of **50** under analogous conditions using 1-methoxynapthalene as the arene coupling partner (not shown in scheme) gave the 4-napthalene bislactim alkylated product in similar yield (71%) and *dr* (95%), respectively. Reaction of **50** with anisole (also not shown in scheme) did not produce a regiochemically pure product, providing a 1:1 mixture of *ortho*/*para*-coupled isomers. Arenes lacking electron-rich methoxy substituents were also not successful coupling partners in this reaction.[40]

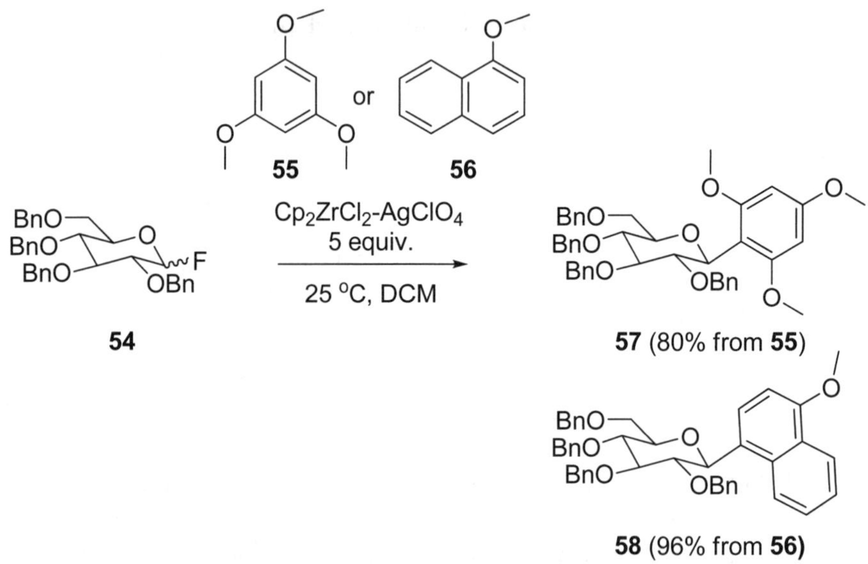

A particularly interesting example of regio- and stereoselective of *C*-arylglycosides, important biologically as antitumor agents, was reported by Matsumoto and co-workers in 1988. Treatment of glycosyl fluoride **54** with Lewis acid Cp_2ZrCl_2–$AgClO_4$ complex with either arene **55** or **56**, provided the equatorial (β-anomer) of *C*-aryl glycoside **57** and **58** in 80% and 96% yields, respectively. The formation of the β-anomer presumably resulted from capture of the oxonium intermediate under conditions of thermodynamic control (5 equiv catalyst). Use of smaller amounts of catalyst (0.2–0.5 equiv) in the majority of cases resulted also in formation of the α-anomer (kinetic product) in addition to the β-anomer. All of the arenes reported contained at least one methoxy substituent.[41]

61a (X,Y=H,H) **61d** (X,Y=Br,H)
61b (X,Y=H,Me) **61e** (X,Y=CN,H)
61c (X,Y=MeO,H) **61f** (X,Y=NO_2,H)

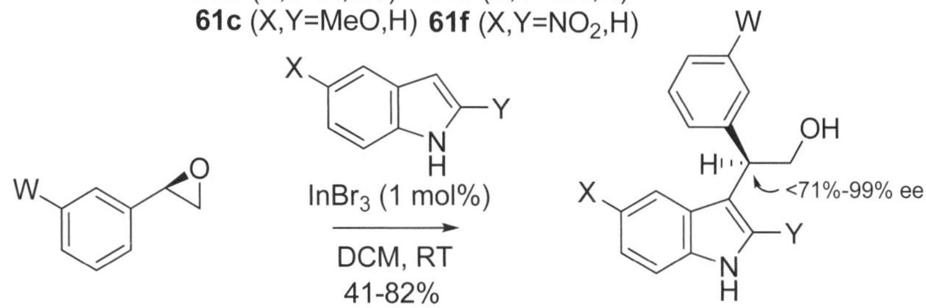

59, 60 (W=H, Cl) (*10 mol% of $InBr_3$ used) **62-69**

Product	W	X	Y	Z	%Yield (ee)
62	H	H	H	H	70 (99)
62*	H	H	H	H	60 (75)
63	H	H	Me	H	79 (99)
64	Cl	H	H	H	65 (99)
65	Cl	H	Me	H	82 (99)
66	H	OMe	H	H	54 (99)
67	H	Br	H	H	54 (99)
68	H	CN	H	H	41 (71)
69	H	NO_2	H	H	24 (ND)

Kotsuki in 1996 reported mild conditions for the regioselective and stereoselective F–C alkylation of indoles with chiral arene oxides. Treatment of (*R*)-styrene oxide **59** with indole (**61a**) gave the corresponding chiral β-3-indolyl alcohol **62** with inversion of configuration (88–92% *ee*, results not shown in scheme) using high pressure (10 kbar) or silica gel as the catalyst (7 days).[42] Umani-Ronchi et al. (2002) have reported impressive results in the alkylation of arene oxides (**59**/**60**) with indoles of type **61** using catalytic $InBr_3$ (1 mol %) in DCM to afford the corresponding alkylated products (**62**–**69**) with essentially complete regiocontrol and stereocontrol (41–82% yield, > 99% *ee*). It is interesting that the use of greater than 5–10 mol % of $InBr_3$ catalyst led to a considerably lower level of stereocontrol (75% *ee* vs. 99% *ee*, respectively) than 1 mol % in the coupling of **59** with **61a.**[43]

Incorporation of a chloro moiety at the 3-position of the styrene oxide (**60**, W = Cl) or a methyl moiety at the 2-position of the indole (**61b**, Y = Me) had no effect on yield or stereocontrol (compare **62**–**65**). Use of either electronically neutral, electron-donating or moderately electron-withdrawing groups at the 5-position of the indole (compare **61a,c,d** respectively where X= H, OMe, Br) gave good yields and stereospecificity of **62**, **66** and **67** (99% *ee*, 54–82% yield). Use of a strongly electron-withdrawing groups at the 5-position (**68-9**; X = CN, NO_2), however, had a deleterious effect on both yield (41% and 24%, respectively) and reaction stereocontrol with each

< 70% *ee*. Substitution of styrene oxide with propylene oxide, an aliphatic oxirane, under these reaction conditions led to complete nonregioselective opening of the aliphatic epoxide (~ 1:1 mixture not shown). Thus the combination of the coupling of arene oxide with a suitably electron-rich arenes is a critical factor in the success of this reaction to achieve a balance of both regiocontrol and stereocontrol.[43]

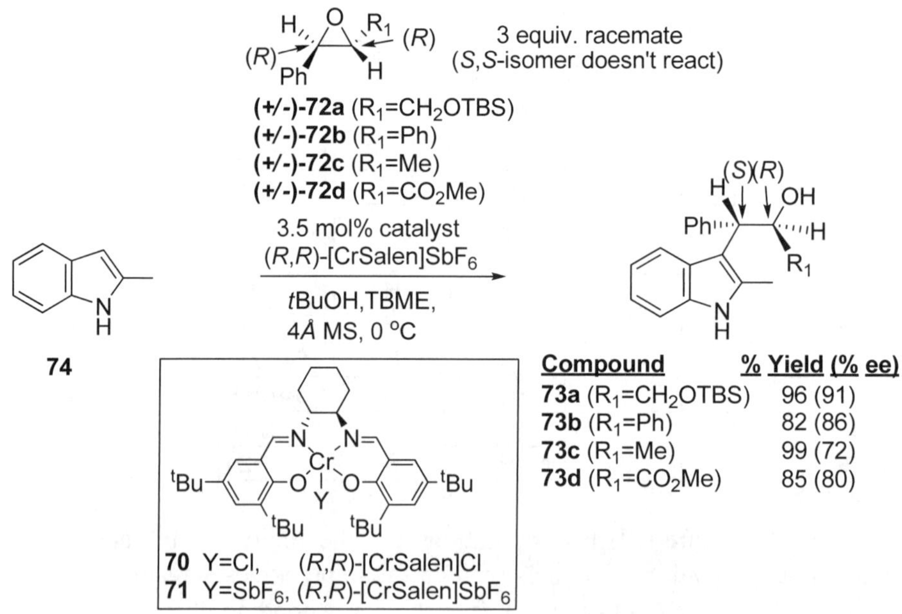

Compound	% Yield (% ee)
73a (R_1=CH_2OTBS)	96 (91)
73b (R_1=Ph)	82 (86)
73c (R_1=Me)	99 (72)
73d (R_1=CO_2Me)	85 (80)

In 2004, Umani-Ronchi et al.[44] discovered that chiral (*R,R*)-[Cr(salen)]Y complexes (Jacobsen catalysts **70**/**71**) could effectively undergo asymmetric F–C alkylation via kinetic resolution of racemic disubstituted epoxides. Treatment of 2-methylindole **74** with 3 equiv *trans*-1,2-disubstituted epoxides (**72a**–**d**) and 3.5 mol % of (*R,R*)-[Cr(salen)]SbF_6 catalyst (**71**) provided excellent yields (82–99%) of the corresponding chiral β-indolyl alcohols (**73a**–**d**) with good optical purity (72–91% *ee*). With this catalyst, only the (*S,S*)-isomer of **72a**–**d** reacts to form the (*R,S*)-isomer of **73a**–**d**, respectively.

As expected, the analogous enantiomerically pure *cis*-epoxide isomer of **72c**, provides a diastereomer of **73c** that has the oposite stereochemistry at the benzylic stereocenter (***R*** instead of ***S***, not shown). During the course of the resolution, the unreacted (*R,R*)-enantiomers of epoxides **73a**–**d** are enantioenriched and can be isolated in excellent optical purity (91–99% *ee*) using only a slight modification of the reaction conditions (data not provided). Thus the F–C alkylation reaction can also be used to effect the kinetic resolution of racemic epoxides. The detailed mechanism and stereochemical model of these reactions was not discussed.[44]

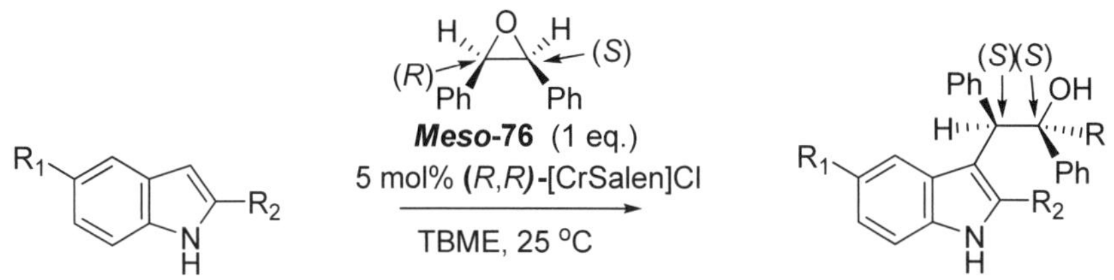

75a-c (1.5 eq)

Compound	% Yield (% ee)
(+/-)-77a (R_1, R_2=H)	98 (93)
(+/-)-77b (R_1=OMe, R_2=H)	95 (90)
(+/-)-77c (R_1=H, R_2=Me)	98 (98)

The (*R*,*R*)-[Cr(salen)]X catalyzed addition of indoles of type **75** was also applied by Umani-Ronchi (2004)[44] to the asymmetric ring opening of *meso*-stilbene oxide (**76**). Treatment of indoles **75a–c** (1.5 equiv) and *meso*-stilbene oxide (1.0 equiv **76**) in TBME with 5 mol% of (*R*,*R*)-[Cr(salen)]Cl afforded the corresponding chiral (*S*,*R*)-β-indolyl alcohols (**77a–c**) in superb yield (95–98%) and enantiomeric purity (90–98% *ee*). Incorporation of a 2-Me moiety or a 5-MeO moiety into the framework of the indole nucleus in place of H improves or reduces the optical purity of the alkylated product [compare **77c** (98% *ee*) and **77b** (90% *ee*) versus **77a** (93% *ee*), respectively].

B Part 1. Chiral Lewis Acid/Brønsted Acid Alkylations of Carbonyl Compounds

The previous investigations of the F–C alkylation of chiral alkylating agents by electron-rich arenes have thus provided a huge role historically in understanding the fundamental mechanistic complexicities of achieving high levels of both regiocontrol and stereocontrol. A selected historical evolution of key F–C asymmetric bond homologation of arenes with electrophiles other than alkylating agents under conditions of chiral catalysis will next be presented. Several interesting review articles[45,46] have recently appeared on this subject.

Fine-tuning of the substrate and reaction conditions is also critical for the asymmetric F–C alkylation of electron-rich aromatic compounds with functionalized aldehydes, ketones and imines. Both the regioselectivity and enantioselectivity of these alkylation reactions can be mediated through use of a complexing chiral catalyst. The catalyst may be either a Lewis or Brønsted base.

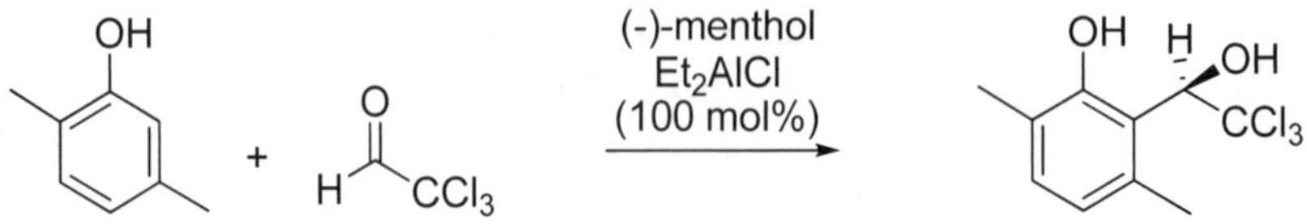

78 **79** **80,** 55% (80%ee)

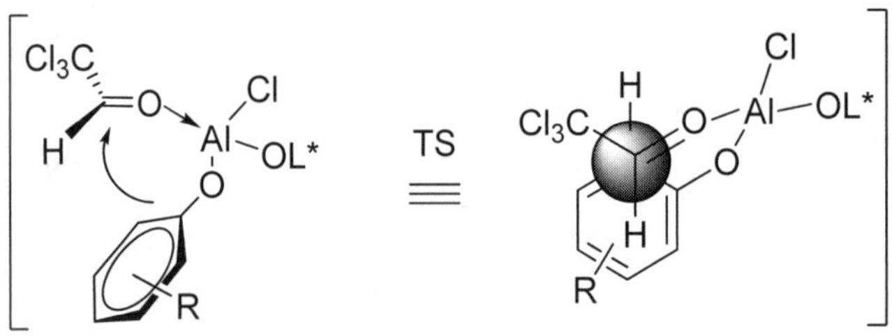

In 1985, Casiraghi reported the asymmetric *ortho*-hydroxylation of substituted phenols such as **78**, an electron-rich arene, with chloral **79** in the presence of a chiral alkoxyaluminum chloride, prepared from stoichiometric amounts of Et_2AlCl and a readily available chiral alcohol (–)-menthol, afforded the corresponding chiral (*R*)-trichlorobenzylic alcohol **80** in 55% yield and 80% *ee*.[47] A clear limitation of this reaction was the requirement of a stoichiometric amount of the catalyst. The proposed stereoinduction model was explained via a chelated 1,5-complex of the chiral Lewis acid with the phenol oxygen and the carbonyl of the aldehyde substrate.

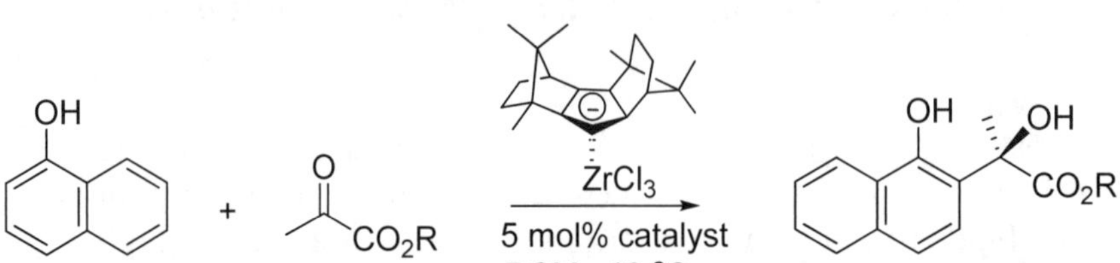

81 (**82a**, R=Et) (**82b**, R=Me) **83a,** R=Et, 56% (84% ee) **83b,** R=Me, 59%

An interesting asymmetric addition of 1-napthol **81** to pyruvic esters **82** using a catalytic amount of a chiral *(1R,4S,1'R,4'S)* dibornacyclopentadienyl-zirconocene Lewis acid was reported by Erker in 1990. Treatment of 1-napthol (**81**, 1 equiv) to pyruvic esters (**82a** and **82b**, 5 equiv) using 5 mol % of $ZrCl_3$–dibornacyclopentadienyl catalyst complex at –10 °C in DCM and water (27 mol %) afforded the corresponding substituted α-hydroxy esters **83a** and **83b**, respectively, in yields of 56–59%. The reported optical purity of **83a** (84–89% *ee*) depended on the conversion of **82a** (70–90%). The presence of a small amount of water appeared to increase the enantio-selectivity of the reaction on the order of 10% *ee*, even though it was reported the catalyst gradually loses the chiral dibornacyclopentadienyl ligand to form an achiral catalytically active species. A bidentate interaction of the phenol hydroxyl and the carbonyl moiety was speculated to be the cause of the exclusive *ortho*-selectivity. The optical purity of **83b** (prepared in 59% yield from **82b**) was not provided. This chiral catalyst was prepared efficiently in just a few steps from (+)-camphor.[48]

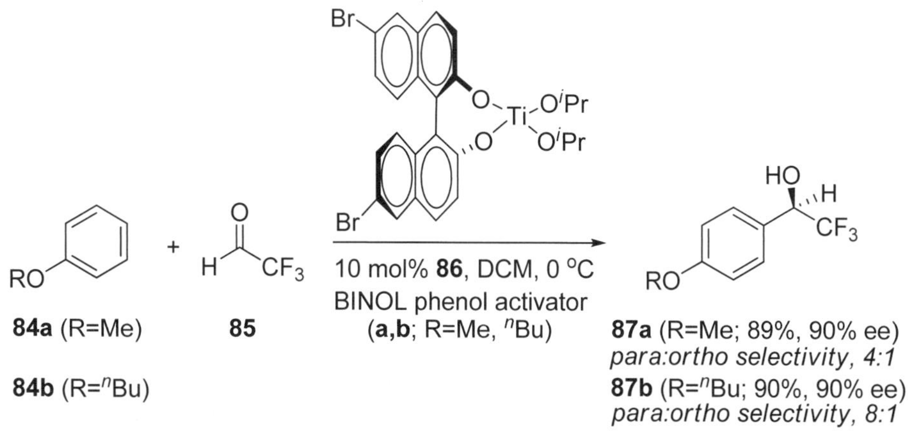

In 2000, Mikami reported the asymmetric addition of phenyl ethers **84a/84b** to fluoral (**85**) using 10 mol % of catalyst Ti(O^{i}Pr)$_2$-6,6′-(*R,R*)-Br$_2$-BINOL **86** with 1:1 addition of matched 6,6′-(*R*)-Br$_2$-BINOL biphenol activator (Brønsted acid of **86**) to afford the corresponding (*R*)-1-aryl-2,2,2-trifluoromethyl alcohols **87a/87b** in 89–90% yield and 90% *ee*.[49] The *para/ortho*-selectivity varied from 4 : 1 to 8 : 1 with just a simple change of R = Me to R = ^{n}Bu, demonstrating another complication in achieving complete selective regiochemical control. The *para*-selectivity suggests that co-complexation of the phenol oxygen of the arene and carbonyl moiety of the substrate with the Lewis acid does not play a significant role in the stereoinduction model. It is interesting that the enantioselectivity of **87a/87b** was reduced ~ 10% to 79% and 83%, respectively, without addition of the 6,6′-(*R,R*)-Br$_2$-BINOL Brønsted acid coactivator (bisphenol Brønsted acid precursor of **86**). Incorporation of the electron-withdrawing 6,6′-Br$_2$ substituents in place of hydrogen into the BINOL scaffold increased Lewis acidity of the catalyst. This proved to be a major factor in optimizing the enantioselectivity of the reaction (results not provided) by modulation of the reactivity of the catalyst.

C Control of Indole Polyalkylation Side Reaction in the Asymmetric Alkylation of Indoles with Carbonyl Compounds (Section B, Part 1) and Imines (Subsequent Section B, Part 2)

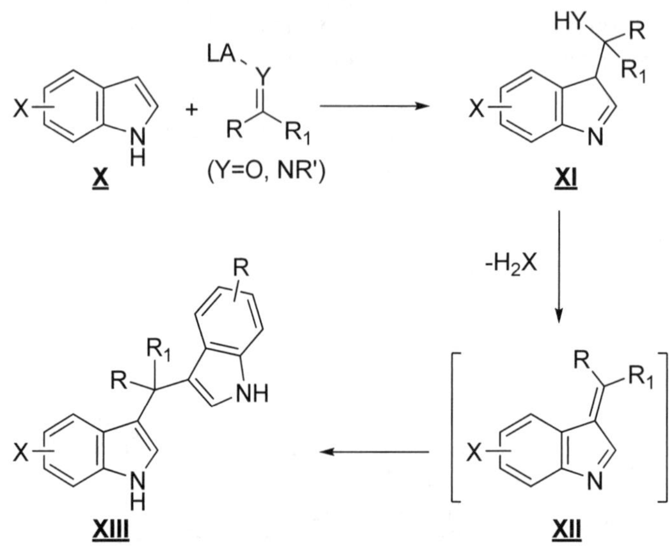

In the asymmetric alkylation of electron-rich heterocycles such as indoles with aldehydes, ketones and the corresponding imines under conditions of homogeneous and heterogeneous catalysis, additional regiochemical complications can arise such as polyalkylation.[46,50] For example, the regioselective *C*-3 alkylation of indoles of type **X** produces an unstable *C*-3 aminomethyl or hydroxymethyl alkylated product (**XI**) which can eliminate H_2X, creating another electrophile **XII** which alkylates again to produce the bisindolylmethylene alkylated product of type **XIII**. The reactivity of these side reactions can be modulated through the judicious matching of reaction conditions with a suitably reactive complexing bulky chiral catalyst. Choice of solvent can play a critical role in the formation or minimization of this side product. The byproduct can be favored under polar protic solvent conditions and can be a significant issue in lowering yield under Bronsted acid catalysis.

Dong (2007) has recently extended this reaction to 5-substituted indoles of type **88** using glyoxal as substrate in the presence of 10 mol % of $Ti(O^iPr)_2$–(*S*)-BINOL (**90**) in Et_2O to afford the corresponding 3-substituted (*S*)-α-hydroxyester indoles of type **89** in 64–88% yield and 62–96% *ee*.[51] As expected, electron-releasing substituents on the R_1 position of the nucleus gave the best overall results in terms of yield and enantioselectivity of the reaction (results not provided). The choice of ether as solvent at < 0 °C was important in minimizing formation of the bisindolylmethylene ester alkylated side adduct of type **91**.

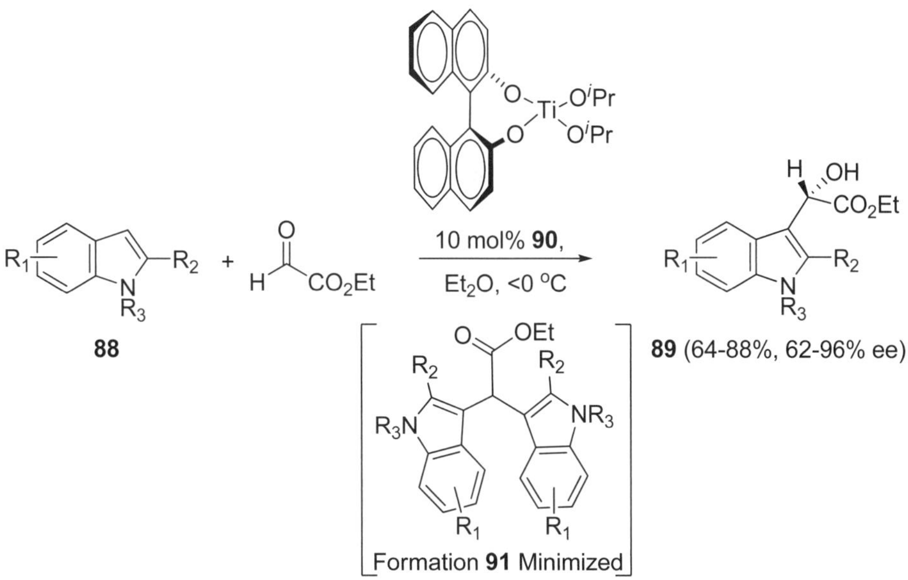

88

89 (64-88%, 62-96% ee)

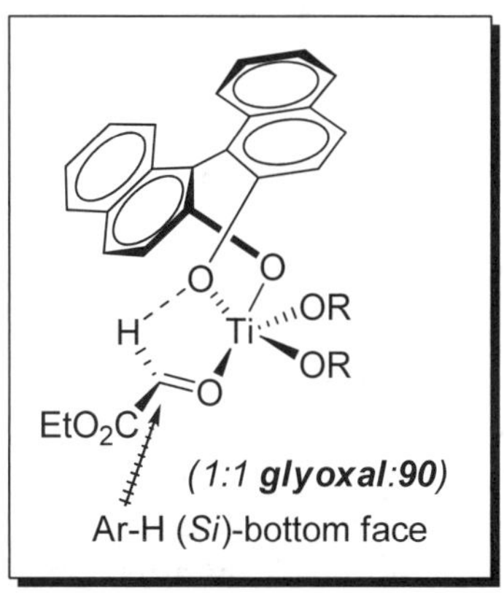

The proposed *Si*-face stereoinduction model involves stabilization of the Lewis acid–carbonyl complex with a formyl *H*-bond to the BINOL ether oxygen (1:1 complex of ethyl glyoxal:**90**). Replacement of the formyl hydrogen with an alkyl group provided little of the expected product with the bisindolyl alkylated side adduct (**91**) predominating and minimal stereocontrol for the associated adduct (not shown in scheme, ~ 10% *ee*).

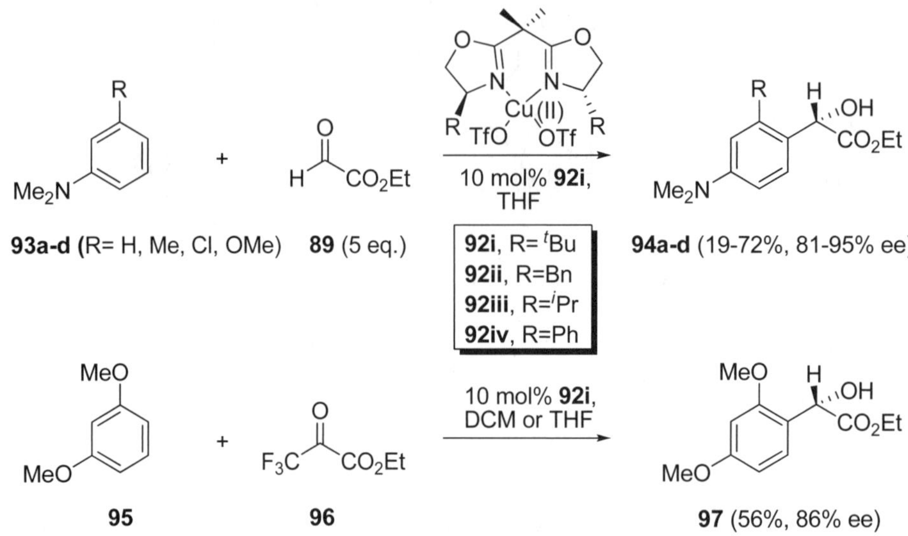

In 2000, Jorgenson[52] reported the catalytic asymmetric addition of electron-rich arenes to glyoxylates and trifluoromethylpyruvates using 10 mol % (*S*)-Cu(OTf)$_2$-^{t}Bu-BOX catalyst (**92i**), one of a number chiral cationic Cu^{+2} bisoxazoline (BOX) bistriflate complexes developed by Evans, which has been recently reviewed (some examples shown in scheme, **92i–iv**, [Cu-BOX](X)$_2$).[53a,b] Treatment of unsubstituted (**93a**) or 3-substituted anilines containing activating or deactivating groups (**93b–d**; R = H, Me, halo, OMe) to ethyl glyoxylate (**89**) using 10 mol % Cu(OTf)$_2$-^{t}Bu-BOX catalyst (**92i**) gave the corresponding (*S*)-chiral hydroxyesters **94a–d** in 19–72% yield and 81–95% *ee*. The lowest yield (19%) was obtained with **94d**, containing the most electron-releasing substituent (OMe). Under the same conditions, treatment of 1,3-dimethoxybenzene (**95**) with ethyl trifluoropyruvate (**96**) as the electrophile, gave a 56% yield of the (*S*)-hydroxyester **97** in 86% *ee*.

Using almost identical reaction conditions, Jorgenson (2001) explored the scope of the reaction of trifluoroethyl pyruvate (**96**) with other electron-rich heterocycles, such as furans (**98a/98b**), thiophene (**98c**), and pyrroles (**98d,f**), to afford the corresponding 2-substituted (*S*)-α-hydroxyester heterocycles (**99a–f**) in 15–80% yield and 79–89% *ee*.[54] Treatment of unsubstituted, 2- and 4-substituted indoles **100a–e** with **96** gave the corresponding (*S*)-3-substituted indolyl α-hydroxy esters **101a–e** in 61–94% yields and 83–94% *ee*. Substitution of the heterocyclic scafffold did not affect stereocontrol but greatly influenced regiocontrol.

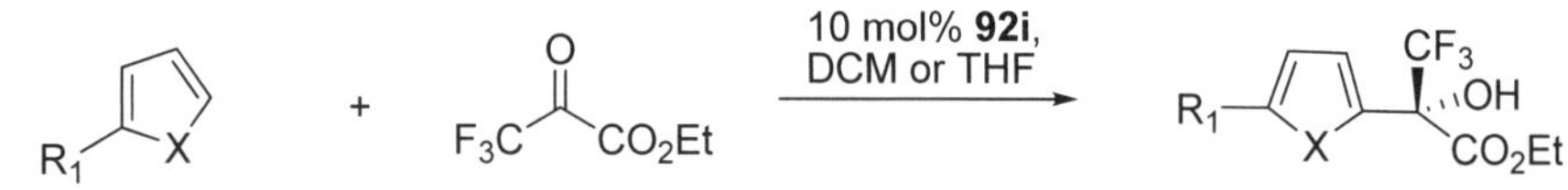

98a/b (R_1=H/Me, X=O) **96** **99a/b** (15/65%, 81/93% ee)
98c (R_1=Me, X=S) **99c** (16%, 79% ee)
98d/e (R_1=H/X=NH,NMe **99d/e** (80%/42%, 83/93% ee)
98f (R_1=COMe/X=NMe **99f** (69%, 89% ee)

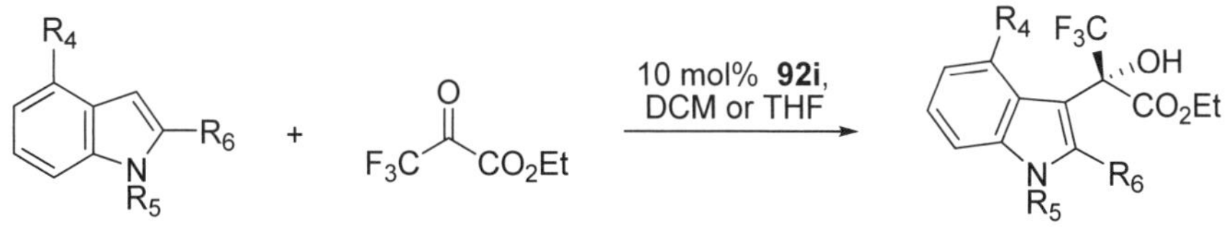

100a (R_4,R_5,R_6=H,H,H) **96** **101a** (93%, 83% ee)
100b (R_4,R_5,R_6=H,Me,H) **101b** (94%, 89% ee)
100c (R_4=Cl,R_5,R_6=H,H) **101c** (70%, 89% ee)
100d (R_4,R_5=H,H,R_6=Me **101d** (88%, 94% ee)
100e (R_4,R_5=H,H,R_6=Ph **101e** (61%, 87% ee)

Considerably lower yields were observed in formation of heterocycles of type **99** lacking a 5-methyl substituent (2-position of precursors **98**, compare furans **98a/98b** vs. pyrroles **98d/98f** in respective yields of 15%/65% and 42%/69%), presumably due to a competitive 2,5-dialkylation pathway. Replacement of the pyrrole *N*–H moiety of **98d** with an *N*–Me moiety (**98e**), resulted in a substantial yield decrease of **99d** and **99e**, respectively (from 80% to 42%). This trend was not observed with the indole scaffold (**101a** vs **101b** obtained in 93–94% yield from **100a/b**, respectively) perhaps due to a greater distance of the substituent from the reacting center. A reduction in yield (70% vs. 93%) with improved stereocontrol (89% vs. 83% *ee*) was observed 4-Cl indole substitution (compare **101c** vs **101a**). Comparison of effects of 2-substitution show amelioration of yield with 2-phenyl substitution (61% vs. 93% of **101e** vs. **101a**, respectively) with no significant effect on stereocontrol (87% *ee* vs. 83% *ee*).[54]

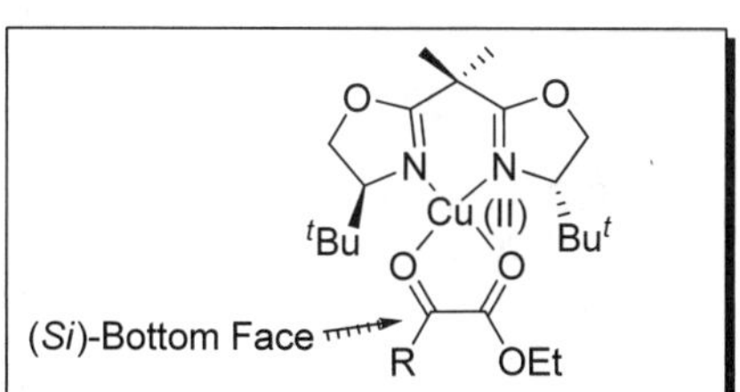

Proposed Binding Model

The originally proposed binding model is consistent with coordination of the divalent copper complex to the 1,2-dicarbonyl moieties in bidentate fashion, making a square planar copper intermediate resulting in the shielding of the *Re*-face of the carbonyl moiety by the *tert*-butyl substituent. When the R group does not contain π-bond (sp^3 hybridized), *Si*-face 1,2-attack of the α-keto carbonyl moiety is the major alkylation pathway forming the chiral (*S*)-α-hydroxy ester, otherwise conjugare 1,4-addition occurs instead (see section E for examples of asymmetric 1,4-Michael additions).[54]

This catalyst system is well designed to have an excellent coordinating metal and easily tunable BOX ligand. Among the divalent ions in the first transition series, Cu(II) forms the most stable ligand–metal complexes (Mn < Fe < Co < Ni < Cu > Zn) acccording to the Irving–Williams series rankings. The BOX catalysts are of type [Cu-BOX]$(X)_2$, where X is a weak or noncoordinating ligand (OTf^- or SbF_6^-), and coordination of a bidentate substrate is thus favored in the equatorial plane. Jahn–Teller distortion in the d^9 complex elongates the remaining apical sites where X may or may not reside promoting a very fast X ligand exchange rate. Thus the binding geometry can vary with substrate and nature of BOX ligand.[53b]

The use of chiral Brønsted acids (organocatalyst class) in asymmetric F–C reactions has been extensively reviewed.[55] They can be used instead of oxophilic chiral Lewis acids for the asymmetric coupling of indoles to less reactive ketone substrates such as trihalopyruvates. The use by Mikami (2000)[49] of a chiral phenol cocatalyst in boosting enantioselectivity (~ 10%) perhaps provided the first clue to such a possibility.

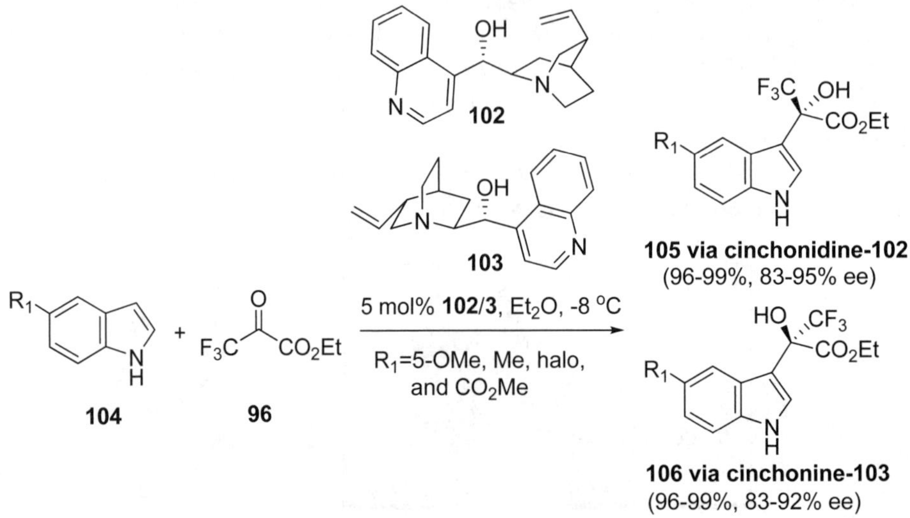

Török (2005) reported the use of chiral cinchona alkaloid Brønsted acid organocatalysts (5 mol %), cinchonidine (**102**) and cinchonine (**103**), in the enantioselective addition of 5-substituted indoles of type **104** (where R_1 = OMe, Me, halo, CO_2Me) to ethyl 3,3,3-trifluoropyruvate **96** to afford the corresponding (*S*)- and (*R*)-3-indolyl hydroxy esters, **105** and **106**, respectively, in 96–99% yield and 83–95% *ee*.[56a,b] Not shown in the scheme is the coupling of the analogous 6-methylsubstituted analogues of **104**, which afforded the corresponding 6-substituted adducts of **105** and **106** in similar yield (97 and 98%, respectively) and enantiomeric purity (95% *ee* and 90% *ee*, respectively) using 5 mol % of **102** and **103**, respectively. In the case of cinchodine catalyst (**102**), the rates of hydroxyalkylation were two orders of magnitude higher than the uncatalyzed process, indicating the enantiodifferentiation was a kinetic phenomenom. The stereoinduction model was not elaborated upon but clearly involves *H*-bonding. Evidence for this is that considerable racemization is observed with catalysts bearing a free *C*-9 hydroxyl, an unquaternized nitrogen, and when *N*-1 position of indole is not alkylated.[56a,b]

B Part 2. Chiral Lewis Acid/Brønsted Acid Alkylations of Imines

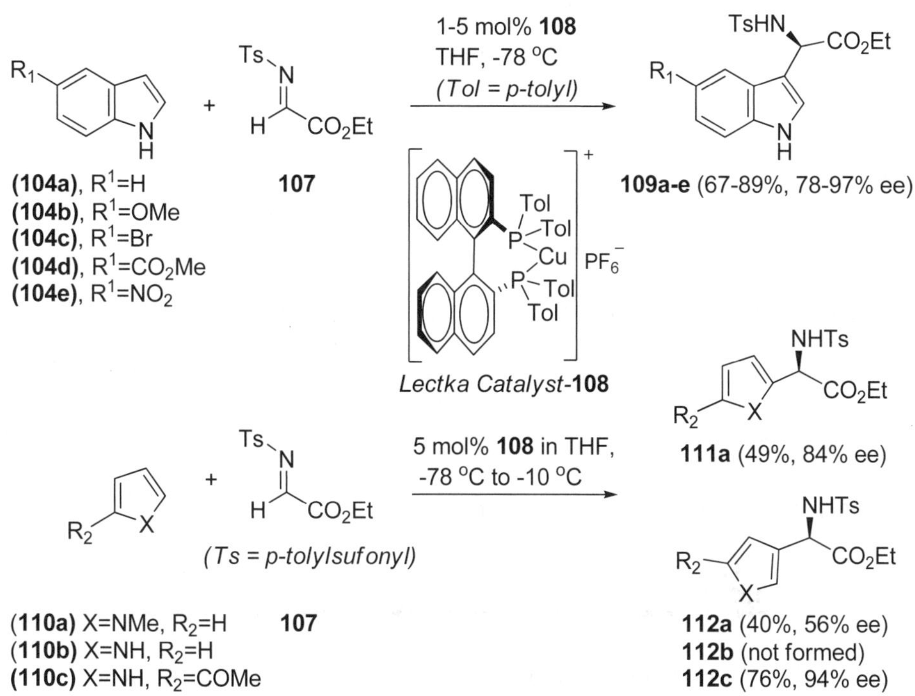

The first general method for the asymmetric aza-F–C alkylation of a wide variety electron-rich heterocycles with electron-deficient aldimines using the chiral Lectka catalyst was first reported by Johannsen (1999),[57a] but later modified by Jorgenson (2000–2002) to improve synthetic practicality and scope.[57b,c] Johannsen[57a] reported that treatment of a variety of a 5-substituted indoles (**104a–e**) with *N*-Ts imines of glyoxylic esters (**107**) in THF at –78 °C, catalyzed by 1–5 mol % Lectka catalyst (**108**), (*R*)-bis[2,2′-di-*p*-tolylphosphanyl)-1,1′-napthyl-$CuPF_6$ complex, gave the unnatural (*R*)-amino acid esters **109a-e** in good yield (67–89%) and enantiomeric purity (78–97% *ee*). The parent indole as well as electron-releasing and electron-withdrawing groups substituted on the 5-position (**104a–d**; R = H, OMe, bromo, CO_2Me) were well tolerated with the expected adducts (**109a–d**) formed in very good yield (67–89%) and optical purity (88–97% *ee*), except in the case where R = NO_2 (**109e** from **104e**, 71% yield, 78% *ee*). The majority of the products could be obtained enantiomerically pure through use of a single recrystallization.

The reaction with pyruvate **107** was extended to *N*-methyl pyrrole (**110a** where R_2 = H) as substrate but was not regioselective, affording 49% (84% *ee*) of the 2-substituted product (**111a**) and 40% yield (56% *ee*) of the 3-substituted pyrrole adduct (**112a**). None of the desired adduct (**112b**) was formed with *N*-1-unsubstituted pyrrole (**110b**), the free NH forming an aminal with imine **107** instead (product not shown). The reaction of 2-acetyl pyrrole (**110c**), however, gave none of the 5-substituted analogue (**111c**), affording instead the 4-substituted adduct (**112c**) in 76% yield and 94% *ee*.[57a]

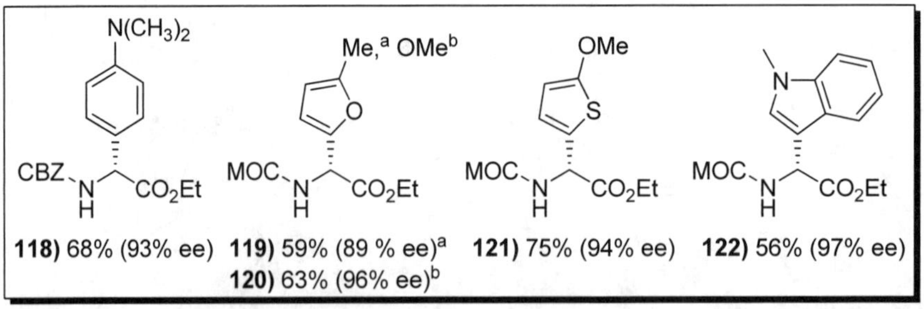

Jorgenson (2000–2002) demonstrated the effective coupling of a host arenes and heteroarenes (**113a–c**) with *N*-MOC and *N*-CBZ carbamate-protected-aldiminoesters (**114** and **115**, respectively) in the presence of catalyst **108** (5–10%) to afford the corresponding *R*-amino acid derivatives (**117a–c**) in good yield and optical purity (vide supra).[57b,c] The carbamate protected iminoesters (**114–116**) were chosen over the *N*-Ts analogues due to the ease of their removal in a subsequent *N*-deprotection step. A mixed solvent system (9:1 toluene/DCM) was employed to allow synthesis of iminoesters **114–116** via a aza-Wittig coupling sequence from the corresponding glyoxylate esters (not shown in scheme), followed by the F–C coupling all in one pot (for representative examples, see **118–122**). The reported yields are for the overall one-pot reaction sequence.

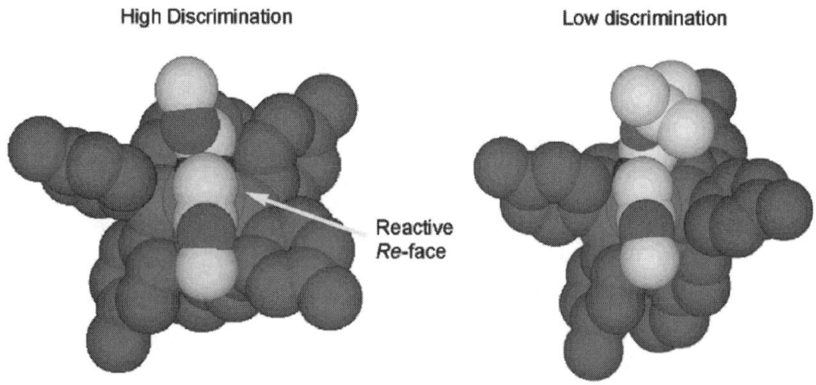

(*R*)-Tol-BINAP-Cu(I)
Complex With 114 (N-MOC)

(*R*)-Tol-BINAP-Cu(I)
Complex With 116 (N-BOC)

The proposed stereoinduction model involves a *para*-tolyl ligand of the catalyst blocking the bottom *Si*-face of the α-iminoester leaving the exposed top *Re*-face open for reaction with the arene moiety. The coupling of *N*-MOC and *N*-CBZ analogues **114–115** gave predictable high enantioselectivity. Coupling of the *N*-BOC moiety (**116**) occurred in modest yields but suffered from poor stereoselectivity leading to either the (*R*) or (*S*)-enantiomer (< 60% *ee*) of **117**, depending on reaction conditions. This result was predictable based on modelling (density functional theory—(DFT) minimization calculations) of the catalyst complexes with **114** and **116**. Poorer facial discrimination of the larger *tert*-butyl moiety of **116** versus the methyl moiety of **114** was predicted due to greater steric crowding on the *Re*-face.[57]

Terada in 2004 explored the coupling of 2-methoxy substituted furan **123** with electron-rich and electron-poor *N*-BOC aryl aldimines **124a–j** in the presence of 2 mol % binapthol phosphoric acid (*R*)-**125**, a chiral Brønsted acid, to afford adducts **126a–j** in high yield (80–96%) and optical purity (86–97%). This reaction was performed on gram scale using even lower catalyst loading (0.5 mol %) with the added benefit the catalyst could be easily recovered and reused. The researchers demonstrated the synthetic utility of the transformation by elaborating the furan product **126b** to γ-butenolide **127b** via a high-yielding two-step sequence (sequence not shown).[58]

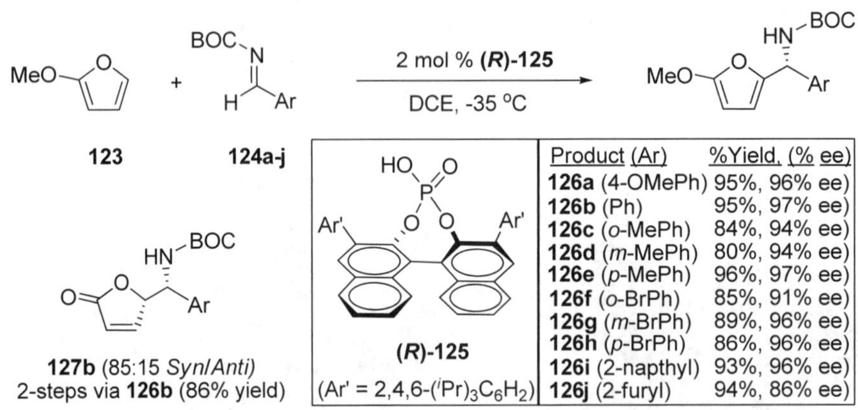

Product (Ar)	%Yield, (% ee)
126a (4-OMePh)	95%, 96% ee)
126b (Ph)	95%, 97% ee)
126c (*o*-MePh)	84%, 94% ee)
126d (*m*-MePh)	80%, 94% ee)
126e (*p*-MePh)	96%, 97% ee)
126f (*o*-BrPh)	85%, 91% ee)
126g (*m*-BrPh)	89%, 96% ee)
126h (*p*-BrPh)	86%, 96% ee)
126i (2-napthyl)	93%, 96% ee)
126j (2-furyl)	94%, 86% ee)

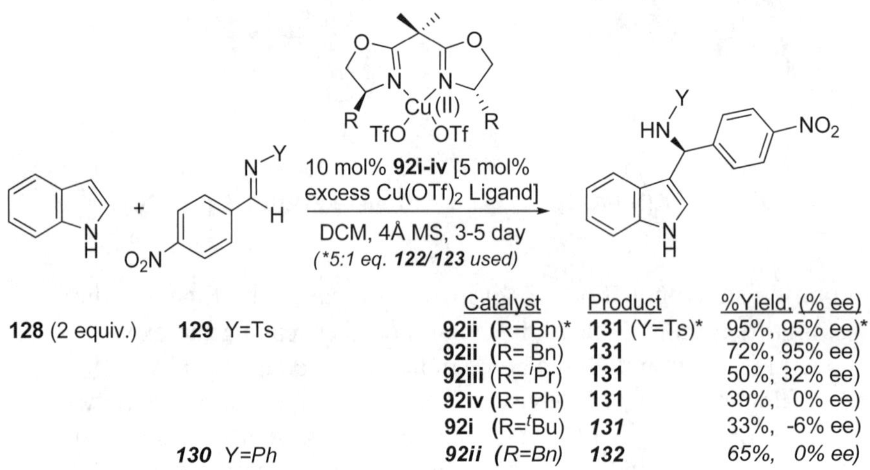

Catalyst	Product	%Yield, (% ee)
92ii (R= Bn)*	**131** (Y=Ts)*	95%, 95% ee)*
92ii (R= Bn)	**131**	72%, 95% ee)
92iii (R= ^{i}Pr)	**131**	50%, 32% ee)
92iv (R= Ph)	**131**	39%, 0% ee)
92i (R=^{t}Bu)	***131***	33%, -6% ee)
92ii *(R=Bn)*	***132***	*65%, 0% ee)*

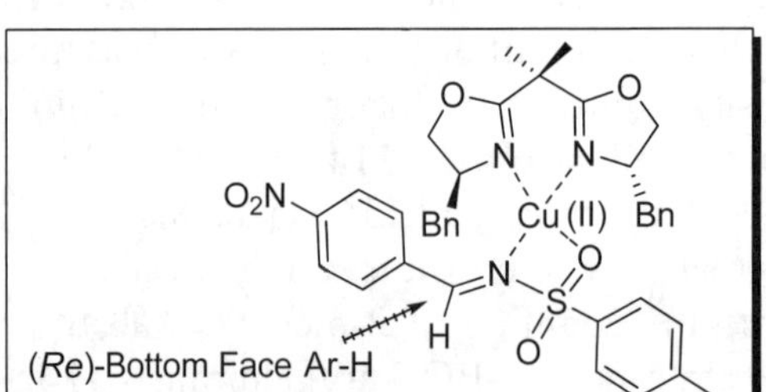

Proposed Binding Model of Complex 92ii/129

days) using a 5:1 stoichiometry of indole **128** to imine **129**. The coupling conditions involved the use of 10 mol % of (*S*)-Cu(OTf)$_2$-Bn-BOX catalyst (**92ii**, R = Bn) at 25 °C in DCM and 4 Å MS to afford a 95% yield of (*S*)-3-indolylarylmethanamine **131** in 95% *ee*. Treatment of *N*-phenyl imine **130**, the desulfonylated analogue of imine **129**, to these same reaction conditions, gave only the racemic adduct **132** in 65% yield. With a decreased 2:1 stoichiometry of indole **128** to *N*-arylsulfonyl aldimine **129**, the adduct **131** was obtained in 72% yield with the same optical purity (95% *ee*). Under the same reaction conditions, poorer conversions (33–50% yield) and enantioselectivities were observed in the formation of **131** using other Cu^{+2} BOX bistriflate catalysts bearing larger isoxazoline ligand substituents (**92iii**, **92iv**, and **92i**, where R = ^{i}Pr, Ph, and ^{t}Bu, respectively) in place of catalyst **92ii** (R = Bn).[59]

The proposed binding model involves 1,3-coordination of the copper to the sulfonyl and imine heteroatoms resulting in the shielding of the *Si*-face (top face as shown) of the imine moiety of **129** by the benzyl substituent of **92ii** and attack of the arene on the exposed *Re*-face (bottom face shown). The drop of enantioselectivity observed with catalysts **92iii/93iv/92i** is presumably due to increased $A^{1,3}$-strain between the arene of the imine and the larger R group of the ligand bound to the catalyst resulting in twisting of the bound conformation.[59]

Using the optimized reaction conditions, a 5:1 stoichiometry of indole **128** with *N*-nosyl arylaldimine **133a–d**, the electronic effects of the imine moiety on yield and enantioselectivity on the formation of the corresponding adducts (**134a–d**) was next examined. The *N*-nosyl protecting group was chosen due to ease of removal in a subsequent deprotection step. The effect of 4-substituention on the reaction was marginal as imines **133a–c** (where R = H, Me, F) gave comparable yields (68–86%) and enantioselectivities (92–94% *ee*) of **134a–c**. A lower yield (47%) of **134d** (R = OMe, strong electron-donating group) was observed, but with comparable enantiocontrol (88% *ee*).

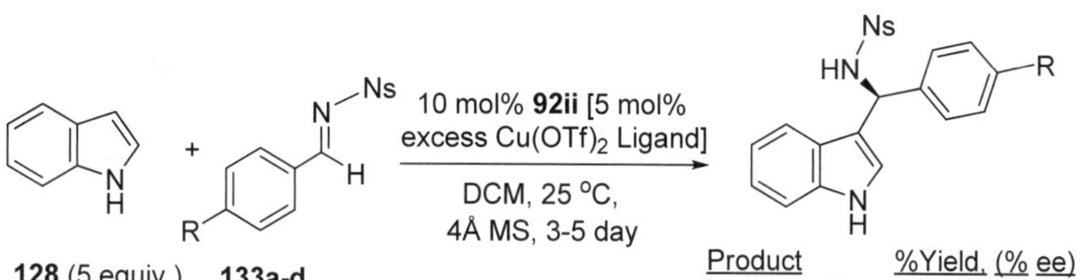

Product	%Yield, (% ee)
134a (R= H)	86%, 94% ee)
134b (R= Me)	84%, 94% ee)
134c (R= F)	68%, 92% ee)
134d (R= OMe)	47%, 88% ee)

The reaction proved to be remarkably tolerant of substitution on the indole scaffold. Further exploration of indole substituents will not be shown in the scheme but will be discussed. Comparison of the coupling of a 5-OMe analogue versus the unsubstituted analogue of indole **128** with unsubstituted imine **133a** (R = H) afforded similar results as the 5-OMe analogue of **133a** was obtained in 92% yield (94% *ee*). Introduction of a 5-Br substituent into

128, however, lowered the yield (50%) and enantioselectivity (77% *ee*) in the coupling of **133a**.[59]

The concept of tunable chiral thiourea based organocatalysts, useful for a wide variety of asymmetric reactions, was invented by Jacobsen in 1999 and has been extensively reviewed over the last few years.[60] They have been particularly useful in the F–C alkylations of imines. In 2006,[61] Deng discovered thiourea catalyst **135** to be an effective general catalyst in promoting effective asymmetric F–C reactions of substituted indoles (**136**) containing either electron-donating or withdrawing groups with aromatic or aliphatic *N*-arylsulfonylimines (**137** and **138** respectively) to afford the corresponding adducts **139** and **140** in very good yield (53–98%) and stereocontrol (83–96% *ee*).[61] Treatment of a variety of a 4-, 5- and 6-substituted indoles (**136**) with *N*-Ts or *N*-Bs imines of aromatic imines (**137**) in EtOAc at 50 °C and 10 mol % thiourea organocatalyst **135** in EtOAc at 50 °C, gave the unnatural (*S*)-3-indolylarylmethanamine **139** in 86–98% yield and 86–98% *ee*. Remarkably, the reaction of indole (**128**) with aliphatic *N*-Ts imines **138** under the same reaction conditions promoted formation of the corresponding indolylalkylmethanamine **140** in 53–86% yield and 94–96% *ee*. Since Jorgensen[57] and Zhou's comparable methods[59] are limited to use of more activated imine substrates such as aromatic imine and imine esters, this method has proven to be more versatile in terms of scope. The mechanism of activation of thiourea organocatalysis will be presented in greater scope in the next section.

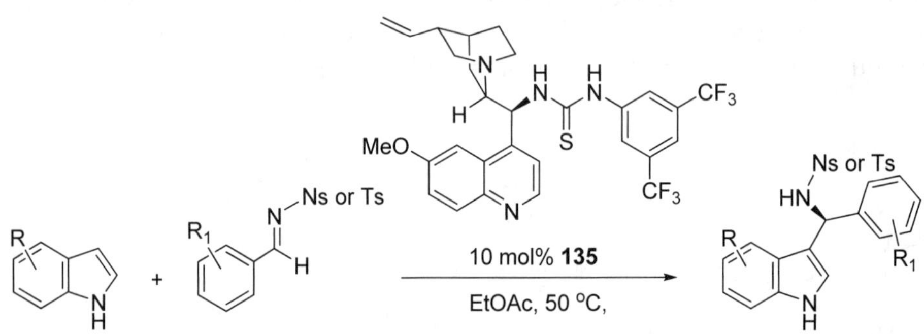

136 (2 equiv.) **137** (1 equiv.)

R = 4-Cl, 3-OMe, 2-Br, 4-CF_3, 4-Me, H

R_1 = 6-Cl, 6-Br, 6-OMe 5-Me, 4-OMe

139 86-98% (86-96% ee)

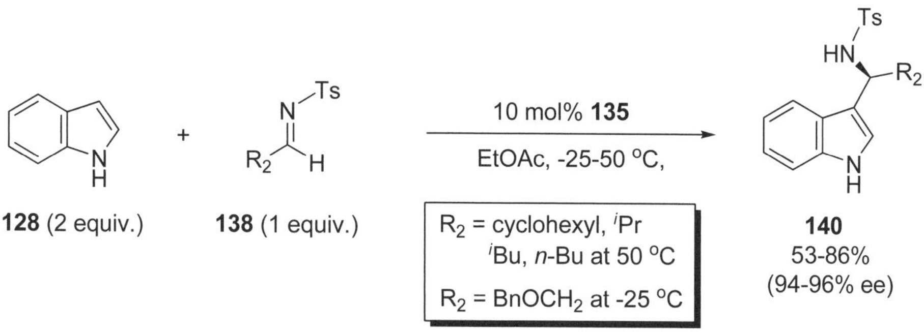

128 (2 equiv.) **138** (1 equiv.) **140** 53-86% (94-96% ee)

The use of aliphatic imines in aza-Friedel–Crafts reactions has been limited by stability to the reaction conditions, particularly with enecarbamates or enamides. Terada (2007)[62] demonstrated that substituted indoles of type **136** containing either electron-donating or electron-withdrawing groups at the 5- or 6-positions could be reacted with *E*- or *Z*-*N*-Boc-protected enecarbamates **141** in the presence of 5 mol % (*R*)-BINOL-phosphoric acid (chiral Brønsted acid **125**) in CH_3CN at 0 °C to afford the Friedel–Crafts adducts **142** in 63–98% yield and 90–94% *ee*. The (*Z*)- and (*E*)-enecarbamates of **141** afforded adducts **142** in identical enantioselectivities but with different rates. It was speculated that the rate-determining step involved formation of a common imine intermediate by protonation of the enecarbamates by the phosphoric acid catalyst. In addition, a strong solvent effect was observed with polar protophobic solvent shown to greatly improve catalytic efficiency.

Similar methodology for asymmetric construction of quaternary asymmetric centers was introduced by Zhou in 2007.[63a,b] Treatment of α-aryl enamides **143** with indole **136** in the presence of 10 mol % (*S*)-BINOL-phosphoric acid catalyst **125** gave the resulting Friedel–Crafts adducts **144** in 73–97% yield and 73–97% *ee*. Presumably, isomerization of enamine **143** to imine **143a** under the influence of the catalyst occurred before alkylation by indole **136**. Alkylation of either of the free *N*–H moieties of the indole and/or enamide led to a complete loss of reactivity. This finding supports the proposed stereoinduction model where chiral phosphoric acid catalyst **125** activates both the indole and ketamine (**143a**) via two *H*-bonding interactions of the trimolecular complex **136/143/125**. The protonation of ketimine **143a**, formed by isomerization of enamide **143**, then activates it for nucleophilic attack of the indole on the *Re*-face, resulting in the observed adduct **144**.

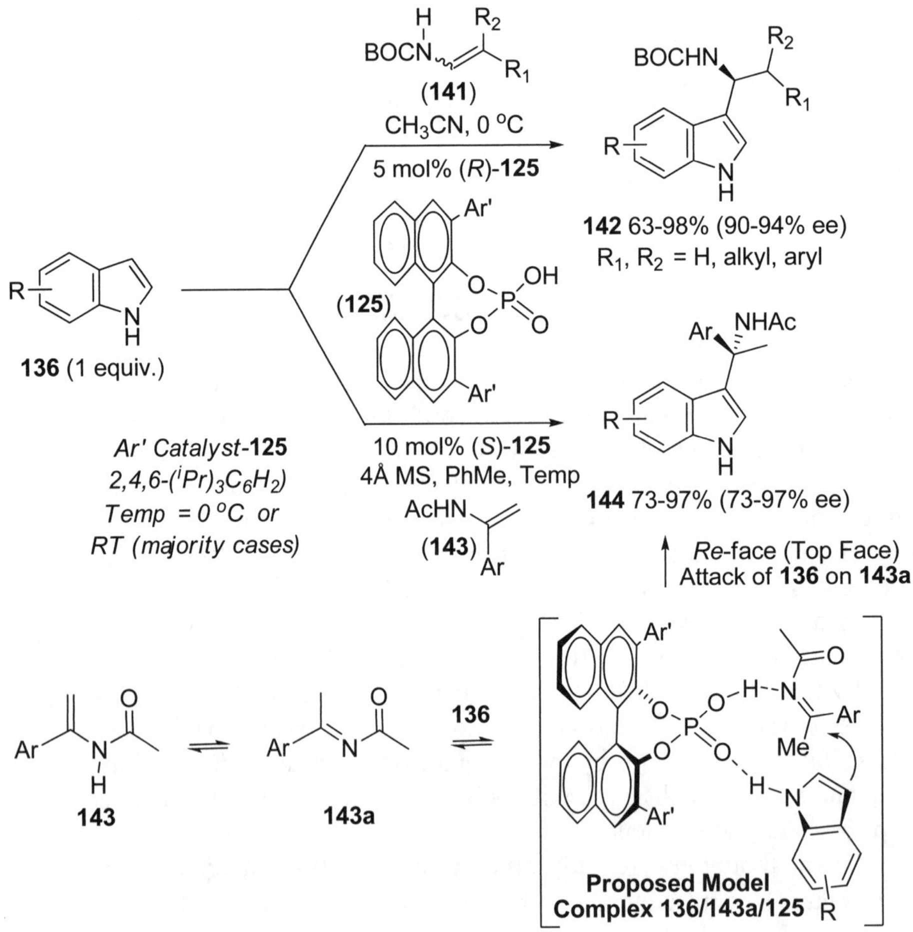

D *Asymmetric Pictet–Spengler and Related N-Acyliminium Cyclizations*

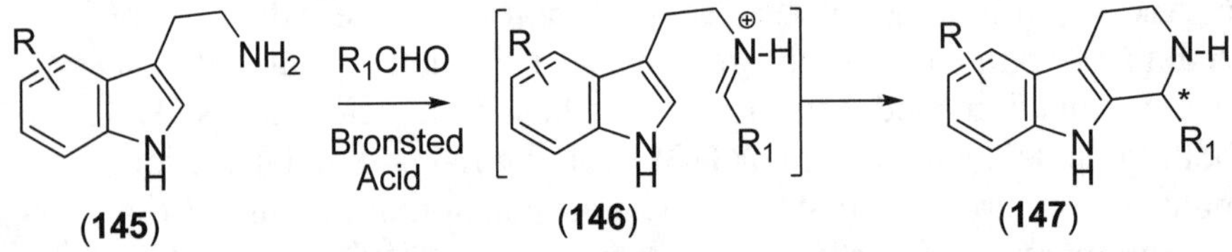

The Pictet–Spengler reaction is an important reaction involving condensation of an unsubstituted or substituted tryptamine **145** with an aldehyde R_1CHO, usually in the presence of a Brønsted acid to afford an iminium ion **146** which undergoes intramolecular cyclization to afford biologically important tetrahydro-β-carboline (**147**). Although the cyclization affords the *C*-2 adduct, the initial F–C product is the spirocyclic *C*-3 adduct, which undergoes rearrangement to the more stable *C*-2 adduct (not shown). In addition other related analogues may be also be accessed such as the tetrahydroisoquinoline scaffolds (not shown).[64] In this section, discussion of the asymmetric Pictet–Spengler reaction will be limited to iminium ion and related cyclizations using Lewis and Brønsted acid chiral catalysts.

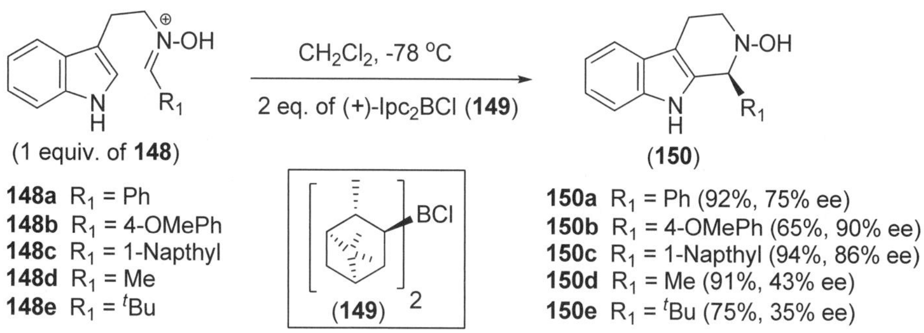

Nakagawa and co-workers in 1996, investigated the first asymmetric Pictet-Spengler reaction of N_b-hydroxytryptamines (nitrone analogues **148a–e**) using a super stoichiometric amount (2 equiv) of a chiral Lewis acid (diisopinylcampheyl-chloroborane, Ipc_2BCl, **149**) to afford the corresponding (*S*)-*Nb*-hydroxytetrahydro-β-carbolines (**150a–e**) in 75–94% yield and 35-90% *ee*.[65] Good yields (65–94%) and enantiomeric purity (75–90% *ee*) were observed when R_1 was aromatic (**150a–c** via **148a–c**), however, poor stereocontrol was observed when R_1 was aliphatic (**150d,e** in 35%–43% *ee* via **148d,e**). The requirement of a superstoichiometric amount of the chiral catalyst is the likely result of catalyst inhibition by the Lewis basic product.

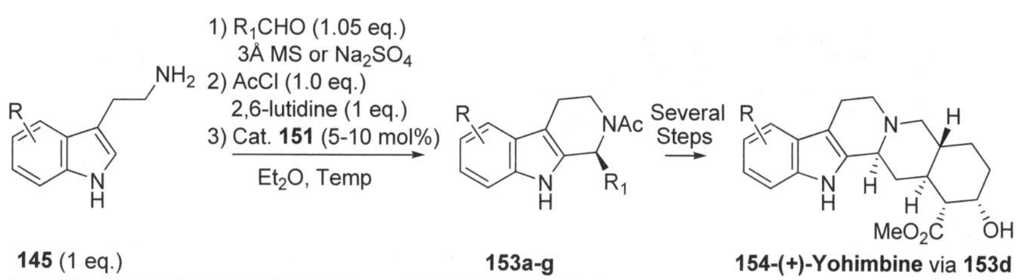

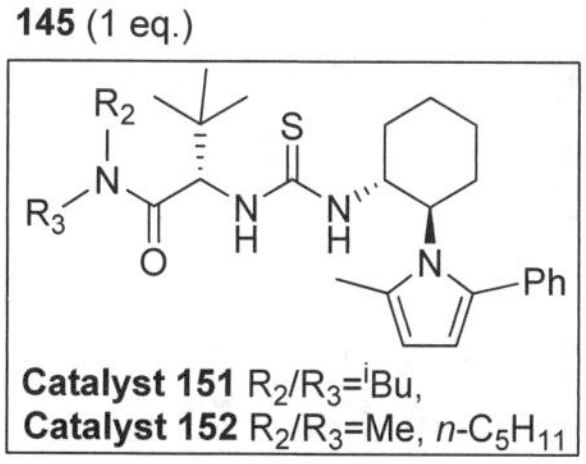

153a-g	R	R_1	Temp	%Yield, (% ee)
153a)	H	$CH(Et)_2$	-30 °C	65%, (93% ee)
153b)	H	$CH(Me)_2$	-40 °C	67%, (85% ee)
153c)	H	C_5H_{11}	-60 °C	65%, (95% ee)
153d)	H	$(CH_2)_2OTBDPS$	-60 °C	77%, (90% ee)
153e)	H	*i*Bu	-60 °C	75%, (93% ee)
153f)	5-OMe	$CH(Et)_2$	-40 °C	81%, (93% ee)
153g)	6-OMe	$CH(Et)_2$	-40 °C	76%, (86% ee)

Greater success has been achieved with chiral Brønsted acid or chiral hydrogen bond donor-catalyzed Pictet–Spengler-type reactions involving the use of *N*-acyliminium or *N*-sulfenyliminium ions as the electrophilic component. The use of tunable thiourea organocatalysts (i.e., **151** or **152**), championed by Jacobsen has played a key role.[60] Jacobsen in 2004 published an enantioselective one-pot three-step method for the synthesis of intermediates for indole alkaloids, tetrahydro-β-carbolines (**153a–g**), involving a Pictet–Spengler *N*-acyliminium ion-cyclization reaction as the key step.[66a] The key *N*-acyliminium ion-intermediate is formed via a two-step dehydration-acylation sequence: starting from tryptamine **145** (1.0 equiv) and aliphatic aldehyde R_1CHO (1.05 equiv), then cyclized using 5–10 mol % of Jacobsen's thiourea catalyst **151** in Et_2O at low temperature to generate the corresponding (*S*)-tetrahydro-β-carbolines (**153a–g**) in 65–81% yield and 85–93% *ee*. There was no significant amelioration of yield or optical purity with methoxy substitution at the the 5- or 6-positions when R_1 was aliphatic (compare **153f,g** versus **153a**) Jacobsen completed the enantioselective total synthesis of the indole alkaloid (+)-yohimbine (**154**) from chiral tetrahydro-β-carboline intermediate **153d** in 2008.[66b]

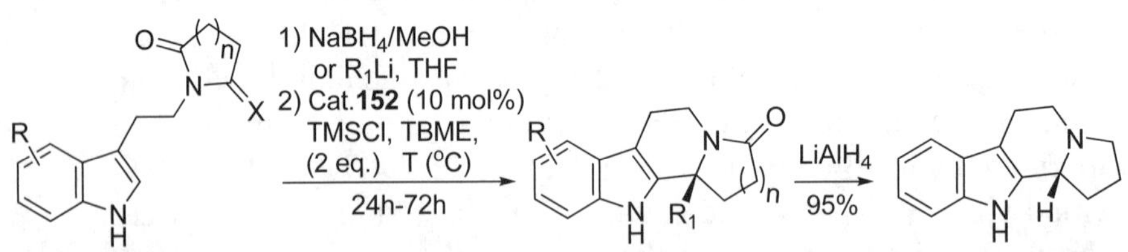

156a-g (X=R1, OH, 1 eq.)
155a-g (X=O)

157a-g
(63-92%, 91-98% ee)

158 (+)-Hamicine
via lactam 157a

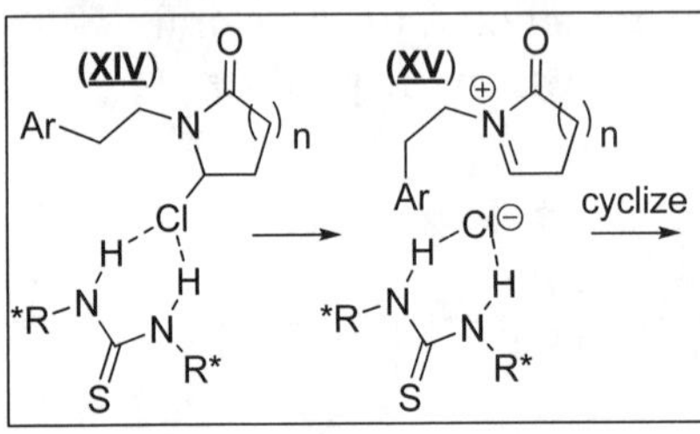

157	R	n	R1	Temp	%Yield, (% ee)
a)	H	1	H	-55 °C	90%, (97% ee)
b)	H	1	Me	-78 °C	92%, (96% ee)
c)	H	1	*n*-Bu	-78 °C	74%, (98% ee)
d)	5-Br	1	H	-55 °C	88%, (96% ee)
e)	5-OMe	1	H	-55 °C	86%, (95% ee)
f)	5-OMe	1	Me	-78 °C	84%, (91% ee)
g)	H	2	Me	-78 °C	63%, (92% ee)

This methodology has been extended to other related types of *N*-acylimminium ion cyclizations. No specific thiourea catalyst has been found to work for all transformations within this class. Hence each specific reaction must be carefully screened. Jacobsen and co-workers (2007) have also shown that γ-hydroxy-2-pyrrolidinones or γ-hydroxy-2-piperidinone 3-substituted indoles (**156a–g**) activated by use of 2 equiv TMSCl as dehydrating agent in the presence of 10 mol % of Jacobsen chiral thiourea organocatalyst **152** in TBME, can undergo asymmetric Pictet–Spengler cyclization to afford either indolizidinones or quinazolidinones (**157a–g**) in high yield (63–92%) and enantiomeric purity (63–98% *ee*). The effect of indole substituents on the cyclization reaction was next investigated.[67]

Substitution at the 5-, 6- and 7-position of indole **156** with either electron-donating or withdrawing groups was also well tolerated (6- and 7-substitution results are not shown in scheme). The reaction of unsubstituted and 5-substituted indole hydroxylactams, **156a–g**, gave the corresponding chiral indolizidinones **157a–f** in 74–92% yield and 91–98% *ee*, regardless of the electronics of substitution. Increase of the size of the R_1 moiety had little effect on yield and stereocontrol (see **157a–c**). The reaction of γ-hydroxypiperidine (**156g**) fared equally well, affording quinazolidinone **157g** in 63% yield and 92% *ee*. The key γ-hydroxylactam precursors (**156a–g**) were prepared by a two-step sequence involving condensation of the corresponding tryptamines **145** with succinic anhydride or glutaric anhydride in toluene/AcOH (tryptamine condensation step with corresponding anhydrides not shown in Scheme), followed by $NaBH_4$/MeOH or organolithium mediated addition to the carbonyl of the resulting intermediate imides **155a–g**. The enantioselective total synthesis of the indole alkaloid (+)-hamicine (**158**) was accomplished in just four steps in 60% overall yield from tryptamine **145a**. The final step of the synthesis involved a $LiAlH_4$ amide reduction of indolizidinone **157a** to afford (+)-hamicine (**158**) in impressive 95% yield.[67]

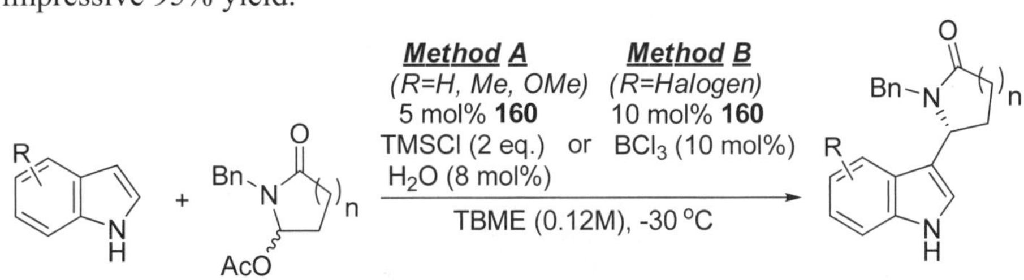

(1 equiv) **128** (R = H) **136** (R = Sub.)

(1 equiv) **159i** (n=1) **159ii** (n=2)

Method A: **161a-f** 70-93% (80-94% ee)
Method B: **161g-k** 47-81% (92-97% ee)

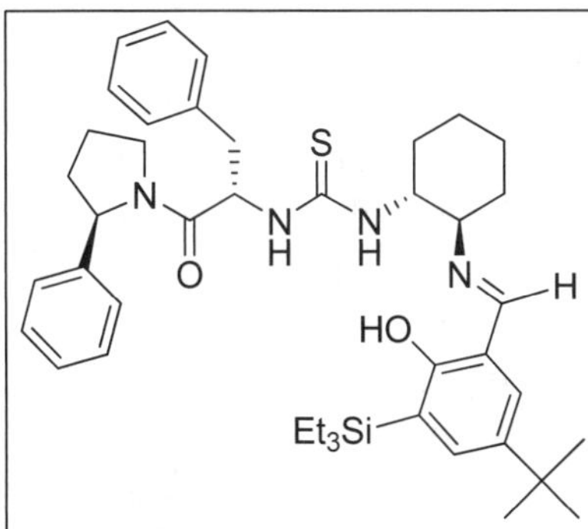

Jacobsen Catalyst 160

161	R	n	Method	% Yield, (% ee)
a)	H	1	A	90, (93% ee)
b)	H	2	A	70, (93% ee)
c)	5-Me	1	A	79, (90% ee)
d)	5-Me	2	A	93, (94% ee)
e)	6-OMe	1	A	80, (80% ee)
f)	6-OMe	2	A	76, (88% ee)
g)	4-F	1	B	57, (96% ee)
h)	4-Br	1	B	47, (97% ee)
i)	5-Br	1	B	80, (93% ee)
j)	6-Cl	1	B	81, (96% ee)
k)	6-Cl	2	B	40, (92% ee)

It was proposed that the reactivity and catalytic mechanism for chiral H-bond catalysis stems from dual H-bonding interactions of the thiourea with the chloride of **XIV** promoting its loss to form reactive *N*-acyliminium ion (**XV**).

Recently, Jacobsen and co-workers (2009) have explored the intermolecular version of this reaction.[68] They have found that *N*-benzyl substituted γ-acetoxy-2-pyrrolidinones (**159i**) or γ-acetoxy-2-piperidinones (**159ii**), activated by use of 2 equiv TMSCl as dehydrating agent in the presence of 10 mol% of Jacobsen chiral thiourea catalyst **160** in TBME at –30 °C, can undergo asymmetric *N*-acyliminium ion cyclization with unsubstituted (**128**) or electron-rich indoles (general formula **136**) to afford either (*R*)-3-substituted γ-substitued pyrrolidinones or piperidinones (general formula **161a–f**) in good yield (70–93%) and enantiomeric purity (80–94% *ee*). No significant difference was observed in formation of pyrrolidinone versus piperidinone analogues was observed in yield or stereocontrol (**161a,c,e,** versus **161b,d,f**, respectively).

This method (A) worked well with indole **128** or indole analogues **136** with electron-donating groups at the 4-, 5-, and 6-positions (Me, vinyl or OMe, not all results shown in scheme). Less reactive halo-substituted indole analogues (electron-withdrawing groups typically gave poor conversion but high enantioselectivity, results not shown in scheme. Use of the α-acetoxy lactam susbstrate for activation over the α-hydroxy lactam was necessary due to the higher solubility of the corresponding iminium ion intermediate. Activation of **159i/159ii** (n = 1 and n = 2 respectively) with 10 mol % of BCl_3 and treatment of the respective 4-, 5- and 6-halo substituted analogues of **136** (0.12 M in TBME) with 10 mol % **160**, afforded the corresponding pyrrolidinone and piperidine analogies (**161g–k**) in 47–81% yield and 92–97% *ee*. Using this method (B), substantially higher yields were observed in the formation of pyrrolidinone versus piperidinone analogues (compare 6-Cl indole substitution affording yields of **161j/161k** in yields of 81% and 40%, respectively).

The proposed catalytic mechanism accounting for the reactivity (methods A and B as well as the related intramolecular cyclizations) involves generation of a α-chlorolactam intermediate **162** formed in situ from the corresponding α-acetoxylactam **159i/159ii** which is trapped by indoles **128/136** through an S_N1-type anion-binding mechanism. In method A, the synergistic effect of TMSCl (2 equiv) and catalytic H_2O (8 mol %) suggests that α-acetoxylactam **159i** reacts with HCl generated in situ to form a key chlorolactam intermediate **162**. Since the enantioselectivity depends on the chloride ion concentration, the chlorolactam is believed to be a key intermediate that reacts with indole (**128** or **136**) by a S_N1-type anion binding mechanism via the intermediacy of the resulting *N*-acyl iminium ion. Method B requires the use of more reactive BCl_3 to generate the same intermediate.

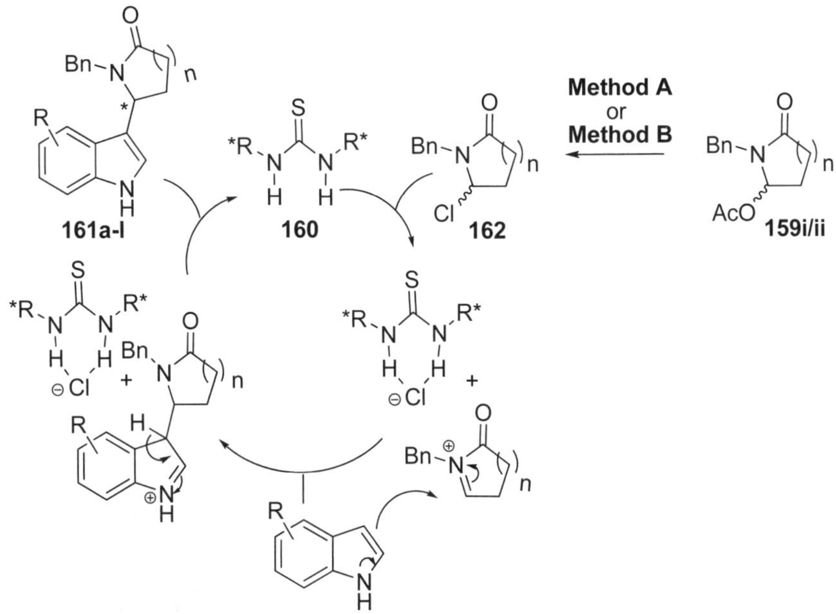

In 2008, Jacobsen reported a clever pyrrole asymmetric Pictet–Spengler like *N*-acyliminium cyclization strategy of 3-substituted β-pyrroloethyl hydroxylactams **163** prepared via substituted succinimide precursor **164**, allowing access to either chiral tricyclic pyrroloindolizidone (**165**–**167**) and pyrroloquinolizidine scaffolds (**168**, **169**), depending on whether precursor **163** nitrogen is *N*-protected.[69] The key cyclization intermediates of general class **163** were prepared via reduction ($NaBH_4$/MeOH) or organolithium addition (RLi/THF/–78 °C) of the carbonyl moiety of the corresponding β-pyrroloethyl succinamide and glutarimide precursors **164** (see Ref. 69).

Cyclization of the *N*-TIPS protected pyrrolohydroxylactam **163** (where $n = 1$ and R_1 = alkyl, H), activated by 2 equiv TMSCl in TBME at –55 °C to –30 °C, gave a racemic 1.5-3:1 mixture of the *C*-4 (**165**) versus *C*-2 cyclization product (**168**), regardless of whether R_1 was substituted with H or alkyl. Cyclization at *C*-2 was unexpected since the *N*-TIPS moiety effectively shields the *N*-1, *C*-2 and *C*-5 positions. Coaddition of 20 mol % of Jacobsen chiral thiourea catalyst **152** to **163** under the same reaction condition, afforded the desired *C*-4 adduct **168** in good yield (49–77%) and excellent enantiomeric purity (92–96% *ee*). Evidently co-complexation of the organocatalyst to the *N*-acylimminium ion derived from **163**, provided an effective discriminating spatial environment.

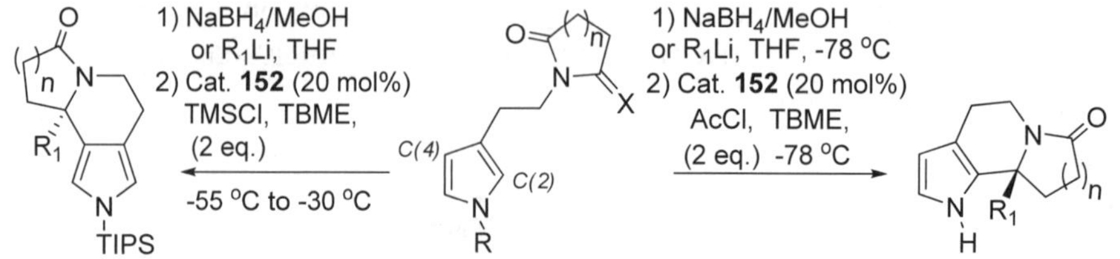

165 (n=1, 49-77%, 92-96% ee) (R$_1$=H, Me, ^{n}Pr, ^{n}Bu, ^{i}Bu, Ph)

166 (n=2, 70-75%, 93-97% ee (R$_1$=Me, Bu)

167 (n=2; R$_1$=H; 76%, 70% ee

a **163** (X=R$_1$, OH; R=H, TIPS) **164** (X=O)

^{a}*see ref. 69 for details*

168 (n=1, 51-86%, 88-93% ee) (R$_1$=H, Me, ^{n}Pr, ^{n}Bu, ^{i}Bu)

169 (n=2, 54-71%, 52-65% ee) (R$_1$=H, Me, ^{n}Bu)

Treatment of the *N*-TIPS protected pyrrolohydroxylactams **163** (where n = 2, R = alkyl) gave equally good results (70–75% yield, 93–97% *ee*); however, replacement of R_1 moiety with H gave a dramatic decrease in stereocontrol of **167** (76% yield, 70% *ee*).[69] Cyclization of the unprotected *N*-H pyrrolohydroxylactam **163** (where n = 1 and R_1 = alkyl, H), activated by 2 equiv AcCl in TBME at –78 °C, gave the expected *C*-2 cyclized adduct **168** in 51–86% yield and 88–93% *ee*. Due to the higher reactivity of the *C*-2 cyclization mode, activation using AcCl at –78 °C proved sufficient. Treatment of *N*-H pyrrolohydroxylactam **163** (n = 2, R = alkyl and H) to the same reaction conditions, gave good yields (54–71%) of **169** but with considerably reduced stereocontrol (52–65%). The reason for this was not clear. These authors also showed that these *C*-2 and *C*-4 cyclized chiral pyrroloindolizidone and pyrroloquinolizidine scaffolds could be converted to many clever and highly functionalized chemotypes (transformation not shown in scheme).[69.]

One practical limitation for implementation of these methods on pilot plant scale is the reliance of achieving high enantioselectivity of these reactions at very low temperatures (< –30 °C). A giant stride toward achieving goal this has recently been reported. In 2009, Jacobsen reinvestigated the asymmetric Pictet–Spengler reaction of 6-OMe substituted and unsubstituted tryptamines (type **145**) with a wide variety of aromatic and aliphatic aldehydes (type R_1CHO) in Toluene at room temperature using 20 mol % of a chiral thiourea organocatalyst **(170a)** in combination with a weak Brønsted acid, benzoic acid (20 mol %), to afford the tetrahydro-β-carboline adducts **147a–g** in excellent yield (45–94%) and enantiomeric purity (85–95% *ee*). Matching of the catalyst to the reaction again proved critical as Jacobsen catalyst **170a** (R_4 = ^{i}Pr) proved superior to **170b** (R_4 = ^{t}Bu), details not provided in scheme.[70]

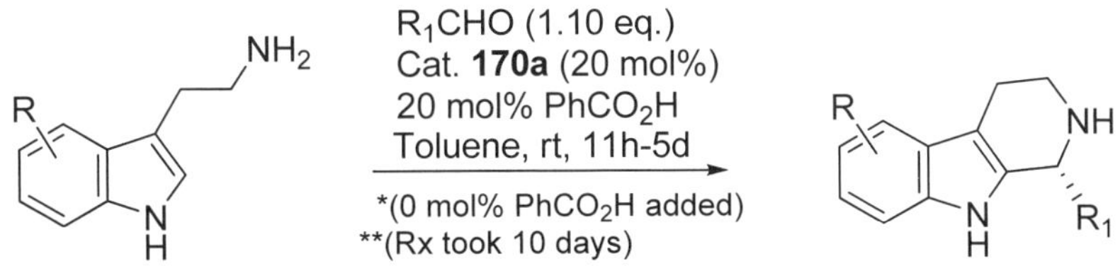

145 (1 equiv.) **147a-g**

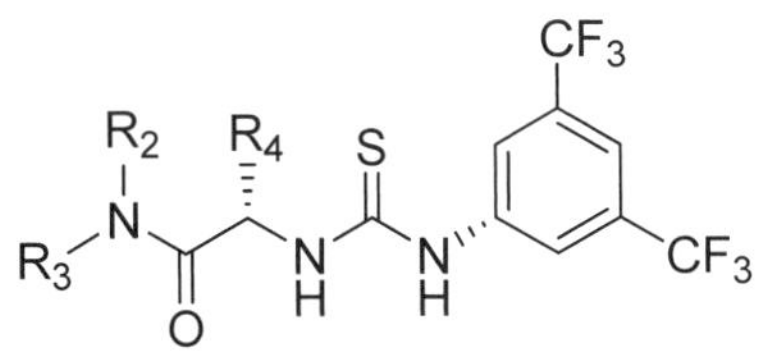

170a R_2, R_3=Me, Bn and R_4=^{i}Pr
170b R_2, R_3=Me, Bn and R_4=^{t}Bu

147	R	R_1	%Yield, (% ee)
a)	6-OMe	4-OMe(Ph)	78%, (85% ee)
b)	6-OMe	Ph	94%, (86% ee)
c)	6-OMe	4-Br(Ph)	79%, (94% ee)
d)	6-OMe	2-Br(Ph)	74%, (95% ee)
e)**	H	2-Br(Ph)	45%, (95% ee)
f)	6-OMe	$CH(Me)_2$	60%, (88% ee)
f)*	6-OMe	$CH(Me)_2$	90%, (94% ee)
g)*	6-OMe	*n*-pentyl	74%, (86% ee)

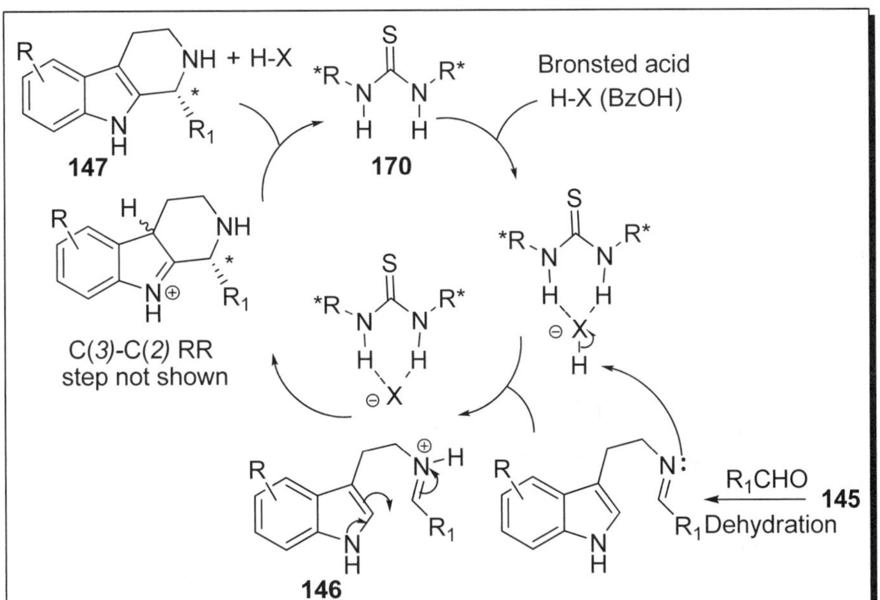

As expected, the unsubstituted tryptamine **145** (where R = H) coupled with 2-bromobenzaldehyde afforded **147** in the lowest observed yield (45%, 10-day reaction) but maintained excellent stereocontrol (95%). Both 5- and 6-methoxysubstituted tryptamines **145** coupled with electron-poor arene and aliphatic aldehydes in superior yield and optical purities (see **147c,d** data). The combination of the reactive electron-rich 6-OMe substituted tryptamine **145** with either *i*-butyraldehyde or *n*-hexanal (highly reactive aliphatic aldehydes) required no weak Brønsted cocatalysis (none $PhCO_2H$ added) but longer reaction times (data not provided) and still providing the expected adducts **147f,g** in a respectable 74–90% yield and 86–94% *ee*.

The greater scope of this reaction was attributed to the dual cyclic Brønsted acid/H-Bond donar cocatalysis mechanism. The catalytic cycle initially involves imine protonation by the chiral thiourea catalyst **170** associated via H-bonding to the conjugate base (X^-) of a weak Brønsted acid (H–X, benzoic acid in this case) additive. Intramolecular cyclization of the protonated iminium ion **146**, followed by rearomatization regenerates the Brønsted acid cocatlayst (benzoic acid).[70a] Note for brevity, the plausible rearrangement (RR) step of the inital C*(3)*-spiroalkylated adduct[70b] to the final tetrahydrohydroisoquinoline scaffold **147** is not shown.

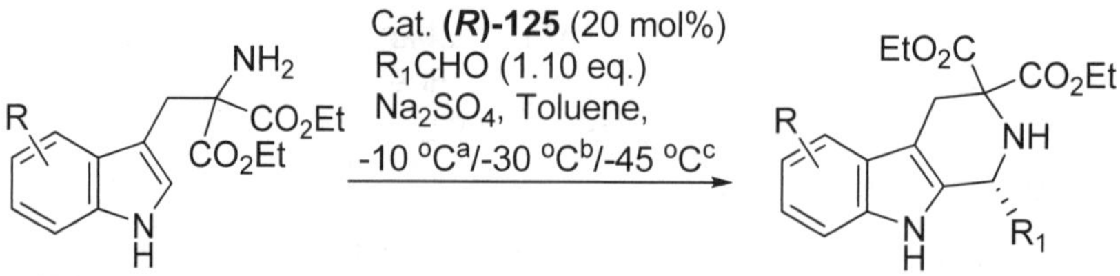

171 (1 equiv.) **172a-j**

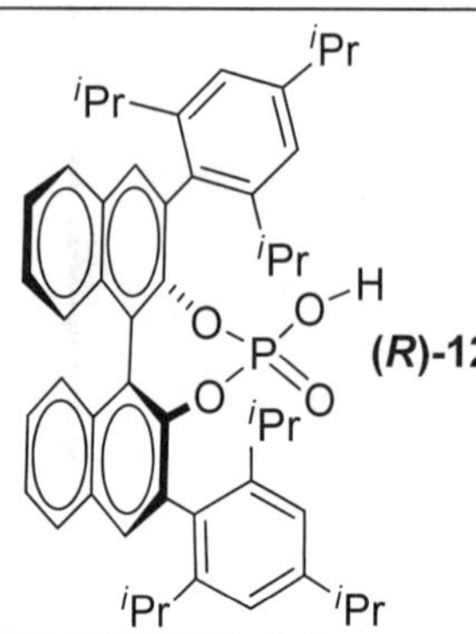

(*R*)-125

172	R	R_1	%Yield, (% ee)
a)[a]	H	Et	76%, (88% ee)
b)[a]	5-OMe	Et	96%, (90% ee)
c)[a]	6-OMe	Et	94%, (86% ee)
d)[b]	5-OMe	*n*-Bu	90%, (87% ee)
e)[b]	5-OMe	^{i}Bu	96%, (80% ee)
f)[c]	5-OMe	cyclohexyl	64%, (94% ee)
g)[a]	5-OMe	Ph	82%, (62% ee)
h)[a]	5-OMe	Ph(4-CN)	60%, (80% ee)
i)[a]	H	Ph(4-CN)	40%, (89% ee)
j)[a]	5-OMe	Ph(4-NO_2)	98%, (96% ee)

Stronger H-bond donors such as chiral Brønsted acids have also been used in the asymmetric Pictet-Spengler reaction. List (2007) has reported the chiral Brønsted acid-catalyzed asymmetric Pictet–Spengler cyclization of conformationally biased unsubstituted, 5-, and 6-methoxy substituted tryptamines **171** with aliphatic and aromatic aldehydes R_1CHO.[71] The tryptamines **171** were conformationally conformationally biased by α,α-amino geminal diester mode of substitution (Thorpe–Ingold effect). Thus treatment of tryptamines **171** with aldehydes R_1CHO in the presence of Na_2SO_4 dehydrating agent and 20 mol % of chiral (*R*)-phosphoric acid catalyst (**125**) in toluene at low temperature (10 °C to –45 °C) afforded the corresponding gem diester substituted (*R*)-β-carbolines (**172a–j**) in 40–98% yield and 80–96% *ee*. Simple tryptamine, or phenethylamine-derived imines lacking gem disubstitution (i.e., **145**) do not undergo cyclization, limiting the reaction scope. Instead, products are formed resulting from sequential homo-aldol condensation and imine formation reactions (not shown). These known side reactions, observed with strong Brønsted acid-mediated Pictet–Spengler reactions of tryptamines with aliphatic aldehydes, can significantly lower the yield of the reaction.

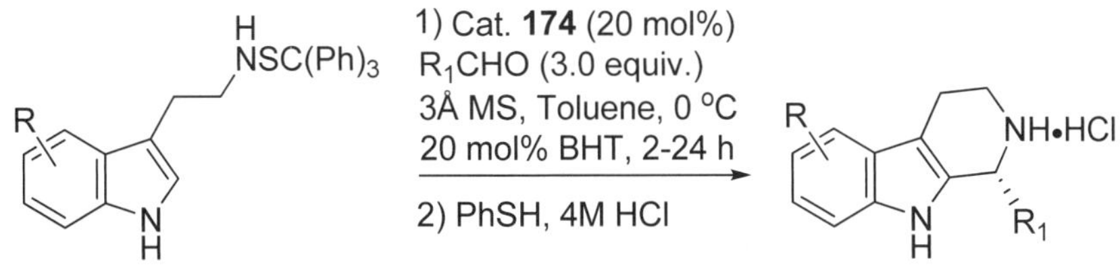

173 (1 equiv.) **175a-h**

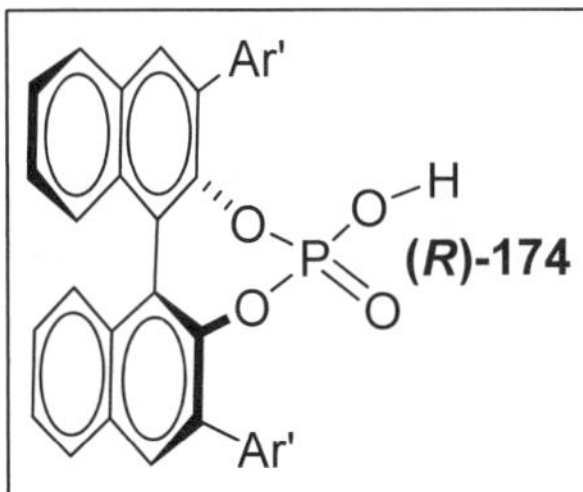

Ar' = 3,5-$(CF_3)_2C_6H_3$

175	R	%Yield, (% ee)
a)	Me	88%, (30% ee)
b)	*n*-pentyl	87%, (84% ee)
c)	2-ethyl(Ph)	88%, (76% ee)
d)	Bn	90%, (87% ee)
e)	iPr	77%, (78% ee)
f)	cyclohexyl	81%, (72% ee)
g)	Ph	70%, (82% ee)
h)	Ph(4-NO_2)	78%, (82% ee)

Hiemstra (2007) eliminated the competing enamine alkylation pathway by cyclization of unsubstituted *N*-tritylsulfenyltryptamines **173** with aliphatic and aromatic aldehydes (R_1CHO) in the presence (20 mol %) of a related chiral phosphoric Brønsted acid catalyst (*R*)-**174**.[72] The yields (70–90%) and optical purities (30–87% *ee*) reported for (*R*)-β-carbolines hydrochlorides (**175a**–**h**) are for a one-pot operation that includes the Pictet-Spengler cyclization step and removal of the nitrogen protecting group. The key cyclization step involved treatment of tryptamine **173** (1 equiv) with either an aliphatic or aromatic aldehyde (3 equiv R_1CHO), 3 Å MS and 20 mol % BHT in toluene at 0 °C using 20 mol% of phosphoric acid catalyst ***R*** **174**. The deprotection conditions involve treatment of the crude Pictet-Spengler adduct reaction mixture with 1.2 equiv thiophenol, followed by 4 M HCl/dioxanes to isolate the enantioenriched β-carbolines (**175a**–**h**) as HCl salts. The addition of 3,5-di-(*tert*-butyl)-4-hydroxytoluene (BHT) to the reaction mixture as a radical scavenger prevented loss of the *S*-tritylsulfenyl *N*-protecting group occurring via free-radical-mediated homolytic *N*-S bond cleavage.

E Asymmetric Friedel Crafts Michael-Type Alkylations using Chiral Lewis acids and Organocatalysts

In section B, many examples reported by Jorgensen in 2001 of successful asymmetric 1,2-attack of electron-rich arenes onto the carbonyl moiety of an α-ketoester catalyzed by 5–10 mol % (*S*)-Cu(OTf)$_2$-^{t}Bu-BOX catalysts (**92i-iv**) were shown.[57b,c] During the same time period. Jorgensen also reported many examples of asymmetric thermodynamic conjugate addition of

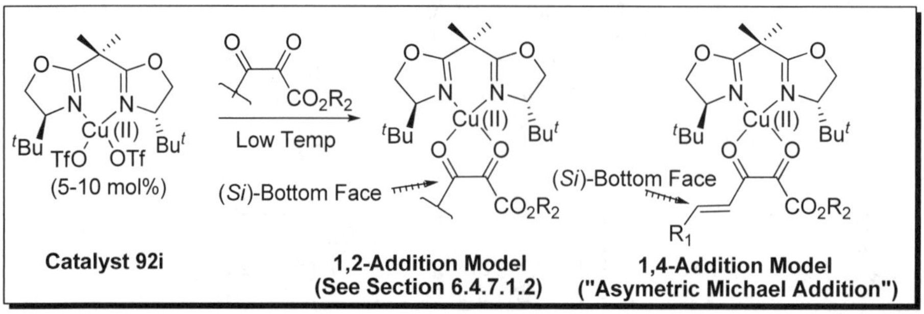

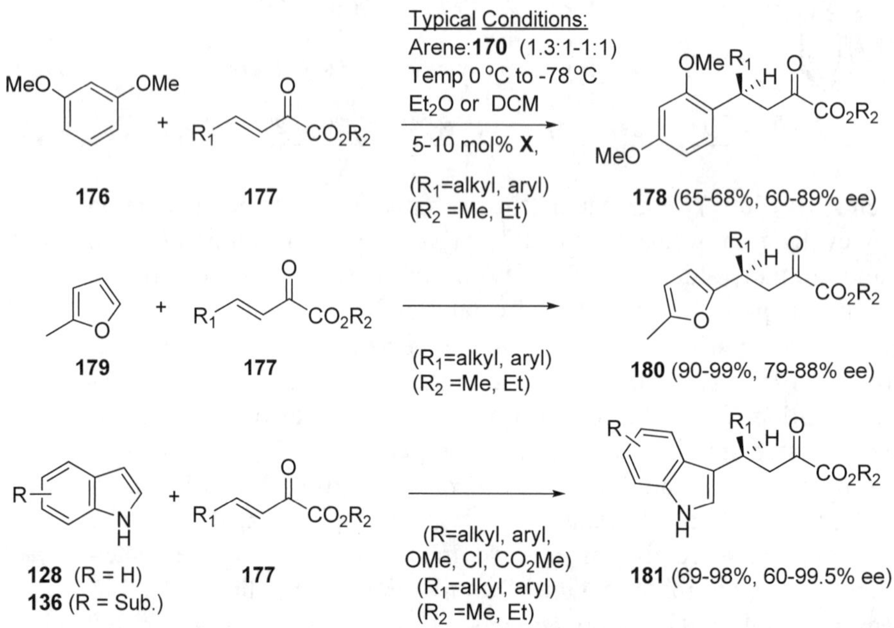

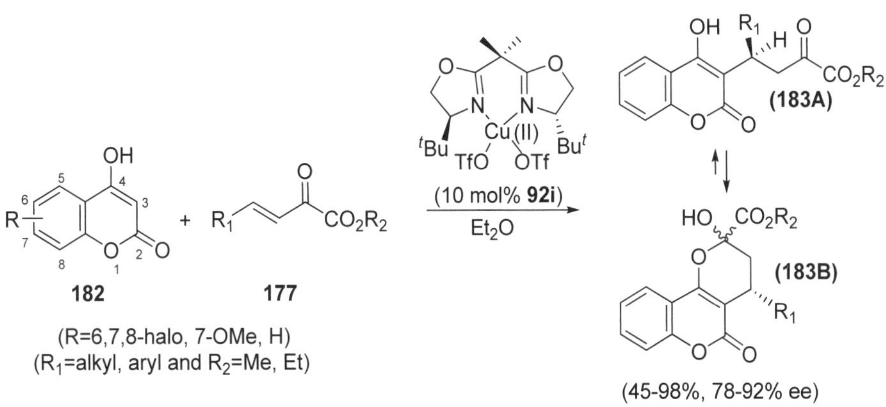

β-substituted α,β-unsaturated ketoesters (1,4-attack to the alkene), under virtually identical reaction conditions, via the same attack trajectory (vide supra).[73] Thus the 1,4-addition of β,γ-unsaturated α-keto esters (R_1 = alkyl, aryl and R_2 = Me, Et) with a wide variety of arene scaffolds (**176**, **179**, and **128/136** respectively) using 5–10 mol % (*S*)-$Cu(OTf)_2$-*t*Bu-BOX catalyst (**92i**) gave the corresponding (3*S*)-3-arylsubstituted Michael adducts **178**, **180** and **181** in good yield with moderate to excellent enantiocontrol (see above scheme). The following general observations are not shown in the scheme. The use of the aerobically stable (*S*)-$Zn(OTf)_2$-^{t}Bu-BOX as catalyst gave comparable levels of stereocontrol to $Cu(OTf)_2$-^{t}Bu-BOX catalyst **92i**. The use of α,β-unsaturated α-keto ethyl ester **177** (R_2 = Et) versus α-keto methyl esters (R_2 = Me) as Michael acceptors gave higher levels of stereocontrol. Enantiocontrol of substituted indole adducts (**181**) was remarkably intolerant of the electronics of substitution at the 5- and 6-positions (halogen, alkoxy, etc.).[73]

Jorgensen in 2003 applied this methodology to the synthesis of chiral 3-substituted 4-hydroxycoumarins, important biologically active chemotypes as anticoagulant and antibiotics.[74a] Coupling of a variety of substituted 4-hydroxycoumarins **182** with methyl/ethyl 3-alkyl and 3-phenyl substituted α-ketoesters of type **177** in the presence of 10 mol % of **92i** catalyst in Et_2O, gave the corresponding Michael adduct (**183A**), which underwent intramolecular cyclization to exist primarily as tricycle **183B** in 45–98% yield in 78–92% *ee*. A similar strategy enabled the total synthesis of Warfarin.[74b]

Jorgensen (2001) also investigated the asymmetric alkylation of unsubstituted (**128**) or substituted indoles (**136**) with β-aryl substitituted alkylidene malonates (**184**), under the same reaction conditions, but met with only modest success (< 70% *ee* achieved).[75] Treatment of a wide variety of unsubstituted (**128**) or substitituted indoles **136** with β-aryl substitituted alkylidene malonates (**184**) as the substrate with the same catalyst (10 mol %

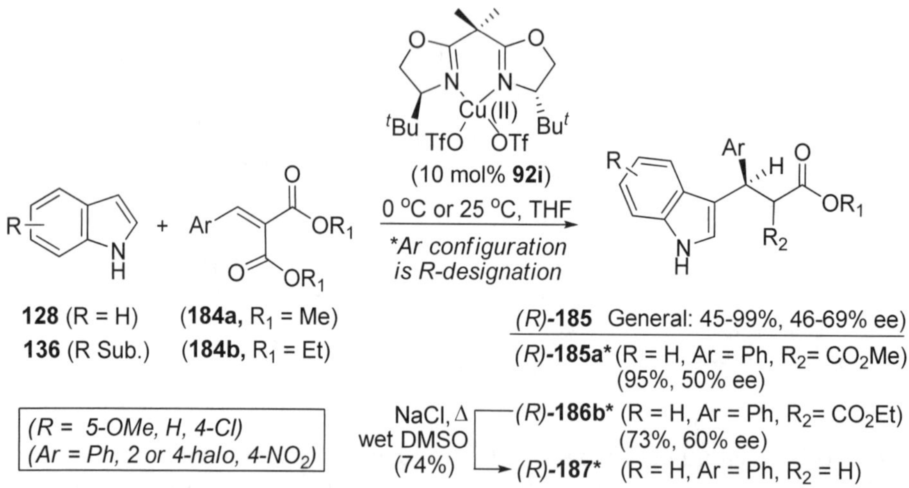

of (*S*)-Cu(OTf)$_2$-^{t}Bu-BOX catalyst **92i**) in THF at 0 °C gave the corresponding Michael adducts (**185** in the *R*-configuration, designated as (*R*)**-185**) in 45–99% yield and 46–69% *ee*. The reactions of methyl and ethyl-β-phenyl-substitituted alkylidene malonate esters (**184a/184b,** where Ar = Ph, R_1 = Me and Et, respectively) with indole **128** under the reaction conditions afforded the methyl ester adduct (*R*)**-185a** in higher yield (95%) but lower enantioselectivity (50% *ee*) than the corresponding ethyl ester (*R*)-**186b** (73% yield, 60% *ee*). Subjection of (*R*)**-186b** to Krapcho decarboxylation conditions (NaCl/wet DMSO/160 °C) afforded ethyl (3*R*)-3-(3-indolyl) 3-phenylpropionate (*R*)**-187** (R = H, R_1 = Et, Ar = Ph) in 74% yield. The diminished stereocontrol observed with this substrate relative to β-substituted α,β-unsaturated ketoesters was presumably because the alkylidene malonate bidentate catalyst complex would place the reacting olefinic center on the ligand C_2-axis, farther from the chiral center than the reacting carbonyl of the Cu(BOX)-α-ketoester alkylidene bidendate complex.

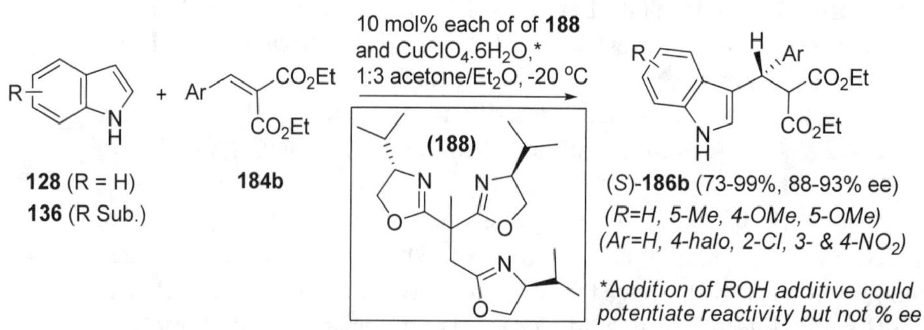

Tang from 2002 to 2004 published many papers[76a,b] demonstrating efficient asymmetric addition of unsubstituted (**128**) and substituted indoles **136** and β-aryl substituted alkylidene malonates **184b** using a novel copper catalyst (10 mol %) prepared either from a pseudo C_3-symmetric functionalized triazoline (TOX) ligand **188** (all *S*-stereocenters) or from several modified BOX ligands (not shown).[76c,d] The initial coupling reaction was conducted in 1:3 acetone/Et_2O at –20 °C using a copper catalyst complex derived from 10 mol % of TOX ligand **188** and $Cu(ClO_4)_2{\bullet}6H_2O$ in 1:3 acetone/Et_2O, to afford the corresponding (*S*)-adduct, (*S*)-**186b**, in high yield (73–99%) and enantioselectivity (88–93% *ee*).[76a] The effect of reaction temperature (–20 °C) proved critical for optimal enantioselectivity. In addition, the reaction proved relatively insensitive to the electronics of indoles **128/136** (R = H, 5-Me and 4- and 5-OMe) and the Ar moiety of the β-aryl alkylidene ethyl malonate **184b** (Ar = 4-H, 4-halo, 2-Cl, 3- and 4-NO_2). The TOX ligands of the catalyst were designed to be capable of a tridentate interaction with the metal center in the proposed catalytic cycle (see below).

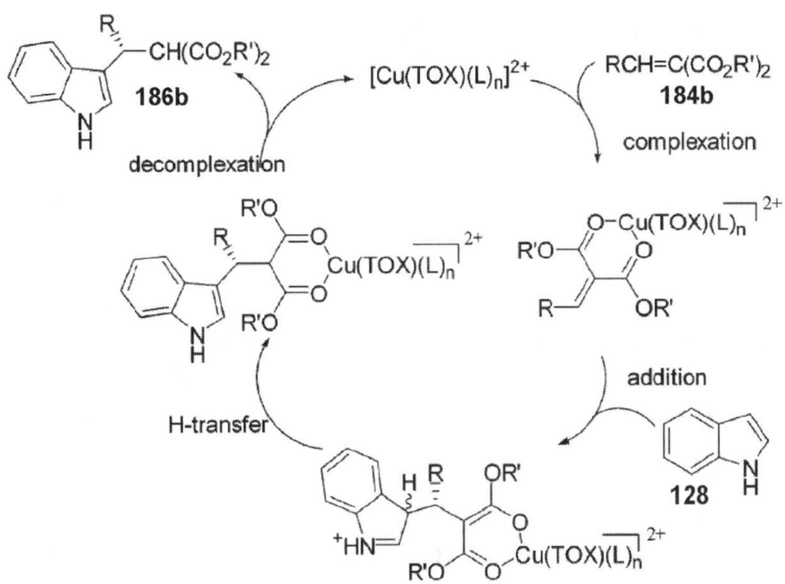

The following results are not shown in the schemes but will be discussed. The reaction of unsubstituted indole **128** with benzylidene ethyl malonate **184b** (where Ar = Ph) afforded a 50% yield and 85% *ee* of (*S*)-**186b** (where Ar = Ph) at 0 °C, but with coaddition of 2 equiv $(CF_3)_2CHOH$, a clearly faster reaction with a higher yield (99%) and improved enantiocontrol (85% *ee*) was observed. The lower reaction temperature, however, appeared to be responsible for the improved stereocontrol. When the same reaction conditions using same additive was conducted at even lower temperature (–25 °C), a modest compromise of conversion (84% yield) with improved stereocontrol (89% *ee*) of (*S*)-**186b** was realized. The enantioselectivity dropped considerably to 60% *ee* when the β-aryl alkylidene substituent of (*S*)-**186b** was replaced with a β-ethyl alkylidene moiety (where Ar replaced by Et) and reacted with **128** the same reaction conditions. Thus this method proved limited to the use of benzylidene substituted malonates as substrates.[76b]

A thorough investigation of the effect of solvent, variation in substrate stoichiometry, ratio of ligand/copper in complex, catalyst loading, reaction temperature, and chiral ligand was then undertaken. The optimal solvent system involved use of branched alcohol additive (2 equiv $(CF_3)_2CHOH$, ^{i}PrOH, and ^{i}BuOH), to speed up the reaction and the optimal trisoxazoline ligand containing 3(*S*)-*iso*-propyl substituents at the three isoxoazoline stereocenters. The addition of an alcohol as a coadditive (2 equiv) at low temperature (> –20 °C) was shown to accelerate both the reaction rate and enantioselectivity, which optimally led to its use as solvent. The use of bulkier branched alcohols as solvent gave an increase in stereoselectivity. The addition of > 200 equiv water apparently impaired catalytic activity but not enantioselectivity.[76b]

These results suggest possible coordination of the alcohol moiety to the catalyst active site (possibly H_2O from catalyst $Cu(ClO_4)_2{\bullet}6H_2O$ in initial scheme). The greater overall water tolerance and catalytic stability of the active intermediates of the trisoxazoline (TOX) catalyst complex over the (*S*)-$Cu(OTf)_2$-^{t}Bu bisoxazoline (BOX) catalyst system was initially attributed to the greater number of donor atoms involved. Deuterium labeling of the alcoholic solvent (ROD) showed 83% incorporation of the deuterium at the acidic α-position of the malonate of **184b**, showing the alcohol was involved in an H-transfer step (see catalytic cycle). The undeuterated product (*S*)-**186b** remained unchanged when subjected to similar conditions and was consistent with the proposed catalytic cycle.[76b]

Remarkable solvent effects were observed on the stereocontrol of this reaction. The use of either $Cu(ClO_4)_2{\bullet}6H_2O$ and $Cu(OTf)_2$ as sources of divalent copper ion were comparable in branched protic solvents, with the latter affording slightly better stereocontrol (~ 5% *ee*) at 15 °C. If the solvent was changed to a nonpolar aprotic solvent, a change to the opposite stereo orientation occurred [(*S*)-**186b** to (*R*)-**186b**)]. This change was greatest when1,1,2,2-tetrachloroethane (1,1,2,2,-DCE) was used as solvent. Treatment of a 10 mol % of ligand with 15 mol % $Cu(OTf)_2$ in this solvent at 15 °C afforded the opposite enantiomer (*R*)-**186b** in in 66% yield and 74% *ee*. Thus either enantiomer of adduct **186b** could thus be obtained in reasonable enantiomeric purity just by change of solvent.[76b,76e]

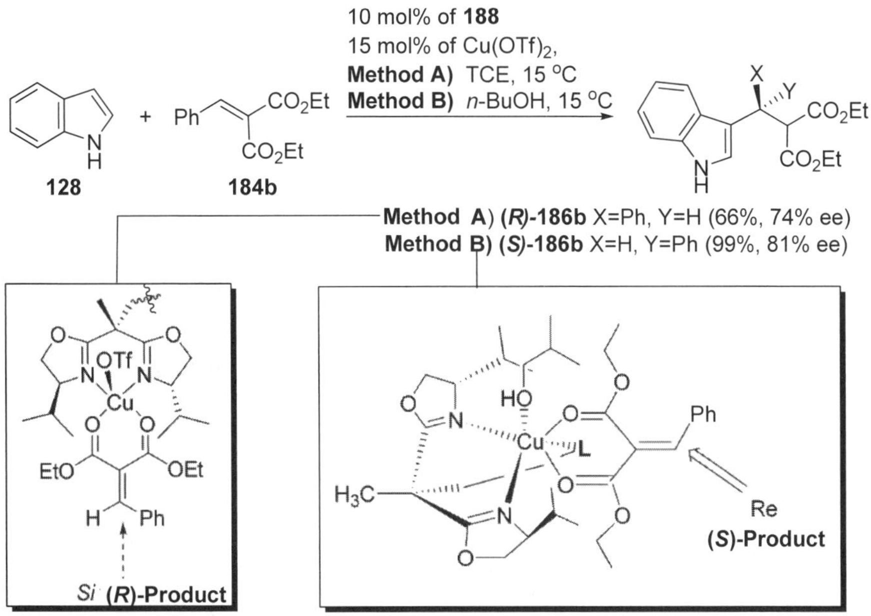

The proposed stereoinduction model has the substrate/catalyst complex in a square pyramidal geometry with the slight excess of triflate displacing a pendant isoxazoline, resulting in *Si*-face attack and formation of (*R*)-**186b**. In fact, an *X*-ray structure previously reported by Evans of catalyst $[Cu((S,S)\text{-}^{t}Bu\text{-}BOX)(H_2O)_2](OTf)_2$, shows quite clear square-pyramidal geometry.[77] Change to a catalyst coordinating solvent such as *n*-BuOH under the same reaction conditions, however, provided a 99% yield of (*S*)-**186b** in 81% *ee*. The change in sense of stereoinduction in alcohol solvent assumes a distorted octahedral geometry with one of the sites occupied by the alcohol site with the *Re*-face exposed allowing selective formation of (*S*)-**186b**.[76b]

A popular notion in asymmetric catalyses that an excess of a chiral ligand with respect to the metal improves enantioselectivity, because a background reaction by free metal is suppressed, proved not to be applicable in the next example. At the end of 2006, Reiser[78] discovered that 5 mol % modified (*S*)-azabis(oxazoline) copper(II)-complex **189a** (catalyst R = ^{i}Pr with *bis*-OTf counterion) or **189b** (catalyst where R = Ph with *bis*-ClO_4 counterion) could promote the addition of indole **128** to benzylidene malonates **184b** to afford the (*R*)-enantiomer adduct, (*R*)-**186b**, in 90–98% yield with up to > 99% *ee*, provided that excess of chiral ligand is avoided (see entries 1–5). The data suggest that one of the oxazole moieties of excess ligand could bind to the catalyst in the resting state (**A**). This would result in up to three nitrogen ligands being bound to copper in **A**. To reach the catalytically active species, one of the competing isoxazole ligands

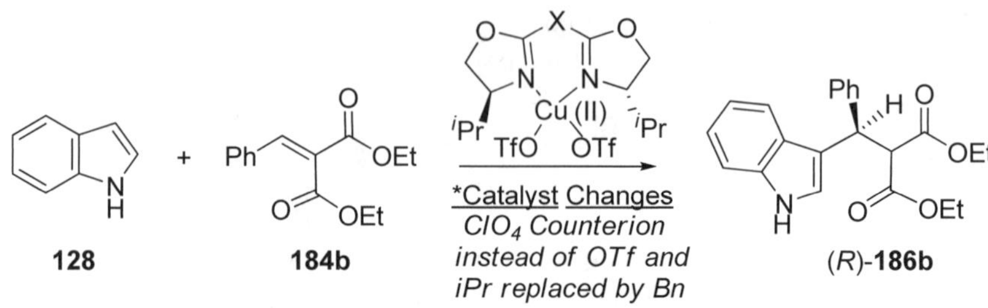

Entry	Catalyst Used	Solvent	Cu:Ligand (mol%)	Yield-**186b** (%ee)
1)	**189a** (X=NH)	EtOH	1.0:1.30 (3.8:5.0)	98% (81% ee)
2)	**189a** (X=NH)	EtOH	1.0:1.10 (4.5:5.0)	93% (85% ee)
3)	**189a** (X=NH)	EtOH	1.0:1.04 (4.8:5.0)	97% (>99% ee)
4)	**189b** (X=NH)*	EtOH	1.0:1.30 (3.8:5.0)	90% (87% ee)
5)	**189b** (X=NH)*	EtOH	1.0:1.04 (4.8:5.0)	96% (95% ee)
6)	**92iii** ($X=CMe)_2$	EtOH	1.0:1.04 (4.8:5.0)	89% (99% ee)

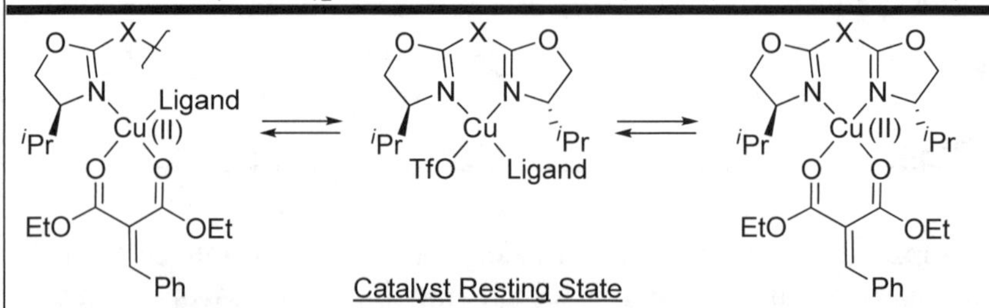

B-Low Enantioselectivity **A**-Catalyst/Ligand **C**-High Enantioselectivity

has to dissociate off first to bind the two carbonyl moieties of the substrate. If subsequent dissociation of one of the chelating oxazoline moieties occurs to create a species such as **B**. The likelihood of this type of displacement would be higher in the presence of a higher than 1:1 stoichiometry of ligand-catalyst complex. The use of a ratio closer to 1:1 would have a greater likelihood of retaining the chelated bisoxazoline moiety coordinated to the copper and result in higher enantioselectivity (active complex **C**). All of these complexes may exist in overall square pyramidal geometry with the fifth position occupied by a triflate or alcohol ligand (omitted for clarity).[78a]

The most significant aspect of this work was that variation of the ligand:catalyst ratio should be investigated in cases where low enantioselectivity is observed. For example, use of (*S*)-$Cu(OTf)_2$-^{t}Bu-BOX catalyst **92iii** in the same coupling reaction under the optimized reaction conditions (4.8/5.0 mol %, 1.0:1.04 Cu/ligand ratio) gave an impressive 89% yield of (*R*)-**186b** in 99% *ee* (entry 6). Increase of ligand:catalyst ratio to 1.0 : 1.20 under similar reaction conditions lowered the enantioselectivity of this conversion to 79% *ee* (results not shown in scheme). Thus the lower enantioselectivity (60% *ee*) reported for the same transformation by Jorgensen in 2001 using **92i** (R = ^{t}Bu) as catalyst may have resulted from differences in reaction conditions employed and not due to a greater proximity of the chiral ligand complex to the reacting olefinic center as had been originally speculated.[78a]

Other significant articles from 2003 to 2005 not appearing in this chapter important to the development of the field include (1) Umani-Ronchi advances of the asymmetric F–C Michael-type reaction on the use of α,β-unsaturated thioesters with a chiral Pd(II)-(Tol-binap) catalyst with the design of a single-point binding catalyst for this important transformation[79a–c] and (2) Ricci's use of chiral Jacobsen-type catalysts containing key requisite thiourea and hydroxyl moiety binding elements capable of bifunctional activation in the asymmetric F–C addition of nitroalkenes to unsubstituted indoles (**128**) (containing a few *N*-1 moiety) in preparation of chiral β-alkyl and aryl functionalized tryptamine precursors.[79d]

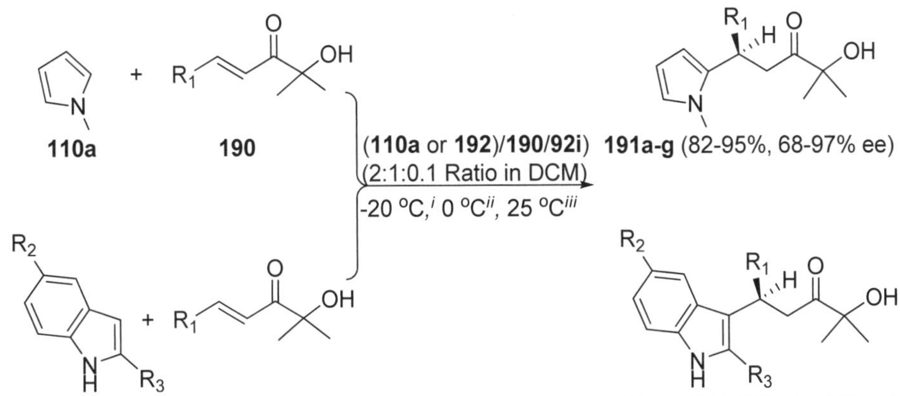

191	R1	%Yield (% ee)
a)i	Et	88% (94% ee)
b)i	tBu	86% (94% ee)
c)iii	2-ethyl(Ph)	86% (92% ee)
d)i	*n*-hexyl	82% (96% ee)
e)ii	iPr	86% (95% ee)
f)iii	cyclohexyl	84% (97% ee)
g)iii	Ph	95% (68% ee)

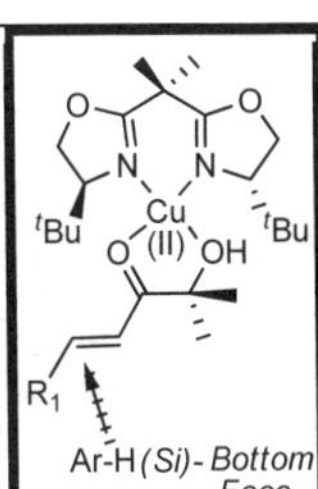

193	R1	R2	R3	%Yield (% ee)
a)i	2-ethyl(Ph)	H	H	85% (94% ee)
b)i	2-ethyl(Ph)	H	Me	89%,(93% ee)
c)i	2-ethyl(Ph)	MeO	H	96% (97% ee)
d)i	*n*-hexyl	H	H	85% (96% ee)
e)ii	iPr	H	H	44% (85% ee)
f)i	cyclohexyl	H	H	32% (85% ee)
h)i	4-Cl(Ph)	H	H	95% (83% ee)

Palomo in 2005 demonstrated that *N*-methyl pyrrole **110a** in the presence of 10 mol % of (*S*)-Cu(OTf)$_2$-^{t}Bu-BOX catalyst **92i** in DCM reacted with β-alkyl- or aryl-substituted α-hydroxyenones **190** to afford the corresponding optically active Friedel–Crafts pyrrole adducts (**191a–g**) in high yields (82–95%) and excellent enantioselectivities (68–97% *ee*).[80] All enones of type **190** containing sterically hindered 2° alkyl R_1 substituents required higher temperatures (25 °C) for conversion than the corresponding 1° alkyl R_1 moieties (–20 °C to 0 °C). The lowest enantioselectivity (68% *ee*) was obtained for a β-substituted aromatic hydroxyenone (R_1 = Ph, **191g**).

Indoles of type **192** also coupled well with **190**, affording the corresponding adducts (**193a–g** in 32–96% yield and 83–97% *ee*. The authors also demonstrated that it was possible to transform the Friedel–Crafts alkylation products **191a–g** and **193a–g** into the corresponding optically enriched aldehydes, carboxylic acids, and ketones via standard elaboration of the α-hydroxycarbonyl moiety (transformation steps not provided in the scheme).[80]

Evans, in a series of papers dated from 2003 to 2007, has reported the use of cationic Sc(III)-bis(oxazolino)pyridine catalysts (Scandium pybox) for the asymmetric Michael addition of electron-rich arenes to α,β-unsaturated 2-acylphosphonates and 2-acylimidazoles to afford the corresponding Michael adducts.[81a–e] These catalysts are one of the most versatile in terms of matching high enantioselectivity and yield in substrate compatibility. The usefulness of this reaction is further enhanced by the synthetic flexibility in the conversion of these "active ester" adducts to new scaffolds. For example, the corresponding 2-acylphosphonate moiety of these adducts can be elaborated to the corresponding chiral β-substituted acid, ester, and amide via a high yield nucleophilic displacement reaction. Adducts containing the more robust 2-acylimidazole moiety, however, offer even greater synthetic flexibility in that they can be converted to the corresponding chiral β-substituted aldehyde, ketone, acid, ester, and amide moieties using a one-pot two or three reaction sequence (ester/amide transformations shown).

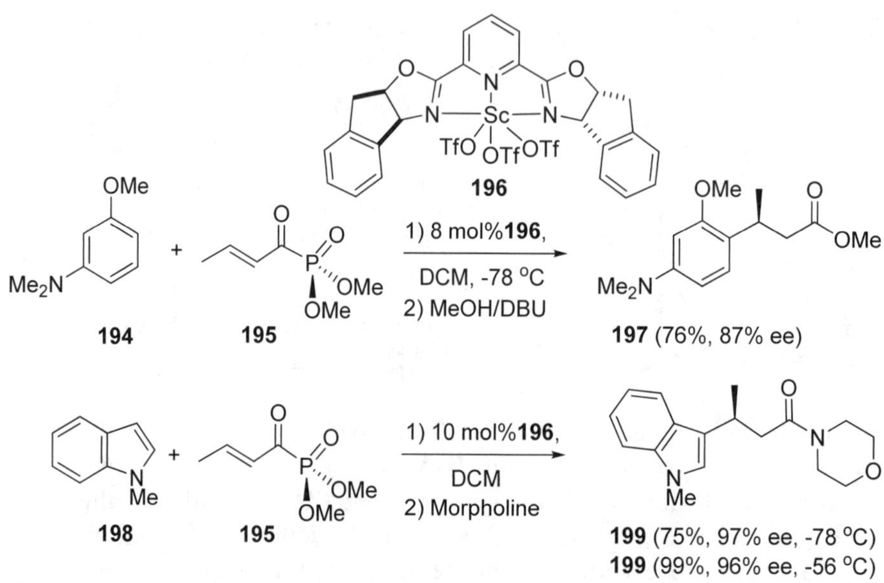

Evans started with α,β-unsaturated 2-acylphosphonate **195** with β-methyl substitution. Treatment of 3-*N*,*N*-dimethylaminoanisole (**194**) with α,β-unsaturated 2-acylphosphonate **195** in the presence of 8 mol % of (*S*,*S*)-

Sc(III)-Inda-pybox triflate complex **196** in DCM at –78 °C, followed by methanolysis of the resulting acylphosphonate Michael adduct using MeOH/DBU, gave the corresponding 3(*S*)-methyl substituted β-aryl propionate methyl ester **197** in 76% yield and 87% *ee*. The reaction of 1-methylindole (**198**, 0.2 M) with α,β-unsaturated 2-acylphosphonate **195** in the presence of 10 mol % catalyst **196** in DCM at –78 °C, followed by amination of the resulting acylphosphonate Michael adduct using morpholine, gave the corresponding morpholine 3(*S*)-methyl substituted β-aryl propionamide **199** in 75–78% yield and 96–97% *ee* (0.2M). At –56 °C, the yield of **199** was improved to 99% a identical enantiocontrol (96%).[81a,d]

Variation of the 1-, 2- and 5-positions of the indole and β-substituent of the α,β-unsaturated 2-acylphosphonate under identical reaction conditions (–78 °C, DCM, 10 mol% **196**) are not shown in the scheme but will be discussed. Replacement of the *N*-1-methyl moiety of indole **198** with *N*-benzyl, *N*-allyl and *N*-Me gave almost identical results in terms of yield of reaction (76–85%) and enantioselectivity (96–99%), but replacement of the indole *N*-1-alkyl moiety with H significantly lowered the enantioselectivity (83% *ee*) but not yield (83%). Replacement of the hydrogen at the 2-position of **198** with a 2-Me moiety led to a dramatic decrease in enantiocontrol (86%) but not yield (94%). It should also be noted that for *N*-1-Bn indoles, the 5-position tolerated both electron-withdrawing and donating groups (CO_2Me, halogen and OMe) relative to the parent *N*-1-Bn-5-*H*-indole with yields (65–85%) and enatioselectivities ranging from 97–99% *ee*. Increase of the size of the β-methyl (alkyl) asubstituent of **195** led to comparable yields and enantioselectivity (57% yield; 94% *ee* vs. 82% yield; 99% *ee*), where this substituent was replaced by $CH_2OTBDPS$ and ^{i}Pr, respectively. A 3-phenyl (3-aryl) analogue, however, was not well tolerated (85% yield, 80% *ee*).[81a,d]

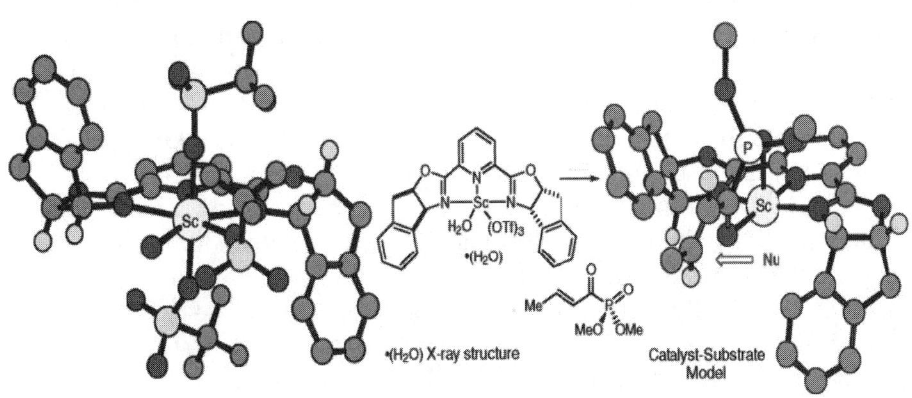

Several drawbacks in the use of the α,β-unsaturated 2-acylphosphonates in acylphosphonates in the asymmetric Michael alkylation, including low substrate tolerance (only β-alkyl substituted enones gave *ee*'s > 90%) and, that for high levels of stereoselectivity (> 90% *ee*), the reaction had to be conducted at temperatures below –50 °C. In addition, the α,β-unsaturated 2-acylphosphonates proved quite labile and could be prepared only in modest yields (20–47%). The pentagonal bipyramid geometry observed in the *X*-ray structure of this catalyst (**196**) led to the rationale that the sense of asymmetric induction is driven by placement of the sterically demanding phosphonate in the less demanding apical position orientating the oxygen toward the ligand plane. Thus addition of nucleophiles is believed to be directed from the indicated *S*-cis enoate diastereoface favored to minimize non-bonded interactions.

Evans et al. then investigated use of more robust and easily prepared Michael acceptor substrates, such as *N*-methyl (R = Me) α,β-unsaturated 2-acylimidazole derivative **200a** containing a small β-methyl (R_1 = Me) substituent.[81b–d] The asymmetric coupling of this substrate allowed great variation in the nature and size of both the R and the R_1 substituent and flexible in the coupling of a wide range of arenes and related heterocycles. In addition, these reactions show great potential for pilot plant scale reactions that could be conducted at temperatures as high as 0 °C with < 2 mol % of the catalyst with very good enantiocontrol. For example, treatment of 2-methoxyfuran **123** and *N*-H pyrrole **110b** with **200a** in the presence of 2 mol% of catalyst **196** in CH_3CN 4 Å MS at 0 °C afforded the respective 3(*R*)-methyl substituted 1,4-adducts **201a** (99% yield, 88% *ee*) and **202a** (82%, 88% *ee*) in very good yield and enantioselectivity, respectively. The coupling of furan **123** and pyrrole **110b** with **201b** having a larger *N*-*i*-Pr substituent (R = ^{i}Pr) under the same reaction conditions gave similar yield and improved enantioselectivities of the corresponding adducts **201b** (85% yield, 89% *ee*) and **202b** (90%, 93% *ee*). Evans completed the synthesis of (+)-heliotridine **205b** from **202b** using a using a two-step triflate-activated intramolecular amide cyclization of the pyrrole *N*-H moiety onto the activated acyl imidazoium ion of **202b** in the presence of Hunigs base to afford **203b** as the key step to afford **203b**. Hydrogenation of the pyrrole moiety of **203b** and $LiAlH_4$ reduction of the resulting bicyclic tertiary amide **204b** gave **205b**.

Coupling of the even larger *N*-Ph imidazole-substituted Michael acceptor (**200c**, R = Ph) with pyrrole **110b** under the coupling conditions led to a higher yield (98%) of **202c** but resulted in no significant improvement of enantioselectivity (94%). Reduction of the reaction temperature to –40 °C in the coupling of furan **123** with **200a** gave a reduced yield (65%) of **201a** with improved enantiocontrol (98% *ee*).[81b] No significant improvement in enantiocontrol of **202b** was observed in the coupling of pyrrole **110b** with either *N*-methyl or the sterically more encumbered *N*-*i*-Pr substituted α,β-unsaturated 2-acylimidazole **200a** and **200b**, respectively, when conducted under similar conditions (5 mol % of catalyst **196** at –40 °C versus 2 mol % of **196** at 0 °C).[81b–d]

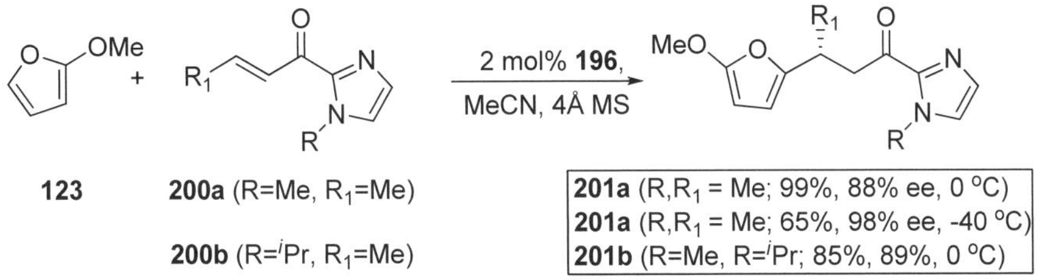

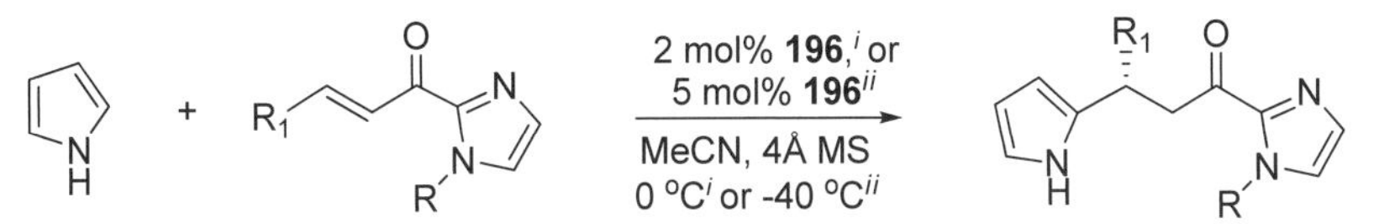

110b

200a (R=Me, R_1=Me)
200b (R=^{i}Pr, R_1=Me)
200c (R=Ph, R_1=Me)
200d (R=^{i}Pr, R_1=^{i}Pr)
200e (R=^{i}Pr, R_1=CO_2Et)
200f (R=^{i}Pr, R_1=4-CO_2Me(Ph))
200g (R=^{i}Pr, R_1=4-MeO(Ph))

202	R	R_1	%Yield, (% ee)
a)^i	Me	Me	82%, (88% ee)
a)^{ii}	Me	Me	69%, (87% ee)
b)^i	^{i}Pr	Me	90%, (93% ee)
b)^{ii}	^{i}Pr	Me	91%, (94% ee)
c)^i	Ph	Me	99%, (94% ee)
d)^i	^{i}Pr	^{i}Pr	90%, (91% ee)
e)^i	^{i}Pr	CO_2Et	99%, (84% ee)
f)^i	^{i}Pr	4-CO_2Me(Ph)	99%, (96% ee)
g)^i	^{i}Pr	4-OMe(Ph)	98%, (92% ee)

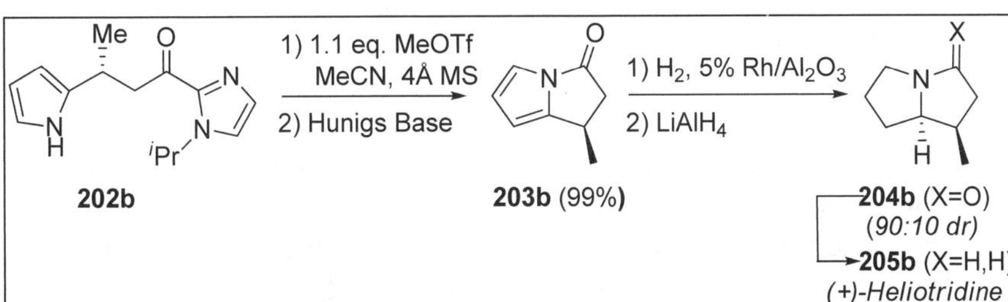

A subsequent detailed study of the reaction has revealed an inverse correlation of higher catalyst loading with lower enantioselectivity, which may have offset any increase in stereoselectivity by lowering of the reaction temperature. In addition, Evans has also discovered that the temperature-dependent control of enantioselectivity of α,β-unsaturated 2-acylimidazole of type **200** was remarkably flat relative to the 2-acylphosphonate analogue **195**.[81d] Investigation of pyrrole **110b**, various β-substituted (R_1-substituted) *N*-^{i}Pr-substituted α,β-unsaturated 2-acylimidazole **200d–g** (R = ^{i}Pr), afforded the corresponding adducts **202d–g** with a remarkable consistency of yield (90–99%) and enantioselectivity (84–96% *ee*). This is remarkable considering that the coupling of α,β-unsaturated 2-acylphosphonate **195**

worked well only with β-alkyl substitution and required remarkably reduced temperatures (< –40 °C) to obtain good enantiocontrol. The following observations are not shown in the scheme but will be discussed. Replacement of the *N*-H pyrrole **110b** with the *N*-methyl pyrrole **110a** in the coupling of **200b** (R = ^{i}Pr) led to a significant loss of enantiocontrol (69% yield, 77% *ee*, 5 mol % of catalyst **196**) even at low temperature (–40 °C).

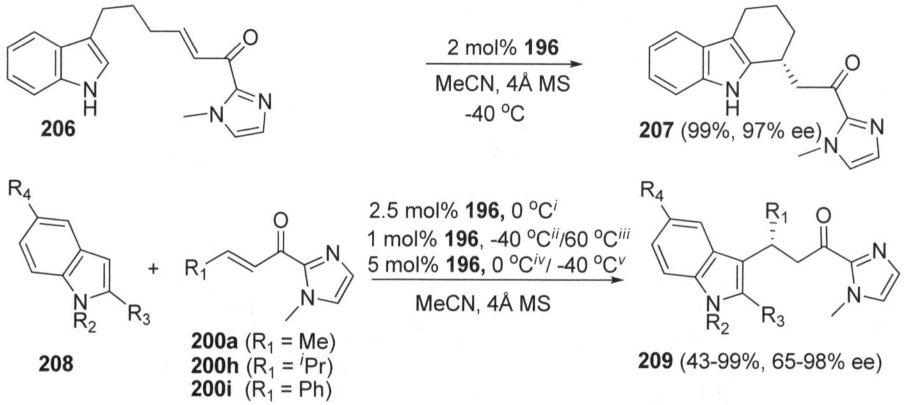

Intramolecular cyclization of tethered indole **206** under the same catalytic conditions (2 mol % of **196**) at –40 °C, gave an impressive 99% yield of **207** in 97% *ee.* To enantioselectively couple *B*-substituted α,β-unsaturated 2-acylimidazole of type **200** intermolecularly with indoles of type **208** under the standard coupling conditions (2.5 mol % **196** at 0 °C), capping of the free indole NH (R_2 = H) was required (compare adducts **209A** versus **209B** obtained 65 and 93% *ee*, respectively, formed in the reaction of **200a** (R_1 = Me) with indole of type **208**). Use of even less of the catalyst (1 mol % **196**), at lower temperature (–40 °C) gave an improved 97% yield and 98% *ee* of **209B**. Increase of the reaction temperature to 65 °C (minimum of 1 mol % of catalyst **196**) afforded **209C** in a surprisingly high 99% yield and 90% *ee*, highlighting the relatively flat temperature dependance of stereocontrol in the reaction.[81b]

Substitution of the indole at *C*-2 with Me with a larger Ph moiety led to a significant drop in yield and stereoselectivity of **200** (88%, 91% *ee*) and **209E** 43%, 66% *ee*).[81d] Alkylation of *N*-Bn indole (**208**) with **200a** under the reaction conditions afforded **209F** in 90% yield and 98% *ee*. There was no loss of stereocontrol of the reaction with replacement of the 5-H substituent of **208** with 5-halo or alkoxy substituent in the coupling of **200a** to afford **209G** in 70% (95% *ee*) and **209H** in 99% yield (97% *ee*), respectively. Variation of the R_1 β-substituent of **200** from methyl (**200a**) to ^{i}Pr (**200g**) or Ph (**200j**) in the coupling of *N*-Me indole (**208**) showed no significant decrease of stereocontrol (compare **209b** formed in 93% *ee* versus **209i** and **209j** formed in 94% and 91% *ee*, respectively).

Overall, this coupling substrate showed the greatest flexibility in the coupling of arene heterocycles with a wide variety of β-substituted (R_1-substituted) *N*-^{i}Pr-substituted α,β-unsaturated 2-acylimidazoles. The resulting acyl imidazolides could be converted in straightforward fashion to either aldehyde, ketone, acid, ester, or amide moieties using a one-pot two- or three-reaction sequence (not shown in scheme).[81b–e]

The proposed stereoinduction model is complex and can vary, depending of the catalyst and active ester substrate used. The binding of the 2-acylimidazole substrate of type **200a** to catalyst **196** is believed to lead to lead to a seven-coordinate pentagonal bipyramid 1:1:1 product/substrate/catalyst complex at lower catalyst loading, but at higher catalyst loading, a pentagonal bipyramid 1:1 product substrate complex is believed to form. Both binding modes lead to the same stereochemical outcome in this case. The imidazole group is planar, and two of these units can occupy the equatorial positions of the catalyst, with one being the substrate and the other being the product molecule. This would place the carbonyl groups at the apical positions with the enone of the substrate oriented for nucleophilic attack on the *Re*-face, in accord with a sense of stereoinduction observed in the reaction according to Evans.[81d.]

In 2009, the general asymmetric coupling of substituted pyrroles (**110**) and indoles (**208**) with 2-acylimidazole substrate of type **200** was extended to other catalyst complexes such as 0.15–0.3 mol % of copper 4,4'-dimethyl-2,2'-bipyridine [Cu(dmbpy)]/DNA complexes [1.4–7 ug/mL DNA base pairs] in aqueous media. Good enantioselectivities (83–93%) of the corresponding adducts **202** and **209,** respectively, were observed using sources of DNA such as st-DNA (salmon testes-DNA) and DNA-1 with the self–complimentary sequence of d(TCAGGGCCCTGA)$_2$. A key limitation of this approach was the requirement of at least 5:1 ratio of arene to **200** to achieve good conversion to the adducts (no scheme provided)[81f]

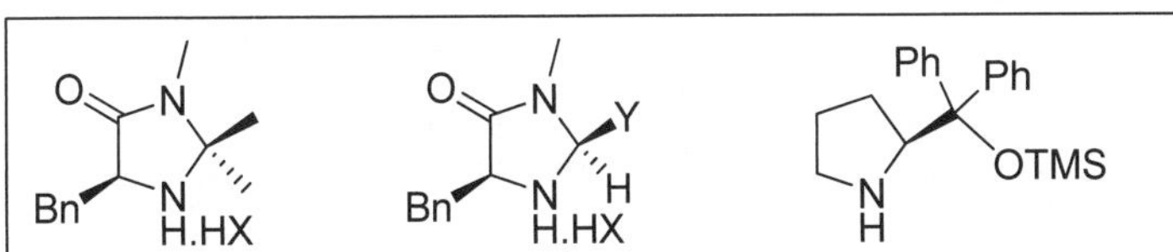

210 **211A** (Y= ^{t}Bu) **211B** (Y= 5-Me-2-furyl) **212** (No HX Needed)

*(For catalysts **210-211A, 211B**, HX usually TFA or HCl salt)*

The final portion of section E will focus on the use of covalent binding secondary amine organocatalysts such **210–212** for the asymmetric F–C Michael Type alkylations of arenes with α,β-unsaturated aldehydes of type **213**, first pioneered by MacMillan back in 2000.[82a–d] The initial organocatalysts used (i.e., **210**, **211A**) were prepared in just a few synthetic steps on pilot plant scale from inexpensive commercially available amino acids (not shown in scheme). Catalyst loading for these type of covalent organocatalysts is typically much higher (~ 20 mol %) than for the aforementioned chiral thiourea and Brønsted acid organocatalysts (typically < 5 mol %), which involve noncovalent H-bonding interactions essential for catalyst activity. Thus the ease of preparation on scale is important to the practicality of the methodology.

During the period 2000–2002, MacMillan's group reported the asymmetric coupling of substituted electron-rich arenes (anilines **214**), pyrroles (**216**) and indoles (**218**) with α,β-unsaturated aldehydes (**213**) in the presence of 20 mol % organocatalyst **210** or **211A** to afford the corresponding adducts (**215**, **217**, and **219**, respectively), in very good yields and enantioselectivities.[82a–c] Thus secondary amine organocatalysts forming reversible chiral iminium ions in situ which increase reactivity of the electrophile by LUMO activation as well as provide a spatially discriminating environment for enantiofacial (*Re*-face selectivity of the alkene moiety) and regioselective 1,4-addition by the arene moiety.

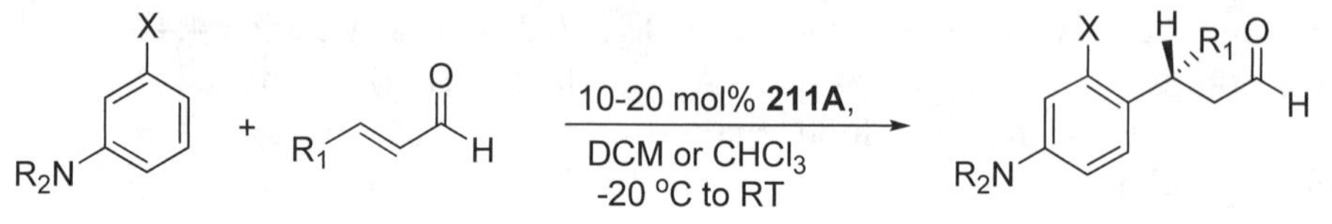

214 **213** *(R=alkyl and X=OMe, H, Cl, SMe)* *(R_1=alkyl, aryl, CO_2Me, CH_2OBn)* **215** (>70%, 72-97% ee)

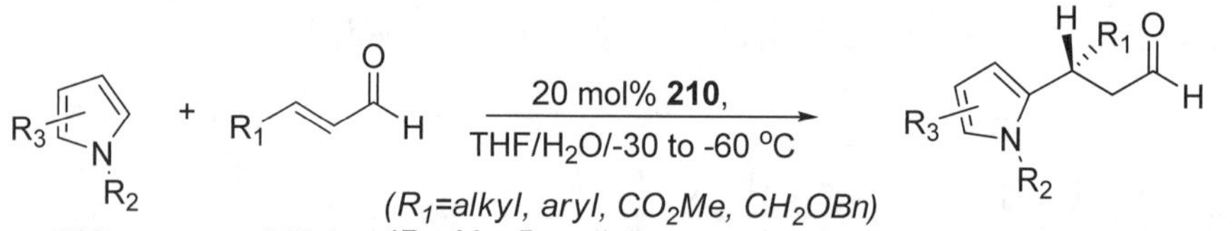

216 **213** *(R_1=alkyl, aryl, CO_2Me, CH_2OBn)* *(R_3=Me, Bn, allyl)* **217** (>73%, 86-99% ee)

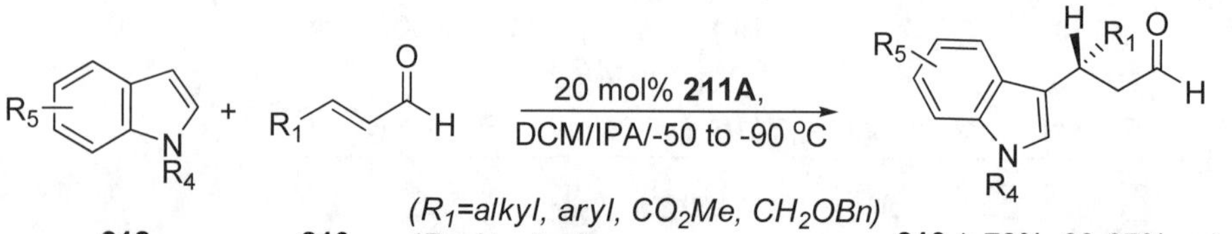

218 **213** *(R_1=alkyl, aryl, CO_2Me, CH_2OBn)* *(R_4=H, alkyl)* *(R_5=Me, Cl, OMe)* **219** (>72%, 89-97% ee)

Overall, **211A** is a better better designed catalyst than **210** though both may be adequate if the arene and Michael acceptor **213** are of appropriate reactivity. The only difference in the catalysts is that the geminal dimethyl position of the aminal moiety of **210** is replaced by the β-chiral ^{t}Bu moiety (**211A**). Several control elements are at work. The lone pair of the secondary amino moiety of catalyst **210** required for formation of the iminium ion is sterically less accessible (vicinal geminal dimethyl groups occupies space on both faces) and is less reactive than with catalyst **211A** where a hydrogen is syn to the lone pair and the ^{t}Bu moiety occupies the opposite face. The catalyst-activated iminium ion complex of **211A**:**213** was anticipated to more optimally populate the (*E*)-isomer to avoid nonbonding interactions between the substrate olefin and the ^{t}Bu moiety. As a result, the benzyl group on the catalyst framework will effectively shield the *Si*-face of the activated olefin, leaving the *Re*-face exposed to indole addition, leading to the (*R*)-adducts with high stereoselectivity. The same steric influences of the iminium ion complex of both catalysts preclude arene 1,2-addition.

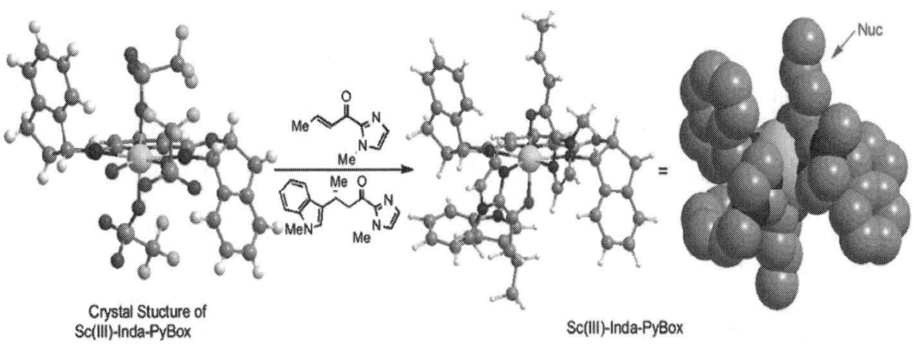

Recently, Wang (2009) demonstrated a practical improvement of this methodology through the use of Lewis base–Lewis base bifunctional catalysis instead of the traditional MacMillan base catalyst with acid cocatalysts of formula HX.[82e] It was shown that coupling of a wide variety of *N*-1 unsubstituted R_5-substituted indoles of type **218** (R_4 = H, 1.20 equiv) and β-substituted α,β-unsaturated aldehydes (**213**, 1.0 equiv) in the presence of diphenyl prolinol TMS ether catalyst **212** (5–20 mol %) in ethereal solvents such as TBME at 25 °C to –20 °C could afford adducts **219** in very good yields (66–95%) and enantioselectivities (92–98%).

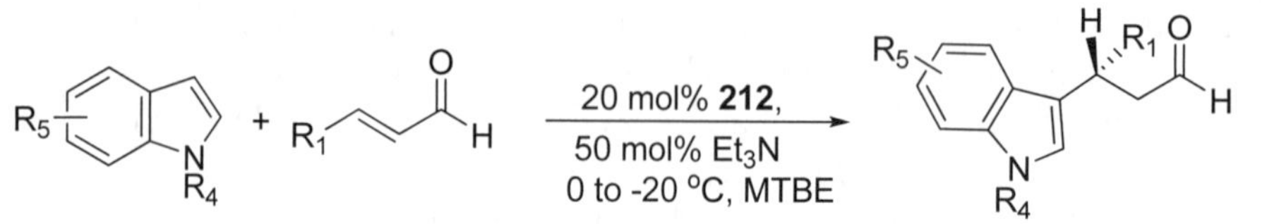

218 (1.2 eq.) **213** (1 eq.) *(R_1 = alkyl, aryl, and R_4 = H)* **219**

(R_5=2-Me, 5-OMe, 5-Br, 7-Me) (66-95%, 92-98% ee)

In this mode of catalysis, it is believed that chiral amine Lewis base catalyst **212** was used to activate the α,β-unsaturated aldehyde **213** by lowering the energy of the LUMO of the electrophile and induction of chirality of the reaction through covalent formation of the intermediate iminium ion complex as proposed by MacMillan.[82a–c] The difference with this bifunctional catalysis mechanism is that addition of a second Lewis base (i.e., triethylamine) acts as a cocatalyst to activate the nucleophilic reagent (indole *N*-H moiety) by raising the energy of the HOMO of the nucleophile by deprotonation or hydrogen-bond interaction (lowering the resulting LUMO–HOMO gap),m increasing the efficiency and practicality of the process. The elimination of the potentially corrosive acid cocatalyst and use of higher reaction temperature make this method attractive for optimization on pilot plant scale. This method does have the limitation that the nucleophile **218** (*N*-H indoles, where R_4 = H) must be capable of activation by the base cocatlyst (Et_3N or $^i Pr_2NEt$).[82e]

The following results are not shown in the scheme but will be discussed. The reaction proved equally effective in other ether based solvents such as THF, ether, dioxane, and DME but gave no conversion to the product in polar aprotic solvents such as DMF or CH_3CN. It is interesting that use nonpolar solvents such as toluene and DCM gave adequate conversion very poor stereocontrol (< 40% *ee*) in formation of the adduct. It was also demonstrated that as little as 5 mol % of catalyst **212** could be used, though a detailed study of the effect of lowering the catalyst loading wasn't reported.[82e]

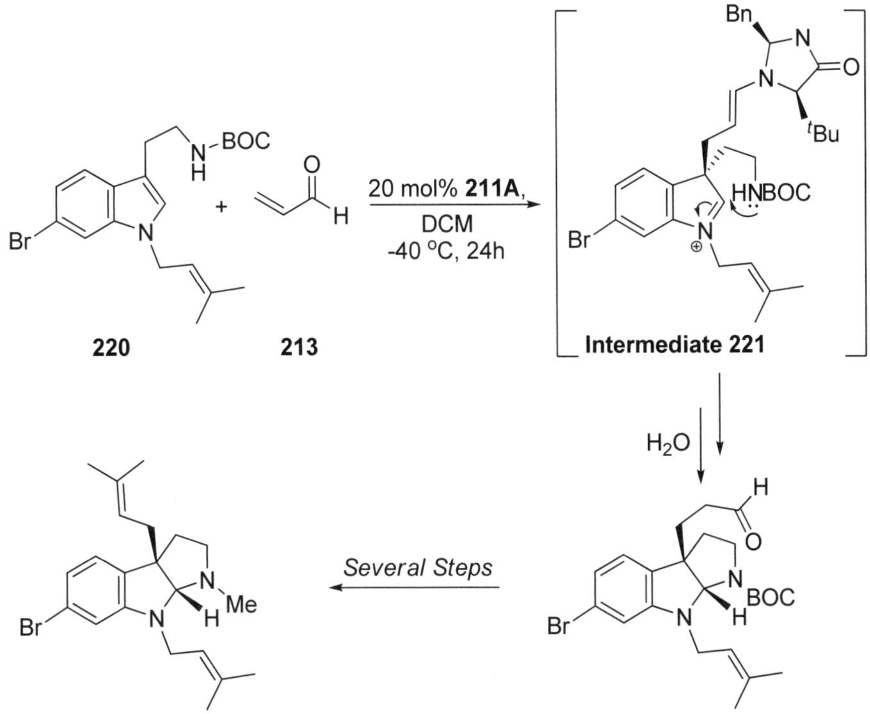

223 (-)-Flustramine B **222** (78%, 90% ee)

Several medicinally important active agents were reported to be synthesized using the MacMillan coupling methodology as the key step in the syntheses. In 2004, MacMillan completed an elegant synthesis of (–)-Flustramine, a pyrrolidinone marine alkaloid (K-channel blocker) using a cascade-cyclization strategy,[82f] whereas workers at Bristol-Myers Squibb in 2005 synthesized a serotonin reuptake inhibitor (SRI) for potential use as an antidepressant via the use of a modified MacMillan catalyst **211B**.[82g] Impressive examples of highly enantioselective intramolecular indole cyclizations were reported by Xiao in 2007 (scheme and results not shown).[83]

One limitation of this method was that it can be applied only enantioselectively to less reactive *α,β*-unsaturated aldehydes and not the corresponding ketones. Many successful strategies for the asymmetric Friedel–Crafts alkylation of indoles with *α,β*-unsaturated enones catalyzed by Lewis acids[79b,c,84a,b] or organocatalysts[85a–e] have also recently been reported. Two of these methods published in 2007 involve use of noncovalent organocatalysts using H-bonding interactions by creation of a catalytic salt by combination of an achiral[85b] and chiral Brønsted acid[85c] with a Brønsted base derived from a primary amine modified cinchona alkaloids. They were applied to afford the corresponding indole adducts with stereocontrol up to 88–96% *ee*, respectively.

6.2.8 *Experimental*

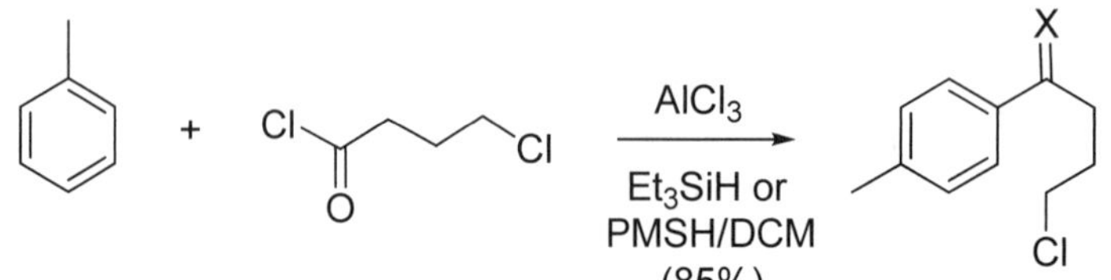

13 **14** **15** (X = O) **16** (X = H,H)

One-Pot two-step generalized procedure for the preparation of 1-(4-chlorobutyl)-4-methylbenzene (16) **from 4-chlorobutanoyl chloride** (14) **and toluene** (13)

A solution of 5.12g (36.3 mmol) 4-chlorobutanoyl chloride (**14**) was stirred in 30 mL CH_2Cl_2 and 4.4 g (33.0 mmol) anhydrous $AlC1_3$ was slowly added. The reaction mixture was stirred until reaction completion. To this mixture at 25 °C was added a solurtion of 2.76 g (30.0 mmol) toluene in 10 mL of CH_2Cl_2, followed by addition of neat 9.25 g, (80 mmol) neat Et_3SiH (80.0 mmol). The reaction mixture was stirred until reaction completion then subjected to traditional aqueous workup with purification by chromatography to afford 3.81 g (86%) **16**: b.p. 97–100 °C/0.015 mm Hg).

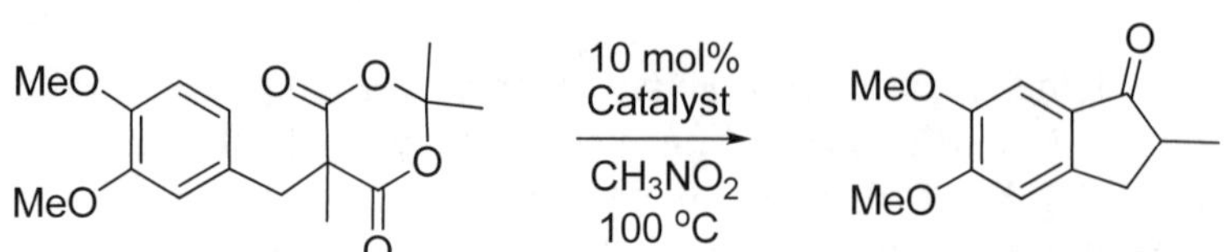

23

24 (Catalyst, Yield)

BF_3•OEt_2	90%
$Sc(OTf)_3$	85%
TfOH	86%

General preparation of 5,6-dimethoxy-2-methyl-2,3-dihydro-1*H*-inden-1-one (24) from 23[33f,h]

To a flame-dried two-necked flask equipped with reflux condensor preheated to 100 °C was added 5 mL of a 0.095 M CH_3NO_2 solution of Meldrum's acid derivative **23** (0.475 mmol) under a dry N_2 atmosphere. The solution was heated 5 min; after which, 10 mol% (0.0475 mmol) of the catalysts (9 μL BF_3•OEt_2, 5 μL TfOH or 25 mg $Sc(OTf)_2$) wer e quickly added. The solution was heated to reflux for 20 min. The mixture was cooled to rrom temperature and concentrated, and the residue was purified by silica gel flash chromatography, eluting with 2:1 ethyl acetate/hexane to afford **24** as a white solid (m.p. 131–132 °C) in the yields indicated in the table.

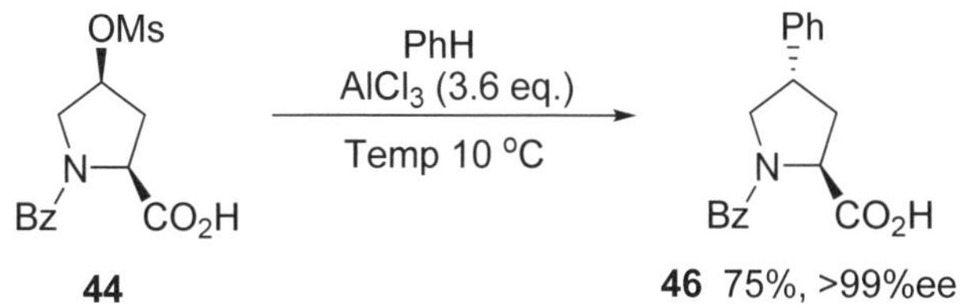

46 75%, >99%ee

Preparation of *trans*-1-Benzoyl-4-Phenyl-L-Proline (46) from 44[39b]

To a dry t,hree-necked, 2-L Morton flask equipped with overhead stirrer, temperature probe and N_2 inlet was added 124.2 g (3.6 equiv, 0.93 mol) $AlCl_3$ in 810 mL benzene. The mixture was cooled as necessary in a dry ice/acetone bath with the slow addition of 81.0 g (0.26 mol) **44** in small solid portions at a rate to maintain the internal temperature at 6 °C (removal of bath as necessary). The mixture was stirred 5.5 h, allowing the internal temperature to rise from 7 to 10 °C as necessary by cooling. The resulting homogeneous mixture was hydrolyzed by the slow addition of 990 mL (2.97 mol) 3 M aqueous HCl solution so the internal temperature did not exceed 30 °C. The solution was diluted with 180 mL brine and stirred at room temperature overnight, the resulting precipitated crude product containing **46** was filtered. The resulting solid was washed with 390 mL (0.39 mol) 1 N HCl and water (4 × 500 mL). The resulting crude product was dissolved in 240 mL hot *n*-butyl acetate, dried (anhydrous Na_2SO_4), cooled to room temperature and seeded to promote recrystallization. A total of 57.4 g (75%) enantiomerically pure **46** was isolated after suitable drying as a white solid [m.p. 137–138.5 °C), $[\alpha]_D$ –62.3 (MeOH, c = 1)].

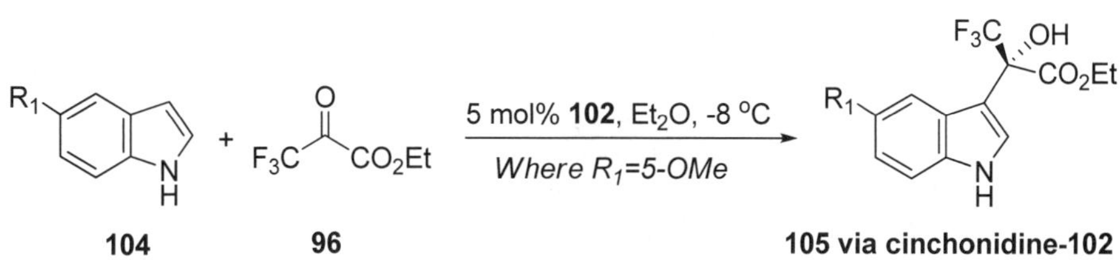

105 via cinchonidine-102
(98%, 93% ee)

Preparation of (*S*)-3,3,3-trifluoro-2-hydroxy-2-(5-methyl-3-indolyl)-propionic acid ethyl ester (105, R_1 = Me) from 102 (R_1 = Me), and 96 via cinchonidine alkaloid catalyst 102[56b]

To a sealable glass reaction vessel under inert atmosphere containing 5-methylindole (0.5 mmol) and cinchonidine (**102**, 0.0375 mmol) was added 3 mL anhydrous Et_2O. The solution was stirred at –8 °C (salt-ice cooling bath) for 30 min. Then 0.75 mmol ethyl 3,3,3-trifluoropyruvate was added, and the mixture was stirred at –8 °C (salt-ice cooling bath) for 3 h. The mixture was concentrated with removal of the solvent and excess ethyl trifluoropyruvate by evaporation. The mixture then was dissolved in ether and the catalyst was removed by use of 500 mg scavenging agent K-10 montmorillonite (a solid acid). The cinchonidine-K-10 complex was removed by filtration and the solvent evaporated to afford **105** (R_1 = Me) in 98% yield and 93% *ee* (chiral HPLC): mp 75.2–76.5 °C

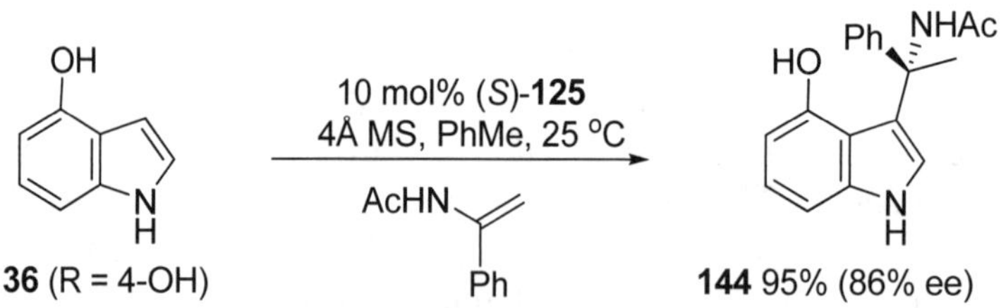

136 (R = 4-OH) (**143**, Ar = Ph) **144** 95% (86% ee)

Preparation of (*S*)-*N*-(1-(4-hydroxy-1*H*-indol-3-yl)-1-phenylethyl)acetamide (144, Ar = Ph, R = 4-OH) from 4-hydroxyindole (136, R = 4-OH), enamide (143, Ar = Ph) and chiral phosphoric acid catalyst (125)[63b]

To an oven-dried Schlenk tube was added phosphoric acid **(*S*)-125** (7.5 mg, 0.01 mmol), 4-hydroxyindole **136** (0.14 mmol), *N*-(1-phenylvinyl)acetamide (enamide **143**, Ar = Ph, 0.1 mmol), and 90 mg 4 Å molecular sieves. The mixture was degassed, placed under a N_2 atmosphere dissolved in toluene (1.5 mL) and stirred at RT until TLC indicated reaction completion. The solvent was removed under vacuum and the residue purified by flash column chromatography on silica gel (eluted with 1:1 ethyl acetate/petroleum ether to afford the (*S*)-adduct **144** (R = 4-OH) in 95% yield and 86% *ee*: m.p. 107–110 °C; $[\alpha]_D$ = –62.3 (*c* 0.7, acetone).

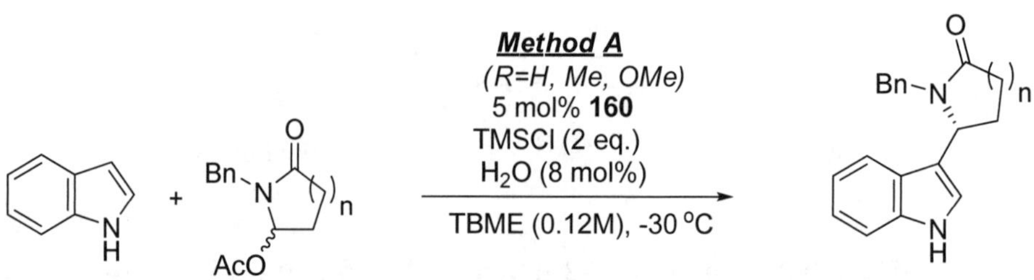

(1 equiv) **128** (1 equiv) **159i** (n=1) **161a** (n = 1) 90, (93% ee)* *99% ee (Et_2O)

Preparation of (*R*)-1-benzyl-5-(1*H*-indol-3-yl)pyrrolidin-2-one (161a) from Indole (128), 1-benzyl-5-oxopyrrolidin-2-yl acetate (159i, *n* = 1): and thiourea catalyst (160) using method A[68]

The thiourea catalyst **160** (0.110 g, 0.156 mmol) and indole (0.367 g, 3.15 mmol, **128**) were each placed in a 100-mL flame-dried round-bottomed flask, which was sealed with a rubber septum. The flask was flushed with N_2, and anhydrous TBME (14.7 mL) was added. The resulting yellow solution was cooled to –78 °C, and a solution of acetoxylactam **159i** in TBME (6.0 mL, 6.34 mmol, 0.53 M) was added. Next, solutions in TBME of TMSCl (3.95 mL, 6.34 mmol, 1.6 M) and H_2O (1.84 mL, 0.276 mmol, 0.15 M) were added sequentially, and the mixture was warmed to –30 °C and stirred for 24 h. The heterogeneous reaction mixture was quenched by the addition of a solution of NaOEt in EtOH (1.84 mL, 21 wt %), followed by the immediate addition of water (9.2 mL). The mixture was allowed to warm to room temperature and was diluted with EtOAc until all solids had dissolved (the amount of EtOAc added (90 mL). The layers were separated, and the organic layer was dried (Na_2SO_4) and concentrated in vacuo. The crude product (93% *ee*) was purified by trituration from Et_2O (3 × 10 mL) to yield **161a** (0.854 g, 90% yield, 99% *ee*) as a colorless crystalline solid: $[\alpha]_D = -49°$ (c = 1.3, MeOH).

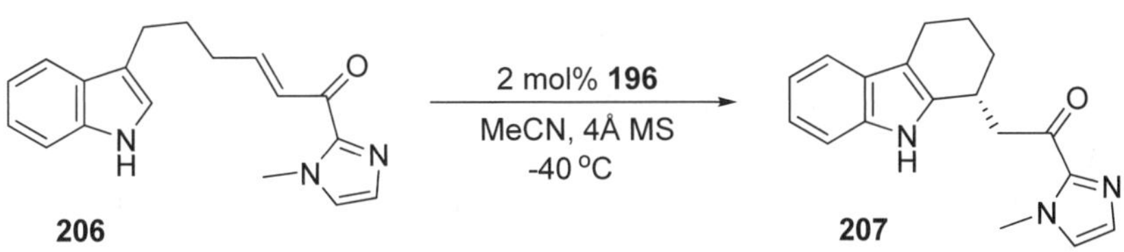

206 **207** (99%, 97% ee)

Preparation of (*R*)-1-(1-methyl-1*H*-imidazol-2-yl)-2-(2,3,4,9-tetrahydro-1*H*-carbazol-1-yl)ethanone (207) from (E)-6-(1*H*-indol-3-yl)-1-(1-methyl-1*H*-imidazol-2-yl)hex-2-en-1-one (206) and catalyst complex (196)[81b,d]

To a dried 2-dram vial in a drybox was added an appropriate amount scandium(III) triflate, 4 Å MS (15 mg/0.13 mmol of substrate), and 1.2 equiv (*S*)-Indapybox ligand. The vial was capped with a septum and purged with 1 mL dichloromethane. The catalyst was allowed to age for 2 h at rt. The dichloromethane was removed by a steady stream of N_2 to afford catalyst complex **196**. To 2 mol % (0.0021 mmol) **196** was added 1 mL of acetonitrile. The reaction was cooled to –40 °C for 15 min before 30 mg (0.103 mmol, 1 equiv) (*E*)-6-(1*H*-indol-3-yl)-1-(1-methyl-1*H*-imidazol-2-yl)hex-2-en-1-one **206** was added to the vial. After 18 h of moderate stirring at –40 °C, the title compound was purified by flash silica chromatography (R_f = 0.31, 50% EtOAc/hexanes) to afford 30 mg (99% yield, 97% *ee*) **207**.

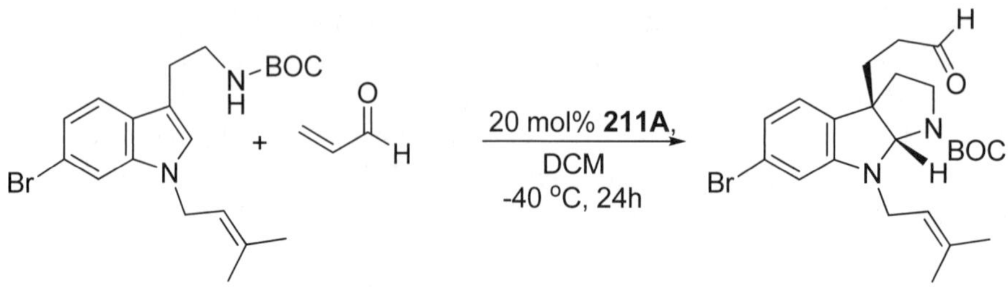

220 **213** **222** (78%, 90% ee)

Preparation of (*S*)-6-bromo-8-(3-prenyl)-3a-(3-oxo-propyl)-3,3a,8,8a-tetrahydro-2*H*-pyrrolo[2,3-*b*]indole-1-carboxylic acid *tert*-butyl ester (222) **from *N*-10-BOC-1-prenyl-6-bromotryptamine** (220)**, acrolein** (213)**, and MacMillan catalyst** 211A[82f]

To an amber 2-dram vial equipped with a magnetic stir bar was added (2*S*,5*S*)-5-benzyl-2-*tert*-butyl-3-methyl-imidazolidin-4-one (catalyst **211A**) in 4.2 mL CH_2Cl_2 cooled to −84°C, 258 mg acid (9.8 μL, 0.13 mmol) was added, followed by trifluoroacetic acid (0.64 mmol) and *N*-10-BOC-1-prenyl-6-bromotryptamine (**220**). The solution was stirred for 5 min before addition of 0.17 mL (2.56 mmol) acrolein **213** and then stirred for 72 h. The resulting suspension was stirred at constant temperature until complete consumption of the indole was observed as determined by TLC. The reaction mixture was then treated with 20 mL pH 7.0 buffer and extracted with diethyl ether (2 × 25 mL) and concentrated in vacuo. The resulting residue was purified by silica gel chromatography (solvents noted) to afford **222** as a colorless oil (231 mg, 78% yield, 90% *ee*) after silica gel chromatography in 10% EtOAc/hexanes as a colorless, viscous oil. $[\alpha]_D = -218.9$ ($c = 1.0$, $CHCl_3$).

6.2.9 References

1. [R] Olah, G. A.; Krishnamurti, R.; Prakash, G. K. S. in *Friedel–Crafts Alkylation in Comprehensive Organic Synthesis, Vol. 3,* Trost B. M., Fleming I. eds.; Pergamen Press, Oxford, **1991**, 293.
2. Miethchen R.; Kroger C.-F. *Z. Chem.* **1975**, *15*, 135.
3. [R] Olah G. A., Dear R. R. *Friedel–Crafts and Related Reactions*, Wiley-Interscience, New York, **1963-1965**.
4. [R] Roberts, R. M.; Khalaf, A. *Friedel Crafts Alkylation Chemistry: A Century of Discovery,* Marcel Dekker, New York, **1984**.
5. [R] Olah G. A. *Friedel–Crafts Chemistry*, Wiley-Interscience, New York, **1973**.
6. [R] Ashdown, A. *Ind. Eng. Chem.* **1927**, 1063.
7. Crafts, J. M.; Friedel C. *Compt. Rend.* **1877**, *84*, 1450. See also Crafts, J. M.; Friedel C. *Compt. Rend.* **1877**, *85*, 74 and 673.
8. [R] Olah G. A.; Krishnamurphy, R.; Prakash, G. K. S. In *Kirk-Ohmer Encyclopedia of Chemical Technology*, 5th Ed **2005**, *12*, 159. For a detailed list of relative Lewis acid activities in F–C reactions, see Table 1, page 1071.
9. [R] Price, C. C. In *Organic Reactions*, Wiley, New York, **1946**, *3*, 2.

10. [R] Laue, T.; Plagens, A. *Named Organic Reactions* Wiley, New York, **2005**, p. 120.
11. Brown, H. C. *J. Am. Chem. Soc.* **1953**, *75*, 6285.
12. Olah, G. A.; Kuhn, S. J. *J. Am. Chem. Soc.* **1958**, *80*, 6541.
13. Effenberger, F. *Chem. Unserer Zeit* **1979**, *13*, 879.
14. [R] Smith, M. B.; March, J. *March's Advanced Organic Chemistry* Wiley, New York, **2001**, p. 711.
15. (a) Nakajima, T.; Suga, S.; Sugita, T.; Ichikawa, K. K. *Tetrahedron* **1969**, *25*, 1807. (b)See also Nakajima, T.; Nakamoto Y.; Suga, S. *Bull. Chem. Soc. Jpn.* **1975**, *48*, 960. (c) For cases of almost 100% inversion, see Piccolo, O.; Azzena, U.; Melloni, G.; Delogu, G.; Valoti, E.; *J. Org. Chem.* **1991**, *56*, 183
16. Brown. H. C.; Jungk, H. *J. Am. Chem. Soc.* **1956**, *78*, 2182.
17. [R] Barton, D.; Ollis, W.D. "*Comprehensive Organic Chemistry*" Pergamon Press, New York, **1979**, *1*, 268–269.
18. Olah, G. A.; Olah, J. A. *J. Am. Chem. Soc.* **1976**, *98*, 1839.
19. Tagematsu, A.; Sugita, K.; Nakane R. *Bull. Chem. Soc. Jpn.* **1978**, *51*, 2082.
20. (a) Olah, G. A.; Kobayashi, S.; Tashiro, M. *J. Am. Chem. Soc.* **1972**, *94*, 7448. See also Asaoka T.; Shimasaki, C.; Taki, K.; Funayama, M.; Sakano M.; Kamimura, Y. *Yuki Gosei Kagaku Kyokai Shi*, **1969**, *27*, 783. For earlier study, see Russell, G.; *J. Am. Chem. Soc.* **1959**, *81*, 4834, [R] (b) For an impressive review of Lewis acid in homogenerous and heterogeneous catalysis, see Corma, A.; Garcia, H. *Chem. Rev.* **2003**, 103, 4307.
21. Mine, N.; Fujiwara, Y.; Taniguchi, H. *Chem. Lett.* **1986**, 357.
22. Nenitzescu, C. D.; Cantuniari, E. I. P. *Chem. Ber.* **1933**, *66*, 1097.
23. Allen, R. H.; Yats, L. D. *J. Am. Chem. Soc.* **1961**, *83*, 2799.
24. Jacobsen O. *Chem. Ber.* **1885**, *18*, 338.
25. Bun-Hoi, N. G.; Cagneunt, P. *Bull. Soc. Chim. Fr.* **1942**, *9*, 887.
26. Ipatieff, V. N.; Pines, H.; Schmerling, L. *J. Org. Chem.* **1940**, *5*, 253.
27. Tashiro, M. *Synthesis*, **1979**, 921. (b) Tashiro, M.; Itoh, T.; Yoshiya, H.; Fukata, G. *Org. Prep. Proc. Intl.* **1984**, *16*, 155.
28. McOmie, J. F. W.; Saleh, S. A. *Tetrahedron* **1973**, *29*, 4003.
29. Lewis, N.; Morgan, I. *Synth. Commun.* **1988**, *18*, 1783.
30. [R] (a) See Francis, A. W.; Reid, E. E. *Ind. Eng. Chem.* **1946**, *38*, 1194. and Weisgermel, K.; Arpe, H. J. *Industrial Organic Chemistry*, Verlag Chemie, Berlin, **1978**. (b) Green, M. M.; Witcoff, H. A. *Organic Chemistry Principles and Industrial Practice* Wiley-VCH, Weinheim. **2003**. pp. 10–21, (c) [R] For an excellent review on the use of clays in synthesis including F–C Reactions, see Dasgupta, S.; Török, B. *Org. Prep. Proc. Intl.* **2008**, *40*, 1.
31. Calloway, N. O. *J. Am. Chem. Soc.* **1937**, *59*, 1474, Brown H. C.; Junck, H. J. Am. Chem. Soc. **1955**, *77*, 5584. For selective monoalkylation of dihaloalkanes containing different halo groups, see Olah, G. A.; Kuhn, S. J. *J. Org. Chem.* **1964**, 39, 2317.
32. [R] Carey, F. A.; Sundberg R. J. *Advanced Organic Chemistry. Part B: Reaction and Synthesis* Plenum Press, New York, **1980**, p. 267.
33. (a) Eisenbraun, E. J.; Hinman, C. W.; Springer, J. M.; Burnham, J. W.; Chou, T. S.; Flanagan, P. W.; Hamming, M. C. *J. Org. Chem.* **1971**, *36*, 2480. [R] (b) For a review of the Haworth reaction sequence, see Agranaut, I.; Shih, Y. *J. Chem. Educ.* **1976**, 53, 488. (c) Jaxa-Chamiec, A.; Shah, V. P.; Kruse, L. I. *J. Chem. Soc. Perkin Trans. I* **1989**, 1705. (d) Coppi, L.; Ricci, A.; Taddai, M. *J. Org. Chem.* **1988**, *53*, 911. (e) Aggarwala, V. P.; Gopal, R.; Garg, S. P. *J. Org. Chem.* **1973**, *38*, 1247. (f) Fillion, E.; Fishlock, D. *Tetrahedron* **2009**, *65,* 6682. (g) Fillion, E.; Fishlock, D. Wilsilly, A. Golf, J. M. *J. Org. Chem.* **2005**, *70*, 1316. (h) Fillion E.; Fishlock, D. *Org. Lett.* **2003**, *5*, 4653. (i) Li, W.; Lai, H.; Ge, Z.; Ding, C.; Zhou, Y. *Synth. Commun.* **2007**, *37*, 1595.
34. (a) Brown, H. C.; Marino, G.; Stock, L. M.; *J. Am. Chem. Soc.* **1959**, *81*, 3310. (b) Brown, H. C.; Marino, G. *J. Am. Chem. Soc.* **1959**, *81*, 5610. (c) Olah G. A.; Moffatt, M. E.; Kuhn, S. J.; Hardie, B. A. *J. Am. Chem. Soc.* **1964**, *86*, 2198.
35. (a) Olah, G. A.; Kobayashi, S. *J. Am. Chem. Soc.* **1971**, *93*, 6964. (b) Teranishi, K.; Hayashi, S.; Nakatsuka, S.; Goto, T.; *Synthesis* **1995**, 506. [R] (c) For an excellent review on green homogeneous and heterogeneous catalysis describing advances, including the modern uses of supercritical CO_2 and ionic liquids to improve reactivity in F–C reactions, see Como, A.; Garcia, H. *Chem Rev.* **2003**, *103*, 4307.

36. Brauman, J. I.; Solladie-Cavallo, A. *J. Chem. Soc., Chem. Commun.* **1968**, 1124.
37. Brauman, J. I.; Pandell, A. J. *J. Am. Chem. Soc.* **1967**, *89*, 5421.
38. Masuda, S.; Nakajima, T.; Suga, S. *Bull. Chem. Soc. Jpn.* **1983**, *56*, 1089.
39. (a) Kronenthal, D. R.; Mueller, R. H.; Kuester, P. L.; Kissick, T. P.; Johnson, E. J. *Tetrahedron Lett.* **1990**, *31*, 1241. (b) For procedural details, see US Pat. 4,912,231.
40. Scholkopf, U.; Grutter, S.; Anderskewitz, R.; Egert, E.; Dyrbusch, M. *Angew Chem., Int. Ed. Engl.* **1987**, *26*, 683.
41. Matsumoto, T.; Maeta, M.; Suzuki, K.; Tsuchihashi, G. *Tetrahedron Lett.* **1989**, *30*, 833.
42. Kotsuki, H.; Hayashida, T.; Shimanouchi, T.; Nishizawa, H. *J. Org. Chem.* **1996**, *61*, 984.
43. Bandini, M.; Cozzi, P. G.; Melchiorre, P.; Umani-Ranchi, A. *J. Org. Chem.* **2002**, *67*, 5386.
44. Bandini, M.; Cozzi, P. G.; Melchiorre, P.; Umani-Ranchi, A. *Angew. Chem., Int. Ed.* **2004**, *43*, 84.
45. [R] (a) Y.Wang, K. Ding, L. Dai, *Chemtracts* **2001**, *14*, 610. [R] (b) Bandini, M.; Cozzi, P. G.; Melchiorre, P.; Umani-Ranchi, A. *Angew. Chem., Int. Ed.* **2004**, *43*, 550
46. [R] Poulson, T. B.; Jorgensen, K. A. *Chem. Rev.***2008**, *108*(8), 2903.
47. F. Bigi, G. Casiraghi, G. Casnati, G. Sartori, *J. Org. Chem.***1985**, *50*, 5018.
48. G. Erker, A. A. H. van der Zeijden; *Angew. Chem., Int. Ed. Engl.***1990**, *29*, 512.
49. A. Ishii,V. A. Soloshonok, K. Mikami, *J. Org. Chem.* **2000**, *65*, 1597.
50. See reference 13 of reference 46 cited within (*vide supra*) and for examples of homogeneous catalysis, see: (a) Yadav, J. S.; Subba Reddy, B. V.; Murthy, C. V. S. R.; Mahesh Kumar, G.; Madan, C. *Synthesis* **2001**, 783. (b) Hao, J.; Taktak, S.; Aikawa, K.; Yusa, Y.; Hatano, M.; Mikami, K. *Synlett* **2001**, 1443. (c) For examples of heterogeneous catalysis, see Ramesh, C.; Banerjee, J.; Pal, R.; Das, B. *Adv. Synth. Catal.* **2003**, 345, 557 .
51. Dong, H.-M.; Lu, H.-H.; Lu, L.-Q.; Chen, C.-B.; Xiao, W.-J. *Adv. Synth. Catal.* **2007**, *349*, 1597.
52. Gathergood, N.; Zhuang, W.; Jørgensen, K. A. *J. Am. Chem.Soc.* **2000**, *122*, 12517.
53. [R] (a) For a review of the use of chiral bisoxazoline-Lewis acid complexes in asymmetric synthesis, see Desimoni, G.; Faita, G.;Jørgensen, K. A. *Chem. Rev.* **2006**, *106*, 3561. [R] (b) Johnson, J. S.; Evans, D. A. *Acc. Chem. Res.* **2000**, *33*, 325..
54. Zhuang, W.; Gathergood, N.; Hazell, R. G.; Jørgensen, K. A. *J. Org. Chem.* **2001**, *66*, 1009.
55. [R] For recent reviews on catalysis by chiral Brønsted acids in asymmetric F–C reactions, see (a) Connon, S. J. *Angew. Chem., Int. Ed.* **2006**, *45*, 3909. (b) Akiyama, T.; Itoh, J.; Fuchibe, K. *Adv. Synth. Catal.* **2006**, *348*, 999. (c) Taylor, M. S.; Jacobsen, E. N. *Angew. Chem.* **2006**, *118*, 1550. (d) Taylor, M. S.; Jacobsen, E. N. *Angew. Chem., Int. Ed.* **2006**, *45*, 1520.
56. (a) Török, B.; Abid, M.; London, G.; Esquibel, J.; Török, M.; Mhadgut, S. C.; Yan, P.; Prakesh, G. K. S. *Angew. Chem., Int. Ed.* **2005**, *44*, 3086. (b) Török, B.; Abid, M.; London, G.; Esquibel, J.; Török, M.; Mhadgut, S. C.; Yan, P.; Prakesh, G. K. S. *Angew. Chem.* **2005**, *117*, 3146.
57. (a) Johannsen, M. *Chem. Commun.* **1999**, 2233. b) Saaby, S.; Fang, X.; Gathergood, N.; Jørgensen, K. A. *Angew. Chem. Int. Ed.* **2000**, *39*, 4114. (c) Saaby, S.; BayLn, P.; Aburel, P. S.; Jørgensen, K. A., *J. Org. Chem.* **2002**, *67*, 4352.
58. Terada, M.; Sorimachi, K.; Uraguchi, D. *J. Am. Chem. Soc.* **2004**, *126*, 11805.
59. Jia, Y.-X.; Xie, J.-H.; Duan, H.-F.; Wang, L.-X.; Zhou, Q.-L. *Org. Lett.* **2006**, *8*, 1621.
60. For a general review of chiral H-bond donars in asymmetric catalysis including Jacobsen type thiourea catalyzed reactions, see the following articles and references cited within [R] (a) Doyle, A. G.; Jacobsen, E. N. *Chem. Rev.* **2007**, *107*, 5713-5743. (b) See also references cited within Taylor, M. S.; Tokunaga, N.; Jacobsen, E. N. *Angew. Chem., Int. Ed.* **2005**, *44*, 6700.
61. Li, H.; Wang, Y.-Q.; Deng, L. *Org. Lett.* **2006**, *8*, 4063.
62. Terada, M.; Sorimachi, K. *J. Am. Chem. Soc.* **2007**, *129*, 292.
63. (a) Jia, Y.-X.; Zhong, J.; Zhu, S.-F.; Zhang, C.-M.; Zhou, Q.-L. *Angew. Chem., Int. Ed.* **2007**, *46*, 5565. (b) Jia, Y.-X.; Zhong, J.; Zhu, S.-F.; Zhang, C.-M.; Zhou, Q.-L. *Angew.Chem.* **2007**, *119*, 5661.
64. [R] (a) Whaley, W. M.; Govindachari, T. R. *Organic Reactions* Vol. 6, Wiley, New York, **1951**, p. 151. (b) Cox, E. D.; Cook, J. M. *Chem. Rev.* **1995**, *95*, 1797. (c) Pictet, A.; Spengler, T. *Ber. Dtsch. Chem. Ges.* **1911**, *44*, 2030. d) Tatsui, G. *J. Pharm. Soc. Jpn.* **1928,** *48*, 453. [R] (e) Cox, E. D.; Cook, J. M. *Chem. Rev.* **1995**, *95*, 1797. (f) Kaufmann, T. *New Methods*

in the Asymmetric Synthesis of Nitrogen Heterocycles Research SignPost, Trivandrum, India, **2005**, Ch. 4, 99.
65. Kawate, T.; Yamada, H.; Soe, T.; Nakagawa, M. *Tetrahedron: Asymmetry* **1996**, *7(5)*, 1249.
66. (a) Taylor, M. S.; Jacobsen, E. N. *J. Am. Chem. Soc.* **2004**, *126*, 10558. (b) Mergott, D. J.; Suend, S. L.; Jacobsen, E. N. *Org. Lett.* **2008**, *10*, 745.
67. Raheem, I. T.; Thiara, P. S.; Peterson, E. A.; Jacobsen, E. N. *J. Am. Chem. Soc.* **2007**, *129*, 13404.
68. Petersen, E.; A.; Jacobsen, E. N. *Angew. Chem., Int. Ed.* **2009**, *48*, 6328.
69. Raheem, I. T.; Thiara, P. S.; Jacobsen, E. N. *Org. Lett.* **2008**, *10*, 1577.
70. (a) Klausen, R. E.; Jacobsen, E. N. *Org.Lett.* **2009**, *11*, 887. (b) Gansesan, A.; Heathcock, C. H. *Tetrahedron Lett.* **1993**, *3*, 439.
71. Seayad, J.; Seayad, A. M.; List, B. *J. Am. Chem. Soc.* **2006**, *128*, 1086.
72. Wanner, M. J.; van der Haas, R. N. S.; de Cuba, K. R.; van Maarseveen, J. H.; Hiemstra, H. *Angew. Chem., Int. Ed.* **2007**, *46*, 7485.
73. Jensen, K. B.; Thorhauge, J.; Hazell, R. G.; Jørgensen, K. A. *Angew.Chem., Int. Ed.* **2001**, *40*, 160.
74. (a) Halland, N.; Velgaard, T.; Jørgensen, K. A. *J. Org. Chem.* **2003**, *68*, 5067. (b) Halland, T.; Hansen, T.; Jørgensen, K. A. *Angew. Chem., Int. Ed.* **2003**, *42*, 4955.
75. Zhuang, W.; Hansen, T.; Jørgensen, K. A. *Chem. Commun.* **2001**, 347.
76. (a) Zhou, J.; Tang, Y. *J. Am. Chem. Soc.* **2002**, *124*, 9030. (b) Zhou, J.; Ye, M.-C.; Huang, Z.; Tang, Y. *J. Org. Chem.* **2004**, *69*, 1309. (c) Zhou, J.; Tang, Y. *Chem. Commun.* **2004**, 432. (d) Ye, M.-C.; Li, B.;Zhou, J.; Sun, X.-L.; Tang, Y. *J. Org. Chem.* **2005**, *70*, 6108. (e) Similar results have just been reported by Fu using an optimized BOX catalyst bearing a β-aryl-substituted alkylidene linker on the ligand, see Sung, Y. J.; Li, N.; Zheng, Z.-B.; Liu, L.; Yu, Y. B.; Qin, Z.-H.; Fu, B. *Adv. Synth. Catal.* **2009**, *351*, 3113.
77. Evans, D. A.; Rovis, T.; Kozlowski, M. C.; Tedrow, J. S. *J. Am. Chem. Soc.* **1999,** *121,* 1994.
78. Rasappan, R.; Hager, M.; Gissibl, A.; Reiser, O. *Org. Lett.* **2006**, *8*, 6099.
79. (a) Bandini, M.; Melloni, A.; Tommasi, S.; Umani-Ronchi, A. *Helv. Chem. Acta* **2003**, *86*, 3753–3763. (b) Bandini, M.; Fagioli, M.; Melchiorre, A. M.; Umani-Ronchi, A. *Tetrahedron Lett.* **2003**, *44*, 5843–5846. (c) Bandini,M.; Fagioli, M.; Garavelli, M.; Melloni, A.; Trigari, V.; Umani-Ronchi, A. *J. Org. Chem.* **2004**, *69*, 7511–7518. (d) Herrera, R. P.; Sgarzani, V.; Bernardi, L.; Ricci, A. *Angew. Chem., Int. Ed.* **2005**, *44*, 6576.
80. Palomo, C.; Oiarbide, M.; Kardak, B. G.; Garcia, J. M.; Linden, A. *J. Am. Chem. Soc.* **2005**, *127*, 4154.
81. (a) Evans, D. A.; Scheidt, K. A.; Fandrick, K. R.; Lam, H. W.; Wu, J. *J. Am. Chem. Soc.* **2003**, *125*, 10780. (b) Evans, D. A.; Fandrick, K. R.; Song, H.-J. *J. Am. Chem. Soc.* **2005**, *127*, 8942. (c) Evans, D. A.; Fandrick, K. R. *Org. Lett.* **2006**, *8*, 2249. (d) Evans, D. A.;Fandrick, K. R.; Song, H.-J.; Scheidt, K. A.; Xu, R. *J. Am. Chem. Soc.* **2007**, *129*, 10029. (e) Evans has also shown this methodology to applicable to the preparation of substituted 2-substituted indoles; see reference 81d. (f) Boersma, A. J.; Feringa B. L.;Roelfes, G. *Angew. Chem., Int. Ed.* **2009**, *48*, 3346.
82. (a) Paras, N. A.; MacMillan, D. W. C. *J. Am. Chem. Soc.* **2001**, *123*, 4370. (b) Austin, J. F.; MacMillan, D. W. C. *J. Am. Chem. Soc.* **2002**, *124*, 1172. (c) Paras, N. A.; MacMillan, D. W. C. *J. Am. Chem. Soc.* **2002**, *124*, 7894. (d) Huang, Y.; Walji, A. M.; Larsen, C. H.; MacMillan, D. W. C. *J. Am.Chem. Soc.* **2005**, *127*, 15051. (e) Hong, L.;Wang, L.; Chen, C.; Zhang , B.; Wang , R. *Adv. Synth. Catal.* **2009**, *351*, 772.
(f) Austin, J. F.; Kim, S.-G.; Sinz, C. J.; Xiao, W.-J.; MacMillan, D. W. C. *Proc. Natl. Acad. Sci.* **2004**, *101*, 5482. (g) King, H. D.; Meng, Z.; Denhart, D.; Mattson, R.; Kimura, R.; Wu, D.; Gao, Q.; Macor, J. E. *Org. Lett.* **2005**, *7*, 3437. (h) Li, C.-F.; Liu, H.; Liao, J.; Cao, Y.-J.; Liu, X.-P.; Xiao, W.-J. *Org. Lett.* **2007**, *9*, 1847.
83. Li, C.-F.; Liu, H.; Liao, J.; Cao, Y.-J.; Liu, X.-P.; Xiao, W.-J. *Org.Lett.* **2007**, *9*, 1847.
84. (a) Blay, G.; Fernandez, I.; Pedro, J. R. *Adv. Synth. Catal.* **2009**, *351*, 2433. (b) Blay, G.; Fernandez, I.; Pedro, J. R.; Vila, C. *Org. Lett.* **2007**, *9*, 2601.
85. (a) Tang, H.-Y.; Lu, A.-D.; Zhou, Z.-H.; Zhao, G.-F.; He, L.-N.; Tang, C. C. *Eur. J. Org. Chem.* **2008**, 1406. (b) Chen, W.; Du, W.; Yue, L.; Li, R.; Wu, Y.; Ding, L.-S.; Chen, Y.-C. *Org. Biomol. Chem.* **2007**, *5*, 816. (c) Bartoli, G.; Bosco, M.; Carlone, A.; Pesciaoli, F.; Sambri, L.; Melchiorre, P. *Org. Lett.* **2007**, *9*, 1403. (d) Li, D.-P.; Guo, Y.-C.; Ding, Y.;

Xiao, W.-J. *Chem. Commun.* **2006**, 799. (e) Zhou, W.; Xu, L.-W.; Li, L.;Yang, L.;.Xia C.-G. Eur. *J. Org. Chem.* **2006**, 5225.

6.3 Houben–Hoesch Reaction

Richard J. Mullins and Matthew C. O'Reilly

6.3.1 Description

The electrophilic substitution of an activated nitrile onto an aromatic ring is known as the Houben–Hoesch reaction. The resulting imine is immediately hydrolyzed to yield the corresponding ketone. The reaction requires a Lewis acid and/or a protic acid, and results in arylketone products analogous to those which might be obtained under related Friedel–Crafts acylation conditions.

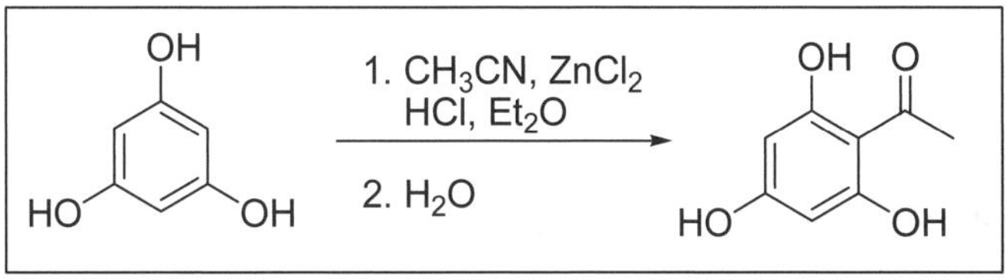

6.3.2 Historical Perspective

The Friedel–Crafts reaction, one of the more important and useful organic transformations, was discovered in 1877 by Charles Friedel and James Crafts. This class of reactions is generally thought to include all electrophilic alkylations and acylations of aromatic rings promoted by Lewis acids (traditionally $AlCl_3$ or $FeCl_3$). Due to their synthetic utility, Friedel–Crafts reactions have been extensively studied and used across a broad and diverse area of chemical research.

In 1898, Ludwig Gatterman reported the Lewis acid promoted reaction of hydrocyanic acid and benzene to produce aromatic aldehydes in a reaction that now bears his name.[1] The related electrophilic substitution of an activated nitrile onto an electron-rich aromatic ring was first reported in 1915 by the German chemist Kurt Hoesch,[2] who later served as the biographer of the legendary chemist Emil Fischer.[3] The original paper details the reaction of phloroglucinol with a number of alkyl and aryl nitriles, the products of which are still being used in modern synthetic efforts. In 1926, Josef Houben, following a stint in the military during which he served as head of the war laboratory[4] and extended and generalized the reaction discovered by Hoesch while at the Biologische Reichsanstalt in Berlin.[5] Known mostly for his contribution to the *Houben–Weyl Methods of Organic Chemistry* reference series, his work on the title reaction earned him the honor of having it named after him. For a more thorough discussion of the history of the Houben–Hoesch reaction, the reader is directed to an excellent review.[6]

6.3.3 Mechanism

The Houben–Hoesch reaction proceeds via a straightforward electrophilic aromatic substitution mechanism. After protonation or Lewis acid activation of the alkyl nitrile, nucleophilic attack by the electron-rich aromatic ring produces the resonance stabilized intermediate **1**. Elimination of H^+ regenerates the aromatic ring, resulting in the imine **2**, which is rapidly hydrolyzed to produce the aryl ketone **3**.[7]

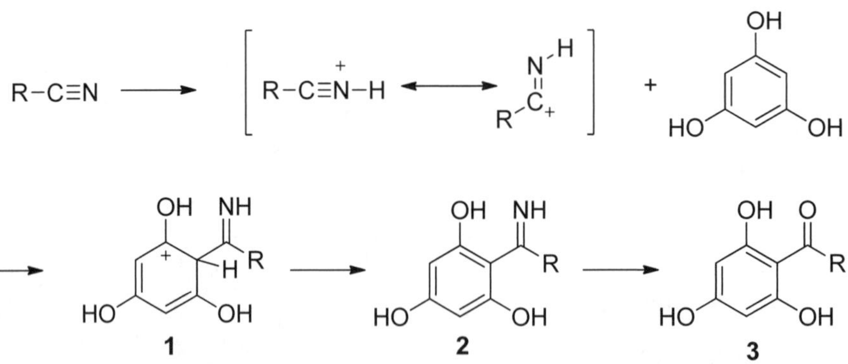

Although the mechanism above is considered generally correct, there have been multiple studies, both theoretical and experimental, which have focused on the more subtle aspects of the mechanism.[8–16] These studies focus mainly on the exact identity of the electrophilic species that is attacked in the first step. The structure of the reactive intermediate depends highly on the conditions utilized in the reaction as well as the particular aromatic species undergoing the transformation. For example, with an electron-rich species, it is generally thought that activation by Lewis acid or protic acid to produce the intermediate cation is sufficient to allow the reaction to occur.[16] An alternate species **5** is suggested to be the electrophile when a hydrated Lewis acid is used in the reaction.[16] Other authors have suggested the initial formation of a 1,3,5-triazinium salt, such as **4**, which then acts as the active electrophile when the Houben–Hoesch reaction is run using triflic acid as catalyst.[11] Finally, when benzene itself is used in the Houben–Hoesch, or the related Gattermann reaction, it is thought that the protonated nitrile is insufficiently electrophilic for the reaction to occur and that a dicationic species is in fact, the electrophilic species in the reaction.[12,13]

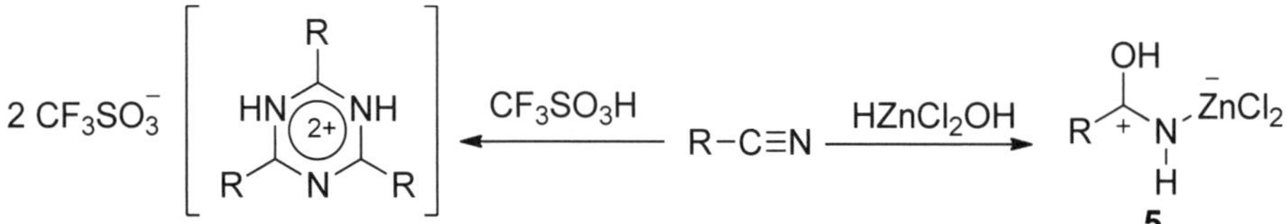

4

6.3.4 *Variations and Improvement*

2-Aminophenyl ketones have found utility as starting materials for the synthesis of 1,4-benzodiazepines, as well as several other important classes of drugs. Seeking an efficient and regioselective method for synthesis of this class of molecule, Sugasawa and co-workers used BCl_3 to direct the *ortho*-acylation of aniline, overcoming the typical difficulties associated with this particular Friedel–Crafts/Houben–Hoesch reaction.[17] Specifically, the high reactivity of anilines, as well as their propensity to direct electrophilic substitution at the *para*-position, are both remedied using this technique. As shown, precomplexation of *N*-methyl aniline (**6**) and BCl_3 results in **7**, which upon treatment with benzonitrile provides **9** after subsequent hydrolysis in high yield, gave exclusively as the *ortho*-regioisomer. The regioselectivity is rationalized as arising from the cyclic transition state **8**.[17–19] This reaction has found expanded utility in recent synthetic endeavors, using a variety of electrophilic coupling partners in addition to nitriles, and is generally known as the Sugasawa reaction.

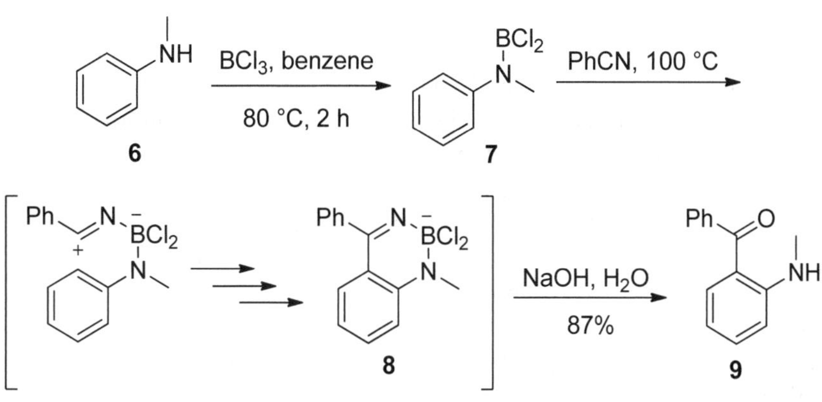

6.3.5 *Synthetic Utility*

For a complete description of synthetic utility of the Houben–Hoesch reaction before 1970 as well as some mechanistic discussions, the reader is directed to an excellent review on the subject.[6] The Houben–Hoesch reaction has been widely used for the synthesis of a number of interesting compounds.[20–25] Because of the relatively low electrophilicity of nitriles as compared to other carboxylic acid derivatives, the majority of useful Houben–Hoesch reactions are conducted with very electron-rich aromatic rings. In addition, due to the strong conditions required by the reaction, it is most often utilized in the early stages of a synthesis. As such, the utility of the Houben–Hoesch is not exclusively in the product of the reaction itself, but the manner in which the reaction is used to achieve the synthesis of a more complex target. For instance, a key step in the synthesis of flavonoid **12** is a Houben–Hoesch reaction.[26] In one of the more common applications of this reaction, upon treatment of phloroglucinol (**10**) and acetonitrile with $ZnCl_2$ and HCl, aryl ketone **11** is produced in good yield.

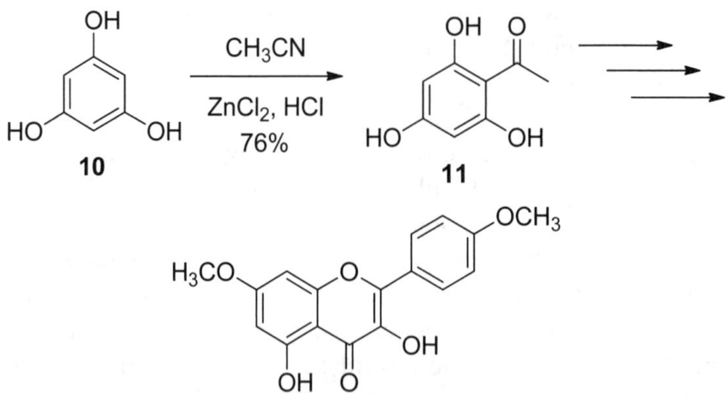

10 **11** **12**

Owing to the high nucleophilicity of phloroglucinol (**10**), it has been been widely used as a coupling partner in the Houben–Hoesch reaction.[27–38] An example of this is illustrated in Wandless and co-workers' synthesis of flavonol derivatives to be used as probes of biological processes.[33] After dissolving phloroglucinol (**10**) and benzoyloxyacetonitrile in diethyl ether, treatment with HCl and subsequent aqueous workup resulted in the preparation of **13** in high yield. Similar processes using the related electron-rich aromatics resorcinol and orcinol have been used in syntheses of modified flavonoids[39] and benzoxanthones.[40] The microwave-promoted Houben–Hoesch reaction in an ionic liquid has also recently been reported, using phloroglucinol as the aromatic species.[41]

10 **13**

The Houben–Hoesch reaction of phloroglucinol has been elegantly used by Rama Rao and co-workers in a concise synthesis of the chromanol moiety of the antiHIV agent, calanolide A.[42] As shown below, the chiral, non-racemic nitrile **14** is reacted with phloroglucinol (**10**) under standard Houben–Hoesch conditions to yield the ketone **15**. Although the yield was modest (25%), the efficiency of this one-pot process offers significant advantages over previous methods for making similar compounds.

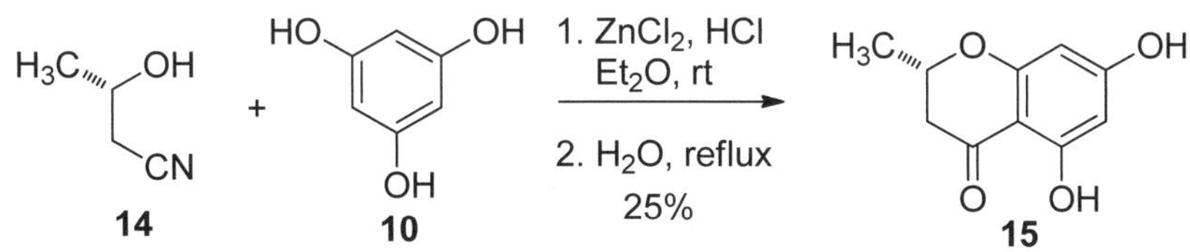

In the same work,[42] an intramolecular Houben–Hoesch was efficiently carried out to produce **16**. Notable in this instance is the improved yield of this ring-forming transformation.

As long as the nucleophilic partner is substituted with electron-donating substituents, the Houben–Hoesch reaction has broad scope. This is most directly illustrated by the work of Parmar and co-workers who used Houben–Hoesch conditions for the synthesis of a large number of benzyl phenyl ketones.[43]

The Houben–Hoesch reaction has found synthetic utility not only with substituted benzene derivatives but also with other π-excessive heterocycles. An impressive example is provided in efforts directed toward the synthesis of novel 2-[5-aroylpyrrolo]alkanoic acids, for evaluation of their potential analgesic and antiinflammatory activities.[44] Treatment of substituted pyrrole **17** and 3-cyanopyridine (**18**) with acid in dry chloroform resulted in the preparation of **19** in good yield.

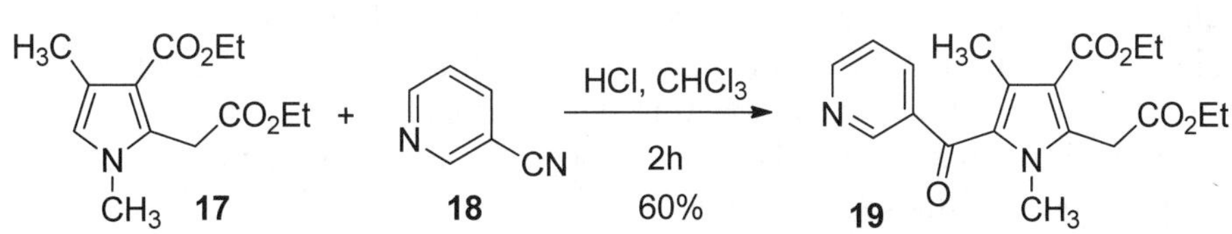

The Houben–Hoesch has also been used in reactions with indoles.[45,46] As a representative example, dimethoxyindole **20** is reacted with benzylcyanide under acidic conditions to produce the indole **21** which has been acylated at the 7-position.[47] The $SnCl_4$-mediated Houben–Hoesch acylation of indole at the 3-position has also been reported to proceed in high yield and with exclusive regioselectivity.[48]

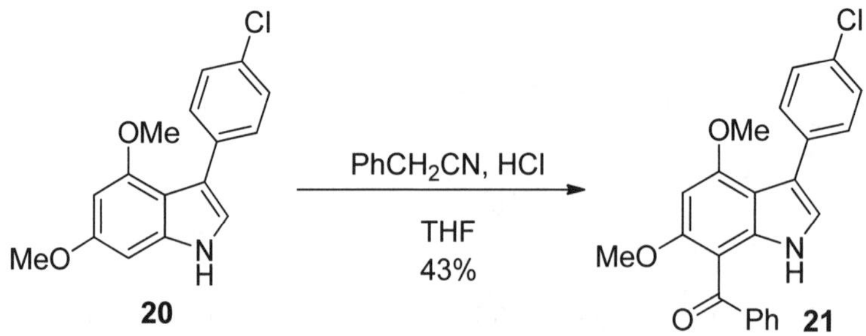

Issues of regioselectivity in the Houben–Hoesch and related Friedel–Crafts reactions have been studied extensively. As is common with a majority of electrophilic aromatic substitution reactions, substitution typically occurs *ortho* or *para* to electron-donating substituents, with issues of steric strain playing a role in the relative ratio of *ortho* and *para* products. In many of the Houben–Hoesch reactions discussed thus far, a single major regioisomer was produced, whether because of some particular electronic or steric effect, which directed substitution in that way or because the aromatic partner in the reaction was such that substitution at one or more positions would result in the same product. Although *para*-substituted products are typically easier to produce via these reactions, there has been substantial interest in achieving exclusive *ortho*-substitution, while maintaining high yields.

One of the first observations of competitive *ortho*-acylation in the Houben–Hoesch reaction came from the Johnston laboratories, where the reaction of phenol and acetonitrile with $AlCl_3$ resulted in approximately equivalent amounts of the *ortho*- and *para*-isomers of the acetophenone product. While the yield of both isomers was low, perhaps owing to the diminished reactivity of the aromatic ring possessing only one electron-donating substituent, a useful strategy for directing substitution to the *ortho*-position emerged from these studies.[49] Coordination of the Lewis acid with the phenolic oxygen and the nitrile nitrogen give complex **22**, which directs substitution to the *ortho*-position. It was this same rationale that has been so effectively applied in what has become known as the Sugasawa reaction for the *ortho*-acylation of anilines and phenols.[17–19,50]

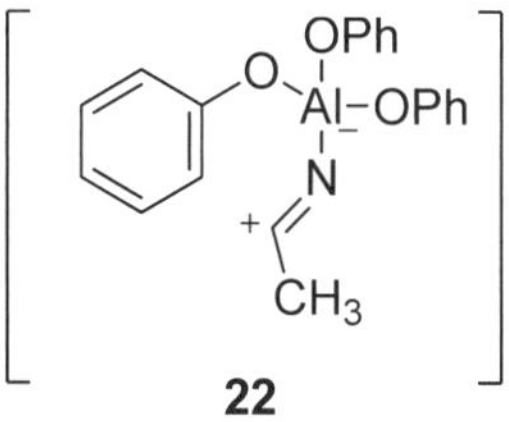

22

The Sugasawa modification has been effectively used in the synthesis of a large number of biologically interesting molecules.[51–53] In particular, the 2-aminophenyl ketone products of the Sugasawa reaction can be readily transformed into indoles via a method, which was also developed by Sugasawa and co-workers.[54–56] This method was impressively demonstrated in the total synthesis of (±)-dragmacidin.[57] Specifically, dibromoaniline **23** and chloroacetonitrile (**24**) were coupled using BCl_3 and $TiCl_4$ to produce ketone **25** in very high yield. Conversion to indole **26** was subsequently effected by treatment of **25** with $NaBH_4$ in refluxing dioxane. In related work, the synthesis of 2-substituted indoles featured a Sugasawa modified Houben–Hoesch reaction for preparation of the substituted 2′-amino-2-chloroacetophenone starting materials.[58,59]

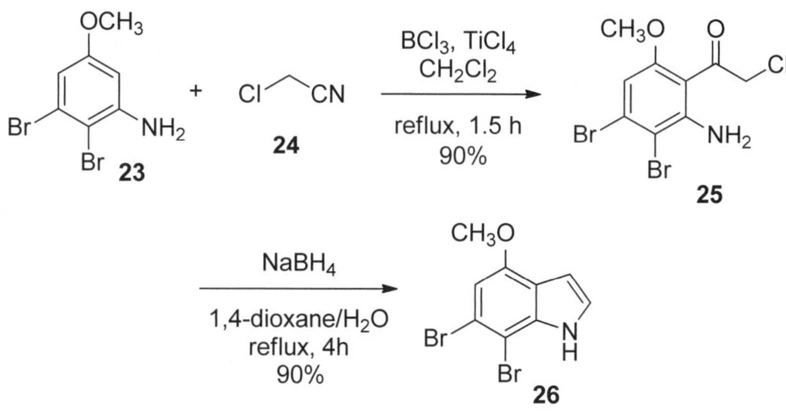

The Sugasawa modification has been used by Houpis and co-workers for the synthesis of the novel reverse transcriptase inhibitor **29**.[60] In this example, cyclopropyl nitrile was coupled with *p*-chloroaniline (**27**) using BCl_3 and $GaCl_3$ as the second Lewis acid. Following an acidic workup, ketone **28** was isolated in 74% yield, along with a small amount (~ 7%) of a product arising from cyclopropane opening.

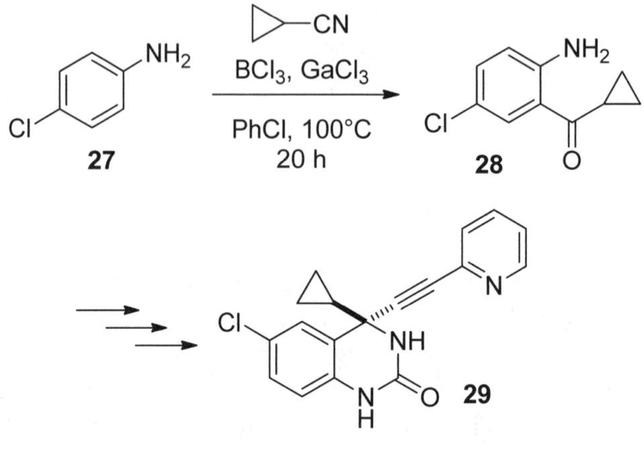

The use of $GaCl_3$ in lieu of the more standard Lewis acid $AlCl_3$, was a result of previous studies by Houpis and co-workers, which delineated the role of the second Lewis acid in the Sugasawa modification.[19] NMR studies suggest the presence of supercomplex **30** upon mixing of the respective aniline, nitrile, BCl_3 and a second Lewis acid, such as $GaCl_3$. The role of the second Lewis acid seems to be to stabilize, and therefore make more favorable, the formation of the supercomplex. As such, Lewis acids with higher affinities for chloride ion should shift the equilibrium to the supercomplex **30**, enabling more efficient *ortho*-acylation. In all cases screened, the use of $GaCl_3$ under milder conditions resulted in higher yields than $AlCl_3$.[19]

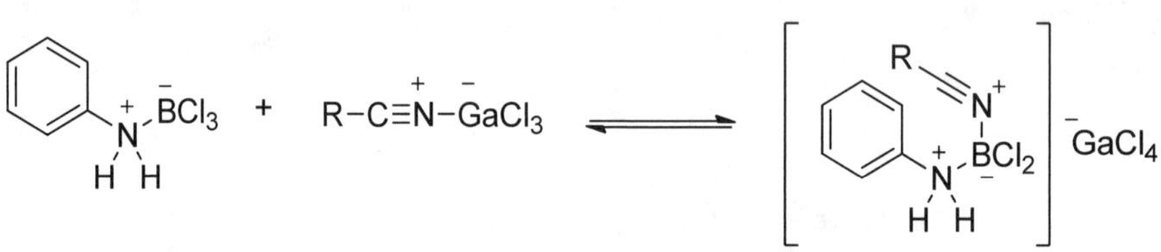

30

An alkylnitrilium variant of the Houben–Hoesch reaction was used in the synthesis of 11-hydroxy-*O*-methylsterigmatocystin (**34**).[61] The preparation of **34** was sought to investigate the role of cytochrome P-450 in the biosynthesis of aflatoxin, a widespread food contaminant and environmental carcinogen. As demonstrated below, nitrilium salt **31** is coupled with **32** to produce **33** in high yield.

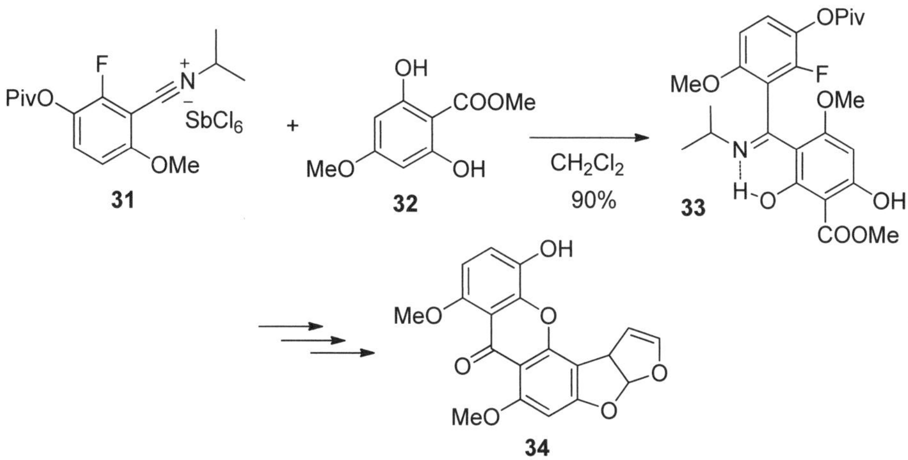

A similar and creative use of an in situ generated alkylnitrilium was demonstrated in the simple, one-pot synthesis of 2-benzazapine derivatives, of interest due to their presence in many pharmaceutically active agents.[62] Occurring by way of simultaneous Ritter and Houben–Hoesch reactions, the transformation proceeds to form new C–N and C–C bonds, respectively. Mechanistically, resonance-stabilized carbocation **36**, formed by protonation of the allylic alcohol in **35**, is attacked by propionitrile to produce nitrilium cation **37**. The Houben–Hoesch reaction then occurs with the nitrilium species to complete the formation of the benzazapine **38**, in high yield. The reaction has proven general for preparation of a number of 2-benzazepine derivatives in moderate to good yields.

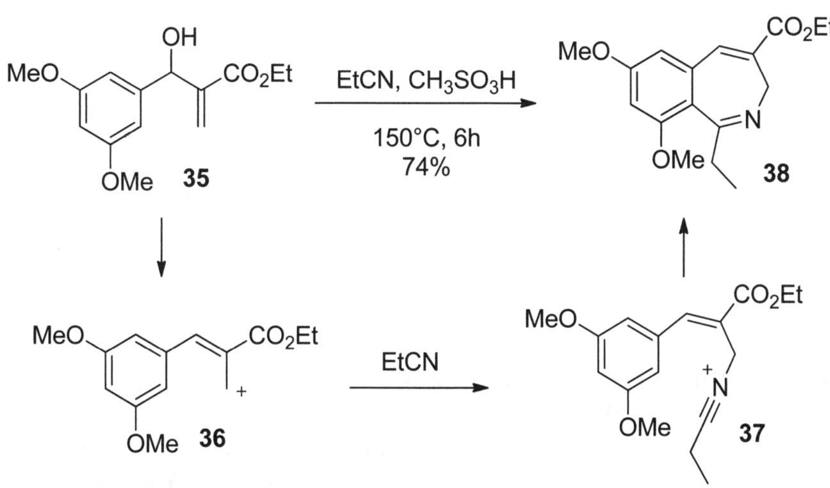

The intramolecular attack of a nitrilium ion is a key step in the Meerwein quinazoline synthesis.[63] A recent example of this reaction can be seen in the synthesis of 4-(*N,N*-dimethylamino)-2-arylquinazolines, as demonstrated below.[64] Treatment of imidoyl chloride **39** with *N,N*-dimethylcyanamide (**40**) followed by $TiCl_4$ results in the formation of intermediate **41**. Mechanistically, after the production of the likely intermediate nitrilium ion **41**, the Houben–Hoesch reaction occurs to give quinazoline **42**. A related procedure has been utilized for the synthesis of phenanthridines.[65]

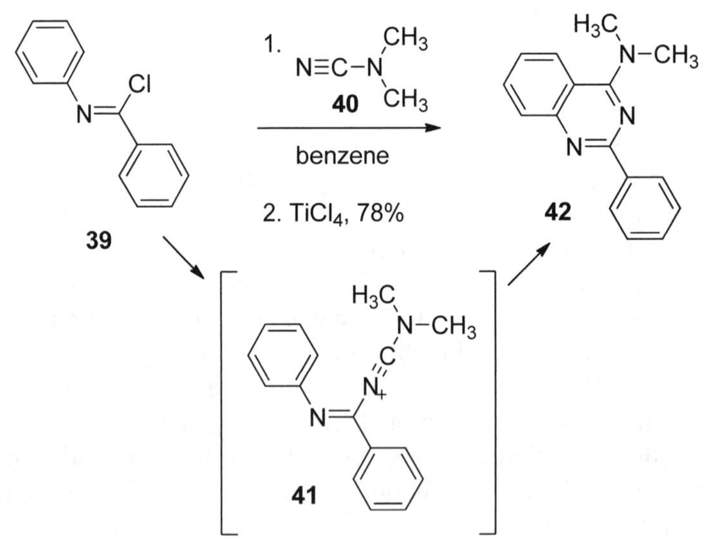

6.3.6 *Experimental*

Houben–Hoesch reaction

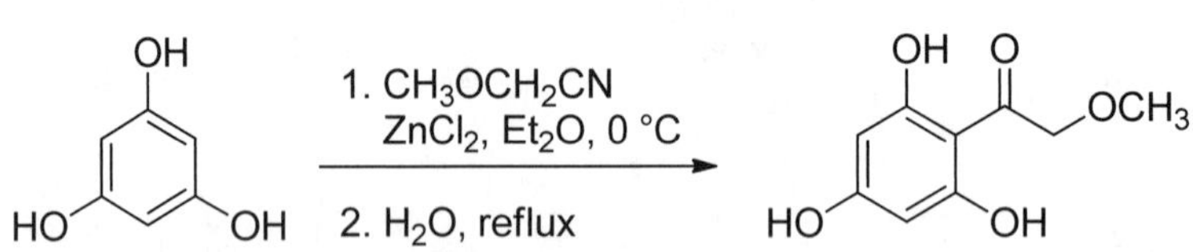

10 43

2′4′6′-Trihydroxy-2-methoxyacetophenone (43)[29]

Anhydrous phloroglucinol (**10**) (6.3 g, 50 mmol) was dissolved in dry ether (60 mL). Anhydrous $ZnCl_2$ (1.8 g, 13 mmol) and methoxyacetonitrile (3.7 mL, 50 mmol) were added. The flask was placed in an icebath. Anhydrous HCl gas was bubbled through the solution for 2 h under stirring. The ketimine hydrochloride which precipitated from the solution was filtered off, washed twice with a small amount of dry ether, dissolved in water and refluxed for 30 min. After cooling, a pink crystalline product precipitated which was recrystallized from water. Yield: 80%.

Sugasawa modification

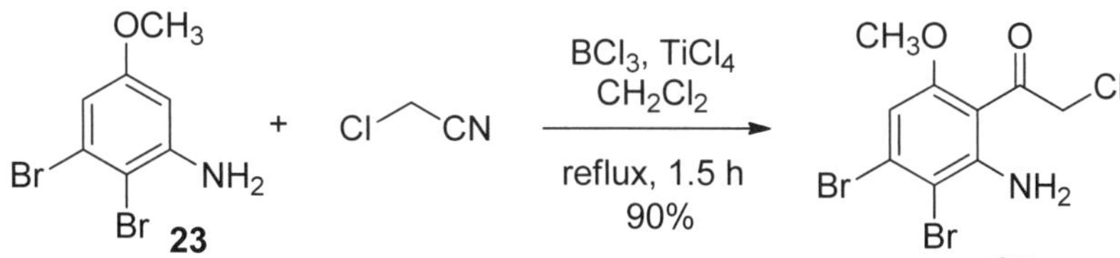

2-Amino-3,4-dibromo-6-methoxy-α-chloroacetophenone (25)[58]
To a stirred solution of **23** (32 g, 0.17 mol) in CH_2Cl_2 (300 mL) cooled in an ice bath was added dropwise successively, boron trichloride (1 M in CH_2Cl_2, 180 mL, 0.18 mol), chloroacetonitrile (14.3 g, 0.19 mol), and titanium tetrachloride (1 M in CH_2Cl_2, 190 ml, 0.19 mol). The resulting mixture was refluxed for 1.5 h. After being cooled to room temperature, the mixture was carefully poured into a mixture of ice and 20% HC1 (700 mL). The organic solvent was distilled. The residue was heated on a water bath (90 °C) for 30 min. After the solution was cooled to room temperature, the solid was filtered off and partitioned between ether (1.4 L) and 1 N NaOH (500 mL). The organic layer was separated and washed with brine, dried over Na_2SO_4, and concentrated. The resulting solid was recrystallized from ethanol to afford **25** (55 g) in 90% yield.

6.3.6 References

1. Gatterman, W.; Berchelmann, W. *Ber.* **1898**, *31*, 1765–1769.
2. Hoesch, K. *Ber.* **1915**, *48*, 1122–1133.
3. [R] Fischer, H. O. L. *Annu. Rev. Biochem.* **1960**, *29*, 1–14.
4. [R] Ronge, G. In *Neue Deutsche Biographie*, Dunker & Humblot, Berlin, **1972**, p. 659.
5. Houben, J. *Ber.* **1926**, *59*, 2878–2891.
6. [R] Ruske, W. In *Friedel–Crafts and Related Reactions*, Olah, G. A. ed.; Interscience, New York, **1964**, 383–497.
7. [R] Li, Jie J. *Name Reactions: A Collection of Detailed Reaction Mechanisms*, Springer-Verlag, Telos, N.Y., **2002**, p. 200.
8. Alagona, G.; Tomasi, J. *J. Mol. Struct.* **1983**, *91*, 263–288.
9. Jeffrey, E. A.; Satchell, D. P. N. *J. Chem. Soc. B.* **1966**, 579–586.
10. Fodor, G.; Nagubandi, S. *Tetrahedron* **1979**, *36*, 1279–1300.
11. Amer, M. I.; Booth, B. L.; Noori, G. F. M.; Proença, M. F. J. R. P. *J. Chem. Soc., Perkin Trans. 1* **1983**, 1075–1082.
12. Yato, M.; Ohwada, T.; Shudo, K. *J. Am. Chem. Soc.* **1991**, *113*, 691–692.
13. Sato, Y.; Yato, M.; Ohwada, T.; Saito S.; Shudo, K. *J. Am. Chem. Soc.* **1995**, *117*, 3037–3043.
14. Booth, B. L.; Noori, G. F. M. *J. Chem. Soc., Perkin Trans. 1* **1980**, 2894–2900.

15. Stang, P. J.; Anderson, A. G. *J. Am. Chem. Soc.* **1978**, *100*, 1520–1525.
16. Li, S.; Li, Y.; Zheng, H. *Eur. J. Mass Spectrom.* **2005**, *11*, 389–392.
17. Sugasawa, T.; Toyoda, T.; Adachi, M.; Sasakura, K. *J. Am. Chem. Soc.* **1978**, *100*, 4842–4852.
18. Toyoda, T.; Sasakura, K.; Sugasawa, T. *J. Org. Chem.* **1981**, *46*, 189–191.
19. Douglas, A. W.; Abramson, N. L.; Houpis, I. N.; Karady, S.; Molina, A.; Xavier, L. C.; Yasuda, N. *Tetrahedron Lett.* **1994**, *35*, 6807–6810.
20. Mák, M.; Nógrádi, M.; Szöllősy, A. *Tetrahedron* **2006**, *62*, 8425–8429.
21. Beck, G.; Bergmann, A.; Kebeler, A.; Wess, G. *Tetrahedron Lett.* **1990**, *31*, 7293–7296.
22. Shokol, T. V.; Turov, V. A.; Turov, A. V.; Krivokhizha, N. V.; Semenyuchenko, V. V.; Khilya, V. P. *Chem. Heterocycl. Compd.,* **2005**, *41*, 1411–1418.
23. Ahluwalia, V. K.; Kumar, D.; Gupta, M. C. *Indian J. Chem., Sect B* **1978**, *16B*, 574–578.
24. Cameron, D. W.; Deutscher, K. R.; Feutrill, G. I.; Hunt, D. E. *Aust. J. Chem.* **1982**, *35*, 1451–1461.
25. Tamiya, J.; Sorensen, E. J. *Tetrahedron* **2003**, 59, 6921–6932.
26. Chengmin, Z.; Fuchu, L.; Hongyou, Z. *Chinese Journal of Appl. Chem.* **1998**, *15*, 68–69.
27. Friščić, T.; Drab, D. M.; MacGillivray, L. R. *Org. Lett.* **2004**, *6*, 4647–4650.
28. Van der Schyf, C. J.; Dekker, T. G.; Fourie, T. G.; Snyckers, F. O. *Antimicrob. Agents Chemother.* **1986**, *30*, 375–381.
29. Boers, F.; Deng, B.-L.; Lemière, G.; Lepoivre, J.; De Groot, A.; Dommisse, R.; Vlietinck, A. *Arch. Pharm. Pharm. Med. Chem.* **1997**, *330*, 313–316.
30. Bass, R. J. *J. Chem. Soc., Chem. Comm.* **1976**, 78–79.
31. Mustafa, K. A.; Kjaergaard, H. G.; Perry, N. B.; Weavers, R. T. *Tetrahedron* **2003**, *59*, 6113–6120.
32. Zheng, X.; Meng, W.-D.; Qing, F.-L. *Tetrahedron Lett.* **2004**, *45*, 8083–8085.
33. Tanaka, H.; Stohlmeyer, M. M.; Wandless, T. J.; Taylor, L. P. *Tetrahedron Lett.* **2000**, *41*, 9735–9739.
34. Howells, H. P.; Little, J. G. *J. Am. Chem. Soc.* **1932**, *54*, 2451–2453.
35. Shriner, R. L.; Grosser, F. *J. Am. Chem. Soc.* **1942**, *64*, 382–384.
36. Urgaonkar, S.; Shaw, J. T. *J. Org. Chem.* **2007**, *72*, 4582–4585.
37. Hastings, J. M.; Hadden, M. K.; Blagg, B. S. J. *J. Org. Chem.* **2008**, *73*, 369–373.
38. Beney, C.; Mariotte, A.; Boumendjel, A. *Heterocycles* **2001**, *55*, 967–972.
39. Arkhipov, V. V.; Smirnov, M. N.; Khilya, V. P. *Chem. Heterocycl. Compd.* **1997**, *33*, 515–519.
40. Kjaer, D.; Kjaer, A.; Risbjerg, E. *J. Chem. Soc., Perkin Trans. 1* **1983**, 2815–2820.
41. Hakala, U.; Wähälä, K. *Tetrahedron Lett.* **2006**, *47*, 8375–8378.
42. Rama Rao, A. V.; Gaitonde, A. S.; Prakash, K. R. C.; Rao, S. P. *Tetrahedron Lett.* **1994**, *35*, 6347–6350.
43. Parmar, V. S.; Pati, H. N.; Azim, A.; Kumar, R.; Himanshu, K. S. B..; Prasad, A. K.; Errington, W. *Bioorg. Med. Chem.* **1998**, *6*, 109–118.
44. Chang, M. N.; Biftu, T.; Boulton, D. A.; Finke, P. E.; Hammond, M. L.; Pessolano, A. A.; Zambias, R. A.; Bailey, P.; Goldenberg, M.; Rackham, A. *Eur. J. Med. Chem, Chim. Ther.* **1986**, *21*, 363–369.
45. Albrecht, R.; Heindl, J.; Loge, O. *Eur. J. Med. Chem.* **1985**, *20*, 57–60.
46. Norcini, G.; Allievi, L.; Bertolini, G.; Casagrande, C.; Miragoli, G.; Santangelo, F.; Semeraro, C. *Eur. J. Med. Chem.* **1993**, *28*, 505–511.
47. Black, D. S.; Kumar, N.; Wahyuningsih, T. D. *ARKIVOC* **2008**, *6*, 42–51.
48. Ottoni, O.; Neder, A., Dias, A. K. B.; Cruz, R. P. A.; Aquino, L. B. *Org. Lett.* **2001**, *3*, 1005–1007.
49. Johnston, H. W. *J. Org. Chem.* **1959**, *25*, 454–455.
50. Bigi, F.; Maggi, R.; Sartori, G.; Casnati, G.; Bocelli, G. *Gazz. Chim. Ital.* **1992**, *122*, 283–289.
51. Nourry, A.; Legoupy, S.; Huet, F. *Tetrahedron* **2008**, *64*, 2241–2250.
52. Atechian, S.; Nock, N.; Norcross, R. D.; Ratni, H.; Thomas, A. W.; Verron, J.; Masciadri, R. *Tetrahedron* **2007**, *63*, 2811–2823.
53. Merlic, C. A.; Motamed, S.; Quinn, B. *J. Org. Chem.* **1995**, *60*, 3365–3369.
54. Sugasawa, T.; Adachi, M.; Sasakura, K.; Kitagawa, A. *J. Org. Chem.* **1979**, *44*, 578–586.

55. Wager, C. A. B.; Miller, S. A. *J. Label. Compd. Radiopharm.* **2006**, *49*, 615–622.
56. Nimtz, M.; Häfelinger, G. *Liebigs Ann. Chem.* **1987**, *9*, 765–770.
57. Jiang, B.; Smallheer, J. M.; Amaral-Ly, C.; Wuonola, M. A. *J. Org. Chem.* **1994**, *59*, 6823–6827.
58. Pei, T.; Chen, C.-y.; Dormer, P. G.; Davies, I. W. *Angew. Chem. Int. Ed.* **2008**, *47*, 4231–4233.
69. Pei, T.; Tellers, D. M.; Streckfuss, E. C.; Chen, C.-y.; Davies, I. W. *Tetrahedron* **2009**, *65*, 3285–3291.
60. Houpis, I. N.; Molina, A.; Douglas, A. W.; Xavier, L.; Lynch, J.; Volante, R. P.; Reider, P. J. *Tetrahedron Lett.* **1994**, *35*, 6811–6814.
61. Udwary, D. W.; Casillas, L. K.; Townsend, C. A. *J. Am. Chem. Soc.* **2002**, *124*, 5294–5303.
62. Basavaiah, D.; Satyanarayana, T. *Chem. Commun.* **2004**, 32–33.
63. Meerwein, H.; Laasch, P.; Mersch, R.; Netwig, J. *Chem. Ber.* **1956**, *89*, 224–238.
64. Zieliński, W.; Kudelko, A.; Holt, E. M. *Heterocycles* **1996**, *43*, 1201–1209.
65. Petterson, R. C.; Bennett, J. T.; Lankin, D. C.; Lin, G. W.; Mykytka, J. P.; Troendle, T. G. *J. Org. Chem.* **1974**, *39*, 1841–1845.

6.4 Kolbe–Schmitt Reaction

Martin E. Hayes

6.4.1 Description

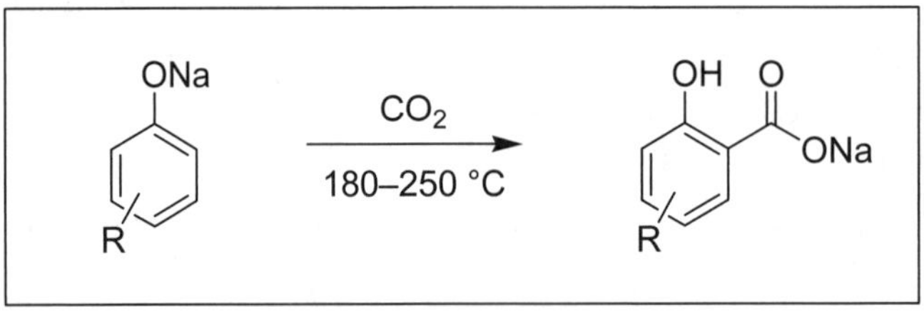

The Kolbe–Schmitt reaction is the carboxylation of phenolic salts, traditionally using carbon dioxide gas at elevated temperatures and pressures. Salicylic acids derived from *ortho*-carboxylation are most commonly obtained while the *para*-carboxylation and di-carboxylation products can also be prepared through judicious choice of counterions and reaction medium.[1]

6.4.2 Historical Perspective

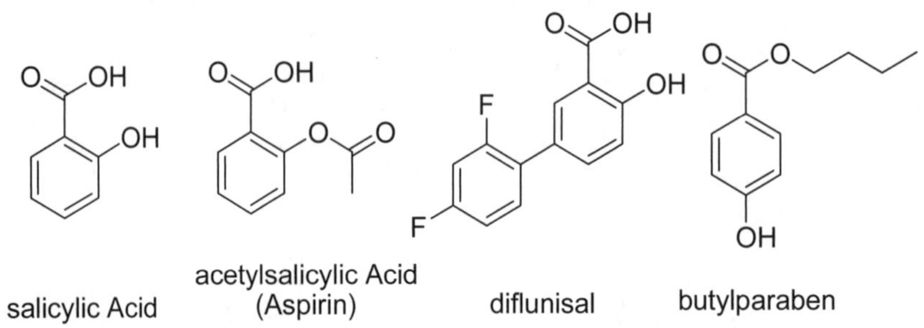

Kolbe was a student of Wöhler and later assistant to Bunsen,[2] and he is credited with coining the term *synthesis* after the first preparation of acetic acid from carbon disulfide.[3] He also made substantial contributions to the establishment of the modern theory of molecular structure during the mid-1800s, including predicting the existence of secondary and tertiary alcohols prior to their first preparations.[2] The Kolbe–Schmitt reaction, not to be confused with the Kolbe electrolysis reaction, remains the most direct method for preparing aromatic hydroxylacids and has been used extensively in the industrial synthesis of analgesic salicylates, including aspirin and diflunisal, and biocidic parabens, such as butylparaben. Rudolph Schmitt, a doctoral student of Kolbe's, introduced the use of a sealed pressure reactor, which allows for more reproducible and scalable yields of hydroxy acids.[3]

6.4.3 Mechanism

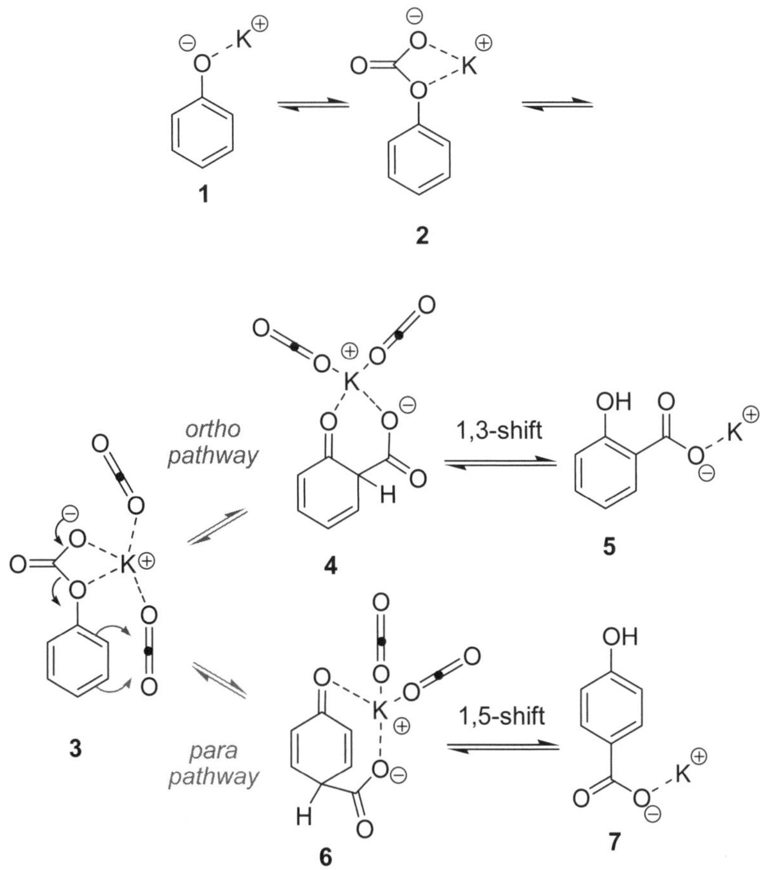

The mechanism has been extensively studied, including isotopic labeling,[4] spectroscopy,[5] and ab initio calculations.[6] There are a wide variety of conditions under which the Kolbe–Schmitt reaction can be executed along with several mechanistic subtleties,[7] including the reversibility of the reaction pathway.[8] However, the classical conditions[9] involving combination of gaseous carbon dioxide with the potassium salt of phenol at temperatures up to 250 °C are believed to proceed via the addition of phenoxide to carbon dioxide to give a phenylcarbonate salt **2**. This intermediate is then solvated by additional molecules of CO_2 at elevated pressure.[10] The solvated CO_2 is attacked by the π-system and, depending on the size of the counterion,[11] gives either *ortho*-substitution (**4**) or *para*-substitution (**6**). The intermediates **4** and **6**, can then tautomerize to the resulting salicylate (**5**) or paraben (**7**) salts.[12]

Alternative mechanisms, including the combination of two molar equivalents of phenoxide with one mole of carbon dioxide,[1] have been proposed, however spectroscopic and empirical observations along with ab initio calculations are consistent with the activation of CO_2 via a phenylcarbonate salt such as **2**. Recently developed modifications using super-critical CO_2[13] and ionic liquids,[14] however, may involve alternative mechanistic pathways as such reaction conditions have not yet been carefully examined.

6.4.4 Variations and Improvements

The original conditions[15] employed by Kolbe involved the formation of sodium phenoxide through evaporation of a molar equivalent mixture of phenol and aqueous sodium hydroxide. The hygroscopic sodium phenoxide is then heated at 180 °C while a stream of carbon dioxide is passed over the molten salt. The mixture is then heated at 220–250 °C to give the dianion of salicylic acid along with carbon dioxide and phenol, both of which distill away from the reaction mixture. Under these conditions, only a 50% theoretical yield of salicylic acid can be realized and often less is isolated.

The Schmitt modification[16] introduces a pressure reactor, typically an autoclave, whereby excess phenol and carbon dioxide do not escape and are efficiently converted to the hydroxy acid product. Using anhydrous sodium phenoxide, obtained by treatment of phenol with sodium metal, allows for a nearly quantitative isolation of salicylic acid after acidification. While the reaction is sensitive to water, it can be run in a variety of solvents, including aromatics, dioxane, pyridine, and DMF or under solvent-free conditions.[6]

A significantly more convenient method for the preparation of hydroxy acids was disclosed by Marasse[17] in which a phenol is combined with an excess of an anhydrous carbonate salt at elevated temperature and pressure. The reaction is limited to potassium, rubidium, and cesium carbonate, yet avoids the stepwise preparation of anhydrous phenoxide salts. Recently, aqueous conditions have also been disclosed, which obviate the need for anhydrous carbonate reagents in some cases.[18]

Methods using microreactors[19] and microwave heating[20] have been described, which provide for significant increases in throughput[21] for the preparation of simple aromatic hydroxy acids. Alternative reaction media such a supercritical CO_2 and ionic liquids have also been incorporated as both solvents and reagents, which, if recycled, may provide further efficiencies in the flow process.[22]

6.4.5 *Synthetic Utility*

General Utility

The regiochemical outcome of the reaction largely depends on the counterion[23] and reaction media[6] and has been extensively reviewed.[1] In general, smaller counterions, such as sodium and lithium, favor *ortho*-substitution of the carboxy group, while larger counterions, like potassium, preferentially lead to *para*-substitution.[24]

The reaction can be extended to naphthalene ring systems to prepare *o*-hydroxynapthanoic acids in good yields. For example, the potassium salt of 2-napthanol (**8**) can be converted to 2-hydroxy-1-napthanoic acid (**9**) in 52% yield at 50 °C under 725 psi CO_2.[4] At higher temperatures, usually above 200 °C, isomerization to give various mixtures of 2-hydroxy-3-napthanoic acid (10) and 2-hydroxy-6-napthanoic acid (**11**) is known to occur.[23,25]

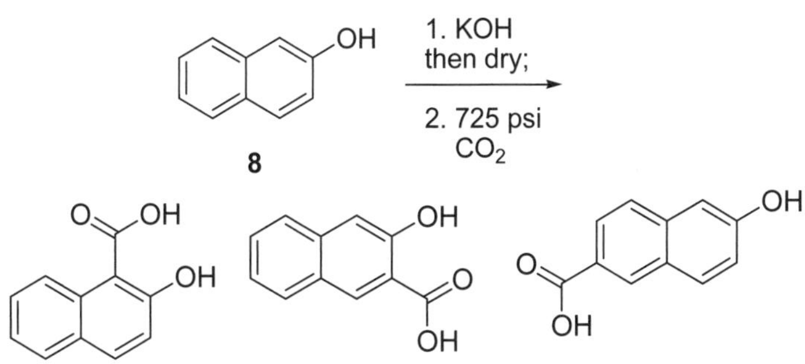

Temperature	**9**	**10**	**11**
20 °C	52%	0%	0%
230 °C	22%	33%	9%

The regiochemical outcome for substituted aromatics is also strongly influenced by field effects, with donating groups favoring *ortho*- and *para*-substitution relative to the hydroxyl. For example, 3-aminophenol provides *para*-amino salicylic acid (**13**) as the major product along with variable amounts of the corresponding isophthalic acid resulting from dicarboxylation.[26] Similarly electron-withdrawing groups inhibit the overall reaction rate, as in the case of 3-nitrophenol.[27] The rate of reaction is also subject to steric effects with alkyl substitutions *meta* to the hydroxyl acting as blocking groups for carboxyl substitution *ortho* to the hydroxyl.[27,28] The reaction is also sensitive to the presence of oxygen with competitive dimerization to give the corresponding biaryl under aerobic reaction conditions.[29]

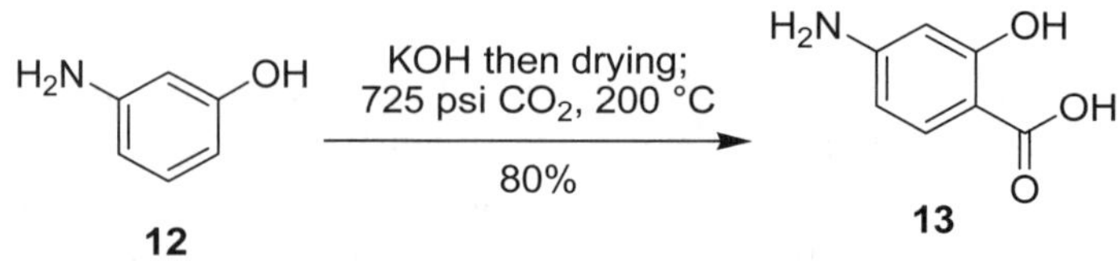

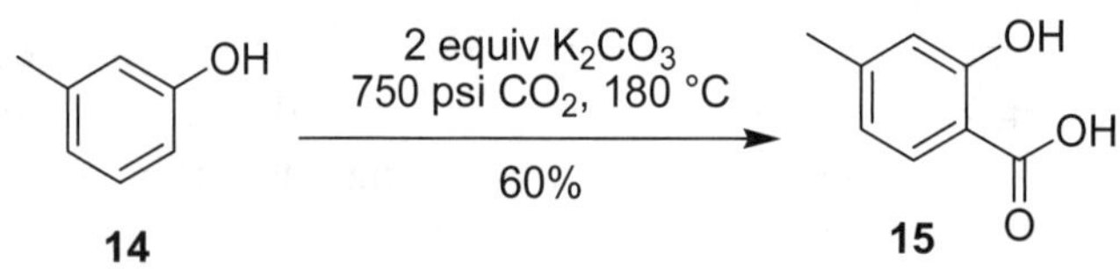

The reaction has been extended to a number of heterocycles, including hydroxy-pyridines where the 4-position relative to the ring nitrogen can be favored for carboxylation.[1] Also the direct addition of carbon dioxide has been reported for several azoles devoid of a hydroxyl group. The azole nitrogen presumably forms an intermediate carbamate salt, which is subject to ring carboxylation and tautomerization to give carboxyazoles, such as 2-pyrrolecarboxylic acid (**17**).[30] This has also been demonstrated on imidazoles and benzimidazoles where the 2-position is selectively carboxylated.[31] It is nteresting that the Kolbe–Schmitt reaction with indoles provides direct access to 3-carboxyindole (**19**),[32] while organolithium-mediated carboxylation affords the 2-carboxyindole (**20**),[33] exclusively.

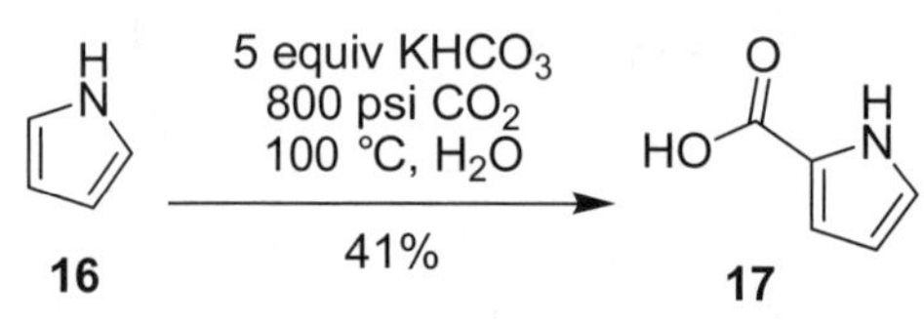

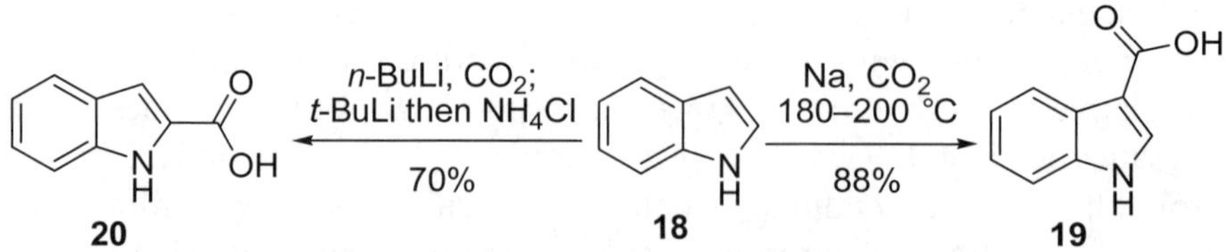

Applications in the total synthesis of natural products

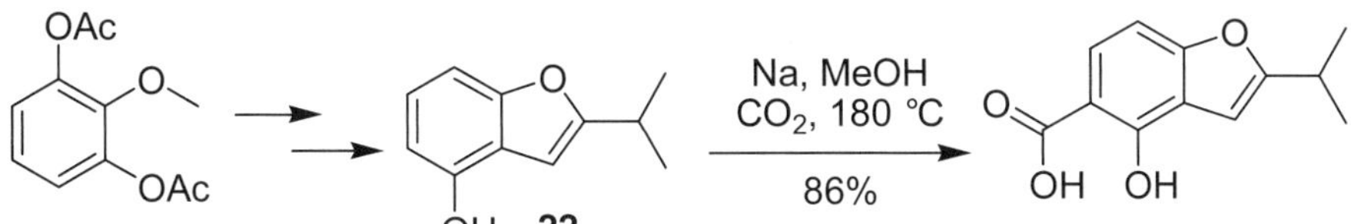

21 **22** **23**

The Kolbe–Schmitt reaction has been used as a key step in natural product synthesis. The early syntheses of isotubaic acid (Rotenic acid) used a Kolbe–Schmitt reaction in the final step.[34] Shriner and co-workers converted isotubanol (**22**) to its sodium salt under anhydrous conditions, then heated the phenoxide with solid carbon dioxide at 180 °C in a sealed reaction vessel to give an 86% yield of the natural product.[35] Sheehan and co-workers also used the Kolbe–Schmitt reaction to prepare an early intermediate in the synthesis of gossypol where thymol is efficiently converted to *o*-thymotic acid in 65% yield.[36]

Applications in the synthesis of pharmaceuticals

The Kolbe–Schmitt reaction made possible the first industrial synthesis of salicylic acid and then acetyl salicylic acid, as two of the most important and widely used analgesics in history.[3] Schmitt is credited with establishing a scalable protocol for salicylic acid manufacture, which provided a cornerstone for the burgeoning German fine chemicals industry in the late 19th century.

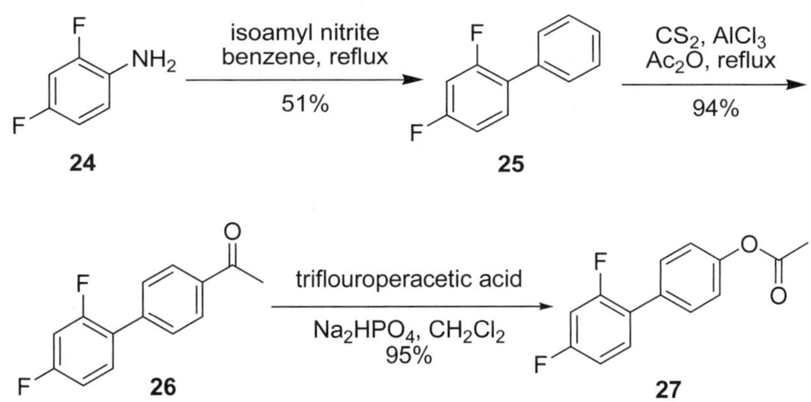

24 **25** **26** **27**

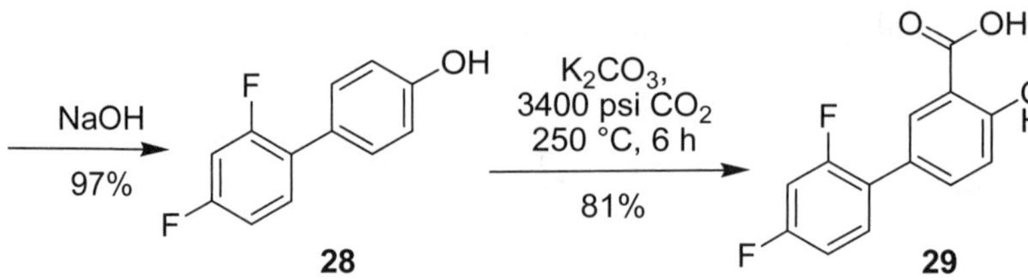

Extensive research by T. Y. Shen and co-workers around salicylic acid analogs[37,38] at Merck, Sharp & Dohme led to more potent and longer lasting salicylates, such as Diflunisal. The commercial synthesis of Diflunisal involves a Marasse modification of the Kolbe–Schmitt reaction using biaryl phenol **28** at 250 °C with 3400 psi CO_2 for 6 h to give **29** in 81% yield. The Kolbe–Schmitt reaction has also been used to prepare HIV integrase inhibitors[39] where 8-hydroxyquinaldine (**30**) is converted to its corresponding quinalidic acid (**31**) in the course of cellular structure-activity-relationship studies.

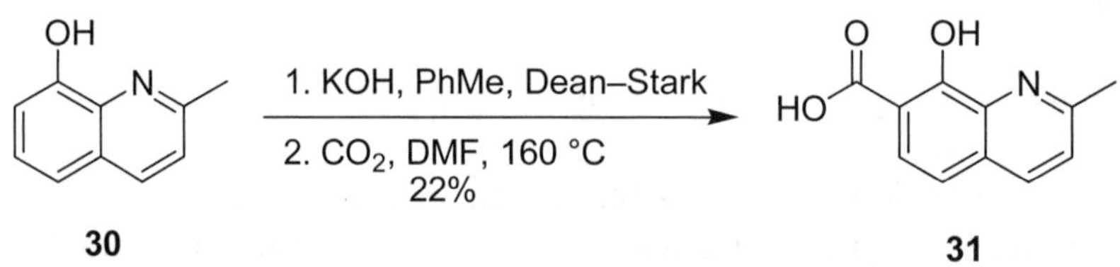

The use of alkyl parabens as biocides in food, drug, and cosmetics manufacture is widespread, despite controversial claims[40] of links to certain cancers.[41] Industrial syntheses of paraben esters use the Kolbe–Schmitt reaction, principally via the corresponding potassium phenoxide, which favors *para*-carboxylation. A modification that strongly favors *para*-subtitution involves potassium ethylcarbonate as a base, which gives a 93% yield of 4-hydroxybenzoic acid (**33**) from phenol at 225 °C under an atmosphere of 25 atm CO_2.[42] Additional protocols using carboxylase enzymes under an atmosphere of CO_2[43] and with supercritical CO_2[44] have been reported to furnish *para*-carboxylation exclusively and in high yields.

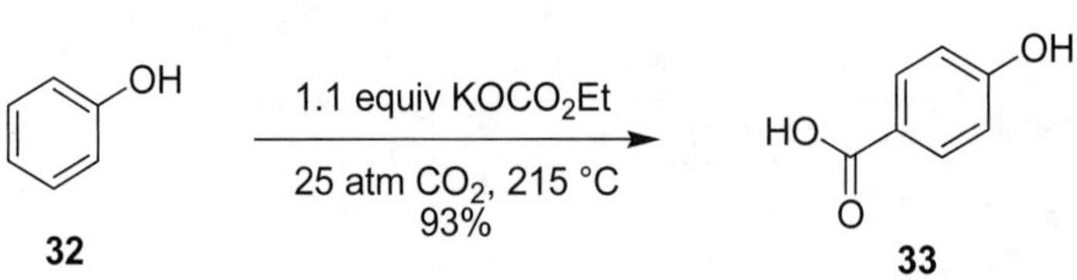

6.4.6 *Experimental*

Preparation of 8-Hydroxy-7-quinaldic Acid (31)[39]

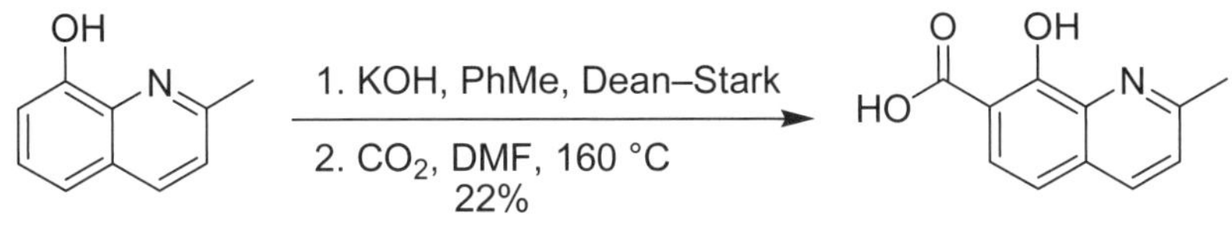

To a suspension of 8-hydroxyquinaldine (**30**, 29 g, 0.18 mol) in toluene (130 mL) was added potassium hydroxide (11.3 g, 0.20 mol). The stirred mixture was heated under reflux for 24 h, collecting the water of the reaction in a Dean–Stark trap. After the mixture cooled at 20 °C, DMF (100 mL) was added, and the Dean–Stark trap was replaced with a distillation column. The reaction mixture was progressively heated until most of the toluene had been distilled. When the temperature reached 140 °C, a stream of CO_2 was passed into the solution and continued throughout the reaction. The distillation of the solvents was continued while the temperature was gradually raised to 160 °C. The reaction mixture was heated for 2 h at this temperature and then cooled to 20 °C. The stream of CO_2 was stopped, and water (250 mL) was added. The solution was acidified to pH = 7 with concentrated HCl and the mixture was extracted with ethyl acetate. The aqueous phase was acidified to pH = 4.2, and the precipitate was filtered, washed with water, and dried in vacuo. The crude acid was recrystallized in 2-propanol to give acid **31** (8.0 g, 22%) as yellow crystals: m.p. 206–208 °C.

Preparation of *para*-hydroxybenzoic acid (33)[42]

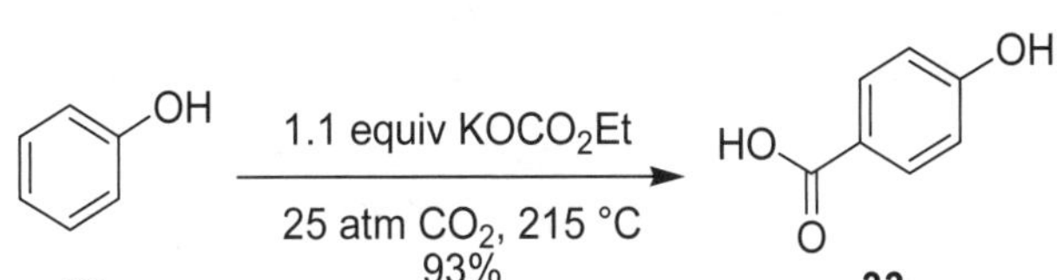

A glass reactor placed in a steel autoclave was charged with phenol (2.35 g, 0.025 mol) and potassium ethyl carbonate (3.46 g, 0.027 mol). The autoclave was closed, and carbon dioxide was fed to a pressure of 25 atm in two steps. The reaction mixture was stirred at this pressure, and its temperature was raised for 6 h to 215 °C (heating rate 32 °C/h) and then kept at this temperature for 1 h. After cooling to room temperature, the reaction mixture was treated with water. The aqueous solution was treated with toluene to remove unreacted phenol and then acidified to give *para*-hydroxybenzoic acid (3.21 g, 93% yield), m.p. = 214–216 °C.

6.4.7 References

1. [R] Lindsey, A. S.; Jeskey, H. *Chem. Rev.* **1957**, *57*, 583–620.
2. [R] Meyer, E. V.; McGowan, G. *A History of Chemistry from Earliest Times to the Present Day Being also an Introduction to the Study of the Science*; MacMillan, London, 1906, pp. 323–332.
3. [R] Rocke, A. J. *The Quiet Revolution: Hermann Kolbe and the Science of Organic Chemistry*, University of California Press, Berkeley, 1993, pp. 291–296, 304–309.
4. Kosugi, Y.; Imaoka, Y.; Gotoh, F.; Rahim, M.; Matsui, Y.; Sakanishi, K. *Org. Biomol. Chem.* **2003**, *1*, 817–821.
5. Hales, J. L.; Jones, J.; Lindsey, A. *J. Chem. Soc.* **1954**, 3145–3151.
6. Stanescu, I.; Achenie, L.E. *Chem. Eng. Sci.* **2006**, *61*, 6199–6212.
7. Kosugi, Y.; Takahashi, K.; Imaoka, Y. *J. Chem. Res. (S)* **1999**, 114–115.
8. Rahim, M. A.; Matsui, Y.; Matsuyama, T.; Kosugi, Y. *Bull. Chem. Soc. Jpn.* **2003**, *76*, 2191–2195.
9. Kolbe, H. *J. Prakt. Chem.* **1874**, *10*, 89–112.
10. Marković, Z.; Marković, S.; Manojlović, N.; Predojević-Simović, J. *J. Chem. Inf. Model.* **2007**, *47*, 1520–1525.
11. Marković, S.; Marković, Z.; Begović, N.; Manojlović, N. *Russ. J. Phys. Chem. A.* **2007**, *81*, 1392–1397.
12. Marković, Z.; Marković, S. *J. Chem. Inf. Model.* **2008**, *48*, 143–147.
13. Iijima, T.; Yamaguchi, T. *Tetrahedron Lett.* **2007**, *48*, 5309–5311.
14. Stark, A.; Huebschmann, S.; Sellin, M.; Kralisch, D.; Trotzki, R.; Onuschka, B. *Chem. Eng. Tech.* **2009**, *32*, 1730–1738.
15. Kolbe, H.; Lautemann, E. *Justus Liebigs Ann. Chem.* **1860**, *115*, 157–206.
16. Schmitt, R. *J. Prakt. Chem.* **1885**, *31*, 397–411.
17. Baine, O.; Adamson, G. F.; Barton, J. W.; Fitch, J. L.; Swayampati, D. R.; Jeskey, H. *J. Org. Chem.* **1954**, *19*, 510–514.
18. Morinaga, N.; Uchigashima, M.; Rahim, M. A.; Onishi, K.; Takahashi, K.; Kosugi, Y. *J. Chem. Soc. Jpn.* **2002**, 467.
19. Hessel, V.; Hofmann, C.; Lob, P.; Lohndorf, J.; Lowe, H.; Ziogas, A. *Org. Proc. Res. Dev.* **2005**, *9*, 479–489.
20. Benaskar, F.; Hessel, V.; Krtschil, U.; Löb, P.; Stark, A. *Org. Proc. Res. Dev.* **2009**, *13*, 970–982.
21. Krtschil, U.; Hessel, V.; Reinhard, D.; Stark, A. *Chem. Eng. Tech.* **2009**, *32*, 1774–1789.
22. Huebschmann, S.; Kralisch, D.; Hessel, V.; Krtschil, U.; Kompter, C. *Chem. Eng. Tech.* **2009**, *32*, 1757–1765.
23. Rahim, M. A.; Matsui, Y.; Kosugi, Y. *Bull. Chem. Soc. Jpn.* **2002**, *75*, 619–622.
24. Marković, Z.; Marković, S.; Begović, N. *J. Chem. Inf. Model* **2006**, *46*, 1957–1964.
25. Marković, Z.; Marković, S.; Đurović, I. *Monatsh. Chem.* **2008**, *139*, 329–335.
26. Doub, L.; Schaeffer, J. A.; Stevenson, O. L.; Walker, C. T.; Vanderbelt, J. M. *J. Org. Chem.* **1958**, *23*, 1422–1424.
27. Wessely, F.; Benedikt, K.; Benger, H.; Prillinger, G. F. *Monatsh. Chem.* **1950**, *81*, 1071–1091.
28. De, A. U.; Saha, B. P. *J. Med. Chem.* **1971**, *14*, 265–266.
29. Chidambaram, M.; Sorenson, J. *J. Pharm. Sci.* **1991**, *80*, 810–811.
30. Scott, J. W.; Focella, A.; Hengartner, U. O.; Parrish, D. R.; Valentine, D., Jr. *Syn. Comm.* **1980**, *10*, 529–540.
31. Kempe, U.; Dockner, T.; Koehler, H. Patent Application EP0328955, 1989-08-23.
32. Binder, H.; Koch, O. Pat. Appl. DE1193946, 1965-06-03.
33. Katritzky, A. R.; Akutagawa, K. *Tetrahedron Lett.* **1985**, *26*, 5935–5938.
34. Reichstein, T.; Hirt, R. *Helv. Chim. Acta.* **1933**, *16*, 121–129.

35. Shriner, R. L.; Witte, M. *J. Am. Chem. Soc.* **1941**, *63*, 1108–1110.
36. Shirley, D. A.; Brody, S. S.; Sheehan, W. C. *J. Org. Chem.* **1957**, *22*, 495–497.
37. Jones, H.; Fordice, M. W.; Greenwald, R. B.; Hannah, J.; Jacobs, A.; Ruyle, W. V.; Walford, G. L.; Shen, T. Y. *J. Med. Chem.* **1978**, *21*, 1100–1104.
38. Hannah, J.; Ruyle, W. V.; Jones, H.; Matzuk, A. R.; Kelly, K. W.; Witzel, B. E.; Holtz, W. J.; Houser, R. A.; Shen, T. Y. *J. Med. Chem.* **1978**, *21*, 1093–1100.
39. Mekouar, K.; Mouscadet, J. F.; Desmaele, D.; Subra, F.; Leh, H.; Savoure, D.; Auclair, C.; d'Angelo, J. *J. Med. Chem.* **1998**, *41*, 2846–2857.
40. Darbre, P.; Harvey, P. *J. Appl. Toxicol.* **2008**, *28*, 561–578.
41. Anderson, F. A. *Int. J. Toxicol.* **2008**, *27 Suppl. 4,* 1–82.
42. Suerbaev, K. A.; Akhmetova, G. B.; Shalmagambetov, K. M. *Russ. J. Gen. Chem.* **2005**, *75*, 1498–1499.
43. Aresta, M.; Quaranta, E.; Liberio, R.; Dileo, C.; Tommasi, I. *Tetrahedron* **1998**, *54*, 8841–8846.
44. Dibenedetto, A.; Noce, R. L.; Pastore, C.; Aresta, M.; Fragale, C. *Envir. Chem. Lett.* **2006**, *3*, 145–148.

6.5 Vilsmeier–Haack Reaction

Brian Goess

6.5.1 Description

When a dialkyl amide is treated with phosphorus oxychloride a Vilsmeier–Haack reagent (hereafter Vilsmeier reagent) is formed. Subsequent treatment with an electron-rich π system leads to substitution and formation of an intermediate iminium ion, which can be hydrolyzed to yield an aldehyde. The overall process is known as a Vilsmeier–Haack reaction (hereafter Vilsmeier reaction).[1]

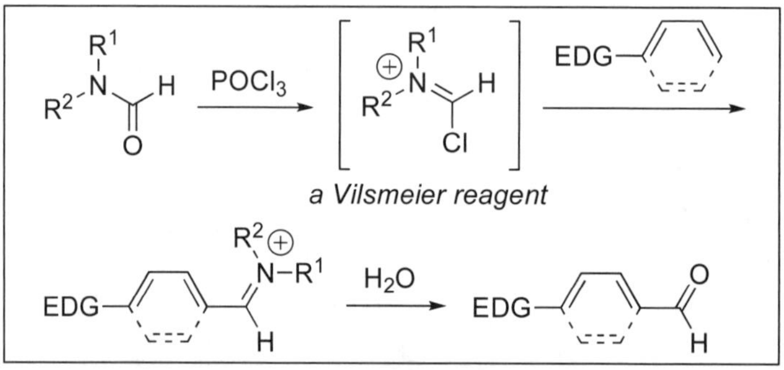

6.5.2 Historical Perspective

As early as 1902 it was recognized that the combination of formanilide and phosphoryl chloride will formylate 1,3-dihydroxybenzene.[2] Though the scope of the reaction was limited, this seminal result foreshadowed the 1925 discovery by Vilsmeier and colleagues that the reaction of *N*-methylacetanilide and phosphoryl chloride yielded, among other things, 4-chloro-1,2-dimethylquinolinium chloride.[3]

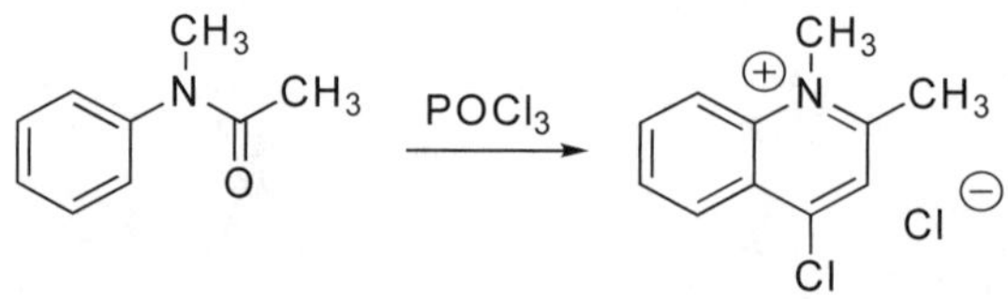

Subsequent investigations indicated that a key intermediate in this reaction was an iminium salt that reacts with aromatic compounds, such as *N*,*N*-dimethylaniline, to yield products of aromatic substitution.[4] At that time there was little definitive evidence for the identity of X in the following scheme.

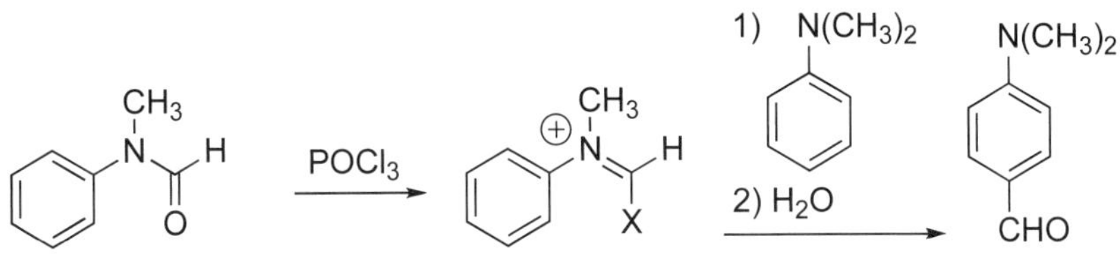

It was then quickly discovered that a number of acid chlorides, including thionyl chloride, oxalyl chloride, phosgene, and carbonyl chloride were capable of reacting with a *N,N*-dialkylformamides to form a weakly electrophilic chloromethyliminium salt, now known as a Vilsmeier reagent. These salts can be isolated before introduction of a nucleophile and have been found to react well with electron-rich aromatic and heteroaromatic compounds,[1c] as well as certain types of alkenes.[1b] The initially formed iminium intermediate is usually hydrolyzed to produce an aldehyde. In modern synthesis applications, the most common reagents used for the formation of Vilsmeier reagents are DMF and $POCl_3$.

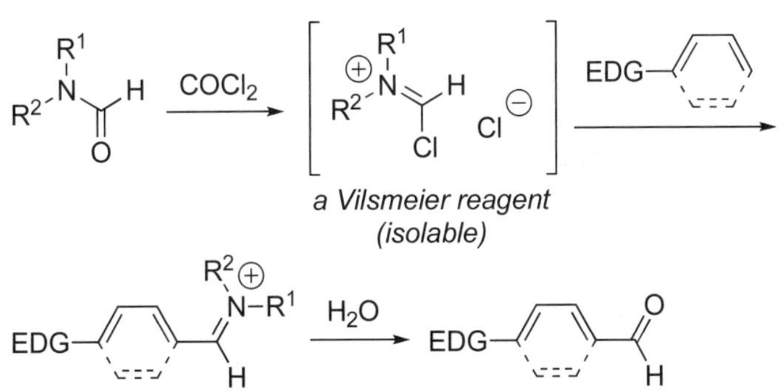

Much of the research on Vilsmeier reactions in the past 50 years has focused on expanding the scope of this transformation. Whereas the reaction was once limited to electron-rich aromatics and heteroaromatics, aliphatic substrates have increasingly been found to react with Vilsmeier reagents. Furthermore, alternate transformations of the iminium intermediate to form products with functional groups other than aldehyde have been developed. Finally, access to a diverse range of heterocycles is now possible due to discovery of substrate classes that are capable of undergoing intramolecular annulation reactions on the iminium intermediate.

6.5.3 Mechanism

When DMF is treated with carbonyl chloride, a sequence of two simple addition-elimination reactions takes place to yield a Vilsmeier reagent.

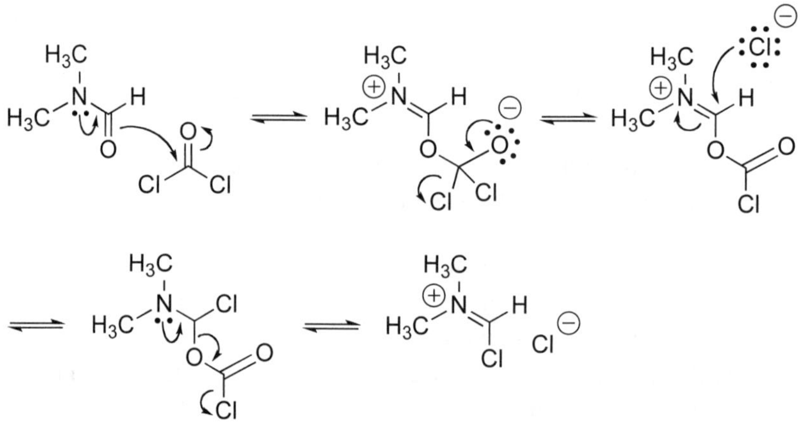

When DMF is treated with $POCl_3$, a more complex equilibrium mixture of iminium salts of varying electrophilicities is produced. A similar sequence of addition-elimination reactions initially takes place to yield iminium salts **1** and **2**, which can react further with DMF to produce **3** and **4**.[5] Although each salt below is capable of undergoing a subsequent acylation reaction, salt **2** is generally considered to be the active Vilsmeier reagent.

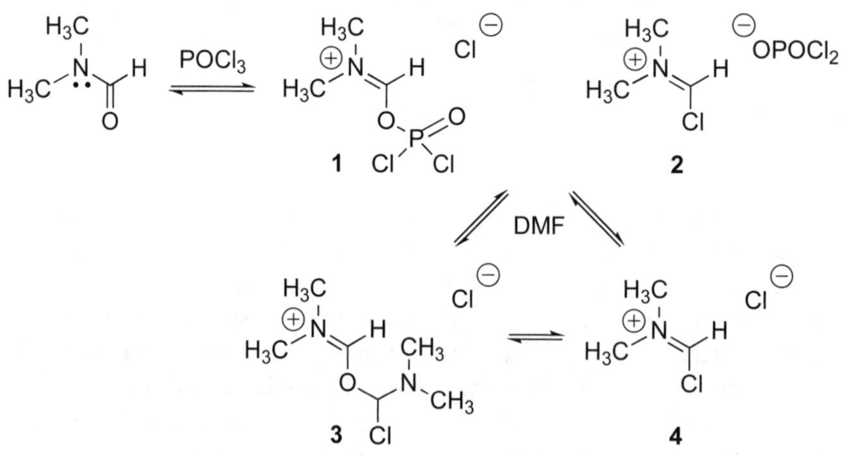

Upon introduction of an aromatic nucleophile, an electrophilic aromatic substitution takes place to yield an intermediate iminium ion (**5**), often with good regioselectivity in favor of the isomer with less steric strain.[6]

5 then undergoes a final addition-elimination reaction with a nucleophile, usually water, to yield the product aldehyde (**6**). Aqueous workup conditions vary considerably but are usually basic. Whenever an imine is hydrolyzed in the following examples, water will be listed as the reagent in a separate step.

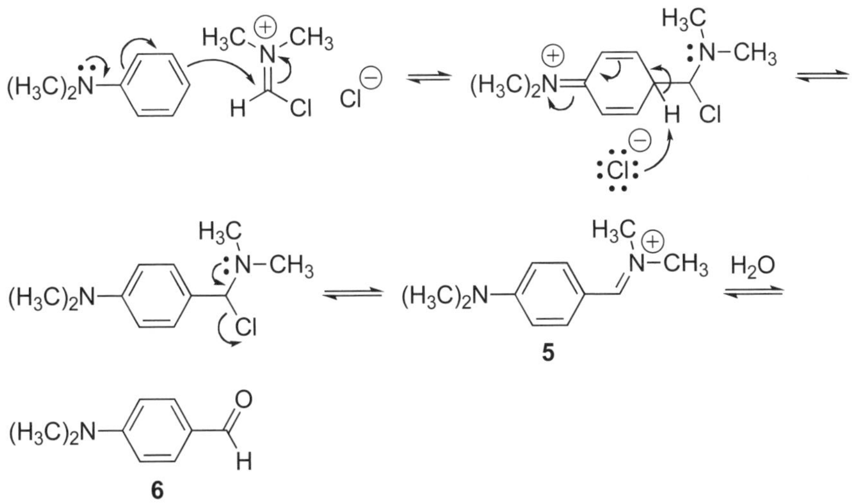

6.5.4 *Variations, Improvements, and Modifications*

The Vilsmeier reaction is used to this day in largely the same way as it was originally conceived, though its substrate scope continues to expand. One important variation, known as the Bischler–Napieralski isoquinoline synthesis, involves an intramolecular Fiedel–Crafts acylation reaction where the acylating agent resembles a Vilsmeier electrophile. If the acylated phenethylamine substrate contains an α-hydroxyl group, a subsequent dehydration yields an isoquinoline.[7]

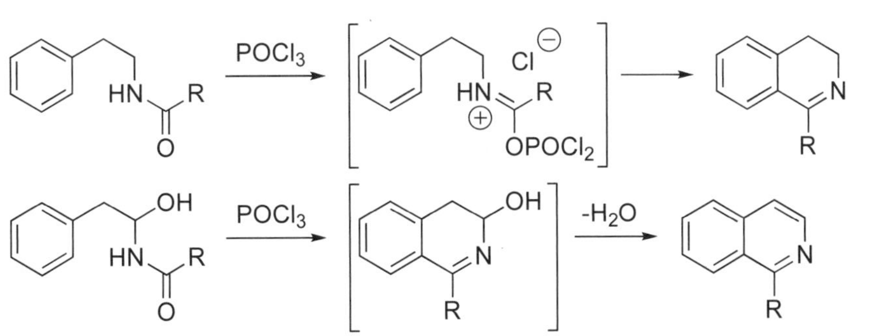

Each component of the standard Vilsmeier conditions can be modified to increase the variety of products that may be generated within this reaction manifold. Acetylation is the most common variant, which requires the use of dimethylacetamide (DMA) in place of DMF (**7**→**8**).[8] Though less common, more sophisticated acylations can be achieved with this methodology (**9**→**10**).[9] Furthermore, when the initially formed iminium salt is sufficiently stable, it can be isolated without hydrolysis (**11**→**12**).[10]

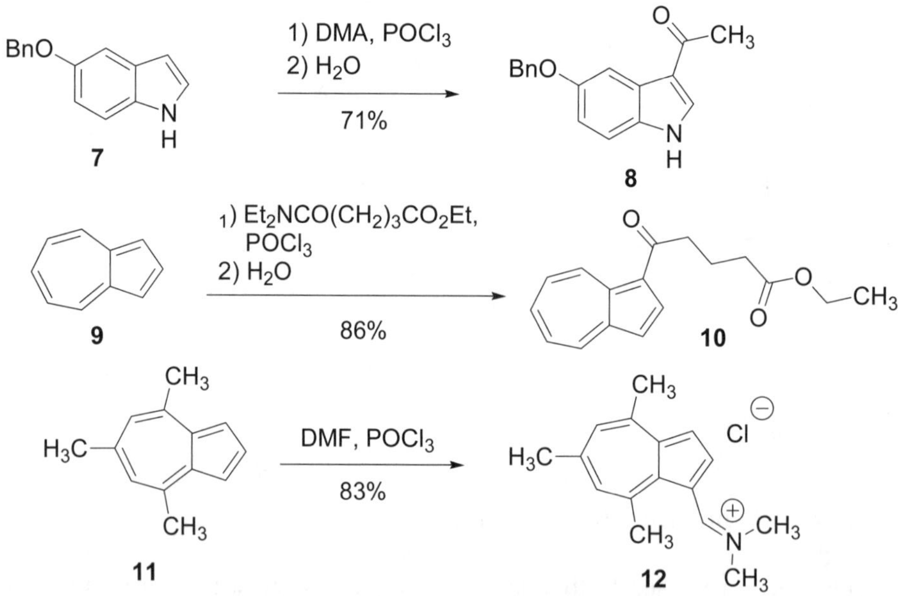

7 **8** **9** **10** **11** **12**

The intermediate iminium salts can also be transformed into functional groups other than carbonyls. Three common variants are quenching with hydrogen sulfide to yields thioaldehydes (**13**→**14**),[11] oxime formation with hydroxylamine followed by dehydration to yield nitriles (**15**→**16**),[12] and hydride reduction to yield dialkylamines (**17**→**18**).[13]

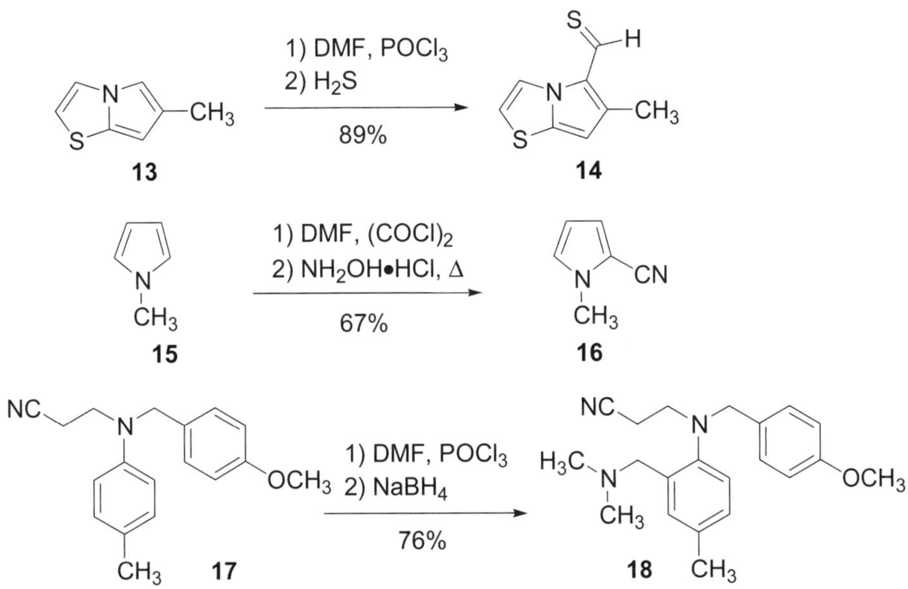

Finally, the extension of the Vilsmeier reaction to nonaromatic substrates has greatly expanded its scope. Most aliphatic alkenes are unreactive toward Vilsmeier reagents, which are only modestly electrophilic. Accordingly, alkenes bearing π-donating substituents show increased reactivity, but only a limited set of functional groups generate simple formylated products. For example, enol ethers are formylated to yield β-ketoaldehydes (**19**→**20**),[14] styrenes yield cinnamaldehydes (**21**→**22**),[15] and conjugated dienes, often formed in situ from the dehydration of allylic alcohols, yield conjugated dienals (**23**→**24**).[16]

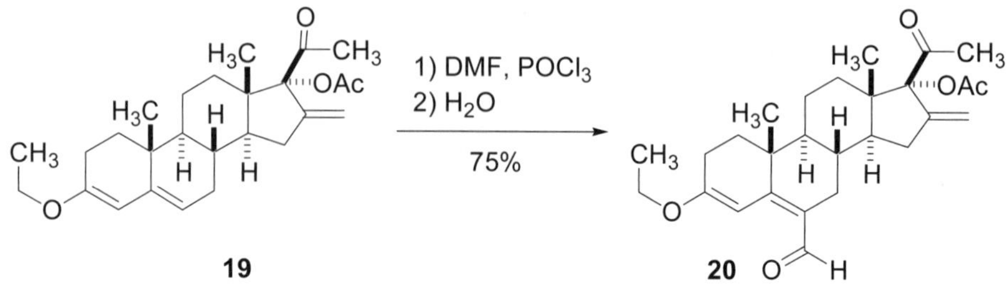

19 20

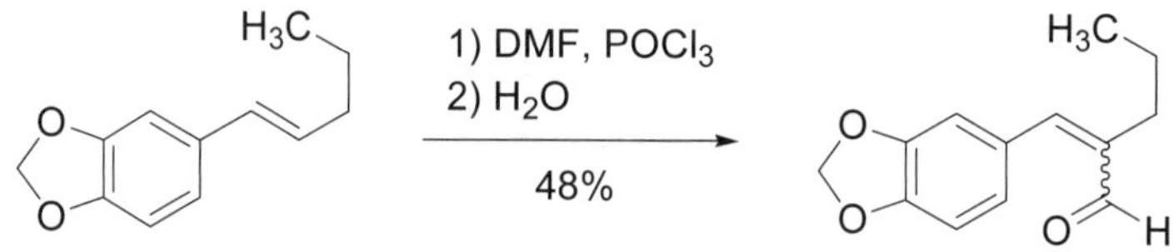

21 22

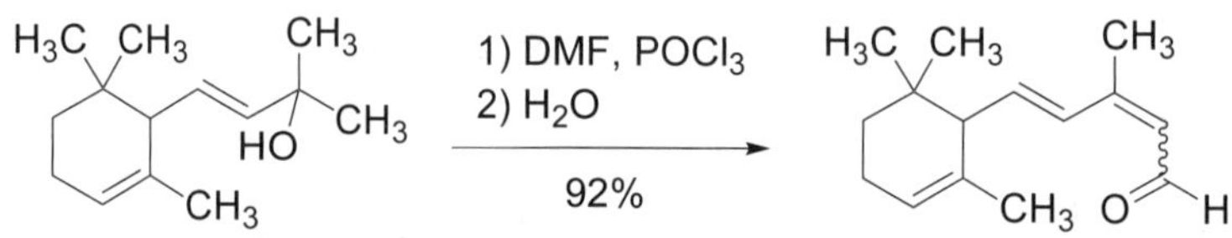

23 24

However, the reactions of most non-aromatic π nucleophiles with Vilsmeier reagents yield more complicated products.[1b] The reaction of the following ketone, via its enol tautomer, is illustrative.[17]

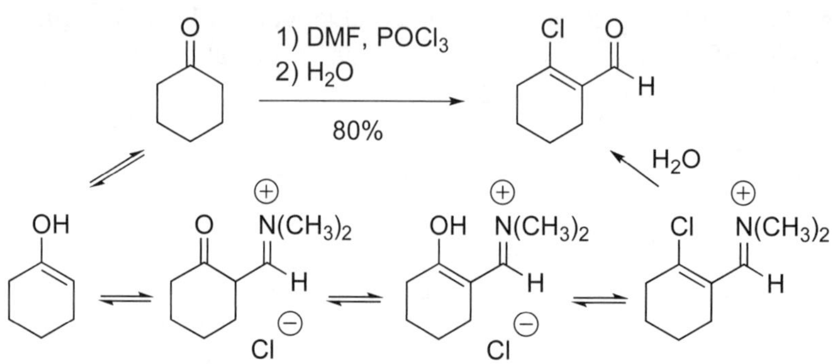

6.5.5 *Synthetic Utility*

This section describes prototypical reactions of Vilsmeier salts with electron-rich aromatic rings, an area where the Vilsmeier reaction has proven particularly valuable. Reactions are grouped into three categories: formylation of common heteroaromatic compounds with a focus on regioselectivity, tandem formylation-cyclization sequences that annulate aromatic compounds, and applications of Vilsmeier reactions in natural product synthesis.

Formylation of common heteroaromatic compounds

On treatment with Vilsmeier salts, *N*-alkylpyrroles formylate in the expected 2-position unless the alkyl group is bulky, in which case formylation at the 3-position is preferred.[18] The 2-substituted pyrroles tend to formylate at the 5-position, and 3-substituted pyrroles formylate at the 2- and 5-position with low regioselectivity.

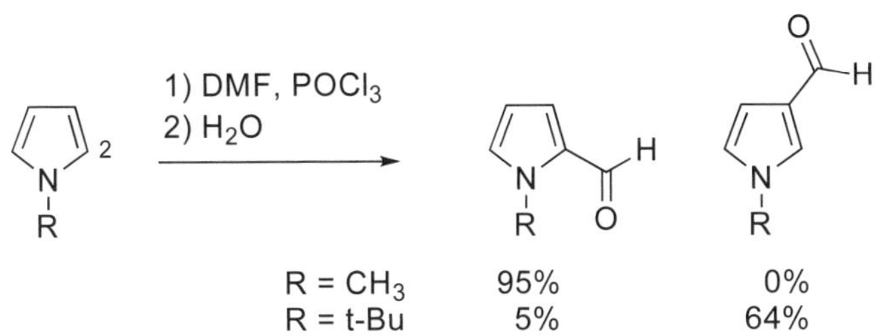

Furans give similar regioselectivities; however, contrasteric regioselectivity is observed in the formylation of 3-methylfuran.[19]

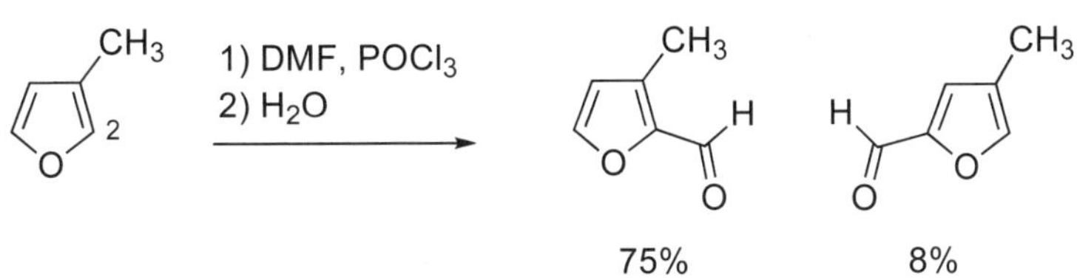

Indoles formylate in the 3-position,[20] but indoles already substituted at the 3-position results in predominantly *N*-formylation with a lesser amount of 2-formylation.[21] Benzofurmans fomylate selectively at the 2-position.[22]

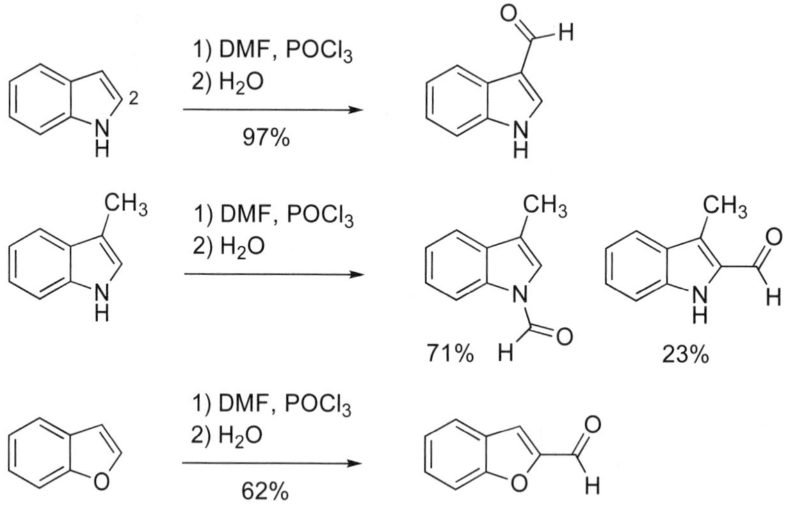

Formylation-cyclization sequences that annulate heteroaromatic compounds

The Bishler–Napieralski reaction, described above, is a named reaction in which Vilsmeier formylation results in a molecule that can participate in a subsequent cyclization reaction, resulting in a net annulation. There are numerous other examples of similar tandem sequences, most of which are not named reactions, and they generally give only modest yields.[1e] For instance, when ketone **25** is treated with a Vilsmeier reagent, isoflavone **26** is produced in nearly quantitative yield.[23]

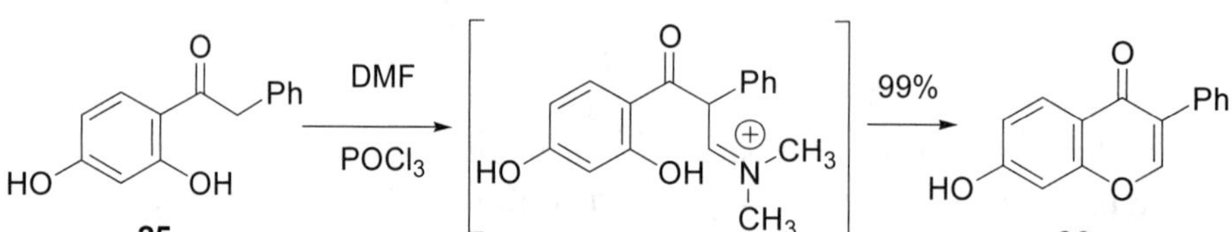

25 **26**

Benzothiophenes can be prepared from the corresponding phenyl ethers in a similar manner. In this case, the yield increased drastically when a more electron-donating amine substituent was present on the substrate.[24]

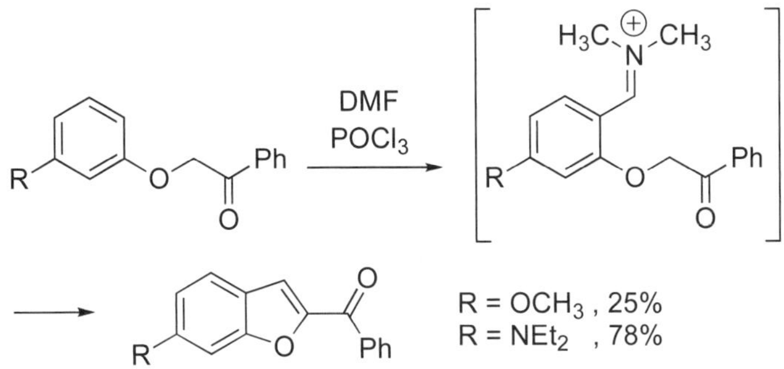

Pyrimidine derivatives also are formylated under Vilsmeier conditions. In the following example, a phenyl susbtituent is ideally placed for a subsequent annulation.[25]

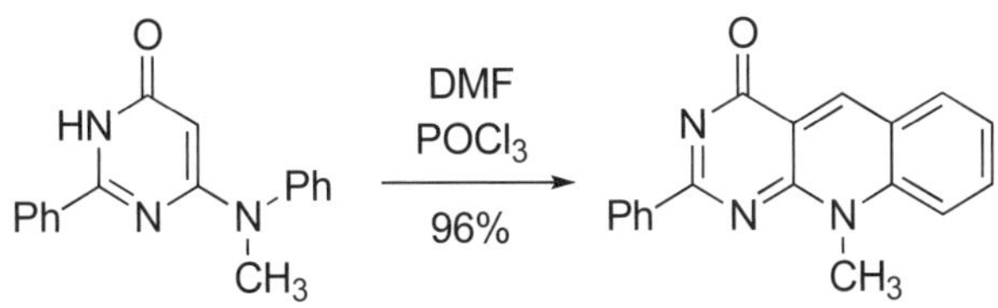

Applications of Vilsmeier reactions in natural product synthesis

A standard Vilsmeier formylation was used as the final step in a synthesis of homofascaplysin C (**28**).[26]

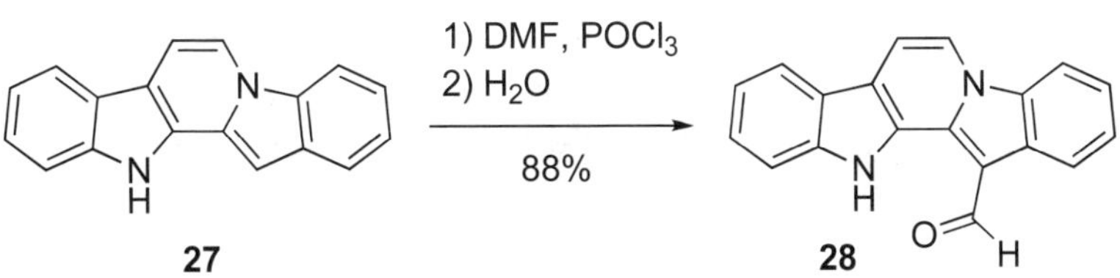

Vilsmeier formylation of enol ether **29** was an early step in the preparation of illudin C.[27] In this case, a concomitant bromination also took place to form **30**.

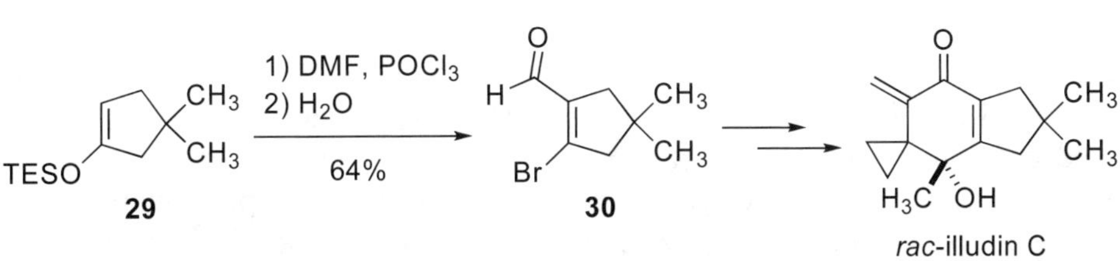

Finally, an adaptation of Vilsmeier conditions was used successfully in a synthesis of tashiromine. In this case, the amide was activated by triflic anhydride.[28]

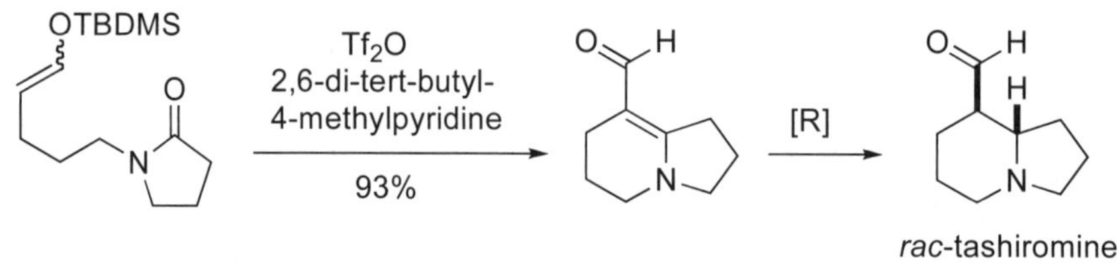

6.5.6 *Experimental*

Compound 28[26]

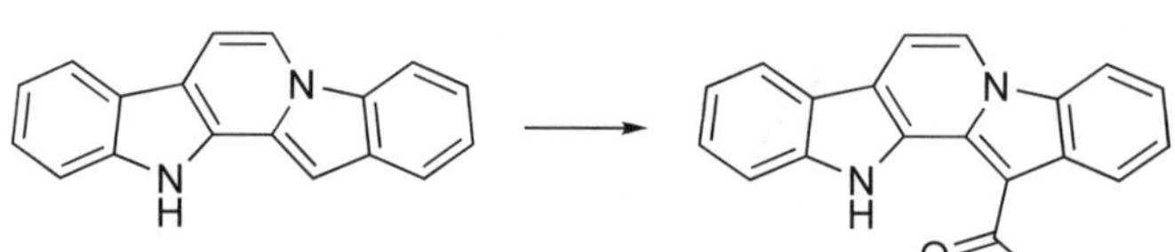

27 **28**

Phosphorus oxychloride (0.2 mL, 2.2 mmol) was added slowly via syringe to ice-cold, dry DMF (5 mL). The resulting solution, protected from moisture with a drying tube, was stirred for 15 min at room temperature and then recooled. A solution of **27** (0.53 g, 2.0 mmol) in DMF (10 mL) was then added over 2 min, and the solution was then stirred at room temperature for 3 h and poured into ice water (50 mL). The yellow mixture was made alkaline with 2 M sodium hydroxide, and the resulting precipitate was collected by filtration, washed thoroughly with water, and dried (50 °C, ~1 torr) to give **28** (0.50 g, 88%) as a yellow solid.

Compound 30[27]

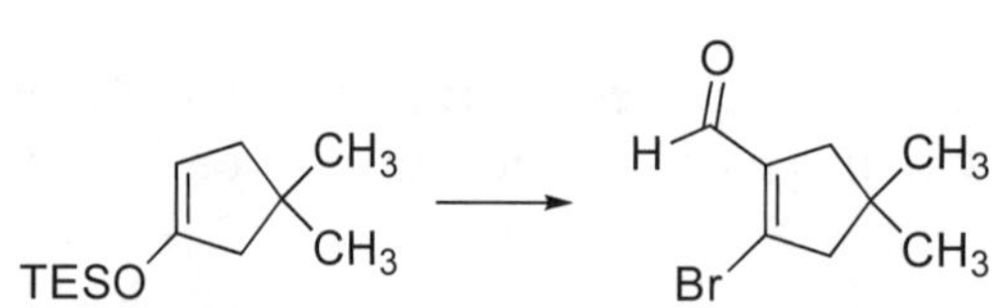

29 **30**

To a solution of DMF (513 mL, 6.63 mmol) in methylene chloride (10 mL) was added phosphorus oxybromide (1.58 g, 5.52 mmol), and the resulting solution was stirred at room temperature for 1 hr as a white precipitate formed. A solution of **29** (500 mg, 2.21 mmol) in methylene chloride (2 mL) was added to the mixture, and the resultant slurry was stirred at room temperature for 72 h and poured onto ice (5 g). The solution was neutralized with sodium bicarbonate and extracted with hexane/ether (9:1). The combined organic layers were washed with saturated aqueous sodium bicarbonate, dried over sodium sulfate, and concentrated. Purification by silica get chromatography (ethyl acetate/hexane, 3:97, silica gel deactivated with 10% triethylamine) gave exclusively one regioisomer as a colorless oil (289 mg, 64%).

6.5.7 References

1. [R] (a) Perumal, P. T. *Indian J. Heterocycl. Chem.* **2001** *11*, 1–8. (b) Jones, G.; Stanforth, S. P. *Org. React.* **2000**, *56*, 355–659. (c) Jones, G.; Stanforth, S. P. *Org. React.* **1997**, *49*, 1–330. (d) Meth-Cohn, O.; Stanforth, S. P. *Comp. Org. Syn. Vol. 2*, Trost, B. M.; Fleming, I., eds. Pergamon Press, Oxford, **1991**, pp. 777–794. (e) Meth-Cohn, O.; Tarnowski, B. *Adv. Heterocycl. Chem.* **1982**, *31*, 207–236. (f) Jutz, C. *Adv. Org. Chem.* **1976**, *9*, 225–342. (g) Seshadri, S. *J. Sci. Ind. Res.* **1973**, *32*, 128–149. (h) Burn, D. *Chem. Ind.* **1973**, 870–873.
2. Dimroth, O.; Zoeppritz, R. *Ber. Dtsch. Chem. Ges.* **1902**, *35*, 995–996.
3. Fischer, O.; Müller, A.; Vilsmeier, A. *J. Prakt. Chem.* **1925**, 69–87.
4. Vilsmeier, A.; Haack, A. *Ber.* **1927**, *60B*, 119–122.
5. Martin, G. J.; Poignant, S. *J. Chem. Soc., Perkin Trans. 2* **1974**, 642–646.
6. Downie, I. M.; Earle, M. J.; Heaney, H.; Shuhaibar, K. F. *Tetrahedron* **1993**, *49*, 4015–4034.
7. [R] Whaley, W. M.; Govindachari, T. R. *Org. React.* **1951**, *6*, 74-150.
8. Anthony, W. C. *J. Org. Chem.* **1960**, *25*, 2049–2053.
9. Anderson, Jr. A. G.; Breazeale, R. D. *J. Org. Chem.* **1969**, *34*, 2375–2384.
10. Hafner, K.; Schneider, J. *Justus Liebigs Ann. Chem.* **1959**, *624*, 37–47.
11. McKenzie, S.; Molloy, B. B.; Reid, D. H. *J. Chem. Soc. C* **1966**, 1908–1913.
12. Barnett, G. H.; Anderson, H. J.; Loader, C. E. *Can. J. Chem.* **1980**, *58*, 409–411.
13. Shawcross, A. P.; Stanforth, S. P. *Tetrahedron* **1989**, *45*, 7063–7076.
14. Popper, T. L.; Faro, H. P.; Carlon, F. E.; Herzog, H. L. *J. Med. Chem.* **1972**, *15*, 555–556.
15. Witiak, D. T.; Williams, D. R.; Kakodkar, S. V.; Hite, G.; Shen, M.-S. *J. Org. Chem.* **1974**, *39*, 1242–1247.
16. Reddy, M. P.; Rao, G. S. K. *Ind. J. Chem., Sec. B* **1982**, *21B*, 757–758.
17. Paquette, L. A.; Johnson, B. A.; Hinga, F. M. *Org. Synth.* **1966**, *46*, 18–20.
18. Jones, R. A.; Candy, C. F.; Wright, P. H. *J. Chem. Soc., C.* **1970**, 2563–2567.
19. Chadwick, D. J.; Chambers, J.; Hargraves, H. E.; Meakins, G. D.; Snowden, R. L. *J. Chem. Soc. Perkin Trans. 1* **1973**, 2327–2332.
20. Klohr, S. E.; Cassady, J. M. *Synth. Commun.* **1988**, *18*, 671–674.
21. Chatterjee, A.; Biswas, K. M. *J. Org. Chem.* **1973**, *38*, 4002–4004.
22. Suu, V. T.; Buu-Hoi, N. P.; Xuong, N. D. *Bull. Soc. Chim. Fr.* **1962**, 1875–1877.
23. Kagal, S. A.; Nair, P. M.; Ventkataraman, K. *Tetraheron Lett.* **1962**, 593–597.
24. Hirota, T.; Fujita, H.; Sasaki, K.; Namba, T.; Hayakawa, S. *Heterocycles* **1986**, *24*, 771–776.
25. Yoneda, F.; Mori, K.; Ono, M.; Kadokawa, Y.; Nagao, E.; Yamaguchi, H. *Chem. Pharm. Bull.* **1980**, *28*, 3514–3520.
26. Gribble, G. W.; Pelcman, B. *J. Org. Chem.* **1992**, *57*, 3636–3642.
27. Aungst, Jr., R. A.; Chan, C.; Funk, R. L. *Org. Lett.* **2001**, *3*, 2611–2613.
28. Bélanger, G.; Larouche-Gauthier, R.; Ménard, F.; Nantel, M.; Barabé, F. *J. Org. Chem.* **2006**, *71*, 704–712.

6.6 von Richter Reaction

Martin E. Hayes

6.6.1 Description

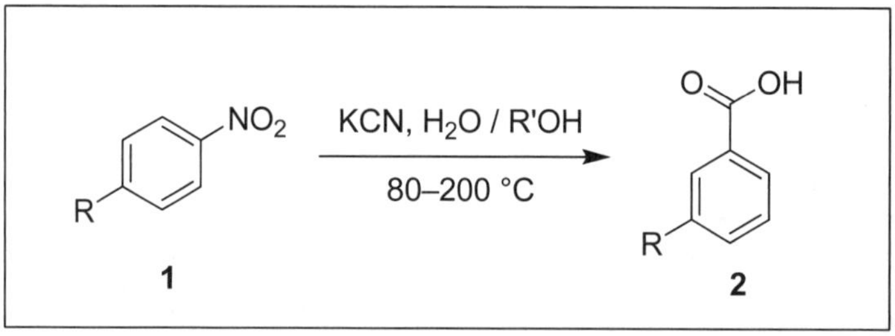

The von Richter reaction is the nucleophilic aromatic substitution of a nitroarene (**1**) with potassium cyanide to give the *cine*-substituted benzoic acid (**2**). The reaction is characterized by the observed regiochemistry of the product where the carboxyl group occupies a position *ortho* to the nitro group that is lost.

6.6.2 Historical Perspective

Victor von Richter was a German chemist[1] whose support of Mendeleev's theory of periodicity,[2] later formalized into the modern Periodic Table of the Elements, played an important role in its adoption throughout western Europe. He was widely regarded for his textbooks on organic[3] and inorganic chemistry[4] that were used throughout Europe and the United States at the turn of the 20th century. Among his research accomplishments is the first synthesis of racemic BINOL[5] and the first synthesis of cinnoline.[6]

The von Richter reaction, also known as the von Richter rearrangement, was first disclosed in 1871[7–9] and the results were revised[10] in 1875 to reflect the correct regiochemistry of the benzoic acid products. The reaction was largely ignored until the mid-20th century[11] when it was further examined in a series of critical experiments that lead to a fuller understanding of the scope and ultimate elucidation of the mechanism. The reaction has been of limited synthetic value due to the low isolated yield of the products.

6.6.3 Mechanism

The mechanism of the von Richter reaction was a subject of active investigation during the mid-20th century and accounts of the debate surrounding it have been detailed previously.[12] The reaction is a classic example of an aromatic *cine*-substitution[13] and one of only a handful of such reactions that do not proceed via the intermediacy of an aryne.[14] The mechanism has been probed through a series of critical studies using ^{15}N-labeled ammonia[15] and substrate, ^{2}H-labeled solvent and substrate,[16] and ^{18}O-labeled water[17] along with careful analysis of the reaction byproducts.

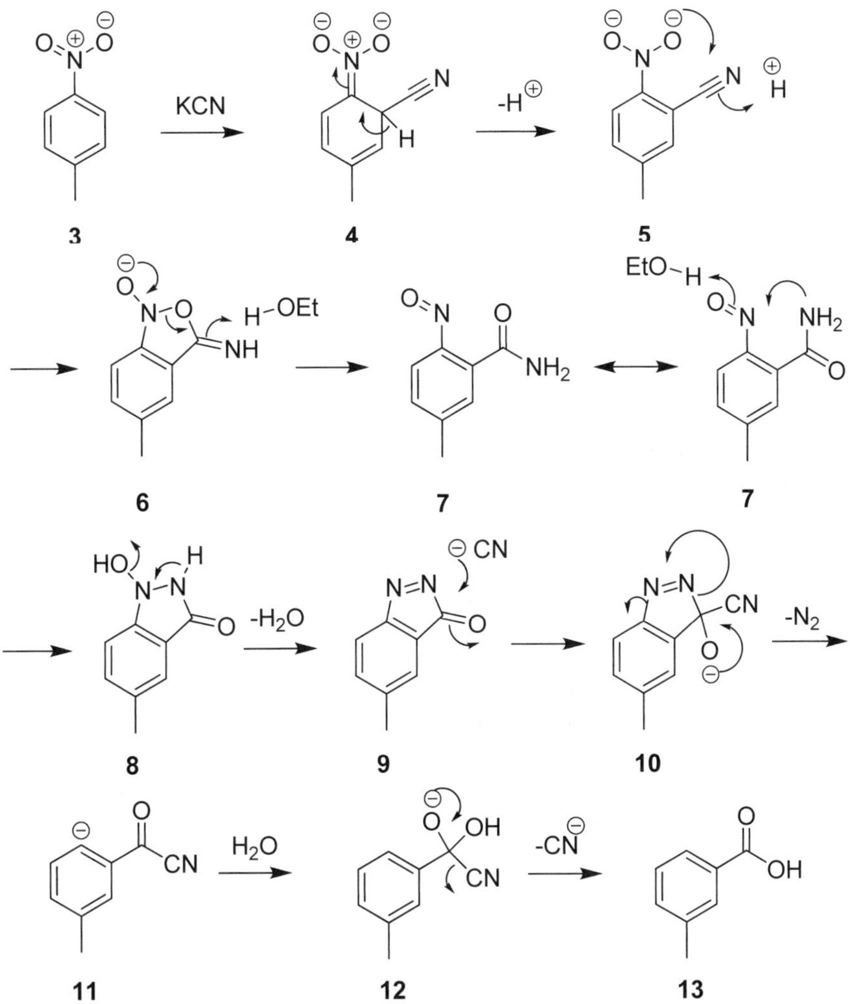

The currently accepted mechanism involves the addition of cyanide *ortho*- to the nitro group to give intermediate **5**. The nitrile is then subject to intramolecular attack to generate the bicyclic intermediate **6**. Previous mechanistic hypotheses[18] have invoked this same intermediate (**6**), yet its fate was a matter of contention until 1960 when Rosenblum,[15] in a series of elegant ^{15}N-labeling experiments coupled with careful interpretation of existing results, proposed that the heterocyclic ring ionizes to give an aryl nitroso amide (**7**). This nitroso intermediate was independently synthesized and demonstrated to provide the observed product when exposed to the von Richter reaction conditions.[19]

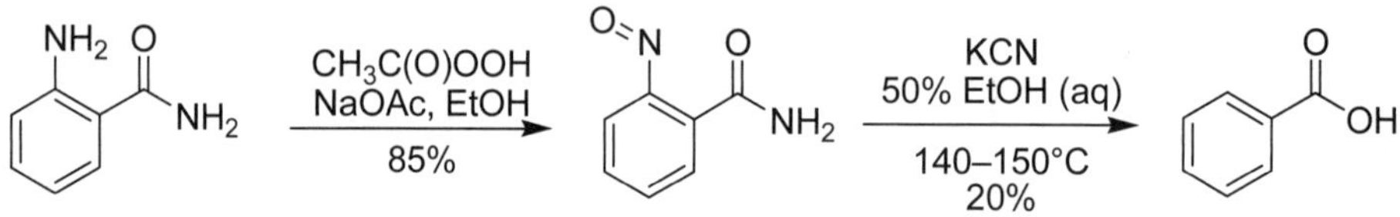

14 **15** **16**

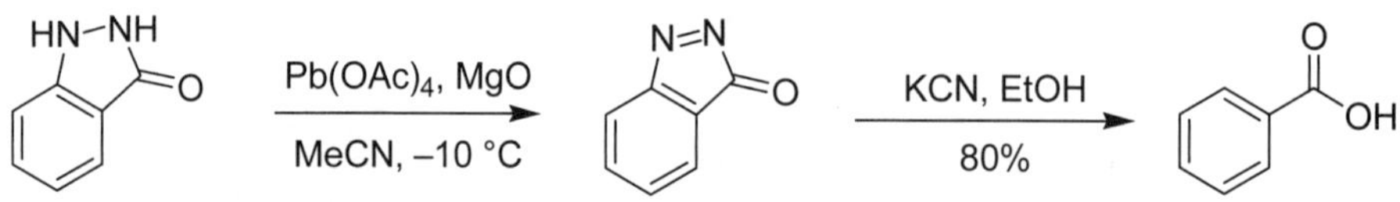

17 **18** **16**

Intermediate **7** is then proposed to undergo recombination of the nitroso and *ortho*-amide groups to form an indazolonone **9**. The unstable indazolonone then decomposes to nitrogen gas and the *cine*-substituted benzoic acid (**13**) via the benzoylnitrile (**11**). The proposed indazolonone intermediate has also been independently prepared and shown to readily decompose to the observed benzoic acid product.[20] Alternative proposed intermediates, such as a benzonitrile **20**, have been observed to hydrolyze at a rate that is demonstrably slower than the overall reaction rate,[21] conclusively eliminating it as a competent intermediate.

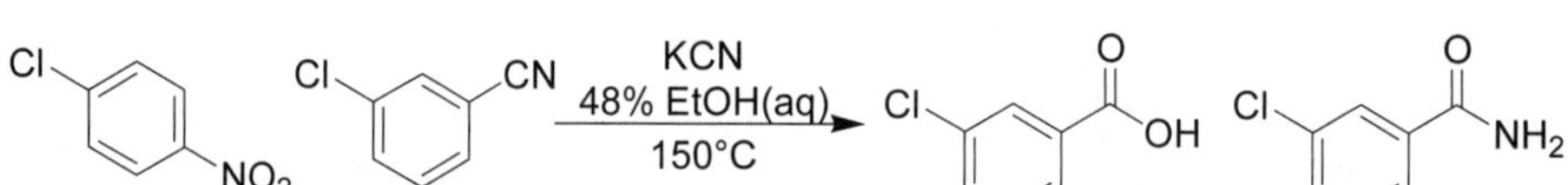

19 **20** **21** **22**

40% 30%

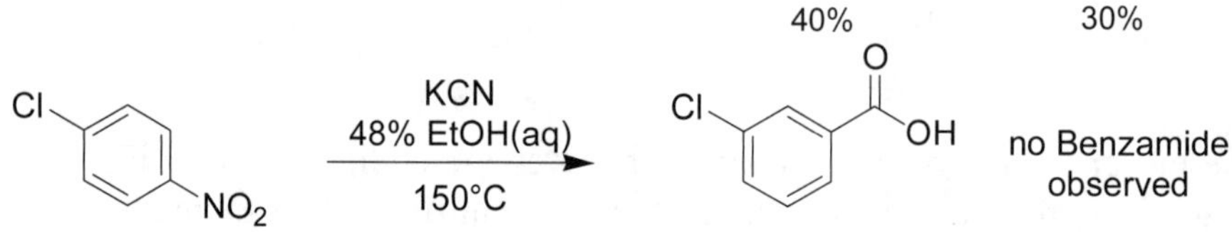

19 **21**

31%

6.6.4 Variations and Improvements

The scope of the von Richter reaction has been examined with respect to substituent effects where the sum of the arene field effects plays an important role in determining optimal conversion to the product. Bunnett[18] reported in 1956 that appreciable conversion was observed only with substituents having summed sigma values, excluding the nitro group, between –0.2 and 0.6 where optimal conversion occurs with sigma values between 0.2 and 0.5. The reaction is inhibited by substituents *ortho-* to the nitro group which prevent efficient orbital overlap of the pi-system due to steric interactions.

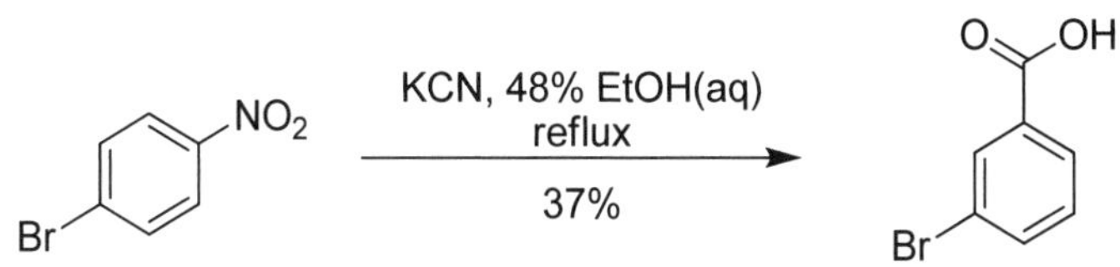

23 **24**

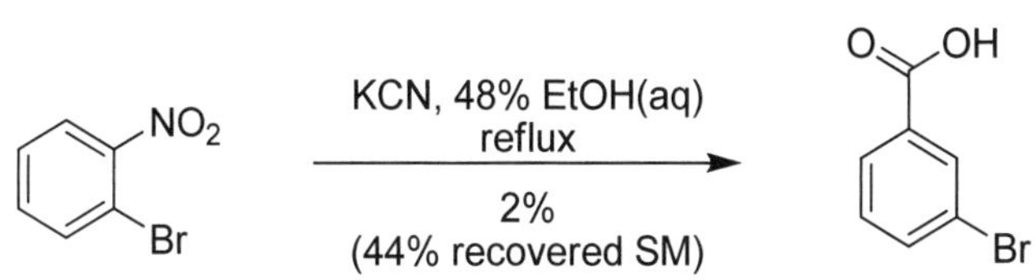

25 **24**

Reactions of di-nitro compounds typically do not proceed via the von Richter pathway but rather give products of *ipso*-substitution at one of the nitro centers. Similarly, activated heterocycles such as 6-nitroquinoline, are reported to proceed via *ipso*-substitution instead of the von Richter pathway.[22]

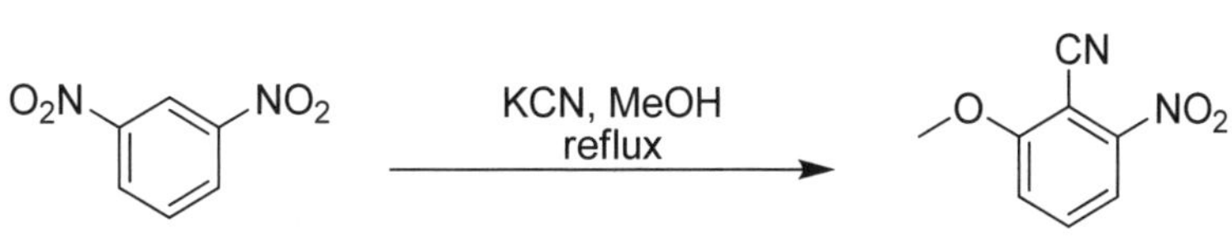

26 **27**

Many examples of the von Richter reaction have been conducted in sealed reaction vessels, however, a more convenient method involves refluxing the reaction mixture, a modification that has been shown to give comparable yields.[16] The reaction proceeds optimally in aqueous alcohol solvents including aqueous mixtures of 48–95% methanol, ethanol, 2-ethoxyethanol, or ethylene glycol. The latter solvents have been employed, owing to their higher boiling points, while the standard reaction medium is 48% aqueous ethanol.[22] Limited examples of other solvents have been disclosed, most notably the use of DMSO has been reported to give several byproducts not previously disclosed.[23]

6.6.5 Synthetic Utility

The synthetic utility of the von Richter reaction has been limited due to the low isolated yields of the benzoic acid products. The reported yields starting from nitroarenes range from 0.5–50%, though in some cases unreacted starting material can also be recovered.[22] The low overall conversion is believed to be due, at least in part, to competing hydrolysis of the cyanide reagent in the reaction medium and often improved conversion is observed with greater excess of cyanide salts.[16]

Published examples of the von Richter reaction have been limited to nitrobenzene and nitronapthalene derivatives. In the case of 2-nitronapthalene, treatment with potassium cyanide in refluxing 48% ethanol for 4 h provides a 13% yield of 1-carboxynapthalene.[18]

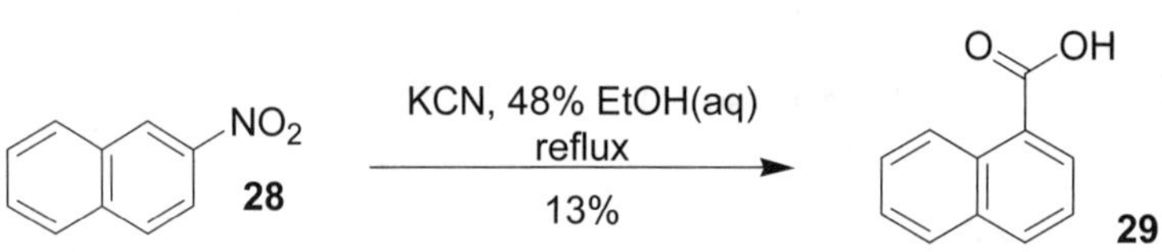

Recently, a variation of the von Richter reaction was disclosed involving *N*-arylnitrones as surrogate nitroarenes that also are proposed to undergo a *cine*-substitution reaction with potassium cyanide.[24] The reaction with potassium cyanide in refluxing 33% ethanol for 2 h provides a 52% yield of benzoic acid and is one of the highest yielding examples of a von Richter reaction.

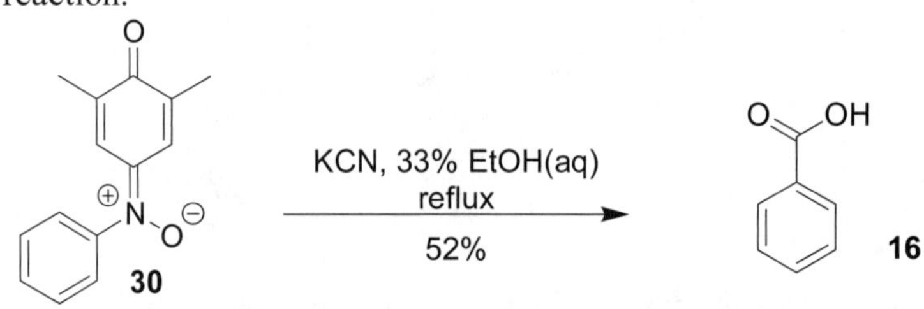

6.6.6 Experimental

Preparation of 3-Chlorobenzoic Acid (21)[22]

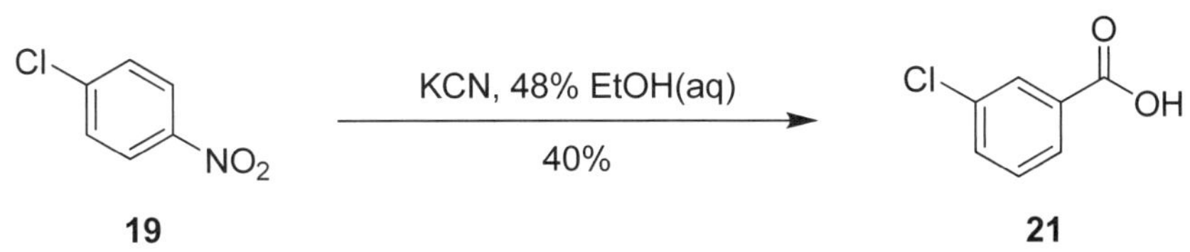

A flask was charged with 4-chloronitrobenzene (3.0 g, 0.019 mol) and KCN (6.45 g, 0.099 mol) in 26 mL of 48% aqueous ethanol, and the reaction mixture was refluxed for 48h. The mixture was allowed to cool to rt and transferred into a flask with the aid of water, then the solution was made basic with the addition of NaOH. Organic solvent and unreacted starting material were removed by steam distillation and then the residue was acidified. Steam was again passed through the acidified solution until 1.5 L of distillate was collected. The pH of the distillate was adjusted with $NaHCO_3$, and the volume of the solution was reduced via distillation to approximately 200 mL. Then the residue was acidified. The resulting precipitate was collected by filtration to give 1.2 g (0.0076 mol) 3-chlorobenzoic acid.

6.6.7 References

1. Pransnitz, G. *Chem. Ber.* **1891**, *24*, 1123–1130.
2. Kaji, M. *Foundations of Chemistry.* **2003**, *5*, 189–214.
3. v. Richter, V. Smith, E., Ed. *Victor von Richter's Organic Chemistry*, 2 ed. Blakiston, Philadelphia, 1900.
4. v. Richter, V. Klinger, H.; Smith, E., Ed. *Victor von Richter's Textbook of Inorganic Chemistry*, 5 ed. P. Blakiston, Philadelphia, 1900.
5. Brunel, J. *Chem. Rev.* **2005**, *105*, 857–989.
6. v. Richter, V. *Chem. Ber.* **1883**, *16*, 677–683.
7. v. Richter, V. *Chem. Ber.* **1871**, *4*, 21–22.
8. v. Richter, V. *Chem. Ber.* **1871**, *4*, 459–468.
9. v. Richter, V. *Chem. Ber.* **1871**, *4*, 553–555.
10. v. Richter, V. *Chem. Ber.* **1875**, *8*, 1418–1425.
11. [R] Huisgen, J. S. *Angew. Chem.* **1960**, *72*, 294–315.
12. [R] Shine, H. Shine, H. J. ed. *Aromatic Rearrangements*, Elsevier, New York, 1967, pp. 326–335.
13. [R] Suwinski, J. S.; Swierczek, K. *Tetrahedron* **2001**, *57*, 1639–1662.
14. [R] Bunnett, J. *Q. Rev. Chem. Soc.* **1958**, *122*, 1–16.
15. Rosenblum, M. *J. Am. Chem. Soc.* **1960**, *82*, 3796–3798.
16. Bunnett, J. F.; Rauhut, M. M.; Knutson, D.; Bussell, G. E. *J. Am. Chem. Soc.* **1954**, *76*, 5755–5761.
17. Samuel, D. *J. Chem. Soc.* **1960**, 1318–1320.
18. Bunnett, J. F.; Rauhut, M. M. *J. Org. Chem.* **1956**, *21*, 934–938.

19. Ibne-Rasa, K. M.; Koubek, E. *J. Org. Chem.* **1963**, *28*, 3240–3241.
20. Ullman, E.; Bartkus, E. *Chem. Ind.* **1962**, 93–94.
21. Bunnett, J. F.; Rauhut, M. M. *J. Org. Chem.* **1956**, *21*, 944–948.
22. Bunnett, J. F.; Cormack, J. F.; McKay, F. C. *J. Org. Chem.* **1950**, *15*, 481–490.
23. Rogers, G. T.; Ulbricht, T. L. V. *Tetrahedron Lett.* **1968**, *9*, 1029–1032.

Appendixes

Appendix 1
Table of Contents for Volume 1: ***Name Reactions in Heterocyclic Chemistry***
Published in 2005

Appendix 2
Table of Contents for Volume 2:
Name Reactions for Functional Group Transformations

Appendix 3
Table of Contents for Volume 3: *Name Reactions for Homologations-Part I*
Published in 2009

Appendix 4
Table of Contents for Volume 3: *Name Reactions for Homologations-Part II*
Published in 2009

Appendix 5
Table of Contents for Volume 6: *Name Reactions in Heterocyclic Chemistry-II*
Due in 2011

Chapter 10 Miscellaneous reactions in heterocyclic chemistry

Appendixes

Subject index

Index